The Cambridge Encyclopedia of Earth Sciences

Editorial Advisory Board

The Cambridge Encyclopedia of Earth Sciences

Editor-in-Chief David G. Smith PhD
Department of Earth Sciences, The Open University, 1979–1981

CAMBRIDGE UNIVERSITY PRESS
Cambridge
London New York New Rochelle Melbourne Sydney

Contributors

Editorial director: James R. Clark
Managing editor: Barbara Horn
Executive editor: Judith Ravenscroft
Designers: Juanita Grout, Karen Osborne
Design assistants: Richard Glydden, Carol McCleeve
Diagram visualizers: Michael Fry, Wolfgang Mezger
Picture Researcher: Mary Fane
Indexer: Richard Raper

First published in 1982 by the Press Syndicate of the University of Cambridge, The Pitt Building, Trumpington Street, Cambridge CB2 1RP
32 East 57th Street, New York, N.Y. 10022, USA
296 Beaconsfield Parade, Middle Park, Melbourne 3206, Australia

First published in the United States in 1981 by Crown Publishers Inc. and the Press Syndicate of the University of Cambridge

First published in Canada in 1981 by Prentice-Hall Canada Inc. and the Press Syndicate of the University of Cambridge

British Library Cataloguing in Publication Data

The Cambridge encyclopedia of earth sciences
1. Earth sciences
I. Smith, David G.
550 QE26.2 81-3313

ISBN 0 521 23900 1

Created, designed and produced by Trewin Copplestone Books Limited
Advance House, 101–109 Ladbroke Grove, London W11

Printed and bound
by Graficromo s.a., Cordoba, Spain

Dr Timothy C. Atkinson
University of East Anglia

Dr Stanley M. Awramik
University of California, Santa Barbara

Dr Peter Brimblecombe
University of East Anglia

Dr Geoffrey Brown
The Open University

Dr Simon Conway Morris
The Open University

Dr Trevor D. Davies
University of East Anglia

Dr Stephen A. Drury
The Open University

Dr Graham Evans
University of London

Dr Peter W. Francis
The Open University

Professor William S. Fyfe
University of Western Ontario

Dr Christopher J. Hawkesworth
The Open University

Professor Gordon L. Herries Davies
University of Dublin

Professor John A. Jacobs
University of Cambridge

Dr P. M. Kelly
University of East Anglia

Dr R. Kerrich
University of Western Ontario

Professor Kurt Lambeck
Australian National University

Dr Michael R. Leeder
University of Leeds

Dr Simon Mitton
University of Cambridge

Dr Roberto Peveraro
British National Oil Corporation

Professor Hans-Ulrich Schmincke
Institut für Mineralogie
Ruhr-Universität, Bochum

Dr Andrew C. Scott
University of London

Dr Robert M. Shackleton
The Open University

Dr David G. Smith
The Open University

Dr Peter J. Smith
The Open University

Dr Sandra Smith
The Open University

Dr Donald H. Tarling
University of Newcastle-upon-Tyne

Dr Richard S. Thorpe
The Open University

Dr Christopher E. Vincent
University of East Anglia

Dr Antony C. Waltham
Trent Polytechnic

Dr David Wood
Phillips Petroleum Company

Dr Nigel H. Woodcock
University of Cambridge

Dr John B. Wright
The Open University

Contents

Foreword

It is very seldom that those working in any area of scientific research find themselves caught up in an episode of conceptual change so rapid and so fundamental that external observers of the field begin to use terms such as 'scientific revolution' to describe the activity they perceive. Yet this is precisely what has happened in the Earth sciences over the last two decades.

The fires of revolt were kindled in Cambridge in the early 1960s by a visitor from Princeton, the late Harry Hess, who argued that the ocean basins must be ephemeral features of the Earth's surface, that continents were able to move apart by the generation of new ocean floor between them and to come together by the destruction of the intervening ocean by the process now known as subduction. It was in Cambridge, too, that a few years later the flames were fanned by Drummond Matthews and Fred Vine who recognized the significance of the previously incomprehensible magnetic lineations of the ocean floor and used them both to show that Hess's ideas were correct, and to establish the rates at which new ocean floor grew.

At first these ideas were received with great scepticism. After all, the ideas that continents had moved with respect to each other during geologic time were hardly new. Wegener, Taylor and others in the first decades of this century had thoroughly explored these ideas and failed to make any universal or lasting impact on geologists or geophysicists.

During the late 1960s, however, the new ideas gained further ground. Morgan in Princeton, Parker in the University of California, McKenzie in Cambridge, Oliver and his collaborators at Columbia University, and Le Pichon at Brest, all nearly simultaneously made dramatic new discoveries, which is many cases partly overlapped with each other. This was a breathless and exciting time in the subject. Not only did the new observations confirm the ideas of Hess, Vine and Matthews, but they showed that these ideas could be extended to formulate the group of concepts now known as plate tectonics, a unifying theory for the mechanical behaviour of the outer part of the Earth which provided for the first time a coherent explanation for the distribution of ocean basins, mountain ranges, earthquakes and volcanoes.

There had, of course, been attempts to explain these phenomena previously but the hypothesis of plate tectonics had one enormous advantage over its predecessors: it was testable. It predicted the character of earthquakes in the known seismic zones in different parts of the world and it predicted the age of the ocean floor.

This latter prediction led to the largest and most ambitious collaborative international experiment ever undertaken in the Earth sciences, the Deep Sea Drilling Project, begun in 1969. With a group of universities in the United States taking the lead, the technology was developed to drill holes and recover cores from the floors of the deep oceans to test by direct observation the age prediction of the new theory. This drilling programme was successful beyond all reasonable expectation; not only were the predictions verified but, almost more important, knowledge of the floor of the deep ocean and the processes which control its development was enormously increased. At the time of writing this programme is in its final phase.

Geologists, geophysicists and geochemists were faced, therefore, with a whole new conceptual framework against which to reconsider the accumulated observations and conclusions of half a century. Many of the basic dogmas to be found in the standard texts were at best half-truths and at worst quite wrong. The texts simply had to be rewritten.

For this reason, if for no other, it is a highly opportune time for this new Encyclopedia of Earth Sciences to make its appearance. The aim is both to provide ready access to the wealth of geological information which has been gleaned over the years and to show how it fits into the framework of plate tectonics.

It would be misleading if this foreword were to conclude without a word of warning. Without question plate tectonics represents the most important single conceptual advance Earth science has known; but that does not mean that it is the last. The subject is still developing rapidly with the steady growth, testing and pruning of ideas. There are still important aspects of the Earth's behaviour which we cannot begin to understand. Some of these no doubt will provide the substance for future editions of this encyclopedia.

2 June 1981

E. R. Oxburgh
Department of Earth Sciences
Cambridge

Introduction

Today the term plate tectonics is not only familiar to scientists, it is entering everyday language and finding its way into dictionaries. The resulting transformation of the way in which we regard the Earth has brought about a reappraisal of the vast accumulation of information about the surface of the Earth collected by geographers, geologists, oceanographers and others over the centuries, reviewing and reinterpreting that information in the comprehensive new framework provided by the new plate tectonic concept. But more than that, this revolution has brought about a change of approach in the scientific study of the Earth. No longer is geology a second-rate descriptive science—it has joined forces with the physical sciences in a new and more cohesive whole to which the term Earth sciences is now so much more appropriate.

This Encyclopedia is designed to be used as a work of reference, through its comprehensive index, but it is also a book that is designed to be picked up and read, as much by the interested general reader as by students of the Earth sciences using it as a reader.

We start by putting the Earth and its study into the perspective of history and its place in the universe. Part Two tackles some challenging aspects of the physical and chemical constitution and behaviour of the Earth as a whole, and introduces the materials of which it is made. As well as providing insight into the inaccessible depths of the Earth, and back into its immensely long history, these topics provide much information necessary to the understanding of the Earth's crust, which is dealt with in Part Three. Here we consider the activity of the outermost skin of the Earth, the only part of the solid Earth that we can observe directly. Most of that activity is concentrated along narrow zones, the boundaries between the plates; it is these plate margins that are the focus of our attention, together with the effect of the enormous forces involved in plate movements both on solid rocks and on their redistribution as recycled waste.

In Part Four we are treading on rather more familiar ground in dealing with those outer zones of the Earth that form our own environment, so it is perhaps surprising that so much is still being learnt about the atmosphere, the oceans and their interaction with the solid Earth. Also of highly topical interest is the study of the origins of life. Its subsequent evolution provides, in the fossil record, a time-scale for the study of the history of the Earth's surface environments. Part Five brings in human interaction with the Earth—a vast topic of which we can deal with only a limited selection here. Part Six takes us back into space again, to see how what we have learnt about the Earth can be applied to our nearest planetary neighbours, and conversely how the space programme (which happened to start at about the same time as the plate tectonic theory was being formulated) has brought better information on those oldest parts of Earth history which are not recorded here on Earth. Finally there is a glossary. Deriving from its lengthy history as a descriptive science dealing with an extraordinary range of phenomena, Earth science has accumulated a particularly rich vocabulary of its own—after all, if you come across something new which you do not fully understand, it always helps to give it a name. This terminology can be an enormous help in summing up complex concepts in a single word or phrase, but it can also be a barrier to understanding. We have therefore tried to keep the technical terms to those with which anyone seriously interested in the subject should be familiar.

The Cambridge Encyclopedia of Earth Sciences was conceived, written and published within less than three years. Its authors are all active in research in the fields in which they have written, and the standing of the editorial panel can be judged from the appearance in the text of several of their names, many of which are now firmly associated with important discoveries.

David G. Smith

Units and abbreviations

Units

Earth scientists use some of the standard international (SI) system of scientific units, some units that predate the SI system, and a few units that are particular to their own field of study.

The base units of the SI system include the metre (length), kilogram (mass), second (time), ampere (electric current) and kelvin (temperature). Other units can be expressed as mathematical combinations of these.

Multiples of SI units come in steps of one thousand. To economize on space and to make comparison between numbers easier, scientists use a form of mathematical shorthand called the powers of ten, or index notation. By this means it is possible to express all numbers as the multiple of a number between 1 and 10 (the mantissa), and a power of ten (the index); many pocket calculators use this notation for numbers which would otherwise be too large or too small for the display. The power or index is a number written as a superscript (eg, in 10^3, 3 is the power) that tells us by how many places to move the decimal point to the right. For numbers smaller than 1 the power is written as a negative number and tells us by how many steps to move the decimal point to the left. Thus $10^3 = 1000$, $10^6 = 1\,000\,000$, $10^{-2} = 0.01$, and $10^{-5} = 0.00001$. Similarly, $2.4 \times 10^4 = 24\,000$, and $5.36 \times 10^{-3} = 0.00536$.

The prefixes used to extend the magnitude of the basic units are:

giga (G) $= 10^9$
mega (M) $= 10^6$
kilo (k) $= 10^3$
milli (m) $= 10^{-3}$
micro (μ) $= 10^{-6}$
nano (n) $= 10^{-9}$
pico (p) $= 10^{-12}$

Abbreviations

A = ampere
a = year
C = Celsius or centigrade
cal = calorie
cm = centimetre
dyn = dyne
G = giga
hfu = heat flow unit
Hz = hertz
J = joule
K = kelvin
k = kilo-
kb = kilobar
kg = kilogramme
km = kilometre
kWh = kilowatt-hour
l = litre
M = mega-
m = metre
m = milli-
Ma = million years
mb = millibar
mgal = milligal
mm = millimetre
mW = milliwatt
N = newton
n = nano-
p = pico-
Pa = pascal
rad = radian
s = second
T = tesla
t = tonne
V = volt
v = velocity
W = watt
μm = micrometre (micron)

Conversion factors

length
1 metre (m) = 1.094 yards
1 centimetre (cm) = 0.394 inches
1 kilometre (km) = 0.621 miles
1 light year $= 9.4605 \times 10^{15}$ m
$= 5.88 \times 10^{12}$ miles

volume
1 litre (l) = 1000.028 cubic centimetres
= 0.220 UK gallons
= 0.264 US gallons
1 cubic metre (m^3) = 1.308 cubic yards

mass
1 kilogramme (kg) = 2.205 pounds
1 tonne (t) = 1000 kilogrammes

temperature
1 degree on the centigrade (C) scale = 1 degree on the kelvin (K) scale
0°C = 273.15 K
1 degree on the centigrade scale = 9/5 degrees on the Fahrenheit (F) scale
0°C = 32°F

acceleration
1 milligal (mgal) $= 10^{-3}$ m/s/s (metres per second per second)

force
1 newton (N) = 1 kg m/s^2 = 0.225 pounds-force (lbf)
1 dyne (dyn) = 10 micronewtons

energy
1 joule (J) = 1 newton metre = 0.239 calorie
1 kilowatt hour (kWh) $= 3.6 \times 10^6$ joules

power
1 watt (W) = 1 joule per second
1 kilowatt (kW) $= 10^3$ watts = 1.341 horsepower

heatflow
1 heatflow unit (hfu) = 0.042 W/m^2 = 1 $\mu cal/cm^2/s$

pressure or stress
1 pascal (Pa) = 1 N/m^2
= 0.000145 lbf/in^2
1 kilobar (kb) $= 10^8$ N/m^2
$= 1.45 \times 10^4$ lbf/in^2
1 millibar (mb) = 100 N/m^2
1 atmosphere $= 1.01325 \times 10^5$ Pa
= 1013.25 millibars
= 14.69 lbf/in^2

viscosity
1 poise = 0.1 Ns/m^2

frequency
1 hertz (Hz) = 1 cycle/s

angular measure
1 arc second = 1/3600 degree
1 radian (rad) = 180/π degrees

magnetic induction
1 tesla (T) = 10 000 gauss
1 gamma $= 10^{-9}$ T

Chemical elements

Ac	actinium	N	nitrogen
Ag	silver	Na	sodium
Al	aluminium	Nb	niobium
Ar	argon	Nd	neodymium
As	arsenic	Ne	neon
At	astatine	Ni	nickel
Au	gold	O	oxygen
B	boron	Os	osmium
Ba	barium	P	phosphorus
Be	beryllium	Pa	protactinium
Bi	bismuth	Pb	lead
Br	bromine	Pd	palladium
C	carbon	Pm	promethium
Ca	calcium	Po	polonium
Cd	cadmium	Pr	praseodymium
Ce	cerium	Pt	platinum
Cl	chlorine	Ra	radium
Co	cobalt	Rb	rubidium
Cr	chromium	Re	rhenium
Cs	caesium	Rh	rhodium
Cu	copper	Rn	radon
Dy	dysprosium	Ru	ruthenium
Er	erbium	S	sulphur
Eu	europium	Sb	antimony
F	fluorine	Sc	scandium
Fe	iron	Se	selenium
Fr	francium	Si	silicon
Ga	gallium	Sm	samarium
Gd	gadolinium	Sn	tin
Ge	germanium	Sr	strontium
H	hydrogen	Ta	tantalum
He	helium	Tb	terbium
Hf	hafnium	Tc	technetium
Hg	mercury	Te	tellurium
Ho	holmium	Th	thorium
I	iodine	Ti	titanium
In	indium	Tl	thallium
Ir	iridium	Tm	thulium
K	potassium	U	uranium
Kr	krypton	V	vanadium
La	lanthanum	W	tungsten
Li	lithium	Xe	xeron
Lu	lutetium	Y	yttrium
Mg	magnesium	Yb	ytterbium
Mn	manganese	Zn	zinc
Mo	molybdenum	Zr	zirconium

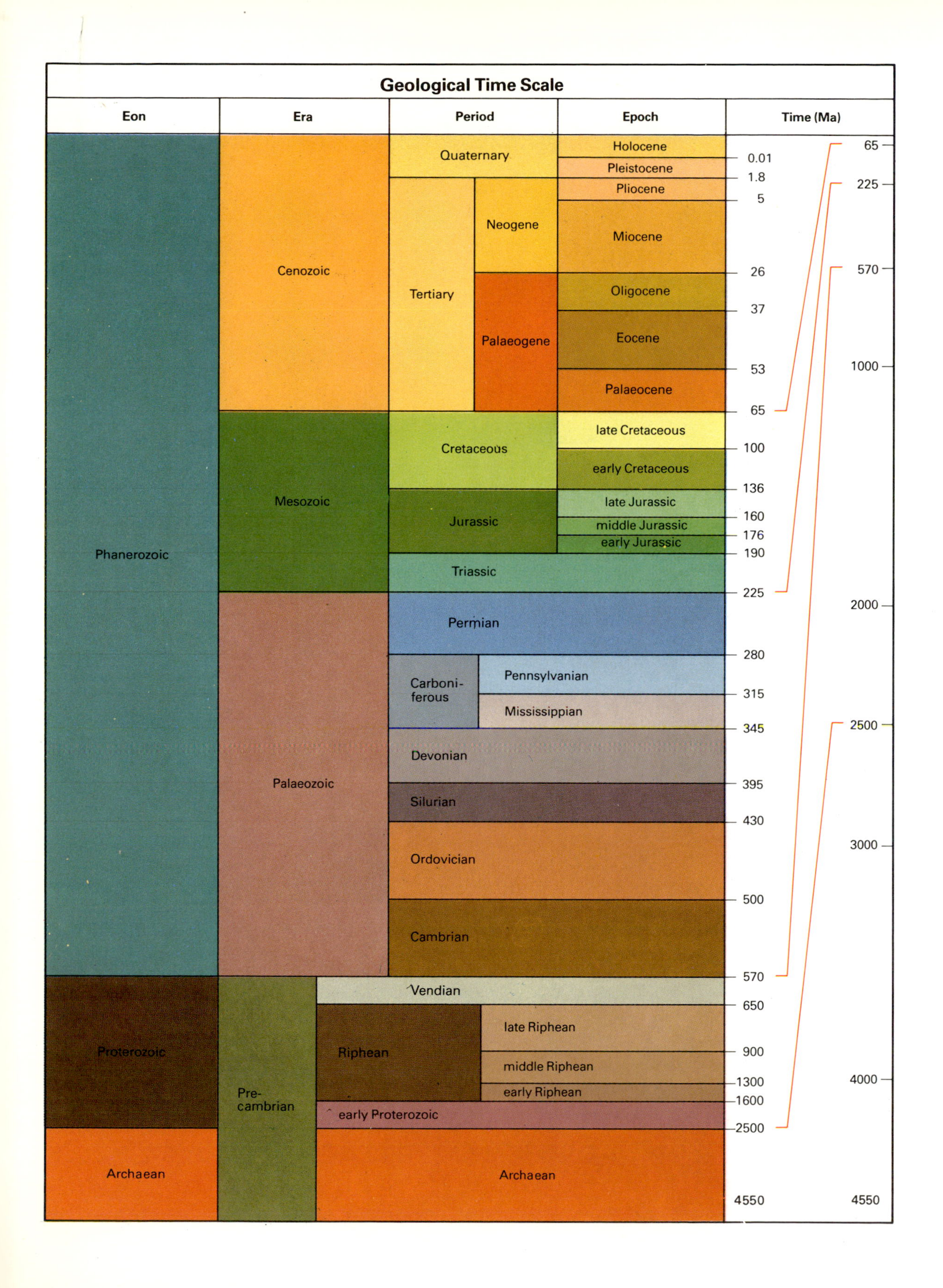

Geological Time Scale
Eon
Era
Period
Epoch
Time (Ma)
Phanerozoic
Proterozoic
Archaean
Cenozoic
Mesozoic
Palaeozoic
Pre-cambrian
Quaternary
Tertiary
Neogene
Palaeogene
Cretaceous
Jurassic
Triassic
Permian
Carboni-ferous
Pennsylvanian
Mississippian
Devonian
Silurian
Ordovician
Cambrian
Vendian
Riphean
late Riphean
middle Riphean
early Riphean
early Proterozoic
Archaean
Holocene
Pleistocene
Pliocene
Miocene
Oligocene
Eocene
Palaeocene
late Cretaceous
early Cretaceous
late Jurassic
middle Jurassic
early Jurassic
0.01
1.8
5
26
37
53
65
100
136
160
176
190
225
280
315
345
395
430
500
570
650
900
1300
1600
2500
4550
65
225
570
1000
2000
2500
3000
4000
4550

SCENO
SYSTEMATIS
PTOLE:
GRAPHIA
MVNDANI
MAICI.
SEPTENTRIO
MERIDIES
OCCIDENS
ORIENS
TROPICVS ÆSTIVVS SEV TROPICVS CANCRI
TROPICVS HYBERNVS SIVE TROPICVS CAPRICORNI
CIRCVLVS ARCTICVS
CIRCVLVS ANTARCTICVS
POLVS ARCTICVS
POLVS ANTARCTICVS
SATVRNVS
IVPITER
SOL
MARS
LVNA
VENVS
MERCVRIVS

Part One: The Earth Sciences in Perspective

A feature of all scientific endeavour is that the better we come to understand our subject matter, the less significant becomes our own small world and our place in it. It is therefore fitting to start with reminders both of the place of present-day Earth science in the history of the interpretation of geological phenomena, and of the place of the Earth in the unimaginably wider context of our Galaxy and of the universe as a whole. Geology had scarcely been born as a science before its practitioners began to delve into its history, but it is relatively recently that this interest has developed into a science of its own. The aim here is to view developments in understanding the Earth in the historical context of the general state of human knowledge and of the contemporary social and even political climate. We can see the revolutionary acceptance of continental drift as only one in a succession of upheavals, starting with the overthrow of the geocentric concept of the universe. Will future generations ridicule twentieth-century reluctance to accept the mobility of the Earth's crust in the way that we have tended to pour scorn on the dependence of eighteenth-century geologists on the biblical Flood?

An eminent scientist claimed earlier this century that there are but two areas of scientific research, comprising between them the study of everything on the Earth (geology) and everything outside it (astronomy). The boundaries between these two have been substantially eroded, particularly since the coming of the era of space exploration. At the end of the book we return to the direct application of Earth science and its methods to the study of the other bodies of the Solar System, but in this section an astronomer tells us about the nature and history of the universe, leading to the formation of the planet that forms the subject of the bulk of this book.

The Ptolemaic system of the universe, from Atlas Coelestis, Harmonia Macrocosmica, *by Andreas Cellarius, Amsterdam, 1660.*

1 The history of the Earth sciences

Early theories of the Earth

Mankind has long taken interest in the nature and origin of the rocks forming the Earth's surface. Among the ancient Greeks, Herodotus (*c* 485–425 BC) recorded the presence of marine shells far inland in Egypt and concluded that they had been left there by the retreating waters of some earlier sea. Pythagoras (sixth century BC) is said to have taken the presence of fossil marine creatures high in the mountains of Greece as evidence of the elevation of a former sea bed, and Strabo (*c* 63 BC–AD 23) certainly accepted such a view. As the Mediterranean lies in one of the world's volcanic and seismic regions many Greek and Roman authors were fascinated by the phenomena of volcanoes and earthquakes. Strabo, in common with many later writers, saw volcanoes as natural safety valves designed to permit the escape of dangerous terrestrial vapours, and the Roman, Seneca (*c* AD 3–65), regarded earthquakes as the result of those same vapours becoming turbulent within the Earth's supposedly cavernous interior. In China, in 132, Chang Hêng (78–139) invented an 'earthquake weathercock', the first seismograph ever constructed. Rather later, Avicenna (980–1037), the father of the Earth sciences in the Arab world, discussed not only earthquakes but a wide variety of other geological phenomena. He also displayed a sound understanding of many Earth processes at a time when, in western Europe, Christian scholars were more interested in saints and seraphim than in sandstones and sapphires. Some of the ideas then current in western Europe must now strike us as bizarre. Gems, for instance, were commonly believed to form as a result of some celestial energy penetrating into the Earth, the energy impinging less obliquely upon the tropics than upon higher latitudes with a result that precious stones were thought to be most abundant in the Earth's warmest climes. Yet, despite such fanciful early notions, it was in western Europe rather than in the Middle or Far East that, in the centuries following 1500, the modern Earth sciences originated.

To some extent Europeans were drawn to the Earth sciences in the aftermath of the Renaissance because of the obvious practical applications of those sciences during an age in which the demand for minerals increased. Not least, the replacement of the medieval feudal economy by capitalism resulted in a pressing need for various precious metals for coinage purposes. *De re metallica* by Georgius Agricola (1494–1555), published in Basle in 1556, offers among much else a detailed discussion of the field occurrence of mineral veins and is a superb early example of applied geology (Figure 1.1). Important though such practical matters undoubtedly were, it was theoretical questions

1.1: *One of the many illustrations by Georgius Agricola depicting mineral veins: a 'principal vein' (A), a 'transverse vein' (B) and a 'vein cutting principal one obliquely' (C). From Agricola's* De re metallica *(Basle, 1556).*

that mostly engaged the attention of scholars interested in the Earth sciences. Here one problem above all others commanded their attention: what was the origin of the 'figured stones', or fossils, found so widely in the Earth's rocks? Some scholars dismissed fossils as meaningless sports of nature. Others saw them as growths formed within rocks as a result of the Earth's *vis plastica* (moulding force). Yet others, such as Leonardo da Vinci (1452–1519), correctly interpreted fossils as the remains of former marine creatures which had become entombed within the rocks forming on some ancient sea bed, the strata and their fossils then being elevated into their present terrestrial environments (Figure 1.2).

During the seventeenth century it was accepted increasingly that fossils could only be organic in origin and there developed the parallel notion that the necessary exchange of ancient sea bed into modern dry land must have occurred during that universal catastrophe described in the Bible—the Noachian Deluge. In the Christian world it was generally accepted that the Bible offered a divinely inspired account of the Earth's early history, and as a result the Flood became a favourite device with most students of the Earth sciences. Even down to the early years of the nineteenth century the Flood was still being invoked by thoroughly reputable scientists in explanation of phenomena ranging from the drift mantle of the continents to the configuration of the north Atlantic. Belief in the Flood was by no means the only scripturally based idea to influence the Earth sciences; there was the even more important belief in the Earth's recent origin. The various genealogical tables of the Old Testament were studied by chronologers in order to arrive at a date for the Creation. The most renowned of these chronologers was the Irish prelate James Ussher (1581–1656), who in 1650 published his conclusion that the Creation had taken place in the year 4004 BC, with the Flood following in the year 2349 BC. Such dates obtained wide currency, but they posed a problem. How could so recent a date for the origin of the Earth be reconciled with the increasing evidence, derived from field observation, that since the time of its formation the Earth's surface had been the scene of many vicissitudes. How, for instance, in the short period that seemed to be available, could great thicknesses of sedimentary strata accumulate, mountain ranges upheave, and valleys be excavated? Clearly, it seemed, reconciliation could be effected only by supposing all those many changes to have been the result of almost instantaneous diluvial, seismic and volcanic catastrophes. In this manner originated the interpretation of the history of the Earth as having consisted of periods of quiescence interrupted by violent cataclysms—an interpretation that came to be known as Catastrophism.

During the seventeenth century scholars began to formulate what were then termed 'theories of the Earth'. These were attempts to devise convincing and fairly detailed histories of the globe; basically their authors sought to employ science in the amplification of the somewhat bald Genesis account of the Earth's origin and early development. The theorists all made extensive use of catastrophic events, but while their efforts now seem naive they do deserve our attention as pioneer essays in historical geology. The French philosopher René Descartes (1596–1650) was an early influence in this new genre, but geologically the most important of the theories was that published in Florence in 1669 by Nicolaus Steno (1638–86), a Dane resident in Tuscany. Steno illustrated his theory by means of a series of diagrams (Figure 1.3) representing successive stages in the evolution of the Earth. He displays an understanding of many important geological concepts. He accepts the organic origin of fossils; he perceives the true nature of sedimentary strata; he appreciates the work of a marine transgression; he implicitly formulates the law of order of superposition (in any normal sequence the upper strata are younger than the lower); and he illustrates what we now know as a geological unconformity. His theory thus stands as one of the great classics of geological literature.

1.2: *Early representations of fossils. From Robert Plot's* The Natural History of Oxfordshire, Being an Essay towards the Natural History of England *(second edition, Oxford, 1705).*

The Neptunian and Plutonic theories

The theories devised by the seventeenth-century cosmogonists continued to excite much attention during the following century, but slowly it came to be appreciated that such theories were nothing more than imaginative exercises. It began to be seen that any true history of the Earth's surface must be founded upon the painstaking field investigation of the actual rocks themselves. It was during the eighteenth century that the first such detailed local studies began to be made. In 1756 Johann Gottlob Lehmann (1700–67) published a careful study of the rocks of the Harz Mountains and the Erzgebirge; at about the same time Giovanni Arduino (1713–95) divided the rocks of northern Italy into the categories of Primitive (schists), Secondary (limestones and other fossiliferous stratified rocks), Tertiary (less well-consolidated fossiliferous strata) and Volcanic (lavas and tuffs); Torbern Olof Bergman (1735–84) explored the stratigraphy of southern Sweden; George Christian Füchsel (1722–73) investigated the rocks of Thuringia; Nicolas Desmarest (1725–1815) explored the extinct volcanoes of central France; and Sir William Hamilton (1730–1803) devoted many years to a study of Vesuvius in all its moods.

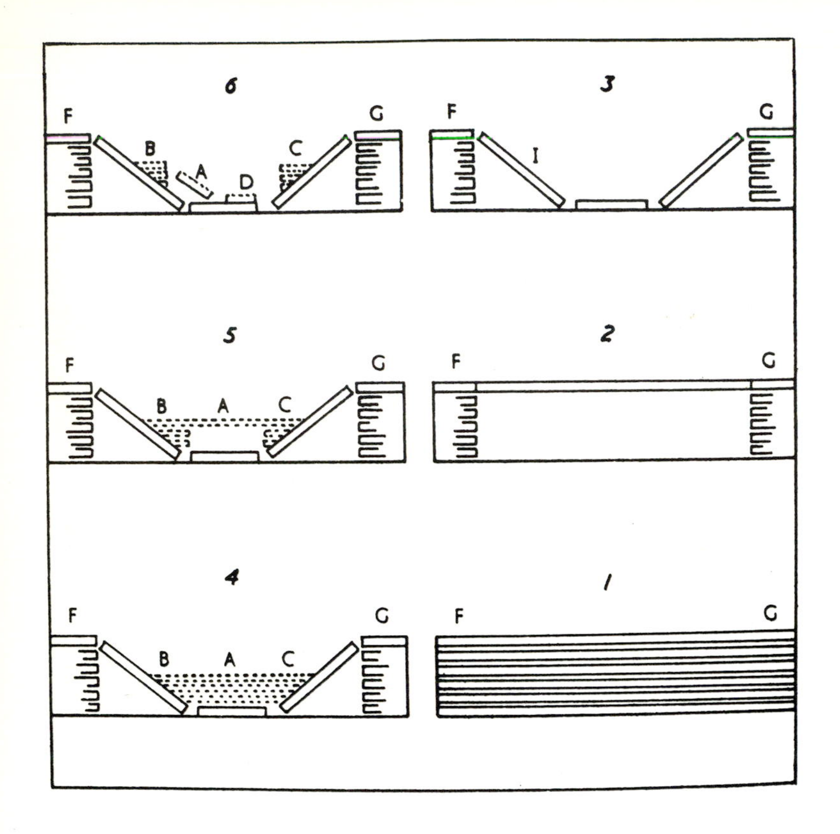

1.3: *Nicolaus Steno's interpretation of the history of the Tuscan landscapes. Redrawn from* Nicolai Stenonis de solido intra solidum naturaliter contento dissertationis prodromus *(Florence, 1669).* 1 *During the Creation the globe is submerged beneath a universal fluid from which precipitation takes place to form horizontal, superimposed beds of sedimentary rock (F–G). In the absence of continents, plants and animals the individual strata are uniform in composition and free from fossils and all other heterogeneous material.* 2 *The newly formed strata have emerged from the fluid to become the Earth's pristine continents. Beneath the continents water is sapping the continental foundations, eating away the strata, and opening up great caverns.* 3 *Large areas of the continents have now been undermined, and the roofs of the great caverns collapse (I). The resulting valleys are drowned by the sea and this explains the Flood.* 4 *In the Flood's waters new sedimentary strata are formed (B–A–C). These younger rocks incorporate plant and animal remains from the antediluvial world and are therefore fossiliferous.* 5 *These new rocks emerge at the close of the Flood and are themselves undermined as a second generation of caverns develops.* 6 *A period of renewed crustal collapse reduces the youngest strata to their present disordered state (A–D) and forms the Earth's modern topography.*

Concurrently with these detailed field investigations attempts were made to incorporate the fresh discoveries into a new generation of more securely grounded theories of the Earth. One of those theories proved to be very influential in its own day—the theory devised by Abraham Gottlob Werner (1749–1817), who was Professor of Mining and Mineralogy in the Saxon Mining Academy at Freiberg. Werner believed that the Earth originally consisted of a primeval nucleus with an uneven surface completely enveloped in the waters of a universal liquid. The oldest of the Earth's rocks, he claimed, were formed in this liquid as chemical precipitates, their bedding either reflecting the configuration of the nucleus itself or else being a result of the slumping of the deposits while they were still in a plastic condition. Most bedding structures were thus regarded as original to the stratum in question, and since there were as yet neither living organisms nor continents, all these ancient rocks were held to be devoid of both fossils and clastic material. Such ancient deposits Werner termed 'Primitive rocks' (*Urgebirge*), and they included rocks such as granite, gneiss, schist and quartzite. Next, as a result of a gradual lowering in the level of the fluid, the Earth's continents began to emerge and were soon colonized by the first plants and animals. Some of the rocks formed in the liquid at this time thus contain both fossils and clastic debris, but the continents were still small in extent and most of the rocks dating from this stage are therefore still chemical precipitates. The rocks belonging to this stage, Werner claimed, included greywackes, slates and some limestone and, since they were supposed to be partly chemical in origin and partly clastic, he termed them 'Transition rocks' (*Übergangs-gebirge*). His final major class of rocks was the Floetz (*Flötz-gebirge*). By the time of their deposition the level of the fluid had subsided still further and the continents had become much more extensive. Thus the Floetz rocks—rocks such as limestone, mudstone and sandstone—are almost entirely clastic in origin and replete with fossils, but, since the fluid now occupied a much diminished area, the Floetz rocks themselves are much more restricted in extent than the earlier Primitive and Transition rocks. The final Floetz rocks consisted of deposits such as alluvium, shingle and tufa, and after their deposition the fluid sank to its present level to become the waters of the world's modern oceans. According to the theory all veins and dykes were fissures which had been filled by percolating solutions seeping down from the fluid above. The fluid played so important a role in this entire system that the theory came to be known as the Neptunian theory.

The theory clearly had its deficiencies. How, for instance, could the fluid once have held virtually all of the Earth's rocks in solution? What brought about temporal variation in the type of material being precipitated? Above all, what caused the fluid gradually to disappear? Contemporary critics of the theory were not slow to raise such problems, and later historians have tended to ridicule the entire Neptunian system. The theory is nevertheless important for three reasons. First, it was a pioneer attempt to establish a universal stratigraphical sequence, because the various formations within the Primitive, Transition and Floetz categories were supposed to occur in the same order the world over. Second, by pretending to offer a universally valid key to stratigraphy, the theory encouraged its adherents to make detailed studies of the rocks in localities throughout the world in an effort to ascertain the position of those rocks within the Neptunian hierarchy. Finally, the theory stimulated a major controversy over the nature of basalt, and efforts to resolve that issue led to a quickening of interest in the Earth sciences generally. That controversy itself merits more than just a mere mention. According to the Neptunian theory basalt was a chemical precipitate formed in the universal liquid, but many of Werner's con-

1.4: *Columnar basalts at the Giant's Causeway, County Antrim, Ireland. A painting executed by Susanna Drury in the 1730s.*

temporaries had already accepted Desmarest's view that basalt was an igneous rock extruded from ancient volcanoes. One of the sites that became crucial in the debate was the finely displayed basalt at the Giant's Causeway in northern Ireland (Figure 1.4). When the Vulcanists, as Desmarest's supporters came to be called, urged that basalt, such as that at the Causeway, was identical in character to some of the material extruded from modern volcanoes, the Neptunists responded that in these cases normal 'sedimentary' basalt lying at depth had been melted by the subterranean fires associated with the volcano and then extruded in the molten state before finally cooling and regaining its original 'sedimentary' form. Not until the early years of the nineteenth century did the Vulcanists finally emerge victorious.

Werner's theory, like all its predecessors, offered a linear view of Earth history. It envisaged both a clear beginning to Earth history, with the commencement of precipitation from the universal liquid, and an equally clear end, with the conversion of the final remnants of the liquid into the world's modern oceans. Since the fluid was now gone, there clearly could be no further precipitation of rocks, but behind this fact there lurked a major problem. The continents are clearly undergoing steady reduction under the ceaseless attack of the forces of denudation, and obviously the continents are doomed to eventual destruction. This was an unpalatable fact in an age when an anthropocentric view of nature predominated. If the continents were destroyed, then where would man live? Strangely, it was an attempt to reconcile the anthropocentric view of nature with a belief in denudation—what has been termed the denudation dilemma—that stimulated the formulation of another theory of the Earth—a theory that gave modern geology its dynamic framework. The author of that theory was James Hutton (1726–97) of Edinburgh, where a modern plaque upon his grave now hails him as 'The Founder of Modern Geology'.

Hutton first propounded his theory publicly in 1785, although it was not published in its final form until 1795. Unlike previous theories of the Earth, Hutton's theory was in no way concerned with the origin of the Earth. His theory is merely an account of the operation of the Earth-machine and not a discussion of its original construction. Fundamental to the theory was Hutton's conception of geological time. He dismissed the scriptural time scale of most earlier theorists and instead argued that the Earth was immensely old. He concluded his essay of 1785 with the now famous phrase that to Earth history 'we find no vestige of a beginning,—no prospect of an end'. Hutton fully accepted the reality of denudation. He regarded all the continental topography as the product of rain and river action, and he admitted that incessant and prolonged denudation could only result in the eventual destruction of the present continents. But he resolved the denudation dilemma by noting that the Earth machine possessed a built-in mechanism for self-repair. The debris worn from the continents is transported by rivers into the sea and there, he observed, the debris accumulates upon the sea bed layer by layer. Gradually, he argued, this material is compacted and lithified under the influence of the terrestrial heat, thus bringing new sedimentary strata into being. Eventually, as the old continents become dank lowlands following long-continued denudation, the new rocks are elevated from the sea bed to form the fresh generation of continents necessary to ensure a continuity of human habitat. This uplift he attributed to the expansive power of the terrestrial heat, and he believed that during the uplift the rocks had acquired most of those contortions so widely evident in the continental strata today.

In contrast to the linear Neptunian theory, Hutton's theory was cyclical: the sequence of denudation, transportation, sedimentation, lithification, uplift and renewed denudation can be repeated for as long as the Earth exists. It is a theory that, in its broad outline, has proved to be valid. As the geostrophic cycle, it forms the basis of present-day geology. But Hutton did more than provide geology with its conceptual framework; inherent in his theory are three other ideas of great importance. First, there are his views on granite. The Neptunists had claimed granite as a 'Primitive rock', part of the Earth's primeval surface, but in the Huttonian system, with its conception of continent succeeding continent in endless succession, there was no place for primitive rocks dating back to the Earth's origins. Hutton therefore claimed that granite was an intrusive rock consolidated from an igneous melt and that it was thus younger, not older, than the adjacent strata. Likewise, he claimed that all dykes and mineral veins had been formed by molten material rising from the Earth's hot interior. Indeed, terrestrial heat played so important a part in Hutton's theory that it soon came to be known as the Plutonic theory. Second, he believed that there should exist what he termed 'compound masses'—places where the rocks of one continent had been later submerged, coated with fresh strata and then re-elevated to form a new landmass. In such cases, Hutton argued, the older of the two rock sequences must have been transformed as a result of its being subjected to the influence of the terrestrial heat not once but twice. His theory thus contained the germ of the idea of rock metamorphism. The identification of actual examples of such 'compound masses', containing what today would be termed an unconformity, was for Hutton a matter of importance as offering significant confirmation of his theory. It was with much satisfaction that he identified three such sites in Scotland during 1787–8 (Figure 1.6). Finally, having set his theory against a background of limitless time, Hutton had no need to invoke sudden catastrophes in explanation of natural phenomena. In his eyes all the major transformations of the Earth's surface, from the excavation of valleys to the elevation of continents, were a cumulative response to innumerable trivial changes occurring through an eon of time. He held that the processes that had acted in the past differed neither in kind nor degree from those processes active in the world today. Hutton thus replaced the catastrophism of his contemporaries by a new philosophy of nature, which was shortly to be entitled Uniformitarianism.

1.5: *James Hutton '. . . rather astonished at the shapes which his favourite rocks have suddenly taken'. From* A Series of Original Portraits and Caricature Etchings, by the late John Kay *(Edinburgh, 1838).*

The development of geological cartography and stratigraphy

Several decades elapsed before Hutton's theory earned general acceptance, but these were the decades during which geology began to be recognized as a fully fledged and independent subject. It was around 1800 that the science came to be known by its modern name of geology and that it began to be institutionalized. The world's first geological society—the Geological Society of London—was founded in 1807, and journals devoted to the publication of research in the discipline came into being. For example, the *Transactions of the Geological Society of London* began publication in 1811 and *The American Journal of Science* in 1818. Geology was added to the curriculum of many European and American universities, and the first generation of geological textbooks appeared—a generation that included such classics as Robert Bakewell's *An Introduction to Geology* (London 1813), Parker Cleaveland's *An Elementary Treatise on Mineralogy and Geology* (Boston 1816) and

1.6: *James Hutton's unconformity at Siccar Point, Berwickshire, Scotland. Upper Old Red Sandstone (late Devonian), dipping gently, overlies vertical slaty mudstones and greywacke sandstones of Silurian age.*

Carl Friedrich Naumann's *Lehrbuch der Geognosie* (Leipzig 1850–4). Nowhere was the new status of geology more evident than in the British Isles where something approaching a geological mania swept through the country between 1800 and 1860. By the latter date geological societies had been founded in Britain not only in London, but in Cornwall, Dublin, Edinburgh, Yorkshire, Manchester, Dudley, Glasgow and Liverpool. Collectors were prepared to pay high prices for good geological specimens, while some geological books were among the literary best sellers of the period. To some extent the popularity of the science in Britain was related to the Industrial Revolution, to which geology was of fundamental importance. The advent of steam-powered industry created a vigorous demand for coal and iron. Large-scale urban expansion resulted in enormously increased requirements for building stones, brick clays and glass sands. The swollen populations of the new towns and cities needed water supplies and, eventually, huge new cemeteries located upon suitable geological formations. The laying out of the new lines of communication—canals, roads and then railways—created an urgent demand for geological information. Their construction laid bare, in cuttings and tunnels, immense geological sections such as had never before been seen. During the course of the nineteenth century, industrialization occurred in the remainder of Europe and in North America.

The recognition that geology was a key to industrialization, and thus to national prosperity, encouraged many governments to finance the geological surveys necessary to provide inventories of local geological resources. The earliest of all national geological surveys was planned in France in 1822 by the Corps Royal des Mines. Field work started in 1825 and led to the publication in 1841 of a *Carte géologique de la France* in six sheets at a scale of 1:500 000. In the British Isles the Geological Survey of England and Wales was established in 1835, under Sir Henry Thomas De La Beche (1796–1855), although ten years earlier the Ordnance Survey had embarked upon an abortive geological survey of Ireland. Across the Atlantic, in 1818, Governor DeWitt Clinton invited Amos Eaton (1776–1842) to give a course of lectures on chemistry and geology to the members of the New York State legislature, and the first actual state survey to be undertaken at public expense was that conducted by Denison Olmsted (1791–1859) in North Carolina in 1823. Within only sixteen years similar surveys had been undertaken in fifteen other states of the Union. Later, the Federal government itself financed geological surveying expeditions, such as those of Ferdinand Vandeveer Hayden (1829–87), but it was not until 1879 that the United States Geological Survey was finally established. By 1879 many other national geological surveys had been founded, including the Geological Survey of Canada (1841), the survey of India (1850) and the survey of Sweden (1858).

For all these surveys the production of geological maps was a matter of prime importance. Indeed, geologists the world over now accepted that careful field mapping of the Earth's rocks was the most fundamental task of their science. Fossils were collected and used to identify and date the various formations; the formational limits were traced out in the field and plotted upon topographical base maps; and finally the task was completed by adding colour washes to the map to differentiate the various rock types. This was the great age of both stratigraphy and geological cartography, and the pioneers in these two areas have always been held in high esteem by the geological fraternity. The most renowned of these pioneers was the English land surveyor and canal engineer William Smith (1769–1839), who has long been known as 'the father of English geology'. In the 1790s Smith discovered that the various strata around Bath could be distinguished by their fossil assemblages, and he soon began to use his discovery to assist him in the construction of a geological map of the whole of England and Wales. After some delay, the map was finally published in

1815, at a scale of 1:316 800, and today it is one of the most highly prized classics of the science. In France similar stratigraphical studies were undertaken in the Paris Basin by Georges Cuvier (1769–1832) and Alexandre Brongniart (1770–1847), their memoir and accompanying map being published in 1811. In the USA the pioneer of geological cartography was William Maclure (1767–1840) whose map depicting the eastern USA was first published in 1809 (Figure 1.9).

The work of the stratigraphers led gradually to the emergence of the stratigraphical column as known to modern geology. Naturally it was the sequence of the younger rocks that was deciphered first; those rocks contain most fossils, are least disturbed and, since they rarely form rugged terrain, are freely accessible. By the 1830s a great deal was known about the Upper Palaeozoic and younger rocks, but little sense had as yet been made of the Lower Palaeozoic and older rocks. They were still just thrown together into an omnibus group styled 'greywackes'. The problem of the greywackes was first tackled seriously by two British geologists: Sir Roderick Impey Murchison (1792–1871) and Adam Sedgwick (1785–1873). They commenced operations in 1831, with Murchison working down the geological succession on the borders of England and Wales, while Sedgwick tried to work up the succession in North Wales. In 1835 Murchison proudly consigned his rocks to what he called 'the Silurian system' while Sedgwick, with equal pride, placed his rocks into what he designated 'the Cambrian system'. Further field study revealed that the lower members of Murchison's Silurian system were identical with the highest components of Sedgwick's Cambrian system. Neither geologist was willing to surrender a part of his system to the other, and there ensued between the two men one of the most bitter of all geological controversies. Not until 1879, after the deaths of the two chief protagonists, was a solution found as a result of a suggestion by Charles Lapworth (1842–1920) that the lower Silurian and the upper Cambrian should both be removed from their respective systems to form a new system, the Ordovician. Lapworth's proposal completed the basic stratigraphical column as it is now known.

1.7: *The stratigraphical column.*

Period	Author	Date
Recent	Charles Lyell	1833
Pleistocene	Charles Lyell	1839
Pliocene	Charles Lyell	1833
Miocene	Charles Lyell	1833
Oligocene	Heinrich Ernst Beyrich	1854
Eocene	Charles Lyell	1833
Palaeocene	Wilhelm Philipp Schimper	1874
Cretaceous	Jean Baptiste Julien d'Omalius d'Halloy	1822
Jurassic	Alexander von Humboldt	1799
Triassic	Friedrich August von Alberti	1834
Permian	Roderick Impey Murchison	1841
Carboniferous	William Daniel Conybeare and William Phillips	1822
Devonian	Roderick Impey Murchison and Adam Sedgwick	1839
Silurian	Roderick Impey Murchison	1835
Ordovician	Charles Lapworth	1879
Cambrian	Adam Sedgwick	1835

1.8: *Collecting fossils in Cherry Hinton chalk pit, Cambridgeshire. From the Rev. Isaac Taylor's* Scenes in England *(London, 1822). This locality, to which Professor Adam Sedgwick used to take his students on horseback, is still visited by students of geology.*

1.9: *William Maclure's geological map of the USA first published in 1809. This is the map as it appeared in* Journal de physique, de chimie, d'histoire naturelle et des arts, *Volume 72 (Paris, 1811).*

Time in geology

Most geologists of the first half of the nineteenth century accepted that the Earth must be very old. As far back as 1779 Georges Louis Leclerc, Comte de Buffon (1707–88), had argued that 75 000 years must have elapsed since the Creation; in Russia Mikhail Vasilievich Lomonosov (1711–65) had proposed that the Earth must be hundreds of thousands of years old; in 1785 Hutton had hoped for time unlimited; and by the 1830s geologists were thinking of the age of the Earth in terms of millions of years. To some extent this development was a result of diminished respect for the Scriptures and their chronology, but far more important in bringing about the change were two discoveries arising from stratigraphical studies. First, such studies had revealed the enormous total thickness of the world's sedimentary pile. In *On the Origin of Species* (1859), for instance, Charles Darwin (1809–82) quoted $13\frac{3}{4}$ miles (22 km) as the cumulative total of the British stratigraphical column. Clearly the formation of so great a thickness of sedimentary strata must have occupied an enormous length of time. Second, within such sedimentary rocks an amazing variety of fossils, many of them representing now extinct life forms, was discovered. Again, it seemed obvious that the development, existence and ultimate extinction of so many species must have occupied an eon of time. Thus from the 1830s onwards geologists really felt few temporal constraints upon their speculations, although in the middle decades of the century the physicist Lord Kelvin (1824–1907) did cause some consternation in the geological camp when he employed physical arguments in an effort to restrict the age of the Earth to less than fifty million years. His reasoning was based upon the supposed rate of the Earth's cooling from its presumed origin as a gaseous ball, but his logic was later undermined by the discovery that the release of radioactive energy was constantly augmenting the Earth's heat balance. This left the geologists free to return to their belief in an enormously ancient Earth, and today it is thought that the true age of the Earth must be around 4600 million years.

It was Sir Charles Lyell (1797–1875) who made the geological world fully aware of the possibilities inherent in the new, extended time scale. The first volume of his classic *Principles of Geology* was published in London in 1830, and achieved enormous popularity, reaching its twelfth English edition in 1875. It was Lyell's object to demonstrate the validity of Hutton's thesis that slow-acting natural processes, such as those operating around us today, are entirely adequate to explain all geological phenomena, provided sufficient time is allowed for their operation. There was no need to invoke catastrophic events, no need even to suppose that present processes had operated with greater intensity in former years. In short, Lyell taught that the present is the key to the past; he became the foremost advocate of the uniformitarian philosophy. One site held particular significance for Lyell and was regularly illustrated in the *Principles*: the so-called Temple of Jupiter Serapis lying near Pozzuoli on the Bay of Naples (Figure 1.10). The temple is known to have stood above sea level in the second century AD, but the presence in the marble columns of marine bivalve borings up to a height of 7 m above high-water mark shows that at some time since the second century the temple has been submerged to at least that depth. It then rose again so that in Lyell's day the platform of the temple stood 300 mm below high-water mark. All this, Lyell observed, had happened as a result of Earth movements so gentle that some of the temple's slender columns had survived without being toppled.

1.10: *The so-called Temple of Jupiter Serapis near Pozzuoli on the Bay of Naples. From Sir Charles Lyell's* Principles of Geology *(seventh edition, London, 1847).*

Nineteenth-century advances

The Huttonian theory had suggested that Earth history consisted of an endless succession of interchanges of land and sea. This sequence of submergences and emergences was essentially unravelled by the nineteenth-century stratigraphers through their studies of the Earth's strata, some formations being of terrestrial origin and others marine. But they were to discover that the story of the Earth was far more complex than Hutton could ever have supposed. In three areas—palaeontology, palaeoclimatology and tectonics—the unexpected complexities proved to be quite astonishing in their character.

Long before the nineteenth century it had been recognized that many fossils were representatives of now extinct species. It had even become clear that extinction was by no means a fate reserved exclusively for small creatures such as corals and molluscs; Pierre Simon Pallas (1741–1811), for instance, had in the 1770s described the remains of the extinct mammoth and woolly rhinoceros which are so widespread in the Russian Arctic. The geological world was nevertheless entirely unprepared for the discovery of the remains of the great Mesozoic reptiles—the dinosaurs. The first dinosaur fossils to be identified—they were parts of an iguanodon—were unearthed in the 1820s by Gideon Algernon Mantell (1790–1852) at Tilgate Forest in the Cretaceous rocks of the English Weald. Equally famous were the remains of the ichthyosaurs and plesiosaurs retrieved by Mary Anning (1799–1847) from the Liassic clays at Lyme Regis on the south coast of England. By the middle of the century a wide range of massive Mesozoic creatures was known; they immediately captured the public interest and this they have ever since retained (Figure 1.11). Many of the most remarkable nineteenth-century discoveries of extinct reptilian and other vertebrate life forms were made between the 1860s and the 1880s in the Mesozoic and Tertiary beds of the American West, ranging from New Mexico northwards into Wyoming. In that region a fierce rivalry developed between Edward Drinker Cope (1840–97) and Othniel Charles Marsh (1831–99), each man determined to disregard both natural hazards and the Indian menace in an effort to outdo the other in the unearthing of further and even stranger new species. It was these palaeontological discoveries, the outlandish and the commonplace, which provided the vital historical underpinning for that most famed of nineteenth-century scientific theories—Darwin's theory of the origin of species by means of natural selection.

Geology revealed that dramatic changes had occurred in the Earth's flora and fauna, and remarkable fluctuations in the Earth's climates were also discovered. In particular it became clear that recently there had been a glacial period during which large portions of the Earth's surface had been shrouded by ice sheets such as those that today survive only in Antarctica and Greenland. The glacial theory began to be developed during the 1820s in Switzerland following the realization that erratics, moraines, striations and other phenomena lying downstream of the snouts of the present glaciers could most conveniently be

1.11: *A giant iguanadon on show in the Brussels Museum. From* Le Nature *(Paris, 27 October 1883).*

explained by supposing those glaciers once to have been more extensive than they are today. In 1836 the idea was taken up by Louis John Rodolph Agassiz (1807–73) and in the following year he began to use the term the Ice Age (*die Eiszeit*) and to voice the somewhat exaggerated view that the whole of Europe as far south as the Mediterranean had recently been engulfed by ice flowing southwards from the North Pole (Figure 1.12). To many, this idea of a glacial 'deluge' seemed dangerously like a reversion to Catastrophism, particularly since there seemed to be no mechanism available to account for the supposed refrigeration. But gradually, during the 1840s and 1850s, it became increasingly clear that the ice of former glaciers did offer the best explanation for the vast spreads of sand and gravel mantling so much of the northern hemisphere. Indeed, it was soon perceived that the Pleistocene glaciation was but the most recent of a number of such events; it was in 1856 that William Thomas Blanford (1832–1905) ascribed a glacial origin to the Talchir conglomerates of India, thus taking a first step towards the recognition of the widespread Permo-Carboniferous glaciation of lands that were once located in the southern hemisphere.

The third of the three startling discoveries was that our continental rocks have been subjected to tectonic movements upon a gigantic scale. In some places vast slices of rock were found to have been forced over their neighbours for distances of tens of kilometres, while in other places detailed mapping of the rocks showed them to have been bent into huge recumbent folds or nappes. It was in 1841 that Arnold Escher von der Linth (1807–72) first described such structures in the Alps, and his work there was later developed by Albert Heim (1849–1937) (Figure 1.13). In the 1880s Lapworth identified similar structures in the north-west Highlands of Scotland where large-scale movement has occurred along the Moine thrust. Looking at the cross-sections of the region arising from his field studies, Lapworth remarked: 'Such sections are to a certain extent astounding, yet they do occur'. His amazement is understandable, but such overthrusts are now known to be a commonplace phenomenon.

The first half of the nineteenth century saw the Earth sciences take rapid strides forward; the second half of the century was a period of consolidation. One sign of the increasing maturity of the Earth sciences in the years after 1850 was the establishment of specialist societies such as the Mineralogical Society of Great Britain and Ireland (1876) or the Seismological Society of Japan (1880). Geologists were also feeling the need for closer international cooperation, and, following an American initiative, the International Union of Geological Sciences was founded. Its first meeting was held in Paris in 1878, since when the Union has reconvened every few years in cities the world over.

One of the most significant developments in the Earth sciences during the second half of the nineteenth century was the development of a new and vitally important technique: the examination of rocks in thin section using the petrological microscope. Its originator was William Nicol (*c* 1768–1851) of Edinburgh. Anxious to study the microstructures in fossil woods, in about 1830 he ground down some wood

1.12: *Glacially polished rocks and morainic debris at the edge of the Zermatt glacier. From Louis Agassiz's* Études sur les glaciers *(Neuchâtel, 1840).*

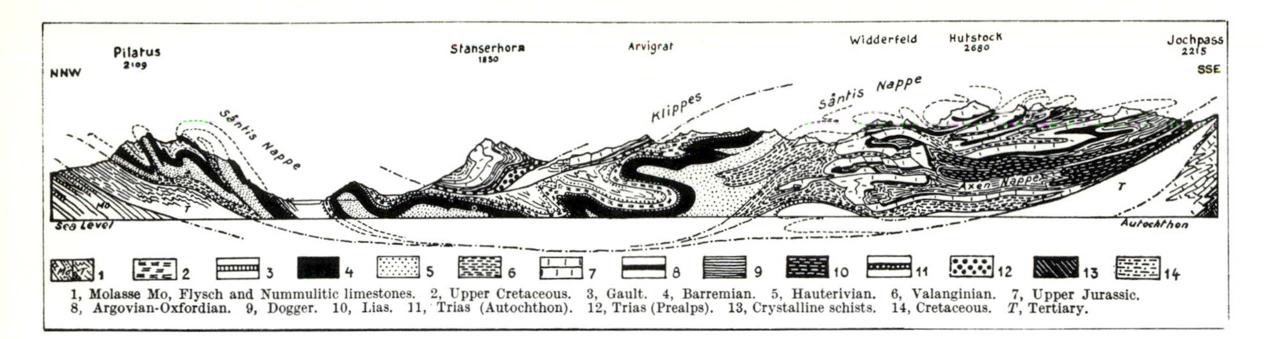

1.13: *A section across 'the High Calcareous Alps' of central Switzerland compiled by Albert Heim. From Léon William Collet's* The Structure of the Alps *(London, 1927).*

specimens until they were transparent enough to be studied beneath a microscope. Initially there was little interest in his idea, but in the 1850s Henry Clifton Sorby (1826–1908) learned of Nicol's technique and applied it to the production of thin sections of actual rocks. Again geologists were slow in appreciating the full significance of the work—some thought it absurd to study the structure of a mountain through a microscope—but in the 1870s the achievements of Nicol and Sorby at last made their impact when Ferdinand Zirkel (1838–1912) in Leipzig adopted the new technique enthusiastically. From that time onwards the rock in thin section and the petrological microscope have been two of geologists' most valued aids (Figure 1.14).

One other development of the second half of the nineteenth cencentury deserves mention because, although trivial in itself, it held enormous portents for the future. In 1859 at Titusville, Pennsylvania, the Drake well, the world's first oil well, was drilled. Until then geologists had thought about neither the nature of oil and natural gas nor the possibility of large-scale exploitation of the world's oil resources. This was soon to change, for from that small beginning in 1859 has grown the enormous modern oil- and gas-producing industries, in which geology and geologists have played a vital role.

The twentieth century

During the twentieth century the skills of the geologist have been in constant demand and the geological profession has in consequence expanded enormously. The technology basic to modern society lays strenuous claim upon all manner of materials locked into the Earth's crust, and the discovery of those materials has become one of the prime tasks of the Earth scientist. As a result of such explorations in all corners of the world, understanding of the Earth's structure has recently grown apace. The acquisition of this new knowledge of the Earth has been greatly assisted by the modern ability to view the Earth's surface from afar. First, aerial photography gave geologists fresh perspectives on the terrestrial configuration, and more recently satellite imagery has revealed patterns in the Earth's structure that would never have been discovered by ground examination. Another important development is the realization that some of the Earth's rocks contain built-in 'natural chronometers'. It was Lord Rutherford (1871–1937) who in 1905 suggested that the newly discovered radioactive properties of certain elements could be employed as a key to the measurement of

The 'London' Petrological Microscope.

"Dick" Model. No. 1321. Large Model.

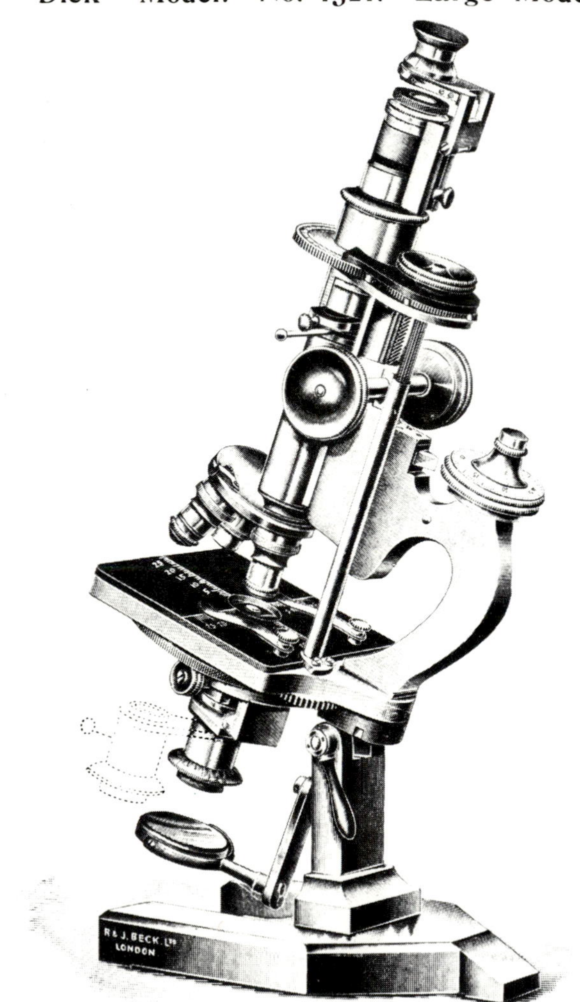

1.14: *An early petrological microscope. From a catalogue of scientific instruments issued by R. & J. Beck Ltd (London, c 1910).*

1.15: *Alfred Wegener's maps illustrating his conception of continental drift. The stippled areas are shallow seas. From Alfred Wegener's* The Origin of Continents and Oceans *(London, 1924).*

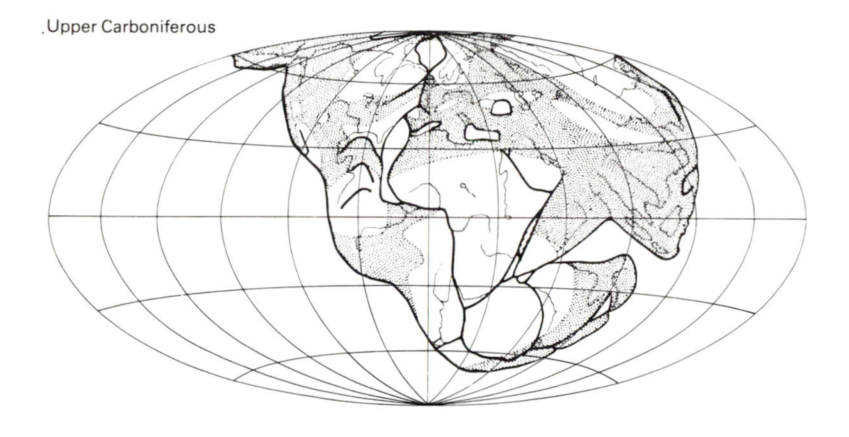

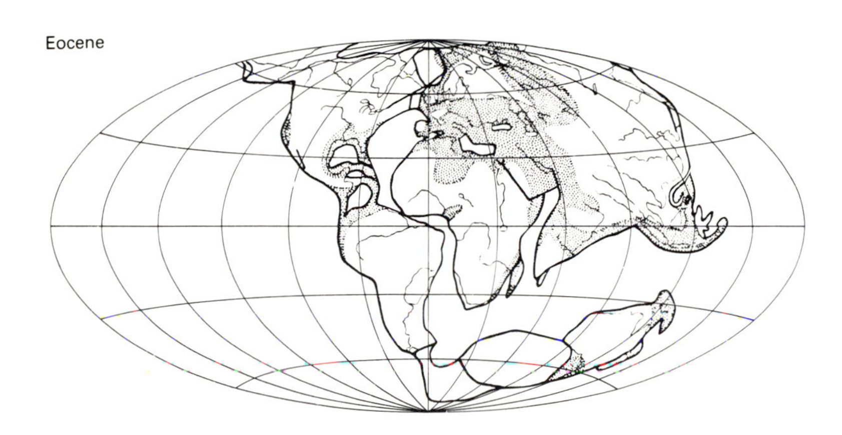

geological time in quantitative terms. The earliest efforts in this field of radiometric dating involved using the known decay rates of uranium and thorium into helium and lead, but today a variety of other methods is employed, allowing a remarkable precision in the dating of many of the Earth's rocks (see Chapter 9).

The twentieth century has certainly seen many exciting developments within the Earth sciences, but there is one that transcends all others in its importance: the advent of the plate tectonic theory. The theory's origins go back to the early years of the century. In 1912 Alfred Lothar Wegener (1880–1930) suggested that during the Palaeozoic all the Earth's present continents had been grouped together to form one large supercontinent which he termed Pangaea. During Mesozoic times, he claimed, Pangaea began to break up and the fragments drifted apart to become the world's present continents (Figure 1.15). In support of his theory of continental drift he offered a wide variety of evidence, but few geologists found his thesis convincing. The chief objection was that the geophysicists of the day regarded the Earth as a rigid body and there seemed to be no plausible mechanism that could result in the continents being dragged from their moorings and sent careering off across the face of the globe. Arthur Holmes (1890–1965) did suggest that an appropriate mechanism might be found in convection currents operating within the Earth's mantle, but little attention was paid to his ideas. Attitudes towards continental movement nevertheless began to change in the 1950s when studies in the palaeomagnetism of rocks first suggested that the continents had indeed shifted their positions. This discovery paved the way for the development in the 1960s of the modern concept of the Earth's surface as consisting of a series of mobile plates (see Chapter 10). This new and exciting conceptual model has reinvigorated the Earth sciences and encouraged fundamental re-appraisals of a type not witnessed in the Earth sciences since the early years of the last century. Strangely, modern plate tectonic theory might well have appealed to the cosmogonists of the late seventeenth century. The conception of mobile plates colliding to form mountain ranges, grinding past each other to trigger earthquakes and being subducted to bring oceanic trenches into being seems strangely reminiscent of those catastrophic theories of the Earth of three hundred years ago. But there is a major difference. Plate movement takes place with infinite slowness against the vast time scale of modern geology and not with a cataclysmic suddenness such as had to be invoked in those distant days when the whole of a complex Earth history had to be crammed into the few millennia permitted by the Old Testament chronology.

2 The Earth in space

The planet Earth is insignificant when compared to the universe at large. From the distance of the nearest stars it would be impossible, using techniques currently available to us, to see our planetary system. The planets comprise only 0.13 per cent of the mass of the solar system, and most of that planetary mass is due to the planet Jupiter. This chapter reviews the properties of the Earth from an astronomical standpoint, relating its history to that of the Sun, the stars and the far universe. This is of importance in understanding the history of the planet, because the geological processes that have occurred in the past 4600 million years (Ma) were largely predetermined by the initial mass, composition and location in space of the Earth.

To compare the Earth with other objects in the universe it is convenient to refer to Figure 2.1 which gives the dimensions and masses of a range of astronomical objects.

The most striking comparison is between the Earth and the universe: according to the astronomical evidence, the universe is about 10^{28} times more massive than the Earth, and their radii differ by twenty powers of ten. This leads to an enormous difference in their relative densities: the Earth has the density normally associated with solid rock, whereas the average density of the universe amounts to around one hydrogen atom per cubic metre. Furthermore the age of the Earth is only about one-quarter the age of the universe.

The universe

Observations of the distant galaxies indicate that the universe as a whole is expanding, with the remotest objects receding at speeds approaching the velocity of light. This suggests that the density of the universe is decreasing, and that it is still expanding from an earlier, very dense phase. In 1963 radio astronomers detected weak extra-terrestrial radiation uniformly distributed across the sky. This has been interpreted as a 'fossil relic' of the 'big bang', which marked the beginning of the universe as we know it. The background radiation is a last remnant of the fireball from which the universe has expanded. Today this radiation is so dilute, so thinly spread around the vast radius of the universe, that the average temperature far away from any star is a mere 2.7 degrees above absolute zero. In the first few seconds of the primeval fireball the subsequent history of the universe at large was determined: its expansion rate, composition and ability to form galaxies, stars and planets.

Within the first few minutes the universe had settled down to the chemical composition it still possesses: nearly 75 per cent hydroȩ and 25 per cent helium by mass. The degree of turbulence in this f maelstrom has carried over into the rotational motions of the galaxies, stars and even the spin of the Earth on its axis. The elements heavier than hydrogen and helium did not exist in the early universe but were created subsequently in exploding stars.

By about a million years after the big bang the universe had expanded sufficiently for the formation of clumps of matter which gave rise to galaxies and galaxy clusters (Figure 2.2). Ever since then these galaxies have been moving apart in a universal expansion. Astronomers are not able to decide whether this expansion will continue for ever, or whether the gravitational attractions between the galaxies might eventually reverse it. The answer is of philosophical rather than scientific importance perhaps, since a positive result favouring the

Object	**Mass** (kg)	**Radius** (m)	**Mean density** (kg/m^3)	**Effective temperature**	**Age** (years)
Moon	7.4×10^{22}	1.7×10^{6}	3.3×10^{3}	−150°C +200°C (surface)	4.6×10^{9}
Earth	6.0×10^{24}	6.4×10^{6}	5.5×10^{3}	14°C	4.6×10^{9}
Jupiter	1.9×10^{27}	7.1×10^{7}	1.3×10^{3}	120 K	4.6×10^{9}
Sun	2.0×10^{30}	7.0×10^{8}	1.4×10^{3}	6 000 K (surface) 1.5×10^{7} K (centre)	5.0×10^{9}
neutron star	2.0×10^{30}	$\sim 1 \times 10^{4}$	$\sim 1 \times 10^{18}$		
Galaxy	2.0×10^{41}	$\sim 1 \times 10^{18}$	$\sim 2 \times 10^{-11}$		$\sim 1 \times 10^{10}$
galaxy clusters	$\sim 10^{42} - 10^{44}$	$\sim 10^{19} - 10^{20}$	$\sim 1 \times 10^{-12}$		
universe	$\sim 1 \times 10^{53}$	$\sim 1 \times 10^{26}$	$\sim 1 \times 10^{-26}$	2.7 K	$\sim 2 \times 10^{10}$

2.1: *Comparative values for the physical properties of astronomical objects. In some cases the values are not known with great precision.*

2.2: *The nearby galaxy cluster in the constellation Perseus. Many of its members seem to form a chain of galaxies. This cluster has spiral, elliptical and irregular galaxies, as well as a strong X-ray galaxy.*

second possibility might indicate that the universe runs through an infinite cycle of expansion followed by collapse, followed again by re-birth in a cosmic fireball. The present rate of expansion indicates that from thirteen to twenty billion years have elapsed since the big bang.

Galaxies

Galaxies are the major building blocks of the universe. A galaxy is a giant family of many millions of stars, and it is held together by its own gravitational field. Most of the material universe is organized into galaxies of stars, together with gas and dust.

There are three main types of galaxy: spiral, elliptical and irregular. Our own Milky Way is a spiral galaxy: a flattish disc of stars with two spiral arms emerging from its central nucleus (Figure 2.3). About one-quarter of all galaxies have this shape. Spiral galaxies are well supplied with the interstellar gas in which new stars form; as the rotating spiral pattern sweeps around the galaxy it compresses gas and dust, triggering the formation of bright young stars in its arms. The elliptical galaxies have a symmetrical elliptical or spheroidal shape with no obvious structure. Most of their member stars are very old (10 000 Ma) and since ellipticals are devoid of interstellar gas, no new stars are forming in them. The biggest and brightest galaxies in the universe are ellipticals with masses of about 10^{13} times that of our Sun; these giants may frequently be sources of strong radio emission, in which case they are called radio galaxies. About two-thirds of all galaxies are elliptical. Irregular galaxies comprise about one-tenth of all galaxies and they come in many subclasses.

Some terrestrial length scales can be expressed as intervals of time: the time to fly from one continent to another or the time it takes to drive to work, for example. By comparison with these familiar yard-sticks the distances to the galaxies are incomprehensibly large, but they too are made more manageable by using a time calibration, in this case the distance that light goes in one year (9.46×10^{15} m). On such a scale the nearest galaxies to our own are 180 000 light years away, and the nearest giant spiral galaxy, the Andromeda galaxy, is two million light years away (Figure 2.4). The most distant luminous objects seen by our telescopes are probably ten thousand million light years away. Their light was already halfway here before the Earth even formed. The light from the nearby Virgo galaxy cluster set out when reptiles still dominated the animal world.

2.3: *A highly schematic cross-section of our Galaxy. Most of the stars and dust are concentrated into the main spiral disc. This is surrounded by a halo containing very hot gas of extremely low density together with hundreds of globular star clusters, which are the oldest surviving star groups from the early formation of our Galaxy.*

Our Galaxy

On a clear night, when there is no Moon, a band of faint light can be seen crossing the heavens from horizon to horizon. This is the Milky Way, which Galileo's telescope resolved into a myriad of faint stars, stars that make up our own enormous galactic family. In the southern hemisphere the Milky Way is generally more prominent, although it is obscured in parts by dense clouds of interstellar dust.

Radio and optical astronomers have found that our Galaxy has the shape of a flat disc with a central bulge. Its diameter is roughly a hundred thousand light years; in the nucleus the thickness reaches ten thousand light years, whereas in the disc it is five hundred to two thousand light years thick. Within this pancake we find stars, gas and dust, and superimposed on the general distribution of stars is the pair of spiral arms. We do not know exactly how far the Sun is from the centre, but it is conventionally taken to be thirty-three thousand light years away. Thus our solar system is relatively far from the galactic centre.

Observations of star and cloud velocities within the Milky Way show that the Galaxy is rotating. This is necessary since the angular momentum prevents the contents of the Galaxy simply falling under gravity into the galactic centre. Presumably the angular momenta of the galaxies reflect turbulences in the early universe. At the centre of the Galaxy there is a swirling mass of gas and stars which takes only a million years or so to circuit the centre. Out at the Sun's distance everything is more leisurely, the rotation speed being a mere 250 km per second and the period 250 million years for one orbit. Since its formation the Earth has travelled around the Galaxy about twenty times, a distance of roughly 10^{19} km.

Surrounding the disc of our Galaxy there is a spherical halo that contains the star families termed globular clusters. These ancient clusters mark out the shape that the Galaxy had when it first formed some 13 000 to 20 000 Ma ago. As the gas distribution in the Galaxy condensed under its own gravitational force to the disc, these clusters were left stranded—a cosmic fossil to remind us of the Galaxy's original shape.

Scattered through the universe are uncounted billions of galaxies. An important feature of our home Galaxy is its sheer size, for it is one of the largest spirals known. It contains roughly 10^{11} stars, whose total rate of radiating energy is about 10^{37} watts.

Stars

The stars appear in our telescopes as points of light, so it might seem that rather little can be learned about them. Among the facts that can be determined about a star are its brightness, which is expressible as the total energy output of the star; its distance, by a variety of techniques; and, from the spectrum of its radiation, its surface temperature and chemical composition.

2.4: *The central part of the great spiral galaxy in Andromeda, showing the dense, starry nucleus containing billions of stars. This galaxy is only two million light years from the solar system.*

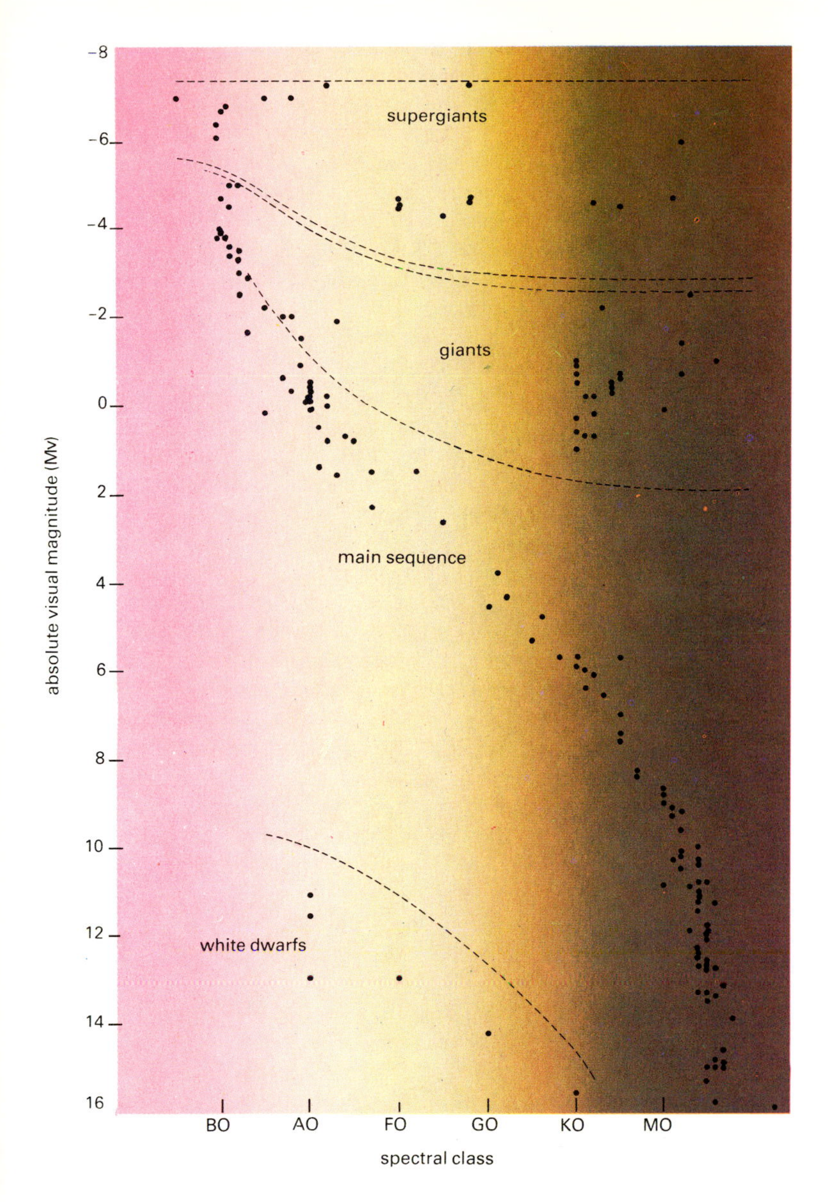

2.5: *The Hertzsprung-Russell diagram, showing the empirical relationship of the rate at which a star emits energy (its absolute magnitude) and its spectral class (indicating surface temperature from hot to cold).*

2.6: *A region of star formation (Horsehead Nebula), showing the dark clouds from which new stars condense, and bright young stars above it.*

Two of the above properties, the energy radiated and the surface temperature, can be plotted on a graph (Figure 2.5). When this is done for a large sample of stars, parts of the diagram are densely crowded with points and others are empty. This shows that the surface temperature and rate of energy output (or luminosity) are not independent of each other. In the temperature–luminosity graph there is a broad band from upper left to lower right, known as the main sequence. Our Sun is located on this band, at the part corresponding to a temperature of 6000 degrees. Above and to the right of the main sequence are located very luminous giant and supergiant stars. Although these are cooler than the Sun they have much larger surface areas and therefore a higher luminosity. Below the main sequence are small, faint stars known as white dwarfs. The general features of this temperature–luminosity diagram can be accounted for in the following manner.

Stars are being formed all the time in the Galaxy. They condense under gravity from dense clouds of interstellar gas (Figure 2.6). The early phases of star formation are poorly understood, although it is clear that gravitational forces must somehow gain the upper hand in a section of a cloud, as a result of which it starts to collapse on itself. As this process proceeds, the cloud breaks up into smaller units which then progressively contract as well. Within a shrinking cloudlet the gravitational forces increase as the radius decreases. This process ceases only when the internal gas pressure of the cloudlet becomes large enough to balance the gravitational force. At this stage the fragment is already radiating in the infrared as a protostar.

The warm protostar continues to contract, and consequently the temperature and pressure at its centre rise. Eventually the central temperature reaches about 10^7 degrees, and at this point nuclear reactions begin. The release of energy in these reactions then stabilizes the collapse, because the heat being radiated from the surface is matched by energy released at the centre.

Other fragments in the interstellar cloud go through the same process, with the end result that a cluster of very young stars is formed (Figure 2.7). Not all the gas and dust ends up inside the stars. Some will disperse back into interstellar space and some may condense into planetary systems.

When a new star has settled down it has a luminosity and temperature that position it on the main sequence. Its precise location depends only on its mass. At the upper end of the main sequence we find stars twenty, thirty, or even fifty times the mass of the Sun, while at the lower end they may be only half a solar mass.

The energy release mechanism for all main sequence stars is the conversion of hydrogen into helium through a series of nuclear reactions (Figure 2.8). The precise details are less important than the main features of the process. It is of course only the central part of the atom, the nucleus, that is involved in the reaction. In fact, the pressure and temperature in the core of a star are so high that the normal atomic structure of matter has instead been replaced by a mixture of electrons and nuclei. At a temperature of 10^7 degrees the hydrogen nuclei collide sufficiently hard to overcome electrostatic repulsion, and they can then fuse to form the next stable element, helium. This is a reaction that releases prodigious quantities of energy, because the helium nucleus is 0.65 per cent less massive than the four hydrogen nuclei (protons) that are needed to make it. The 'missing' mass (m) is converted to electromagnetic energy (E), in accordance with the Einstein relation $E = mc^2$, where c represents the speed of light. Every second, in a star like the Sun, 655 million tonnes of hydrogen are converted to 650 million tonnes of helium: five million tonnes of matter are annihilated to release energy.

2.7: *All stars were probably members of star clusters originally. This is Kappa Crucis, a beautiful jewel-like cluster in the southern sky.*

From our personal viewpoint a crucial property of stars is their long-term stability, which astrophysicists recognize as being due to a number of lucky 'accidents'. For example, if the di-proton (two-proton nucleus) were stable, stars would race through their life cycle in millions rather than billions of years. The reactions that take place are actually rather improbable, and this is why a star like the Sun can spin out the fuel supply for ten billion years. If the reactions had a higher probability, stellar evolution would proceed much faster. One of the major findings of stratigraphic geology has been the discovery of the huge time span, measured in billions of years, that preceded the emergence of hominids on this planet. Our presence in the universe is made possible only by the long time scale associated with stellar evolution.

The rate at which a star evolves depends strongly on its mass. Stars a few times more massive than the Sun have significantly higher central temperatures. Consequently they use their fuel faster and reach the later stages of stellar evolution sooner than solar-mass stars.

When the hydrogen fuel is depleted, the end point of a star's life is determined mainly by its mass. Low-mass stars start to shrink as the temperature falls. This can ignite hydrogen-burning reactions in the unprocessed outer layers. The contracting core of the star becomes hot enough to sustain further reactions: it burns helium to form carbon and oxygen, carbon to form neon and magnesium, oxygen to form silicon, and silicon is burnt to make iron. Meanwhile the burning in the outer layers can lift these layers away to form a planetary nebula (so called because its appearance in an astronomical telescope resembles the disc presented by a planet, in contrast with the point image of a star; there is no suggestion that this gas subsequently forms planets). The remnant core finally cools down and collapses until forces between electrons are high enough to stave off further gravitational contraction. At this stage it has become a white dwarf star, with a density some million times greater than ordinary matter at 10^9 kg/m^3. Ultimately it cools to an inert mass that plays no further part in the life of the Galaxy.

Stars that are somewhat more massive than the Sun probably finish their evolution with a stupendous explosion. The dense reactor core runs through a series of reactions that form elements as far as the iron group in the periodic table (see page 53). Eventually the contraction causes the internal pressure of this core to reach a stage in which its electrons combine with its protons to form neutrons. This change removes the electron pressure support at the centre and instantaneously triggers a catastrophic implosion. The outer layers of the star suddenly find that the core has collapsed beneath them: they, in turn, rain down towards the centre. Temperatures now rise, out of control, and in less than a second runaway nuclear reactions are detonated in the unspent fuel of the outer layers. A gigantic explosion

2.8: *The beautiful Orion Nebula, 1600 light years away, is the most prominent gaseous nebula in the sky. The glowing red gas is hydrogen, the major constituent of stars and the commonest element in the universe.*

2.9: *Vela supernova remnant is a blast wave of debris cast into space when a star exploded 100 000 years ago. Eventually these glowing ashes will merge with the interstellar medium and become the matter that will form a future star.*

throws most of the stellar material back into interstellar space. Supernova explosions are thought to be the major site of element synthesis in the universe, the means by which the elements beyond hydrogen and helium have been created since the primeval big bang.

The ejecta from a supernova outburst form a supernova remnant (Figure 2.9). Initially the remnant may expand at a good fraction of the velocity of light, but ultimately this slows to the ambient velocity as the matter ploughs into the interstellar medium. At the site of the explosion is the collapsed core of the original star. For a stellar relic with a mass between 0.1 and 1.4 solar masses the final stable configuration is a neutron star, when the object has the density associated with nuclear matter, some 10^{18} kg/m^3. It is then supported by neutron pressure against any further collapse beyond its radius of 10 km or so. When they are made in supernova explosions, these neutron stars rotate rapidly, preserving the angular momentum of the pre-supernova star, which had a far larger radius. They are observed by radio astronomers as pulsars. If the central relic exceeds 1.4 solar masses it collapses indefinitely and forms a black hole, a region of space from which nothing, not even light, can escape.

The origin and distribution of the elements

An important goal of astrophysics, and of fundamental importance to the Earth sciences, is to account for the origin and distribution of the chemical elements. Simply discovering the present composition of the Earth, the planets, the Sun, stars and the Galaxy is a major occupation.

The actual distribution of the chemical elements in the universe can be found either by the direct analysis of samples in the laboratory or by the detailed analysis of the spectra of distant bodies. The former method can be applied at present only to the Earth, the Moon and meteorites. All the remaining data on the relative abundances of the elements have come from spectroscopy. The optical spectra of the Sun and stars have in them thousands of dark absorption lines which are caused by the various atoms in the cool upper atmospheres of the stars. The relative strengths of these lines can be measured and interpreted to yield the relative abundances of the various elements in a particular star (Figure 2.10).

It is clearly the case that the inner planets of our solar system, such as Mercury and the Earth, have very little of the lightest and most volatile elements left; after all, they are rocky planets. The outer giant planets, such as Jupiter and Saturn, are probably nearer to possessing the original composition of the solar system. The composition of the outer layers of the Sun itself is probably the best clue to the nature of the material from which the solar system condensed; the Sun has a strong gravitational field and it cannot have lost its lighter gases selectively in the way that the Earth has. The outer layers have not been contaminated by the helium manufactured at the centre. The solar abundances show that, by mass, its composition is approximately three-quarters hydrogen and one-quarter helium, a faithful reflection of the elemental distribution in the early universe. This is not the whole story, however: about 1 per cent of the Sun's mass is contributed by the so-called heavier elements, that is elements other than hydrogen and helium. The relative abundances of these elements show some interesting trends. There is a general decrease in abundance as we go to heavier elements: lead and gold are far scarcer cosmically than silicon and sulphur, and a handful of the lightest elements, lithium, beryllium and boron, are rare relative to their neighbours in the periodic table of the elements. These light elements are destroyed early in the evolution of protostars. There is also a bunching of abundant elements around iron.

The general features of the Sun's composition today can be taken to be typical of the material from which the planetary system formed. It would be unreasonable to suppose that the planets condensed from cloud fragments with a fundamentally different composition to the proto-Sun. It is also reassuring that the relative abundances of the

2.11: *The strong resemblance of Mercury, shown here, to the Moon. Hummocky terrain and faults can be seen in this cratered area.*

2.10: *The relative abundances of the elements in the solar system plotted against atomic mass. Astrophysicists have shown how the main features of this curve arise. All elements other than hydrogen and helium have been manufactured in cosmic nuclear furnaces and explosions.*

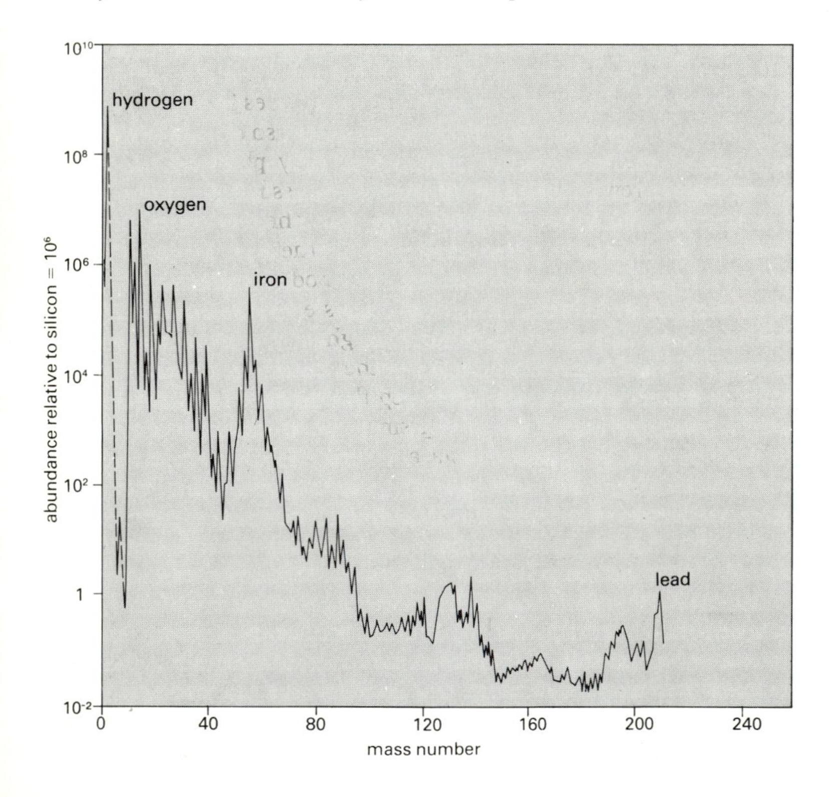

heavier elements in the Earth's crust, the lunar rocks and meteorites, are entirely consistent with the solar abundance distribution.

Most stars have a composition similar to that of the Sun, if exotic kinds such as neutron stars are excluded. It can, therefore, be taken that our Sun and solar system are not the products of any rare astronomical circumstances.

It used to be supposed that the universe has always had its present chemical composition, but it is now clear that the products of stellar evolution, and especially of supernovae, have modified this assumption. Astrophysicists are in reasonable agreement that the cosmic ubiquity of hydrogen and helium indicates that these at least are primeval, and their relative proportions were essentially fixed by the time the universe was ten minutes old. The remaining elements have been synthesized subsequently, by controlled and explosive burning in stars. Astrophysical theory is able to give a reasonable explanation for the abundances that are actually seen. It can explain why iron-group elements are common; why the post-iron-group elements (those of higher atomic number than iron) are rare; why lithium, beryllium and boron are scarce; and why nuclei and isotopes that have a nucleon number that is an exact multiple of four are commoner than those that are not exactly divisible by four. These successes indicate that astronomers have been able to give a coherent account of the history of matter.

It can now begin to be understood why the Sun and the Earth have arrived on the astronomical stage relatively late. The Earth is overwhelmingly composed of heavier elements, those that simply did not exist in the early universe. It was not until earlier generations of stars had exploded as supernovae, contaminating the interstellar medium with the products of nuclear burning, that stars with heavier elements could be born. Out of these cosmically enriched clouds smaller bodies containing elements other than hydrogen and helium could condense also. Practically all the nuclei in all the atoms on the Earth (except the hydrogen in the water) have been manufactured in massive stars that were born, lived and exploded long before the Earth itself could form. The carbon nuclei in the black ink on this page were cooked up ten billion years ago in a stellar furnace with a temperature of 10^8 degrees centigrade.

The origin of the terrestrial planets

The golden age of planetary exploration in the 1960s and 1970s provided new insights into the seemingly intractable problem of accounting for the planets that orbit the Sun (see Chapter 27). On Earth the oldest rocks have an age of 3800 Ma, but they do not provide useful information on its earliest history. The lunar rocks do give us a snapshot of the first 700 Ma of solar system history. The rocks of the dark maria regions on the Moon are datable at 3300 Ma to 3800 Ma. In the lunar highlands they are as old as 4200 Ma, and rocks that are even more ancient show that the Moon was very hot some 4400 Ma ago. This evidence for a hot Moon implies that the primitive Earth, being larger, must have been hotter still, and therefore volcanic activity and outgassing of the lighter gases must have been commonplace in its early history (see Chapter 16).

The solar system planets that do not possess thick atmospheres, Mercury and Mars, and also the Moon, suffered intensive bombardment in the past (Figure 2.11). The most ancient lunar rocks are not lumps of lava; rather they are fragments welded together into rocky masses, having the appearance of collision debris. On Mercury the cratering is at saturation density, no part of the planet having escaped repeated barrages. There is no way in which the Earth can have escaped a similar epoch of cratering, but eons of tectonic activity and weathering processes have removed all traces of it. The lunar rocks indicate that the intense bombardment ended about 4000 Ma ago, by which time most of the large rocks in interplanetary space had already crashed into planets and their moons. From the cratered surfaces it can be deduced that one of the final phases of planetary formation was the accretion of fragments onto the planets. It is instructive to trace this argument yet further back in time to see if accretion processes can give a reasonable account of the origin of the planets.

There is no one picture of planetary formation accepted by all planetary scientists. The tentative outline given here is a broad indication of what might have happened after the Sun had condensed from its interstellar cloud. Rotational energy in the collapsing proto-Sun would cause cloud fragments to be left stranded, spinning in orbit round the central massive object. Angular momentum would force this material to orbit in a flat disc, which is why the planetary orbits are almost confined to a plane.

Within this disc protoplanets then condensed out, each with its own gravitational field. Huge numbers of these small protoplanets might have formed, and they would gradually have settled into larger accretions by gravitational coalescence. Certain meteorites, the chondrites, provide us with the best evidence of what probably went on at this stage in the evolution of the solar system. These meteorites are generally thought to be typical samples of the clumps of particulate matter that condensed from dust and grains in the interstellar cloud. One of the stumbling blocks is explaining how these smallish rocks contrived to form planetary nuclei a few kilometres in size. Once such nuclei formed they swept up further rocks and also collided and joined with each other. In such a manner the rocky planets of the inner solar system grew. The later stages of the process might have involved glancing collisions, which could explain why the rotation axes of the planets are inclined at a variety of angles to their orbital planes.

An important feature of the accretion model is the kinetic energy contributed by the in-falling fragments. It is estimated that this could have raised the surface temperature as high as 10 000 degrees on the Earth. Another source of heating was the formation of a dense core. Radioactive heating continues to keep the interior of the Earth fluid (see Chapter 9). The Moon, being small, soon lost its heat energy; tectonic activity ceased 4000 Ma ago and vulcanism 3200 Ma ago (see Chapter 27).

The origin of the Jovian planets

The outer planets, Jupiter, Saturn, Uranus and Neptune, are fundamentally different from the rocky masses of the inner solar system. They are gas-rich planets with lower densities than the terrestrial planets. The explanation may be that the inner planets had gaseous envelopes too, but that these have been ripped away by atmospheric

2.12: *The swirling clouds of Jupiter's upper atmosphere mark out several prominent circulation belts. Note the numerous small eddies trapped in these belts. The Great Red Spot is the largest of these whirlwinds. The two satellites are Io (left) and Europa. The scan lines are produced by the way the photograph was taken, rather like a television picture. This photograph was taken on 5 February 1979 by Voyager 1.*

tides raised by the Sun. Another possibility is that the ices of water, ammonia and methane condensed only in the cooler outer regions of the solar nebula. Snowflakes of these substances would readily stick together on contact, so that accretion may have been much swifter in the outer solar system. Eventually Jupiter and Saturn had grown sufficiently massive for their gravitational fields to capture even the lightest gases, which are scarce in the inner solar system today.

The outer planets have a profusion of natural satellites. Some may be captured asteroids. Others may have formed in the vicinity of their parent bodies in analogous manner to the accretion of the planets proper.

The minor planets, or asteroids, are probably planetesimals left stranded by chance during the great phase of accretion. Their combined mass is too small (less than 0.02 per cent of the Earth's mass) to make a proper planet. Comets may be icy planetesimals from the furthest outposts of the solar system.

Dynamics of the solar system

The solar system has now settled into a stable arrangement in which nine planets follow orbits that are widely spaced and nearly circular. If this were not the case it is obvious that collisions would occur.

The Earth's orbit has a fairly small eccentricity. Its mean distance is around 150 million km from the Sun, and the greatest and least distances are 153 million and 147 million km. The mean distances increase by roughly 75 per cent from one planet to the next, a remarkably regular arrangement first discovered in the eighteenth century.

The Earth's spin axis is inclined to its orbital plane by $23\frac{1}{2}°$. Over long periods of time, the combined gravitational pull of the Sun and Moon can alter this slightly, and the angle is known to have changed by about half a degree in the past four thousand years. The direction in space to which the spin axis points also changes gradually, through precession, moving full circuit in about 26 000 years. The complex behaviour of the angle and direction of the rotational axis is important to studies of data from rock magnetism.

Astronomical events and the Earth

The Earth is still influenced by certain astronomical phenomena, although the influence of some of these is still speculative. The Earth is affected by matter falling onto it and by the type of radiation, and its intensity, shining onto it.

Interplanetary space still contains stony rubble which crashes onto

2.13: *A meteorite crater in Arizona, USA, formed about 50 Ma ago.*

the Earth's upper atmosphere at velocities of tens of kilometres per second. Fragments with a mass exceeding 1 kg reach the Earth's surface, and about twenty of these meteorites are recorded each year. Very large meteorites (of 100 or more tonnes) are not significantly braked by the atmosphere and they cause impact craters. The Arizona meteorite crater, 1.2 km in diameter and 100 m deep, is the best preserved of these, and it is a few thousand years old. The total amount of debris that the Earth sweeps up in a year is 10 000 tonnes, most of which comprises small particles weighing 1 g or less. The Tunguska event of 1906, which flattened a huge area of Siberia, is thought to have been a collision of the Earth with a comet, an event that is unlikely to happen more than once in a billion years.

Our Sun displays highly energetic phenomena, principally flares, in its surface layers. These inject clouds of electrically charged particles into space. As these brush past the Earth they can interact with the upper atmosphere and enhance the aurora. At the same time the ionosphere is temporarily destroyed, which inhibits long-distance radio-wave propagation.

The Sun's average energy production may also vary over millions of years. Any evidence favouring this is almost non-existent. It is an attractive idea for explaining fluctuations in the Earth's climate, but extremely difficult to investigate properly. On short time scales, roughly eleven years, the sunspot density reaches a maximum. At this time the aurora is prominent, on account of the increased production of charged particles from active regions on the Sun. Perhaps this cycle subtly modulates the Earth's climate and weather; real evidence for this is, at best, weak (see Chapter 18). It is of course possible that the amount of solar radiation reaching the Earth could vary even though the Sun itself remains constant. This would be the case if our solar system were to pass through an interstellar cloud, and it is not impossible that this has happened at some stage.

In the realms of speculative catastrophism it can be asked: what would happen if the Earth were close to a supernova or if it were hit by a meteorite? This question may be relevant to the sudden disappearance of perhaps as many as 75 per cent of all the species on this planet about 65 Ma ago. This major extinction event occurred within less than a few hundred thousand years, and possibly in a mere thousand years. Clearly the biosphere was subject to major stress at that time. It may have been a purely terrestrial event caused by, for example, a great increase in volcanic activity. One candidate for extraterrestrial influence is a nearby supernova. One within thirty light years would bathe the Earth in X-rays and high-energy particles. This would grossly modify the atmosphere and cause a dramatic increase in radiation diseases, a drop in temperature and a worldwide drought. The Earth could get near to such a supernova when the solar system passes through one of the spiral arms. Another possibility is a collision with an asteroid or very large meteorite. There is evidence in ocean sediments that something odd happened 65 Ma ago. Elements that are more abundant in extraterrestrial matter than in the Earth's crust are commoner in this layer. This is consistent with a huge collision that would also have clouded the atmosphere with dust, eliminating most of the sunlight. Ideas of this type are, of their nature, hard to confirm, but cosmic catastrophes are attractive as explanations of extinctions such as that of the giant reptiles (see Chapter 22).

2.14: *Saturn's ring system, captured in detail by the Voyager 1 camera, in 1980.*

Other planetary systems

The possible existence of worlds other than our own is an ancient astronomical problem, given a new twist as a theory for the origin of our own system is pieced together. In discussing this problem it is important to realize that our Sun is a solitary star, and such stars are in a minority. Most stars are members of binary or multiple systems, and planets would not survive for long in such an environment. When we survey the single stars it is not possible to see any planets nor could we expect to do so with present techniques. In the future it may be possible to search for such planets directly by using large telescopes in space, or by detecting tiny but regular changes in the velocity of a star, which would indicate that unseen companions were in orbit around it. The theoretical approach suggests that planetary systems may be common around single solar-type stars. Nothing in the present theories assumes that our Sun has been subjected to any rare or unique process. Hence planets such as the Earth, Jupiter, Mercury, and perhaps even unlike any local examples, may be plentiful in the Galaxy and the universe at large. This conclusion has encouraged elaborate plans for searching for extraterrestrial intelligence, although none of these projects has received significant funding. For the present the Earth remains the only planet available for detailed scientific investigation.

Part Two: Physics and Chemistry of the Earth

The geologist is trained to observe and describe the rocks, minerals and fossils of a particular area, quarry, mine or borehole. The geochemist and geophysicist can take a more experimental approach to the study of the Earth in keeping with the scientific nature of their parent disciplines of physics and chemistry. They can propose hypotheses or models, and test them against available data, or devise new techniques for collecting data. Physicists and chemists had earlier made occasional interjections into geological debate, sometimes with later contradictory results, such as the 'proof' (by Kelvin) of the 100 Ma age of the Earth, and (by Jeffreys) of the impossibility of continental drift. Without the assistance of geophysics and geochemistry, it would be impossible to answer the questions such as 'what is the Earth made of', and 'how does it work?' Conversely, physicists and chemists willing to turn their attention from the workbench to the larger laboratory of the Earth itself have found data to interpret and problems to solve on the grand scale.

Geophysics and geochemistry are by no means separate disciplines of the Earth sciences; indeed they are closely interdependent. The seismologist cannot interpret the passage of earthquake waves through the Earth without some knowledge of its possible range of chemical compositions, and the geochemist studying the derivation of the rocks of the crust from those of the mantle needs an appreciation of the physical phenomena of heat flow and convection. The vast bulk of the Earth is completely inaccessible to the geologist with his hammer and microscope, and only the measurement of physical variables and the study of extraterrestrial chemistry can answer some of the most fundamental questions about our planet.

Archaean gneiss, approximately 2800 Ma old, at Nensen Bugt, Umivik, Greenland.

3 Seismology and the deep structure of the Earth

Seismology is the study of earthquakes. Earthquakes can cause severe damage to buildings and loss of life, although many of the largest earthquakes occur under the oceans or in uninhabited parts of the Earth with little ill effect on human beings. On the other hand some quite small earthquakes cause heavy casualties and widespread destruction because they occur near the surface beneath cities whose buildings have not been designed to withstand the shaking of the Earth (see Chapter 26). Considerable effort has gone into earthquake prediction and even earthquake control—so far without too much success.

More positively, observations of the records of elastic waves generated by earthquakes or man-made explosions have provided the main source of our information about the Earth's deep interior. Moreover the pattern of earthquakes observed on the Earth's surface has played a major part in the development of the concept of plate tectonics—earthquake belts marking plate boundaries (see Chapter 10).

Elastic waves

The Earth is continually undergoing deformation due to stresses that are set up within it. If the stresses continue to build up over a long time fracture may take place, resulting in an earthquake. This involves a sudden release of energy, part of which takes the form of elastic waves which travel through the Earth. These waves emanate from a confined region below the surface, called the focus. The point on the surface of the Earth vertically above the focus is called the epicentre.

There are three basic types of elastic wave: two body waves, ie waves that propagate within a body of rock, and surface waves, ie waves whose motion is restricted to near the surface, their amplitude decreasing with depth (Figure 3.1). The faster of the body waves is called the primary, or P, wave. It is a longitudinal wave, ie as it spreads out it alternately pushes (compresses) and pulls (dilates) the rock. P waves, in common with sound waves, can travel through both solid

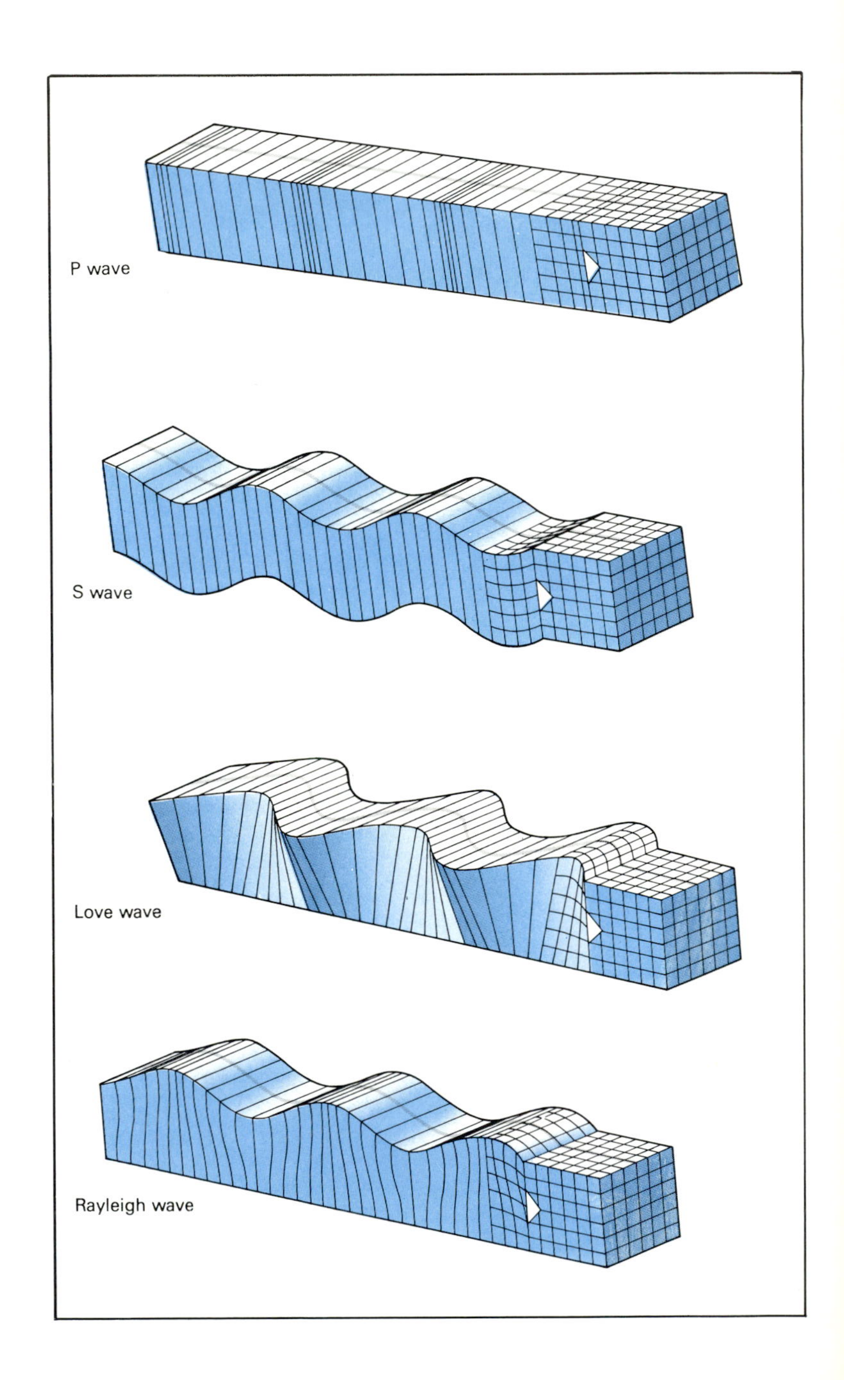

3.1: *The form of ground motion near the surface of the Earth for four types of earthquake waves:* P wave, S wave, Love wave and Rayleigh wave. *The motion of Love waves is essentially the same as that of S waves with no vertical displacement. The ground moves from side to side in a horizontal plane parallel to the Earth's surface at right angles to the direction of propagation. Rayleigh waves are like rolling ocean waves, the material disturbed by the wave moving both vertically and horizontally in a vertical plane in the direction in which the waves are travelling.*

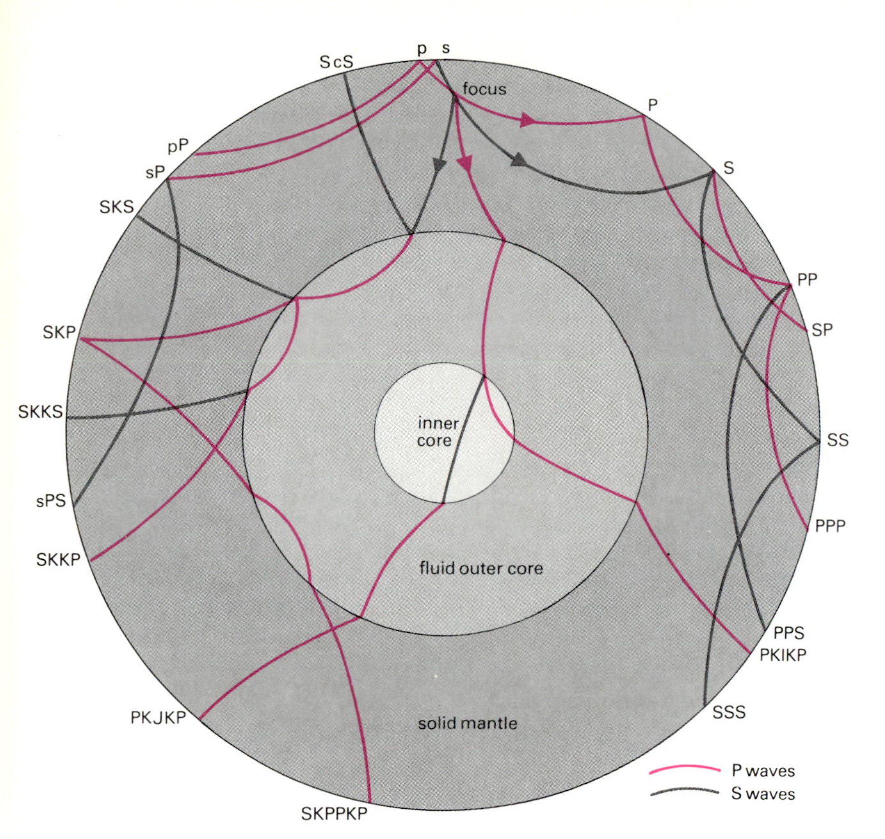

3.2: *Some of the many possible reflections and refractions of elastic waves at discontinuities within the Earth. P and S waves reaching the surface directly without refraction are labelled* p *and* s. *A P wave reflected from the Earth's surface gives rise to both a P wave and an S wave (called PP and PS waves, respectively). Likewise an S wave reflected from the surface can give both P and S waves (called SP and SS waves respectively). The letters* c *and* i *denote reflections from the outer and inner core boundaries respectively, and K and I are P waves that have come through the outer and inner core. Thus ScS is an S wave that has travelled down to the core boundary and been reflected as an S wave. Similarly SKP is an S wave that has been refracted into the outer core (necessarily as a P wave) and refracted back into the mantle as a P wave.*

rock and liquid material, such as volcanic magma or the water of the oceans. The slower body wave is called the secondary, or S, wave, and as it propagates it shears the rock sideways, at right angles to the direction of travel. P waves involve change of volume, whereas S waves involve distortion without change of volume. S waves, in common with light waves, are transverse and so are polarized. Vertical (SV) polarization is distinguished from horizontal (SH) polarization.

It can be shown that the velocity of P waves is given by:

$$V_p = \sqrt{\frac{k + \frac{4}{3}\mu}{\rho}}$$ and that of S waves by: $$V_s = \sqrt{\frac{\mu}{\rho}}$$

where ρ is the density, k the adiabatic incompressibility and μ the rigidity. Both P and S wave velocities depend only on the elastic constants and the density of the medium. In particular, if the rigidity is zero, V_s is zero; ie shear waves cannot be transmitted through a material of zero rigidity, such as the oceans.

If a stone is thrown into a pond, or a strong wind blows across an open stretch of water, the surface of the water is agitated and waves that travel away from the source of disturbance are set up. Such waves are called surface waves and are controlled by gravity. They can also occur in a solid, when they are controlled by elasticity. There are two types of surface waves in solids: Love waves and Rayleigh waves. Rayleigh waves can be generated on a uniform solid, whereas Love waves are possible only if the material is non-uniform. Love waves require a surface layer in which the velocity of S waves is less than that below. The velocity of Love waves also depends on the ratio of the wavelength to the thickness of the layer, so that unlike P and S waves Love waves show dispersion. If the velocity is found by observation for a number of wavelengths it is possible to determine the thickness of the layer. Surface waves travel more slowly than body waves and, of the two surface waves, Love waves generally travel faster than Rayleigh waves. Thus as the waves spread out into the Earth's crust from the earthquake source the different types separate out from one another in a predictable pattern (see Figure 3.5).

When an elastic wave meets a sharp boundary between two media of different properties part of it will be reflected and part refracted, and laws of reflection and refraction apply, rather like those for light waves. The case of elastic waves, however, is more complicated since waves of both P and S type may be reflected and refracted. The theory is based on Fermat's principle, according to which an elastic wave takes the quickest path between any two points. This does not imply that there is only one path. There may be a number of alternative paths, but each must involve a minimum transit time.

The basic principle of refraction in optics is Snell's law, which is equally applicable to seismic waves; the angles of reflection and refraction are related to the angle of incidence by the wave velocities V_p and V_s of P and S waves in the two media. There are two major discontinuities in the Earth, one just below the surface, separating the crust from the mantle, and the other at a depth of about 2900 km, separating the mantle from the core. The core may also be divided into an inner core and an outer core. All these discontinuities may variously reflect or refract seismic waves, leading to a great variety of potential pathways (Figure 3.2).

Distribution of earthquakes

Earthquakes have been classified rather arbitrarily as shallow (0–70 km), intermediate (70–300 km) or deep (below 300 km) according to their depth of focus. The deepest earthquakes occur at a depth of about 700 km. Figure 3.3 shows the global distribution of earthquakes in the depth ranges 0–100 km and 100–700 km. Deep earthquakes are found in the South American Andes, the Tonga Islands, Samoa, the New Hebrides chain, the Japan Sea, Indonesia and the Caribbean Antilles: each of these regions is also associated with a deep ocean trench. On average the frequency of earthquakes in these regions decreases

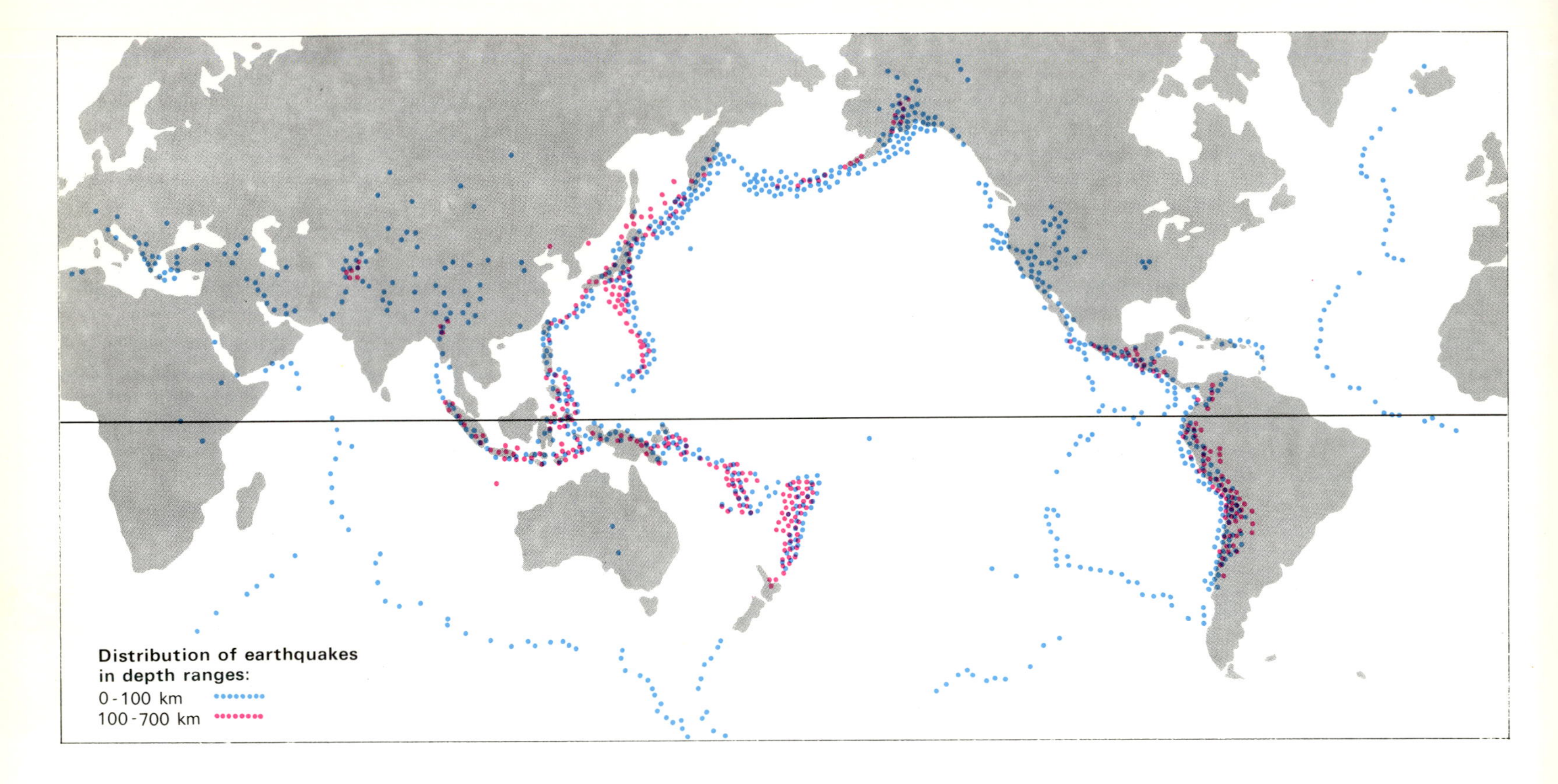

3.3: *The distribution of earthquakes in two depth ranges for 1961–7. Note the continuity of the belts of shallower earthquakes that outline large, stable blocks, and the narrowness of the belts.*

rapidly below a depth of about 200 km. Some intermediate and deep-focus earthquakes are located away from the Pacific region, in the Hindu Kush, Romania, the Aegean Sea and Spain.

Clearly defined belts of seismic activity separate large oceanic and continental regions almost devoid of earthquake centres. Other concentrations of earthquakes occur in oceanic areas along the centre of the Atlantic and Indian Oceans, which are the sites of immense submarine ridges. Dense concentrations of earthquakes coincide also with the island arcs such as those in the Pacific and eastern Caribbean. There is a well-defined pattern in the depths of focuses of earthquakes near such island arcs. The focuses lie in a narrow zone which dips from near the trench beneath the island arc at an angle of about 45°. The significance of this dipping seismic zone (the Benioff zone) is discussed in Chapter 10. On the other side of the Pacific the whole western coast of Central and South America is marked by strong seismic activity. In contrast the eastern part of South America is almost entirely free from earthquakes, and they almost never occur in the large central and northern areas of Canada, much of Siberia, west Africa and Australia. There is, however, a long trans-Asiatic zone of high seismicity running approximately east–west from Burma through the Himalaya Mountains and central Asia to the Caucasus and Mediterranean. In Europe earthquake activity is quite widespread, in particular in Turkey, Greece, Yugoslavia, Italy, Spain and Portugal.

Most moderate to large shallow earthquakes are followed in the same area for a few hours (and even for several months) afterwards by numerous smaller earthquakes called aftershocks. The great Rat Island earthquake in the Aleutian Islands on 4 February 1965 was followed in the next twenty-four days by more than 750 aftershocks large enough to be recorded by distant seismographs. A few earthquakes are preceded by smaller 'foreshocks' in the source area.

The detection and mechanism of earthquakes

Earthquakes are recorded by seismographs which measure movement of the ground. A mass is loosely coupled to the Earth by means of a pendulum or by suspending it from a spring. When the ground moves the mass tends to remain stationary because of its inertia, and the displacement of the Earth relative to it can be recorded. This motion is magnified electronically and modern instruments can detect ground displacements as small as 0.1 nanometre (nm). This is more sensitivity than can be used since the ground is continually being disturbed by winds, waves and man-made noise. The signal is usually recorded on a sheet mounted on a drum, which rotates at a fixed rate and moves sideways as it rotates. A continuous record is obtained, which, when the sheet is removed from the drum, appears as a number of parallel lines (Figure 3.4).

A moving system such as that described will have a natural period of vibration. If it is carefully designed it may have about the same response over a considerable range of periods. However, seismic waves can have periods ranging from a few tenths of a second to many minutes, and no vibrating system can give an equal and adequate response over this range. It is therefore necessary to design instruments for specific purposes, and a well-equipped seismograph station will have more than one set of instruments. It is common practice to have one set that will respond well to short-period waves with periods from 0.2–2 seconds and another set to respond to a long-period range of 15–100 seconds.

The travel time of a seismic wave from its source to any point on the Earth's surface is a direct measure of the distance between the two points. Seismologists have been able to determine by trial and error the average travel time of P and S waves for any given distance. The times have been printed in tables and graphs as a function of the distance. These expected travel times can then be compared with actual measurements of the distance between an earthquake source and a seismic observatory, and the distance between the observatory

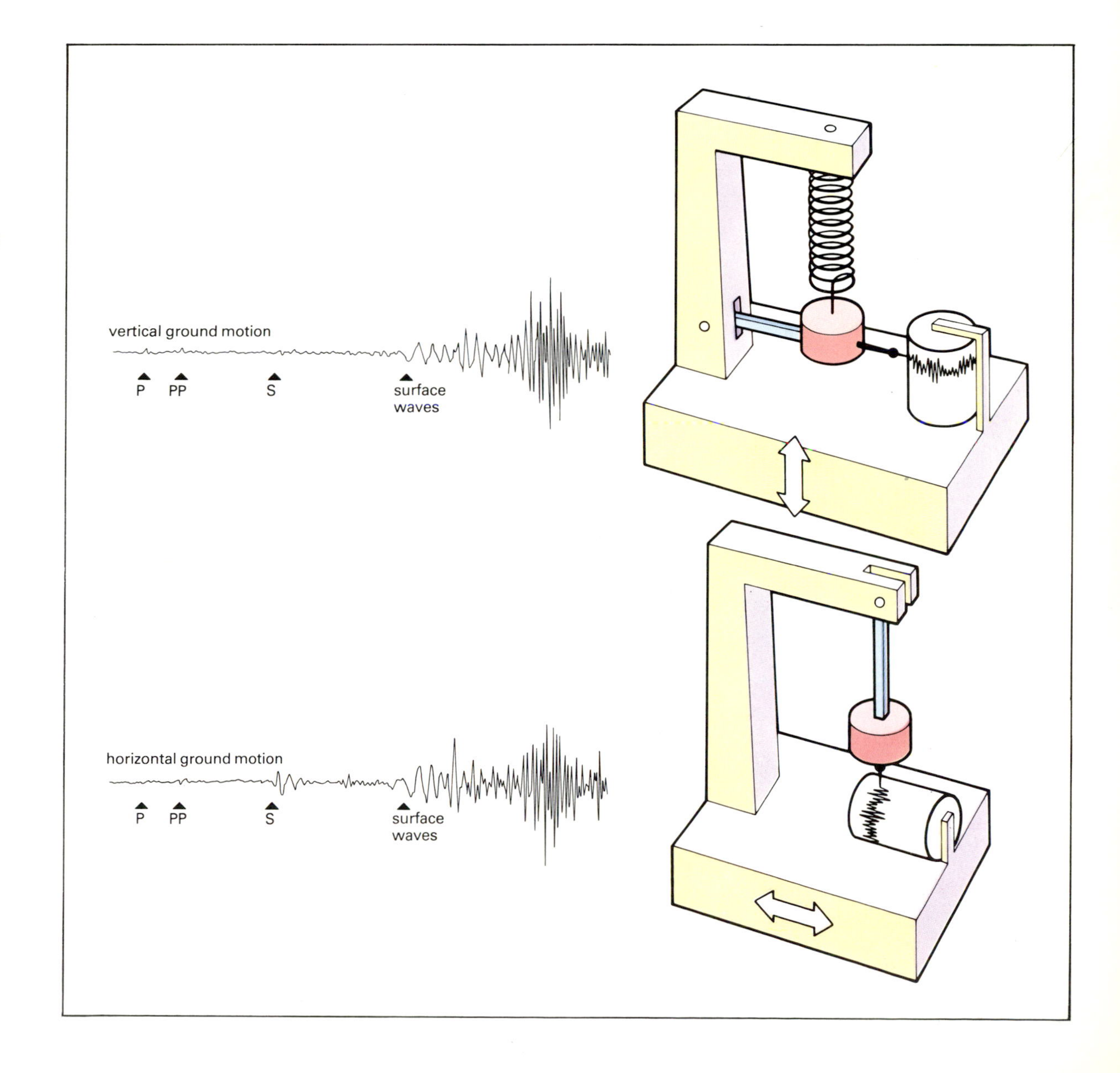

3.4: *Seismometers are designed to measure either vertical or horizontal ground motions. A set of seismometers always includes one for recording vertical movements and two, at right angles to one another, for horizontal motions. One of each kind is shown here, with examples of the signals received from a single earthquake event. Successive arrivals (from left to right) of four different types of seismic wave are shown.*

and epicentre can be read from the tables. Alternatively the difference in travel times of P and S waves can be used to determine the distance graphically to the epicentre (Figure 3.5). If arrival times at only one observatory are available then only the distance of the earthquake source from that observatory can be determined and not its location. If arrival times at three observatories are available then the epicentre and time of occurrence of the earthquake can be determined by triangulation. In practice, readings from many observatories are used. A moderate-sized earthquake on the mid-Atlantic ridge, for example, might be located using readings from sixty or more seismic stations around the world. The calculation is carried out by high-speed computers.

The energy and magnitude of earthquakes

About 100 joules of energy is released from each cubic metre of rock at the time of an earthquake. This may not seem very much. However, a large earthquake associated with a fault 1000 km long, extending 100 km downwards, and distorting the ground as far as 50 km on either side of the fault amounts to a strained volume of 10^{16} cubic metres which gives a total of 10^{18} joules. This is equivalent to about a thousand nuclear explosions, each with a strength of 1 megaton of TNT. Although the energy released gives a measure of the size of an earthquake it is not easy to determine the fault dimensions, slip and other factors needed to compute it. Seismologists have therefore adopted the Richter magnitude scale which relates the local magnitude to the logarithm of the maximum amplitude recorded by seismographs (an empirical factor takes into account the attenuation of seismic waves as they spread out from the focus). Because magnitudes are based on a logarithmic scale, an increase in magnitude of one unit corresponds to an increase in the amount of energy released by a factor of about thirty. The actual range of magnitudes is enormous, the largest events representing motions a hundred million times (9 magnitudes) as large as those of small events. The scale is open at both ends. It does not have a maximum value of ten, as is often reported. Two or three earthquakes this century have had Richter magnitudes of 8.9. The 1906 San Francisco earthquake had a Richter magnitude of 8.25, and the great Chilean earthquake of 22 May 1960, one of 8.5. At the other extreme earthquakes with a magnitude of less than −2 are routinely observed by seismologists studying micro-earthquakes. Most earthquakes are small, and each year about 800 000 small tremors recorded by instruments are not felt by human beings. Generally speaking shallow earthquakes have to reach a Richter magnitude of more than 5.5 before significant damage occurs near the source. Great earthquakes with magnitudes exceeding 8 occur about once every five to ten years.

The current practice at seismic observatories is to use two magnitude scales both of which are different from the original Richter scale. The reason for using two scales is that earthquakes having deep focuses give very different seismic records from those having shallow focuses, even though the amount of energy released in each event might be the same. In particular deep-focus earthquakes produce only small or insignificant trains of surface waves.

A new measure of earthquake strength called seismic moment (defined as the product of the rigidity of the rock, the area of faulting and the amount of slip) has recently been suggested. This measure of earthquake size has indicated the need for some revision of previous estimates of the relative magnitudes of great earthquakes. Another scale—the intensity—has been used to describe the 'bigness' of an earthquake. The intensity scale is a measure of the observed effects of earthquake damage at a particular place and is not based on actual measurements of any kind (see Chapter 26).

At some distance (more than about 150 km) from an earthquake the shorter-period vibrations tend to be damped out, resulting in ground periods of more than a second. Tall buildings often have periods of this order, and resonant vibrations may be set up, such as the 'pounding' of adjacent tall buildings in Los Angeles at the time of the Kern County earthquake. This effect has been particularly pronounced in Mexico City. The business section of this city is built on very soft soil, which shows a predominant period of vibration in earthquakes of about 2.5 seconds. Many tall buildings have been constructed in this district, and their periods of vibration are very nearly the same. When, in 1957, an earthquake of magnitude 7.5 occurred off the coast, 350 km from the city, many of these tall buildings were thrown violently into their natural periods of vibration and suffered major damage. Smaller buildings, on the other hand, many of them old, came through the earthquake unharmed.

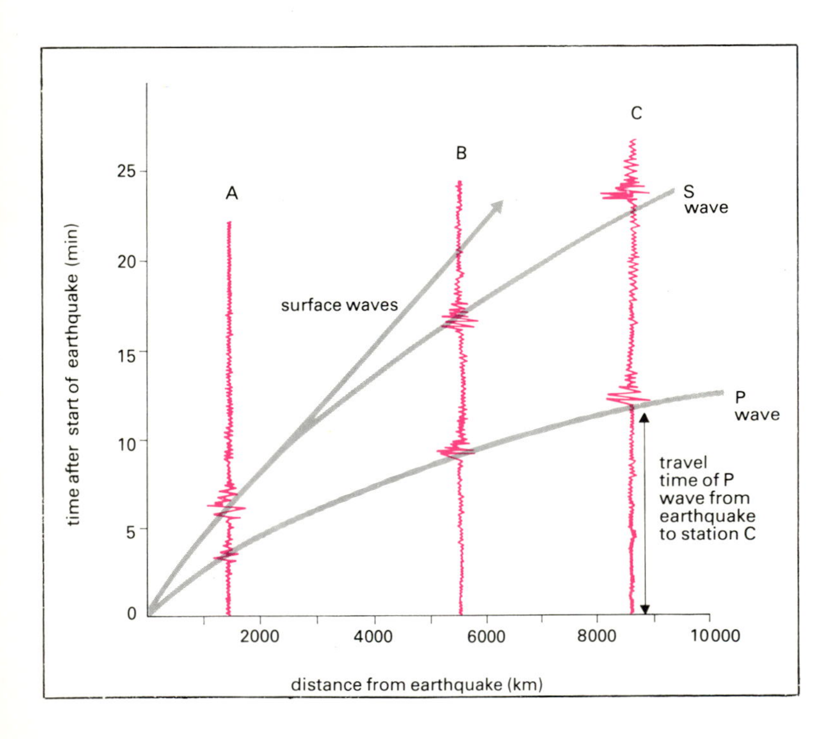

3.5: *Travel time curves for P, S and surface waves. To locate an earthquake epicentre, the time interval observed at a given station is matched against the travel time curves for P and S waves until the distance is found at which the separation between the curves agrees with the observed S–P time difference. If the distance from three stations A, B and C is known, the epicentre can be located.*

3.6: *Earthquake damage in Los Angeles, California. Originating in the San Andreas fault system, the earthquake took place on 9 February 1971.*

The distinction between earthquakes and underground explosions

Earthquakes are associated with large fractures or faults in the Earth's outer surface. One of the most widely accepted explanations of shallow earthquakes is Reid's (1911) elastic-rebound theory, which attributes earthquakes to the accumulation of strain energy in tectonic regions and the sudden release of this energy by faulting when the fracture strength is exceeded. In great earthquakes the offset across the fault may be as much as 15 m. Once the frictional bond is broken the elastic strain energy, which has accumulated slowly for tens or hundreds of years, is suddenly released and the resulting rupture will travel at a speed of about 3.5 km per second, continuing in some cases for as much as 1000 km. The effect of underground nuclear explosions is quite different, consisting of outwardly radiating P waves. An important application of seismology is the detection of such explosions, and their discrimination from natural earthquakes. An earthquake, being associated with a mechanical couple (a pair of forces acting in parallel but opposite directions), should produce alternate phases of compression and dilation in four quadrants related to the fault plane. Seismographs in two quadrants will therefore record a compression as the first motion detected, and those in the other two will record an initial dilatation. Almost all earthquakes give this quadrant distribution of compressions and dilatations from which the direction of the fault plane can be discovered. An explosion, on the other hand, in which energy is radiating outwards from a single point, should cause compression everywhere, so that all seismograph stations record an initial compression. Consequently it was initially hoped that seismology would be able to distinguish between nuclear explosions and earthquakes, but uncertainty arises when there are not stations close to an earthquake's epicentre. Since there are many parts of the world, such as ocean bottoms and isolated land masses, in which it is impossible to place stations, earthquakes occasionally give initial compressions at all available stations.

Earthquake control

The concept of fracture followed by frictional sliding across a fault face is plausible only for shallow shocks. At depths below a few kilometres the overburden pressure exceeds the shear strength of the rock, and dry frictional sliding is impossible. If a pore fluid is present under pressure the effective normal stress across the fault is reduced. This has the effect of lubricating the fault, allowing sliding to occur at greater depths. In 1966 a series of small earthquakes in Denver, Colorado, was attributed by some seismologists to the injection of fluid into a waste-disposal well. This interpretation was questioned on the grounds that the area was seismically active in any case and that earthquakes continued to occur after injection ceased. The correlation between the amount of water injected and the number of earthquakes was, however, quite strong, and this chance discovery was followed up in 1969 by a planned field experiment by the United States Geological Survey at the Rangely Oil Field in western Colorado. A large number of oil wells were available at the site, and water could be regularly injected into the wells or pumped out of them. The results of the experiments showed an excellent correlation between the quantity of fluid injected and local earthquake activity. When the pore fluid pressure reached a critical level earthquake activity increased. When the pressure dropped as a result of water withdrawal seismic activity decreased. Again it should be pointed out that the wells at Rangely penetrated pre-existing faults and that the crust in the region was already under some tectonic strain, as indicated by the occurrence of small local earthquakes over the previous years.

The Denver and Rangely investigations suggest the possibility of earthquake control. One suggestion is to pump water through deep boreholes into faults in a region where natural earthquakes might be particularly hazardous. An outbreak of small earthquakes might thus be induced, reducing the amount of strain energy stored in the crust in the vicinity and reducing the probability of a large earthquake. Tampering with the forces of nature in this way is, however, dangerous. If control was attempted along a major active fault the consequences might be especially damaging. On the other hand 'de-straining' crustal rocks by water injection at the site of a large structure, such as a dam, might be feasible.

Earthquake prediction

Many methods have been used to try to predict earthquakes. These have ranged from attempts to identify typical 'earthquake weather' to observations of strange animal behaviour prior to an earthquake. Since the 1960s much scientific effort has been directed to the problem of prediction, particularly in the USA, the Soviet Union, Japan and China. It must be confessed, however, that little success has been achieved. The absence of any obvious precursors to the moderately large earthquake (magnitude 5.7) that occurred in August 1979 near Coyote Lake, 100 km south-east of San Francisco, suggests that some earthquakes may be extremely difficult or even impossible to predict. The most publicized lack of forewarning was the tragic earthquake which on 27 July 1976 almost razed Tangshan, an industrial city in China of one million people. Unofficial reports estimated a death toll of about 650 000, including about a hundred people in Peking, some 150 km to the west. It is estimated that an additional 780 000 people were injured.

Historical studies of world seismicity patterns have made it possible to predict the probable place at which a damaging earthquake can be expected to occur. However, such records do not enable us to forecast a precise time of occurrence. Even in China, where between 500 and 1000 destructive earthquakes have occurred within the past 2700 years, statistical studies have not revealed any clear periodicities between great earthquakes, although they do indicate that long periods of quiescence can elapse between them.

Many different forces, including severe weather conditions, volcanic activity and the gravitational pull of the Moon, Sun and planets, have been suggested as earthquake triggers. Numerous catalogues of earthquakes, including complete lists for California, have been searched for such precipitators without any convincing results. Some of the more promising lines of research are the detection of strain in crustal rocks by geodetic surveys (of the shape of the Earth), the identification of suspicious gaps in the regular occurrence of earthquakes in both time and space, and the observation of foreshocks. In recent years the major earthquake prediction effort has been directed to more precise measurements of fluctuations in the physical properties of crustal rocks in seismically active continental areas. Special sensing devices have been installed to observe long-term changes in the rocks, but the number of measurements is still limited, and results have thus far been conflicting. In some, unusual behaviour was seen before a local earthquake, while in others nothing significant occurred before the event or, alternatively variations were observed that were not associated with earthquakes.

If some properties of rocks change before an earthquake, then the speed of seismic waves may also vary. Some of the first information on precursory changes in the travel time of waves in moderate earthquakes was obtained in 1962 in the Tadzhikistan region of the Soviet Union. Measurements there suggested that P velocities changed by about 10–15 per cent before the occurrence of local earthquakes. Fieldwork since then, in the Soviet Union and elsewhere, has indicated that in some cases the velocity of P waves in the focal region decreases by about 10 per cent for a time and then, just before the main shock occurs, increases again to a more normal value. More detailed checks have now been made in a number of countries, with mixed results. In the USA, in 1971, researchers at the Lamont Doherty Geological Observatory, working on quite small earthquakes in the Adirondacks in New York, detected increases in the travel time of P waves before three small earthquakes. In contrast similar tests carried out in central California have shown that fluctuations in travel times before a number of small to moderate earthquakes along the San Andreas fault were not significant. This negative result indicates that any precursory changes in the velocity of P waves before small to moderate earthquakes are probably highly localized around the region.

A second variable that has been used for prediction is precursory changes in ground level, such as ground tilts in earthquake-prone regions. A third possible variable is the release into the atmosphere of the inert gas radon, particularly from deep wells along active fault zones. It has been claimed that significantly increased concentrations of radon have been detected just before earthquakes in some parts of the Soviet Union. However, because so few measurements in different geological circumstances are available it is impossible to say whether the observed increases are exceptional rather than normal variations in radon concentration.

A fourth variable, to which a good deal of attention has been paid, is the electrical conductivity of the rocks in an earthquake zone. It is known from laboratory experiments on rock samples that the electrical resistance of water-saturated rocks such as granite changes drastically just before the rocks fracture at high pressure. A few field experiments to check this result have been made in fault zones in the Soviet Union, China, Japan and the USA, and decreases in electrical resistance before earthquakes have been reported.

Part of the difficulty in predicting earthquakes is the lack of a proper understanding of how the two faces of a fault slide by one another. Unlike a theoretical model in which a fault is thought of as two surfaces flush to one another, real faults tend to be rough and irregular. Such irregularities extend several kilometres below the surface at which earthquakes actually occur. These irregularities may be bends and short offsets in the direction of the fault, bumps or rough spots on the faces of the fault or differences in the material caught up between the fault faces. Any or all of these irregularities could hinder movement along the fault. One result would be that instead of stress being evenly distributed along the fault it would tend to be concentrated at the irregularities. When the stresses reach a critical level the fault might rupture at one of these stress concentrations before most of the fault is ready to break. Most of the 'jolt' of the earthquake could come from the small rupture, but the stress released there could also cause slippage elsewhere on the fault.

The relationship between depth and seismic velocity

The times at which signals from the same earthquake arrive at different seismic observatories can be recorded, and thus it is possible to determine the travel time of the disturbance in terms of distance. Figure 3.5

3.7: *A trace of the San Andreas fault in Carrizo Plain, California. Offset gulleys clearly show the sense of movement on the fault.*

illustrates travel time curves for P, S and surface waves plotted against distance from the earthquake. The graph for the first arrival of surface waves is a straight line as it depends simply on geographical distance. The graphs for P and S waves, however, are curves showing that travel times do not increase proportionately with distance. The explanation is that their velocity is greater at depth within the Earth. The travel time depends on how the velocities of the different seismic waves change as they pass through materials having different elastic properties. From the travel time curves it is thus possible to obtain velocity–depth curves (Figure 3.9) which provide the basic information for estimating the physical properties of the Earth's interior, starting with the density. Although velocity generally increases with depth it is possible to predict the effects of a zone in which the velocity decreases with depth. This will cause refraction, as shown in Figure 3.10, and

3.8: *Effects in Anchorage of the Alaska earthquake of 27 March 1964. The epicentre was 169 km to the east, near the entrance to Prince William Sound.*

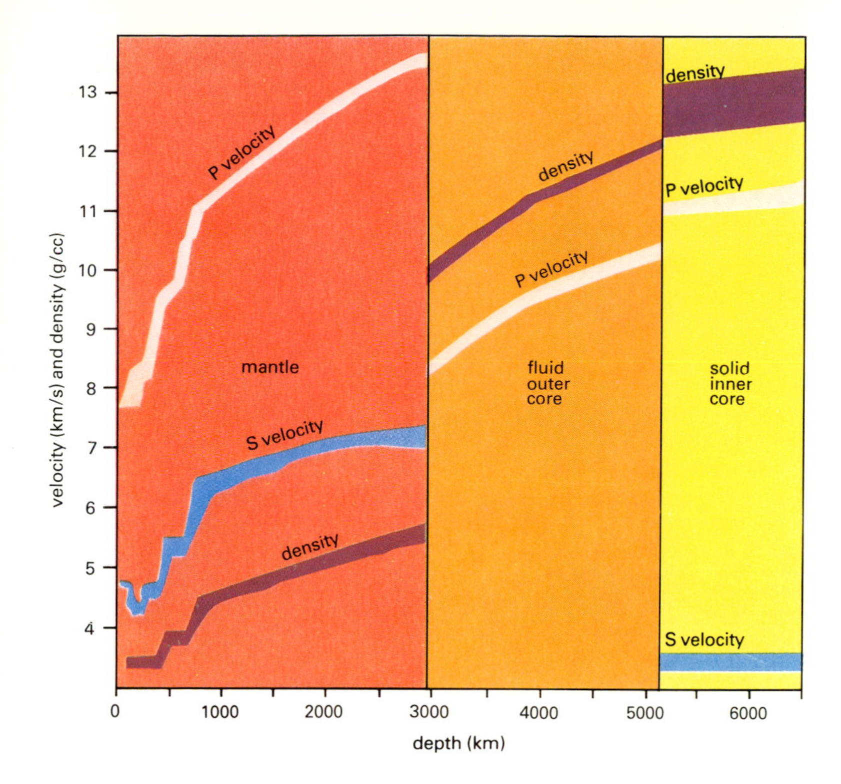

3.9: *The variation of* P *and* S *wave velocity and density within the Earth. The width of the narrow bands indicate the uncertainty in the data.*

there will be a range of distances from the source of waves in which arrivals are not observed (a shadow zone). There is a discontinuous drop in the velocity of P waves across the core–mantle boundary leading to a shadow zone, which was predicted by Oldham in 1906 and verified by Gutenberg in 1912. The shadow zone applies to P waves in the range 105° to 143° (ie 11 600 km to 16 000 km) away from the source. However, some waves are recorded in this range, and so it cannot be regarded as a true shadow zone. The amplitudes of such waves are, nevertheless, much reduced and for many years their presence was attributed to diffraction around the boundary of the core. Lehmann suggested in 1935 that the waves recorded in the shadow zone had passed through an inner core in which the P velocity is significantly greater than that in the outer core. Later work has corroborated her hypothesis and the existence of a solid inner core is now well established.

Physical properties of the Earth's interior

A primary application of seismology is the determination of the structure and constitution of the Earth's interior from measurements made at the surface. Fundamental to this problem is the determination of the changes in density with depth. The density depends on the pressure, temperature and an indefinite number of variables specifying the chemical composition. Since the stress in the Earth's interior is essentially equivalent to a hydrostatic pressure, the variation in pressure with depth is a straightforward mathematical function. If further simplifying assumptions are made concerning temperature and chemical homogeneity it is possible to derive another equation linking density with depth. This equation was first obtained by Adams and Williamson in 1923 and may be integrated numerically to obtain the density distribution in those regions of the Earth where chemical and other variations may be neglected.

Additional information on the physical properties of the Earth's mantle and core may be obtained from an analysis of the free oscillations of the Earth. If a bell is struck with a hammer it rings. The Earth also rings when it is disturbed by a great earthquake and the entire globe vibrates like a bell for as long as several weeks. The tones of the Earth's vibrations are pitched too low for the human ear to hear, but modern seismographs are sensitive enough to detect these low-frequency oscillations. There are two classes of such oscillations, torsional (or toroidal) and spheroidal (Figure 3.11). In the simplest torsional mode there is no displacement at the equator, and the two hemispheres oscillate in antiphase. The fundamental spheroidal oscillation is an alternating compression and rarefaction of the whole Earth. The next higher mode is the 'football' mode. The nodes are two parallels of latitude dividing the surface into three zones. As the sphere oscillates it is distorted alternately into an oblate and prolate spheroid. Its period is 53 minutes, corresponding to E flat in the twentieth octave below middle C. There are higher modes with an increasing number of subdivided zonal distributions, and each of these modes also has overtones with internal nodal surfaces.

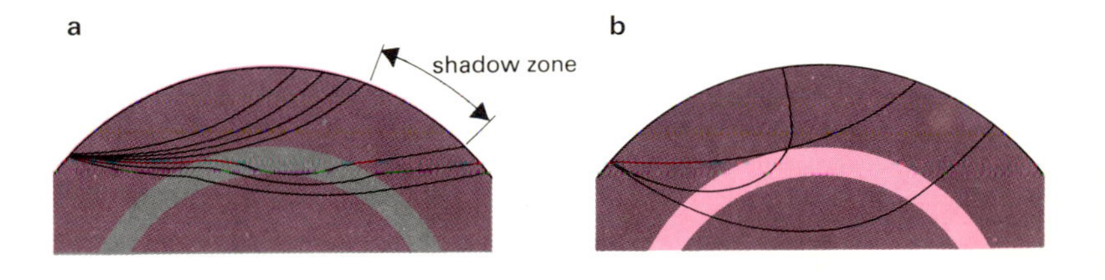

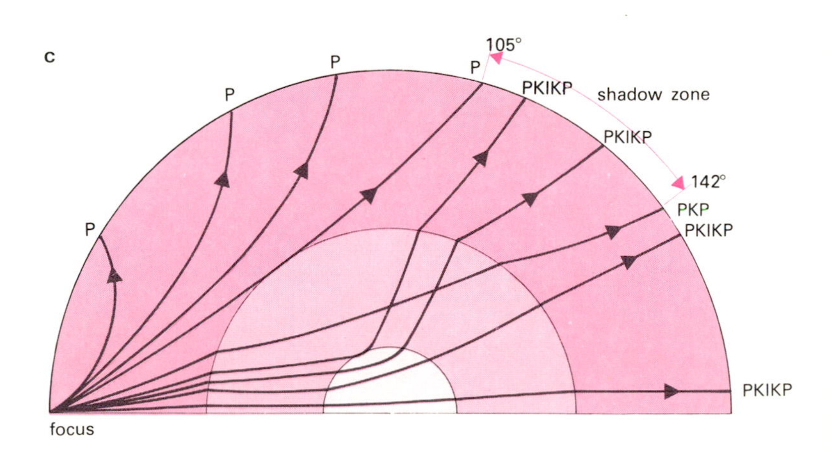

3.10: *The effect seismic waves of a layer in which the velocity* (**a**) *decreases with depth and* (**b**) *increases with depth in an otherwise 'normal' Earth.* (**c**) P, PKP *and* PKIKP *waves and the shadow zone.*

Free oscillations of the Earth were first observed in 1952, although it was not until the great Chilean earthquake of 22 May 1960 that improved instrumentation enabled detailed measurements to be made. Additional results were obtained from the great Alaska earthquake of 28 March 1964. Although the free oscillations of a uniform elastic sphere were first investigated more than a hundred years ago, calculations for realistic Earth models, taking into account detailed structure, gravitational forces and rotation, were not undertaken until fairly recently. Such calculations would have been impossible without the aid of modern computers. The free oscillations excited by a major earthquake last for several days, but their amplitude diminishes because the Earth is not a perfectly elastic body. The precise way in which it diminishes can be determined for each mode of oscillation and information on the elastic properties of the Earth's interior obtained.

It should be noted, however, that the 'average' Earth to which the average periods of free oscillation modes relate is not necessarily the same as an Earth model to which the currently available average travel times of seismic body waves apply. This is because earthquake epicentres and recording stations are not randomly distributed over the Earth's surface. It must also be stressed that since the variations in the physical properties of the Earth are in general continuous, and the amount of data available is finite, there is no unique solution to the problem, which is further compounded by the fact that the data themselves contain errors. However, it has proved possible to place bounds on possible solutions, and certain properties of the Earth are known to a high degree of accuracy.

For example, the pressure distribution may be obtained once the density distribution has been determined. Since the density is used only to determine the pressure gradient the pressure distribution is insensitive to small changes in the density distribution and may be determined quite accurately. The variation with depth of the gravitational acceleration (g) can also be calculated. Its value does not differ by more than 1 per cent from 9.9 m per second per second (its value at the Earth's surface; see Chapter 6) until a depth of over 2400 km is reached. On the other hand values of g deep within the Earth are highly dependent on density, and calculated values for depths below 4000 km may be in error by as much as 5 per cent. From a knowledge of the density distribution, values of the elastic constants can be computed.

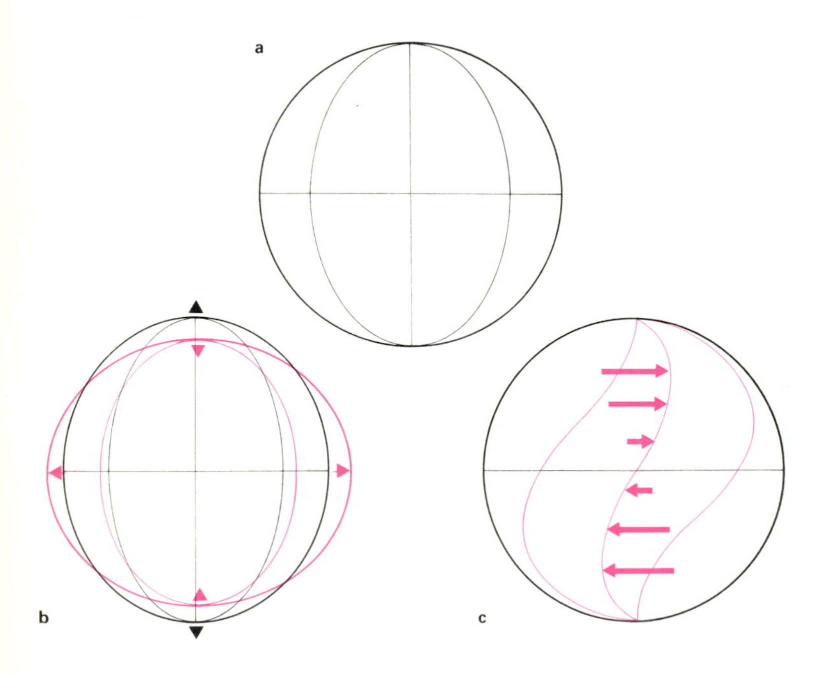

3.11: *The static Earth* (**a**) *and two classes of free oscillation* (**b** *and* **c**).

Even if the variation of density and other physical properties through the Earth were known in detail, laboratory experiments on rocks and minerals are necessary to identify the materials in the Earth's interior. Studies of wave propagation measurements in rocks show that the velocity of compressional waves depends principally upon density and mean atomic weight. However, most common rocks have mean atomic weights close to 21 or 22 regardless of composition, and rocks or minerals of very different compositions may have the same densities and seismic velocities. Thus it is not easy to infer chemical composition from seismic data alone, and laboratory experiments conducted in the conditions of pressure and temperature that exist deep within the Earth are necessary.

The pioneering experimental work of Bridgman, beginning in the 1920s, was undertaken at pressures up to 100 kilobars, which corresponds to a depth of only 300 km within the Earth. However, in the past few years, dynamic determinations of the compressibility of minerals and rocks have been made by a number of researchers up to pressures in excess of those at the centre of the Earth. Such high pressures are created for very short time intervals behind the front of a strong shock wave set up by an explosive charge. They are an order of magnitude greater than those that can be obtained by static methods. Although the pressures are maintained for intervals of usually less than a microsecond this is long enough to allow measurements to be made of some geophysically important quantities. More recently, sustained high pressures and temperatures have been made possible by the refinement of the diamond-anvil pressure cell, an instrument in which two tiny diamond faces are squeezed together with a solid sample in between. Previously, high pressures were obtained by applying very large forces to relatively large surface areas. In the new technique high pressures are generated by applying modest forces to very small areas (a few hundred micrometres in diameter). A problem in this approach is the possible failure of the diamonds themselves. Even carefully selected and shaped diamonds will crack and shatter into diamond dust if not properly aligned and supported. Pressures of 1.7 megabars, equivalent to that just below the core–mantle boundary, have been obtained. At these pressures the diamonds (the hardest substance known) began to deform.

The composition of the Earth

Geological sampling and laboratory measurements of the properties of rocks and minerals show that P wave velocities may be associated with composition. The continental crust consists mainly of granitic rocks

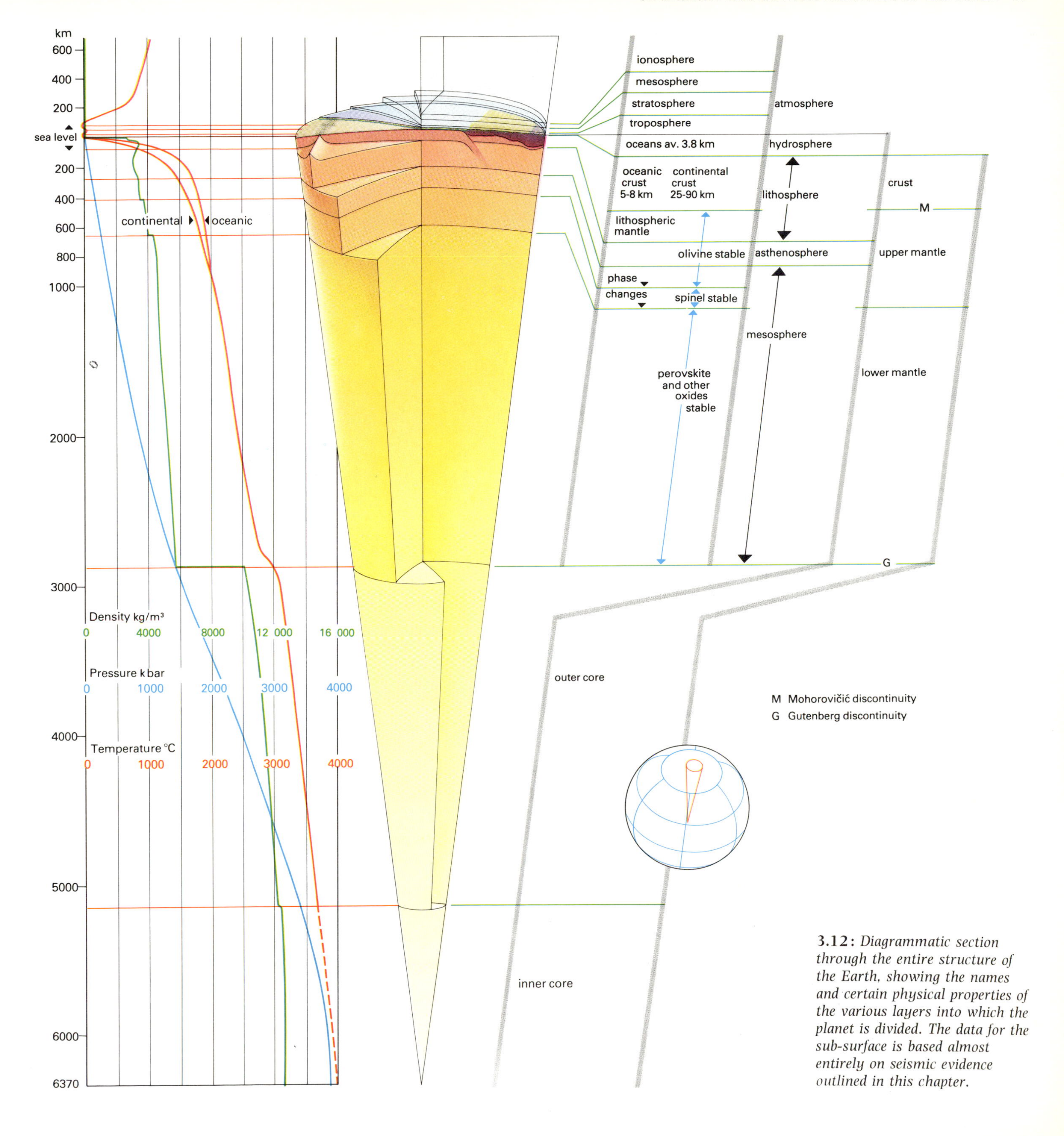

3.12: *Diagrammatic section through the entire structure of the Earth, showing the names and certain physical properties of the various layers into which the planet is divided. The data for the sub-surface is based almost entirely on seismic evidence outlined in this chapter.*

with gabbro near the bottom. There is no granite on the floors of the deep oceans, the crust there being entirely basalt and gabbro. The mantle below the Mohorovičić discontinuity consists primarily of the dense ultrabasic rock peridotite. The thickness of the crust varies from about 33 km under continents (even more under mountains) to about 5 km under ocean floors.

The structure of the outermost 700 km of the Earth as revealed by the velocity of S waves is shown in Figure 3.13. The uppermost region is called the lithosphere, a slab about 70 km thick in which the continents are embedded. Its lower boundary is marked by an abrupt decrease in shear wave velocity. The lithosphere is characterized by high velocity and efficient propagation of seismic waves, implying solidity and strength. Below the lithosphere is the asthenosphere (or zone of weakness). It is also a low velocity zone where seismic waves are attenuated more strongly than anywhere else in the Earth. These features of the asthenosphere may indicate partial melting to the extent of 1 to 10 per cent. The velocity and density in both the lithosphere and asthenosphere would be satisfied by a peridotitic composition (ie predominantly of the mineral olivine), the boundary at 70 km being the phase change separating the solid lithosphere from the weaker asthenosphere.

The asthenosphere extends to a depth of about 250 km, where the rocks again become solid. There is also a slight increase of velocity with depth because of increasing pressure. There is then a very rapid increase in velocity and density through a narrow zone at a depth of about 400 km. This increase is too rapid to be accounted for by a composition change; a change of phase (ie a closer packing on the atomic level) is required. This theoretical explanation was verified in 1969 independently by two researchers who subjected olivine to high pressures and found that at a critical pressure and temperature its atoms take up a more compact arrangement, changing into the spinel structure. These critical conditions of temperature and pressure occur at a depth of about 400 km in the Earth. There is another region of rapid change at around 650 km following a region of small gradual changes with depth. This zone was first indicated by shock wave experiments, and in 1974 workers in Japan and the USA reached these pressures in static compression tests and found that olivine breaks down into dense, simple oxides of iron, silicon and magnesium. Recent diamond-anvil experiments at the Australian National University, Canberra, and at the Carnegie Institution of Washington have helped establish the perovskite type of crystal structure as a major form for mantle rock above 0.3 megabar, equivalent to a depth of about 650 km. Geochemical and geophysical evidence indicates that the mantle is composed largely of oxygen, silicon, magnesium and iron, but there is less agreement about how these elements are grouped to form minerals and how the mineral structures adjust to the different pressures and temperatures at different depths. A likely combination of mantle minerals is the three silicates, olivine, pyroxene and garnet. It has been suggested that the sudden large change in density at 650 km in part results from the atoms in the olivine crystal adjusting to the pressure

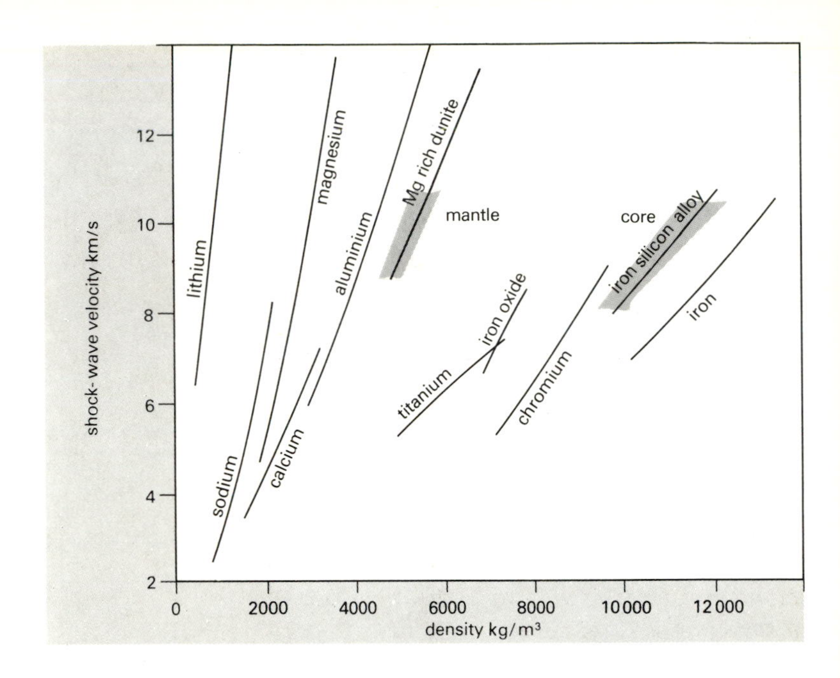

3.14: *Shock wave data for certain materials under varying degrees of compression for testing hypotheses of the composition of the Earth's mantle and core. Seismic evidence limits the properties of mantle and core to the shaded areas. The dunite, consisting almost entirely of magnesium-rich olivine, most closely approximates the properties of the mantle, while an iron–silicon alloy comes closest to matching those of the core.*

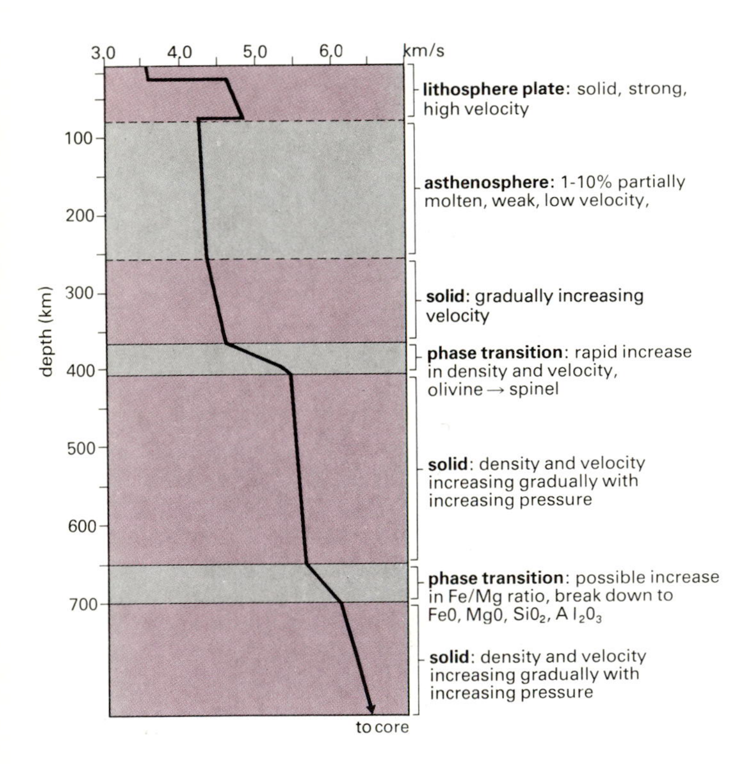

3.13: *The structure of the outermost 700 km of the Earth illustrated by a plot of S-wave velocity against depth.*

by changing into the denser perovskite form, in which the oxygen atoms are closer to each other. The lower mantle extending from about 650 km to the core at a depth of about 2885 km shows little change in either composition or phase, and velocity and density increase fairly smoothly with increasing pressure.

Shear waves have never been observed in the outer part of the core, which is thus considered to be liquid. However, evidence for the existence of an inner core, which is solid, has steadily increased. In particular it has been shown that the free oscillation data demand a solid inner core in which the average shear velocity is about 3.6 km per second. The observation of shear waves in the inner core would establish its rigidity. In 1972 shear waves were claimed to have been seen on a seismograph, and a value for their velocity of about 2.95 km per second was deduced. Unfortunately this cannot be reconciled with the value of 3.6 km per second from the free oscillation data. It is not clear exactly what phase had been identified on the seismograph.

All evidence indicates that the inner core consists primarily of iron and nickel and the outer core of molten iron alloyed with 8–20 per cent of a light element, giving an average atomic number of about 23. Shock wave data for iron indicate that both densities and bulk moduli in the outer core are less than those of iron under equivalent conditions (Figure 3.14), although their gradients through the outer core are consistent with gross chemical homogeneity (ie uniform intermixing of iron with a lighter, more compressible element or compound). Both densities and bulk moduli for the inner core are compatible with those of iron, suggesting that the inner/outer core boundary is likely to be a compositional as well as a phase boundary. There is no firm evidence as to the identity of the light element in the outer core. It must be reasonably abundant, miscible with iron and possess chemical properties that would allow it to enter the core. The prime candidates are silicon and sulphur; more recently oxygen has been suggested.

Seismology only gives us a 'snapshot' of the interior of the Earth as it is today and gives no information about its structure in the past or its evolution. The existence of a liquid outer core and solid inner core raises the question of the Earth's origin. Has the Earth always had such a structure, or has it evolved over geological time? This question cannot be divorced from the much broader question of the origin of the Earth and solar system (see Chapter 2). It is usually assumed that the embryo Earth was homogeneous and later differentiated into crust, mantle and core. Inhomogeneous models have also been suggested in which the (mainly iron) core formed first, the silicate mantle being deposited upon it later. There is one constraint, however, that can be imposed upon the possible evolution of the Earth's core. It is generally believed that the Earth's magnetic field is due to some form of electromagnetic induction in the fluid, electrically conducting outer core (see Chapter 7). Rocks as old as about three thousand million years have been found to possess remanent magnetization, and so the Earth must have had a molten outer core comparable to its present size at least that long ago. A major problem in understanding the evolution of the Earth is the means by which it could have heated up sufficiently to lead to a (predominantly iron) molten outer core that long ago.

Fairly short periods (about twenty million years after accretion) are now favoured as being the time required to form the core. An upper limit is about five hundred million years so that core formation should have been essentially complete during the first tenth of the Earth's history. Deducing the subsequent evolution from an approximately homogeneous all-fluid core into the present state of a solid (mainly iron) inner core and fluid outer core containing some light alloying element is another problem. This transformation may have proceeded much more slowly and still may not be complete. It should be noted that gravitational energy would be released by the growth of the inner core and this may provide the main source of energy for driving the geodynamo (see Chapter 9).

4 The chemistry of the Earth

Present-day ideas on the chemistry of the Earth are derived from many sources because there is no direct method of sampling the interior of our planet. Data from the chemistry of objects in space (the Sun, the interstellar medium and meteorites) have been vital to the construction of current chemical models. The basic physical properties of our planet, as indicated by its moment of inertia and mass and, particularly, by the nature of seismic wave propagation, have provided critical information. Finally any suggestions as to the chemistry of the Earth must be compatible with knowledge of the physical state of materials in the internal pressure–temperature regime of our planet. Thus any discussion of the Earth's chemistry must begin with a consideration of the chemistry of all matter in the universe.

The cosmic abundance of the elements

Analysis of the chemical composition of every type of object in the universe, including distant stars, the Sun, planets, the Moon and meteorites, is a major undertaking for scientists. The quantity of an element in a star is determined by examining the spectra of emitted radiation. Direct chemical analysis is so far possible only for meteorites and the surfaces of the Earth, the Moon and Mars. Chemical studies of rocks in the laboratory by means of highly sensitive techniques such as mass spectrometry permit analysis of all the chemical elements and their isotopes, even those present at the one part per billion level, equivalent to one gram in a thousand tonnes (see also Chapter 8).

4.1: *A photomicrograph (× 5, crossed polars) of a thin section through the Barwell meteorite, which fell on Christmas Eve, 1965, at Leicester, England. The circular areas are chondrules.*

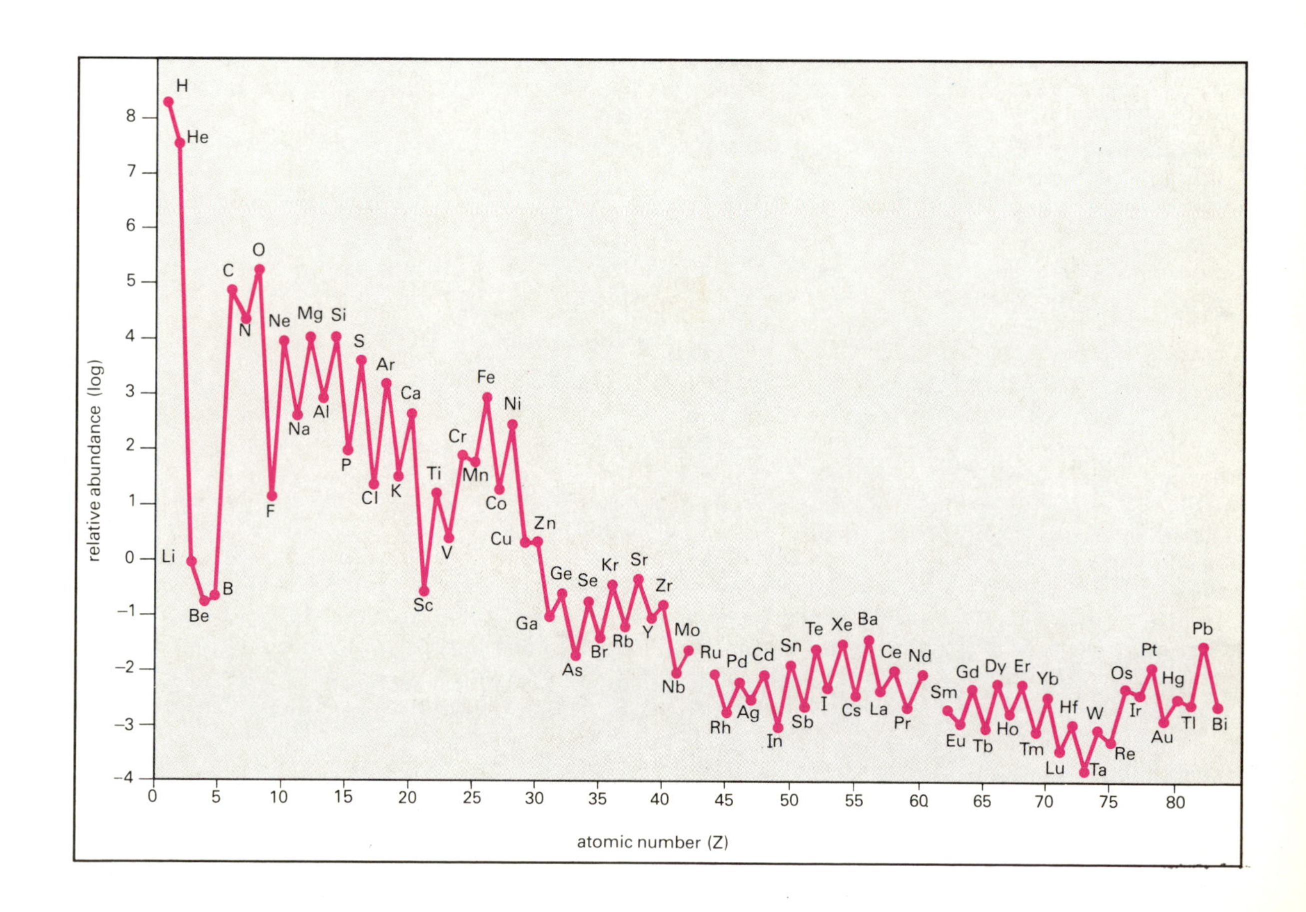

4.2: *Cosmic abundance of the chemical elements. Note the logarithmic vertical scale. Hydrogen* (H) *and helium* (He) *are about a thousand times (3 units) more plentiful than most other elements. Abundances of all the elements have been related to that of silicon* (Si), *which is given a value of 10 000 (4 on the log scale). Note the pronounced peak at iron* (Fe)*; the rarity of the light elements lithium* (Li), *beryllium* (Be) *and boron* (B)*; and the overall odd-even effect for the abundances of neighbouring elements.* H *and* He *were produced shortly after the 'big bang'; whereas all other elements have been synthesized since by nuclear fusion reactions in the interiors of stars.*

4.3: *Abundances of the chemical elements in the solar system, adjusted relative to silicon, so that silicon is one million* (1.0×10^6).

Element	Chemical symbol	Atomic number (Z)	Atomic mass (A)	Abundance
hydrogen	H	1	1.008	4.8×10^{10}
helium	He	2	4.003	3.9×10^{9}
lithium	Li	3	6.939	16.0
beryllium	Be	4	9.012	0.81
boron	B	5	10.81	6.2
carbon	C	6	12.01	1.7×10^{7}
nitrogen	N	7	14.01	4.6×10^{6}
oxygen	O	8	16.00	4.4×10^{7}
fluorine	F	9	19.00	2.5×10^{3}
neon	Ne	10	20.18	4.4×10^{6}
sodium	Na	11	22.99	3.5×10^{4}
magnesium	Mg	12	24.31	1.04×10^{6}
aluminium	Al	13	26.98	8.4×10^{4}
silicon	Si	14	28.09	1.0×10^{6}
phosphorus	P	15	30.97	8.1×10^{3}
sulphur	S	16	32.06	8.0×10^{5}
chlorine	Cl	17	35.45	2.1×10^{3}
argon	Ar	18	39.95	3.4×10^{5}
potassium	K	19	39.10	2.1×10^{3}
calcium	Ca	20	40.08	7.2×10^{4}
scandium	Sc	21	44.96	35.0
titanium	Ti	22	47.90	2.4×10^{3}
vanadium	V	23	50.94	5.9×10^{2}
chromium	Cr	24	52.00	1.24×10^{4}
manganese	Mn	25	54.94	6.2×10^{3}
iron	Fe	26	55.85	2.5×10^{5}
cobalt	Co	27	58.93	1.9×10^{3}
nickel	Ni	28	58.71	4.5×10^{4}
copper	Cu	29	63.54	4.2×10^{2}
zinc	Zn	30	65.37	6.3×10^{2}
gallium	Ga	31	69.72	28.0
germanium	Ge	32	72.59	76.0
arsenic	As	33	74.92	3.8
selenium	Se	34	78.96	2.7×10^{2}
bromine	Br	35	79.91	5.4
krypton	Kr	36	83.80	25.0
rubidium	Rb	37	85.47	4.1
strontium	Sr	38	87.62	25.0
yttrium	Y	39	88.91	4.7
zirconium	Zr	40	91.22	23.0
niobium	Nb	41	92.91	0.9
molybdenum	Mo	42	95.94	2.5
technetium	Tc	43	99.00	–
ruthenium	Ru	44	101.0	1.83
rhodium	Rh	45	102.9	0.33
palladium	Pd	46	106.4	1.33
silver	Ag	47	107.9	0.33
cadmium	Cd	48	112.4	1.2
indium	In	49	114.8	0.1
tin	Sn	50	118.7	1.7
antimony	Sb	51	121.8	0.2
tellurium	Te	52	127.6	3.1
iodine	I	53	126.9	0.41
xenon	Xe	54	131.3	3.0
cesium	Cs	55	132.9	0.21
barium	Ba	56	137.4	5.0
lanthanum	La	57	138.9	0.47
cerium	Ce	58	140.1	1.38
praseodymium	Pr	59	140.9	0.19
neodymium	Nd	60	144.2	0.88
promethium	Pm	61	147.0	–
samarium	Sm	62	150.4	0.28
europium	Eu	63	152.0	0.10
gadolinium	Gd	64	157.3	0.43
terbium	Tb	65	158.9	0.061
dysprosium	Dy	66	162.5	0.45
holmium	Ho	67	164.9	0.093
erbium	Er	68	167.3	0.28
thulium	Tm	69	168.9	0.041
ytterbium	Yb	70	173.0	0.22
lutetium	Lu	71	175.0	0.036
hafnium	Hf	72	178.5	0.31
tantalum	Ta	73	181.0	0.019
tungsten	W	74	183.9	0.16
rhenium	Re	75	186.2	0.059
osmium	Os	76	190.2	0.86
iridium	Ir	77	192.2	0.96
platinum	Pt	78	195.1	1.4
gold	Au	79	197.0	0.18
mercury	Hg	80	200.6	0.60
thallium	Tl	81	204.4	0.13
lead	Pb	82	207.2	1.3
bismuth	Bi	83	209.0	0.19
thorium	Th	90	232.0	0.04
uranium	U	92	238.0	0.01

Given the basic knowledge of nucleosynthetic reactions, together with the chemical composition of the Sun, it is now possible to provide a clear interpretation of the measured cosmic abundance of the elements (Figures 4.2 and 4.3). Hydrogen (75.4 per cent) and helium (23.1 per cent) are overwhelmingly the most plentiful elements. All the remaining elements together make up only 1.5 per cent of the total. There are eight other elements with significant abundances, including the life-forming elements carbon (C), oxygen (O), nitrogen (N) and phosphorus (P). A pronounced peak in abundance occurs for elements of atomic number close to iron, due to the maximum binding energy per nucleon. Lithium, beryllium and boron are unusually rare compared to their neighbours (helium, carbon, nitrogen and oxygen). Elements of even atomic number are more plentiful than adjacent elements of odd atomic number.

The chemistry of the interstellar medium

Space between stars is not empty but is sparsely occupied by diffuse matter, both gaseous and particulate. Locally this matter is gathered into clouds, and in our galaxy such clouds create dark patches in the Milky Way by obscuring starlight.

This interstellar medium is composed of primordial gaseous hydrogen and helium synthesized in the big bang, together with a mix of heavier elements, created in the envelopes of red giant stars and blown out into space by the pressure of the solar wind, and others spewed into space during periodic supernovae. This is the material out of which new stars such as our Sun and planets are born when the clouds undergo local condensation into a hot, denser cloud, or solar nebula.

Knowledge of the physical and chemical properties of the interstellar medium comes from spectral studies of starlight which is modified by both absorption and scattering in its passage through the clouds. Hydrogen and helium are the predominant gaseous elements. Gaseous carbon, oxygen and nitrogen do not occur in the same abundance as observed in the stars, but instead are believed to be present as solid chemical compounds. Thus the particles of interstellar dust are composed of mineral cores about 0.1 micrometre (μm) in size, surrounded by a mantle of dirty ice 0.3 μm in diameter. Minerals that have been tentatively identified are quartz, magnetite, and magnesium silicate; water, ice, metallic iron and graphite are also present. There is a growing list of more complex molecules that have been identified, especially organic molecules composed of the life-forming elements carbon, oxygen and nitrogen. For instance ammonia (NH_3), carbon monoxide (CO), thioformaldehyde (H_2CS) and even complex species like HC_7N (cyanotriacetylene) have been detected.

Stars born early in the evolution of the galaxy (called first generation stars) are composed principally of primordial hydrogen and helium. However, second and later generation stars condense out of interstellar gas enriched in all the heavier elements and compounds mentioned above, due to the course of nucleosynthesis. Compared to the universe as a whole, the solar system, and particularly the Earth, is enriched in heavier elements and life-essential elements (C, O, N, P). All the heavier elements on Earth were synthesized inside a generation of stars that evolved and died prior to the birth of the Sun and planets from a second generation solar nebula.

The origin and chemical evolution of the solar system

There is general agreement that the Sun and planets formed out of a rotating disc of interstellar gas and dust termed the solar nebula, some 4600 Ma ago. The bulk composition of the Sun provides the best

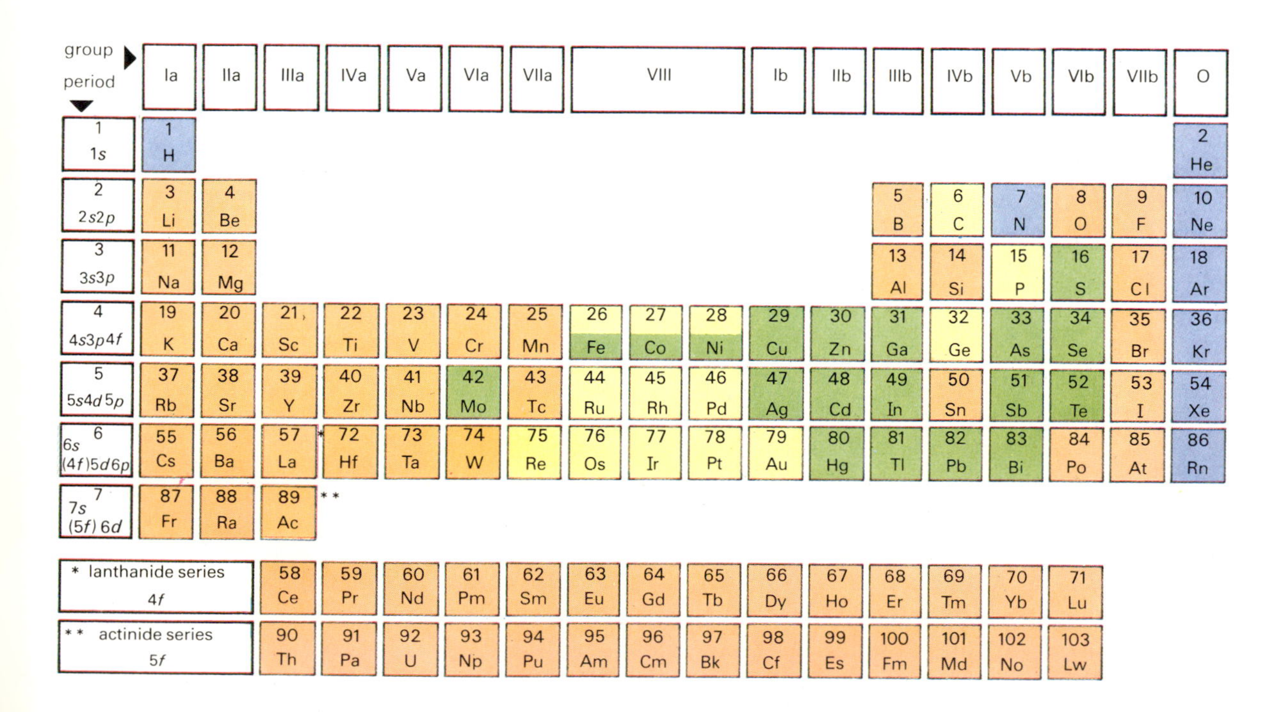

4.4: *The periodic table of the elements, showing their geochemical classification.*

indication of the overall chemical nature of interstellar matter from which the Sun and planets condensed (Figure 4.6).

Whereas nuclear reactions at high temperatures in stellar interiors govern the synthesis and distribution of elements in the universe, as discussed above, chemical forces determine the variety of chemical compounds formed at much lower ambient temperatures in the solar nebula. Initial temperatures within the solar nebula are believed to have been about 2000 degrees in proximity to the Sun, diminishing to less than 20 degrees at the periphery, in the present location of Pluto. Conditions of temperature exert a strong control on the nature of compounds and minerals that condense out of the solar nebula gas; the sequence of condensation with respect to diminishing temperature is given in Figure 4.5, and this is likely to have resulted in a chemical zonation from the inner to the outer reaches of the solar system. From these considerations it is clear that the temperature conditions under which a planet accreted will depend on its distance from the Sun, with temperature in turn dictating the nature of material from which the planet coalesced.

The planets may be classified into an inner group of terrestrial planets (Mercury, Venus, Earth, Mars) possessing low mass but high density, and an outer group of giant planets (Jupiter, Saturn, Uranus, Neptune) having high mass but low density. Chemical zonation within the solar nebula accounts for the broad chemical compositions of the two different planetary groups. The terrestrial planets condensed close to the Sun, receiving a major complement of high-temperature and high-density silicate together with nickel–iron condensates, with a correspondingly small quantity of low-temperature condensates. On the other hand the giant outer planets accreted from cool material

4.5: *Simplified sequence for the condensation of minerals and compounds from the solar nebula, and approximate positions of planetary bodies in the sequence.*

Condensation temperature	Planet	Class of material	Minerals and compounds	Minor elements co-condensing	
1800–1500		high temperature condensates (principally oxides of calcium and aluminium)	corundum–Al_2O_3 calcium oxide–CaO melilite–$Ca(Al,Mg)(Si,Al_2)O_7$ perovskite–$CaTiO_3$ spinel–$MgAl_2O_4$	platinum metals tungsten (W) molybdenum (Mo) tantalum (Ta) zirconium (Zr) rare earth elements	
1500–1000	Mercury	intermediate temperature condensates (principally metallic iron plus magnesium, calcium and aluminium silicates)	metallic nickel–iron alloy diopside–$CaMg(SiO_3)_2$ forsterite–Mg_2SiO_4 enstatite–$MgSiO_3$ anorthite–$CaAl_2Si_2O_8$	chromium (Cr) phosphorus (P) gold (Au) lithium (Li)	arsenic (As) copper (Cu) gallium (Ga)
1000–500	Venus	low to intermediate temperature condensates (principally calcium, sodium, magnesium and iron silicates)	plagioclase–$(CaAl,NaSi)AlSi_2O_8$ olivine–$(Mg,Fe)_2SiO_4$ pyroxene–$(Mg,Fe)SiO_3$ troilite–FeS	silver (Ag) antimony (Sb) germanium (Ge) fluorine (F) tin (Sn) zinc (Zn)	selenium (Se) tellurium (Te) cadmium (Ca) sulphur (S) lead (Pb)
	Earth				
500–200	Mars asteroids	low temperature condensates (sulphates, carbonates, silicates combined with water as (OH) and carbon compounds)	magnetite–Fe_3O_4 sulphate–eg $CaSO_4$ carbonate–eg $CaCO_3$ hydrous silicates carbonaceous compounds	bismuth (Bi) indium (In) thallium (Tl)	
less than 200	satellites of Jupiter	ices	water ice–H_2O ammonia ice–NH_3 methane ice–CH_4		
less than 20		condensation of gases to liquids	neon liquid hydrogen liquid helium liquid		

Note Minerals formed at higher temperatures may react with remaining nebula gases to lower temperature condensates as temperatures diminish. Approximate temperatures at which the terrestrial planets (Mercury, Venus, Earth, Mars), asteroids and satellites of Jupiter accreted from condensates according to the equilibrium condensation model are given. The terrestrial planets are endowed with a greater proportion of 'rocky' components, whereas the satellites of Jupiter have more 'icy' compounds.

4.6: *Abundances of the chemical elements in the solar atmosphere relative to type C1 carbonaceous chondrites. The one-to-one relationship of abundance for most elements indicates that the Sun and carbonaceous chondrites both condensed out of chemically similar interstellar material constituting the early solar nebula. All abundances are in terms of atoms per one atom of silicon. The scale is logarithmic.*

further out in the solar nebula. They thus acquired a relatively small quantity of rocky, low-temperature hydrous silicate materials, which now reside in the planetary cores (and their moons), but a large endowment of the low-density volatile elements (hydrogen, helium, carbon, oxygen, nitrogen), present as liquids, gases and ices.

Within the group of terrestrial planets Mercury and Earth have high densities (5440 and 5220 kg/m^3 respectively), but Mars (3940) possesses lower bulk density. These observations could be explained by the former two planetary bodies having received a larger ratio of dense nickel–iron relative to silicate (and therefore more massive cores) during accretion. Inasmuch as all the terrestrial planets accreted in the same general region of the solar nebula, there are difficulties encountered in explaining major variations in the metal to silicate ratio. A more probable alternative is that all the terrestrial planets accreted from material with comparable quantities of most chemical elements, but that greater availability of oxygen in Venus and Mars converted more metallic iron to its oxidized state of iron oxide (density 3.78 inside a planet), while Earth and Mercury retained a greater proportion of iron in the reduced metallic state (density 3.99 inside a planet).

Meteorites

Meteorites are solid bodies travelling from space which impact on the terrestrial surface when captured by the Earth's gravitational field. Most meteorites are believed to have originated within the solar system, deriving specifically from asteroids, rocky objects moving around the Sun in a wide variety of orbits but principally confined to swarms marshalled in a belt between Mars and Jupiter. The parent bodies of some asteroids were small planets, termed planetesimals, perhaps a few hundred kilometres in diameter, which once resided in the region now occupied by the asteroid belt. Planetesimals accreted by coalescence of small objects, including solar nebula dust, at the time the solar system formed. They underwent intense heating to a molten state, such that a core, mantle and crust differentiated. Subsequently they were fragmented by mutual collisions to form the asteroids.

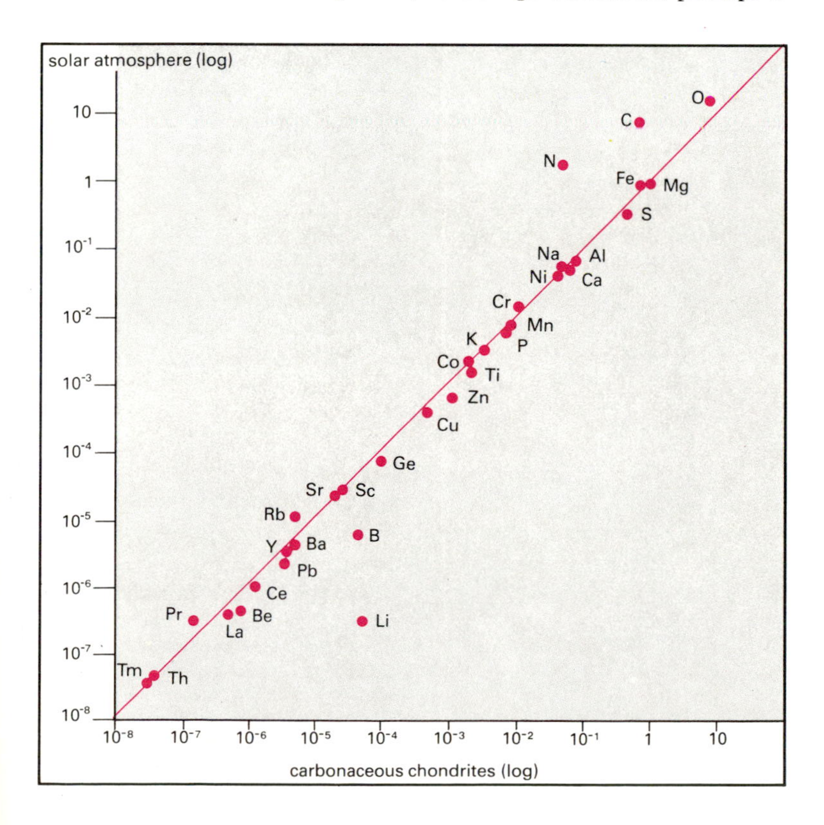

Studies of meteorite chemistry have far-reaching consequences, beyond the intrinsic interest of these rocky objects from space. In particular, meteorites provide special insight into the nature and chemical composition of planetary interiors, since the different classes appear to represent the core, mantle and crust respectively of their fragmented parent planetesimals. By contrast the Earth's interior is not accessible to observation, so we have no direct knowledge of its internal chemistry.

Meteorites may be categorized into five broad classes on the basis of chemistry and mineral content as follows.

1 Chondrites are mostly composed of silicate minerals; they represent a planetesimal mantle. Included in this class is the important group of carbonaceous chondrites.

2 Achondrites are mostly composed of silicate minerals, and represent the mantle and crust of a planetesimal.

3 Irons consist of 98 per cent metal on average, being derived from the cores of planetesimals.

4 Stony irons consist of 50 per cent metal and 50 per cent silicate on average.

5 Tektites are composed of a silica-rich glass.

Carbonaceous chondrites are of special interest because of their unusual chemistry. They are composed of spherical particles termed chondrules, embedded in a fine-grained mineral matrix. Chondrules consist of the iron–magnesium silicate minerals olivine and pyroxene, with minor nickel–iron alloy together with the mineral troilite (FeS). The coexistence of metallic iron with the iron-bearing silicates olivine and pyroxene attests to an oxygen-depleted environment of formation in the solar nebula. By contrast in the terrestrial crust and mantle all iron is combined with oxygen, entering minerals as iron oxide (FeO), and so no independent metallic iron is present. Carbonaceous chondrites also contain calcium–aluminium inclusions composed of the minerals spinel ($MgAl_2O_4$), melilite ($Ca_2Al_2SiO_7$–$Ca_2MgSi_2O_7$), garnet ($Ca_3Al_2(SiO_4)_3$), perovskite ($CaTiO_3$) and others. High contents of calcium and aluminium in these minerals reflect elevated condensation temperatures, as indicated in Figure 4.5. The fine matrix consists of low-temperature condensates, such as the minerals serpentine and clay which contain bonded water, together with about 5 per cent volatile organic compounds.

Carbonaceous chondrites have a special significance, their chemical composition being close to that of the Sun (Figure 4.6). This chemical correspondence suggests that carbonaceous chondrites are samples of material formed in the solar nebula at the same time as the Sun, but largely unmodified by melting as occurred in the planets and planetesimals. The coexistence of both high- and low-temperature minerals in single bodies may indicate mixing of high- and low-temperature condensates in the solar nebula.

Chondrites have the same overall proportions of iron and silicate minerals as carbonaceous chondrites, but exhibit pronounced depletion of volatile trace elements compared to solar abundances, and are devoid of water and volatile organic compounds present in carbonaceous chondrites. For instance the highly volatile elements bismuth, thallium and indium are depleted by a factor of a hundred to a thousand relative to their abundance in carbonaceous chondrites. Removal of these volatile elements is believed to have arisen both from melting of the parent planetesimals and from the subsequent collisions.

Achondrite meteorites resemble rocks from the terrestrial mantle and crust (basalts), thus providing evidence for early volcanic activity on the parent planetesimals prior to total cooling. In common with chondrites, they exhibit extreme depletion of volatile trace elements and water compared to terrestrial rocks.

Iron meteorites consist essentially of nickel–iron alloy (4–20 per cent Ni), with minor troilite, cobalt, phosphorus and graphite. Stony iron meteorites contain nickel–iron alloy and silicate minerals in approximately equal proportions. Iron and stony iron meteorites provide direct evidence of the behaviour of siderophile elements (those with an affinity for iron) and lithophile elements (those having an affinity for silicate minerals) during melting and internal differentiation of their parent planetesimals into an iron core and silicate mantle. Direct analysis of the iron and stony silicate components reveals that the key siderophile trace metals cobalt, nickel, iridium, gold and tungsten are preferentially enriched in the iron core. This knowledge of the siderophile element distribution between iron and silicate is vital to interpreting the extent to which the bulk Earth was depleted in siderophile elements by scavenging into the iron core during core formation. Furthermore the low siderophile element content of the Moon, given that the Moon has only a small core, requires that it was derived from a planetary body that had already undergone internal differentiation and core formation. Generalized chemical compositions for the various classes of meteorites are given in Figure 4.7.

Meteorites have retained an important chemical and isotopic record of processes involved in the earliest stages of solar system formation. Chondritic meteorites give the oldest dates of any rocky material yet analysed: their age of 4600 million years is taken to represent the time of formation of the solar system. Planetesimals (from which the meteorites are derived) accreted, melted, differentiated and cooled within a few million years; thus their isotopic clocks were 'set' from the moment of cooling and have not subsequently been disturbed. On the Earth, however, internal heating, mixing and volcanic activity has continually reset isotopic clocks in rocks, so that no very old ages are preserved.

Independent evidence for rapid accretion, melting and cooling of planetesimals in the early solar nebula is retained in the form of exotic isotopic anomalies within some meteorites. The anomalous isotopes arose as decay products of short-lived radioactive nuclides such as iodine–29 (half life of 16×10^6 years), plutonium–244 (82×10^6 years) and aluminium–26 (10^6 years). These exotic parent nuclides are believed to have been synthesized in a supernova nearby in the Galaxy, expelled into space, captured by the solar nebula and incorporated into some chondrites, all within a few million years before the radioactive parents had decayed to insignificant levels. If the exotic nuclides were indigenous to the solar nebula, rather than introduced from a supernova, they would be uniformly distributed among all meteorites, but this is not the case. For instance, the notable Allende meteorite has an excess of magnesium–26 over 'normal' levels, due to the radio-

4.7: *The percentage chemical composition of meteorites, with their metal and silicate fractions, compared to the Earth's crust.*

	Metal[1] (from irons)	Metal[1] (from chondrites)	Silicate (from chondrites)	Average chondrite	Carbonaceous chondrite[2] (type 1)	Tektite[3] (average)	Earth's crust[4]
oxygen			43.7	33.2		47.8	46.6
iron	90.8	90.7	9.88	27.2	27.34	3.73	5.0
silicon			22.5	17.1	31.12	34.3	27.7
magnesium			18.8	14.3	32.48	1.33	2.09
sulphur				1.93	5.65		0.026
nickel	8.59	8.80		1.64	1.37		0.075
calcium			1.67	1.27	1.81	1.64	3.63
aluminium			1.60	1.22	2.68	6.8	8.13
sodium			0.84	0.64	1.97	1.97	2.83
chromium			0.51	0.39	0.40		0.01
manganese			0.33	0.25	0.22	0.09	0.09
phosphorus			0.14	0.11	0.33		0.11
cobalt	0.63	0.48		0.09	0.07		0.025
potassium			1.11	0.08	0.12	1.77	2.59
titanium			0.08	0.06	0.09	0.48	0.44
Volatile species							
H_2O				0.27	19.7		0.68
C				0.15	2.99		
organic compounds					6.71		

Notes

1 Iron meteorites and the iron fraction of chondrites are remarkably similar in composition and are relatively enriched in the siderophile element cobalt.

2 Carbonaceous chondrites are extremely volatile-rich (the volatile species are given separately).

3 Tektites are chemically unlike other meteorites or the Earth's crust.

4 The Earth's crust has greater abundances of silicon, aluminium and oxygen relative to average chondrite, together with elevated levels of the lithophile elements sodium, potassium, calcium and titanium, but lower than average chondrite levels of the siderophile elements sulphur, nickel, chromium, manganese, phosphorus and cobalt, due to their selective incorporation into the Earth's metallic core.

active decay of aluminium-26, and also an excess of oxygen-16 above the 'normal' terrestrial proportion of this isotope. It is possible that the supernova responsible for the exotic nuclides actually triggered contraction of an interstellar gas cloud into the early solar nebula.

Tektites are small (200–300 g) disc-shaped objects that result from melting and ablation by aerodynamic frictional heating during rapid passage through the atmosphere. Unlike meteorites, tektites occur in specific geographical locations known as strewn fields, within which they are contemporaneous, due to the fact that they fall in intermittent showers. Tektites also lack the signs of bombardment by solar cosmic rays, which produce ^{3}He, ^{20}Ne, ^{26}Ne, ^{36}Ar etc in all other meteorites; this is taken to indicate a short residence time in space. Tektites of all ages, even Archaean, are known.

Chemically tektites are unlike any terrestrial volcanic rocks or lunar surface rocks, but in some ways resemble terrestrial sediments termed greywackes and certain lunar granites. They are characterized

4.8: *Below left, iron meteorite (×1), which was part of the Henbury Fall, Northern Territory, Australia. The fall is believed to have occurred in prehistoric times, the craters being comparatively fresh. Above right, a polished section of a surface cut through an iron meteorite which fell in Arizona, USA. The internal pattern is called Widmanstätten structure, which is produced as a result of the redistribution of the nickel and iron into bands of different minerals during cooling. Below right, a stony-iron meteorite (×1) which fell at Krasnoyarsk in southern Siberia, USSR. The large rounded crystals are olivine.*

by high SiO_2, Al_2O_3, K_2O and CaO and low MgO, Fe_2O_3 and Na_2O; they contain low abundances of the volatile trace metals lead (Pb), thallium (Th), copper (Cu) and zinc (Zn), and there is almost total depletion of water and carbon dioxide. Tektites are probably not related to terrestrial volcanism, nor derived as splashes of molten rock from giant meteorite impacts. They may be the product of a lunar granitic volcano firing debris at the Earth, but their origin remains a subject of controversy.

The chemical composition of the Moon

The overall chemical composition of the Moon, the Earth's only satellite, is now known in broad terms, but ideas on its internal structure and detailed chemistry are still conjectural. Hypotheses for the origin of the Moon are numerous, but considerations of lunar chemistry as well as of celestial dynamics strongly limit the possibilities. The low mean density of the Moon (3340 kg/m^3) indicates that, relative to the Earth, the Moon is depleted in iron (the most plentiful heavy element in the solar system), and this is reflected in the presence of only a small metallic iron core, independently deduced from evidence of a lunar magnetic field.

4.9: *A selection of tektites to show the range of aerodynamically moulded forms.*

Samples of rock returned from the Moon during the Apollo missions may be separated into two classes on the basis of their mineral content and chemistry. In the first group is the rock type called anorthosite, forming a lunar crust that is 40 km thick; the second group comprises basalts which occupy the lunar maria (plains).

The lunar anorthosite crust separated from a magma ocean perhaps 400 km thick about 4400 million years ago, by upward flotation of plagioclase feldspar ($CaAl_2Si_2O_8$). Because of the preponderance of plagioclase, the lunar crust has a notably high content of calcium and aluminium. Lunar anorthosites have chemical characteristics unlike those of comparable terrestrial counterparts; in particular the calcium content of plagioclase is unusually high, and the abundances of certain volatile elements (Br, Cd, Ce, Sb, Te, Zn etc) are elevated relative to mare basalts. Furthermore, the entire lunar crust has been subjected to intense meteorite bombardment, contaminating surface rocks with material of a chondrite type of chemistry.

There also exists a group of crustal rocks termed norites, which are notable for either high or low contents of certain 'incompatible' elements, including potassium (K), the rare earth elements (REE) and phosphorus (P), to which the acronym KREEP is applied. Incompatible elements are so termed because during partial melting they are selectively concentrated from the solid into the melt fraction. KREEP norites appear to have formed by local partial melting of the anorthositic crust.

Lunar basalts resemble terrestrial basalts in overall chemistry, each having been derived by partial melting of the mantles of their respective parent bodies. However, two major differences exist in lunar basalts: a wide variation in abundance of titanium and certain other elements in some mare basalts, coupled with extreme depletion of the volatile trace elements, water and carbon dioxide in all basalts sampled. Depletion of the most volatile major elements, silicon and sodium, is also evident in lunar basalts relative to their terrestrial counterparts. It is believed that low-titanium basalts were derived by simple partial melting of the lunar mantle, whereas the high-titanium basalts were contaminated by mixing with the residual heavy mineral 'sink' left over beneath the crust, including the titanium mineral ilmenite ($FeTiO_3$).

Lunar basalts possess abundances of non-volatile siderophile elements (Ni, Co, W, Os, Ir, P) which closely resemble those in the Earth's mantle. Given that the low terrestrial mantle abundances are the result of scavenging of these elements into the metallic iron core, due to their affinity with iron, it follows that the comparable lunar abundances also require that the lunar mantle equilibrated chemically with a metallic iron phase. Yet the Moon possesses only a small core. These and other considerations have led to the idea that the Moon split off from the Earth. It is postulated that during the Earth's melting the metallic iron segregated into the core, sweeping the siderophile elements out of the mantle. Transfer of mass towards the Earth's centre led to an increased rate of spin, such that part of the Earth's mantle was 'thrown off' by centrifugal force into a disc orbiting the Earth. The Moon formed by coalescence of the disc, which accounts for certain geochemical similarities to the terrestrial mantle and for loss of volatiles in the pre-lunar disc during fission. In this context it is significant that the volatile siderophile elements (Cu, Ga, Ce, As, Ag, Sb, Au) are depleted in the Moon relative to the Earth precisely because of this devolatilization episode.

The chemical composition of the Earth

A visitor from space, approaching the planet Earth, would observe that there is a significant atmosphere, that about 70 per cent of the surface is covered by oceans and that 30 per cent of the surface is above sea level. On the basis of observations of the rocky surface, both above and below sea level, and from the increasingly sophisticated observations of modern seismology, the Earth can be regarded as consisting of a series of layers increasing in density toward the centre. The Earth can be divided into a series of well-defined geospheres: atmosphere, hydrosphere, crust, mantle and core. Each part may be further subdivided: the water on continents can be distinguished from that of the oceans, the continental crust from the oceanic crust, the upper and lower mantles, the outer liquid core and the inner solid core (Figure 3.12). This section summarizes knowledge of their chemistry.

Since the outer shells can be sampled directly, knowledge of their chemical composition is as good as the sampling and the methods used in chemical description. As depth increases, direct sampling rapidly becomes more difficult, and indirect approaches to ascertain chemical composition become increasingly important. Seismic data and data from cosmic chemistry, solar chemistry and meteorite chemistry—all of which provide information on the inner physical properties of the planet—must be used to obtain models of deep Earth chemistry.

The basic control of the Earth's geochemical structure

Given the composition of the various parts and their masses (Figure 4.10) an overall composition of the Earth might be like that shown in Figure 4.11. From this composition, dominated by iron, oxygen, silicon, magnesium, aluminium and the alkali metals, it is not difficult to explain the gross features of the planet. First, if all the atomic components were to be mixed in a container of appropriate volume, competition would arise between the metallic elements and non-metallic elements (O, S) to form oxides and sulphides. When the affinity of the elements with oxygen is considered (based on the free energy of formation of the oxides) it is clear that species such as MgO, SiO_2, Al_2O_3, Na_2O and CaO would form in preference to FeO. Thus as the bulk composition is oxygen-deficient, most of the iron and nickel would be left in the unoxidized state. Of course, the separate oxide phases would combine to form silicates (eg $MgO + SiO_2 \rightarrow MgSiO_3$ (pyroxene) or olivine (Mg_2SiO_4)).

If the condition of gravitational equilibrium were now to be imposed the most dense materials would be at the centre and the mixture would stratify, forming the dense inner metallic core and the outer silicate mantle. Atomic species at low concentrations would tend to separate in a sympathetic manner. Thus a noble metal such as gold would tend to be associated with metallic iron, and an oxyphile element such as uranium with the lighter silicate fraction.

Next it may be assumed that the Earth is heated until it is totally or partially molten. The source of this energy may be energy of accumulation, energy of core separation or radioactive heating. Given such heating and partial fusion, volatile species with very low melting temperatures (H_2O, CO_2, N_2, Ar etc) will rapidly boil off, to form the atmosphere and hydrosphere systems. In the mantle the most fusible light liquids will rise to the surface, enriching the upper layers, in such components as SiO_2, Al_2O_3, Na_2O and K_2O, while the heavy fusible components will move to the core (Fe, Ni, FeS etc).

Early geochemists recognized this basic pattern of separation of the elements and classified elements as atmophile (preferring to be in the atmosphere), lithophile (in silicates and oxides), chalcophile (in sulphides) and siderophile (in iron or metals).

Given a hot early Earth rapid separation into the layers described above would be expected. The separation is essentially a result of the volatility, fusibility and density of the most stable chemical units (metals, oxides, silicates etc). However, it should be noted that, given the very high pressures and temperatures within the Earth, compounds may form in structures quite different from those that are familiar at the Earth's surface. Thus while SiO_2 occurs at the surface as the common mineral quartz, of density 2650 kg/m^3, at depth the common mineral will be the polymorph stishovite of density 4290 kg/m^3. All present evidence shows that the Earth had reached something like its present structure about 4000 million years ago. The oldest samples of crust, dating from about 3800 million years ago, contain rocks quite similar to those forming today, and most were formed in a submarine environment (see Chapter 16).

4.10: *Mass relations of major parts of the Earth (kg).*

atmosphere	5.3×10^{18}
hydrosphere	1.4×10^{21}
biomass	1.0×10^{15}
oceanic crust	7.0×10^{21}
continental crust	1.6×10^{22}
mantle	4.1×10^{24}
core	1.9×10^{24}

4.11: *A possible composition of the Earth (per cent by weight).*

Fe	35	S	1.9	Mn	0.22
O	30	Ca	1.1	Co	0.13
Si	15	Al	1.1	P	0.10
Mg	13	Na	0.57	K	0.08
Ni	2.4	Cr	0.26	Ti	0.05

The atmosphere

The atmosphere of Earth is unique in the solar system. At sea level atmospheric pressure is about 1 atmosphere, ie 1×10^5 pascals (Pa). The total mass of the atmosphere is 5.3×10^{18} kg (see Figure 4.10), a very small part of the total mass of the planet, which is 5.973×10^{24} kg. The nature of the atmosphere is a result of gravitational stratification and the very complex photochemical interactions that take place between solar radiation and the molecular species present (see Chapter 17).

At sea level the typical composition is as shown in Figure 4.12. Except for N_2, O_2 and Ar, all gases are present in very small amounts. The thermal structure of the atmosphere is a complex result of pressure change, radiative balances and photochemical processes. Thus temperatures decrease upwards from the surface to about 12 km, increase from 12 to 50 km, decrease again to about 100 km and then steadily increase. At high altitudes photodissociation processes are extreme, and single atoms or ionized atoms dominate the tenuous atmosphere; in the upper layers of the atmosphere a large fraction of light atoms may reach the escape velocity, and it is estimated that about two-thirds of all hydrogen atoms present will escape in about a thousand years. There may also be some loss of helium, but for all heavier species escape is negligible. As all hydrogen compounds are subject to photochemical dissociation, their lifetime in the upper atmosphere is short.

The hydrosphere

The mass of liquid water on our planet is about 1.4×10^{21} kg, about 2 per cent of the mass being in streams, lakes and ice sheets, while 98 per cent is in the oceans. Continental (fresh) waters are relatively dilute solutions, while ocean waters are more saline (Figure 4.13). Inspection of the appropriate data shows that ocean water does not simply result from concentration of inflowing river waters, but that interactions with the solid crust must play a major part in its chemical evolution.

The crust

The outer layers of the solid Earth, the crust, consist mainly of materials with density below 3000 kg/m^3. This crust can be divided into continental crust (mass 1.6×10^{22} kg), with an average density around 2700 kg/m^3, and ocean floor crust (mass 7.0×10^{21} kg), with density around 3000 kg/m^3. The continental crust averages about 30 km in thickness, while the oceanic crust is in the range of 5 to 10 km in thickness. Knowledge of the chemical composition of the crust results from sampling the various rock types accessible at the surface, and from mining and drilling beneath the surface. Only recently has ship-based drilling made it possible to sample the ocean floor crust. The rock types most likely to be dominant at greater depth must be assessed from knowledge of structural geology or from seismic studies. Clearly, there are many uncertainties in deciding on the average composition given in Figure 4.14.

4.12: *The composition of dry air at sea level.*

Species	Per cent by volume
N_2	78.08
O_2	20.95
Ar	0.93
CO_2	0.031
Ne	0.001 8
He	0.000 52
Kr	0.000 11
Xe	0.000 008 7
H_2	0.000 05
CH_4	0.000 2
NO	0.000 05
O_3	0.000 007 (summer)
	0.000 002 (winter)

4.13: *Major ionic components in river water and seawater.*

River water (parts per million)

HCO_3^-	58.4	Na^+	6.3	Ca^{2+}	15.0
SO_4^{2-}	11.2	K^+	2.3	Fe^{2+}	0.67
Cl^-	7.8	Mg^{2+}	4.1	SiO_2	13.1

Seawater (g/kg)

Cl^-	18.98	Mg^{2+}	1.27	HCO_3^-	0.14
Na^+	10.54	Ca^{2+}	0.40	Br^-	0.06
SO_4^{2-}	2.46	K^+	0.38	H_3BO_3	0.02

4.14: *The composition of the Earth's crust (parts per million).*

H	1 520	Cu	68	Pr	9.1
Li	18	Zn	76	Nd	39.6
Be	2	Ga	19	Sm	7.0
B	9	Ge	1.5	Eu	2.1
C	180	As	1.8	Gd	6.1
N	19	Se	0.05	Tb	1.18
O	456 000	Br	2.5	Ho	1.26
F	544	Rb	78	Er	3.46
Na	22 700	Sr	384	Tm	0.5
Mg	27 640	Y	31	Yb	3.1
Al	83 600	Zr	162	Hf	2.8
Si	273 000	Nb	20	Ta	1.7
P	1 120	Mo	1.2	W	1.2
S	340	Pd	0.015	Re	0.000 7
Cl	126	Ag	0.08	Os	0.005
K	18 400	Cd	0.16	Ir	0.001
Ca	46 600	In	0.24	Pt	0.01
Sc	25	Sn	2.1	Au	0.002
Ti	6 320	Sb	0.2	Hg	0.08
V	136	I	0.46	Tl	0.7
Cr	122	Cs	2.6	Pb	13.0
Mn	1 060	Ba	390	Bi	0.008
Fe	62 200	La	34.6	Th	8.1
Co	29	Ce	66.4	U	2.3
Ni	99				

4.15: *A model composition of the upper mantle (per cent).*

SiO_2	45.16	K_2O	0.13
MgO	37.49	Cr_2O_3	0.43
FeO	8.04	NiO	0.2
Fe_2O_3	0.46	CoO	0.01
Al_2O_3	3.54	TiO_2	0.71
CaO	3.08	MnO	0.14
Na_2O	0.57	P_2O_5	0.06

The mantle

It is known from seismic studies that the mantle, extending from 10 to 90 km beneath the surface, to the core boundary at 3486 km from the Earth's centre, is made of materials considerably more dense than those of the crust. This region of the Earth cannot be sampled directly and, as noted above, ideas on its chemistry are based on cosmic chemistry, meteorite chemistry, seismic properties etc. Further constraints are imposed by what is known of the overall heat production of the Earth and, of great importance, by the fact that the material must be capable of producing the volcanic magmas that are known to rise from the deep interior at oceanic ridges and volcanic islands such as Hawaii. Knowledge of the chemistry of the mantle, particularly in its deepest parts, is uncertain. The typical mantle composition shown in Figure 4.15 will allow production of basalt liquids and will produce mineral phases of appropriate seismic properties at depth. Modern evidence of element and isotopic abundances clearly shows that the mantle is not uniform in composition. The abundances of siderophile elements such as Pt, Pd and Au indicate that these must have been largely partitioned into the metallic core during core formation.

The core

Seismic studies and the total mass of the Earth clearly show that the core (total mass 1.9×10^{24} kg, radius 3485.7 km) must have an average density of 11 000 kg/m^3. The inner core is solid while the outer core is liquid. Further, the core must have properties appropriate to the generation of the magnetic field. Given the density constraints, and the seismic constraint that the mean atomic number of the core is a little lower than for pure iron, most researchers assume that the core is dominated by iron and nickel (as in some metallic meteorites), but with some dilution by light elements. Candidates for this dilution include C, S, Si, H and perhaps O. At present the exact composition is still a matter of debate. As discussed in the paragraph above, the core probably contains most of the Earth's siderophile elements.

The biosphere

While the total mass of living matter is small (10^{15} kg), the biomass is important in concentrating and transporting certain species in the hydrosphere–atmosphere system. Most biological species require a wide range of elements, and concentrations in living organisms can lead to remarkable enrichment of some metals. These processes are often reflected in the metal content of fossil carbon deposits, such as coal. Thus there are elements that can be considered as biophile. Biological activity is very important near the surface, for it can produce highly reducing conditions, eg in subaqueous environments or, in the regions of photosynthesis, oxidizing conditions.

The present geochemical cycle

Over the past few decades there has been a spectacular accumulation of evidence to show that the Earth is cooling by a convective as well as by a conductive process. The convective process controls most of the major surface processes, such as continental drift, ocean floor spreading, mountain building and volcanism. Convection is a mixing process and, in recent years, geochemists have become aware that this phenomenon must be considered when models of Earth chemistry are posed. Not all the aspects of the convective process have yet been quantified. This section discusses general chemical processes associated with present Earth dynamics, for a knowledge of present processes is vital to our understanding of those of the past. Ultimately models of the distribution of elements must take into account these cooling and mixing processes. Clearly the present distribution of elements in the major layers of the Earth has not necessarily always existed.

The geochemical cycle, Stage 1: rising convection cells

Radioactive heat production in the mantle leads to the formation of hot mantle plumes, which rise at widely spaced centres to produce the volcanic phenomena observed at the great ocean ridge mountain ranges (see Chapter 13). At these ridges new light crust is formed from magmas, which crystallize to rocks of the basalt–peridotite group. This new crust forms at a rate of about 4×10^{13} kg/a, and the process carries almost half the energy produced in the interior to these highly localized sites. Most of the phenomena occur in a submarine environment. The new crust is formed, from both extrusive volcanic rocks and their intrusive equivalents, beneath the lava cover. The process is rapid, and has formed about 70 per cent of the Earth's crust in the past 200 million years, ie 5 per cent of the Earth's history.

Most of the new magmas arriving at or near the surface cool and crystallize in a submarine environment. As magma cools it contracts, cracks and becomes porous and permeable. It has been directly observed and shown by all theoretical treatments that, where there are hot permeable rocks under seawater, cooling must involve convective circulation of the latter. At the ocean ridges almost half the thermal energy in the new hot crust is transferred to seawater. Observers in submersibles have recorded impressive discharges of hot water near submarine volcanic sites, with hot water reaching temperatures in excess of 300°C.

The interaction of cold seawater with hot basaltic rocks produces profound chemical changes in the rocks and modifies the seawater. Thus an exchange process occurs between the hydrosphere and new mantle-derived crust. The scale of the exchange is related to the thermal energy available at the ridges, and the entire ocean mass is circulated through the ridge zones every few million years.

Recent studies of ocean floor rocks, both *in situ* and where they appear on land (notable occurrences can be found in Cyprus and Oman), show the profound chemical changes that occur. For example H_2O is fixed in hydrated minerals; O_2 dissolved in seawater oxidizes the upper layers and produces minerals such as haematite and magnetite (Fe_2O_3 and Fe_3O_4); bicarbonate in ocean water is fixed in carbonate minerals; sulphate in seawater may be precipitated as calcium sulphate, or reduced to form pyrite (FeS_2) by reaction with iron in the basalts; sodium ions in seawater exchange with cations in basalt silicates to form the mineral albite ($NaAlSi_3O_8$); and potassium ions in seawater are fixed in complex clay minerals formed by low temperature alteration.

As the cold seawater convects down into the basaltic crust, it becomes heated, and eventually discharges back to the surface. On account of the exchange reactions the fluid discharged is quite different in chemistry. It becomes enriched in gaseous hydrogen and especially in trace metals, resulting from the sodium–potassium fixation processes. Almost all the transition metals are enriched, and metals, present as chlorides, are returned to the ocean–rock interface; they include Cu, Mn, Fe, Cr, Ni, Zn, Ag and Au. Some of these metals may be precipitated as the ascending fluids cool, some may be precipitated as sulphides at the marine interface and some may be dispersed into the overlying seawater to contribute to metal-rich muddy sediments. Such fluids may also contribute to the elements that form the widely distributed manganese nodules rich in Mn, Fe, Co, Ni and Cu. At the present time these phenomena are the subject of intensive research, and already submersibles have directly observed and sampled new metallic ores formed by such processes. The altered rocks are chemically complex, as is to be expected for a convective process, where the pressure, temperature and chemistry of the fluid all change along the flow path.

At the great ocean ridge systems of our planet new crust is formed as a product of partial fusion of rising hot mantle. This rock is modified by cooling, through seawater exchange. Components from the atmosphere and hydrosphere are added to the solid crust, and the oceans are enriched in some components from the mantle. Many of the great sulphide and oxide ore deposits are formed by this process.

The geochemical cycle, Stage 2: sinking convection cells

It is now well established that the new crust and lithosphere, formed and chemically modified at ocean ridges, spreads away from the ridges

4.16: *A manganese nodule dredged from the floor of the Pacific Ocean. The maximum diameter is 70 mm. The average composition of the nodules is 20 per cent manganese, 15 per cent iron, varying amounts of clay, calcium carbonate, and volcanic fragments, and traces of copper, cobalt and nickel. This material is built up by accretion around a nucleus, which can be a pebble, volcanic fragment, organic particle, or even a shark's tooth.*

4.17: *A broken surface through the same nodule displaying the characteristic internal layering which is built up by accretion at a rate in the order of 10 mm per 5 million years.*

and is almost completely returned to the mantle near the great trench systems of the Earth, in what are termed subduction zones (see Chapter 14). In this process all the chemical substances fixed in the altered ocean crust are returned to the mantle, to depths perhaps as great as 700 km, as seismic studies have shown. Thus atmosphere–hydrosphere components are recycled deep into the mantle.

Before the older, cooler and heavier crust reaches the subduction zone it is covered by a sediment layer. In the subduction process these sediments may be scraped off to build continental crust, or subducted. The exact balance of these two possible processes is not well known at the present time, but recent studies have shown that, where the angle of descent of the lithosphere is steep, sediments appear to be subducted. As typical deep ocean sediments are rich in SiO_2, Al_2O_3, potassium-bearing clay minerals and even uranium, these species may be returned to the mantle. There is no doubt that much research in the next decade will be directed at quantifying exactly what materials are subducted. The process has profound implications for mantle chemistry and models of crustal evolution.

One feature of this process is clear. Given the present knowledge of what is subducted there must be a most efficient process of return flow. Consider just one species, water. At present water is returned to the mantle at a rate of about 1.5×10^{12} kg/a (assuming 5 per cent water in the ocean crust). This rate might be doubled if pelagic sediments were subducted. As the ocean mass is 1.4×10^{21} kg, this entire mass would be removed in a thousand million years (or half this time, if sediments are subducted). Since there are still oceans now this water must be returned to the surface.

The return flux is not well quantified. Some fluids must be released by metamorphic dehydration processes as the crust descends and becomes hotter. But the most obvious return process involves the volcanism that is associated with the regions above subduction zones. As the descending crust becomes sufficiently heated dehydration processes may cause melting of the materials and melting in the hotter overlying mantle (Figure 4.18). These rising andesite melts have exactly the compositions that might be expected. They are generally similar to the basalts formed at the ridges, but have higher contents of elements that are involved in the modification of the primary ridge materials, with a possible contribution from sediments. Thus the andesite magmas are often rich in gases (H_2O, CO_2) and elements such as potassium, sodium, silicon and uranium. The exact balance between what goes down and what comes back to the surface is not quantified, but is a subject of intense research at the present time.

The magmas that rise above subduction zones carry heat to the overlying crust. In many regions this crust is continental, and it responds to the heating from below by partial melting. The process leads to the production of magmas of the granite family, and these rocks form a major part of the crust above subduction zones. Granitic magmas tend to rise in giant (often 500 km^3) bubbles and, where they penetrate to near the surface, they also cool, crack and can be water-cooled by typical continental ground waters. Again the fluid convective processes associated with the rise of these hot materials lead to the formation of major ore deposits, particularly involving elements such as copper, tin, tungsten, molybdenum and gold. (The processes of mineralization at active subduction zones are discussed in greater detail in Chapter 14.)

The geochemical cycle, Stage 3: erosion

The convective processes outlined above produce the major surface features of the Earth and, in particular, continuously generate high elevations, which are then subject to the various forms of erosion by particle and solution transport. These erosional processes, involving the atmosphere and continental hydrosphere and biosphere, close the geochemical cycle.

Rain water dissolves rocks, and carries the dissolved species to the oceans. The primary rocks of the crust are formed by igneous processes; the minerals in these rocks are unstable in the oxygenated aqueous surface environment. Thus typical feldspar minerals ($NaAlSi_3O_8$, $KAlSi_3O_8$, $CaAl_2Si_2O_8$), which make up about 50 per cent of igneous rocks, react with water, producing a residue of complex clays ($Al_2Si_2O_5(OH)_4$):

$$\text{feldspar} + H_2O \rightarrow (Na^+, K^+, Ca^{2+} + OH^- + Si(OH)_4 \text{ in solution}) + \text{clay}$$

If the process continues in very stable regions, eventually all the Na, K and Ca moves to the oceans, and a second process

$$Al_2Si_2O_5(OH)_4 + H_2O \rightarrow Al_2O_3.3H_2O + (Si(OH)_4 \text{ in solution})$$

leads to the formation of a residue rich in Al_2O_3, a 'soil' termed bauxite, a potential ore of aluminium.

Reactions of the iron–magnesium minerals of igneous rocks (see Chapter 5) may be illustrated by the mineral olivine as follows:

$$(MgFe)_2SiO_4 + \tfrac{1}{2}O_2 + H_2O \rightarrow (Si(OH)_4 + Mg^{++} + 2OH^- \text{ in solution}) + Fe_2O_3$$

In this case, the end product of leaching is a residue highly enriched in ferric oxide, typical of red tropical soils. Often, carbon dioxide dissolved in the water controls the solution of minerals; carbonate minerals are dissolved as bicarbonates:

$$CaCO_3 + CO_2 + H_2O \rightarrow (Ca^{++} + 2HCO_3^- \text{ in solution})$$

The present erosion rate of the continents is about 10^{14} kg/a. As the total mass of continental crust is 1.6×10^{22} kg, it could theoretically be entirely reworked in a few hundred million years, although the continued existence of much older crust shows that this recycling is not happening globally. This process recharges the dissolved chemicals of the oceans, some of which modify the chemistry of the igneous crust of the ocean floor, which eventually returns to the mantle.

The geochemical evolution of the Earth

One of the great objectives of the Earth sciences is to understand the history of our planet, starting from its birth some 4550 million years ago. We may also wish to predict the directions of future change. At present the problems far outweigh the facts, and every time new information comes from exploration of other members of our solar system new surprises appear and models change with the new observations. This section discusses some of the present ideas, but it should be stressed that the available data are still few and imperfect. Improvements will undoubtedly come as the techniques of observation become increasingly sophisticated.

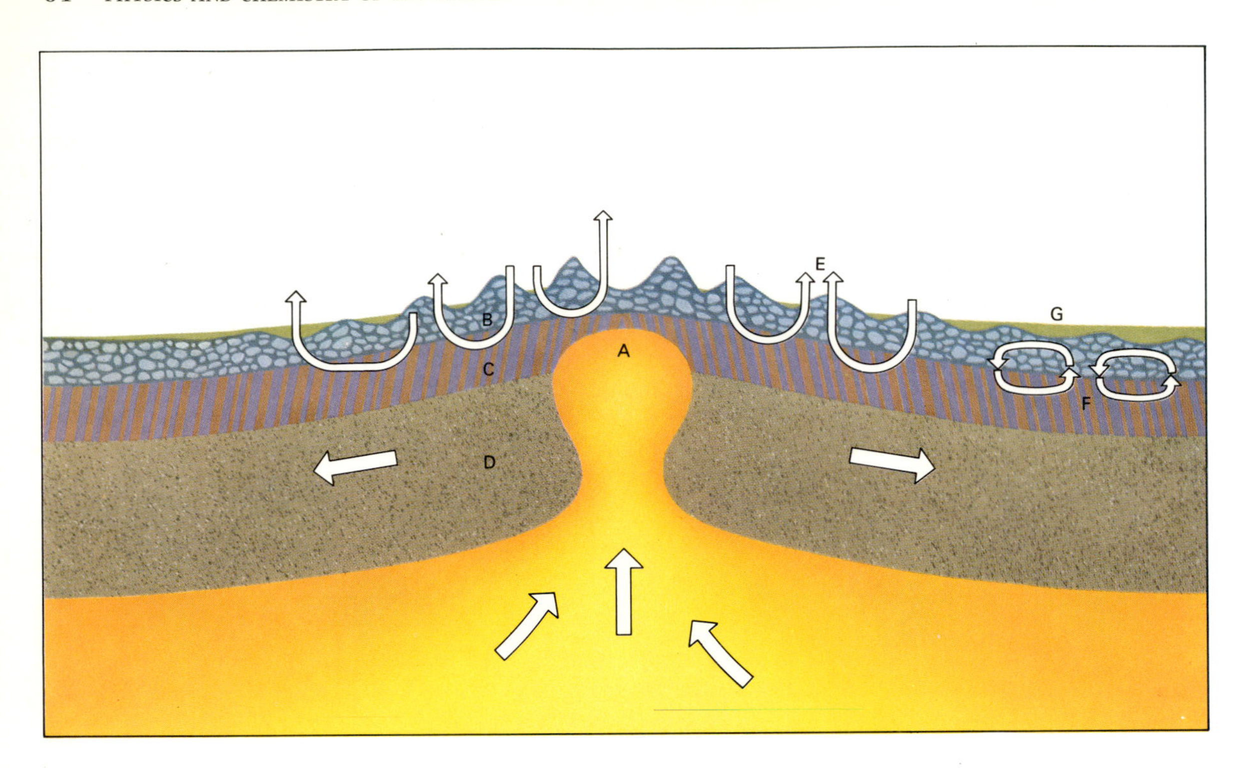

4.18: *Left: the ocean ridge where hot mantle materials rise and create new oceanic crust. The surface B consists of basaltic lava flows and pillow lavas; layer C below is made up of magma-filled cracks or dykes, and layer D of gabbro-peridotite. A large, basaltic magma chamber is shown at the ridge. The crust cools by conductive heat flow and by driving seawater convection cells, which vent to the ocean near the ridge (E). In favourable cases sulphide ore deposits may form from the discharge of metals in solution* (Cu, Zn, Fe, Ag). *The new crust moves away from the ridge and becomes sediment-covered (G). Fluid convection becomes confined in the basaltic layer beneath the impermeable sediments. Below left: a subduction zone where sediment-covered ocean crust and lithosphere (about 100 km thick) descends back to the mantle near a trench environment (A). Sediments may be scraped off, adding to the continent, but some may be dragged back to the mantle. At depth (D) water released from the descending slab may trigger melting and production of andesite melts which rise to form volcanoes. The thermal flux may lead to crustal fusion and the creation of large granite plutons (B), which may be cooled by convecting groundwaters (C), again leading to metal concentrations and ore deposits* (Cu, Mo, W, Sn, *etc*).

Accretion and early modification

As discussed earlier in this chapter modern observations on objects in the solar system point to rapid accumulation of the major bodies in a primitive dust cloud surrounding our Sun. Evidence from all the major solid bodies in the solar system shows that during the first five hundred million years all larger objects were subjected to intense bombardment by meteorites, some of very large sizes. Melting phenomena in small meteorite bodies suggest that there may have been heating by short-lived radioisotopes, such as ^{26}Al, while some researchers have suggested that super-heavy nuclides may have been present. For large bodies a massive contribution to heating must have come from bombardment and energy of gravitational accumulation. Evidence, particularly from the structure of the Moon, points to a high degree of melting of the early planets, with what are now called 'magma oceans' forming the outer layers.

Given high degrees of melting at the early stages, certain processes must follow, and would obliterate evidence as to whether the planet accumulated homogeneously or heterogeneously. For the Earth: the heavy metallic core would rapidly settle out, carrying with it the necessary light mixed elements (O, S, Si, H etc); the mantle would crystallize, with light feldspathic materials floating and the dense iron–magnesium silicates sinking; minor elements would be distributed in the major regions according to their oxygen affinity and density; volatile gases (H_2O, CO_2, N_2 etc) would rapidly form a heavy atmosphere in solution equilibrium with possible magma oceans beneath the outer shell; atmospheric photochemical processes and high-level hydrogen escape would rapidly limit or eliminate light gases such as methane, ammonia and hydrogen in the atmosphere, and produce low levels of free oxygen and ozone; and the early surface would be catastrophically reworked by meteorite bombardment.

The geological record begins

The oldest rocks preserved on Earth were formed approximately 3800 Ma ago and are surprisingly normal. Such rocks at Isua, Greenland, were formed in a submarine environment, and submarine basalts and sediments, including iron formations, form part of the succession. The record, shown in almost all of the small ancient remnants, is dominated by rocks formed in a submarine environment. There is no doubt that the Earth had a well-developed ocean and crust 3800 Ma ago (see Chapter 16).

Archaean sediments and igneous rocks are less potassic than their modern counterparts (Figure 4.19). Biological structures and biomolecules were present 3500 Ma ago. Metamorphic rocks show evidence for slightly, but not drastically, higher thermal gradients in the crust. Submarine hydrothermal ore deposits were common, and water-cooling processes were similar to those for the modern Earth. The greater heat production in the mantle was reflected in higher crustal gradients and in the eruption of some magma types with melting temperatures of 1600–1700°C, greater than present magma temperatures (1200°C). There is a surprising lack of features that might be correlated with meteoric phenomena. The oxygen isotope ratios of silica in sedimentary cherts suggest that ocean temperatures might have been much higher than at present; there may have been a 'greenhouse effect' on the Earth (a similar phenomenon has been suggested for Mars). There is little definitive evidence that the atmosphere was significantly different from that of today. All evidence on the structure of the ancient crust points to a much closer spacing of mantle magma penetration sites, and convection cells appear to have been smaller than for the present Earth. Finally there is good evidence that the average composition of the crust, ocean water, and sediments became more like the present at the Archaean–Proterozoic break, 2500 Ma ago; long linear structural features similar to modern mountain belts appear in terrains of this age.

In summary, there is surprisingly little evidence for a hotter early Earth. The efficiency of erosional processes, due to the hydrosphere, has eliminated much of the evidence of our early history. It is certain that convective mixing has always occurred, and was perhaps extremely vigorous in the early Earth. Early vigorous convective mixing probably resulted in a less clear distinction between continental and oceanic crust than is the case for the present-day Earth. The modern type of convection, and evolution of the modern structure, may have

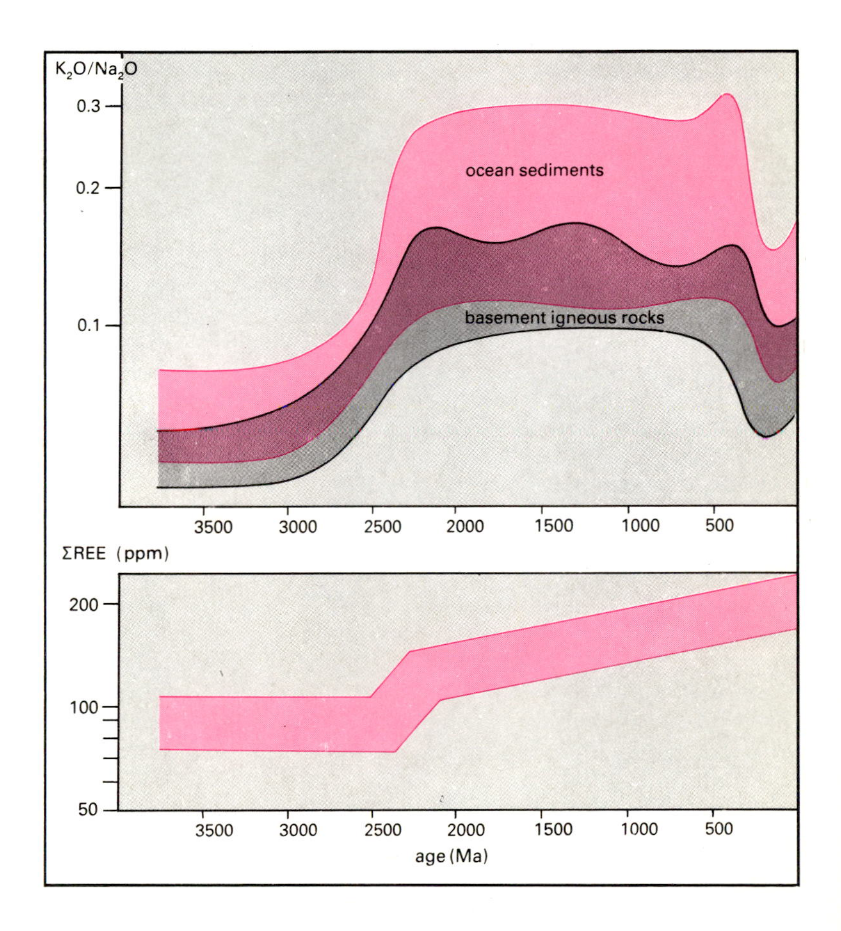

4.19: (**a**) *The trend in the potassium/sodium ratio of ocean sediments and basement igneous rocks through time.* (**b**) *The enrichment of rare earth elements in sediments. The exact causes of such patterns are not known, but there is evidence that the modern style of convection commenced about 2500 Ma ago.*

been initiated about 2500 Ma ago. The record of life appears to be as old as could possibly be expected from the preservation of appropriate rocks (see Chapter 21).

What of the future? If the Earth slowly cools, and if it continues to mix by the subduction process, there should be changes. Our present geochemical arrangement is a complex function of gravitational equilibrium and thermal effects, which drive light fusible and volatile species to the surface. For a cool Earth, more water might be fixed in hydrated phases in the mantle; silica-rich rocks, so characteristic of the continental crust, could return to the mantle and exist more stably in mantle Mg–Fe-rich phases. Convective mechanisms are now beginning to be understood in relation to the present time for a convecting Earth that is slowly cooling. Other planets show different stages in this evolutionary pattern. There is no doubt that many of our planet's unique features are a result of the chemical transport processes which depend on the presence of the hydrosphere.

Changes in the Earth's composition

Escape of the primitive atmosphere

The Earth and other planets accreted by coalescence of planetesimals with a wide variety of compositions, including the volatile-rich carbonaceous chondrites. Initial accretion was under cool conditions, but by the time 0.1 of the Earth's mass had accumulated gravity would have been so strong that incoming planetesimals arrived at a velocity fast enough to induce vaporization on impact. Volatile elements were released to form a primitive atmosphere, whereas non-volatile elements condensed to form an outer layer of molten rock. It is therefore likely that the early Earth was composed of a cool, volatile-rich, solid interior, surrounded by an ocean of molten rock, a magma ocean more than 1000 km thick. Sinking of heavy molten iron through the magma ocean, the process of separating dense metallic iron towards the Earth's centre to form a core, released an additional tremendous quantity of gravitational energy as heat. Heating from the combined sources of impacting meteorites and core separation caused outgassing of volatile elements trapped in the Earth's interior to form an early atmosphere, chiefly of carbon monoxide (CO), hydrogen sulphide (H_2S), nitrogen (N_2), hydrogen (H_2) and water together with a proportion of the more volatile trace elements such as chromium, manganese and sodium. This primitive atmosphere probably totalled about 10 per cent of the Earth's mass, but it was steadily blown off during the first billion years as outgassing proceeded by a combination of intense solar wind from the young Sun, rapid rotation of the Earth and turbulent mixing with the solar nebula.

The Earth's present atmosphere of nitrogen, oxygen, argon and carbon dioxide is essentially secondary, derived by outgassing of the Earth's hot interior while the mantle solidified. The dominant gaseous species of C, N and S were reduced in the primitive atmosphere, as dictated by the presence of reduced metallic iron throughout the Earth's volume. Once the dense metallic iron had been swept by gravity into the core, oxidized gaseous species of C, N and S were produced in reactions mediated by the more oxidized mantle. Further strong modification of the early atmosphere arose from biological photosynthesis and fermentation.

That the Earth's present atmosphere is secondary can be deduced from the low content of the noble gases neon (Ne), krypton (Kr) and xenon (Xe) relative to their abundance in the Sun and material from which the Earth accreted. For instance, the solar nitrogen/neon ratio is about 0.8, whereas the extant terrestrial atmospheric ratio is about 40 000. Noble gases are relatively immune to losses, so that the current excess of nitrogen is due to secondary generation of this gas.

The Earth's atmosphere is now relatively stable, with minor losses induced by specialized processes. Only hydrogen and to a lesser extent helium can escape from the upper atmosphere into space, a process known as thermal evaporation. Hydrodynamical loss involves the escape of gases when local supersonic velocities are attained. Photodissociation, the break-up of gas molecules by incident solar radiation, may enhance the loss of hydrogen by converting ammonia (NH_3) into its constituent nitrogen and hydrogen molecules. Combined losses are not at present appreciable.

The influx of meteorites

Heavy cratering on the surfaces of the Moon, Mars, Mercury, the Jovian moon Callisto and the Saturnian moon Dione attests to the in-

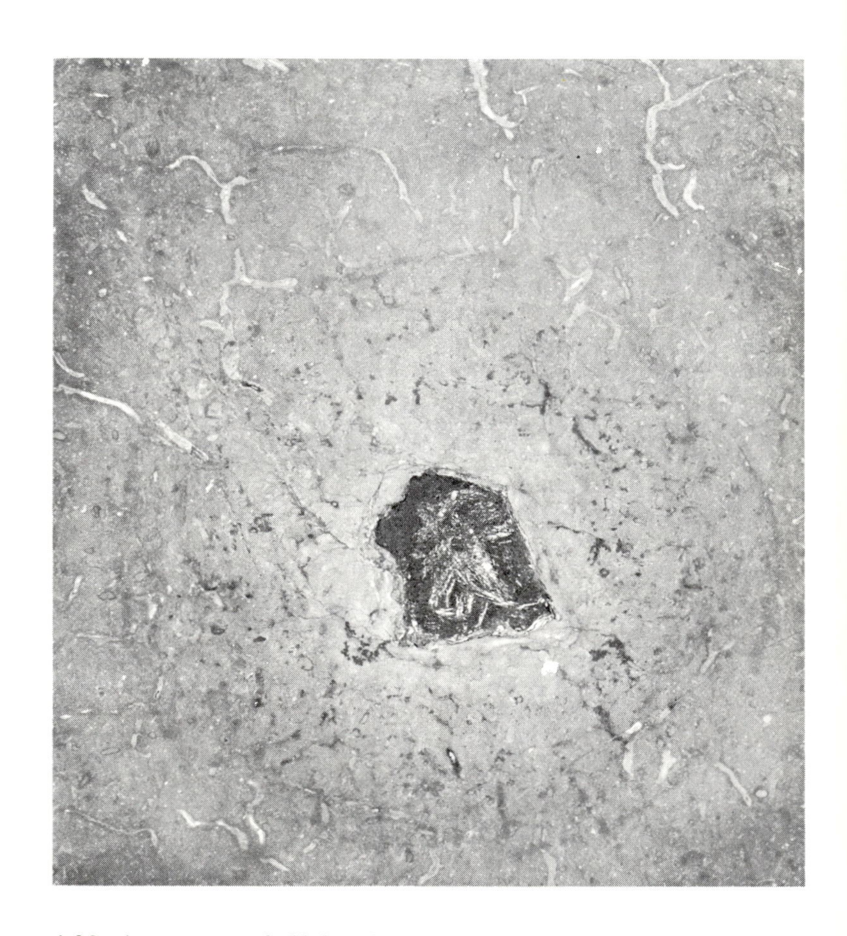

4.20: *A meteorite embedded in the shell of a straight shelled nautiloid (Mollusca) in a 463-million-year-old Ordovician limestone. It seems likely that the meteorite was the cause of death.*

tensity of meteorite bombardment of the planets and their satellites. Examination of the density of craters and their ages on the Moon reveals a tremendous rate of influx about 4400 Ma ago, rapidly diminishing by 3800 Ma ago and gradually tailing off subsequently.

The cumulative masses of meteorites swept up by the gravitational fields of the Moon and Earth from 4400 Ma ago to the present are estimated to be about 10^{19} and 20×10^{19} kg respectively, representing about 0.003 Earth mass. Taking into account the volatile content of meteorites (about 10 per cent for carbonaceous chondrites), the capture process would account for less than 0.1 per cent of the present mass of water in the world's oceans and only a small fraction of the Earth's total inventory of carbon or other volatile elements.

The fiery trail of meteorites heated to incandescence by atmospheric friction is vivid evidence for the infall of rocky material from space. Such visible falls are comparatively rare, however, and by far the largest mass of debris captured from space by the Earth's gravity consists of a steady rain of microscopic meteorites or dust particles which are too small to be observed directly. Measurement of their total influx requires elaborate chemical techniques based on exotic nuclides or anomalous quantities of the rare metal iridium.

Meteorites and dust particles in space are constantly bombarded by cosmic rays emitted in a powerful stream from the Sun. Cosmic rays produce exotic nuclides such as ^{26}Al, ^{53}Mn and ^{59}Ni in collisions with dust particles, but the Earth's atmosphere effectively shields surface rocks from exposure. Thus the presence of ^{59}Ni in ocean sediments can be attributed entirely to microscopic meteorites and is found to be about 2 per cent of the total nickel present. Meteorites contain a 'normal' cosmic abundance of iridium, but the Earth's surface rocks are strongly depleted in this metal because of its siderophile character (preference for the metallic iron core rather than the rocky crust). Hence the quantity of iridium in surface rocks provides a measure of the influx of meteorites. A major iridium anomaly has been detected in rocks at the Cretaceous–Palaeocene boundary (65 Ma ago), which coincides with the extinction of the dinosaurs. It has been suggested from this evidence that the impact of a large meteorite, or a nearby supernova, severely modified surface terrestrial conditions.

Changes due to radioactivity

Finally radioactive decay involves conversion of a parent nuclide into a daughter nuclide, accompanied by the release of energy. Of the various long-lived radioactive elements present in rocks, only two daughter nuclides are produced in sufficient quantity to increase appreciably the Earth's inventory. First, potassium–40 undergoes decay to argon–40, producing a progressively larger content of this noble gas in the Earth's atmosphere. Second, uranium and thorium both undergo decay to lead plus helium: new daughter lead represents only a few per cent of the Earth's intrinsic complement of lead, but the extra helium production is significant.

5.1: *Rock composed of crystals of calcite (white), pyrite (gold), sphalerite (black), and galena (silver).*

5 Earth materials: minerals and rocks

Minerals are the basic chemical compounds of the solid Earth. About two thousand minerals are known, and between them they contain all the chemical elements of which the Earth is made. Rocks are physical mixtures of minerals, usually in the form of interlocking crystals or of grains bound together by a natural cement. An enormous variety of rocks make up the Earth, especially its outer crust (about 18 to 50 km thick), but the vast majority are composed of not more than a few dozen of the two thousand minerals that are known.

The physical properties of minerals

Nearly all minerals are crystalline with compositions that are fixed or vary within specified limits. This means that they have well-defined physical and chemical properties which give each mineral a separate identity. Under appropriate physical and chemical conditions minerals can develop crystal forms that are among the most beautiful of natural shapes. Well-crystallized minerals are bounded by regular arrange-

a

b

5.2: (**a**) *Crystals of pyrite* (FeS_2) *showing features of cubic symmetry.*

(**b**) *Crystals of quartz* (SiO_2) *showing prismatic and pyramidal faces.*

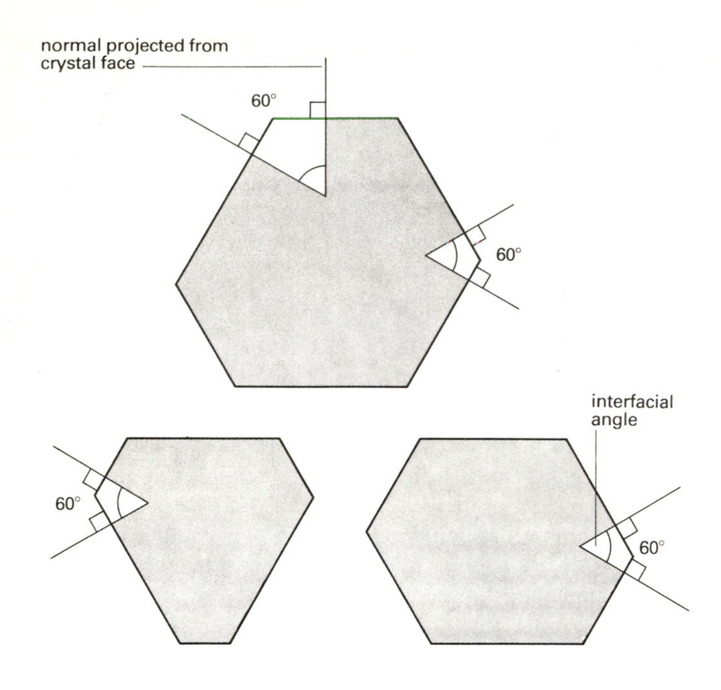

5.3: *Diagrammatic cross-sections through three quartz crystals showing how the angles between the columnar faces are always the same. Interfacial angles are conventionally defined as those between the normals to the faces (60° in this case).*

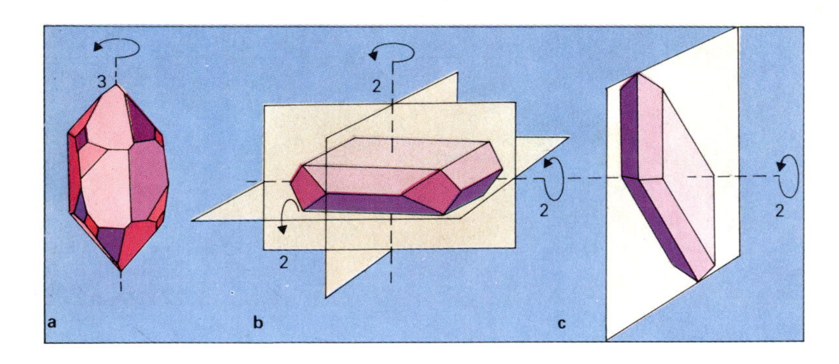

5.4: (a) *An idealized quartz crystal with a three-fold rotation axis. Quartz crystallizes in the trigonal system, despite the apparent hexagonal symmetry of its prismatic and pyramidal faces (see Figure 5.3)—note the arrangement of smaller faces.* (b) *An idealized barite crystal with three two-fold axes and three mirror planes. Barite crystallizes in the orthorhombic system.* (c) *An idealized gypsum crystal with a single two-fold axis and one mirror plane. Gypsum crystallizes in the monoclinic system. (An n-fold symmetry axis means that the crystal presents the same aspect n times on a 360° rotation about that axis; a mirror plane of symmetry means that the two halves of the crystal are mirror images on either side of the plane.)*

ments of plane surfaces, called crystal faces, which are characteristic for each mineral and independent of size.

All crystals possess symmetry, though it is not always immediately apparent. The cubes of pyrite and the hexagonal columns of quartz in Figure 5.2 are obvious enough, but the symmetry of others is more elusive. The underlying symmetry of a crystal can be further obscured when the sizes and shapes of its faces are not uniform, as frequently happens. The quartz crystals in Figure 5.2, for example, all have both columnar and pyramidal faces, but the latter have developed differently in each crystal. However, the angles between the columnar faces (and the pyramidal faces) are the same in these crystals and on all other quartz crystals, wherever they occur (Figure 5.3).

The constancy of interfacial angles, which underlies the patterns of symmetry among crystals, was first established by Nicolaus Steno (1638–86), a Danish monk, in the seventeenth century. Systematic measurements of interfacial angles on countless specimens of both natural and artificial crystals have enabled mineralogists to establish that all crystalline substances can be assigned to one of seven symmetry systems, each characterized by specific combinations of symmetry elements of which the principal ones are axes, planes and centres of symmetry.

Crystals of the same mineral can vary widely in appearance because crystal faces develop in different ways depending upon the conditions of crystallization. The common mineral calcite is a good example, and two of its numerous crystal forms are portrayed in Figure 5.6(a).

A great many minerals break along planes of weakness called cleavage planes, so that broken fragments are bounded by more or less well-developed plane surfaces. These cleavage surfaces are sometimes so smooth that they can be mistaken for crystal faces. The direction of cleavage in minerals is commonly parallel to crystal faces, though equally commonly it is not. Cleavage planes bear precise geometrical

5.5: *The seven symmetry systems and their principal elements.*

Symmetry system[1]	Symmetry elements[2]
cubic	four three-fold axes (through cube corners)
tetragonal	one four-fold axis
hexagonal	one six-fold axis
trigonal	one three-fold axis
orthorhombic	three two-fold axes
monoclinic	one two-fold axis
triclinic	no axes (or mirror planes)

Notes

1 The seven symmetry systems can be further subdivided into a total of thirty-two symmetry classes according to the presence or absence of further symmetry elements (planes, centres, axes).

2 Only the principal axes of rotation are given.

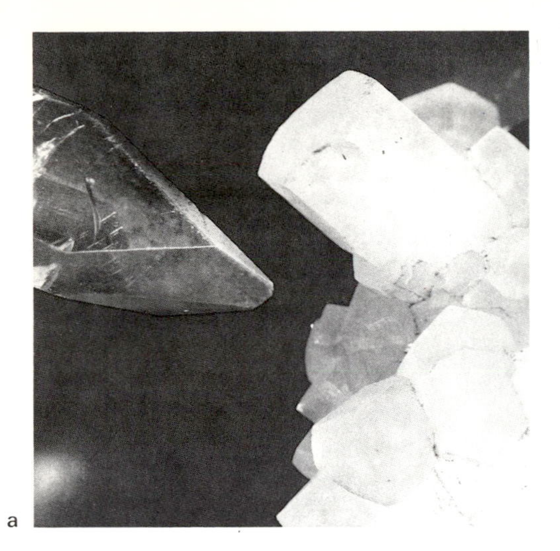

a

b

c

5.6: *Crystals can develop in a very wide variety of forms, according not only to their chemical composition and hence their internal structure and symmetry but also to the site and environment in which they grew.* (**a**) *Two contrasted crystal forms of calcite* ($CaCO_3$)*; that on the left is commonly known as dog-tooth spar and that on the right as nail-head spar. The contrast arises from the development in each of different crystal faces.* (**b**) *Rosettes of needlelike stibnite crystals* (Sb_2S_3). (**c**) *A section of the concentric structure formed by minutely crystalline malachite* ($CuCO_3.Cu(OH)_2$) *which rarely occurs as well-formed crystals. The tiny crystals are arranged perpendicular to the layers.*

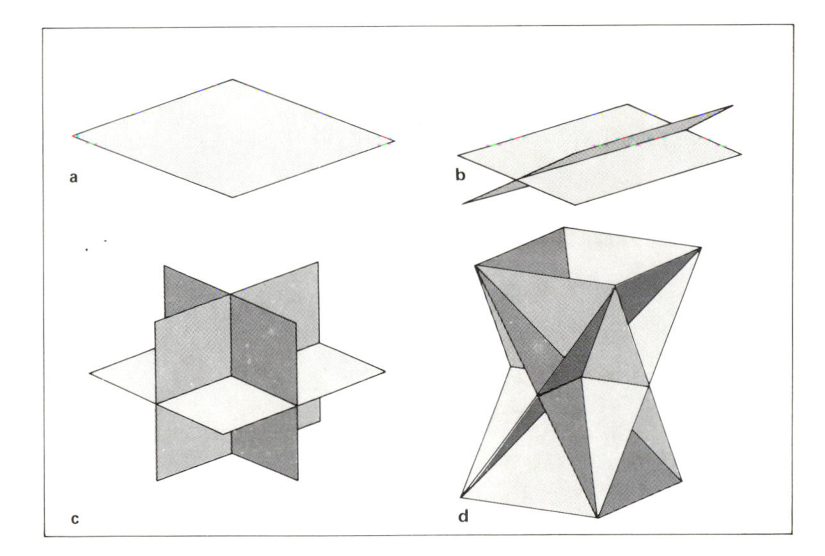

e

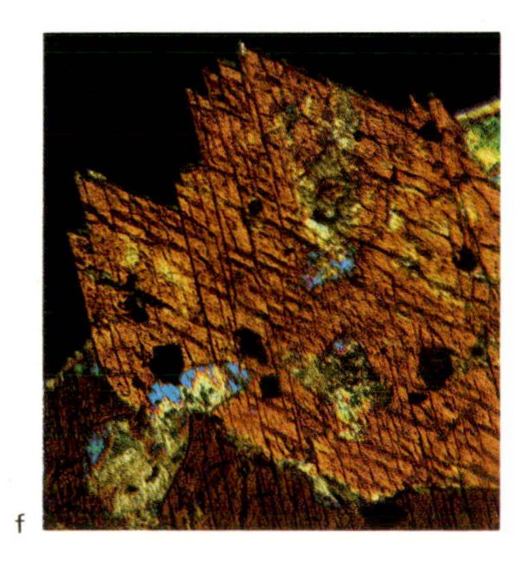

f

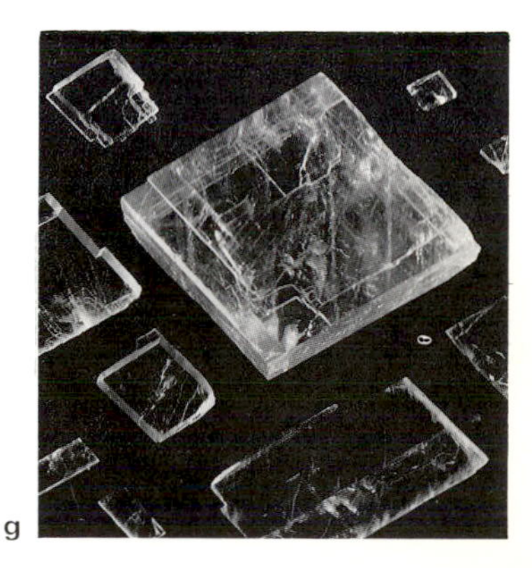

g

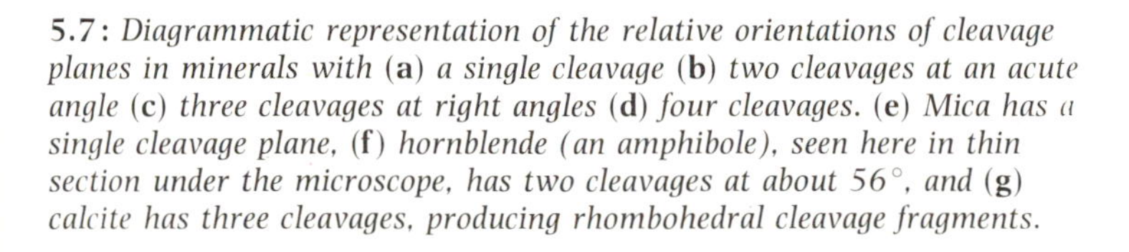

5.7: *Diagrammatic representation of the relative orientations of cleavage planes in minerals with* (**a**) *a single cleavage* (**b**) *two cleavages at an acute angle* (**c**) *three cleavages at right angles* (**d**) *four cleavages.* (**e**) *Mica has a single cleavage plane,* (**f**) *hornblende (an amphibole), seen here in thin section under the microscope, has two cleavages at about 56°, and* (**g**) *calcite has three cleavages, producing rhombohedral cleavage fragments.*

a

b

c

d

5.8: *Examples of twinning in minerals.* (**a**) *A characteristic 'swallow-tail' twin of gypsum.* (**b**) *Carlsbad twinning in orthoclase feldspar, a form of rotation twinning.* (**c**) *A twinned crystal of aragonite, another form of calcium carbonate* ($CaCO_3$) *which, unlike calcite, crystallizes in the orthorhombic crystallographic system.* (**d**) *Multiple twinning (sometimes called lamellar twinning) as seen in a photomicrograph of a thin section; each of the 'stripes' represents an individual twin.*

relationships to one another as well as to crystal faces, and these relationships are characteristic for each mineral: intercleavage angles are constant, just as interfacial angles are.

A particular mineral may develop several different combinations of crystal faces, depending on the environment of crystallization, but the number and orientation of its cleavage planes are always the same. Cleavages represent planes of weakness in the mineral. Crystal faces are planes of growth.

Crystal growth is rarely uniform; crystal faces frequently develop unevenly, and some minerals have a tendency to form twinned crystals. In these a crystal face becomes a twin plane, on either side of which the crystal develops in a different orientation, forming in effect two (or more) crystals joined together in a definite geometrical relationship.

In the most straightforward type of twinning two halves of a crystal are mirror images of each other, but more complex twins involve relative rotations of individuals on either side of the twin plane. Some minerals are more prone to twinning than others, and some of these develop multiple twins, made up of several twinned subindividuals within a single crystal.

The classification of minerals

The most convenient and widely used way of classifying minerals is on the basis of their chemical composition. Figure 5.9 lists the principal mineral groups with some common examples of the minerals in each. Within each group the minerals have generally similar physical properties: many sulphides are grey or yellow with a metallic lustre; oxides are hard, dense and generally dark coloured, while hydroxides occur commonly as earthy-looking, non-crystalline (amorphous) masses of low density; and most chlorides, sulphates and carbonates are translucent or glassy (vitreous) and quite soft (easily scratched).

This community of physical properties can include similarity of crystal form and cleavage. When crystalline substances with analogous chemical formulas exhibit similar crystal forms they are said to be isomorphous. The carbonates of iron, magnesium and calcium (and manganese), for example, all crystallize in the trigonal system and have rhombohedral cleavage. Similarity of chemical composition is not always accompanied by isomorphism, however. The minerals sylvite (KCl) and halite (NaCl), for instance, crystallize in the orthorhombic and cubic systems respectively.

Conversely some minerals have quite different crystal forms and symmetries, but identical compositions; they are polymorphous. Probably the best-known examples are the polymorphs of carbon (diamond and graphite) and of calcium carbonate (calcite and aragonite). As a general rule the denser polymorphs will form under conditions of higher pressure and/or lower temperature than the less dense polymorphs. Thus diamond can only form at pressures equivalent to depths of around 150 km below the Earth's surface, which accounts for the scarcity of natural diamonds: they are brought to the surface by volcanic activity that has to originate at considerable depths and is therefore rather exceptional. Diamond is metastable at the Earth's surface, but (fortunately for industry and jewellers), its rate of inversion to the low-density polymorph, graphite, is infinitesimal. Aragonite, however, the dense polymorph of calcium carbonate, inverts to calcite over measurable periods of geological time; aragonite is rare in rocks older than about fifty million years.

Most of the mineral groups listed in Figure 5.9 can be considered as products of simple chemical reactions, and that is how mineralogists thought of them until well into the present century. Oxides and hydroxides are easily understood in these terms, and the chemical formulas of many of the other minerals look like those of salts resulting from acid–base reactions (eg NaCl—halite, and $CaSO_4$—anhydrite). However, the variety of compositions within and between the silicate subgroups is too complex for them to be satisfactorily categorized as the products of reactions with recognizable silicic acids. It required the development of X-ray crystallography before a proper understanding of the silicate minerals could be achieved.

5.9: *The classification of some important minerals by chemical composition (with symbols or formulas).*

Note
The chemical compositions of many minerals, notably the silicates, are given in simplified form. Hydroxides are defined as those minerals in which the principal anion component is the hydroxyl ion, OH^-. Some other mineral groups, especially the silicates, have hydroxyl ions as a component of complex polyatomic anions, and these are sometimes also called hydrous minerals.

Mineral	Formula
Halides	
halite	NaCl
sylvite	KCl
fluorite	CaF_2
Sulphides	
galena	PbS
chalcopyrite	$CuFeS_2$
pyrite	FeS_2
pyrrhotite	FeS
Oxides	
rutile	TiO_2
haematite	Fe_2O_3
magnetite	Fe_3O_4
ilmenite	$FeTiO_3$
chromite	$(Mg,Fe)Cr_2O_4$
Hydroxides	
goethite	$Fe_2O_3.H_2O$
gibbsite	$Al(OH)_3$
Carbonates	
calcite	$CaCO_3$
aragonite	$CaCO_3$
siderite	$FeCO_3$
magnesite	$MgCO_3$
dolomite	$CaMg(CO_3)_2$
Sulphates	
gypsum	$CaSO_4.2H_2O$
anhydrite	$CaSO_4$
barite	$BaSO_4$
Phosphates	
apatite	$Ca_5(PO_4)_3(F,Cl,OH)$
Silicates	
olivine	$(Mg,Fe)_2SiO_4$
garnet group	
pyrope	$Mg_3Al_2(SiO_4)_3$
almandine	$Fe_3Al_2(SiO_4)_3$
zircon	$ZrSiO_4$
sillimanite, kyanite, andalusite	polymorphs of Al_2SiO_5
sphene	$CaTiSiO_5$
pyroxene group	
augite	$Ca(Mg,Fe)Si_2O_6$ (clinopyroxene)
enstatite	$(Mg,Fe)SiO_3$ (orthopyroxene)
amphibole group	
hornblende	$Ca_2(Mg,Fe)_5Si_8O_{22}(OH)_2$
mica group	
biotite	$K(Fe,Mg)_3(AlSi_3O_{10})(OH)_2$
phlogopite	$KMg_3(AlSi_3O_{10})(OH)_2$
muscovite	$KAl_2(AlSi_3O_{10})(OH)_2$
chlorite	$(Mg,Fe,Al)_6(Si,Al)_4O_{10}(OH)_8$
clay minerals	
kaolinite	$Al_4Si_4O_{10}(OH)_8$
feldspar group	
orthoclase	$KAlSi_3O_8$
plagioclase series	
albite	$NaAlSi_3O_8$
anorthite	$CaAl_2Si_2O_8$
quartz	SiO_2
feldspathoid group	
leucite	$KAlSi_2O_6$
nepheline	$NaAlSiO_4$

The internal structure of minerals

The perfection of form and symmetry possessed by crystals and cleavage fragments of minerals must be a manifestation of internal structures, in which planes of atoms or molecules control the formation of crystal faces and cleavage directions. The French mineralogist René Haüy was among the first to arrive at this conclusion, early in the nineteenth century, but experimental verification had to wait another hundred years or so, until the discovery in 1912 that X-rays are diffracted by crystals. The wavelength of X-rays is comparable with the spacing between atoms and ions in crystals—of the order of a few hundred picometres (pm).

The internal structure of any mineral can be described in general terms as a three-dimensional array of positively and negatively charged ions (cations and anions respectively), built up by the regular repetition of a basic configuration, the unit cell. The symmetry and form of a crystal is determined by the shape of its unit cell, which is unique for each mineral and is its basic 'building block' (Figure 5.11). Measurement of the spacing between planes of ions in the structure enables the shape and dimensions of the unit cell and the sizes of the ions in the structure to be calculated. Crystal structures can be represented either by lattice models which portray the spatial relationships between constituent ions, or as close-packed models, which are made up of spheres of different sizes and approximate more closely the real state of affairs inside crystals (see Figure 5.12 for examples).

The chemical formula of a mineral expresses the relative proportions of its constituent elements; thus in halite (NaCl) there is a 1:1 ratio of sodium to chloride ions. The unit cell, the repeat unit of the structure, generally contains small whole multiples of the chemical formula; for example, that of halite contains 4NaCl.

The simple form of halite demonstrates how the inferences the early mineralogists drew about crystal structure from their study of morphology and other properties have been substantially confirmed. Crystal faces and cleavages develop parallel to lattice planes. A variety of lattice planes may be available for crystal growth, but the positions of planes of weakness—lattice planes between which the bonding is weakest—can never change. In most minerals cleavage directions are parallel to some of the crystal faces: halite and mica are good examples (see Figure 5.7).

The sizes of ions

Anions are formed when atoms gain one or more electrons, and cations are formed when atoms lose one or more electrons. Partly for this reason, and partly because the atoms of anion-forming elements tend to be larger than those of cation-forming elements, anions are generally larger than cations. Crystal structures can therefore conveniently be envisaged as geometrical arrangements of large spheres (anions) packed around small spheres (cations) in such a way that the spaces between them are minimized and the positive and negative charges cancel each other out. The resulting geometrical configurations will depend upon the relative sizes of the ions: the larger the cation, the more anions can be fitted around it. Accordingly, other things being equal, crystal structures in which cations are large will be less dense than those in which cations are small. The density of sylvite (KCl), for example, is less than half that of pyrrhotite (FeS), though their 'molecular' weights are not very different (75 and 88 respectively). This does not always work: the lead sulphide, galena (PbS), is very dense, despite the relatively large size of the lead ion (120 pm), simply because lead has a very high atomic mass. But whatever the packing configuration ions are always arranged in clearly defined planes within the

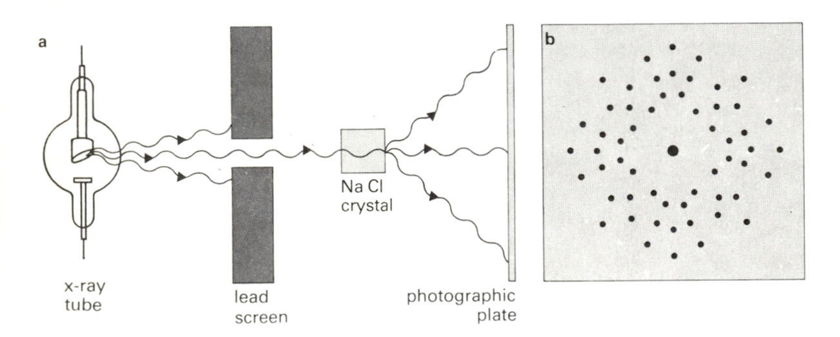

5.10: *The principle of X-ray diffraction by crystals.* (**a**) *A parallel beam of X-rays of a single wavelength (ie monochromatic) impinges on parallel planes of atoms or ions at an angle* θ. *The beam interacts with the atoms or ions and is diffracted. Some of the diffracted rays will emerge from the crystal at the same angle* θ *as the incident rays. For certain values of* θ, *the path difference will be a whole number (n) of wavelengths. When this happens the diffracted rays are intensified by constructive interference, and* (**b**) *are recorded on the photographic film. By simple trigonometry the wavelength* λ *(known) and the angle* θ *(measured) are related to the interplanar spacing, d, by the relationship:* $n\lambda = 2d \sin \theta$, *known as the Bragg equation after the founder of X-ray crystallography Sir Lawrence Bragg. The pattern shown here is for halite* (NaCl).

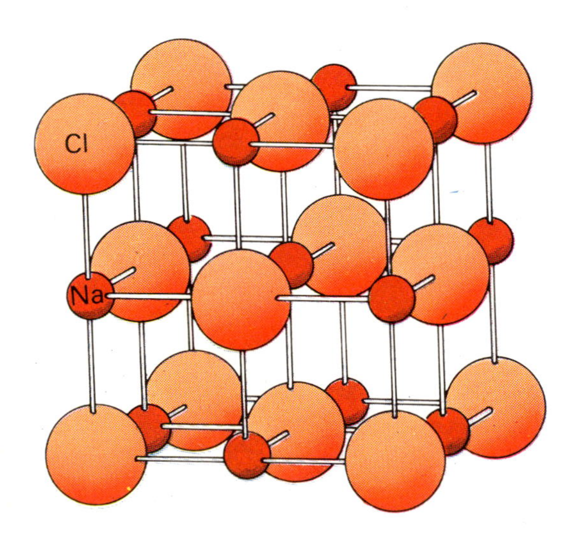

5.11: *Lattice model of the structure of halite* (NaCl), *the first mineral to be investigated using X-rays. It is a three-dimensional array of alternating smaller sodium* (Na^+) *and larger chloride* (Cl^-) *ions. The whole array can be extended indefinitely by repetitions of the cubic unit cell, which determines the symmetry and form of halite crystals. The planes parallel to the sides of the cube define the three perfect cubic cleavage planes of halite.*

structure, which find expression either as crystal faces or as cleavage planes, or as both.

Cations are more variable in size than anions, which accounts for the occurrence of isomorphism between some minerals with analogous chemical formulas but not between others. Thus, calcite ($CaCO_3$), magnesite ($MgCO_3$), siderite ($FeCO_3$) and rhodochrosite ($MnCO_3$) are isomorphous because the sizes of their cations are such that they can all form rhombohedral (trigonal) structures with the carbonate (CO_3^{2-}) anion. On the other hand, the potassium ion is about one and a half times the size of the sodium ion, and so halite (NaCl) and sylvite (KCl) are not isomorphous.

It is very useful to regard crystals as ionic solids made up of neatly packed solid spheres, but such a representation is also something of a simplification. Ions are not rigid spheres, and it is better to regard them as slightly elastic, with radii that adjust slightly to different packing configurations. For instance, in minerals containing calcium (Ca) where Ca^{2+} is identified as being surrounded by six oxygens in the structure, the ionic radii of Ca^{2+} and O^{2-} are respectively 108 pm and 132 pm. In structures in which each calcium ion is surrounded by eight oxygens, on the other hand, the radii increase to 120 pm and 134 pm respectively. This is the result of changes in the interaction between the electrons of the central cation and the surrounding anions —the electron shells of both cation and anions all tend to be 'pulled out' more. Nor are all crystal structures wholly ionic, although it remains convenient to treat them as such. In many minerals the bonding is partly or wholly covalent, notably among the native elements, some sulphides (eg pyrite, FeS_2) and also within those anions that are polyatomic complexes, such as the carbonate (CO_3^{2-}) and silicate (SiO_4^{4-}) anions.

X-ray crystallography has greatly extended the scope of mineralogical investigations, particularly when crystals are very small, as in cryptocrystalline materials like chalcedony or in the rare minerals that form minute crystals of submillimetre dimensions. But the most striking advances have been in understanding the structures of silicate minerals.

The silicate minerals

Oxygen and silicon are by far the most abundant elements in the crust and mantle, which together form about four-fifths of the Earth's volume. The silicates are thus overwhelmingly the dominant group of minerals. They are also the most diverse, encompassing a rich and wide variety, from sheetlike mica to fibrous asbestos, from hard quartz to soft talc, from deep-red garnet to deep-blue lapis lazuli. The reason for this great diversity of minerals within a single major group is the versatile behaviour of the silicate anion (SiO_4^{4-}), the so-called silicate tetrahedron. What sets it apart from other complex anions, such as the carbonate (CO_3^{2-}) and the sulphate (SO_4^{2-}) ions, is its ability to 'polymerize' and form giant polyanions. This is achieved by the sharing of oxygens between tetrahedrons, to build rings, chains, bands, sheets and framework structures. Each of these variously linked arrangements forms the 'skeleton' for a particular group of silicate minerals, the giant polyanions being held together by cations of appropriate size and charge, which mostly fit into spaces that are surrounded by

a

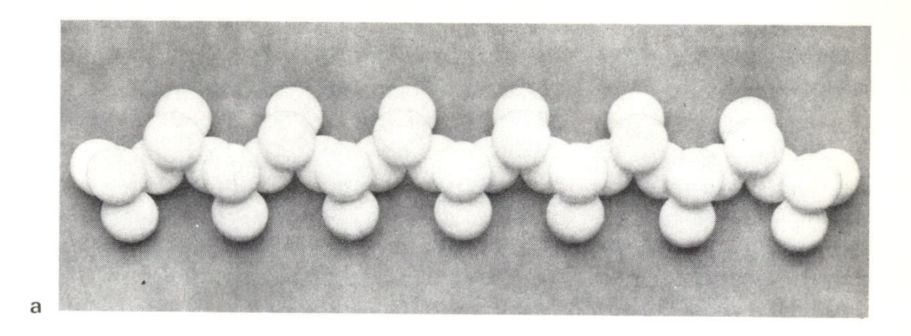

b

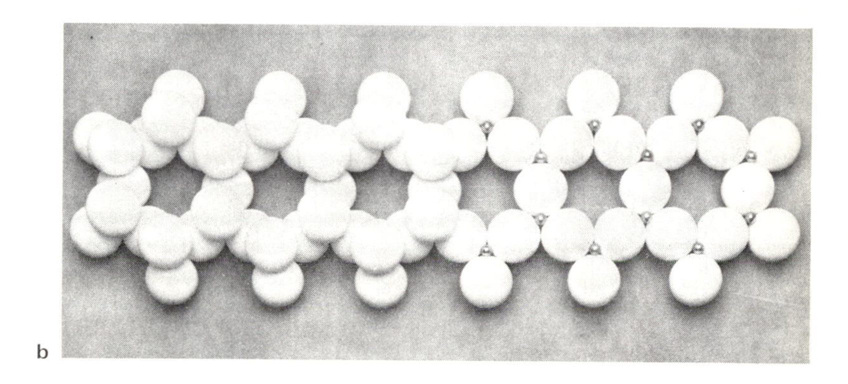

c

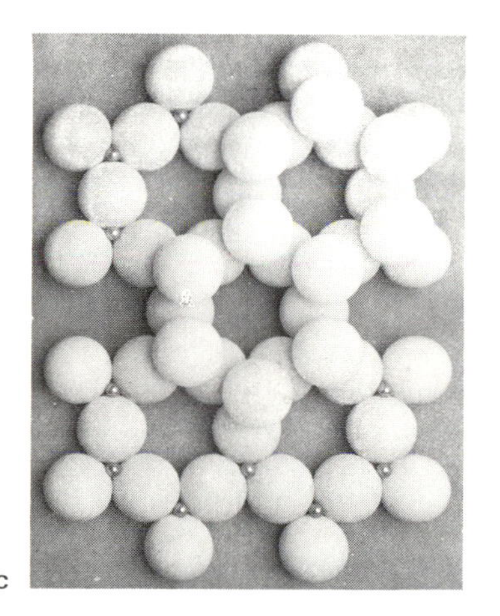

d

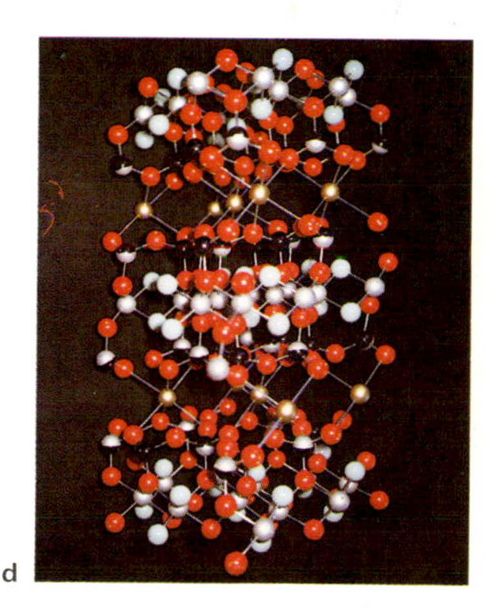

5.12: *Close-packed models showing the basic structures of three groups of silicate minerals formed by oxygen-sharing between silicate tetrahedrons; oxygen atoms are represented by ping-pong balls and silicons by ball-bearings.* (**a**) *Chains of tetrahedrons form the structural basis of the pyroxene group of minerals.* (**b**) *Bands formed by oxygen-sharing between chains constitute the structural basis of the amphibole group.* (**c**) *Sheets of tetrahedrons are formed when three oxygens of each tetrahedron are shared. This is the basic structure of the mica group of minerals.* (**d**) *In the mica group, the sheets of silicate tetrahedrons are held together by large positive ions, such as potassium (the golden balls in this lattice model). This structure gives rise to the single perfect cleavage characteristic of these minerals.*

unshared oxygens at tetrahedral apexes.

The silicate tetrahedron can also behave like the other complex anions and pack together as independent SiO_4^{4-} units, along with smaller cations, forming compact structures and dense minerals such as olivines and garnets.

There is a general decrease in density of silicate minerals as the number of shared oxygens in the silicate framework increases, and larger spaces become available within and between the polyanions to accommodate bigger cations. Moreover, the hexagonal holes in the bands and sheets of amphiboles and micas are occupied by large light hydroxyl (OH^-) ions. Amphiboles are therefore less dense than pyroxenes, but they are denser than micas which contain proportionately more OH^-.

The partly covalent Si–O bond within the silicate polyanions is stronger than the ionic bonds between the silicate polyanions and the cations, so cleavage and crystal morphology in silicates is controlled chiefly by the form of the silicate framework.

Substitution and solid solution

Ions of similar size can substitute readily for one another in crystal structures. An analogy has been made with a bricklayer, who can use bricks of different colours or materials when building a wall; so long

5.13: *The structure and physical properties of the silicates.*

Structure type	Formula of silicon–oxygen unit	Number of oxygens shared	Mineral example	Characteristics	Specific gravity (water has sg = 1)	Hardness on Moh scale (1–10)	
tetrahedra	SiO_4	0	olivines garnets	Dense, hard compact structures; crystals tend to have equi-dimensional shape without cleavage.	3.5–4.3	6–7.5	single tetrahedron
ring silicates	Si_6O_{18}	2	tourmaline beryl	Fairly dense, hard and compact; crystals tend to be elongate and columnar (called prismatic); cleavage is across the columns—between the stacked rings.	2.7–3.2	7–8	ring
chain silicates	SiO_3	2	pyroxenes	Generally almost equi-dimensional, and as dense and hard as olivines and garnets; cleavage is not always well developed due to the tight binding of the chains by cations.	3–4	5–6	chains
band silicates	Si_4O_{11}	$2\frac{1}{2}$	amphiboles	Less compact and less dense than pyroxenes; the individual double-chain bands are less strongly held by cations than are the single chains of pyroxenes; cleavage is much better developed in amphiboles.	2.8–3.6	5–6	double chains
sheet silicates	Si_2O_5	3	talc micas serpentine clay minerals	Less compact and dense than other minerals; low density, perfect cleavage and low hardness, due to weak bonding between the internally strong polysilicate anionic sheets.	2.6–3.3	1–3	sheets
framework silicates	SiO_2	4	quartz feldspars	Compact structures which are rather open and thus less dense; great hardness due to strong bonds in three dimensions; cleavage is variable, absent in quartz but quite well developed in feldspars.	2.6–2.8	6–7	three dimensional networks

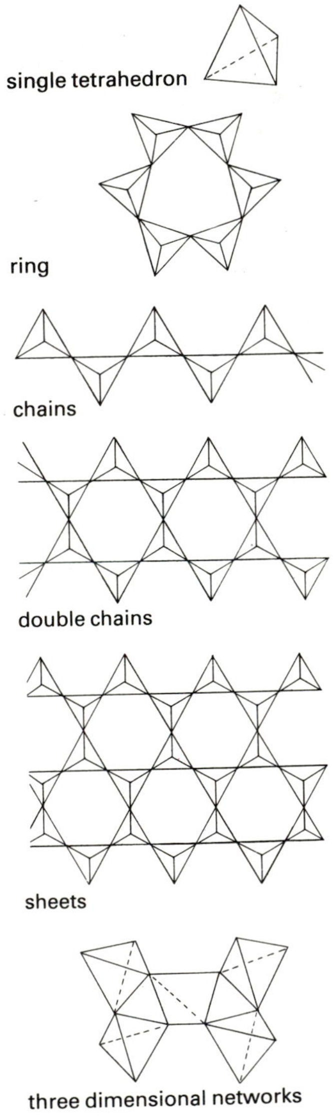

as the bricks are all of the same size the shape of the structure is unaffected, and only its composition changes.

Ionic substitution is more widespread among the silicates than other minerals, chiefly because of the silicates' more complex structures. A common example is the interchangeability of Mg^{2+} and Fe^{2+}, whose ionic radii are very similar (80 pm and 71 pm respectively). In the olivines, for example, there is a complete range of compositions (mixed crystals) from fayalite (Fe_2SiO_4) at one end to forsterite (Mg_2SiO_4) at the other, ie complete solid solution exists between these two end-member compositions. Most natural olivines have Mg-rich compositions near that of forsterite, but compositions throughout almost the whole range have been recorded. One of the most important ionic substitutions in silicate minerals is that of Al^{3+} (47 pm) for Si^{4+} (34 pm) in the silicate tetrahedron. The $AlO_4{}^{5-}$ aluminate ion is tetrahedral, and its insertion in no way impairs the formation of large polyanions by oxygen-sharing. Thus the geometry of silicate structures remains unaffected, but the aluminate ion has an extra negative charge and is responsible for much of the diversity among silicate minerals. It plays a key role in the framework silicates, which include quartz and the feldspars, the most abundant and widespread minerals of the Earth's crust.

In framework silicates all the tetrahedral apexes are linked by oxygen-sharing, producing very rigid structures in which each silicon is effectively bonded to only half an oxygen ion. In the absence of other ions the structure is electrically neutral, and its formula is SiO_2—that of the mineral quartz, the least destructible of all the common minerals.

In the feldspars some of the Si^{4+} in the framework is replaced by Al^{3+}, and the charge deficiency is made up by positive ions. The available spaces in framework structures can accommodate relatively large positive ions, and so feldspars do not have high density. However, they are denser than the micas because the latter have the extra spaces to accommodate very light OH^- ions.

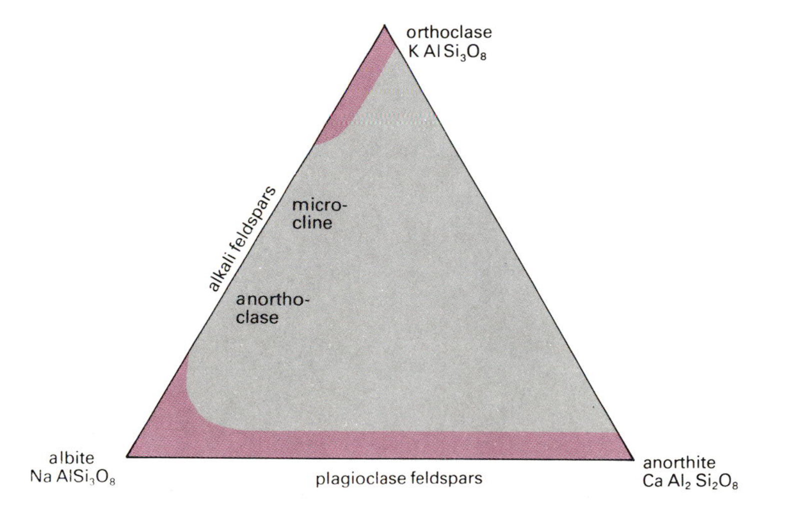

5.14: *Relationships among the feldspars. The pink areas indicate approximate limits of solid solution among the feldspars. The similar ionic radii of* Na^+ *(124 pm) and* Ca^{2+} *(120 pm) ensure complete solid solution between albite and anorthite—the plagioclase series. Because of the charge difference between* Na^+ *and* Ca^{2+} *some of the silicons are replaced by aluminium* (Al^{3+})*. Although solid solution between sodium and potassium feldspars occurs only at high temperatures because of the large size of the potassium ion, these minerals are collectively termed the alkali feldspars.*

These examples illustrate a central principle of ionic substitution and emphasize the brick-laying analogy: the chief criteria for substitution are, first of all, size and, second, charge. There is little obvious correlation with the chemical affinities implicit in the ordering of elements in the periodic table (see page 53). For example magnesium is a Group IIA alkaline earth element, whereas iron is a Group VIII transition metal, and their chemical properties are very different. Yet their ions substitute for one another in all proportions in mineral structures. Such contrasts are commonplace throughout the mineral kingdom.

Isomorphism is commonly associated with extensive ionic substitution, but one is not a necessary consequence of the other. In the garnet group, for instance, there is complete solid solution between pyrope ($Mg_3Al_2(SiO_4)_3$) and almandine ($Fe_3Al_2(SiO_4)_3$), but not between them and the calcium analogue (it should be noted that aluminium (Al) is present in these structures as an ordinary cation and is not substituting for silicon). Yet all garnets are isomorphous, crystallizing as distinctive dodecahedrons. Again, among the carbonates there is complete solid solution between magnesite ($MgCO_3$) and siderite ($FeCO_3$) (and rhodochrosite ($MnCO_3$) as well), but not between them and calcite ($CaCO_3$). Yet they are all isomorphous, forming rhombohedral (trigonal) crystals. On the other hand, there is substantial, if limited, substitution of iron (Fe) in the structure of sphalerite (ZnS), though ZnS and FeS are not isomorphous.

Trace elements

All minerals contain small amounts of less abundant elements, substituting for similar-sized cations of major elements in their structures. For instance, nickel replaces some of the magnesium and iron in olivines; barium occurs in potassium feldspars, along with a concomitant replacement of some aluminium for silicon; and appreciable concentrations of strontium are found in calcium-rich minerals such as plagioclase feldspars. The variety of trace elements is greater in the silicates than in other minerals because their structural complexity makes many more 'holes' of appropriate size available for substitution.

Trace elements are often responsible for the variation in colour between different specimens of the same mineral, and for this reason colour is not a good diagnostic property for mineral identification. Only in very exceptional circumstances are trace elements found forming minerals in their own right. The principal reason for this lies in their typically low availability.

Under the physical and chemical conditions appropriate to the formation of a particular mineral, its composition (and hence its structure) will to a great extent be determined by the relative abundance of different elements, ie by their availability. Potassium is more abundant than barium, and so potassium feldspars are widespread while barium feldspars are not. Calcium is more abundant than strontium, and so calcite ($CaCO_3$) is more common than celestine ($SrCO_3$). Magnesium is more abundant than nickel, and so most olivines are magnesium-rich, and nickel-rich olivine does not occur naturally. On the other hand, when availability, size and charge are comparable, then chemical behaviour becomes important. Thus high temperatures favour the formation of magnesium-rich rather than iron-rich olivines, of calcium-rich rather than sodium-rich feldspars. This applies particularly to the ten or so elements that between them make up over 99 per cent of the minerals of which the Earth's crust and mantle are composed. Most abundant of them all is oxygen. Almost all the minerals in Figure 5.9 are little more than close-packed arrangements of oxygen with cations of various sizes in the spaces between them. The Earth's crust is therefore composed principally of oxygen (no less than 95 per cent by volume, on account of the large size of the oxygen anion in comparison with other common ions). The crust and mantle are sometimes collectively termed the oxysphere in consequence, and the predominant cation to be found in the interstices of the oxysphere is Si^{4+}.

Silicate minerals form the crust and the mantle of the Earth down to 2900 km or so. However, their structures at these depths are very different from those displayed by the same minerals at the surface. Olivine and pyroxene are the predominant minerals of the mantle, and experiments have shown that at pressures corresponding to depths of only a few hundred kilometres they undergo polymorphic transitions to more compact structures of higher density, whose ionic configurations are similar to those of a group of oxide minerals known as spinels.

Our understanding of silicate structures is comparatively recent, as X-ray techniques were not applied systematically to the silicates until well into the 1920s. In view of the structural complexities that have been revealed—a consequence partly of the polymerizing tendencies of SiO_4^{4-} and partly of the ability of the resulting structures to accept a multitude of ions of diverse size and charge—it is small wonder that the early mineralogists had difficulty in classifying silicates as salts of simple acids.

The analysis of minerals and rocks

Early mineralogists and petrologists identified minerals by their more obvious physical properties (hardness, density, cleavage, crystal form and so on), and they determined the chemical compositions of minerals and rocks by so-called 'wet' gravimetric methods, which involved dissolution of the sample followed by selective precipitation and weighing of individual constituents.

With the discovery of the polarization of light early in the nineteenth century the observation of physical properties of minerals could be extended to a study of their behaviour in polarized light. The application of X-rays to analysis of crystal structure in the early twentieth century was followed by the development of several indirect and generally non-destructive methods of chemical analysis. These new techniques have supplemented rather than supplanted the earlier well-tried methods, which are still widely used. All students of mineralogy learn how to identify minerals on the basis of their 'external' properties, and gravimetric analysis, though somewhat laborious despite modern improvements, still provides the standards against which indirect physical methods of analysis are calibrated.

Three main kinds of physical method exist for the determination of the nature and chemical composition of minerals and rocks: (1) optical methods, involving the use of polarizing microscopes; (2) diffraction methods using either X-rays (see above) or electron beams to elucidate the internal structure of minerals; and (3) analysis of emission or absorption spectra, which reflect the relative abundances of elements. Of these three, the first is the longest established. It requires the simplest and least costly equipment and it remains the most powerful tool for the laboratory study of minerals and rocks.

The great advantage of the polarizing microscope is its versatility. All transparent minerals, save those crystallizing in the cubic system, are themselves capable of polarizing light, and they produce complicated but characteristic combinations of absorption and interference colours when polarized light is passed through thin sections of rocks or mineral grains. The identification of minerals can also be refined by the measurement of their refractive indexes using carefully calibrated immersion oils. Some minerals are opaque, however, even in thin sections, and these can be examined by polishing their surfaces and examining them in reflected polarized light.

Because the resolution that can be obtained in any microscope system is limited by the wavelength of the radiation used, there is a lower limit to the size of objects that can be distinguished by optical microscopy. Under the most favourable conditions optical microscopes cannot resolve objects less than half a micron or so in size (and few petrological microscopes are capable of this). The electron microscope is capable of far greater resolution.

Although electrons are generally thought of as charged particles they also possess wavelike properties, and electron beams can be reflected, refracted and diffracted, just as beams of light can. However, their wavelength is much less (in the order of a few pm (1 pm = 10^{-12} m) which is smaller than atomic dimensions) than that of light (which is around 10^{-7} m). Moreover, because electrons are charged particles electron beams can be focused using electromagnetic 'lenses', the magnification of which can be controlled by changing the strength of the current in the lenses.

The electrons are produced by a heated filament, and the beam is controlled by the various lenses in the system, which are analogous to the condenser, objective and ocular lenses in optical microscopes. The whole system must, however, be maintained under high-vacuum conditions, and the preparation of specimens is a complicated process, because it is not possible simply to transmit electrons through, or reflect them from, the object to be studied.

Optical emission and X-ray fluorescence (XRF) spectrometry are methods for determining the chemical composition of minerals and rocks. They therefore contrast with the techniques just described, which are concerned with physical properties and characteristics. Optical emission spectrography depends upon spectral analysis of radiation produced by the excitation of atoms by strong heating. When the radiation is passed through a prism, it is refracted and dispersed according to wavelength. The optical emission spectrum of an element consists of a series of lines, corresponding to radiation of different specific and characteristic wavelengths. The intensity of the lines in the spectrum can be measured, and it is related to the concentration of the corresponding element in the sample.

XRF spectrography combines the principles of X-ray diffraction and emission spectrography. It provides a rapid method of analysis and is nowadays widely used. The atoms are excited by a primary X-ray beam, rather than by heating, and the greater energy input causes them to emit energy at X-ray wavelengths (a few tens of pm).

Just as with optical emission spectrography, the X-radiation for each element is of specific and characteristic wavelength, and the intensity of the radiation can be measured and is proportional to the concentration of the element in the sample.

The electron microprobe, whose development was pioneered in the 1950s by J. V. P. Long of Cambridge, England, embodies perhaps the most important new technique of any for the chemical analysis of minerals. Previously, it was only possible to measure the bulk composition of minerals separated from rocks, a method that could not always eliminate microscopic impurities adhering to or enclosed within mineral grains. The electron microprobe enables the composition of minerals to be measured *in situ* within the rock, and furthermore it enables variations of composition within individual mineral grains and between different grains of the same mineral in a rock to be determined.

The technique is based upon principles applied in reflected light and electron microscopy and XRF analysis: a high-energy beam of electrons is focused upon the polished and carbon-coated specimen surface. The electrons excite the atoms in the sample to energy levels at which they emit their characteristic X-radiation. The X-rays from the sample are then diffracted and detected in just the same way as they are for XRF analysis. Because the specimen is polished, it can be viewed in the specimen chamber by an optical (reflecting) microscope system, and grains or parts of grains can be selected for analysis; and because the electron beam can be focused down to about 1 micron diameter, selected spot analyses or traverses can be made across individual grains. In addition, whole grains can be scanned to determine the overall distribution within them.

Most of these methods (the notable exception is transmitted light microscopy) require sophisticated equipment, power sources with stabilized voltage, cooling systems of constant water pressure and (for electron microscopy and microprobe) high-vacuum chambers. Developments and extensions of the methods are constantly being made and are applied in a wide variety of fields, not just to those of mineralogy and petrology. It is well known, for instance, that the electron microscope is widely used in biological studies, and XRF and the electron microprobe has obvious applications in the study of alloys and ceramics.

The physical properties of rocks

Rocks are physical mixtures rather than chemical compounds, without the fixed attributes of form, density, hardness and chemical composition that make, say, crystals of quartz or cleavage rhombs of calcite instantly recognizable wherever they are found. No two rocks are alike. They may be given the same name, but there will always be differences in the relative sizes, shapes and proportions of the mineral grains of which they are composed. There are, however, limits to the possible combinations of these variables, especially the relative proportions of minerals among the common rocks.

The texture of a rock, which can be defined as the physical relationship between its constituent minerals, provides much information about the physical environment in which the rock was formed. On the basis of textural criteria alone it is usually possible to assign a rock to one of the three main groups, each of which is discussed more fully later in this chapter. Igneous rocks are mostly characterized by an interlocking crystalline texture resulting from solidification of magma, which is molten rock material originating within the crust or the upper mantle. Smaller grain sizes are usually indicative of more rapid cooling (eg volcanic lavas). Most sedimentary rocks are aggregates of more or less rounded mineral grains, the spaces between them filled by finer material. They represent accumulations of material eroded from pre-existing rocks and deposited in successive layers, usually beneath the sea. Finer-grain sizes are indicative of quieter depositional conditions (eg lagoon silts and muds). Metamorphic rocks are also crystalline, but often display a banding or preferred alignment of minerals. They are products of solid state recrystallization that occurs when pre-existing rocks are subjected to the combined effects of heat and pressure, often deep within the crust. Finer-grain sizes can be broadly correlated with less intense conditions of metamorphism (eg slates).

Texture can often provide more immediate information about the origin of a rock than its mineralogical composition can. Figure 5.15 illustrates the basic textural characteristics of the three main rock groups and shows how a few minerals can be combined together in rocks of very different appearance and origin.

a

b

c

d

5.15: *The textures of rocks:* (**a**) *Igneous rock (diorite) formed of interlocking crystals of feldspar, quartz, hornblende and biotite mica.* (**b**) *Sedimentary rock (sandstone) formed of rounded grains of quartz.* (**c**) *Contact metamorphosed rock (spotted hornfels) in which clumps of new minerals have grown at high temperature.* (**d**) *Metamorphic rock (schist) formed mainly of crystals of mica in subparallel alignment, having grown under conditions of high temperature and pressure (regional metamorphism).*

The mineralogical and chemical composition of rocks

All the common rocks are made of silicate minerals, with the exception of limestones, which consist of calcium carbonate (calcite, $CaCO_3$); but rock-forming minerals tend to occur in fairly well-defined associations. Quartz, feldspars and micas are common constituents in all three main rock groups, whereas amphiboles and pyroxenes are rare in sedimentary rocks but common among both igneous and metamorphic rocks. Olivines and garnets are practically confined to igneous and metamorphic rocks respectively, and calcite is widespread among sedimentary and metamorphic rocks but is not a major constituent of igneous rocks.

When a rock specimen is crushed to a homogeneous powder and chemically analysed each of the minerals contributes its quota of elements to the bulk chemical composition of the rock. Chemical analyses of rocks are normally expressed in terms of the oxides of the major constituent elements: this has no mineralogical significance but simply expresses the overwhelming abundance of oxygen. As a result of the predominance of silicon over all the remaining elements the principal component in chemical analyses of common rocks, save limestones, is silica. Rocks rich in quartz and alkali feldspars (eg granites) will contain a higher proportion of silica (SiO_2), potassium (K_2O) and sodium (Na_2O) in their analyses, and less of the basic oxides (MgO, FeO, CaO) than will rocks rich in plagioclase feldspar and pyroxene (eg basalts). Figure 5.16 lists mineralogical and chemical correlations for some common rocks.

While metamorphic rocks have chemical compositions similar to igneous and sedimentary rocks they may have quite different mineralogical compositions (the main exceptions are marbles and quartzites which are metamorphosed limestones and sandstones). Metamorphism is primarily a recrystallization process, whereby elements are redistributed into new minerals; there is only very limited movement into or out of the rock, and so the chemical composition remains essentially that of the original rock.

5.16: *Composition of some common rocks in terms of their chemistry (left) and mineralogy (right).*

Oxide	Granite (igneous rocks)	Basalt (igneous rocks)	Amphibolite[1] (metamorphic rocks)	Schist[2] (metamorphic rocks)	Shale (sedimentary rocks)	Sandstone (sedimentary rocks)	Limestone (sedimentary rocks)
SiO_2	70.8	49.0	49.3	63.3	62.4	94.4	5.2
TiO_2	0.4	1.0	1.2	1.4	1.1	0.1	0.1
Al_2O_3	14.5	18.2	16.9	17.9	16.6	1.1	0.8
Fe_2O_3	1.6	3.2	3.6	3.6	3.2	0.4	0.3
FeO	1.8	6.0	6.8	2.6	2.1	0.2	0.2
MgO	0.9	7.6	7.0	1.6	2.5	0.1	7.9
CaO	1.8	11.0	9.5	1.9	1.7	1.6	42.6
Na_2O	3.3	2.5	2.9	1.3	0.9	0.1	0.1
K_2O	4.0	0.9	1.1	3.1	3.0	0.2	0.3
H_2O (water[3])	0.8	0.4	1.5	2.6	5.2	0.3	0.7
CO_2 (carbon dioxide[3])					1.0	1.1	41.6
	99.9	99.8	99.8	99.3	99.7	99.6	99.8

Mineral	Granite (igneous rocks)	Basalt (igneous rocks)	Amphibolite[1] (metamorphic rocks)	Schist[2] (metamorphic rocks)	Shale (sedimentary rocks)	Sandstone (sedimentary rocks)	Limestone (sedimentary rocks)
quartz	30			32	17	97	3
alkali feldspar	60	5				1	1
plagioclase	5	45	42	18			
pyroxene		40					
amphibole			50				
olivine		5					
biotite	4		5	7			
muscovite				38	1	1	
magnetite	1	5	3	3	1	1	1
staurolite				2			
clay minerals					80		1
calcite					1		94
	100	100	100	100	100	100	100

Notes

Chemical analyses are in weight per cent oxides; mineral compositions are in volume per cent and are approximate. Chemical analyses do not add up to exactly 100 per cent because of the small errors inherent in the analytical methods. Several different techniques are used, some being appropriate to only one element. Analyses totalling between 99.5 and 100.5 are usually considered to be acceptable.

1 The amphibolite is probably a metamorphosed basalt.

2 The schist is probably a metamorphosed shale.

3 Metamorphism generally reduces the amount of H_2O and CO_2 in rocks, especially sedimentary rocks. H_2O increases in amphibolites, because it is taken up from wet sediments interlayered with basalts and goes into amphiboles.

The classification of rocks

Rocks are generally classified on the basis of their textural features, especially grain size, and mineralogical composition. Figures 5.17, 5.20 and 5.22 provide classification schemes for the more important igneous, sedimentary and metamorphic rocks. Because rocks are mixtures of minerals there are no hard and fast boundaries between the different rock types named in these figures. Only the principal minerals are shown in each figure; all rocks have small amounts of minerals in addition to those listed.

Igneous rocks

Igneous rocks are formed when magma cools and solidifies. All magmas contain small amounts of dissolved gases, chiefly water vapour and CO_2, which help to reduce their viscosity and enable them to flow more easily. They frequently carry crystals that have begun to form as the magma moves from the site of generation up towards the surface. Magmas that reach the surface form volcanic, or extrusive, rocks, and those that are emplaced within the crust form intrusive rock bodies of various shapes and sizes.

An important chemical subdivision of igneous rocks is into acid, intermediate, basic (or mafic) and ultrabasic (ultramafic) categories. The terms acid and basic are inappropriate but stem from the idea that silicate minerals were salts of 'silicic acid'. The more silica in the rock, the more 'acid' its parent liquid. Acid igneous rocks contain sufficient silica for at least 10 per cent of the rock to be quartz. Intermediate rocks have up to 10 per cent quartz and basic rocks typically have no quartz; all the silica is taken up in feldspars and other silicates. In ultrabasic rocks there is normally too little silica even to form feldspars, and the basic (ferromagnesian) silicates predominate.

Granite and basalt are the two principal kinds of igneous rock in the Earth's crust and their magmas have quite different characteristics. Basaltic magmas have temperatures in the approximate range 1200°C to 900°C and relatively low viscosity. They form the spectacular rivers of lava that are such well-known features of Hawaiian and Icelandic volcanoes. Basaltic lava flows are usually thin and extensive, and basaltic volcanoes generally have gentle shield-like profiles. Rapid escape of the dissolved gases, as confining pressures are released during ascent to the surface, leads to more explosive eruptions, with fire-fountaining and the formation of volcanic bombs, scoria and volcanic ash. These pyroclastic materials either pile up round the erupting vent or are spread out over wide areas as layers among the lava flows; when consolidated into rock they are called volcanic agglomerate.

Bubbles of gas are also released as the lavas cool and crystallize, resulting in the vesicular texture commonly seen on the surface of lava flows. Vesicles filled by minerals deposited from groundwater solutions percolating through old lavas are called amygdales.

Basaltic magmas erupted under water adopt globular or cylindrical forms a metre or so across; they are termed pillow lavas on account of their characteristic shapes in cross-section, which reveal a

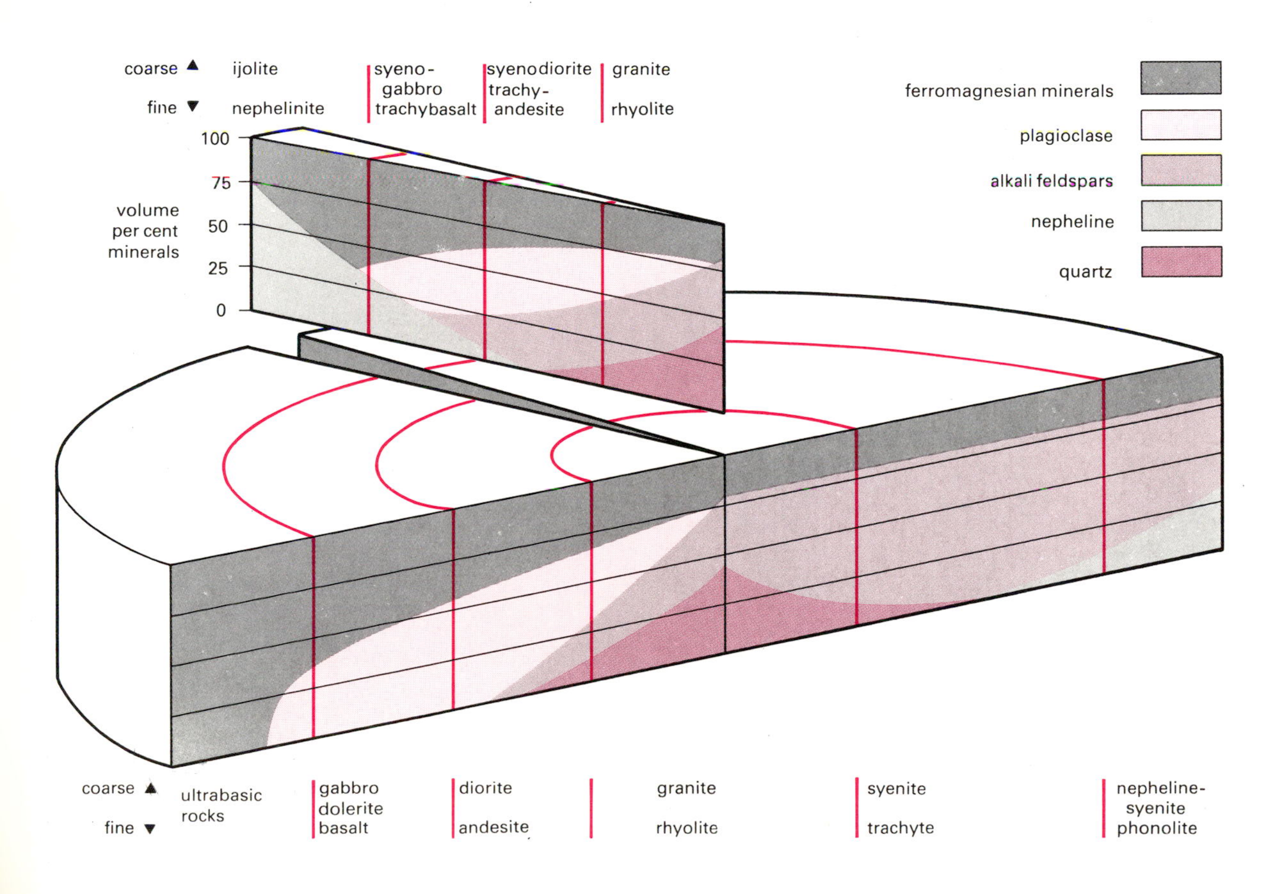

5.17: *Classification of major igneous rock types according to their content of essential minerals. There is continuous variation among them and many intermediate types are given names not shown here; there are also more extreme types consisting almost entirely of a single mineral. All igneous rocks also contain small percentages of 'accessory' minerals whose presence or absence does not affect the classification.*

a

b

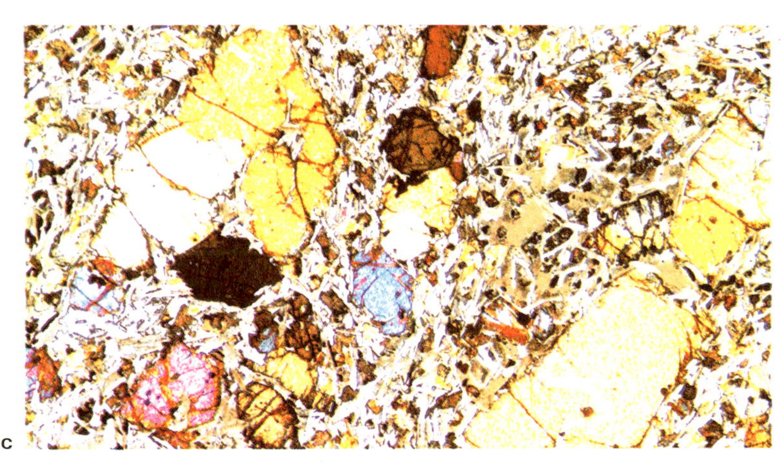

c

d

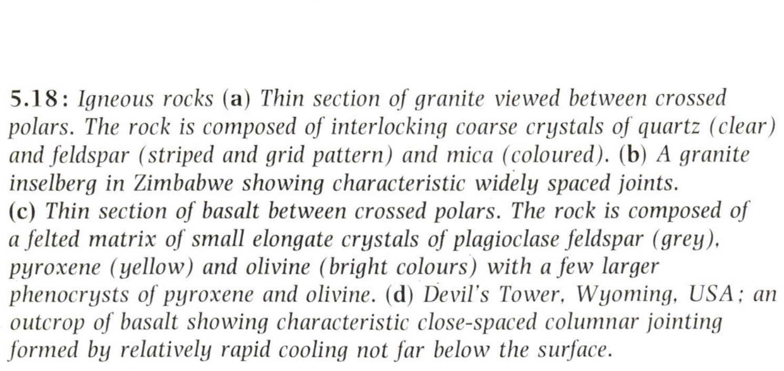

5.18: *Igneous rocks* (**a**) *Thin section of granite viewed between crossed polars. The rock is composed of interlocking coarse crystals of quartz (clear) and feldspar (striped and grid pattern) and mica (coloured).* (**b**) *A granite inselberg in Zimbabwe showing characteristic widely spaced joints.* (**c**) *Thin section of basalt between crossed polars. The rock is composed of a felted matrix of small elongate crystals of plagioclase feldspar (grey), pyroxene (yellow) and olivine (bright colours) with a few larger phenocrysts of pyroxene and olivine.* (**d**) *Devil's Tower, Wyoming, USA; an outcrop of basalt showing characteristic close-spaced columnar jointing formed by relatively rapid cooling not far below the surface.*

fine-grained or glassy rind surrounding coarser-grained interiors with a radiating fracture pattern produced on cooling. The topmost kilometre or so of the oceanic crust (beneath the cover of sediments) is formed almost entirely of pillow lavas.

Magmas of granitic composition have temperatures in the approximate range 800°C to 600°C. They are a great deal more viscous than basaltic magmas and do not reach the surface so often. When they do, the lavas are pasty masses extruded slowly as thick flows or bulbous domes of rhyolite or obsidian. Magmas of granitic (rhyolitic) composition typically erupt with enormous violence, however, as the dissolved gases are explosively released from the viscous mass to produce a voluminous incandescent froth of gas and rock particles, as in the sudden release of pressure from a bottle of beer that has been shaken. The incandescent clouds travel at great speeds and may cover areas of hundreds of square kilometres to thicknesses of tens or even hundreds of metres. When the mass finally comes to rest, the particles in the lower layers are often hot enough to weld together to form pyroclastic rocks called welded tuffs, or ignimbrites. The overlying layers of less consolidated tuffs commonly contain fragments of the highly vesicular, frothy-looking rock known as pumice.

Granitic (rhyolitic) magmas are more viscous than basaltic magmas partly on account of their lower temperatures and partly because of their greater silica contents (70–75 per cent as against 45–50 per cent). Silicate tetrahedrons tend to link together in the magma even before crystallization starts. The linkages offer resistance to flow, and the higher the silica content, the more viscous the magma. As a result obsidian (glassy rhyolite) is commoner than basaltic glass: the viscous rhyolitic melt, with its high proportion of randomly linked tetrahedrons, is readily chilled to a more or less structureless glass, and there is little chance for the constituents to become ordered into crystals.

Andesitic magmas are intermediate in composition between those of granite (rhyolite) and basalt and have intermediate characteristics. More viscous than basalt, and less viscous than rhyolite, they characteristically erupt roughly equal amounts of thick and rather rubbly lava flows and pyroclastic deposits, to build the high steep-sided cones that are commonly regarded as the 'typical' volcano shape.

Magmas that fail to reach the surface penetrate fractures and other zones of weakness in the crust and crystallize as igneous intrusions. They may be approximately tabular (sills, dykes, laccoliths and lopoliths) or roughly cylindrical (volcanic plugs or necks, stocks and bosses). Some larger intrusions may represent the subsurface magma chambers that lay beneath ancient volcanoes. Granitic magmas also form the enormous, elongate intrusions called batholiths that occur in the deeper parts of major orogenic (mountain-building) belts and have dimensions measurable in tens to hundreds of kilometres.

Evidence to account for what has happened to the rock that formerly filled the space now occupied by the vast granite batholiths is inconclusive. However, it is generally agreed that there are three ways in which the batholiths could have formed. The granite may have 'domed' the overlying rocks, displacing them upwards and sideways; it may have 'stoped' its way upwards, by breaking off blocks of overlying rocks, which then sank through the less dense granitic magma; or it may have been formed largely *in situ* by a combination of partial melting (see below) and very complex geochemical processes, whereby continental crustal rocks of suitable composition are transformed into granites. These three processes are not mutually exclusive, and indeed many large batholiths bear evidence of all three having played some part in their formation.

The textures of intrusive igneous rocks are generally coarser than those of extrusive (volcanic) rocks, because they have cooled more slowly; and textures in shallow-level (hypabyssal) intrusions, especially the smaller sills and dykes, are finer grained than in large deep-seated (plutonic) intrusions. The margins of intrusive rock bodies are finer grained than their interiors, because the magma is chilled against surrounding rocks (which are sometimes referred to as the 'country' rocks).

Porphyritic texture is common among both volcanic and hypabyssal intrusive rocks and indicates an initial phase of slow cooling and crystallization under plutonic conditions, followed by more rapid cooling at the surface or at shallow depth. It provides a clear demonstration that magmas not only crystallize over a range of temperatures but are also commonly emplaced or erupted as more or less mushy mixtures of liquid and early-formed crystals. Porphyries have been widely used for statuary and ornamental stones throughout history.

Crystallization and fractionation of magmas

It has been recognized for a long time from a study of textures of igneous rocks, particularly porphyritic textures, that there is a sequence of crystallization of minerals from magma. Ferromagnesian minerals tend to dominate among early-formed minerals, while feldspars and micas (and quartz) crystallize later. Charles Darwin was among the first to realize that the sinking or floating of early-formed crystals (phenocrysts) in a large, slowly cooling body of magma would change the composition of the remaining melt, which he suggested was a cause of variation among igneous rocks. Large, slowly cooled masses of basaltic magma (their original composition preserved in the chilled margins) exemplify this process admirably. Well-studied examples include the Palisades sill of New Jersey and the Skaergaard intrusion of east Greenland. The intrusions are layered because after they were emplaced the large body of still fluid basaltic magma cooled slowly and minerals sank to the bottom as they crystallized and accumulated in layers. The lower layers are ultrabasic, rich in olivine and pyroxene crystals which formed at the early stages of cooling when temperatures were still quite high. They sank through the slowly cooling magma and settled at the base of the intrusion, just as sedimentary particles accumulate under water (hence the term cumulate for rocks formed in this way; see Figure 8.6). Higher in the intrusion the mineral composition becomes that of intermediate and even acid rocks: olivine disappears, the plagioclase feldspar becomes more sodic and some quartz may appear. This happens because the segregation of olivine and pyroxene has depleted the magma in magnesium (and iron) and thus relatively enriched it in the other constituents. An originally basaltic magma can thus be fractionated or differentiated by the removal of successive crops of crystals to produce a series of rocks ranging from ultrabasic to acid in composition.

Basaltic volcanoes commonly have variable amounts of intermediate to acid volcanic rocks associated with them; these are usually attributed to the eruption of magmas of appropriate composition produced by the fractionation of a basaltic parent magma at depth.

A more quantitative approach to the study of crystal fractionation and magmatic differentiation was begun in about 1915 by N. L. Bowen at the Geophysical Laboratory in Washington, DC, which remains a principal centre for experimental petrology. This science consists in essence of subjecting natural rocks or mixtures of specially synthesized minerals to melting and crystallization experiments under carefully controlled conditions of temperature and pressure.

By combining information from experimental petrology and observations of igneous rocks in the field, a very generalized diagram can be drawn to show the appropriate order of crystallization of minerals in common igneous rocks. Melting and crystallization behaviour among rock-forming silicates is very complex, and only the salient features are shown in Figure 5.19, which cannot be taken as more than an approximate guide; for instance, many basalts contain no olivine and many granites contain no muscovite.

Magmatic differentiation leads to a progressive increase in silica

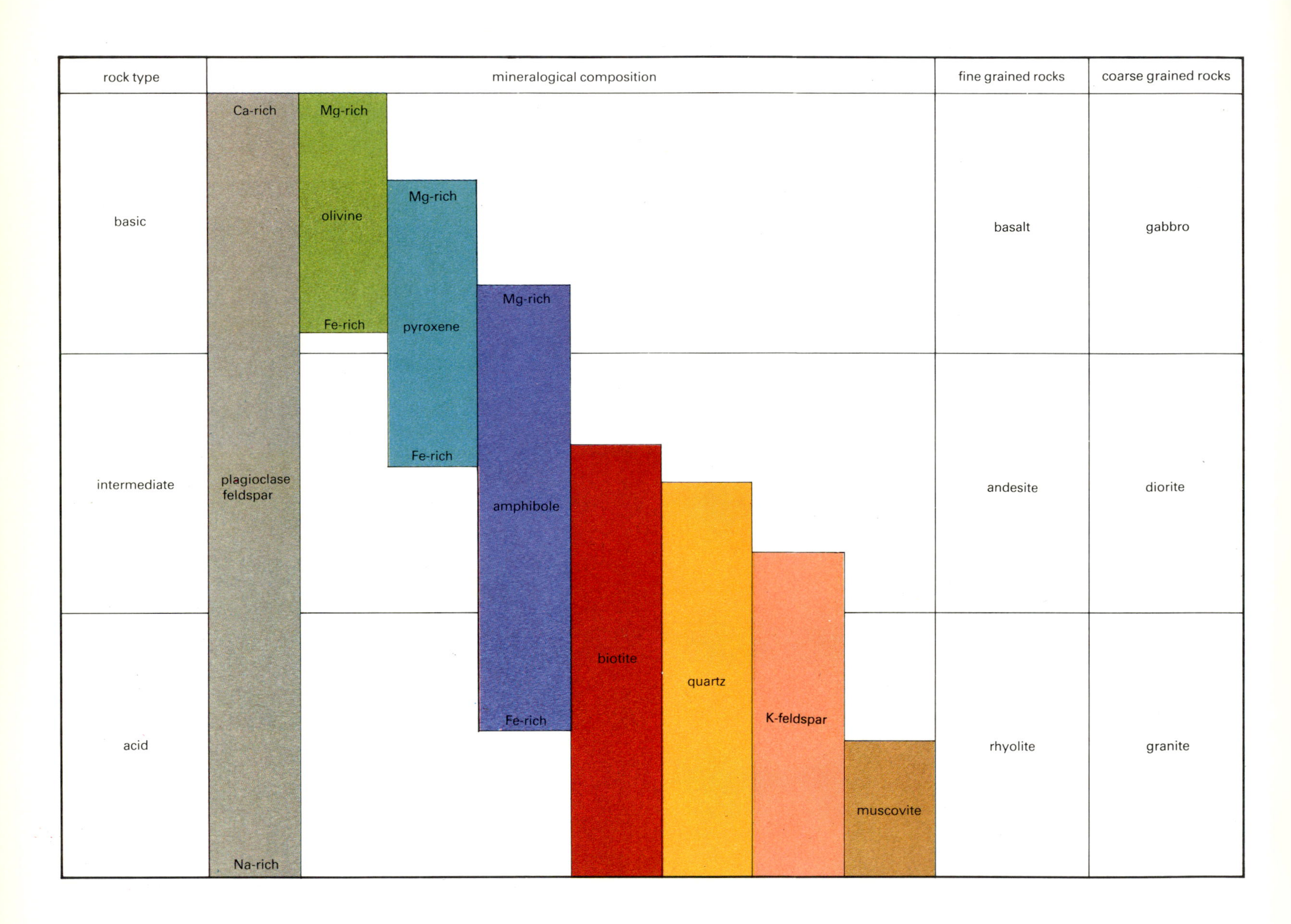

5.19: *The approximate order of crystallization of minerals in common igneous rocks. There is a decrease of density of minerals with falling temperature, together with a trend from* Mg *to* Fe *among ferromagnesian mineral groups and from* Ca *to* Na *among plagioclase feldspars. Thus there is a compositional control on temperature of crystallization:* Mg-rich *olivine and* Ca-rich *plagioclase, for instance, crystallize at higher temperatures than* Fe-rich *olivine and* Na-rich *plagioclase.*

and alkalis, and thus rocks of granitic composition are one obvious end product. However, syenites and nepheline syenites (and their fine-grained equivalents trachyte and phonolite) can also be produced in this way as a result of differences in the composition of the basaltic parent or in the conditions of fractionation, or in both. A whole spectrum of fractionation trends is thus possible. Where a group, or suite, of igneous rocks can be shown to be related in space and time it can almost always also be referred to one of these fractionation trends. Such associations commonly define what are sometimes called petrographic provinces.

Generation of magmas

Magmas crystallize over a range of temperatures. The first minerals to form have compositions that are different from the overall bulk composition of the magma. In the same way rocks melt over a range of temperatures; and the first liquid to form has a different composition from the rock. Magma generation is a partial melting process, and the ultimate source of all magmas is the Earth's mantle, which consists of peridotite. It is composed principally of olivine, together with some pyroxene, and small but variable amounts of other minerals, including plagioclase, chromite and phlogopite. At depths of around one hundred to two hundred kilometres the low-velocity zone marks the boundary between the lithosphere, of which the crust is the topmost part, and the asthenosphere, which extends downwards to the base of the upper mantle. Pressure and temperature conditions in the asthenosphere are such that a small increase in temperature or decrease in pressure can induce partial melting of peridotite (about 10–30 per cent).

Basaltic magmas are generated by direct partial melting of mantle peridotite, and they are most abundantly generated beneath active oceanic ridges, the sites of formation of lithospheric plates. The whole of the oceanic crust is basaltic in composition, consisting of volcanic rocks (pillow lavas) erupted on the ocean floor and underlain by basaltic dykes, representing the fissures through which the lavas erupted; these in turn are underlain by gabbros, with cumulate layers in their lower parts, representing the magma chambers in which the magma accumulated prior to eruption (see Chapter 13).

Andesitic magmas are generated above the subduction zones, where lithospheric plates converge. Here the partial melting of mantle peridotite is complicated by the presence of the wet sediments and basalts of the oceanic crust on top of the descending lithospheric plate, and the result is a more silica-rich magma. Andesitic magmas are responsible for the formation of new continental crust, which grows by progressive accumulation of andesitic volcanics above and dioritic intrusions below (see Chapter 14).

Large amounts of granitic magmas are also generated above subduction zones, both by partial melting at the base of the andesitic crust and by fractionation of andesitic or even basaltic magmas modified by some mixing with crustal material.

The only true primary magmas are those of basaltic composition, in the sense that they can be generated only by partial melting of mantle peridotite. In contrast andesite and granite can be formed either by the fractionation of basaltic magma or by direct partial melting in the mantle or crust. The same can be said of almost all other igneous rocks, and the vast range of igneous rock compositions results from a complex interplay of these two fundamental processes.

Sedimentary rocks

Sediments and sedimentary rocks cover most of the ocean floor and about three-quarters of the continental land area, forming a veneer rarely more than a few kilometres thick upon the predominantly metamorphic and igneous continental crust. Where there has been major subsidence, sediment accumulations can reach thicknesses of 20–30 km. Most abundant are greywackes, sandstones, shales and limestones, of which the essential minerals are few in number: quartz, feldspars, clay minerals and calcite.

The agents of mechanical weathering physically disrupt solid rocks exposed at the Earth's surface. Such agents include frost action in mountainous areas, pounding by waves at cliffs, differential expansion and contraction in deserts and the wedging action of growing plant roots. Further physical breakdown accompanies transportation by rivers, by wind or by waves and currents along coasts. Individual particles become smaller and more rounded and minerals with good cleavage are rapidly reduced in size by repeated fracturing.

These physical processes are largely responsible for determining the textural features of sedimentary rocks. Their mineralogical composition is largely the result of chemical weathering, which decomposes the minerals of igneous and metamorphic rocks and forms new

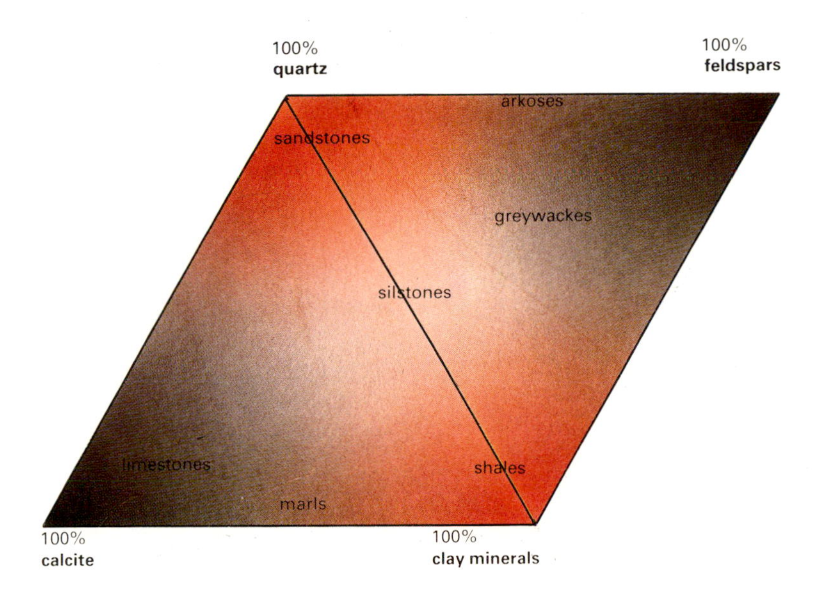

5.20: *Classification of the sedimentary rocks. Of the four major mineral constituents shown, quartz, feldspar and clay minerals are detrital (clastic or terrigenous) components, and calcium carbonate is chemically precipitated, usually by the activities of shell-building marine organisms. Other common sedimentary rocks include conglomerates (large, rounded rock fragments, up to boulder size); breccias (large, angular rock fragments); chert (biogenically precipitated silica); and evaporites (mostly halite and gypsum chemically precipitated by the evaporation of seawater).*

a

b

c

d

e

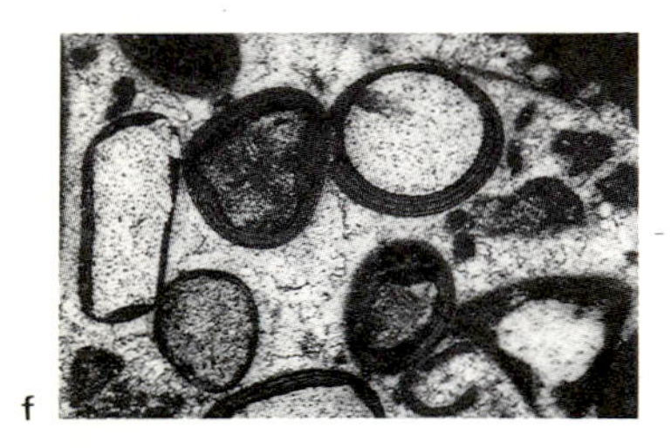
f

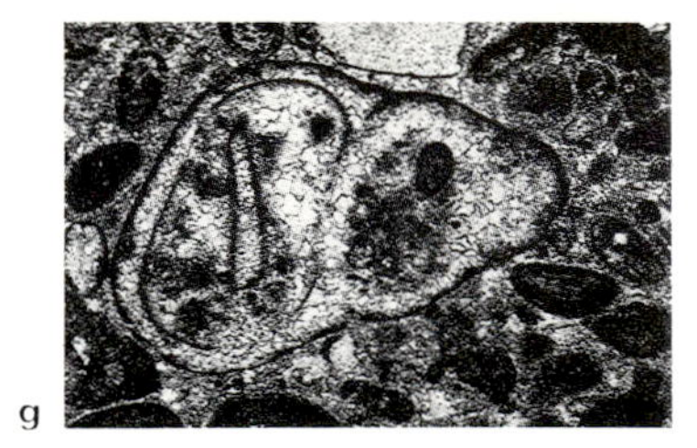
g

5.21: *Sedimentary rocks.* (**a**) *Thin section of a sandstone in polarized light. The individual grains, mainly quartz, are well-rounded, showing both mineralogical and textural maturity.* (**b**) *Thin section of a greywacke in polarized light. The grains are very variable in size, shape and composition, including angular particles of quartz, feldspar and rock fragments with finer grained particles including clay minerals filling the spaces between the grains. The textural and mineralogical immaturity suggest quite a different history from the sandstone (a).* (**c**) *Shales are the most abundant of sedimentary rocks, but the most difficult to study because their fine grain-size makes identification of individual minerals impossible without X-ray and other analytical techniques. Late Triassic shales with interbedded, more resistant, sandstones outcrop on steep cliffs on the island of Hopen in the Spitsbergen archipelago.* (**d**) *Cross-bedded sandstones of the Navajo Sandstone Formation, Zion National Park, Utah. These well-sorted sands were deposited on the leeward slopes of sand dunes in a desert in the Jurassic period. Migration of dunes as the deposit built up resulted in partial erosion of existing dunes so that each set of deposits cuts across the top of an older set. A wind-blown (aeolian) sand such as this would consist more purely of quartz than the sandstone in (a).* (**e**) *Outcrops of a limestone bed composed largely of fragments of the stems of crinoids (stalked echinoderms).* (**f**) *Thin section of an oolitic limestone. The grains are rounded and well sorted for size like the quartz grains in (a). The individual grains are each formed of several layers of finely crystalline calcite; this oolitic texture would have originated from gentle current action in warm shallow waters saturated in calcium carbonate. The spaces between the grains are entirely filled with clear calcite crystals which grew after the grains had settled into place.* (**g**) *Thin section of a micritic (muddy) limestone. The particles, including a small gastropod (snail) shell, are poorly sorted and the partly recrystallized matrix evidently included much fine grained calcite mud which would have been washed away under the conditions of formation of the oolitic limestone (f). Diagenetic recrystallization has also affected parts of the gastropod shell.*

ones. Igneous and metamorphic rocks are formed at high temperatures and commonly also at high pressures. When cooled and exposed at the Earth's surface, they are in a very different environment from that in which they originally crystallized. They are not in equilibrium under the new conditions and are no longer chemically stable. Chemical decomposition is slow in cold dry air, however, because water is necessary for chemical weathering, and so rates of mineral decomposition are highest under warm and humid conditions.

The silicate minerals of igneous and metamorphic rocks are broken down by hydration reactions, which come under the general heading of hydrolysis. The reactions are aided by small amounts of dissolved CO_2 in rainwater, which also increases the solubility of calcium carbonate and accounts for the rapid chemical weathering of limestones.

There is a broad inverse correlation between the temperature of magmatic crystallization of minerals (Figure 5.19) and their relative stabilities in the weathering environment. Olivine and pyroxene, for example, formed at highest temperatures, break down most readily, whereas muscovite is only very slowly decomposed, and quartz is hardly affected at all, except by slow dissolution.

As sedimentary particles are transported to the site of ultimate deposition they are subject to sorting into different fractions according to size or density; individual particles will also become more rounded with time and the decomposition of unstable minerals continues. Mature sediments, generally speaking, are those in which the constituents have spent a long time in the weathering environment. Examples are a pure quartz sandstone with well-rounded grains, accumulated on a beach or a sand dune, the finer material winnowed away by waves or wind, and a pure shale, consisting entirely of clay mineral particles accumulated in still water. A typical immature sediment is greywacke, consisting of angular fragments of quartz, feldspars, probably some ferromagnesian minerals and a good deal of clay.

An economically important product of sediment sorting is the concentration of dense, resistant minerals, such as zircon, magnetite, ilmenite and rutile, as well as gold and diamonds. These alluvial, or placer, deposits are what is left behind when less dense mineral fragments are washed away by strong currents or by wave action. The occurrence of heavy mineral fractions in sediments can provide a useful guide to the source, or provenance, of the sediments.

The sorting of sediments is a process of fractionation, somewhat analogous to the fractionation found in igneous rocks: minerals are segregated into different fractions, generally on the basis of density. When a granite is weathered and disaggregated, for example, the quartz and feldspar which make up about 90 per cent of the original rock will go their separate ways and may eventually be completely segregated from one another, to accumulate as deposits of pure quartz sand and pure clays.

The textural characteristics and mineral content of sedimentary rocks reflect their depositional environment and history. Grain size analysis of fragmental rocks, for example, gives an indication of the degree of sorting to which the particles have been subjected, and provides information about the conditions under which the sediment was transported and deposited. Other indications are provided by various sedimentary structures. They are generally seen in outcrops of sedimentary rocks and show the way in which the sediments were transported over the surface and then deposited. River and dune sands, for example, frequently exhibit cross-bedding; beach deposits are commonly ripple-marked; bedding laminations are characteristic of deposits accumulated in very quiet water; and graded bedding indicates that there were fluctuating currents in the area of deposition. Wind is also a powerful agent of sediment transport and deposition. It behaves in much the same way as water, and produces similar sedimentary structures (see Chapter 20).

Chemical processes

Most of the aluminium in the minerals of the original rock goes into clay minerals and much of the silicon is retained in quartz, but the other principal constituents are mainly removed in solution, though they do not always travel far. Iron is normally in the Fe^{2+} state in minerals; on weathering it rapidly oxidizes to very insoluble Fe^{3+} and is precipitated in soils near the site of weathering as hydrated iron oxides. Under prolonged tropical weathering thick iron-rich laterite layers can develop over wide areas. Even clay minerals can be decomposed under these conditions, silica being removed in solution to leave deposits enriched in insoluble aluminium hydroxides (bauxite) provided the original rock was not rich in iron.

Magnesium, calcium, sodium and potassium usually stay in solution and end up in the oceans. However, variable amounts of all of these elements enter the structures of new clay minerals: potassium, in a variety called illite, and to a lesser extent calcium and magnesium (and even sodium), in another family of clay minerals, which is called the montmorillonites.

Calcium is precipitated from seawater mainly by the action of marine organisms. It is precipitated mostly as calcium carbonate in the shells and skeletons of invertebrates (corals, bivalves, planktonic animals, etc), but also as calcium phosphate in the bony skeletons of fishes. In warm water it may also be inorganically precipitated, frequently as a fine-grained calcite mud. In shallow warm current-swept areas free of terrigenous material, subspherical grains with a concentric structure, called ooliths, are a characteristic product of carbonate precipitation. When calcium carbonate is precipitated either organically or inorganically from seawater, it is initially in the form of the aragonite polymorph, which subsequently inverts to calcite. The precipitation of calcium carbonate, either organically or inorganically, is the first stage in the development of new limestones. Small amounts of silica also enter the sea in solution and are used by a minority of organisms, chiefly planktonic varieties, to build skeletons which may subsequently accumulate in layers on the sea floor and are lithified to chert (of which flint is a variety).

When seawater is evaporated a whole variety of salts is precipitated as crystals to form evaporite deposits, which are composed principally of halite (NaCl) and calcium sulphate (either as anhydrite or gypsum), along with smaller amounts of potassium and magnesium salts. These are inorganic precipitates and have interlocking crystalline textures, superficially similar to those of some igneous rocks.

Diagenesis and authigenesis are processes that affect sediments after they have been deposited. Diagenesis encompasses the conversion of sediment to rock (lithification) and involves mainly the cementation of grains by material deposited from migrating solutions as they are driven out during compaction and a certain amount of recrystallization, particularly of very fine clay mineral grains and of calcite. Authigenesis involves the growth of new minerals in sediments as part of the diagenetic process. Good examples are the development of pyrite, gypsum or feldspar crystals in various kinds of shale, as a result of redistribution of elements within the sediments by the migrating solutions.

Metamorphic rocks

Metamorphism can be said to begin when temperature and/or pressure has risen enough to effect obvious changes of original igneous or sedimentary textures or mineralogy, but there is no hard and fast boundary between diagenesis and metamorphism.

Metamorphism is a solid-state recrystallization process, but the presence of water is crucial. It may be water still trapped in pore spaces or combined as OH^- ions in hydrous minerals such as amphiboles or micas or clay minerals, and it is present only in small amounts (never more than a small percentage). It acts both as a catalyst for metamorphic reactions and as a kind of intergranular lubricant between mineral grains, providing the transporting medium in which ions are exchanged and rearranged into new mineral structures.

With increases of temperature and pressure the aqueous fluid phase becomes more mobile and is progressively driven out of the rocks or locked up again in new hydrous minerals. New textures and mineral assemblages develop in the rocks in response to the changing conditions of temperature and/or pressure. The textures and mineral assemblages of metamorphic rocks can be used to determine the conditions of temperature and pressure under which the rocks were recrystallized. In this context it is customary to speak of the metamorphic grade of a rock; the higher the grade, the more extreme the conditions of temperature and/or pressure during its formation.

There are two principal kinds of metamorphism: contact, or thermal, metamorphism, in which temperature is the principal agent of change; and regional metamorphism, in which the effects of increased pressure and shearing stress are combined with those of temperature.

Rocks in the vicinity of igneous intrusions are literally baked by the heat given off by the cooling magma, and igneous intrusions are surrounded by metamorphic aureoles, in which the metamorphic grade decreases away from the boundary of the intrusion. The textures of contact metamorphic rocks are generally granular and unfoliated, with randomly oriented crystals, because of the absence of directed pressures. The very hard and splintery rocks formed by contact metamorphism of fine-grained sediments and some lavas are called hornfels, the mineral composition of which will depend upon the nature of the original rock and the temperature of metamorphism. Biotite is a common constituent of hornfelsed clay-rich rocks and may be accompanied by large crystals (porphyroblasts) of aluminosilicate minerals such as andalusite (chiastolite) and cordierite, giving the rock a typically 'spotted' appearance (Figure 5.15). However, the contact metamorphism of originally high temperature rocks such as basalt results simply in the development of new textures and not of new minerals.

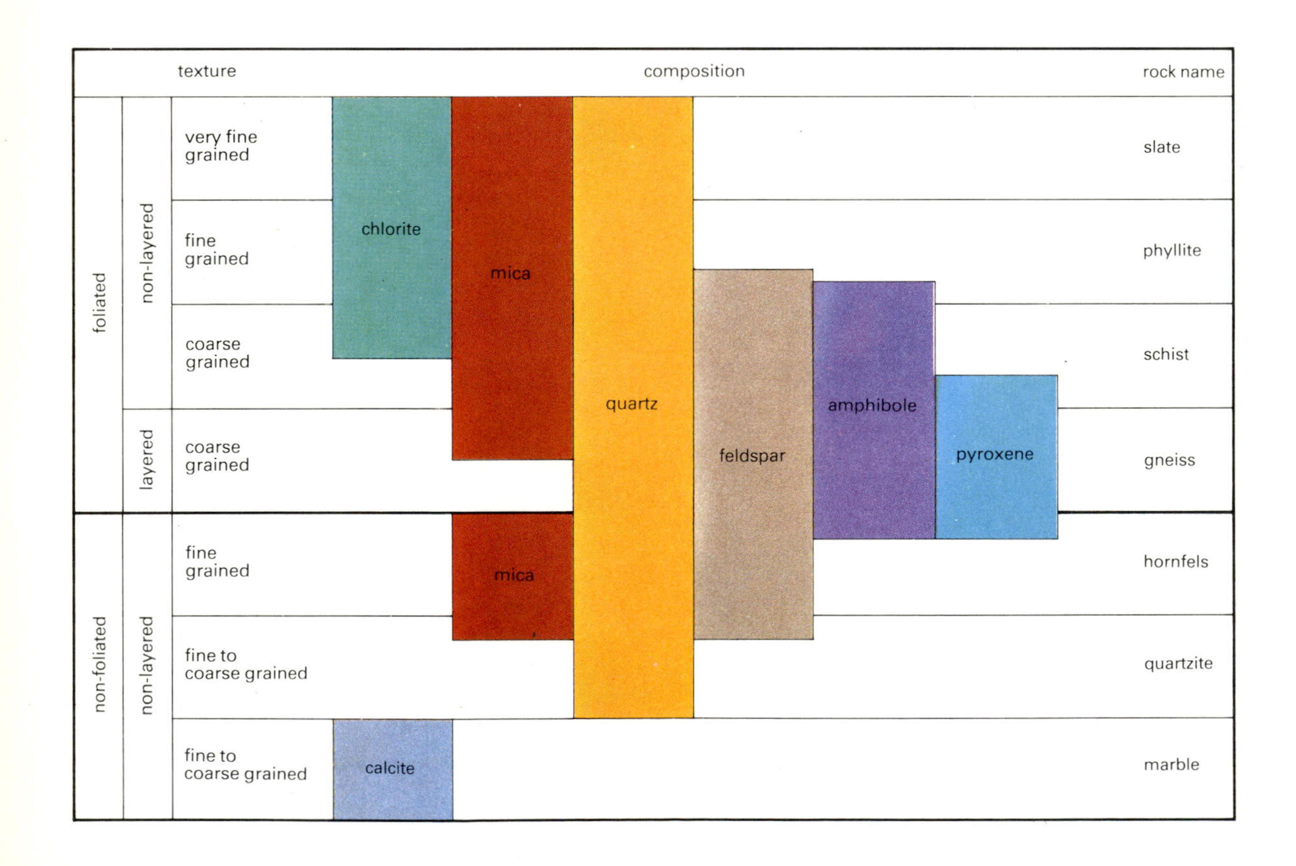

5.22: *Classification of the metamorphic rocks. Foliated rocks are those possessing a definite planar structure. The diagram shows the minerals that may be present in a particular metamorphic rock, the mineral assemblage depending partly on the composition of the original rock and partly on the conditions of metamorphism.*

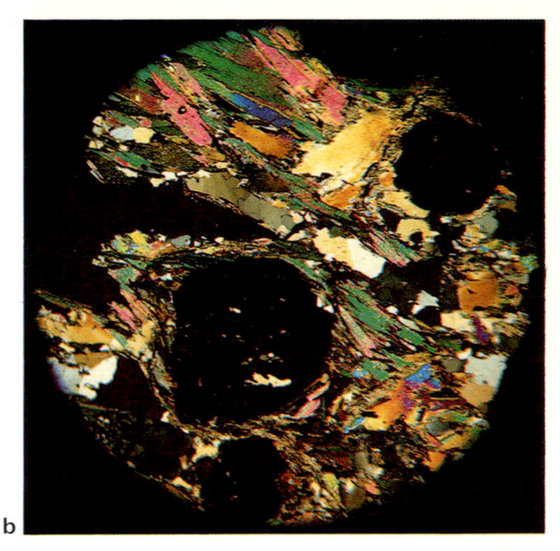

5.23: *Metamorphic rocks.* (**a**) *Thin section of hornfels (compare with Figure 5.15(c); this very fine grained rock was produced by contact metamorphism of a basalt.* (**b**) *Thin section of mica schist (compare Figure 5.15(d)).* (**c**) *Outcrop of gneiss of Precambrian age in East Greenland.*

Regionally metamorphosed rocks underlie huge areas of the Earth's crust formed in the deeper parts of orogenic belts. They are the products of high temperatures and pressures. Most regional metamorphism occurs where lithospheric plates converge, particularly at collision zones between continental masses (as in the Alps and Himalayas) where great thicknesses of sedimentary and igneous rocks are buried and internally deformed and heated. The rocks are almost always foliated, possessing a directional texture defined by the alignment of minerals such as micas and amphiboles at right angles to the principal stress. There is a general increase of grain size with increasing metamorphic grade: gneisses and schists are coarser grained than phyllites, which in turn are coarser than slates.

Metamorphic zones and facies

The study of metamorphic rocks is complicated by the fact that the same conditions of metamorphism may affect rocks of widely differing origins and compositions, leading to the development of totally dissimilar mineral assemblages. In order to define zones in which conditions of metamorphism have been approximately the same, it is necessary to compare rocks of similar original composition. Such studies were initiated by George Barrow in the south-eastern Highlands of Scotland in the 1880s. Barrow concentrated on pelitic rocks —those formed by metamorphism of originally clay-rich sediments (as distinct from psammitic rocks, formed from sandstones)—to ensure generally similar bulk compositions. He distinguished successive zones of progressively increasing metamorphic grade, each zone marked by the appearance of a distinctive metamorphic mineral. Boundaries between zones are called isograds; ideally, all points on an isograd surface reached the same conditions of temperature and pressure during metamorphism. The rocks of highest grade were metamorphosed under the highest temperatures and pressures, ie at the greatest depths.

The type of mineral assemblage developed in successive zones of regional metamorphism depends upon the thermal gradient, ie on the rate of increase of temperature with depth (and hence pressure). Other zonal sequences are possible. The zonal sequence developed in the Abukuma region of Japan or the Buchan region of north-east Scotland, for example, is the result of regional metamorphism under somewhat lower pressures than those that prevailed at similar temperatures in Barrovian zones, ie the thermal gradient was somewhat steeper. Among the most useful indices of pressure and temperature conditions during metamorphism are the three polymorphs of Al_2SiO_5: kyanite, sillimanite and andalusite.

A complete classification of metamorphic terrains should take into account variations in the bulk composition of rocks as well as of temperature and pressure. The concept of metamorphic facies, first proposed by the Finnish petrologist Penti Eskola in about 1915, is that of a group of rocks of varying composition, which have all been metamorphosed under similar conditions. Each facies covers a large range of conditions, and the nomenclature is based on the mineral assemblages that are developed in metamorphosed basalts (Figure 5.26). Facies can be divided into subfacies according to different bulk compositions, and facies series can be used to deduce the thermal gradient that prevailed during metamorphism.

A definite hierarchy of metamorphic minerals and textures is developed in rocks of a particular composition, in response to rising temperatures and pressures. The great majority of metamorphic rocks retain the features imposed during prograde metamorphism, which is why it is possible to deduce the conditions under which they formed.

The reactions that form new minerals take place relatively rapidly as temperature rises, as energy is added to the system and in the presence of an aqueous phase. Once past the metamorphic climax, when temperatures and pressures begin to fall again, the mineral assemblages might be expected to revert to phases stable under the changing conditions. However, the rocks have been 'dried out' by expulsion of the aqueous phase, and re-equilibration reactions are infinitely slow when temperature is falling and energy is leaving the system. Retrograde metamorphism is therefore rare, unless the rocks are maintained at some intermediate temperature for prolonged periods or water is somehow introduced into them, as along a major fault or fracture.

5.24: *Typical mineral assemblages developed in pelitic (clay-rich) rocks. The metamorphic mineral assemblages are for progressive metamorphism as identified by successive index minerals characterizing zones of rising pressure and temperature conditions. The zones are grouped into facies, which define broad regions of temperature and pressure within which the metamorphic mineral assemblages that characterize groups of zones are formed. Index minerals defining zones are based on progressive (prograde) metamorphism of rocks of pelitic composition. The facies classification is based on progressive metamorphism of rocks of basaltic composition.*

Barrovian zones

subfacies	mineral assemblages	facies
chlorite	quartz-muscovite-chlorite-albite	
biotite	quartz-muscovite-chlorite-biotite-albite	greenschist
garnet	quartz-muscovite-chlorite-biotite-garnet-albite or oligoclase	
staurolite	quartz-muscovite-biotite-garnet-oligoclase-staurolite	
kyanite	quartz-muscovite-biotite-garnet-oligoclase-kyanite	amphibolite
sillimanite	quartz-muscovite-biotite-garnet-oligoclase-sillimanite	

Buchan zones

subfacies	mineral assemblages	facies
chlorite and biotite	quartz-muscovite-biotite-chlorite-albite	greenschist
andalusite	quartz-muscovite-biotite-chlorite-andalusite-oligoclase	
cordierite	quartz-muscovite-biotite-oligoclase-cordierite-garnet	amphibolite
sillimanite	quartz-muscovite-biotite-oligoclase-sillimante-garnet	

Metasomatism

Although metamorphism can be considered for theoretical purposes to be an isochemical process, with negligible change in bulk composition, in practice it is generally accompanied by small changes in chemistry, most notably by loss of volatiles including water. When there is significant addition of new elements by migration into the rock the process is known as metasomatism, of which there are various kinds. The most widespread occurrence of metasomatism is in oceanic crust which is subject to a form of regional metamorphism to greenschist and amphibolite facies rocks under a very shallow thermal gradient (high temperatures and low pressures). New oceanic crust formed at active spreading ridges is rapidly fractured and penetrated to depths

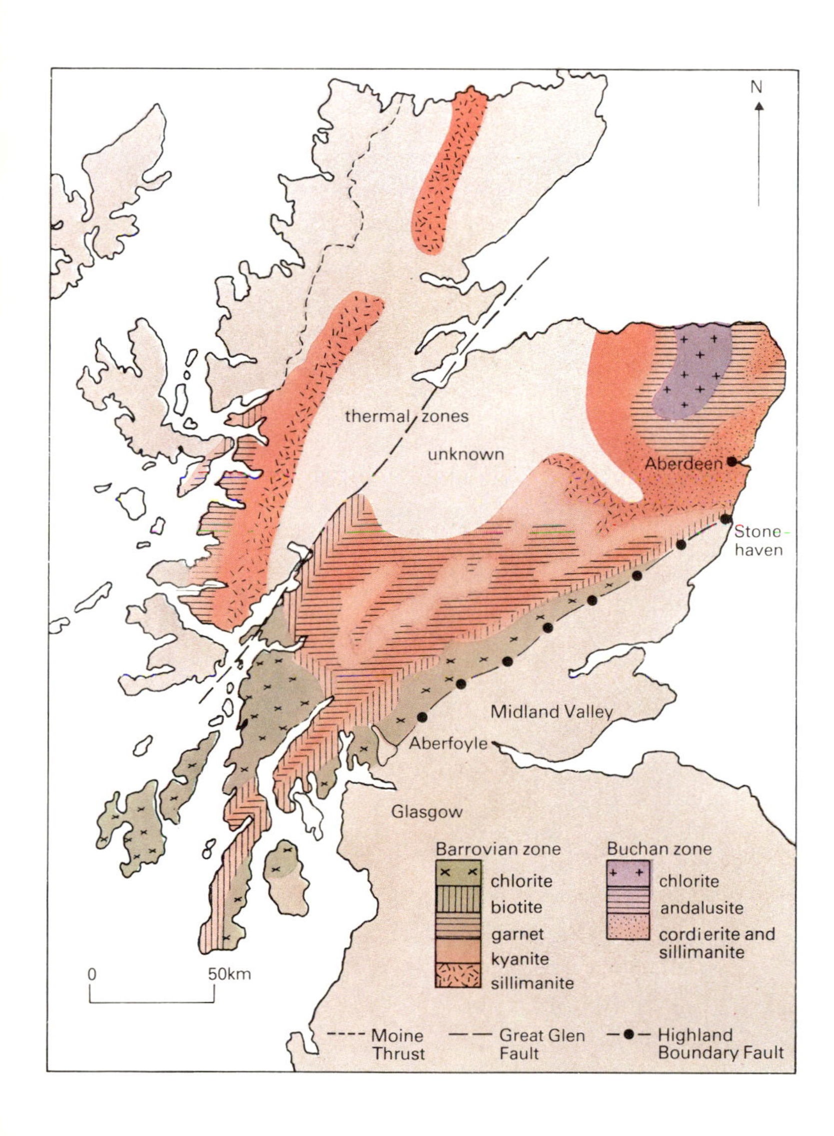

5.25: *Metamorphic zones in the south-east Highlands of Scotland, the area of Barrow's pioneer work.*

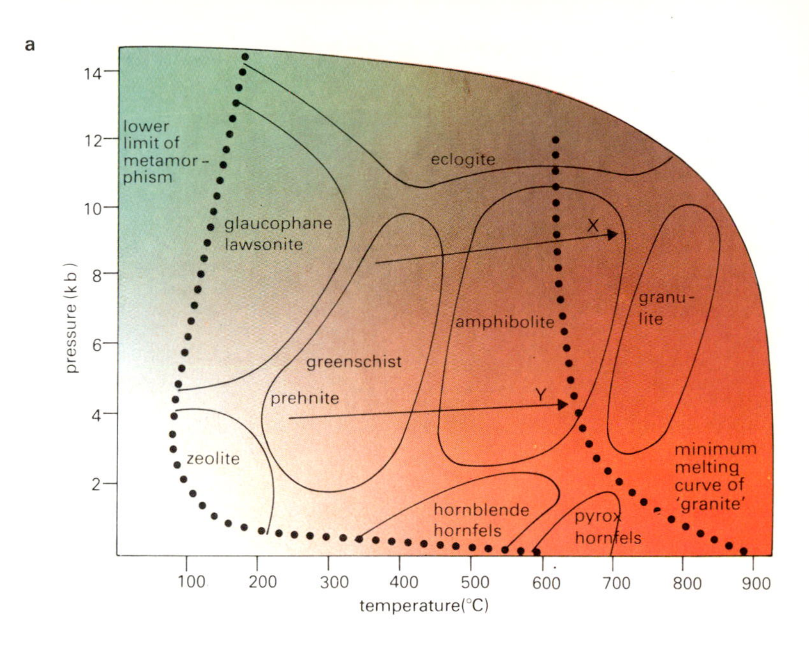

b

facies	environment	pressure/temperature conditions
zeolite	Deeply buried sediments	low T low P
prehnite/pumpellyite	More deeply buried sediments	low T moderate P
glaucophane/lawsonite	Still more deeply buried sediments, notably in circum-Pacific belt; probably occupies deepest levels of ocean trenches	low T high P
greenschist	Extremely widespread facies, low grade regional metamorphism in orogenic belts also typical of vast areas of Precambrian cratons and of oceanic crust	moderate T moderate P
amphibolite	Also extremely abundant; high grade regional metamorphism in orogenic belts and also typical of vast areas of Precambrian cratons	high T high P
hornblende hornfels	Contact aureoles	high T low P
pyroxene hornfels	Very hot, inner zones of aureoles	very high T low P
granulite	Highest grade of regional metamorphism, occuring as small parts of Precambrian cratons	low T low P
eclogite	Not well understood; circum-Pacific glaucophane schist areas	high T very high P

5.26: (**a**) *The pressure/temperature fields and* (**b**) *the occurrence of the major metamorphic facies. Metamorphism at higher pressure and lower temperature than the normal geothermal gradient will be in the glaucophane schist facies, and at lower pressures and higher temperatures metamorphism is in the hornfels facies. Rocks metamorphosed in the eclogite facies are rather rare. The arrows indicate the geothermal gradients represented by (X) Barrovian and (Y) Buchan zones (Figure 5.25).*

of a few kilometres by seawater, which reacts with the hot rocks. The basaltic rocks are metamorphosed and metasomatized. Hydrous minerals, notably chlorite and amphiboles, are formed, and the sodium content of the rocks increases two- or three-fold, with a corresponding decrease in calcium: other constituents are affected to some degree. The oceanic crust that reaches a subduction zone is therefore rather different in composition from that generated at the spreading ridge. Hydrothermal alteration of ultrabasic rocks can transform olivines and pyroxenes into serpentine ($Mg_6Si_4O_{10}(OH)_8$) by the addition of water, a process which produces a volume increase and a lowering of density.

Metasomatism also occurs in contact metamorphosed rocks, particularly near granites, where highly mobile water-rich (hydrothermal) fluids formed at a late stage of crystallization have permeated the country rocks for short distances. Metasomatic introduction of new elements into limestones produces distinctive mineral assemblages known as skarns, containing calcium-rich garnets, pyroxenes and other minerals. It is possible that some iron ores and sulphide deposits occurring along the contacts of large intrusions have a metasomatic origin, particularly where the deposits are zoned, ie those formed at higher temperatures are found nearer the intrusion while lower temperature minerals are found further away from it. Hydrothermal deposits form the principal sources of a number of important minerals: wolframite ($FeWO_4$), cassiterite (SnO_2) and molybdenite (MoS_2) are high temperature examples, while pyrite (FeS_2), cinnabar (HgS) and stibnite (Sb_2S_3) are low temperature examples. Many important sulphide deposits are not obviously associated with intrusions, however, and the chloride-rich, sulphur-deficient hydrothermal solutions in which the metals became concentrated could have had other origins, such as the connate water originally trapped in sedimentary rocks at the time of their deposition. Such water, expelled at depth and heated geothermally, could leach metal ions from rocks through which it passed, depositing them on contact with a sulphur-rich stratum; this may be the origin of the economically important strata-bound sulphide deposits, explaining why they are so often confined to particular sedimentary layers. Another economically important effect of hydrothermal metasomatism is the kaolinization of feldspars, typically in granite intrusions. Kaolinite ($Al_4Si_4O_{10}(OH)_8$), a clay mineral, is produced by normal weathering of feldspars, but can be generated in much larger quantities hydrothermally. Addition of water is clearly necessary as feldspar contains none, while addition of carbon dioxide removes the anions (Na^+, K^+, Ca^{2+}) as soluble carbonates.

Metasomatism in which the agent of chemical change is gaseous rather than aqueous is referred to as pneumatolysis; the chemically active fluids in this case include fluorine-bearing compounds. Pneumatolysis is especially characteristic of the late stages of cooling of some granites and it may affect both the contact metamorphic aureole of the intrusion and the igneous mass itself; tourmaline is commonly formed in this way when boron is also present in the fluids.

5.27: *The beginnings of melting in Archaean gneisses, southern India. The 3400 Ma old gneisses show sheets and more diffuse zones of granitic material cutting across the original structure of the gneiss. The granitic material is interpreted as resulting from the melting of gneiss at high temperature and on introduction of water vapour about 2600 Ma ago.*

Migmatites and melting

With progressive increase in temperature and pressure the conditions of regional metamorphism pass into those of magma generation. Granitic magmas can form from rocks of appropriate composition in the presence of a little water, at temperatures as low as 600°C, provided the pressures are high enough (Figure 5.26). These conditions are met in the deeper zones of mountain belts (collision orogenic zones), where there are all gradations from high grade metamorphism to melting and magma generation. A combination of the two may result in migmatites, composite rocks in which high grade gneisses are injected by granitic magma derived from partial melting of nearby rocks. Plastic deformation is common in migmatites, which thus become difficult to interpret. The greater the degree of magma generation and deformation, the greater the loss of original texture, until anatexis finally destroys all trace of pre-existing structure, and metamorphism grades into igneous processes. There is a complete cycle of rock change between the three main categories of igneous, sedimentary and metamorphic; the processes involved are discussed in subsequent chapters.

6 The gravitational mechanics of the Earth

Physicists have deduced that there are four types of force that operate between material bodies: the electromagnetic force, the gravitational force and the strong and weak nuclear forces. The second of these, gravity, is the most important in astronomy and the Earth sciences. It is well known that Isaac Newton (1643–1727) gave a correct mathematical description of the laws of gravity while he was in his early twenties. Newton's law of gravitation states that, for two point masses, the force attracting them is proportional to the product of their masses, and inversely proportional to the square of the distance of their separation. Although Newton's law has been superseded by Einstein's general theory of relativity it is completely adequate for all practical purposes in geophysics and much of astrophysics.

In order to write down Newton's law F can represent the force of attraction, m_1 and m_2 the pair of point masses and r the distance between them. The law is then written thus: $F = G\, m_1 m_2 / r^2$. The quantity G, the gravitational constant, is found experimentally and is difficult to measure with high precision. In the international metric system its value is 6.670×10^{-11} N m^2/kg^2. This small value for the constant reflects the fact that gravity is an intrinsically weak force. In everyday life we are only aware of the gravitational attraction of the Earth itself, which has a large mass ($M = 5.98 \times 10^{24}$ kg); we experience this force as 'weight'. Geophysics and astrophysics deal with very large masses—planets, moons, stars and galaxies—as well as enormous distances. Under such conditions it is gravity that is the dominant natural force.

The law of gravity is central to the study of the physics of the Earth as it determines the general shape of a planet, influences many tectonic processes that shape the planetary surface and controls the orbital and rotational motions. Much of the discussion in this chapter will deal with the far-reaching consequences of this simple law.

Gravity and the solar system

When we consider the orbits in the solar system, such as the Earth's path round the Sun, we have to take into account the gravitational attractions of many masses, not just the two that appear in Newton's law. The total attraction experienced by any one planet is found by taking the vector sum of the individual attractions. This was once a formidable computational problem which had to be laboriously worked out, but it is now relatively easy to solve by computer. Knowledge of the planetary orbits has been greatly increased through space missions and accurate high-speed computing.

Orbital and spin motions

As Copernicus (1473–1543) suggested, the Earth moves around the Sun in an elliptical orbit, the orientation of which is fixed in space. When the motion is averaged over a long time interval the mean orbital plane defines a useful frame of reference for describing the orbital motions of the Moon and other planets, as well as for describing the orientation of the rotation axes of planets and satellites in the solar system: this plane is called the ecliptic. Most solar system bodies do not move far out of this plane. The Earth's instantaneous rotation or spin is about an axis that is inclined to the ecliptic by about $66\frac{1}{2}°$, an angle that remains more or less constant. The Earth's equator is therefore inclined at an equal angle to the ecliptic and this inclination, known as the obliquity of the equator, is responsible for the annual seasonal motion of the Sun's path in the sky as seen from the Earth.

The Moon moves about the Earth, also in a slightly eccentric orbit, in a plane that maintains an almost constant inclination, about 5°, to the ecliptic. Because of the obliquity, the inclination of the lunar orbit upon the equator varies periodically with the period of the lunar month, about twenty-eight days, the time it takes for the Moon to make one revolution about the Earth. The values of the constants in these motions partly reflect the conditions at the time of formation of the planets: for example the length of the year and the eccentricity of the Earth's orbit. Other 'constants', such as the period of the lunar orbit and the length of the Earth's day, are consequences of the subsequent dynamic evolution of the solar system.

The motions of the Earth–Moon–Sun system cannot be fully treated by discussing only the attractions between two bodies at a time, although this simplification does explain the dominant characteristics of the motions. Thus the elliptical motion of the Earth about the Sun is principally due to the interaction of the gravity fields of these two bodies, while the motion of the Moon about the Earth is, in the first approximation, a consequence of the gravitational interactions between these two bodies only. In a complete discussion the problem must be treated as one of mutual gravitational interactions between the three bodies. Thus, because of the Sun's attraction of the Moon, the latter's motion will oscillate in orientation and size by small amounts about a mean orbit that is almost elliptical.

If the three bodies, Sun, Moon and Earth, could be considered as point masses or as spherically symmetric bodies the resulting orbital and spin motions could be readily evaluated from Newton's laws. But the Earth and the Moon are not perfectly spherical and this causes considerable complexity in solving the equations governing the motion of the Earth and Moon.

Motions of the spin axes

The most significant departure from the spherical shape of the Earth is its flattening at the poles. Our planet's shape can be better approximated by an oblate spheroid, symmetrical about the spin axis. The flattening, which is a direct consequence of the planet's rotation, can be described in several ways but, most simply, it means that the polar diameter is about 43 km shorter than the equatorial diameter. This departure from radial symmetry causes the Sun and Moon to exert additional forces (torques) on the Earth, so inducing shifts in the position of the rotational axis in space. These shifts are the precession and nutation discussed in the following section. Neither is the Moon a strictly radially symmetrical body and the Earth's and Sun's attraction will similarly cause its spin axis to precess in space. This is referred to as the lunar libration.

6.1: *'A philosopher delivering a lecture on the orrery', a painting by Joseph Wright of Derby. An orrery is a clockwork model of the planetary system named after Charles Boyle, 4th Earl of Orrery (1676–1731), for whom one of the first was made.*

The rotation of the Earth

Precession and nutation

Viewed by an observer on the Earth and near the north (or south) pole, stars appear to trace out concentric circles whose centre defines the celestial north (or south) pole, the extension of the Earth's rotational axis. The celestial north pole currently lies close to the star Polaris. Careful observations over many years have revealed that the position of this celestial pole changes relative to the stars, as was noted by Hipparchus about 120 BC. The rotation axis is observed slowly to trace out a cone, with a half-angle of $23\frac{1}{2}°$, about the pole of the ecliptic. It takes about 26 000 years to go full circle around the heavens. In about 3000 BC the pole star was Alpha Draconis; Alpha Cephei will be near the pole in AD 7500. This steady motion of the rotation axis in space is termed the precession of the Earth.

As noted above, the major axis of the oblate Earth is inclined to the ecliptic. In consequence the net gravitational force on the Earth due to the Sun does not pass through the centre of mass of the Earth. This results in a torque being exerted about the centre. The torque attempts to draw the equator into the plane of the ecliptic but the spinning Earth resists this. Instead the torque achieves a motion of the spin axis about the pole of the ecliptic.

The Moon acts on the Earth in a similar way and the observed precession is the sum of the polar and lunar torques plus a rather minor contribution arising from the other planets. The Earth's orbit about the Sun is somewhat eccentric and twice a year the Sun passes over the equator where it is aligned with the Earth's bulge; consequently the solar torque varies periodically, as does the Moon's torque. The net result is that the secular precessional motion of the rotational axis is perturbed by small oscillations or 'nodding' motions called the forced nutations. The principal nutation term has a period of nineteen years, and the size of the nodding motion is nine seconds of arc. This arises from a nineteen-year periodicity in the inclination of the Moon's orbit. It was first detected and explained by the astronomer James Bradley (1692–1762) in 1747.

Although the precessional and nutational motions are mainly of astronomical interest, they are also of some geophysical consequence in that their amplitudes depend on the oblateness of the Earth and therefore provide some information on the internal structure of the Earth. This is discussed below.

Polar motion

If an observer at either pole recorded photographically the small concentric circles traced out by the circumpolar stars, with time the centre of the circles would be seen to be slightly displaced on the photographic plate, and this centre would itself trace out a small circle over about a one-year period. This reflects a motion of the rotational axis relative to an Earth-fixed reference frame and is referred to as polar motion. It is quite distinct from precession and nutation. If the motion is observed from space the Earth as a whole appears to wobble about its axis. This motion was predicted by L. Euler (1707–83) in 1765, but it was not observed until the end of the nineteenth century.

The motion of the pole is deduced from careful measurements of the positions of the stars. Figure 6.2 illustrates the pole path as observed over a two-year period—a meandering motion that is typical of the polar path observed over the past eighty years. An analysis of such observations indicates that the motion is mainly made up of two periodic oscillations, one with a fourteen-month period, the other with a twelve-month period, both with amplitudes of the order of 0.1 arc second. Hence the wobble of the rotational axis at the north and south poles, relative to the Earth's surface, is merely a few metres, and precise astronomical measurements are needed to observe it. Yet these observations yield considerable information on the Earth, which will now be discussed.

To understand the fourteen-month oscillation the Earth may be regarded as a rigid oblate spheroid as in the previous discussion of the precession and nutation. If the rotational axis of a rigid body is initially aligned with its principal axis of maximum inertia then that body will rotate uniformly with the two axes always remaining parallel; however, if for any reason the two axes are tilted relative to each other, Euler's theory predicts that the rotational axis appears to wobble about the principal axis. For the Earth this theory predicts a 305-day wobble. But the observed period is not 305 but 430 days, as first noted by S. C. Chandler in 1892. This increase is explained by the Earth not responding to the rotation as a rigid body. The theory must be extended to the rotation of a deformable Earth in which the mantle's elastic deformation must be considered as well as movements in the oceans and in the liquid core. The theory is relatively complicated but one aspect of the Earth's non-rigidity can be mentioned here.

Consider that the Earth has a liquid core filling a spherical cavity. If the fluid had no viscosity then there would be no frictional forces linking motions in the core to the mantle, and vice versa. The outer mantle would be able to slide round the inner core without disturbing the fluid motions within. If this simplification described the real situation then it would reduce the theoretical period of the wobble by about thirty days. The elasticity of the mantle on the other hand increases the period by about 120–130 days and the oceans lengthen it by a further thirty days, so that the computed period is now close to the observed 430 days. From this discussion it can be seen that observations of polar motion assist the deduction of the Earth's structure.

The observed wobble, now referred to as the Chandler wobble, being a free oscillation, should ultimately be damped since no free oscillation will persist indefinitely in physics. Yet the observations suggest that the oscillation has persisted for nearly a century, albeit with considerable fluctuation in amplitude and phase, and this suggests that there is some mechanism exciting this motion. One mechanism (there are several) is the excitation of the wobble by large earthquakes. This

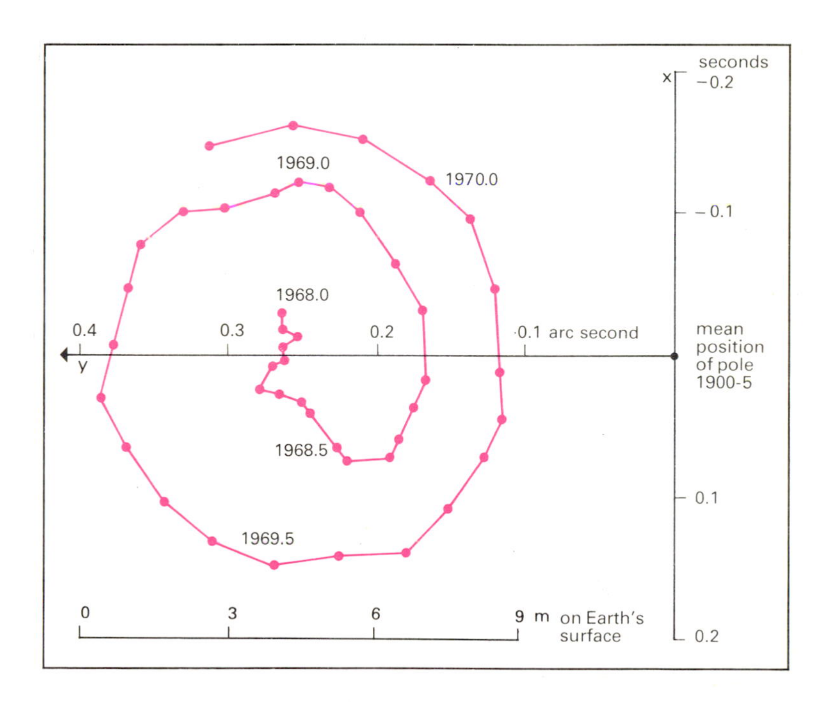

6.2: *The pole path from 1968 to 1970. The pole positions are given at intervals of 0.05 year. The x axis is directed towards the Greenwich meridian. The origin corresponds to the mean position of the pole for the years 1900–5 and the pole now appears to rotate about a mean position that has shifted a total of about 0.25 arc second in a direction about 90° west of the Greenwich meridian.*

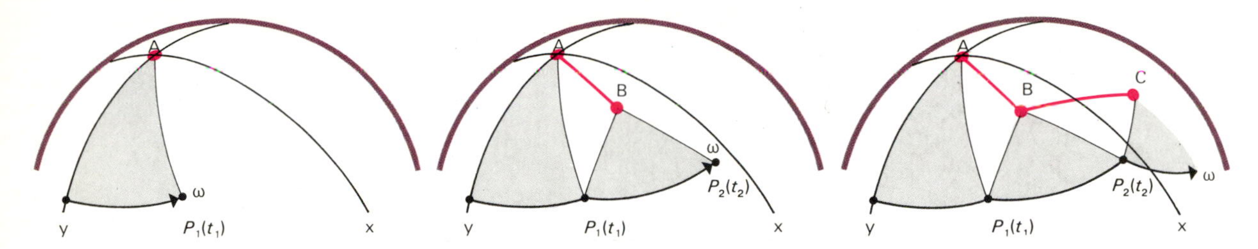

6.3: *Motion of the instantaneous rotational axis ω about the principal axis. Until time t_1, when the rotational axis is at P_1, the pole has rotated about the principal axis at A. At time t_1, an earthquake occurs and modifies the mass distribution such that the principal axis is now at B. The pole now moves about this new position until the next earthquake occurs at time t_2 when the pole is at P_2. The principal axis then jumps to C.*

is illustrated in Figure 6.3. Suppose that while the rotational axis wobbles about the principal axis at A, a large earthquake occurs at time t_1. This seismic event, if large enough, changes the mass distribution within the Earth and shifts the position of the principal axis from A to B. The rotational axis now moves about this new position of the principal axis leaving behind it a kink in the pole path. At some later time t_2, a second earthquake occurs and a second kink forms. In this way the Chandler wobble can be maintained as long as sufficiently large earthquakes occur at irregular but frequent intervals. This points to the main problem with this model, for while some earthquakes appear to be sufficiently large to shift the principal axis by the requisite amount, there do not appear to have been enough of them to maintain the wobble throughout the past hundred or so years. It is more likely to be a combination of the catastrophic seismic shock and slower deformations preceding or following the main shock that displaces the principal axis.

The other, annual, oscillation in the polar motion is a consequence of the Earth's mass distribution being periodically modified by seasonal redistribution of mass within and between the atmosphere, oceans and surface and ground water.

In addition to these two periodic motions the pole path observations exhibit a secular drift in a direction roughly along a meridian 70° west of Greenwich and at a rate of about 0.002–0.003 arc second per year. This is apparently a consequence of an exchange of water mass between the polar ice sheets and the oceans and of the Earth's response to the changing load on its surface.

Changes in the length of day

The third aspect of the Earth's rotation concerns the angular velocity about the instantaneous rotational axis. Astronomers observe the times of transit of a star across their meridian using precise atomic clocks to establish the time scale, and this provides a measure of the length of the day. These observations indicate that the interval between successive transits varies perceptibly, and that the Earth is usually either ahead or behind the time kept by the clocks. Typically the length of the day fluctuates by 1 part in 10^8, equivalent to about 10^{-3} second, a small but observable amount. Observations of the length of day have been made regularly since about 1820, but only with the introduction

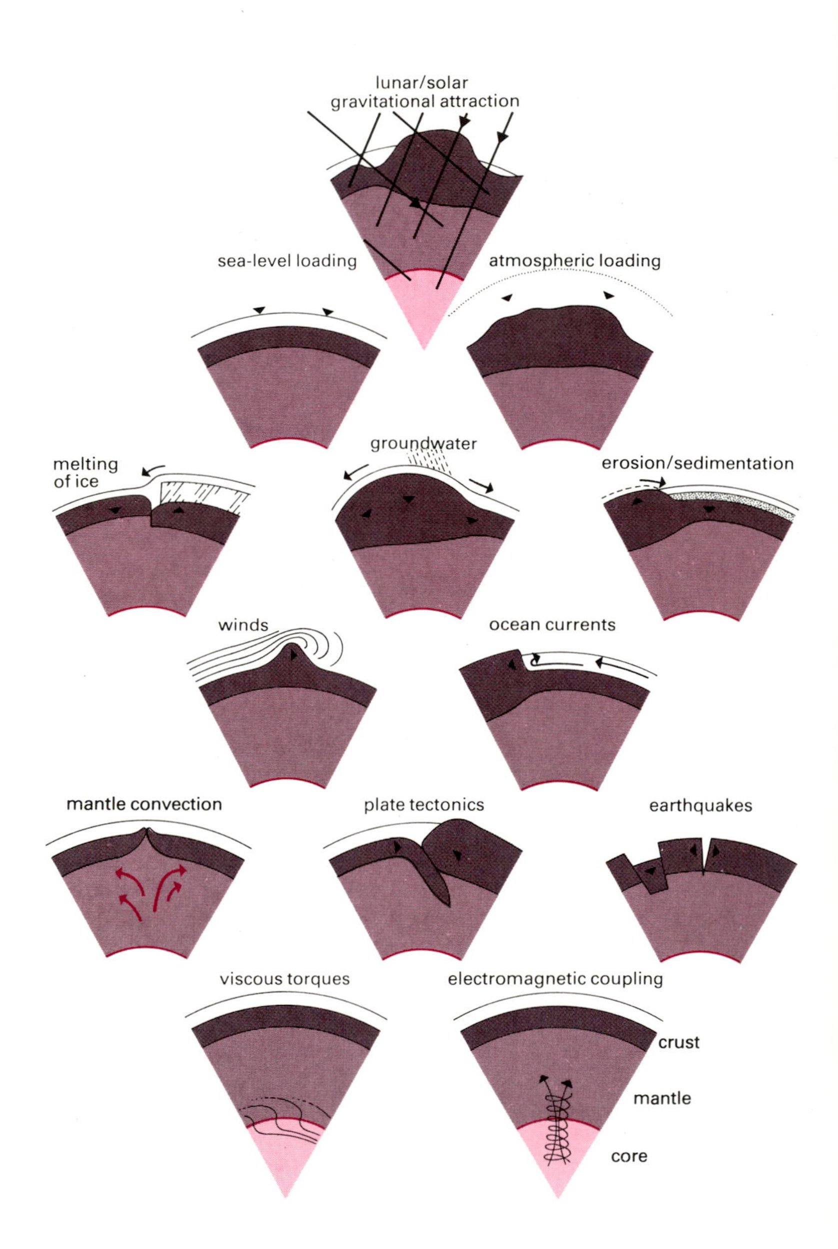

6.4: *A schematic representation of the forces acting on the Earth that will perturb the rotational motion.*

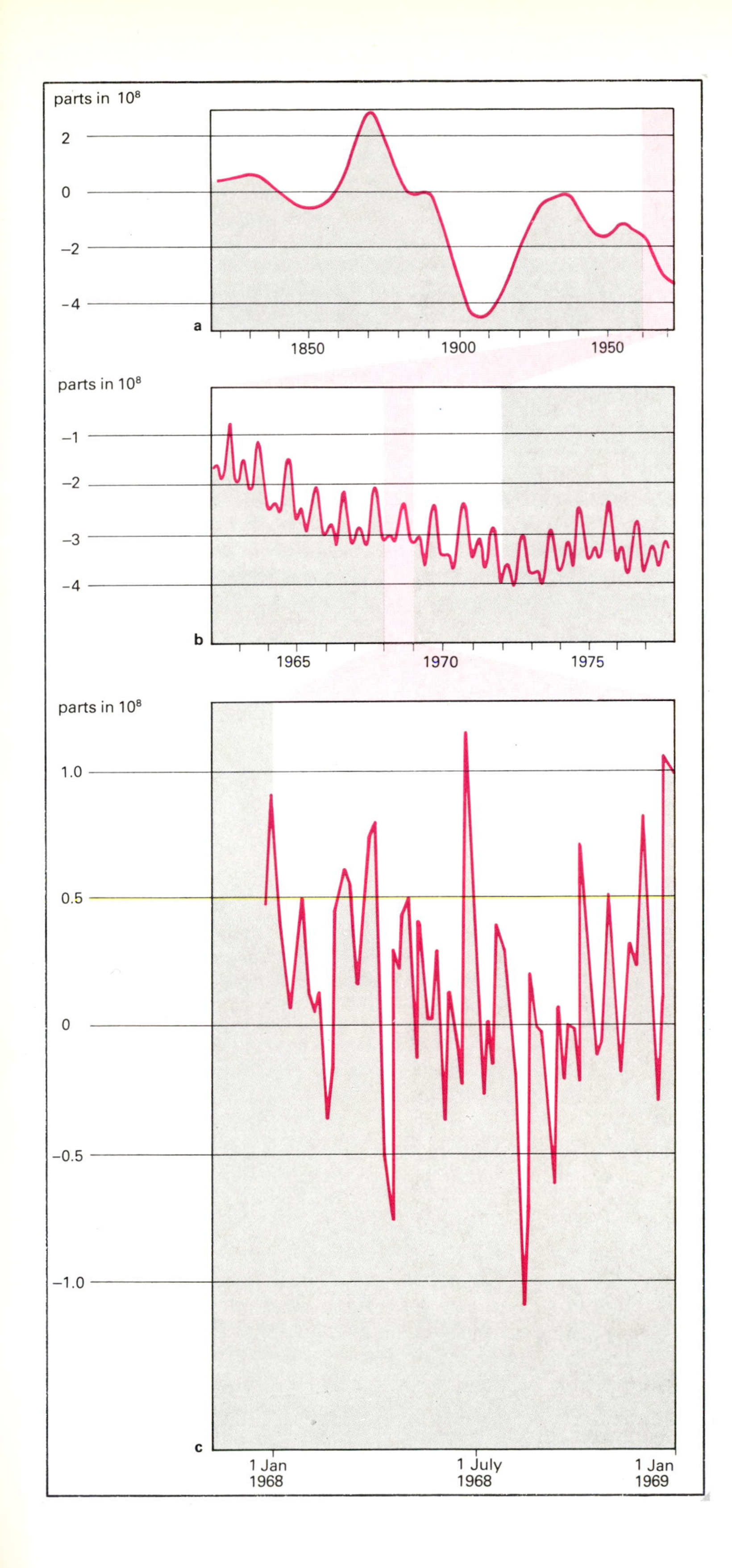

6.5: *Proportional changes in the length of day as observed on three different time scales.* (**a**) *From 1820 to 1970. Only fluctuations that persist for about ten years or longer are evident.* (**b**) *From 1962 to 1977. Seasonal variations are clearly evident.* (**c**) *Upon removal of the seasonal effects, highly irregular variations are seen. Some of these may represent 'noise' in the astronomical data while some are due to a rapid exchange of angular momentum between the Earth and atmosphere.*

of atomic time in about 1955 have the observations been sufficiently precise to establish a comprehensive picture of these changes. Prior to this time only changes that persisted for five to ten years or longer could be seen but now changes on a week-to-week basis are detectable.

Figure 6.5 illustrates some results. In Figure 6.5(a) only the variations occurring on a 'decade' time scale are indicated and most notable is the sudden decrease in spin velocity from about 1870 to 1900, during which period the length of day increased by nearly 10 milliseconds. This is followed by a reversal in trend from 1900 to about 1930 and a second period of deceleration continuing into the 1970s. The origin of these changes has remained obscure but only the core contains sufficient mass and mobility to explain these fluctuations. The most likely explanation is that they are caused by interactions at the core–mantle boundary between the magnetic field (in and moving with the core fluid) and the electrically conducting lower mantle. This results in a variation in the extent to which the core moves with the mantle and in a concomitant change in the mantle spin. The theory is complex and not readily verifiable since the magnetic field responsible is largely shielded from the observer by the low electrical conductivity of the upper mantle.

Of the higher frequency fluctuations observed clearly since 1955 the annual and semi-annual behaviour is mainly caused by an exchange of angular momentum in the east–west circulation of the atmosphere with the angular momentum of the solid Earth. In simple terms, as the westerly winds speed up so the Earth slows down, and vice versa. In addition to these seasonal terms many irregular changes in rotation rate occur in the frequency range of about 0.3 cycle per year to perhaps as high as 10–20 cycles per year and these are also largely due to irregular fluctuations in the atmospheric circulation.

The astronomical observations of the Earth's rotation have provided very fascinating data whose interpretations impinge upon many aspects of the Earth sciences. Briefly, any force that exerts a torque on the Earth's crust, or that results in a redistribution of mass within the Earth, is a candidate for perturbing the Earth's rotation. There is a wide range of phenomena that do this, including the secular tidal torques, to be discussed below, mantle convection, fluctuations in the magnetic field, relative motions in the core, oceans and atmosphere, the direct attraction of the Sun and Moon and the concomitant tidal deformations.

The motion of the Moon

Orbital motion

The complex problem of determining the Moon's orbital motion occupied many of the great mathematicians of the eighteenth century,

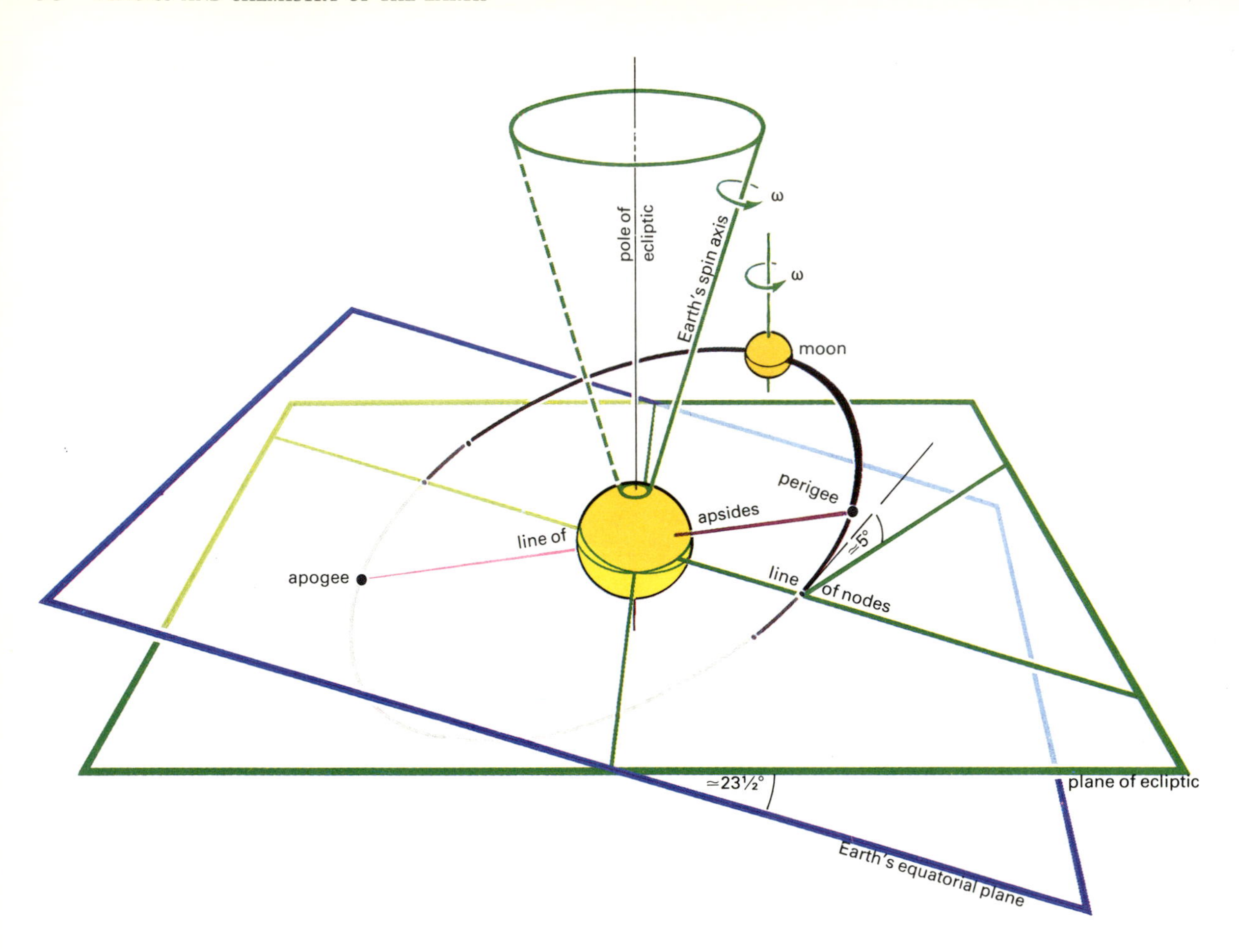

6.6: *Orbital geometry of the Earth–Moon system. The Earth's average orbital motion around the Sun defines the ecliptic and the Earth's spin axis rotates about the pole of the ecliptic once every 26 000 years because of the precession. The Earth's equinox, the intersection of the equator and ecliptic, moves along the ecliptic at the same rate. The lunar orbit intercepts the ecliptic along the line of nodes which moves around the ecliptic because of the solar attraction. For the same reason the line of apsides precesses in the orbital plane. The Moon's spin axis remains normal to the ecliptic.*

including Euler, Clairaut, D'Alembert, Lagrange and Laplace. To the intrinsic interest of the problem was added the desire to test Newton's theory. A further incentive was the substantial financial rewards that governments and scientific societies offered to those mathematicians who could provide the precise tables of the Moon's motion essential for navigation at sea.

The simplest approximation to the motion of the Moon is an ellipse that remains fixed in size, shape and orientation. This ellipse has a mean radius of about 380 000 km and an eccentricity of about 0.055, and is inclined by about 5° to the ecliptic. The orbital period of the Moon moving along this simplified ellipse would be about twenty-seven days for motion observed relative to the background stars. This is referred to as the sidereal month. During one sidereal month the Sun moves eastwards by nearly 30° and the time between successive full Moons —the synodic month—which occurs when the three bodies are aligned, is in consequence longer than the sidereal month by nearly two days.

The simple motion assumed above is much perturbed by the gravitational attraction of the Sun and to a lesser degree by the Earth's departure from a point mass as well as by the attractions of the other planets. The main consequence of the solar attraction is to precess the orbit in its plane, such that the line of apsides—the line joining the point at which the Moon is nearest to the Earth (perigee) to the point at which the Moon is furthest away (apogee)—rotates in the plane of the orbit with a period of about 8.9 years (Figure 6.6). A second consequence is that the orbital plane itself precesses about the ecliptic, such that the line of nodes—the intersection of the orbital plane and the ecliptic—makes one revolution in about 18.7 years, in a direction opposite to the Moon's orbital motion. Superimposed upon these secular motions of the orbit are a number of periodic perturbations due to the variation of the solar attraction with the continual changing Earth–Moon–Sun geometry.

Librations

The Moon always shows the same face to the Earth. However, this represents only an average state, for although the lunar spin velocity is constant, the Moon's orbital velocity along the elliptic path about the Earth is not. Thus there are some times when the spin is behind its orbital motion and other times when it is ahead; at such times it becomes possible to see a little beyond the limits of the average disc directed towards the Earth. Furthermore, since the inclination of the Moon's orbit on the ecliptic is 5.2° and the inclination of the Moon's equator on the ecliptic is about 1.5°, it becomes possible to see about 6.7° of latitude beyond the two poles. Together these circumstances are referred to as the optical librations. They permitted nearly 60 per cent of the Moon's surface to be photographed and investigated from the Earth long before the lunar orbiter programme provided nearly complete photographic coverage. The librations have also permitted the lunar topography to be estimated for the limb regions, observations that continue to provide control on lunar mapping.

Of geophysical interest are the much smaller physical librations, analogous to the Earth's precession and nutation. These are a consequence of the Earth exerting a torque on the asymmetrical mass distribution of the Moon. Their magnitudes do not exceed a small fraction of a degree, and while their existence was predicted by Newton they were not observed until 1839. The amplitudes of these oscillations depend on the distribution of mass within the Moon, and in common with the precession observations of the Earth their observation provides some information on the shape and mass distribution of the planet. The Moon's spin being only about one twenty-seventh of that of the Earth, the lunar polar flattening is much less than that of the Earth. The Moon is subject to a solid tidal deformation in the same way that the Earth is tidally deformed (see below) but with the difference that the lunar tidal bulge is predominantly and permanently oriented towards the Earth. Thus the Moon possesses an equatorial bulge and the Moon's shape can be described approximately as a triaxial ellipsoid with the major axis directed towards the Earth.

Tides

Ocean tides

Perhaps a more familiar consequence of the Moon's gravitational attraction than precession and nutation are the Earth's tides. This phenomenon is perhaps most readily understood by viewing the Earth as a rigid sphere covered by an ocean layer of uniform depth. The Moon's gravitational attraction on the Earth is slightly greater on the side of the Earth towards it than elsewhere and this causes a small bulge in the ocean layer that is directed towards the Moon. On the opposite side of the Earth, away from the Moon, the gravitational attraction is a minimum; however, the lunar attraction on the solid Earth exceeds that on the water since the former is closer to the Moon. The solid Earth is pulled more towards the Moon than is the water which actually appears to be pulled away from the Earth as a second bulge. During the daily rotation of the Earth underneath the Moon, the tidal bulge moves with the Moon and therefore around the surface of the Earth with one bulge always facing the Moon and the second bulge directly opposite. The time between successive transits of the Moon across the observer's meridian is about twenty-five hours. At any point on the Earth's surface the successive lunar tides pass at intervals of about twelve and a half hours.

Likewise, the gravitational attraction of the Sun also raises two tidal bulges of about half the amplitude of the lunar tide and these travel around the Earth in nearly twenty-four hours. When the Sun and Moon are aligned the two tides reinforce each other, producing the spring tides; this occurs every two weeks at full or new Moon. When the Sun and Moon are separated in longitude by 90° the two tides partly cancel and the combined tide has a minimum amplitude. These are the neap tides. The actual pattern of tidal periodicities is much more complicated than this since the intensity and direction of the force of gravity varies with the Moon's orbital motion along its inclined and eccentric path. An analysis of a long series of tidal observations reveals a large number of oscillations whose periods cluster around twelve and twenty-four hours as well as a number of much longer periods up to 18.6 years.

Observed tidal patterns

The actual spatial pattern is complicated by the ocean–land distribution, by the variable depth of the ocean and by frictional forces along the sea floor. Figure 6.7 illustrates the tidal pattern of the principal semidiurnal lunar tide around the British Isles, where the tides are particularly complex. The lines of equal phase, called cotidal lines, join

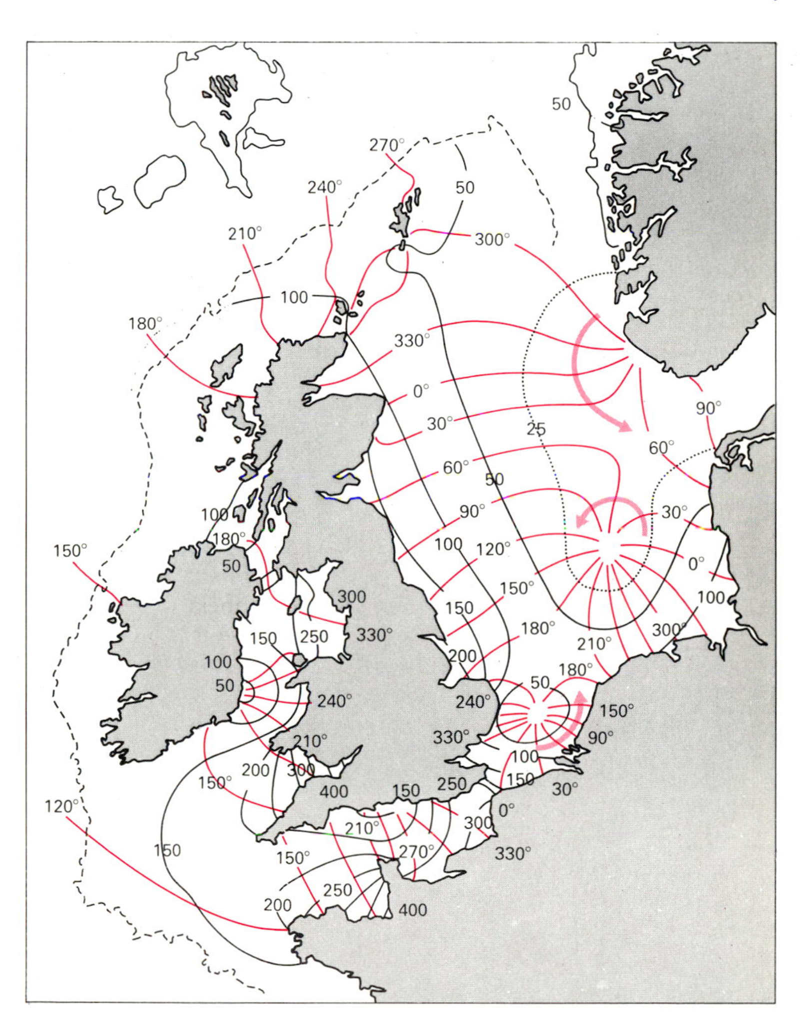

6.7: *The principal lunar semidiurnal tide around the British Isles. The co-range (black) lines show the occurrence of equal amplitude in centimetres. The equal phase (red) lines give the phase of the tide relative to the lunar transit across the Greenwich meridian. The arrows denote the sense of rotation at the amphidromes (points of zero tidal amplitude).*

points at which high water occurs at the same time. Of note are the regions where the tidal amplitudes are zero. These are the amphidromic points about which tidal currents rotate without there being a change in amplitude. One such point occurs in the North Sea between England and Denmark and the tidal currents rotate about it in an anticlockwise manner. Thus along the western boundary of the North Sea the maximum tidal amplitude occurs progressively later from the Shetlands to the Strait of Dover. Figure 6.8 illustrates a global representation of the same tide. These global representations do not illustrate the detail of the local tidal patterns along the coasts where the geometry of the coast line and of the sea floor can result in tidal amplitudes exceeding 12 m, as in the Bay of Fundy, Canada.

The ocean tides and the associated tidal currents are of importance in navigation in coastal and shallow seas. They also influence the coastal landform by forming sand bars or barrier islands that are subsequently stabilized by vegetation and may become a permanent feature of the local geography.

Tidal energy

One frequently mentioned use of the ocean tide is as a source of energy, by using the tidal currents to drive turbines. The total energy stored in one cycle of the ocean tide is of the order of 10^{18} joules. This is comparable with the geothermal loss of heat but four orders of magnitude less than the solar energy received by the Earth during the same period. For much of the oceans the energy density is extremely low and only in some coastal areas, where the tidal currents are sufficiently enhanced by local conditions, may the extraction of this energy be worthwhile (see also Chapter 9).

Solid tides

Sensitive instruments capable of measuring small changes in the local gravity or small tilts of the Earth's crust indicate that the solid Earth is also subject to tides, implying that the Earth does not respond as a rigid body to the gravitational attraction of the Sun and Moon. Typically, an observer on the Earth's surface will move up and down in

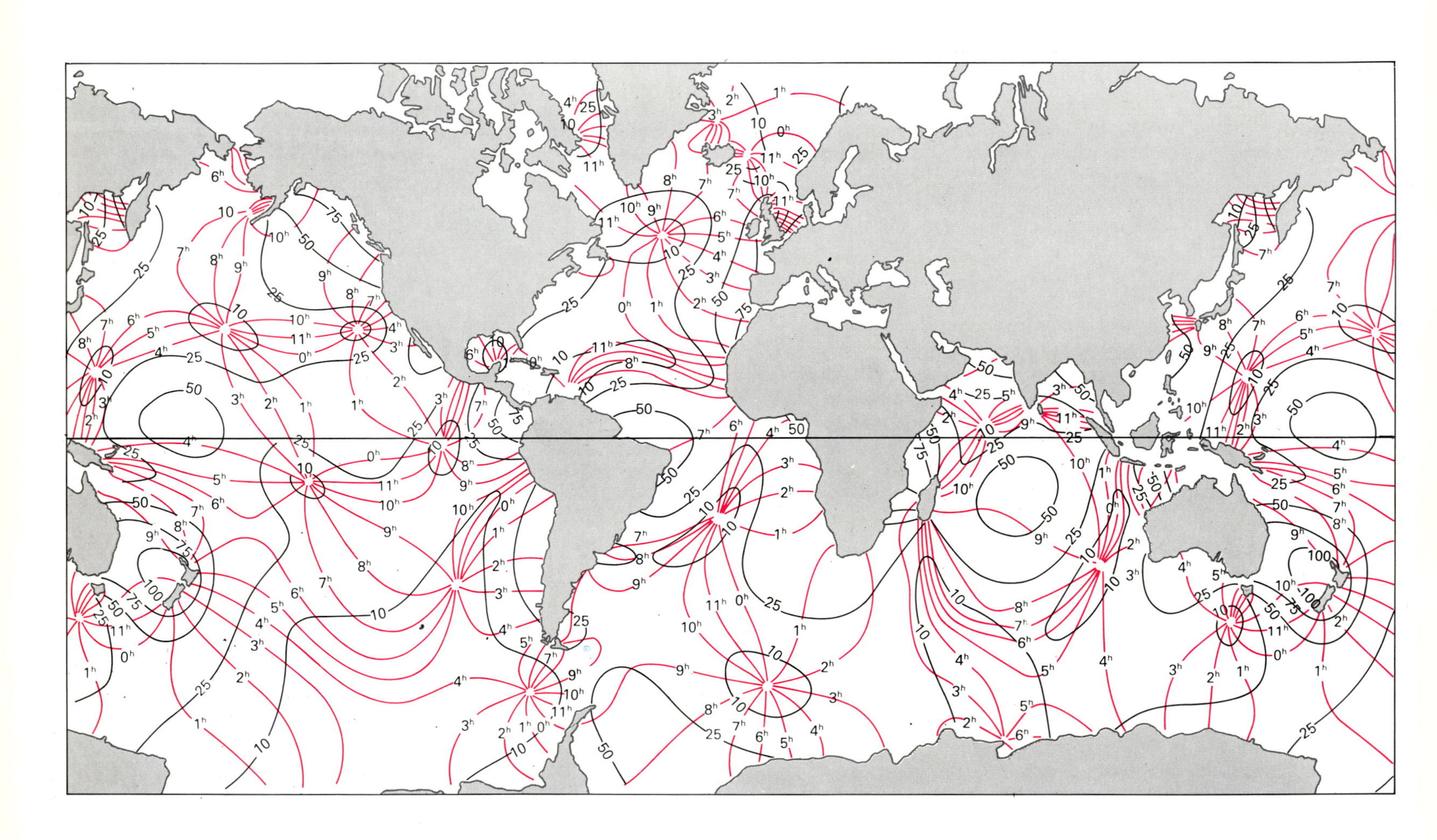

6.8: *The global lunar semidiurnal tide. The co-range or equal amplitude lines are black and the cotidal or equal phase lines (in lunar hours) are red.*

twelve hours by as much as 500 mm. At the same time gravity on the surface will change fractionally. This is the solid or elastic tide. Its amplitude is a function not only of the magnitude of the tide-raising force but also of the elastic properties of the Earth. The Earth's response to this force is not instantaneous since the planet is not a pure elastic body and the tidal bulge lags behind the applied force by a small angle that is unlikely to exceed a degree. The observations of the amplitude and lag of the tidal response are of considerable geophysical interest in that they reflect elastic and anelastic properties of a periodically stressed planet at relatively low frequencies. As such they are complementary to seismic studies of the planet's response to the high frequency waves excited by earthquakes and to rotation studies which measure the response at much lower frequencies.

The most reliable observations of the tide are obtained with precise gravimeters located on the Earth's surface. During the tidal cycle gravity varies because of the direct attraction by the Sun or Moon, the variable distance of the tidally deformed surface from the centre of mass and the redistribution of mass inside the body. The combined effect of these last two factors is about 15 per cent of the direct attraction. The gravimeter cannot distinguish between the solid tide and the attraction of nearby ocean water, a contribution that will also vary periodically because of the ocean tide. A simpler remedy is to observe the solid tides far from coastlines. Figure 6.9 gives an example of a record taken at a station in central Australia.

Tidal friction

If, for the moment, we consider only the solid tide, the bulge will be aligned with the Earth–Moon axis if the tidal response is that of an elastic body. But in a more realistic model the deformation is subject to frictional dissipation and the response is somewhat delayed. During the 'delay' time the Moon will have moved through a small angle along its orbit. Thus the bulge appears to be ahead of the Moon, as shown in Figure 6.10. The lunar attraction on the nearest misaligned bulge exceeds that on the farside bulge and a torque is exerted which does not vanish when averaged over an orbital period of the Moon. The consequence of this torque is a change in the Earth's angular momentum or, equivalently, a decrease in the spin of the Earth such that the length of day is increasing, at present by about 0.001 seconds in a hundred years. At the same time the bulge exerts an equal but opposite torque on the Moon, slows the Moon down in its orbital motion and leads to an increase in the Earth–Moon distance. This secular change is of the order of a few centimetres per year. Both changes are small but when integrated over longer time intervals the consequences become very significant. For example, after about two thousand years the Earth is misorientated relative to stars by some 10–15° in longitude and the position at which a solar eclipse is observed is displaced by this amount from a position computed on the basis of a uniform rotation of the Earth. Thus the theory of the Earth's tidal acceleration can be tested by predicting, on the assumption of uniform rotation, places and times of eclipses or of other astronomical configurations and comparing them with observations of these events as recorded in the literature and history of older civilizations. Conversely, once the acceleration has been established such comparisons can be used for dating purposes.

Over longer time periods the consequence of the small but persistent tidal acceleration becomes even more dramatic. Four hundred million years ago the length of day was about twenty-two hours and the year consisted of about four hundred days. Curiously enough this can be tested against the records of tidal and diurnal cycles contained in the fossils of certain corals and bivalves. The growth of these organisms is controlled by the daylight and tidal cycles which influence the biological processes that deposit the thin incremental layers of calcium carbonate in their skeletons. The available results confirm that the tidal acceleration has continued over at least the last 500 Ma. The fossil remains of one of these early 'astronomers' is seen in Figure 6.11.

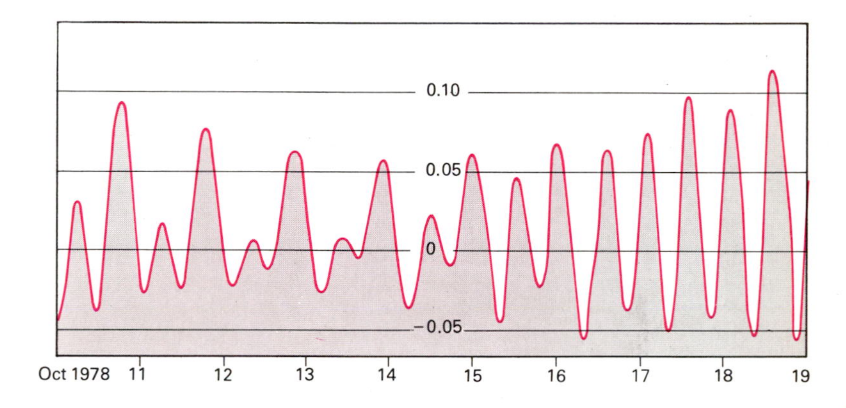

6.9: *Gravity variations in central Australia due to the solid tides over a nine-day period in October 1978.*

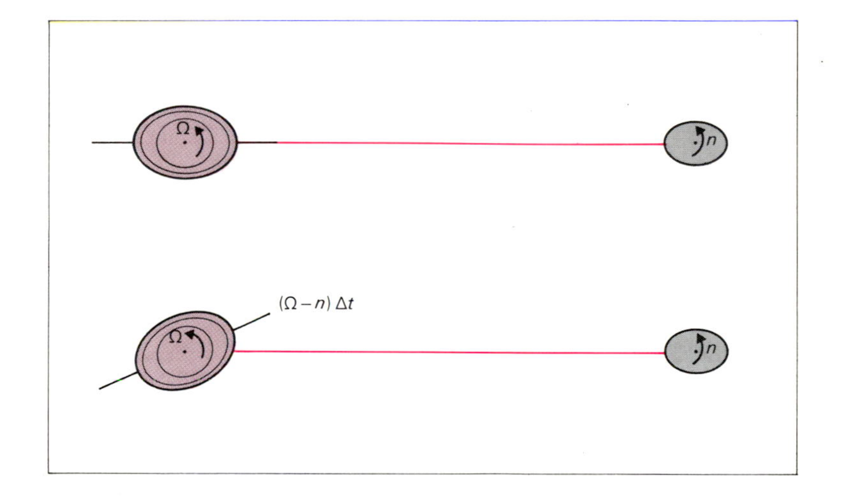

6.10: *Because of the lunar attraction the elastic Earth deforms into an ellipsoid whose major axis is aligned with the Moon (top). If there is a delay Δt in the response due to the Earth not being a perfectly elastic body, the bulge will have been rotated through a small angle ΩΔt. The Moon will also have moved through a much smaller angle nΔt and the bulge will no longer be aligned with the Earth–Moon axis. The torque exerted by the Moon on the bulge slows the Earth down and the torque exerted by the bulge on the Moon slows the Moon down in its orbit.*

Gravity

The shape of the Earth

Gravity varies from place to place on the Earth's surface, reflecting the asymmetrical distribution of mass in the Earth's crust and mantle. A major departure from symmetry is the Earth's oblateness. Because of diurnal rotation any element of mass within the Earth is subject to the gravitational attraction by the remainder of the planet and by the centrifugal force acting in a direction perpendicular to the rotational axis. This latter force is at a maximum at the equator and acts in the opposite direction to the within-planet attraction. At the poles the centrifugal force vanishes. If the Earth responded as a fluid to these latitude-dependent forces its equilibrium form would be an oblate ellipsoid. The observed flattening is close to that expected for a fluid body, suggesting that the Earth responds essentially as a fluid to forces that act over very long time periods.

A more precise description of the shape necessitates the introduction of the term 'geoid', a level surface that closely approximates the ocean surface. Because of the anomalous mass distribution within the Earth the geoid departs from the ideal fluid ellipsoid, in some places rising above it and in other places falling below it by up to 100 m (Figure 6.12). When scientists talk about the shape of the Earth they usually mean the geometric form of the geoid. Only at sea is the geoid directly accessible and its form can be determined from gravity observations taken at sea level. On land the geoid is deduced from gravity observations taken at the surface.

6.11: *The epitheca of a Devonian fossil coral. The epitheca forms a part of the coral skeleton that records the growth rhythms in the form of a fine structure of ridges. This example illustrates thirteen bands, each consisting of about thirty ridges. These are interpreted as daily growth lines modulated by a monthly influence in the growth.*

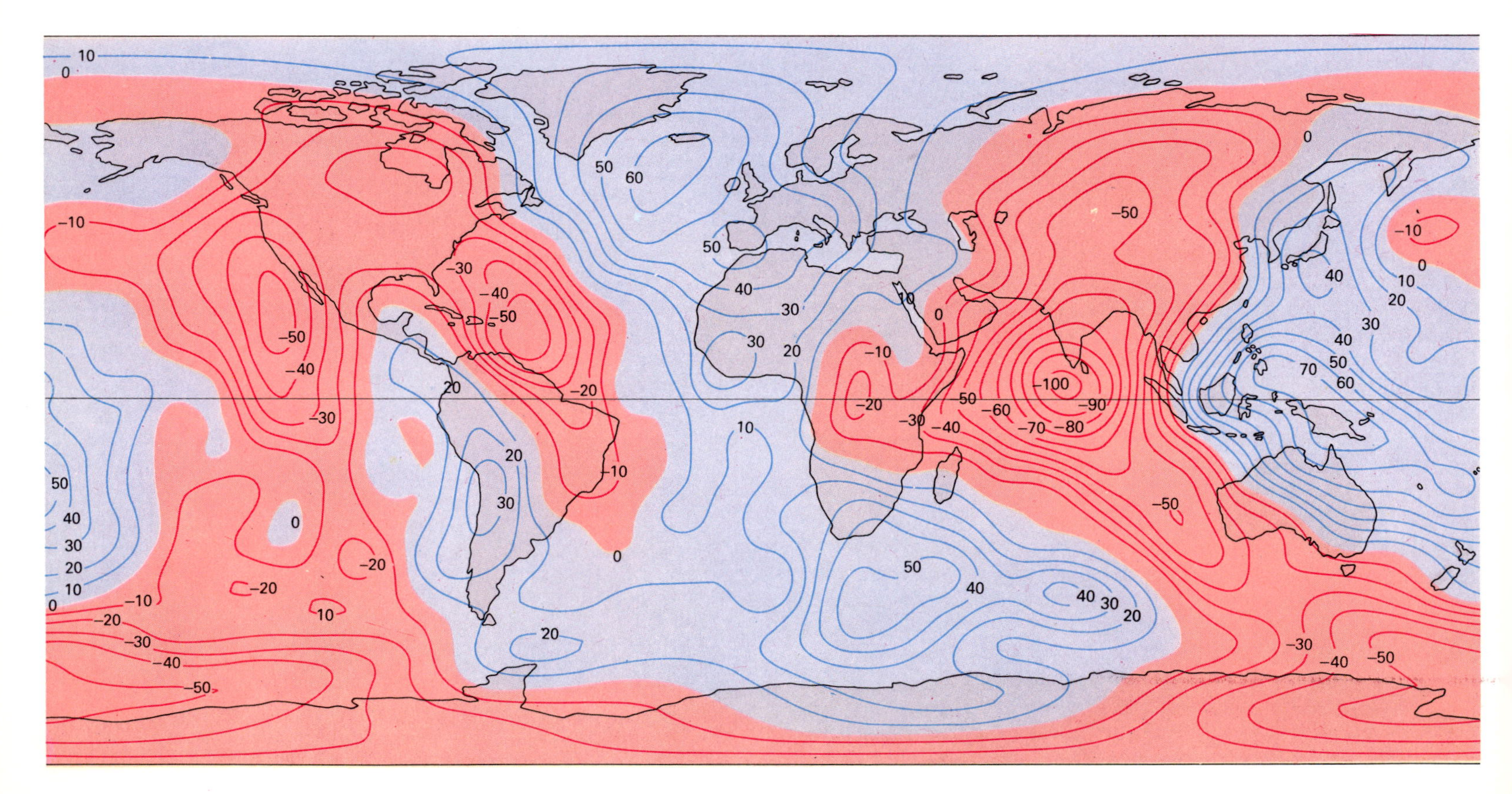

6.12: *The global geoid as deduced from an analysis of satellite orbit perturbations and from surface gravity. The heights (m) relate to the best-fitting ellipsoid. Negative regions, where the geoid lies below this ellipsoid, are red.*

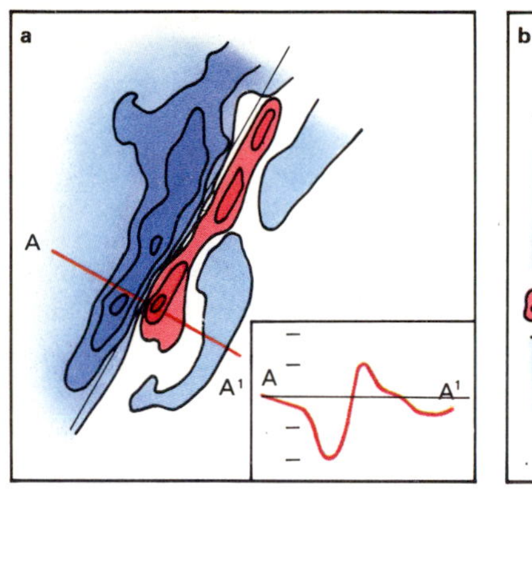

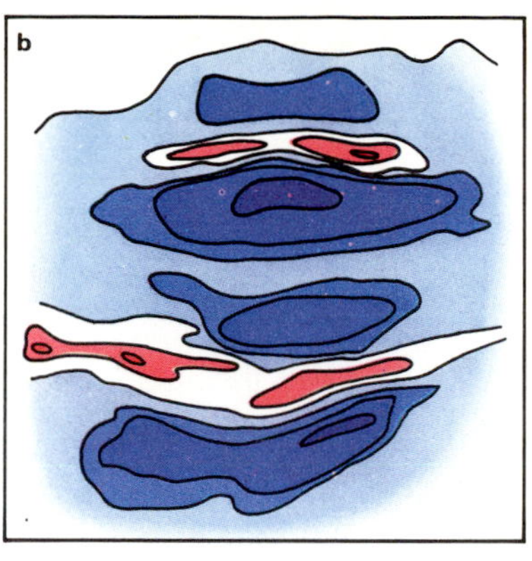

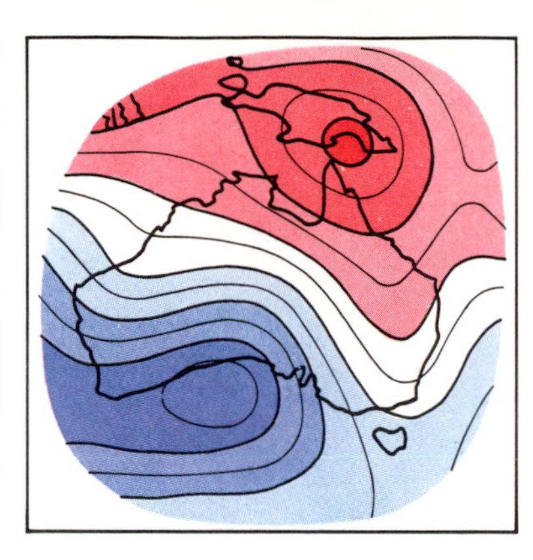

−100 −80 −60 −40 −20 0 20 40 60
scale units: mgals

−200 −120 -160 −80 −40 0 40 80

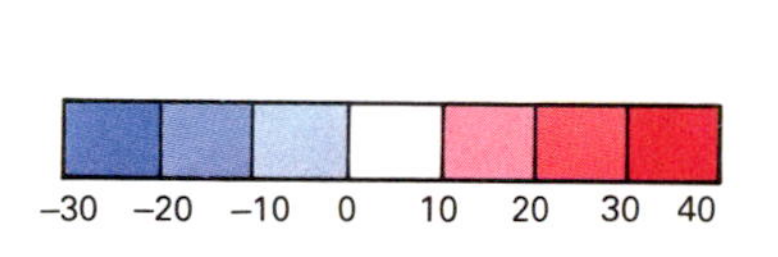

6.13: *Gravity anomalies at different scales.* (**a**) *Bouguer anomaly in W. Australia over the faulted boundary between Archaean and Proterozoic rocks.* (**b**) *Linear negative Bouguer anomalies occurring over sedimentary basins in central Australia; the size of the anomaly implies considerable crustal thickening under the basins and this implies a significant departure from isostatic equilibrium.* (**c**) *Regional gravity anomalies over Australia; negative over the sedimentary basins of the south-west, and positive over the island arcs and subduction zones of New Guinea.*

Gravity anomalies

Gravity can be measured with a pendulum apparatus since the period of a pendulum depends only on its length and on gravity. Gravity is therefore essentially a measurement of a time interval and of length—the period of the pendulum and its effective length. Thus, if the pendulum length is kept constant, differences in gravity will be reflected in differences in the period. Gravity can also be measured using gravimeters, in which a weight is suspended from a coiled spring whose length changes in proportion to a change in gravity. Whereas pendulum instruments can measure absolute gravity, gravimeters can measure only changes in gravity, but the latter are more sensitive to small changes as well as simpler and faster to operate. Most gravity measurements are now made using gravimeters that measure gravity relative to a base station at which absolute gravity has been measured. Pendulum measurements typically have an accuracy of one part in 10^6, while gravimeters operated on land may be three or four orders of magnitude more precise. A major development in the past two decades has been the measurement of gravity at sea with one part in 10^6 precision, despite the disturbances caused by the accelerations of the ship. Thus gravity measurements are now available for many regions of the world's oceans as well as on land. In recent years absolute gravity has also been measured with great precision using 'free-fall' instruments in which the time interval required for a mass to fall through a known distance in a vacuum is measured. With these instruments such precision has been attained as to raise the possibility that secular and long-period changes in absolute gravity can now be measured.

Gravity measurements taken on the Earth's surface vary from place to place because of the Earth's oblateness, because of the variable distance from the centre of mass and because of an asymmetrical mass distribution within the planet. For geological and geophysical studies only the last of these effects is of real interest and it is convenient to correct the observed gravity for the first two factors (the free-air and Bouguer corrections, see page 106).

The difference between a corrected gravity measurement and the theoretical value for an idealized ellipsoidal Earth is termed the gravity anomaly. Gravity anomalies of a short wavelength or extending over a small area reflect near-surface density anomalies whereas anomalies persisting over large areas generally reflect mantle anomalies, a distinction that is a consequence of the attenuation of the force of gravity with distance. Figure 6.13 illustrates several examples.

The small- and intermediate-sized anomalies are readily surveyed with gravimeters but the regional and global anomalies are more difficult to determine in this way, not only because the ground survey approach is time-consuming but also because the long wavelength fluctuations are difficult to separate from the shorter wavelengths and the instrumental drifts that make gravimeters unsuitable for measuring the long wavelength anomalies, particularly at sea. However, satellites have provided a much more direct way of measuring the global and regional gravity field.

Measurement by satellite

An artificial satellite in an orbit about and close to the Earth would follow an elliptical orbit like that described for the Moon if the Earth were a radially symmetrical sphere. The better approximation of the Earth as an oblate spheroid requires that the Earth exert a torque on the satellite that tries to rotate the orbital plane into the equator. But because the angular momentum of the satellite motion is conserved the orbital plane precesses about the Earth's symmetry axis. Hence the effect of the Earth's flattening is primarily to cause the satellite's orbit to precess, both about the rotational axis and within the orbital plane, at a rate that depends on the exact form of the flattening. Typically the rates of precession are a few degrees per day and are readily deduced from a series of observations of the satellite's position from tracking stations on Earth.

Precise observations of an artificial satellite's motion show that the orbit is periodically perturbed from the precessing elliptical motion due to the departures in the mass distribution from the oblate spheroid approximation. As the satellite passes over a density anomaly it is accelerated in its orbit, and while the density anomalies may be relatively small the cumulative accelerations of the satellite during successive passes become measurable.

The study of the Earth's gravity field using satellites involves celestial mechanics, satellite geodesy and geophysics. The problem of celestial mechanics is to describe the motion of the satellite not only because of the Earth's gravity but also because of a variety of other forces. The satellite is attracted by the Sun and Moon. Furthermore the tidal deformations of the Earth result in a time-dependent attraction by the Earth on the satellite. Other forces of a non-gravitational origin include the drag forces experienced by the satellite moving in the tenuous atmosphere that exists even at altitudes well above 2000 km, and the force that radiation emitted by the Sun exerts on the satellite. The problem of satellite geodesy is to compare the motion described by the equations of celestial mechanics with observations of the satellite's position and to deduce the various parameters—gravitational or otherwise—entering into the celestial mechanics theory, describing the anomalous gravity field or defining the atmospheric drag coefficient or the intensity of the solar radiation. The problem for geophysicists is the interpretation of these results, to which we return below.

6.14: *Left, simultaneous range observations from the three laser stations on Continent A fix the satellite's position* S_1*. The observation from Continent B fixes this fourth station on a sphere centred at* S_1*. A second set of observations when the satellite is at* S_2 *determines the station along the intersection of two spheres while a third set fixes the station on B relative to those on A. This is a purely geometric mode of position determination in which no orbital information is required. Repeating the observations at different times would determine the displacements of A relative to B due to ocean-floor spreading. Right, in the orbital mode the satellite is observed from a number of stations on Continent A. The orbit is determined from these observations and the motion is extrapolated to the time when the satellite is observed from Continent B (or from a ship). The position of this tracking station can then be determined relative to the orbit.*

Tracking methods

Observations of satellite positions can be made in several ways by optical or electronic methods. A much used technique has been to photograph a satellite at night against a star background while the satellite is still illuminated by the Sun. These observations can give satellite positions accurate to about 10 m. Another much used method is the Doppler tracking of a satellite that emits a continuous radio signal at a constant frequency. Because of the motion of the source relative to the observer, the received signal is shifted to a lower frequency by an amount that varies during the spacecraft's passage over the tracking station. The comparison of the observed frequency with the standard frequency contains information on the satellite's motion relative to the observer and on its position in its orbit with a precision that ranges from about 10 m to 1 or 2 m.

During the past decade one of the most precise tracking methods to be developed has been the determination of the distance to the satellite by measuring the travel time of a very short laser pulse transmitted from the tracking station, reflected by an array of reflectors and received back at the station. Accuracies of a few centimetres are now possible in this way, which means that spacecraft accelerations due to quite small forces can be determined with considerable precision.

The satellite's motion is about the Earth's centre of mass and a logical choice for the origin of a terrestrial reference system is about this point. The positions of the tracking stations in this geocentric system will generally not be known with an accuracy that is compatible with that of the observations and it becomes necessary to determine simultaneously the orbital and force-field parameters and the

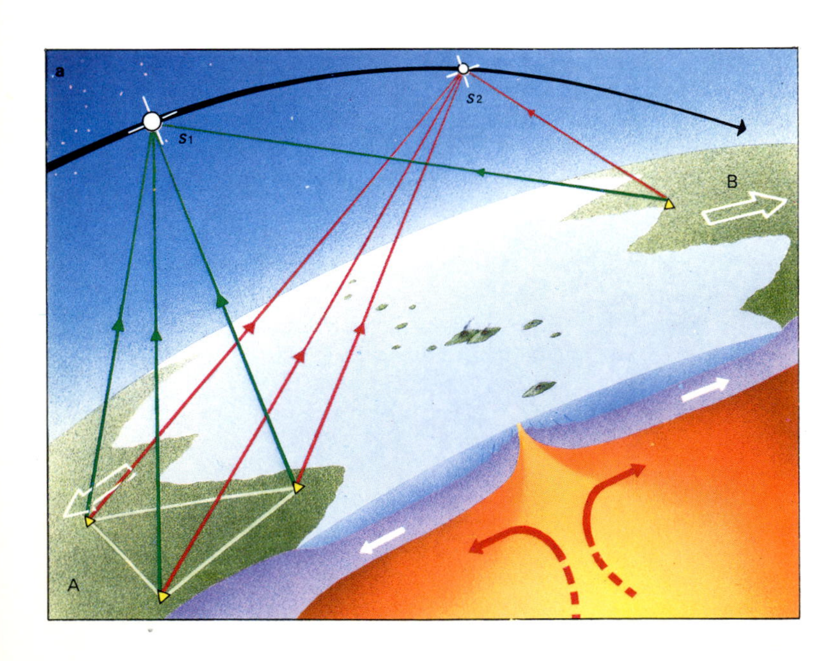

station positions. The latter can be determined in several ways.

A simple geometric procedure exists in which the satellite is used as a target that can be observed simultaneously from a number of stations. Assume, for example, that laser range measurements are made simultaneously from three stations whose relative positions are known from conventional terrestrial geodetic measurements. These observations determine the satellite's location as the intersection of three spheres with radii equal to the observed ranges. A simultaneous observation from a fourth station, of unknown position, fixes this station on a sphere centred at the satellite (Figure 6.14). Repeating the observations when the satellite is at different points in its orbit fixes the position of the fourth station relative to the other three. It is possible to measure relative station positions over continental and intercontinental distances with accuracies that approach those of the measurements themselves. This leads to the possibility that plate tectonic motions can be measured directly rather than inferred from less direct geophysical evidence.

While this geometric approach is simple in concept and does not require information on the forces that act on the satellite, the method is not widely used because the condition that the satellite be visible from many stations is restrictive. Instead, a procedure is usually adopted in which the station positions and the unknown parameters quantifying the forces are both determined simultaneously. Once these parameters have been determined with a satisfactory precision the motion of the spacecraft can be regarded as known and predictable and any additional observations can be used to determine the station's position relative to the orbit. This procedure is widely adopted, particularly when rapid position determination is required, as for navigation using the Doppler tracking of satellites.

Satellites used in these studies are varied. Almost any satellite can be photographed, whereas laser ranging requires that the satellite be fitted with cube-corner reflectors which reflect the incident laser pulse back to the station. Doppler observations require an on-board continuous transmitter of a constant frequency radiowave. To achieve high accuracies in describing the orbital motions, it is important that the air drag and radiation pressure forces are reduced. The simplest way of ensuring this is to have spherical satellites, free of protruding solar panels and antennae, with uniform reflectance properties and made of dense material. Geodetic requirements led to the design of two very dense, spherical satellites which are covered with cube-corner reflectors and are tracked entirely with lasers. The first satellite (Figure 6.15) was launched by the European Space Agency in 1975 into an orbit at an altitude of about 800 km. This satellite is used mainly in the determination of the Earth's gravity field and in the study of other forces on the satellite, principally those due to the periodic tidal deformations of the Earth. The second satellite, launched as part of the US space programme, is considerably larger and heavier and has been placed in a much higher orbit at an altitude of about 5000 km. Being further from the Earth, it is less perturbed by the Earth's gravity field than are lower spacecraft and this makes the satellite most useful for the precise determination of positions of points on the Earth and for measuring continental drift.

For gravity studies the lower the satellite the more sensitive it is to the Earth's anomalous density structure and the more useful it is, geophysically speaking. However, the air drag force also becomes more important and the drag perturbations begin to dominate the gravity field perturbations. Also, the lower the satellite, the more difficult it becomes to track it regularly from the ground. Together these factors limit the heights of geodetically useful satellites to about 700–800 km, meaning that only gravity anomalies of an extent exceeding about 2000 km can be detected using the methods outlined above.

6.15: *The satellite Starlette launched by the European Space Agency in February 1975. The satellite is a sphere of 250 mm diameter and has sixty reflectors distributed over its surface. Its core is made mainly of uranium giving it a weight of 35 kg and a density of about 18 kg/m*^{3}. *The satellite is tracked by lasers as a means of determining the Earth's gravity field and tidal deformation.*

An important development has been the use of radar altimeters to measure directly the height of the satellite over the oceans. The ability to track satellites from the ground with high precision means that the orbits are now well known and that it is possible accurately to compute the satellite's position within a reference framework whose origin lies at the centre of mass of the Earth. If the height of the satellite is measured using an on-board radar then the position of the reflecting surface can be determined relative to the orbit. Thus, for measurements made over the oceans the geoid can be measured directly with considerable precision and spatial resolution. Figure 6.16 illustrates some examples of these measurements obtained from the Geos 3 satellite launched in 1975. Results such as these are of considerable interest in understanding aspects of the oceanic crust and upper mantle and most of the world's ocean areas have now been surveyed in this way by the Geos 3 and Seasat satellites.

Interpretation of gravity

With the results from satellite and gravimeter observations gravity is a better-known quantity over the Earth's surface than nearly all other geophysical quantitites. Yet the interpretation of gravity is fraught with difficulty because a given gravity anomaly on the surface can be modelled by an infinite number of different density distributions. The geophysical problem is to separate the plausible from the improbable models. This inherent ambiguity can be illustrated by the simple case of gravity on a sphere of radius R and mass m. If all the mass m is concentrated at the centre of the sphere, gravity will be Gm/R^2. Likewise, if the mass is distributed in an infinitely thin layer just below the surface of the sphere gravity will also be Gm/R^2 and any intermediate radially symmetric density distribution between these two extremes will give the same value. In consequence, gravity observations on their own are not always very useful and are best interpreted together with other geophysical measurements and geological considerations. Examples of this will occur in later chapters.

Gravity anomalies

Despite this negative aspect gravity observations have provided some very useful results. One is that the variations in gravity are less than would be expected if they were due to topography alone. An example of this is illustrated in Figure 6.17, in which the result of gravity measurements at sea is plotted over a mid-ocean ridge in the form of gravity anomalies. A second example is given by the gravity anomalies over the European Alps. The observed gravity is first reduced to the geoid, the free-air correction, and then corrected for the density of the rock between the two surfaces, the Bouguer correction. A third example is given by gravity across the continental margins of south-east Australia: the corrections for the attraction of the topography and ocean water only increase the anomaly. All three examples indicate that the fully corrected gravity anomaly is a function of elevation. Over highly elevated terrain the anomaly is markedly negative while over the oceans it becomes strongly positive, increasing in magnitude with

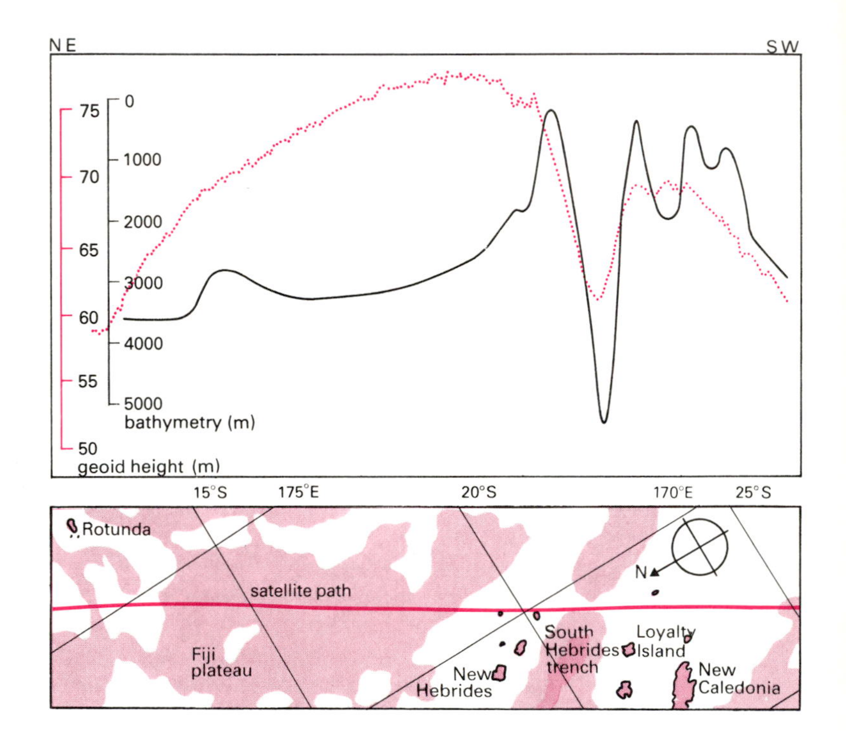

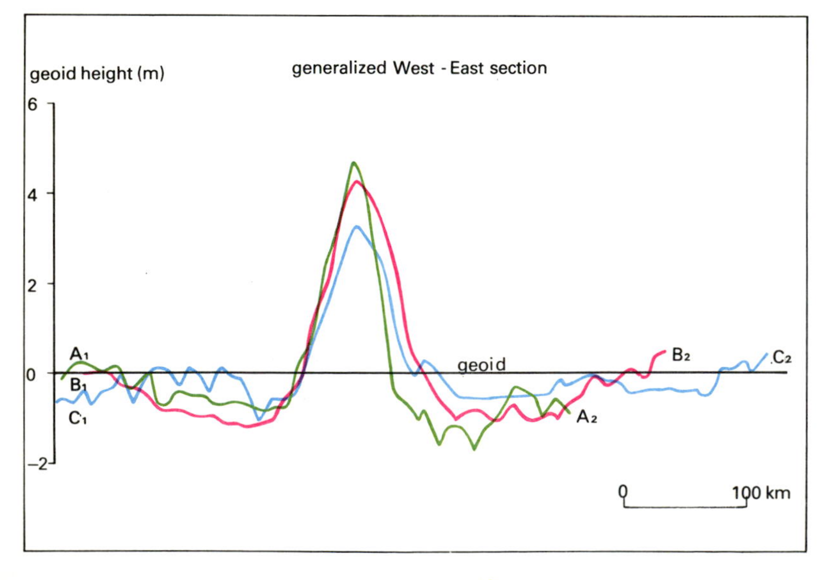

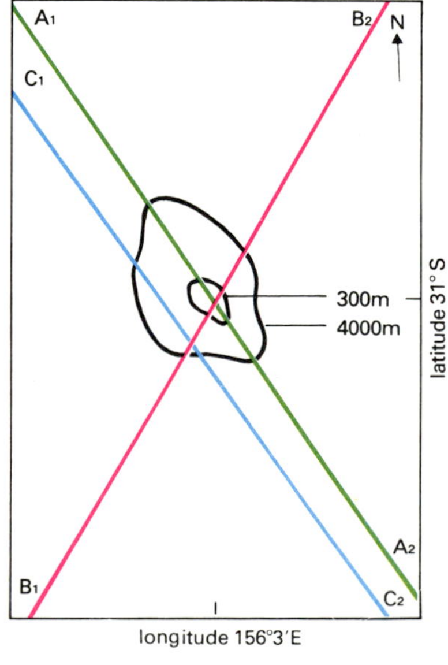

6.16: *Above, Geos 3 altimeter profile in the south-west Pacific and the ocean bathymetry. The geoid shows a pronounced negative anomaly over the South Hebrides trench. Left, Geos 3 altimeter profiles over the Derwent–Hunter seamount in the Tasman Sea. The geoid has pronounced positive anomalies over the centre of the seamount and negative anomalies over the flanks. This suggests that the load of the seamount is supported regionally rather than locally.*

increasing depth. This suggests that there is much more to gravity than just topography. In particular, it suggests that regions elevated above the geoid are associated with a mass deficit somewhere in the crust below the mountain while the ocean basins are associated with a mass excess in their crust, density anomalies which in both cases tend to compensate for the surface load.

The seismic evidence indicates that the Earth's crust varies considerably in thickness, being thin under the oceans compared with its thickness under the continents. The seismic evidence also points to an increase in density across the crust–mantle boundary, from about 2800–2900 kg/m^3 to 3200–3300 kg/m^3. Thus one explanation of the gravity anomalies is in terms of a crust of variable thickness; under highly elevated regions the crust is thicker and tends to compensate for the extra surface load while under the oceans the thinner crust results in the dense mantle material being closer to the surface. The crust generally behaves as a rigid layer capable of supporting stress differences but at some depth below the crust the mantle begins to behave more as a plastic material, deforming when subject to stress differences. Evidence for this response comes from laboratory measurements, from observations of post-glacial uplift and from the observation that the Earth's oblateness is very similar to what it would be if the Earth as a whole responded as a fluid to the centrifugal force. Furthermore the plate tectonic hypothesis (see Chapter 10) requires that within the upper mantle there is a region that deforms when subject to stress over a geological time scale so as to permit the lithospheric plates to move with respect to the deeper mantle.

Isostasy

Together, the gravity observations, the seismic evidence and the rheological response of the upper mantle support the models of isostasy proposed more than a hundred years ago. These were offered as an explanation of geodetic observations in India which suggested that the attraction of the Himalayas was less than it would be if this mountain range was simply on top of a radially homogeneous crust and mantle. The isostatic model assumes that the crust can be represented by blocks floating on a fluid-like and denser mantle, with each block moving vertically and independently of its neighbour. Figure 6.18(a) illustrates the model proposed by G. B. Airy (1801–92) in 1851 to explain the Indian observations. All blocks are of equal density but different heights, and they float, the longest blocks extending deepest into the mantle. At a certain depth, equal to or greater than the thickest crust, the pressure is everywhere constant and below it the mantle is in a state of hydrostatic equilibrium. The gravity anomaly according to this model consists of two contributions. One is from the visible parts of the blocks, the topography, and the second is from the 'roots', the parts of the blocks that have displaced the more dense mantle material. The two contributions are of opposite sign, and the resulting gravity anomaly is considerably smaller than if it were due to the topography alone (see Figure 6.17). At the same time as Airy expounded his model of isostatic compensation, J. H. Pratt proposed the same isostatic principle, of constant pressure at some constant depth below the surface, but with a model in which the crust is assumed to be of a variable density but with a constant depth such that under elevated areas the density is less than under low-lying areas (Figure 6.18(b)). Airy's model was based on the assumption that the topographic load stresses the Earth's crust beyond its strength-bearing capacity so that failure occurs by normal faulting, and that the crust below the load is depressed until the isostatic condition is attained. Pratt, on the other

6.17: *Three examples of the relationship between gravity anomalies and topography. Note the different scales used.* (**a**) *Mid-Atlantic ridge. The free-air anomaly is small and variable across the ridge. When 'corrected' for the Bouguer effect the Bouguer anomaly is strongly positive and the deeper the ocean the larger the anomaly.* (**b**) *The European Alps. The free-air anomaly is very variable particularly near the high peaks. The Bouguer anomaly is negative and increases in magnitude with altitude.* (**c**) *The continental margin of south-east Australia. The free-air anomaly is positive over land and negative over the ocean but the Bouguer anomaly is of opposite sign and more pronounced.*

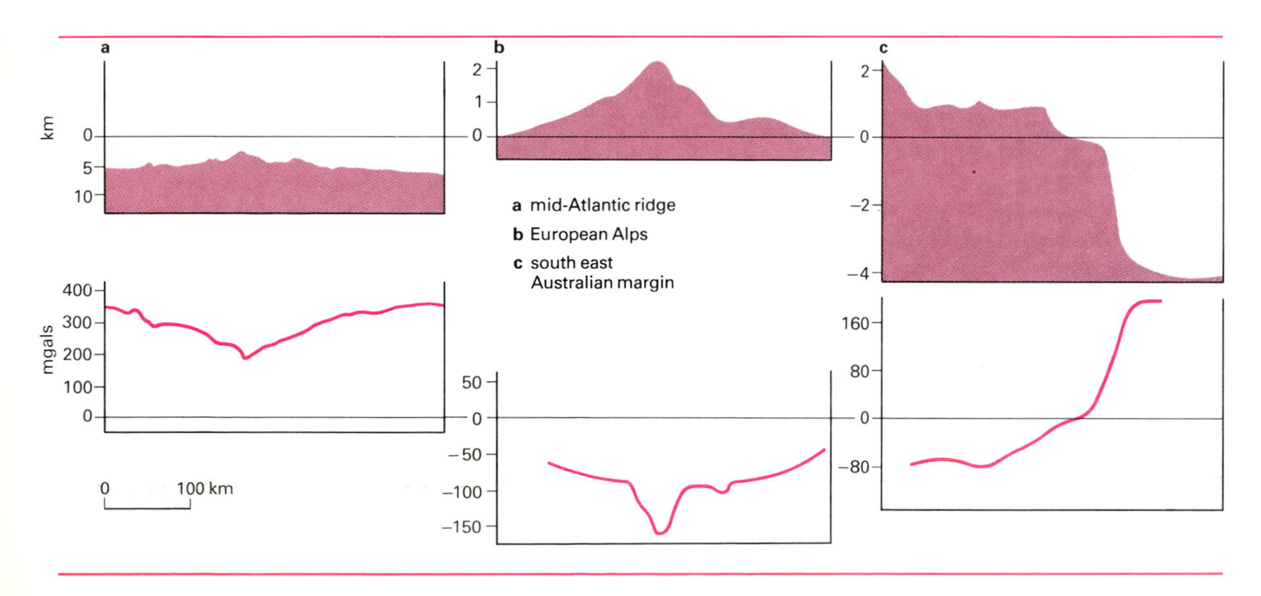

hand, assumed that the mountains were the result of a thermal expansion of the crust as a result of a heat source in the crust itself or in the mantle below.

While both models are based on the apparently unrealistic assumption that the mantle behaves as a fluid and the equally unrealistic assumption that the response of one block is independent of neighbouring blocks, both describe surprisingly well the gravity observed over regions of variable terrain and indicate that the principle of isostasy is obeyed over much of the world.

Global anomalies

One advantage of the universality and simplicity of the isostasy hypothesis is that it is relatively simple to correct gravity measurements to take account of topography elevations and their roots. The gravity so corrected is the isostatic anomaly. If the traditional isostatic models were complete then these anomalies would be everywhere zero. Yet this is not so. Classical isostasy is not sufficient to explain the global gravity anomalies and the only satisfactory explanation is that they are due to density anomalies below the crust and not directly associated with the topography. This is perhaps the most important result that has come from satellite geodesy to date. From it two quite contradictory conclusions can be drawn about the Earth's mantle. One is that the mantle is sufficiently rigid for it to be able to support density anomalies elastically so that the present gravity field reflects conditions in the Earth at some time in its remote geological past. The concomitant stress differences in the mantle are of the order of 500 bars and the general consensus of Earth scientists is that this is excessive, that mantle materials will flow when subject to such stresses at the temperatures and pressures characteristic of the mantle. The most convincing evidence that flow would occur is seen in the postglacial rebound phenomenon in which the removal of late Pleistocene ice loads has resulted in a slow rebound of the originally depressed crust and in a flow of the mantle in response to the changes in stress which do not exceed a few tens of bars.

The alternative interpretation is that the gravity anomalies are associated with mantle convection and that the density anomalies are a consequence of temperature at a given depth not being everywhere the same. This interpretation is now widely accepted by geophysicists and is reinforced by the rather remarkable correlation that exists between the gravity anomalies and the surface expression of plate tectonics, the plate boundaries. The major subduction zones around the Pacific Ocean are all associated with rather broad gravity anomalies. The collision zones of the African and Indian plates with the Eurasian plate are also associated with positive but milder gravity anomalies as are parts of the ocean ridges, particularly Greenland and the Azores in the north Atlantic. The negative anomalies lie mostly over ocean basins and over old continental shields. These gravity anomalies are not fully understood but they do indicate that there is a relation between them and mantle observations. The question now is not so much whether convection occurs, but concerns such matters as the scale of the cells, the depth to which convection occurs, what drives it and the periodicity of the motion. The answers to these questions are still largely speculative but the gravity anomalies may make a useful contribution towards reaching an understanding.

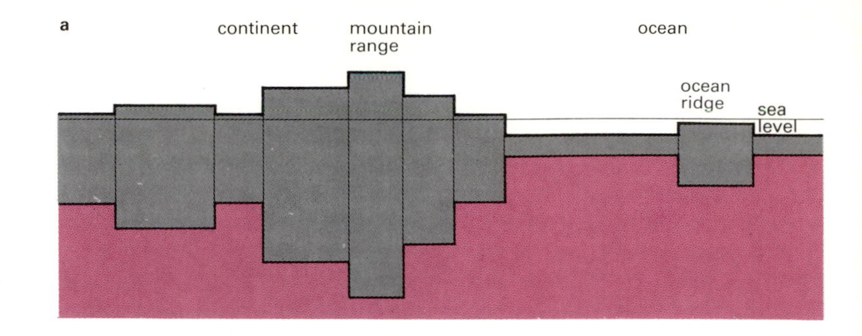

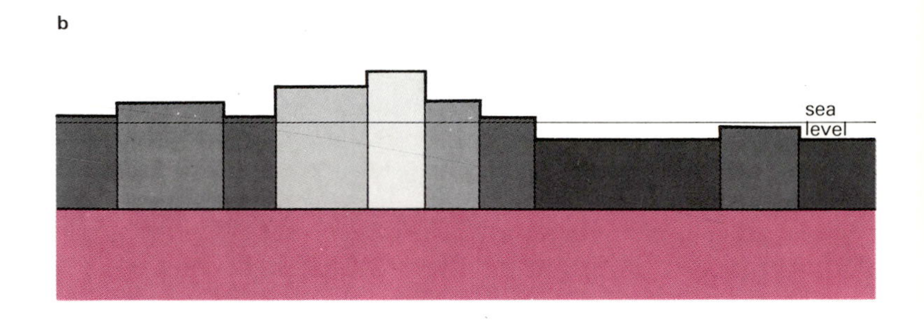

6.18: *Two interpretations of the same topographic profile according to the isostatic hypothesis of Airy and of Pratt.* (**a**) *Airy's model, in which the crust is of variable thickness but of constant density and is thicker under elevated terrain than under depressions such as the oceans. The depth of the underlying 'root' is related to the height of the overlying topography.* (**b**) *Pratt's model, in which the crust is of variable density but its base is at a constant depth below sea level. Topographic height is related in this model to crustal density at that point. In both models the pressure due to the weight of the overlying crust is constant at a certain depth, which is always greater than the deepest crust.*

7 The Earth as a magnet

Most people are familiar with the fact that the earth has a magnetic field which influences a magnetic compass. But although this is now one of the most common pieces of basic scientific knowledge among non-scientists, the existence of the geomagnetic field came to be recognized only a few hundred years ago. One of the consequences of the field, on the other hand, was familiar to natural philosophers thousands of years earlier. Because the geomagnetic field exists, many rocks upon formation acquire a magnetization which they usually retain until they are broken down, heated to a very high temperature or re-assimilated into the Earth's interior. Most of the rocks now accessible at the Earth's surface are so weakly magnetic that their magnetizations can only be detected and measured by sensitive magnetometers developed during the twentieth century. There is one exception, however, in lodestone, a naturally occurring form of almost pure magnetite (Fe_3O_4) with a magnetization whose strength is noticeable without measuring instruments. The Greeks were familiar with lodestone by at least 600 BC and thus with an indirect effect of the geomagnetic field. But although they saw that lodestone exerts an influence on other bits of lodestone at a distance, they never came to appreciate that the Earth exerts a similar influence. They never discovered polarity (the north and south poles of a lodestone magnet) and they never knew the compass, the one invention that might have led them to conclude that the Earth itself acts as a magnet.

The discovery of the directional properties of magnetic material was left to the Chinese who, in about the first century BC, constructed the first compass in the form of a lodestone spoon carefully balanced upon a smooth polished board. A modern reconstruction of this remarkable instrument is shown in Figure 7.1. When lodestone is cut into such an elongate form it automatically becomes polarized with its extreme ends having a north and south pole, respectively. If free to rotate, therefore, it will act as a compass needle and thus come to rest in a roughly north–south direction under the influence of the Earth's magnetic field. Not that the Chinese realized the Earth was having an influence, at least in the sense of physical cause and effect. Neither at the time the compass was invented nor for many centuries afterwards was a scientific connection made between the Earth and the behaviour of the spoon. It was used instead for geomancy, the divination of the future course of events. By the fourteenth century the Chinese had found that they could float a compass in a bowl of water and use it as a navigational aid across open oceans. Towards the end of the sixteenth century European scientists began to recognize that magnetism is one of the Earth's intrinsic properties. At about that time William Gilbert (1540–1603), an English physician, showed that the behaviour of a compass needle at the Earth's surface was almost identical to that of an iron needle placed on the surface of a lodestone sphere, from which it was but a short step to concluding that the Earth itself acts as though it is a huge magnetic sphere.

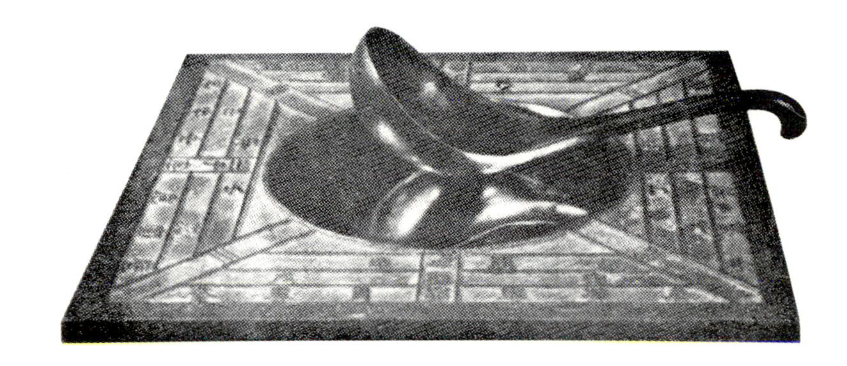

7.1: *A replica of a lodestone spoon balanced upon a polished bronze plate, constructed several decades ago from an ancient Chinese pattern. This was the earliest form of the magnetic compass and is known from Chinese literature to have been in use by at least the first century* BC. *It works because a piece of lodestone cut into an elongated form automatically becomes polarized with opposing magnetic poles* (N *and* S) *at opposite ends. The variety of signs on the base plate indicates its ancient use for geomancy.*

The Earth's present-day magnetic field

A magnetic field is a vector quantity, which means that it has both magnitude and direction. As the Chinese discovered without realizing it, a compass needle free to rotate at a particular point will come to rest in the direction of the field at that point; or, to put it another way, the compass may be used to measure the field direction. If used to determine the field directions in the Earth's immediate vicinity, the result looks like that illustrated in Figure 7.2. The most obvious conclusion to be drawn from this is that the Earth's field exhibits a pattern; it is not just random. Indeed, the field shape appears to be identical to that of a bar magnet, the simplest field (known as a dipole field) that can exist.

Moreover, the field pattern is symmetrical about the Earth's centre, although not about the rotational axis. What this regularity and symmetry suggest above all else is that the magnetic field near the Earth's surface is mainly of internal origin, for no field produced externally would be likely to have such a close relationship with the Earth's geometry.

To say anything more than that, however, requires a closer examination of the geomagnetic field than Figure 7.2 allows; and to make such an examination it is necessary first to define the quantities by which a field may be specified and the frame of reference within which those quantities may be given universal significance. If at a point O the magnetic field is F, then, as Figure 7.3 shows, the field may be represented by a line whose length represents the strength of the field at O and whose direction indicates the field's direction. The coordinate system within which this direction is then specified is the rectangular coordinate system defined by geographic north, geographic east and the vertical (downwards). In general, the field F will lie at an angle to this frame of reference. In mathematical terms, however, F is equivalent to the three components X, Y and Z at right angles, ie along the directions that define the frame of reference.

A small compass needle at O, free to rotate in any direction whatsoever, will set along the line F, ie in the field direction. This is the sort of situation represented by Figure 7.2 which, though drawn in two dimensions, should be imagined as a three-dimensional pattern completely surrounding the Earth. But although Figure 7.2 was constructed from data obtained from a completely free compass needle, this is not the sort of freedom an observer usually has. The common compass needle used to measure the field at the Earth's surface, for example, is normally mounted on a vertical pivot and is thus constrained to rotate in a horizontal plane only. If used at a point such as O, therefore, it will come to rest not along F but along B, the component of F in the horizontal plane.

The angle (D) that B makes with the geographic (true) north is known as the angle of declination, or simply declination (the nautical term is 'variation'). Declination is usually measured in degrees east of true north (ETN), although when the value of declination approaches 360°ETN it is sometimes more convenient to specify it in degrees west of true north (eg 351°ETN = 9°WTN). The angle (I) between B and F (which can be measured using a compass needle mounted on a horizontal pivot and thus constrained to move only in the vertical plane) is known as the angle of inclination, or simply inclination. It is regarded as positive if F lies below the horizontal and negative if it lies above.

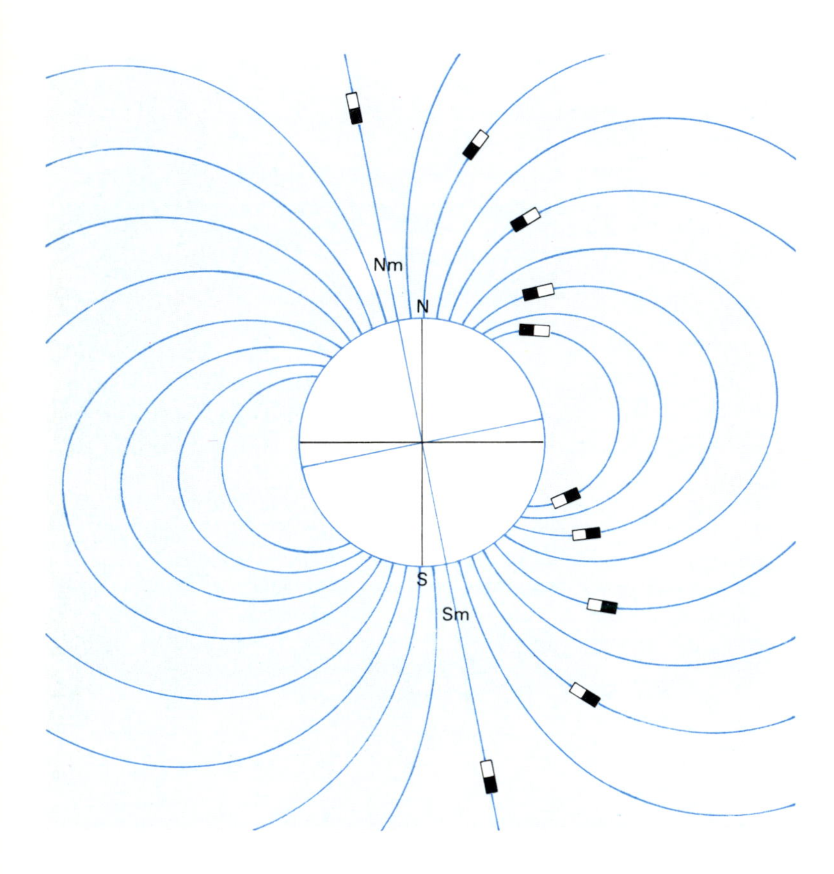

7.2: *The broad pattern of the magnetic field around the Earth as delineated by a compass needle free to rotate in any direction. The dark halves of the compass needles are the north poles and point in the field direction. The field, which should be imagined in three dimensions, is identical to that of a short bar magnet at the Earth's centre and inclined at 11° to the Earth's rotational axis. At the geomagnetic equator (which is inclined at 11° to the geographic equator) the needle is aligned with the Earth's surface; at the north* (Nm) *and south* (Sm) *geomagnetic poles (which lie 11° of latitude away from the corresponding geographic poles,* N *and* S*) it points vertically downwards and upwards, respectively.*

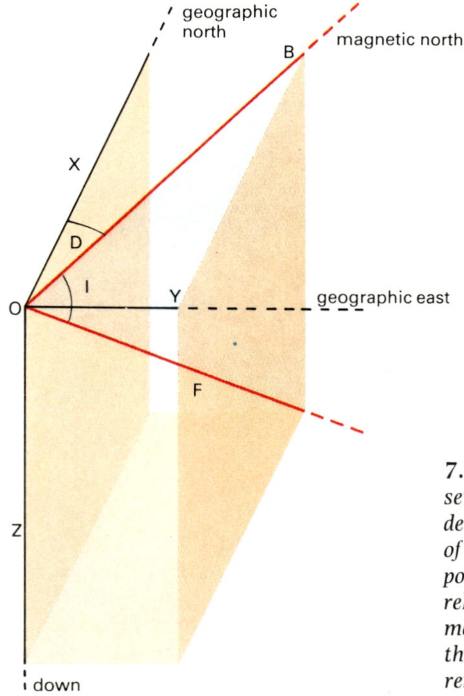

7.3: *The magnetic elements—the seven quantities used to define and describe the direction and strength of the magnetic field acting at point* O. *The seven elements are related trigonometrically, which means that knowledge of only three elements enables the remaining four to be calculated.*

The vertical plane through B (and hence F) is known as the magnetic meridian at O, just as the vertical plane through O and geographic north is the geographic meridian.

X, Y, Z, B, F, D and I are known as the magnetic elements. They are used to specify and/or describe the Earth's magnetic field, although to specify the field completely it is not necessary to determine all seven elements. There are trigonometric relationships between them which make it possible to derive the four remaining elements from three known ones. Once the elements have been obtained at a number of points it is usual to plot them on a world map, not directly but as 'contours' linking points at which a given element has the same value. Such maps are known generally as isomagnetic charts. Lines joining points at which the declination values are the same are known as isogonic lines, or simply isogonics, and give rise to an isogonic chart. Similarly, lines of inclination are known as isoclinic lines, or isoclinics, and give rise to an isoclinic chart. Isomagnetic lines and charts for the other elements do not have special names.

Figures 7.4, 7.5 and 7.6 show, respectively, the isogonic chart, the isoclinic chart and the isomagnetic chart of the total field strength, for the year 1975. Although there are patterns of sorts, the maps suggest

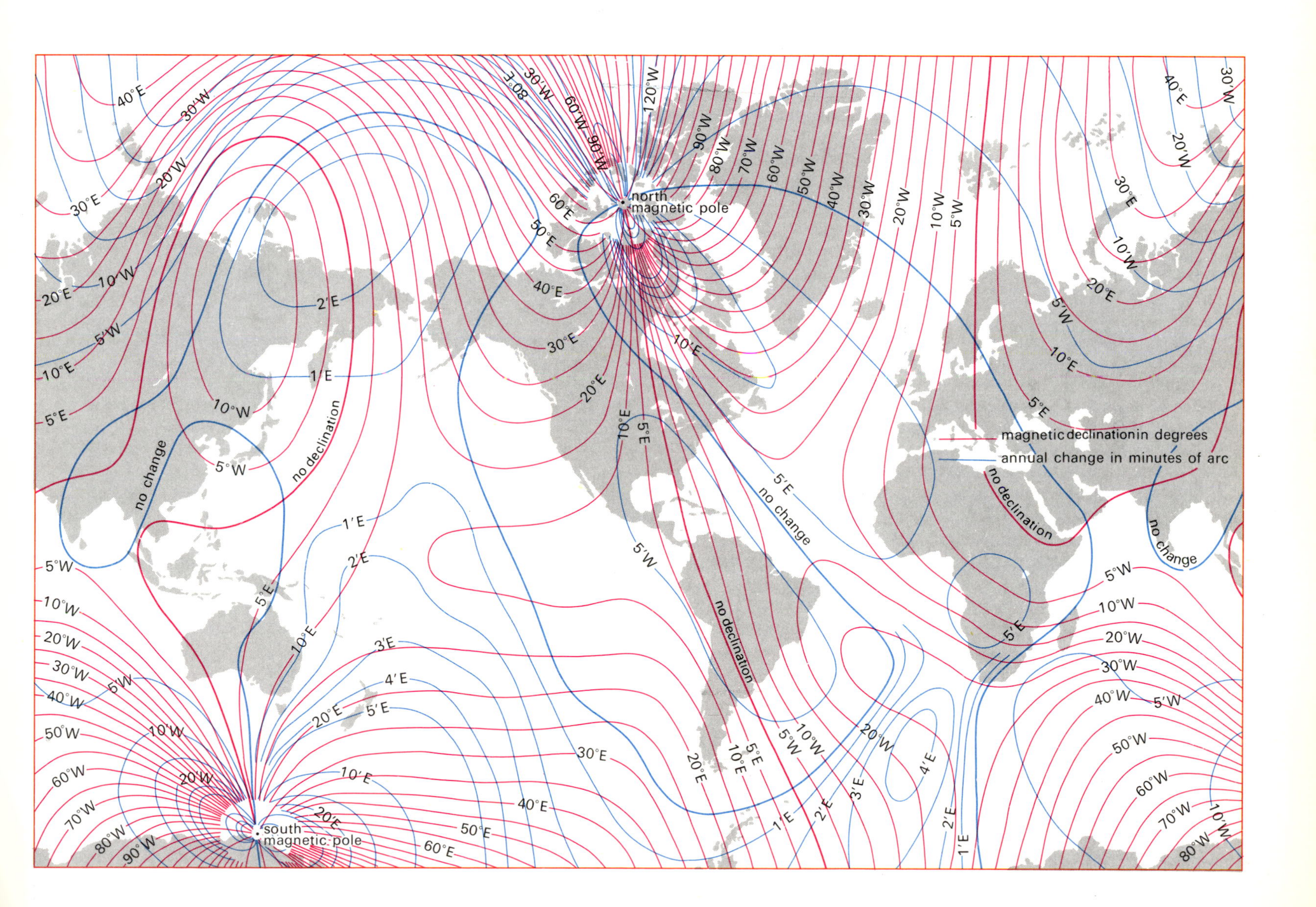

7.4: *The isogonic chart for 1975—the map of world declination. The red contours are lines of equal declination, where the declination is specified as east or west of true north (eg lines marked 20°E, 10°W). The blue lines are lines of equal rate of change of declination where the rate of change is expressed as minutes of arc per year to the east or west (eg lines marked 5′E, 2′W). The magnetic poles shown here and on Figures 7.5 and 7.6 are the magnetic dip poles, not the geomagnetic poles.*

that the geomagnetic field is quite complicated—much more so than the simple dipole pattern of Figure 7.2 would seem to imply. If the geomagnetic field were simply that of a dipole lying along the rotational axis, for example, a horizontal compass needle at the Earth's surface would everywhere point towards geographic north, the declination would therefore be everywhere zero and there would be no isogonic lines as such. The isoclinic lines, on the other hand, would be a series of lines parallel to lines of geographic latitude in Figure 7.5, as would the lines of equal field strength in Figure 7.6. As far as the real Earth is concerned, things would not be quite as simple as this, for the dipole is inclined to the rotational axis; but whatever the slope of the dipole it would certainly not alone lead to field distortion to the extent demonstrated by Figures 7.4, 7.5 and 7.6.

In fact the complexity of the isomagnetic charts is a little deceptive. The Earth's field is largely that of a dipole, but superimposed upon it there is a more or less randomly directed field known as the non-dipole field. The non-dipole components are generally much smaller than the main dipole in strength but they nevertheless appear to have a considerable distorting effect when viewed in chart form. Their true significance is best demonstrated not in pictures but in mathematical analysis (specifically spherical harmonic analysis), the results of which are important and quite easily stated.

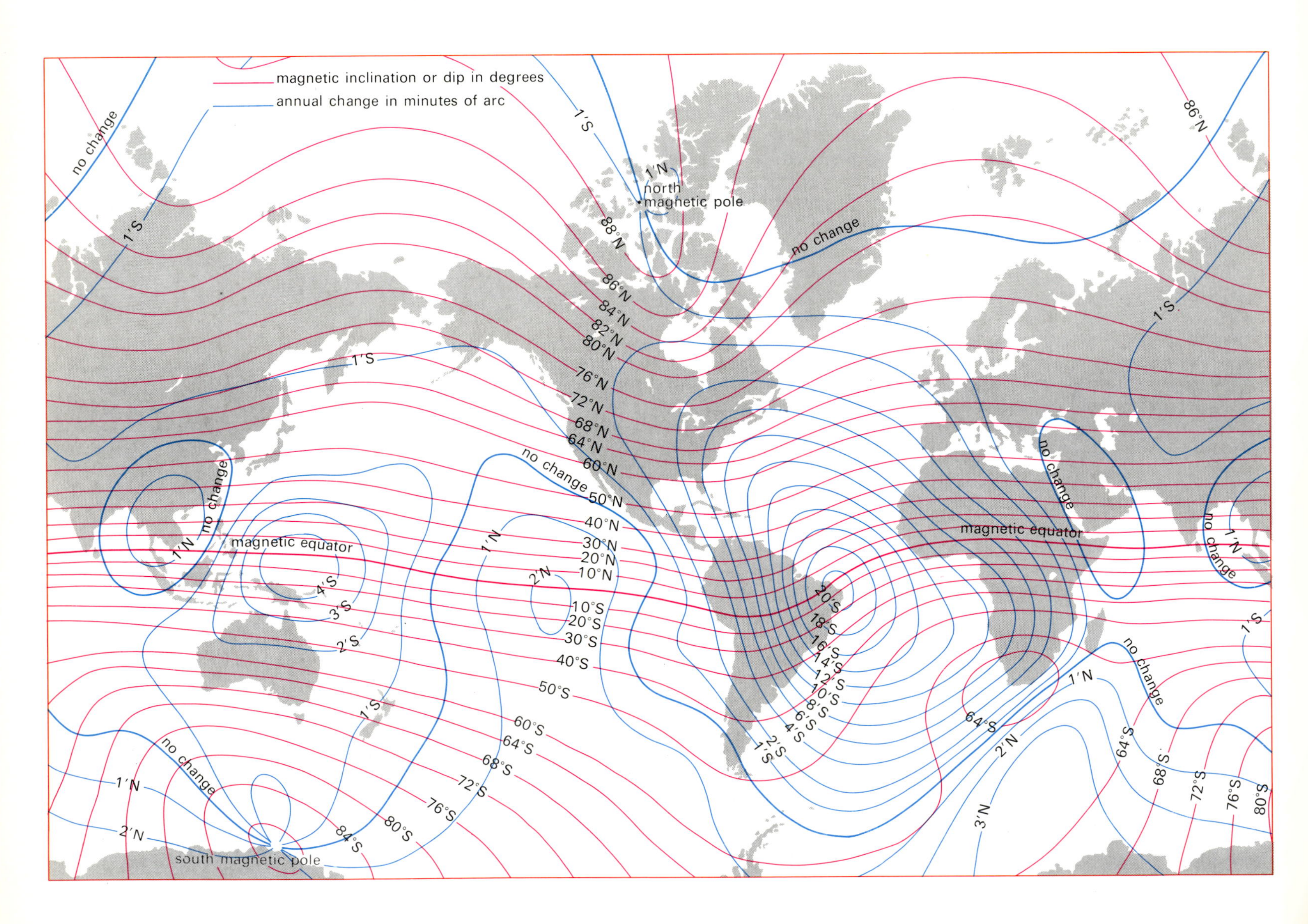

7.5: *The isoclinic chart for 1975—the map of world inclination. The red contours are lines of equal inclination, and/or equal geomagnetic latitude. Geomagnetic latitude (λ) is related to inclination (I) by* $\tan I = 2 \tan \lambda$. *The contours are the same but the labels are different. In this diagram the lines are marked as geomagnetic latitude (eg 30° N = geomagnetic latitude 30° N ≡ inclination + 49°). The blue lines are equal rates of change of geomagnetic latitude where the rate of change is expressed as minutes of arc (latitude) to the north or south (eg lines marked 2′N, 2′S).*

1 The geomagnetic field can be regarded as the combined effect of a series of separate components, namely, dipole (two magnetic poles), quadrupole (four poles), octupole (eight poles) etc. Of these, by far the strongest is the dipole component, whose field resembles that of a bar magnet or a solenoid. This does not mean that the geomagnetic field is actually produced by a bar magnet in the Earth but only that, whatever the origin of the field, the shape of the field is very similar to that from a bar magnet. In fact the source of the field is complex and difficult to visualize, for which reason it is often convenient to imagine a dipole at the Earth's centre (the geomagnetic dipole). All components other than the dipole are known collectively as the non-dipole field. This division of the geomagnetic field into dipole and non-dipole was first derived mathematically by the German mathematician K. F. Gauss (1777–1855) in 1839 and is in essence a mathematical fiction. As will be seen later, however, it so happens that the dipole and non-dipole fields also have rather different physical origins.

2 All but a small percentage of the field (dipole and non-dipole) observed at the Earth's surface is produced by processes inside the Earth, ie most of the field is of internal origin. The small remainder of the field is produced externally, probably by electromagnetic phenomena in the upper atmosphere. The external field is not very important but it does account for the very rapid (hour-to-hour or even minute-to-minute) but small changes observed in the Earth's surface field.

3 The geomagnetic dipole lies at the centre of the Earth (ie it is geocentric) but not along the rotational axis. In fact it slopes at 11° to

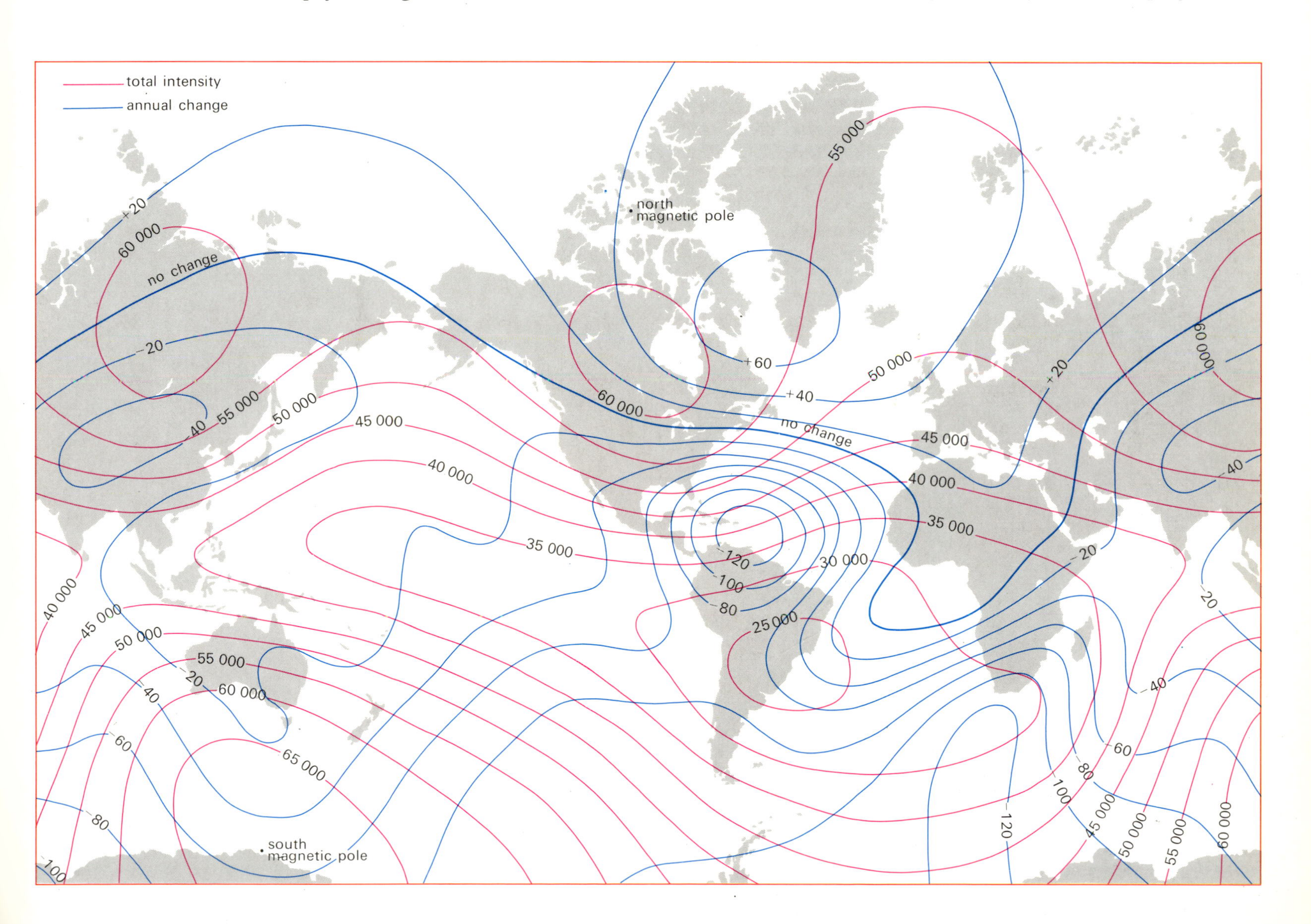

7.6: *The map of world field strength for 1975. The red contours are lines of equal field strength in units of* 10^{-9} *tesla. The blue contours are lines of equal rate of change of field strength in units of* $\pm 10^{-9}$ *tesla per year.*

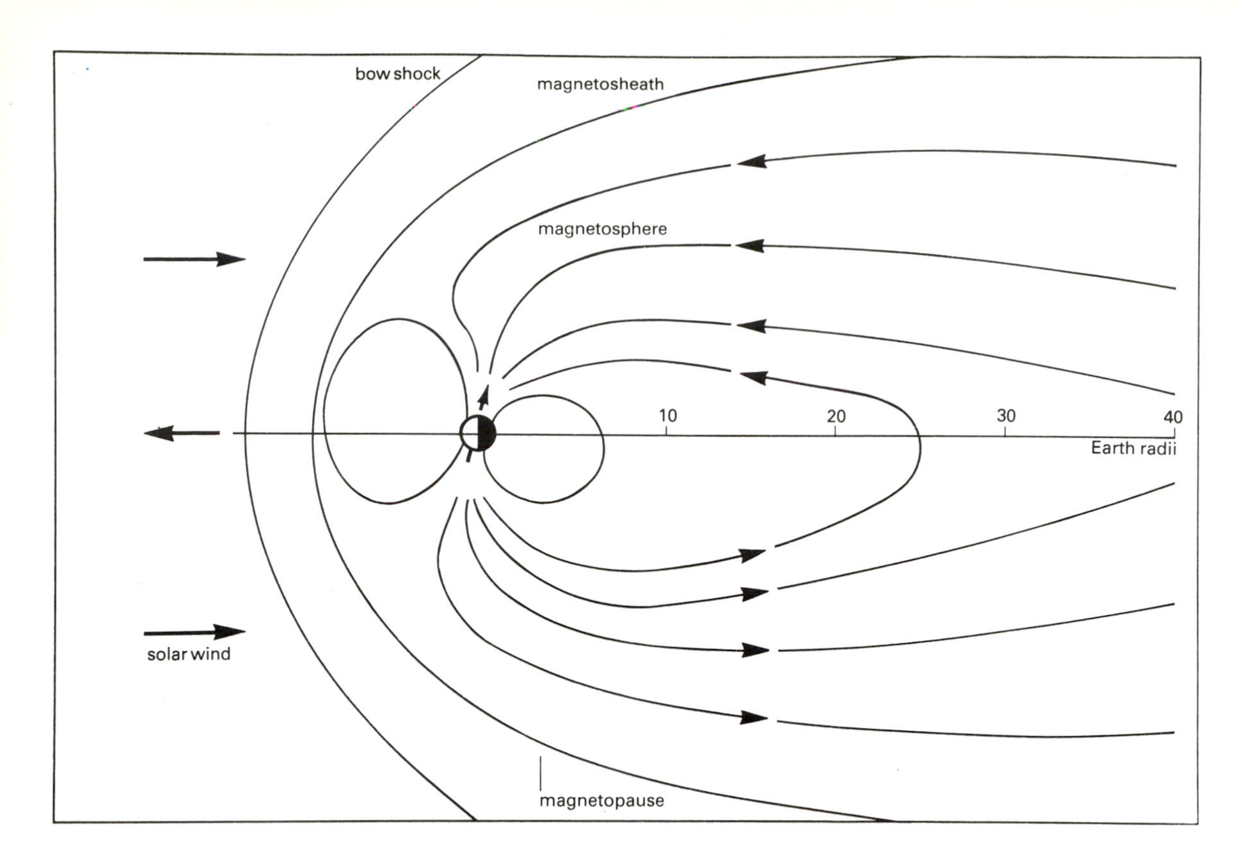

7.7: Left, the charged solar wind particles streaming out from the Sun interact with the Earth's magnetic field, compressing the field on the forward side and stretching it on the trailing side. The intrinsic symmetry of the field thus disappears when viewed from 10 Earth radii or more from the Earth's centre. Below left, the corona of the Sun, its outer atmosphere, in a computer enhancement of a coronagraph taken from Skylab 3 in 1973; the Sun itself is obscured by a disc in the telescope. The thin layers of gas forming the corona are trapped by the Sun's magnetic field, the shape of which is thus picked out by the brightness contours in this image.

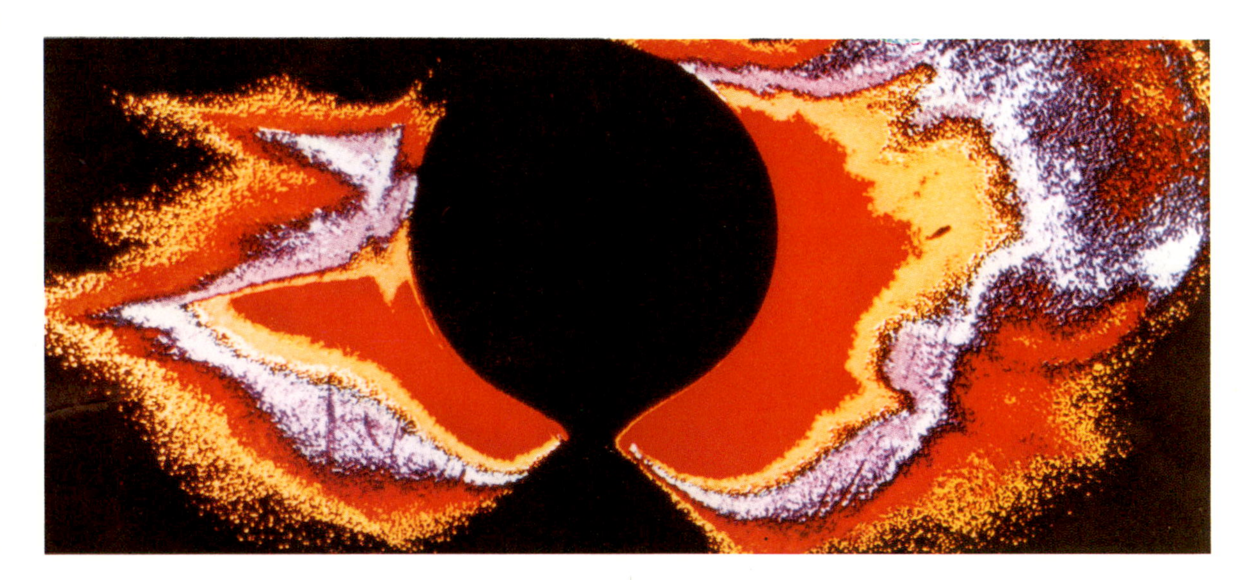

the rotational axis, giving geomagnetic poles at 79°N, 70°W (north geomagnetic pole) and 79°S, 110°E (south geomagnetic pole); in other words, the geomagnetic poles and geographic poles do not coincide. If the geomagnetic field were solely dipolar, as shown in Figure 7.2, the north pole of a compass needle would point vertically downwards (inclination = +90°) at the north geomagnetic pole and vertically upwards (inclination = −90°) at the south geomagnetic pole. (It should be noted that as by a fundamental law of magnetism like poles repel and unlike poles attract the south pole of the geomagnetic dipole must be pointing north; in other words, the north geomagnetic pole corresponds to the south pole of the geomagnetic dipole itself, and vice versa.)

4 In practice the geomagnetic field is not solely dipolar; there are irregular non-dipole components present as well. This means that at the geomagnetic poles there is an extra field which changes the inclination from +90° and −90°, respectively. The points at which the inclination is actually +90° and −90° are elsewhere, at places known as the magnetic dip poles. The north magnetic dip pole is now at about 75°N, 101°W and the south magnetic dip pole is at about 67°S, 143°E. Because the non-dipole field is irregular and thus different in each hemisphere, it does not have the same effect in each polar region. The magnetic dip poles are thus not antipodal. Moreover, as will be seen later, the non-dipole field changes fairly rapidly and so, therefore, do the positions of the magnetic dip poles.

5 If, again, the geomagnetic field were solely dipolar, the strength of the field would be 0.62×10^{-4} tesla at the geomagnetic poles and 0.31×10^{-4} tesla at the geomagnetic equator. Again, however, there are deviations from these values at the real geomagnetic poles because of the presence of non-zero, non-dipole components. On average, the non-dipole field strength is only about 5 per cent of that of the dipole field, but at odd individual spots it can reach 30–40 per cent of the dipole field.

So far in this section it has been assumed that the Earth's main field, though produced internally, is allowed to remain undisturbed by external influences. Close to the Earth itself this is indeed more or less so, which explains why the field pattern in Figure 7.2 remains symmetrical. Much further out, however, there is an interaction between the geomagnetic field and the solar wind, the plasma (stream of charged particles) that emanates from the Sun. On the side of the Earth towards the Sun the solar wind compresses the geomagnetic field, whereas on the opposite side of the Earth the field is greatly elongate (Figure 7.7). The field is thus confined to a zone known as the magnetosphere, the boundary of which is called the magnetopause. The position of the magnetopause varies slightly as the intensity of the solar wind changes, but in the solar direction it lies on average about 10 Earth radii from the Earth's centre, while away from the Sun it extends out to very large distances of at least 60 Earth radii. As the solar particles are travelling at speeds of about 1000 km per second when they encounter the outer edges of the geomagnetic field, a shock wave is produced some distance before the magnetopause, giving, between the magnetopause and the wave front, a transition zone known as the magnetosheath.

Geomagnetic field changes: direct observation

There are odd observations of declination and inclination going back to the sixteenth century, although systematic measurements of the magnetic elements have only been made during the past 140 years or so—and the earlier of even these leave much to be desired so far as accuracy and general trustworthiness are concerned. Surprisingly, perhaps, time changes in the field, known usually as secular variations, were revealed by some of the very earliest observations. In 1635, for example, Henry Gellibrand reported his discovery that between 1580 and 1634 the declination at London had changed from 11.3°E to 4.1°E. Much more recently it has become clear that all the magnetic elements are continuously changing; and the systematic field measurements of the nineteenth and twentieth centuries have provided a fairly detailed picture of the nature and extent of these changes.

Analysis shows that the geomagnetic field has been predominantly dipolar throughout the whole of the period covered by direct observation, although the axis of the dipole has not remained entirely stationary and the strength of the dipole has decreased at the rate of about 5 per cent a century (Figure 7.8). As far as the position of the dipole axis and its poles are concerned, for the past 140 years the latitude of the north geomagnetic pole has remained at about 79°N, but the longitude has changed from about 64°W to 70°W. Thus the angle between the geomagnetic (dipole) and geographic (rotational) axes has remained at about 11°, but the polar longitude has moved at an average rate of about 0.04° longitude a year.

It would be tempting to deduce from this that the geomagnetic dipole always lies at an angle of 11° from the rotational axis; but it would be a mistake to do so, for 140 years is really a very short time upon which to base a conclusion about long-term field behaviour. Moreover, there is circumstantial evidence against the validity of the perpetual 11° angle. It is reasonable to suppose that on average the dipole axis will lie along the rotational axis, for the rotational axis is the only unique axis in the Earth; apart from the rotational axis all straight lines through the Earth's centre are indistinguishable from each other. On the other hand, it is a clear observational fact that throughout the period of direct observation the geomagnetic and geographical axes have definitely not coincided. The only way of reconciling fact and hypothesis is to suppose that the dipole 'wobbles' about the rotational axis—so that the two are coincident when averaged over a certain period of time but not necessarily coincident at any given time. From the rate at which the geomagnetic pole has been moving in recent years, it would appear that the 'certain period of time' must be at least ten thousand years.

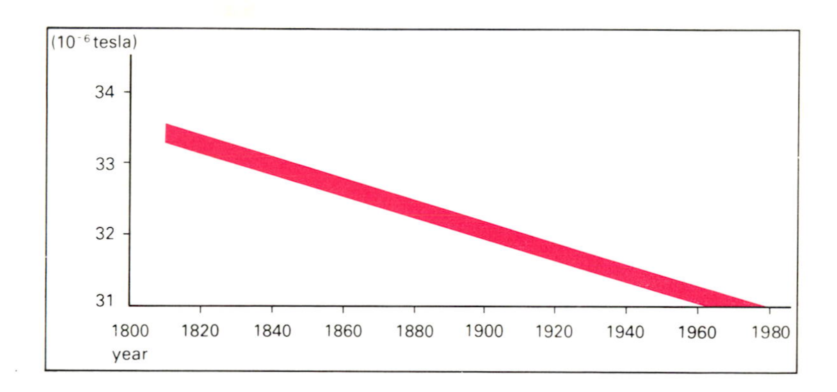

7.8: *The strength of the geomagnetic dipole has decreased strikingly over the period of direct field observation. The numbers on the vertical axis actually refer to the dipole field strength at the geomagnetic equator. The points are average values in the fifty-year intervals 1815–65, 1865–1915, 1915–65, respectively.*

The non-dipole field, by contrast, has been changing much more rapidly, on typical time scales of the order of a thousand years or even a hundred years. This is well illustrated by comparing Figures 7.9 and 7.10, which show the isomagnetic maps of the Z (vertical) component of the non-dipole field (only) for, respectively, 1835 and 1965. The two charts are broadly similar but have significant detail differences. During the 130 years the feature marked A, for example, has drifted westwards and about doubled its maximum strength. The feature marked K has also drifted westwards, albeit not by as much as A, and has also increased in strength, albeit not at such a great rate as A. The feature marked R disappeared altogether between 1835 and 1965.

Over the past 140 years there has been a general tendency for the non-dipole field to drift westwards at an average rate of 0.2° longitude a year. This would seem to imply that if the field were to persist long enough it would move right around the Earth in less than two thousand years; but in fact it does not remain in the same form for that long. The features of the non-dipole field are continuously changing–growing, diminishing, expanding, contracting, disappearing and reappearing–on time scales of a hundred to a thousand years, which indicate variation at least an order of magnitude more rapid than the rate of change of the dipole field.

Geomagnetic field changes: indirect measurements

The Earth's magnetic field has been observed directly for a tantalizingly short period. What was happening to the field before the nineteenth century? To what extent are the trends observed over the past 140 years typical of field behaviour? Are there any field variations with characteristic time scales greater than a thousand to a hundred thousand years? Has the geomagnetic field always been dipolar? Has there always been a field, for that matter? These are important geophysical questions; and fortunately they can now be answered, at least in part, thanks to the remarkable phenomenon mentioned briefly at the beginning of this chapter–the magnetization of rocks. The study of such magnetization is called palaeomagnetism.

Almost all rocks contain very small quantities of iron compounds capable of acquiring a magnetization. As far as palaeomagnetism is concerned the most important magnetic minerals are the titanomagnetites, a series of compounds based on magnetite (Fe_3O_4) in which some of the iron has been replaced by titanium, and titanohaematite, the haematite (Fe_2O_3) series with similar, but generally less, titanium replacement. Palaeomagnetic studies have most often made use of basalts, of which the chief magnetic constituent is titanomagnetite, and sandstone, in which the chief magnetic mineral is titanohematite, although many other rock types have also been used. In some of the less common rocks studied palaeomagnetically the crucial magnetic component has not been an oxide at all but iron sulphide.

Igneous rocks acquire their magnetization, known as thermoremanent magnetization (TRM), as they gradually cool; the atomic-scale basic magnets within the magnetic grains become aligned in the ambient field direction as the rocks cool. Sedimentary rocks, by contrast, often acquire a detrital remanent magnetization (DRM); as the small magnetic particles settle through water they gradually orient themselves under the influence of the field. The real surprise, however, is not that rocks acquire magnetization in the first place but that they are able to retain it over tens, hundreds or even thousands of millions of years. The basic reason for this is that many rocks contain very small magnetic grains with very high coercive forces (the reverse magnetic fields needed to demagnetize the material)–high enough indeed

7.9: The strength of the vertical (Z) component of the non-dipole field for 1835, positive contours, black; negative, red. The units are 10^{-7} tesla.

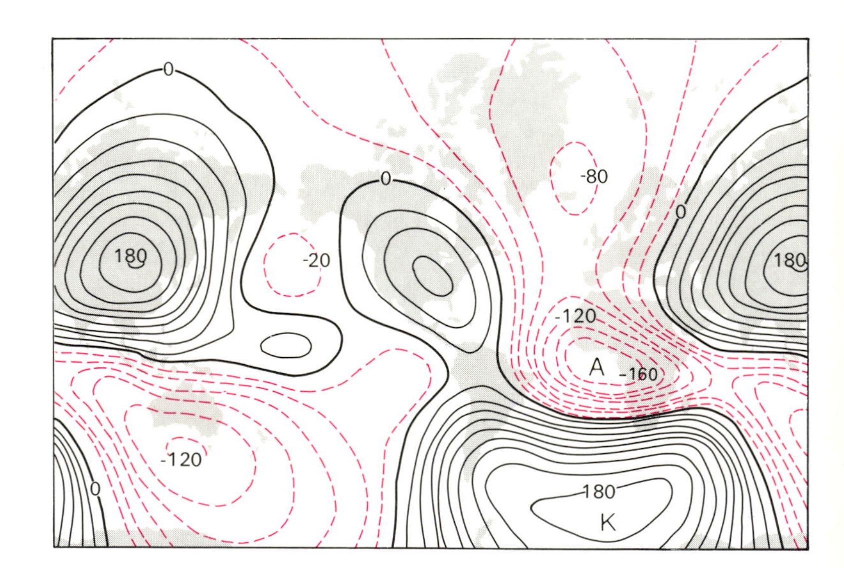

7.10: The strength of the vertical (Z) component of the non-dipole field for 1965. The units are 10^{-7} tesla. (Compare Figure 7.9.)

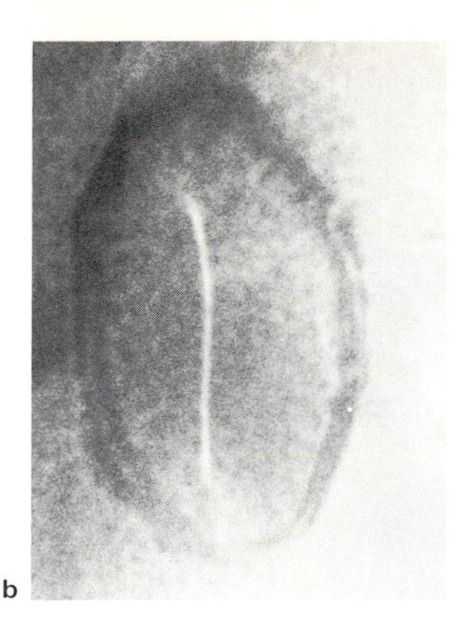

7.11: *The most common magnetic mineral in rocks is titanomagnetite, magnetite* (Fe_3O_4) *in which some of the iron* (Fe) *has been replaced by titanium* (Ti). (**a**) *A conventional electron microscope image of a grain in focus; the grain is about 2 μm long.* (**b**) *The same grain using a new technique in which the image is slightly defocused. The defocusing reveals that the grain is divided internally into sections known as 'domains' separated by 'domain walls' (the bright line).*

The material surrounding the magnetic grain is the non-magnetic mineral garnet, and the rock is a metadolerite from Greenland. Metadolerites are not frequently used for palaeomagnetic work, but one is shown here because it has given one of the best pictorial results so far using the new defocusing technique.

7.12: *Magnetic grains in igneous rock used for palaeomagnetic purposes vary widely in appearance. In this one, which is fairly highly oxidized, the titanomagnetite has been altered to titanohaemetite (bright yellow) and pseudobrookite (the yellowish brown elongated section). What remains of the original titanomagnetite (also yellowish brown) contains dark rods of the mineral spinel.*

to protect the original magnetization directions against most of the vicissitudes of subsequent geological history. Rock magnetizations may be very weak but they are also often very stable. They remain today, therefore, available to be measured and hence to divulge information on the directions (and in fewer instances the strengths) of the Earth's magnetic field in times past.

The oldest rock known to possess a magnetization dating from the time the rock was formed is almost three thousand million years old. Unfortunately, rocks of that age and older are relatively rare, and many of those that do remain are so highly altered that their magnetizations are suspect; but there is still a chance that even older original magnetizations will be found. In the meantime, it can be said with confidence that the Earth has had a magnetic field for at least two-thirds of its life of 4600 million years.

But has the field been predominantly dipolar for all that time? When it became possible, earlier this century, to detect and measure the weak magnetizations of rocks this was just the sort of question the new subject of palaeomagnetism set out to answer. The discovery that rocks have recorded ancient magnetic fields seemed to present an ideal opportunity to study the history of the field throughout the period covered by the rock record. However, matters turned out to be rather more complicated than that because of an unforeseen phenomenon, although it has nevertheless proved possible to discover something about the ancient field.

From declination and inclination of the ancient geomagnetic field at a particular time, as measured from orientated samples of rock, it is possible to determine the position of the corresponding ancient geomagnetic pole, or palaeomagnetic pole position. (Both poles may be determined, but for simplicity in discussion it is usual to refer only to the north geomagnetic/palaeomagnetic pole.) A set of pole positions thus derived from rocks less than seven thousand years old is shown in Figure 7.13. Although the pole positions are quite scattered, they are nevertheless clearly seen to be grouped around the geographic pole rather than the present geomagnetic pole. This is highly significant, for it seems to confirm the previous supposition that when averaged over a suitable time period the geomagnetic and geographic poles coincide. The average of the eleven palaeomagnetic poles lies very close to the geographic pole but bears no relation to the position of the present geomagnetic pole. The reason that the individual poles give such a great scatter is that no single one of them was derived from samples covering a sufficient period of time to allow the dipole wobble to be averaged out. In other words, each of the pole positions reflects the fact that over the short time interval covered by the rocks concerned the geomagnetic pole lay some distance from the geographic pole. Taken together, however, the eleven poles, covering a period of thousands of years, do more or less average out the dipole wobble.

Considering that the rocks involved came from a number of widely separated sites, the results in Figure 7.13 are an impressive demonstration that for the past seven thousand years at least the Earth's magnetic field has been mainly dipolar. Admittedly there is a slight trick involved here in so far as, in order to calculate pole positions from declination and inclination measurements, it is necessary to assume dipolarity of the field in the first place. But this is not as serious as it sounds. The fact that making the dipolar assumption leads to a close grouping of palaeomagnetic poles proves the point on the grounds of

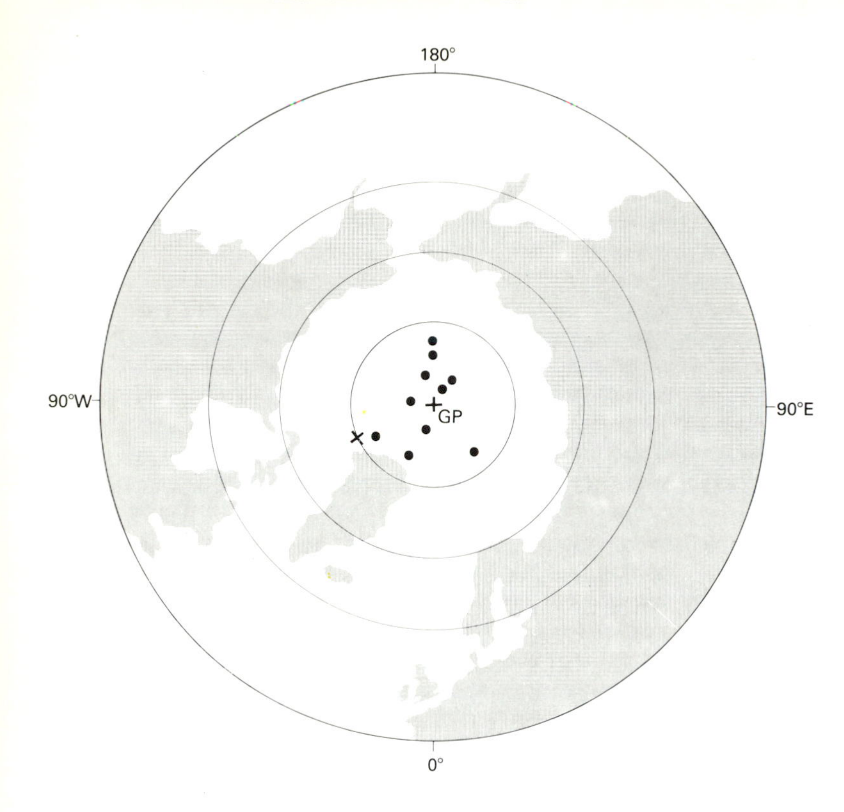

7.13: *Palaeomagnetic north poles obtained from rocks less than seven thousand years old. GP is the north geographic pole and X is the present north geomagnetic pole. Each palaeomagnetic pole shown is the average of poles obtained from a number of rock samples covering a period of at least several hundred years; so most, if not all, of the scatter that may have occurred from non-dipole field effects should have been averaged out. However, none of the rock sets covers a period sufficiently long to average out the effects of dipole wobble.*

consistency. If the field had not been mainly dipolar over the past seven thousand years, the calculated poles would have been scattered all over the globe.

A similar result is obtained from rocks up to several million years old, and the same conclusion may be drawn. Beyond that, however, something quite different begins to happen as the palaeomagnetic poles obtained from rocks deviate further and further away from the geographic pole. So far as a single continent was concerned, this phenomenon surprised, but did not disturb, the early palaeomagnetists. For a single continent the palaeomagnetic pole positions for rocks in order of decreasing age lie along a path that ends at the geographic pole. From this it would appear that over the past few hundred million years or so the geomagnetic pole has gradually moved to its present position from some distance away, a phenomenon given the name of polar wander. But, as has been seen, there is good reason to suppose that the geomagnetic pole should always have coincided more or less with the geographic pole. There appeared to be a conflict, therefore, between observation and theory, although the pioneers of palaeomagnetism were apparently not too upset at having to reject the theory.

What they could not live with so easily, however, was the subsequent realization that the rocks from different continents gave different polar wander paths, as Figure 7.14 shows. There can only be one north geomagnetic pole at any given time, assuming the field to be dipolar; and so if rocks of the same age from n different continents give n different pole positions then either polar wander cannot be the right explanation or for much of the past few hundred million years the geomagnetic field has been nothing like dipolar. It is possible to debate this particular conundrum for a considerable time; but, to cut a long story short, the simplest explanation consistent with all the data is that (a) the geomagnetic field has always been predominantly dipolar and (b) the continents have moved with respect to each other. The geomagnetic pole has remained fixed, give or take a little wobble, and the continents have drifted (Figure 7.15). It was the palaeomagnetic data

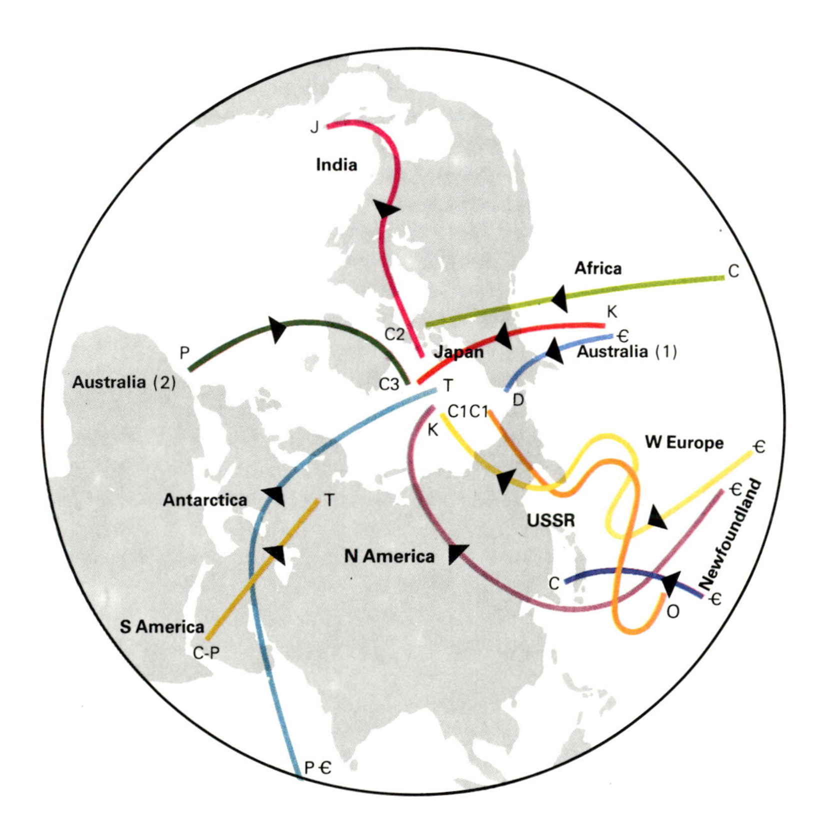

7.14: *Apparent (north) polar wander curves for rocks from different continents. The term 'apparent' is used because, taken together, the curves demonstrate that little, if any, polar wander has occurred; the chief phenomenon responsible for the lack of coincidence is continental drift. The following letters designate age: C3 (late Cenozoic); C2 (middle Cenozoic); C1 (early Cenozoic); K (Cretaceous); J (Jurassic); Tr (Triassic); P (Permian); C (Carboniferous); D (Devonian); O (Ordovician); € (Cambrian); P€ (Precambrian).*

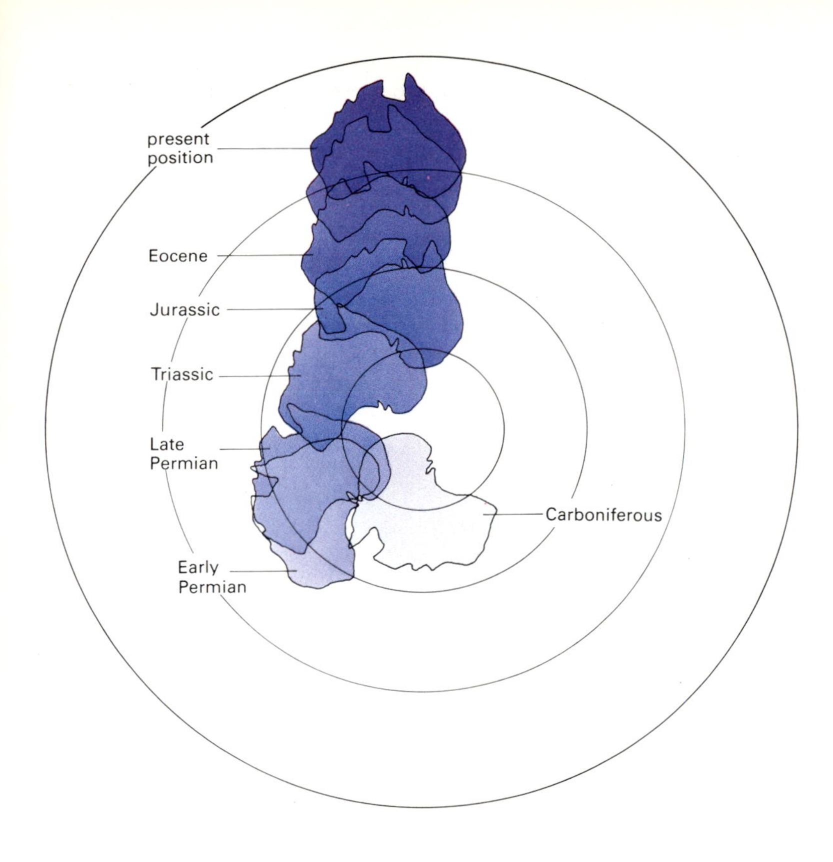

7.15: *The movement of Australia relative to the South Pole from Carboniferous times onwards (inferred from palaeomagnetic data).*

that finally convinced most Earth scientists that the continents have indeed drifted, thus bringing to an end a controversy that had lasted about forty years. Magnetic data are not the only support for drift, however, as Chapter 10 makes clear.

In the meantime there is more to be learned about the Earth's magnetic field as such. The field may have been essentially dipolar for at least several hundred million years and the dipole axis may have coincided on average with the rotational axis, but what about the strength of the dipole over that time? Unfortunately it is much more difficult to determine the strength of the ancient geomagnetic field than the direction—and for a very simple reason. When a rock is formed it acquires a magnetization in the direction of the Earth's field at the rock site and with a strength proportional to the strength of the field; or, to put it another way, magnetization is a vector quantity having both direction and magnitude. Now if a little, but unknown, amount of that original magnetization is subsequently lost for one reason or another (the rock may have been heated slightly, for example, thereby erasing part of the magnetization), the chances of being able to determine its original magnitude have gone for ever and with them the chances of being able to determine the strength of the field that produced it. On the other hand, the magnetization remaining will still lie in its original direction, a direction that can still be measured. The direction will be lost only if the original magnetization disappears completely; the total strength will be lost if only a bit of it does.

For this reason there are very few rocks from which reliable ancient field strengths may be obtained. On the other hand, it is rather easier to obtain field strengths from archaeological samples, such as kilns that have been baked to a high temperature, if only because they are relatively young and have thus had less time to become altered. Figure 7.16 shows the results from such materials covering roughly the past nine thousand years. The chief conclusion to be drawn is that, far from remaining constant, the geomagnetic field strength has been fluctuating with a typical period of the order of ten thousand years. It is not possible to extend these results much further back in time because well-dated materials closely spaced in time are not available; but what little evidence there is suggests that fluctuations in dipole strength have been occurring throughout the period covered by the rock record.

The discovery of field strength fluctuations was unexpected. A far more important, though equally unexpected, discovery was that of field reversals. The early palaeomagnetists soon found that some rocks (about 50 per cent, it became evident later) were magnetized in precisely the opposite direction than that to be expected on the basis of formation in the present geomagnetic field (which can, as will become clear, usefully be called the 'normal' field). Their first reaction was to suppose that these rocks actually acquired their magnetizations in the direction opposite to that of the ancient field or that, having acquired a magnet-

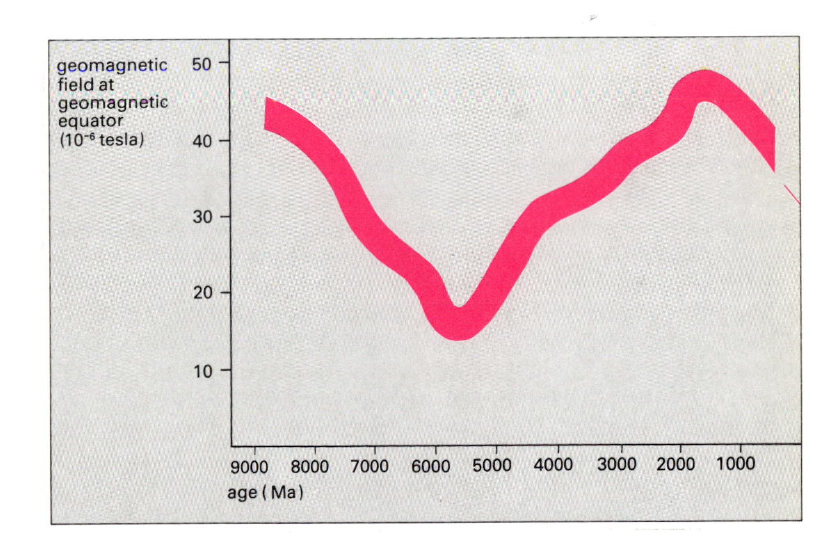

7.16: *Change in the strength of the geomagnetic dipole over the past eight to nine thousand years, based on values obtained from archaeological specimens. The numbers on the vertical axis actually refer to the dipole field strength at the geomagnetic equator. The short sloping line on the extreme right is the directly observed dipole strength decrease as shown in Figure 7.8.*

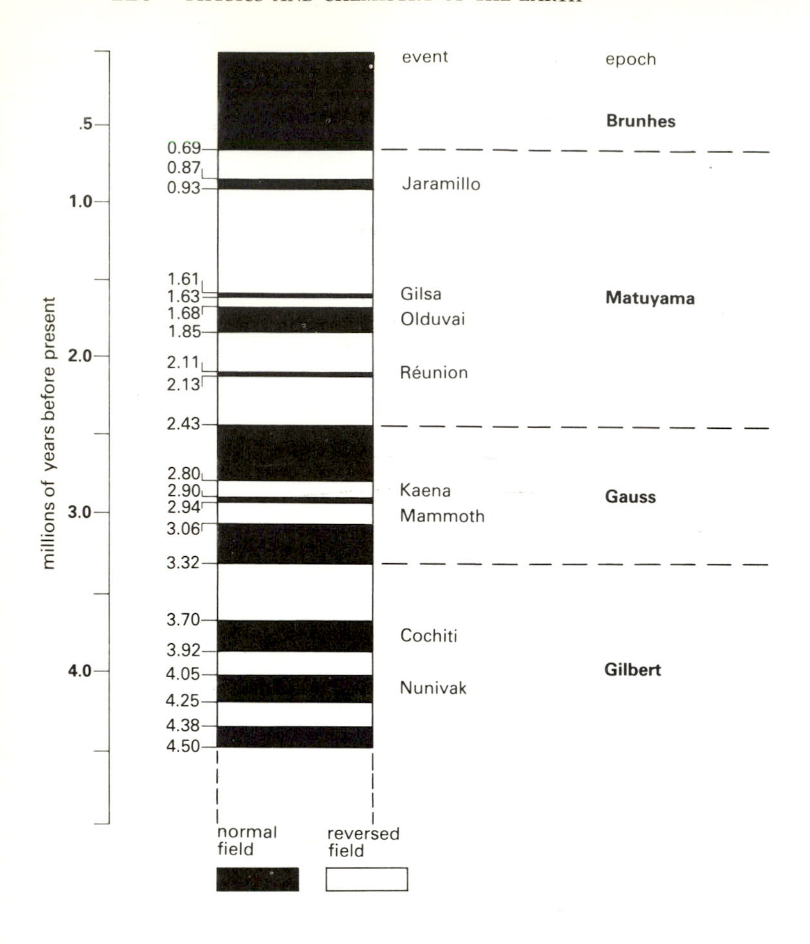

7.17: *The polarity-time scale for the past four to five million years, based on continental igneous rocks. The scale has been divided into epochs and events, which have been given names. The scale may not be complete; there may be many very short events yet to be discovered.*

ization in the same direction as the field, they subsequently underwent some process (chemical change, perhaps) whereby the magnetization reversed itself. These hypotheses together came to be known as self-reversal; and a number of quite plausible mechanisms were proposed by which self-reversal might take place. But despite exhaustive tests no convincing evidence for them could be found. There was, on the other hand, a quite different possible explanation for the existence of reversely magnetized rocks, namely that at the time the 'reversed' samples acquired their magnetizations the Earth's magnetic field was actually reversed in direction. In other words, it was suggested that, whereas the geomagnetic dipole now (in its 'normal' state) has its south pole in the northern hemisphere (corresponding to the so-called north geomagnetic pole) and its north pole in the southern hemisphere, there have been times in the past when the dipole pointed the opposite way (the 'reversed' state). This supposed phenomenon came to be known as field reversal.

When field reversal was first proposed it seemed less likely than self-reversal, although support for the idea grew as the inability to detect self-reversal continued. In any case it soon became clear that there was a simple test to decide the issue. The self-reversal process must depend upon the rock or rocks concerned; the ability of a rock to self-reverse its magnetization, if indeed it has such an ability, must be due to some property, possibly chemical, of the rock itself. It is reasonable to suppose, therefore, that in a collection of rocks of the same age some will have this property and some will not. If field reversal is valid, however, rocks of a given age must be either all normally magnetized or all reversely magnetized, for the field can never have been both normal and reversed at the same time. All rocks of the same age, irrespective of their chemistry or mineralogy, should thus have the same polarity.

And they do. This was demonstrated in a brilliant piece of work carried out largely by Allan Cox and Richard R. Doell of the US Geological Survey during the mid-1960s. Cox and Doell collected as many young continental igneous rocks as they could, measured their magnetic polarities (normal or reversed) and dated them accurately by potassium-argon dating (see Chapter 8). The results, updated somewhat, are shown in Figure 7.17. Because all rocks of the same age do indeed have the same polarity, it proved possible to build up a polarity-time scale showing precisely when the geomagnetic field was normal and when it was reversed, at least for the past 4.5 million years or so. For older rocks, unfortunately, the standard errors on the ages obtained begin to become significant in relation to the lengths of some of the shorter normal and reversed periods, which means that the older epochs and events cannot be defined with any precision. However, reversely magnetized rocks are known throughout the rock record; therefore it can be said with confidence that field reversal is a long-term fundamental property of the geomagnetic field and not just a recent aberration.

Moreover, although the polarity time scale cannot be extended further back using well-dated continental rocks, some progress can be made with the help of ocean sediments. When sediments are laid down on the ocean floor, they too become magnetized in the Earth's field direction, forming a series of superimposed layers of material that is normally and reversely magnetized alternately, reflecting the constant switching of the field between polarities. A vertical core drilled through the sediments will therefore contain a series of normally and reversely magnetized sections with the most recent field reversals recorded at the top of the core.

Unfortunately there are limits to what can be done with such material. For one thing sediments cannot generally be dated radiometrically, and even if they could the calculation of ages of rocks older than about 4.5 million years would suffer from the same precision problems as continental rocks. In the case of sediments fossil dating is often possible, of course, although it is seldom precise enough to enable reversal boundaries to be delineated with any accuracy. On the other hand, sediments do have the advantage that the relative chronology is clear, whereas continental rocks have to be collected from geographically random sites around the world. Sedimentary cores can also represent very long periods of time, although care must be taken to spot hiatuses in sedimentation.

Assuming no breaks in sedimentation in the crucial region towards the top of a core, comparison of the polarity pattern in the core with the continental polarity-time scale permits calculation of the average sedimentation rate and even some of its variations, if any, over the past four to five million years. To take a very simple example, if the distance from the top of a sediment core to the first reversal boundary is 1.38 m, this length corresponds to the 0.69 million years represented by the Brunhes normal epoch (Figure 7.17). The average sedimentation rate over that epoch in that particular core was therefore 1380/690 000 = 0.002 mm per year. Then, if the sedimentation rate is assumed to have always been constant, it is possible to extend the dating to many older reversals deeper in the core. On this basis, for example, a reversal taking place at a distance of 20 m from the top of the core could be dated at ten million years ago. Of course, assuming a constant sedimentation rate is dangerous in itself, but sometimes there are local geological factors by which the reality, or otherwise, of uniform sedimentation may be judged.

Reversals as recorded in sediments form the basis of magnetostratigraphy, the science of which will be developed in greater detail in Chapter 23. In the meantime this chapter passes on to another, and in fact much more important, type of field reversal record.

Magnetic anomalies

A magnetized rock is a magnet, albeit a weak one, and as such produces a magnetic field around itself. If a rock is close enough to the Earth's surface its magnetic field should be observable at the Earth's surface, where it will combine with the Earth's magnetic field. In other words the magnetic field observed in the area concerned will be a two-part field comprising (a) the general geomagnetic field in the area and (b) the local field from the nearby magnetized rock.

The field from the rock is known as an anomalous field, or magnetic anomaly, which simply means that it is a magnetic field present in addition to the general geomagnetic field. The word 'addition' may be misleading, however. Because fields are vectors, the combined field referred to above may be greater or smaller than the geomagnetic field acting alone. If the field from the magnetized body lies more or less in the same direction as the Earth's magnetic field at the site, the two fields will reinforce each other, and the total field will be greater than the Earth's magnetic field alone. The anomaly in this case is said to be positive. If, on the other hand, the two basic fields are opposite in direction, they will tend to cancel each other, and the total field will be smaller than the Earth's field alone. The anomaly is then negative.

The detection of a magnetic anomaly—in other words the discovery of a total magnetic field at the Earth's surface which differs from that to be expected from the geomagnetic field alone—is a sign that there is magnetized material below. In the seventeenth century, within only a few decades of William Gilbert's proposal of the existence of a general geomagnetic field, magnetic anomalies were already being used in Sweden to detect buried bodies of iron ore. But it was not until the twentieth century that anomalies became widely used to investigate much more weakly magnetized features such as buried hills, igneous intrusions, salt domes and concealed meteorites, using techniques that have also been adapted to detect buried artefacts such as pipelines. Magnetic anomaly investigations now play an important part in prospecting and exploration, especially in respect of local and comparatively small-scale features.

7.18: *The pattern of oceanic magnetic anomalies over part of the Reykjanes ridge to the south of Iceland. The black zones are positive anomalies, thought to be produced by normally magnetized rocks beneath. The white areas are negative anomalies, thought to be produced by reversely magnetized rocks beneath. The central anomaly is positive because the rocks now rising at the ridge axis, cooling and acquiring magnetization are being magnetized in the Earth's present magnetic field.*

During the 1960s, however, magnetic anomalies on a much larger scale also became important, playing a vital role in the Earth sciences revolution from which arose a widespread belief in the mobility of continents. The fact is that the magnetism of rocks not only provided the crucial evidence in favour of continental drift, it also enabled the drift of continents to be placed in the wider context of plate tectonics by confirming the theory of ocean floor spreading. Anticipating Chapter 10, it is necessary to envisage here the early stages of ocean floor spreading. The world's ocean floors are host to a linked system of active oceanic ridges with a total length of tens of thousands of kilometres. At these ridges molten material from the Earth's interior rises, cools and solidifies. It then splits and moves laterally away from the ridges in both directions as new material follows on from below, thus forming the spreading ocean floor.

When the molten material cools and solidifies it acquires a thermo-remanent magnetization in the direction of the Earth's field at the time, which means that sometimes the magnetization is normal and sometimes it is reversed. The consequence is that the ocean floors comprise an alternating sequence of normally and reversely magnetized linear blocks parallel to the ridges. These blocks, in turn, give rise to an alternating system of linear positive and negative magnetic anomalies above. An example of such anomalies is shown in Figure 7.18. It was the discovery and analysis of such anomalies during the mid-1960s that convinced most Earth scientists that the hypothesis of ocean floor spreading, proposed several years before on quite different grounds, was valid.

Further, just as the pattern of reversals in an oceanic sediment core may be used to determine the average sedimentation rate, so may the pattern of reversals observed above the oceans be used to determine the average rate of ocean floor spreading. Again to take a very simple example, if the distance from a ridge axis to the first reversal boundary revealed by the anomalies is 13.8 km, this distance corresponds to the 0.69 million years represented by the Brunhes normal epoch (see Figure 7.17). The average spreading rate over that epoch in that particular ocean was therefore 13 800 000/690 000 = 20 mm a year. Assuming a constant spreading rate, much older reversals may again be dated, as they were with ocean sediments supposedly forming at a uniform rate.

Magnetic anomalies have now been observed in all the world's major oceans, just as magnetized sediment cores have, and just as magnetized rocks have been found on all continents. Moreover, these indications of past geomagnetic fields are perpetually being joined by permanent records of present fields in the form of new magnetized igneous and sedimentary rocks. But where does all this magnetism originate? What causes the Earth's magnetic field in the first place?

Origin of the geomagnetic field

Any viable theory for the origin of the Earth's magnetic field must obviously be able to explain all the known characteristics of the field. However, it must also be consistent with what is known about the structure and properties of the Earth from other sources. This places severe constraints on any model of field origin. When all the constraints are taken into account it perhaps comes as little surprise that, although some effects of the geomagnetic field were known to the Greeks more than two thousand years ago, Earth scientists had little idea of the cause of the field as recently as the 1940s. Since then, moreover, there have been new discoveries of phenomena (field reversals, for example) which must be added to the list of field characteristics to be accounted for.

It would perhaps be useful to summarize what these characteristics are. The Earth's magnetic field has existed for at least three thousand million years. As far as we know, the field has been predominantly dipolar throughout that period and on average has been axial (along the Earth's rotational axis). The average conceals short-term effects, however, for the dipole axis has apparently 'wobbled' slightly (perhaps up to a maximum of about 11°) about the rotational axis in such a way as to average out to an axial direction only over a time of at least ten thousand years. The strength of the dipole also varies, exhibiting fluctuations on a similar time scale. On a much longer time scale (a hundred thousand to a million years), on the other hand, are reversals of the dipole, a phenomenon apparently characteristic of the geomagnetic field throughout the whole of its life. Equally persistent has been the much smaller non-dipole field, the components of which vary much more rapidly with time scales of a hundred to a thousand years.

These facts show why it is so difficult to envisage a field production mechanism. For example, the simplest possible explanation for the geomagnetic field is that the Earth is a huge permanent magnet. This theory was proposed by William Gilbert in AD 1600 and had no rival until very recently. Its great attraction was based on the experimental facts that the shape of the magnetic field produced by the Earth is essentially the same as that produced by a uniformly magnetized sphere and that the latter is identical to the field produced by a small magnetic dipole at the centre of the sphere. In other words, a permanently magnetic Earth would explain the main feature of the geomagnetic field—its dipolarity.

But it would explain little else. For example, how could permanent magnetism explain the variations in the dipole's strength and direction? And while it is just possible to envisage the non-dipole field as the results of anomalous magnetic bodies embedded in the Earth, how could permanently magnetic bodies of whatever size account for the non-dipole field variations, which are even more rapid than the dipole variations? The upheavals that the Earth would have to undergo if solid permanent magnets were to move to produce field variations on time scales of a hundred to a thousand years would be such as to shatter the planet. Furthermore, even forgetting the variations, it is not possible for magnetic rocks in the Earth to account for the magnitude of the observed field. Every magnetic material possesses a characteristic temperature, known as the Curie temperature or Curie point, above which it is no longer capable of possessing magnetization. For iron-bearing minerals the Curie points are all lower than 800°C. Yet temperature increases with depth in the Earth at an average rate of about 30°C per kilometre. In other words, within a few tens of kilometres of the Earth's surface, no material could possibly be magnetic. If the geomagnetic field is the result of permanent magnetism, the magnetism must reside in a thin outer shell of the Earth. In theory this does not present a difficulty, for such a shell would also give a dipole-shaped field, but the practical fact is that the rocks in the Earth's outer shell are not magnetized strongly enough.

With the failure of the idea of permanent magnetism Earth scientists proposed a number of other theories, none of which proved successful. In 1947, to take but one of the more important examples,

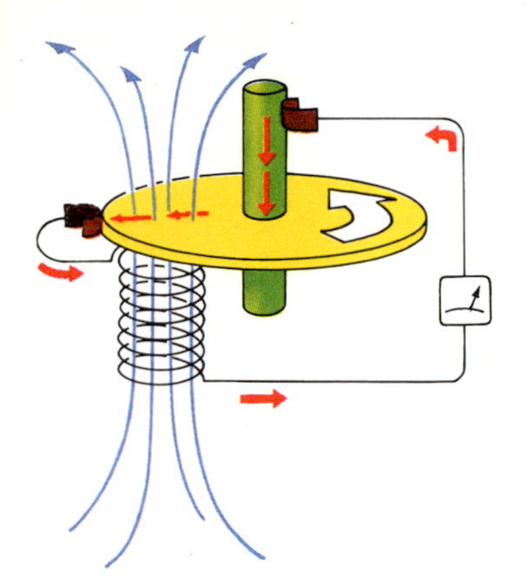

7.19: *A mechanical model of the self-exciting dynamo. The copper disc is kept rotating about the vertical shaft. The shaft and the outer edge of the disc are connected via brushes to make a complete circuit including a coil. As long as the disc is kept rotating in the direction shown current will flow in the direction indicated by the straight arrows, producing a magnetic field in the coil as represented by the curved arrows.*

Professor P. M. S. Blackett suggested that magnetism is simply a fundamental property of rotating bodies, in other words that rotating bodies may produce magnetic fields just because they are rotating. This idea was inspired in part by the negative factor that plausible theories of field origin were proving difficult to imagine and in part by the more positive point that certain rotating stars demonstrably produce magnetic fields. But it did not work. For one thing, very sensitive experiments designed to detect magnetic fields from rotating gold cylinders proved negative. Moreover, Blackett's hypothesis predicted that the horizontal component of the geomagnetic field should decrease with depth in the Earth whereas observation shows it to increase. Nor, finally, would rotational magnetism be any more successful than conventional permanent magnetism in accounting for rapid variations in the Earth's magnetic field.

It is easy to be wise after the event, but the rapid variations in the Earth's dipole and non-dipole fields should have given a clue to the seat of the geomagnetic field much sooner than they did. For the only conceivable source of such variations is a fluid part of the Earth, and that must mean the liquid outer core. Moreover, the outer core is known to be largely iron, or at least a good conductor; and it has long been known that moving conductors, electric currents and magnetic fields are no strangers to one another. The circumstances in the fluid outer core are thus promising, although that is far from suggesting that it is an easy task to envisage what actually occurs there to generate the geomagnetic field.

In fact it may never be known precisely how the geomagnetic field is produced. What we can do, however, is to demonstrate that there are possible field generation processes; and in this connection it is useful to think in terms of a mechanical analogy such as that shown schematically in Figure 7.19. The large copper disc is set rotating in an initial magnetic field. Since, when a conductor is moved in a magnetic field, an electric current is generated in the conductor, a current is thus produced in the copper disc. This current is then fed through a circuit which includes a coil. But the current in the coil generates a magnetic field which, because of the coil's position, cuts the rotating disc. Thus even more current is generated in the disc, the new current also passes through the coil, yet more magnetic field is produced, and so on. In short, so long as the disc is kept rotating (and so long as there was an initial magnetic field just to set the process off), the system will keep generating a magnetic field for ever. This is not a perpetual motion machine, because a constant energy input is required to keep the disc rotating. It is a dynamo which produces its own magnetic field; in other words, it is not just a simple dynamo but a self-exciting dynamo.

Of course, there are no rotating copper discs, coils or connecting wires in the Earth's outer core. But there are equivalents. The conductor is the iron of the core itself; and the iron is almost certainly in motion, probably in the form of convection generated by the release of radioactive heat. And once in motion the core will contain its own current loops, threaded by magnetic lines of force. The situation in the core must be extremely complex; and it is difficult, if not impossible, to describe its field generation mechanism in mathematical terms. Nevertheless it has been possible to show mathematically in a general way that there do exist motions in a spherical fluid conductor that could give rise to self-exciting dynamo action. This does not mean that these particular motions are the ones actually occurring within the Earth; but they do show that field generation in the core is at least feasible.

Whatever the form of the circulation in the core the motions are likely to be influenced by the Earth's rotation, which makes it quite reasonable that the geomagnetic dipole axis and the rotational axis should coincide on average. Moreover, there seems to be no reason why the magnetic field should be able to distinguish between the two (north-pointing and south-pointing) directions along the rotational axis; and so it is equally reasonable that the dipole should point in each direction for about 50 per cent of the time. Finally, the motions in the body of the core are likely to be modified where they come into contact with the stationary mantle above, giving rise to eddies. It is quite easy to see in these rapid and random perimeter movements the source of the non-dipole field and its variations.

In short, although the details may never be known, the fluid outer core seems to be the only part of the Earth able to account in principle for all the geomagnetic field characteristics.

8 Trace element and isotope geochemistry

Over eighty naturally occurring elements together make up less than one weight per cent of the planet Earth. Yet many of these so-called trace elements have an importance that far outweighs their abundance. Their very scarcity gives them value, and considerable time and money have been invested in studying the natural processes by which they have become concentrated into ore bodies. Organisms have adapted to their low concentrations so that minute amounts may be necessary for life (iodine and cobalt), while comparatively small increases in concentration (arsenic and lead) can be poisonous. A few emit radiation and thus play vital roles in being responsible for the heat within the Earth (see Chapter 9) and in fulfilling our energy requirements; some are now used as radiometric 'clocks', thus enabling us to date geological events throughout Earth history. Finally, much has been learnt from the study of trace elements because of their sensitivity to their physical and chemical environments. Unlike major elements, trace elements rarely form minerals of their own, and are therefore forced to seek 'sites' in slightly alien crystal structures. However, it is precisely because of this need for compromise between the nature of the trace element and the proffered sites in the available minerals that the study of trace elements is so invaluable. Their distribution between minerals and any fluid (eg seawater, freshwater, magma) is extremely sensitive to parameters such as the composition of the fluid, oxidation state, temperature and to a lesser extent pressure, in addition to the chemistry of the minerals. Some of the trace elements most widely used in geology are listed in Figure 8.2.

Analytical techniques

Rapid developments in technology over the past twenty years are reflected in the wide range of analytical techniques available to Earth scientists. X-ray fluorescence spectrometry has already been discussed (Chapter 5) in the context of major element analyses. It is also widely used to determine trace element abundances; the sample powder is not diluted with a flux but is simply compressed into a disc and loaded into the spectrometer. Most elements which are present in more than five parts per million (ppm) may be analysed and the precision is controlled primarily by the length of time over which the operator is prepared to measure the intensities of the secondary (or fluorescent) X-rays.

Two other physical techniques that are becoming increasingly popular in the search for more precise trace element analyses are neutron activation and mass spectrometry.

Neutron activation analysis

This technique depends on the 'activation' of the elements of interest by bombardment of the sample with fast-moving neutrons in the core of a nuclear reactor. Neutrons are added to the nucleus of stable isotopes, to produce new and unstable radionuclides, which then decay emitting particles with characteristic energies which can be discriminated and measured with a scintillation counter. For example, the concentration of uranium may be determined by activating the common isotope ^{238}U to ^{239}U and then measuring its decay, thus:

$$^{238}_{92}U + 1 \text{ neutron} \rightarrow {}^{239}_{92}U \xrightarrow{\beta 1 \text{ decay}} {}^{239}_{93}Np \xrightarrow{\beta 1 \text{ decay}} {}^{239}_{94}Pu$$

The two β^- decays of ^{239}U and ^{239}Np are time-dependent: they have half-lives (ie the time taken for half the original number of atoms to decay) of 23.5 minutes and 2.35 days respectively, and so measurements must be made soon after activation.

In practice many samples may be packed into a single reactor can. They are loaded, together with standards, in a specific orientation thus permitting any variations in the neutron flux across the can to be determined accurately. The technique is more time-consuming and less flexible than X-ray fluorescence, but it is extremely sensitive (detection limits are often less than 1 ppm) and is widely used to analyse for elements such as barium (Ba), hafnium (Hf), scandium (Sc), uranium (U) and thorium (Th), and the rare earth elements.

Mass spectrometry

A mass spectrometer (Figure 8.1) separates charged atoms and molecules on the basis of their masses. It consists of a source of positively charged ions, a magnetic analyser and an ion collector all of which are evacuated to pressures of the order of 10^{-9} to 10^{-12} atm. Both gaseous and solid samples can be analysed. A gaseous sample is allowed to leak into the source through a small orifice whereupon the molecules are ionized by bombardment with electrons. Solid samples are deposited as a salt on a filament (usually of tantalum (Ta) or rhenium (Re)) which is then heated electrically to temperatures sufficient to ionize the element to be analysed, perhaps 1500–2000°C. In both cases the resulting ions are accelerated by an adjustable voltage and are collimated into an ion beam by a number of suitably spaced slit plates.

The unresolved ion beam enters a magnetic field (generated by an electromagnet) which deflects the ions into circular paths whose radii are proportional to the masses of the isotopes, the heavier ions being deflected less than the lighter ones. The separated ion beams then continue down the analyser tube to the collector, a metallic cup positioned behind a slit plate. By adjusting the accelerating voltage in the source and/or the magnetic field, one ion beam at a time can be focused on the collector cup where it is neutralized by electrons which flow from ground through a resistor (10^{10} to 10^{12} ohms). The voltage difference generated across the terminals of the resistor is magnified, then measured by an extremely sensitive voltmeter whose output may be fed to a strip chart recorder or converted into digital output.

Mass analysis of an element (or compound) consisting of several isotopes (isotopic masses) is usually obtained by varying the magnetic field so that the separated ion beams are successively focused into the collector. The resulting signal, traced for example on a chart recorder, consists of a series of peaks and valleys which form the element's mass spectrum. Each peak represents a discrete isotope whose abundance is proportional to its relative height. If the output from the collector is digitized then the magnetic field will be switched in a series of jumps, rather than continuously, so that the intensity of the ion beam can be measured for a fixed period of time.

In the Earth sciences mass spectrometers have two main uses. First, they can measure precisely differences in isotope ratios of an element caused by radioactive decay or mass fractionation. High precision is obtained by having a very stable ion beam and then repeatedly cycling around the mass spectrum so that many isotope ratios are measured for each sample. For example strontium (Sr) and neodymium (Nd) isotope ratios can be determined to ± 0.005 per cent, although such an analysis might take three to seven hours. Second, they enable the concentration of an element to be determined by a technique called isotope dilution. If a sample whose isotope ratios are known is mixed with a spike which has the same isotopes but in different proportions, then the isotope ratios of the mixture depend on the concentrations of the element in the sample and in the spike. Provided that the concentration in the spike is known then that in the sample may be calculated. This technique is extremely sensitive since only 10^{-7} to 10^{-8} g of the element needs to be loaded on the filament and it is potentially available for any element having more than one isotope. Trace elements routinely measured by this technique include potassium (K), rubidium (Rb), barium (Ba), strontium (Sr), uranium (U), lead (Pb) and the rare earths lanthanum (La), cerium (Ce), neodymium (Nd), samarium (Sm), europium (Eu), gadolinium (Gd), dysprosium (Dy), erbium (Er), ytterbium (Yb) and lutetium (Lu). Errors can vary from 0.1 to 3 per cent depending on the element being measured.

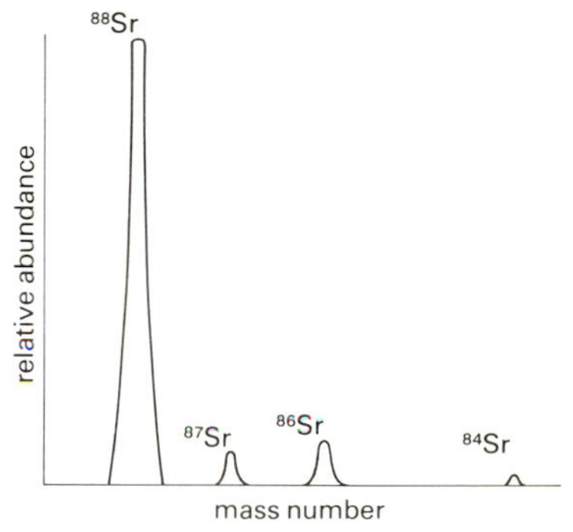

8.1: *Left, a solid-source mass-spectrometer. The sample is placed in the cylindrical chamber (centre) from where the ion beam is deflected by the large magnet towards the collector (nearest the camera). The instrument is controlled and the data are analysed by the electronics in front of the operator. The spectrum, above, appears on the plotter below his left hand. The example shows a typical analysis of Sr isotopes.*

Igneous processes

The behaviour of trace elements in igneous processes is usually discussed in terms of their distribution between the different minerals, the melt or magma, and any volatile-rich phase that may be present. More formally the distribution of a trace element between two co-existing phases is described by its partition coefficient K_D, where

$$K_D = \frac{\text{concentration of element in phase 1}}{\text{concentration of element in phase 2}}$$

and is constant for any particular equilibrium. For a mineral in equilibrium with a liquid, partition coefficients are conventionally written so that the concentration of the element in the mineral is divided by that in the liquid: thus the size of the partition coefficient reflects the ease with which an element can substitute into a particular mineral.

Factors controlling trace element substitution

The trace elements that find it easiest to obtain sites in a crystallizing mineral are those that have ions of similar size and charge to one of the major elements in the crystal structure. Thus Rb^+ (147 pm) substitutes for K^+ (133 pm), and Sr^{2+} (112 pm) substitutes for Ca^{2+} (99 pm) in a wide range of silicate minerals. However, the details of the relationship between ease of substitution (as reflected in the K_D), ionic charge and ionic radius may be most simply illustrated on a diagram of partition coefficient plotted against ionic radius.

Figure 8.3 shows the distribution of a number of elements between pyroxene phenocrysts and groundmass crystals (believed to represent the liquid) in a basalt from Tahiti. The pyroxene structure obviously finds it easiest to incorporate divalent ions of ionic radii 79 and 101 pm, which are slightly smaller than its major lattice-forming cations, namely Mg^{2+} and Ca^{2+}.

Since different minerals have different crystal structures, different sites are available for trace element substitution. Curves of K_D against ionic radius for all the common minerals would each have a characteristic pattern and show a distinctive imprint on any liquids with which they have been in equilibrium. The majority of trace elements in Figure 8.2 tend to have very low partition coefficients between common minerals and basic (low silica) magmas. Exceptions for which $K_D > 1$ include Ni in olivine, Y in garnet, and both Sr and Eu in plagioclase feldspar, and since these are rare they are often diagnostic of the particular mineral. Elements with $K_D > 1$ are often described as 'compatible', whereas those with $K_D < 1$ are called 'incompatible' because they find it difficult to fit into the crystal structure.

Partial melting and fractional crystallization

Igneous rocks typically consist of several different minerals, each with its own melting point. The rocks therefore melt and crystallize over a range of temperature and pressure in which the coexisting crystals and liquid interact. The distribution of particular trace elements will depend largely on which minerals are present and the nature of the 'sites' available therein.

8.2: *Trace element abundances in common rocks and minerals (parts per million; dashes indicate abundances of less than 0.01 ppm).*

	Rubidium (Rb)	Strontium (Sr)	Nickel (Ni)	Chromium (Cr)	Zirconium (Zr)	Yttrium (Y)	Neodymium (Nd)	Samarium (Sm)	Lead (Pb)	Uranium (U)
Rocks										
peridotite	0.2	10	1 800	2 500	10	3	0.5	0.2	1	0.05
basalt										
mid-ocean ridge	1	100	200	450	100	30	10	3	10	0.5
intraplate	40	550	50	40	200	35	50	8	35	6
andesite	35	400	5	15	150	30	30	7	6	1
granite	100	250	9	5	100	15	50	9	20	5
shale	140	300	60	90	160	26	30	6	20	4
sandstone	60	20	2	35	220	40	15	3	7	0.5
limestone	3	700	20	11	20	30	1	0.2	9	2
Minerals										
olivine	–	0.1	2 000	1 000	1	0.3	–	–	–	–
pyroxene	–	5	200	4 500	5	10	2	0.08	0.1	–
garnet	–	0.1	20	450	30	60	2	0.9	1	–
plagioclase feldspar	0.07	220	0.1	0.1	1	0.3	0.02	0.01	1	–

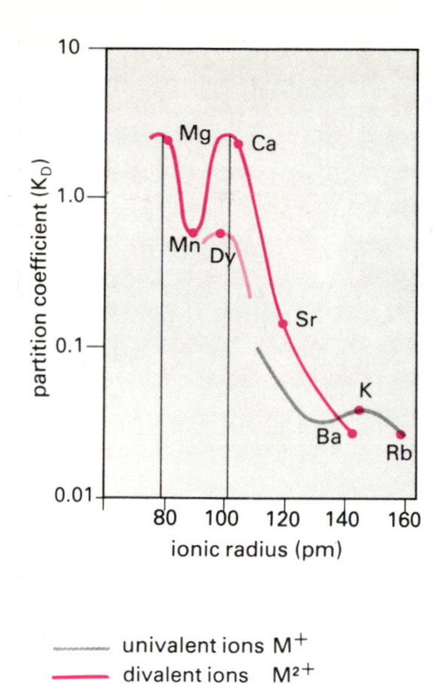

8.3: A graph of partition coefficient (K_D) against ionic radius for a number of ions distributed between pyroxene crystals and the groundmass in a basalt from Tahiti. The size of K_D reflects the ease with which a particular element can substitute into the pyroxene crystal lattice. The curves include two maxima at about 79 and 101 pm which represent the optimum ionic sizes for substitution in the pyroxene crystal structure. Since the curve for divalent ions (M^{2+}) is higher than that for trivalent ions (M^{3+}), the former are more readily accepted into pyroxene crystals. The size of the ion can be more important than its charge: eg Sr^{2+} finds it more difficult (ie has a lower K_D) than Dy^{3+} whose ionic radius is much closer to the optimum value of 101 pm.

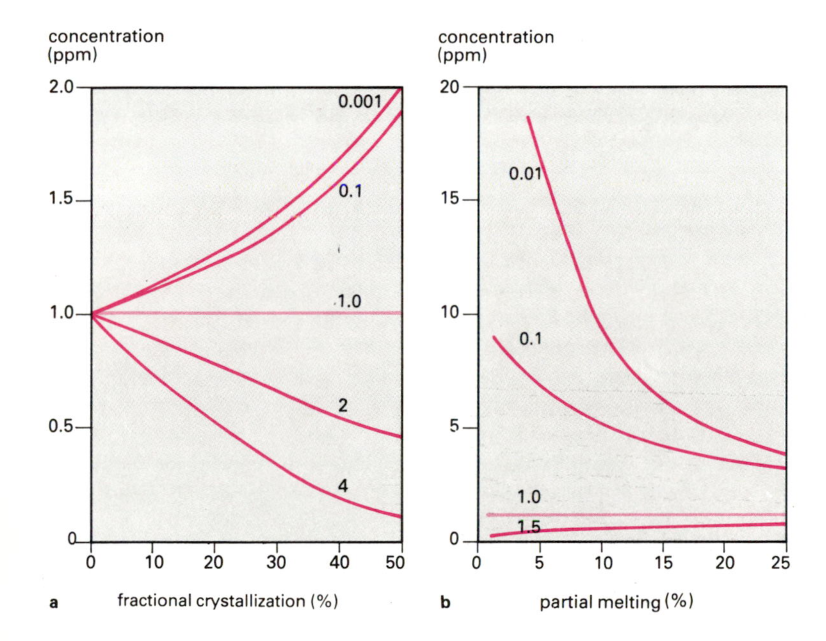

8.4: Calculated relative concentrations of trace elements in a liquid after different degrees of partial melting and fractional crystallization. The original solid and liquid respectively contained 1 ppm, and different curves reflect particular partition coefficients, K_D. For both partial melting and fractional crystallization, elements with small K_D are concentrated in the liquid, whereas those with large K_D (>1) remain in the crystals and are thus depleted in the liquid.

Figure 8.4 illustrates the calculated concentrations of trace elements (with different partition coefficients) in the liquid after various amounts of partial melting and fractional crystallization. In both cases elements with $K_D < 1$ are concentrated in the liquid; those concentrations are very sensitive to the amount of fractional crystallization or partial melting, and become more dilute as the volume of liquid increases. Conversely elements with $K_D > 1$ will tend to remain in the crystals (while they are present) and so will have correspondingly low concentrations in the liquid.

For many problems it is also useful to consider how the ratio of two trace elements varies during fractional crystallization and partial melting. Incompatible elements ($K_D < 1$) tend to have very similar distribution patterns in the liquid, even though their partition coefficients may vary by a couple of orders of magnitude: compare the curves for $K_D = 0.001$ and 0.1. Thus whereas the concentrations of incompatible elements change dramatically as partial melting or fractional crystallization proceeds, the ratio of two such elements remains relatively constant. In general provided that the volume of liquid exceeds about 10 per cent, the ratio of two incompatible elements is similar to that in the original magma in the case of fractional crystallization, or its source rock in the case of partial melting. It follows that we may estimate the ratio of two incompatible elements in the source of an igneous rock with much more confidence than their concentrations can be estimated.

Many igneous rocks become altered and/or metamorphosed after they have crystallized and thus it is prudent to consider elements that are little affected by such secondary processes, for example zirconium (Zr) and yttrium (Y). Figure 8.5 illustrates the variation of Zr/Y and Zr (ppm) observed in recent basalts erupted in oceanic areas. Those of Zr/Y in particular are too great to be attributed to differences in the amount of partial melting and fractional crystallization during the formation of these basalts, and indicate that the upper mantle source rocks must have significantly different Zr/Y ratios. The important conclusion, first demonstrated by Professor Paul Gast in the 1960s, is that the upper mantle is not chemically homogeneous, but contains variations, particularly in trace element abundances and their ratios.

With time some of these trace element differences result in differences in isotope composition (see below). The nature of such variations, how they are generated and how long they have existed are vital questions in our efforts to understand the evolution of the outer regions of the Earth. They might reflect material left behind when new crust is formed, or the partial reassimilation of crustal material when it returns to the mantle along destructive plate margins. How long chemical variations can persist will then depend on the mixing efficiency of convection in the upper mantle.

In summary, igneous processes are primarily responsible for the present distribution of elements within the Earth. In particular, since melts migrate to the outer portions of the Earth the crust is comparatively enriched in these incompatible elements (eg K, Rb, Sr, Ce) but depleted in compatible elements (eg Ni, Cr) which tend to remain behind in residual crystals in the upper mantle.

Ore processes

Our industrial requirements now include most of the minor and trace elements. Although present in almost every rock, they usually occur in such small amounts (perhaps one gram in a tonne of rock) that their extraction is clearly impractical. It is therefore important to study how and where they have become concentrated by the natural processes known as ore genesis.

An ore is simply a rock in which the concentrations of certain elements are high enough for it to be a worthwhile mining proposition. These elements are concentrated into particular minerals (ore minerals). It is these that are mined and whose formation is important in the study of ore genesis. The ore minerals may be igneous, metamorphic or sedimentary: they may crystallize from unusual fluids or be concentrated by mechanical processes as are grains of sand on a river bed.

Igneous processes are primarily responsible for the distribution of elements within the Earth and trace elements tend to be concentrated either in the higher-temperature, early-crystallizing minerals, or in the coexisting liquid. Ore bodies may form both by mechanically concentrating early formed minerals which contain economic compatible elements, and at the place where for some physicochemical reason the liquid rich in incompatible elements finally crystallizes. One of the best examples of the former is in the Bushveld complex in southern Africa. About two thousand million years (Ma) ago a large volume of basic magma intruded older sedimentary rocks in the form of a huge saucer-shaped magma chamber. As the magma cooled, a mixture of silicate and oxide minerals crystallized which, being dense, accumulated at the bottom of the chamber. Today they occur as layers of minerals (Figure 8.6) some of which may be traced for many miles. One of the minerals of economic importance is chromite ($FeCr_2O_4$), the source of most of the world's chromium.

The second process of trace element concentration occurs as crystallization proceeds and incompatible elements remain in the liquid, their concentration increasing as the volume of liquid shrinks. One such element is hydrogen, and since it readily combines with oxygen to form water, water may also be thought of as incompatible. Thus granites contain more water than the basalts from which they are often derived. Eventually, when the concentration of water is too great for it all to dissolve in the silicate magma, it separates out into a water-rich fluid. This upsets the pre-existing distribution of trace elements between the silicate magma and the crystallizing minerals, since many elements prefer the new conditions offered in the water-rich fluid. Such a fluid is also more mobile than a silicate magma and can be injected into fine cracks in the surrounding rocks, where it solidifies as pegmatite (Figure 8.7). Most pegmatites are not economic, but many contain

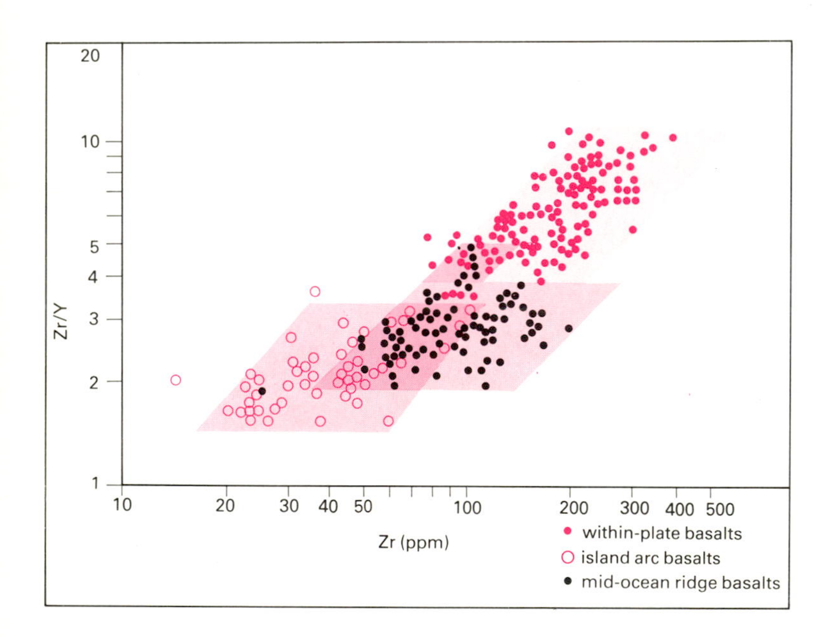

8.5: *Variations in Zr/Y and Zr contents in young basaltic rocks erupted in oceanic areas. The Zr contents reflect both the concentrations in their source rocks and the amount of partial melting and fractional crystallization, but the latter have little effect on the Zr/Y ratios which are therefore similar to those in their source rocks. Differences in Zr/Y, in particular, are attributed to variations in the upper mantle source regions of these basic volcanic rocks; and rocks erupted in different tectonic environments appear to have been derived from source rocks with different trace element characteristics.*

8.6: *Layers of minerals including chromite in the Bushveld Complex, Dwarb River, north Transvaal, South Africa.*

unusual minerals, and they are the sole source of some rare elements.

A similar but much larger-scale process is the build-up of magmatic fluids in areas of high magmatic activity. We know from the gases released and the minerals formed around volcanic vents (Figure 8.8) that magmas contain many volatile materials (eg carbon dioxide, sulphur and halogens, in addition to water) which are released spontaneously when the magma nears the surface and the pressure is reduced. The build-up and movement of such fluids at depth in the crust can concentrate particular elements, forcibly invade and break up the surrounding rocks and precipitate often large ore bodies. Probably the best examples are the sizable copper deposits of the South American Andes.

Similar, at least in principle, to those deposits formed by magmatic waters are those concentrated by hot hydrothermal fluids. Usually the water starts at or near the Earth's surface; it might be seawater, or water derived directly from the atmosphere (meteoric water). It is heated up, concentrates certain elements as it passes through the crust and then deposits them in economic ore bodies. Unlike the processes discussed previously this entails scavenging elements from pre-existing rock, and it is therefore a secondary process.

The spectacular geysers and hot springs of areas such as the western USA and northern New Zealand are ample evidence that hydrothermal fluids do occur. Beneath the oceans too they have been

8.7: *Pegmatites in outcrop and in hand specimen. Above, sheet of acid (silica-rich) pegmatite intruded into metamorphosed diorite, Guernsey, English Channel. Right, hand specimen of pegmatite from Portsoy, Banff, Scotland. The large crystals are of pink feldspar, white quartz, shiny yellow mica, black tourmaline and brownish sphene.*

reported from many mid-ocean ridges, since the first discovery of hot brines on the floor of the Red Sea in 1963. More recently the hot springs of the Galápagos rise have received special attention because of the unique ecological community they support (Figure 19.13). The hot water being expelled is seawater modified by its passage through the oceanic crust; some of the elements it picks up are redeposited as sulphides within the crust, others are clearly added to the oceans and hence help to determine the trace element composition of seawater itself. Much of the evidence underlying this description comes from a study of stable isotopes (see Figure 8.11).

Sedimentary ore deposits are also secondary, involving the reworking and redistribution of pre-existing material. This may take place chemically when elements soluble in near-surface waters are leached out, and has obviously occurred when rock is weathered and crumbling. However, extremely mobile elements such as uranium have often been leached from rocks which appear fresh and unaltered. To form an ore body conditions must then change so that the soluble element becomes insoluble and precipitates. For example, although uranium is extremely mobile under oxidizing conditions, it quickly becomes insoluble and drops out of solution when the conditions become more reducing (Figure 8.10).

Sedimentary processes may also concentrate material mechanically into ore deposits. The action of transporting grains in water sorts them on the basis of size, weight and shape, a process utilized by the small-time prospector 'panning' for gold. Economic deposits formed by this sorting process are known as placer deposits.

8.8: *Evidence of hydrothermal activity in presently active volcanic areas. Left, hot gases emerging from hydrothermal vents, Noboribetsu volcano, Japan. Above, sulphur crystals forming a chimney around a small fumarole, Merapi volcano, Java.*

8.9: *Underwater photograph taken by* Alvin *at 2500 m depth on the crest of the East Pacific Rise at 21°N. Water at 350°C emerges as a clear solution from orifices 15 cm in diameter at a few metres per second. Rapid mixing with the local seawater induces precipitation of minerals such as sulphides of iron, zinc and copper, forming a black smoky cloud and building the conduit which rises above the pillow basalts.*

Stable isotopes

Many elements consist of more than one isotope. Some are unstable and decay, emitting radiation, whereas those which are not measurably radioactive are termed stable isotopes. Isotopes of a particular element have the same electron configuration and differ essentially in the mass of their nuclei. In general, chemical characteristics reflect the former, and physical properties are more sensitive to changes in mass. Thus there are small but significant differences in the physical and physicochemical properties of different isotopes, even though their purely chemical behaviour is broadly similar. This effect is most marked in isotopes having the greatest mass difference, for example in hydrogen isotopes, where one, deuterium (D), has twice the mass of the other. It is very much less significant in the isotopes of an element like strontium (Sr) where the relative mass difference between ^{87}Sr and ^{86}Sr is comparatively small.

The stable isotopes most widely used in the Earth sciences are those of light elements such as hydrogen, carbon, oxygen and sulphur. Figure 8.12 gives details of hydrogen and oxygen isotopes, which will be used to illustrate the basic principles.

The hydrosphere

The difference in mass between isotopes of a particular element means that molecules containing lighter isotopes have a greater internal vibrational energy than those containing heavier isotopes. The former therefore require slightly less energy to vaporize which is why, for example, the boiling point of H_2O (100°C) is lower than that of D_2O (101.4°C) at the same pressure. Similarly, evaporation of natural waters, which contain different isotopes of hydrogen and oxygen, results in molecules containing the lighter isotopes being vaporized more readily (Figure 8.11). Thus water vapour in clouds is isotopically lighter (having lower δD and δ^{18}O values) than water in the sea below. Moreover isotopically heavier molecules condense more readily so that rain water has higher δD and δ^{18}O values than the clouds from which it came (the delta notation is explained in Figure 8.12).

The difference in the isotope ratios of coexisting water and water vapour has been determined experimentally and appears to be constant at any particular temperature, eg $\delta^{18}O_{\text{water}} - \delta^{18}O_{\text{water vapour}}$ is approximately ten at 20°C. This difference, generally referred to as

8.10: *Uranium mineralization in Namibia, south-west Africa. The uranium, leached from nearby granite, has been concentrated in adjacent calcareous sediments, largely as the mineral uranophane.*

the amount of isotope fractionation, is greater at low and much less at high temperatures. Thus rain or snow under the colder conditions of winter, or high altitudes, tends to have lower $\delta^{18}O$ and δD values. Similarly the isotopic compositions of precipitation in the polar regions is much lighter than that in the tropics where it is still close to the composition of seawater.

This relationship between $\delta^{18}O$ and temperature has also facilitated the study of climatic variations in the recent geological past. The isotope record preserved in calcareous fossil shells (usually foraminifera) reflects both changes in temperature and in the isotope composition of seawater. Since the latter is sensitive to the amount and the isotope ratios of water removed during, for example, the glacial periods, oxygen isotope studies have revealed detailed records of climatic evolution (see Chapters 18 and 23).

The lithosphere

The use of stable isotopes in the study of palaeoclimates is primarily as a geothermometer. It relies on the observation that the amount of oxygen isotope fractionation varies with temperature. Since oxygen isotopes are also fractionated between coexisting minerals several mineral pairs have been investigated experimentally and then used as geothermometers in both igneous and metamorphic rocks. Common examples include quartz–magnetite, quartz–mica and olivine–feldspar, which cover temperatures in the range 1300–300°C.

Another consequence of the decrease in stable isotope fractionation with increasing temperature is that most mantle-derived rocks have a very restricted range in hydrogen and oxygen isotope ratios (Figure 8.12). To observe large stable isotope variations in rocks it is necessary to look at lower-temperature environments, and particularly those in which interaction with water can occur. Most sediments (and hence many metamorphic rocks) tend to have heavy oxygen isotope ratios, with the result that particularly the upper levels of the continental crust appear to be characterized by high $\delta^{18}O$ values. Since many granitic rocks also have high $^{18}O/^{16}O$ ratios it suggests that they are either derived by partial melting of pre-existing continental crust, or that they have at least assimilated some crustal material. Thus a widespread application of stable isotopes is to assess whether a particular igneous rock originated in the crust or the mantle, or perhaps represents a mixture of both.

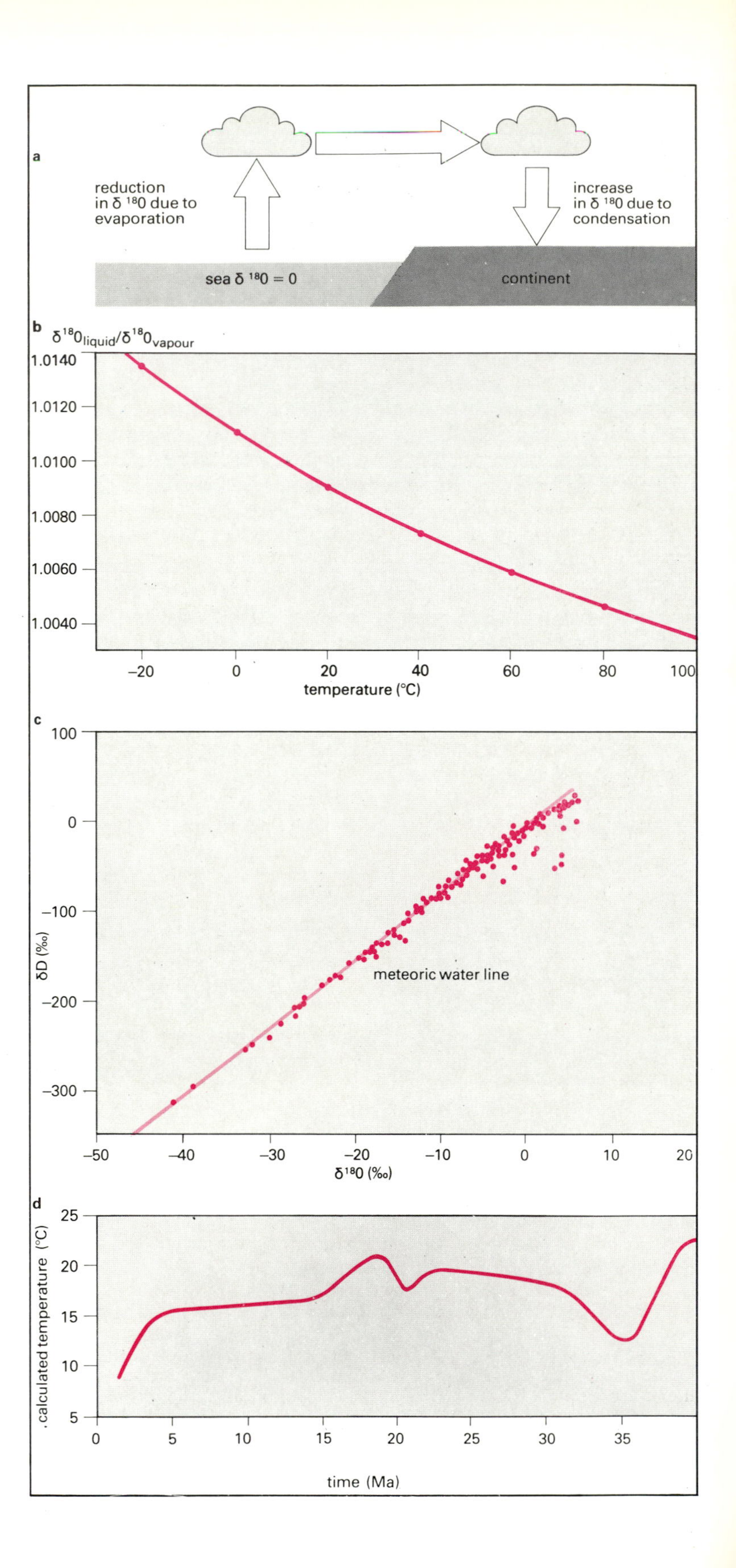

8.11: *Hydrogen and oxygen isotopes in the hydrosphere.* (**a**) *Variation in oxygen isotope values generated during the meteorological cycle.* (**b**) *Oxygen isotope fractionation between water and water vapour at different temperatures. Similar curves have been determined experimentally for the fractionation of oxygen and hydrogen isotopes between minerals and water, and between coexisting minerals. All are sensitive to temperature, and the latter used widely as geothermometers.* (**c**) *Variation in oxygen and hydrogen isotopes in meteoric waters. This correlation is known as the meteoric water line which reflects different amounts of isotope fractionation taking place during evaporation and condensation at different temperatures.* (**d**) *A typical palaeotemperature curve showing variation of water temperature over the past 40 Ma. Temperatures are calculated from the* $\delta^{18}O$ *values of foraminifera.*

However, arguably the greatest contribution of stable isotope studies has been in the field of water–rock interaction, an area in which many ore deposits are formed. A surprising result of combined oxygen and hydrogen isotope studies is that several igneous intrusive complexes have been shown to have very low $\delta^{18}O$ and δD values, eg on the island of Skye in Scotland. Since meteoric waters appear to be the only likely source of such light hydrogen and oxygen isotope ratios, the low values in the rocks are attributed to exchange with considerable volumes of meteoric water. Mass balance considerations suggest that each cubic kilometre of rock must have interacted with a similar volume of water (total 2000 km^3), and one of the unresolved questions of great concern to geologists studying such igneous rocks is the effect of all this water on, in particular, their trace element composition.

Finally let us consider the effect of water–rock interaction in the oceanic crust. Figure 8.9 clearly shows hot water being expelled from hydrothermal vents in the ocean floor. In other areas, however, ocean floor rocks have been tectonically emplaced above present sea level as ophiolites where they now provide both spectacular geological sections through fragments of oceanic crust and important economic deposits of sulphides (eg copper). By analysing the stable isotope ratios of both rocks and minerals which have interacted with and been deposited from such hydrothermal fluids it is possible to determine both their origins and in what quantity they may have passed through the rocks.

The calculated 'fields' for water in equilibrium with both visibly altered ocean floor basalt and minerals from veins through which the fluids passed are illustrated in Figure 8.13. For temperatures in excess of about 300°C those fields are similar to that of ocean water, demonstrating that the fluids themselves were from seawater. It is a spectacular piece of evidence that at moderate temperatures close to the site of basalt generation along mid-ocean ridges considerable quantities of ocean water circulate through the oceanic crust, probably to a depth of 3 to 4 km. One important consequence is the deposition of sizable copper sulphide ore bodies, such as those in Cyprus.

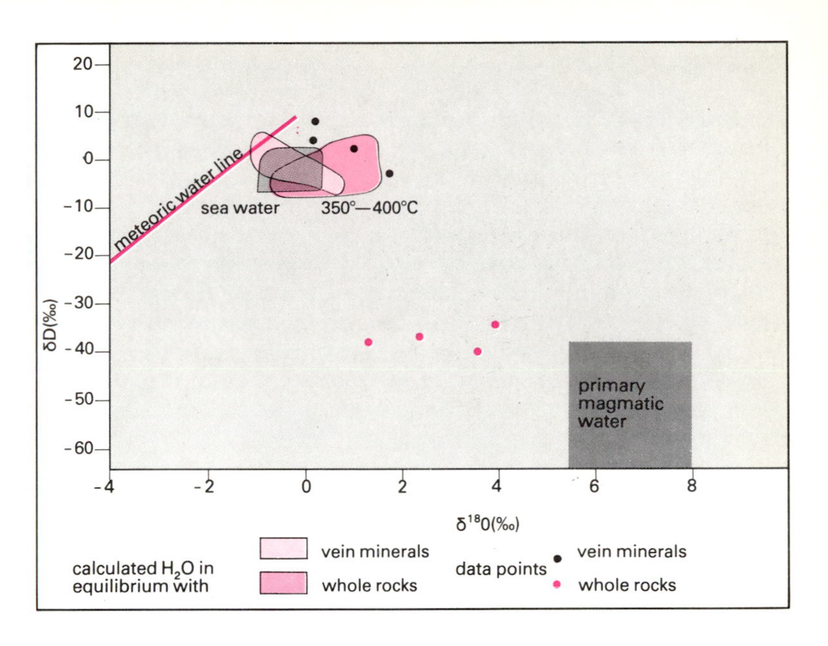

8.13: *A plot of hydrogen (δD) and oxygen ($\delta^{18}O$) isotopes illustrating effects of water–rock interaction associated with sulphide ore deposits in a fragment of old ocean crust on Cyprus. The meteoric water line is as in Figure 8.11(c), and the field for ocean water is slightly displaced to the right. Temperatures are estimated from observed mineral assemblages, and isotope ratios of water in equilibrium with the rocks and minerals are then calculated from experimentally determined fractionation curves similar to that in Figure 8.11(b). Calculated fields for the waters plot close to that for ocean water and a long way from that for primary magmatic water (ie from the mantle). Thus the fluids that pervaded these ocean floor rocks, and from which the sulphides were deposited, originated from seawater.*

8.12: *Stable isotopes of oxygen and hydrogen.*

Elements	Natural isotopes		Measured ratios	Peridotites and basalts	Andesites and granites	Sediments	Metamorphic rocks
oxygen	^{16}O	99.76%	$^{18}O/^{16}O$,	+5 to +7.5	+12 to −6	+5 to +25	+3 to +20
	^{17}O	0.04%	expressed as				
	^{18}O	0.20%	$\delta^{18}O$ ($^0/_{00}$)				
hydrogen	^{1}H	99.984%	D/H,	−50 to −90	−60 to −180	−40 to −100	−60 to −120
	^{2}H	0.016%	expressed as				
			δD ($^0/_{00}$)				

Note So that results from different laboratories may be compared with one another the isotope ratios are reported as delta values calculated relative to an international standard. For oxygen and hydrogen that standard is Standard Mean Ocean Water (SMOW) and the delta values are calculated as follows:

$$\delta D = \left(\frac{D/H_{sample}}{D/H_{SMOW}} - 1\right) \times 1000\ ^0/_{00}$$

and $$\delta^{18}O = \left(\frac{^{18}O/^{16}O_{sample}}{^{18}O/^{16}O_{SMOW}} - 1\right) \times 1000\ ^0/_{00}$$

The units are in parts per thousand ($^0/_{00}$). The δD and $\delta^{18}O$ values of SMOW are zero. Samples that are isotopically lighter (have a greater concentration of the lighter isotope) than SMOW have negative delta values, and samples that are isotopically heaviei than SMOW have positive delta values.

Radiogenic isotopes

Isotopes that are the product of radioactivity are termed radiogenic. Often they are themselves stable, but they have accumulated through time by the process of radioactive decay. Radioactivity itself has only been recognized for about a hundred years and the first attempts to harness it as a radiometric time-piece occurred at the beginning of this century. Inevitably the early work relied on the more obvious radioactivity of uranium, but since then several trace elements have been shown to be radioactive (Figure 8.14) and their decay schemes have become widely used to determine the ages of both rocks and minerals. In addition the isotope composition of a rock at the time of its formation often reflects both how it was formed and the evolution of its source region.

Geochronology: the dating of rocks and minerals

The amount of a radiogenic isotope that has accumulated in a sample since its formation may be used to determine its age. The process of radioactive decay is exponential, ie the number of atoms that disintegrate in a particular period of time depends on the number that are present. More formally the law of radioactive decay states that the number of atoms disintegrating per unit time ($-dN/dt$) is proportional to the total number of radioactive atoms (N) which are present. Thus $-dN/dt = \lambda N$, where λ is the decay constant and has a characteristic value for each radioactive isotope. It represents the probability that an atom disintegrates in unit time, which may also be expressed by the simpler concept of a half-life.

From the previous equation, $N_O = Ne^{\lambda t}$, where N_O is the number of radioactive atoms at the formation of the sample, N is the number present today, and e is the base of the natural logarithms. On taking logarithms, $t = (1/\lambda) \ln (N_O/N)$, which is the basic equation for calculating the age of a sample formed at time t. In practice it is easier to measure radiogenic isotopes, and since the number of radiogenic isotopes (N_R) was produced by the disintegration of the same number of radioactive isotopes, $N_R = N_O - N$. Substituting into the equation for determining t gives

$$t = \frac{1}{\lambda} \ln \frac{N + N_R}{N}$$

Several varieties of this basic equation are used widely in geological age calculations. The rate of decay of an atom may also be expressed in terms of its half-life ($T_{\frac{1}{2}}$), the time required for half a given number of atoms of a radioactive isotope to decay. Alternatively, since the disintegration of one radioactive isotope produces one radiogenic isotope, it is the time at which the number of radiogenic isotopes produced equals the number of radioactive isotopes remaining. Thus if $N_R = N$ it follows from the last equation that the half-life may be simply related to the decay constant, $T_{\frac{1}{2}} = \ln 2/\lambda$. By using the general equation for determining t in the form $N_R = N(e^{\lambda t} - 1)$, two further points may be illustrated by the decay scheme ^{87}Rb to ^{87}Sr. ^{87}Sr is the radiogenic isotope, and so $^{87}Sr = {}^{87}Rb(e^{\lambda t} - 1)$. However, mass spectrometers are more simply used to measure ratios of isotopes than their absolute abundance and it is therefore convenient to rewrite that equation in the form $^{87}Sr/^{86}Sr = (^{87}Rb/^{86}Sr)(e^{\lambda t} - 1)$.

^{86}Sr is chosen because it is neither radioactive nor radiogenic and so its overall abundance may be assumed to be constant. The age t could therefore be arrived at by comparing the ratio of those two isotopes of strontium with the ratio of ^{87}Rb to ^{86}Sr, on the assumption that all the ^{87}Sr has been produced by decay of ^{87}Rb since the formation of the sample, t million years ago. However, in the case of strontium some ^{87}Sr is not radiogenic and even some of the radiogenic ^{87}Sr will have been produced before time t, ie before the formation of the rock or mineral that is to be dated. Thus it is necessary to subtract the ^{87}Sr which was already present at time t from that which is measured today to estimate the amount of ^{87}Sr actually produced by the decay of ^{87}Rb since time t. Thus

$$(^{87}Sr/^{86}Sr) - (^{87}Sr/^{86}Sr)_O = (^{87}Rb/^{86}Sr)(e^{\lambda t} - 1), \text{ or}$$
$$(^{87}Sr/^{86}Sr) = (^{87}Rb/^{86}Sr)(e^{\lambda t} - 1) + (^{87}Sr/^{86}Sr)_O,$$

where $(^{87}Sr/^{86}Sr)_O$ is the strontium isotope ratio which existed at the

8.14: *Radioactive decay schemes used in geochronology and isotope geochemistry.*

Radioactive isotope	Natural abundance (atoms %)	Decay type	Decay constant λ ($\times 10^{-11}$/year)	Radiogenic isotope	Approximate abundance (atoms %)	Ratio measured	
^{40}K	0.0118	K capture	5.81	^{40}Ar	99.7	$^{40}Ar/^{36}Ar$	1
^{87}Rb	27.85	beta	1.42	^{87}Sr	7	$^{87}Sr/^{86}Sr$	2
^{147}Sm	14.97	alpha	0.654	^{143}Nd	12	$^{143}Nd/^{144}Nd$	3
^{232}Th	100	chain	4.95	^{208}Pb	52	$^{208}Pb/^{204}Pb$	4
^{235}U	0.72	chain	98.485	^{207}Pb	22	$^{207}Pb/^{204}Pb$	4
^{238}U	99.28	chain	15.5125	^{206}Pb	25	$^{206}Pb/^{204}Pb$	4

Notes

1 Argon is readily mobilized, and thus this scheme usually records the last thermal event in any area.

2 The most widely used decay scheme in geochronology.

3 Highly resistant to secondary processes of alteration and metamorphism, but with a small decay constant.

4 The mobility of U in near-surface rocks and the toxicity of ^{232}Th means that many studies only report Pb-isotope results (plotted on Pb-Pb diagrams, eg Figures 8.16 and 8.17).

formation of the sample, t million years ago, and is generally referred to as the initial strontium ratio. This equation is in such a form ($y = mx + c$) that a graph of $^{87}Sr/^{86}Sr$ against $^{87}Rb/^{86}Sr$ should be a straight line with a slope of $(e^{\lambda t} - 1)$ intercepting the $^{87}Sr/^{86}Sr$ axis at a point corresponding to $(^{87}Sr/^{86}Sr)_O$. Such a line is called an isochron, since it connects points of equal age, and diagrams of this form are most widely used for the decay schemes of $^{87}Rb \rightarrow {}^{87}Sr$ and $^{147}Sm \rightarrow {}^{143}Nd$. Obviously at least two samples must be analysed for an isochron line to be drawn, and in practice the more samples that are analysed and the greater the spread in isotope ratios, the more precisely will the age t be determined.

Samples plotted on an isochron diagram may be either of different minerals separated from the same rock or, perhaps more commonly, of whole rock samples which have been crushed to a homogeneous powder before analysis (Figure 8.15). In all cases, however, it is extremely important that the samples had the same initial $^{87}Sr/^{86}Sr$ ratio at the time t that is being determined. Experience shows that this condition is most frequently met by igneous rocks crystallizing from the same liquid magma; it is sometimes true for metamorphic rocks if metamorphism was sufficiently pervasive to homogenize the $^{87}Sr/^{86}Sr$ ratios, but it is much less likely to occur among sediments with variable amounts of detritus from different sources.

A further assumption underlying the interpretation of isochrons is that the trace element ratio has not changed since time t, except by the process of radioactive decay. Provided that the samples are fresh and unaltered such an assumption appears to be justified for Rb/Sr and Sm/Nd. Uranium, however, is readily oxidized to a highly mobile state so that the U/Pb ratios measured for igneous rocks are rarely the same

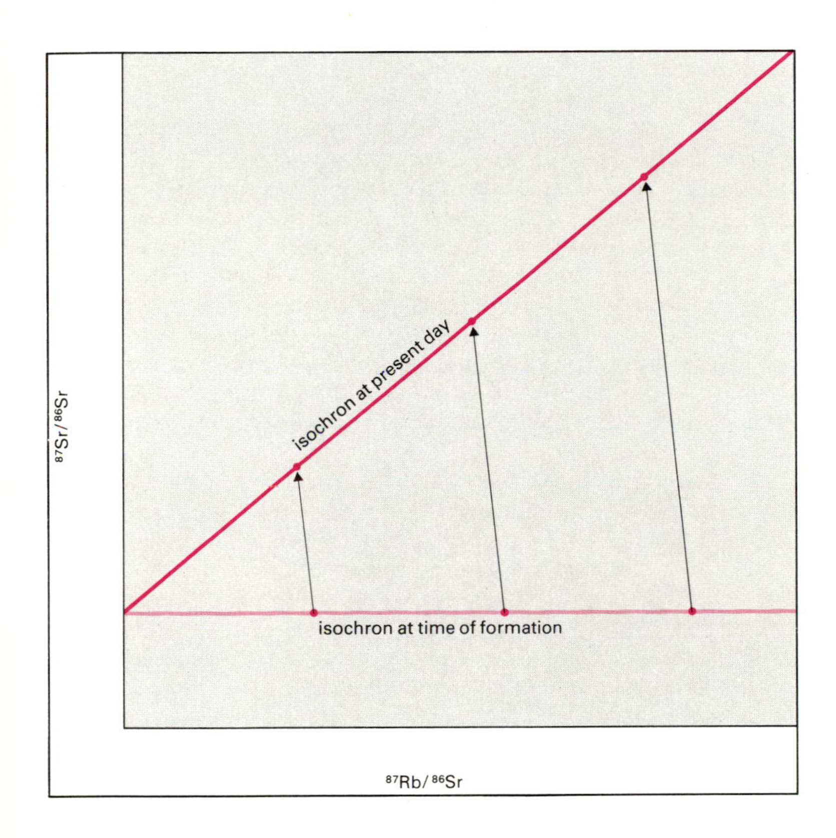

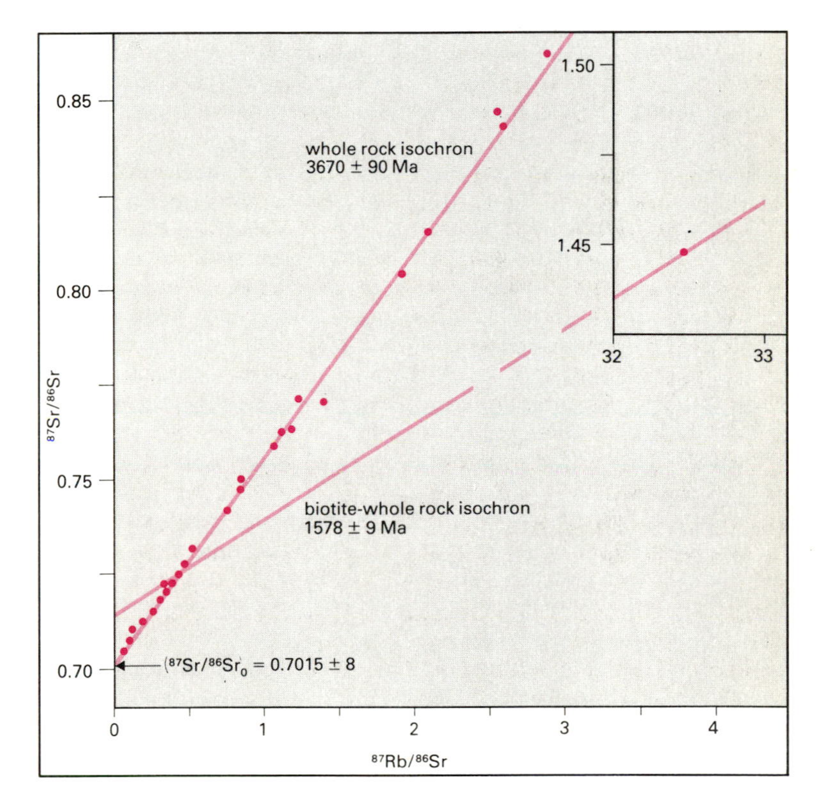

8.15: *Above, an Rb/Sr isochron diagram illustrating the evolution of isotope ratios with time after crystallization from a homogeneous magma. At the time of crystallization all four samples have the same* $^{87}Sr/^{86}Sr$ *ratio but different* $^{87}Rb/^{86}Sr$ ($^{87}Rb/^{86}Sr \simeq 2.9 \times Rb/Sr$). *The size of the subsequent increase in* $^{87}Sr/^{86}Sr$ *due to the decay of* ^{87}Rb *then depends on the ratio of* $^{87}Rb/^{86}Sr$ *of the particular sample, so that samples evolve as shown and now lie on a single straight line of slope* $(e^{\lambda t} - 1)$. *Since* λ *is known (Figure 8.14), t, the age, may be calculated. Right, results on whole rock and mineral samples from Narssaq near Godthaab in west Greenland. The twenty-five whole rock samples of these gneissic rocks lie on an isochron whose slope corresponds to an age of 3670 ± 90 Ma (errors quoted at ninety-five per cent confidence limits). This is interpreted as the time of formation of these rocks which are consequently among the oldest known on Earth. In contrast the mineral biotite, which was separated physically from one of the whole rock samples, plots to the right of the whole rock isochron, and the age of the two-point biotite–whole rock isochron is 1578 ± 9 Ma. Thus the biotite, and presumably the other minerals in equilibrium with it, are much younger than the host rock: their isotopic clocks have been reset by some heating event 1600 Ma ago. The power of this technique is that different events may be recognized and dated in geologically complex terrains. Note the change of scale into the inset box.*

as those acquired during crystallization. This problem may be avoided by combining equations for the two U–Pb decay schemes, forming a relationship between $^{206}Pb/^{204}Pb$ and $^{207}Pb/^{204}Pb$ in which the uranium contents need not be known.

For $^{238}U \rightarrow {}^{206}Pb$,

$$\frac{^{206}Pb}{^{204}Pb} = \left[\frac{^{206}Pb}{^{204}Pb}\right]_o + \frac{^{238}U}{^{204}Pb}(e^{\lambda_1 t} - 1), \text{ and}$$

for $^{235}U \rightarrow {}^{207}Pb$,

$$\frac{^{207}Pb}{^{204}Pb} = \left[\frac{^{207}Pb}{^{204}Pb}\right]_o + \frac{^{235}U}{^{204}Pb}(e^{\lambda_2 t} - 1)$$

As before the subscript $_o$ signifies the initial Pb isotope ratio, ie at time t. Combining these two equations,

$$\frac{\left(\frac{^{207}Pb}{^{204}Pb}\right) - \left(\frac{^{207}Pb}{^{204}Pb}\right)_o}{\left(\frac{^{206}Pb}{^{204}Pb}\right) - \left(\frac{^{206}Pb}{^{204}Pb}\right)_o} = \frac{^{235}U\, e^{\lambda_2 t} - 1}{^{238}U\, e^{\lambda_1 t} - 1}$$

The ratio $^{235}U/^{238}U$ is a constant equal to 1/137.8 for all uranium of normal composition in the Earth, Moon and meteorites at the present time. Note also that on a diagram of $^{207}Pb/^{204}Pb$ against $^{206}Pb/^{204}Pb$ a suite of related rocks and minerals (ie of the same age and with the same initial Pb isotope ratios) should plot on a straight line of slope m:

$$m = \frac{1}{137.8}\frac{e^{\lambda_2 t} - 1}{e^{\lambda_1 t} - 1}$$

Thus a straight line on these Pb–Pb diagrams may be interpreted as an isochron, in which case the time t can be calculated from its slope. This technique has made a unique contribution in enabling scientists to determine the age of the Earth and many meteorites as approximately 4550 million years (Ma) (Figure 8.16). However, it must be emphasized that whether or not straight lines on such diagrams are isochrons is still a matter of interpretation.

Evolution of lead isotopes

The generation of radiogenic isotopes, such as those of lead, also enables geochemists to study how different portions of the Earth, and those parts of the solar system from which samples are available, have evolved through time. This is done by analysing old rocks and minerals and calculating the isotope and trace element ratios at the time of their formation, and by studying the variations of isotope ratios in young rocks and estimating how much time was needed to generate those isotopic differences.

For lead isotopes minerals containing lead and little or no uranium provide an invaluable record of the Pb isotope ratios present at the time of formation. Moreover, since most of the U-free samples analysed are actually Pb ores (eg galena (PbS)), the isotope ratios will reflect the average Pb from the large volume of rock scavenged during the formation of the ore minerals.

Lead isotope ratios have now been determined on a large number of ore minerals and many lie on or close to a smooth curve in the Pb–Pb diagram (Figure 8.17). Significantly the age of the minerals analysed varies along this curve which is therefore believed to represent the evolution of that portion of the Earth that is sampled when Pb is concentrated into ore bodies. Its identity has been hotly debated, but the present consensus suggests that this Pb is derived from the continental crust and that the curve in Figure 8.17, known as the conformable Pb ore growth curve, represents a striking record of how Pb isotope ratios have evolved within it. This interpretation is supported by the observation that Pb in oceanic sediments (which is almost entirely derived from the continents and thus represents a reasonable estimate of average crust as sampled by weathering and erosion) also lies on the end of the Pb ore growth curve.

Lead from the upper mantle, as sampled by volcanism, has significantly different isotope ratios from that in the crust. Those from mid-ocean ridge basalt (MORB), for example, have lower $^{207}Pb/^{204}Pb$ and $^{206}Pb/^{204}Pb$ ratios, while in rocks from oceanic islands both ratios tend to be higher. The low $^{207}Pb/^{204}Pb$ ratios of MORB are significant

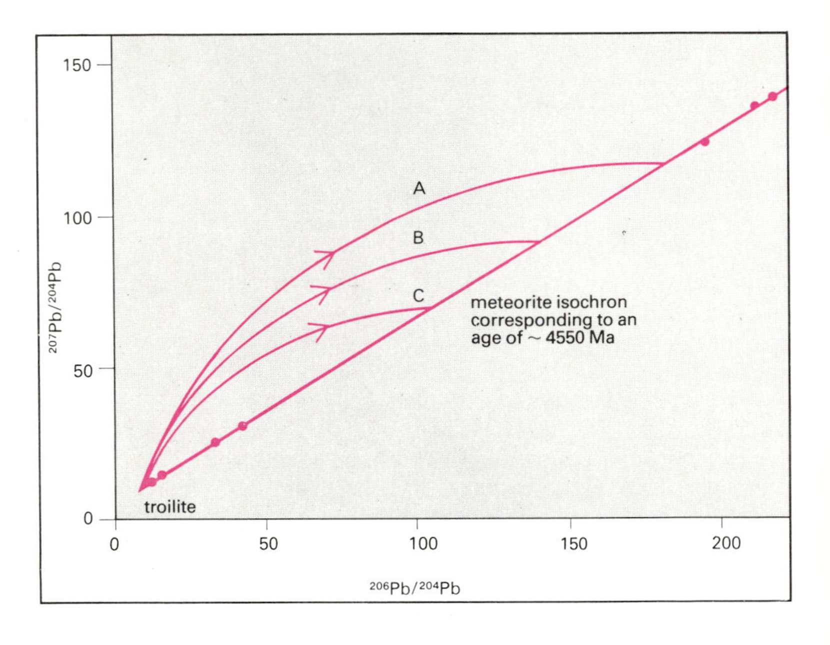

8.16: *Lead isotope ratios from meteorites plotted on a diagram of* $^{207}Pb/^{204}Pb$ *against* $^{206}Pb/^{204}Pb$. *The slope of the straight line corresponds to an age of 4500 Ma which is interpreted as that of the meteorites and of the Earth. The mineral troilite contains no uranium and thus its lead isotope ratios are the same now as when it formed, giving the best available estimate of the initial* Pb *isotope ratios of meteorites and the Earth. Curves A, B and C are schematic evolution paths illustrating how the* Pb *isotope ratios of three samples with different* U/Pb *ratios changed. The* U/Pb *ratio is highest in A and lowest in C. The shape of these curves reflect changes in the* $^{235}U/^{238}U$ *ratio with time: early in Earth history there was relatively more* ^{235}U *so that more* ^{207}Pb *was generated and the curves were steeper. Later on, when most of the* ^{235}U *had decayed, most of the* Pb *produced was* ^{206}Pb *from* ^{238}U *(Figure 8.14). Thus the curves, or 'evolution paths', flatten out.*

because they indicate that the source regions for these basalts had low U/Pb ratios very early in Earth history when more ^{207}Pb was being generated. The question of how early such chemical variations existed is extremely controversial, but central to a great many studies in isotope geology.

Interpretations of the Pb isotope data may be illustrated by two extreme end-member models. The first assumes that the straight line of results from oceanic volcanics is an isochron; its slope corresponds to an age of nearly 2000 Ma and, if valid, this date presumably represents some worldwide event in the upper mantle. At the very least it demonstrates that chemical variations in the upper mantle have persisted for more than 2000 Ma. The alternative school of thought is that such straight lines need not have any age significance, but rather may be generated by mixing processes which have operated throughout Earth history. This has the advantage of being much more in tune with current ideas about convection in the upper mantle (see Chapter 9).

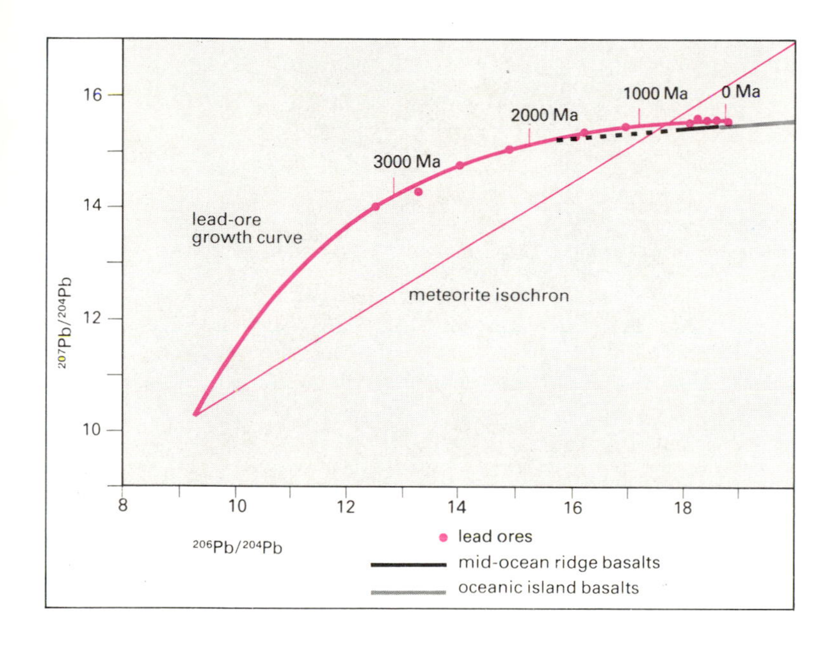

8.17: *Variations of Pb isotope ratios in Pb ores of different ages, the Pb ore growth curve and young mantle-derived volcanic rocks. The Pb ore growth curve reflects the evolution of the crustal source regions of Pb ores formed from just before 3000 Ma to the present day. Volcanic rocks lie on a straight line which, if interpreted as an isochron, has a slope corresponding to an age of almost 2000 Ma: it intersects the growth curve about 2000 Ma ago. Even if this straight line is not interpreted as an isochron, variations in U/Pb ratios must have been present in the crust and the upper mantle before 2000 Ma ago.*

Strontium and neodymium isotopes in magmatic rocks

One popular model for the origin of the Earth suggests that it formed from a dust cloud, and that the proto-Earth was homogeneous, similar in composition to many meteorites. If so, then present-day differences in major and minor elements and isotope ratios (Figures 8.17 and 8.18) must have been generated during the history of the Earth. How and when did this happen? How did the composition of the mantle change when the crust was derived from it? Has the rate of formation of the continental crust always been the same? Once chemical variations are produced in the upper mantle, do they persist? Or does convection cause such efficient mixing that chemical differences become homogenized? Questions such as these have been at least partially responsible for the recent boom in the study of radiogenic isotopes.

Considerations of the trace element geochemistry of recently formed basalts suggest that they could not all be derived from similar source regions and that significant trace element variations must be present in the upper mantle. Since rocks of different Rb/Sr (or Sm/Nd) ratios will with time develop different $^{87}Sr/^{86}Sr$ (or $^{143}Nd/^{144}Nd$) ratios it is to be expected that, as with Pb, isotopes of Sr and Nd are

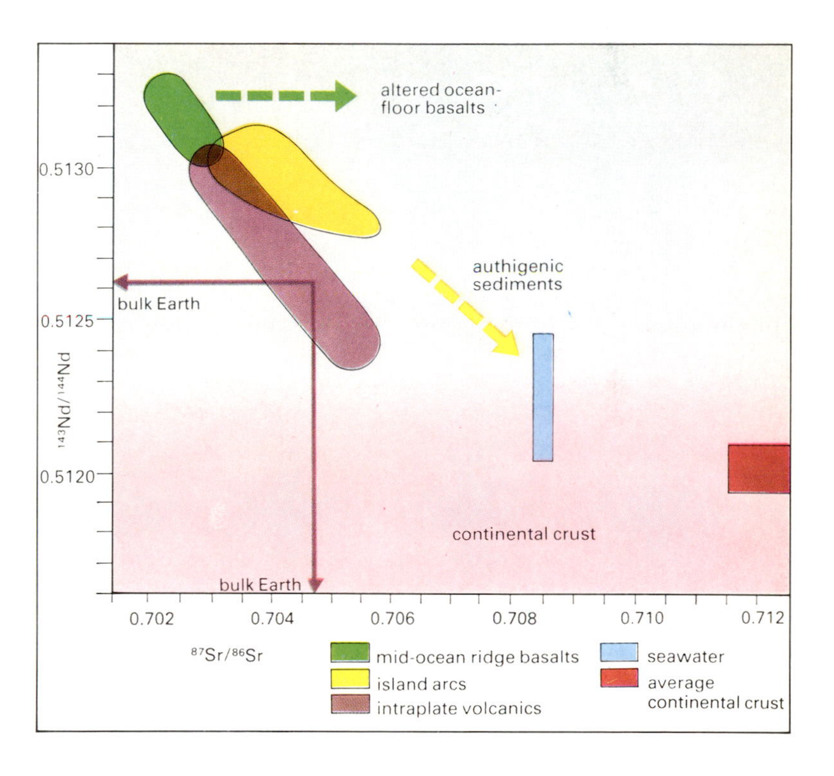

8.18: *Present-day variations in Nd and Sr isotopes in young volcanic rocks, sediments and continental crust. The composition for the (model) bulk Earth is calculated assuming initial isotope and Sm/Nd ratios similar to those of chondritic meteorites. Much of the Sr and Nd in seawater is derived from the continents, and the relatively high $^{87}Sr/^{86}Sr$ ratios of many island arc rocks indicate that some seawater, and hence continental Sr, is being recycled.*

not uniformly distributed in the upper mantle. This is confirmed by the available analyses of recent magmatic rocks believed to have been derived from the mantle for which $^{87}Sr/^{86}Sr = 0.702–0.707$ and $^{143}Nd/^{144}Nd = 0.5133–0.5124$. Moreover since the continental crust tends to have higher Rb/Sr and lower Sm/Nd ratios than the upper mantle, magmas formed in the crust, or which have assimilated crustal material, should have higher $^{87}Sr/^{86}Sr$ and lower $^{143}Nd/^{144}Nd$ ratios than those from the mantle.

Variations in Sr and Nd isotopes in recent rocks are often illustrated on a diagram of $^{143}Nd/^{144}Nd$ against $^{87}Sr/^{86}Sr$ from which we can estimate average ratios for the Earth as a whole. The majority of results from mantle-derived material plot on a single broad trend and the bulk Earth ratio is assumed to plot on that trend. The Earth is believed to have had initial Sm/Nd and $^{143}Nd/^{144}Nd$ ratios like those in chondritic meteorites and if these are used in the appropriate isochron equation the present-day (ie $t = 4550$ Ma) bulk Earth $^{143}Nd/^{144}Nd$ ratio may be calculated to be 0.51262. At $^{143}Nd/^{144}Nd = 0.51262$, $^{87}Sr/^{86}Sr = 0.7045–0.7050$ which is therefore taken as a reasonable estimate of the present-day $^{87}Sr/^{86}Sr$ ratio of the bulk Earth. Adopting an initial $^{87}Sr/^{86}Sr$ ratio of the Earth of 0.69898 (Figure 8.19) indicates that its Rb/Sr ratio is 0.03, significantly lower than chondrites. (Note the very high degree of experimental precision needed to pick out these differences.)

This argument suggests that any piece of the primitive Earth having its original Rb/Sr and Sm/Nd ratios and having remained undisturbed throughout its entire history (4550 Ma) would have present-day $^{87}Sr/^{86}Sr$ and $^{143}Nd/^{144}Nd$ ratios of 0.7045–0.7050 and about 0.5126 respectively. Whether such a sample could ever survive undisturbed is clearly debatable, but the lack of such samples certainly does not diminish the significance of the model bulk Earth as a concept. It permits comparison both with other planets and with estimates of cosmic abundances and it is also widely used as a convenient reference for geochemical comparison between terrestrial rocks. For example, since the majority of igneous rocks in the oceans have lower $^{87}Sr/^{86}Sr$ and higher $^{143}Nd/^{144}Nd$ ratios than the model bulk Earth, it is implied that their source regions had relatively low Rb/Sr and high Sm/Nd ratios for much of their history. However, a sizable proportion of the oceanic volcanic rocks analysed actually contain high Rb/Sr and low Sm/Nd ratios and since these probably reflect similar ratios in their source rocks it suggests that such ratios have only been present in the source rocks for a comparatively short period, ie the particular trace element ratios cannot have been there for long enough to have had a significant effect on the isotope ratios.

This evidence is one indication that at least the upper mantle has experienced a complex history. Another is that the main correlation between Nd and Sr isotopes cannot be generated by a model in which there is just one widespread mantle event, perhaps 2000 Ma ago as suggested by the isochron interpretation of the Pb-isotope data. Rather, a multi-stage (ie many-event) history is required indicating that, as with the Pb data, the way that $^{143}Nd/^{144}Nd$ and $^{87}Sr/^{86}Sr$ vary together may primarily reflect long-term mixing processes in a convecting upper mantle.

One group of volcanic rocks that appears not to plot on the main trend of Nd and Sr isotopes in Figure 8.18 is those erupted along destructive plate margins (see Chapter 14). These tend to be displaced to relatively high $^{87}Sr/^{86}Sr$ ratios (ie they have more radiogenic Sr). Since both seawater (and thus oceanic sediment) and altered ocean floor basalts (see Figure 8.9) also exhibit comparatively radiogenic Sr, this is strong evidence that island arc volcanic rocks contain a component of material released from the subducted oceanic crust. Whether it is released by melting or merely dehydration of the down-going crust is debatable, but in either case these results demonstrate that at least some of the trace elements being added to the crust along destructive plate margins are being recycled. Erosion and dissolution transfers them to seawater where they reside, often for long periods, before being removed by interaction with the ocean crust, transported along to subduction zones and released once again into the overriding crust. The relative amounts of an element which are recycled depends on its chemical characteristics: a mobile element such as potassium will tend to be more readily recycled than more stable elements such as Zr, Y and the rare earths. This inevitably affects their present distribution between the crust and the mantle and indeed potassium is relatively more abundant in crustal rocks.

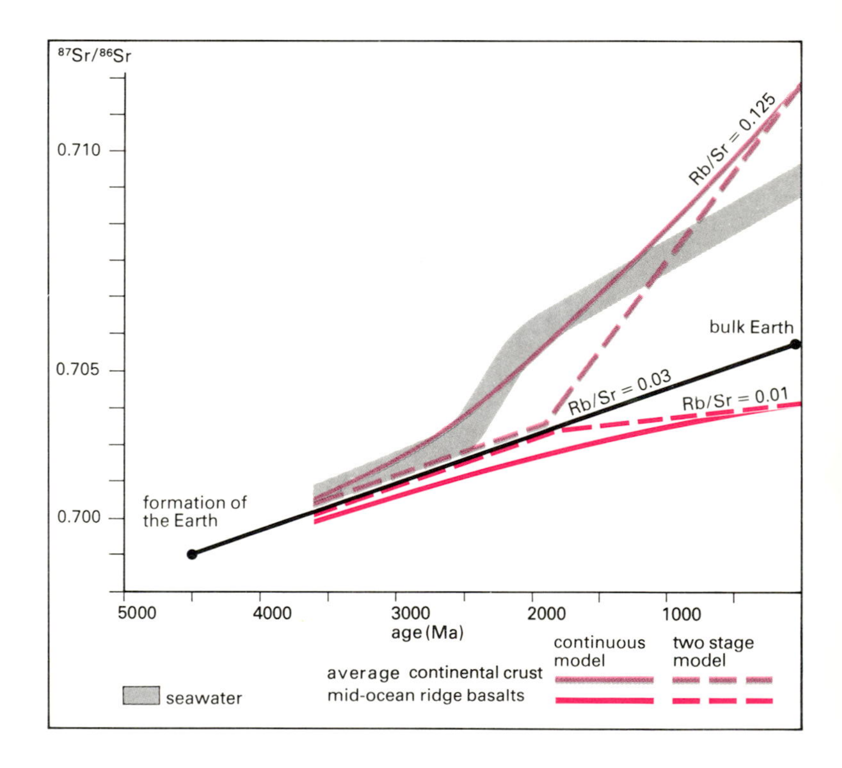

8.19: *Evolution of* $^{87}Sr/^{86}Sr$ *ratios from the formation of the Earth to the present day. The slope of a line depends on the* Rb/Sr *ratio of the particular sample, or segment of the Earth. The Sr-isotope composition of ancient seawater is inferred from that of contemporaneous limestones and the step in its evolution path is believed to reflect the stabilization of high* Rb/Sr *continental crust.*

Isotopic evolution of the Earth

Radiogenic isotopes provide the best chance of evaluating how the composition of particular portions of the Earth, as reflected in their trace element ratios have changed with time, and such changes may be most clearly observed on an isotope evolution diagram. Isotope geochemistry is currently trying to discover how the composition of the crust and that of the mantle (as sampled by MORB) have diverged from the bulk Earth composition; Figure 8.19 thus presents some initial $^{87}Sr/^{86}Sr$ ratios from rocks of different ages. A similar diagram could also be drawn for initial Nd isotope ratios, but fewer data are available.

On an isotope evolution diagram the slope of any positively inclined straight line is the rate of increase through geological time of the isotope ratio, in this case $^{87}Sr/^{86}Sr$, and from the equations discussed earlier it depends on the trace element ratio Rb/Sr. Thus the evolution of the model bulk Earth is represented by a straight line from 0.699 at 4550 Ma to 0.7048 at the present time, corresponding to a ratio of 0.03. Crustal rocks tend to have higher Rb/Sr ratios and will consequently evolve along steeper trajectories, while since many mantle rocks have Rb/Sr ratios similar to or less than that of the bulk Earth they will tend to evolve along shallower trends.

One estimate of the average composition of the crust is that it has a present-day $^{87}Sr/^{86}Sr$ ratio of 0.712 and an Rb/Sr ratio of 0.125. Projecting its evolution back in time on Figure 8.19, it intersects the evolution of the model bulk Earth at about 2000 Ma. If the same calculation is carried out for average MORB (which represents by far the largest volume of upper mantle sampled by volcanism) then its evolution also intersects that of the bulk Earth close to 2000 Ma and, remarkably, this is the same age as that obtained by interpreting the Pb isotope results from oceanic volcanic rocks in terms of an isochron. Moreover, this coincidence of ages deduced from these different arguments encouraged the early suggestion that the evolution of at least the outer portions of the Earth might be described in terms of a two-stage history. It was envisaged that from 4550 to about 2000 Ma the crust and mantle had similar ratios to the bulk Earth and thus evolved together along the path of the bulk Earth. Then around 2000 Ma ago the onset of stage two was heralded by some widespread, presumably catastrophic, event which resulted in different parts of the crust and upper mantle having different trace element ratios. From 2000 Ma ago to the present day these different ratios of Rb/Sr, Sm/Nd, U/Pb etc resulted in the variation in isotope ratios which we observe today.

However, such a two-stage model was always known to be over simplistic. Some objections have already been discussed in the context of combined Nd and Sr isotopes (Figure 8.18). Others are more obvious on Figure 8.19. At the present time, most of the strontium in seawater is derived from the continents, and thus the changing $^{87}Sr/^{86}Sr$ ratio of seawater through time is an indication of the composition of the crust being sampled by the combined forces of weathering and erosion. Early in Earth history the Sr isotope evolution path is very similar to that of the bulk Earth suggesting that there was little high $^{87}Sr/^{86}Sr$ continental crust contributing Sr to those ancient oceans. Then at ~2500 Ma there is a sharp increase in ^{87}Sr believed to reflect the stabilization of high Rb/Sr, and, so with time high $^{87}Sr/^{86}Sr$ continental crust well before 2000 Ma ago. Secondly, magmatic rocks generated as long ago as 2700 Ma often have initial Sr isotope ratios which are lower than the $^{87}Sr/^{86}Sr$ ratio of the bulk Earth at that time. This suggests that their upper mantle source regions had had lower Rb/Sr ratios than the bulk Earth (0.03) for long periods even before 2700 Ma ago.

The available evidence indicates that significant variations in Rb/Sr and $^{87}Sr/^{86}Sr$ were present in both the crust and upper mantle early in Earth history, and it is also clear that the range of initial $^{87}Sr/^{86}Sr$ ratios observed in old rocks is very much less than that in young rocks. Thus the crust and upper mantle have become increasingly heterogeneous with respect to isotope ratios such as $^{87}Sr/^{86}Sr$ and also $^{143}Nd/^{144}Nd$, $^{206}Pb/^{204}Pb$ etc. Once again we are forced away from simplistic two-stage models towards those models capable of reproducing these semicontinuous changes in isotope composition.

If it is assumed that the low Rb/Sr and hence $^{87}Sr/^{86}Sr$ ratios of the mantle are due to the formation of the crust (characterized as it is by relatively high Rb/Sr) then the models are sensitive to changes in composition and rate of formation of stable continental crust with time. This is partly responsible for the recent upsurge in geological interest in those few areas where very old rocks are still preserved, but despite considerable efforts no clear consensus has yet emerged. The increase in range of isotope ratios through time must reflect the progressive stabilization of more crustal material with high Rb/Sr ratios and this is consistent with our knowledge that the Earth is getting steadily cooler. The debate concerns whether this should be interpreted to mean either that the volume of continental crust has increased with time, even though its rate of formation had decreased, or that its volume has remained constant, but in the hotter early evolution of the Earth it was more easily destroyed so that material was recycled too rapidly for significantly different isotope ratios to be generated. In either case the volume of stable crust appears to have increased not least because the destruction of continental crust became more difficult as the Earth cooled.

9 The energy budget of the Earth

On the geological time scale the human race has temporarily inhabited the mobile, changing surface of a vigorously active planet: inconceivably vast amounts of energy are continually interacting with both the interior and the surface of the Earth. If the incident energy of the Sun's rays (Figure 9.1) falling on the Earth for just one day could be stored it would be enough to supply world energy demands for the next century. If harnessed, even the internally generated heat energy flowing through the Earth's surface could supply our energy needs. Human activities and experiences are dominated by the regular cyclic daily and annual patterns of solar heating and, perhaps, of ocean tides, but we are rarely aware of the energetic state of the Earth's interior except when observing natural phenomena such as earthquakes, volcanic eruptions and hot springs. Volcanoes provide ample evidence that the Earth's interior is hot and therefore contains an energy source. However, the quantities of energy involved in these spectacular events are less by about two orders of magnitude than the integrated heat energy conducted through the Earth's surface, which can be detected only by using sophisticated instruments. The energy 'released' from the solid Earth by these natural phenomena is usually transferred directly to the atmosphere as heat which joins, but is totally swamped by, the effects of the solar heating cycle. Incoming solar radiation exceeds the combined effects of these forms of terrestrial radiation by at least three orders of magnitude, but its influence on the Earth's interior is negligible because rocks are such poor conductors of heat that the surface of the Earth acts as an energy barrier to solar radiation. It has therefore been possible to learn much about the Earth's internal energetics by studying patterns of surface heat flow, volcanoes, earthquakes and so on.

This chapter describes the measurement and effects of surface and internal energy transfers based on the energy 'balance sheet' for a single year in the contemporary life of the Earth. It also discusses the extent to which the various different processes involved can be harnessed for human use.

The first attempt to account for the Earth's internal heat was published (1864) by William Thomson (Lord Kelvin). He observed the near-surface temperature gradient in mines, obtained results (an increase with depth of 20–40°C per kilometre) remarkably close to modern values and then calculated how long it would have taken for the Earth to cool to these conditions from an initially molten state by loss of heat from the solid surface. (It was assumed that as the Earth accreted, some of the kinetic energy of the impacting particles would be converted into heat energy, causing internal melting.) Kelvin's maximum figure of 100 Ma for the cooling of the Earth led to a major controversy at the time with, for example, Sir Charles Lyell and Charles Darwin, who saw the need for much greater time intervals to account for the evolution of rock sequences and life forms. These difficulties were not resolved until the turn of the century when, with the discovery of radioactivity by Henri Becquerel, it was realized that the processes of radioactive decay produce heat continuously inside the Earth. A time scale of hundreds of millions of years was soon accepted and has since been refined (by radioactive dating methods; see Chapter 8) to give the current estimate of the Earth's age as 4550 million years (Ma). Although there is some doubt about the exact concentrations of radioactive isotopes, most of the heat flowing out of the Earth is radiogenic, with small contributions from heating due to gravitational energy sources, and even from the Earth's original heat.

9.1: *About 5 × 10^{24} J of solar energy reaches the Earth every year. This is more than 1000 times the energy involved in any other natural process affecting the Earth.*

External energy sources

The principal sources of energy external to the Earth are solar radiation and gravitational energy due to the forces of the Sun and Moon on the rotating Earth. In common with solar energy, gravitational energy is most effective at the Earth's surface where it causes ocean tides, but the income from it is much less than from solar radiation which is responsible for the atmospheric pressure differences which cause winds (and hence ocean waves), evaporation and reprecipitation of water vapour in the hydrologic cycle and photosynthetic activities among organisms. These solar-induced effects have controlled the evolution of surface environments in which life on Earth originated, evolved and became fossilized within sedimentary rock sequences, themselves the product of erosion and deposition processes driven by solar energy (see Chapters 20–23). The origin, potential use and ultimate fate of these two external energy sources are summarized in Figure 9.4.

The origin of solar radiation

In common with all stars, the Sun is a vast nuclear powerhouse, giving clues to the ultimate origin of most forms of energy and even of the chemical elements. Its thermal energy is released by nuclear fusion

9.2: *The Earth's energy budget: chief sources and consumption.*

Energy income[1] (J/a)		Energy expenditure[1] (J/a)	
solar radiation	5.4×10^{24}	direct reflection of solar income	1.9×10^{24}
		solar income absorbed by atmosphere, hydrosphere and lithosphere; ultimately reradiated	3.5×10^{24}
		solar income converted into fossil fuels[2]	10^{17}–10^{18}
radioactive decay of long-lived radio-isotopes (usually assumed to match surface heat flow)	$\sim 10^{21}$	heat flowing through the solid surface by conduction	$\sim 10^{21}$
		heat losses at the solid surface by convection: volcanoes and hot springs	$\sim 10^{19}$
		energy released by earthquakes	10^{18}–10^{19}
gravitational torques that raise tides	$\sim 10^{20}$	tidal dissipation in the oceans	8.0×10^{19}
gravitational potential energy released by crust formation (various estimates)	10^{18}–10^{20}	tidal dissipation within the Earth (various estimates)	10^{15}–10^{19}
gravitational potential energy released by inner core formation, up to	10^{19}	energy requirements of the geomagnetic dynamo	10^{17}–10^{18}

Notes

1 The headings 'income' and 'expenditure' should not be understood too rigorously because energy is never consumed irrevocably but is merely converted into some other form.

2 Potentially recoverable fossil fuel reserves are estimated at 2×10^{23} J; human beings are consuming energy resources at $\sim 3 \times 10^{20}$ J/a.

according to the Einstein relation $E = mc^2$, where E is the energy output resulting from a loss of mass m, and c is the speed of light. The Sun is losing mass at the rate of over 4 million tonnes a second as the following three-stage reaction takes place at a temperature of 2×10^7 °C:

$$^1_1H + ^1_1H \longrightarrow ^2_1H + e^+ \quad \text{(a positron)}$$
$$^2_1H + ^2_1H \longrightarrow ^3_2He + n \quad \text{(a neutron)}$$
$$^3_2He + ^3_2He \longrightarrow ^4_2He + ^1_1H + ^1_1H$$

The net result is that four hydrogen nuclei, each consisting of a single proton (^1_1H), are converted into one helium nucleus, comprising two protons and two neutrons (^4_2He). Since the helium nucleus is slightly less massive than four protons, energy is released and is the origin of sunlight. A loss of 4 million tonnes of mass every second is equivalent to a continuous energy output of 3.6×10^{26} joules per second (J/s or watts), from the Einstein relation, or 1.14×10^{34} J per year (a). Thus only a tiny fraction (about 2 parts in 10^9) of the Sun's energy output is incident on the Earth. The Sun is estimated to contain 70 per cent hydrogen (by mass) and 28 per cent helium, most of which has been produced since it formed. Because the Sun is a relatively small star, it is unlikely to reach sufficiently high temperatures for other fusion reactions to occur, producing elements heavier than helium. Since 2 per cent of the Sun's mass consists of those heavier elements they must have been there when the Sun and planets were first formed. These elements must have resulted from previous nuclear

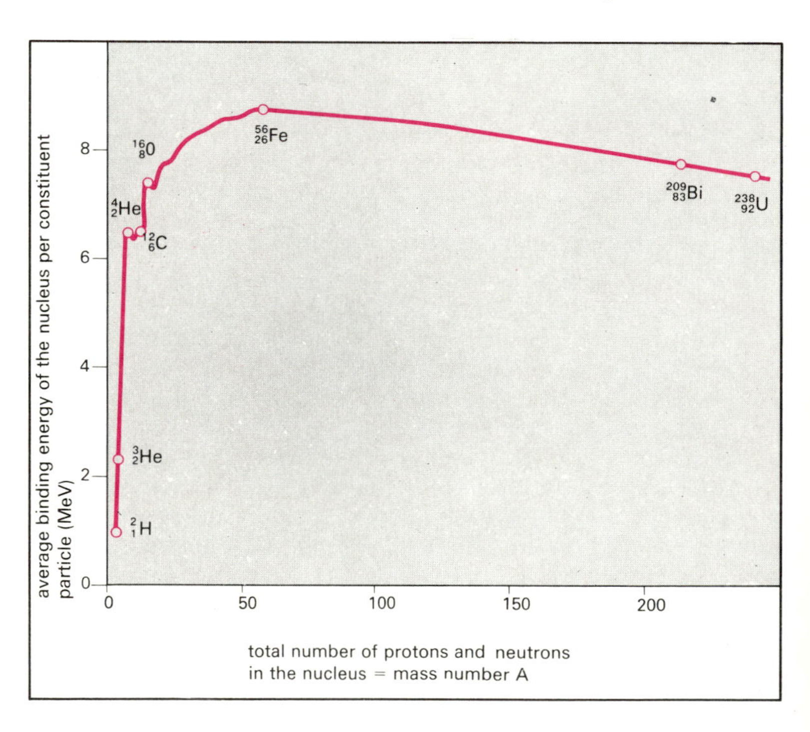

9.3: *The nuclear binding-energy curve: a plot of the energy required to break apart the nuclei of different isotopes (per constituent nuclear particle) against mass number. The higher up the curve a nucleus plots, the more stable it is. The energy is measured in mega-electron volts (MeV).*

fusion reactions in other stars more massive than the Sun.

The nuclear binding energy per constituent nuclear particle (protons and neutrons, called nucleons) increases progressively from hydrogen to iron. Energy equivalent to the difference in binding energy between the reactants and products is released in all nuclear fusion reactions which produce progressively heavier nuclei, up to and including iron. The most common of these reactions in stars involve fusion and addition of helium nuclei to make heavier nuclei having mass numbers that are multiples of four. This explains the importance of isotopes such as $^{12}_{6}C$, $^{16}_{8}O$, $^{20}_{14}Ne$, $^{24}_{12}Mg$, $^{28}_{14}Si$, $^{32}_{16}S$ and $^{56}_{26}Fe$ among the heavier elements in the Sun, the planets and in particular the Earth (compare the analysis in Chapter 4 of the Earth's composition). These elements are thought to have originated by nuclear fusion processes (nucleosyntheses) in relatively large stars whose lives terminated in supernovae explosions, in which prodigious quantities of incandescent gas and dust were ejected at high velocity into interstellar space, so providing the raw materials from which new stars and planets eventually formed. During a supernova explosion conditions are right for elements heavier than iron to be fused even though this requires an energy input (ie the reactions are endothermic). Thus isotopes such as $^{238}_{92}U$ and $^{235}_{92}U$, which are naturally fissile (ie spontaneously split to release energy) and radioactive, are generated within the ejected material. The Earth's internal heat production from radioactive sources depends on the incorporation during its formation of such elements fused in ancient stars.

Solar energy available at the Earth's surface

The Earth intercepts 1.72×10^{17} watts of solar energy (5.42×10^{24} J/a) of which about 35 per cent is reflected by clouds and land surfaces or is back-scattered by dust particles in the upper atmosphere, and 65 per cent is absorbed in the atmosphere or at the Earth's surface. The Earth's curvature causes a greater heating effect in equatorial latitudes than at the poles and this differential heating of the atmosphere combines with the Earth's rotation to produce belts of high and low pressure (see Chapters 17 and 18) between which winds blow. It has been estimated that the kinetic energy of winds over the whole globe is about 3×10^{23} J/a (equivalent to about 10 per cent of the absorbed solar energy), most of which becomes dissipated through the atmosphere as heat. However, only a minute fraction of this total power is potentially available to be harnessed because most atmospheric circu-

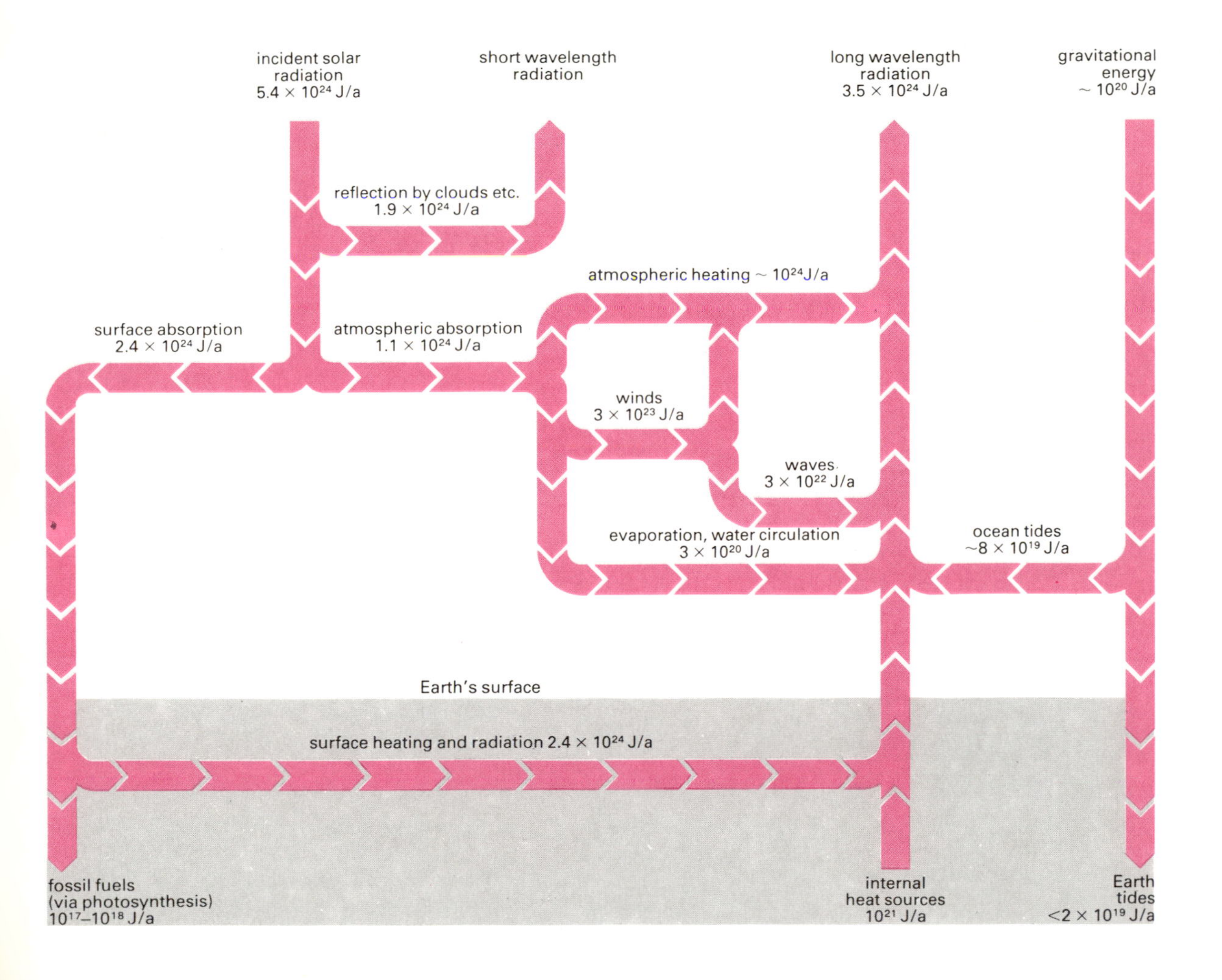

9.4: *A flow diagram summarizing the main sources and directions of energy flow across the Earth's surface and within its oceans and atmosphere at the present day, measured in joules per year (J/a).*

lation takes place at inaccessible altitudes above the Earth's surface. Also, windmills for electrical power generation have low efficiencies. Realistic estimates of available energy from wind power are only about 10^{18} J/a (in comparison with present world energy demands of about 3×10^{20} J/a).

One important consequence of atmospheric circulation is the creation of ocean waves by friction at the sea surface. As much as 10 per cent of the kinetic energy of winds may be converted into water waves although there are major practical and technical problems to be overcome before this energy can be harnessed.

Another small portion ($\sim 3 \times 10^{20}$ J/a) of the absorbed solar energy income is used in evaporating water from the oceans and lakes. This water is returned as rain so that the net effect of evaporation and reprecipitation is to convert solar energy into a gravitational form. Ultimately this energy is also dissipated as heat, through the kinetic energy of water flowing by rivers to the sea. This water flow provides a concentrated and useful form of energy. Although only 5×10^{18} J/a are produced today it is estimated that with appropriate capital investment, about one-third of the available 'hydropower' (ie 10^{20} J/a) might be harnessed.

Wind, wave and hydropower, all natural forms of converted solar energy, together account for a little more than 10 per cent of the absorbed solar energy (3.5×10^{24} J/a). Although most wind energy is mainly added to the heat content of the atmosphere, there is also a direct atmospheric heating effect, so that the atmosphere removes about one-third (1.1×10^{24} J/a) of the absorbed solar energy income. Apart from the small amount used in biological processes (see below) the solid surface of the Earth absorbs the remaining two-thirds (2.4×10^{24} J/a). However, because rocks are poor conductors, heat absorbed at the surface does not penetrate to any great depth, and the incident energy must be reradiated, particularly at night. Since only small amounts of solar energy are converted into energy forms that can be stored relatively permanently by the Earth (mainly fossil fuels; see below), there must be a delicate balance between solar income and terrestrial radiation or the Earth's surface would heat up (a discussion of the 'greenhouse effect' is contained in Chapter 18).

The sharp increases in the world price of fossil fuels during the 1970s have increased efforts to trap even a small part of the Earth's vast solar energy income either by generating electricity directly or by storing the energy in the form of synthetic fuels such as hydrogen and organic compounds. Photoelectric cells at present lack the efficiency to be commercially viable, but heliostat systems, such as that opened in Sicily during 1980, seem to be an interesting prospect. An array of sun-tracking mirrors, or heliostats, focus the Sun's rays onto a solar furnace at the top of a high central tower. Superheated steam is generated and forced at high pressure through a conventional steam turbine which generates electricity; the total efficiency exceeds 15 per cent averaged over long operating periods.

In the longer term endothermic thermochemical reactions, which use solar energy to generate a chemical fuel, may also prove viable. Altogether, about 0.01 per cent of the solar energy reaching the Earth's surface could be used in one form or another, which would exceed the world's present energy production from fossil fuels.

9.5: *Work proceeding at the top of one of the world's largest windmills yet built, at Tvind in Jutland, Denmark. The tower is 52 m high and the three blades are each 27 m long. The windmill was commissioned in 1978 and is designed to generate 4 million kilowatt-hours per year.*

9.6: *Summary of solar energy expenditure and its likely maximum potential as an energy resource.*

Form of energy expenditure[1]	Total energy involved (J/a)	Maximum potential use (J/a)
direct reflection	1.9×10^{24}	
atmospheric circulation (winds and waves)	3.0×10^{23}	$\sim 1.0 \times 10^{18}$
atmospheric heating (includes dissipation of wind energy)	1.1×10^{24}	
evaporation and reprecipitation	3.0×10^{20}	$\sim 1.0 \times 10^{20}$
surface heating	2.4×10^{24}	$\sim 2.4 \times 10^{20}$
biological fuel (via photosynthesis)	1.0×10^{22}	$\sim 1.0 \times 10^{20}$

Note

1 Total solar energy income (5.4×10^{24} J/a) is balanced by three main forms of energy expenditure: direct reflection, atmospheric heating and surface heating $(1.9 + 1.1 + 2.4 = 5.4) \times 10^{24}$ J/a.

9.7: *The solar power plant at Adrano, Sicily.*

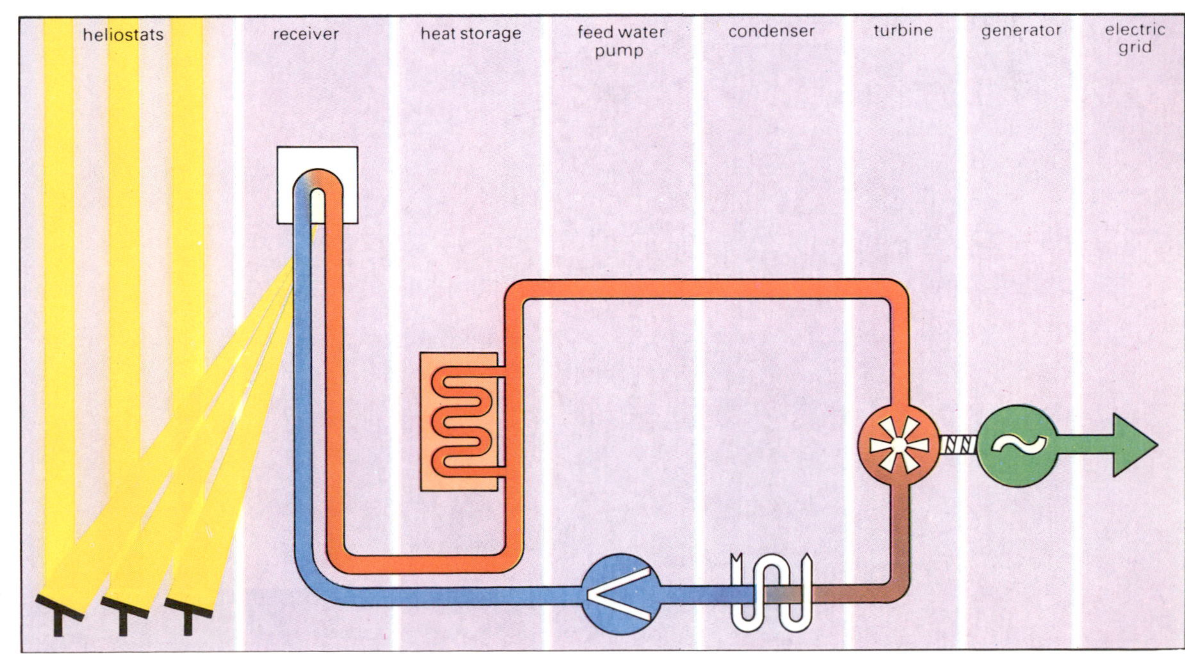

9.8: *A schematic diagram showing the operation of the solar-powered electricity generating plant at Adrano, Sicily. An array of Sun-tracking heliostats focuses the Sun's rays onto a receiver, known as a solar furnace, which generates superheated steam. Steam is forced at high pressure through a generating turbine, then condensed back to liquid form and recycled through the solar furnace.*

Photosynthesis and fossil fuels

All living things on the Earth's surface depend on the trapping of solar radiation by plants. This is true even of higher vertebrates which eat either plants or animals that have eaten plants. The creation of carbohydrate matter in land plants and phytoplankton in the oceans results from the simple chemical reaction:

$$6CO_2 + 6H_2O + 2.8 \times 10^6 \text{ J (solar energy)} \rightarrow C_6H_{12}O_6 + 6O_2$$

In this endothermic reaction the basic raw materials, carbon dioxide and water, produce complex carbohydrates and release oxygen, a process known as photosynthesis. The resulting carbohydrate stores energy until it is broken down by combustion, metabolic activity or some other means. Although at present about 1 per cent of the Earth's available solar energy is used in photosynthesis, much is liberated almost immediately as heat. The remainder, about 1.0×10^{22} J/a, is the new growth component in the form of plants which may be regarded as an addition to the biological fuel (or food) stock, known as the biomass. Thus photosynthesis continuously creates a biomass with an energy content some thirty times greater than the total energy demand of mankind (3×10^{20} J/a). There are many difficulties, however, in burning enough of this fuel to supply our energy needs. The entire animal kingdom depends on plant life for its existence, but as this fuel is consumed and passed on through the food chain it contributes progressively less to the growth of each animal at successively higher nutritional (or trophic) levels. This greatly reduces the amount available as a combustible energy resource. There is also the problem of harvesting a sufficient yield and transporting it to the area of greatest need for combustion. Nevertheless if only 1 per cent of the biomass created each year were burned with reasonable efficiency about one-third of the world's energy demands would be met.

Most of the biological fuel that is not consumed decays naturally in an oxidizing environment to yield heat (eg in compost or a forest fire), but a significant fraction does not. This is preserved in oxygen-poor environments (eg swamps, bogs and marshes) or trapped in marine sediments where it is protected from combustion but undergoes chemical reduction reactions, due mainly to the work of anaerobic bacteria which convert carbohydrates into hydrocarbons. The energy content per unit mass of the material increases and the net result is coal (from plant debris) or oil (from planktonic debris), both of which may be associated with hydrocarbon gases such as methane. The rate of fossil fuel production is difficult to assess from studies of rock sequences because hydrocarbon concentrations vary from near 100 per cent in coal seams down to a few parts per million in other rocks. However, the production rate is probably between 10^{17} and 10^{18} J/a. Very little of the hydrocarbon produced is ever in deposits that are economically extractable. Although the estimated total hydrocarbon content of rocks is about 3×10^{26} J, the energy from that able to be extracted is only about 2×10^{23} J. Of this, 90 per cent is in the form of coal, and oil and natural gas each comprise 5 per cent. Man's use of these fuels is quite disproportionate in comparison to their availability and thus the roughly equal world demand for coal and oil at the present day is a prime contributory factor in the so-called 'oil crisis'. Simple statistics (eg Figure 9.2) suggest that fossil fuels, which have taken several hundreds of millions of years to form, might last several hundred years at present exploitation rates, but this can be true only for coal reserves. Also, electrical energy production from fossil fuels is only 30 per cent efficient so that the production of 2×10^{20} J/a from fossil fuels requires a reserves input equivalent to over 6×10^{20} J/a.

9.9: *Coal seams in an opencast mine at Muswellbrook, New South Wales, Australia. The coal, formed by biochemical degradation and compression of plant debris, is interbedded with sandstones and shales representing the delta on which the plants grew.*

The gravitational influence of the Sun and Moon

The gravitational interactions of the Earth with the Sun and Moon are the source of tidal energy both within the oceans and inside the solid Earth. Because gravitational torques and tensions vary with M/R^3 (M = mass, R = distance) only the Sun and Moon are, respectively, sufficiently massive or close to the Earth to cause significant tidal effects. Chapter 6 explained that gravitational tensions within the oceans tend to cause water to pile up on opposite sides of the rotating Earth such that any point on shore experiences two high and two low tides every day. The relative positions of the Sun and Moon are responsible for the cyclic variation in tidal magnitude. The amount of frictional energy associated with tides can be calculated, from a knowledge of the rotational energy of the Earth, as about 10^{20} J/a, many orders of magnitude less than the solar energy input. About 8×10^{19} J/a appears as the kinetic energy of water motion in the oceans and is ultimately converted into heat, mainly through friction between water and land. Some of the energy of tides can be exploited along coastlines where the natural tidal range (about 1 m) is accentuated by geographical factors so that an exploitable head of water (about 10 m) can be developed across a dam as in the La Rance estuary in France. However, the energy potentially available worldwide is only about 10^{18} J/a.

A small proportion of the energy of tidal tensions is not dissipated through the oceans but is associated with the Earth itself as 'Earth tides'. It is difficult to determine the amount of energy involved as the effect on fluid layers, such as the outer core, is likely to exceed the small distortions that occur at the surface. The amount is unlikely to exceed 2×10^{19} J/a at present, and even though this will be dissipated as heat inside the Earth, the heat energy will be much smaller than the estimated 10^{21} J/a from radioactive sources (see below). However, the relative importance of gravitational and radioactive heating might have changed during the Earth's history.

The solid Earth also experiences a gravitational torque or acceleration, related to its equatorial bulge, which induces precessional movements of the rotational axis (like the cone traced by the axis of a spinning top). But this force is small compared with the combined effects of different forms of tidal friction which act as retarding forces, apparently causing the Earth's period of axial rotation to increase. The day length should therefore increase because the net effect is to transfer angular (rotational) momentum from the Earth to the Moon, which experiences an acceleration. The Moon should therefore orbit more rapidly and move away from the Earth. Interpretations of the geological history of the Earth have suggested that the Moon was at some time much closer to the Earth and must have raised larger tides involving more energy than today. Thus internal heating due to external gravitational forces may then have been more important. While there is some evidence for the truth of this argument the picture is not necessarily quite so simple (see below).

Manifestations of the Earth's internal heat

Volcanoes, hot springs and earthquakes

The occurrence of both volcanoes and hot springs demonstrates that the heat of the interior of the Earth can produce molten rock at temperatures of up to 1500°C and superheated steam. These phenomena are mainly confined to several narrow and elongate zones along certain continental margins and ocean-ridge zones, the currently active boundaries of the mobile tectonic 'plates'. New oceanic crust is created by melting processes beneath ocean-ridge zones (Chapter 13) whereas the rather different melting processes that create new continental crust occur beneath active continental margins (Chapter 14), where oceanic crust is resorbed into the mantle. Both types of melt zone are strongly associated with the lateral movements taking place

9.10: *Explosive contact of lava with seawater during a volcanic eruption of Eldfjell on the island of Heimaey, Iceland, in April 1973.*

within the outer rigid skin or lithosphere of the Earth, which consists of crust and uppermost mantle. However, lateral movements of the lithospheric plates and the occurrence of volcanoes are not simply cause and effect—both must have some deeper underlying cause and the movement of the plates gives further evidence of the heat of the Earth's interior. Thermal convection in the mantle beneath the lithosphere (>100 km depth) is probably responsible for plate motions, and volcanoes are mainly confined to the thermally most disturbed regions, the plate boundaries.

In any year there are innumerable volcanic eruptions, but few of these are either large enough or in sufficiently populated regions to become newsworthy. However, volcanologists keep a close watch and have produced several estimates, all close to 10^{19} J/a, of the amount of energy involved in eruptions. This usually includes the thermal and kinetic energy released at the surface by lava flows, ash eruptions and steam venting (hot springs). The kinetic energy of plate motions is determined simply by combining an average plate velocity (50 mm/a) with the mass of the lithosphere (about 2×10^{23} kg) to give nearly 10^{13} J/a, a relatively small amount of energy compared with that from volcanoes, earthquakes and the heat conducted through the plates. Because the plates are rigid and move in different directions, large amounts of strain energy may be built up and stored, with little plastic deformation along their margins especially where two plates move towards or past each other. Ultimately a rupture (or fault) develops and the strain energy is released as sudden relative movement across the fault plane and in the form of seismic waves which propagate through the Earth away from the fault zone (see Chapter 3): this is an earthquake. Through these motions, the strain energy is dissipated as heat, due to frictional shear along the fault and to vibrations of particles within the material of the Earth.

Earthquakes are confined to the same narrow zones along plate margins as volcanoes and hot springs; many are therefore caused by the relative movements between the lithospheric plates. Others are caused by the volume expansions and contractions associated with

9.11: *A geyser on the mid-Atlantic ridge at Strokkur, Iceland, liberating heat from the Earth's interior in the form of steam.*

the production and crystallization of molten rock. In energy terms a single large earthquake may release about 10^{17} J (eg the 1906 San Francisco earthquake which had a magnitude of 8.2 on the Richter scale). Although such large earthquakes are exceptional the average year brings several magnitude 7 shocks, many of them in relatively unpopulated regions, and a large number of small shocks whose detection is increasing. The total annual release of energy from earthquakes is between 10^{18} and 10^{19} J, which is almost the same as that from volcanoes, but together these only account for a few per cent of the internal energy being 'expended' (in the terms of Figure 9.2) by the Earth. Generally there is little volcanic or seismic activity in the interiors of the plates, yet enormous amounts of energy are transmitted through them as conducted heat. Although the manifestation of this energy is not itself spectacular, it can be measured and these measurements do lead to impressive conclusions about the distribution of energy sources inside the Earth.

Terrestrial heat flow

Measurement

The rigid lithosphere of the Earth conducts heat to the surface from the deep convecting interior, and heat is therefore flowing from the Earth's surface. Since heat flow varies from place to place, some simple means of making accurate measurements is needed. One way of doing this might be to place a saucepan of water on the ground and to observe the rate at which it warms up. But terrestrial heat flow is so small that it would take about two months for the water temperature to rise by 1°C, by which time surface temperatures may have ranged through 20–30°C as a result of solar heating. It is necessary to make measurements at depth in mines or boreholes and this raises the question as to what thermal quantities are being measured.

If heat is flowing from the hot interior to the cool surface, temperature should increase with depth. Figure 9.12 shows a set of temperature measurements made with sensitive electronic thermometers, called thermistors, down a borehole 1 km deep. As expected, temperature does increase with depth but in a series of steps, each characterized by a uniform rate. This rate, the thermal gradient, is actually the inverse of the gradient measured on each step of the graph in Figure 9.12. It is obtained by dividing the difference in temperature at either end of a step ΔT by the depth z of that step. Thus the thermal gradients are 20, 47 and 21°C/km for steps 1–3 respectively. It is not immediately obvious why the rate of temperature increase should vary until the changes of rock type down the borehole and variations in the ease with which different rocks transmit heat are considered. Thus some, such as the shale in step 2, act as more effective insulators of heat than others, such as the limestone and granite in steps 1 and 3. A useful quantity which describes the ability of a material to transmit heat is thermal conductivity k; most metals have high thermal conductivities, whereas most rocks have low values. Limestone and granite have higher thermal conductivities than shale and this causes the variations in thermal gradient. Thermal conductivity is measured in watts per metre per degree (W/m/°C) and describes the energy flowing through a metre of rock across the ends of which there is a temperature difference of 1°. This quantity can be combined with the thermal gradient (°C/m) and the product, in watts per square metre, gives the heat flow q. Thus $q = k(\Delta T/z)$ W/m^2. With this equation and the data of Figure 9.12 the heat flowing through this part of the Earth's crust can be calculated. Remembering that $(\Delta T/z)$ must be converted into units of °C/m, heat flow has almost the same value for each step: 0.07 W/m^2, or 70 mW/m^2 (milliwatts per square metre).

In oceanic areas, where there are soft sea-bottom sediments, a long cylindrical probe carrying thermistors can be plunged into the sediments in order to measure the thermal gradient over a depth of several metres. The probe usually contains a heating element which can be operated remotely, and by registering the rate of temperature increase for a known heat input a measure of thermal conductivity is obtained. This method is simple because the sediments are soft, and because temperature fluctuations on the sea floor are small so that meaningful results can be obtained with small depths of penetration. Thus, of 5000 reported terrestrial heat-flow measurements, nearly 70 per cent are from oceanic areas.

Reliable results are much more difficult and time-consuming to obtain in continental areas because boreholes at least 100 m deep are

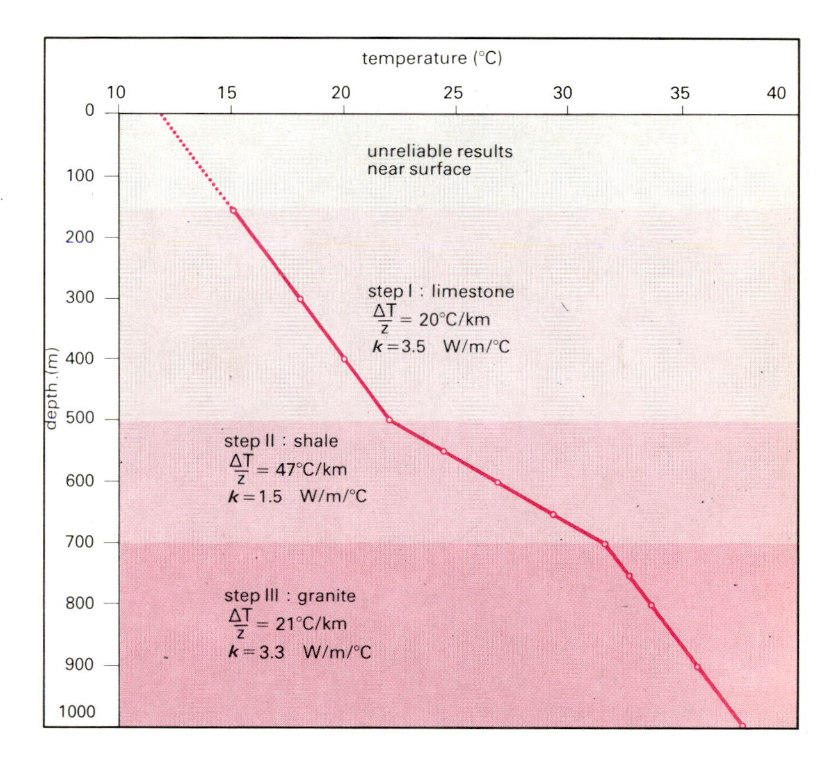

9.12: *A temperature–depth profile down a hypothetical borehole through rock strata with differing thermal conductivities as indicated. Heat flow is constant throughout the column; therefore the lower the conductivity, the greater the increase of temperature with depth, ie the steeper the thermal gradient.*

required in order to escape the effects of daily and annual temperature fluctuations at the surface. Heat-flow studies are therefore usually carried out in boreholes that have been drilled for some other purpose. The temperature profile is again obtained by use of thermistors but, for hard rocks, laboratory studies of core samples are required to obtain thermal conductivity. By maintaining a constant temperature difference between the two ends of a divided-bar apparatus (Figure 9.13) and ensuring good insulation, the temperature differences across known lengths of the rock and a standard material, such as silica, are found. Since the heat flowing through the silica and rock is the same (as in the borehole in Figure 9.12), then, using the previous equation, $k_R(\Delta T/z)_R = k_S(\Delta T/z)_S$, where $_R$ refers to the rock specimen and $_S$ to the silica. Since k_S is known k_R may now be calculated and combined with the value of $\Delta T/z$ obtained from the borehole to give q.

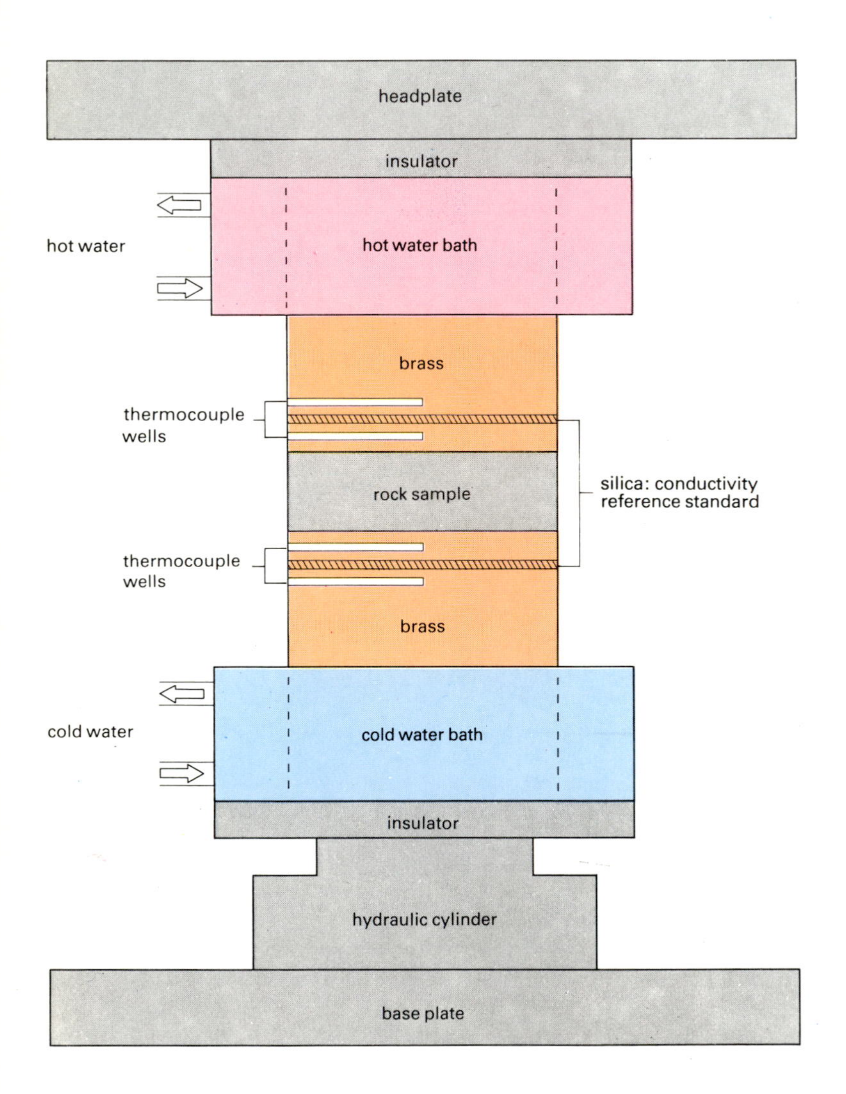

9.13: *A divided-bar apparatus for measuring thermal conductivity in rock samples. The apparatus, shown here in cross-section, is cylindrical. Its centre section comprises lengths of silica and rock across which temperature differences are compared using sensitive thermocouples. Temperature is maintained at constant values at each end of the apparatus using water baths; thus, heat flow through the silica and rock is constant and the temperature gradients measured are inversely related to the thermal conductivities of the two materials.*

Distribution and interpretation

From a statistical survey of 5000 terrestrial heat-flow measurements a world heat-flow map has been drawn (Figure 9.20). Measured values vary from about 30 to over 200 mW/m^2, but the global mean is estimated at 60 mW/m^2, giving an integrated heat flow through the Earth's surface of 10^{21} J/a. The most important zone of high heat flow, centred over the East Pacific rise, coincides with the most rapid rates of present-day sea-floor spreading. Less striking increases are observed over the relatively slow-spreading ridges of the Atlantic and Indian Oceans. Rather more than half the Earth's heat loss takes place through ocean-ridge zones. In continental areas the zones of highest heat flow occupy the same linear belts as those defined by volcanic and earthquake activity. Thus the youngest magmatically active zones of both oceans and continents have the highest heat flow which decreases in both with increasing age. However, there the similarity ends, for the patterns of heat flow reflect different underlying causes related largely to differences in crustal thickness and composition.

Figure 9.14(a) shows the systematic decrease of heat flow with increasing age for the Pacific Ocean; values for young ocean floor, close to the ridge, are high and variable (indicated by the large standard deviation error bars) whereas those for relatively old oceanic lithosphere (>60 Ma) are low and fairly constant. In Figure 9.14(b) the same kind of data from both the north Pacific and north Atlantic are combined and plotted on logarithmic scales to show that different ocean basins have the same pattern of heat flow. Figure 9.14(b) shows that $\log q$ is proportional to $-0.5 \log t$, where t is the age of ocean floor in millions of years; thus heat flow is inversely proportional to the square root of the age of the oceanic lithosphere, ie $q \propto t^{-\frac{1}{2}}$. This relationship holds for all but the oldest parts of the ocean basins.

The reasons why heat flow should decrease in this way are related to the formation of oceanic plates from hot mantle-derived material at ocean ridges. This is followed by mechanical spreading and cooling as the plates age and migrate away from the ridges, eventually to be resorbed into the mantle at destructive plate margins. This surface plate cycle results from some kind of convection at depth in the mantle, as yet not clearly understood. As an oceanic plate moves away from the ridge zone and cools, it thickens progressively because underlying partially molten material from the asthenospheric mantle freezes to its base. The temperature difference between its upper and lower surfaces remains roughly constant, however, so the temperature gradient and therefore the heat flow fall. However, the rate of cooling and thickening decrease with time, thus accounting for the curve of Figure 9.14(a) which expresses the inverse square-root relationship $q \propto t^{-\frac{1}{2}}$. This observation is confirmed by bathymetric and gravity measurements over ocean-ridge zones (see Chapter 13) which indicate that ridges occur because the material beneath is thermally expanded, or less dense, than elsewhere beneath the ocean basins. Since most of the heat escaping by conduction through the oceanic lithosphere is derived from the hot mantle beneath, and little is generated by radioactive sources within the lithosphere, the sources of continental and oceanic heat flow are quite different.

The continental lithosphere is known to vary considerably in both composition and thickness; both these and the temperatures within the lithosphere apparently depend upon the 'age' of the region, ie the time since it was last involved in orogenic or metamorphic processes. This is not necessarily the time of its formation because old metamorphic rocks may have experienced repeated episodes of metamorphism and may have young radiometric ages (see Chapter 8). Figure 9.16 summarizes the decrease with age of temperatures and hence of heat flow within the continental lithosphere for several continental areas. Whereas oceanic heat flow (Figure 9.14(a)) decreases from over 200 to 40 mW/m^2 in about 100 Ma, a similar decrease in continental areas from 70 to 40 mW/m^2 takes about 1000 Ma. Thus, although both curves flatten off at about 40 mW/m^2, the decrease takes much longer in continental areas, again reflecting the contrasting heat sources beneath the oceans and continents. This evidence for the

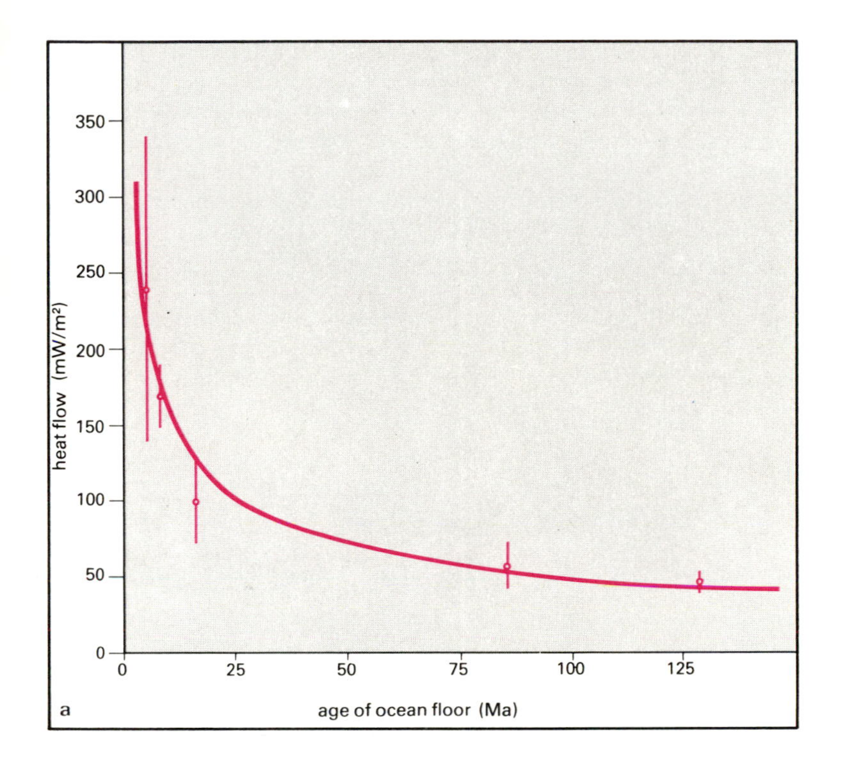

9.14: *The decrease of heat flow with age in the ocean basins,* (**a**) *using linear scales, where a smooth curve has been fitted to averaged values of measured heat flow in the five age zones of the Pacific Ocean floor, and* (**b**) *using logarithmic scales on both axes and data from both the north Atlantic and Pacific Oceans. The error bars show the standard deviations about the mean values.*

9.15: *The concentrations of elements with long-lived radioisotopes and resultant heat productivities for the Earth's crust and mantle.*

	Uranium (ppm)	Thorium (ppm)	Potassium (%)	Heat production (μW/m³)	Volume (m³)	Total heat production (W)	(J/a)
average continental crust	2.3–4.7	9–20	2.5–4.2	1.5–3.0 ×	(5.0 × 10^{18}) =	1.1 × 10^{13} =	3.5 × 10^{20}
average oceanic crust	0.9	2.7	0.4	0.4 ×	(2.5 × 10^{18}) =	1.0 × 10^{12} =	0.3 × 10^{20}
fresh, unmelted mantle	0.015	0.08	0.1	0.02 ×	(9.0 × 10^{20}) =	1.8 × 10^{13} =	5.7 × 10^{20}
total heat production, crust and mantle							= 9.5 × 10^{20}

continental lithosphere containing a heat source is borne out by the average concentration of long-lived radioactive heat-producing isotopes in the upper continental crust (of granodioritic composition) which is about five times greater than in basaltic oceanic crust. The critical isotopes are ^{40}K, ^{232}Th, ^{235}U and ^{238}U, which have radioactive decay half-lives ranging from 700 to 50 000 Ma. This means that their heat output reduces to half its former value on a time scale of the order of 10^9 years. This decay partly explains the pattern of continental heat flow.

Convincing proof that radioactive heat production A in the upper part of the continental lithosphere contributes to continental heat flow (q) is derived from a graph of these quantities (Figure 9.17). Heat flow correlates linearly with surface heat production A_0 in most continental regions, the lowest q and A_0 values being for ancient continental shields and the highest for young mountain belts. The relationship is described by the equation $q = q^* + A_0 h$, where the intercept q^* (also known as reduced heat flow) is the heat flow coming from beneath the radioactive layer whose thickness is a function of the gradient h. Detailed observations have shown that variations of q^* from province to province are relatively small. The decrease of continental heat flow with age is not simply a function of radioactive decay because the upper (relatively radioactive) parts of the continental crust are subject to erosion. It is generally agreed that to maintain the relation discussed above, heat production in the continental crust must decrease logarithmically with depth. Thus, continental heat flow consists of two main components: q^*, a relatively constant input from deep sources, and a quantity $A_0 h$ which decreases with time as heat production A_0 decays and erosion takes place. In addition, as for the oceanic lithosphere, a small transient contribution to the heat flowing from young continental lithosphere may result from its residual magmatic 'heat of formation'.

Figure 9.18 summarizes estimates of the average surface heat flow and the contribution from beneath the Mohorovičić discontinuity. The similarity of average oceanic and continental heat flow is coincidental: whereas only a small fraction of oceanic heat flow is due to crustal (above the Mohorovičić discontinuity) sources, about half the continental heat flow is generated within the continental crust. Heat flow from subcontinental mantle sources must therefore be lower than from suboceanic mantle sources. There are two possible reasons. First, the lower parts of the continental lithosphere may contain less radioactive heat sources and, second, the conductive layer, or lithosphere, is thicker beneath the continents (on average about 200 km) than beneath the oceans (about 50–100 km). Assuming roughly similar temperatures at the bases of the two lithospheres, the average continental temperature gradient and therefore associated heat flow contribution from the deep mantle must be less than from beneath the ocean basins (as in Figure 9.18).

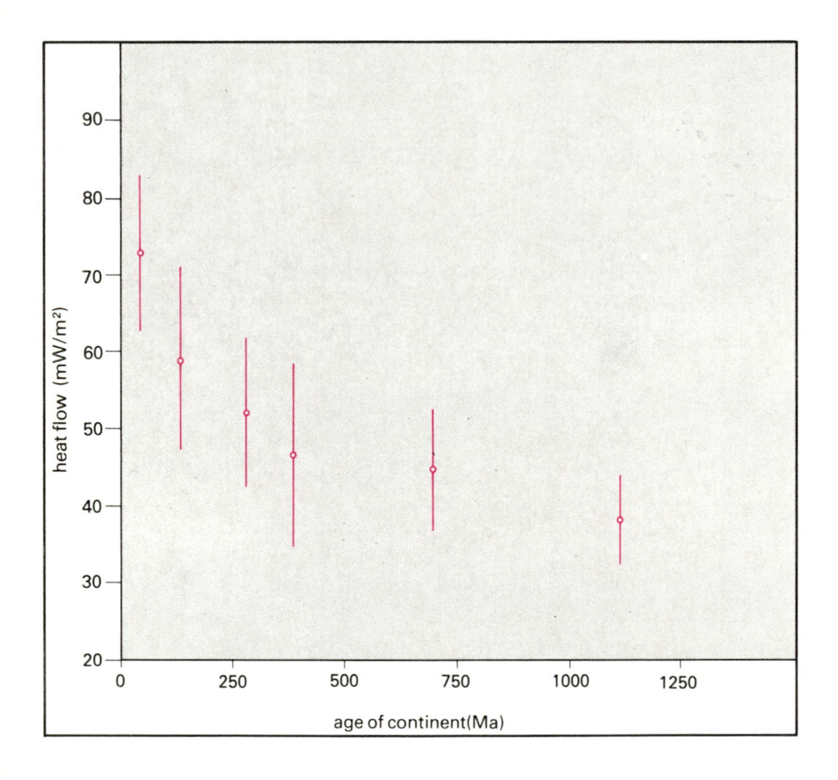

9.16: *The decrease of heat flow with age in continental areas using data from all continents. Notice that continental heat flow decreases to an 'equilibrium' value in about 1000 Ma compared with 100 Ma for oceanic heat flow. Error bars indicate standard deviation about the mean values.*

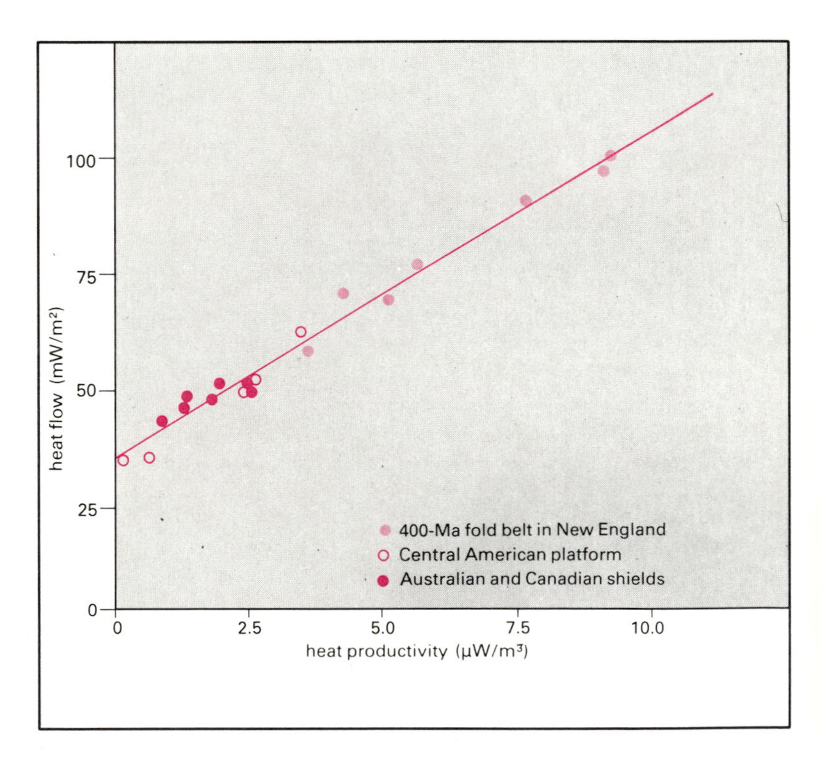

9.17: *The relationship between heat flow and surface heat production for various continental areas. Continental heat flow has two components: heat produced within the upper radioactive layer and heat flowing from deeper within the Earth.*

9.18: *Heat flow averages for oceanic and continental regions.*

	Continental average (W/m^2)	Oceanic average (W/m^2)	Total (W)
through the Earth's surface	53×10^{-3}	62×10^{-3}	3.2×10^{13}
through the Mohorovičić discontinuity	28×10^{-3}	57×10^{-3}	2.4×10^{13}

The Earth's magnetic field: heat flow from the core

The variation in the magnetic field on all imaginable scales of time was shown in Chapter 7 to imply a dynamo-like origin within the Earth's outer core which would require an energy source (see Figure 9.2). Much recent research into field generation mechanisms has centred on the alternative theories of thermally induced convection requiring a heat source in the outer core, or gravitationally induced convection, whereby the fluid outer core continually produces metallic grains which sink, inducing outer core movements under gravity, and then accrete to the surface of the inner core.

In the first case it has been suggested that small quantities of radioisotopes, particularly ^{40}K, may have been incorporated into the outer core when the Earth first separated into layers. It is assumed that the radioactive source is still present and despite the exponential nature of radioactive decay it must still produce enough heat to keep the outer core in motion. The outer core must therefore be above its estimated melting point (of between 2000 and 3000°C). Gravitationally induced convection would require the outer core to be at its melting temperature for iron-rich particles to be crystallizing slowly. The gravitational potential energy of these dense particles would then be exploited to produce convective movements in the outer core. This theory assumes sulphur to be the light element with iron in the outer core (see Chapter 4) and that, as some grains of iron crystallized, the melting (or crystallization) temperature of the remaining, relatively sulphur-enriched, liquid would become lower because of the change in composition. The temperature of the outer core would be falling slowly and continuously, as more and more grains of iron were produced. At the lowest temperature possible for liquid iron–sulphur without the entire mass becoming crystalline (at about 1900°C—the eutectic point—Figure 9.25) the inner core would be 25 times its present volume.

An important consequence of both theories is that the core must be losing heat, either because of continued radioactive heat production, or because it is crystallizing. There is a finite flow of heat from the core to the mantle in the range 0.4 to 1.6×10^{13} W or 25 to 100×10^{-3} W/m^2. This is about the same as the range of variation of heat flow from the surface of the Earth, but the comparison is deceptive since the surface area of the core is less than one-third that of the Earth. However, as much as one-third of the Earth's surface heat flow could be generated from within the core.

Use of the Earth's internal heat: geothermal energy

The close association between hot springs and volcanically active regions such as the boundaries of the tectonic plates was noted earlier. These are areas of exceptionally high heat flow where hot water circulatory systems, usually driven by the thermal energy of shallow magma bodies, are set up near the Earth's surface. This volcanic and hydrothermal activity provides an important source of geothermal energy which has been harnessed by power plants such as those in New Zealand (Figure 9.19), Italy, Iceland, Japan and western North and Central America. High pressure steam at temperatures up to 250°C may be ejected from the ground at velocities up to

9.19: *Wairakei geothermal power station, North Island, New Zealand. It is one of the world's most powerful geothermal plants with an output of 175 MW, generating about 10 per cent of New Zealand's electricity.*

500 m/s, but its use is not necessarily to drive turbines and generate electricity. In Iceland, for example, steam and hot water circulated through a piping network provide domestic and industrial heating. Research into the extraction of heat from the ground is going on in other areas where the manifestation of geothermal heat is less spectacularly developed than along plate boundaries.

Figure 9.18 shows that the average surface continental heat flow is 53×10^{-3} W/m^2 and an average thermal conductivity for near-surface crustal rocks might be 2.5 W/m/°C. These data give an average continental thermal, or geothermal, gradient of 0.021°C/m or 21°C/km increase of temperature with depth. The surface gradient of 21°C/km is only an average which conceals considerable regional differences when variations in heat flow and thermal conductivity are taken into account. The gradient must decrease with increasing depth in the crust because of the logarithmic decrease of heat production mentioned earlier; all other things being equal, the rate of temperature increase with depth (geothermal gradient) is linearly correlated with heat flow, which decreases to about half its surface value at the Mohorovičić discontinuity (Figures 9.17 and 9.18).

To be useful for geothermal purposes, water or steam must be at temperatures in the range 60–100°C for domestic and industrial heating, or 160–200°C for electricity generation. Given the average geothermal gradient just calculated, drilling to uneconomic depths of 5 or 10 km respectively would normally be needed to find such water. However, outside plate boundary zones there do exist many 'semi-thermal' regions where the temperature gradient is sufficiently high for economic exploitation. Figure 9.21 is a subsurface temperature distribution map of western Europe at 1.5 km depth. Away from the Italian plate margin, site of the earliest geothermal power station at Larderello, many areas have elevated gradients. Two diverse examples are in the Paris area and south-west England.

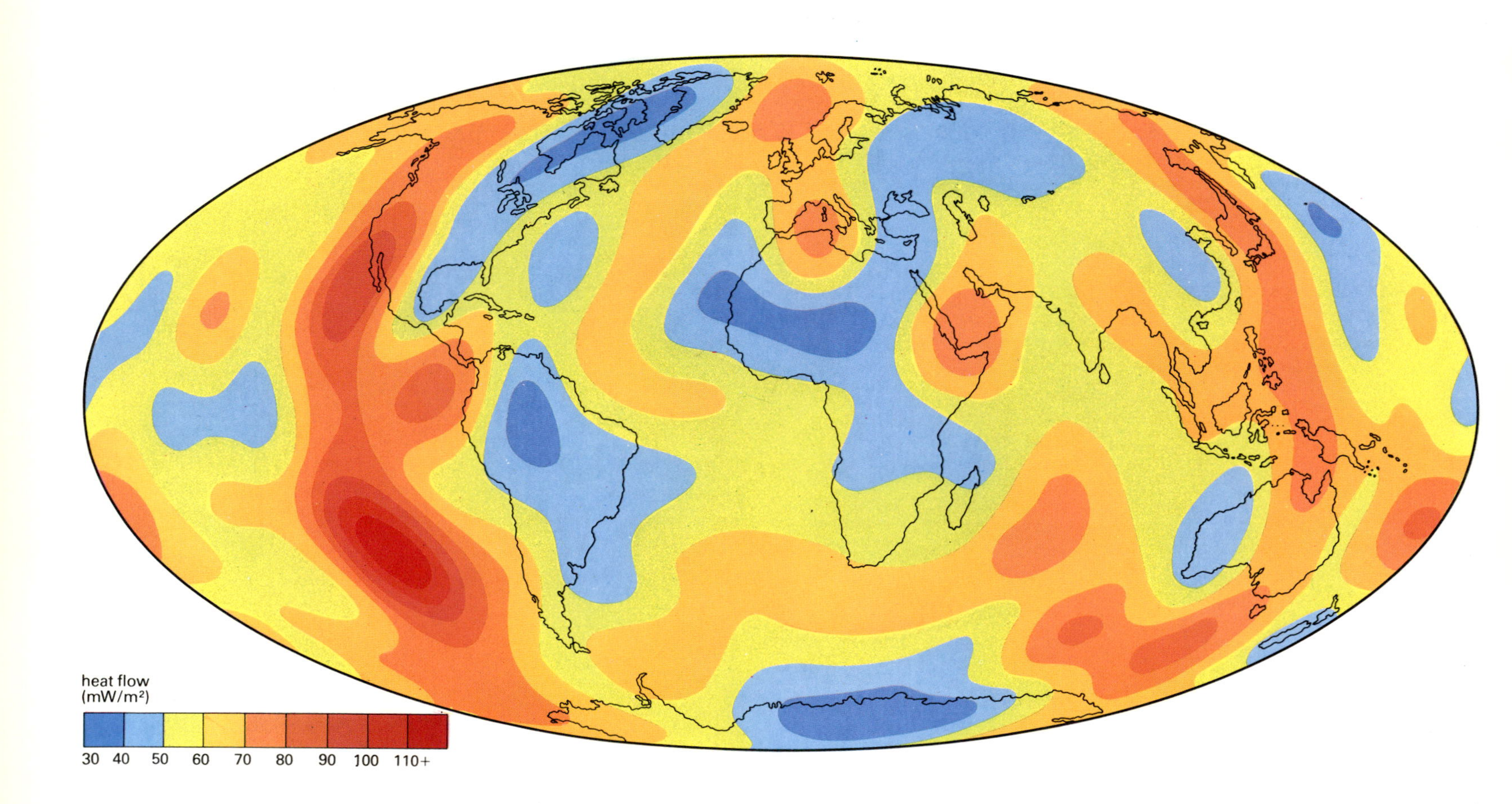

9.20: *Heat-flow map of the world. The map is based on 5000 terrestrial measurements of heat flow, with contours every 10 milliwatts per square metre. The map shows that the zones of highest heat flow occur over active ocean ridges and the zones of lowest heat flow occur within geologically ancient continental interiors and in the older parts of the ocean basins remote from the ocean ridges.*

9.21: *Subsurface temperature distribution at 1500 m depth beneath central and western Europe based on data from deep boreholes and extrapolations using near-surface heat flow/thermal conductivity measurements. Notable high-temperature zones are the active volcanic areas of Italy, the Massif Central and the Rhine Valley; the deep sedimentary basin centred on Paris, and the radioisotope enriched crystalline massif of south-west England.*

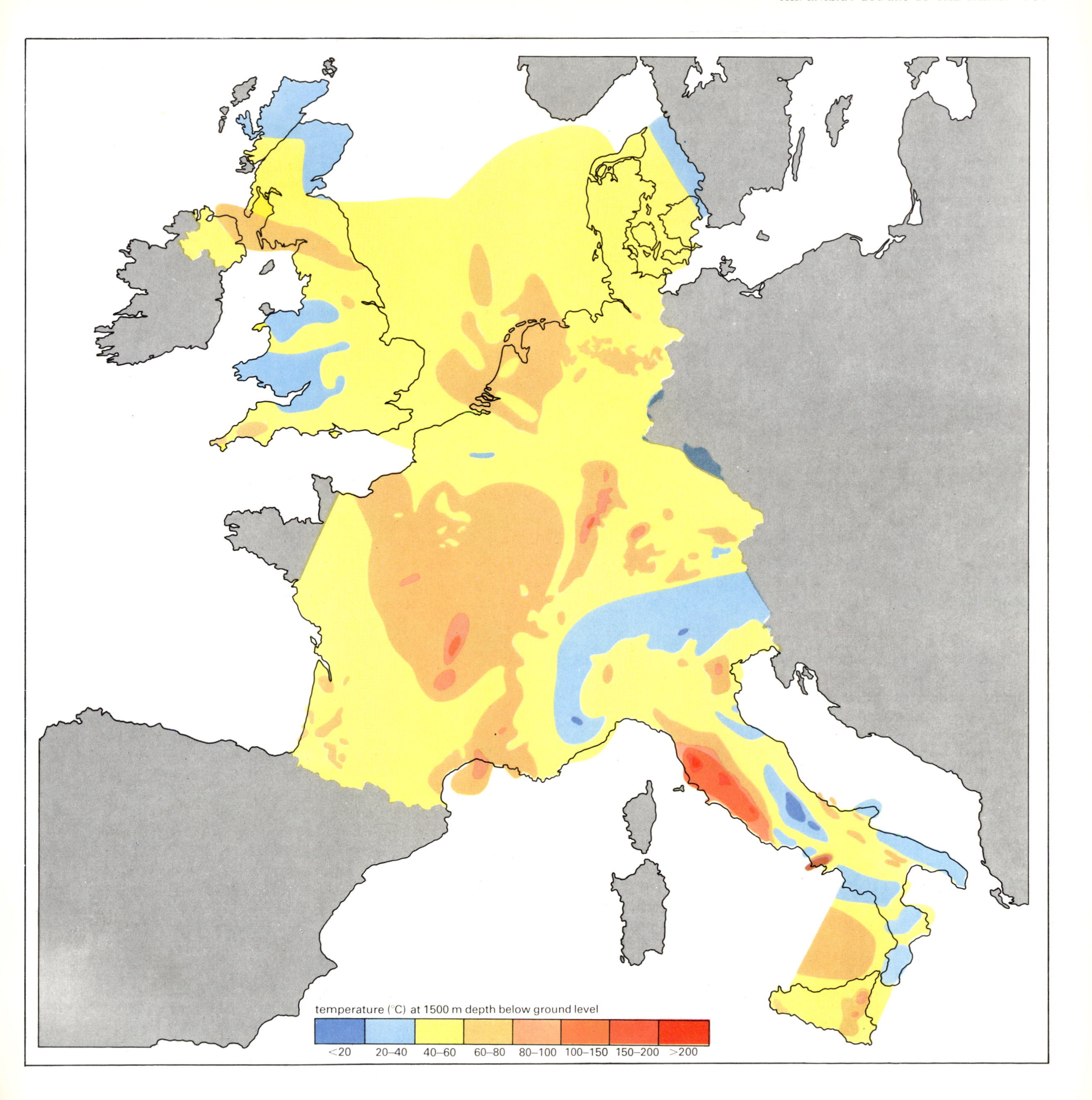
temperature (°C) at 1500 m depth below ground level
<20
20–40
40–60
60–80
80–100
100–150
150–200
>200

Beneath the city of Paris lies a large basin of sedimentary rocks, permeable sandstones and limestones making up a sandwich with layers of impermeable clay. The sedimentary sequence has a low thermal conductivity compared with an equivalent thickness of crystalline rocks and, therefore, the temperature gradient down through it is above average. The water which can migrate freely through the permeable sedimentary units has been tapped by drilling to depths of 1000–2000 m and brought to the surface at 60–100°C to provide space and water heating for several large apartment complexes. A disadvantage of this type of system is its finite lifetime because heat is extracted faster than it can be replenished through low conductivity strata.

In contrast with the Paris basin high subsurface temperatures in south-west England are associated with a large crystalline massif composed of granitic rocks with high concentrations of radioactive isotopes. Both heat production and surface heat flow are high and may remain so down to several kilometres depth. Although the thermal conductivity of granite is also high compared with that of clay-rich sedimentary rocks, elevated subsurface temperatures are thought to characterize such areas and there is a good chance that even temperatures of 160–200°C may be encountered at 4–5 km. Since granite is crystalline and has very limited permeability the exploitation of geothermal heat is a more formidable engineering task than in sedimentary basins. Techniques of energy extraction from hot crystalline rocks are being developed at the Los Alamos Scientific Laboratory (LASL) in New Mexico, and by the Camborne School of Mines in England. The hot dry rock (HDR) technique consists of drilling twin deviated holes to a depth at which appropriate rock temperatures are

9.22: *Adaptation of the LASL hot dry rock geothermal scheme being developed and tested by the Camborne School of Mines in south-west England. (Granite shown here has a cover of sedimentary rocks—see Figure 9.24—but experiments are being conducted first where granite comprises the entire section.) Hydraulic fracturing techniques are used to create a heat exchange surface at depth between the injection and production wells (see arrows) which can be exploited for steam production and the generation of electrical power.*

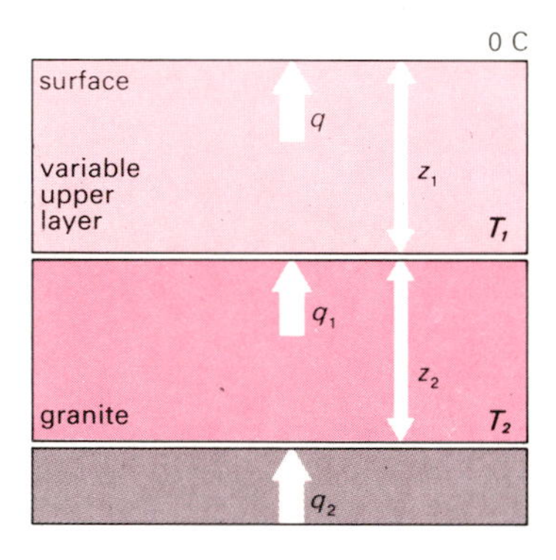

9.23: *A schematic diagram of two rock layers illustrating the calculations in Figure 9.24. Symbols are q = heat flow, z = thickness, T = temperature and k = thermal conductivity.*

9.24: *Projected bottom-hole temperatures at 4 km depth for different crustal successions.*

Nature of succession[1]	Thermal conductivity of top 2 km (W/m/°C)	Temperature at 4 km depth[2] (°C)
all granite to 4 km	3.3	121
2 km limestone, 2 km granite	3.5	118
2 km sandstone, 2 km granite	3.4	119
1 km limestone, 1 km mudstone, 2 km granite	2.5	141
2 km mudstone, 2 km granite	1.5	194
2 km shale with 10% coal, 2 km granite	1.3	215

Notes

1 The successions comprise an upper layer 2 km thick which varies in rock type, overlying a buried granite. Heat flow is constant through each column at 100 mW/m^2 (ie $q = q_1 = q_2$ in Figure 9.23).

2 Calculated from: $q = k_1(\Delta T/z)_1 = k_2(\Delta T/z)_2$ and since $z_1 = z_2 = 2$ km, it follows that $(\Delta T)_1 + (\Delta T)_2 = 2[(q/k_1) + (q/k_2)]$. Surface temperature is assumed to be 0°C.

reached. Hydraulic fracturing enhances the natural fracture system at depth through which water is circulated and recovered for energy extraction through heat exchangers.

Although these systems are experimental their results open up immense worldwide potential for direct energy extraction from crystalline rocks. Ideally, the combined geothermal attributes of low conductivity sedimentary rocks and highly heat productive crystalline rocks might both be exploited if buried granites can be detected successfully. Clearly the right kind of low conductivity sedimentary layer can effectively insulate the buried granite, further enhancing its geothermal potential and obviating the need for excessively deep drilling. Thus, although only about 5×10^{16} J/a of energy are produced today from geothermal sources, this may increase by between one or even two orders of magnitude by the year 2000.

The origin, transport and variation of internal heat

Earlier in this chapter it was implicitly assumed that the heat flowing from the Earth's surface is almost entirely due to radioactive heat sources (see Figure 9.2). Although it has been shown that a significant proportion of continental heat flow does arise in this way, to prove that this is true in general requires much more knowledge about the concentrations of heat-producing elements deep inside the Earth, the way in which heat is transported in the Earth and the relative importance of different transport mechanisms at different stages in the Earth's history. Although it has proved possible to estimate the radioelement content of the crust and mantle from samples (Figure 9.15), these estimates must be treated with caution because they are based entirely on material from the top 200 km of the Earth, leaving 90 per cent unsampled. However, if these concentrations are typical, then the Earth is producing almost enough radiogenic heat to account for its surface heat flow (about 10^{21} J/a). Figure 9.15 shows that the continental crust alone produces about a third of the Earth's radiogenic heat, but it would be imprudent to conclude that the Earth has no other internal heat sources, eg probable heat flow from the core due to radioactive or gravitational sources.

Alternative heat sources in the Earth

When the Earth formed by the accretion of small particles it would have been heated by several mechanisms whose combined effect produced what is known as its original, or primordial, heat. First, the kinetic energy of the particles impacting on the growing surface of the Earth and released as heat has been calculated as 38×10^6 J/kg which if it happened rapidly enough could have raised the temperature of the whole Earth by about 20 000°C. Just how much of this energy was retained rather than being lost back into space by radiation is an open question because there are two major unknowns: the size of the accreting particles and the rate of planetary growth. If the particles were of dust size then accretion would have had to be accomplished in 10 000 years for core-forming temperatures to be achieved (2000–4000°C at the range of Earth pressures, depending on the composition of the core; Figure 9.25). If the particles were much larger, up to 0.1 per cent of the Earth's mass, then their kinetic energy would have been transferred to the deep Earth more efficiently by shock waves and mechanical stirring; accretion with the same result could then have taken several millions of years. As there is good geochemical evidence that the core did form during and after accretion from chemically homogeneous starting material, rather than forming before the silicate mantle (known as heterogeneous accretion), we may take it that the thermal conditions were provided largely by accretional impact energy.

Two other heating mechanisms produced progressively higher temperatures towards the Earth's centre: adiabatic heating and core formation energy. If a volume of any material, such as air in a bicycle pump, is compressed its temperature increases. This is adiabatic heating. Pressure due to the overlying mass of accreted material on the Earth compressed the material at depth and caused its temperature to rise. The rate of adiabatic temperature increase with depth (or adiabatic temperature gradient) in the Earth is about 0.15°C/km, which could have raised the temperature at the centre by about 900°C immediately after accretion. If there had been no additional heat source, the central temperature would immediately have started to fall by conduction of heat to the surface. In any case, the Earth's core could not have formed through adiabatic heating alone because the increase in melting point with depth for possible core materials is as great (Fe–S mixtures) or greater (pure Fe) than the adiabatic temperature increase with depth.

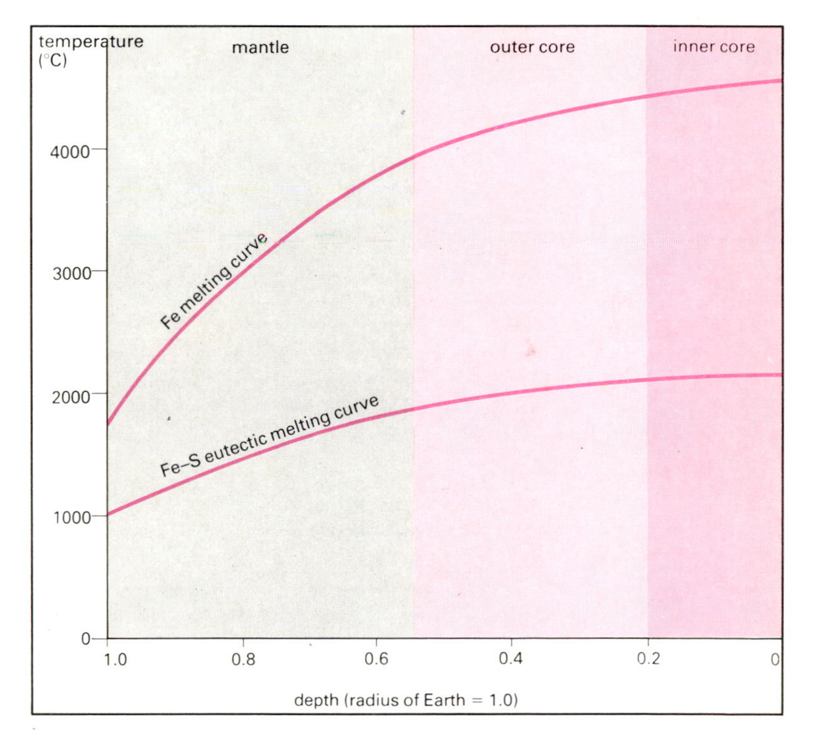

9.25: *The increase of melting temperature with depth in the Earth for pure iron and for iron–sulphur eutectic mixtures (ie the mixture having the lowest melting temperature). Although both curves show higher melting temperatures with increasing depth, that for iron–sulphur mixtures is less steep than that for pure iron. The temperature of the outer core is thought to lie between the two curves.*

Core formation involved the sinking of relatively dense iron-rich materials which displaced lighter mantle silicates upwards, releasing gravitational potential energy by friction and viscous drag, as heat. This core formation energy must have been an important source of heat within the early Earth for two reasons. First, there is good evidence from U–Pb isotopic data (see Chapter 8) that the core formed rapidly, at most within a hundred million years of the Earth's accretion, implying that there was insufficient time for the resultant heat to be lost by transfer to the surface. Second, core formation energy was released throughout the Earth's interior, again implying that the resultant heat was retained more efficiently. The temperature of the entire Earth could have been raised by 1500°C during core formation. However, as with adiabatic heating, core formation energy could not be released until the necessary temperatures for core formation were provided by some other process, such as through accretional impacts.

To the three main sources of primordial heat (in order of decreasing importance: accretional impact energy, core formation energy and adiabatic heating) may be added a fourth: that due to short-lived radioactive isotopes possibly incorporated into the early Earth. For example lunar and meteorite studies show that ^{26}Al (half-life 720 000 years) was incorporated into the solar system, but it is not known whether the Earth acquired a significant quantity of this potentially potent heat source. Timing is crucial because isotopes of this kind, manufactured only during supernovae, decay very rapidly, even on the time scale of planetary accretion. Thus, although the former presence of these isotopes in the solar system indicates that a supernova must have enriched the reservoir of pre-planetary particles just before the planets formed, it is not known to what extent short-lived radio-isotopes caused post-accretion heating of the Earth.

Unlike long-lived radioactive decay, which still produces heat, the Earth's primordial heating was a unique event which took place about 4550 Ma ago. Much of this original heat remains, and cannot be distinguished from that due to radioactive sources, since it is not known whether the Earth is cooling faster than the rate at which modern heat input is declining. However, if heat were to be transported only by conduction from the Earth's centre, it would take 200 000 Ma to be felt at the surface: much longer than the time available. Although it is now virtually certain that heat is transported much more rapidly because of convection in the mantle, the heat flowing from the surface of the Earth today was not produced yesterday, but represents a small proportion of that produced by all sources since the Earth formed. The calculation of heat being produced in the Earth today, based on the radio-element concentration values in Figure 9.15, may therefore have been slightly overestimated.

Although it is by far the most important, long-lived radioactive decay is not the only source of heat produced in the Earth today. The gravitational energy of tidal tensions dissipated within the Earth is unlikely to exceed 2×10^{19} J/a (equivalent to about 2 per cent of the heat derived from radioactive sources) though it may have been much more important in the distant geological past when the Moon was closer to the Earth. The continental crust has provided gravitational potential energy by the upward segregation of a light layer, a similar, but much smaller, contribution to that from core formation. If the continental crust is still growing (Chapter 14) then there may still be a tiny contribution to surface heat flow from this source.

Heat transport and mantle convection

The movement of the tectonic plates about the Earth's surface requires a system of horizontal forces that must be related to movements of the deep mantle of the Earth which is commonly supposed to be convecting in order to transmit the necessary forces to the plates. Heat flow has been discussed in terms of conduction because the lithospheric plates are regarded as the relatively rigid outer skin of the Earth (100–200 km thick) through which heat is indeed transported by convection. Whether or not the deeper mantle beneath the plates is convecting is dependent on two things: the need for the material to deform and the nature of the deforming forces.

The rock material of the Earth's mantle might seem too rigid for the fluid behaviour required for it to convect, but it must be remembered that solids exposed to deforming forces for long enough undergo inelastic deformation. In other words they suffer permanent deformation (as do folded rocks), and on a crystalline scale different parts of an atomic structure undergo relative movement, or creep. Some mantle material must continually undergo inelastic deformation in order to convect slowly. A measure of a material's ability to withstand forces leading to this kind of deformation is its viscosity; for example water has a low viscosity (10^{-3} kg/m/s) and mantle rocks have very high viscosities (10^{21}–10^{27} kg/m/s). The viscosity of the mantle has been measured by gravitational and geomorphological techniques

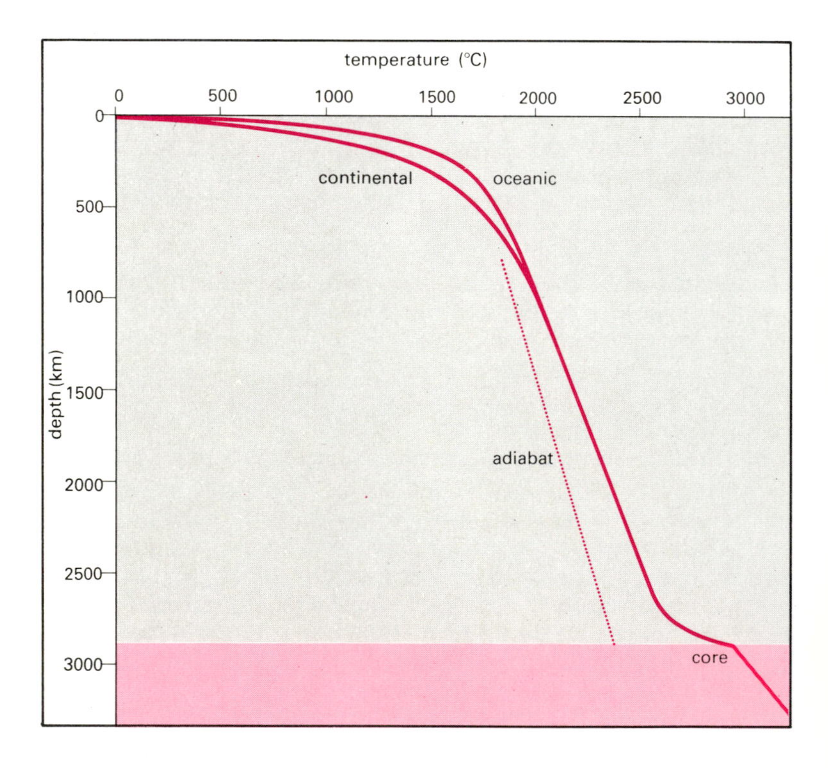

9.26: *Probable temperature profile in the Earth's mantle illustrating the rapid increase of temperature with depth through the lithosphere (conductive heat transport) and the near-adiabatic temperature increase in the deep mantle (convective heat transport).*

from the rate at which it deformed as the stress due to ice-loading during the last glaciation was added and then removed, a process known as isostatic readjustment. Calculations have revealed a three-layer structure: a relatively undeformable, rigid lithosphere overlying an asthenosphere 75 km thick (viscosity 10^{21} kg/m/s), which is only a little less viscous than the rest of the mantle beneath it (10^{21}–10^{22} kg/m/s). The similarity in viscosities means that convection may occur throughout the sublithospheric mantle, depending on the nature of the deforming forces.

Heated material will always convect if the buoyancy forces due to the low density of the hot material near the base of the system can overcome the viscous resisting forces. For example, as long as a pan of jam is being heated, the material at the base becomes thermally expanded and less dense, and rises, to be replaced by colder, denser material from above. The Earth's mantle is much more viscous than jam and much of its heat is generated from within rather than from below, but Lord Rayleigh's general principles of convective behaviour in fluids still hold.

One of the most important factors controlling convection is the temperature difference between the top and bottom of the layer; the greater this difference the greater the difference in densities, and hence the more vigorous the convection. However, this will be the temperature difference in excess of the adiabatic difference (defined earlier) because, under adiabatic conditions, any tendency for material to expand and become buoyant at depth is controlled by the increased pressure there. Once extra heat is added at the base the material there becomes buoyant and will try to convect in order to re-establish stable adiabatic conditions. Lord Rayleigh used a dimensionless constant, the Rayleigh number R_a, to define the mode of convection: R_a = volume coefficient of thermal expansion × temperature difference above adiabatic × density × depth^3 × gravitational acceleration/viscosity × thermal diffusivity. As expected an increase in temperature difference leads to higher values of R_a. Convection is promoted also by a high expansion coefficient, high density and a thick (deep) layer but inhibited by high viscosity and high thermal diffusivity, which leads to greater conductive heat transfer. The Rayleigh number for the Earth's mantle turns out to be $20\,000 \times \Delta T_a$, where ΔT_a is the temperature difference above the adiabatic.

The Rayleigh number for convection to begin in a homogeneous liquid heated at its base is about 2000. Below this value all the heat can be transported by conduction, and above it simple convection cells as wide as they are deep (see Figure 9.27) will appear. As R_a increases, convection becomes more rapid but remains regular until above about 10^6, when the cells break down and turbulence occurs. Thus, if ΔT_a across the sublithospheric mantle exceeds 0.1°C, there will be convection ($R_a > 2000$) for a viscosity of 10^{21} kg/m/s (or $\Delta T_a > 1$°C for convection in a lower mantle of viscosity 10^{22} kg/m/s). The best available estimate is that ΔT_a may well be several hundred degrees, giving R_a just about 10^6, in turn implying that the deep mantle is undergoing vigorous convection.

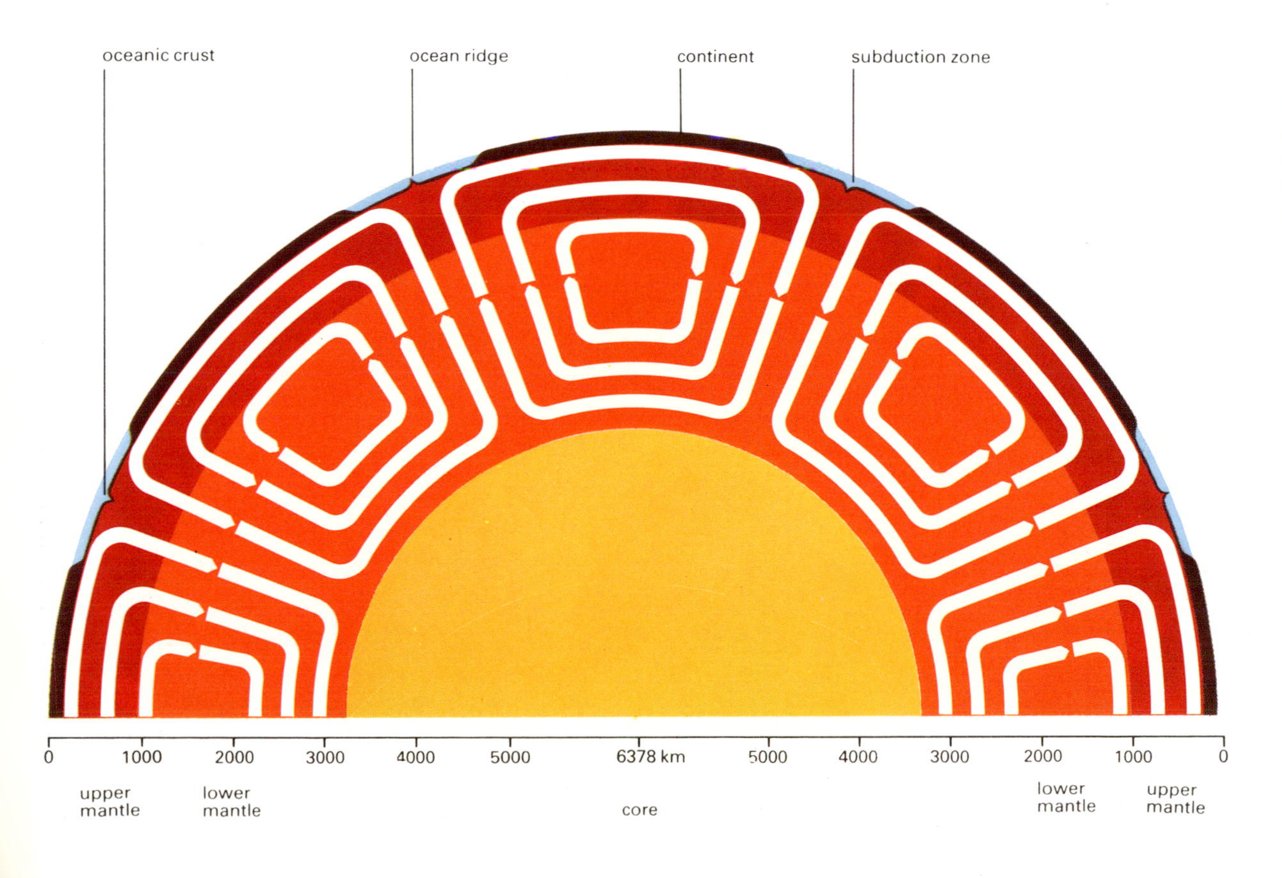

9.27: *A schematic cross-section through the Earth illustrating the concept of whole-mantle convection and its relationship to oceanic lithosphere generation (zones of upwelling) and destruction (zones of down-welling).*

There are many implications of this result. For example the temperature gradient across the lower mantle is self-limiting: if it exceeded the adiabatic by more than the amount shown in Figure 9.26 the increased convection rate developed would remove the excess heat. The temperature distribution of the Earth's interior can therefore be estimated more precisely. There is a relatively high temperature gradient through the rigid lithosphere beneath the continents and oceans where heat is transported by conduction. The heat production of the continental lithosphere is concentrated mostly in the crust, and so temperatures in the upper mantle are slightly higher beneath the continents than beneath the oceans. However, in the deeper levels and from sublithosphere temperatures of about 1600°C (at about 100–200 km depth), the temperature rises to only about 2600°C at the base of the mantle (2885 km). There may be a small region of high-temperature gradient at the core–mantle boundary induced by heat flowing from the core, but the temperature of the outer core, which is convecting independently, is probably quite close to 3000°C (ie between the two melting curves in Figure 9.25).

Another consequence of vigorous convection throughout most of the mantle is that heat transport should be relatively fast and efficient compared with the hundred billion years it would take for heat to be conducted through it. The heat detected flowing from the Earth's surface today will balance that produced during its recent history more

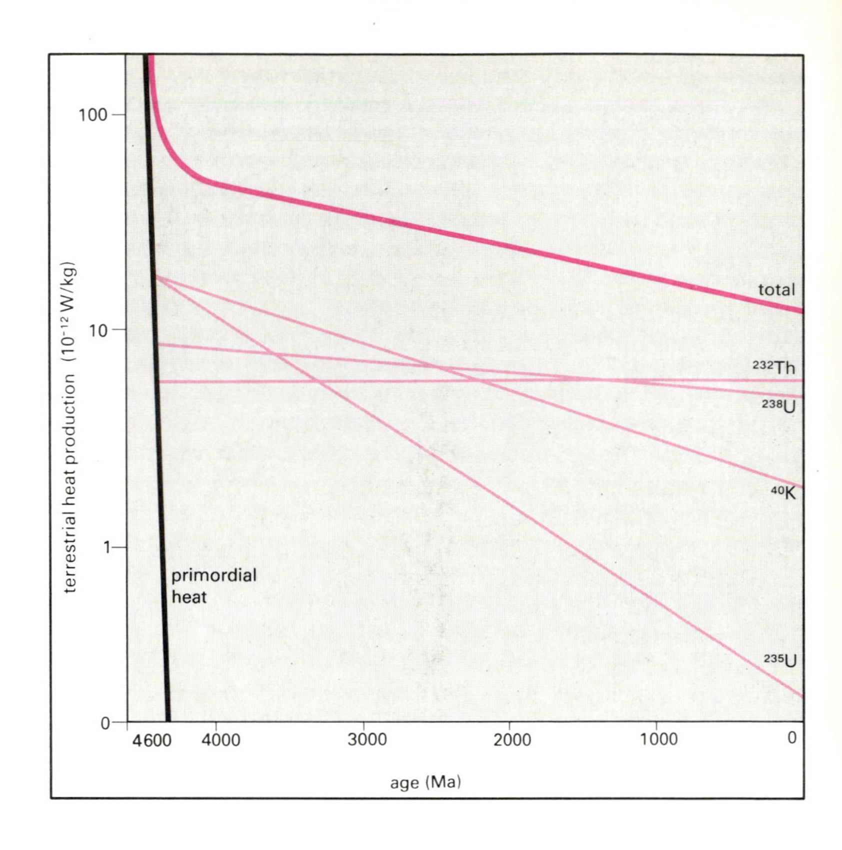

9.28: *Heat production throughout the Earth's history indicating the respective contributions from primordial and long-lived radioactive sources. The heat production scale is logarithmic.*

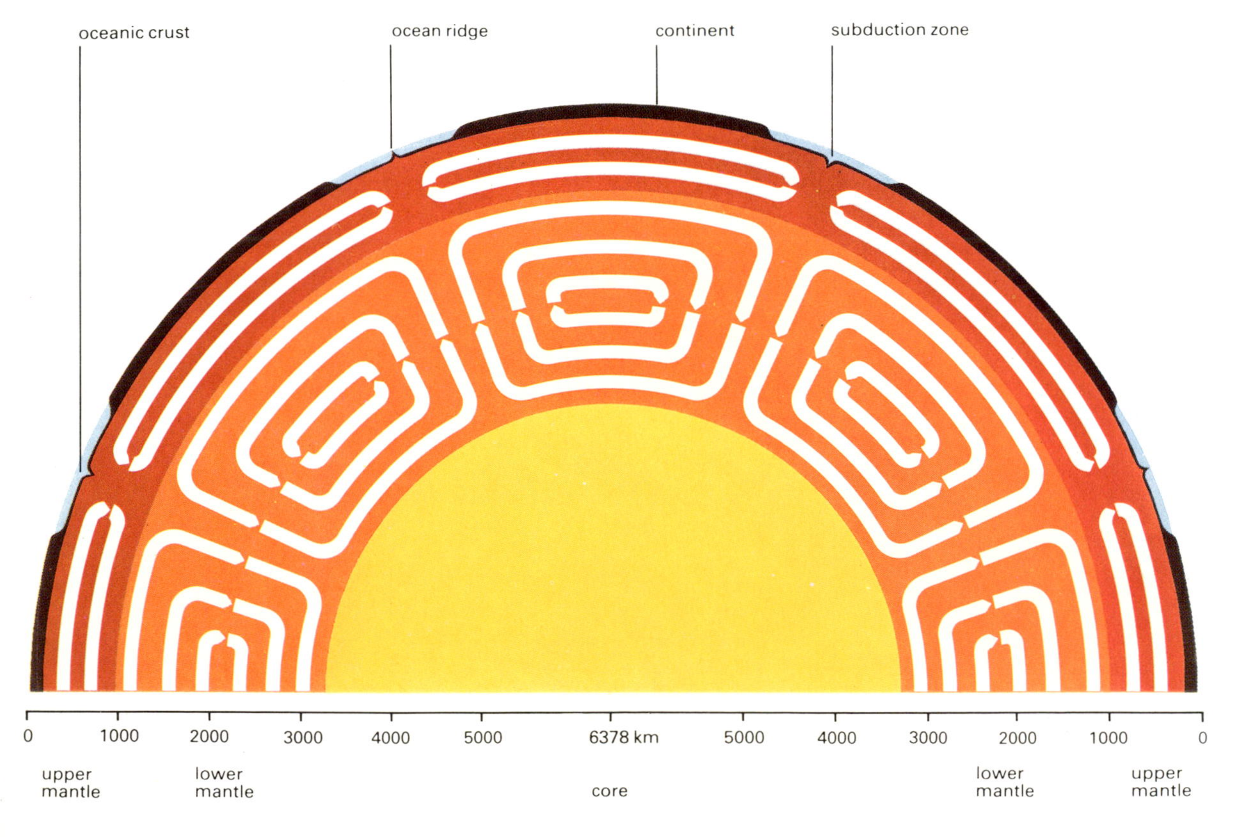

9.29: *A schematic cross-section through the Earth illustrating the concept of layered mantle convection, in contrast with Figure 9.27.*

closely than if primordial heat were still insulated deep within. The Earth must have cooled since its primordial heating, and although it may still be cooling very slowly, heat production and losses today are almost balanced.

A recent suggestion based on isotopic modelling is that convection may involve a kind of layered structure, heat being transported efficiently within and between the layers. Both this and whole-mantle convection are equally consistent with the physical evidence of the Earth's present internal state, but the evidence for layered convection depends on the recognition that the geochemical and isotopic components of the Earth's crust have probably been derived at the expense of only one-third to one-half of the mantle.

A summary of the thermal history of the Earth

The thermal history of the Earth has been dominated by a massive input of primordial heat during and after accretion 4550 Ma ago, followed since by heating mainly from long-lived radioactive sources. The extent of primordial heating is unknown but is likely to have produced temperatures of several thousand degrees at the centre of the Earth. Subsequently, heat produced from the four most important long-lived radioactive sources, and lesser contributions from poorly defined gravitational sources, has declined exponentially and is now smaller by a factor of about 7 than when the Earth first formed. Although the cooling of the Earth's interior may have been progressive, it must have been slow because the nature of the lithosphere–convecting deep mantle system has always adjusted to balance the heat being produced. For example, to dispose of higher internal heat production soon after the Earth's formation, mantle convection and lithospheric plate motions may have been more vigorous than today. Also the lithosphere may have been thinner so that thermal gradients and heat flow were higher. An important feature of exponential decay is that the rate of change decreases with time, hence the logarithmic scale in Figure 9.28. Geological patterns in ancient crustal rocks have shown that major changes took place in tectonic styles in the distant past which may have been linked to major changes in convective patterns inside the Earth. The present dynamic configuration of the Earth's interior will undoubtedly persist long into the future and no major changes are likely to occur in heat flow, heat production and plate motions. Ultimately, after another 10^9–10^{10} years, the rate of mantle convection and plate motions will slow and inevitably grind to a halt, whereupon the remaining heat in the Earth will be lost entirely by conduction. The energy balance sheet of Figure 9.2 provides a summary of the energy interactions within and upon the Earth which, except under circumstances of human intervention, are unlikely to change in the foreseeable future.

Part Three: Crustal Processes and Evolution

The Earth's crust is different in density and composition, and probably in its structural complexity, from the remaining 99.5 per cent of the planet's bulk. It is, apparently, almost entirely younger than the surface layers of the Moon, Mercury and Mars, and unlike all of them, it is in motion. It is the mobile nature of the Earth's crust (more properly its lithosphere) that makes its study so interesting. The final and overdue realization by Earth scientists of the reality of continental drift, and the easily grasped underlying principles of sea-floor spreading and subduction at the margins of otherwise rigid plates, has let loose an orgy of reappraisal of crustal processes. The revolutionary new direction of this subject was possible, however, only because of the already great accumulation of knowledge of the crust against which the new theories could be tested. The recycling of the materials of the crust through mountain-building, erosion and redeposition is a concept two centuries old; plate tectonic theory has added the continued creation and destruction of crustal materials to the range of recycling processes which we know about.

Plate tectonics is far from being an instant answer to all outstanding geological problems, nor indeed is it fully understood. After all, much of the activity takes place several kilometres below sea level, in an environment that scientists are only just beginning to explore. Rare instances of exposure of ocean ridges above sea level demonstrate the rate of activity there and even allow measurement of the extension of the crust across them. But a great deal more information awaits the intrepid few who descend to the submarine ridges in submersibles. How long will it be before the depths of even the trenches are explored? The challenge of plate tectonics is still very much with us: this part of the book must therefore leave open as many questions as it answers.

Aerial view of a fault at Thingvelli, Iceland, where the mid-Atlantic ridge rises above sea level.

10 The crust of the Earth

The crust of the Earth can be simply defined as those rocks that overlie the mantle. The mantle–crust boundary is generally a very sharp chemical and density discontinuity and a very good seismic reflector. It is thus readily identifiable using seismological techniques. The density of mantle rocks, 3100 to 3300 kg per m^3 (the density of water is 1000 kg/m^3), means that they can be easily identified by the appearance of fast seismic arrivals with P wave velocities of some 8.1 ± 0.2 km per second. The fact that the density of the Earth's surface is lower than that of deeper rocks is also shown by comparing the mean density of the Earth, 5517 kg/m^3, with that of the average density of surface rocks, some 2670 kg/m^3.

There are two major kinds of terrestrial crust—oceanic and continental—which each have radically different compositions, ages and origins. The most obvious distinction between them is that the top of the oceanic crust is generally some 3 to 4 km lower-lying than the continental and also contains the deepest parts of the Earth's surface, while the highest elevations are confined to the continental crust (see Figure 10.1). The distribution of elevations on the Earth's surface also emphasizes that, while the oceanic crust is almost entirely submarine, the continental crust is by no means entirely emergent. The coastlines are not, therefore, usually the limits of the extent of continental crust.

The evidence for the nature of the different crusts derives mostly from geophysical studies as surface exposures are necessarily restricted almost entirely to the continents and can only relate to the uppermost part of the crust. However, some samples of the deeper crust can be obtained where molten rocks (magmas), passing through it, have brought broken-off pieces of the deeper crust (xenoliths) to the surface. Drilling is currently under way in some areas to penetrate the lower continental crust, but the areas chosen are necessarily anomalous as cost and technology tend to require that sites are selected where the upper crust is thinnest, and these are usually where the total crust is abnormally thin. In the oceans technological limitations and cost have largely prevented significant drill penetration into the crustal rocks beneath the sediments. The evidence for the nature of the oceanic crust is therefore largely geophysical, but supplemented by dredged samples, limited submersible studies and investigations of a few isolated outcrops of rocks that appear to have the same characteristics as the oceanic crust, such as the Troodos complex of Cyprus, the Vourinos complex in Greece, and similar rocks in Papua New Guinea. Such areas are also anomalous in that their number is few, and special conditions must have existed for their formation and preservation.

This chapter describes the major features of both oceanic and continental crust and the zones where they meet, leading to the unifying concept of plate tectonics within which these characteristics are most satisfactorily explained.

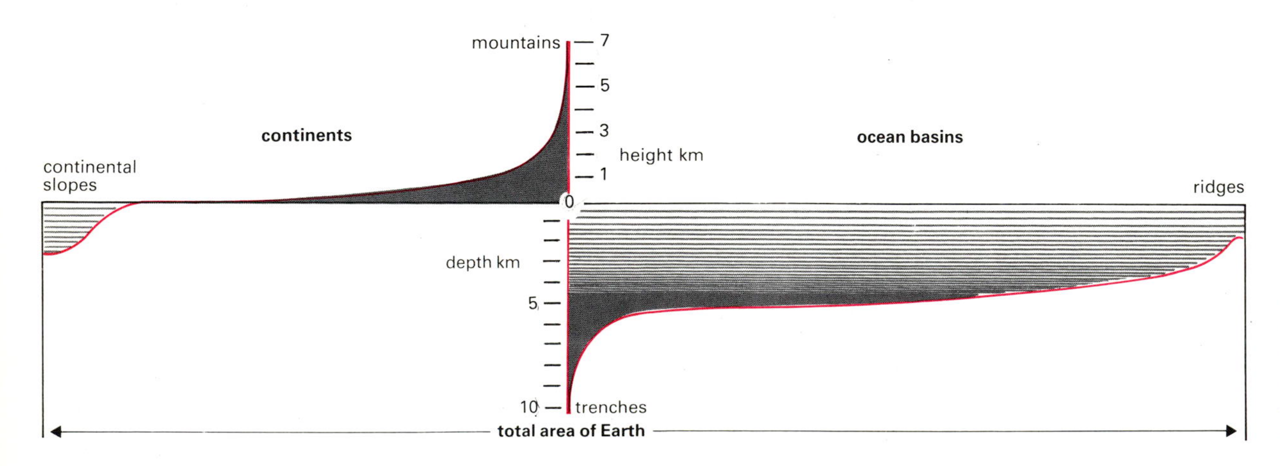

10.1: *A modified hypsometric curve. This is essentially a graph of the height of the Earth's surface against its area, but is modified to illustrate the general distribution of elevation within oceans and continents.*

The oceanic crust

Oceanic ridges

Although the presence of a mid-Atlantic ridge had been known for over two hundred years it was not until 1954 that the ridges, the main topographical feature of the ocean basins, were known to be part of the same continuous feature extending for some 80 000 km through all the oceans. The oceanic ridge system is thus a major element of the Earth's topography, being some 2 to 3 km higher than its bordering oceanic basins. In the Atlantic and Indian Oceans this ridge system is equidistant between the adjacent continents, and in the Arctic it runs midway between the submerged continental Lomonosov ridge (which passes through the North Pole) and the Asian continent. In the Pacific the ridge system is highly asymmetric with respect to the continents as, having passed midway between Antarctica and Australia, it continues through the southern Pacific, into the eastern Pacific and thence northwards. In the Indian Ocean one branch enters the Gulf of Aden and links with the Gulf of Tadjura and the Red Sea. In the Pacific two branches pass from the east Pacific rise towards South America, while the east Pacific rise itself passes into the Gulf of California. The ridge system is presumed to extend beneath the continental crust of the Basin and Range province of the western USA, and then to re-emerge in the Pacific Ocean as the Juan de Fuca ridge, which eventually passes beneath Alaska.

Although almost entirely submarine the ridge does rise above sea level in a few places, which are always associated with geologically young or present-day volcanic activity, as in Iceland, Tristan da Cunha, St Helena and Ascension Island in the Atlantic and the Galápagos in the Pacific. The ridge is also active seismically, with tensional earthquakes of low to moderate strength (generally less than magnitude 6 on the Richter scale) and apparently restricted to the upper 10–12 km and to within a few kilometres of the ridge's central line. (The lack of adequate seismological observatories in oceanic areas means that these determinations are over-estimates and the earthquakes are probably restricted to a zone 2 to 3 km wide in the uppermost 10 km.) The crest of the ridge is, in the Atlantic and Indian Oceans, characterized by a central rift valley which is generally some 2 to 3 km deep and 20 to 30 km wide. In the Indian Ocean the central rift valley is often accompanied by similar, nearly parallel rift valleys which make the central one difficult to identify. Volcanism, where studied, appears to occur along the central line of the rift valley which is also where seismic activity occurs. In the Pacific, earthquake activity appears to be confined to a similarly narrow zone but volcanism seems to have been much greater and has filled the central rift valley so that the ridge appears to be smooth.

Fracture zones

Although clearly a continuous ridge when viewed on a scale of a few hundred kilometres or more, in detail the ridge system is frequently offset by fracture zones (see for example the mid-Atlantic ridge in Figure 10.17). These fracture zones are cliffed areas which vary in width from a few kilometres to 40 to 50 km, but can be traced laterally for hundreds or thousands of kilometres. They appear to be almost entirely confined to the oceanic areas, with rare exceptions where they most closely approach continental edges. The fracture zones are seismically active only along that portion between the offset ridge crests. The parts of the fracture zones extending away from the crests, towards the continents, are seismically very quiet. The earthquakes between the crests are associated with transcurrent motions, implying that the fracture zones are faults in which each side moves horizontally but in opposite directions. The earthquakes away from the interridge segment are very weak and appear to be associated mainly with small vertical displacements between the opposite sides of the fracture zone.

10.2: *The active volcano of Eldfjell, above Heimaey, Iceland, on the mid-Atlantic ridge, throwing up a cloud of water vapour, gases and ash.*

The structure and composition of the oceanic crust

One of the most important characteristics of the structure of the oceanic crust is its remarkable uniformity, with sediments overlying igneous rocks which form three distinct layers (Figure 10.3). The sediments, known to seismologists as layer 1, vary considerably in thickness and in composition; shells and debris from marine plants and animals are particularly abundant near the equator, and detritus from the continents is thickest nearer the continental edges from which it comes. However, the ridge crests are generally free of sediment and the thickness of sediment generally increases to 3 to 4 km as the continental edges are approached.

The three igneous layers are each of very uniform composition and thickness. The upper of these (layer 2) has been drilled and is known to be composed of basaltic lavas, 1 to 2.5 km thick. The basal layer directly overlying the mantle (layer 4) is thin, probably less than

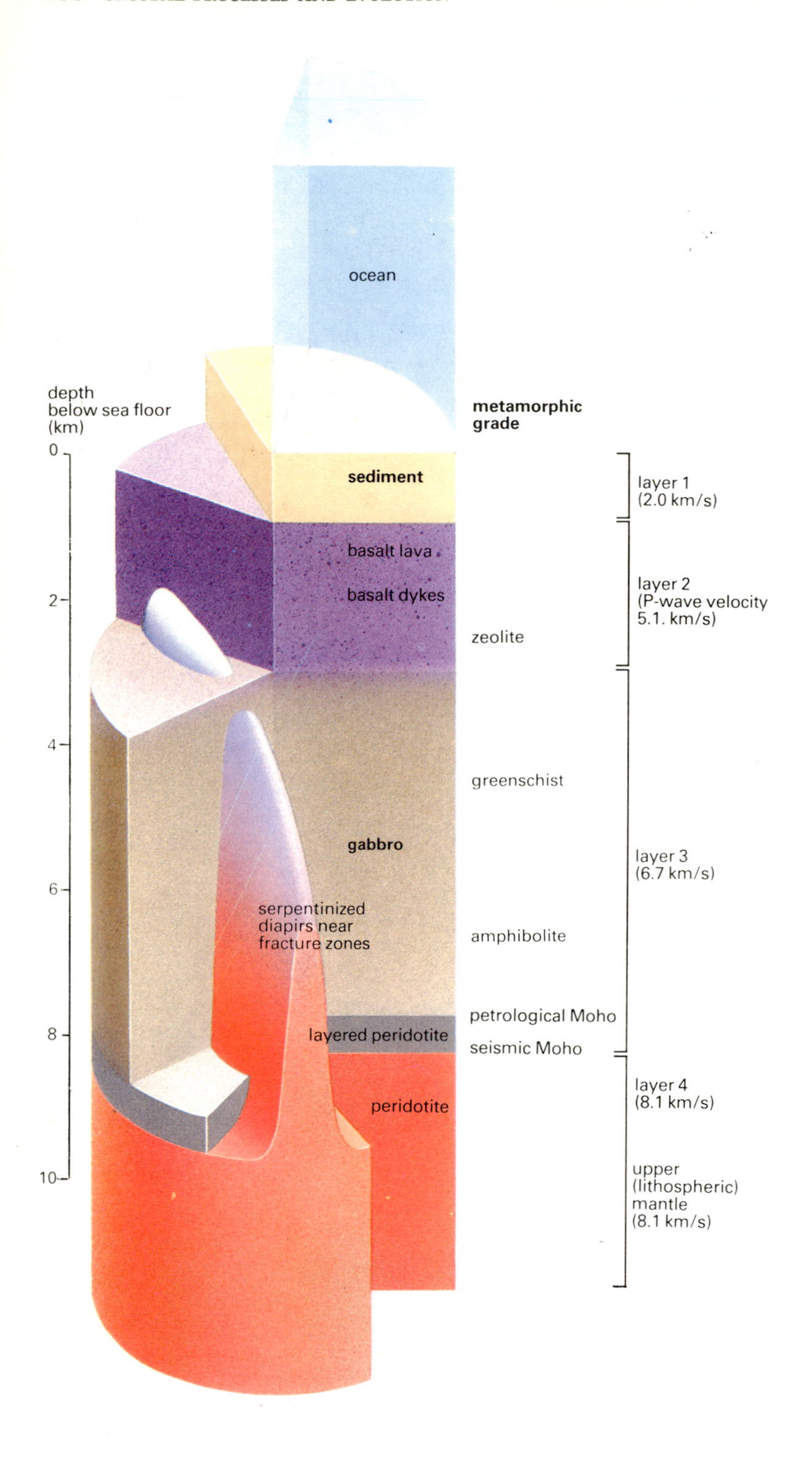

10.3: *A hypothetical section through ocean crust, based on seismic velocity interpretations, evidence from dredged samples and comparisons with outcrops of rocks thought to have once been parts of ocean floors.*

0.5 km thick, and formed of rocks with a density of some 3000 kg/m^3. The main layer (layer 3) is some 5 km thick, and its seismic velocity is consistent with a gabbroic composition. The igneous oceanic crust is therefore composed of basalts, overlying a thick gabbro layer and a thin, probably ultrabasic, olivine-rich layer.

These compositions appear, at first sight, to conflict with those of samples of igneous rocks dredged from the sea floor which tend to fall into four main categories—basalts, gabbros, amphibolites and serpentinized (ie hydrated) peridotites. The basalts and gabbros can clearly be associated with layers 2 and 3 respectively. The amphibolites therefore seem to be metamorphosed equivalents of the gabbros and basalts. The serpentinized peridotites appear to be anomalous as their seismic velocity is less than that of the oceanic crustal rocks. The large quantities of dredged amphibolites seem to reflect the fact that most igneous samples are obtained from the cliff edges of the fracture zones as elsewhere the igneous crust is largely or entirely obscured by sediments. It appears that serpentine mineralization is similarly associated with the fracture zones, which may provide channels for the circulation of hydrothermal systems, in which the water could be derived from either the mantle or the ocean. It is thus generally thought that layer 2 comprises basaltic lavas and dykes, the upper parts of which are in a very low (zeolite) grade of metamorphism and the lower parts, which would also be predominantly dykes, are in a rather higher (greenschist) grade of metamorphism. This uppermost igneous layer would be more brittle than the underlying rocks of layer 3 which may exhibit plastic deformation to prolonged stresses. Locally it seems probable that the gabbro of this layer is in an even higher (amphibolite) grade of metamorphism, particularly so near fracture zones, and that serpentinization may well have been concentrated in such areas. Because of their low density the serpentines may also rise diapirically if they have been formed by hydration from mantle rocks. Such localities are thus also most likely to occur in association with fracture zones.

Linear volcanic island chains

Although oceanic volcanism is mostly confined to the seismically active ridges it does occur, with minor associated seismic activity, well away from the ridges, particularly in the Pacific. The most notable of these areas is Hawaii, where present-day volcanic activity occurs in the main island of Hawaii. Increasingly older volcanicity, active some twenty-seven million years ago, can be traced north-westwards along the Hawaiian chain as far as Midway Island, from where the age of volcanism increases northwards along the Emperor seamounts until they disappear at the Aleutian trench. Similar unidirectional increase in volcanism characterizes, for example, the Society, Austral, Samoan and Cook Islands. These linear volcanic island chains are all essentially aseismic, with present-day volcanic and seismic activity taking place only at their south-eastern extremities.

Similar aseismic linear ridges extend also across the oceanic basins from areas of particularly strong present-day volcanic activity on the mid-ocean ridges, the age of volcanism increasing in both directions away from the present-day activity on the mid-ocean ridge crest. In the Atlantic, the Rio Grande and Walvis rises extend to either side of Tristan da Cunha while the Iceland–Faeroes and Greenland ridges are similarly disposed with respect to Iceland. Other examples can be found in the Pacific Ocean.

Oceanic heat flow and gravity

Measurements of the heat flow from the Earth's interior are made in the oceans using probes dropped into sediments. The temperatures are taken at different depths and the thermal conductivity of the sediments is either measured from core samples or determined from tables for the types of sediment present. These observations have proved critical in understanding the development of the oceanic crust, as they indicate very clearly the high heat flow associated with the seismically active ridges. Statistical analyses suggest that the extremely high heat flow values along the crest of the ridge (1.82 $\pm$ 1.56 heat flow units (hfu)) are paralleled by a steady decrease to normal ocean basin values of 1.28 $\pm$ 0.53 hfu, except for some minor increases some 200 km on either side. The mean heat flow therefore decreases steadily away from the ridges and very closely parallels the decrease in elevation of the sea floor. In contrast, measurements at the ocean surface indicate very uniform values for gravity, irrespective of the topography of the ridges. The only areas where distinct gravity anomalies appear are exactly on the ridge crest, mainly within the rift valley itself. Such observations, when combined with the evidence for a very uniform thickness and density of the oceanic crust, must mean that the topography of the ridge is directly related to the presence of lower density mantle rocks beneath so that the net density of the mantle rocks increases exactly in parallel with the decrease in the elevation of the ridge.

Magnetism and ocean floor spreading

Perhaps the most important geophysical property of the uppermost igneous oceanic crust is its magnetism (see Chapter 7). For many decades ocean surface magnetometers on oceanic voyages monitored the changing strength of the geomagnetic field, but little interpretation of these magnetic anomalies was attempted. It was soon recognized that the crests of the oceanic ridges were characterized by a stronger magnetic anomaly than occurred elsewhere. This allowed rapid identification of the median rift valley in the Indian Ocean and also delineated the crest of the east Pacific rise (which lacks a rift valley). However, the main use made of these oceanic records was to assess the extent to which some of the isolated oceanic islands or submerged peaks could be explained in terms of coral growth on top of a previously volcanic cone, as proposed originally by Charles Darwin, or conversely the extent to which these features were essentially volcanic with little associated coral growth. The islands with a volcanic cone are magnetic and therefore create magnetic anomalies, while purely coral islands would be non-magnetic.

In the late 1950s tests of oceanic equipment were undertaken by the Scripps Institute of Oceanography along a series of closely spaced navigation lines off California. Comparison of these lines indicated a very great similarity in the sequence of magnetic anomalies along each of the profiles (Figure 10.4). It also showed that identical patterns could be traced east–west for several hundred kilometres, but that when traced north–south, the patterns changed drastically across fracture zones. None the less the anomalies could still be matched on either side of a fracture zone when the total pattern on one side was displaced by several hundred kilometres. This provided the first strong evidence for transcurrent movements of several hundred kilometres; displacements of 1160 km were indicated between the two sides of the Mendocino fracture zone and of 265 km along the Pioneer fracture zone. This contrasted with the previous evidence for maximum transcurrent displacements of a few tens of kilometres, for example along the Great Glen fault in Scotland. The anomaly correlations were clear and convincing, but the origin of the anomaly pattern was still unknown and was largely attributed to magnetic susceptibility differences. For example, it was suggested that fault lines, running north–south, had resulted in the formation of sequences of north–south grabens filled with sediment or other materials having a low magnetic susceptibility.

In 1928 Arthur Holmes suggested that the continental crustal blocks separated from each other as a result of convective currents driven by radiogenic heat in the Earth's mantle. In order for this separation to take place Holmes proposed that new oceanic rocks were forming everywhere throughout the ocean basins, although predominantly at the oceanic ridges (which were not then known to be a continuous system). In the early 1960s Harry Hess (and later Robert Dietz) proposed that the formation of new oceanic rocks between the separating continents could be restricted to very localized areas along the crest of the oceanic ridges. As the continents separated new oceanic rocks were added, cooled down and magnetized in the ambient direction of the geomagnetic field.

It was already known that the geomagnetic field spontaneously changes polarity (the magnetic north pole becoming the magnetic south pole, and vice versa), with approximately three such polarity changes having taken place every million years during the past sixty million years. It was thus conceivable that the spreading of the oceanic igneous crust away from the ridges would provide magnetic 'tape-recordings' of the changes of the geomagnetic field. In 1963 the magnetic records of changes on east–west lines passing over the Reykjanes ridge, just south of Iceland, were shown by Fred Vine and Drummond Matthews to be attributable to the alternately normal and reversed magnetization of the upper igneous rocks. The sequence of alternations was identical to the known polarity changes of the geomagnetic field during the past three million years. Furthermore the sequences of polarity changes on either flank of the ridge were mirror images of each other. Subsequent studies of earlier records for magnetic anomalies on oceanic ridges in all other oceans showed that the same sequences and mirror images existed (Figure 10.5). The sequences were more extended on some ridges than on others, presumably because the oceanic floor migrated away from the ridge more quickly in these areas. As further magnetic records were analysed and collected the sequence of polarity changes recorded by the uppermost oceanic igneous rocks was extended further back in time and the correlation with known polarity changes, determined from palaeomagnetic studies of dated continental rock sequences, continued to be confirmed. The degree of correlation between the lateral sequences of magnetic polarity changes in the oceanic rocks and the known time sequence of polarity changes of the Earth's magnetic field was so great that there was very little subsequent hesitation in accepting Harry Hess's model of ocean floor spreading for the creation of the oceanic igneous rocks.

When analyses of the oceanic magnetic anomalies were first undertaken the polarity changes of the geomagnetic field were known with an acceptable degree of precision only for the past four or five million years. The initial dating of the earlier oceanic anomalies was therefore based mainly on an assumption of constant spreading rate. It was known, however, that there were changes in the rates and

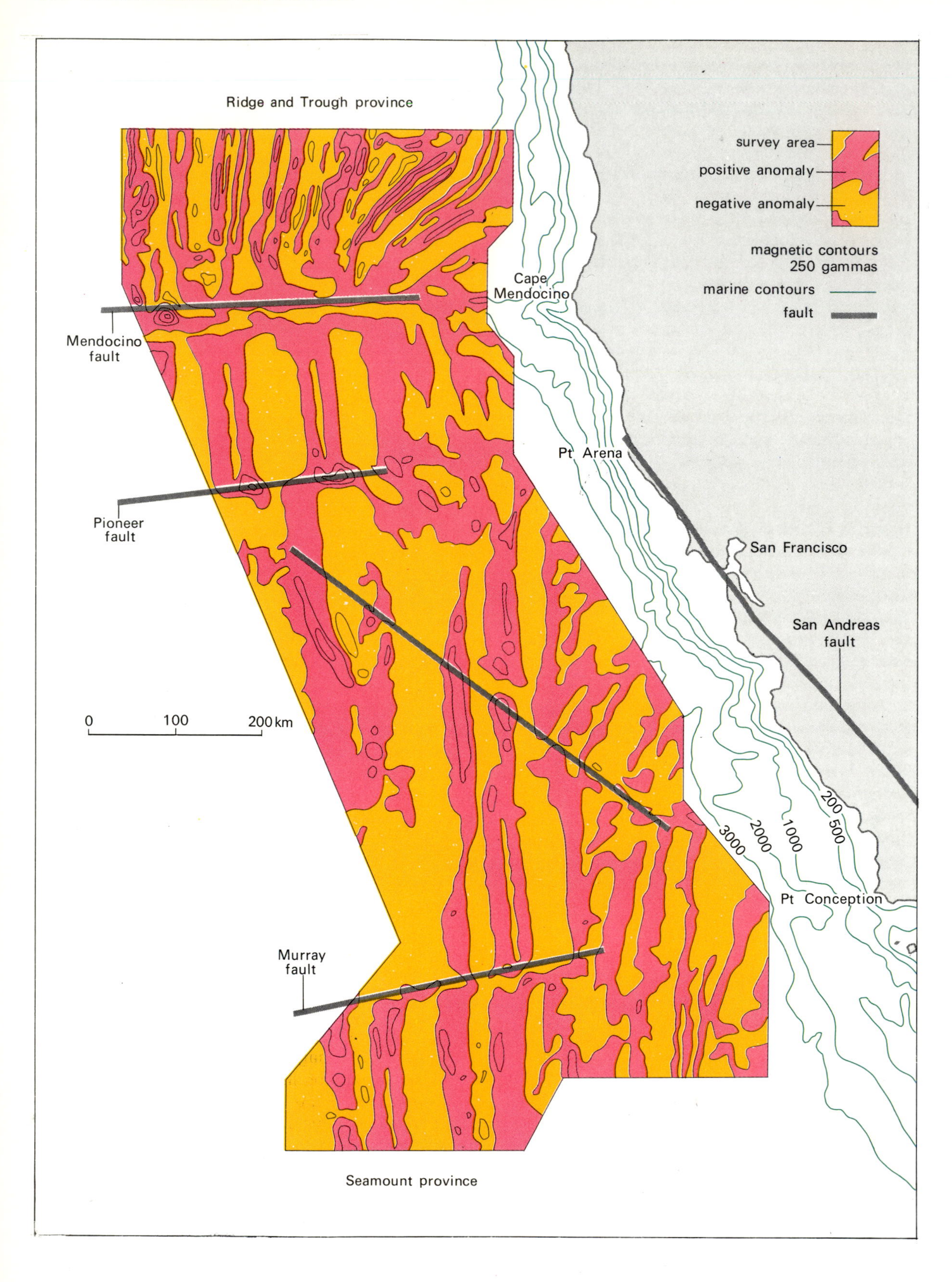

10.4: *Oceanic magnetic anomalies near California. Faults of known locations such as the Mendocino and Pioneer fracture zones can be seen to offset the pattern of positive and negative anomalies.*

directions of spreading in different oceans. After Jim Heirtzler proposed that the polarity sequence in the south Atlantic apparently showed few changes in spreading rate during its evolution, this sequence formed the standard scale for geomagnetic polarity dating for the past 75 million years (Figure 10.6). It was known that the basic assumption of a constant spreading rate was almost certainly invalid, but the choice has proved largely justified. It is only recently that adjustments —of no more than 20 per cent—have been proposed for the age of the older anomalies, on the basis of the stratigraphic age of sediments immediately overlying them.

This model of ocean-floor spreading is clearly supported by the lack of sediments on the crests of the ridges and their gradual increase in thickness towards the continents. However, its main confirmation comes from the dating of the sedimentary layers and of the anomalies far from the ridge crests.

Oceanic sediments and the age of the ocean basins

The increase in thickness of the sediments away from the oceanic ridges has been known for several decades as their base can be readily determined by seismic techniques. The point of contact between the sediments and underlying igneous rocks is irregular, with the essentially horizontally bedded sediments covering, to varying extents, volcanic projections of the igneous basement. The lower sediments show signs of a greater degree of compaction and two seismic layers can often be distinguished. The lower one frequently shows gentle, very large-scale deformation, while the overlying younger sediments are more uniform. Within the sedimentary sequence very distinct seismic reflectors can be traced laterally under the oceans. These reflectors intersect with the igneous basement at different distances from the ridge (Figure 10.7). The lower, and thus older, reflectors intersect the basement at greater distances from the ridge than the younger reflectors. Thus, before the actual age of the sediments could be determined from drill core samples, the relationships between seismic reflectors and the basement suggested that the age of the base of the sediments increased away from the ridges. When samples of the sediments forming the reflectors were obtained from deep-sea drilling cores, they were found to be mostly chert horizons. In the earlier attempts they prevented further drilling because the drill bits became rapidly worn out while trying to penetrate them. Some seismic reflections had already been dated from samples taken more simply from the few areas where these older horizons had been exposed by recent erosion or faulting. It was, however, the Deep Sea Drilling Project (DSDP) that provided the main incontrovertible evidence for the ages of the sediments in the ocean basins.

This project, still active in 1981 as the International Phase of Ocean Drilling (IPOD), began life in 1969 as the Joint Oceanographic Institutions Deep Earth Sampling (JOIDES) project. Since then, some hundreds of borehole cores have been taken from the ocean floor, under several kilometres of seawater, using techniques originally developed for the abortive Mohole Project which was to have drilled right through oceanic crust and into the mantle. That ambition is now unlikely to be achieved, but the results of the DSDP have undoubtedly been of far greater benefit to science.

The technological advances for drilling through the sediments of the deep oceans are comparable in magnitude to those for the lunar landing project, and the results have been similarly outstanding. The sediments in all oceans except the Arctic have now been sampled, with penetration often reaching into the igneous basement. The sediments can then be dated by the fossils they contain. This is by no means a simple process as the perpetual rain of debris from organic life in the upper levels of the ocean is continually filtered by solution in seawater, so that only siliceous skeletal matter persists to depths greater than about 4 km. This is the calcium carbonate ($CaCO_3$) compensation level, below which $CaCO_3$ is completely dissolved in seawater; the precise depth of this level may have changed during geological time. Skeletal matter in the deep oceans accumulates only very slowly. Reworking by bottom currents and burrowing animals and further dissolution are common.

None the less the early cores could be reliably dated and it was clear that, although they rarely penetrated to the basement, the predicted stratigraphical age of the base of the sediments was close to that predicted by the age of the oceanic magnetic anomalies. The age of magnetic anomaly 24 (Figure 10.8) is critical as this is the oldest recorded anomaly between Greenland and Europe and thus dates the age of the opening of this oceanic basin. The original (Heirtzler) estimates dated this event at some 60 million years ago. The dating of fossils in the sediments immediately overlying the anomaly show that they were deposited at the Palaeocene–Eocene boundary, which is variously dated at between 49 and 54 Ma ago. The precise age of many of the oceanic magnetic anomalies still remains uncertain,

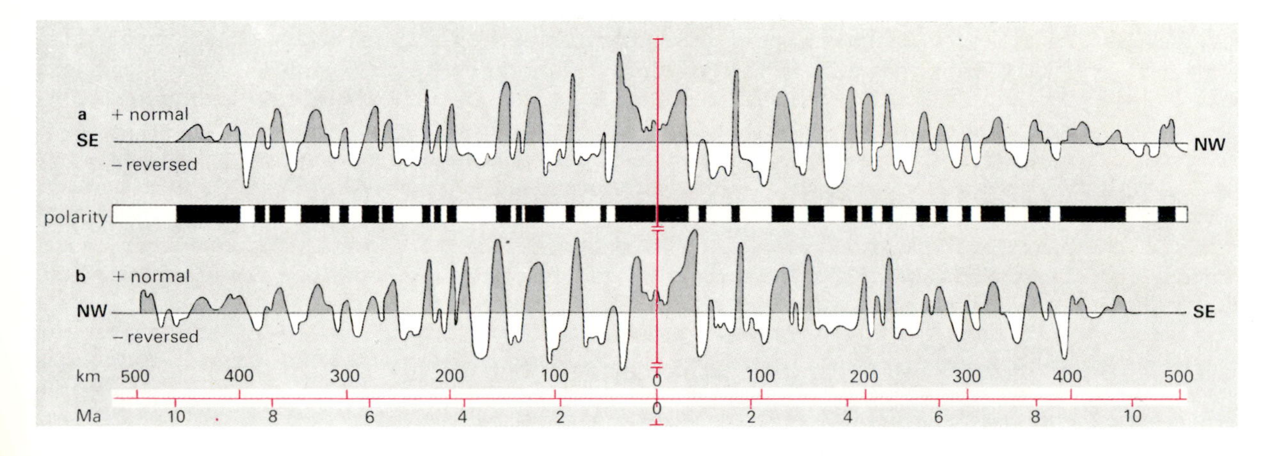

10.5: *Polarity sequences away from oceanic ridge crests. The magnetic anomaly sequences are almost identical on either side of a ridge. If the same profile is plotted as its mirror image (a is the mirror image of b) then the profiles are almost identical.*

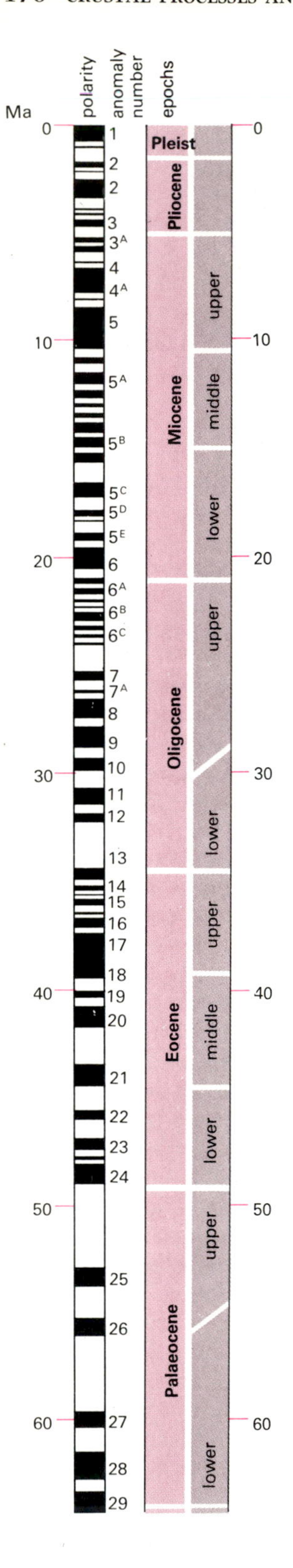

10.6: *The polarity time scale for the past 75 Ma. It is thought that the actual dating of any particular reversal is correct only to ±1 to 2 per cent and that many polarity changes of shorter duration have still not been adequately defined. None the less the polarity scale provides a rapid and cheap method for dating and for time correlations on a global scale.*

within some 2 or 3 Ma, but it is absolutely certain that none of the present oceanic basins is floored by crust that is older than 250 Ma.

The fact that the oceanic basins can be dated by means of their magnetic anomalies has many important consequences. For example, the systematic decrease of both elevation and heat flow away from an oceanic ridge is now known to relate to the square root of the age of the oceanic floor. Conversely the age of the oceanic basins can also be calculated, to a first approximation, simply from their depths. Areas with a very high sedimentation rate cannot be dated so accurately by this method as they will appear shallower and therefore younger than their real age. Nor can it be applied to the major oceanic deep trenches —perhaps the most intriguing features of the Earth's surface.

Oceanic trenches

Although the oceanic basins are generally some 3 to 5 km deep, long narrow trenches occur where water depths can exceed 10 km. These trenches are almost entirely confined to the edges of the Pacific, although one major trench, the Java trench, occurs in the Indian Ocean, two small ones, the Scotia and Antilles trenches, occur in the Atlantic and an even smaller one, the Cretan trench, occurs in the Mediterranean immediately south of Greece. The trenches are generally long, often exceeding 1000 km. They have an asymmetric profile (Figure 10.9) and the 'oceanic' side has a gentler slope than the 'continental' side, although even this slopes only at some 8 to 20° but is often strongly exaggerated in cross-sections. The trenches have varying amounts of sedimentary infill: most show horizontally bedded

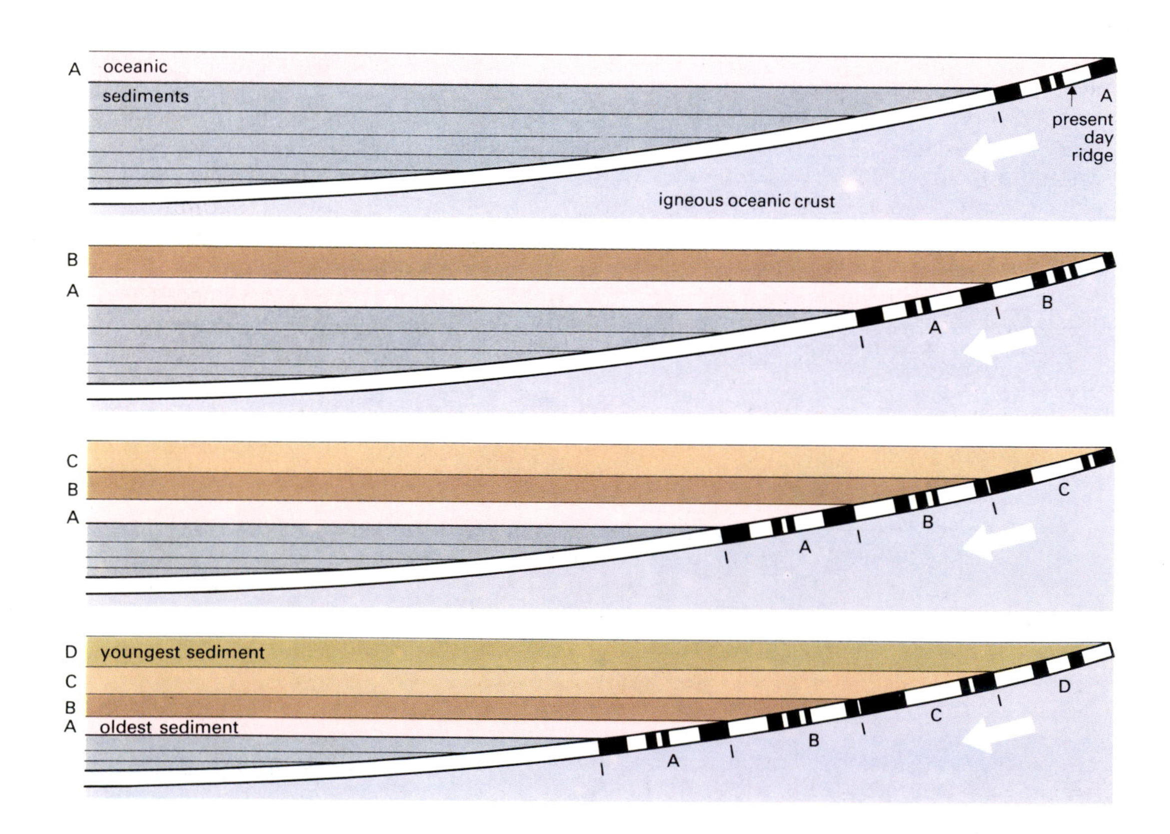

10.7: *Dating oceanic sediments and anomalies.*

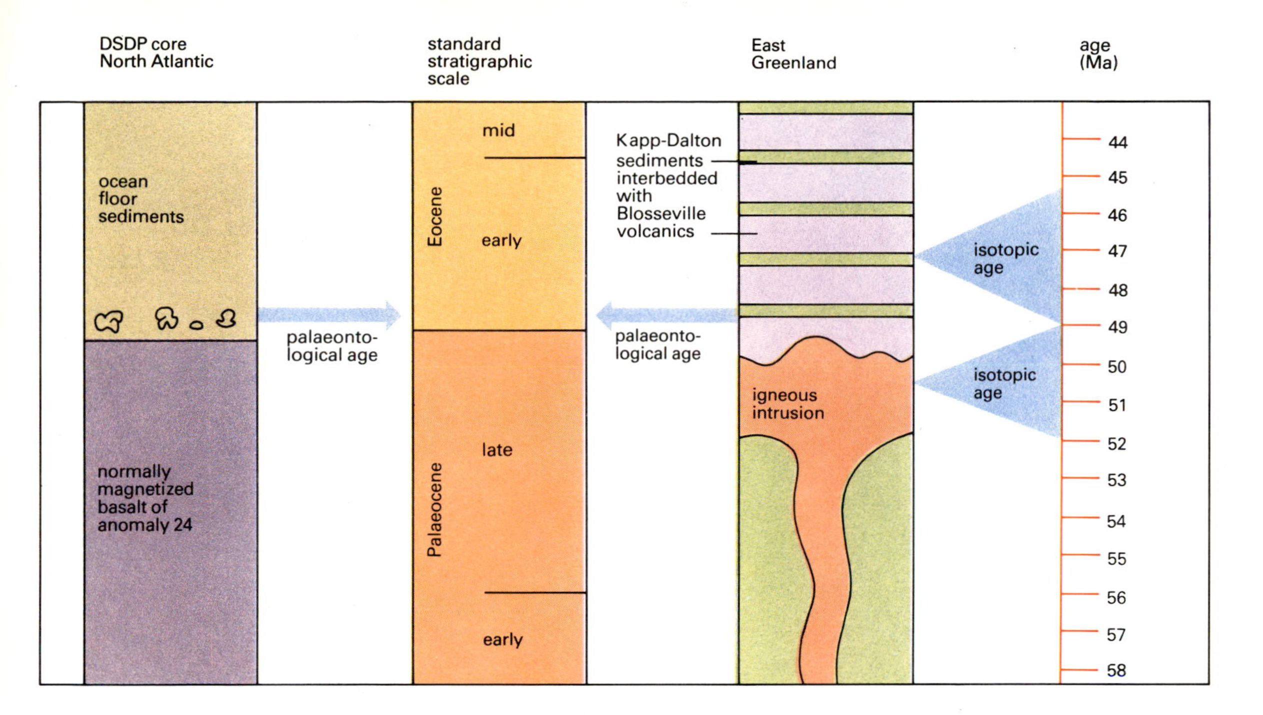

10.8: *Dating magnetic anomaly 24. Fossils of earliest Eocene age are found in sediments overlying ocean floor basalts magnetized during anomaly 24. In Greenland sediments with Palaeocene fossils are intruded and overlain by volcanic rocks with radioactive dates around 48–49 Ma, which in turn are overlain by Eocene sediments. Thus the volcanic rocks were formed at about the Palaeocene–Eocene boundary and thus anomaly 24 can be dated at about 49 Ma.*

sediments at their bottom, but some are filled with sediments. The edges of the trenches are usually marked by faulted blocks on which small amounts of sediment have lodged. Where filled with sediments the trenches can most readily be located by means of their strong earthquake activity, volcanism and long, narrow gravitational and magnetic anomalies. These anomalies lie mostly to one side of the trench, in most cases on the continental side. Some, however, such as the New Hebrides, Solomons and New Guinea trenches, have their associated anomalies on the Pacific side. Similarly, generalizations about the often arcuate curvature of the trenches are dangerous as all degrees of curvature are found. The Kermadec–Tonga trench, for example, is essentially linear, and there are complex curvatures of the trenches in the Celebes. The negative gravity anomaly is always located immediately adjacent to the deepest part of the trench. This anomaly was interpreted in the 1930s as being caused by a low-density crustal 'root' beneath the trench. Subsequent seismic studies have shown that there are no such roots to the trench zone—the oceanic crust simply bends downwards underneath the bordering plate—and no precise cause of these anomalies has been discovered. However, the seismic evidence in some areas appears to be consistent with the emplacement of a wedge of low-density sediments immediately beneath the rocks on the continental side of the trench. This area is one of great complexity with a thick, overlying sequence of sediments derived from the immediately adjacent volcanic arc. In some areas, such as along the Peru–Chile trench, the volcanic activity occurs within the continent, generally some 100–200 km from the trench, while in oceanic areas the volcanoes form an independent island arc.

The composition of the volcanic rocks in such arcs differs in detail with distance from the trench (see Figure 10.11), but their average composition is granodiorite, similar to that of average continental crustal rocks. Robert Coats pointed out in 1962 that the volcanic arc of the western Aleutians is bordered on both sides by oceanic rocks, so that the continental composition of the volcanic rocks in this region can only have been generated from oceanic materials and not from re-worked continental materials. Thus the processes operating in the region of the world's trenches are capable of converting oceanic crust to crust of continental composition.

Subduction zones

The complexity of subduction zones with their combination of igneous, metamorphic and tectonic activity (more fully described in Chapter 14) makes them important in studying ancient 'orogenic belts', linear zones of similar complexity which form mountain belts. The subduction zones not only mark the site of lithosphere consumption but they also appear to be the areas where new continental materials are being generated. Subduction of oceanic rocks subjects them to greater pressures and temperatures at which they lose volatiles, which rise and cause partial melting of the overlying rocks. The resulting rise of magma increases the temperatures nearer the surface, causing metamorphism. The parts of the volcanic arc or continental margin farthest from the trench and overlying the subduction zone are metamorphosed at a higher grade, being usually hydrous amphibolites generated by high temperature and moderate pressure. Nearer the trench magmas are not generated as the descending oceanic plate has not yet reached depths at which volatiles would be given off. The near-surface temperatures are thus lower so the interaction between the two plates causes high-pressure but low-temperature metamorphism of the prehnite–pumpellyite (blueschist) grade.

The recognition of such pairs of metamorphic belts in older conti-

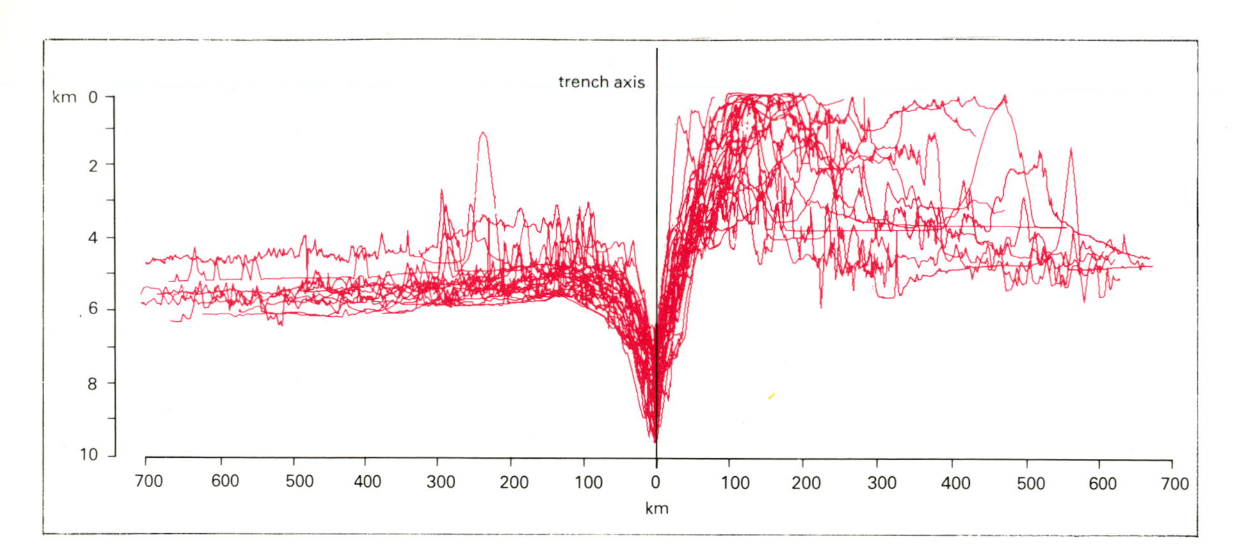

10.9: *Thirty-five topographic sections across different oceanic trenches, all superimposed, centred on their trench axes and with the oceanic crust to the left. The oceanic side of the trench is clearly less steep than the continental side and there is much greater variation in topography over the continental edge than within the ocean basins.*

nental rocks is of major importance in the determination of the direction of subduction in the past (Figure 10.12). At the present day the paired belts in the Japanese arc are clearly consistent with the subduction of Pacific oceanic rocks, while the Alps show similar paired metamorphism for Cretaceous times, indicating the previous subduction of oceanic rocks towards the south some 100 million years ago. This simple pattern becomes more complex on smaller scales and there are areas where the direction of subduction appears to have changed with time, for example in the New Hebrides and Papua New Guinea areas. Under such circumstances extremely complex geochemical and metamorphic patterns emerge that make it difficult to establish the previous history of the region.

The continental crust

Perhaps the major reason why many Earth scientists were initially unwilling to accept the concept of plate tectonics was the obvious complexity of the continental crust, particularly when compared to the extremely simple structure of the oceanic crust. The oceanic crust shows regular, simple age sequences, discrete, uniform layering in both its igneous and sedimentary rocks and uniform behaviour in its physical characteristics such as gravity, heat flow and topography. The age of the oldest known continental crustal rocks is very much greater, almost 4000 Ma, in contrast with an absolute maximum of 250 Ma for the oldest *in situ* oceanic crust, and virtually all geophysical and geological properties of the continental crust change over very short distances. However, while extremely old crustal units are present, the age of even the oldest terrestrial rocks is some 600 Ma younger than the age of the formation of the Earth itself and some 300 to 400 Ma younger than the crusts of other terrestrial planets (Mars, Mercury and possibly Venus) and of the Moon. Clearly the processes involved in crustal formation must have been different on these planets during their earliest histories, a topic which is taken up in greater detail in Chapter 27.

The seismic structure of the continental crust

The base of the continental crust in south-eastern Europe was discovered in 1909 by Andrija Mohorovičić who found that seismic waves were refracted along higher-velocity mantle rocks just below this boundary and so arrived before waves travelling along the surface of the continental crust at surface detecting stations 200 km or more away from the location of the earthquake. Until recently it was generally believed that this Mohorovičić discontinuity marked a very sharp chemical boundary between the continental and mantle rocks, and such a chemical change indeed conforms with the density difference between mantle and crustal rocks. However, recent work has demonstrated that the boundary is locally more complex than previously thought, for seismic reflections indicate varying degrees of penetration of mantle rocks up into the basal crust. None the less the extent of penetration appears to be no more than about a kilometre, and such small irregularities will probably show marked lateral variations reflecting the changes that have occurred in the relationship between mantle and crust as the system has evolved.

The thickness of the continental crust shows wide variations (Figure 10.12) within a range of about 10 to 50 km. There appears to be some correlation between the thickness of the crust and the age of the last orogenic event recorded at its surface. The crust beneath present-day mountain chains is characteristically very thick, with the thickest crust occurring beneath the Himalayas and northern Andes. The crust in older orogenic belts, such as the Appalachian–Caledonian and Variscan Mountains of Europe and North America (see Chapter 16) is generally somewhat thicker than in areas of Proterozoic or older continental crust. However, the thinnest continental crust appears to be associated not only with the oldest Archaean blocks but also with areas where the mantle appears to be active, as along the East African rift valleys and beneath the Basin and Range province of the western USA.

Although seismic records for earthquake waves passing through the crust are extremely complex it is frequently possible to distinguish two main crustal zones, sometimes separated by a velocity discon-

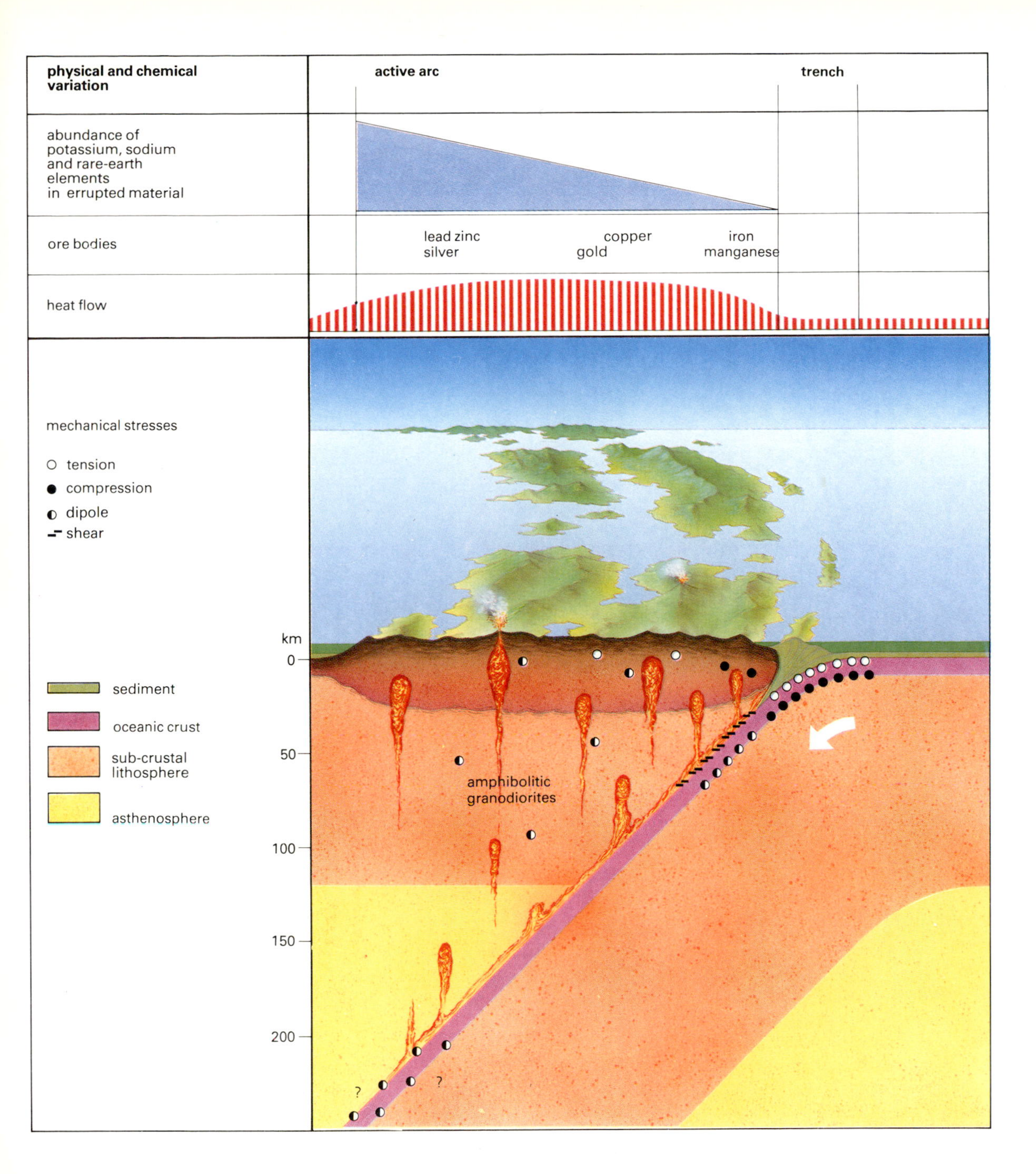

10.10: *The variation in the composition of erupted rocks bordering a subduction zone. As the depth of seismic activity increases the ratio of potassium to sodium of the lavas increases and the rare earth elements (REE) also change. Ore deposits range from iron and manganese near the trench to copper and gold overlying earthquake activity at depths of 150–200 km and to silver, lead and zinc over deeper seismicity. The nature of the seismicity also varies, with tension (T), compressional (C) and dipole (D) source mechanisms having specific distributions.*

tinuity determined by V. Conrad in 1925 at a depth of some 8 to 10 km in Europe. Combined seismic and gravity studies indicate that the upper layer has essentially 'granitic' properties, with an average density of about 2670 kg/m^3, while the lower crust has 'gabbroic' properties and a density of about 3000 kg/m^3. However, xenoliths carried up in volcanic pipes suggest that the lower crust comprises granodioritic-type rocks in a high (garnet/granulite) grade of metamorphism. The density of the lower crust is slightly greater than would be expected if its composition was the same as the upper crustal rocks but in a higher grade of metamorphism. It seems probable, therefore, that the lower crust also includes somewhat more basic or ultrabasic intrusions than the upper crust. None the less it must be emphasized that the seismic records are very complex, often with zones of low seismic velocity related to the Conrad discontinuity. Similarly generalizations about the nature of the lower crust become even less reliable near to active or recently active orogenic belts. The Alps, for example, have a generally thick crust, but local anomalies are common there, and the crust appears to be only a few kilometres thick in, for example, the Ivrea zone. In particular, instead of being in the anhydrous granulitic phase of metamorphism, the metamorphosed continental rocks appear to be in the hydrous amphibolitic grade.

Continental heat flow

In the same way that generalizations concerning the general seismic structure of continental crusts are dangerous, so also are statements about the distribution of heat flow. However, there does appear to be some correlation between the rate of heat flow and the age of the last orogenic event to which the crust was subjected. As in the oceans, the rate of heat flow appears to decrease according to the square root of age, but in the continents it is the age of the last orogenic event affecting the basement rocks. Furthermore the time scale for the decrease in heat flow is very much greater for the continents than for the oceans, the rate of decrease being measured in hundreds of millions of years, rather than in tens of millions. This correlation, although not yet adequately established in all continents, seems to be at odds with the surface distribution of elements producing radiogenic heat. Potassium-rich granites, which are commonly enriched also in heat-generating uranium and thorium, are characteristically of Proterozoic and later age and seem to be able to account for most of the observed continental heat flow without any significant contribution from either lower crustal or mantle sources. The heat-flow observations thus indicate that the continental crust has, at some time, become very strongly vertically differentiated in its heat-producing elements. The geochemical compatibilities of the other lithophile elements and volatiles must have become similarly concentrated in the uppermost parts of the continental crust (Figure 10.13). The time and cause of this differentiation comprise one of the major puzzles in any attempt to understand the evolution of the Earth—yet they have been largely ignored.

The temperature of the lower continental crust

As radiogenic heat appears to be confined to the highest parts of the continental crust the temperatures in the lower crust are difficult to evaluate. They clearly lie below the melting point of the constituent materials, but this temperature is variable, depending especially on the presence or absence of volatiles such as water, carbon dioxide and sulphur. Most estimates place the temperature in the lower crust at between 400 and 800°C, although temperatures of some 900° may occur where volatiles are largely absent. The main evidence for the actual temperatures is based on geochemical relationships in minerals brought to the surface in volcanic pipes. Unfortunately the mere occurrence of such pipes must be regarded as indicating possibly abnormal lower-crustal conditions. Many kimberlites appear to have originated from depths of some 160 to 200 km, but their xenolithic content may reflect anomalous conditions in the upper mantle rather than anomalous crustal properties. In general it seems likely that lower-crustal temperatures are relatively low, possibly 600–700°C, but with anomalously higher and lower temperatures occurring in specific areas.

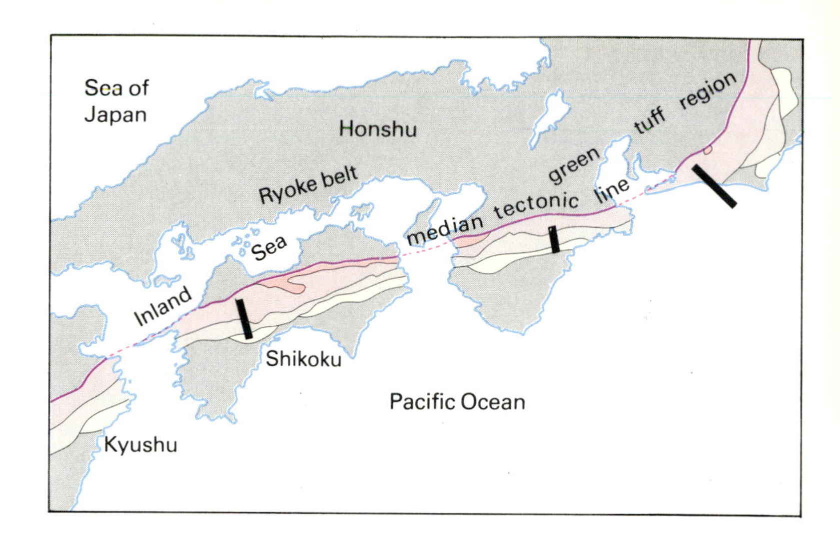

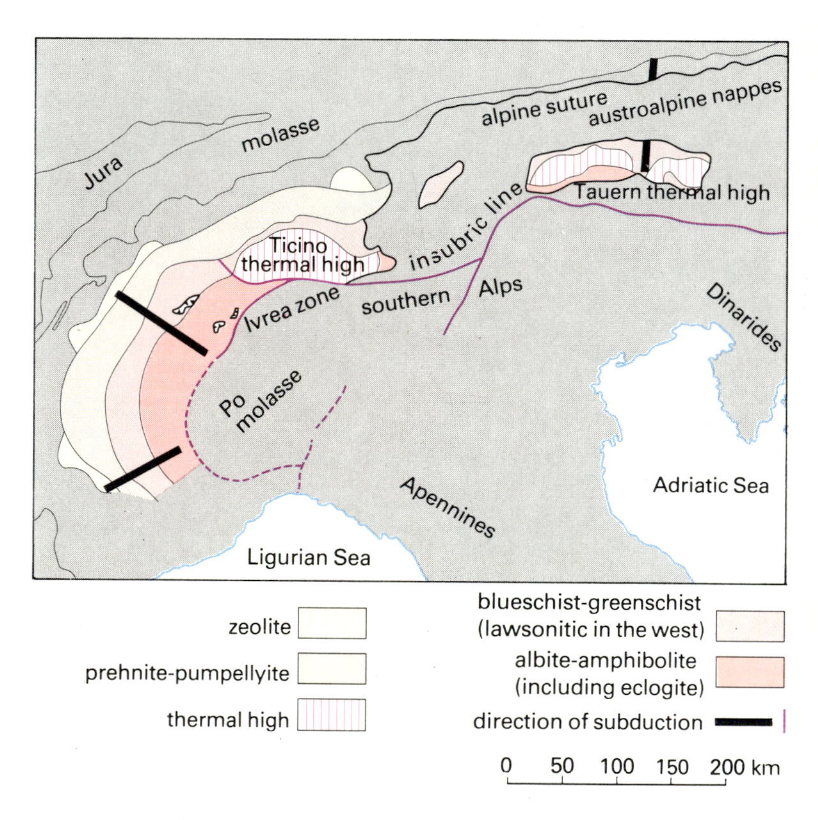

10.11: *Paired metamorphic belts in modern and old subduction zones. Low-grade metamorphic belts lie close to and parallel with the Japan trench today* (**a**), *with higher grade metamorphism at greater distances from the trench. In the Alps a similar gradation of metamorphism can be seen* (**b**) *which can be interpreted as indicating that an ocean, with subduction, lay to the north of the Alps at the time these rocks were metamorphosed some hundred million years ago.*

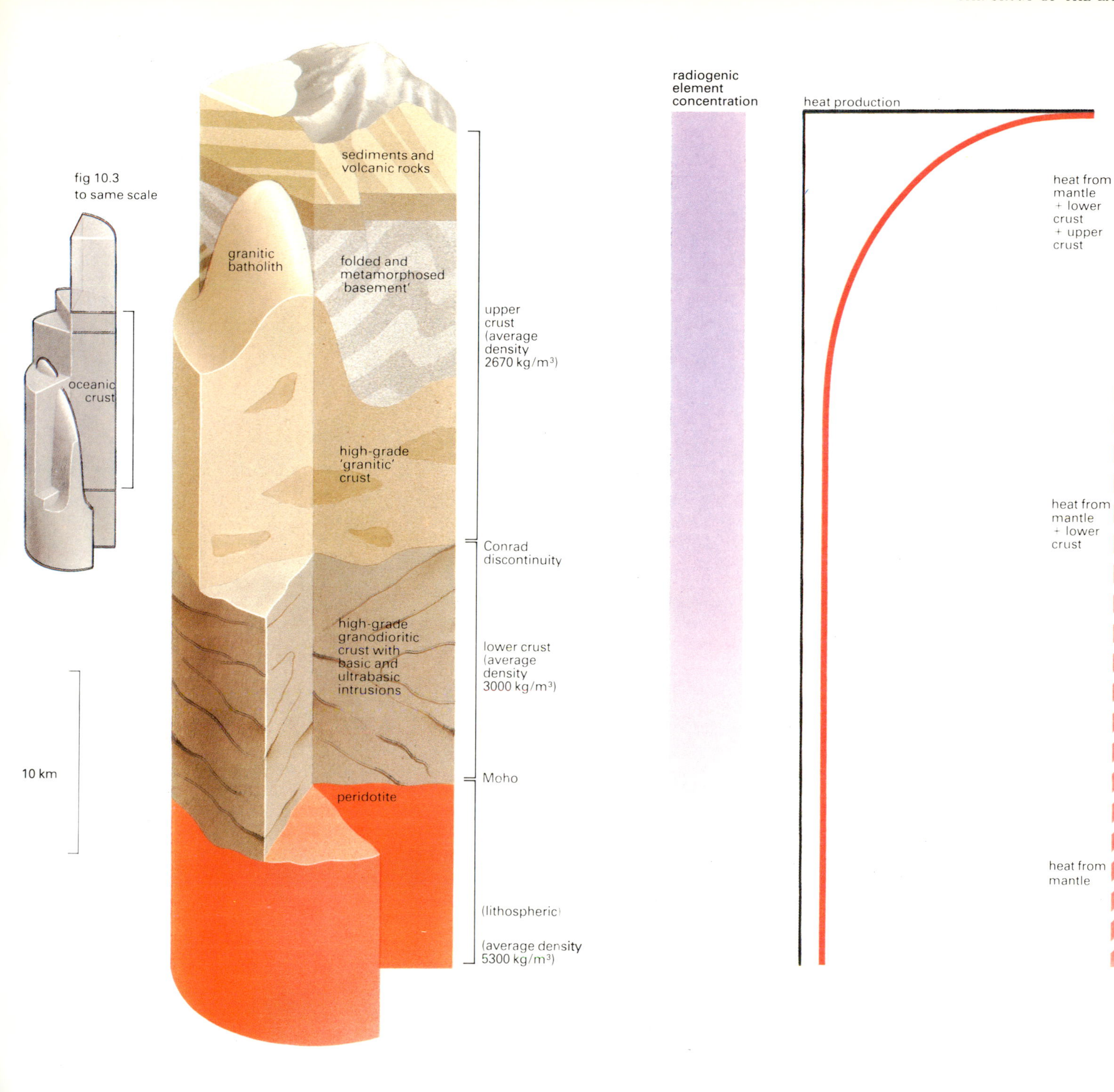

10.12: *A hypothetical section through continental crust, based on seismic data as well as on outcrop information. The oceanic crustal section of Figure 10.3 is redrawn to the same scale for comparison.*

10.13: *Radiogenic heat concentration in the continental crust. The flow of heat from the mantle is only slightly supplemented by radiogenic heat generated within the lower crust; most of the heat flow observed at the surface of ancient cratonic shield areas of the continents is derived from the upper crustal zone.*

Continental crustal evolution

Geological studies of rocks exposed at the Earth's surface indicate that there have been drastic changes in the nature of processes operating within the continental crust (Chapter 16). In brief the main features are an absence of any continental or oceanic rocks during the first 600 Ma of the Earth's history. The Archaean eon, c 3900 to 2500 Ma ago, is represented predominantly by two types of terrain. The gneisses and granodiorites, often in a high (granulitic) grade of metamorphism, form some 80 per cent of the Archaean continental crust and the remainder comprise greenstones and granites, often in an amphibolitic grade of metamorphism. The degree of metamorphism varies, although the bulk of Archaean rocks are now found in at least a greenschist grade of metamorphism.

There is dispute concerning the origin of these different types of terrain. The gneiss-granodiorites are often regarded as 'continental' and the greenstones as 'oceanic'. However, the contact, where visible, between the greenstones and gneisses seems to suggest that the greenstones were erupted onto the gneisses in a way similar to that of the flood basalts accompanying the opening of the Indian and south Atlantic Oceans during Mesozoic times. The net crustal composition is granodioritic—similar to that observed today in modern orogenic belts and suggesting that similar processes were operating during the Archaean. However, there is some detailed evidence for an evolution of the geotectonic processes. The outcrops of greenstones appear to get less arcuate and longer with time, the composition of granites intruding the sequences appears to get more alkaline and the Archaean possibly terminated with a worldwide potassium-rich intrusive episode.

The Proterozoic is characterized mainly by very great continental stability between 2500 and about 1000 Ma ago. The evidence for, or against, continental splitting is poor, but seems to be consistent with the existence of large supercontinents which were able to support sedimentary thicknesses of some 10 km or more. The onset of 'modern' plate tectonics during this period is uncertain. The Grenvillean orogenic episode in North America about 1000 Ma ago seems to provide the earliest reasonably convincing evidence for collisions between continental blocks. However, this evidence is still poor compared with that for the Phanerozoic period, the past 570 Ma during which the Appalachian–Caledonian and Variscan orogenies in North America and Europe can be explained in terms of continental collisions. Similarly the Urals appear to have formed by successive collisions between Europe and Siberia during the past 250 Ma.

Transitions from continental to oceanic crust

The fundamental differences between continental and oceanic crust clearly make it important to understand the transition between the two. This normally occurs near the margin of the continental shelf rather than at the coastline. It has been long recognized that two fundamental types of continental margin exist. Those associated with orogenic and volcanic activity, the active margins, are marked by the deep oceanic trenches (see above) and occur where the oceanic lithosphere is subducted beneath the continental crust. Those in which there is little or no tectonic activity, the passive margins, occur typically on the continental edges bordering the Atlantic and Indian Oceans. They are characteristically bounded by faults whose nature indicates one of two types of passive margin (Figure 10.14). Continental edges that are approximately parallel to the oceanic ridges are distinguished by normal, tensional faulting, while those that are perpendicular to the ridges show evidence of earlier transcurrent motions which can often be correlated with old fracture zones in the oceanic crust. The

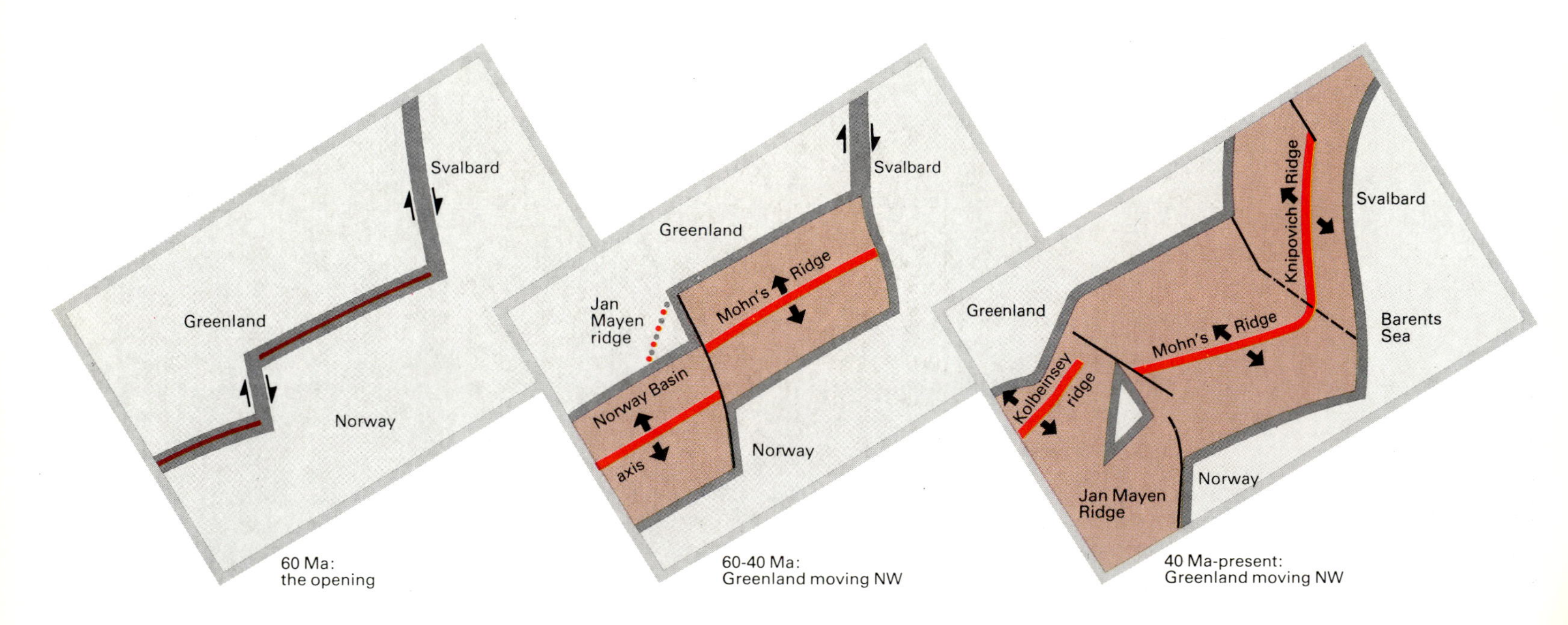

10.14: *Continental edges in the northern north Atlantic, formed either as a transcurrent fault zone or (where parallel to the mid-ocean ridge) by rifting, marked by normal faults.*

distinction between these two types of passive margin is fundamental in assessing their origins. It is also important in estimating their potential for hydrocarbon reserves, since the tension-faulted areas are most likely to contain traps and reservoirs.

The precise nature of passive continental margins can differ significantly in other ways (Figure 10.15). The normally faulted edges are often accompanied by a series of parallel faults where isolated blocks of continental material separate from the original continental block during the early period of ocean formation. The intervening oceanic areas may be filled in by later sediments. In lower latitudes coral reefs sometimes form on the outer edges, enhancing the effectiveness of this sediment trap, for example off the southern Atlantic seaboard of North America. The edge may be further modified by the deposition of sediments derived from the continental areas behind.

In such areas, the continental edge, though modified by later sediments, is geophysically clearly defined by a rapid change in the density and magnetization of the rocks. Such rapid changes demonstrate that the continent–ocean transition is both sharp and deep, which seems to be true of most areas bordering the Atlantic. However, there are areas elsewhere where the transition is much more gradual. Off eastern Africa, for example, the distinction between continental and oceanic crust is much more difficult to define as the one appears to merge gradually into the other (Figure 10.16). The crust in other areas of the world also shows properties that make it difficult to establish whether it is truly oceanic or continental. The Mediterranean Sea, for example, has an undoubtedly oceanic crust in the west, but in the east the crust may be regarded as either thin continental or thick oceanic. Similarly, most of the rises in the south-western Pacific (the Lord Howe rise, the Lau rise, etc) have structures that appear to be intermediate between continental and oceanic. In the south-western Pacific these areas appear to represent subduction zones whose crust-forming activity was interrupted.

The crust between island arc systems and nearby continental blocks appears to be of fundamentally oceanic composition and structure. None the less these marginal basins differ from oceanic basins in several important respects. In most cases there are no clear magnetic anomaly patterns (see Chapter 6), suggesting that their formation is less systematic than that of the main oceanic basins. They are also characterized by a generally high geothermal gradient and a thick sedimentary sequence. The former apparently indicates the widespread (rather than linear) formation of new oceanic crust while the latter indicates the presence of plentiful supplies of sediments from the continents on the one side and active volcanic arcs on the other. In general, however, such marginal basins may be regarded as essentially oceanic.

Plate tectonics

Seismic zones

The division of the Earth's crust into continental and oceanic is obvious when considering their different composition, structure and history. It is surprising therefore that, in terms of tectonic activity, the most important division of the surface rocks cuts across such chemical and structural differences. Present-day tectonic activity can be inferred from the distribution of earthquakes (see Chapter 3), which are confined to extremely narrow but continuous zones. The narrowest and shallowest zones occur along the oceanic ridges and within the fracture zones that lie between offset oceanic ridges. These earthquakes are tensional at the ridges and transcurrent along the fracture zones (see above), and are confined to belts that are at most a few kilometres wide and some 10 km deep. There are also earthquakes along the ridges which are associated with volcanic activity and which may extend to depths of some 40 km. These appear to be generated by the rise of magmas from deeper in the mantle and are thus only indirectly related to the tensional forces operating at the ridges.

Other, often more violent, earthquakes occur along less well-defined zones, particularly those associated with the oceanic trenches, where the detailed picture is complex. Although the surface characteristics of the earthquakes associated with oceanic trenches and island arcs are varied, the majority of such earthquakes appear to be confined to a narrow, dipping zone, which has become known as the Benioff zone. The Benioff zone dips, on average, at some 45°, usually away from the ocean and towards the continent. In detail it is often less steeply inclined at shallow depths and then becomes steeper, eventually reaching depths that may be as great as 710 km but are generally less than some 600 km. The Benioff zone appears to be characterized by tensional earthquakes on the oceanic side of the oceanic trench, with compressional earthquakes sometimes occurring some 10 to 15 km below. At the top of the Benioff zone the earthquakes are attributable mostly to a dip-slip motion between the adjacent sides. This type of activity persists to depths of some 100 km, below which seismic activity appears to be partially suppressed, but at depths greater than 200 km the earthquakes have more complex source mechanisms.

Although most seismic activity is restricted to the Benioff zone more irregular activity occurs in the overlying mantle and crustal rocks. Some of this activity is tensional, especially in the upper few tens of kilometres, but a variety of sources appears to be involved, and most of the earthquakes away from the Benioff zone are thought to be associated with the rise or generation of magma which eventually feeds the volcanic island arc or mountain chain usually bordering the oceanic trenches.

Plates and plate boundaries

The fact that some 80 per cent of the world's earthquakes occur in specific continuous, narrow zones, and have similar source mechanisms within each, led Jason Morgan, in 1968, to propose that the Earth's surface could be regarded as comprising a few large plates, or shells. If each plate is assumed to have a high torsional rigidity (ie, in common with a piece of paper floating on water, it can be moved about on the surface without distorting) earthquake activity can be explained simply in terms of the difference in motion between adjacent plates. At the oceanic ridges the plates are separating, the ridges thus being in tension, while opposing sides of the fracture zones offset between ridge crests will be moving in opposite directions, thereby generating transcurrent seismic activity. As the Earth is not expanding overall, plates are believed to be consumed at the oceanic trenches. Here, the plates are descending, generating earthquakes as they are subducted under the bordering plate. Morgan identified eight major plates, and numer-

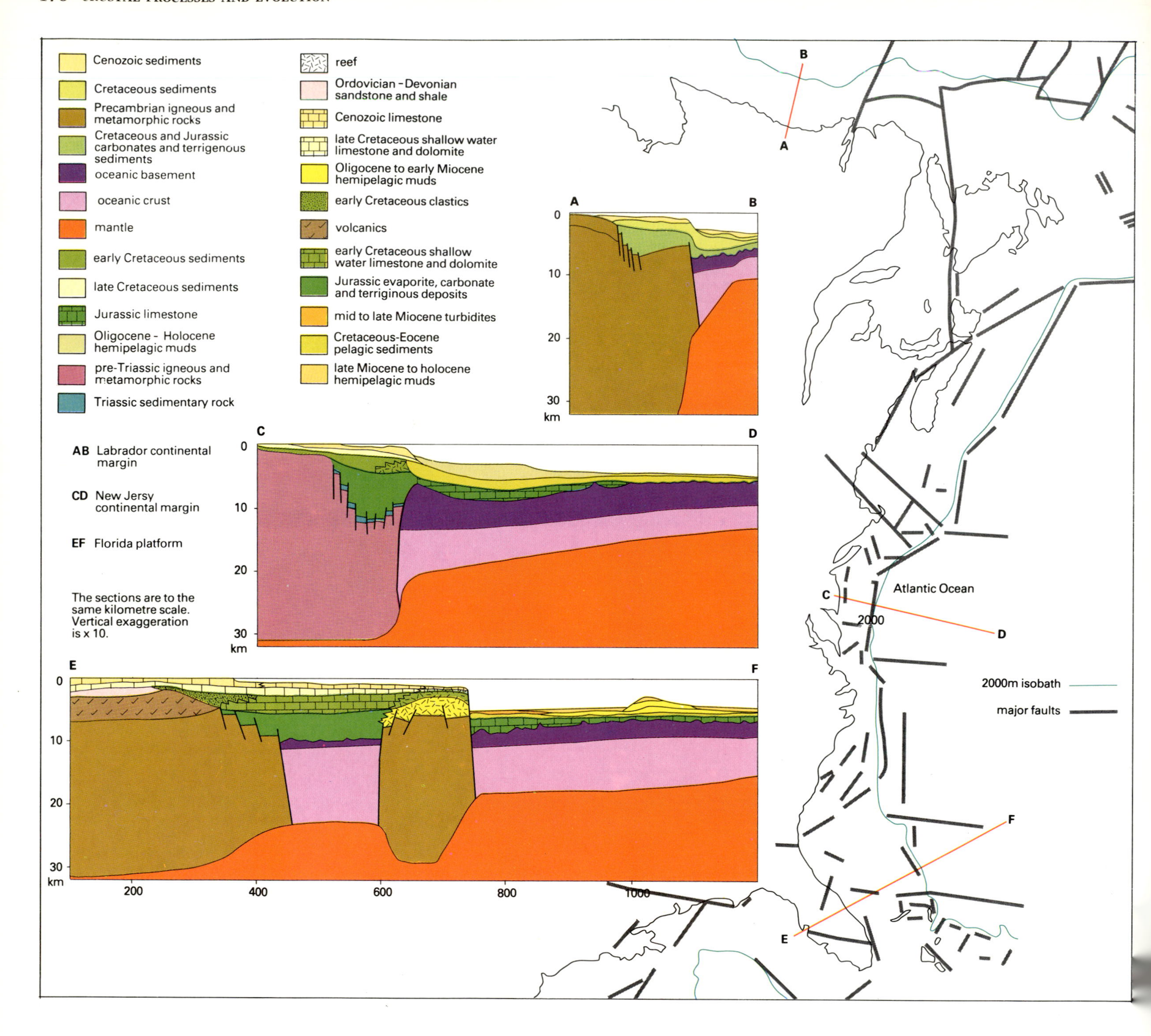

10.15: *The eastern North American continental edge is of fractured origin but differs in its precise nature along its entire length. Off Labrador the faulted edge is overlain by thick sediments but the topographic edge still corresponds with the continental edge. Off New Jersey there is a similar correspondence, but a much thicker sequence of continental sediments which include evaporites because of the lower latitude. Off Florida the edge is complicated by the fragment of continental crust that broke off during separation and formed a nucleus for reefs etc. This resulted in the infilling of the oceanic trough between this block and the main continental crustal block.*

ous smaller plates or microplates have subsequently been recognized in complex areas such as the Mediterranean (Figure 10.17).

It is an important aspect of the plate concept that plate boundaries are not always coincident with the junction between continental and oceanic crust; they may parallel this junction (eg the mid-Atlantic ridge) or meet it at an angle (east Pacific rise). Thus many plates include both continental and oceanic crust. Furthermore, the plates are thicker than the crust alone, and they include that part of the mantle that belongs, with the crust, to the lithosphere.

It is important to emphasize that while this model can explain most of the world's earthquakes, at least one-fifth of known earthquakes do not occur in association with specific continuous boundaries. It should also be emphasized that while the plates have a very strong torsional rigidity, they do not necessarily have a strong flexural rigidity or lateral strength (ie, again in common with a piece of paper on the surface of a pool of water, they can be easily flexed vertically). Thus the plates can move as coherent units, yet still flex under different stresses, such as those required to maintain long-term isostatic equilibrium over both small and large areas.

According to the hypothesis of plate tectonics, plate boundaries may be one of three types, ridge, transform fault or subduction zone, each with particular seismic, tectonic and volcanic characteristics. The three boundary types can usefully be thought of as constructive, conservative and destructive respectively. Junctions between boundaries are known as triple junctions which may comprise one of ten possible combinations of the three types of boundary (Figure 10.18). Some of these combinations are stable in that they will not change their geometry unless there is a fundamental change in the controlling forces. However, if different rates of motion are involved or at least one of the boundaries is a trench, then the geometry of the three junctions must evolve. This means that the nature of the tectonics associated with any one part of a plate margin can be subjected to a widely varying history of stress fields and types of tectonism as associated triple junctions migrate past it. Many of the complexities in the structural evolution of continental areas may thus be explained in terms of the evolution of triple junctions.

The main difficulty in interpreting the past evolution of plate margins arises from the disappearance of the crust (usually oceanic) at subduction zones, thereby destroying at least half of the history of the evolutionary sequence. The occasional preservation of slices of oceanic-type crust in so-called ophiolite complexes may assist in the recognition of a vanished ocean along the sutured collision zone of two continental masses, but these do not reveal the former width of the ocean, whether tens or thousands of kilometres. Some ophiolites may indeed be remnants of flood basalts erupted onto continental rocks with no associated ocean at all.

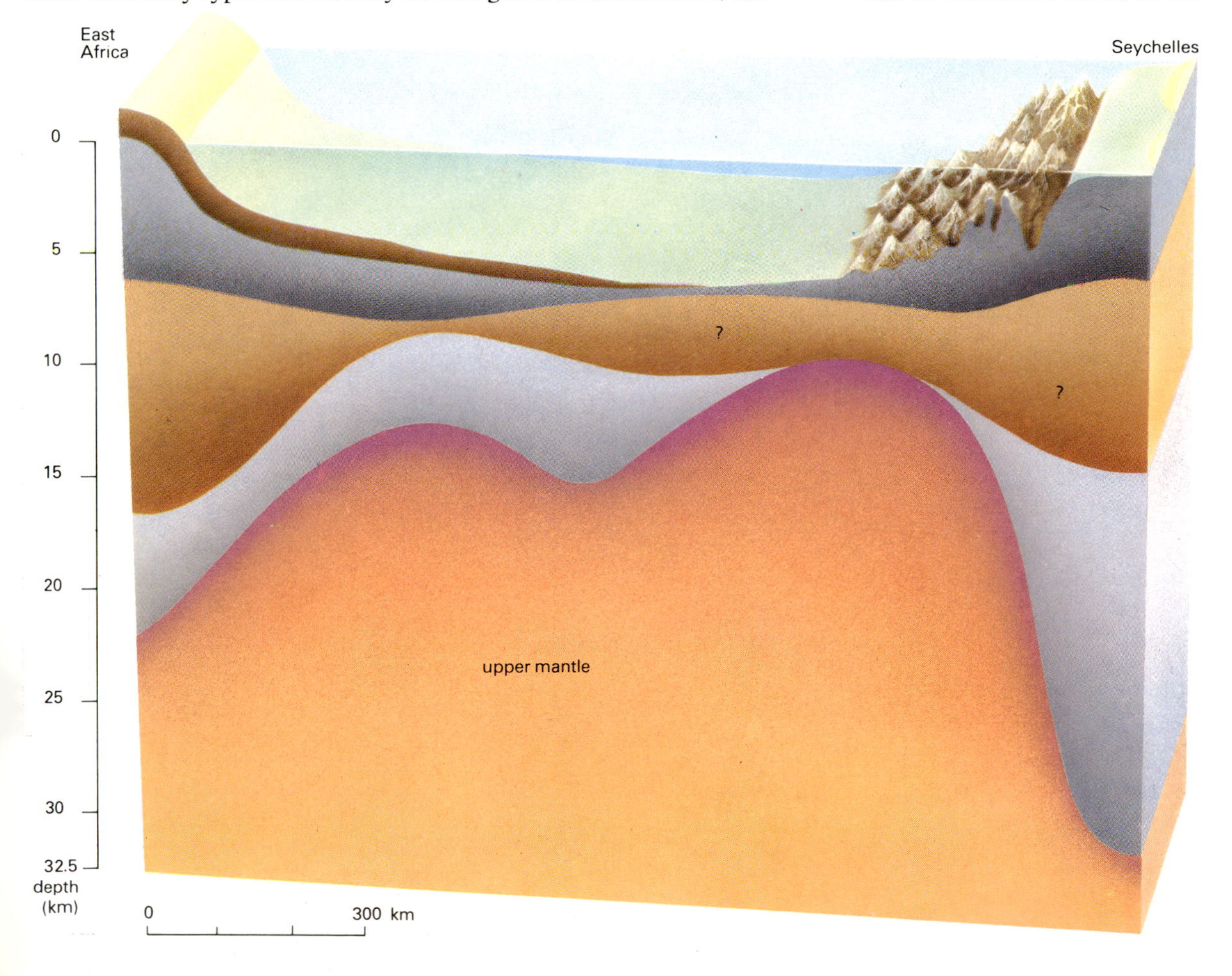

10.16: *A section of anomalous continental-oceanic interface, based on seismic and gravity data obtained between East Africa and the Seychelles. In this area it is difficult to be certain where 'continental' crust ends and 'oceanic' crust commences. This differs drastically from other areas, such as the eastern North American seaboard (Figure 10.15) where the boundary is sharp and distinct.*

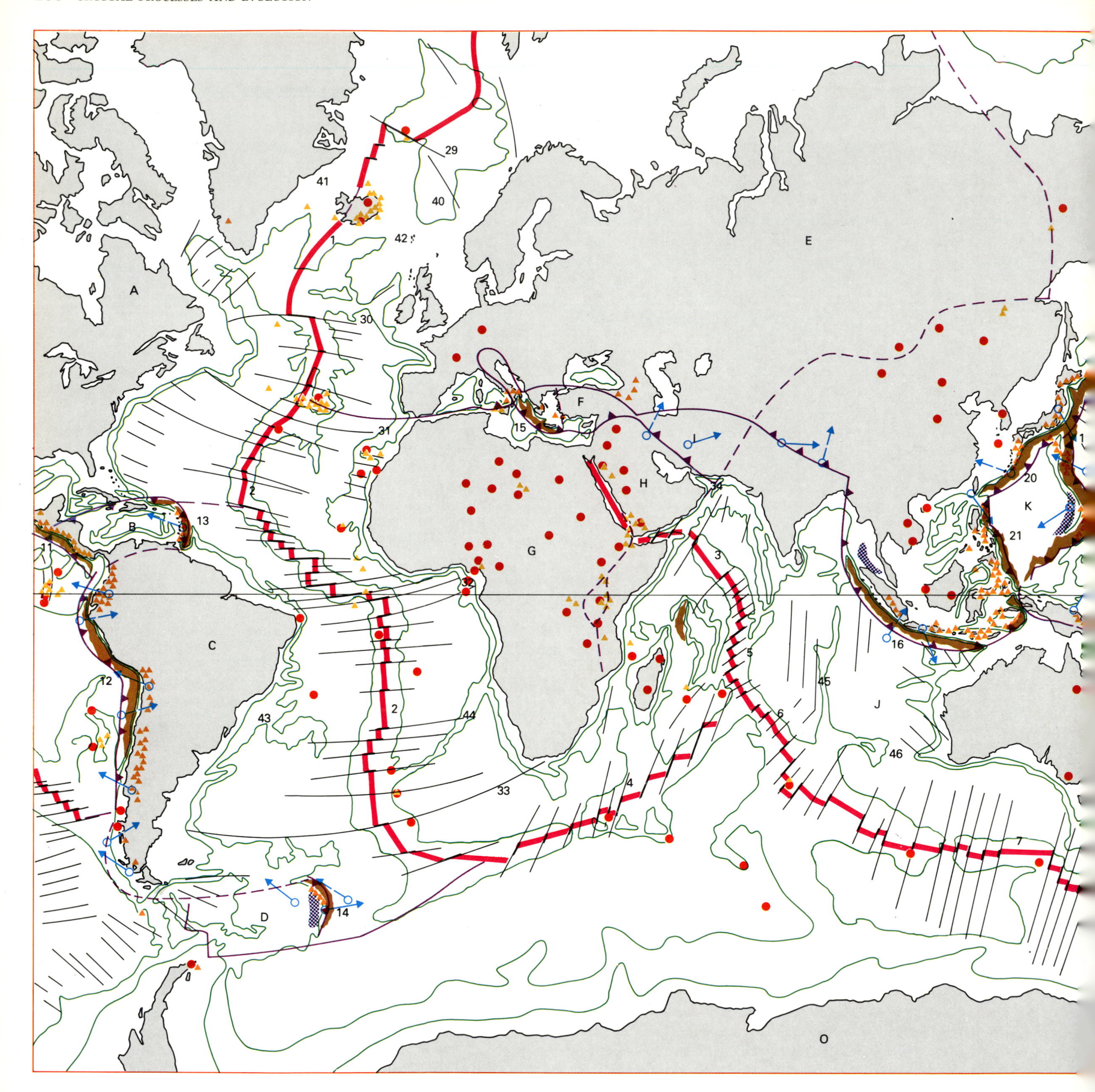

A
B
C
D
E
F
G
H
I
J
K
O
1
2
3
4
5
6
7
11
12
13
14
15
16
20
21
29
30
31
32
33
34
40
41
42
43
44
45
46

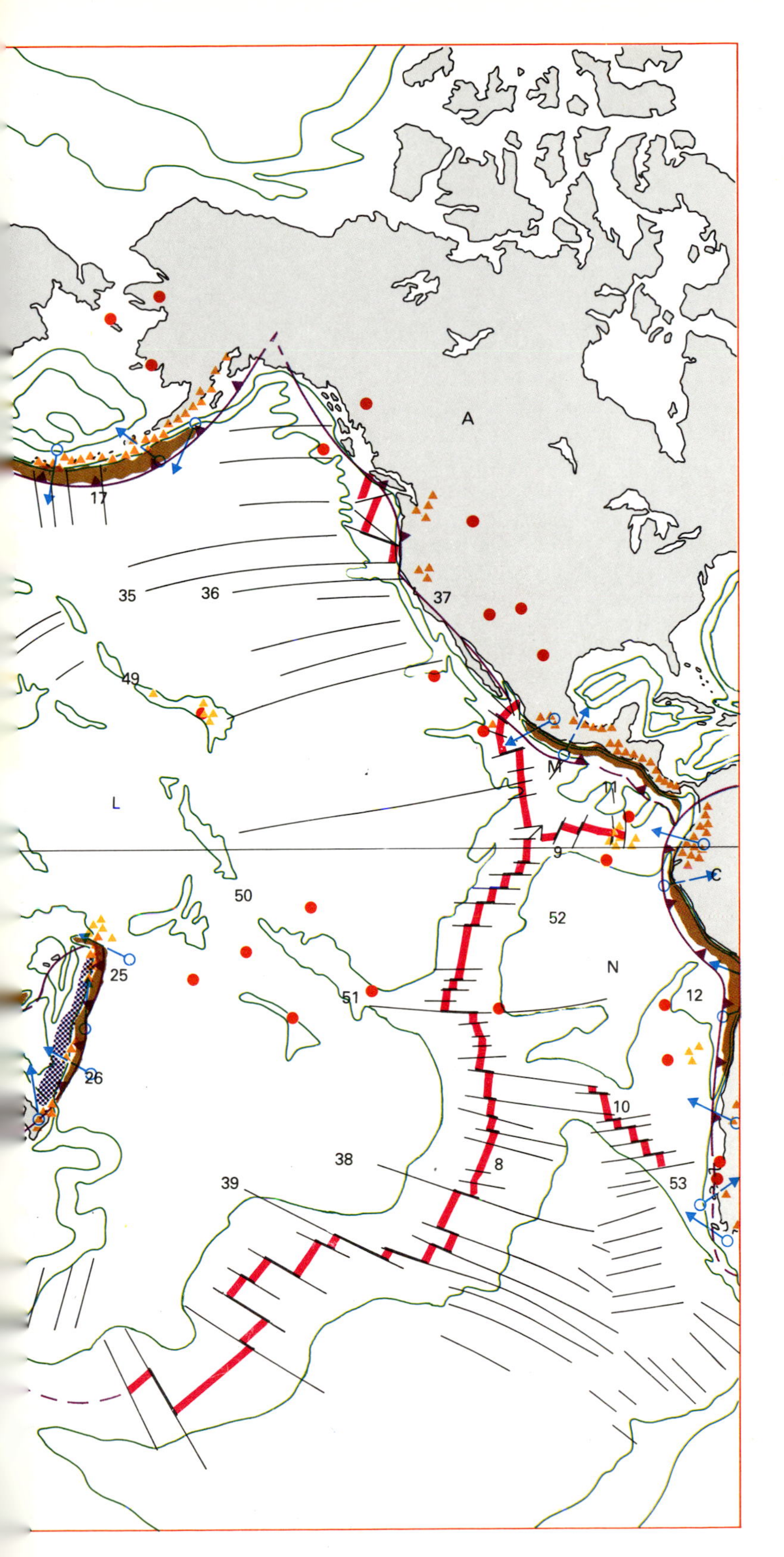

10.17: *A map of the major tectonic features of the Earth's surface.*

submarine contours at 2000m intervals
effusive volcanoes
explosive volcanoes
oceanic trenches
fracture zone

Plate boundaries

transform faults
destructive
constructive
undifferentiated
uncertain and incipient

areas of rapid marginal basin (back-arc) sea floor spreading
'hot spots'
directions of movement at destructive margins of overriding ( ) and downgoing ( ) plates, relative to the reference frame provided by the hot spots

Major lithospheric plates

A	North America	F	Hellenic	K	Phillippine
B	Caribbean	G	African	L	Pacific
C	South America	H	Arabian	M	Cocos
D	Scotia	I	Iranian	N	Nazca
E	Eurasian	J	Indo-Australian	O	Antarctic

Active spreading ridges

1	Reykjanes	5	Mid-Indian	8	East Pacific rise
2	Mid-Atlantic	6	South-east Indian	9	Cocos
3	Carlsberg	7	Antarctic-Pacific	10	Nazca
4	South-west Indian				

Submarine trenches

11	Central American	17	Aleutian	23	New Guinea
12	Peru-Chile	18	Kuril	24	New Hebrides
13	Puerto Rico	19	Japan	25	Tonga
14	South Sandwich	20	Ryukyu	26	Kermadec
15	Cretan	21	Philippine	27	Puysegur
16	Java	22	Pariana	28	Hjort

Fracture zones and transform faults

29	Jan Mayen	33	Falkland	37	San Andreas
30	Gibbs	34	Owen	38	Menard
31	Oceanographer	35	Mendocino	39	Eltanin
32	Romanche	36	Pioneer		

Other major features

40	Jan Mayen ridge	45	Ninety-east ridge	50	Line Islands
41	Greenland ridge	46	Broken ridge	51	Tuamotu Islands
42	Faeroes ridge	47	Lord Howe rise	52	Galapagos ridge
43	Rio Grande rise	48	Emperor seamounts	53	Chile rise
44	Valvis ridge	49	Hawaiian ridge		

The geological record of oceanic ridges will obviously be almost entirely destroyed as the ridge is subducted. In some specific instances, however, the high heat flow associated with the ridge has not only not prevented its subduction but has had identifiable results within the continent after being subducted. In some cases the ridge appears to have initiated the formation of marginal oceanic basins between the volcanic arc associated with the trench and the bordering continent. This appears to have happened on at least two successive occasions in the region of the Scotia plate between Patagonia and the Antarctic peninsula during the past 60 Ma, while the volcanic and tectonic activity in the Basin and Range province of the western USA appears similarly to have been initiated following the subduction of part of the actively spreading east Pacific rise in Miocene times.

Palaeogeographic reconstructions

In view of the clear evidence for the relative motions of different parts of the Earth's crust it is of interest to extrapolate those motions back in time. Furthermore, any understanding of the geological evolution of its surface and fauna and flora requires an assessment of the previous relationships of the continental blocks and of land–sea margins, oceanic circulation and so forth. There are clearly no seismic data for prehistoric periods and it is thus necessary to use geological and geophysical criteria.

Oceanic anomalies and palaeomagnetism

The patterns of ocean-floor magnetic anomalies relate to the location of the active spreading ridge at the time they were formed (see Chapter 6). It is therefore possible to replot adjacent plates so that a specified anomaly on one plate lies adjacent to the anomaly of the same age on the other plate (Figure 10.19). Such a reconstruction must then be true for the continents forming part of each plate for that time, within the precision of the dating and the accuracy of matching the anomalies. Although some anomalies are difficult to distinguish one from another, the sequence of polarity changes means that certain groups of anomalies are readily recognizable. Furthermore, although there are some uncertainties about the precise age of some of the anomalies, the error thus introduced into the replotted separation of continental blocks is small when rates of motion rarely exceed 50 mm per year. On this basis it is relatively easy to make reconstructions between continental blocks separated by a spreading ridge during the past 200 or so Ma. It must be emphasized, however, that such reconstructions establish only the position of the continental crustal blocks relative to each other. The blocks could also have a common motion quite independent of the oceanic anomaly patterns. For example, as the Atlantic Ocean opened, the magnetic anomalies formed parallel to the spreading ridge, but independent results show that the whole Atlantic basin was probably moving northwards at the same time.

Latitudinal changes can be estimated directly from palaeomagnetic studies of continental rocks of known age. When igneous rocks cool, or sediments are deposited, any ferromagnetic particles of the order of one micron in diameter are capable of retaining a record of the ambient direction of magnetization (see Chapter 6). As the average geomagnetic field corresponds closely with the model of a bar magnet at the Earth's centre aligned with the Earth's spin axis, it is thus possible to determine the latitude and orientation of rock samples relative to the past axis of rotation at the time of their formation (Figure 10.20). This means that this remanent magnetization can be used to define the palaeolatitudinal relationships of continental blocks while the oceanic anomalies define their spatial relationships. Thus, when both lines of magnetic evidence are available, reasonably precise palaeogeographical reconstructions can be made. (This is the basis for the construction of the maps in Chapter 23.)

Unfortunately many critical areas of the oceans have still not been studied magnetically and palaeomagnetic data from continental blocks are also sparse. In the oceans the reconstructions are hampered by the lack of polarity changes at certain times. The critical opening phases of the Indian and Atlantic Oceans are problematical because of the lack of readily identifiable anomalies of those ages (Jurassic–Cretaceous). In particular, the magnetic field during most of the Cretaceous period was of normal polarity, as was also the case for much of the Jurassic. In earlier times the Permian period was one of almost entirely reversed polarity, but as no crust has yet been identified of this age in the present oceans the sequence of polarity changes in the Palaeozoic is not of relevance to reconstructions of past continental positions. None the less, while there are numerous uncertain details, there is a general agreement between most Earth scientists on the probable continental relationships during the Mesozoic and Cenozoic eras.

Palaeozoic and earlier palaeogeographies

For the Palaeozoic era and even earlier times, the data base for any reconstruction is extremely poor. No oceanic anomaly data are available and the palaeomagnetic record is often unreliable because of uncertainties about either the ages of the rocks themselves or of the time at which the rocks last became magnetized. Many Precambrian and Palaeozoic rocks have been subjected to later thermal or chemical changes, so they carry a superimposed magnetization that may overprint the originally acquired orientation. Although some data are available and the situation for Palaeozoic times is becoming better known, the main evidence is still from floral and faunal distributions, even though such arguments are often subjective and difficult to evaluate in terms of precise palaeogeographical relationships. One particular problem is that climatic belts and biological provinces do not always follow lines of latitude. For instance, environments on the east and west coasts at the same latitude are usually quite different. Other problems come with the preservation and identification of fossil flora and fauna, and different scientists' interpretations of the same data yield quite different Palaeozoic palaeogeographies.

This problem can be illustrated with a brief consideration of the evidence from Silurian–Devonian 'freshwater' fish. The remains of such fish are normally fragmentary. The evidence shows fairly convincingly that most were living in freshwater rivers and lakes in their mature forms, though it is conceivable that the younger larval stages may have been tolerant of brackish or even marine water. The occurrence of similar species of fish in Australia and northern Canada could be interpreted as indicating that these two land masses were contiguous in Devonian times. The palaeomagnetic data suggest that

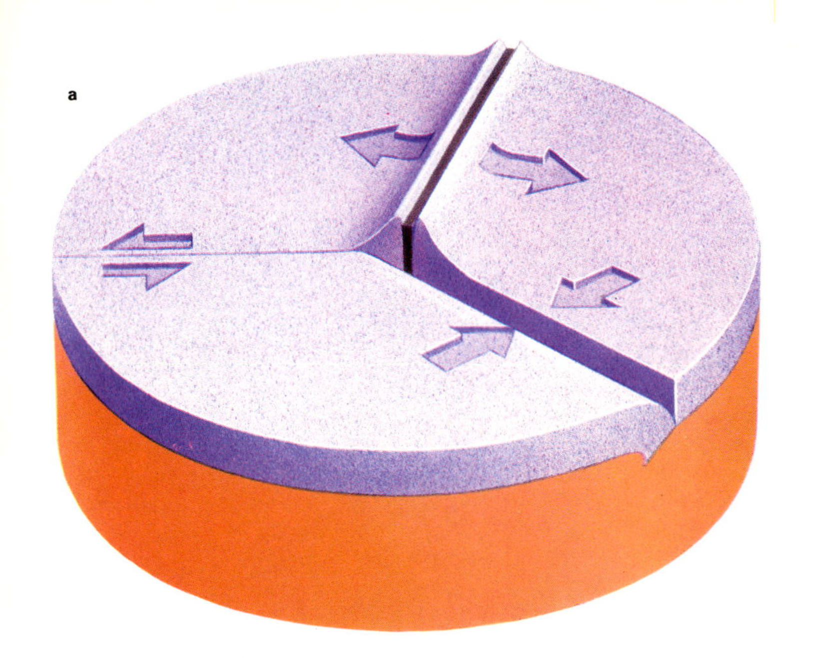

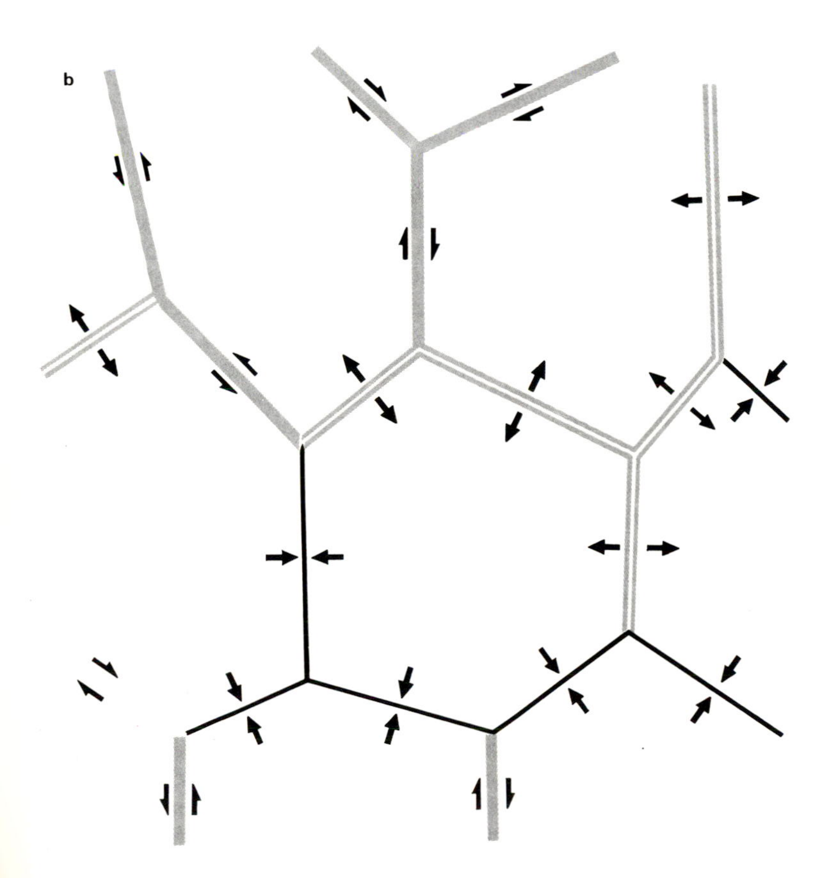

10.18: (**a**) *A triple junction is formed where plate boundaries meet.* (**b**) *Ten possible configurations of the three types of boundary.*

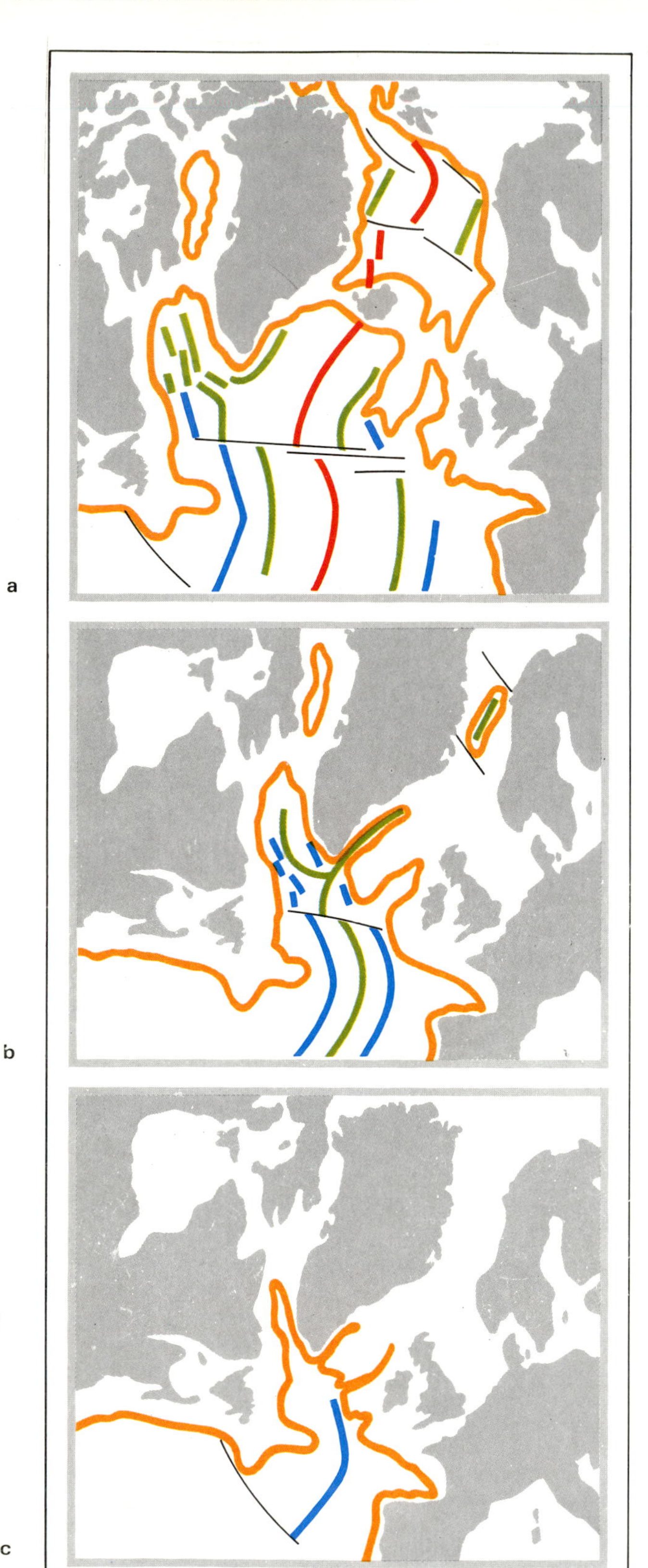

both areas were lying in the southern tropics at that time, which would not be inconsistent with the palaeontology. On the other hand, Australia definitely formed part of the Gondwana continent (South America, Africa, India, Australia, Antarctica and New Zealand) and northern Canada was part of Laurentia (North America, Greenland and northern Europe) in the Devonian (see Figure 23.22). As Africa and North America had only recently been in collision to form the Caledonian Mountains and appear to have remained close during the Carboniferous and Permian, it is clearly impossible for Australia and Canada to have been contiguous in the 'Pacific' area. Thus it is difficult either to satisfy both types of data or to know which to give more weight to.

The situation is clearly even more complex for Precambrian times as there is no fossil evidence available, with the exception of stromatolites which are possible but not proven indicators of low palaeolatitude. The only data are therefore palaeomagnetic and palaeoclimatic, although clearly any reconstructions for this time must also be consistent with the structural development of the areas concerned. Most of the data available for Proterozoic times indicate that the continents were in motion, as palaeolatitude changes are shown consistently by both palaeomagnetic and palaeoclimatic indicators. It appears as if only two major continental blocks were involved during this prolonged era. For Archaean times the different tectonic regime indicates drastically different conditions (see Chapter 16) and palaeomagnetic data may never be sufficient to yield a unique reconstruction for any given time interval. An understanding of the nature of the tectonic processes in the earliest periods must therefore depend on structural analyses and hypothetical models.

Problems in plate tectonics

There are two fundamental problems in the existing concept of plate tectonics. The first is that while the lateral extent of a modern plate can readily be defined by earthquake activity its thickness is still not physically defined. This lack of definition makes it difficult to understand how the plates move. The second major problem is to assess whether plate tectonic activity occurred also in the distant past and, if so, whether its manifestation then was significantly different from that in the present day. Although the past geological history of the Earth (see Chapter 16) indicates the possible nature of its past tectonics, only when the driving forces behind the motion of plates are known can a proper evaluation of past plate movement be made. What is certain, however, is that tectonic activity occurred throughout that part of Earth history recorded in continental rocks, and in this respect the Earth differs from the other inner planets of the solar system.

10.19: *Magnetic anomaly reconstructions in the north Atlantic.* (**a**) *Anomalies up to 50 Ma old (anomaly 24, green) occur at present between Greenland and Europe (red line anomaly 1, present spreading axis; orange line, continental shelf edge).* (**b**) *Closing the ocean between these anomalies restores the continents to their positions of 50 Ma ago.* (**c**) *Similarly, matching anomalies 31/32 (blue) demonstrates the non-existence of the southern north Atlantic some 75–80 Ma ago.*

The physical properties of plates

The surface rocks of the Earth have high flexural rigidities (of about 10^{28} poise) and are thus unable to deform except in response to very strong or extremely prolonged forces. The mantle beneath the continents and oceans has a much lower rigidity (of some 10^{21-22} poise) which is none the less comparable with that of sheet glass at normal room temperatures.

Determinations of rigidity are not easy to undertake and their interpretation can be ambiguous. The most effective long-term measurements are based on the rate of uplift of, for example, parts of Scandinavia following the retreat of the last of the Quaternary ice sheets where 5000-year-old shorelines are now at an elevation of 100 m. However, the area affected by this uplift extends over most of northern Europe and may also have been influenced, to an extent difficult to assess, by other factors. For example changing sea levels in the Baltic and North Sea and Alpine mountain-building processes make it difficult to sort out the rate of uplift due to isostatic recovery alone. Shorter-term relaxation times can also be determined from man-made loading, such as that caused by the waters impounded behind the Kariba Dam in northern Zimbabwe. Similar estimates can also be made using the yielding of crustal rocks in the oceans resulting from loading by volcanic chains such as Hawaii or even from tidal loading of ocean waters on the continental shelves.

Seismic waves passing through the Earth also provide information on the average density of the rocks through which they pass and, by comparison of P and S wave velocities, on some other physical properties such as the elastic constants (see Chapter 3). The physical properties obtained in this way relate to the instantaneous, rather than long-term, response of the mantle rocks as it takes only some twenty minutes for such body waves to pass through the entire Earth. The velocity of seismic waves in mantle rocks in the oceanic areas usually shows a distinct decrease at depths between about 80 and 120 km and this low-velocity zone extends to depths of some 200–250 km, although its lower boundary is usually much more poorly defined than its upper boundary. Within some 50 km of the ridge the top of the low-velocity zone is encountered at even shallower depths and it appears to be very close to the surface at the ridge crest. As volcanism occurs at the crest, the low-velocity zone is conventionally interpreted as being a zone within which partial melting (of about 1 per cent) occurs (see Chapter 3). The presence of a partial melt would cause the phenomenon of a greater slowing and absorption of S than of P waves. Accordingly, the rocks overlying the seismic low-velocity zone are interpreted as being 'rigid' and are termed the lithosphere, while the rocks of the low-velocity zone, being partially molten, will deform readily and are termed the asthenosphere.

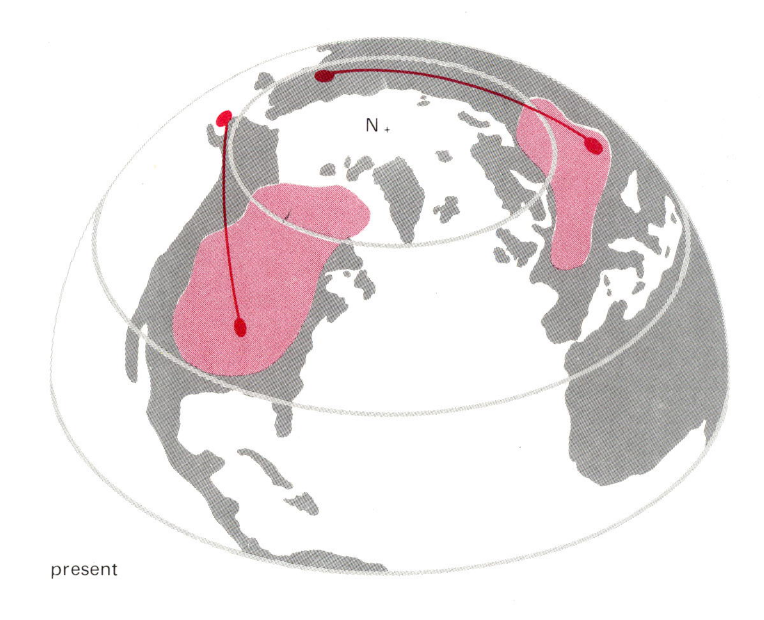

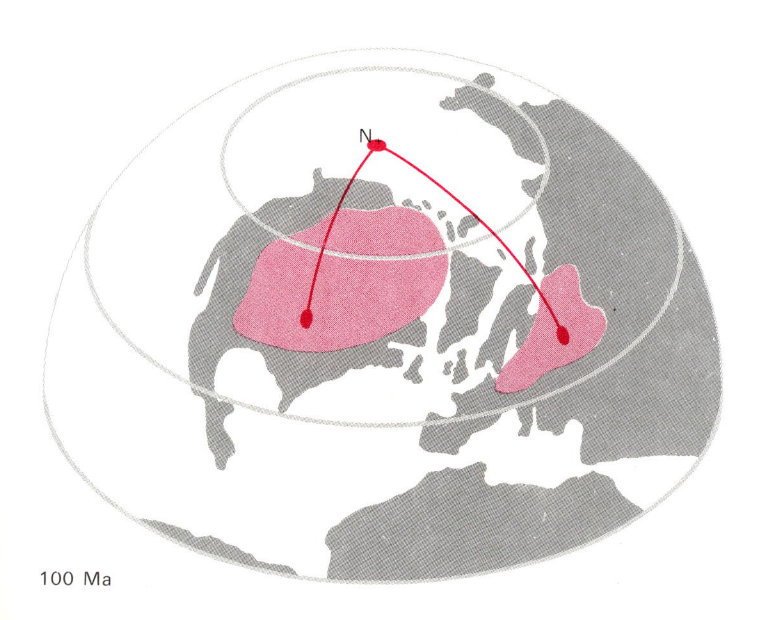

10.20: *The determination of palaeomagnetic pole positions. Rocks that have been magnetized during the past million years or so have an average magnetic direction that points due north, and the inclination of the direction from horizontal changes systematically from zero at the equator to vertical at the geographic (rotational) poles. This suggests that the Earth's magnetic field corresponds to a bar magnet at the centre of the Earth, aligned along its axis. When much older rocks are sampled their average magnetic directions differ from those of the past million years. The location of the average magnetic pole can then be determined from their magnetization as the magnetization obviously points towards the magnetic pole and its inclination indicates the distance of the pole from the sampling site. For rocks some hundred million years old in North America and in Europe, the magnetic vector would point west of north in America and east of north in Europe, but restoring the relative positions of Europe and North America rotates both the palaeomagnetic vectors to north again.*

Such a model is consistent with the gravitational concept of isostasy: rocks of the crust and upper mantle adjust their elevation so that the total mass of rocks above a certain depth is constant (in much the same way as an iceberg floats with eight-ninths of its mass below sea level). This compensation level in the mantle, 111.3 km, was determined in the nineteenth century and it is almost identical with the average thickness of the oceanic lithosphere. Thus the thickness of rocks moving as a plate is substantially greater than that of crustal rocks alone, and the processes occurring within the crust can be understood only in terms of the behaviour of the lithosphere as a whole.

The driving forces behind plate motions

Detailed analyses of within-plate earthquakes indicate very uniform stress directions within individual plates. This has been interpreted as indicating either that the plates are very strongly coupled to mantle motions or that they are completely decoupled from the mantle. In view of the partial-melt interpretation of the seismic low-velocity zone it is conventional to assume that the plates are mechanically decoupled from the mantle. In other words the driving force for plate motion lies within the plates themselves and any mantle flow is a response to such motions and not itself the driving force. Two such mechanisms have been considered as possible driving forces (Figure 10.21).

First, as the oceanic ridges stand some 2 to 3 km higher than their bordering basins there must be a component of gravity operating within the plane of the plate. The magnitude of this component is, however, only just adequate to account for the energy released in earthquakes, some 6×10^{18} joules per year, with virtually none left to move the plates. This mechanism is therefore largely discounted—except for its local effects which explain some tensional tectonics at the ridge crest itself.

Many Earth scientists now believe that the main driving force for plate tectonics occurs in the subduction zones. According to this view, as the oceanic lithosphere cools its density becomes significantly greater than that of the underlying asthenosphere and it eventually sinks into the mantle. As the lithosphere sinks it enters higher-pressure regions and certain minerals undergo phase changes at specific depths. In particular there is an increase in density of some 9 per cent at a depth of 300 km as olivine converts to a garnet structure, and an increase of about 7.5 per cent as this then converts to the spinel form at a depth of about 530 km. Such density increases are known (from studies of seismic arrival times) to occur in normal mantle at about these depths, although the changes actually are not sharp but extend through some tens of kilometres. Smaller density increases occur at other depths. As the oceanic lithosphere is cooler it is more dense and therefore undergoes these phase changes at higher levels in the subducting lithosphere than in the asthenosphere. The increase in density thus creates a major downward force which, assuming it is coupled to the rest of the oceanic lithosphere, pulls the lithosphere over the asthenosphere in the same way as a table cloth can pull itself off a table when a sufficient weight of it has already fallen. Estimates for the magnitude of this force are very uncertain as the precise mechanisms of the phase change processes are unclear and there is disagreement as to whether such processes would inhibit or assist the downward motion. These uncertain factors aside, it is generally agreed that the force would be adequate to account for the known earthquakes and volcanoes as well as plate motions, providing that there is no friction between the lithosphere and asthenosphere.

A few Earth scientists are highly sceptical of this widely accepted model. One problem arises because there is no evidence for a seismic low-velocity zone beneath the older, cratonic blocks of the continents. Beneath younger continental areas the low-velocity zone appears restricted and poorly defined. The continental lithosphere in such regions also appears to be very much thicker than the oceanic lithosphere. The depletion in radioactivity under the continents, for example, is thought to extend to depths of some 300–400 km. The volcanic rocks known as kimberlites, irrespective of their age, are always located in the same tectonic environments. Yet they originate from depths of some 200 km which implies that their sources, deep within the mantle, travel with the continents. It follows that the continental lithosphere must be some 200 km thick, at the very least, and is underlain by 'solid' mantle which has the same seismic properties as that at similar depths beneath the oceans, ie below the oceanic low-velocity zone. A minimum of 200 km of the Earth's outer layer must therefore move with the continental parts of the plates. This must be compensated by return flow in the mantle at depths well below that of the oceanic low-velocity zone. This return-flow is thus within parts of the mantle that are considered too rigid for plastic deformation by those who wish to restrict the mantle response to within the asthenosphere.

More fundamentally, the 'tablecloth' mechanism can only operate once the ocean and subduction zone have formed. Continental separation, with the initiation of new oceans, cannot possibly be explained by such a model. If oceans such as the Atlantic, Arctic and western Indian, some thousands of kilometres across, can be formed without this driving mechanism it seems unnecessary to postulate it. This is not to say that such forces do not exist in subduction zones, but the seismic evidence suggests that they occur within discrete slabs of descending lithosphere that are not in fact coupled to the rest of the oceanic lithosphere.

An alternative view of the driving force follows essentially that originally outlined by Arthur Holmes in 1927. Radiogenic heat production within the mantle accounts for the 10^{21} joules per year lost from the Earth by heat flow. This is a hundred times more than that required to account for both earthquakes and volcanoes. The need for the mantle to rid itself of this heat causes regular convection. Any heat accumulation results in a decrease in the mantle's viscosity and therefore faster convective flow. In other words the mantle maintains itself in a state of constant viscosity (some 10^{21-22} poise). The plates are thus carried about by mantle motions, rather than causing them.

This view, which is not at the moment widely accepted, raises the question of the nature of the seismic low-velocity zone. The upper surface of the low-velocity zone under the oceans is clearly an isotherm as it relates precisely to the decrease in topography and heat flow away from oceanic ridge crests. However, the seismic observations could equally well be explained if this was an isotherm of some 800°C, well below the melting temperature of 1100°C. At temperatures some 200°C below the melting point the viscosity of mantle materials begins to decrease drastically. At this temperature, also, amphiboles are likely

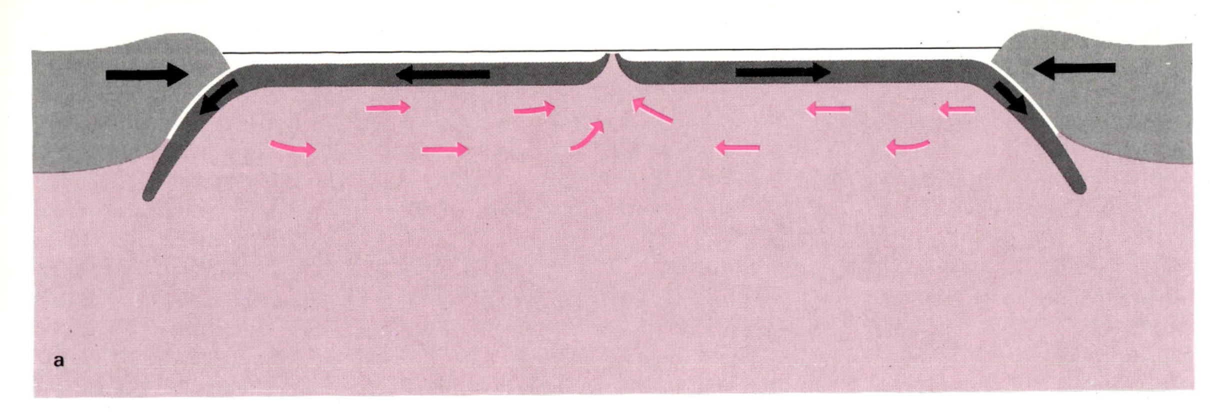

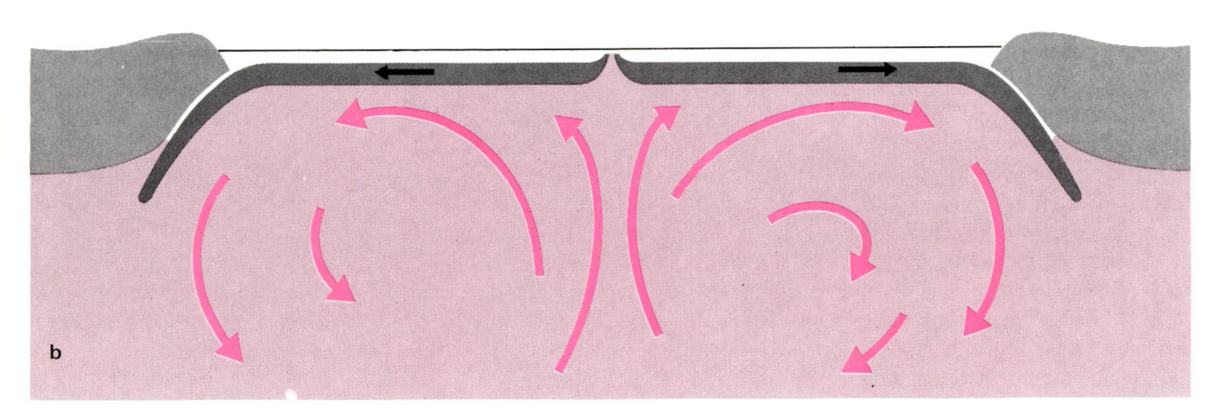

10.21: (**a**) *The conventional plate tectonic model, in which plates are decoupled from the asthenosphere so that mantle flow occurs in response to plate motions and is not the cause of them.* (**b**) *An alternative model, in which mantle convection actually drives the plates.*

to lose water from their structure. The properties of seismic low-velocity zones could thus be explained by the presence of water (or other volatiles) as an intergranular fluid within the rock.

Past plate tectonics

The question whether plate tectonic mechanisms operated as far back in time as the Proterozoic hinges on the geological evidence for changes in the factors which affect present-day motions. If radiogenic heating within the mantle is the prime factor then mantle motions must have been somewhat more active in the past, when radiogenic heat production was correspondingly greater. This would mean that mantle convection rates at the beginning of geological history would have been approximately twice those of today. Irrespective of the driving mechanism the necessarily higher heat flow in the past must have meant that the average thickness of the lithosphere was thinner in the past than today. The density changes necessary to develop a negative buoyancy in the oceanic lithosphere would thus be less likely, possibly inhibiting plate tectonics at least by the 'tablecloth' mechanism.

It is improbable that the increase in average thickness of the lithosphere of any one plate would be regular through time, although the net decrease in radiogenic heat production would be regular. The gradual thickening of the lithosphere would mean that while the temperature of its base remained constant the pressure at that boundary would be increasing. It thus seems conceivable that specific changes in the composition of the lithosphere could take place at specific times. One suggestion is that at the end of the Archaean the pressure–temperature conditions at the base of the continental lithosphere became such that devolatilization occurred, with the differentiation of the continental crust as a result. Such a geochemical change would also drastically alter the physical properties of the continental lithosphere, providing large-scale tectonic stability with little or no continental splitting (see Chapter 16). However, such models are, to say the least, provisional and controversial.

Other problems

There are many other areas of contention. One problem is posed, for example, by the lack of exact symmetry of the aseismic oceanic ridges even when the spreading of the ocean floor was symmetric. Neither the Iceland–Faeroes and Greenland nor the Walvis–Rio Grande ridges are anywhere near as symmetric as would be predicted by application of the spreading model. The mode of subduction of an oceanic ridge is also puzzling. If the mechanism is one of subductive drag, then the descending slab would cease to be coupled when the ridge reached the subduction zone, and hence the far side of the ridge would not be subducted. Similarly it is difficult to see how a ridge over the rising limb of a mantle convection cell could also be carried down on the descending limb of the same cell.

Many, many details remain obscure. For example, volcanism along the Andes, bordering the Peru–Chile trench, is laterally discontinuous, possibly as a result of the varying thicknesses of the subducted oceanic lithosphere from place to place, but the discontinuities remain unexplained. The western USA shows an extremely complex history during the past 50 Ma, with the possibility of two subduction zones being intersected by a subducted ridge. Other evidence indicates discrete motions between different units within the western Cordillera that are not consistent with movements on those subduction zones. There is evidence for oceanic ridge 'jumps', sometimes into continental crust, such as must have occurred to separate Rockall from Europe or the Blake plateau from North America. A newly formed rifted ocean must be very much weaker than the surrounding lithosphere and would therefore be expected to continue to form rather than for the spreading axis to jump to a new area. Sedimentary basins within continents show records of persistent sinking, irregular sinking, and phases of sinking alternating with phases of stability. Such events are obviously related to mantle processes, yet are not explicable in terms of current understanding of plate tectonics. The formation of kimberlites is also fairly well understood in principle, but the reasons for the timing of their emplacement are just not known. These and numerous other examples cannot be explained until more is known of mantle–lithosphere processes, but there seems to be every chance that most features of the history of the Earth's surface can eventually be explained in terms of its mobility.

None the less, it is not possible to exclude other theories completely. It seems most unlikely that the Earth has been subjected to significant expansion during any part of its history, if only because the other planets show no signs of such changes, as they should if the cause was external to the Earth. The fact that the continents can be fitted on a globe of some 55 per cent of its present size has been used to support a theory of expansion, but the same fact implies that the continental crust itself has not expanded—nor have fragments of ancient mantle brought up in kimberlites. Clearly the ingenuity of Earth scientists will continue to be taxed, but the improved understanding of how the crust forms and changes with time has at least provided a framework within which the questions can be asked. At a practical and economic level the framework of plate tectonics has important implications for the development of earthquake, volcano and resource prediction.

11 The deformation of rocks

At first sight one of the puzzles of the geological record is the evidence that rocks can be deformed within the Earth's crust. In everyday terms rocks symbolize rigidity and strength, and the dramatic results of rock deformation (see Figure 11.1) therefore seem to suggest immense forces active within the Earth. The first task of this chapter is to explain how rocks deform at all. It will be shown that other factors, such as temperature, deformation rate and the presence of fluids, are at least as important as large forces in initiating and controlling rock deformation. Variations of these factors within the crust can alter the rheology of rocks, their physical response to applied forces. Rocks which are strong and rigid at the Earth's surface may become weak and plastic at depth. Changes in rock rheology result from changes in the microscopic mechanisms of deformation and both these mechanisms and the bulk rheology of the rock are reflected in the various geometric forms or structures that are produced in the rocks. This chapter describes these different structures and indicates the origin and significance of each type. Finally it will be shown how different regions of the Earth's surface may expose different suites of structures or structural associations. There seems to be a limited number of such associations, a fact that points the way to explaining rock deformation in terms of the large-scale movements of lithospheric plates described in the following chapters.

Rheology of rocks

In the eighteenth and nineteenth centuries geologists appreciated that types of rock deformation that are impossible in the physical conditions at the Earth's surface might nevertheless operate in the very different conditions expected at depth in the Earth. It was certain that pressure increased downwards in the crust due to the increasing weight of overlying rock, and it could be argued that temperatures must usually increase downwards too. Although analogy with other materials suggested that these conditions might cause rocks to deform

11.1: *Folds formed by shortening of originally horizontal planar beds; above, small scale (note lamp post at foot of cliff) near the San Andreas fault, California, and, right, large scale, Sheep Mountain, Wyoming.*

at depth, it was difficult to test or quantify this idea.

A breakthrough came in the early 1900s when experimental apparatus was constructed which could simulate conditions at depth in the crust. The variations in rock rheology revealed by these experiments provide a convenient starting point in the explanation of rock deformation (Figure 11.2). In the deformation apparatus a small cylinder of rock is squeezed lengthways between two pistons, and is confined around its sides by a fluid at a different pressure from that on the specimen ends. Unequal pressures like these are termed stresses, and are measured as the force per unit area on the rock surface. During a deformation experiment the end stress on the piston is altered and the resulting shape change or strain of the rock cylinder

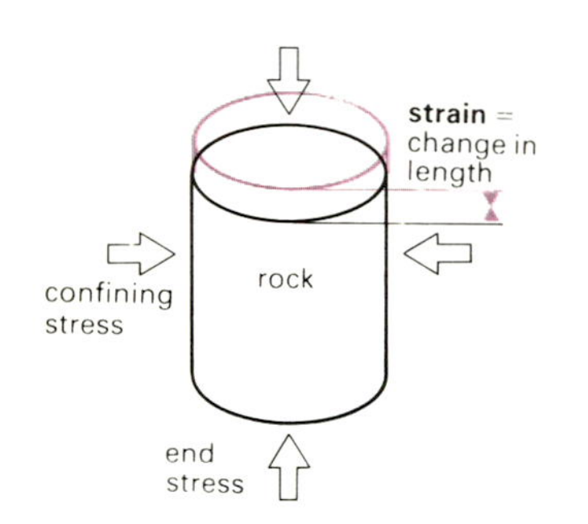

11.2: Experimental compression of rock cylinder showing the relationship (strain) between unstressed and stressed length.

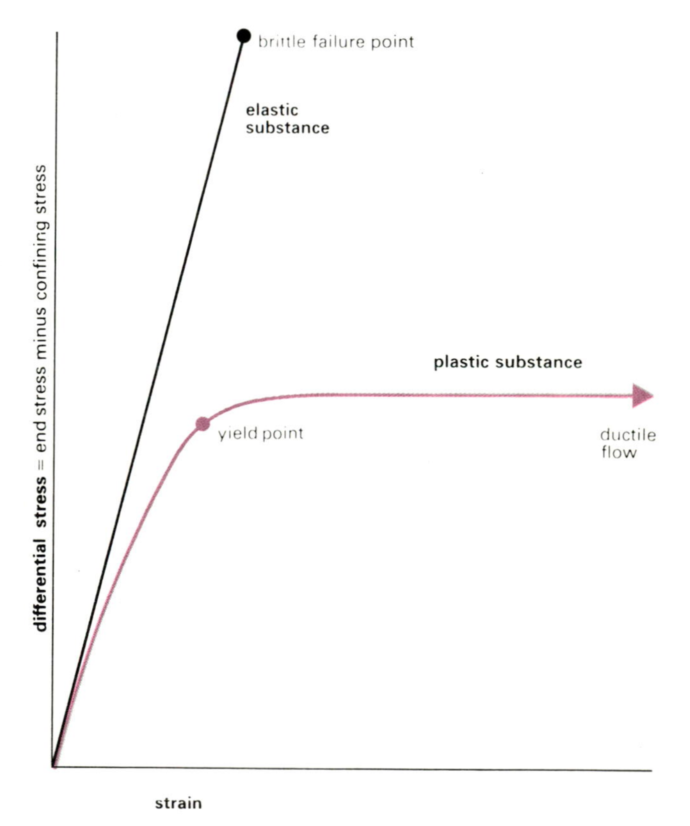

11.3: The results of rock deformation experiments for brittle and ductile materials; the strain (shape change) of the rock specimen is plotted against the imposed stress (force per unit area).

is measured. Natural conditions in the Earth can be simulated by varying both the confining stress and the temperature of the rock specimen. Modern apparatus can generate temperatures exceeding 1000°C and confining stresses of at least 1000 megapascals (MPa). (1 MPa = 1 000 000 pascals (Pa). A stress of 1 Pa is generated by a force of 1 dyne acting over an area of 1 m^2. 1 MPa = about 10 atmospheres.) These pressures correspond to depths of 35 km, near the base of the continental crust. In all conditions the rock strain increases with greater differential stress, the difference between the end stress and confining stress. However there are detailed variations in their stress–strain behaviour which are of great significance. At Earth surface conditions, that is at low temperatures and confining stresses, the amount of rock strain is directly dependent on the applied differential stress. When the applied stresses are released, the specimen quickly reverts to near its original shape—the strain is recoverable. These are features of an elastic rheology and are typical of many familiar materials such as spring steel and rubber. In rocks there is a critical differential stress at which elastic behaviour ceases and the specimen cracks along discrete planes. This type of irrecoverable strain is termed brittle failure (Figure 11.3). It is an important process of permanent rock deformation in the upper levels of the crust.

Conditions deeper in the Earth are simulated in the experiments by raising the confining stress and the temperature. As the differential stress is increased under these new conditions the rock specimen may initially show recoverable elastic behaviour. However, beyond a critical yield stress large strains now occur which are irrecoverable—the specimen does not return to its original shape when the differential stress is removed. This type of permanent deformation, which is taken up more uniformly through the rock than brittle failure, is termed ductile flow. Large strains are possible with little or no increase in differential stress. A rock that behaves in an elastic way at low stresses and in a ductile way beyond the yield stress is called a plastic substance (eg modelling clay, toothpaste and cold porridge).

There are two natural factors other than pressure and temperature that influence whether rocks deform by brittle failure or ductile flow. The most important of these is strain rate—the speed at which the deformation occurs. Experiments suggest that materials that are brittle if deformed quickly may behave in a ductile way if allowed to deform over long periods of time. Natural pitch shows this property at Earth surface conditions. In most rocks it is only important at depth in the crust, where it may be a major control on whether brittle or ductile deformation occurs. The uncertainty arises because experiments cannot reproduce strain rates as low as those possible over the vast period of geological time. Typical geological strain rates are thought to be about 10^{-14} per second, equivalent to a change in length of 1 per cent in 10^{12} seconds (about thirty thousand years). The slowest experimentally possible strain rates are about 10^{-7} per second, ten million times faster than natural rates.

The other important physical factor influencing rock rheology is the presence and abundance of a fluid, most commonly water, within a deforming rock. Fluids can weaken some rock-forming minerals dramatically and wet rocks are therefore more easily deformed than dry rocks, especially at high pressures and temperatures. This effect may be greater if the fluid is at a higher pressure than the rock itself.

The most important lesson to be learned from the studies of rock

11.4: *Above, a fault, and, right, two intersecting sets of joints make a lozenge-shaped pattern on a rock bedding surface.*

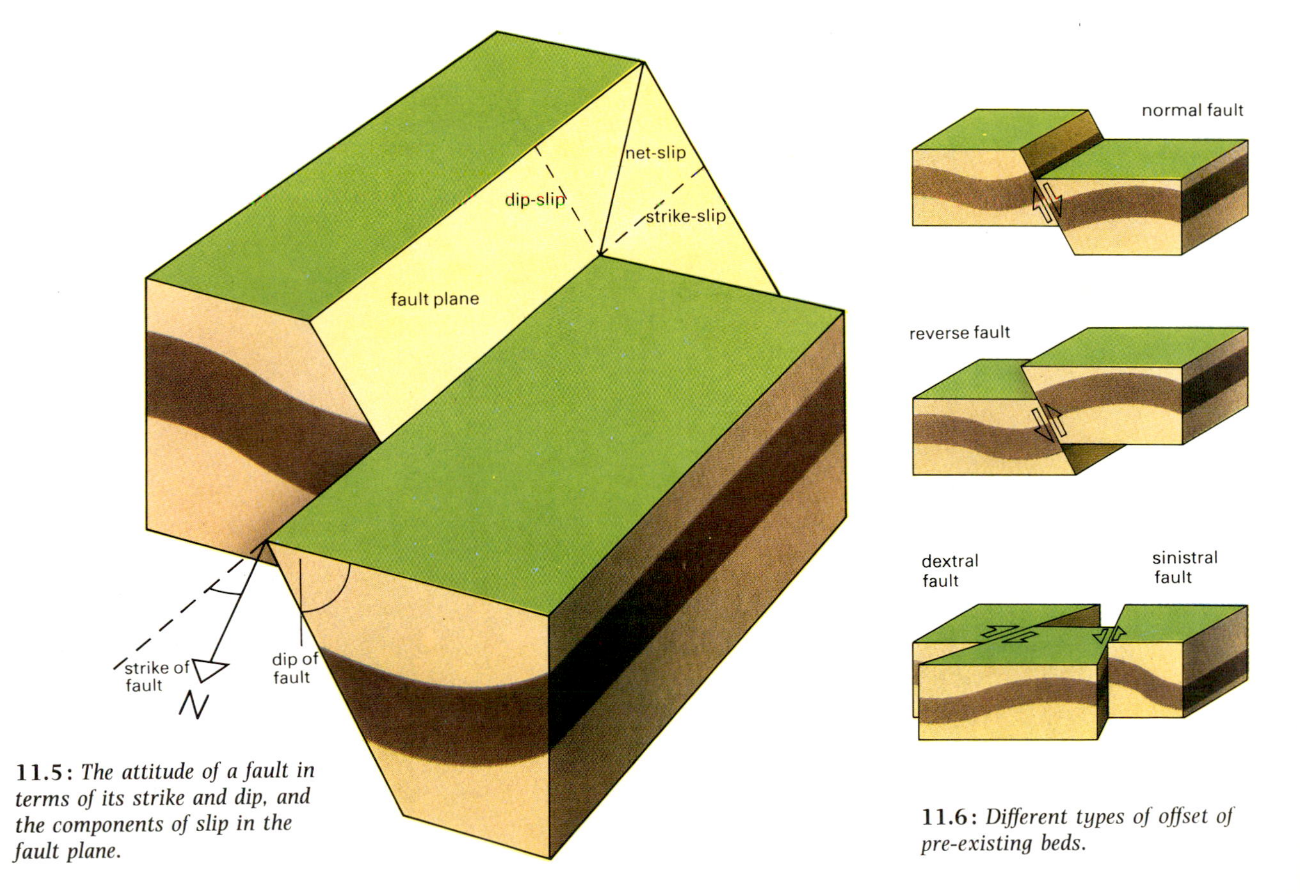

11.5: *The attitude of a fault in terms of its strike and dip, and the components of slip in the fault plane.*

11.6: *Different types of offset of pre-existing beds.*

rheology is the distinction between brittle and ductile behaviour. In general brittle deformation occurs at low temperature and confining pressure, that is in the upper levels of the crust, and is favoured by high strain rates. Ductile deformation characteristically occurs at higher temperatures and pressures, ie lower in the crust, and is promoted by low strain rates. This brittle–ductile distinction can now be used as the framework for describing the deformation structures produced in rocks. The rocks now seen at the Earth's surface have usually been deformed at depth in the crust and later uplifted and exposed by erosion of the overlying rocks.

Brittle structures

Only with brittle failure do rocks with elastic rheology undergo the irrecoverable strain that produces permanent deformation structures. In nature as in experiments the products of brittle failure are discrete fractures. These fractures are termed faults if the rocks on either side have been offset, and joints if no offset is visible to the unaided eye (Figure 11.4).

Faults

Faults, together with joints, can provide valuable information about the stresses that act in the upper levels of the crust. This information is held in the geometrical relationship of faults with their surrounding rocks (Figure 11.5). The attitude of the fault, its three-dimensional orientation in space, can be described by its strike and dip. The strike is the compass direction of a horizonal line within the fault plane; the dip is the angle that the fault makes with the horizontal. Strike and dip are useful measures of the attitude of any planar surface. The displacement of one side of the fault with respect to the other can be measured by joining two points which were originally contiguous across the fault. The length of this join is the net slip on the fault. The amount of slip measured parallel to the fault strike is the strike-slip, and the amount down the line of greatest slope in the fault is the dip-slip. These measures allow a common classification of faults into dip-slip faults where dip-slip displacement is dominant and strike-slip faults with predominant strike-slip displacement. Oblique-slip faults have significant strike-slip and dip-slip components.

Other terms describe the sense of offset by the fault of originally contiguous features (Figure 11.6). As viewed on a vertical plane through the fault, the offset is normal if the fault dips towards the downthrown side of the fault, and reverse if it dips towards the upthrown side. In plan view the offset is dextral if an observer facing the fault sees features on the opposite side of the fault displaced to the right and sinistral if such features are displaced to the left.

When this rather cumbersome terminology is applied to natural faults, certain important generalizations emerge. First, many faults near the Earth's surface show either nearly pure dip-slip or pure strike-slip. Oblique-slip faults are more rare than might be expected. Second, a number of approximately parallel faults of the same type, for example dextral strike-slip faults, often occur together in the same region. Third, such a set of faults is commonly associated with one other non-parallel set. The two sets often make a consistent angle of about 60° with each other.

These relationships can be explained by further experimental results from apparatus where stresses can be varied independently in three directions at right angles (Figure 11.7). Three stress regimes and corresponding fault patterns occur depending on which principal stress direction is vertical. The gravity regime (maximum stress vertical) will produce dip-slip faults with normal displacement and the thrust regime

11.7: *Three different fault patterns, each produced by three perpendicular principal stresses. The relative sizes of the arrows represent the relative sizes of the stresses. Faults tend to form along two planes intersecting parallel to the intermediate stress and making an angle of about 30° with the maximum stress. A different principal stress is vertical for each fault pattern.*

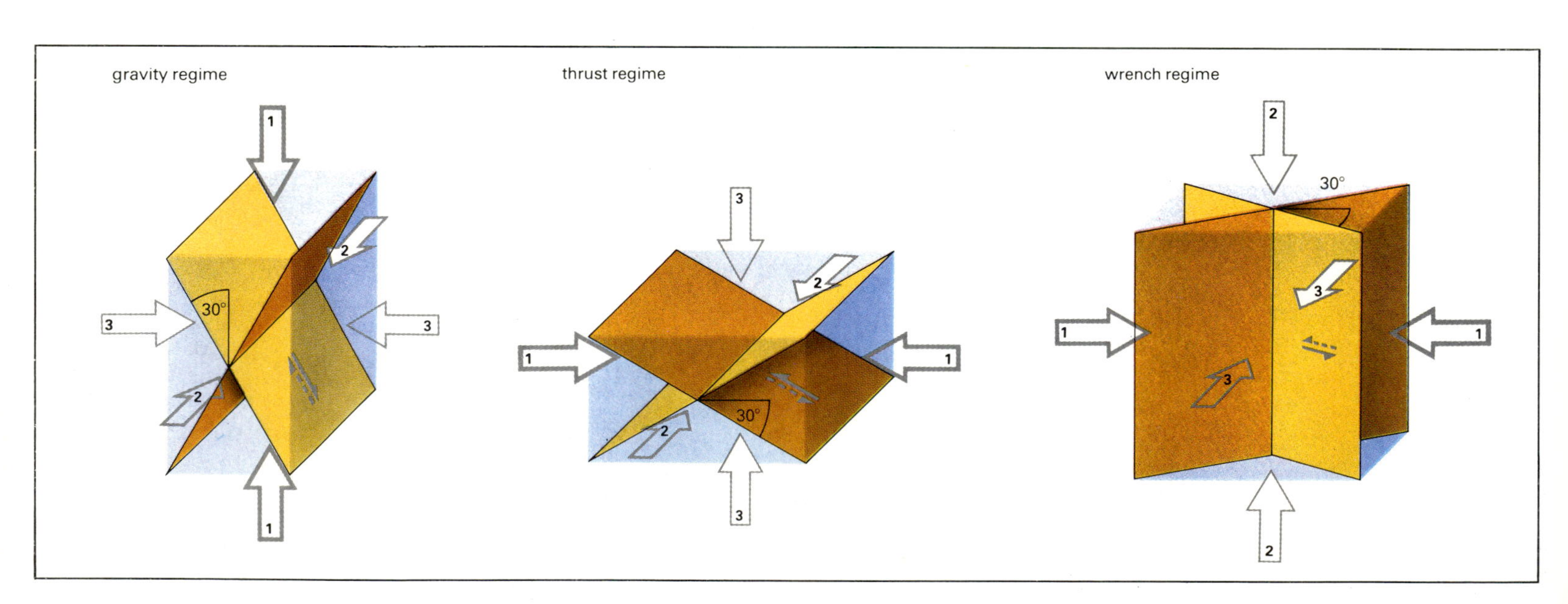

(minimum stress vertical) dip-slip reverse faults. The wrench regime (intermediate stress vertical) will generate strike-slip faults.

In practice the geologist uses observed geometrical features of faults to deduce the stress regime that operated in the crust at the time of their formation. This in turn may allow deductions to be made about the relationship of that part of the crust to movements taking place on a much larger scale.

Joints

Joints are the most commonly seen rock deformation structure. In common with faults they form sets of approximately parallel fractures, and several sets may occur with a consistent angular relationship to form a joint system (see Figure 11.4). Unlike faults, joints within a set are spaced rather closely together. Typical joint spacings are between 10 mm and 1 m, whereas faults are commonly spaced hundreds of metres or some kilometres apart. By definition joints differ from faults in having no offset visible to the unaided eye. However, microscopic examination suggests that some joints do in fact show a small but significant offset.

The mechanical explanation of some joints (shear joints) is rather similar to that of faults—they form at acute angles to the maximum principal stress. These may show small offsets. However, other joints (extension joints) commonly form perpendicular to the minimum stress. These show no offset—the two walls of the fracture merely move apart slightly. Joint systems with more than two joint sets can be produced by superimposed shear and extension joints. The major contrast in spacing of joints and faults is more difficult to explain. The reason seems to be that, whereas the stresses that cause faulting are actively applied from outside the deforming rock body, joint-producing stresses derive from within the rock itself. The closest analogy is with the cracks that form in drying mud. Because the stresses are similar at every point a large number of small fractures form. Once the fracture has formed, the stresses around it are largely dissipated and no further displacement can occur across the fracture. The stresses that produce joints are stored in the elastic strain of grains in the rock, and this strain has been acquired at depth. Joints form when the rocks are uplifted and the elastic strains are released.

Fault and joint fillings

Sometimes fractures remain as simple cracks in the rocks. More commonly they are marked by infilling of various types.

A fault breccia is mainly composed of angular fragments of rocks broken from the fault walls during its displacement. These fragments are surrounded either by fine-grained material also derived from the wall rocks, or by crystalline quartz or calcite. Sometimes no coarse fragments are present—the fine-grained material alone is called fault gouge. Both gouge and breccia are particularly common in faults at a high crustal level. Lower down in the crust extreme fragmentation of rocks along faults may produce mylonite, a hard fine-grained fault rock. Extremely rapid movement on large faults at depth may produce enough frictional heat to melt a thin film of wall rock which solidifies to a glassy rock called pseudotachylite.

The walls of faults often show linear patterns, called slickensides, aligned in the direction of slip on the fault. Sometimes these are scratches caused by the abrasion of one wall against the other. Often they are parallel growths of elongate quartz or calcite crystals. These grow in the direction of dilation or opening of small gaps produced along the moving fault. Crystalline infillings or veins are common along joints as well as faults.

Ductile structures

Whereas brittle deformation produces fractures that may surround almost undeformed volumes of rock, ductile deformation may result in permanent shape changes throughout the deformed mass. The structures produced are geometrically varied, depending on the pressure and temperature conditions, the rock composition and the presence and orientation of pre-existing rock structures. They can be grouped under four headings. Foliations are planar structures directly produced by deformation. Lineations are deformation-induced linear structures. Folds are wavelike deformations or the isolated bending or shearing of pre-existing layers. Boudinage is the elongation and fragmentation of pre-existing layers.

Foliations

Most sedimentary and some igneous rocks contain planar structures (eg bedding) which are a primary product of the rock-forming processes. However, some planar structures clearly intersect or are superimposed on these primary structures. These foliations are usually the product of later deformation and perhaps metamorphism of the rocks. If the rock tends to break along a foliation, the structure is also called a cleavage. A workable descriptive classification is based on whether the foliation is defined by surfaces that have a definite separation visible to the unaided eye (spaced cleavage) or whether the planar structure pervades the whole rock at this scale (continuous cleavage).

Continuous cleavages have a different appearance depending on the metamorphic grade of the rocks during their formation. Slaty cleavage is defined by alignment of very fine-grained minerals and is characteristic of low pressure and temperature conditions. Phyllitic foliation is defined by larger grains, grown at higher pressure and temperature conditions, which give a sheen to foliation surfaces. Schistosity forms at still higher grades and contains new mineral grains easily visible to the unaided eye. Rocks deformed at highest metamorphic grades may show gneissic foliation which has both a continuous element and a spaced compositional banding.

Continuous cleavages result from preferential alignment of mineral grains. Both tabular and elongate grains tend to have their longest dimensions aligned parallel to the foliation. Grains may be aligned by three mechanisms. First, pre-existing grains may be bodily rotated. Second, pre-existing grains may be modified in shape by internal deformation or overgrowths of new material. Third, new minerals may grow in parallel alignment. Experimental and theoretical evidence suggests that in each case the resulting foliation orientation is related to the pattern of strain in the rock. Three mutually perpendicular principal strains can be defined: X is the direction in which lines are most elongate after deformation, Z is the direction of maximum shortening and Y has an intermediate length change. The foliation usually develops perpendicular to Z, the maximum shortening, whichever mechanism of mineral alignment operates. The

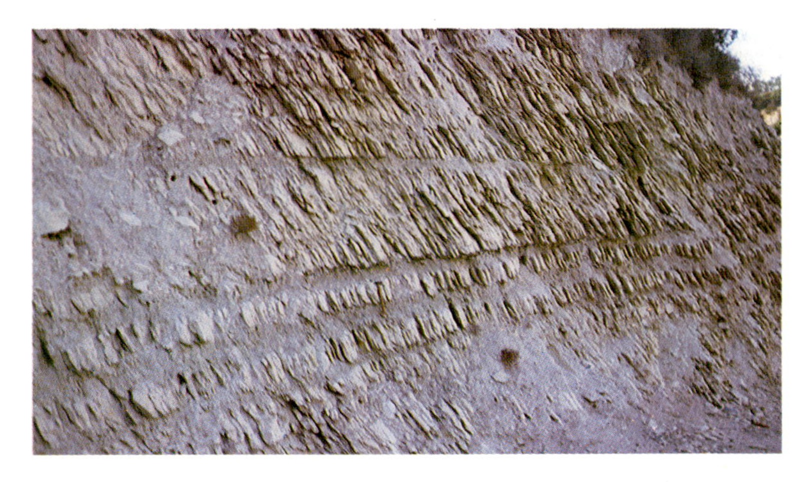

11.8: *Top, primary bedding in sedimentary rocks (horizontal) is cut by a steep foliation (cleavage) produced by deformation. Centre, a cleavage formed by alignment of minerals. This thin section of rock shows a schistosity in a metamorphic rock. Bottom, a spaced cleavage, here formed by the removal of soluble minerals along bands in the rock to produce dark stripes cutting pre-existing structures.*

principal strain directions can therefore be deduced directly from the orientation of continuous foliations. Moreover it may be possible to estimate the magnitude of strain if the deformed rock contains objects whose probable original shape is known. Fossils, some sedimentary structures, pebbles and other sedimentary grains can be used in this way.

Spaced cleavages take three forms. Small-scale folding of a pre-existing planar structure, often a sedimentary lamination or continuous cleavage, produces crenulation cleavage. Selective solution of more soluble minerals from discrete planar zones produces solution cleavage. Finally a close-spaced array of parallel brittle fractures may produce a foliation (fracture cleavage) which is not the result of ductile deformation at all. The separation of spaced cleavage planes is usually less than a few centimetres. Of these three types the process of brittle failure involved in fracture cleavage has already been explained, and crenulation cleavage will be treated along with folds. The remaining process of pressure solution depends on the property of many minerals, particularly quartz and calcite, to dissolve more readily when stressed. These minerals may be dissolved in water at sites of high stress and either be redeposited at sites of lower stress in the same rock or be transported out of the deforming rock to higher levels of the crust, there to form veins. The insoluble residue from pressure solution is darker than the original rock, which often develops alternate light and dark stripes perpendicular to the Z direction.

Lineations

The term lineation is used for both primary and secondary linear structures. Here we are concerned with secondary lineations produced by rock deformation. As with foliations these can be classed as either continuous or spaced.

Continuous lineations and continuous foliations have a closely similar origin. Both are due to alignment of minerals by similar mechanisms, hence the common term mineral lineation. The different structures are produced by different relative magnitudes of the principal strains. If X and Y are roughly similar but Z is considerably smaller, ie the deformation is essentially a shortening along Z, then a foliation will develop. However if X is much larger than either Y or Z, essentially an elongation along X, a lineation will develop parallel to X. Continuous foliation and lineation are end members of a spectrum of mixed linear–planar mineral alignments that develops in intermediate strain conditions.

Spaced lineations can result from the intersection of spaced cleavages with other surfaces, for example sedimentary bedding.

Folds

Consideration of foliations and lineations has shown that ductile deformation may produce either continuous or spaced structures in rocks. Folds are larger-scale spaced structures defined by more or less periodic waveforms in a pre-existing bedding or foliation.

Most fold systems have alternations of antiforms (arch-shaped folds) and synforms (trough-shaped folds) as in Figure 11.9. The terms anticline and syncline are alternatives as long as the stack has younger rocks at the top. The most important of the various geometrical elements of folds are the hinge lines—the positions of maximum surface curvature—and the axial surfaces, which pass through

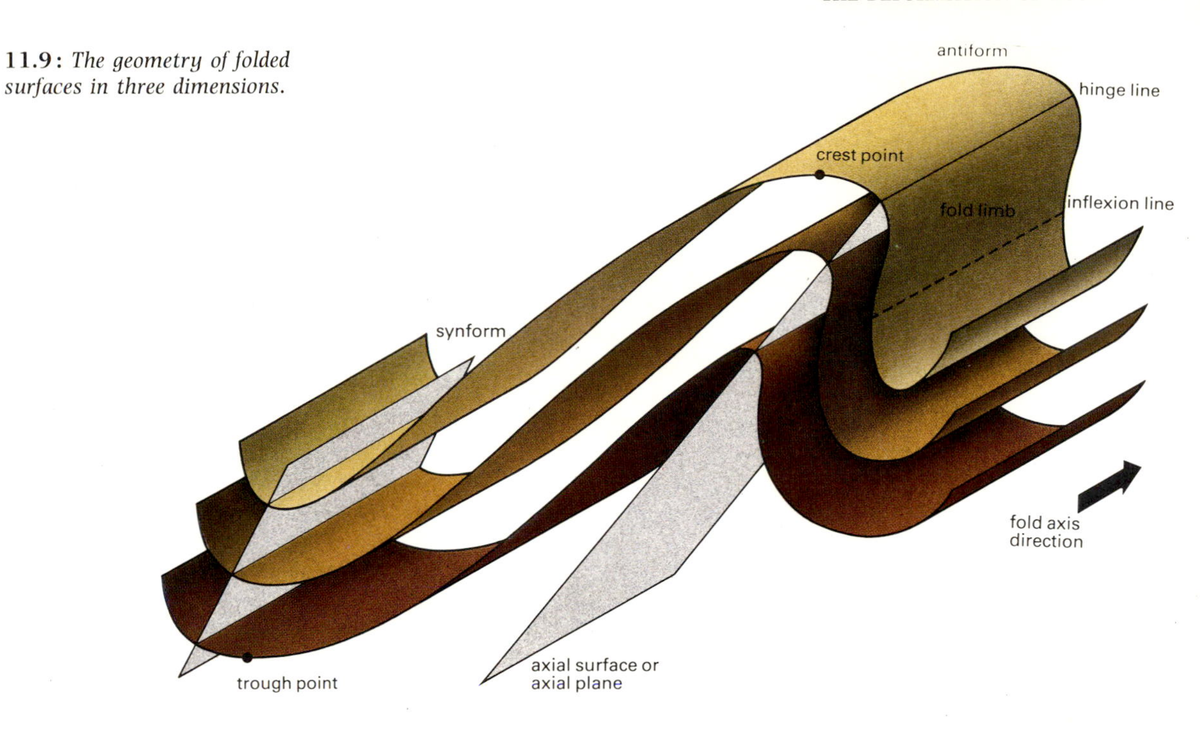

11.9: *The geometry of folded surfaces in three dimensions.*

11.10: *The geometry of folds in sections perpendicular to the fold axis.*

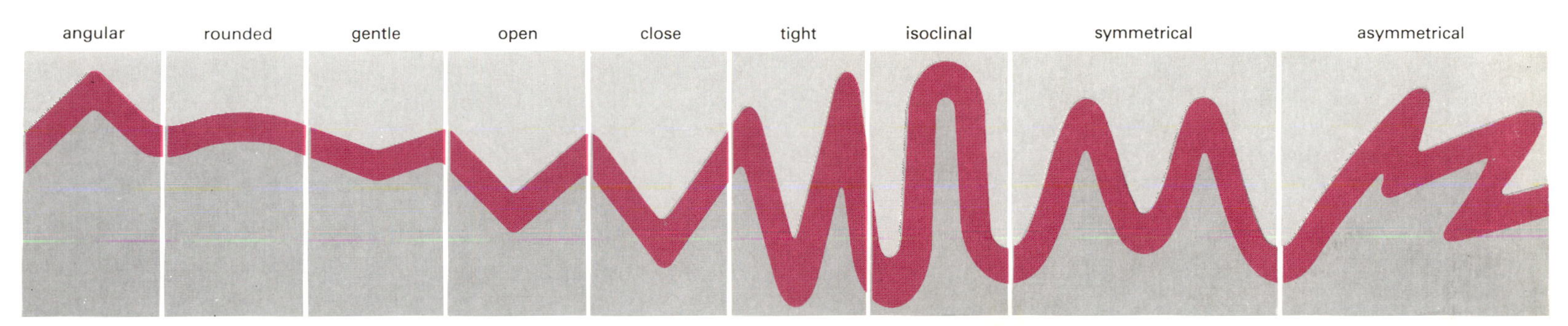

the hinge lines on successive folded surfaces. The orientation of a fold may be specified by the orientation of its hinge line and axial surface. Some folds show similar shapes on any section cut perpendicular to the hinge lines. These are cylindrical folds, and in such folds the direction of the fold hinges is termed the fold axis.

Natural folds encompass every geometrical variant in Figure 11.10 and range in size from a few centimetres to many kilometres. However, there are very few known fold-forming processes and it appears that the natural variety results more from the range of physical properties of the deformed rocks.

Most folds probably form due to shortening of pre-existing layers. Isolated single layers may start to buckle at weak or irregular points, and the resulting disturbance propagates along the layer as a train of folds. For this to happen the isolated layer must be stiffer, perhaps more viscous, than the surrounding rocks. The wavelength of the resulting folds increases with the layer thickness and the relative stiffness. If the direction of maximum shortening (Z) in the rock is parallel to the original layer the folds tend to be symmetrical, that is they have their axial surface perpendicular to the original layer orientation. Asymmetrical folds develop if the maximum shortening was oblique to the layer. Most single-layer buckle folds are parallel folds, that is the layer retains the same thickness around the folds.

Many natural rocks are composed of a sequence of parallel layers. These multilayer sequences behave differently from single layers during shortening. The layering may again buckle at weak points but the disturbance spreads across the layering rather than along it. After small amounts of shortening, bands of deformation, called kink bands, are visible making an angle of about 60° with the layering. With further shortening more bands initiate and older bands propagate further through the rock. Eventually a set of angular, approximately symmetrical, folds called chevron folds may form. These will tend to be asymmetric if the maximum shortening is oblique to the layering. Small-scale multilayer folds are often classed as crenulation cleavage.

Folds can theoretically be produced without shortening of the layering, and some folds certainly form by a shear folding mechanism. The layering is moved passively by flow of rock oblique to the layering, like deforming a picture drawn on the edges of a pack of cards. The resulting structures are similar folds where each folded surface has exactly the same shape as its neighbour, though the layer thickness varies around each fold.

Boudinage

If rock layering is elongate rather than shortened parallel to its length, the stiffer layers may separate into a series of sausage-shaped fragments termed boudins. The long axes of the boudins will tend to be perpendicular to the maximum elongation direction in the rock. As with folds, boudins may occur in isolated single layers or in multilayers, where complex interlocking arrays of boudins may form.

Deformation analysis

In describing rock rheology and deformation structures some ways have been indicated in which crustal conditions in the geological past might be determined. This ultimate aim of deformation analysis involves deduction from the present geometry of structures and the application of relevant experimental and theoretical results. Two particular features of natural deformation complicate the analysis—the fact that one episode of deformation produces a number of related structures and the tendency for the same rocks to be deformed more than once during geological time. An idealized suite of structures formed by a single deformation episode is shown in Figure 11.11.

Most deformed rocks would not show all these structures. Indeed the main value of such a model is that potential geometric relationships may be predicted from a more restricted range of structures. Fold orientations can be estimated just from an assumed axial planar foliation with an intersection lineation or boudins. Major folds, too large to be seen in one exposure, might be located by mapping the variation in the sense of asymmetry of minor folds. The possible position and orientation of veins, perhaps containing economic minerals, can be estimated. This reconstruction of the total geometry of a deformed rock body, often from rather fragmentary evidence, is an essential initial step in the structural analysis of an area. Sometimes geometrical reconstruction is an end in itself, as for instance where economically valuable horizons such as coal or ironstones must be located in a folded area or where the structures themselves control the location of oil or ore deposits.

A second aspect of structural analysis takes the present geometry as a starting point in reconstructing the development of the structures through time, and the relationship of deformation events with other igneous, sedimentary or metamorphic events. Even in an apparently simple geometric situation, such as Figure 11.11, there may be evidence that different structures developed at different times. For instance joints may cut across and therefore be later than other structures such

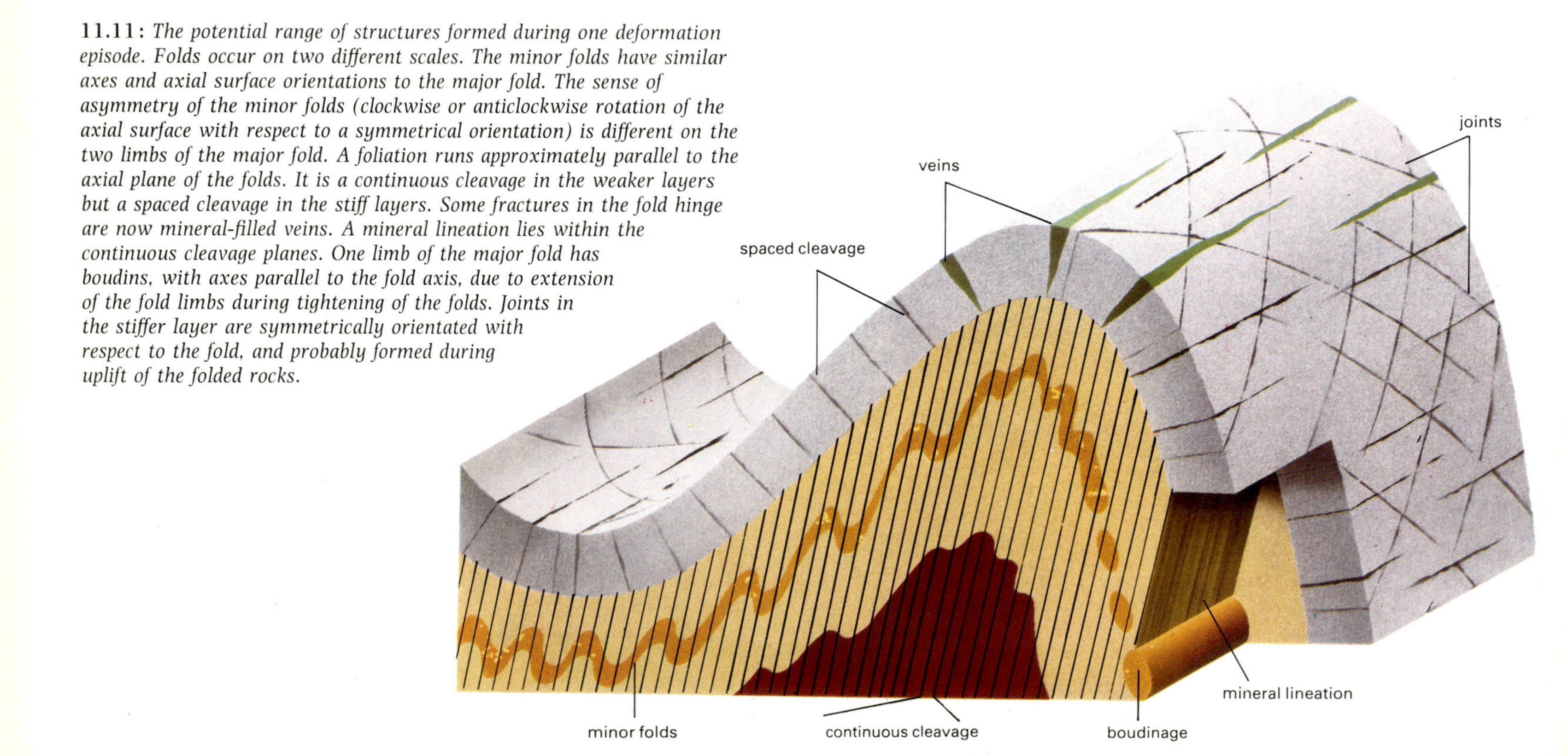

11.11: *The potential range of structures formed during one deformation episode. Folds occur on two different scales. The minor folds have similar axes and axial surface orientations to the major fold. The sense of asymmetry of the minor folds (clockwise or anticlockwise rotation of the axial surface with respect to a symmetrical orientation) is different on the two limbs of the major fold. A foliation runs approximately parallel to the axial plane of the folds. It is a continuous cleavage in the weaker layers but a spaced cleavage in the stiff layers. Some fractures in the fold hinge are now mineral-filled veins. A mineral lineation lies within the continuous cleavage planes. One limb of the major fold has boudins, with axes parallel to the fold axis, due to extension of the fold limbs during tightening of the folds. Joints in the stiffer layer are symmetrically orientated with respect to the fold, and probably formed during uplift of the folded rocks.*

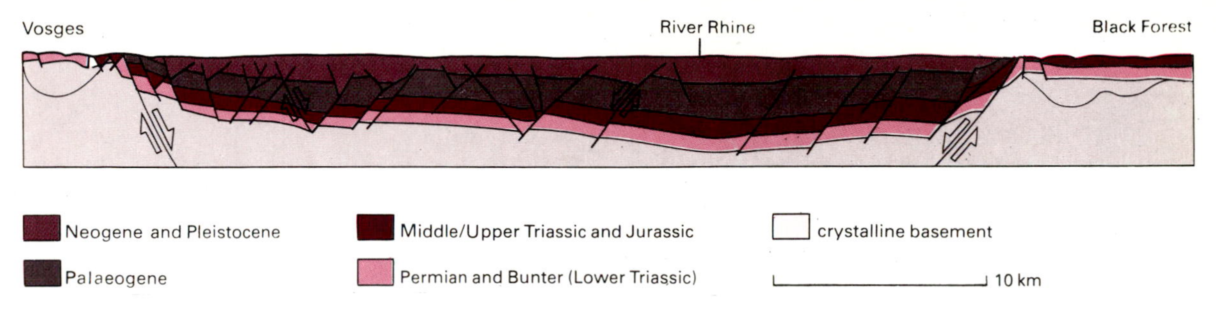

11.12: *A vertical section through the Rhine valley graben north of Karlsruhe. The normal faults indicate crustal extension.*

as foliations or boudins. Veins must be later than, or at least coeval with, the fractures that they fill. Boudins may be folded, or alternatively minor folds may undergo boudinage. These types of observation enable a time sequence of structural events to be reconstructed. Further complications arise when already deformed rocks suffer subsequent deformation with a different geometry, perhaps millions or even hundreds of millions of years later. Such polyphase or superposed deformation can be recognized by a number of structural complications unlikely to have developed during one episode of deformation alone. Foliations, lineations, boudins, fold axial planes or veins may be bent around new folds. A foliation may be finely folded to give a crenulation cleavage. New foliations may cut pre-existing folds. The resulting structural geometry can be very complex.

If the structural history of the rocks through time can be fully unravelled this may give some guide to the corresponding history of crustal conditions. However, fuller understanding of the stresses, temperatures, strain rates and fluid pressures which have acted progressively on the deformed rocks can only come from integration of the structural evidence with that from sedimentary, metamorphic and igneous events. For instance, metamorphic mineral assemblages may give a good estimate of crustal pressure–temperature conditions. To be structurally relevant this information must be slotted into the deformation sequence.

The limitation of structural analysis is that every increment of deformation history is rarely, if ever, recorded in the rocks, and long blank periods of geological time remain. Because of the wide variety of ways in which any particular final rock geometry can be produced, there is always a measure of ambiguity in piecing together structural time sequences.

Structural associations

So far we have been viewing structures on a fairly small scale, from the smallest visible sizes up to perhaps a kilometre or two. However, most structures develop as parts of deformed belts, orogenic belts, which may extend for hundreds or thousands of kilometres. In view of the possible combinations of structures, each with widely variable geometries, an infinite variety of structural style ought to occur in orogenic belts. This is not the case. Certain suites of structures, or structural associations, have consistently recurred in different geographic areas and at different periods of geological time. A whole orogenic belt may comprise only one structural association or a small number of related associations. These associations provide the essential link between the study of small-scale structures in the field and theories of orogenic belt formation which will be described in later chapters. This chapter will concentrate on the geometrical characteristics of some common associations and on their direct interpretation.

Flat-lying sedimentary areas

Paradoxically, areas of relatively undeformed sedimentary rocks are almost as significant for large-scale tectonic interpretations as are deformed zones. They represent areas of tectonic stability suffering only slow and rather uniform subsidence or uplift. On the continents these relatively stable areas, termed cratons, may persist for hundreds or even thousands of millions of years. Examples are the continental interior of North America and much of eastern Europe. Variable subsidence may produce large gentle folds in the bedded sediments. Tighter folds only form by the draping of the sedimentary cover over rare active faults in the underlying basement. Faults, usually normal dip-slip faults, sometimes cut the sediments themselves in zones of differential subsidence or uplift, but vertical joints are more common structures.

Flat-lying sediments also occur on the igneous basement of the present ocean floors. These stable areas persist for shorter times than do continental cratons. Most sediments on the present ocean floors are of Jurassic and younger age (less than about 190 million years), and are thinner than 1 km.

Rift zones

Rifts are characterized by approximately parallel arrays of faults with large dip-slip displacements. Typically these faults bound elongate topographic depressions termed rift valleys or graben (Figure 11.12). Well known continental examples are in the East African rift system, the Rhine graben in Germany and the Baikal rift in the Soviet Union. Recent rift zones commonly contain active volcanoes. In some continental regions, such as the Basin and Range province in the western USA, or western Turkey, the faults are distributed over a wide area, giving an aligned array of valleys (graben) and hills (horsts). In many rift systems there is doubt about the attitude of the faults at depth, although near the surface they are steep and have normal displacements. Thus many rifts are probably due to crustal extension perpendicular to the rift zone. The large-scale causes of this extension are not fully understood. In some cases the rift may be a brittle crack on the crest of a regional crustal uplift. Such uplift is commonly associated with rifts. Alternatively the uplift itself could be a consequence of stretching and thinning of the whole crust, allowing hot mantle material to intrude upwards. Rifting due to crustal downwarp is another possibility, although there is little supporting evidence.

Rifts are not restricted to the continents. Conspicuous rifts run down the axes of many mid-ocean ridges—the best known is along the mid-Atlantic ridge. The sea floor spreading evidence clearly indicates an overall extensional environment at ridges, although the actual topography across these rifts may be controlled more by vertical isostatic processes than by horizontal stretching. A link between oceanic and continental rifts is provided by areas such as the Red Sea where an oceanic ridge intersects a continental area. The early history of the Red Sea closely matches that of continental rifts. Moreover, some of the most extensive ancient rift systems on Earth underlie present-day passive continental margins (see Chapter 10) such as those of the Atlantic. These observations strongly suggest that most ocean basins are initiated as continental rifts. Conversely, however, not all continental rifts develop into oceanic basins. The Benue trough in west Africa and the North Sea graben in north-west Europe are examples of now inactive rifts containing volcanic rocks but no true oceanic crust.

Despite the evidence that most rifts are zones of extension, the term rift is not restricted to this mode of formation. Some rifts, such as the Dead Sea rift, occur in wrench zones dominated by strike-slip faults.

Wrench zones

Wrench zones are linear belts of dominantly strike-slip displacement. Presently active wrench zones such as the Alpine fault in New Zealand, the San Andreas fault zone in the western USA, the north Anatolian fault in Turkey and the Dead Sea zone are dominated by arrays of steeply dipping strike-slip faults. Many of these faults strike parallel to the wrench zone, but in detail they branch and join each other to form a braided pattern. Less commonly faults may strike at a high angle to the zone. Only some of the faults are active at any one time. Average rates of displacement measured across the wrench zones are up to 100 mm a year and total displacements of hundreds of kilometres may accumulate through time.

Although most strain in the surface levels of active wrench zones is probably by brittle failure, ductile deformation may play an important role. Folds may develop with axes either parallel to the wrench zone or more characteristically *en echelon*, that is systematically slightly oblique to the zone. Folding may be intense where there is a component of shortening across the zone, superimposed on the pure strike-slip deformation. Such shortening may be localized at bends in the strike-slip fault as in the Transverse Ranges on the San Andreas fault. Alternatively local bends or large lengths of a fault zone may have a component of elongation across them. This may provoke large dip-slip normal displacements in the zone and form down-dropped rift valleys as in the Dead Sea rift. These depressions, or pull-apart basins, may fill rapidly with thick sedimentary deposits.

In currently active wrench zones only the surface expression of structures is visible, but they may extend to the base of the crust and beyond. Ancient inactive wrench zones may be progressively eroded to reveal their deeper structure. At deeper crustal levels ductile deformation becomes relatively more important than brittle behaviour.

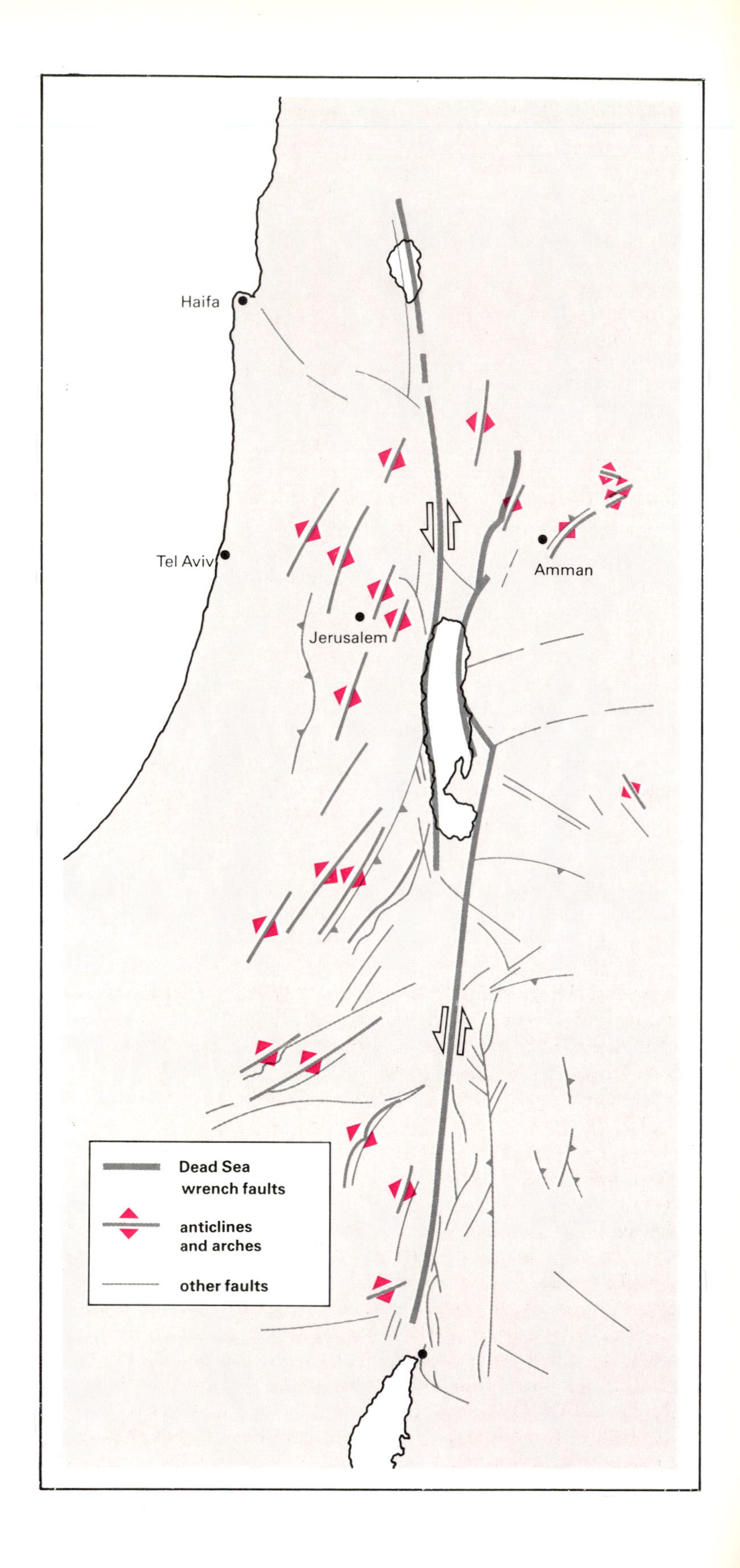

11.13: *The Dead Sea region, a sinistral wrench zone. Strike-slip faults bound the Dead Sea pull-apart basin, and are associated with oblique folds and faults.*

11.14: *Right, a vertical section through the front ranges of the eastern Rockies (Canada) showing numerous thrust faults, low-angle reverse faults. Centre, a vertical section through the European Alps. Fold nappes dominate the structure. Below right, a cross-section through the Jura Mountains (west central Europe). The sedimentary cover, with upright folds, has moved separately over the older crystalline basement.*

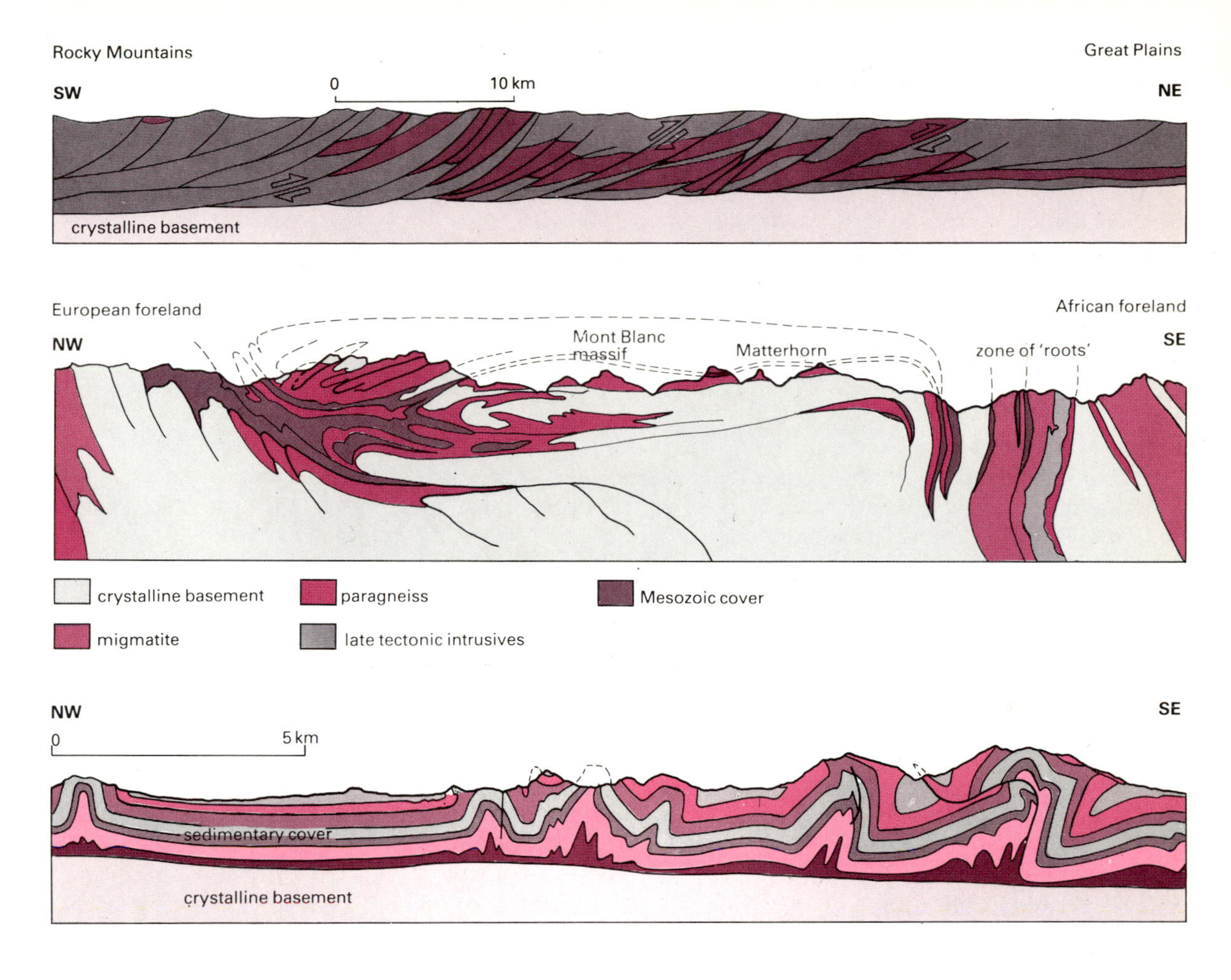

Rather than being taken up on discrete fractures, displacements are taken up across wider ductile shear zones. These occur on all scales from microscopic size to tens of kilometres wide. They are common in deeply eroded ancient crust (eg the late Archaean Limpopo belt, southern Africa, and the early Proterozoic Ikertôq shear belt in western Greenland).

Thrust belts and nappes

So far deformed zones characterized by dip-slip normal faults and by strike-slip faults have been dealt with. The third member of this fault-dominated suite of associations is characterized by dip-slip reverse faults. On a regional scale such faults usually develop with low dips and are termed thrusts. The most spectacular thrust belts are probably the ancient examples of the eastern Rockies and the western Appalachians in North America, showing arrays of parallel thrusts in plan view. In section each thrust has a listric (curved) shape and unites at depth with other thrusts, or with a sole thrust underlying the whole belt. In these examples the attitude of the sole thrust is known from seismic experiments. In other thrust belts it is less certain that the sole thrust has a shallow dip, and some thrusts may cut deep into the basement. As in wrench zones the thrust displacements at depth are often taken up not on discrete faults but on thick shear zones.

All thrust belts must have shortened considerably perpendicular to the thrusts. Moreover sections through many thrust belts imply that individual thrust sheets are allochthonous, that is they have been severed from their basement and transported a considerable distance with respect to the underlying autochthonous basement. In Europe such sheets might be termed nappes or more strictly thrust nappes.

In thrust belts such as the Rockies most of the shortening is accommodated on faults. Folds do occur, especially in thrust sheets that have over-ridden steps in the basement, but these do not dominate the structure. However, there is a complete spectrum of structural styles from these fault-dominated belts to those where displacements are taken up almost totally in ductile folds. The fold nappes of the European Alps typify these ductile belts of shortening. Folds are commonly tight and recumbent. The normal limb of each fold is effectively transported by concentrated deformation on its underlying overturned limb. Some fold nappes deform only sedimentary cover rocks. Others involve crystalline basement, a process that can only occur at fairly high metamorphic grades.

There is considerable uncertainty about the crustal mechanisms that produce shortening in thrust and nappe belts. One obvious possibility is that shortening affects the whole thickness of the crust, or more. However, in belts such as the Rockies, where sedimentary cover is allochthonous over an apparently undeformed basement, the shortening apparently affects only high crustal levels. Either part of the basement has been removed by 'subduction' or shortening has occurred superficially by gravitational forces—downslope gravity sliding of the allochthonous mass or by sideways gravity spreading of a plastic rock mass, like porridge on a plate. Both these mechanisms can operate in moving ice sheets and glaciers. In this and other more detailed ways ice behaves in a rather similar way to rocks, although it deforms much more quickly.

Upright fold belts

This association is dominated by upright folds (steeply dipping axial planes) developed in sedimentary, volcanic or low-grade metamorphic rocks. Individual folds may have subhorizontal hinge lines continuous along strike, or undulating hinge lines which produce elongate domes and basins. A steep axial plane cleavage may or may not be present. In fine-grained lithologies the cleavage may be intense enough to produce slates and the distinctive terrains called slate belts. Examples of upright fold belts are the foothills belt of Iran, the Swiss Jura, the central western fold belt of New South Wales and the Caledonian belt in Wales. Despite their gross similarity, these areas show important differences in structural style, and there is probably no single mode of origin involved. Most involve subhorizontal shortening across the belts, perpendicular to the folds. As with thrust belts this interpretation poses the question of whether the underlying basement too has shortened. Under the Jura it has not, implying a detachment or décollement of the cover from its basement.

Polyphase schist belts

Many low- to medium-grade metamorphic areas show structures that have formed during more than one episode of deformation. Such polyphase or superposed deformation can produce extremely complex structural geometry. (Some of these complications are described in the section on structural analysis.) A common, though by no means general, sequence of structures is as follows: early tight folds with a strong axial plane schistosity are followed by later, less tight folds and associated axial plane crenulation cleavage and finally by angular kink or chevron folds with weak or absent axial plane foliation. The later phases may be developed in restricted areas, or local extra phases may occur. The significance of such time sequences in terms of the formation of schist belts is not fully understood.

The central or internal parts of many orogenic belts have polyphase schist belts. The best known are those of the Caledonian belt of Scandinavia and the British Isles, and of the Appalachian belt of eastern North America.

Gneiss belts

Areas of very high-grade metamorphic rocks may appear structurally less complex than some polyphase schist belts. In fact most gneiss terrains have undergone intense polyphase deformation, and the illusion of relative simplicity can be due to the intense deformation and metamorphism, destroying all evidence of primary sedimentary or igneous structures and also of previous deformation structures. The detailed complexities of gneisses may also be masked by their large grain size, the paucity of the flaky minerals that define foliations and the frequent partial melting of the rock. Partly because of these difficulties, generalizations about sequence of structures in gneisses are not available.

The metamorphic grade in gneiss belts implies that the structures were formed at great depth in the crust. Exposure of such deep crustal levels requires time for many kilometres of uplift and erosion. Therefore the world's most extensive gneiss belts are found in the Precambrian shield areas, for instance, of central Australia and the Baltic and Canadian shields.

Diapiric structures

Diapirs are large buoyant 'bubbles' of rock driven upwards in the crust by their low density compared with surrounding rocks. The most common rocks to behave in this way are, at low temperatures, salt and mud and, at high temperatures, granites and perhaps some gneisses. The geometry associated with a diapir is rather similar whatever the type of rock concerned. The diapir itself is a bulbous or flat-topped cylinder or wall with subvertical sides some kilometres apart. Where the buoyant rock was originally bedded, as with salt, the deformed bedding reveals an extremely complex internal folded structure to the diapir. Rocks overlying the diapir are arched or faulted and surrounding beds are upturned. Diapiric structures are rarely isolated, but occur in arrays sometimes linearly arranged.

Salt diapirs, or salt domes, are particularly important economically in producing structures that may act as oil traps in otherwise undeformed sediments. Examples occur on the Gulf coast of the USA, in the Persian Gulf region and in the North Sea as well as off many passive continental margins. Mud diapirs are particularly common in some large modern deltas such as the Niger and the Mississippi. Granite diapirs may occur as isolated stocks geometrically similar to salt domes, but often coalesce to form extensive terrains of granite intrusions or batholiths. They are common down the western Cordillera of North America and down the Andes in South America.

Mélanges

Extensive terrains in some orogenic belts, for example the Alpine–Himalayan belt, are characterized by apparently chaotically arranged blocks of a number of different rock types floating in a matrix like currants in a cake. The rock body comprising blocks and matrix is termed a mélange. The blocks may range from a few centimetres in size to several kilometres. The matrix may be sedimentary or, more rarely, altered or deformed igneous or metamorphic material. Some mélanges have a strong foliation in either or both the blocks and the matrix, others are undeformed. Two quite distinct modes of mélange origin have been demonstrated. One is by tectonic processes, fragmenting and separating blocks of originally organized rock sequences in large fault zones. The other origin is by purely sedimentary processes. Small blocks are carried down topographic slopes in debris flows supported by a muddy matrix. Some large blocks may slide downslope individually. Distinguishing between the tectonic and sedimentary origin of mélanges is important but often difficult.

12 Volcanic activity away from plate margins

By around the turn of the century it was recognized that volcanic lavas extruded in areas of folding and mountain-building differed in their chemical composition from those in areas of rifting. From the geographical areas believed to be representative of these two types geologists coined the terms 'Pacific' magma types for more calcium-rich (calc-alkaline) and 'Atlantic' magma types for more sodium- and potassium-rich (alkaline) lava compositions. Some seventy years later plate tectonics theory has allowed a rationalization of the relationship between volcanism and the Earth's deformational movements.

Most of the active volcanoes on Earth occur along plate margins, where the bulk of magma that rises from the mantle to form new crust is generated. However, volcanic activity in the interior of continents and oceans, typically in regions of tensional tectonics, is also a major phenomenon, exemplified by the world's most active and best-studied volcano, Kilauea, on the east side of Mauna Loa volcano on the island of Hawaii. Such volcanism is referred to as within-plate, or intraplate, volcanism and is subdivided into continental and oceanic according to its setting. Intraplate volcanoes are generally more accessible to direct observation than those along plate margins. They are more widespread on the continents, in contrast to underwater eruptions along constructive plate margins (see Chapter 13), which are difficult to observe, and to the many active volcanoes above the destructive margins (see Chapter 14) that occur at high elevations and are often snow-covered. Surface volcanoes have three common basic forms and structures: small scoria (cinder) cones with simple structures; low-profile, but generally large, shield volcanoes which in common with scoria cones are usually of basaltic composition; and more complex strato-volcanoes of alternating layers of lava and pyroclastic debris and of more differentiated composition, structurally similar to those of subduction zone volcanoes.

Perhaps the most challenging aspect of the study of intraplate volcanism, especially in the ocean basins, is that it has produced one of the most revolutionary and stimulating concepts in the framework of plate tectonics: the hot spot or plume hypothesis. This hypothesis was developed to describe fundamental dynamic processes in the mantle beneath the plates, perhaps involving the entire mantle. In its present form this model is typical of Earth science theories in general, being based upon observations from many different fields in the Earth sciences, none of which by itself is sufficient to support the hypothesis.

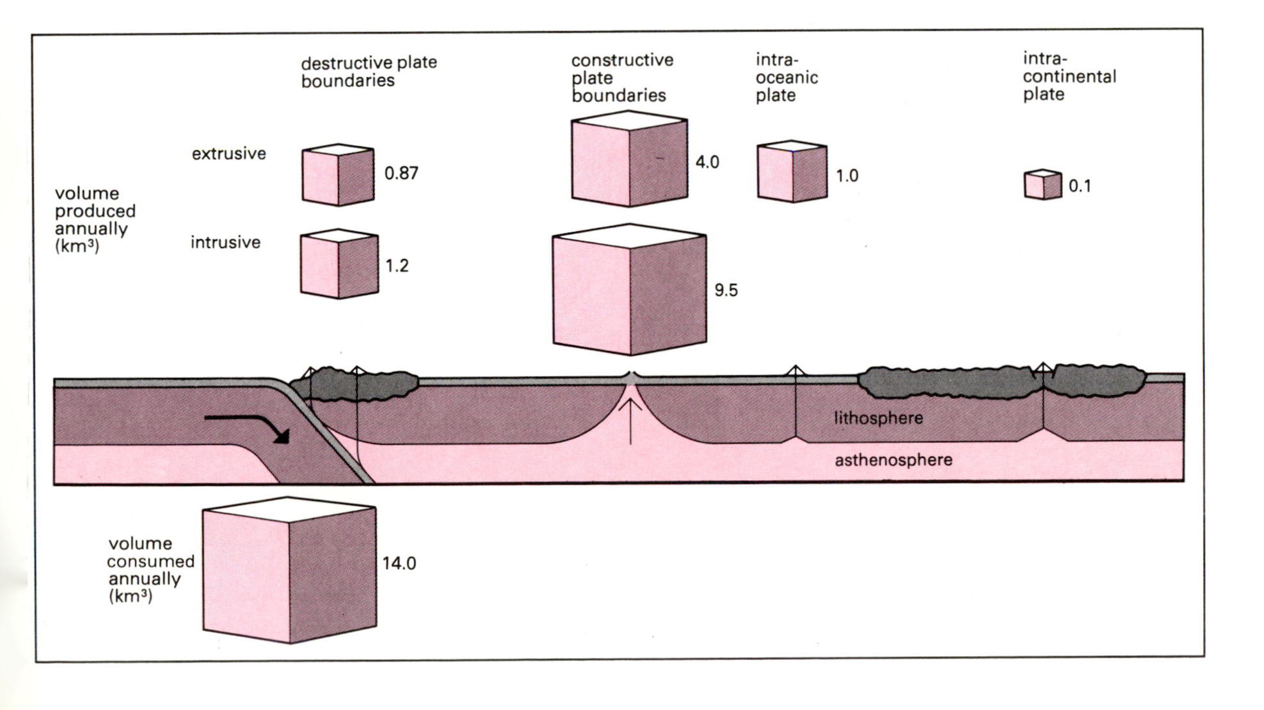

12.1: *Rates of annual magma production in different plate margin settings compared with continental and oceanic igneous activity away from plate margins.*

12.2: *Eruption of basaltic lava, Mauna Ulu, Kilauea, Hawaii, in 1969, showing a lava fountain and cascades of lava, still largely molten.*

Oceanic intraplate volcanism

Many well-known oceanic islands away from the mid-ocean ridges are intraplate volcanoes. Examples are Hawaii, Samoa, Tahiti, Galápagos and Easter Island in the Pacific, the Azores, Madeira, the Canary Islands, St Helena and Ascension in the Atlantic, and Réunion and Mauritius in the Indian Ocean. Almost all these islands are members of larger groups, or archipelagos, or, especially in the Pacific, form parts of very long strings of volcanic islands. Less well known, but more numerous, are the several tens of thousands of seamounts, submarine volcanic mountains whose often circular summits rise up to 2000 m above sea bottom. In the Pacific ocean basin alone there are more than ten thousand with heights above the sea floor exceeding 1000 m.

Those seamounts or submarine volcanoes that have produced enough magma to grow above sea level, forming volcanic islands, are among the highest and biggest mountains on Earth. Mauna Loa in Hawaii, for example, has an elevation of about 10 000 m above the floor of the ocean. Its volume of about 40 000 km^3 is sufficient to cover Switzerland with a layer of basalt lava 1 km thick. Many oceanic intraplate volcanoes exceed 3000 m, contrasting with subduction zone volcanoes, very few of which are higher than 2000 m.

Although the distribution of oceanic intraplate volcanoes does not appear to be random, a convincing explanation for their location has yet to be proposed. For example some islands or island groups occur on or close to mid-ocean ridges, to which their origin must in some way be related. Most prominent among these is Iceland which straddles the mid-Atlantic ridge and is better regarded as an anomalous, elevated part of the ridge rather than a 'true' oceanic island. Others include the Azores, Ascension, the Galápagos and Easter Island. Prominent groups of seamounts whose magma extrusion was insufficient to grow above sea level also occur close to mid-ocean ridges. Other groups, such as the Canary and Madeira Islands and associated large seamounts, have formed close to the boundary between oceanic and continental crust at a passive continental margin. Some such islands may have formed along extensions into oceanic crust of long-lived lines of major crustal weakness in the continents. The occurrence of yet other intraplate volcanoes, such as the Hawaiian chain or St Helena and Tristan da Cunha and Trindade in the Atlantic, is still a mystery.

The structure of oceanic intraplate volcanoes

The structure and composition of seamounts and of the submarine parts of oceanic volcanoes have been studied by a variety of methods: examination of the volcanoes by dredging and by observations from submarines, drilling into submerged volcanoes and finally the study of uplifted, formerly submarine, parts of volcanic islands. These still very scanty data show that submarine volcanoes consist mainly of pillow lavas, intrusive rocks such as dykes and sills and, particularly in the upper parts, increasing amounts of broken-up rocks such as pillow breccias and finer-grained submarine volcanic debris.

The cores of these volcanoes may be made of coarse-grained intrusive rocks such as gabbros, and ultrabasic rocks that represent the former magma chambers. It has been calculated that extrusion rates exceeding one million cubic kilometres per year are needed for a submarine volcano to grow above sea level. Emplacement and extrusive mechanisms change drastically when a volcano grows to within a few tens of metres of the water surface, because decreasing pressure causes magmatic gases to expand so that the volcanism becomes truly explosive. The very large degree of abrasion close to the wave base also impedes the further growth of oceanic volcanoes. Large masses of clastic material may slough off the top and flow for more than 100 km into the surrounding deeper sea, leading to the formation of large aprons of volcanoclastic material around the initial cone. If eruption rates are low volcanoes may stagnate at this stage and become seamounts; if the rates are high what appears to be the major phase of most oceanic volcanoes may now be formed—a shield volcano composed of basaltic lava flows. Structurally many shield volcanoes consist of a main vent area above the major magma chamber and one or more radiating rift zones.

When eruptive rates decline longer-lived magma chambers can be established in the substructure of shield volcanoes. In these, more highly differentiated (trachytic, phonolitic or rhyolitic—see Chapter 5) magmas may form and may be emplaced as plugs, or erupted as lava flows or more explosively. The shield basalts may be succeeded immediately, or after a small time interval, by even more alkaline basaltic lavas. This stage is followed, after erosional periods that may last from about two million years (Ma) in Hawaii to more than 5 Ma

12.3: *Pahoehoe (ropy) lava, Mauna Ulu, Kilauea, Hawaii. The congealed skin became wrinkled by continued flow of the still molten interior.*

12.4: *Montaña de Fuego, Lanzarote, Canary Islands; scoria cones and lava fields dating from the eighteenth century.*

as in Gran Canaria, by the eruption of very low volumes of much more alkaline magmas. Many of these carry chunks of peridotite, evidence of rapid and direct ascent from the mantle. They are commonly erupted from fissures whose orientation is at right angles to the main fissure direction of the shield phase of formation.

Many oceanic islands are fringed by cones of tuff that are due to eruptions caused by the contact of magma with water. They form either when lava flows enter the sea at the shore (littoral cones) or by direct eruption under water or by eruption in nearshore areas where water is present near the surface.

The geochemistry of oceanic intraplate volcanoes

Alkali basalt, the dominant volcanic rock type on oceanic islands, is moderately silica-undersaturated and contains varying proportions of phenocrysts, chiefly clinopyroxene, olivine and plagioclase. Subalkaline or tholeiitic magmas occur mainly in islands with very high production rates but also in some smaller island groups such as Easter Island and the Galápagos. More alkaline and undersaturated lavas (such as basanites, nephelinites and melilite–nephelinites) occur on a number of islands, particularly those with longer magmatic histories. These lavas occur typically during the late post-erosional stage in the evolution of such islands. They are absent from islands such as Iceland, the Azores and Ascension and Easter Islands which are close to mid-ocean ridges. More potassium-rich magmas are rare but are present for example in the Atlantic islands of Tristan da Cunha, Gough, São Miguel and Jan Mayen.

Depending on the composition of the basaltic parent magma, different lineages of magmatic evolution and differentiation are represented on oceanic islands. Most common perhaps are trachytes representing the end stage of differentiation, but more undersaturated, highly differentiated magmas such as phonolites are not rare. Even rhyolites, once thought to be restricted to continental areas or when existing on oceanic islands to be evidence of the presence of continental-type crust beneath them, may occur. They are present in relatively large amounts on Iceland and Gran Canaria. They are typically peralkaline (highly alkaline) in composition. Even carbonatites have been reported in a few volcanic islands such as the Cape Verde Islands. Most recently of all there has been a report of a possible occurrence of kimberlite pipes in Cretaceous oceanic crust in the south-west Pacific; this would mean that every possible magma type had been observed in an oceanic crustal setting.

Detailed studies of trace elements and radiogenic isotopes have led a number of Earth scientists to the conclusion that ocean-island basaltic magmas are generated from very different, probably much deeper, portions of the suboceanic mantle than are the ocean-floor basaltic magmas. However, more recent data show that this picture is oversimplified. It is now realized that the mantle is not homogeneous but may show small- and large-scale lateral changes in its mineralogical and chemical composition.

The age of oceanic intraplate volcanoes

Almost all oceanic volcanic islands are less than about 20 Ma old and a large number have been active in historical times. Although some islands that have become inactive have undoubtedly become eroded to below sea level, it appears that ocean island formation is a relatively recent development in the evolution of igneous oceanic crust. Although oceanic islands are found throughout the ocean basins there is a tendency for the older islands to occur away from the ridges while the younger ones are situated close to the mid-ocean ridges. The islands that are far away from the ridges have resulted from events much later than the formation of the oceanic crust proper, while those close to the mid-ocean ridges may have been formed by processes associated with the evolution of the ridge systems themselves. Some Earth scientists have postulated that oceanic islands close to the ridges show chemical affinities to mid-ocean ridge magmas, while those away from the ridges are generally more alkaline. However there are many exceptions to this pattern, and a more overriding influence seems to be the evolution of ocean island magmas from a less undersaturated shield stage to a highly undersaturated and more alkaline post-erosional phase.

The lifespan of oceanic volcanoes is variable. The shield phase, as indicated above, is generally very short. Some islands, for example in Hawaii, may become inactive after 2 to 3 Ma, while others, such as many of the Canaries, may show a number of volcanic phases over a timespan exceeding 15 Ma. It appears that the active lifespan depends, among other factors, on the spreading rate of the related plate: thus the greater ocean-floor spreading rates in the Pacific lead to short spans of activity while the lower rates in the Atlantic lead to lifespans that are generally longer.

12.5: *Madeira: eroded interior of a large scoria cone 1.5 Ma old, showing bedded cinders, lava flows, and a prominent feeder dyke.*

Continental intraplate volcanism

Continental volcanism appears more variable and complex than oceanic intraplate volcanism. It may occur on both old, cratonic as well as on young, tectonically active crust, but it is not common on very old cratonic shields, indicating that under these areas temperatures may be too low for partial melting of mantle material or that tectonic stability is higher.

At present, several major partly interrelated tectonic settings of continental intraplate volcanism may be distinguished. Perhaps the most classic is that along large intracontinental rifts. These are linear segments, in some cases exceeding 1000 km in length and generally about 30–60 km wide, along which the Earth's crust has subsided by as much as 5 km along the crest of elongate uplifts. Crust and lithospheric mantle are thinner beneath such rifts than under adjacent areas. The East African rift and the Rhine graben are the best-known examples, both showing crustal extension of some 5 km. The Rhine graben subsided chiefly between about 25 and 35 Ma ago, although volcanism within the rift occurred as early as 70 Ma ago. Similarly the East African rift, with its western branch through Uganda and eastern branch through Kenya and Tanzania, has developed chiefly during the past 30 Ma, although earlier stages are recognized in its southern portion. Typical volcanic rocks are alkaline to highly alkaline, ranging from alkali basalt and peralkaline rhyolite associations to more silica-undersaturated lavas, such as basanites and nephelinites and their derivatives such as phonolite. Carbonatites occur in the most alkaline areas. Volcanism is less alkaline in other rift systems, and in some there is a distinct zonation, with a moderately alkaline and even tholeiitic zone in the centre (northern Rio Grande rift, Vogelsberg in the Rhine rift) and an increasing alkalinity with distance from the centre of the depressions.

In the northern part of the East African rift system, the Red Sea and the Afar depression document a transition between continental rifts and oceanic-type constructive plate margins which is also characterized by voluminous, less alkaline (tholeiitic) basaltic volcanism. Most of the East African rifts as well as most other intracontinental rifts do not appear to be mere precursors of oceanic rifts, however, as shown by their long lifespan and by the existence of old, decayed and inactive rift systems such as the Permian Oslo graben. Moreover, the type of continental intraplate volcanism that does appear to herald subsequent oceanic rifting is of a quite different flood basalt type.

Much of the continental intraplate volcanism outside rift zones occurs in uplifted areas such as the Massif Central in France and the Rhenish massif in Germany (forming the western half of a belt of Cenozoic and Quaternary volcanic fields extending eastwards to Silesia), Tibesti and Hoggar in northern Africa, the Yellowstone and Colorado plateaux and the Sierra Nevada in the western USA, and in eastern Australia. Where studied in detail it has been shown that major episodes of volcanic activity are approximately synchronous with periods of significant uplift. For example the Oligocene and Pliocene/Quaternary volcanic phases in the Eifel correspond to major phases of uplift in the Rhenish massif. Similarly the late Miocene and late Pliocene volcanic fields on top of the Sierra Nevada batholith (western USA) were extruded during phases of uplift. In both areas the composition of the younger volcanic rocks is distinctly potassium-

12.6: *Chemical composition (weight %) of oceanic intraplate volcanics, Gran Canaria.*

	1	2	3	4	5	6	7	8
SiO_2	46.2	49.2	71.0	49.4	45.7	38.7	38.6	41.6
TiO_2	3.7	4.4	0.6	3.0	3.8	4.1	3.9	3.6
Al_2O_3	10.9	13.7	14.35	12.7	12.3	9.4	9.7	11.6
Fe_2O_3	4.5	7.6	2.44	5.2	7.2	9.9	4.7	8.4
FeO	8.4	5.8	0.31	7.5	6.7	4.7	8.3	4.1
MnO	0.2	0.2	0.2	0.2	0.2	0.2	0.2	0.2
MgO	10.3	4.4	0.5	8.7	8.7	13.9	14.3	12.7
CaO	11.9	8.8	0.7	9.5	10.6	13.0	13.6	11.5
Na_2O	2.2	3.4	6.1	2.6	2.7	3.1	3.5	3.3
K_2O	1.0	1.6	3.5	0.6	1.1	1.2	1.5	1.6
P_2O_5	0.4	0.7	0.1	0.5	0.9	1.5	1.4	1.0

Notes to columns
Miocene shield series: **1** alkali basalt **2** hawaiite **3** rhyolite
Pliocene shield series: **4** transitional alkali basalt **5** alkali basalt
Post erosional series: **6** nephelinite **7** melilite nephelinite **8** basanite

12.7: *Chemical composition (weight %) of continental intraplate volcanics.*

	1	2	3	4	5
SiO_2	50.1	54.9	42.7	40.9	43.9
TiO_2	1.6	2.2	2.4	2.6	2.6
Al_2O_3	15.5	13.8	11.7	11.2	15.5
Fe_2O_3	–	–	3.0	5.9	4.5
FeO	11.2	12.3	8.2	4.8	6.4
MnO	0.2	0.2	0.2	0.2	0.2
MgO	6.7	3.4	14.1	10.9	8.4
CaO	10.6	7.1	12.1	16.1	12.0
Na_2O	2.9	3.3	3.2	3.2	4.6
K_2O	0.6	1.8	1.6	3.3	1.0
P_2O_5	0.2	0.4	0.9	0.9	0.9

Notes to columns
Flood basalts (Washington, USA): **1** Picture Gorge basalt **2** Yakima basalt
Primitive rift magmas (Eifel region, Germany): **3** basanite
4 melilite nephelinite **5** alkali basalt

rich. In contrast to adjacent zones of rifting, heat flow in these uplifted areas is very low. This low heat flow (if it can be extrapolated to mantle depths) and the potassic nature of the magmas could indicate generation of the magmas by very small degrees of partial melting at depths in excess of 100 km. Some Earth scientists think that the ascent of volatile-rich magmas has caused metasomatism of the uppermost mantle, thus lowering its density and increasing its volume. This causes the uplifts, at some of which the magma erupts at the surface.

The fundamental relationship between uplift and doming, faulting and volcanism recognized some forty years ago by Cloos has since been studied in more detail in Africa. It seems that pauses in the motion of the African lithosphere relative to the mantle led to an accumulation of heat at the base of the lithosphere, generating uplift and partial melting. Thus the onset of major Cenozoic volcanism in Africa about 35 Ma ago may be a consequence of a period of stasis of the African plate from about 45 Ma ago. Differences in the dimensions of domal uplift as well as the volume and chemical composition of erupted magmas may be related to local differences in these thermal anomalies.

However, not all areas of continental intraplate volcanism fit into either the rift or the uplift category. Such an area is the Cenozoic volcanic province of the western USA which, although studied in great detail, is not fully understood. Volcanism in this area falls into two episodes. An earlier phase ranging over a period from about 70 to 20 Ma ago was characterized by dominantly calc-alkaline volcanism of intermediate to silica-rich composition. This was probably related to subduction along the Pacific margin to the west. After only a brief interlude between about 20 and 17 Ma ago most of the area was subjected to pronounced regional extension. For example cumulative tectonic extension exceeded 100 km in the Great Basin. Except in the extreme west, volcanism now changed to either basaltic or alternations of basaltic and rhyolitic in composition. The entire igneous province which extends for about 2000 km north to south and about 1000 km east to west encompasses several major and distinct areas, including the block-faulted Basin and Range province (characterized by high heat flow and a lithosphere only 60 km thick), the Rio Grande rift, the Colorado plateau and, to the north, the Snake River plain, the Yellowstone plateau and the Columbia River flood basalt plateau.

The abrupt westward migration and narrowing of the calc-alkaline belt between 30 and 20 Ma ago is interpreted as an abrupt steepening of the subduction zone prior to final cessation of subduction in the west when it became replaced by a transform fault. Some Earth scientists think that the shallow, downgoing oceanic slab was cut off and left behind as a result of steepening subduction. Its gradual warming and expansion may be the major cause for regional uplift and extension as well as partial melting and magmatism. If this is confirmed, a major zone of intraplate magmatism is thus intimately tied to the late stage in the evolution of a subduction zone.

12.8: *Crater Lake, Oregon, USA, occupying a caldera representing a collapsed volcanic cone which was glaciated at the time of its last eruption.*

Flood basalt provinces

Although this chapter necessarily deals principally with young volcanic areas, mention must be made of a special type of volcanism only one example of which—the Columbia River basalt province—is geologically young. While flood basalt provinces can be examples of intraplate volcanism, most of the older flood basalts were formed just prior to the break-up of continental plates and should thus be considered as an early, continental phase of normally oceanic constructive plate magmatism. Most known flood basalts are Mesozoic or younger; the Karroo in southern Africa is about 180 to 190 Ma old, Paraná in South America about 120 to 130 Ma, the Deccan basalts in India about 65 Ma and the Columbia River basalts in North America chiefly between 14 and 17 Ma. The abundant Mesozoic igneous sills of Tasmania and Antarctica belong to this category, as well as the early Cenozoic flood basalts of eastern Greenland. The Cenozoic igneous province of Britain was apparently formed in a similar tectonic setting but with a slightly different structural evolution, resulting in central volcanic complexes such as Arran, Mull and Skye as well as flood basalts such as those of northern Ireland.

The most impressive aspect of flood basalts is the colossal dimensions both of single lava flows (whose volume may exceed 500 km^3 with areal extents of 40 000 km^2 or more) and of entire lava fields (volumes of the order of 100 000 to a million km^3). The feeder-dyke swarms are of correspondingly gigantic dimensions; the Grande Ronde dyke swarm in Oregon, Washington and Idaho extends for some 200 km and has a width of some 50 km, representing considerable extension of the crust.

Flood basalts are chemically intermediate, with respect to alkali elements, between ocean floor basalts and the more common intracontinental and oceanic island basalts. However, many of the lavas of flood basalt provinces are not very primitive, as shown by their low magnesium content. This presents a major problem when their large volume and lack of highly differentiated rock associations are considered; how and where did such large volumes of magma come to be fractionated? The problem could be explained if the mantle were more iron-rich than is generally assumed and that these magmas were erupted from mantle depths. Alternatively, the magmas could have been erupted from gigantic magma reservoirs at the base of the crust where more primitive mantle-derived magmas were temporarily stored and were able to differentiate prior to eruption (see below).

12.9: *Flood basalts, Columbia River plateau, showing the great thickness of successive tholeiitic lava flows accumulated mainly in Miocene times.*

Mantle nodules, kimberlites and carbonatites

There is little doubt among Earth scientists today that basaltic magmas are formed in the Earth's uppermost mantle, most probably at depths varying between about 20 km below the mid-ocean ridges, where the mantle extends towards the surface, and 200 km below old continental shields, where temperatures are much lower than at similar depths under mid-ocean ridges. The two major lines of evidence for a magma source in the mantle rather than in the crust are that temperatures of erupting basaltic magmas (about 1200°C in Hawaii) are much higher than any in the crust, and that the composition of the crust is either quite different from that of basalt in many continental areas or—where similar—would not generate basaltic magmas upon partial melting.

Although the Earth's mantle is not directly accessible, pieces of olivine-rich rock (peridotite) presumed to be derived from the mantle are brought to the surface in alkaline basaltic lavas. These so-called nodules (xenoliths would be a more accurate term) consist dominantly of the silicates olivine, orthopyroxene (enstatite), clinopyroxene and the oxide spinel. Amphibole (a hydrated mineral) and the magnesian mica phlogopite occur in some, as do garnet and a few other minerals. These mineralogical differences reflect two main classes of mantle: most common are harzburgites, consisting chiefly of olivine and orthopyroxene. These are depleted in those chemical elements (alkalis and the more abundant elements calcium and aluminium) that are major components of basaltic magmas. The magmas bringing these nodules to the surface must thus have traversed 'barren' upper mantle regions which have undergone one or more earlier episodes of partial melting. The less common 'fertile' or undepleted nodules of mantle rock contain more clinopyroxene and garnet, the chief aluminium- and calcium-bearing minerals characteristic of the pressure (ie depth) range in which most basaltic magmas are generated.

Most mantle nodules are not in equilibrium with their host basalts, ie if given time for assimilation they would lose their distinctive character. The conclusion that the magma cannot be derived from mantle regions represented by the nodules is supported by significant differences in isotopic composition between the two. A recent discovery is that many nodules consist of more than one component with different histories. For example, amphibole, increasingly recognized as a common accessory mineral in peridotite nodules, may be the crystallized equivalent of those fluids that are highly enriched in the so-called large-ion lithophile (LIL) elements (or incompatible elements). As was mentioned above, such fluids may also play a role in lowering mantle density and causing uplift in regions of continental intraplate volcanism. Garnet-bearing nodules occur almost exclusively in the rare rock kimberlite and are more abundant in these rocks than the 'depleted' harzburgites.

Kimberlites are intrusive, generally brecciated rocks that occur in pipes, or diatremes, are mostly restricted to the continental lithosphere and have ages from Proterozoic to Cenozoic. They consist of large crystals of olivine, phlogopite, ferromagnesian garnet such as pyrope and almandine, pyroxenes such as enstatite and diopside, and magnesium-rich ilmenite ($FeTiO_3$) set in a fine-grained matrix of serpentine, phlogopite, calcite and other minerals. Kimberlites are the primary source of diamonds, which supports the view that kimberlite magmas form at great depth in the mantle. Kimberlite magmas, being highly enriched in low-melting components as well as in water and carbon dioxide, thus represent one extreme of the range of intraplate alkali basalt compositions. They probably represent only very small degrees of partial melting, perhaps of the order of 1 per cent or less. It has recently been shown by laboratory experiments that such highly alkaline basaltic magmas form at relatively high CO_2/H_2O ratios in the mantle. Carbon dioxide, the second most common volcanic gas phase after water, probably occurs principally in the carbonate mineral dolomite at these mantle depths.

Some very rare magmas, the carbonatites, consist dominantly of carbonates of calcium, magnesium and iron, or of sodium and potassium. They are associated in time and space with highly alkaline basaltic magmas (the nephelinite suite) and locally also with kimberlites. They may represent a liquid that has separated by unmixing from a silicate magma on cooling.

The origin of magmas

Primary basaltic magmas in intraplate and other settings are now generally believed to be formed by partial melting of mantle peridotite. Depths of melting and degree of partial melting vary with composition; a range of 30 to 100 km is a usual depth of origin for most compositions, with up to 30 per cent melting for tholeiitic (subalkaline) basaltic magmas such as the flood basalts and those of Hawaii, up to 15 per cent for the most widespread alkali basalts and less than 10 per cent for the highly alkaline and undersaturated magmas that appear during the late stages in some ocean islands and also in some continental volcanic provinces. Volcanic areas of young, very alkaline volcanism emit especially large quantities of CO_2 and, in rare instances, even erupt carbonatite lavas, as does the active volcano of Oldoinyo Lengai in East Africa.

There are several theories on the formation of the silicic and highly alkaline magmas which give rise to rhyolite, trachyte and phonolite rocks at intraplate volcanoes. There are basically two main ways to generate such magmas. One is by differentiation of the basaltic parent magma during cooling in crustal magma pockets, or chambers. The crystallization of mineral phases such as olivine, pyroxene, amphibole and feldspars and their removal from the magma by gravitational settling leads to more silicic and alkaline residual magma. These less dense liquids migrate to the top of a magma pipe and may be the only erupted products of this process owing to their higher content of volatile compounds such as H_2O, CO_2 and H_2S. The second way to generate such magmas is by partial melting of crustal rocks. The necessary heat would be provided by large volumes of basaltic magmas supplied to the base of the crust or to magma reservoirs within it. One way of distinguishing the two processes is by comparison of major and trace elements and of radiogenic isotope ratios between rocks believed to represent parent and derivative magmas; strong chemical and isotopic similarities indicate differentiation from a single parent magma with no addition of crustal material. Using these results, it seems that partial melting of crustal rocks—a process that at present best explains the vast volumes of rhyolites and granites in subduction zone environments—is increasingly called upon to explain the presence of large volumes of rhyolitic magmas in both continental and oceanic intraplate settings such as Iceland and Gran Canaria, where rhyolitic magmas were generated in relatively large volumes. Thus, with increasing research, the chemical distinctions between volcanic products of quite different tectonic settings begin to blur.

Magmatic and volcanic processes have been studied nowhere in more detail than in intraplate volcanoes, especially Kilauea on Hawaii, where the path of magma from its source area to the surface is now known in broad terms. Earthquake epicentres define a pipelike structure from approximately 40 km beneath the volcano to about 10 km beneath its surface, which is interpreted as the duct up which magma migrates from the mantle into the crust. Between about 10 and 3 km beneath the top of the volcano the pipe becomes a roughly pear-shaped magma chamber. Magma appears to be fed into this holding reservoir at a relatively constant rate.

Magma is periodically released from the chamber either sideways, probably along dykelike conduits into a rift zone, of which the Kilauea east rift zone is especially long (50 km), or upwards to the main Kilauea crater area. It then rises to the surface or into near-surface chambers where the magma may cool slowly and crystallize to form coarse-grained gabbros. Its ascent appears to be due mainly to buoyant forces, while the nucleation and expansion of volatiles into gas bubbles at very low pressures (within the upper few hundreds of metres from the surface) may accelerate the erupting magma to form spectacular lava fountains that may reach heights of several hundred metres.

Hot spots and mantle plumes

There is no better way to monitor the movement of lithospheric plates during the past few tens of millions of years of Earth history than by analysing plate vectors and velocity while assuming that hot spots are fixed in the mantle. Hot spots may signal the sites of major plumes where heat and mantle material and magma are rising from greater depths. Mantle plumes may be instrumental in breaking up and moving continents, and their pathways may be traceable some 200 Ma back in Earth history.

Hot spots and mantle plumes are at present a subject of debate and research which, for convenience, may be divided into the three main fields described below.

Linear volcanic island chains and topographic anomalies

More than a hundred years ago the American geologist Dana noted an increase in age along the Hawaiian island chains from the active island of Hawaii at the south-eastern end of the chain to the strongly eroded island of Niihau north-west of Hawaii. Von Fritsch noted a similar age succession in the Canary Islands in the Atlantic, from the historically most active island of La Palma in the west to the most highly eroded island of Fuerteventura to the east. This principal observation formed the basis for the famous hot spot hypothesis proposed in 1963 by J. T. Wilson. Wilson postulated that the age progression shown by chains of oceanic islands are best explained by movement of the lithosphere over hot spots representing the fixed centres of mantle convection cells. These emit magmas that ooze up into and through the lithosphere to form surface volcanoes whose activity ends once the plate has moved away from the hot spot. The distribution of the hot spots over the lithospheric plates is shown in Figure 10.17.

W.J. Morgan extended these ideas and suggested that hot spots are the expression of mantle material rising from depths as cylindrical bodies ($r \sim 100$ km) driven by thermal convection (mantle plumes). He thought that areas of voluminous intraplate and ridge volcanism were the surface expression of such mantle plumes. Morgan postulated 16 such areas of plume volcanism, later extended by other workers to more than 100.

The Hawaiian–Emperor chain is an alignment of oceanic volcanoes extending for almost 6000 km across the Pacific coast. It comprises some 107 volcanoes representing more than 740 000 km^3 of magma. The Hawaiian chain proper is about 3500 km long and the eight main Hawaiian islands are situated along its south-eastern end. The chain bends about 3500 km north-west of Hawaii and continues as the Emperor chain of seamounts for another 2300 km northwards. It disappears near the intersection of the Kuril and Aleutian trenches.

The Hawaiian–Emperor volcanic chain was formed during the past 70 million years as the Pacific lithospheric plate moved first north, then west relative to a fixed hot spot. A volcano was formed on top of the plate above the hot spot, gradually being cut off from its magma source as the plate continued to move. A new volcano formed on the sea floor along the same trail once enough magma had accumulated to rise through the lithosphere. The rate of migration of volcanism along the Hawaiian–Emperor chain is currently determined as approximately 80 mm per year. Similar, but generally much less well-documented, island chains showing almost identical progression rates occur to the south in the Pacific, notably the Marshall-Austral and Line–Tuamotu chains. Interestingly, they also show a similar configuration, that is a south-eastern branch and a north-trending branch with present activity concentrated along the south-eastern tip. The age of the bend in both volcano chains is identical with that in the Hawaiian–Emperor chain, that is roughly 42 Ma.

Currently available data indicate that similar linear age progressions can also be demonstrated in the Atlantic, but the progression is much less regular than that of the large chains of Pacific islands. One of the major reasons for the irregularity is probably the appreciably lower spreading rate of the Atlantic compared with the Pacific plate.

Sites of intense volcanism are also topographic highs which may extend far beyond the volcanic fields. For example there is a positive topographic anomaly extending for over 1000 km south-east of the Azores hot spot area into lithosphere up to 60 Ma old. Such positive topographic anomalies in the oceans are defined as those areas that are higher than would be expected from subsidence of the lithosphere as it cools while it moves away from the mid-ocean ridge. It has recently been suggested that hot spots and high spots (with or without volcanism) are both due to the same cause, namely disruption of materials and temperatures within the mantle by ascending streams of hotter mantle material. Such high spots have diameters of 200–3000 km at the Earth's surface.

Geophysical characteristics

Many hot spot areas which are expressed at the surface by abundant volcanism and/or unusual topographic elevation are also locally underlain by anomalous mantle to depths of 200–250 km. Most significant are positive gravity anomalies, higher than normal heat flow, thinning of both crust and lithosphere and travel-time delays of seismic waves, which are interpreted as reduction of mantle density by up to 10 per cent, for example below the Yellowstone, Hawaii and Iceland hot spots. These anomalously hot and low-density mantle regions coincide approximately in diameter with the area of the surface volcanic fields and extend through the underlying lithosphere into the asthenosphere. They are thought to represent partially molten mantle material.

Geochemical characteristics

Soon after the mantle plume hypothesis was formulated geochemical arguments were advanced to show that plume magmas could be identified that differed fundamentally from other magmas such as ocean floor basalts. Plume magmas were postulated as deriving from relatively primitive, undepleted mantle (contrasted with the widespread depleted suboceanic mantle). They would thus be enriched in LIL elements such as potassium, rubidium, cesium, strontium and barium, the light rare earth (LRE) elements, and uranium, thorium, lead, zirconium, niobium, yttrium and tantalum and others. They are also enriched in ratios of radiogenic isotopes such as $^{87}Sr/^{86}Sr$, $^{143}Nd/^{144}Nd$ and $^{206}Pb/^{204}Pb$. (In these definitions undepleted would refer to closed-system isotope evolution since Earth formation 4550 Ma ago, while enriched and depleted would refer to systems with parent/daughter ratios of Rb/Sr, Sm/Nd or U/Pb respectively higher or lower than in closed-system evolution.) In some areas, such as the mid-Atlantic ridge south of Iceland (Reykjanes ridge) or south and north of the Azores, values for LIL elements and isotope ratios are found to be intermediate between those of the island hot spots and the 'normal' ocean-floor magma. These gradients are interpreted as due to mixing between undepleted and depleted mantle–magma systems. In some major hot spot areas (Yellowstone, Iceland, Kilauea) mantle-derived helium (^3He) occurs at concentrations fifteen or more times greater than atmospheric—evidence for major mantle degassing. The importance of the geochemical arguments lies in the fact that they may provide more diagnostic criteria than the geophysical data, most of which apply equally well to non-plume magma-generating zones, such as mid-ocean ridges and continental rifts.

Conclusions

During the past decade the most attractive alternative to the hot spot/mantle plume model is that of a propagating fracture, an idea dating back to the 1940s. In terms of plate tectonics such fractures are thought to penetrate the lithosphere and to cause melting by pressure reduction. The fractures might have formed in various ways such as contraction of the cooling lithosphere or lateral stress as it moved.

While each of these mechanisms is theoretically plausible they seem less adequate than the hot spot idea to explain regular linear age progression and the apparent fixed position of some hot spots in the sublithospheric mantle. At present, geochemical arguments have lost some power because both $^{87}Sr/^{86}Sr$ and $^{143}Nd/^{144}Nd$ isotope ratios indicate that hot spot magmas are also derived from depleted—and not primordial—mantle and that their enrichment in LIL elements may be due to geologically recent metasomatism, possibly only a few hundred million years ago. Moreover, geochemical heterogeneities in the north Atlantic suboceanic mantle indicate domains thousands of kilometres in diameter that encompass both volcanic islands and ocean floor.

Geophysical data, on the other hand, provide increasing evidence that anomalous mantle exists directly beneath hot spot areas, although such low-density (partially melted?) mantle is not so far found any deeper than 250 km. It is not clear at present whether the measured geochemical and geophysical hot spot features can be used to infer well-defined hot spots directly or are merely the effects of deeper causes whose exact nature and geometry are still unknown. Recent proposals that hot spot areas are caused by heating of the base of the lithosphere already invoke broader sublithospheric areas than were originally envisaged for mantle plumes.

There is little doubt today that the Earth is convecting, but it is not clear whether convection is restricted to the upper mantle, or whether this can be distinguished from convection in the lower mantle, or whether the entire mantle convects. The rising hotter portions of convection cells may constitute the physical base for hot spots. However, even if melting anomalies are generally related to upwelling mantle, the nature of the presumably wide range of their geometrical configurations is obscured by modifications caused by thick lithospheric plates moving over them. Some systems, such as the Hawaiian–Emperor chain provide excellent evidence for a relatively small melting anomaly fixed in the mantle and stable for some 70 Ma. The majority of intraplate volcanoes, however, show different patterns most of which are related to more linear zones of simultaneous volcanism. An example is the young volcanism of the Basin and Range province in the western USA which is characterized by crustal thinning, a strongly tensional regime and a lateral age progression as on the side of a mid-ocean ridge. Other anomalous areas are those of abundant intraplate volcanism at some passive margins, such as the seamounts and islands off north-west Africa. The hinge zone between the continental and oceanic parts of the African plate formed during continental separation and the increasing sediment loading may influence the triggering of melting anomalies. Finally volcanism in and around continental rifts is less readily interpreted in terms of hot spots except by the propagation of rifts between adjacent hot spots.

The hot spot–plume model owes much of its theoretical attractiveness to the contrast with the linear belts of volcanism that characterize mid-ocean ridges and subduction zones. Both are basic elements of the plate tectonics theory. Since operation of plate tectonic processes prior to about 200 Ma ago is uncertain mid-ocean ridges and subduction zones as understood today may not have existed during early geological history. On the other hand, it is generally agreed that mantle convection was more vigorous during the early history of the Earth, that geothermal gradients were locally much steeper and that much of the early crust was of dominantly submarine volcanic origin, forming stationary plates possibly much smaller than the present plates. If hot spots are understood as localized rather than linear melting anomalies, and their magmatic component as possibly generated by decompression in convectively rising mantle, then hot spots were probably dominating magmatism during the early evolution of the Earth, although the scant record of early rocks prohibits definite conclusions. Such a model of magmatic evolution was called 'hot spot tectonics' by Fyfe in 1978.

Alkali basalts are very rare among Proterozoic and Archaean rocks, most early volcanics being chemically similar to present-day tholeiites and calc-alkaline basic magmas. This might be expected, because decreasing geothermal gradients and increasing plate thickness towards the present would favour increased occurrence of alkaline basaltic magmas resulting from low degrees of partial melting. If hot spots are defined by chemical criteria (eg LIL-element-enriched), then hot spot volcanism would be restricted to the recent geological past. However, as discussed above, it seems more reasonable at present to define hot spots as localized melting anomalies, possibly due to convection, without specifying the chemical composition of the magmas. The study of intraplate volcanism has clearly contributed a surprising amount to the plate tectonics theory and Earth history, and there is undoubtedly a great deal yet to learn.

12.10: *This sequence of stills from a film records the event which resulted in the formation of an individual lava pillow. Lava from an eruption of Mauna Ulu, Hawaii, in 1972 and 1973 flowed into the sea. Here the molten lava flowing under water breaks through the half-solid crust, until the surface in contact with the water cools sufficiently to solidify in a pillow shape. Ultimately the pillow breaks away from the lava mass and falls farther down the slope.*

13 Constructive margin processes

The theory of plate tectonics developed from the proposal in the early 1960s of the process of ocean-floor spreading at the actively volcanic mid-ocean ridge system. The process was convincingly confirmed when the predicted symmetrical distribution of linear magnetic stripes about the mid-ocean ridges was discovered in all the major ocean basins (see Figure 10.5). Magma injected along the axes of the mid-ocean ridges becomes magnetized in the direction of the Earth's prevailing magnetic field as it cools. If the cooled material is split and displaced away from ridge by subsequent magma injection, it forms a pattern, symmetrical about the ridge axis, of stripes of alternately normal and reversed magnetism, as the polarity of the Earth's magnetic field reverses from time to time. In consequence the volcanic portion of the ocean floor becomes progressively older away from the mid-ocean ridge axes.

The recognition of sea floor spreading has stimulated many diverse investigations of oceanic crust and mid-ocean ridges because of their new-found significance as material recently derived from the upper mantle and as constructive plate margins respectively. Most of our present knowledge of ocean floor geology comes from these studies which are still in progress. Only recently have models for the structure of the oceanic crust been integrated with those for the geochemical and geophysical processes occurring at mid-ocean ridges.

General features of the mid-ocean ridge system

Mid-ocean ridges are probably the surface manifestation of advection in mantle convection cells (see Chapter 9). Decompression induces partial melting in the uprising ultrabasic mantle material. The buoyant basaltic magma segregates and rises, accumulating and crystallizing as it cools in magma reservoirs quite close to the surface before being erupted. The depleted ultrabasic residue of the partial melting and the gabbroic crystal cumulates formed in the magma reservoirs constitute important layers in the mantle sections of the lithospheric plates. The basaltic lavas and the dykes that feed them form the upper igneous layers of the oceanic crust. The crust–mantle boundary, determined seismically by a sharp increase in velocity with depth (about 7 to 8 km/s) probably marks the boundary between the gabbroic and ultrabasic cumulates and varies from about 4 to 15 km in depth below the ocean floor. Sediments derived mainly from the hard parts of micro-organisms are deposited throughout the ocean basins and eventually cover the basalt lavas some time after their eruption at ocean ridges.

The fractures that offset the mid-ocean ridges (Figure 10.17) are parallel to the plate motion (generally at right angles to the ridge) and are termed transform faults. The widths of the magnetic stripes are generally proportional to the durations of the normal and reversed polarity periods established in rock sequences of known age. This indicates that the rate of spreading at a particular ridge segment remains approximately constant for substantial periods of time. The calculations of spreading rates indicate that about 4 km^3 of basaltic crust are generated each year at mid-ocean ridges.

Spreading ridges are not necessarily stationary with respect to the rest of the Earth's lithosphere. For example, there is no destructive plate margin between the mid-Atlantic and Carlsberg ridges and as there is no active loss of lithosphere between them, one or both of the ridges must be moving. Movement of the ridge itself may prevent the lithosphere from developing a symmetrical pattern about the ridge.

The ocean crust is elevated at spreading centres but subsides as it cools and becomes more dense. The rate of subsidence, from 1 km in the first ten million years to 20 or 30 m per million years thereafter provides a measure of ocean floor age. The volcanic portion of the crust is thicker at the fast-spreading ridges which generally have broader elevations and more gentle slopes than slow-spreading ridges, many of which may have median rift valleys. The balance between eruption, subsidence and spreading determines the topography of the ridge and the ocean floor. Slow-spreading ridges often show short-term fluctuations in activity leading to a rugged topography.

A minor amount of crust is generated in the fracture zones or transforms. The processes involved are complicated and include crustal uplift, intense tectonization of the rock units and associated breccia zones, minor alkali basalt volcanism of the within-plate type (see Chapter 12) and hydrothermal activity and metallogenesis occurring in material older than in the ridge crest.

Mid-ocean ridge basalts (MORB) have characteristic chemical compositions of a relatively restricted range. They are low-potassium olivine tholeiites rich in calcium and alumina. They are relatively depleted in those trace elements which tend to be concentrated in silicate melts rather than in the silicate minerals found in mantle rocks and crystallizing from basaltic liquids; these are the so-called incompatible or hygromagmatophile (H) elements. The major element composition (weight per cent) of the most magnesian MORB is approximately: SiO_2 = 49.8; TiO_2 = 0.68; Al_2O_3 = 15.0; FeO (total iron) = 8.4; MgO = 10.0; CaO = 13.4; Na_2O = 1.7; K_2O = 0.03. For comparison a MORB that has undergone more extensive crystal fractionation prior to eruption has a composition: SiO_2 = 49.8; TiO_2 = 3.91; Al_2O_3 = 13.0; FeO (total iron) = 14.1; MgO = 5.0; CaO = 9.0; Na_2O = 2.9; K_2O = 0.54.

The majority of MORB lavas have a pillow form, quenched petrographic textures and glass selvages—characteristics of rapid cooling in seawater. Ten per cent of the basalts are coarse-grained and represent intrusive features or slowly cooled lava flows. Below a water depth of 1000 m the basalts contain less than 5 per cent vesicles, because of the water pressure, but those erupted at depths of less than 500 m may have 20 per cent. Ten per cent of the basaltic rocks are glassy breccias with a carbonate matrix and may have been formed by injection of hot magma into wet sediments. When large volumes of lava are erupted onto the sea floor in a single episode, lava ponds may form and result in spectacular morphological features.

Techniques used to study the oceanic crust

A variety of techniques, from direct sampling to geophysical surveys, has been applied to the very difficult problem of studying the oceanic crust, which was far too long ignored by Earth scientists.

Dredging involves dragging a sample collector—usually a chain bag, the dredge—along the ocean floor to pick up loose rock fragments. Basaltic rocks can be recovered only from the mid-ocean ridges or the walls of fracture zones where sediment cover is at a minimum. The recently developed techniques of mapping by side-scan sonar bathymetry and satellite navigation help in the placing of the dredges. However, dredging must be carried out in conjunction with other techniques. The recent development of manned submersibles has enabled geological fieldwork to be carried out under several kilometres of seawater. The accuracy of the sample locations and the detail of the geological structures that can be observed cannot be matched by other methods.

Of the available geophysical techniques magnetic surveys have played an important role in dating different regions of the sea floor and in detecting active and extinct spreading axes. The first marine gravity surveys undertaken in the 1930s showed that high-density rocks must form a major part of the oceanic crust. This was the first indication that the low-density granitic continental crust does not underlie the oceans. Seismic refraction surveys of the 1940s and 1950s established that the oceanic crust was significantly thinner than the continental crust and possessed a well-defined seismic layering with deeper layers having progressively higher compressional wave velocities. These layers were interpreted in terms of a three-layer model with the sediments forming layer 1, basaltic lavas 1 to 2.5 km thick forming layer 2, and a 6-km-thick layer 3 separating the upper crust from the mantle. More recently the refraction data have been interpreted in terms of gradual rather than stepped increases in velocity with depth. Thus the seismic 'boundaries' and the corresponding lithologic 'layers' in the crust are transitional from one to the next.

The abortive Mohole project (1967) was the first attempt to drill through the oceanic crust into the mantle. Although this project failed it helped to stimulate other projects involving the drilling of deep holes into the basaltic portion of the oceanic crust. This was undertaken by the *Glomar Challenger* close to the French-American Mid-Ocean Undersea Study (FAMOUS) area as part of the Deep Sea Drilling Project (DSDP) in 1974. This work was continued from 1976 to 1980 as the International Program of Ocean Drilling (IPOD)—a collaborative project supported by the USA, Britain, France, West Germany, Japan and the Soviet Union. More than five hundred holes have been drilled in the ocean floor by DSDP and IPOD, although only thirty or so have penetrated more than about 30 m into the basaltic basement. Drilling in the uppermost half-kilometre of the north Atlantic oceanic crust has revealed marked episodicity of volcanic activity, ubiquitous low-temperature alteration of the basaltic rocks and evidence of large-scale tectonic disturbance.

Layered series of rocks, comprising deep-sea sediments overlying submarine lavas which give way downwards to a distinctive succession of sheeted basaltic dykes, gabbros and ultrabasic rocks have been discovered in most continents. These sequences are called ophiolite complexes and are generally interpreted as sections of ancient oceanic crust and upper mantle. Limitations to the direct extrapolation from these ophiolite complexes to normal ocean crust include the fact that many of the basalts present are more akin to those erupted at destructive plate margins, suggesting that the complexes originated in back-

13.1: *The top of a hollow pillar of basalt occurring with others in a fossil pool of fluid lava. It was formed around conduits of heated seawater escaping from a lava pool. The lava was subsequently drained, leaving the pillars behind.*

13.2: *The manned submersibles such as* Alvin *with sophisticated navigation, photographic and sampling systems have added a new dimension to the study of sea floor spreading.*

arc seas such as those in the west Pacific Ocean, and that disruption of the original sequence and structure makes it difficult to establish the thickness of each lithologic unit.

Exceptionally, segments of the ocean ridge system are sufficiently elevated to emerge above sea level. Iceland and Afar are the only ones of any size, and this alone makes them anomalous. One of the most striking features of these areas is the episodic nature of the volcanic activity and its patchy distribution along the spreading axis at any one time. The crustal thickness of these areas is generally greater than 10 km (significantly greater than for normal ridge segments) and the basalts have different trace element compositions from normal MORB (being more enriched in the H elements).

The formation of the oceanic crust

The two main processes that control the chemical composition of the igneous portion of the oceanic crust are the magmatic processes at mid-ocean ridges to generate the igneous rocks and the subsequent metamorphism of these igneous rocks when they react with seawater.

Volcanic eruptions are the final stage of a process that begins with the partial melting of ultrabasic rocks in the mantle source region. This volcanic activity at mid-ocean ridges continues for up to several hundred million years and the processes of magma generation and eruption must remain broadly similar throughout this period in order to generate the layered structure of the oceanic crust. Beneath the crust the uppermost layer of the mantle is solid and coupled with the crust to form the lithosphere. However, a layer of the mantle beneath the lithosphere behaves plastically because it is near its melting point and may contain about 1 or 2 per cent of a melt phase. This layer is the seismic low-velocity zone and is called the asthenosphere; its top lies typically about 80 km below the Earth's surface except at mid-ocean ridges where it rises to within a few kilometres of the surface. The oceanic lithosphere rides over the asthenosphere as it moves away from the mid-ocean ridges. Where the asthenosphere rises beneath the ridges (see Figure 4.18) partial melting takes place and the less dense magma generated rises and accumulates at the base of the crust in magma reservoirs or chambers. The denser unmelted mantle residue of the asthenosphere accretes to and thickens the oceanic lithosphere. In these magma reservoirs the composition of the magmas is modified by processes of crystal extraction, crystal accumulation and magma mixing. The magmas are subsequently erupted onto the sea floor or intruded into crustal layer 2 as dykes or sheets. The composition of the erupted basalts is therefore dependent on three main factors—partial melting processes, magma chamber processes and composition of the mantle source.

Partial melting of the mantle

The melting behaviour of peridotite (the main ultrabasic rock of the upper mantle) under different temperature conditions is summarized in Figure 13.3. The magma produced by partial melting is less dense than the solid residue and the two part company when the upward buoyancy of the magma exceeds the frictional forces trying to keep it in place. For very small degrees of partial melting (less than 1 or 2 per cent) it is unlikely that the magma will be able to segregate from the solid residue. However, the shearing forces induced by the motions of convection in the slightly plastic mantle beneath a mid-ocean ridge will cause the magma droplets to coalesce into larger drops. At a critical size the magma drops segregate and rise faster than the mantle residue. The magma accumulates as it rises and eventually feeds a magma reservoir near the base of the crust. MORB are thought to be the result of partial melting of between 10 and 30 per cent.

The concentrations of the incompatible or H elements are particularly sensitive to different conditions of partial melting. The behaviour of trace elements during partial melting is governed by simple equations which are based on the degree of partial melting and a distribution coefficient (ratio of the concentration of an element in the solid phases to the concentration of an element in the equilibrium melt phase), which by definition is less than one for all incompatible elements (see Chapter 8).

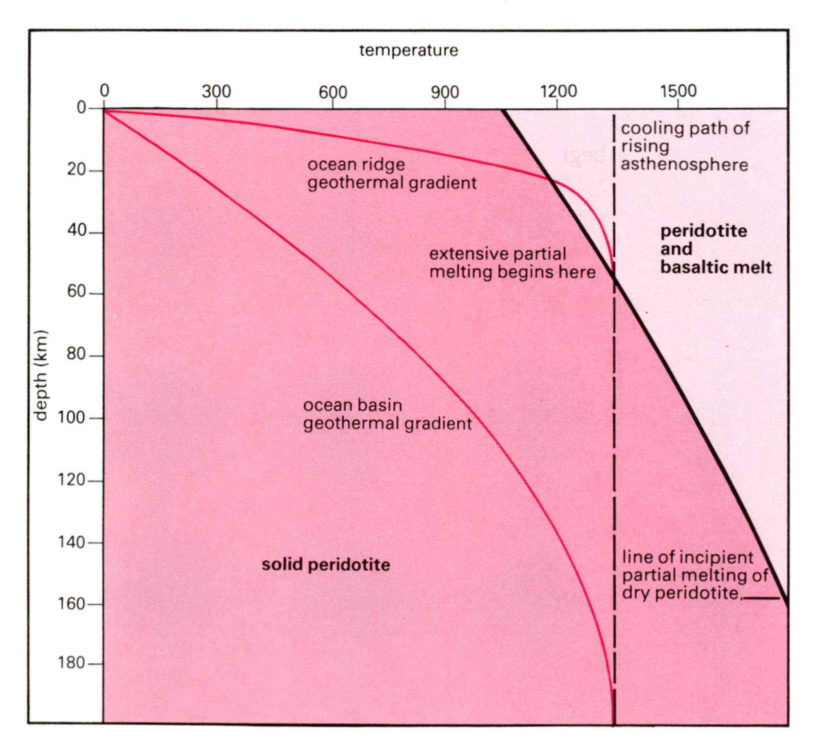

13.3: *Temperature conditions in terms of depth in the oceanic crust and underlying upper mantle. Typical geothermal gradients are shown for the ocean basins and for ocean ridges, where much higher temperatures are found at much shallower depths. Also shown is the melting curve for typical mantle rock (peridotite)—the melting point changes with pressure and hence with depth. As the asthenosphere rises (broken line) it cools only slightly, and at a depth of about 60 km it encounters conditions where it can begin to melt. Assuming a composition similar to peridotite, partial melting of the upper mantle yields a basaltic magma of MORB type if the degree of melting exceeds about 10 per cent (smaller degrees of melting yield alkali basalts or more silica undersaturated magmas).*

Fractionation in magma reservoirs

Crystallization takes place for the most part in the magma reservoir and this prevents a high proportion of the magma from ever ascending to the sea floor. Crystallization commences from the top and the minerals formed generally sink to the floor of the reservoir in a manner similar to the precipitation of salt crystals in a cup of hot salty water as it cools. The minerals with the highest density sink most rapidly and those whose density is less than that of the magma may remain buoyant, as is the case for some plagioclase feldspars. This settling process leads to the formation of mineral layers on the floor of the reservoir and the change in composition of the remaining magma, a process known as crystal fractionation. The residual magma becomes depleted in the elements of which the crystals are composed, and enriched in the elements not contained in the crystal extract, for example the H elements. The minerals crystallizing from magmas as they rise grow slowly and may become quite large (several tens of millimetres) and some are usually brought to the surface in lavas that are erupted. At the Earth's surface, especially on the sea floor, the lavas cool rapidly and most of the crystals that form are small and so the large crystals, or phenocrysts, from the magma chambers can be easily distinguished. The compositions of these phenocrysts therefore help to establish the compositional changes that occurred in the lavas with which they are associated prior to their eruption.

Intermittent convection in the magma reservoir gives rise to changes in the composition of the crystal extract and this compositional layering is more complex and on a larger scale than the mineral layering (Figure 13.4).

Melting experiments on MORB under pressure conditions equivalent to those at about 2 or 3 km below the Earth's surface indicate that heavy olivine and chrome-rich spinel are generally the first minerals to crystallize, forming the basal dunite (a rock with more than 90 per cent olivine) and chromite layers of the crystals precipitated, or cumulates, in the magma reservoirs. This crystal extract has a much higher magnesium, chromium and nickel content than the magma which therefore becomes relatively enriched in silicon, aluminium and calcium. When significant proportions of plagioclase are involved in the crystal extract there is an enrichment of iron, a slight depletion of calcium and aluminium and an approximately constant proportion of silicon in the residual liquid, and the cumulates reach the composition of gabbro. As the magma cools and crystallization proceeds, calcium-rich pyroxene begins to crystallize and its proportion in the crystal extract gradually increases while that of olivine declines. This causes a decrease in the ratio of calcium to aluminium in the residual liquid which cannot be brought about by crystallization of the other minerals.

13.4: *Mineral layering in a granodiorite intrusion in the Sierra Nevada Mountains in California. The dark crystals are hornblende and sphene; the lighter ones are feldspar and quartz.*

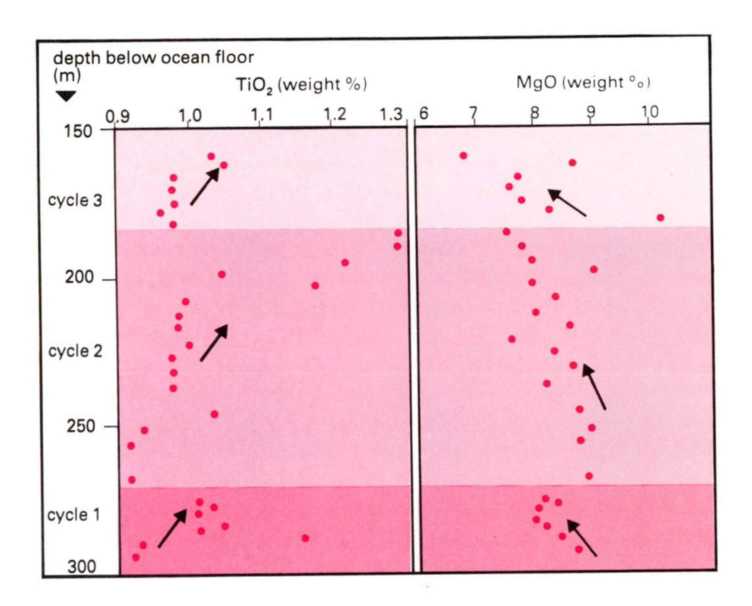

13.5: *IPOD site 412, near the FAMOUS area of the mid-Atlantic ridge between latitudes 36°N and 37°N, penetrated 131 m of basalt comprising twenty-seven separate lava flows. The basalts contain plagioclase, olivine and pyroxene phenocrysts and vary from coarse-grained to microcrystalline. The H element compositions of basalts from this site are consistent with the derivation of all the lavas from a single parent magma. Concentrations of* TiO_2 *and* MgO *have been plotted against depth in the drill hole. The magnesium composition of the basalts varies from about 9 per cent* MgO *to about 7.5 per cent* MgO. *Crystal fractionation must have been episodic in order to explain the cyclic variation in titanium and magnesium. Three cycles are present probably representing the periodic replenishment of the magma reservoir with more magnesium-rich magma.*

Studies of the chemical changes in a vertical sequence of lavas sometimes reveal distinct trends or cycles that monitor the evolution of the magma composition in the reservoir at the time of their eruption (Figure 13.5). There is substantial petrological evidence to suggest that fractionated magma depleted in magnesium is periodically remixed with more 'primitive', magnesium-rich magma which is repeatedly injected into the reservoir. Mixing of magmas in this way helps to explain some chemical variations observed in series of MORB erupted at any one locality.

Magma-mixing tends to maintain a moderately fractionated (magnesium-poor) magma in the reservoir, but results in the magma having higher concentrations of the H elements than could be produced from a single parent magma by simple crystal fractionation processes. Mixing of two magmas, which have been derived from a single mantle melt but have suffered different degrees of crystal fractionation, would produce a magma with the same H element ratios as the original magmas. Processes other than fractional crystallization and mixing of magmas derived from a single mantle melting episode must be involved in the production of MORB with different H element ratios.

There is evidence that fractional crystallization of some MORB may also occur at higher pressures in deeper magma reservoirs. This may be an important process contributing to the overall chemical variations observed in MORB.

Basalts erupted at fast-spreading ridges are generally more fractionated (average MgO about 4 to 6 weight per cent) than those erupted at slow-spreading ridges (average MgO about 7 to 9 weight per cent). The frequent extraction of magma from a large reservoir (Figure 13.6) will minimize the compositional effects of injection of primitive magma and magma-mixing and lead to a more evolved steady state magma composition.

At several drill sites in the Atlantic Ocean two or more primitive magma types with quite different compositions (including H element ratios) occur interlayered (Figure 13.7). Two magma reservoirs must have been involved in these emplacements and they must have been small and not continuous (Figure 13.8).

Geochemical studies on basalt samples from holes drilled at different latitudes in the Atlantic Ocean have shown that consistent variations in some trace element abundances and isotope ratios exist in crust formed at different ridge segments. It seems that such variations reflect differences in the composition of the upper mantle source regions from which the basalts are derived.

Metamorphic processes in the ocean crust

Seawater circulation through oceanic crust at high and low temperatures produces important chemical and mineralogical changes within the rocks. Thermally driven convection of seawater at spreading ridges has been suggested to explain the anomalously low heat flow in young oceanic crust and as providing hydrothermal fluids for exchanging various elements between the crust and the oceanic waters. The hydrothermal activity in Iceland provides direct evidence of the deep circulation of water within the crust as do jets of hot water emerging from vents on the Galápagos spreading centre.

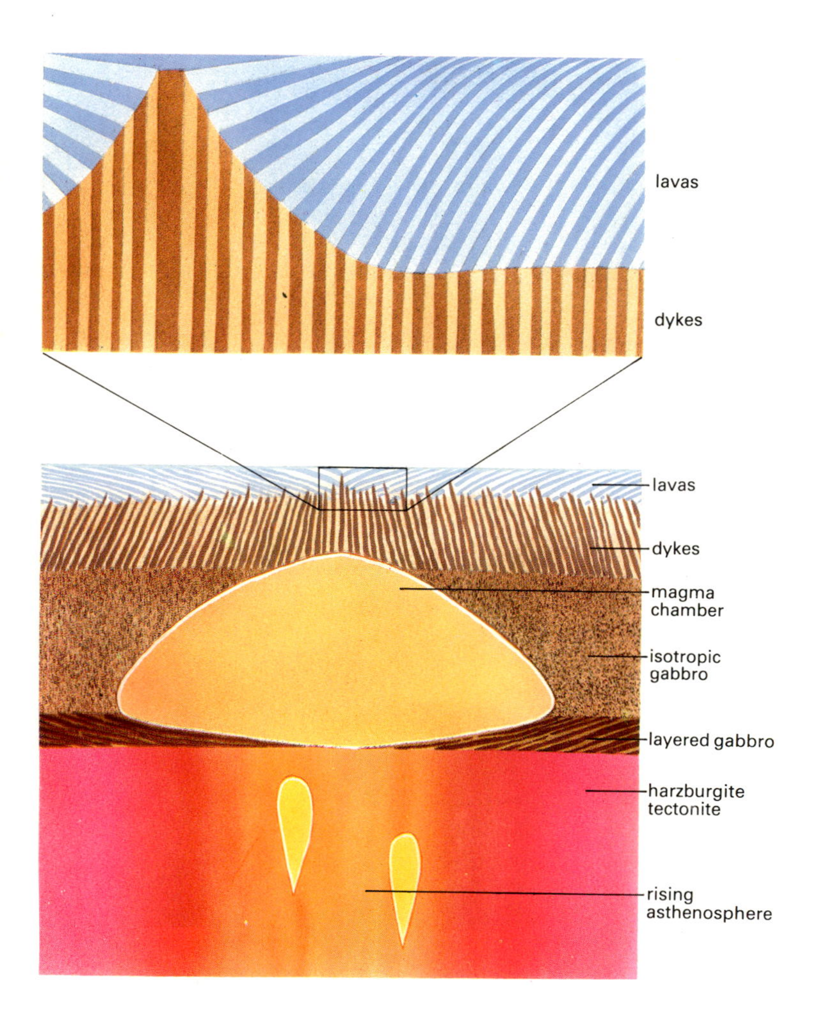

13.6: *A schematic diagram of an idealized model for crustal development during perfect sea floor spreading, to which the oceanic crust produced at fast spreading ridges probably approximates. In this model each successive dyke is intruded up the centre of the preceding one; the central fissure is stationary. The overlying lava pile, above each dyke, thins away from the ridge crest, and so has a triangular cross-section. As the lavas accumulate, and subsidence takes place, there is an overall inward inclination of the lava layers towards the ridge axis.*

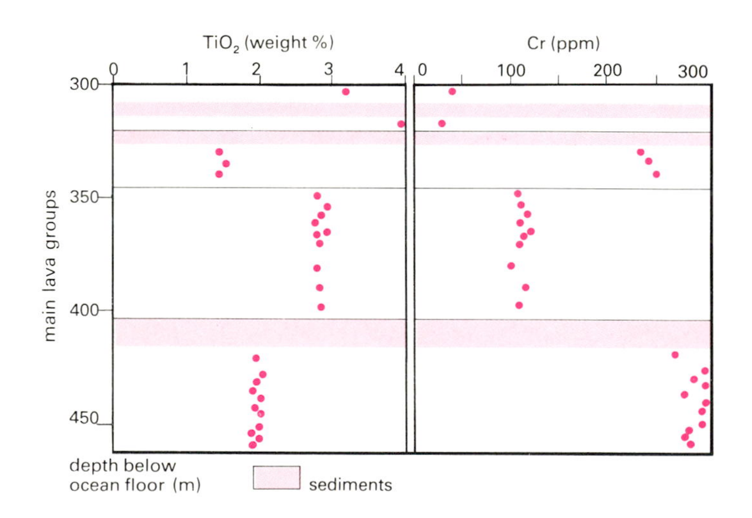

13.7: *Basement stratigraphy in IPOD drill hole 407 on the western flank of the Reykjanes ridge, south of Iceland, with concentrations of titanium and chromium plotted against depth. Sediments (fine-grained chalks and limestones) are commonly interlayered with the extrusive basalt, particularly in the upper 200 m or so. The different compositions of the four groups of lavas cannot all be produced by fractional crystallization and mixing of magmas derived from a single parent magma.*

Hydrothermal convection is so efficient at removing heat that it is likely to be episodic, with periods of active convection alternating with periods of repose while the thermal gradient necessary to restart convection is restored. As the temperature gradient decreases away from the ridge crest the circulation of water decreases. Hydrothermal circulation of seawater through the oceanic crust is an important process for only about five million years after crust formation. Most of the metamorphic alteration of MORB takes place in the first one or two million years.

Evidence from ophiolite complexes suggests that hydrothermal metamorphism has a significant effect only on the layers of the crust above the gabbros. This indicates that strong seawater penetration and circulation occurs in a permeable layer about 2 or 3 km thick. This is accounted for by fissures parallel to the ridge crest, the spacing between lava pillows and intensive fracturing in the basalts. Later, infilling of these spaces with secondary minerals reduces the permeability of older, deeper crust. The persistence of easily altered olivine and basaltic glass in even the oldest basalts indicates that the crust must become sealed from circulating seawater at a relatively early stage.

Almost all metamorphic rocks from the ocean floor preserve their original igneous textures and have had only their mineralogy and chemistry modified (Figure 13.9). According to the maximum conditions of temperature and pressure to which the rocks have been subjected they can be referred to a succession of metamorphic facies. The brownstone facies is characterized by low-temperature (~ 50°C), oxidizing conditions prevailing throughout the oceanic crust. The zeolite facies begins at temperatures above about 50°C and is characterized by the development of crystals of the zeolite group of minerals. At the upper limits of this facies zeolites are no longer stable and are replaced by albite (replacing more calcium-rich plagioclase feldspars) and chlorite (replacing olivine and glass) at about 250°C to 300°C. This is the lower limit of the greenschist facies of metamorphism, named for the greenish colour of the chlorite. Rocks of this metamorphic grade are sometimes known as spilites. Above about 500°C the amphibolite facies succeeds the greenschist facies and is generally associated only with intrusive rocks because these conditions are attained only in the deeper regions of the crust. It is this grade of metamorphism to which the oceanic crust is subjected in subduction zones (see Chapter 14).

In the ultrabasic rocks the mineral serpentine forms by the action of water on olivine. This occurs below 400°C and involves considerable increase in volume and decrease in density. Hence low-densitv serpentinites intrude upwards through oceanic crust as diapirs, often bringing with them fragments of the deeper layers.

The lower oceanic crust formed at a spreading centre has yet to be sampled *in situ* by drilling, thus direct evidence of its structure and composition is lacking. Fresh gabbros which have been pushed up to high levels in the crust by diapirism have seismic characteristics close to the average for the lower crust. Therefore crustal layer 3 is generally thought to consist of gabbro (see Figure 13.6), although the velocity changes from layer 2 values to layer 3 values could equally correspond to a metamorphic change separating greenschist facies above and amphibolite facies below. A more detailed knowledge of the lower crustal material must therefore await direct sampling by deeper drilling.

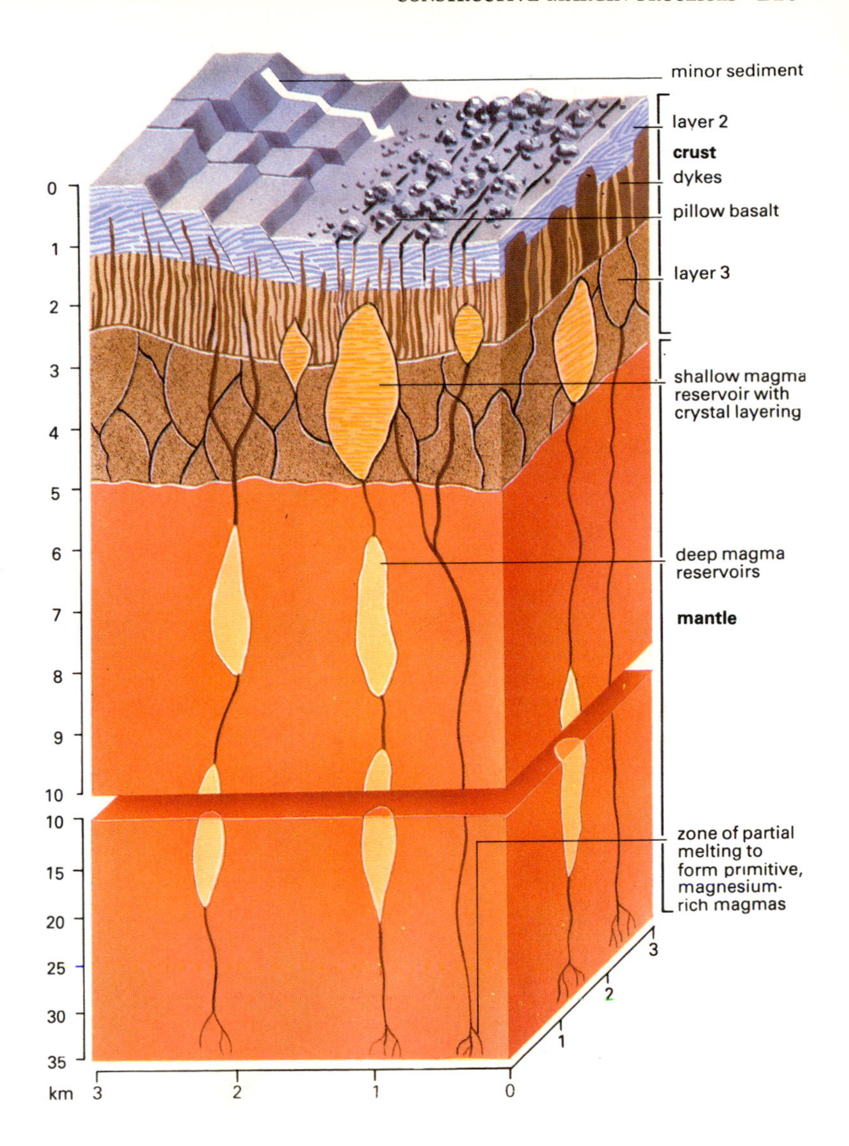

13.8: *This schematic diagram explains the petrological and geochemical variations in basalts erupted in the rift zone of the mid-Atlantic ridge. At slow-spreading ridges magmas are generated by partial melting of sources in more than one region of the upper mantle and rise separately. Some rise directly to high levels, others are held temporarily in deep magma reservoirs. The latter undergo varying degrees of high pressure fractional crystallization. Magma reservoirs are small and periodic lateral shifts occur in volcanic activity at the surface depending on the locations of the reservoirs available to be tapped at a given time. Periodic influxes of primitive magma from depth result in varying degrees of magma mixing. It is during the time of the shifts of magmatic activity that the most primitive and also the most evolved basalts have the best chance of reaching the surface. At other times only moderately evolved basalts are erupted.*

Chemical balances between the oceans and the oceanic crust

Hydrothermal activity associated with the formation of new oceanic crust exerts an important influence on the seawater chemistry. Much magnesium is removed from seawater at the ridges, going to form various clay minerals, as is sulphur which is attracted to certain metallic elements to form massive sulphide deposits. On the other hand the ridges form a primary source for the alkali metals rubidium and lithium. They also support a flux of calcium into seawater equivalent to that derived from the non-carbonate weathering of continents, and amounts of potassium, silicon and barium equal to about one half of the river load. The isotopic composition of oceanic sulphate may also be dominated by this process.

The concentrations of strontium, barium and uranium can be significantly increased in basalts suffering low-temperature metamorphism, whereas these elements are removed from basalts at higher temperatures. The strontium isotope composition of MORB changes significantly during metamorphism towards the composition of seawater. The glass phase in basalts contains most of the H elements present in them and alteration of this phase at low temperatures (sometimes called palagonitization) can lead to significant changes in the concentration of some of these elements which are immobile in more crystalline rocks. Unfortunately with the limited data currently available the relative proportions of the different metamorphic facies present in average oceanic crust are not accurately known.

a

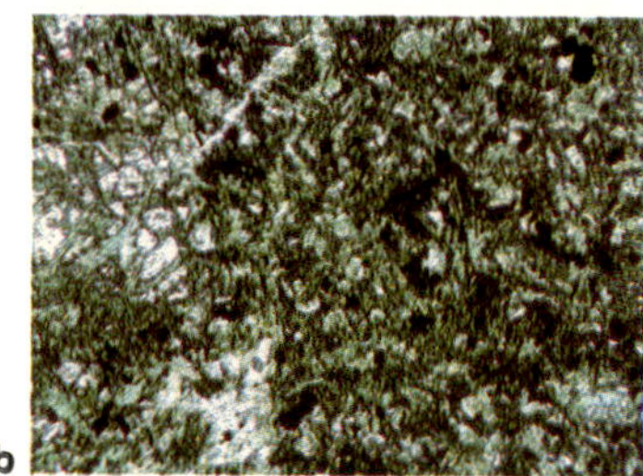

b

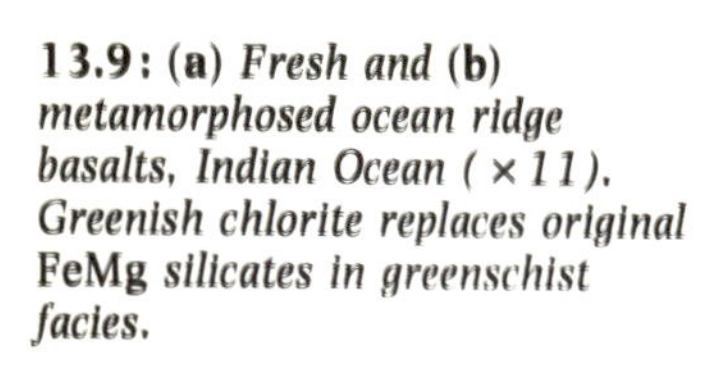

13.9: (**a**) *Fresh and* (**b**) *metamorphosed ocean ridge basalts, Indian Ocean (×11). Greenish chlorite replaces original* FeMg *silicates in greenschist facies.*

13.10: *Massive sulphide deposits recently discovered in steep-sided depressions (20–30 m deep and wide) less than 1 km from the axis of the east Pacific rise.*

Mineralization at ocean ridges

Three characteristic types of metal-rich deposits have been identified in contemporary ridge crest environments. The first is rich in iron and manganese, the second is rich in manganese and the third is rich in iron sulphide and manganese-depleted. The first type is the most common and often forms the basal deposit of oceanic sediments. Indeed it is one of the components (along with ultrabasic rocks, pillow lavas and chert) used to define ophiolite assemblages and to identify ancient oceanic crust. The first and second type of deposit are now becoming economically important sources of manganese. For instance projects are now being considered to recover manganese nodules in commercial quantities from the sea floor surrounding the east Pacific rise (see Chapter 25).

Iron-sulphide-rich, manganese-depleted ore deposits have been found at sites presumed to be the original discharge positions of circulating hydrothermal fluids and where iron and manganese-rich sediments known as umbers overlie the igneous rocks.

In the Troodos ophiolite in Cyprus the distance between sulphide deposits is about 5 km and this is consistent with the idea that they were deposited by hydrothermal convection in the crust. The manganiferous sediments, on the other hand, were deposited on the sea floor after discharge of the thermal solutions.

Features on the east Pacific rise (Figure 13.10) and the Galápagos ridge (see Figure 8.9) suggest that the various metal-rich deposits are different manifestations of a single phenomenon. Heated seawater in the basaltic layer convects and brings up high-temperature (~350°C), acidic, reducing solutions that mix with cooler water percolating downwards. Massive sulphides are deposited first, depleting the solution in copper, nickel, cadmium, zinc, mercury, arsenic, selenium, chromium and uranium. Manganese crusts are formed later from cool oxidizing solutions containing only a few per cent of the original hydrothermal fluid. Sediments rich in both iron and manganese represent an intermediate stage.

Experimental evidence indicates that massive sulphides are formed by the rapid introduction of hot acidic fluids into the sea floor. During slower flow the fluids are oxidized by seawater and the sulphides contain silica (as opal) and native sulphur. Manganese dioxide is deposited by slow, oxidizing solutions. This tends to occur where the surface rock is extensively broken.

Sediments rich in both iron and manganese are particularly associated with fast-spreading ridges. The more intense tectonic and thermal activity in these areas serves to increase the permeability of the crust and hence to encourage more extensive circulation of seawater and dilution of the hydrothermal fluids. The lower temperatures in such systems would not favour the formation of massive sulphides and the higher flow rates would not favour pure MnO_2 deposits. Instead metalliferous sediments rich in both iron and manganese develop.

The structural development of the basaltic layer

Drilling in the north Atlantic Ocean has provided unexpected evidence for widespread tilting and rotation of blocks of oceanic crust. The structural disruption occurs in the median valley during crustal formation rather than during migration to the flanks of the ridge. Magnetic inclinations in drilled basalt sections often deviate systematically from the expected values through considerable depths and no section greater than 250 m is uniformly magnetized. Lateral lithologic and magnetic variability is evident where adjacent holes have been drilled a few hundred metres apart. Also there is surprisingly poor agreement between the sense of effective magnetization in the holes drilled and the associated linear anomalies mapped from the ocean surface. These facts indicate that the upper 600-m section of basaltic crust is not responsible for the linear anomaly pattern. The conclusion is that the formation of crust along the mid-Atlantic ridge is episodic in nature and that there is large-scale rotation of crustal blocks at the ridge crest. The exact location of the source of the linear magnetic anomalies is uncertain at the present time, but the sheeted dyke layer of oceanic crust is a possible source.

In several IPOD drill holes the tilt of the basaltic lavas increases with depth. This suggests that the crust in the inner median valley of the ridge cannot withstand the load of a new pile of volcanic material and it subsides along a series of faults. Other than at transform faults, earthquakes are concentrated at ridge crests in the oceans, implying that most of the faulting and block rotations occur there. Palaeomagnetic studies of drilled basalt sections suggest that not only is the activity at ridge crests episodic but the time for each period of crust formation is short compared with the intervals between activity, when most of the tilting takes place. The occurrence of sediments interlayered with the basalt lavas in the upper part of crustal layer 2 bears this out. Figure 13.11 shows a schematic model for tectonic rotation of crust in the inner rift valley of the mid-Atlantic ridge controlled by movement along listric (concave) normal faults, also known as growth faults. The combination of increased crustal thickness and strength away from the inner median valley and the greater loading of previously formed crust close to the zone of more recent activity causes the trailing edge of the previously formed crustal block to suffer the greatest subsidence. A crustal structure similar to that proposed by this tectonic model occurs in Iceland (Figure 13.12).

Continuing programmes of crustal drilling, dredging, geophysical surveys and direct observations from submersibles, coupled with studies of ophiolite complexes and emergent segments of constructive plate margins will undoubtedly provide us with more details of the processes occurring in this environment which is of such critical importance in the transfer of materials from the mantle to the crust, hydrosphere and atmosphere.

13.11: *Episodic eruption, subsidence and lateral displacement of volcanoes in the median valley of the mid-Atlantic ridge.*

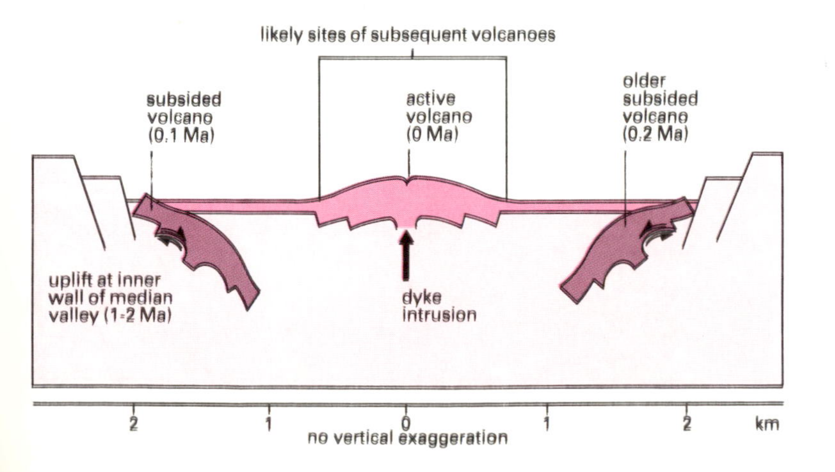

13.12: *A block diagram integrating and generalizing data for the upper crustal structure in Iceland with that for lower crustal and upper mantle structure and composition from ophiolite complexes. The model explains the spatial petrological and geochemical variations in the lavas in terms of episodic magmatic activity and is similar to the models presented in Figures 13.8 and 13.11 for the normal segments of the mid-Atlantic ridge.*

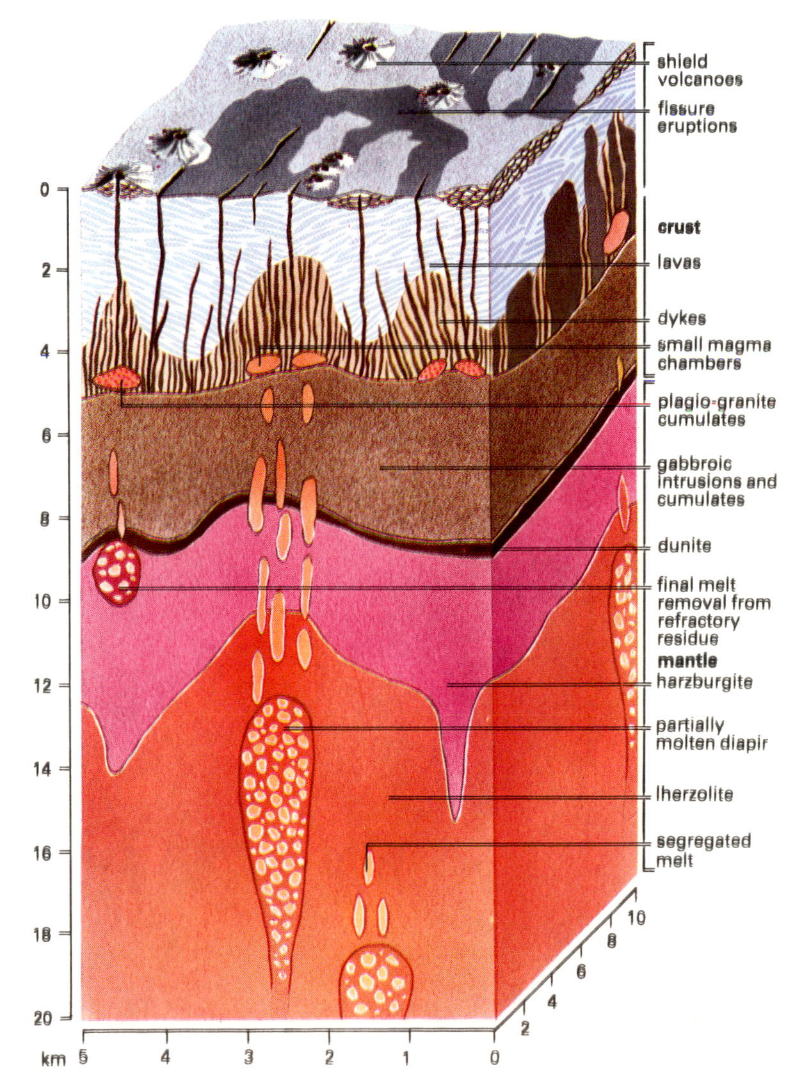

14 Destructive margin processes

Geological and geophysical characteristics

The Earth's crust has formed from the mantle as a result of magmatic activity over at least 3800 million years (Ma). At present such activity occurs dominantly at lithospheric plate boundaries and to a minor degree within plates. Volcanism at constructive margins (oceanic ridges) is dominantly submarine and basaltic, with the products forming the oceanic crust. Destructive plate boundaries form island arcs and active continental margins, where the volcanic products range from basalt through andesite and dacite to rhyolite. This chapter describes the processes active at destructive plate margins and those involved in collisions between island arcs and continents and between continents, ie the processes forming orogenic belts.

The advent of the theory of plate tectonics (see Chapter 10) during the 1960s led to a reappraisal of processes in orogenic zones. Before the plate tectonics era, orogenic activity was considered to conform to a broad pattern or cycle, the earliest stage of which was the eruption of basaltic lavas (spilites) within a flysch-type sedimentary sequence deposited in a linear belt or geosyncline along a continental margin. Then, during the early stages of deformation of the sedimentary-volcanic succession, ultrabasic and basic intrusions were emplaced. The third stage, during and after the main episode of deformation, was the intrusion of batholiths of diorite–granodiorite–granite composition. Finally, following elevation of the folded sedimentary–volcanic pile, volcanic rocks of the orogenic andesite association were erupted.

Before the concept of plate tectonics the major problem was to explain this orogenic cycle. It was assumed that the geosynclinal belts evolved in depressions in the continental crust, possibly located over sinking mantle convection currents. It then followed that deformation could occur when the continental crust below the depression melted and became less rigid, allowing the sedimentary-volcanic pile to be compressed and deformed. When sinking ceased, uplift of the deformed orogenic belt occurred and melting in the mantle or at the base of the thickened pile led to the formation of magmas that were intruded or erupted at the surface. Plate tectonic processes provide an alternative to this 'ensialic' model of the orogenic cycle.

Island arcs, active continental margins and collision zones have been abundant throughout much of geological time. At present most of these features are located around the Pacific (see Figure 10.17). The western Pacific has major arcs extending from the New Zealand–Tonga system, through New Britain–Papua New Guinea and the Mariana–Volcano Islands, to the Japanese–Kuril–Kamchatka system. The Mariana–Volcano system and the Japanese arc are separated from the continent by 'marginal' basins formed, similarly to larger ocean basins, by sea-floor spreading. Thus the Sea of Japan formed as a result of subduction below the Japanese landmass, which caused melting in the mantle and splitting away of the Japanese islands from the Asian mainland. Such splitting explains the origin of marginal or 'back-arc' basins as a result of orogenic processes. Such back-arc basins occur behind the Kamchatka–Kuril arcs (Sea of Okhotsk), the Japanese arc (Sea of Japan), the Philippines (the Philippine Sea) and Taiwan (East China Sea). As these marginal seas have increased in area the arcuate forms of the island arcs bounding them have become more pronounced. Formation of a back-arc basin might split the island arc off an inactive arc on the continent side of the basin and the active arc. Such 'bending' of the eastern Pacific island arcs is a result of the process of back-arc spreading. With the exception of Japan, these arcs appear to lack ancient continental basement.

In contrast to the western Pacific the eastern Pacific has no island arcs but has volcanism on an active continental margin extending from the western USA, through Mexico and Central America, to the western margin of South America. A similar situation exists along the Indonesian margins of south-east Asia, but here the continent is largely submarine. In the Atlantic, the Lesser Antilles and the South Sandwich Islands are young arcs associated with marginal basins. The characteristics of the basins behind the Mediterranean Aeolian and Aegean arcs are, however, more ambiguous. The most extensive active continental collision zone extends from the Alps through Turkey and Iran to the Himalayas.

Some of the major geological and geophysical characteristics of volcanic arcs and active continental margins were described in Chapter 10. The major features are summarized below.

1 Arcuate distribution of islands, or a linear belt of volcanism, with a length of the order of several thousands of kilometres and a relatively narrow width. In many arcs, particularly where they are split by back-arc spreading, the area on the oceanic side of the volcanic belt consists of an outer arc region and an accretionary wedge or subduction complex extending to the trench.

2 A trench on the oceanic side, often between 6000 and 11 000 m deep, and (behind island arcs) a shallow tray-shaped marginal basin sea on the continental side, generally less than 3000 m deep.

14.1: *Landsat image of Kunashir Island in the Kuril island arc in 1973 showing a volcanic plume drifting to the east of the erupting Mt Tiatia at an altitude of up to 4.6 km. The arrow indicates North.*

3 Active volcanism, with an abrupt oceanward boundary of the volcanic zone parallel to, and about 200 km from, the oceanic trench. This is known as the volcanic front. The concentration of volcanoes is greatest here and decreases with increasing distance from the trench.
4 Active seismicity, including shallow, intermediate and deep earthquakes, extending as a well-defined plane from below the trench towards the marginal basin or continental side. Compared with the mantle below the island arc, that below the marginal basin has low seismic velocities and high attenuation of seismic waves.
5 A marked gravity anomaly belt: a negative anomaly of up to 100 milligals is associated with the trench; positive anomalies occur on the arc or continental margin.
6 A marked heat-flow anomaly belt: heat flow is relatively low in the trench area (generally <40 mW/m^2) and higher at the island arc or continental margin (>40 mW/m^2).
7 In some island arcs, notably Japan, there is a distinct zonal arrangement in the composition of volcanic rocks. Volcanoes near the volcanic front erupt tholeiitic basalts, while volcanoes farther away erupt more alkaline basalts. In many, but not all, island arcs and active continental margins the potassium content at a given silica percentage increases towards the back of the arc.

The characteristics of volcanic arcs and active continental margins summarized above may be explained in terms of plate tectonics. Cold, dense lithosphere sinks into the mantle from the oceanic trench along a seismic plane termed the Benioff, or subduction, zone. This explains the negative gravity anomalies and low heat flow associated with oceanic trenches. At a particular depth the descending slab melts or dehydrates, releasing volatiles into the overlying mantle wedge thus causing melting, rise and intrusion of magma and surface volcanism. More extensive island arc orogenesis may lead to more extensive melting and the formation of a marginal basin. The formation of island arcs as outlined above has been termed island-arc-type orogeny.

Island arc orogeny

The sequence of events in the formation of a single volcanic arc may be as follows. The development of a trench may be associated with initiation of plate descent, and simultaneously there may be complex thrusting of wedge-shaped slices of oceanic crust and mantle. During early thrusting, oceanic sediments (including cherts, argillites and carbonates) may slide into the trench by gravity. Such slides may carry blocks of basic and ultrabasic rocks derived from disrupted, faulted and thrust blocks.

The slide deposit may be carried into a deformed sedimentary accretionary wedge beneath and behind the trench. This wedge, a complex mixture, or melange, of varied sedimentary and igneous rocks scraped from the descending plate, may experience strong deformation and metamorphism within the blueschist facies (low temperature–high pressure) in the low heat flow regimes of the trench. Such subduction complexes occur in many western Pacific volcanic arcs.

The evolution of the accretionary prism depends upon the way in which sedimentary material is scraped off the oceanic plate descending below the island arc. Where a thick sediment cover overlies the oceanic plate entering the subduction zone (eg as a result of proximity to a continental landmass) pockets of sediment are sequentially scraped off

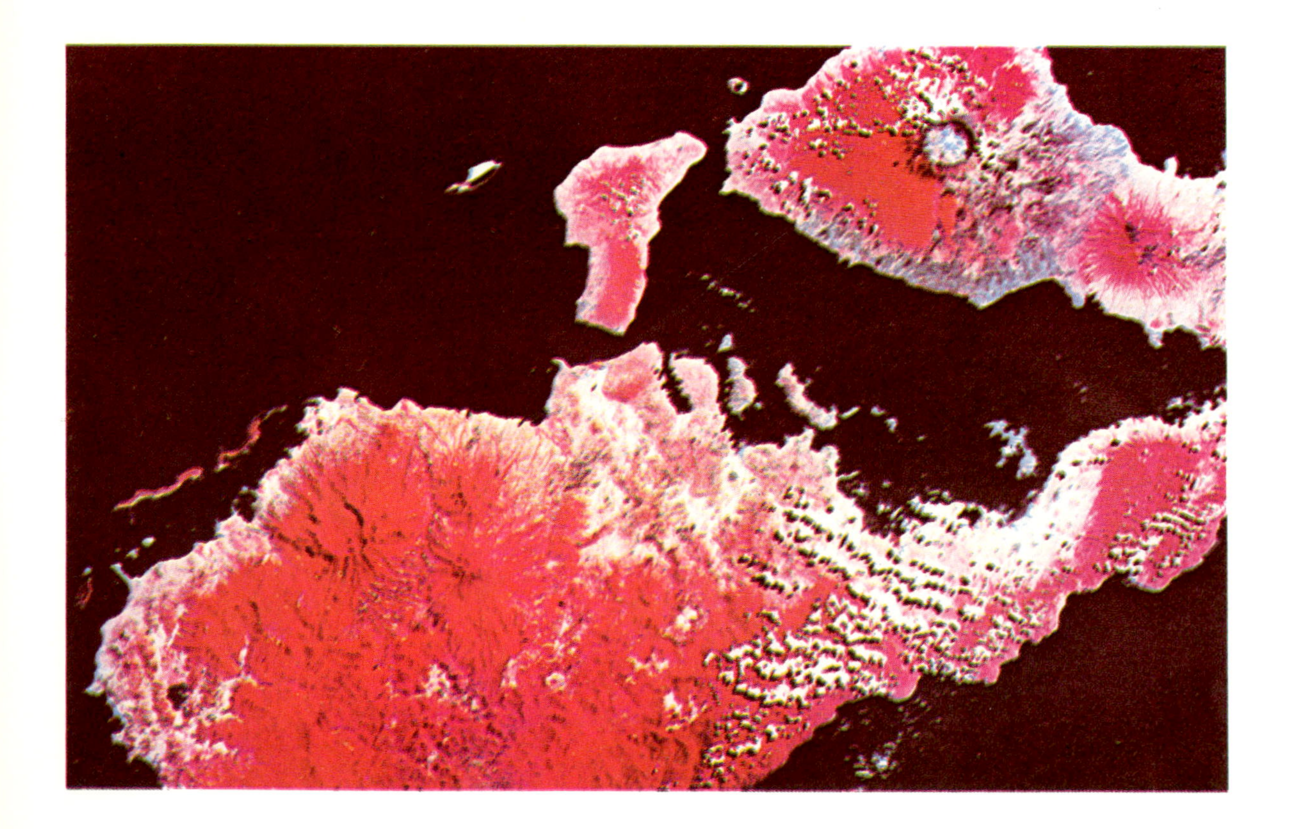

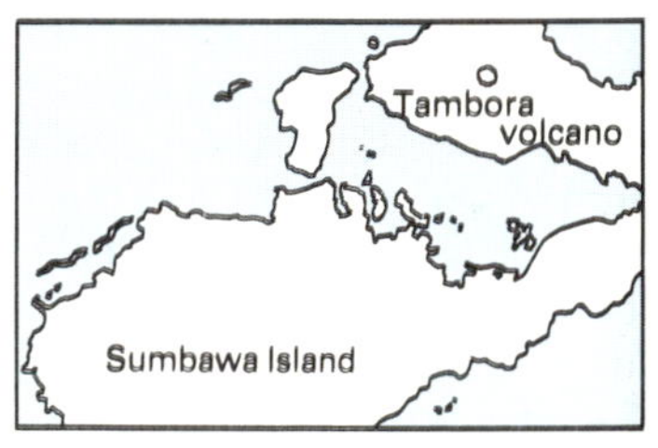

14.2: *Landsat image of Sumbawa Island in the Indonesian island arc. The photograph shows Tambora volcano which gave the largest historic eruption in Indonesia in 1815. In the eruption some 30 km³ of the cone was blown away, depositing a thick ash layer which was about 60 cm in thickness at a distance of 70 km from the volcano.*

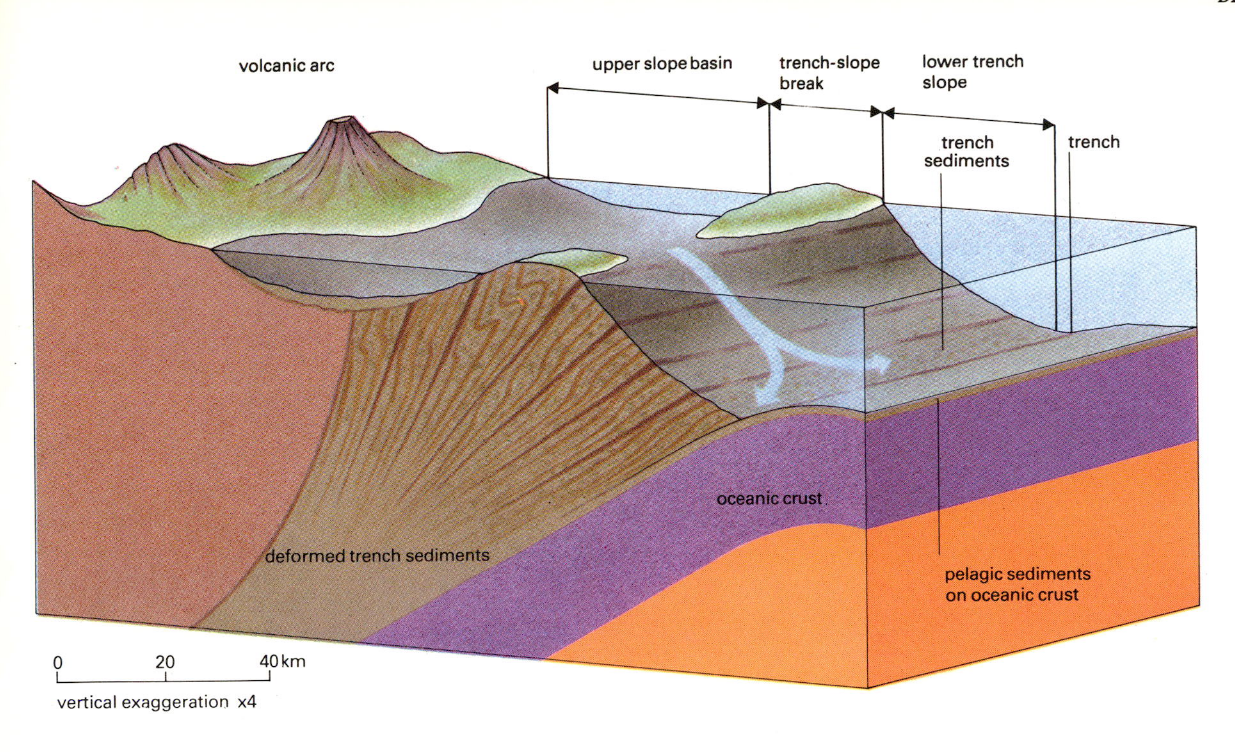

14.3: *Block diagram of a typical subduction zone of island arc type, to show the various subdivisions of an arc system and the structural complexity of the subducted sediments which include both pelagic sediment of oceanic origin and sediment (flysch) derived from the arc itself.*

the descending plate, since they are much less dense than the oceanic lithosphere. When the process occurs for a prolonged period of time an accretionary prism is built up in the outer arc region. The accretionary prism consists of a sequence of layers of sedimentary rocks emplaced above the descending oceanic plate. Each layer may show a sequence from ophiolitic ocean floor rocks, through cherts and/or argillites up into flysch sediments. Hence the beds within each layer get younger towards the island arc but the sequence of layers as a whole youngs towards the ocean. In active island arcs the breaks between the sediment layers are manifested as regularly spaced, large fault planes. These are initially low-angle fault planes but rotate to a high angle as the accretionary prism builds up. Prolonged accretion then gives rise to a ridge feature: the trench-slope break at the top of the tectonically rising lower trench slope. This trench-slope break forms the outer arc terrain, separating a relatively undeformed upper slope basin between the volcanic arc and the break, and the accretionary prism between the break and the trench. Generally the trench-slope break is below sea level in active arcs, but in some cases (eg the Mentawai Islands, west of Sumatra, and in Alaska) parts are emergent and form islands composed of layers of trench sediments separated by the characteristic faulting.

When the descending plate capped with hydrated oceanic crust and/or some oceanic sediment reaches a depth of about 100 km, dehydration releases hydrous fluids into the overlying mantle wedge. This initiates melting, and the parent magmas for the igneous rocks of the volcanic arc are formed. At or following this stage the arc may become split by back-arc spreading as described earlier. In the volcanic arc the rocks are of basic and intermediate composition; gabbro and diorite intrusions are emplaced into the crust, and a volcanic pile of basalt and basaltic andesite begins to accumulate at the surface. Where continued magmatic activity heats and thins the lithosphere, high temperature–high pressure metamorphism of the crust occurs. The crust thickens by intrusion and growth of the volcanic pile, together with continued deformation and accretion. Migration of the igneous axis may occur, yielding inner and outer arcs. During crustal growth an oceanward thickening wedge of flysch sediment, derived from the volcanic belt, develops between the active volcanic front and the trench. At this stage a volcanic arc with paired metamorphic zonation (low temperature–high pressure near the trench, high temperature–low pressure near the active axis) may be developed.

Clearly, therefore, oceanic trenches and their associated sedimentary wedges are likely to be extremely complicated terrains, with complex geometric and time relationships between varied components of the oceanic crust and upper mantle, melange, intrusive and volcanic rocks and derived flysch sediments.

Cordilleran belts

In contrast to the island-arc-type orogeny described above the orogenic belts of the eastern Pacific occur not as island arcs but along continental margins. These are cordilleran-type orogenic belts. Such a situation may originate where an inactive Atlantic-type continental margin becomes active by development of a trench. During and after the formation of a destructive margin near the continental margin, the sequence of events within and behind the newly formed trench may be similar to that described above for island arcs. However, the presence of continental crust provides additional complexities. For example, the sedimentary rocks formed before initiation of the trench may include continental shelf carbonates and continental rise argillites and clastic rocks. The sequence of events may include formation of blueschist melanges and flysch accumulations.

When the subducted oceanic lithosphere reaches depths of about 100 km, dehydration and possibly melting of the descending oceanic crust occurs. The resulting magmas rise, intrusions may thus be emplaced and submarine volcanic rocks may be erupted near the volcanic front. As in island arcs these igneous rocks are relatively basic in composition and include diorites with basalts and basaltic andesite. The igneous axis may migrate towards the continent and then encroach upon older crust. This leads to underplating and intrusion, and therefore to thickening of the continental crust. Uplift will cause subsequent volcanism to be subaerial. At this stage the intrusive rocks belong to the diorite–granodiorite–granite association, and erupted rocks are andesite and dacite lavas and pyroclastic rocks (air-fall and pyroclastic flows, including ignimbrites) containing dacite and rhyolite. Formation of these rocks may involve melting of old continental basement. The intrusive and extrusive rocks may show a compositional polarity analogous to that described for island arcs. The subaerial volcanic belt forms an axis from which sediments are eroded and then transported both into the trench and towards the continent. Subsequent evolution may be dominated by thrusting towards the continental side of the magmatic axis, involving the flysch and sometimes the continental basement.

Active orogenic belts have a range of characteristics between cordilleran- and island arc-type belts. Complex orogenic belts develop from collision of an island arc with either an Atlantic-type continental margin or a cordilleran-type orogenic belt. A further complexity arises from the direction of subduction below the arc; subduction might be oblique to the trench of the arc or might pass into strike-slip faults that traverse some island arcs (eg Sumatra) and cordilleran margins (eg New Zealand). In addition multiple collisions of island arcs as a result of subduction processes or strike-slip faulting may lead to formation of larger continental masses with complex structural characteristics.

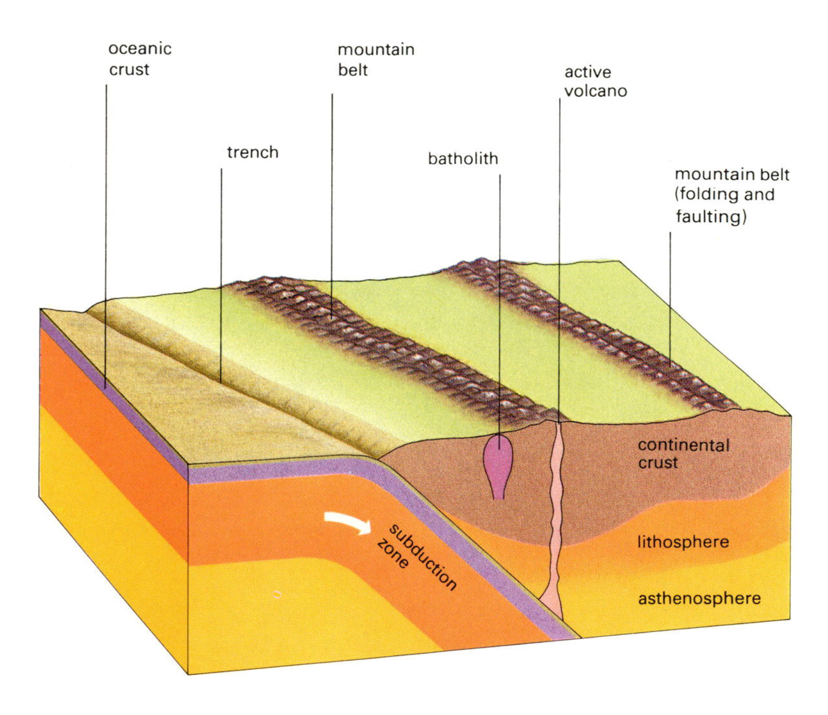

14.4: *Destructive plate margin of cordilleran type. Subduction of oceanic crust along a trench gives rise to mountain-building and intrusive and extrusive igneous activity within the continent.*

Continental collision zones

The most complex situation develops where two continental blocks approach and collide by subduction of the intervening oceanic crust. Such a collision is believed to be responsible for formation of the Alpine–Himalayan orogenic belt. In the simplest case, we may consider the approach of an Atlantic-type continental margin to another continent bounded by a cordilleran-type orogenic belt. As the Atlantic-type margin meets the trench there may be deformation and thrusting of the continental basement, and nappes may eventually form. Oceanic crust, upper mantle, oceanic sediments and flysch deposits are deformed and thrust over earlier thrust-sheets. When the collision is arrested by the buoyancy of the collided continental masses the collision zone becomes an area of thickened crust characterized by complex faulting, including prominent lateral faulting, and uplift.

The structures developed during collision between continents depend upon the natures of the sedimentary sequence and the basement on the Atlantic-type margin, and the shapes of the colliding margins. Where margins are irregular the first continental segments to collide will experience the earliest and most intense deformation. Such segments may be partially subducted or split so that the upper part becomes 'flaked' onto the continent. Alternatively the segments may be extended by lateral faulting, so as to generate a more even collision. Against such regions the trench zone becomes a narrow suture from which fragments of oceanic crust, mantle, oceanic sediments and flysch sediments are transported. These regions are likely to have extensive continentally derived molasse sediments in external troughs. In contrast to collided segments some portions of incidental areas of the continental margins may never collide so that relatively undeformed areas of oceanic crust are preserved within the orogenic belt.

Continent–continent collision zones are clearly likely to be extremely complex. Before collision the ocean may include small island arcs and small continental masses. At the initiation of collision the margins may split, forming a complex pattern of small continental fragments that eventually become sutured along the collision zone.

Petrological characteristics

Volcanic rocks

Whereas the volcanic rocks erupted along the destructive margins can be relatively easily collected and studied, this is not true of the intrusive rocks believed to be present below the volcanic belt. Because such intrusive rocks are exposed by erosion in areas of active or geologically recent subduction they are widely believed to represent magmas that have crystallized below an overlying contemporaneous volcanic belt. However, this is very difficult to prove. Furthermore,

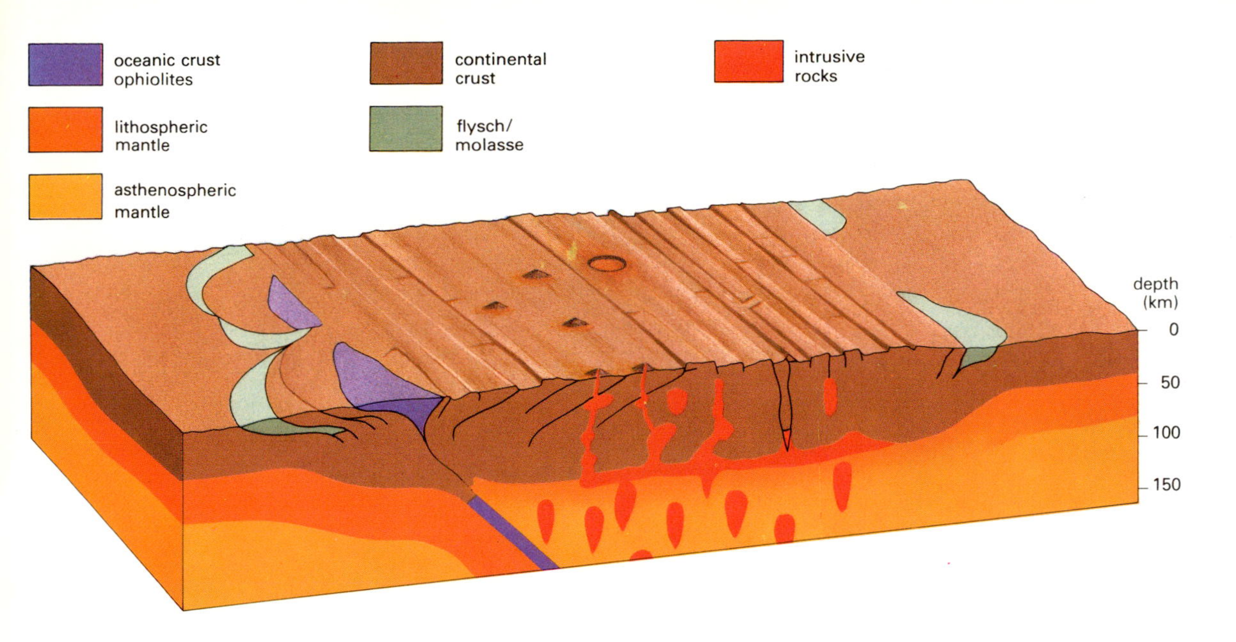

14.5: *Continent–continent collision zone. The intense deformation imposed by the subduction of continental crust is largely accommodated by complex thrusting developed within the leading edges of both plates.*

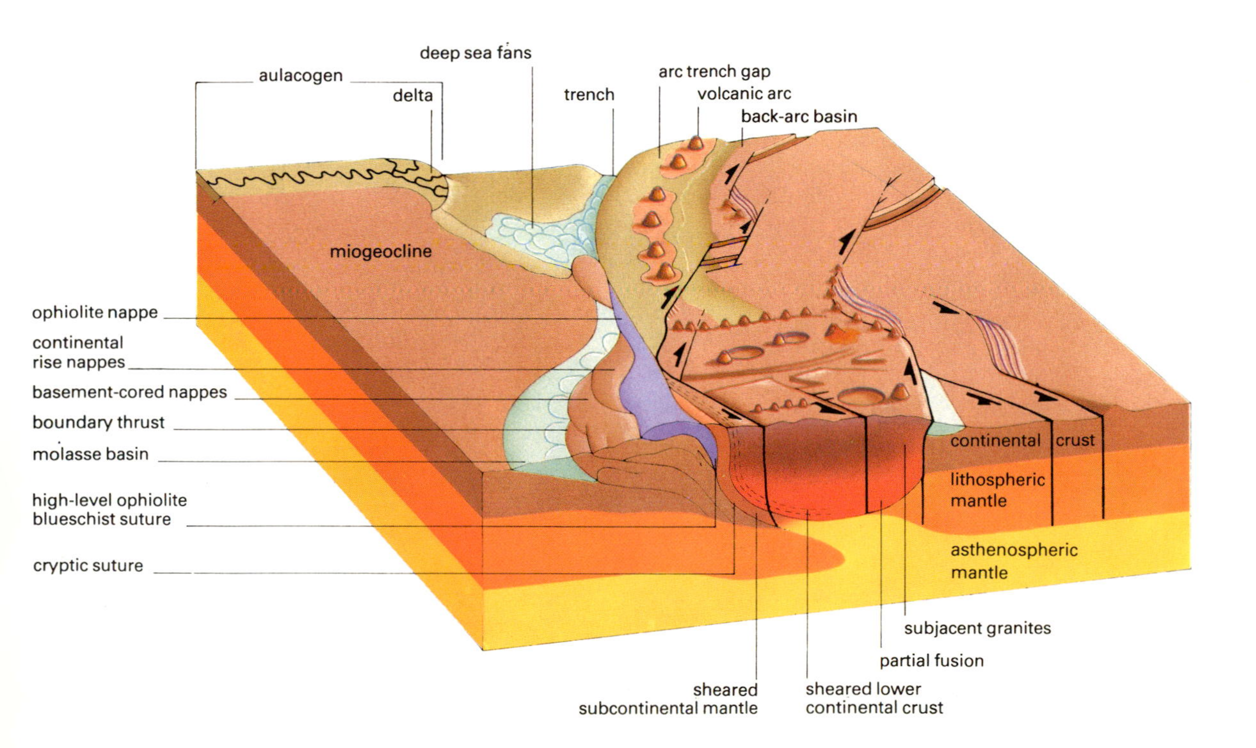

14.6: *Block diagram showing the potential complexities of continent–continent collision between irregular continental margins.*

there are likely to be differences between the compositions of contemporaneous extrusive and intrusive rocks, reflecting their contrasted occurrence and cooling environment. Nevertheless the mineral and chemical compositions of extrusive and intrusive rocks emplaced at destructive margins show clear parallels.

As already explained, the igneous rocks of destructive margins range from basic to acid in chemical composition; the intrusive rocks form the gabbro–diorite–granodiorite–granite association and the volcanic rocks form the basalt–andesite–dacite–rhyolite association. There is continuous mineralogical variation within these series, as shown in Figure 5.17. The basic rocks are composed of Ca-rich plagioclase, pyroxenes and/or amphiboles with minor olivine and

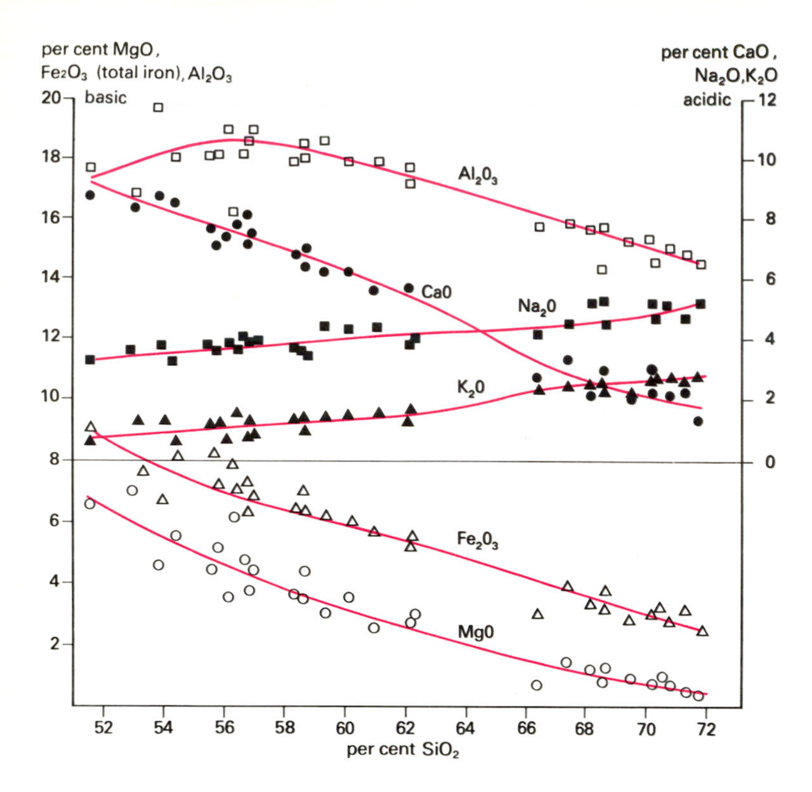

14.7: *Chemical variation diagram showing proportions of important elements (as their oxides) against percentage of silica (ie acidity). Samples were from the Cascades volcanic province, north-west USA.*

Fe–Ti oxide, and grade through intermediate rocks, composed largely of intermediate plagioclase with pyroxenes and/or amphiboles and quartz, to acid rocks composed largely of Na-rich plagioclase, alkali feldspar and quartz. These mineralogical variations are paralleled by continuous chemical variations, as shown on a plot of element oxides against silica (SiO_2) in Figure 14.7. From basic to acid compositions there is a decrease in total FeO, MgO and CaO, and an increase in Na_2O and K_2O. Chemical analyses of representative intrusive and extrusive rocks are given in Figure 14.8.

The volcanic rocks show a progressive increase in K_2O relative to SiO_2 from the trench towards the island arc or continent. The volcanic associations may therefore be further subdivided on a plot of K_2O against SiO_2. Although variation is continuous certain distinctive associations may be recognized and these are low-K, calc-alkaline, high-K and shoshonitic associations. Volcanic rocks of the low-K association have tholeiitic characteristics and are often termed the island arc tholeiite series. This, like other tholeiitic series, shows a trend of enrichment of total FeO relative to MgO in a diagram showing relative variations in $Na_2O + K_2O$ (alkalis), total FeO (FeO_t) and MgO (an AFM diagram). By contrast, the calc-alkaline association does not show such Fe enrichment. The tholeiitic association is dominated by basalts and basaltic andesites, while the calc-alkaline association has greater amounts of andesites. The characteristics of island arc tholeiitic and calc-alkaline associations are summarized in Figure 14.10. It is important to note that there is complete gradation from the island arc tholeiitic series, through low-K, calc-alkaline, to the calc-alkaline association.

The western Pacific island arcs appear to show evolutionary sequence of island arc volcanism. Relatively young ones, such as the Mariana and Tonga–Kermadec arcs and the South Sandwich arc in the south Atlantic, while exhibiting a complete spectrum of volcanic rocks, from basalt to rhyolite, are dominantly composed of basalt and basaltic andesite of the island arc tholeiitic association. Older and hence more evolved arcs such as Japan, Indonesia, Kamchatka and the Aleutians, and continental margins like the Cascades in north-west USA and the central Andes, have a similar range of composition but consist dominantly of calc-alkaline andesites. In some evolved arcs, such as Indonesia, and continental margins, such as the central Andes, high-K calc-alkaline and shoshonitic volcanic rocks are erupted. Thus

14.8: *Representative chemical analyses of volcanic and intrusive rocks from destructive plate margins.*

	1	2	3	4	5	6	7	8
SiO_2	50.3	62.5	67.4	74.2	48.3	59.1	66.2	75.3
TiO_2	1.0	0.6	0.7	0.3	0.6	0.6	0.6	0.2
Al_2O_3	20.3	15.9	15.3	13.3	19.1	17.1	15.5	12.9
Fe_2O_3	3.0	1.3	3.6	0.9	2.8	2.4	1.7	0.3
FeO	5.5	3.3	1.1	0.9	7.0	5.0	2.6	0.8
MnO	0.2	0.1	0.1	0.1	0.2	0.2	0.1	0.1
MgO	4.3	3.5	1.0	0.3	8.0	2.7	1.8	0.4
CaO	11.0	4.6	3.4	1.6	11.2	5.7	3.1	0.8
Na_2O	3.3	4.2	4.6	4.2	2.0	3.1	2.8	3.9
K_2O	0.4	2.8	2.4	3.2	0.2	2.6	4.4	4.4
P_2O_5	0.1	0.2	0.2	0.1	0.1	0.2	0.1	
H_2O	0.7	1.8	0.8	1.0				
Total	100.1	100.8	100.6	100.1	99.5	98.7	98.9	99.1

Notes to columns (terms in parentheses refer to the classification of volcanic rocks as shown in Figure 14.9).

1 Basalt (low-K tholeiite), Mt Misery volcano, St Kitts, Lesser Antilles.

2 Andesite (high-K andesite), San Pedro volcano, north Chile.

3 Dacite (dacite), Maungaongaonga, New Zealand.

4 Rhyolite (average of twenty-five analyses), New Zealand.

5 Gabbro (average of eight analyses), Peruvian coastal batholith.

6 Diorite, Linga superunit, Arequipa segment of Peruvian coastal batholith.

7 Granodiorite, Linga superunit, Arequipa segment of Peruvian coastal batholith.

8 Granite, Pativilca pluton, Lima segment of Peruvian coastal batholith.

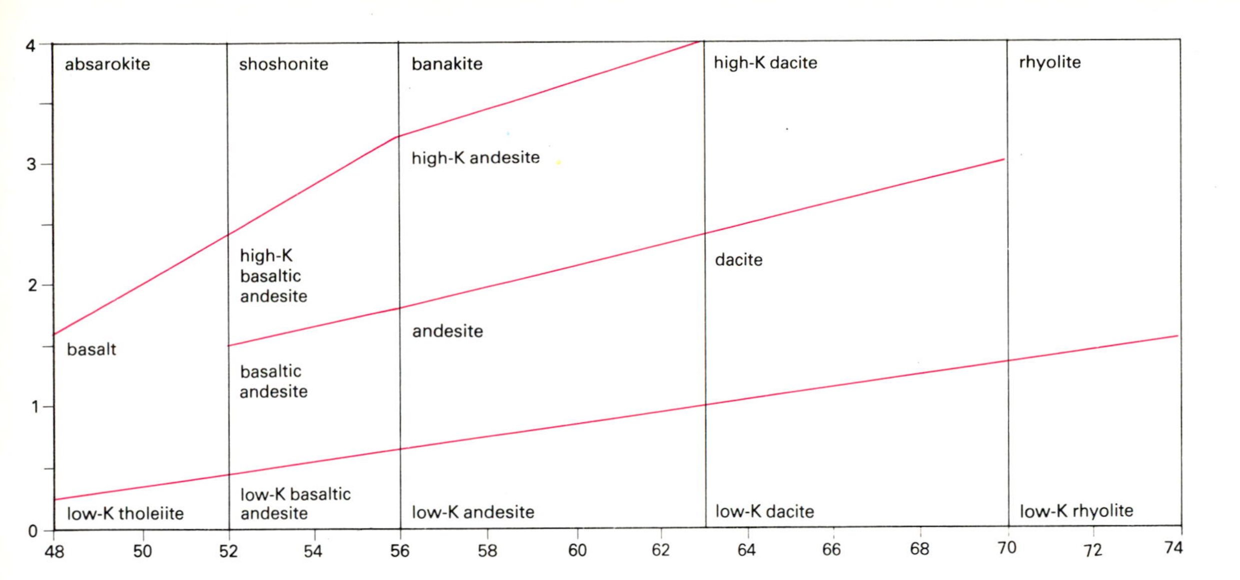

14.9: *Classification of orogenic volcanic associations on a plot of potash (% K_2O) against silica (% SiO_2).*

14.10: *Some characteristics of island arc tholeiitic and calc-alkaline associations.*

	Island arc tholeiitic associations	Calc-alkaline associations
SiO_2 range	45–70%	50–70%
SiO_2 mode	53%	59%
FeO_t/MgO_{55} ratio	3.5–1.5	2–1
FeO_t/MgO_{60} ratio	4.5–2.5	3–1.5
K_2O_{55}	0.4–1.5%	0.8–2%
K_2O_{60}	0.6–1.5%	1–3%
Mineralogy	probably pigeonite groundmass	probably ortho-pyroxene groundmass
Fe–Ti oxides phenocrysts	common	rare
Hydrated minerals	generally absent	more common
Fe(+V,+Ti) enrichment	present	absent
Incompatible elements	10–30 × chondritic	30–100 × chondritic
K/Rb ratio	500–1 300	220–560
Na_2O/K_2O ratio	4–6	2–3
Rare earths	unfractionated	strongly fractionated
La/Yb ratio	0.7–3.5	3.5–20

the lower parts of evolved island arcs consist of tholeiitic rocks, while the upper parts comprise calc-alkaline and shoshonitic volcanic rocks. There is, therefore, a compositional stratification in an evolved island arc, the upper part being richer in SiO_2, K_2O and associated trace elements, and having a lower FeO/MgO ratio, because of the stratigraphic variations described above.

Increase in K_2O with distance from the oceanic trench at a given SiO_2 content was first demonstrated in the Japanese arc where tholeiitic volcanic rocks occur near the trench, followed inland by calc-alkaline and alkaline volcanic rocks. This compositional trend has been demonstrated in many arcs, Indonesia for example, and continental margins (the central Andes), but is not found everywhere. Since the major chemical variation is in the K_2O content at given SiO_2, and since this correlates with depth to the Benioff zone (*h*), it is known as the K–*h* relationship.

Intrusive rocks

The chemical characteristics and the compositional variations with space and time outlined for volcanic rocks above also characterize intrusive rocks of destructive margins. However, the relationship of extrusive and intrusive rocks is not unequivocal: young intrusive rocks are not yet exposed by erosion, and exposed intrusions may not represent magmatic liquid compositions or may be hydrothermally altered because of slow cooling in the presence of meteoric water.

Intrusive rocks emplaced at destructive margins and continent–continent collision zones have a broad spectrum of basic to acid chemical compositions. Intrusive associations are referred to as igneous, or I-type, granites and sedimentary, or S-type, granites, these being end-members of a spectrum of granite types. The characteristics of I- and S-type granites (Figure 14.13) probably indicate a fundamental difference between the source regions referred to in their names. Both I- and S-type granites occur in distinctive tectonic environments which are described later.

The compositional variations in the intrusive rocks of continental margins parallel those in volcanic associations. These intrusive rocks, exposed in relatively young arcs, such as those of the Caribbean and in the Central American margin, and continental margins, such as Panama, are dominantly tholeiitic gabbros and diorites, equivalent to the basic members of the island arc tholeiitic volcanic series. The compositions of the large batholithic intrusions of the western USA and the Andes range from gabbros through diorite and granodiorite to granite. The granites have I-type characteristics. However, the dominant rocks are of intermediate chemical composition, with calc-alkaline characteristics.

In the central Andes, where intrusive rocks with a range of ages occur at one location, the earlier (Jurassic) intrusions consist of more basic gabbros and diorites, while younger (Cenozoic) intrusions are intermediate and acid granodiorites and granites. Thus the time–composition relationship among destructive margin intrusive rocks seems to resemble that described for volcanic rocks. Where large areas of Mesozoic and Cenozoic batholithic intrusions are exposed, as in the western USA, the compositions show an eastward increase in K_2O at a given SiO_2 content. In the west the dominant rock types are diorites and quartz diorites, whereas in the east granodiorites and granites are more abundant. Since there is abundant geological evidence for contemporaneous subduction this regional variation may be interpreted as a K–*h* relationship indicating Mesozoic and Cenozoic plate

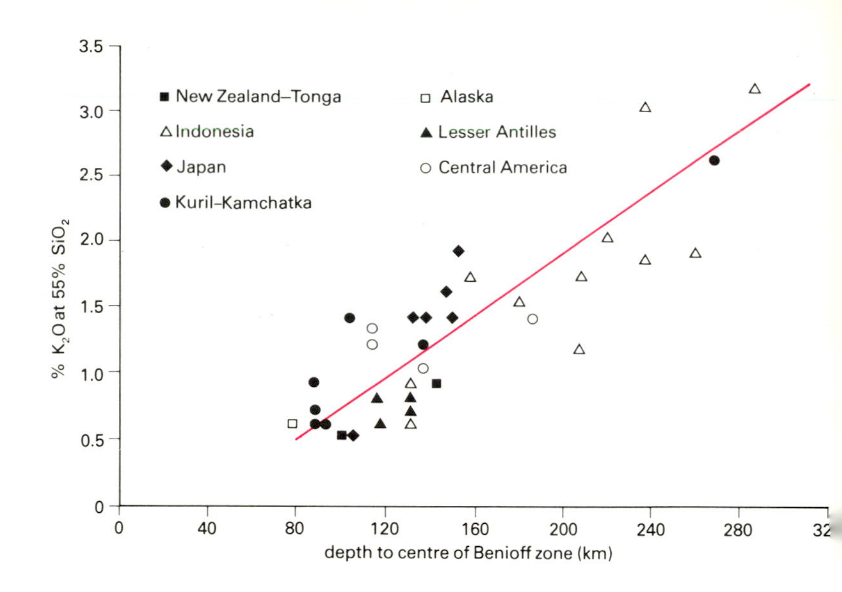

14.12: *Variation in K_2O/SiO_2 ratio with distance above Benioff zone for volcanic rocks from several areas, with a statistically fitted correlation line.*

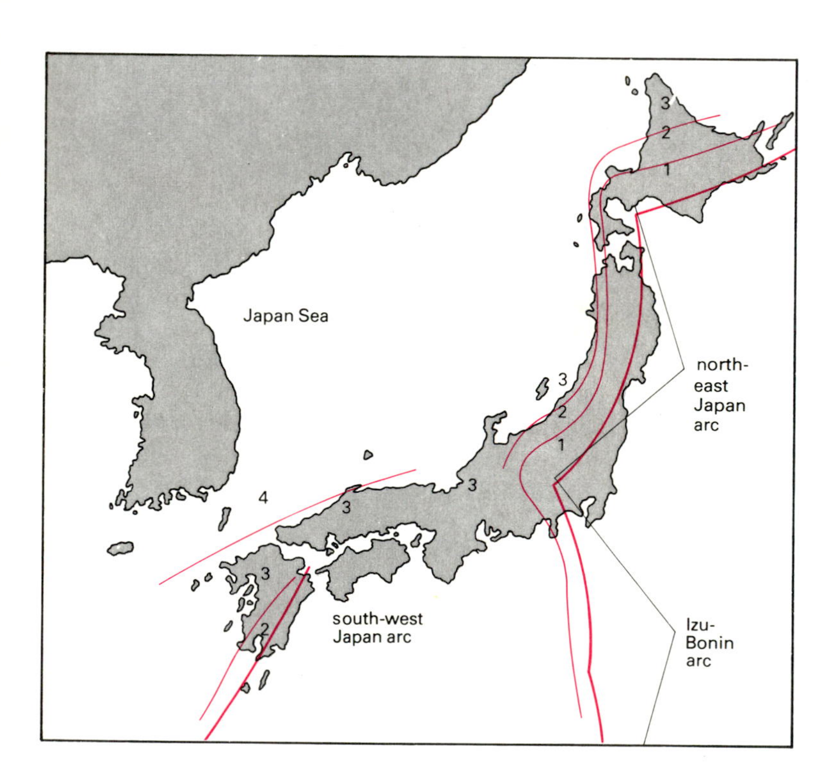

14.11: *Compositional variation of young volcanic rocks across the Japanese arc; 1 toleiitic and calc-alkali, 2 calc-alkali, 3 calc-alkali and sodic alkali, 4 sodic alkali.*

14.13: *Tectonic settings and geochemical characteristics of I-type and S-type granites.*

	I-type	S-type
Chemical composition	Part of a broad spectrum from basic (gabbro) through intermediate (diorite and granodiorite) to acid (granite).	Restricted acid compositional range.
	High Na_2O contents and Na_2O/K_2O ratios (>3.2% Na_2O at 5.0% K_2O).	Low Na_2O contents and Na_2O/K_2O ratios (<3.2% Na_2O at 5.0% K_2O).
Characteristic accessory minerals	Pyroxene, hornblende, biotite, magnetite.	Muscovite, biotite, cordierite, garnet, ilmenite.
Oxygen isotope composition	Relatively low $\delta^{18}O$ values, typically 5–10‰.	Relatively high $\delta^{18}O$ values, typically >10‰.
Strontium isotope composition	Relatively low initial $^{87}Sr/^{86}Sr$ ratios, typically <0.708.	Relatively high $^{87}Sr/^{86}Sr$ values, typically >0.708.
Associated mineral deposits	Porphyry copper and molybdenum deposits.	Tin–tungsten deposits.

subduction from west to east. Therefore, although destructive margin intrusive rocks are more difficult to study than volcanic rocks, the compositional variations and relationships appear to parallel those described for volcanic associations.

The origin of destructive margin magmas

A major problem in the evolution of island arcs and active continental margins concerns the petrogenesis of the orogenic basalt–andesite–dacite–rhyolite volcanic association and its intrusive equivalent. Island arcs are composed of these volcanic rocks, which thus represent the major volcanic components at active margins. Since the continents are close to andesite in composition it seems reasonable to argue that the continents themselves have grown by lateral accretion of island arcs and by addition of rocks of the orogenic andesite association. The petrogenesis of this volcanic association is, therefore, of major significance in the Earth sciences.

Before the development of the plate tectonic theory in the late 1960s the petrogenesis of the orogenic volcanic association was discussed largely in terms of fractional crystallization of mantle-derived basaltic magma and/or melting of pre-existing sialic material. However the existence of a complete spectrum of orogenic volcanic rocks varying in composition from basalt to rhyolite in entirely oceanic island arcs such as the Marianas, Tonga–Kermadec and the western Aleutian arcs indicates that this volcanic association can be derived from the mantle.

Any theory of petrogenesis of the orogenic volcanic association must explain the geological, geophysical and compositional characteristics and their variation in space and time outlined earlier. The geological and geophysical characteristics of orogenic zones indicate subduction of oceanic lithosphere, and consequently the petrogenesis of the association is discussed in terms of possible contributions from subducted oceanic crust and from the overlying wedge of mantle. Furthermore, melting pre-existing continental crust where available might contribute to orogenic magmas.

Subducted ocean crust

Chapter 13 described how basaltic ocean crust formed at a spreading ridge becomes hydrated and chemically altered by interaction with heated ocean water during cooling. When such ocean crust descends into the mantle by subduction it is heated by conduction during passage into hotter mantle and may be frictionally heated along the inclined Benioff zone. At depths of 80–100 km the hydrated oceanic crust dehydrates, causing the conversion of amphibolite to eclogite (pyroxene and garnet) and water. It has been proposed that the more tholeiitic volcanic associations might result from varying degrees of partial melting of amphibolitized oceanic crust, while calc-alkaline associations would result from varying degrees of partial melting of eclogite ocean crust. Although this appears to be consistent with plate tectonic theory several problems are now seen. Most models for temperature variation along subduction zones propose that temperatures along the Benioff zone are not sufficiently high for partial melting to occur. Furthermore, experiments carried out on samples of basaltic composition at pressures equivalent to depths of 80–100 km show that the partial melts formed initially do not correspond to basalts or andesites of the orogenic volcanic association. Partial melting of subducted oceanic crust may not, therefore, contribute significantly to orogenic volcanism.

The mantle wedge

Even though melting of subducted oceanic crust is unlikely to contribute to destructive margin magmas, such hydrated crust must dehydrate at 80–100 km and release water into the overlying wedge of lithospheric mantle. Since the melting temperature of mantle peridotite is lowered by addition of water, this should cause melting and the model for temperature variation around a subducting plate below an island arc shows that dehydration of oceanic crust will indeed cause partial melting in the mantle wedge (Figure 14.14). Experiments indicate that the composition of the melts will range from basaltic to andesitic for higher and lower degrees of partial melting respectively. During dehydration the water driven from the descending oceanic

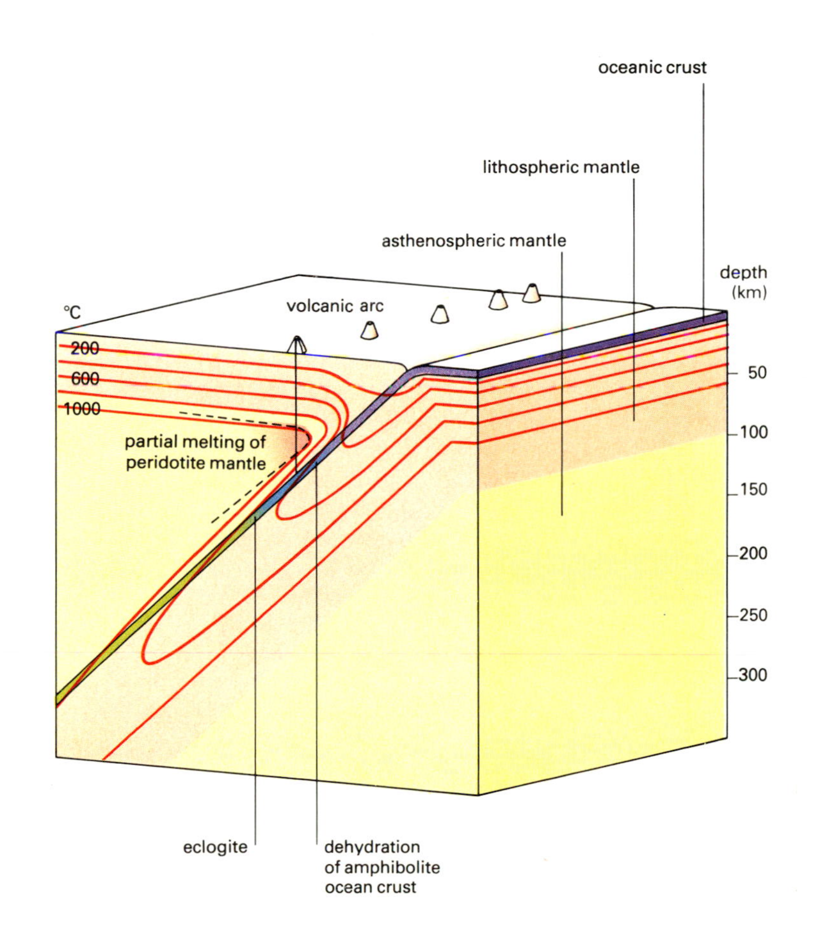

14.14: *Computer-generated model of a temperature variation in a subduction zone, showing the depth of occurrence of processes important for generation of destructive margin magmas.*

crust will transport additional Si, K and associated trace elements into the mantle wedge, and the resulting magmas may be richer in these elements than mantle-derived magmas formed elsewhere. The formation of orogenic volcanic rocks by partial melting of the wedge of enriched mantle peridotite above subduction zones is now the most attractive model for formation of the orogenic volcanic association.

The continental crust

The possible incorporation of a sialic component into the magma from pre-existing continental crust can happen in three different ways.

First, at a subduction zone, the subducted material may include oceanic pelagic sediment or continental flysch-type sedimentary material, as part of the oceanic lithosphere. Such sediments are mainly accreted into a subduction complex, as explained earlier, but the possibility exists that some sediment is subducted and then scraped off at shallow depth ('subcreted'), or further subducted into the zone of dehydration or partial melting, where it could contribute to magmagenesis. Certain evidence from lead isotopes (see Chapter 8) indicates that magmas in some island arcs such as the Aleutians contain a component from subducted oceanic pelagic sediment. Thus sedimentary material, some of it originally derived from continental crust, probably contributes to magmas at destructive margins.

Second, oceanic crust formed at spreading ridges experiences submarine weathering and chemical exchange of elements with circulating seawater in hydrothermal systems (see Chapter 13). The elements dissolved in seawater are derived from weathering of continental and oceanic sources. Elements such as K, Na and Mg, the concentrations of which are relatively high in seawater, become fixed in weathered and hydrothermally altered oceanic crust. Subduction of such crust might take continent-derived elements into the zone of dehydration and partial melting, thus contributing to magma.

Third, where mantle-derived destructive margin magmas pass through thick continental crust as they do in the western USA and the Andes, a further source of continental components becomes available. At such a cordilleran type destructive margin, melting might occur within subducted oceanic crust, the peridotite mantle wedge and/or the lower continental crust. Igneous rocks formed above continental destructive margins might therefore be 'contaminated' by a crustal component, or might even be formed entirely by melting of pre-existing continental crust. Examples might include the dacite–rhyolite ignimbrites erupted in areas of thick continental crust at Andean continental margins and in Alpine–Himalayan belts.

Space–time relationships

The change from the island arc tholeiitic series dominated by basaltic rocks in young island arcs to the calc-alkaline associations with abundant andesites in more evolved island arcs appears to correspond with increasing thickness of crust. Island arc tholeiitic associations occur where the crust is less than 20 km in thickness, whereas intermediate calc-alkaline associations occur where the crust is thicker. This is shown by the correlation with crustal thickness of both mean SiO_2 and the FeO/MgO ratio for volcanic rocks with 55 per cent SiO_2. The thicker the crust, the longer the time of ascent of mantle-derived magmas and the greater the opportunity for fractional crystallization during ascent. The calc-alkaline characteristics of the magma could also be enhanced by melting of solidified basaltic rock within or at the base of the crust as crustal thickness increases.

Space–composition relationships

The K–*h* relationship described earlier was originally taken to indicate that the Benioff zone exerts a controlling influence on the composition of orogenic volcanic rocks. However, if it is accepted that the parent magmas of the volcanic rocks derive from the mantle wedge, the influence of the Benioff zone is indirect. The K–*h* relationship might therefore suggest that the degree of partial melting decreases from the Benioff zone into the wedge. Alternatively, the relationship might reflect the increasing degrees of fractional crystallization experienced by magma derived from just above the subducting slab according to the thickness of mantle wedge through which it had to rise to reach the surface.

From the above it is clear that the petrogenesis of orogenic igneous rocks might involve components from the oceanic crust (partial melt with or without fluids for dehydration), peridotite mantle wedge, or continental crust (subducted sediment, seawater component or crustal melt). The abundance of basaltic rocks and the occurrence of a spectrum from basalt or basaltic andesite to rhyolite suggest that the orogenic volcanic associations are derived from basaltic parent magmas. As in other settings these would be formed primarily by partial melting of mantle peridotite.

The compositional similarities between extrusive and intrusive igneous rocks suggest that the batholithic magmas emplaced in the western USA and the Andes might have an origin similar to that just described for the volcanic rocks. For the I-type granites such features as the association with basic and intermediate intrusive rocks and the relatively low initial $^{87}Sr/^{86}Sr$ ratios are usually considered to be

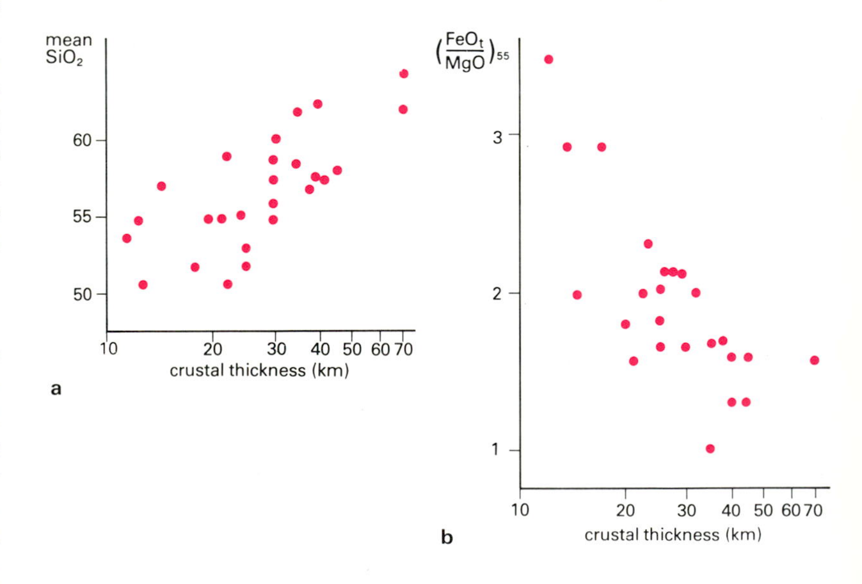

14.15: *Correlation with crustal thickness of* (**a**) *silica and* (**b**) *total* FeO/MgO *at 55 per cent silica, in rocks of the orogenic volcanic association.*

14.16: *Looking west along the south side of the Hunza Valley where the main river valley cuts across the axis of the Karakoram mountain chain. The view is towards the northern slope of Mt Rakaposhi ('Queen of the Snows'), which is the steepest in the world. At its summit the mountain consists of ophiolitic ocean crust, and marks the very northern boundary of what was once an island arc in the Tethys Sea, now squeezed up between India and Asia.*

evidence of formation by fractional crystallization or partial melting of mantle-derived igneous material. In contrast the acid composition and relatively high initial $^{87}Sr/^{86}Sr$ ratios of S-type granites suggest formation by melting of sedimentary continental crust. Their low Na content may indicate formation by melting of sedimentary source rocks from which the Na has been partitioned into seawater. Similarly, their low oxidation state, indicated by the presence of ilmenite rather than magnetite, may reflect the presence of reducing agents such as carbonaceous shales within the sedimentary source materials. However, in view of the problems of correlating extrusive and intrusive rocks, it is more difficult to apply such models to batholithic rocks. Since these rocks are confined to areas of pre-existing continental crust it is frequently argued that they might contain a large component of crustal melt in comparison with contemporaneous volcanic rocks. However, the compositional similarities of contemporaneous extrusive and intrusive igneous rocks suggest that they share a common origin.

Ancient destructive margins

Ancient island arcs

Presently or recently active island arc orogenic belts are transient features, since island arcs are eventually accreted onto continental margins, and active continental margins become inactive when subduction ceases. Deformation and magmatic activity occur during collision, but the process of collision causes convergence, and hence tectonic and magmatic activity cease. There is good evidence for the formation and destruction of the ocean floor only for the past two hundred million years. However, many characteristics of ancient linear belts of metamorphism, magmatism and deformation suggest that they might have formed in island arc destructive margin settings. These characteristics are as follows.

1 A relatively narrow (<100 km) linear belt with distinctive associations of sedimentary and igneous rock types.

2 The occurrence of igneous rocks dominated by intrusive gabbro and diorite and extrusive basalt and basaltic andesite of the orogenic association.

3 Volcanic flysch derived from erosion of volcanic rocks associated with chaotic melanges.

4 No evidence of ancient continental basement, but the presence of oceanic crustal basement and/or ophiolitic fragments of oceanic crust in melanges.

5 Metamorphism of flysch and melanges at low temperature–high pressure (blueschist facies) and high temperature–high pressure (amphibolite and granulite facies) in pairs of parallel belts.

6 Variation in composition of volcanic rocks, eg increase in K at a given SiO_2 concentration, from one side of the belt to the other.

Identification of some or most of these characteristics within a linear continental terrain would indicate formation by island-arc-type orogeny. The patterns of metamorphic facies and chemical variation in volcanic rocks may be used to infer the direction of subduction of oceanic crust below the island arc. These criteria have been used to identify island arcs within numerous ancient terrains, and hence to re-interpret the geology of many areas. Several examples have been identified, but the most complete section through an ancient island arc exists in north-west Pakistan. The newly constructed Karakoram highway exposes a section through a Cretaceous stratigraphic sequence of rocks which was tilted and uplifted during the formation of the Himalayan mountain belt as India collided with Asia. At the base of the section there is a variety of ultrabasic rocks including peridotites, which are overlain by amphibolite and metamorphosed gabbro. Above are gabbroic intrusions overlain by calc-alkaline intermediate and acid intrusive and volcanic rocks. At the top are pillow lavas and volcanic-derived flysch. The sequence lies within a belt up to 150 km in width; its thickness is generally 10–15 km and locally up to 40 km. This sequence of rock types does not correspond to an ophiolite sequence, and the absence of ancient continental basement indicates that this is not simply a section through ordinary continental

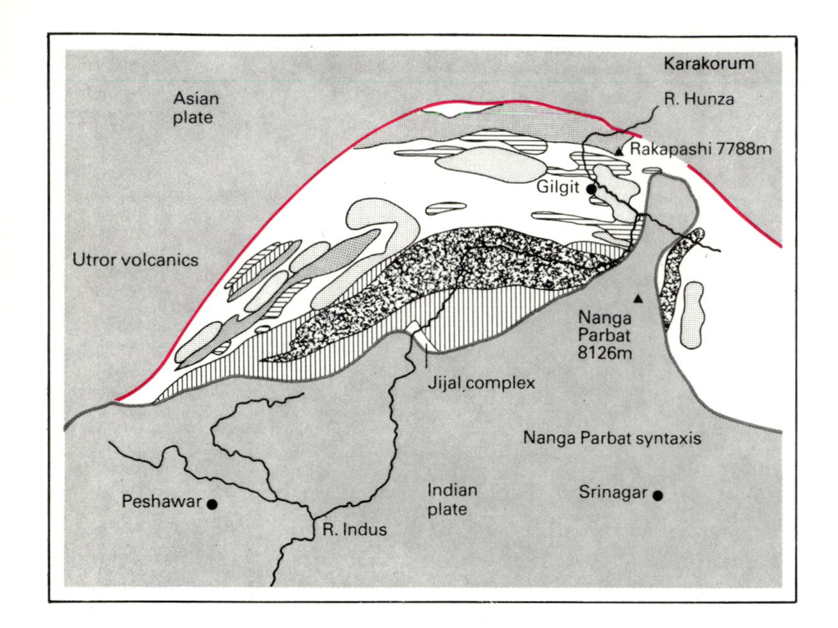

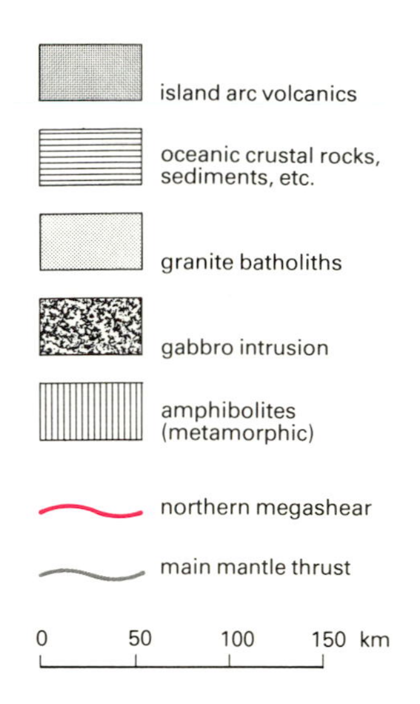

14.17: Simplified geological map of the Kohistan block, an upturned ancient island arc deformed in the Himalayan collision orogeny.

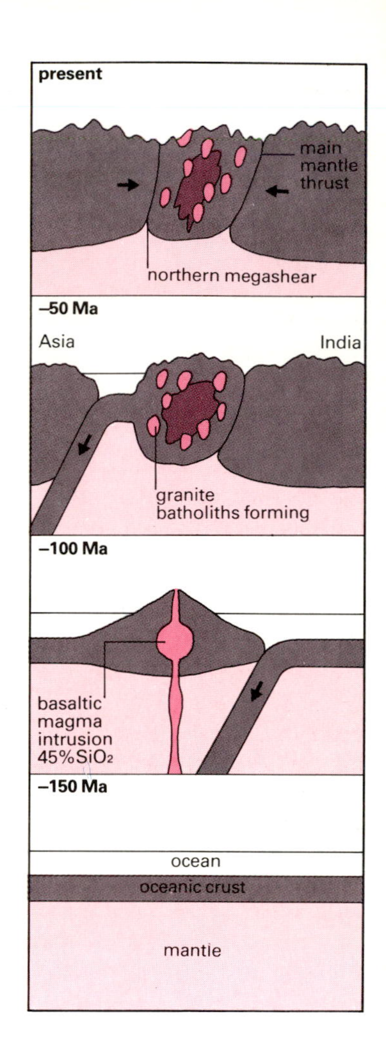

14.18: Diagrammatic cross-sections illustrating the development and eventual destruction of the island arc which now forms the Kohistan block.

crust. The sequence is therefore interpreted as a complete section through a Cretaceous calc-alkaline island arc. According to this hypothesis the ultrabasic rocks are part of the uppermost mantle, and the gabbros and amphibolites constituting the bulk of the sequence correspond to calc-alkaline intrusions and metamorphosed oceanic crust. The lavas and sedimentary rocks at the top of the sequence correspond to the uppermost part of the island arc.

Ancient cordilleran belts

All gradations are possible between island arcs and cordilleran orogenic belts. Nevertheless it is possible to pose criteria for the recognition of orogenic belts formed at cordilleran type destructive margins:

1 The occurrence of igneous rocks dominated by intrusive tonalite, granodiorite and granite and extrusive andesite, dacite and rhyolite of the orogenic association.

2 Platform-type and flysch sedimentary rocks, the latter derived from volcanic and ancient continental crustal sources, and possibly associated with chaotic melanges.

3 The occurrence of ancient continental crustal basement and possibly ophiolitic fragments of oceanic crust in melanges.

4 Widespread high temperature–high pressure (amphibolite- and granulite-facies) metamorphism of ancient crustal basement and possibly partial melting.

The identification of these characteristics within a linear continental terrain would indicate formation at an active continental margin by cordilleran orogeny. The patterns of metamorphic facies and chemical variation in volcanic or intrusive rocks may be used to infer the direction of subduction of oceanic crust below the cordilleran margin.

The best examples of ancient cordilleran belts occur in western Canada, the western USA and western South America. An older example was active in the areas of Scotland and Northern Ireland during north-westward subduction of ocean floor during the Caledonian orogeny (see Chapter 16). Another good example of an older cordilleran margin exists in central Peru. Although the Nazca plate is now being subducted below central Peru, there is no active volcanism such as occurs further north in Ecuador and in south Peru and north Chile. However, during the Mesozoic–Cenozoic (about two hundred to ten million years ago) central Peru was the site of an active cordilleran margin, whose products are now uplifted and exposed to erosion. Central Peru has a basement that includes Precambrian and Palaeozoic crystalline rocks overlain by Palaeozoic sediments and intruded by early Palaeozoic igneous rocks. Mesozoic sedimentary rocks were deposited upon this basement in narrow subsiding belts separated by upstanding horsts. These sedimentary rocks locally reach substantial thicknesses (>7000 km) and they are characterized by rapid lateral facies changes. During the early Cretaceous this sedimentary succession was added to by submarine volcanic rocks; the entire succession then formed the 'country rock' for the orogenic activity which followed.

Between the early Cretaceous and late Cenozoic the Peruvian batholith was emplaced into the sedimentary-volcanic belt. The emplacement was controlled by older linear structures so that, although the component rocks were emplaced as hundreds of relatively small plutons with diameters up to about 20 km, the batholith as a whole is about 1600 km long and about 100 km wide. Enough detail is now known to divide the batholith into units, each unit representing an evolutionary sequence from more basic to more acid compositions. As a whole the batholith shows an evolution from relatively basic (gabbro–diorite) to intermediate and acid units (granodiorite–granite) showing I-type characteristics. The plutonic rocks were all intruded near to the surface of the Earth and are mostly overlain by subaerial volcanic rocks, although the youngest intrusions are emplaced within the volcanics.

The characteristics of the Mesozoic–Cenozoic geological history of Peru are those of a cordilleran plate margin. Of particular importance is the older sialic crust below the orogenic belt and its characteristic influence on sedimentation, structure and magmatism. Furthermore, the absence of flysch, ophiolitic material in melanges and of low temperature–high pressure metamorphism, distinguishes this example from an island-arc-type orogeny.

Ancient collision zones

The identification of ancient island arc–continent and continent–continent collision zones is complicated because, prior to collision, the arc and/or continental margins may have experienced island arc or cordilleran orogenic activity respectively. Collision zones therefore have to be interpreted not only in terms of collision and post-collision processes but also in terms of pre-collision processes such as island-arc-type and cordilleran orogenies. The following characteristics apply specifically to continent–continent collision zones.

1 An extensive (300–1000 km wide) belt, approximately linear on a large scale (>1000 km) with distinctive associations of sedimentary

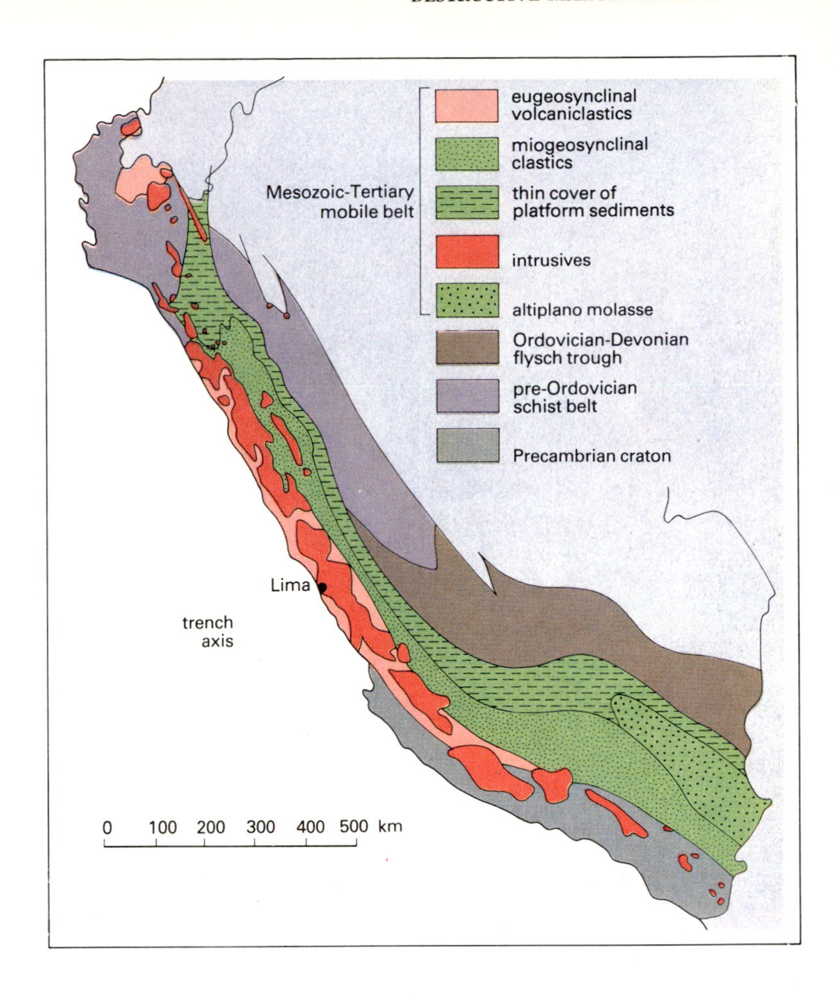

14.19: Geological map of part of the orogenic belt of coastal Peru.

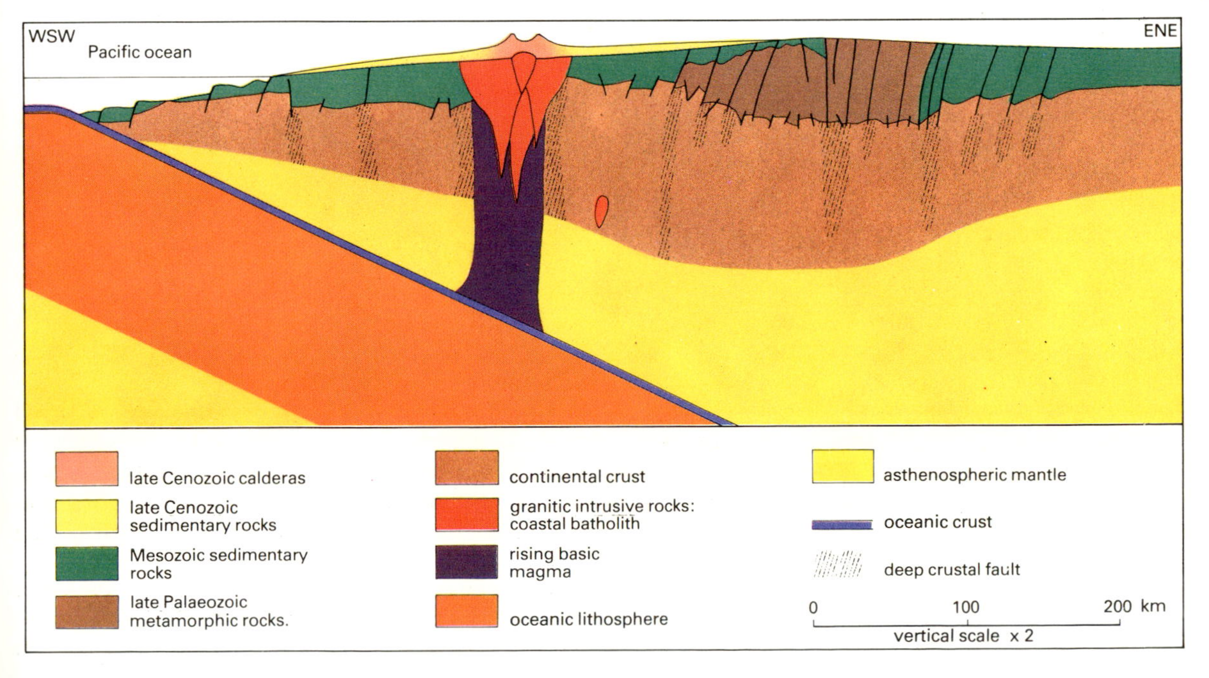

14.20: Cross-section of Peruvian destructive margin. Structures and intrusions developed in the thickened continental lithosphere are typical of a cordilleran orogenic belt.

and igneous rock types.

2 The occurrence of igneous rocks dominated by granodiorites and granites (often muscovite–biotite granites with S-type characteristics), and extrusive dacite and rhyolite.

3 Sedimentary rocks may be thick flysch deposited before collision and molasse derived from an uplifted mountain belt and deposited in basins inside (internal basins) or outside (external basins) the mountain belt.

4 Widespread deformation of pre-collision flysch-type sediments to form complex folded terrains, including nappes.

5 Metamorphism of flysch at low temperature–high pressure (blueschist facies) where subducted near to the suture zone, and at high temperature–high pressure facies away from this zone.

6 Widespread lateral faulting around the uplifted collided area and widespread regional metamorphism of ancient crustal basement at high temperature–high pressure amphibolite- and granulite-facies, and possibly partial melting.

Many, but not all, of these features of ancient continent–continent collision zones, deduced from studies of the Alpine–Himalayan belt, are found in older orogenic belts such as the Variscan belt of western Europe. The Variscan belt has been interpreted in a number of ways (Figure 14.21).

A Permo-Triassic reconstruction of the continental area around the north Atlantic shows an extensive area affected by deformation, metamorphism and magmatic activity during the Carboniferous–Permian. This area may be divided into three parts. The most extensive part, Zone 3, is largely underlain by late Precambrian basement, in which the main phase of deformation occurred during the late Carboniferous. The basement within Zone 3 was variably affected by this activity, in some regions experiencing low-grade deformation, in others being metamorphosed at amphibolite and granulite facies, and locally undergoing partial melting to form migmatite belts. Zone 3 was also intruded by late Carboniferous granites, such as those of Iberia, Cornwall, the Massif Central, the Vosges and the Bohemian massif, which are of comparatively restricted composition, having relatively high K_2O contents and K_2O/Na_2O ratios. They include muscovite–biotite granites which are associated with tin deposits in Cornwall and the Massif Central. These granites therefore have S-type characteristics contrasting with the I-type granites of the western USA and Andean cordilleran margins. The palaeogeography of Zone 3 in the late Carboniferous was that of a widespread segmented plateau with strike-slip faulting, internal clastic and coal swamp basins and widespread subaerial pyroclastic volcanism.

Although there has been much debate about the plate tectonic processes responsible for formation of the Variscan belt, many of the features described above may be explained in terms of continent–continent collision processes. In particular the great width, widespread high temperature–pressure metamorphism, emplacement of S-type granites and widespread reactivation of pre-Variscan basement are readily interpreted in terms of continental collision processes. Hence Zone 2 may be a suture zone of continental collision between northern and southern European continental blocks. In this case Zone 3 would have formed by crustal thickening during and following convergence of northern and southern Europe, as shown schematically in Figure 14.5 (see also Chapter 16).

Continental growth and recycling

The overall chemical composition of the continental crust corresponds to that of an igneous rock intermediate in composition between tonalite (andesite) and granodiorite (dacite). Since such igneous rocks are added to the continental crust from the mantle at destructive plate margins, igneous activity of this kind might have been responsible for the growth of the continental crust. Such growth resulting from formation and accretion of island arcs is termed the andesite model of crustal growth. This is examined below.

Many island arcs show compositional variation from tholeiitic to calc-alkaline, and sometimes shoshonitic, with both time and distance from the oceanic trench. The volume proportions of basalts, andesites and dacites in these island arc associations are shown in Figure 14.22. By assuming that an island arc is 'mature' or 'developed', with well-developed compositional relationships, it is possible to calculate the volume proportions of rock types in it as also shown in Figure 14.22. The chemical composition of a 'developed' island arc is shown in Figure 14.23, together with an average composition of Archaean volcanic rocks and estimates of the overall composition of the continental crust.

The calculated island arc composition in Figure 14.23 is similar to the average composition of Canadian Archaean volcanic rocks, indicating that the processes responsible for formation of volcanic rocks in island arcs may have been similar since the Archaean. The compositions of the developed island arc (and Archaean average) are similar to some estimates for the composition of the continental crust. These similarities have been taken to support the andesite model of crustal

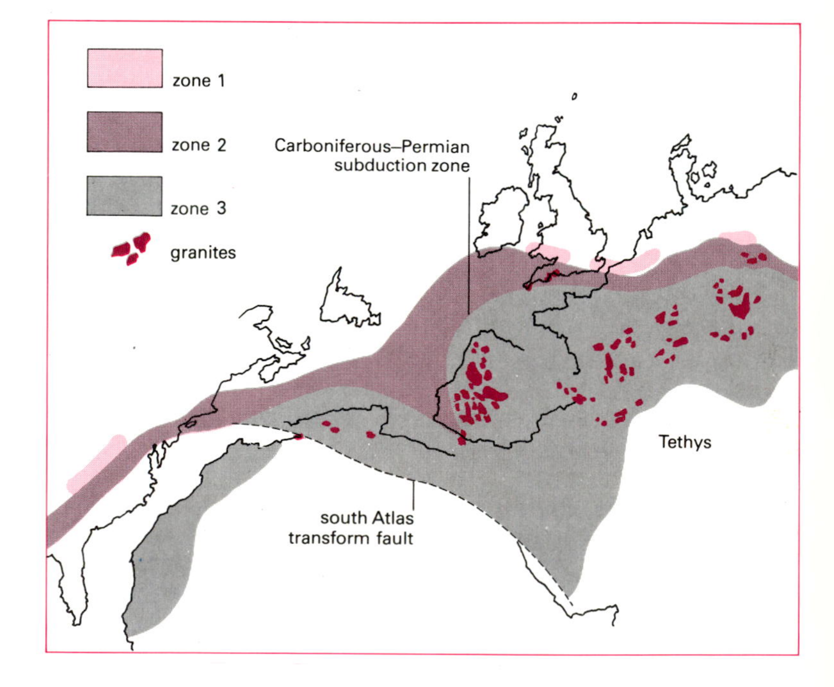

14.21: *Major zones (see text) of the Hercynian–Mauritainide–southern Appalachian fold belt on a Permo–Triassic reconstruction of the continents across the central and north Atlantic.*

growth, and it has been widely accepted that the formation and lateral accretion of island arcs has added to the continental crust throughout geological time. However, this hypothesis poses some problems.

Although the average andesites in Figure 14.23 correspond to some estimates of the composition of the continental crust, many other such estimates have much higher concentrations of potash and related trace elements than the island arc averages. Furthermore, estimates for the composition of the upper crust in shield areas are generally much higher in respect of silica and potash than is true of island arcs. In order to convert island arcs into continental crust, therefore, considerable vertical chemical differentiation is needed to yield an upper crust rich in silica and potash and a complementary, depleted lower crust. Although developed island arcs may be compositionally stratified, with a lower, more tholeiitic crust and an upper calc-alkaline crust, for the andesite model to be valid further differentiation must occur during or following accretion. But ancient accreted island arcs such as the Kohistan sequence in the Karakoram do not seem to have undergone subsequent crustal heating and magmatism, which might have led to further differentiation of the island arc material. It is nevertheless likely that crustal growth occurs by emplacement of mantle-derived igneous rocks below accreted island arc complexes or below cordilleran margins. To distinguish this from lateral accretion of island arcs the process is termed vertical accretion, or underplating.

Whatever the mechanism, if crustal growth results from emplacement of largely mantle-derived destructive margin magmas then the latter play an important role in geochemical cycles. The simplest such cycle is initiated by melting of mantle peridotite at ocean ridges to form oceanic crust. The oceanic lithosphere (crust and uppermost mantle) moves away from the ridge and eventually descends at a subduction zone. The subducted oceanic crust either dehydrates or melts to produce destructive margin magmas. The extrusive and intrusive igneous rocks so formed are accreted to the continental crust. The residual oceanic crust material then descends into the mantle. This simple model implies an irreversible differentiation of elements from the mantle, leading to the generation of continental crust.

The geochemical cycle is not as simple as this, because crustal components may be eroded from the continental crust and either fixed in hydrothermally altered oceanic crust or subducted *en masse* back into the mantle, to take part in petrogenesis of destructive margin magmas. The picture is complicated further since elements behave very differently during the processes of weathering, river transport and precipitation upon or within the ocean crust. Some elements, such as K, Rb, Sr, U and Th, are readily weathered from continental rocks, and high proportions of these in ocean waters may be of continental origin. These elements are also readily fixed in oceanic crust during hydrothermal alteration. Although there are large uncertainties about the quantities involved, it is possible that large proportions recirculate in the geochemical cycle. In this case the elements are not irreversibly differentiated from the mantle, as in the simple model described earlier. However, elements that are not readily weathered and eroded from continents, and whose concentration in ocean water is not therefore contributed to from continental sources, might well become concentrated in the continental crust, by irreversible differentiation from the mantle.

14.22: *Volume proportions of rock types of the three volcanic associations of island arcs.*

	Association	Basalts (<53% SiO_2)	Andesites (53–62% SiO_2)	Dacites (>62% SiO_2)
Volume proportions of rocks of each association	tholeiitic	50	35	15
	calc-alkaline	13	55	32
	shoshonitic	50	40	10
Volume proportions of rock types in the whole island arc[1]	tholeiitic	42.5	29.7	12.7
	calc-alkaline	1.6	7.0	4.0
	shoshonitic	1.3	1.0	0.2

Note

1 It is assumed that island arcs evolved simply in three stages and that the relationships of the various rock associations are the same as those determined for the Quaternary of Japan.

14.23: *Comparison of island arc volcanic rocks with composition of ancient continental crust and theoretical crustal average compositions.*

	1	2	3	4	5	6	7
SiO_2	58.78	56.15	66.0	66.06	66.0	54.0	58.0
TiO_2	0.84	1.01	0.6	0.54	0.6	0.9	0.8
Al_2O_3	15.58	15.63	16.0	16.08	15.3	19.0	18.0
Fe_2O_3	2.63	2.43	4.5*	1.42	1.9	9.0*	7.5*
FeO	5.04	7.70		3.14	3.1		
MnO	0.11	0.18		0.08	0.1		
MgO	4.57	5.24	2.3	2.22	2.4	4.1	3.5
CaO	8.02	7.76	3.5	3.44	3.7	9.5	7.5
Na_2O	3.39	3.00	3.8	3.95	3.2	3.4	3.5
K_2O	0.82	0.68	3.3	2.90	3.5	0.6	1.5
P_2O_5	0.22	0.22		0.16	0.2		
Total	100.00	100.00	100.0	99.99	100.0	100.5	100.3

Notes to columns

1 Composition of 'developed island arc' is calculated from average analyses and rock proportions on an H_2O- and CO_2-free basis.

2 Average Archaean volcanic rock from the Canadian shield.

3 Average composition of the upper continental crust.

4 Composition of the Canadian shield.

5 Composition of the Ukrainian shield.

6 Average composition of the lower continental crust based on an upper crustal to lower crustal ratio of 0.33 to 0.67.

7 Average composition of the bulk continental crust.

*Total Fe as FeO.

The complexities of the geochemical cycle make it difficult to interpret the contribution of destructive margin magmas to continental growth. It was argued earlier that such magmas are largely mantle-derived and that, regardless of whether the continental crust has grown by lateral accretion of island arcs or by vertical accretion below active continental margins, such additions have been a major mechanism of continental growth. Bearing in mind crustal recycling, it is possible to examine the rate of growth of continental crust throughout geological time.

Various models have been proposed for the growth of continental crust. In the first the rate of crustal growth is assumed to be proportional to the decay of the Earth's radioactive heat sources (Chapter 9). The Earth's heat controls the rate of all internal geological processes and would have been greatest early in the planet's history, subsequently declining exponentially. Therefore the rate of crustal growth would have been highest early in the Earth's history and would have declined exponentially to the present day. This, the simplest model, cannot be fully valid since the oldest continental rocks known are only about 3800 Ma in age. It would be necessary to assume that until about 3800 Ma ago mantle convection was too vigorous for crust to remain stable at the Earth's surface without being subducted back into the mantle, or that meteorite bombardment might have obliterated any continental crust formed before that date.

In view of the above problem, a second model has been proposed, according to which continental crust formation became significant only after 4000 Ma ago, and the rate then declined exponentially as predicted in the first. Both of these models are consistent with geological evidence that some 50–70 per cent of the volume of the continental crust formed during the Archaean (up to 2500 Ma ago). However, the latter observation is also consistent with the other two models. The third suggests that the present volume of continental crust was established by 2500 Ma ago and that at this time a steady state was established, so that subducted continent-derived material was balanced by magmatic material supplied by the mantle to the

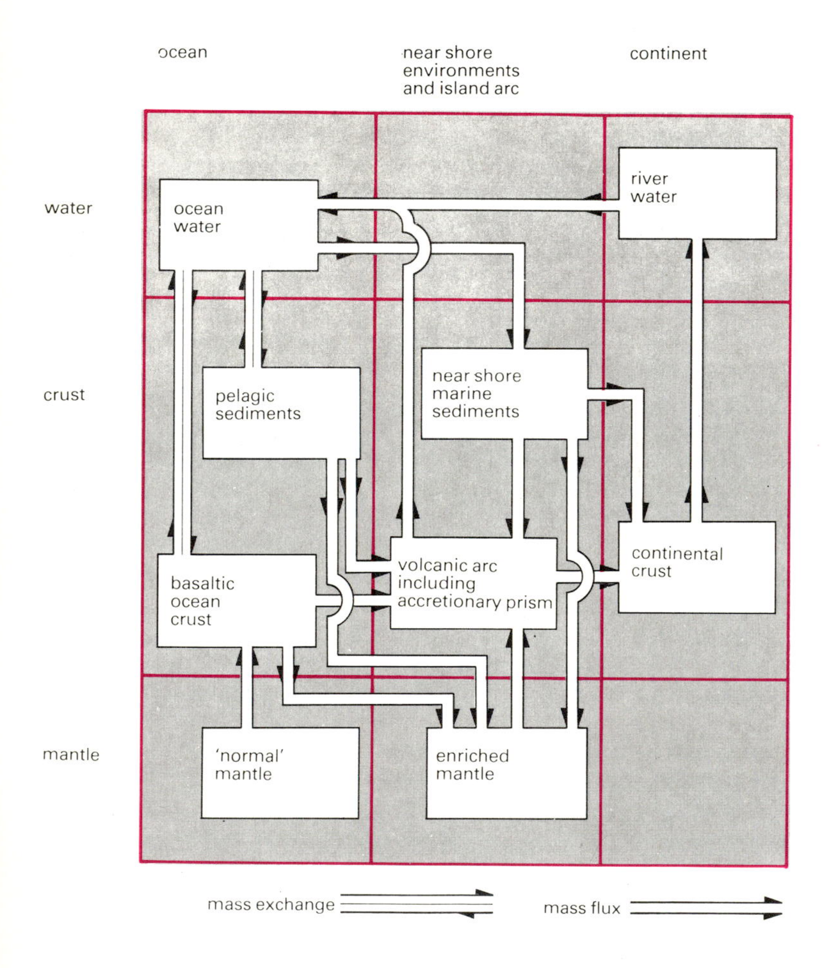

14.24: *Schematic flow diagram showing possible pathways for individual elements through the plate tectonic cycle.*

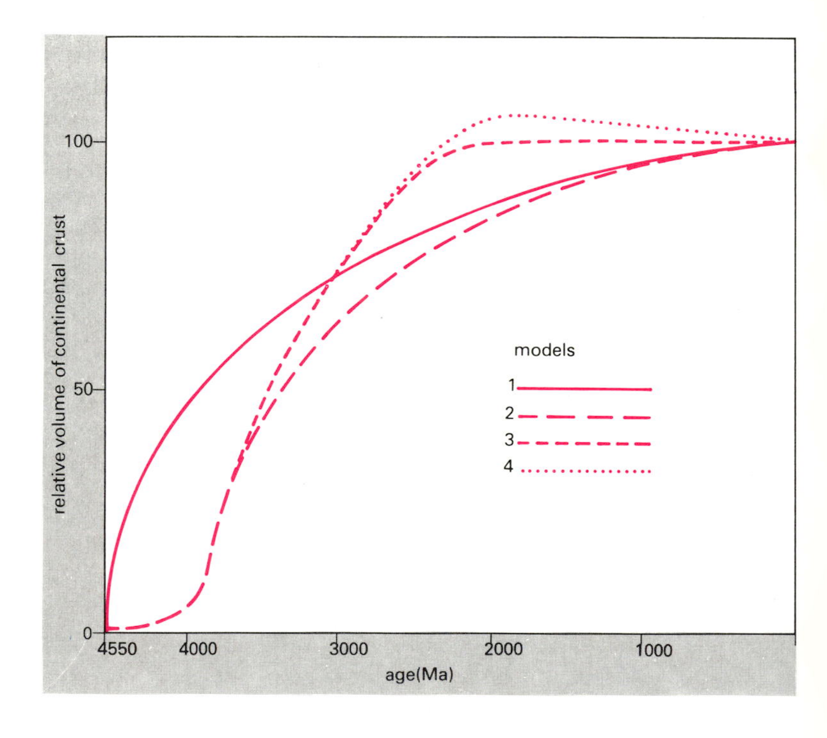

14.25: *Four possible models for the growth in volume of the continental crust through geological time.*

crust. A more radical fourth model proposes that the continental volume was greatest during the Archaean and has since diminished because subduction of continental material was more rapid than continental growth. Although model 2 probably has the greatest support at present, followed by model 3, the mechanism and rate of growth of continental crust is a subject of vigorous debate, and none of the models can be completely excluded on the basis of present knowledge.

Mineralization

Ore deposits are genetically related to their host rocks and to their overall geological setting. Following recognition of the relationships of volcanic and intrusive rocks to island arc, cordilleran and continent collision zones, mineralization at destructive margins can now be discussed in terms of plate tectonics. Such mineralization occurs in four settings: island arc–cordilleran margins, back-arc regions behind both the island arcs and volcanic belts of active continental margins, outer regions of island arcs, and continent–continent collision zones. The characteristics of island arcs and cordilleran margins were described earlier. Back-arc regions lie on the landward side of volcanic belts on active continental margins, up to 700 km from the trench. They may include a fold-thrust belt with thrusts dipping towards the trench. Examples of such back-arc belts are the eastern Cordilleras of Bolivia, and the Cretaceous fold-thrust belt of the North American Cordilleras. Outer arc regions lie on the oceanic side of some island arcs and consist of a wedge of flysch-type sediment forming an accretionary prism as described above. The outer arc region also includes ophiolitic masses of oceanic crust, formed at a spreading centre and tectonically emplaced within the arc. These may have characteristic mineral deposits which, since they are now exposed at destructive plate margins, are briefly reviewed here. Some outer arc regions, for example in south-west Japan, the Mentawi Islands (west of Sumatra) and the Alaska–Aleutians arc are intruded by granitic rocks. Finally collision zones resulting from either island arc–continent or continent–continent collision are sites of mineralization. In all four settings mineralization is associated with both volcanic and intrusive rocks.

Mineralization at island arc–cordilleran margins

At island arc–cordilleran margins the major ores are porphyry copper deposits associated with porphyritic intrusions and Kuroko-type sulphide deposits associated with basic volcanic rocks.

Porphyry copper deposits are of large tonnage and low grade, containing between about 0.4 and 1.0 per cent copper disseminated in largely igneous host rocks, which commonly include porphyritic diorite, granodiorite and/or granite intrusions. The primary ore minerals are chalcopyrite ($CuFeS_2$), bornite (Cu_5FeS_4) and pyrite (FeS_2), but these are frequently altered by hydrothermal solutions and redistributed by downward percolation of groundwater to form subsurface 'blankets' containing oxidized copper minerals in which the grade may be increased to 1–2 per cent copper. Analogous deposits of disseminated molybdenum are called 'porphyry molybdenum deposits'. Over 70 per cent of the world's present copper production and nearly 100 per cent of its molybdenum production are from such porphyry deposits.

Porphyry copper and molybdenum deposits are associated with intrusive rocks, and so cannot be directly related to active island arc or cordilleran margins. However, it is clear from their distribution in belts identified as Mesozoic and Cenozoic island arcs and cordilleran margins that such porphyry deposits are formed in destructive margin settings. Here the source of the metals poses problems similar to those concerning the source of the associated calc-alkaline magmas. Since the metals are essentially trace components of the magmas, their origin is a matter of additional uncertainty: they may be derived from small amounts of subducted metal-rich pelagic clays, melted or dehydrated oceanic crust or the mantle wedge; or they may be concentrated from within the crust during crystallization of the porphyritic intrusions.

The major mineralization associated with volcanic rocks in the island arc setting takes the form of lenslike bodies rich in sulphides and conformable with the stratification of the submarine volcanic host rocks. These are known as Kuroko-type deposits, after mines on Honshu Island, Japan. The main metals, in decreasing order of abundance, are iron, zinc, lead and copper, with minor variable amounts of silver and gold. The grades of Kuroko-type deposits are higher than for porphyry copper deposits, commonly about 7 per cent combined lead and zinc, but because they are less common the world metal production from them is much less.

Kuroko-type deposits occur in many young island arcs and some cordilleran margins, eg in Mexico and Chile. They are believed to result from submarine hot-spring activity related to explosions on the flanks of the submarine dacite or rhyolite domes. The ore fluid that deposited the sulphides is believed to be a mixture of magmatic water, derived from the cooling volcanic dome, and large volumes of heated seawater. The sources of the metals may be the same as for porphyry deposits.

Back-arc mineralization

The world's best-known and most productive tin province is in Bolivia in the back-arc region to the east of the active Andean volcanic belt. The ore is disseminated throughout dacite and rhyolite volcanic rocks (lavas and ignimbrites), sedimentary rocks and porphyritic intrusions. In a similar tectonic setting in Burma and western Thailand, tin (cassiterite) and tungsten (wolframite—$FeWO_4$, and scheelite—$CaWO_4$) are associated with Cretaceous to Palaeocene intrusive rocks including muscovite–biotite–granites, intruded into Carboniferous sedimentary rocks, the host rocks for the mineralization.

Since back-arc regions are closely associated with active volcanic belts, but are more distant from the trench and the underlying subducted oceanic plate, the source of the ore metals is even more uncertain than in island arc–cordilleran mineral deposits. The metals may be derived from subducted oceanic crust or sediment, or may be extracted from continental crust by fluids expelled from the subducted oceanic lithosphere.

Outer arc mineralization

Cenozoic granitic intrusions occur within deformed flysch on the ocean side (outer zone) of Japan, the Mentawi Islands and the Aleutian arc. The Japanese granitic rocks have S-type characteristics and are associated with tin–copper–arsenic deposits, locally containing lead and zinc. The emplacement of granites within flysch behind a trench rather than in the volcanic arc is unusual, implying that the source of the metals was within the flysch rather than the underlying subducted oceanic lithosphere.

The outer arc area may contain ophiolitic masses of oceanic crust, a complete section through which may include pelagic sediment, pillow basalts, dykes, gabbros and harzburgites or dunites, these representing successively deeper layers in the oceanic crust and underlying mantle. These ophiolites have characteristic mineralization. Within the pillow lavas disseminated sulphide deposits include pyrite (FeS_2) together with chalcopyrite ($CuFeS_2$), and sphalerite (ZnS) with minor galena (PbS) and pyrrhotite ($Fe_{0.8-1.0}S$). The occurrence of metal-rich brines and muds in the Red Sea axial trough indicates that such sulphide deposits formed on the ocean floor. A second type of deposit consists of lenses or pods of chromite ($FeCr_2O_4$) or nickel sulphide segregations within the harzburgite or dunite part of the ophiolite. Economic chromite deposits of this type occur in Cuba and in the Philippine island arc which also has nickel sulphide deposits. In some tropical island arcs weathering of harzburgite or dunite can produce laterites enriched in nickel.

Collision-zone mineralization

Collision zone granites are mainly of the S-type, thus differing from the mainly I-type granites of island arc and cordilleran margins (Figure 14.13). Many important tin deposits are associated with granites having these characteristics, and may therefore have formed within continental crust during or following collision with a continent or island arc.

The granites emplaced in the Himalayan continent–continent collision zone form large plutons and inclined sheets intruding Palaeozoic sedimentary rocks. They include muscovite–biotite–granites and some have high initial $^{87}Sr/^{86}Sr$ ratios, interpreted as indicating that they originated by fusion of ancient sialic crust. Some of these granites show extensive mineralization with cassiterite (SnO_2), xenotime (YPO_4) and scheelite ($CaWO_4$) around the margins. In older orogenic belts, tin-bearing granites with similar characteristics include the Hercynian granites of south-west England and the Erzgebirge (Czechoslovakia–East Germany) and late Triassic granites extending from central Thailand through Malaysia into Indonesia.

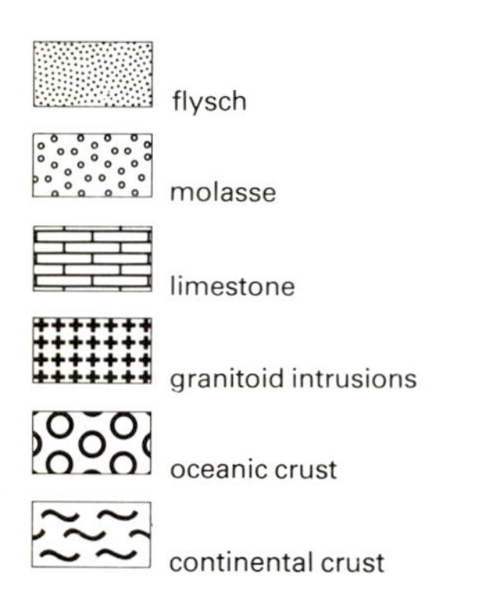

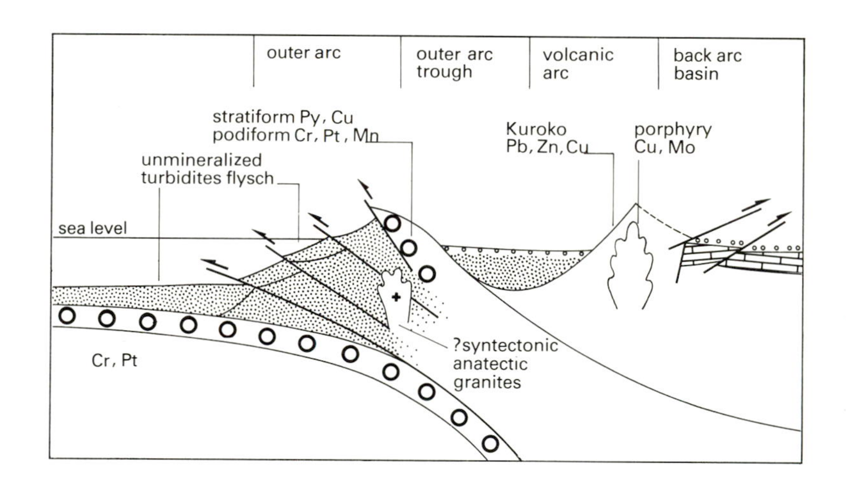

14.26: *Location of different kinds of mineralization in the different parts of an island arc orogenic belt.*

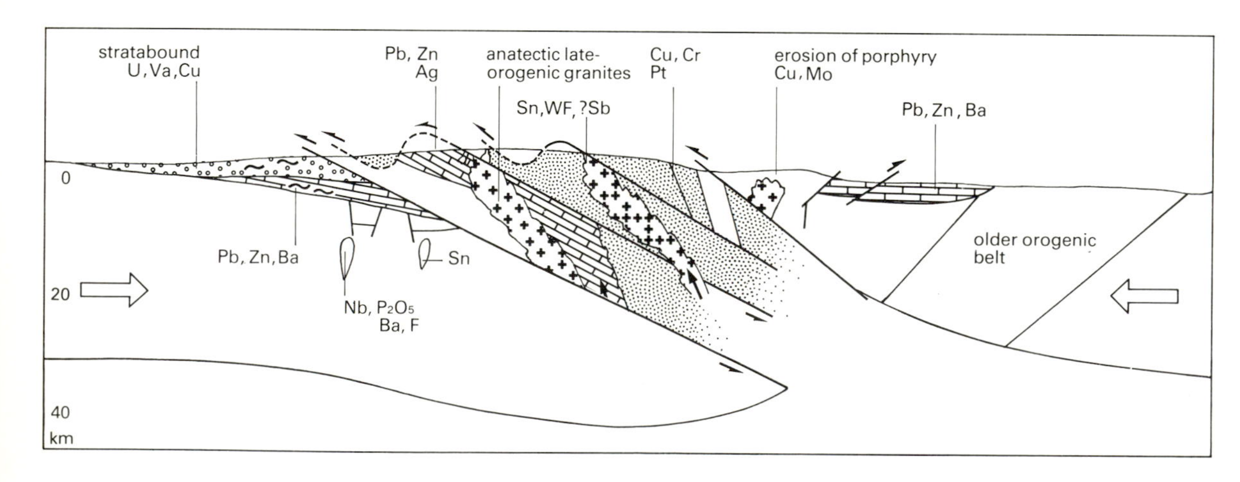

14.27: *Location of different kinds of mineralization in the late stages of a continent–continent collision zone.*

14.28: *Computer enhanced Landsat satellite image of Atacama drylake along the border between Chile and Bolivia. Volcanoes in the Altiplano region are visible along the eastern side of the picture. The Chuquicamata area containing one of the world's largest copper deposits is in the north-west corner.*

Summary

The above account indicates that island arcs and cordilleran margins characterized by volcanic and intrusive rocks of the associations described earlier may be highly mineralized areas of destructive margins. They include porphyry copper and molybdenum deposits (island arcs and cordilleran margins) and Kuroko-type metal sulphide deposits (island arcs). Back-arc regions are characterized by S-type granites with tin, tungsten and uranium deposits; outer arc regions, also with S-type granites, may contain tin and related metals. Collision-zone S-type granites emplaced within continental collision zones contain tin–tungsten mineralization.

There has been much argument about the value of plate tectonic concepts in the search for economic mineral deposits. Plate tectonics provides an elegant framework for the description of igneous activity and the petrogenesis of igneous rocks. At present, however, it seems that the association of different types of mineral deposit with distinctive types of igneous rock remains the best empirical guide to the possible occurrence of mineral deposits, regardless of genetic hypotheses based on lithospheric plate movements. Improvements in prospecting strategies will undoubtedly come with further understanding of the tectonic and geochemical processes at destructive plate margins.

15 Tectonics and sedimentation

The Earth as a recycling system

Since the advent of the plate tectonics theory and the development of more sophisticated mantle convection models it has become clear that the Earth's surface and interior processes are closely linked. The Earth can be regarded as a gigantic machine that creates and destroys lithosphere as it slowly loses internal heat. Some portions of the crust are elevated as a consequence of plate interactions, eg the sedimentary strata with shallow water fossils near the crest of Mount Everest, while others suffer depression, eg the great thicknesses of shallow water strata found at depth in sedimentary basins such as the North Sea or Gulf of Mexico.

Central to this vision of the Earth is the notion of planetary recycling. The agents of weathering and erosion act ceaselessly upon the uplifted portions of the crust around destructive plate margins such as the Andean chain or around areas of plate collision like the Alpine–Himalayan system. The sedimentary particles 'liberated' from pre-existing sedimentary, igneous and metamorphic rocks in these areas are transported by water, wind and mass flow (see Chapter 20) into areas of lower relief either within the continental landmass or at the continental–oceanic boundary. Great thicknesses of sedimentary strata may result because of persistent deposition of sediment in deep water or in areas of the crust undergoing subsidence.

Regions showing thick net accumulation of sediment are termed sedimentary basins. As the sedimentary particles become buried at depth they lose much of their water content, and mineral precipitation transforms the soft, wet sediment into dry, hard sedimentary rock. Geological activity such as plate motions may then cause folding and uplift, for instance as a passive plate margin is changed into an active margin again, and the vast recycling process starts anew.

The potential energy required for transport of sediment particles is ultimately provided by mantle heat by way of the motions of lithospheric plates. This chapter will examine the broad scope of recycling, with particular reference to the rates of processes involved.

Isostasy and Earth surface relief

Cross-sections through the Earth's crust and topmost mantle based on data gained from gravity and seismic studies show a number of interesting features (Figures 15.1 and 15.2). Continental crust is thicker under high mountain ranges than under low-lying terrain. On the other hand the greatest mountain ranges in the world, submarine mid-ocean ridges, are not accompanied by thickened oceanic crust but are underlain by substantially lighter materials.

The term isostasy has been given to the theory that accounts for observed correlations between surface relief and crustal composition and thickness. Isostasy is a simple application of Archimedes' principle and is best understood by analogy with floating icebergs. Whereas the Airy hypothesis achieves isostatic equilibrium by postulating variations in crustal thickness the Pratt hypothesis does so by having lateral variations in crustal or mantle density.

One of the best natural examples of isostasy in action is the continuing 'glacial rebound' of areas like the Canadian Shield and Baltic Shield (see Figure 15.5) which were overlain by great thicknesses of ice in the Quaternary Ice Age. The great weight of accumulated ice caused the Earth's crust to buckle downwards by as much as 700 m, the sag being accommodated by the creep of mantle materials away from the application of the extra load. Melting of the ice between 18 000 and 6000 years ago caused reverse flow of mantle material, giving uplift which continues today at rates of up to 10 mm per year.

Many loads applied onto the Earth's crust can, of course, be borne comfortably without significant isostatic downwarp. These are narrow or locally restricted loads like some volcanic edifices. The mechanisms by which isostatic equilibrium is reached are varied and not entirely

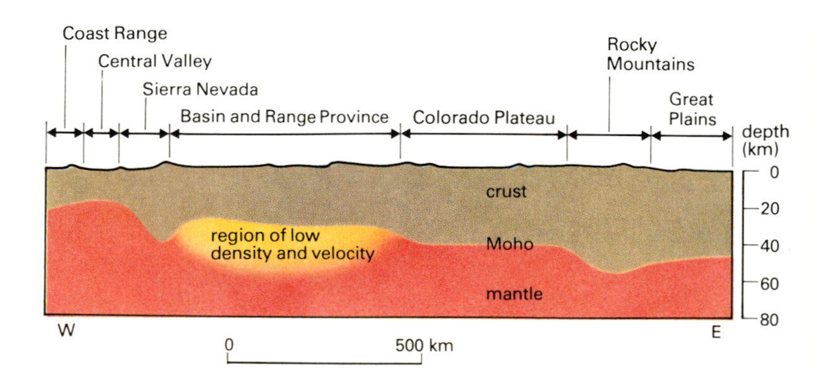

15.1: *Variations in crustal thickness from San Francisco, California, to Lamar, Colorado, as revealed by seismic refraction surveys. Note the thickened crustal root zones under the Sierra Nevada and the Rocky Mountains, compensating for excess surface relief as a form of Airy-type isostatic compensation.*

clear. The establishment of Pratt-type equilibrium under mid-ocean ridges and some rift valleys is caused by mantle partial melting over the ascending limbs of convection currents. A partial melt of mantle amounting to about 10 per cent leads to a significant density decrease, causing uplift and magma extrusion. Measurement of the amount of uplift of the ridges above the mean depth of the ocean floor (2–3 km) agrees well with independent calculations of the degree and likely density of mantle melting.

The origins of the crustal roots that underlie the great mountain belts of the world are more problematical. One possibility is that roots develop because of crustal underthrusting (Figure 15.3), as seen in the Himalayas. Another is that diapiric materials are added onto the base of the crust during subduction, the so-called 'underplating' process. A final possibility is that there is crustal shortening associated with plate

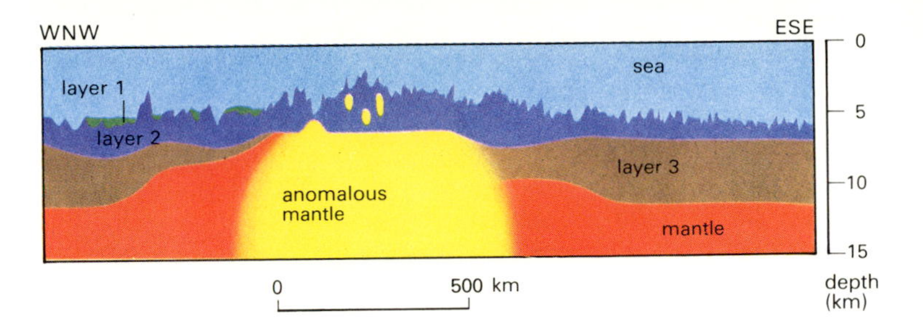

15.2: *Variation in crustal thickness, composition and relief across the mid-Atlantic ridge, from 36.6°N, 48.3°W to 25.5°N, 29.8°W, approximately. Layer 1 rocks are unconsolidated pelagic oceanic sediments present only in intermontane basins. Layer 2 rocks are basaltic in composition, comprising pillow lavas and dykes. Layer 3 rocks are amphibolites and gabbros. The anomalous (low-density) mantle under the mid-ocean ridge shows low seismic velocities due to partial melting above the ascending portion of a convection cell. Gravity measurements show that the mid-ocean ridge is in approximate isotatic equilibrium. This equilibrium is not due to crustal thickening (compare Figure 15.1) but to the presence of low-density upper mantle. The isostatic compensation is thus a form of Pratt-type isostasy.*

15.3: *A diagrammatic cross-section showing that the collision between the Indian and Eurasian plates in the Himalayas has a greatly thickened root due to lithospheric shortening by underthrusting along major faults during the Cenozoic era.*

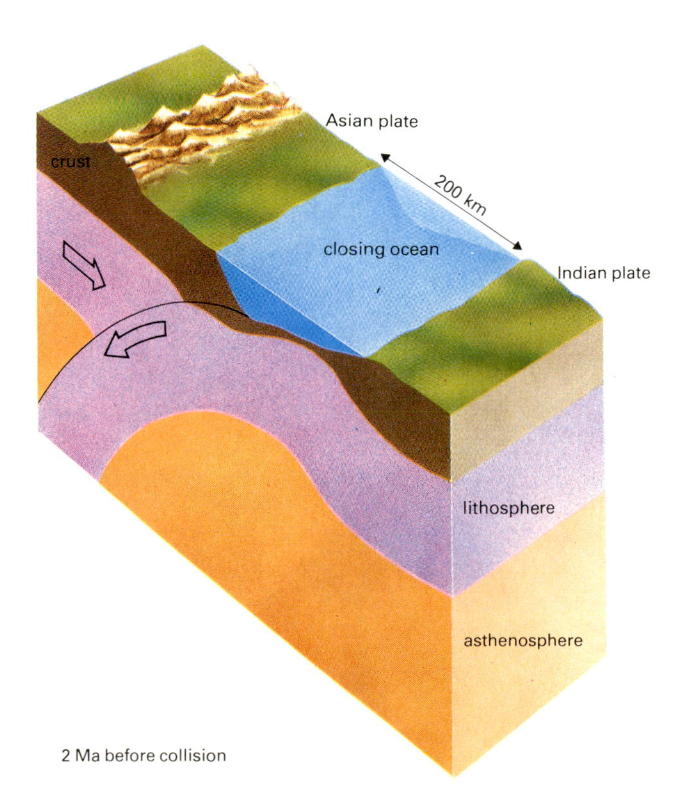

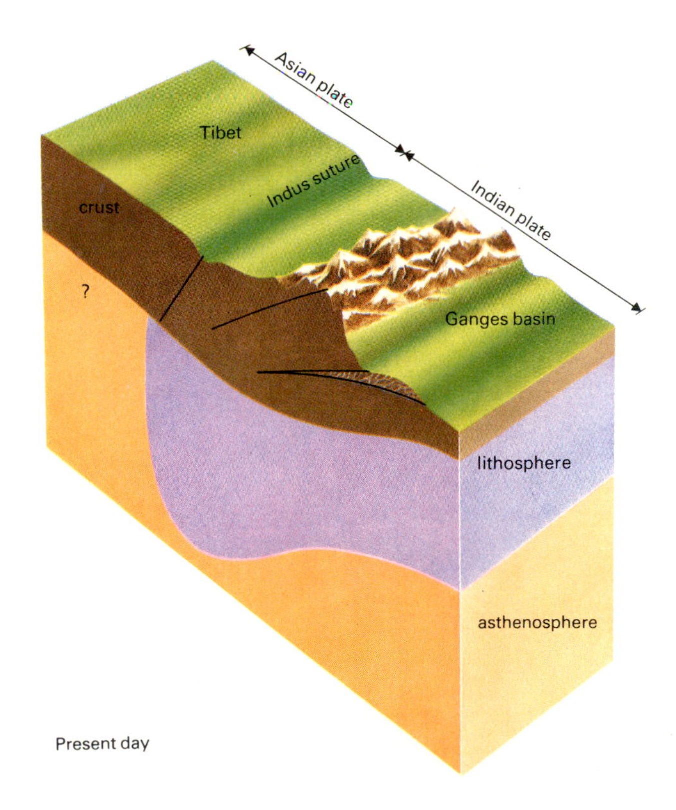

destruction. As mountain belts become denuded by erosion and weathering there must be a corresponding uplift to preserve isostatic equilibrium, until eventually both the mountain belt and its compensatory root disappear.

However, not all parts of the Earth are in isostatic equilibrium. Substantial departures occur in the ocean trenches and island arcs where great stresses associated with plate collisions cause local effects that can far exceed the compensatory stresses associated with isostasy. Once the restraining forces are released equilibrium will nevertheless rapidly occur. An analogy may be made with the removal of a finger pressing on to an ice cube in a drink: buoyancy forces will cause the cube to rise to its equilibrium level.

Vertical crustal movements and uplift rates

Concern with the lateral movement of plates over the Earth's surface (rates of movement of 20–75 mm per year) has tended to divert attention from important vertical crustal movements, both upward and downward, in many areas. This section will concentrate on positive vertical movements.

A broad if rather crude minimum estimate of mean crustal uplift rate may be gained from knowledge of the altitude of fossil layers known to have originated in shallow seas. This method involves the assumption that sea level has remained fairly constant. Thus Mesozoic marine faunas in rocks close to the summit of Everest provide indisputable evidence for uplift exceeding 9 km over about 100 million years.

A second method applicable to relatively recent geological uplift involves the dating of geomorphic surfaces such as raised coral reefs, raised beaches, raised cliff lines and warped planation horizons. Thorium or carbon isotope dating of calcium carbonate shell debris, or other dating methods (including archaeological techniques), may then be used to work out the time taken for the feature to have reached its present height. A disadvantage lies in the assumption that sea level has remained constant in the recent past, something that cannot be accepted (see page 244). After appropriate corrections have been made it has been shown that some oceanic islands have undergone average rates of uplift of between 0.3 and 2 mm per year. The warping of continental planation surfaces is an important gross indicator of uplift, but independent evidence is required for dating the beginning and end of the upwarp period in order to obtain the uplift rate.

Perhaps the most accurate measure of instantaneous uplift rate is provided by repeated, highly accurate levelling using precision instruments such as lasers. A direct indication of change of relief over the time span between the levelling intervals is provided in this way.

Finally use can be made of the tiny tracks left by the radioactive decay of ^{238}U in minerals such as apatite and zircon in the time since they cooled below a certain critical temperature (above which the fission tracks are not preserved) as mountain belts are uplifted after orogeny. The density of these tracks is proportional to the time interval of radioactive decay (Figure 15.4). Good correspondence is found between ages derived from fission tracks and from repeated accurate levelling in areas such as the Alps. Rates of vertical crustal movements thus measured in many present-day orogenic mountain belts are between 0.5 and 1.5 mm per year.

Three main types of regional crustal uplift may be considered. Orogenic uplift (rates vary between 1.5 and 17.5 mm per year) of mountain belts formed above subduction zones or along lithosphere sutures is the most spectacular. Uplift is a response to crustal thickening and leads to Airy-type isostatic compensation. Taphrogenic uplift (rates vary between 0.2 and 5.0 mm per year) occurs along rift valley features such as the continental graben of the Rhine, Baikal and East Africa and along the mid-ocean ridges. Uplift here is due to Pratt-type isostatic imbalance along constructive or incipient constructive plate margins produced by upper mantle partial melting above local, regional or worldwide convection cells. Isostatic rebound uplift occurs in response either to the unloading of the crust of thick continental ice sheets (Figure 15.5) or to continuing uplift of orogenic belts after initial isostatic equilibrium has been established because of unloading by erosion. Uplift rates of the latter type measured in the Alps are typically around 0.5 mm per year.

Rates of erosion versus rates of uplift

As soon as new sedimentary, igneous and metamorphic rocks in orogenic belts are raised above sea level they are weathered and the

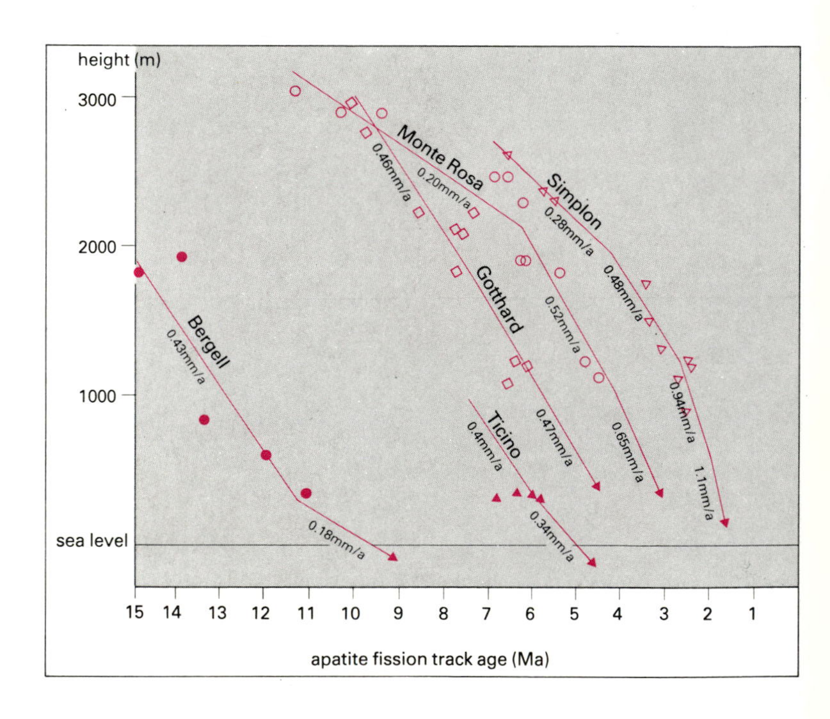

15.4: *Elevation of sampling location versus apatite fission track age in the central Alps. The ages from fission tracks are seen to be greater at higher than at lower altitudes. Since the ages indicate when the rock temperature passed below a critical threshold, the difference in age between samples from different levels gives a good estimate of the mean uplift rate as levels of rock passed below the lower cooling threshold for track production. Mean values of crustal uplift are given alongside the best-fit line for each location.*

resulting detritus is transferred by water, wind, ice and mass movements down the local valley gravity slope (Figure 15.6). The fact that very high mountains can exist at all is proof, however, that rates of erosion must be considerably less than rates of uplift. Measurements of sediment yield from small drainage basins in the western USA enable estimates to be made of the denudation rate, which shows a mean value of about 0.1 mm per year, significantly less than the uplift rates obtained from many orogenic belts. The dependence of the erosional process on climate and relief should be remembered when applying such figures. Increasing surface run-off will tend to cause an increase in sediment yield. Opposing this trend will be the effect of surface vegetation. Arid and semi-arid areas have little vegetation cover. The incoming of grass and forest cover as rainfall increases will tend to reduce drastically the proportion of weathered hill-slope mantle removed by surface run-off. For the geological past before the development of land vegetation it has been suggested that this constraint upon increasing sediment yield with increasing surface run-off was absent. The whole of the Earth's surface would thus have behaved like modern semi-arid areas. Progressive reduction of sediment yields in hinterland areas with rainfall greater than 250 mm would have followed the establishment of the early Mesozoic coniferous forests and the early Cenozoic grasses.

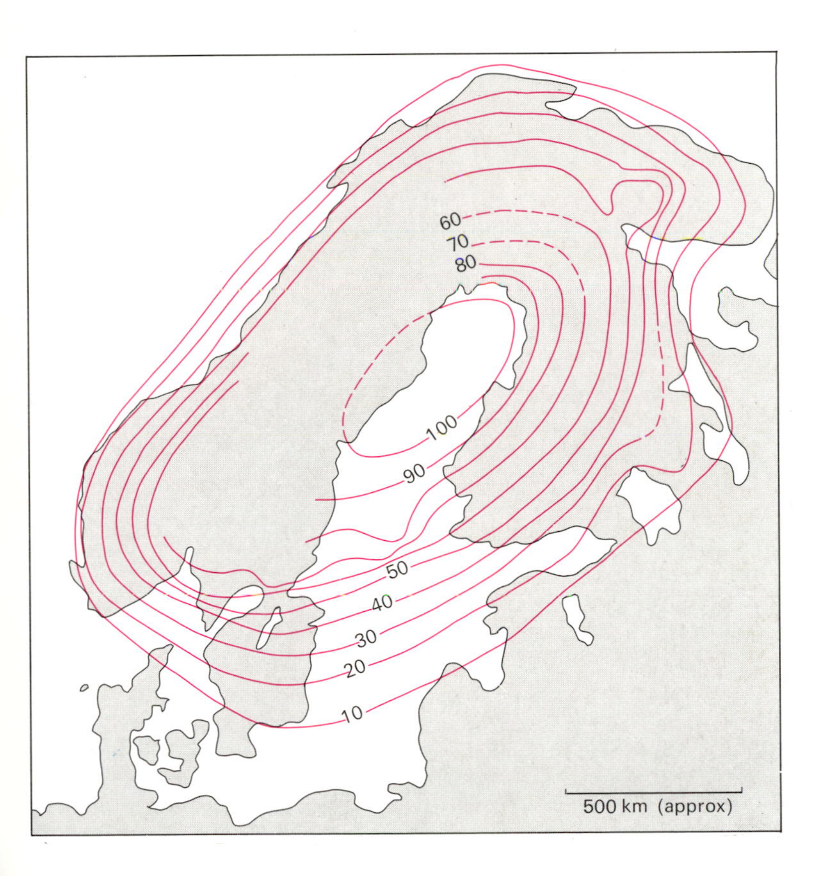

15.5: *A map of the present elevation (m) of the shoreline of 5000 years ago in Scandinavia to illustrate isostatic 'rebound' following deglaciation.*

One important consequence of rapid orogenic uplift is superimposed drainage, in which rivers cut deep gorges through the structural grain and through all types of rock. Superb examples occur in the Alps and Himalayas (Figure 15.7) and illustrate the fact that landscape erosion consists of two parts: the slope processes acting on valley sides (mass movement, avalanching, surface wash), and the downcutting process

15.6: *The braidplain of the River Durance in the French Alps. The river transports detritus from the Alps into the Rhône river system and hence into the Mediterranean. Terraces of Pleistocene river gravels in the background attest to periodic aggradation and degradation during glacial and interglacial epochs.*

of the stream itself. The valley form results from the delicate balance between the two.

A final point regarding erosion and its regional effects is an isostatic one. Just as the addition of new light material beneath an orogenic belt causes orogenic uplift due to root production, so the removal of material by erosion must cause uplift, since after a time the mountain will be overcompensated by the light root. Erosion here will cause periodic uplift through isostatic rebound once a critical threshold is reached. Uplift of areas like the Alps at the present time (~0.5 mm per year) may partly result from this mechanism.

15.7: *The Gorge de Guile in the Briançonnais zone of the French Alps. The gorge cuts vertically through nappes folded and uplifted in the late Cenozoic. It is a good example of superimposed drainage, the river having continued to cut down as the nappe pile was uplifted.*

Areas and rates of deposition and subsidence

Eroded detritus from highlands is deposited as sedimentary strata. Deposition may occur in various sedimentary environments, which may be broadly classified as continental, continental margin and oceanic. The former includes great inland drainage basins, lakes and continental deserts and the latter coastal plains, deltas, shorelines, shelves, continental slopes/rises and ocean basins.

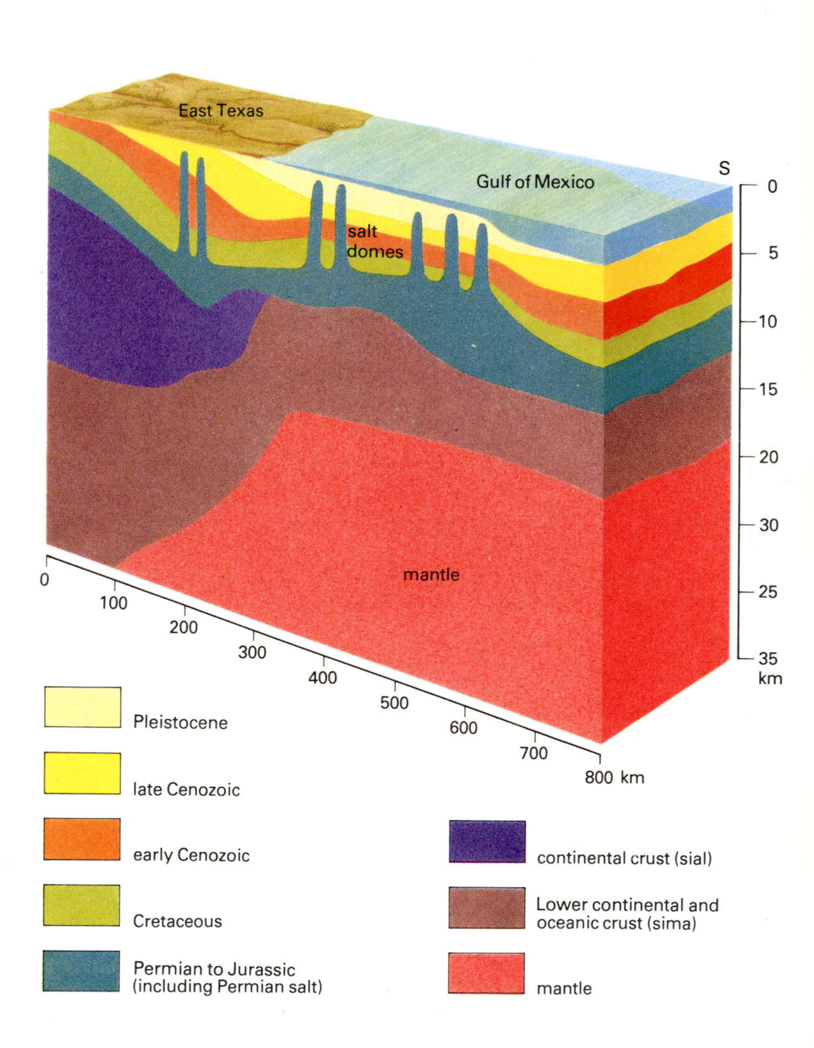

15.8: *North–south section from east Texas to the Gulf of Mexico to show the thick sequence of sediments deposited during the Mesozoic and Cenozoic. The Gulf has been a subsiding continental margin since its formation in the Permian (the salt domes originate from a thick layer of Permian halite). The main centres of deposition, defined by zones of maximum sediment thickness, have moved seawards as the continental shelf has built outwards. Much of the Cenozoic sequence comprises deltaic clastic facies laid down by drainage systems ancestral to the modern Mississippi and Rio Grande rivers and their neighbours.*

Great thicknesses of continental margin deposits, extending back in time to about two hundred million years ago, when the present sea floor spreading cycle began, exist in many areas of the world. For example, more than 8 km of Permian to Recent sediments occur in the Gulf of Mexico (Figure 15.8); more than 4 km in the North Sea (Figure 15.9); and more than 7 km in the Niger delta. The question arises as to how such great thicknesses may accumulate.

One possibility is that of deep basin infill, sediment wedges being deposited by simple outbuilding into relatively deep water. Examples occur in abundance in the world's major deltas (Figure 15.10), where there is rapid delta outbuilding (termed progradation). Other spectacular examples include the greater submarine fans of the ocean basins, often fed by turbidity currents sourced in adjacent shelf or deltaic areas via submarine canyons. Areas such as the Alps and Himalayas are bounded by so-called foredeeps which are depressed areas adjacent to the rising orogenic belts or nappe folds. These foredeeps are infilled with very coarse-grained sediments eroded from the nearby mountains. Tremendous thicknesses of sediment may accumulate in such areas, aided by substantial additional subsidence due to the isostatic effects of the weight of deposited sediment. Consider, for example, a delta building out into a marine basin 2 km deep and filling the basin to sea level with sediment. Sediment of density, say, 2000 kg per m^3 thus displaces water of density 1000 kg per m^3. The extra load acts upon underlying mantle material of density 3300 kg per m^3. Extra subsidence, magnitude $2 \times (3300 - 1000/3300 - 2000) = 3.4$ km, is thus possible at equilibrium.

A second sort of process, in contrast, involves continuing deposition in fairly shallow water. Deep boreholes in such sedimentary basins reveal great thicknesses of shallow-water sedimentary material. A mechanism is thus required for active and prolonged subsidence simultaneous with sedimentation over tens or hundreds of millions of years (additional subsidence will arise due to loading, as noted previously). Such sedimentary basins occur mainly on the passive margins of Atlantic type and along the great zones of continental rift valleys, providing good clues as to their primary origins. Possible mechanisms include cooling of initially hot upper mantle, creep of lower crustal material out towards the ocean basin and lithosphere stretching followed by rise of hot asthenosphere, causing subsidence as the asthenosphere cools down. The basins develop through crustal fragmentation due to tensional stresses generated during and after the rifting that accompanies plate break-up. Many such basins are fault-bounded with active faulted margins, although in the case of the major North Sea basin, a later phase of basinal sagging, originating by creep or thermal decay, was superimposed upon the earlier rifted phase.

A further type of faulted basin characterizes continental margins dominated by great transform faults such as the San Andreas and associated faults of southern California or the Alpine fault of New Zealand. Here the transcurrent motion along irregular boundaries creates local downwarps which subside extremely rapidly and are filled by internal continental deposits of streams, alluvial fans and lake bodies.

To a large degree the rate of subsidence of the Earth's crust controls the extent of preservation of deposited sediment. Direct determination of present-day subsidence rates may be achieved by repeated accurate levelling. Values from sedimentary basins fall between 0.3 and 2.5 mm per year. The indirect determination of subsidence rates in long-lived sedimentary basins is possible by dividing the total sediment thickness by the time interval of deposition. This procedure assumes that the whole sedimentary infill was deposited at a rate comparable with the mean subsidence rate. Large errors may arise in such determinations since long periods of non-deposition or erosion may have occurred and compaction effects, such as the squeezing out of inter-

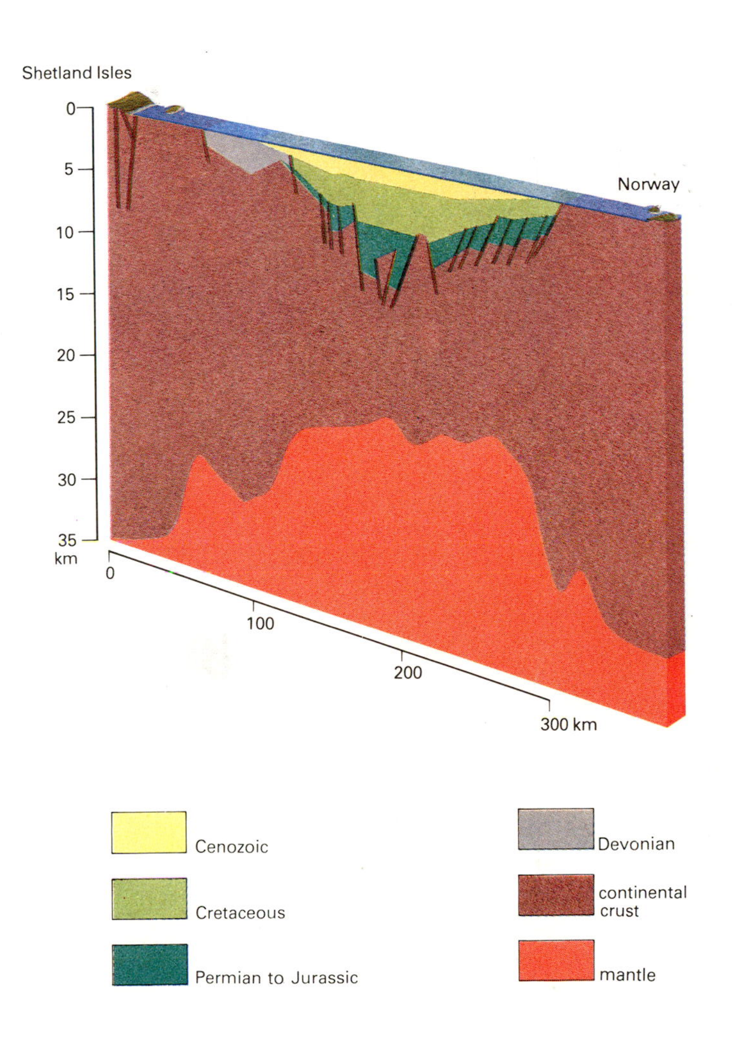

15.9: *A section across the northern North Sea from seismic refraction and reflection data. Note the form of the sedimentary basin, with its change from an early fault-bounded morphology to a later broad 'downsag'. Note also the marked crustal thinning under the basin, possibly due to lithosphere stretching. The relatively light Mesozoic and Cenozoic sediments are isostatically compensated by dense upper mantle above the general level (~30 km) of the mainland Mohorovičić discontinuity.*

stitial water, will progressively decrease the net sediment thickness as deposition and burial continue.

Similar problems arise in the determination of sediment deposition rates. Since some sediment may be partly eroded or reworked within a generally depositional period it becomes necessary to distinguish a net deposition rate, measured over a fairly long time interval, from a local, short-term deposition rate. The net rate takes into account all interruptions to sedimentation over a long time span. Sections of sedimentary rock found in the geological record may contain many gaps in sediment deposition at bedding planes and disconformities. Rates of short-term deposition thus only rarely apply to the whole of thick sedimentary successions. Short-term (Holocene) rates vary from the astonishingly low 0.0001 mm per year for modern pelagic oceanic red clays to rates in excess of 1.5 mm per year for flood sediments of the Mississippi River. In many coastal and coastal plain environments the rate of sediment deposition is roughly equal to the rates of crustal subsidence noted above.

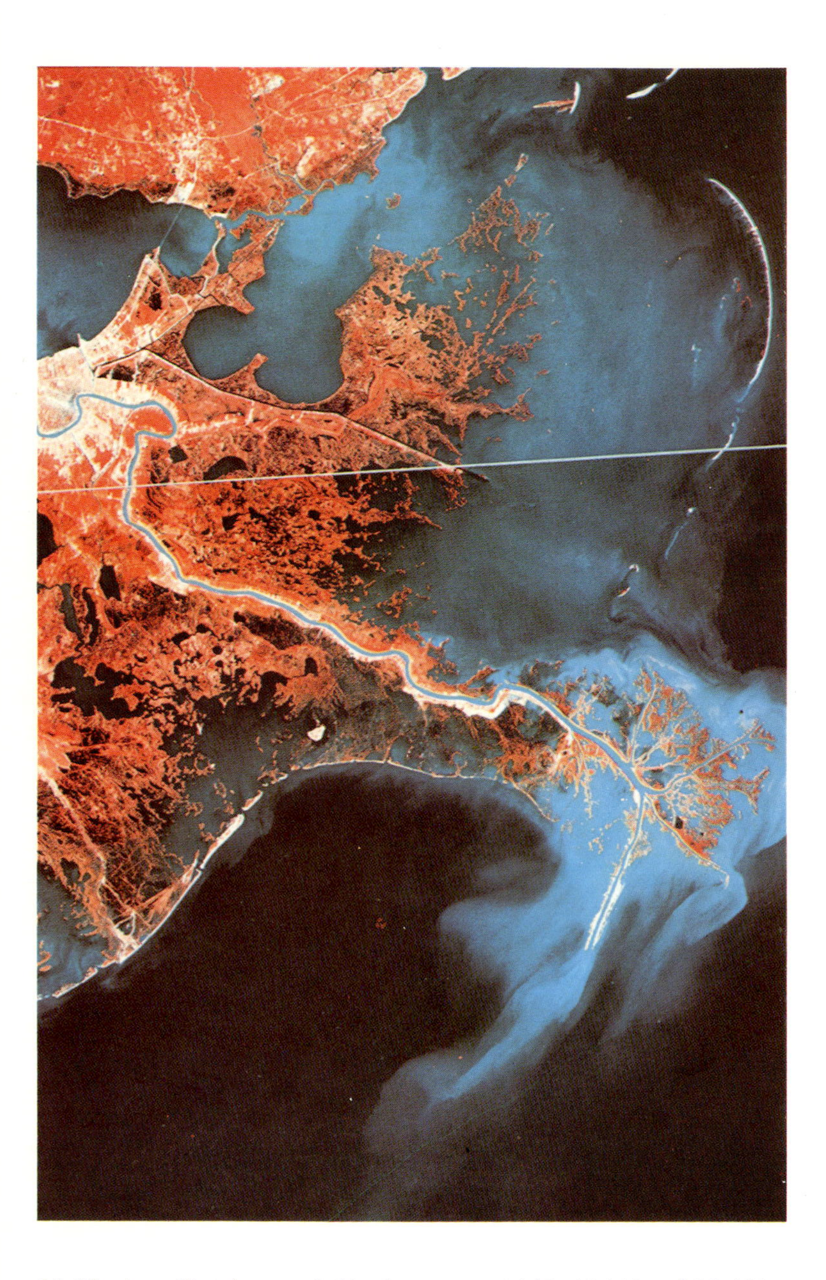

15.10: *A satellite photograph (Landsat images 2443–15 460 and 15 462) to show the Mississippi delta region. Much of the Cenozoic sediment (up to 6 km thick) deposited along the northern continental margin of the Gulf of Mexico has been supplied by precursors of the modern Mississippi and its neighbours.*

The rate of sedimentation is determined by dividing the deposited sediment thickness by the time interval of deposition. The latter quantity may be determined by fossil dating, radioactive decay of certain minerals or by direct measurement in the shorter term (Figure 15.11). One important factor that must be taken into account when measuring present-day deposition (and erosion) rates is the great modifications in the rates of Earth surface processes caused by human activities. This is particularly true of terrestrial environments where, for example, extensive deforestation may greatly increase sediment yields from drainage basins.

A still strident 'catastrophist' school among geologists maintains that such events as extreme river floods and hurricanes must dominate the sedimentary rock record. Students of terrestrial landforms (geomorphologists) have long recognized that the essential concept here is one of magnitude versus frequency. For many Earth surface processes the amount of work (the product of frequency and rate) performed by a natural event such as a river flow or a wave breaking must attain a maximum. The rarer, high-magnitude events ('catastrophes') cannot be held responsible for the majority of work performed in depositing or eroding sediment.

Eustasy, tectonism and sedimentary cycles

Eustasy is the term given to worldwide changes of sea level. The main cause of such changes in the fairly recent past has been the periodic formation and melting of great continental ice caps. Eustatic changes of sea level caused by this process are rapid by geological standards. For example the last great transgression that resulted after melting of the Quaternary ice sheets covering much of northern Europe and northern America is calculated to have caused the sea level to rise by about 7 mm per year. There is, however, good evidence gained from careful study of sediments preserved in sedimentary basins that much slower rises and falls of sea level have produced marked changes of sediment and faunal characters over wide areas. This has been worked out in most detail for Jurassic and Cretaceous times, when ice caps are thought to have been absent. Recognition of transgressional and regressional stratigraphic trends is of vital importance to hydrocarbon exploration, and much of the pioneer work in this field (now termed seismic stratigraphy) was done by oil company geologists with access to sophisticated seismic reflection profiles across continental margins.

The hypothesis now favoured to explain slow synchronous changes of sea level is based upon the periodic uplift and subsidence of mid-ocean ridges. The chief evidence for this hypothesis is the correlation of eustatic sea level falls and rises, as deduced from seismic stratigraphy, and the sedimentary evidence of shallowing and deepening, with periods of less and more rapid sea floor spreading, as deduced from magnetic anomalies of the oceanic crust. Rapid spreading is thought to be accompanied by 'swelling' of the mid-ocean ridge, presumably due to a Pratt-type isostatic compensation as more upper mantle material is melted beneath the ridge by a hotter than average convective overturn. Slow subsidence of the ridge system then follows as the spreading rate decreases. The sedimentary cycles produced in this way thus show deep-to-shallow-to-deep changes in depositional environments. Physical erosion during shallowing phases may lead to a minor unconformity on the shelf margin, which is then overlain by transgressive onlap of sediment deposited during the next rise in sea level.

Major tectono-sedimentary provinces

The sites of deposition of sediment eroded from upland areas and the nature of that sediment bear a direct relation to the major tectonic regimes. Just as distinct magmatic provinces associated with the various types of constructive and destructive plate margins can be defined (see Chapters 12–14), so also can sedimentary provinces which arise because of the various combinations of hinterland relief, composition and tectonic regime. The cycle of plate motions that begins with plate break-up and ocean formation and ends with an orogenic belt is thus reflected in the nature of sediment deposition.

The relationship between tectonics and sedimentation was postulated by geologists long before the basic tenets of plate tectonics were established. It is important to consider the attitudes of geologists to this relationship before the development of the plate tectonics theory. Perhaps the most important concept was that of the geosyncline: deep linear troughs were thought to have been progressively filled with sediment during the pre-folding, folding and post-folding histories of an orogenic belt. Studies of orogenic belts of various ages gave rise to a bewildering number of geosynclinal types that were appropriate to particular belts like the Alps, the Caledonides, the Appalachians and the Hercynides. There was much acrimonious debate concerning the correct or most general geosynclinal model. The rapid growth of the plate tectonics theory gave rise to a more logical approach based upon the classic geological principle of uniformitarianism (see Chapter 1). By studying modern sedimentary processes in relation to large-scale plate tectonic processes, coherent models can be established which may then be used to interpret orogenic belts. The diversity of plate motions and interactions with time is thus contained within a central strand of ordered theory. Many geologists are now actively debating the details of the relations between tectonics and sedimentation in particular orogenic belts. The most vital problem remaining is that of how far into the geological past the plate tectonics theory is applicable. It is doubtful, for example, whether Archaean orogenesis is explicable in terms of the theory.

In the sections that follow the interrelationships between tectonics and sedimentation in each stage of the plate tectonics cycle will be examined, with an indication of how the sedimentary products interrelate with the magmatic and other features already outlined.

Continental rift valleys

Rifts characteristically lie astride great areas of updomed crust and in many cases the graben itself forms a Y-shaped pattern in plan. The low-lying graben floor is separated from the adjacent horst by steeply dipping normal faults. Gravity surveys confirm the presence of these master normal faults and reveal large negative gravity anomalies over the graben, caused by crustal thinning of the downthrown block. Seismic surveys reveal that grabens are underlain by a 'pillow' of low-density mantle material which extends well under the horst blocks on either side. The pattern of rift occurrences over uplifted areas in Africa and their dimensions support the view that they have developed over the upwelling portions of polygonal convection cells under a stationary plate. The thesis here is to regard such rifts as incipient areas of plate break-up. Some rifts are still operative while others are now clearly dormant. Yet other examples lie deeply buried beneath the margins of opening oceans.

On a broad scale the development of a rift complex within a pre-

15.11: *Short to medium term erosion or deposition rates in modern environments may be measured as illustrated on this Scottish tidal flat. A plug of sediment of known length is extracted from the silty muds, a white paste of silica 'flour' and water being substituted in its place. After a period, in this case two years, the plug is dug up and examined. Erosion will reduce its length; deposition will produce layers of sediment above it. Here, deposition has occurred at a rate of about 5 mm per year.*

viously 'inert' plate has great consequences for drainage. River systems may become reversed or truncated and large lake bodies develop both within and adjacent to rifts. Some rifts provide the sites for major river locations, eg the Rhine river valley. Establishment of the graben gives rise to a system of alluvial fans along the fault-bounded margins which may be periodically rejuvenated by fault movement. Sediment is thus transferred from the horsts and graben margins into the graben interior where it is deposited rapidly on alluvial flats or as small deltas leading into lakes. In the lakes, depending upon climatic factors, sediment of biological origin is intermixed with the finest detrital sediments. Rifts in arid climates, such as the Kenya rifts, contain much chemical sediment produced by evaporation of shallow lake bodies. Rifts in humid climates may support an abundant plant cover and thus give rise to peat and, ultimately, coal deposits. Spectacular examples of thick brown coals in the lower Rhine graben are thought to have formed in the warm, humid Cenozoic climate. Interbedded with all these sedimentary strata are voluminous outpourings of lavas and pyroclastic material, cut by intrusive plugs.

Because the central graben can respond to loading, the dumping of sedimentary material may build up to such an extent that much further subsidence can occur in addition to infilling of the initial 'hole-in-the-ground'. Rifts are thus prime areas for preservation of deposited sediment and fossils. Preserved sediments in the Kenyan and Ethiopian rifts contain the earliest remains of fossil Man, who browsed and hunted beside the ancestral lakes of the graben some five million years ago.

Young opening oceans

Initial plate break-up gives rise to new ocean crust forming along a mid-ocean ridge. Two sorts of process seem to be responsible. In some areas such as the Red Sea the young opening ocean is directly preceded by a continental graben phase which may have lasted for several, or several tens of, million years. Portions of rifts abandoned during this process now form active or defunct graben that strike into the continental heartlands. These are termed failed arms. Ocean formation by this model must thus always leave a permanent sedimentary and igneous record in the portion of the pre-existing rift that will now lie astride the new continental margin (Figure 15.13). There are, however, many areas of the world where this record does not occur and where initial plate break-up seems to have occurred quickly by crustal thinning, monoclinal flexuring and coast-parallel dyke intrusion followed by voluminous outpourings of tholeiitic basalts. Examples are the eastern Greenland and South African continental margins.

Turning now to the sedimentary consequences of complete plate break-up, it is very commonly found that this was accompanied by massive development of rock salt precipitated from the marine waters that inundated the young oceans as they developed (Figure 15.12). Thus oil exploration boreholes around continental margins such as the eastern north Atlantic, Gulf of Mexico and Red Sea may show thickness of such salt of up to 3 km. Evidently the initial circulation of marine waters in young oceans in semi-arid climates was poor, allowing evaporation and eventual precipitation of salts. These initial sedimentary deposits of evaporative origin are then succeeded by deposits of more 'normal' coastal, shelf and deep ocean sedimentary material as the ocean connection becomes more permanent and efficient circulation occurs. Thus the Red Sea today is maintained at just above normal salinity via the Hormuz Straits. All along the Red Sea coast coral reefs have developed whose breakdown and attrition contributes biological sediment for future limestone development. Unlike that of more mature oceans the continental shelf of the Red Sea is very narrow. This feature and the low tidal range in the narrow ocean prohibit extensive tidal reworking of the shallow water sediments.

A further consequence of plate break-up is the inpouring of great continental river systems into the ocean margins. Initial opening of the south Atlantic was accompanied by the inflow of the great Niger–Benue river system along the failed arm of the Benue trough. The thick sedimentary fill to this trough has been dominated in the past sixty million years by the progradation of a great thickness of sediments onto the continental margin and ocean crust.

The above simple scheme is complicated by the formation of small subsidiary sedimentary basins floored by thinned continental or oceanic lithosphere separated from the main ocean by microcontinental plates. Such basins result from short-lived or abortive spreading centres.

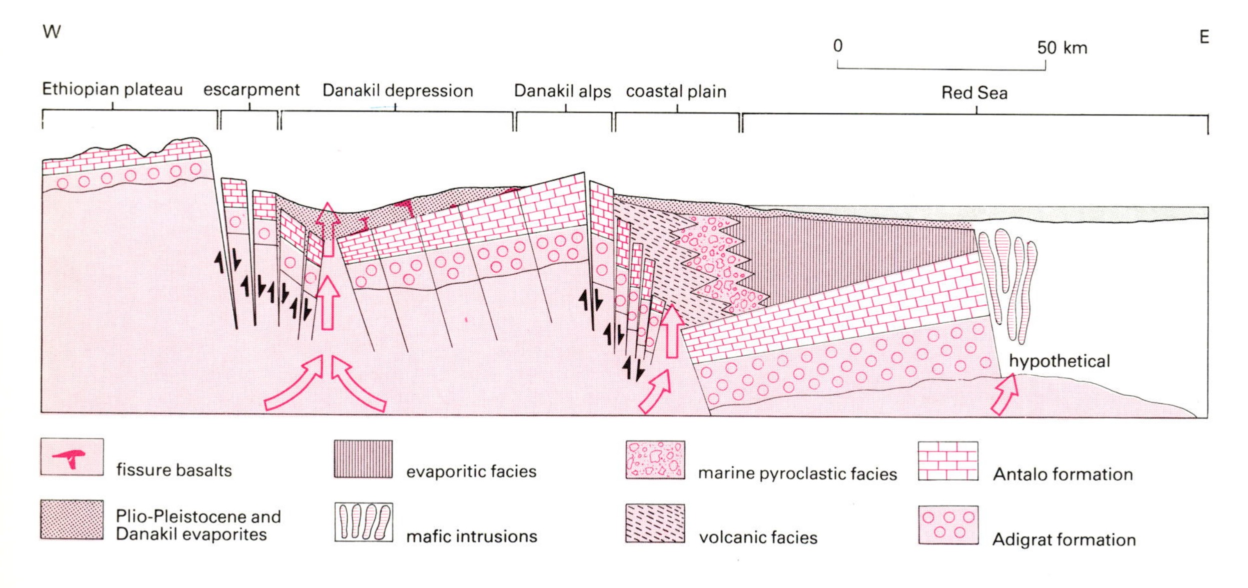

15.12: *A schematic section across the Red Sea and Danakil depression, Ethiopia, to show tilted fault blocks, sites of fissure basalt eruption and extent of subsurface Oligocene–Miocene evaporites formed by marine incursion into the Red Sea rift from the Indian Ocean. Sea floor spreading began in the Pliocene and Quaternary and has continued to the present.*

Many examples exist in the north Atlantic, the best being the Rockall basin and the Bahamas platform. Subsidence of such basins and platforms has led to a thick sequence of sedimentary deposits. If the bank is completely isolated from sources of detrital sediments, as in the Bahamas bank, the deposits consist mainly of calcium carbonate.

Mature opening oceans

The fringes of a mature opening ocean such as the Atlantic are the sites of sedimentary basins as continued crustal subsidence on the shelf keeps pace with sediment input. Deep boreholes drilled in search of oil and seismic studies show that such basins as the North Sea basin and those of the eastern North American seaboard have a sedimentary infill that dates back to the initiation of sea floor spreading in the adjacent ocean (Figures 15.8, 15.9, 15.13).

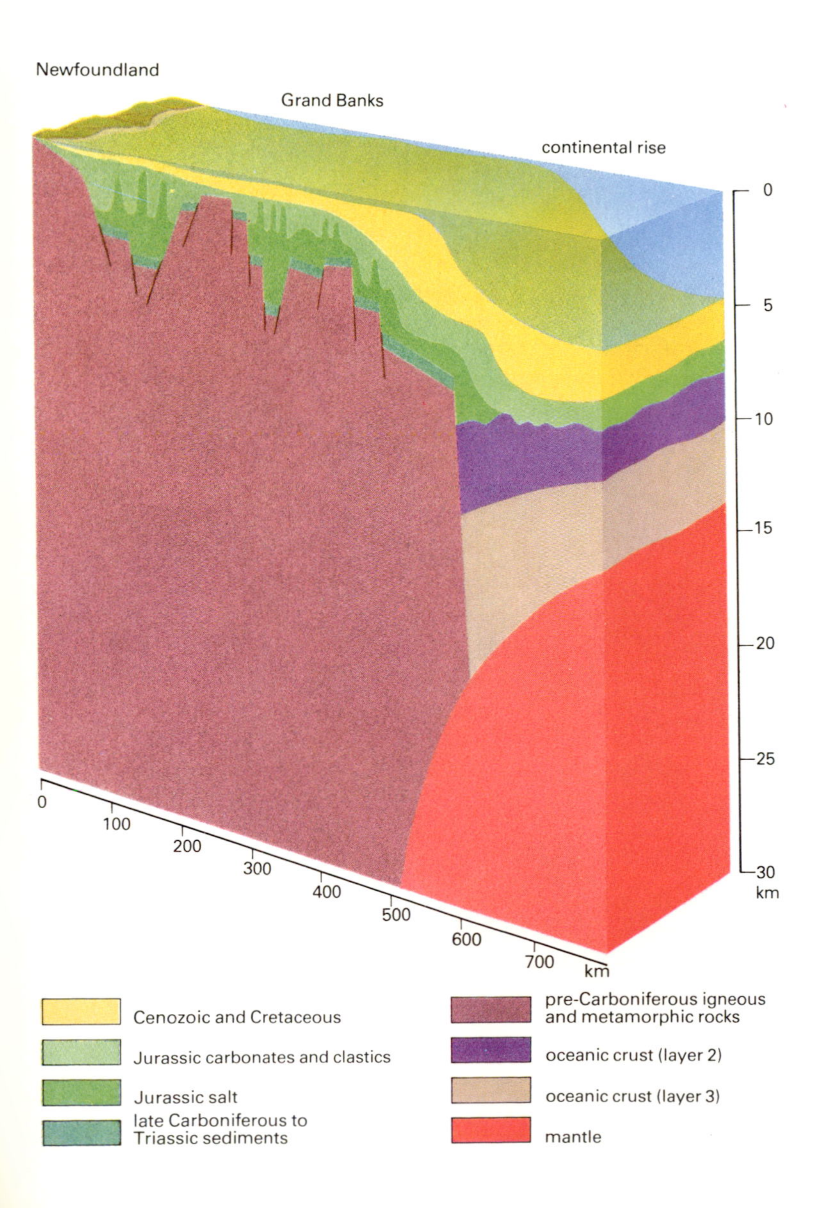

In mature opening oceans the hinterlands for the coastal plain rivers are relatively low-lying and subject to considerable weathering on stable hill slopes. The clastic sedimentary material thus produced tends to be composed of relatively stable minerals such as quartz, particularly if the 'oldlands' in the drainage system are themselves composed of sedimentary rock. Many margins show a broad low-lying coastal plain crossed by rivers if the climate is humid. Such coastal plains of alluviation may also include swamplands of peat, a future source of coal where the plain is part of a sedimentary basin undergoing subsidence. In arid climates the coastal plain sediments will contain evaporite deposits laid down in sabkhas—an Arabic term for coastal salt flats now widely used in geology (see Chapter 20). Major river systems such as those of the Mississippi (Figure 15.10) and the Niger bring down so much sediment in their waters that large deltas, again with extensive swamplands, are established despite the tendency of waves and tides to redistribute it along the coasts. As noted above such major depositional centres have a lengthy history, and the sediments in the deep subsurface of Nigeria, Texas and Louisiana are major oil and gas reservoirs. Outbuilding of a major deltaic wedge into deeper water causes much additional crustal subsidence by loading (Figure 15.8). Should the deltaic wedge be underlain by thick rock salt (precipitated as the oceanic basin first formed), plastic flow of the salt produces spectacular vertical columns termed salt domes. Some near-surface salt domes in the Gulf of Mexico are derived from as deep as 8 km, the salt being of Permian age.

The broad shelves of many ocean margins have been constructed by subsidence of continental crust and by sediment accumulations. The sediment brought into such shelves is usually widely redistributed by tidal and wave currents. Material deposited close to the shelf edges is further distributed by slumping-off down the continental rise as turbidity currents. Submarine canyons cut into the rise may funnel these currents and cause deposition of sediment on submarine fans or on abyssal plains. Many large deltas are fronted by such fans since the delta front slopes are particularly unstable and prone to sediment sliding and turbidity current generation. Some of this sediment brought into the deep oceans may be reworked by cold, dense, ocean floor currents (termed 'contour' or 'thermohaline' currents) that occur as polar waters sink down and travel towards the equator. Further out in the deepest abyssal plains of the oceans, less and less continent-derived sediment is deposited so that the ocean floor retains its rugged topography. The net effect of all these processes acting in the oceans seawards of the shallow shelf is to produce a great wedge of redeposited sediment. The thickness of the wedge off the eastern north Atlantic may reach 3 km.

Continental rifting, ocean crust formation and continued ocean opening combine to produce a fairly predictable sequence of sedimen-

15.13: *A diagrammatic cross-section across the Grand Banks, off Newfoundland. This section, typical of those across mature passive 'trailing edge' margins, shows evidence for down-faulted Palaeozoic basement whose graben and horsts are infilled by Jurassic sediments, including salt, deposited during the earliest phase of north Atlantic opening. Note the later progradation of Cretaceous and Cenozoic sediments out towards the Atlantic Ocean and the formation of a broad continental shelf.*

tary products. Because of great crustal subsidence many of these products are preserved as deep sedimentary sequences. Thus a borehole strategically placed on the margin of the north Atlantic off New York would reveal to the geologist the history of ocean development at that point since Permo-Triassic times. The margins of the great oceans, extending from coastal plain to abyssal plain, are modern geosynclines comprising linear basins parallel to the coast in continental crust having a shelf margin with shallow-water infill and continental rise basins with a deep-water infill.

Closing oceans

Subduction of ocean lithosphere completely transforms the tectonic and sedimentary system established on the continental margins of opening oceans. The sediment deposited on these continental rise and shelf basins is subject to folding and faulting due to compression arising from underthrusting. The massive uplift of these rocks to produce the cordilleran mountain chains that generally bound closing oceans is accompanied by extensive calc-alkaline plutonism and volcanism (Figure 15.14). As noted previously (Figure 15.3), the uplift is probably due to a process of underplating and magmatic addition which leads to the formation of a thickened crustal root zone.

During the process of subduction a new depository for sediment is formed along the deep oceanic trenches. The sediment eroded from rising mountain chains is transferred rapidly to the shelf where slumping or wave action causes downslope movement into the trench zone along submarine fans. Great zones of submarine 'landsliding' occur on certain steep slopes forming chaotic sedimentary deposits known as melange. The active underthrusting usually associated with subduction causes successively younger layers of the previously deposited continental rise sediment to be scraped off along reverse faults and plastered onto the trench wall to form a subduction complex. This process of accreting wedges has great value in the discovery and interpretation of trench positions in ancient orogenic belts. Thus recent work has established the Southern Uplands of Scotland as a subduction complex resulting from closure of the Iapetus Ocean in Ordovician and Silurian times. New sediments being introduced into the trench and its slope may coat such an accretionary wedge or infill an extensive fore-arc basin. If derived from the newly uplifted cordillera the sedimentary particles should include fragments of volcanic and plutonic rocks and minerals derived from the volcanic arc.

In many cases subduction leads to the formation of a volcanically active island arc separated from the continental plate by a broad shallow sea known as a back-arc basin, eg the South China Sea. Sediment derived from the adjacent continent is now dumped into the back-arc basin while active subduction is confined to the oceanward side of the island arc where an accretionary wedge may develop.

The plate dynamics of an ocean like the Pacific are much complicated by plate movements along transform faults such as the San Andreas system and by the juxtaposition of ridges and basins along the faulted continental margin. As noted previously the complex lateral movement of crustal blocks leads to very rapid subsidence and uplift in a complex pattern of relatively small basins. Sediment eroded from uplifted fault blocks is transferred into the beach zone and hence into the mouths of submarine canyons that extend very close inshore, as seen off the Californian coast.

The Mediterranean is an ocean of the greatest complexity, with an astonishing variety of microplates, subduction zones and transform faults. One event for which the proto-Mediterranean has become famous is the Messinian salinity crisis about five million years ago. The connection between north Africa and Spain which currently keeps the ocean at normal oceanic salinity was blocked off by some tectonic event in the Miocene, whereupon the whole ocean evaporated, producing a layer of evaporite salts (halite, anhydrite) up to 1 km thick. Large volumes of evaporite deposits may thus form when a closing ocean becomes isolated as well as when a newly opening ocean first becomes linked to the 'mother' ocean.

15.14: *The Alaska Range in south-eastern Alaska is a typical range of mountains along the edge of a subduction zone.*

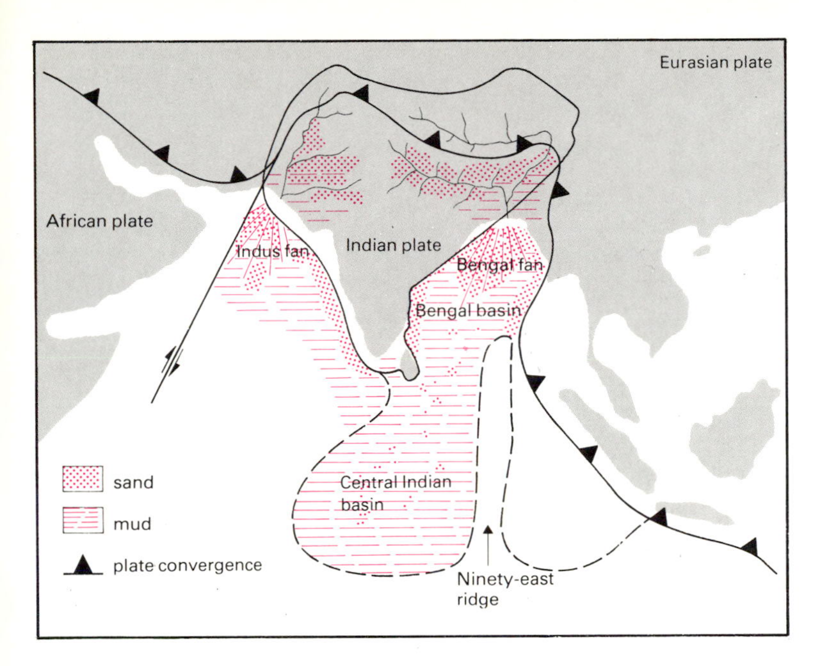

15.15: *Depositional patterns that have developed around the Indian subcontinental plate as a result of collision with the Euro-Asian plate. Himalayan molasse sediment transported by the Indus and Ganges–Brahmaputra river systems is deposited in deltas and gigantic submarine fans. Some 3 km of Cenozoic sediment are recorded in parts of the Bengal fan. The Nicobar fan was abandoned as a site of deposition when the Ninetyeast ridge and the Sunda trench converged, cutting off its supply of sediment.*

Continental plate collisions

Our knowledge of the effects of colliding plates on erosion and sediment deposition come from portions of the great Alpine–Himalayan fold mountain belt which partly resulted from the closure of the Tethyan ocean that had existed as an east–west feature since early in Mesozoic times. The impact of the collision process caused overthrusts and nappe folds to develop in many areas. For example, in the Austrian Alps a great slab of shallow-water-deposited limestones was thrust over the plate junction from the south, and today forms the region known as the Calcareous Alps. In the Himalayan belt, lithosphere thickening has occurred due to overthrusting (Figure 15.3).

The most important sedimentary effect of continental plate collision is the erosion and deposition of uplifted rocks that had previously accumulated along the ancient plate margins. Enormous quantities of sediment have been derived from the Alpine–Himalayan belt since their formation some thirty million years ago. This sediment was, and still is, being dumped in low-lying areas adjacent to the rising mountains as river or lake deposits and is known as molasse. Great thicknesses of molasse accumulated all around the Alpine arc during the Cenozoic period. The total thickness of molasse greatly exceeds that of all previously deposited sediment in some parts of the Alpine belt. For example, some 163 000 km^3 of sediment are preserved in north Alpine molasse basins, representing an average stripping of about 4 km in about the past twenty-five million years. Calculations of denudation rates (0.4–1 mm per year) during this period are $2\frac{1}{2}$ to 5 times greater than the observed stripping rate, implying that much of the eroded sediment bypasses the basins to be deposited elsewhere, eg the Po and Rhine basins, deltas and submarine fans. Along the southern margin of the Himalayan mountain ranges molasse is being laid down as fluvial deposits by the great river systems of the Indus and Ganges–Brahmaputra in a well-defined 'foredeep' basin. Sediment is also transferred into the adjacent oceans as gigantic submarine fans (Figure 15.15). All the major rivers that drain the Alpine–Himalayan area are transferring sediment into adjacent shelf and ocean basins, where deposition results on coastlines, deltas and submarine fans. Here the recycling concept is seen in action again, the deposited sediment forming the infill for the next generation of mountain belts which will inevitably grow as present-day plate motions produce new closing oceans and continental plate collisions.

Conclusion

This chapter has centred on the role of erosion and deposition of sediment in the vast recycling process that occurs as the Earth's lithospheric plates interact with each other. Tectonic and isostatic processes cause uplift and subsidence while geomorphic and sedimentary processes cause erosion and deposition. A logical scheme of birth, destruction and renewal will continue as long as sufficient heat is produced in the Earth's mantle to drive the plates around the surface of the globe. Long ago Hutton, the brilliant Scottish philosopher, conceived of the Earth as a recycling machine whose activity depended on heat from its 'central fire'. This profound deductive conclusion is elegantly confirmed by today's more detailed knowledge of tectonics and sedimentation, accumulated long after Hutton's first pronouncements on geological recycling astonished and outraged the thinking world in the early nineteenth century.

16 The history of the Earth's crust

The Archaean eon and before

For James Hutton in the eighteenth century the geological history of Britain held 'no vestige of a beginning nor any prospect of an end'. Although historical geology has advanced immensely since its early days Hutton's observation still has a ring of truth. The age of the Earth and of most meteorites is known, from painstaking study of isotope systematics (see Chapter 8), to be around 4600 million years (Ma). However, the oldest rocks known to exist on our planet and therefore the starting point for investigating the history of its crust are less than 3900 Ma old. This leaves a 700-Ma gap before the stock in trade of geologists became available. Quite literally, infinitely more is known about the early history of the Moon, Mercury and Mars than of the Earth, a fact for which there is a simple explanation. Major crustal activity on those planetary bodies slowed down to insignificant levels, in comparison with that of the Earth, around 3800 Ma ago, whereas the Earth continued in violent upheaval to the present day. The Earth's crust and its evolution are reflections of these overturns.

Evidence from the Moon, Mars and Mercury suggests that all the inner planets suffered episodic bombardment by smaller planetesimals from their accretion until about 3800 Ma ago (see Chapter 27). This is reflected in their deeply cratered surfaces. It is impossible that the Earth escaped such a battering; indeed, as it is a more massive body, it may have suffered a much greater planetesimal flux early in its history. There are two main factors explaining the lack of evidence for the missing 700 Ma. First, any primordial crust would have been disrupted by impacts, possibly beyond recognition. Second, any part that escaped would have been incorporated into younger materials and masked by their swamping effect. If, as will be suggested, primordial crust was of basic or ultrabasic rock, most of it would have been recycled into the Earth's interior by processes akin to subduction. On no other inner planet is there evidence for such recycling by mechanisms of a plate tectonic kind, and the earliest crust is still widely preserved, though in a much modified form.

The history of the Earth's crust cannot therefore be related from the beginning. The early part of crustal history can only be inferred from the evolution of its components which emerged successively out of the turbulent epoch of extraterrestrial processes. This chapter begins, therefore, by looking at rocks formed in the succeeding eon, the Archaean, and discussing some of the models for their formation.

16.1: *Dawn over Kilauea, Hawaii.*

The Archaean eon

With only a very few exceptions rocks older than 2500 Ma are found only in those areas that have been geologically stable, except for faulting, uplift and erosion, for the past 1000 Ma. These cratons are the nuclei of the major masses of continental crust. Only that part of the Archaean geological cycle that culminated in sialic crust is preserved. Some relics of Archaean oceanic crust may be locked in it,

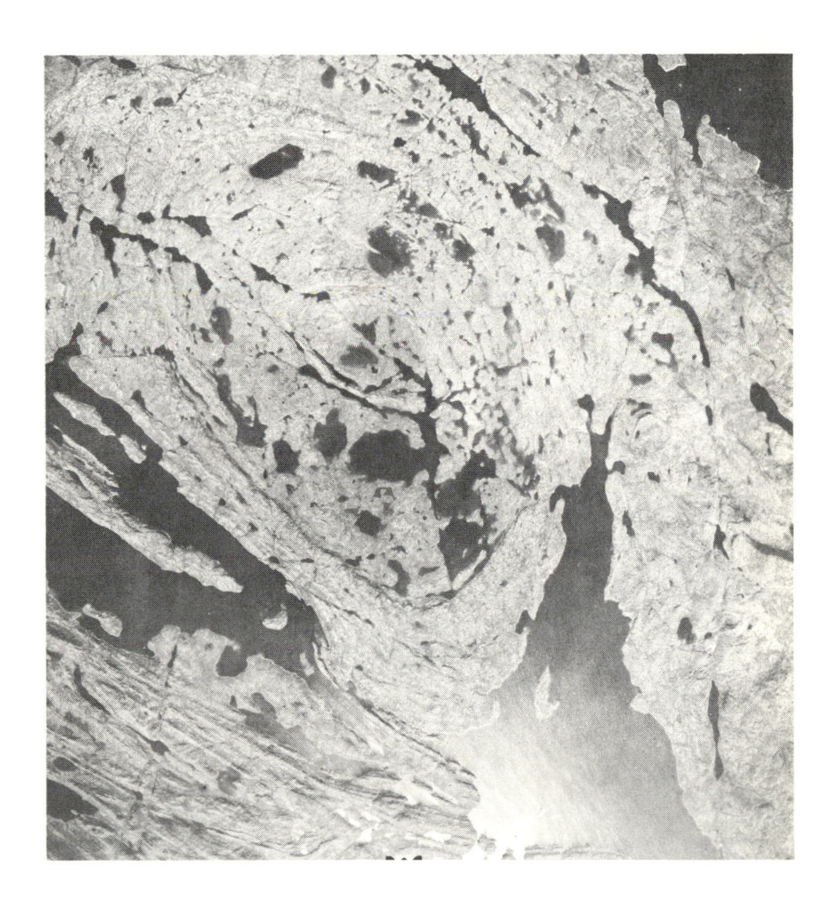

16.2: *An aerial photograph of complexly folded gneisses in the Archaean craton of west Greenland. The strange outcrop shapes reflect repeated refolding after the formation of the gneisses' banding.*

but their existence is by no means certain and such relics would probably be atypical.

Archaean geology appears simple on a gross scale. It is dominated by coarse-grained, sialic rocks which crudely approximate to granites, granodiorites or tonalites and can be loosely termed granitoids. These may be strongly deformed, banded metamorphic rocks called gneisses, which reveal an immensely complex structural history, or groups of homogeneous, clearly intrusive igneous masses. Set in this granitoid 'sea' are smaller areas of rocks which formed on top of the crust. These supracrustal materials comprise a mixture of volcanic and sedimentary associations. In some parts of the cratons the supracrustals are barely deformed and have suffered only slight metamorphism. The volcanic components have a greenish tinge imparted by hydrous metamorphic minerals, and hence they are often called greenstone belts. Such areas are low-grade terrains where the mutual relations between supracrustals and granitoids are fairly easy to see. Other areas have been repeatedly deformed and metamorphosed at temperatures as high as 900°C and as deep as 40 km. In these high-grade terrains the supracrustals have lost their primary textures through recrystallization, their relations to the granitoids have been obliterated and considerable geochemical changes have been induced.

It is in the low-grade terrains that evidence for magmatic, volcanic and surface processes can best be assembled. The high-grade terrains provide the key to understanding the way in which continental crust 'ripens' and becomes stabilized. Only in a very few areas is it possible to see how both types of Archaean terrain relate to one another because they are often separated by later zones of large-scale crustal movements.

16.3: *A false-colour Landsat image of the low-grade Archaean terrain of the Pilbara area, Western Australia. The white-to-yellow masses are large Archaean granitoid intrusions that have domed up the older greenstone belt volcanics and sediments (mainly blue and green) so that their bedding is nearly vertical yet closely follows the margins of the granitoids.*

16.4: *A detailed map of the world's oldest known rocks, the Isua supracrustals, reveals that they are structurally enveloped between slabs of slightly younger Amîtsoq gneisses, due to horizontal tectonics. Both major Archaean units are cut by undeformed basic dykes.*

The Earth's oldest rocks

A retreat by the Greenland ice cap, perhaps in historical times, exposed a horseshoe-shaped outcrop 1–3 km wide of supracrustal rocks at a place in west Greenland called Isua by the Innuit (Eskimo) people. It is bounded on both sides by granitoid gneisses, contains small intrusions of the gneisses and the overall shape is due to folding, thrusting and refolding. Both units and the structures are cut at high angles by basaltic dykes (Figure 16.4).

The whole Archaean complex of west Greenland is high grade. Most direct evidence for the age of one component relative to another has been removed and all units are parallel. In unravelling the area the basic dykes assumed an importance far beyond their volume or their petrogenetic significance. Among the gneisses one suite was intruded by the dykes, as at Isua, or contained deformed remnants of them. The other gneisses were free of dykes. The Greenland Geological Survey used this simple distribution in the late 1960s to divide the gneisses in the area around Godthaab into older, pre-dyke, Amîtsoq gneisses and post-dyke, Nûk gneisses. As all the rocks were known to be Archaean the two sets of gneisses became obvious subjects for radio-

16.5: *A conglomerate from Isua, evidently of sedimentary origin and containing cobbles of volcanic rock.*

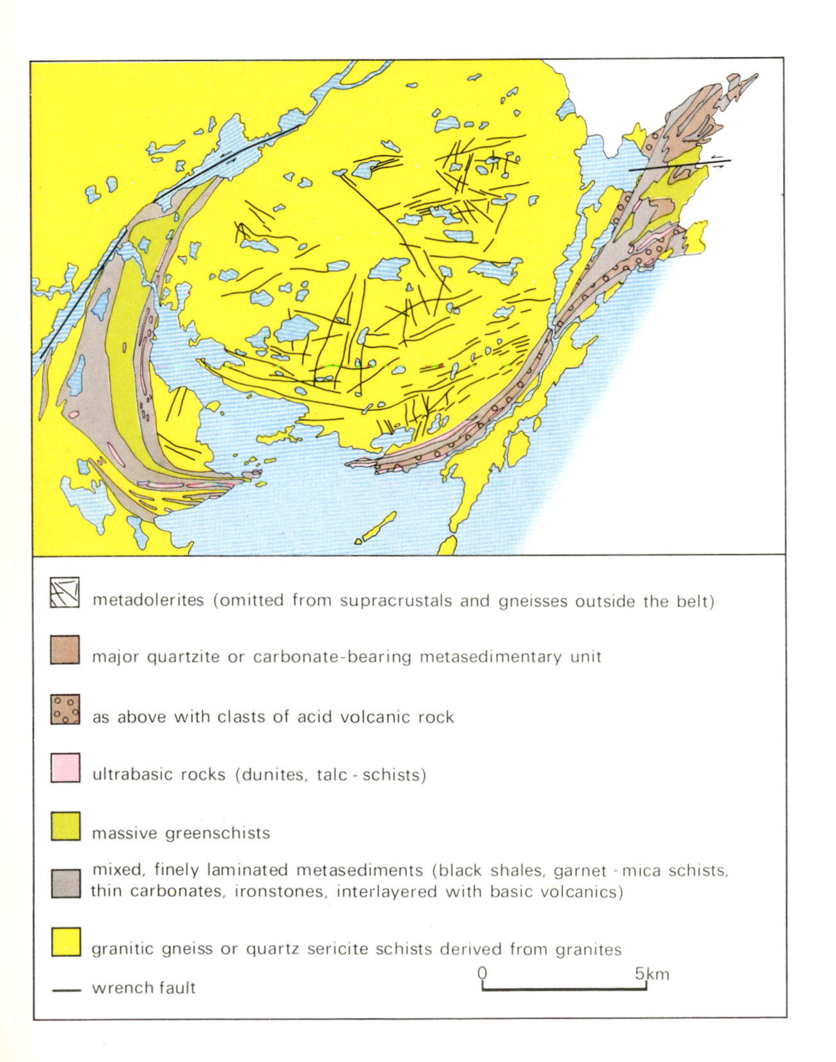

16.6: *An example of Amîtsoq gneisses at Isua in which the younger, intrusive basic igneous dykes have themselves been strongly deformed.*

metric dating. The Amîtsoq gneisses extended known geological time by more than 200 Ma, to around 3700 Ma. Moreover their strontium isotope ratios and geochemistry showed that they had originated from a mantle-depth source and were not derived from a primordial continental crust, which many geologists and geochemists had supposed to have arisen at the time of core formation. The Nûk gneisses, some 600 Ma younger, were formed in an analogous manner and show no connection in their strontium isotopes to the older Amîtsoq protocontinent. Very careful work on the Nûk lead isotopes, however, reveals variable contamination with lead of Amîtsoq age, which has provided some idea of the extent of the first known continental slab. It can have been only a few thousand square kilometres in area, as the vast majority of Nûk gneisses are uncontaminated by older crustal lead.

The metasediments at Isua, and smaller but similar blocks, streaks and lenses in the Amîtsoq gneisses, are clearly pre-Amîtsoq and thus undisputed as the world's oldest known rocks. As well as basic and ultrabasic lavas they contain chemically precipitated ironstones and cherts, rocks once rich in clay minerals and a conglomerate containing rounded cobbles of acid volcanic rocks. The last unit is dated at 3824 ± 10 Ma, but more important than its antiquity is the fact that it was clearly deposited by water, as were the other sedimentary units. Not only was the earliest Archaean surface environment dominated by liquid water, but stable-isotope studies from the ironstones strongly hint at active, though primitive, organic life (see Chapter 21).

The rocks from west Greenland, now being supplemented by rocks from other continents dated at around 3600 Ma, offer two clear hints about early geological processes. First, the characteristic geochemistry of the granitoid Amîtsoq gneisses was derived by partial melting of basic material containing the metamorphic mineral garnet, which suggests some form of subduction and associated magmatism. Second, although different in chemistry from most modern sediments, the Isua supracrustals were laid down in some kind of body of water (Figure 16.5). Both internal and surface processes by around 3800 Ma ago were not totally dissimilar from those prevailing today and, as will be seen, this has encouraged many geologists to adopt a uniformitarian approach to early terrestrial processes.

It is only when very ancient terrains of low metamorphic grade are examined that clear evidence emerges of profound differences between the Archaean and the present.

Low-grade Archaean terrains

Geologists have been active in the Archaean of southern Africa for nearly a century, for one simple reason—gold. The source for the Precambrian sedimentary gold deposits of the Witwatersrand have long been known to be the older, Archaean supracrustals of such places as the Barberton mountain land. The lowermost parts of this greenstone belt have been dated at 3540 ± 30 Ma, by the samarium–neodymium method. Being only slightly deformed and metamorphosed the Barberton area comprises the oldest known low-grade Archaean terrain (Figure 16.7).

As elsewhere, Archaean supracrustals at Barberton are engulfed by granitoids of various kinds and ages. One type, a complex gneiss, is suspected by many to have an age close to that of the Amîtsoq gneisses, though reliable dates are not yet available. The relationship between this ancient gneiss complex and the Barberton supracrustals is obscured by intervening younger granitoid intrusions. Many of these intrusions appear to have risen as blobs, so that the Barberton belt looks as though it has been hit by huge billiard balls. The net result is that it has proved impossible to demonstrate whether the supracrustals were laid down upon an earlier continental foundation or were formed as primitive crust.

The Barberton supracrustals represent a gradual change in surface conditions which may have taken longer than 400 Ma, equivalent to the time from the Devonian to the present. They are subdivided into a lower volcanic-dominated group and two younger sediment-dominated groups. The older of the sedimentary groups comprises turbidites containing detritus from the erosion of thick successions of volcanic rocks. It is only in the uppermost group, which is composed of sandstones and conglomerates, that there is clear evidence that continental crust was in existence and being eroded to supply quartz, feldspar and sometimes fragments of granitoids (Figure 16.8).

The oldest group consists of a series of volcanic–sedimentary cycles, lavas at the base of each cycle being overlain by thin sediments. The sediment units are dominantly cherts, which are either chemical precipitates of silica or formed by replacement of evaporite deposits, such as gypsum, by silica. There is no conclusive evidence that these sediments contain detritus derived by erosion of continental crust, and they most probably arose from fluids emanated from the volcanics or from the interaction of surface water with the volcanics. In the volcanic parts of the cycles there is a gradual upward change from basic and ultrabasic lavas to a basic–intermediate–acid lava association.

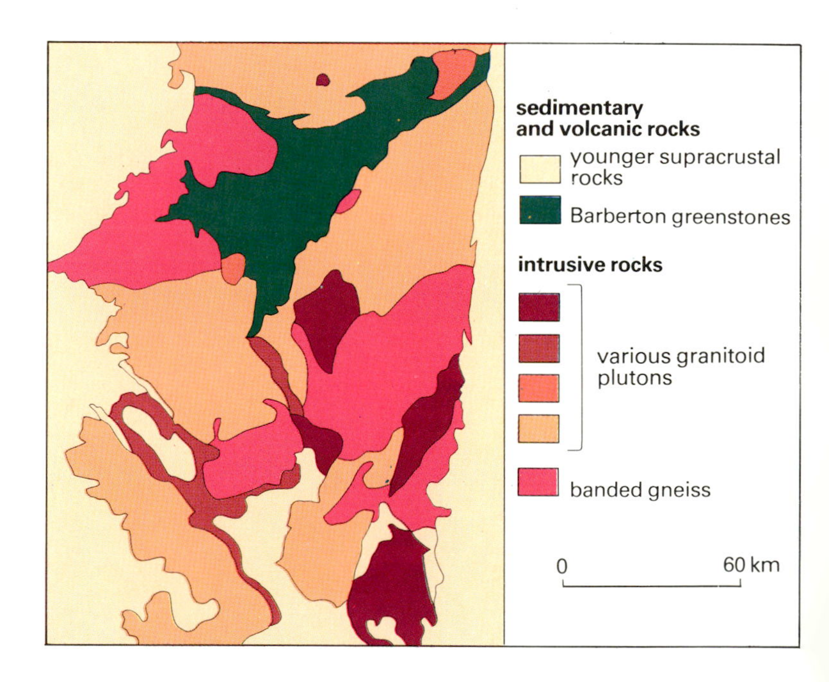

16.7: *The strange shape of the Barberton greenstone belt, 3500 Ma old, reflects the forceful intrusion of a whole series of younger granitoid plutons and is typical of many Archaean low-grade supracrustal belts.*

The ultrabasic lavas are of most interest since, with very few exceptions, they are restricted to the Archaean and are most common in the oldest areas. They have two unmistakably volcanic features. Not only do they contain pillow structures formed by eruption in standing water, but they show curious dendritic networks of the minerals pyroxene and olivine which are called spinifex textures because of their resemblance to a thorny shrub. Spinifex textures are characteristic of sudden cooling, or quenching of high-temperature silicate melts. By experimentally remelting Barberton ultrabasic lavas it has been shown that they were quenched at temperatures of 1600°C, some 400°C hotter than modern basaltic lavas.

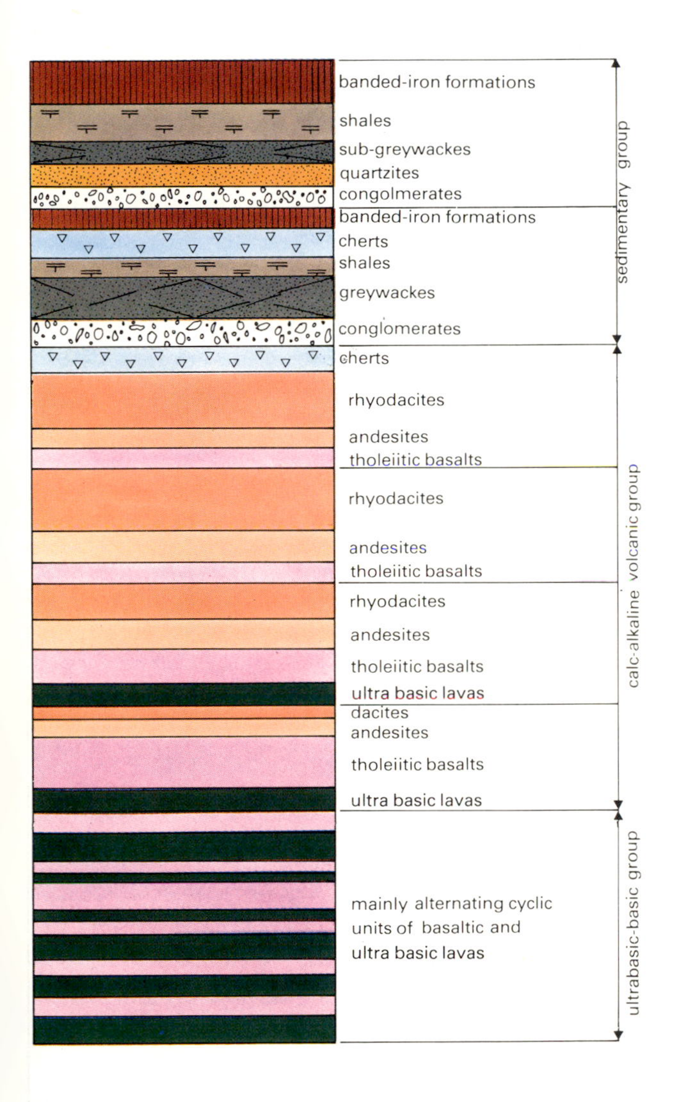

16.8: *The gradually changing cyclicity of deposition in the Barberton belt, revealed by a generalized stratigraphic column, suggests a change from primitive volcanicity without sedimentation from coeval continental crust, through sedimentation in an unstable environment to typical sedimentation of a stable continental area.*

The ultrabasic composition of some Barberton lavas and their extremely high temperatures of eruption suggest that they were derived by much higher degrees of mantle melting than modern lavas. This implies that the Archaean upper mantle was hotter and temperatures increased more rapidly with depth than at present. If this was the case, it has important implications for the way in which the Archaean crust and lithosphere behaved.

Supracrustal sequences dominated by basic–ultrabasic lavas and sediments with no recognizable sialic component, such as those from Barberton, are present as small belts in most Archaean terrains. Those at Isua and some in India, Zimbabwe and Western Australia comprise, together with Barberton, one of three Archaean volcanic–sedimentary associations. This type seems to represent the earliest fragment of crustal history recorded in these areas (older than 3500 Ma) and may be the nearest we shall get to the Earth's primordial crust. Another type of association is also displayed in Archaean low-grade terrains and is much more widespread.

In Canada, Scandinavia, Zimbabwe, India and Australia the low-grade terrains contain large numbers of elongate supracrustal belts which in many cases clearly rest unconformably upon older granitoid gneisses. Strangely they nearly all have ages falling in the period 2800–2600 Ma and their formation was in each case very closely followed by the emplacement of granitoid plutons, further adding to the already established local continental crust. Many of these belts have basal volcanic components dominated by basaltic lavas but also containing both ultrabasic and, more commonly, calc-alkaline and silicic lavas. These volcanics are nearly always blanketed by thick piles of immature sediments such as greywacke turbidites, which in some cases have been proved to derive from both volcanic and granitoid sources. The relationships and primary structures are sometimes so well preserved that it is possible to reconstruct the facies variations in these belts quite precisely (Figure 16.9).

The association of basaltic–andesite–rhyolite volcanics with immature turbidites, together with the geochemical similarity of the volcanics of some belts to those in modern destructive margin environments, initially encouraged the uniformitarian view that the late Archaean was dominated by island arcs. More recent discoveries cast doubt on this view. For instance there are constructive margin and within-plate affinities among the late Archaean lavas. Sedimentary infilling of the basins appears to have occurred from several directions instead of the unidirectional character of modern island-arc sedimentary wedges. One popular compromise is the view that the belts were analogous to modern back-arc basins, which is convenient in that a wide spectrum of magma types can be found there. However, perhaps it is safest to say that the late Archaean Earth probably had just as wide, if not wider, variations of tectonic setting for magmagenesis as does the Earth today, and that many of these settings, together with some peculiar to the Archaean, are represented by greenstone belts.

The factor above all others that conceals the true setting and relations of late Archaean supracrustals (except in one or two cases) is their structural complexity. The structures responsible are quite unlike those found today in orogenic belts. In many examples the later granitoid plutons have domed up the supracrustals leading to dome and basin interference patterns (see Figure 16.3). In Zimbabwe, parts of Canada and India such domal features are superimposed on huge

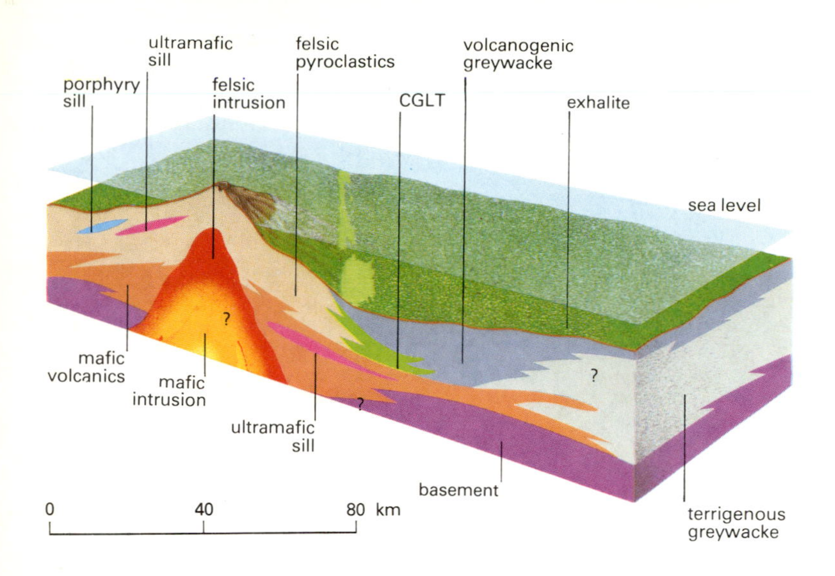

16.9: *A block diagram of the slightly deformed Abitibi belt of Ontario shows the facies variations in a late Archaean greenstone belt, as revealed by detailed geological mapping.*

nappes reflecting earlier horizontal tectonics. Moreover, the Archaean continental crust was sometimes broken up by later wide shear zones or mobile belts such as the Limpopo belt of southern Africa. It may be that these interfering major structures are at least as responsible for the intriguing parallelism of greenstone belts as is the possibility of parallelism of the original basins in which they accumulated.

In the late Archaean low-grade terrains another feature unfamiliar in more modern settings is the swamping of enormous areas by what have been graphically termed 'gregarious' batholiths. Granitoid batholiths are common enough in Alpine orogenic belts but they are restricted to linear arrays, whereas the late Archaean granitoid batholiths occupy areas as broad as they are long. There seems to have been a worldwide event of continental crust formation in the late Archaean, for it is possible to predict, with a fair degree of confidence, an age of between 2700 and 2500 Ma for such bodies, wherever they are found. Their origin must have been linked in some way with the processes governing formation of the enveloped supracrustal belts, as in most cases little more than 50–100 Ma spans the entire evolution from earliest supracrustals to the most highly fractionated granitoids. Isotopic data from the batholiths consistently indicate that they emerged not from the older sialic materials, but from a geochemically more primitive basaltic source, perhaps the Archaean equivalent of modern oceanic crust.

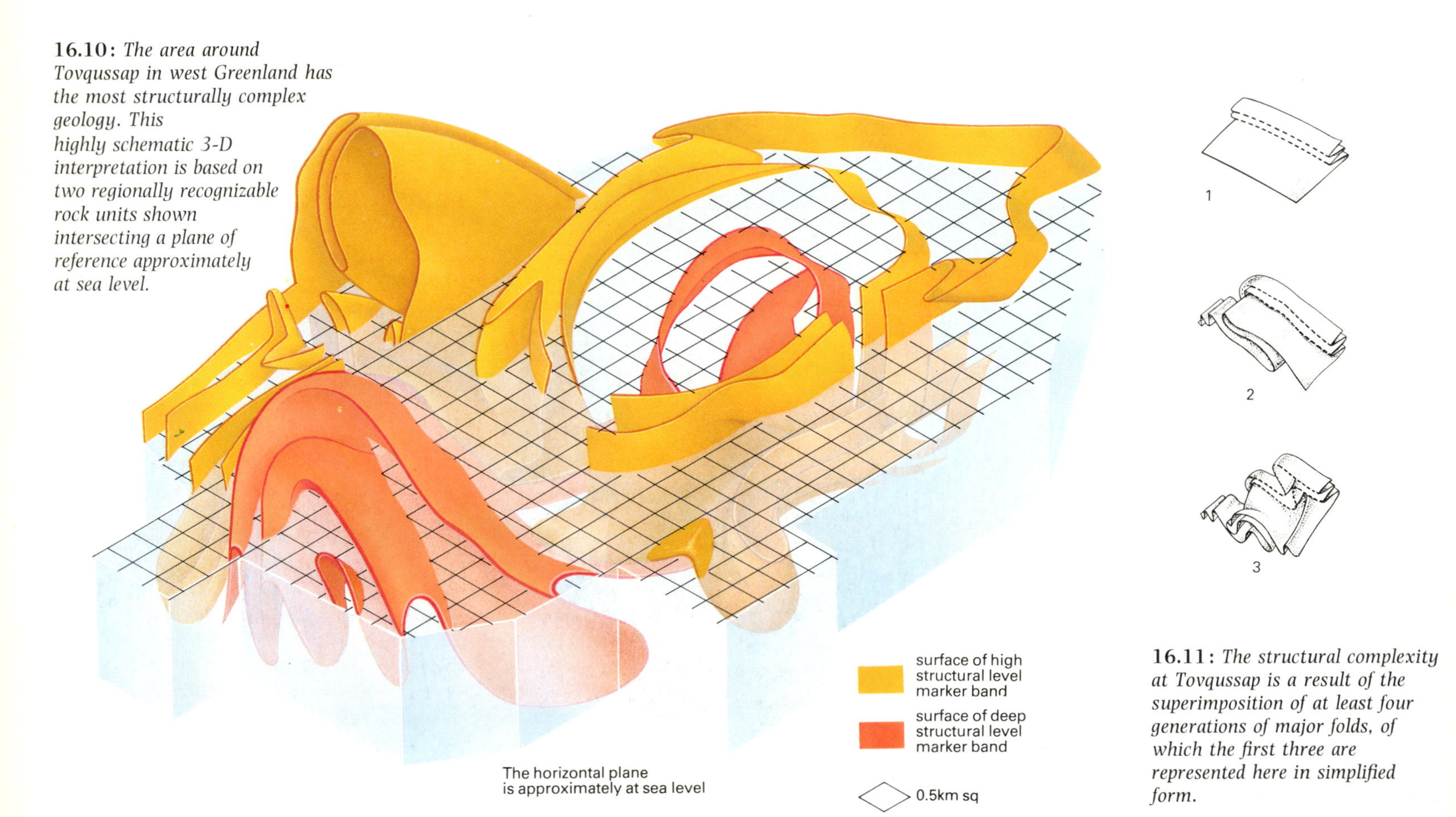

16.10: *The area around Tovqussap in west Greenland has the most structurally complex geology. This highly schematic 3-D interpretation is based on two regionally recognizable rock units shown intersecting a plane of reference approximately at sea level.*

16.11: *The structural complexity at Tovqussap is a result of the superimposition of at least four generations of major folds, of which the first three are represented here in simplified form.*

It has been argued that the single most important event in the evolution of the continental crust was that of the late Archaean. Most, if not nearly all, of the present continental masses may have been present by 2500 Ma ago and this is borne out by the continual discovery of relict Archaean patches in much younger, reworked terrains. There is also evidence from the late Archaean that parts of the continents were beginning to be stabilized as permanent, rigid cratons.

High-grade terrains

From virtually every geological standpoint—structural, geochemical, mineralogical and petrological—the most complex areas of the Earth are those bearing evidence of pressures corresponding to depths of 20 to 50 km in the Archaean continental crust. These are the high-grade terrains. They too are present in every craton and have an intriguingly similar range of ages in the time span 2800–2600 Ma.

The first impression of high-grade Archaean areas is that they are a jumble of a seemingly infinite variety of rock types, dominated by granitoid gneisses but containing deformed dykes, layered igneous intrusions, metamorphosed volcanics and sediments and diverse fragments of uncertain origins. Their unravelling is complicated not only by the bizarre nature of their folding (Figures 16.10 and 16.11), but by the lack of clear relationships due to the parallelism of their components. This was long thought to reflect the recrystallization and folding of an originally layered sequence of sediments and volcanic rocks. However, it is now becoming clear that it is a consequence of the repeated slicing together of older crust, supracrustals and conformable intrusions of granitoid magma, like the shuffling of a deck of cards.

16.12: *A spectacular example of a cumulate texture in anorthosite at Tovqussap, in which balls of plagioclase feldspar have settled into a finer basic groundmass which was formerly basaltic magma.*

It is quite clear from their content of metasedimentary rocks that parts of these high-grade terrains once resided at the Earth's surface. It is in this category of supracrustal assemblage that an apparent contradiction emerges. The immense complexity of these rocks bears witness to the extreme instability of their tectonic setting at one stage in their history. However, the sediments comprise pure quartzites, former clays, banded ironstones and dolomitic limestones, all of which are characteristic of stable continental-shelf deposition. This is in marked contrast to the oldest Archaean sediments which contain little if any evidence of derivation from older sial, and to the late Archaean turbidite-dominated supracrustals which formed in very unstable basins. Further evidence for initial stability in high-grade terrains comes from the layered igneous complexes often adjacent to the supracrustals. They contain spectacular evidence for quiet crystal accumulation and slow differentiation. In one such complex, at Fiskenaesset in west Greenland, a complete igneous stratigraphy revealed by thin chromite layers has survived deformation and must have formed in a thin body that underlay hundreds if not thousands of square kilometres.

Clearly something remarkable happened to all these high-grade areas around 2800 to 2600 Ma ago. In every case the interleaving and complex folding that characterizes them was followed by high-temperature metamorphism at depths down to 40 or more km. This produced a characteristic suite of metamorphic rocks whose mineral assemblages are assigned to the granulite facies (see Chapter 5). They are pyroxene gneisses with the same range of bulk compositions as lower-grade gneisses elsewhere (dioritic, tonalitic, granodioritic and even granitic) but which contain only tiny amounts of hydrous minerals such as amphibole or mica—they are nearly anhydrous rocks. Detailed study of their geochemistry reveals that not only are they depleted in water (H_2O), but they also have very low contents of the heat-producing elements potassium (K), rubidium (Rb), uranium (U) and thorium (Th). Such deficiencies probably resulted from the purging of water, and of elements that can easily be carried in aqueous fluids, during the metamorphism of these rocks. This could be explained by the permeation of the deep crust by vapour rich in carbon dioxide which forced out all water in the system; indeed, there are fluid inclusions in the granulite minerals which contain very high proportions of CO_2. Whatever the process, it removed both the heat producing capacity of the bulk of the rocks and their ability to be melted by normal means. They were destined to become relatively cool and rigid foundations of the continents. Only if later affected by water, heat and shear stress would they again suffer deformation.

It is not known precisely how such a state could have arisen. Such areas at the surface today are underlain by continental crust 30 to 40 km thick; they may have been 70 to 90 km thick when they formed. These thicknesses are matched today only by continent–continent collision zones such as the Himalayas and the Tibetan plateau. Two main theories have been put forward to explain this enormous crustal thickening. One is that it was due to emplacement of layer after layer of silicic magma from below. The other proposes that it resulted from horizontal tectonics which stacked the earlier crust into several repetitions. Quite possibly, a combination of both processes was responsible. Whatever the cause, it was accompanied by the presence of CO_2 vapour that rid the lower crust of water and anything else that was incompatible with the new metamorphic minerals.

Models of Archaean tectonics

One of the outstanding questions in modern geology challenges what is traditionally its most fundamental philosophy: is the present always the key to the past? Particularly, did plate tectonics operate in the Precambrian, as far back as 3800 Ma? Access to definitive features such as oceanic magnetic anomaly patterns is impossible, and palaeomagnetic pole positions have yet to be determined for the Archaean. An approach to the problem must lie in practical geology and geochemistry and in general theories about the way in which rocks ought to have behaved in response to the presumed geothermal conditions of the Archaean.

Geologists are in nearly unanimous agreement that ophiolitic complexes are relics of oceanic crust that failed to be subducted.

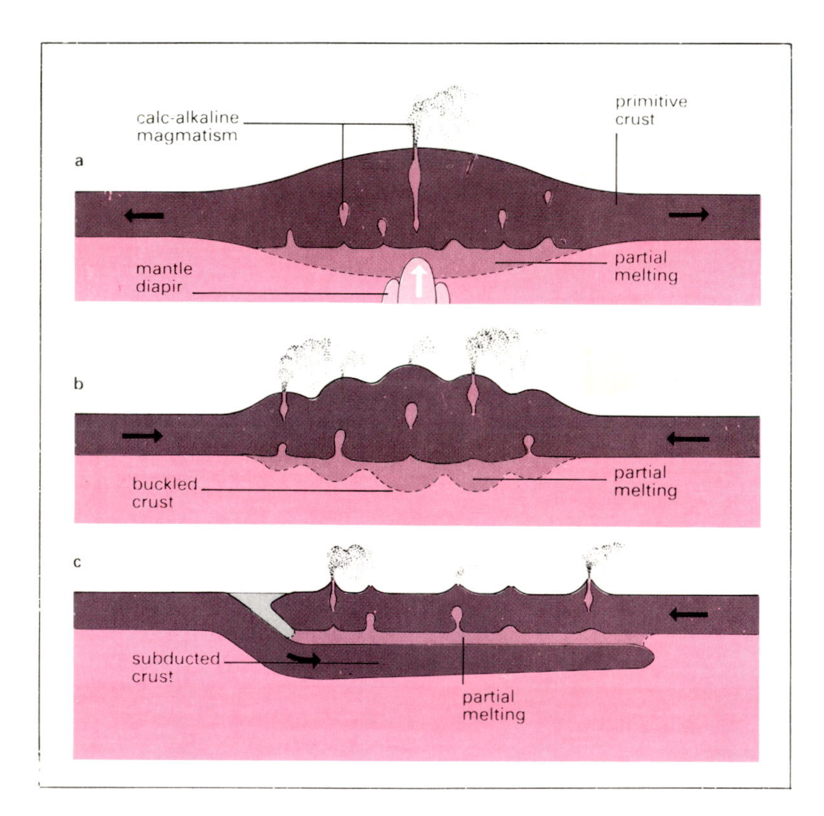

16.13: *Primary garnet-bearing basaltic crust can be induced to remelt at high pressure under a variety of plausible tectonic conditions:* (**a**) *beneath a very large shield volcano,* (**b**) *where crust has been thickened by buckling, or* (**c**) *where it is forced beneath an overriding crustal or lithospheric slab.*

They agree that occurrence of blueschists and eclogites mark the sites of ancient low temperature/high pressure metamorphism which are associated with subduction of relatively cool oceanic lithosphere. They also agree that there are no unequivocal Archaean ophiolites, blueschists or eclogites. However, there are Archaean igneous rocks, both volcanic and plutonic, whose geochemistry is the same as is found in modern plate tectonic settings. But in all cases these modern settings have structural attributes that are not obviously present in their supposed Archaean counterparts. This contradictory picture is hardly surprising given the age, complexity and incompleteness of the Archaean record, and it enables several divergent schools of thought to thrive.

Before discussing these alternatives an important guideline must be established. The geochemistry of modern igneous rocks apparently enables them to be classified in terms of their setting at oceanic rifts, island arcs or continental margins. However, the chemistry merely reflects the specific conditions prevailing in their source regions. These conditions are pressure, temperature, abundance of H_2O and CO_2, the composition of the source material, its mineralogy and the extent to which it melts. The geochemistry of the magma produced can then be altered further during crystallization and interaction with the rocks through which it passes to the surface. Particular combinations of these conditions are present in modern tectonic settings, but they are not unique and could be just as easily mimicked in other settings. For instance the partial melting of garnet-bearing basic rock, so important for the origin of any continental crust, could occur at the subsiding base of huge basaltic volcanoes, by downward buckling of basic crust, just as well as by subduction (Figure 16.13).

There are as many Archaean models as there are established Archaean specialists, but the models fall into three groups which can be termed the uniformitarian, the non-uniformitarian and the inverted uniformitarian.

Uniformitarian models accept that there may have been differences in the Archaean rate of sea floor spreading, amount of partial melting and extent of continental crust, but insist that fundamentally the same tectonic environments were in existence as today. The higher heat flow in the Archaean, due to a higher level of radioactivity, was accommodated by more rapid sea floor spreading, so that geothermal gradients were essentially the same as now. Growth of continental masses, in this model, was accomplished by collision and accretion of numerous island arc systems, underridden by subduction zones (Figure 16.14(a and b)). Those areas now characterized by high-grade terrains were the zones of most intense igneous activity and crustal thickening, and were at the leading edges of plates where earlier marginal basin sediments were caught up in the deformation. In the more quiescent back-arc environment initial rifting and shallow melting produced basal volcanic sequences which were capped by sedimentary detritus from the fore-arc to form greenstone belt associations. When several such arcs had accreted they were further bolstered by granites melted from the deeper regions of the thickened crust (Figure 16.14(b)).

The non-uniformitarian models form a very diverse collection. Some incorporate the possibility of impact-induced phenomena akin to processes associated with the lunar maria, but most concentrate upon the greater mobility of the lithosphere when heat flow was higher

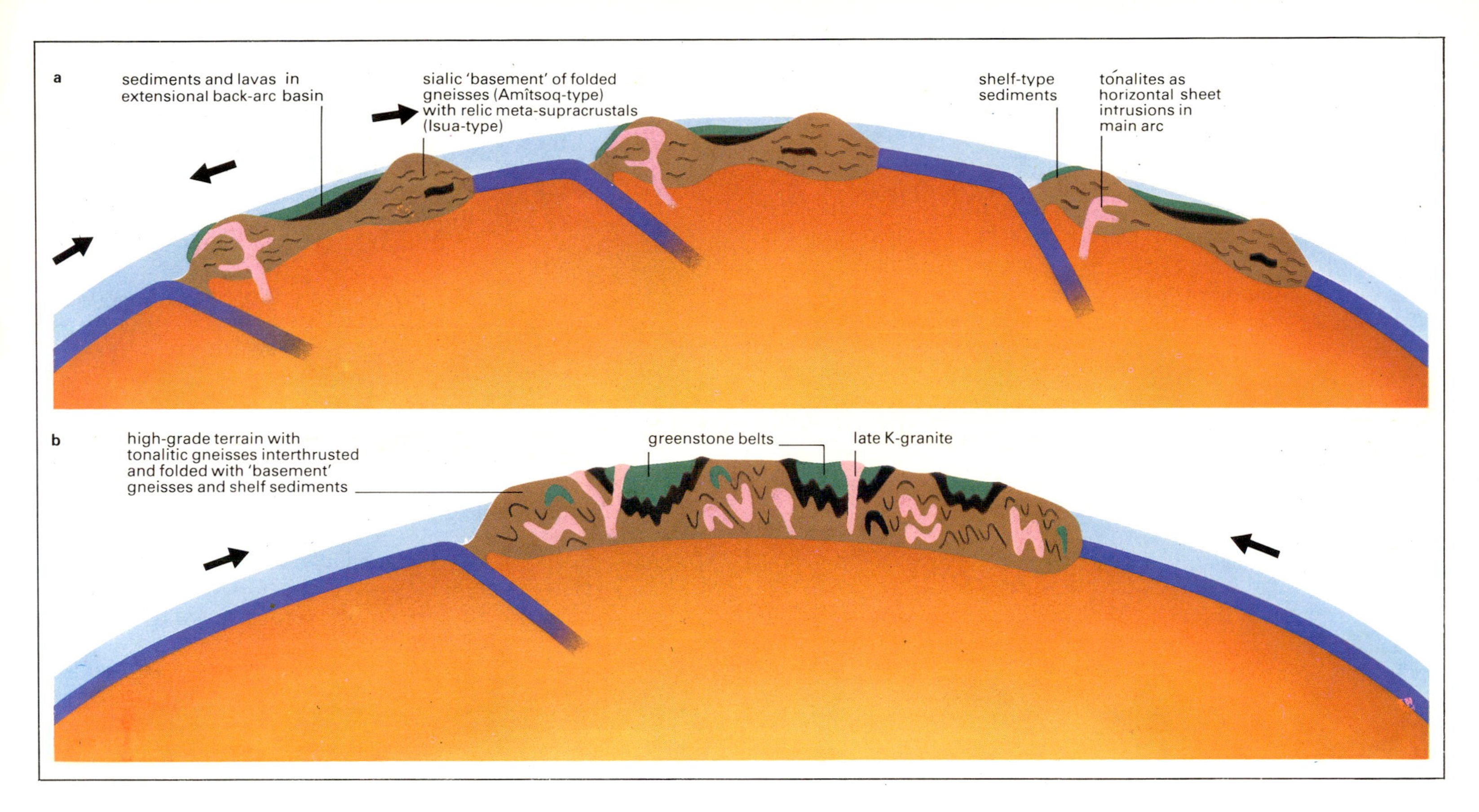

16.14: *A uniformitarian plate tectonic model of continental growth in the Archaean.* (**a**) *Numerous small sialic plates accrete,* (**b**) *metamorphosing the destructive margin rocks to high-grade terrains, and deforming the back-arc basins to form greenstone belts.*

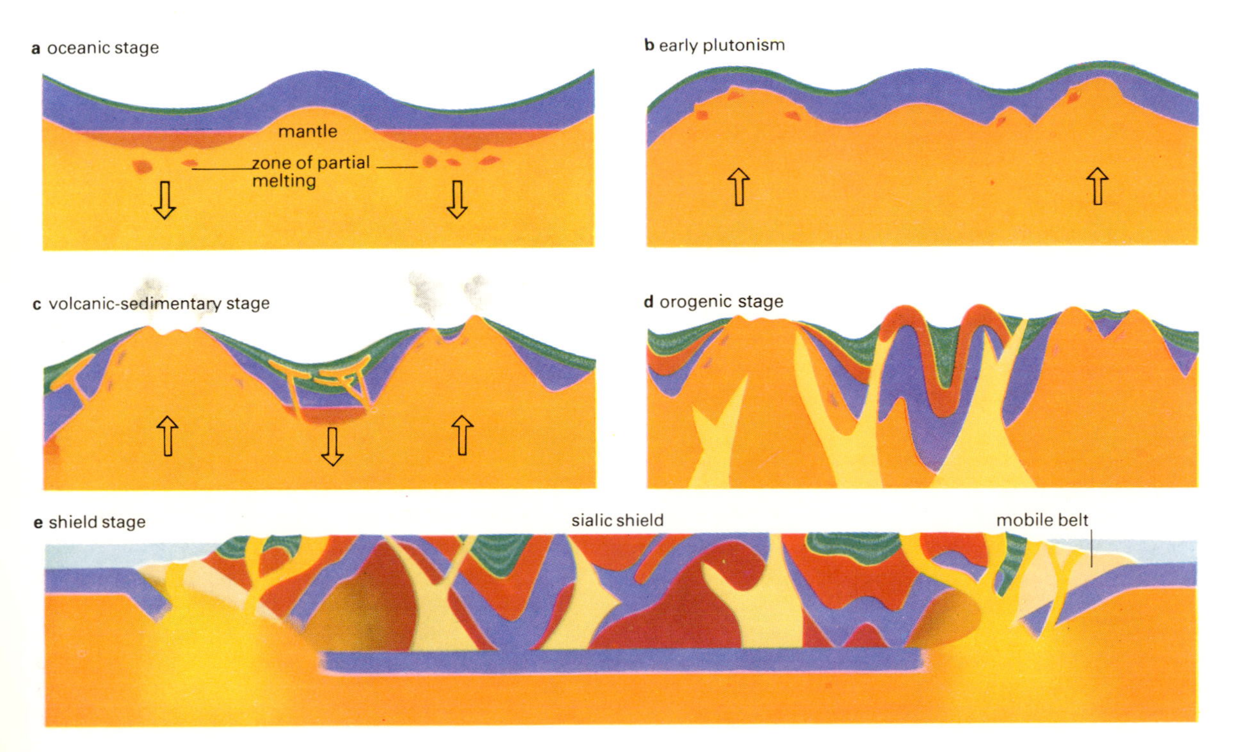

16.15: *A non-uniformitarian model of Archaean tectonics.* (**a**) *Oceanic crustal sags accumulate sediment derived from uplifts.* (**b**) *Partial melting generates* Na-rich *acid magmas.* (**c**) *Subsidence and further partial melting generate calc-alkaline magmas of the greenstone association.* (**d**) *Folding and metamorphism of the greenstone belts. Partial melting of the now thickened crust yields* K-rich *granitic magmas.* (**e**) *The whole assemblage cools to form a sialic shield.*

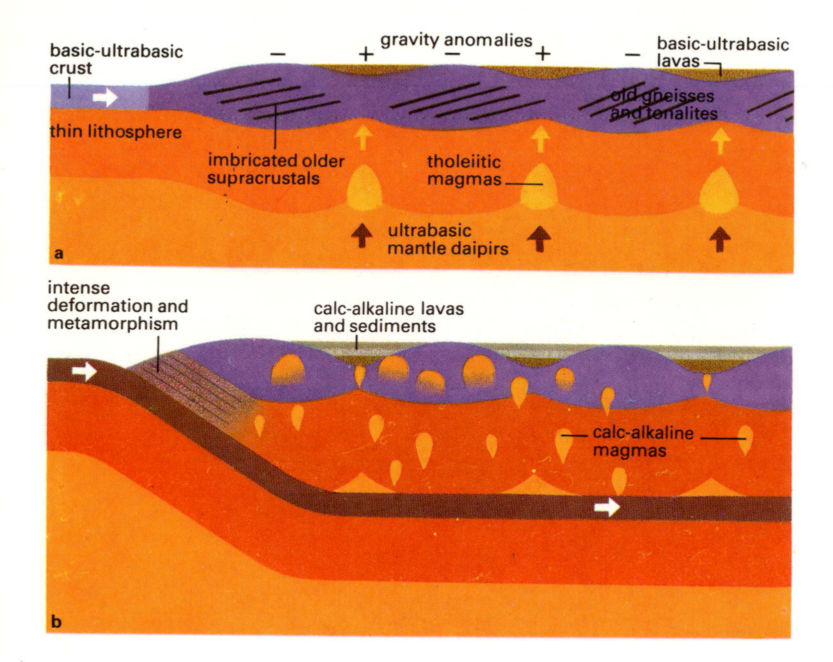

16.16: *An inverted uniformitarian model of Archaean tectonics, dominated by a steep geothermal gradient.* (**a**) *Initial crustal buckling and spreading, perhaps in response to mantle processes, leads to* (**b**) *shallow subduction and consequent widespread generation of silica-rich magma.*

(Figure 16.15(a to e)). Deformation or downsagging of basic crust would have led to partial melting at its base and consequent rise of light acid magmas. Continued diapirism of this kind and isostatic subsidence of the intervening crust created basins into which new magmas were poured out and which were filled by sedimentary debris to form the greenstone belt association. High-grade terrains represent the deeper levels of this system. Tectonic movements in this model are dominantly vertical, though horizontal deformation could have been a contributory factor.

Inverted uniformitarian models are based on an explicit break with the philosophy that the present is the key to the past. They replace it with the hypothesis that plate tectonics arose out of some previous mode of whole Earth activity. This is based on the probable changes in the Earth's thermal behaviour and most particularly on a decrease through time in the geothermal gradient with its consequences for the lithosphere. Had the Earth at some stage had an average geothermal gradient in excess of 15°C/km then basaltic rocks could not have been transformed into eclogite, which is denser than the mantle. The consequence of this is that steep subduction would have been unlikely, but shallow-angled underplating of the lithosphere may have been possible. Further consequences of a higher geothermal gradient would have been a thinner lithosphere, shallower and more thorough partial melting of the mantle, and a greater likelihood of buckling of the lithosphere by crustal spreading processes (Figure 16.16(a and b)). A possible consequence of this sort of behaviour might have been the occurrence of zones of generation of silica-rich magma over much broader areas than modern arc systems. Small, parallel linear basins might have formed along lithospheric downwarps, providing ideal sites for volcanic and sedimentary accumulation. This, together with the multiple slicing together of thin but extensive sial at the leading edges of such systems, gives a close approximation to the patterns seen in Archaean high- and low-grade terrains. Another advantage of models such as this is that they allow for sudden changes in overall behaviour of the lithosphere in response to changes in the geothermal gradient. Such great breaks are thought to be reflected in the Archaean–Proterozoic boundary (2500 Ma), and possibly within the late Proterozoic.

Sparse as information on the earliest Archaean is, the possibilities can now be examined for the missing 700 Ma which culminated in the production of crust 3800 Ma ago.

The earliest history of the Earth

Modern theory suggests that one of the direct outcomes of the original accretion of the material that was to form the Earth was the trapping within it of sufficient energy for its major component with the lowest melting point, iron sulphide and metallic iron, to be partially melted and to sink, thereby forming the core. Gravitational energy released by this process was so huge that, irrespective of its initial temperature, the Earth must have been heated to such a degree that it experienced a complete geochemical upheaval. This completed the formation of the core, drove off much of the Earth's volatile elements, particularly gases, and resulted in at least partial melting of the entire mantle. Only when cooling by radiation at the surface was sufficient to allow a solid skin to form can geological evolution be safely assumed to have begun. But what specific processes were involved?

Undoubtedly volcanism played a primary role, releasing gases that eventually formed a secondary atmosphere, but volcanism is only an outward manifestation of the Earth's internal processes. Convection is the only efficient means of transferring thermal energy from the interior of a planet (see Chapter 9), and theory suggests that convection in the early Earth would have taken the form of small-scale thermal jets superimposed on a larger-scale overturn. Such jets would dominate volcanism, and it can be speculated that volcanoes were mainly concentrated in great centres over the jets, forming huge radially spreading shields of lava. Another feature leading to localized thermal highs could have been the major impacts that also dominated the Moon, Mars and Mercury at this time. In the case of the largest planetesimals, impact energy transformed into seismic waves would have been distributed to great depths. The thermal inertia of the Earth would have ensured that hot spots such as these had long lives. They too would produce central volcanism.

If magma was welling up to the Earth's surface at hot spots, some older cooler material had to descend in compensation. Perhaps such sinking took place along relatively cold zones between hot spots. Radial spreading from all hot spots would ensure a global polygonal network of such zones and it is along such 'cold sinks' that the first light continental material may have formed. Once formed, the incompressibility of sialic material would ensure that it could never defy gravity and return to the mantle, and continent formation can be said to have begun.

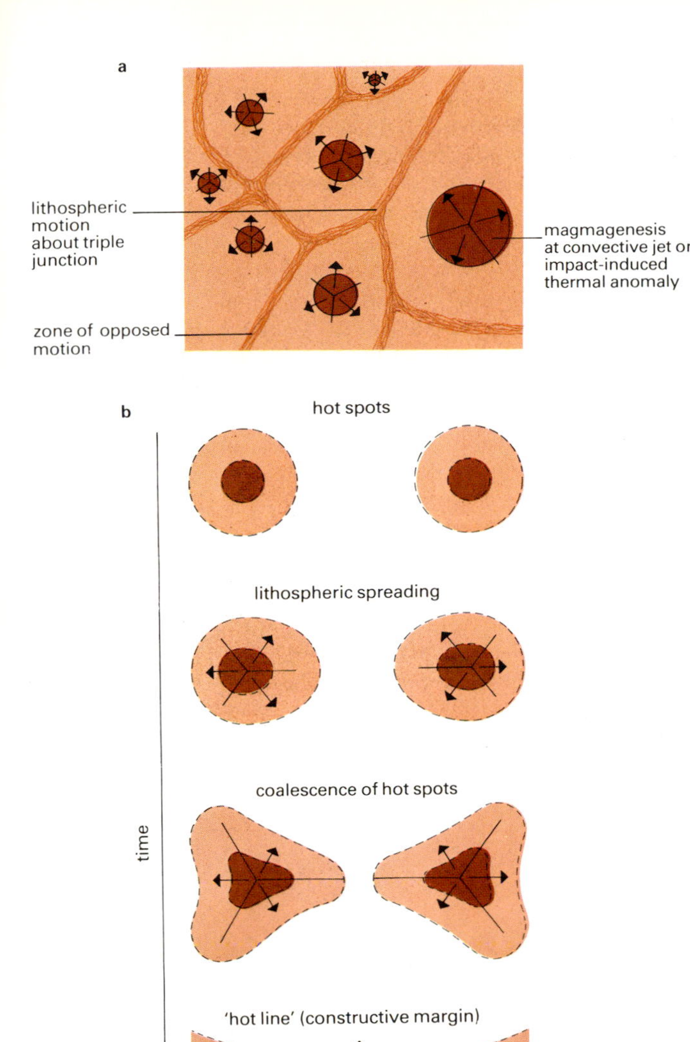

16.17: (**a**) *Concentration of primary basaltic magmatism at centres, over small-scale thermal jets and thermal anomalies induced by large impacts, may have led to radial spreading of the primitive Earth's crust. The radial motion would then have been accommodated by a polygonal network of relatively cool, deformed zones where crust was recycled into the mantle system. Formation of acid, unsubductible magmas might have occurred by partial melting at the base of the shield volcanoes or in the polygonal system of sinking crust.* (**b**) *The tendency of thermal anomalies on a closed spherical surface to coalesce could lead to the gradual appearance of linear spreading zones, once impacts had ceased or thermal jets had become less significant forms of convection. This proposal leads to an evolutionary sequence towards modern plate tectonics.*

A random distribution of hot spots on a closed surface such as the Earth cannot remain stable unless such thermal anomalies are either being continually produced or are migrating rapidly. Sooner, rather than later, they would interfere with one another and join up as a system of hot lines (Figure 16.17(a and b)) with outward lithospheric spreading analogous to that along present linear constructive plate margins. Modern plate tectonic processes may have arisen out of similar stable patterns, through some intermediate stage or stages in the Precambrian as the influence of hot spots waned with the parallel growth to dominance of large-scale convection cells. Modern hot spots (see Chapter 12), which remain stationary relative to lithospheric plate movements and must reflect some deep mantle thermal anomalies, may be relics of the Earth's primitive tectonic behaviour.

Models such as this remain shots in the dark, and their further development requires not only greater knowledge about the oldest Archaean rocks but also the discovery of relics of even earlier materials. Another fruitful line is to develop more detailed theories about the roles played in geochemical evolution by both the Earth's internal effects on its primary constituents and the effects of admixing extraterrestrial matter in major impacts. To a large extent therefore the key area for research is in the inner planets whose evolution was virtually 'frozen' 3800 Ma ago and where the most primitive materials in the solar system reside. Whatever the approach used, the missing 700 Ma of Earth history is the last really major hurdle in developing a framework for planetary science.

The Proterozoic eon

The boundary between the Archaean and Proterozoic eons is conventionally taken at 2500 Ma ago. There was no sudden change in crustal evolution at that time, but there is a widely held view that the rate and scale, and possibly even the nature, of tectonic processes in the Archaean differed from those of the Proterozoic and Phanerozoic. In particular many geologists now consider that the bulk of the continental crust had formed by the close of Archaean times and that tectonic processes since then have been more concerned with recycling continental crust than with adding to it; such recycling implies a plate tectonic regime.

The thermal state of the Earth may also have been different in Archaean times. There is evidence of much steeper geothermal gradients, probably related to the relative thinness of the crust, and radioactive heat production must then have been two or three times greater than today. The most efficient method of ridding the Earth of heat is by volcanism, which should have been more active, perhaps driving the thinner and smaller crustal plates faster than those of today, if Archaean plate tectonics is to be accepted. Modern plate tectonic processes are more widely, though not universally, accepted to explain features of Proterozoic fold belts.

Although greenstone belts are especially characteristic of Archaean terrains, there are Proterozoic and even Phanerozoic equivalents; the

Barberton belt (>3400 Ma old), the Ivory Coast belt (2200 Ma) and the Rocas Verdes complex of the Chilean Andes of Cretaceous age differ only in that the older they are the less linear they are. Likewise, although Archaean terrains apparently show less distinction between mobile belts and stable cratons than Proterozoic regions, there were certainly extensive areas of rigid continental crust already in existence at least in the later Archaean. For example in southern Africa (Transvaal and Swaziland) the Archaean Pongola system, 10 km thick, accumulated on continental crust, and the Limpopo belt, about 2700 Ma old, is a linear orogenic belt comparable in its internal structure to the much younger Grenville belt, 1000 Ma old, in eastern Canada. Thus the change from the typical Archaean regime to the rather more familiar processes of the Proterozoic did not take place sharply or even at the same time everywhere. The main difference is that the further back in time the more fragmentary the record and the more difficult it is to decipher the history.

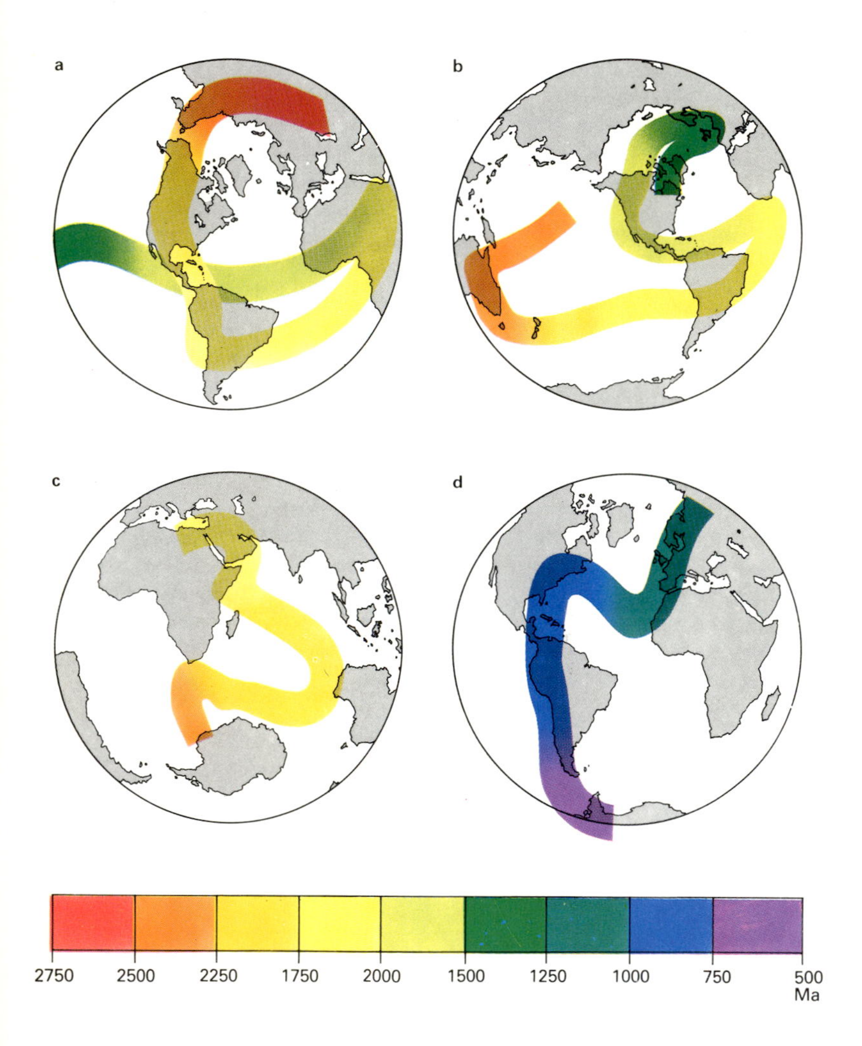

16.18: *Apparent polar wander paths for Proterozoic and Phanerozoic time. Each shows, on a present-day configuration, the changing north magnetic pole positions indicated by successively younger magnetized rocks from each of several continents or parts of continents.*

It seems that by the beginning of the Proterozoic, 2500 Ma ago, most of the present continental crust was already in existence. Later additions, in the form of accreted island arcs in areas underlain by oceanic crust and now incorporated into the continents, make up only a small total area. The thickness of the continental crust was probably about the same as it is now.

While the area and thickness of the continental crust may not have changed dramatically in the past 2500 Ma its geographical distribution then was quite different. This is known from the study of the remanent magnetism of Proterozoic rocks. This evidence is usually displayed in the form of apparent polar wandering curves (see Chapter 7). These curves show that the continental areas were moving, relative to the poles, at rates similar to present plate motions throughout the time (over 3000 Ma) for which there are data. On the other hand the data do not demonstrate motions of the continental plates relative to one another. The opening and closing by subduction of small oceans cannot yet be proved or disproved palaeomagnetically because the methods are not sufficiently precise, and it can always be argued that large oceans may have existed for very short periods. However, if oceans, whether large or small, were formed and then destroyed, it seems that they must often have closed again in such a way as to restore the plates on either side very close to their previous relative positions. This is not by any means improbable as shown by the relatively recent case of the Alps: two plates moved about a thousand kilometres apart and then returned to within about a hundred kilometres of their original relative positions.

One striking feature of the polar wandering curves is that they show a number of sudden changes of direction. These have been called hairpins, and they seem to be connected with phases of intense crustal deformation. It was proposed in the 1960s that these deformation phases were themselves the results of changes in the number and pattern of convection cells in the Earth's mantle as the core increased in diameter. However, subsequent geochemical calculations and the evidence that a strong magnetic field already existed in the Archaean, imply that the core separated early in the Earth's history.

Early Proterozoic continental dyke swarms and basic intrusions

The rigidity and brittle fracturing of large areas of continental crust in early Proterozoic times is strikingly shown by swarms of ancient basic dykes, such as those in the Canadian Shield, Greenland and north-west Scotland, the Great Dyke (2550 Ma old) which extends for 500 km across the Archaean crust of Zimbabwe, and those that cut the Archaean rocks of Mysore in southern India. Comparable systems of basic dykes of Cenozoic age are associated with the opening of the north Atlantic, while the Jurassic Karroo dolerite dyke swarms of southern Africa are associated with the break-up of the southern supercontinent Gondwanaland. It is therefore likely that the early Proterozoic basic dyke swarms indicate similar phases of continental fracturing.

Early Proterozoic fold belts

Major phases of deformation, metamorphism and intrusion of granitic plutons, about 2000 Ma ago, are recognized in all the continents. Several lines of evidence suggest that at least some of these early Proterozoic fold belts are the products of plate tectonic processes. Probable remnants of oceanic crust are represented by ophiolites in south-east Finland, by basic and ultrabasic rocks in the circum-Ungava belt and perhaps by the enormously thick Mt Stanley volcanics on the Ruwenzori in Uganda. Early Proterozoic aulacogens, first recognized in Siberia, and later in north-west Canada, associated with the Coronation geosyncline are wedge-shaped rifts, containing thick sediments, and indicate early but arrested plate separation along rifts propagating from triple junctions. Widespread exposure of high-pressure granulites in some belts, as in the Ubendian of Tanzania and in the Arequipa block in Peru, implies crustal thickening similar to that characterizing the much younger Himalayan and Andean orogenic belts.

Major shear zones developed during the early Proterozoic in the north Atlantic region. They are imprecisely dated and may have been repeatedly reactivated. One of them extends for almost 2500 km in the Nelson River area of Canada. These shear zones may have been transform faults, or possibly slip zones of the type attributed to indentation of one plate into another during collision.

16.19: *Early Proterozoic fold belts.*

Fold belt	**Age** (Ma)
North America:	
Circum-Ungava belts (Labrador trough, Cape Smith and Belcher belts)	1600–1750
Coronation geosyncline and associated aulacogens	1750–2100
Penokean and Hudsonian belts	1950–2200
Eurasia:	
Laxfordian (NW Scotland)	1750
Svecofennian (E Scandinavia)	1700–1900
Africa:	
Eburnean (W Africa)	1800–2200
Ubendian, Ruzizi, Ruwenzorian	1800–2000
Usagaran (Tanzania)	2000
South America:	
Trans-Amazonian	2000
Arequipa block, Peru	1950
Australia:	
Arunta and Gawler provinces	1700–2000

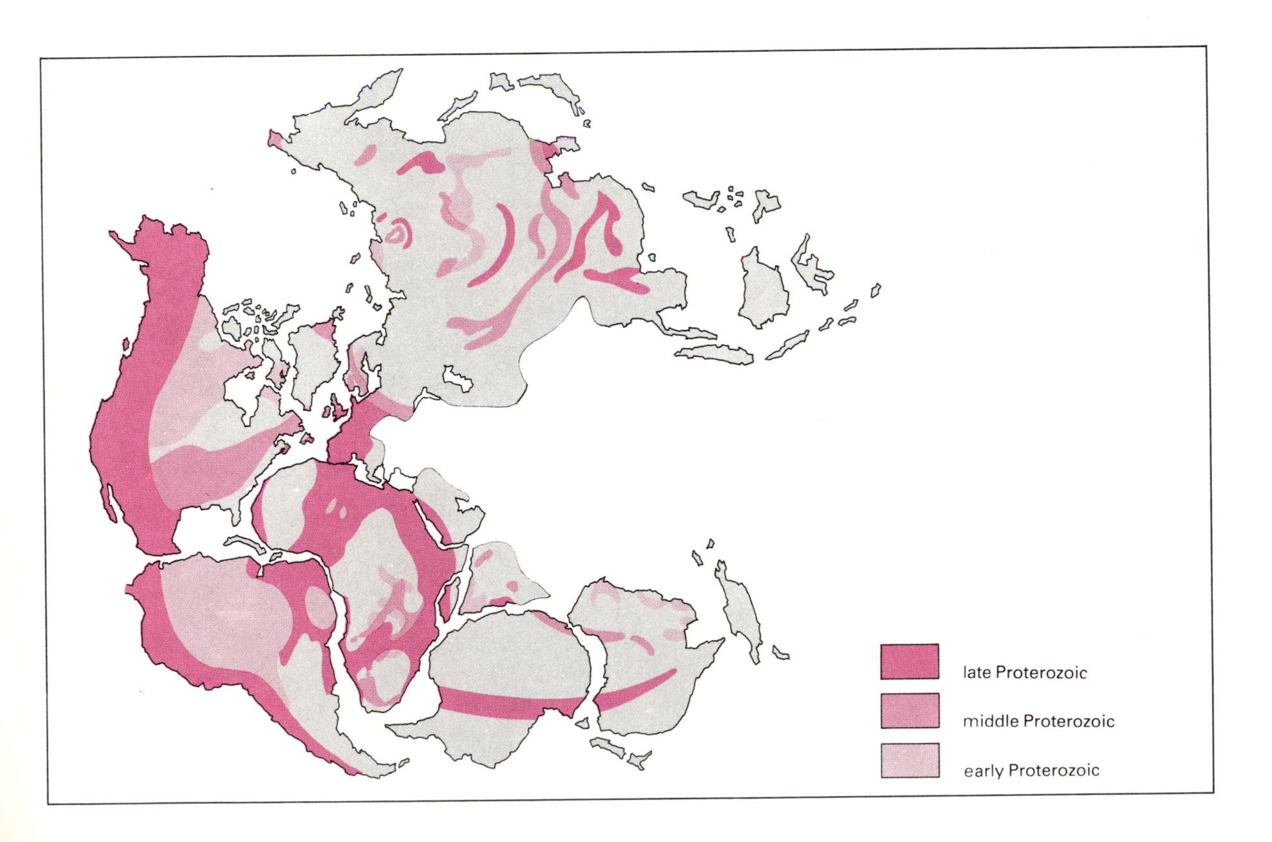

16.20: *A world map of Proterozoic fold belts. The distribution of the continents so long ago is uncertain, and so the distribution of their fold belts is plotted on the Permo–Triassic Pangaea reconstruction, which restores the movements of only the past 200 Ma.*

Mid-Proterozoic crustal evolution

After the intense crustal activity between 2200 and 1700 Ma ago, within-plate magmatism and fracturing occurred in many areas of Proterozoic rocks, as well as deposition of sediments, mainly sandstones. The characteristic igneous rocks of this period are alkali basalts, granites and rhyolites. Two very distinctive rock types are particularly characteristic of a wide zone in the north Atlantic region stretching from north California across Canada through south Greenland, north-west Scotland, to Scandinavia and beyond. Within this mobile belt there are intrusions of anorthosite, which consist largely of plagioclase feldspar, and Rapakivi granites, which have large pink ovoid potash-feldspar crystals rimmed by white plagioclase; acid and basic volcanic rocks also occur. This within-plate magmatism occurred intermittently through the mid-Proterozoic, within the regions previously affected by early Proterozoic metamorphism.

About 1000 to 1200 Ma ago intense deformation and metamorphism affected many parts of the continental crust. The best-known part of this system is the Grenville belt in eastern Canada. This belt continues westwards, mostly below younger rocks, across the USA and eastwards (within later fold belts) through north-west Ireland, the Scottish Highlands and thence into southern Scandinavia. The evolution of the Grenville belt is controversial. The north-west margin is a zone of high shear strain known as the Grenville front. Since various formations and structures, notably the iron ores of the Labrador trough, continue across the Grenville front this line cannot represent a former plate boundary. The structures within the belt are very complex, with interference folds due to superimposed deformations. High-pressure metamorphism implies that the crust was drastically thickened, as if in a collision orogeny, but no suture or ophiolites have been recognized. Palaeomagnetic data of the mid-1970s appeared to indicate a relative separation of two continents by 5000 km about 1150 Ma ago, but later palaeomagnetic work does not support this interpretation. The Grenville rocks recrystallized at high temperatures and the remanent magnetism, imposed over a long period of cooling while the region was moving relative to the geographic poles, is complex. Structures and metamorphism are consistent with a plate collision and no convincing alternative has been proposed. However, the collision model can only be regarded as a working hypothesis.

Such an interpretation is much more likely in the case of the Namaqualand fold belt in southern Africa, also about 1000 Ma old. In Natal a complex of northward-directed thrusts involves basic and ultrabasic rocks which have been interpreted as an ophiolite complex, thought to mark a suture at a destructive plate margin. Related fold belts in central Africa contain many granites with tin mineralization near their contacts.

Plate collisions of a similar age are also indicated much farther to the north-east. In Saudi Arabia the Mecca granites have been interpreted as being related to a series of successively developed subduction zones dipping eastwards beneath island arcs that are separated by sutures marked by ophiolites. This interpretation is supported by an eastward increase in K_2O/Na_2O ratios in the granites.

The Pan-African system

Late Proterozoic fold belts are known in many parts of the world, and many are now interpreted in terms of plate tectonics. However, controversy has surrounded the network of fold belts of this age in Africa, and it is only very recently that intensive study has begun to lead to a clearer understanding of this so-called Pan-African event in crustal history.

The term Pan-African was introduced for all those Precambrian rocks of Africa yielding potassium–argon (K–Ar) ages of between 450 and 550 Ma. These ages were cautiously attributed to a tectono-thermal event, rather than to a mountain-building phase, since there was no clear evidence of classical orogenesis in Africa at that time. It is now realized that K–Ar ages do not usually date the metamorphisms or deformations but rather the stage when the minerals had cooled enough to hold the radiogenic argon gas. The geologically significant events occurred earlier than these cooling ages.

The Pan-African fold belts form a connected network, surrounding and clearly distinguished from a series of cratons which behaved as rigid plates or microplates during the Pan-African events. In north-east and north-west Africa the Pan-African fold belts include well-authenticated fragments of oceanic crust. The importance of these ophiolites is that they show that oceans (or back-arc basins) existed that have now been almost wholly consumed. This implies subduction: hence these late Proterozoic fold belts must be the result of plate tectonic processes. Moreover, many of the ophiolites can be shown to lie along sutures, which are themselves associated with calc-alkaline volcanic and plutonic rocks of island arc types. These are especially well developed in north-east Africa and Arabia where late Proterozoic events as now interpreted involved the accretion of a series of island arcs. Once the plate tectonic interpretation can be shown to apply to north-east and north-west Africa, it follows that it must apply elsewhere. Therefore late Proterozoic plate margins, variously destructive, conservative or constructive, must have formed a continuous network throughout the Pan-African domain.

The plate tectonic interpretation may not explain all the Pan-African belts, and controversies remain as to whether certain parts of the system represent true collision orogenies or only long-lived rifts (aulacogens) without significant relative movement between opposing sides. There is another kind of problem in eastern Africa, in the so-called Mozambique belt. Here erosion has exposed much deeper levels of the fold belts than usual, the grade of metamorphism is very high and the structures complex. Older Proterozoic and even Archaean structures are involved but have not yet been clearly distinguished from the Pan-African structures. There are not enough radiometric dates and it is therefore not yet possible to decide how much is Pan-African, nor whether later Proterozoic subduction zones existed there. It is at least possible that the accreted island arcs that formed the continental crust of Arabia and north-east Africa connect with collision orogenies in central East Africa. The connection between the Pan-African fold belts of eastern and western Africa is regarded as linking two triple junctions, in Namibia and in Mozambique. The connecting fold system includes the Zambian copper belt.

Late Proterozoic fold belts known as the Brasilides continue the Pan-African system in South America, which before 200 Ma ago was attached to Africa. A very extensive, little-explored ophiolite belt is postulated near the western side of the Brasilides in central Brazil. If confirmed, this ophiolite belt would further support the plate tectonic interpretation of the Pan-African fold belts.

The Phanerozoic eon

The palaeontologically critical transition about 570 Ma ago from the Proterozoic to the Phanerozoic eon was unimportant in terms of the evolution of the Earth's crust, although the study of Phanerozoic crustal history is frequently aided by the availability of fossils which facilitate much more refined dating of events than is possible in the Precambrian by radiometric dating. A more important change occurred about 200 Ma ago, when the supercontinent Gondwanaland began to split and move apart. From about that time onwards remnants of oceanic crust are also preserved and allow a far more accurate reconstruction of the relative positions and motions of the continents. The break-up of Gondwanaland coincides roughly with the transition from Palaeozoic to Mesozoic.

Some major Palaeozoic events

The pioneer work on Phanerozoic rocks was carried out in western Europe. A major advance was the recognition in the late 18th century, by James Hutton, of the significance of unconformities. Hutton's discovery soon led to the recognition of two major phases of mountain-building in the Palaeozoic era. These are the Caledonian and the Variscan or Hercynian. Gradually, as periods of folding were more accurately dated and correlated, the idea of a few brief, worldwide

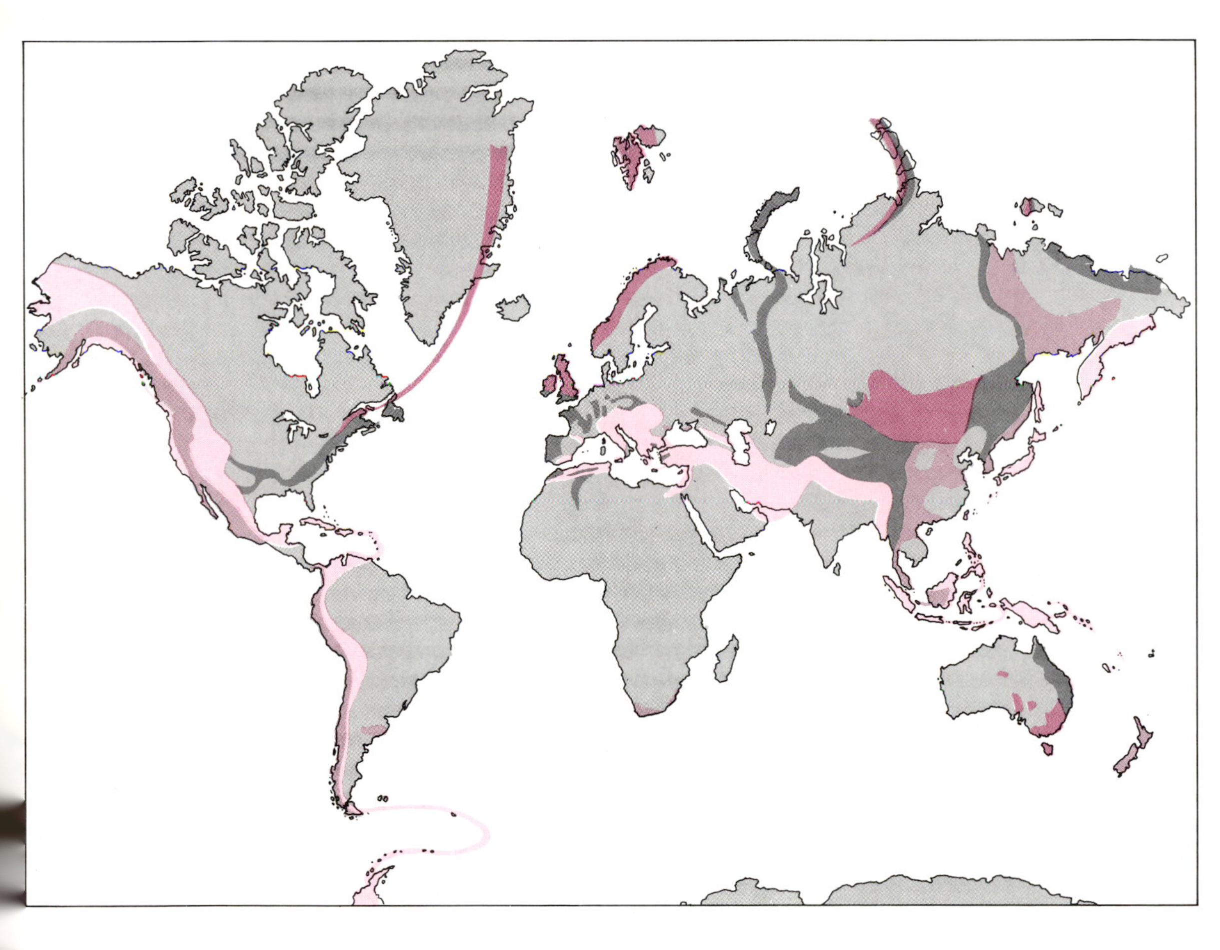

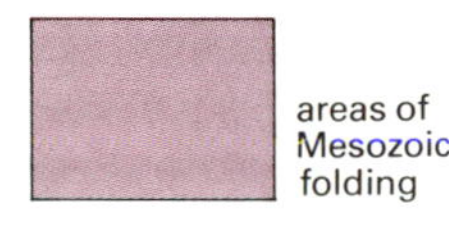
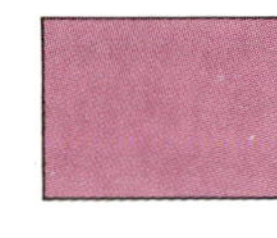

16.21: *Phanerozoic fold belts of the world.*

mountain-building revolutions separated by long quiet intervals gave way to that of numerous distinct phases which, however, were still thought to have worldwide significance. Plate tectonics theory changed these ideas because huge lithospheric plates cannot suddenly stop and start moving. They move continuously, though at varying rates, and the folding, thrusting and other deformations that occur mainly along the destructive plate margins are also continuous, though varying in rate. 'Caledonian' now encompasses all the deformations that occurred between the start of the Cambrian (about 570 Ma ago) and the mid-Devonian (about 380 Ma ago), and 'Variscan' comprises those between the mid-Devonian and the end of the Permian (about 235 Ma ago). Both of these major deformational cycles can now be attributed to complex sequences of events. Both involved closure of more than one ocean and both spanned a considerable length of time because of the irregularities of the approaching continental margins. In each case too the ages of the sediments deformed extend hundreds of millions of years back from the major mountain-building episodes.

The Caledonian deformations

The classic area of Caledonian deformation is Scotland (Roman Caledonia). This has been found to be very complex and the structure is still far from clear (Figure 16.25(a)). The whole Caledonian structure in Britain is strongly asymmetric. The structural grain runs from north-east to south-west, and deformation and metamorphism migrated south-eastwards across it. In the western Scottish Highlands Grenvillean (about 1000 Ma ago) and younger (about 750 Ma ago) folding and metamorphism are recognized. In the southern Highlands the main events of folding and metamorphism occurred about 600 Ma ago. In southern Scotland, northern England and Wales folding (but not metamorphism) occurred about 400 Ma ago. The whole process of building the Caledonides thus took several hundred million years.

The process was not continuous. At least twice, and perhaps more, an ocean opened and was then eliminated by subduction. This is known because small remnants of oceanic crust were preserved, trapped along sutures between collided plates. These ophiolites are seen on Unst in the Shetland Islands, along the Highland Border, and at Ballantrae in Ayrshire. They suggest north-westerly subduction of oceanic crust under the Scottish Highlands, where the complex structures are suggestive of a collision orogeny with considerable crustal thickening. The north-western margin is a thrust zone with evidence of an aggregate 100 km of west-north-westerly transport of metamorphosed late Proterozoic schists over unmetamorphosed late Proterozoic to Ordovician sediments which overlie early Proterozoic and Archaean basement. The southern Highlands are dominated structurally by a huge south-east-facing recumbent fold, or nappe.

A younger and more southerly suture is recognized mainly by contrasts of fossil faunas and of structural style across it; it runs through central Ireland and along the Scottish–English border. This suture, which is not marked by ophiolites, marks the position of a former ocean known as Iapetus. The subduction of the Iapetus Ocean on a Benioff zone dipping south-eastwards would be expected, from our knowledge of present-day plate tectonic processes, to lead to the eruption of calc-alkaline magmas in the overriding plate and the emplacement of large volumes of granite. Such volcanic rocks were abundantly erupted in Ordovician times in the English Lake District, north Wales and south-east Ireland. Granite batholiths were intruded in the Scottish Highlands, northern England and (the largest) in south-east Ireland.

Subduction along the north-western side of the Iapetus Ocean led to the accumulation of a thick stack of north-west-facing slices of Palaeozoic rocks, which can be recognized from equivalents at present-day trenches as an accretionary wedge (forearc prism) along a north-west-dipping Benioff zone. Although many problems remain, it is clear that the British Caledonides were thus built up in a series of plate collisions which accreted successive deformation zones against a continent to the north-west. It is the classic example of an orogenic mountain belt, studied in detail for over a century, to which the plate tectonics theory has brought a framework for its interpretation in terms of present-day processes. The Caledonian belt is continued northwards through western Scandinavia and east Greenland to Spitsbergen, and westwards through Newfoundland into the Appalachians where Caledonian deformation became superimposed by later, Variscan, events (Figure 16.21).

The Variscan deformations in Eurasia

A broad zone of late Palaeozoic deformation stretches across Eurasia, with a triple junction and a major northward branch, the Ural Mountains, in the Soviet Union. As in the Caledonides there were repeated superimposed and migrating deformations. In western Europe the northern limit of the Variscan fold belt is a thrust front in places, such as southern Ireland, south-west England and Belgium. The southern limit, from the South Atlas fault in Morocco to central Tibet, is fairly well defined, but between these limits there are cratons or microplates made of older rocks and unaffected by Variscan deformation, such as the Channel Islands and north Normandy, the Lut block in Iran and the Tsaidam depression in west China. The plate tectonic interpretation of the whole belt is much more controversial than that of the Caledonides.

Ophiolites and high-pressure blueschists are spectacularly developed in the Urals, across which palaeomagnetic data show a large misfit. Here it is clear that a large ocean was eliminated by subduction. In western Europe the evidence for a plate tectonic interpretation is less clear. A possible ophiolite in south-west Britain is thrust northwards over a trench-type melange, and there are blueschists that yield Variscan dates in the Ile de Groix, Brittany. Palaeomagnetic data suggest that an ocean separated the African and European plates before the Variscan closure, but it is not clear whether the main suture runs through south-west England (Lizard) or through south Brittany and the French central plateau, or both.

The Appalachians

The Appalachian Mountains of eastern North America are a direct continuation of the Caledonian and Variscan fold belts of western Europe, but, whereas in western Europe the two are geographically separated, in the Appalachians they are superimposed. The Caledonian part of Appalachian history is one of late Proterozoic continental rifting, accumulation of thick continental shelf and rise sediments in the early Palaeozoic and (in the northern Appalachians) final closure of the Iapetus Ocean in Ordovician to Silurian times. In the southern Appalachians the final deformations occurred later, in Carboniferous to Permian times. Large-scale thrusting with horizontal displacements of up to 125 km, together with mineralization and intrusion of granites, point to a collision orogeny.

This brief account of some of the better-known Palaeozoic fold belts leaves no doubt as to the applicability of plate tectonic processes such as are active today at destructive margins, at least this far back in geological history.

Major Mesozoic and Cenozoic events

The break-up of Gondwanaland

In Alfred Wegener's revolutionary exposition of the theory of continental drift in 1919, through A. L. du Toit's masterly restatement in 1937, to the sudden development of the plate tectonic theory in the 1960s, the break-up of Gondwanaland, and especially the separation of Africa from South America and the formation of the south Atlantic, have constituted crucial evidence. In the evolution of the crust this break-up, which took place largely in Jurassic to Cretaceous times, was a major and perhaps unique event, and it prompts three questions. To what extent did older structures control the positions of the breaks? Can hot spot uplifts from which fractures propagated into continuous rifts be recognized? What forces moved the plates apart and, in particular, carried the Indian plate far into Asia?

The control of pre-existing structures can be seen on two scales. On a large scale the rifts seldom cut through the old cratons but tended to run along the middle of the Pan-African fold belts (Figure 16.20). On a smaller scale the incipient splitting along the African rift system has certainly been guided in many places, most obviously in the Tanganyika rift in south-west Tanzania, by Precambrian mylonite zones. Elsewhere, however, the continental plate was split without regard to previous structure.

Propagation of rift fractures from one hot spot elevation to another is clear in the East African rift system and the Red Sea. The rift between Africa and South America connects a number of topographic domes which were probably related to hot spots.

No plausible proposals have been made as to the nature of the forces which rather suddenly broke up Gondwanaland which had been a unit since early Cambrian times at least. If rifts do indeed develop as fractures propagating between uplifts over hot spots it may be that this can happen only when a continental plate is stationary relative to a system of hot spots for sufficient time. Gondwanaland, from palaeomagnetic evidence, had moved continuously until some time in the Triassic period, perhaps about 50 Ma before break-up began.

Post-Pan-African events in Africa

Within-plate magmatism and deformation are particularly clearly displayed in Africa. The change from destructive margin magmatism to within-plate magmatism can be recognized geochemically by a change from calc-alkaline to alkaline composition of the volcanic and intrusive rocks. Within-plate processes have dominated African tectonics since the end of the Pan-African deformation, more than 500 Ma ago.

Younger magmatic sources appear at first to be randomly distributed and are thought to be plumes of hot material rising from within the mantle. At the surface they are expressed either by ring-intrusions, of which there are hundreds in Africa ranging in age from Precambrian to late Mesozoic and Cenozoic, or by volcanoes. One striking fact of the distribution of the Mesozoic and Cenozoic volcanoes in Africa is that they are almost exclusively confined to the Pan-African domains. The mantle below these former mobile belts must have remained for 500 Ma in a different thermal state from that of the mantle under the cratons.

A particularly distinctive type of magma is kimberlite, famous for the diamonds it contains. This is a peculiar carbonate-rich, alkaline type containing blocks of upper-mantle material brought up from 200 km or more below the surface. Unlike the other volcanic types the diamond-bearing kimberlites are mainly confined to the very old cratons, beneath which the mantle is unusually cool. An unexplained fact is that the majority of the many kimberlites in southern Africa

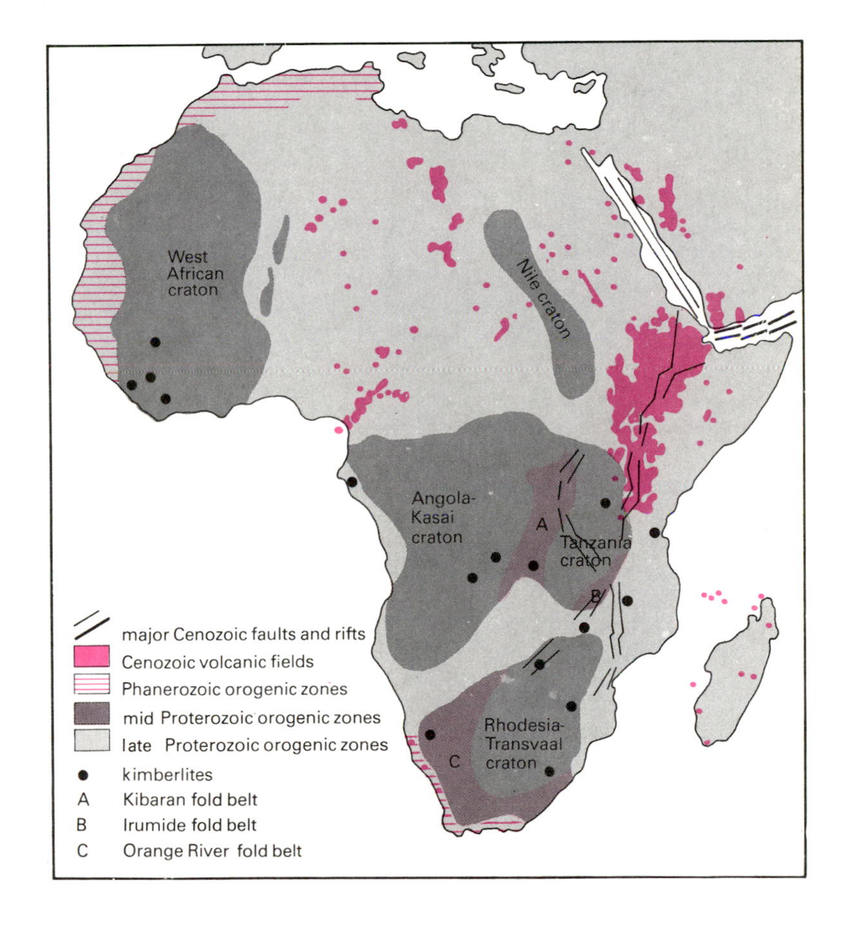

16.22: *Tectonic map of Africa, showing the geological setting of Mesozoic and Cenozoic rifting, volcanism and emplacement of kimberlites. Selected Archaean and Proterozoic geological features are also shown.*

were erupted during quite a short period in the Mesozoic.

The deformation of Africa during the gradual break-up of Gondwanaland in Mesozoic and Cenozoic times produced other extensional fault systems, of which the best known is the East African rift system. Depressions such as the Sudd, the Chad basin, the Congo basin, the Kalahari and Lake Victoria basin were probably formed by way of compensation for elevations, such as the Ethiopian and Kenyan highlands, where there is evidence of hot plumes rising from the mantle.

The evolution of the modern oceans

Before the break-up of Gondwanaland the major oceans were the Pacific, much larger than it is now, and the Tethyan Ocean between Laurasia and Gondwanaland. The Pacific was once thought to mark the scar left by the departure of the Moon, but it is now known that no part of the existing Pacific floor is older than about 200 Ma. This does not mean that the Pacific Ocean itself is young, as there is evidence that its present eastern margin has been near its present position relative to both North and South America for at least 500 Ma. This is shown by the presence of calc-alkaline plutons, parallel to the coast, similar to those formed more recently in the Cordilleras, but dating back to 500 Ma. On the other side of the Pacific the margin has shifted and the Eurasian continent has grown by accretion of island arcs during the past few hundred million years. Very much older structures, about 2000 Ma old, on the Peruvian coast, seem to be cut off at the Pacific margin, suggesting that the ocean formed later than those structures. If so, the lunar hypothesis must be incorrect, the Moon being much older than 2000 Ma.

Unlike the Pacific Ocean, the Indian, Atlantic and Antarctic Oceans are young. They formed as the rifts that split Gondwanaland gradually increased in width. They are still widening, while the Pacific contracts. As for the once wide Tethys, only small remnants remain such as the west Mediterranean, the Black Sea and the Caspian Sea.

Plate motions relative to the mantle (the hot spot frame of reference) can be deduced from the migration of volcanic centres, as in the Hawaiian chain.

Continental rifting: the East African and other rifts

The break-up of Pangaea in the Mesozoic resulted in a great many more rifts than developed into oceans. Understanding of these rifts is important both to the plate tectonic theory and also economically, since some have accumulated thick sedimentary sequences with important hydrocarbon reservoirs.

The movement of the Arabian plate away from the African plate in about the past 30 Ma has opened the Gulf of Aden and the Red Sea, forming wide rifts floored by oceanic crust. Where the Red Sea and the Gulf of Aden meet, another rift branches off, providing a unique example of an active ridge–ridge–ridge triple junction on land, in the Afar region of Ethiopia (Figures 16.23 and 16.24). Southwards from Afar a discontinuous series of rifts stretches 3000 km to southern Malawi. Geophysical studies show that in places the rifts are underlain only a few kilometres below the surface by molten or partially molten material which filters out transverse seismic waves. Seismic reflection surveys in Afar show that the continental crust there has been drastically thinned, partly by extensional faults and perhaps partly by ductile flow. In a very few localities, such as the Ert Ali volcanic range, narrow strips of oceanic-type crust have been formed, and the composition of the lavas in these areas is also of oceanic type. The rest of Afar and the African rifts are underlain entirely by continental crust. A conspicuous feature of the rift system is the presence of large dome-like uplifts such as those that form the Ethiopian plateau and the Kenyan highlands. These swells tend to mark the sites of triple junctions where three rifts join, as in Afar, Kenya and the Mbeya area of Tanzania. These radiating tensional fractures are the natural result of the distension of the crust on the uplifts. The history of these domes can be traced by the successively greater elevation of older erosion surfaces on them. The African plate is splitting, but the fractures have not yet joined into a continuous plate boundary. This would probably involve the development of transform faults between rifts that are offset from one another. The rate of separation diminishes southwards from the Ethiopian rift to Kenya, where the rift has increased in width by 10 km in the past 12 Ma or so. Clearly the African rift system provides a glimpse of a new ocean in the making.

Nearly all the faults bounding the East African rifts are normal dip-slip faults inclined on average at about 65°. The slabs between the faults are rotated, usually dipping away from the upthrown side, so

16.23: *Colour composite photograph from Landsat-1 satellite of the Afar depression at the meeting point of the East African rift system (lower left), the Red Sea rifts (top and right) and the Gulf of Aden rifts (lower right).*

streams drain away from the rifts, while lakes tend to form at the foot of the faults on their downthrown sides. The faults are close together, often irregular, or en echelon, and they have moved repeatedly so that the displacement of older rocks is much larger than that of younger rocks across the same fault. Old fault scarps are eroded back and embayed, while young ones are sharply defined cliff lines. The continuation of rifting to the present day is shown by young faults, which displace Pleistocene deposits containing human artefacts, and by recent earthquakes, which have produced little scarps on the ground surface. Along the rift system earthquakes are frequent and of variable intensity. They are all shallow in origin and tend to recur in particular areas.

Associated with the development of the rift system are volcanoes, lava floods, ignimbrites and many other volcanic phenomena. The volcanic rocks are predominantly alkaline and range from very basic to acid. The larger volcanoes, such as Mt Kenya (now deeply eroded) and Mt Elgon in Kenya, Kilimanjaro in Tanzania and Simien in Ethiopia, are on the flanks of the rift, while others lie within it. Some of the volcanoes, especially in Afar, can be seen to be fed by dykes which fill vertical fissures. Some of these fissures are open at the surface, in common with those in Iceland. Thus the volcanic phenomena also strongly suggest a tensional regime.

In Afar saline water from the Red Sea flowed into the rift during its early history and was so concentrated by evaporation that thick layers of gypsum and salt were deposited. This appears to be typical of the initial rifting of new oceans, as similar evaporites are found under the Red Sea and under the offshore margins of Africa.

Elsewhere in the world other rift systems include the Cenozoic Rhine rift between the Vosges and the Black Forest and the Mesozoic rift system that is the site of the most productive oil fields in the North Sea. In the western USA the Basin and Range province is an area of extensional faulting in some respects similar to the African rifts. The late Cenozoic Rio Grande rift of New Mexico, in common with others, has important mineral deposits associated with it. In Asia, Lake Baikal occupies part of a rift system trending north-east that was formed in Mesozoic times along the line of a late Proterozoic mobile belt.

16.24: *Seismic and volcanic activity in southern Afar in November 1978, in the north-west–south-east trending arm of the rift system, generated new fault scarps and a new volcano. Beyond the 40 m high cone, from which 16 cubic metres of basalt was extruded during this period of activity, there is a young fault scarp.*

The Andes and the North American Cordilleras

The Andes have been built by the prolonged subduction of Pacific oceanic crust under the South American continent, possibly for as long as 1000 Ma (Figure 16.25(c)). The trench, where the Benioff zone reaches the surface, lies about a hundred kilometres offshore and the fore-arc is under water. A trench–arc gap, devoid of volcanoes and virtually without Andean deformation, extends inland in places to include the major areas of coastal plain and deltaic sedimentation. To the east lie the Andean ranges where the crust is thickened by huge batholiths and plutons of granodiorite and allied rocks and by a wide range of basaltic to acid but predominantly andesitic volcanics, including vast sheets of ignimbrites (welded tuffs). The drastic thickening of the crust cannot be attributed to tectonic compression, as sedimentary sequences, showing a long previous history of marine conditions, are only rather weakly folded. It must therefore be due mainly to addition of material from below as intrusive and extrusive magma. Isotopic and geochemical studies show that much of this material came from the mantle rather than by melting of the crust. A spectacular feature of the Andes is the evidence, first noted by Charles Darwin, of very recent and rapid isostatic uplift. Flights of marine terraces rise hundreds of metres above present sea level. On one of these in Peru the remains of a stranded whale can be seen.

In the southern Chilean Andes a back-arc basin developed during the late Mesozoic, on the continental side of the volcanic belt. In this basin there accumulated a thick sequence of mainly basic volcanic rocks including pillow lavas, supposedly representing oceanic crust. It was later compressed to form a synclinal complex—the Rocas Verdes, which has been suggested as a model for the enigmatic Archaean greenstone belts.

The Mesozoic volcanic and sedimentary rocks of the Andes rest on an older foundation of folded Palaeozoic sediments. These in turn are underlain by Precambrian sialic basement which appears to continue under the Andes, to emerge with a cover of nearly flat-lying Devonian and younger strata within coastal regions of Peru. These Precambrian rocks of the Peruvian coast, metamorphosed to granulite facies, are nearly 2000 Ma old. Thus the foundations of the Andes have been of continental character for a very long period.

The North American Cordilleras (Figure 16.25(b)) are more complex than the Andes, from which they differ in several respects. Along the broad coastal zone, notably in California and in Oregon, numerous subduction complexes are exposed. Best known is the Franciscan complex, a gigantic melange with blocks of ocean floor rocks and trench sediments chaotically tumbled together. East of the Franciscan, in the Great Valley, ophiolitic rocks covered by Mesozoic turbidites are believed to represent oceanic crust underlying an outer trench–arc gap.

An important difference from the Andes is that the relative motions

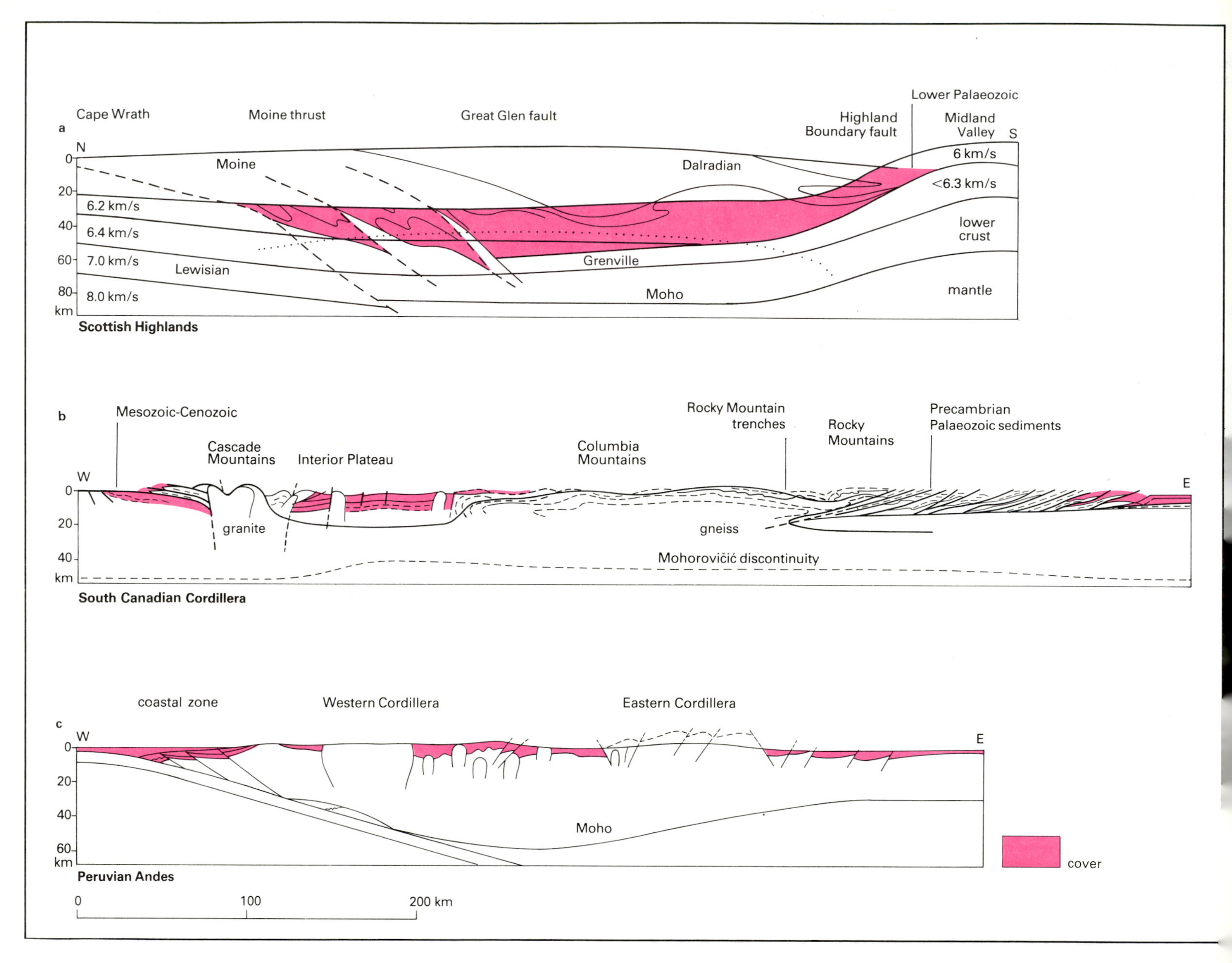

16.25: *Diagrammatic cross-sections through major fold belts, showing relationships between major structural units. All at the same scale, with no vertical exaggeration.* (**a**) *The Scottish Highlands, part of a continent–continent collision zone formed between 1000 and 600 Ma ago; Caledonian orogenic activity to the south of the Highland Boundary Fault continued until 400 Ma ago. This structural section is based on seismic velocity data.* (**b**) *The southern Canadian cordillera. Deformation of the Columbian orogen began in the Mesozoic with blueschist metamorphism suggesting an east-dipping subduction zone.* (**c**) *The Andes of central Peru, a cordilleran orogenic belt at the junction of oceanic and continental crust.*

of three separate plates meeting at an unstable triple junction led to a relative northward motion of plates on the Pacific side of a transform fault, the San Andreas fault. Strike-slip motion on this fault results in repeated earthquakes, including the one that devastated San Francisco in 1906. At the eastern side of the North American Cordilleras, especially in Canada, there is an immense series of thrust slices which are important hydrocarbon traps in the Canadian oil fields. The thrusting may be due to gravity-spreading of the Cordilleran massif as it thickened.

An important feature of both the North and South American Cordilleras is the presence of very large, elongate granitic batholiths. From north to south they include the Coast Range (Canada), Idaho and Sierra Nevada (USA), Baja California (Mexico) and Peruvian and Chilean batholiths. These granites were intruded in the Mesozoic, mainly the Cretaceous period, at depths of 6 kilometres or more beneath the active volcanic region on the continental side of the subduction zone.

It now seems likely that predominantly northward movement of Pacific Ocean floor has plastered a whole series of subduction arcs against the North American margin, from Alaska to Baja California, starting in Jurassic times. Some of those arcs may have travelled hundreds or even thousands of miles relative to North America, which would explain the numerous differences in the Palaeozoic and early Mesozoic sequences to be found in adjacent terrains of the western North American margin. Perhaps there were times in the Palaeozoic and in the Mesozoic when the eastern Pacific, offshore from North America, resembled the western Pacific of today, with its festoons of island arcs.

The island arcs of the west Pacific

The island arcs of the west Pacific are the most regular large-scale geometric features of the Earth's surface. Their geometry is a consequence of the fact that the subducting Pacific plate is a large part of a spherical shell.

While many of the arcs are essentially volcanic and no older than Mesozoic or even Cenozoic, the Japanese arc has had a much longer history. The Chichibu geosyncline comprises up to 20 km of Palaeozoic marine sediments and volcanics, which were intensely folded in Triassic times. There is no evidence of underlying pre-Palaeozoic basement (older dated zircons probably came from mainland Asia). These Palaeozoic rocks are much younger than the oceanic crust under the Sea of Japan. Back-arc spreading displaced the arc away from the Asian mainland and increased its curvature, as shown by palaeomagnetic evidence.

Post-Triassic sediments and volcanic rocks are mostly terrestrial, except in the northern island of Hokkaido. Repeated subduction led to obduction of ophiolites, development of paired metamorphic belts (see Chapter 14) and intrusion of granites, in successive accretions on the oceanic side of the Japanese arcs.

North of Japan, the Kuril arc is much younger, the oldest visible rocks being late Cretaceous, and the western Aleutian arc is entirely Cenozoic in age. To the south of Japan, the Mariana and Tonga island arcs are no older than Cenozoic, but a much longer history of existence at an oceanic margin is recorded in New Zealand. Sedimentary and volcanic rocks there stretch back to the earliest Palaeozoic while later events include a very large dextral shift on the Alpine transform fault and wholesale rotation. In New Caledonia enormous ophiolitic masses with important mineral deposits were forced up during subduction.

The striking difference between the cordilleras on one side of the Pacific and the festoons of island arcs on the other is attributed to the motion of the Pacific plate relative to the underlying mantle. The relative motion is deduced from the hot-spot frame of reference (see Chapter 10). The Pacific floor, which sinks into the mantle at the trenches, is at the same time moving away from the Asian continent but towards the Americas. Relative motions of the plates deduced in this way are shown in Figure 10.17, which also depicts present-day island arcs.

Cenozoic collision orogenies: the Alpine–Himalayan chain

It is no coincidence that the type example of a mountain range formed by collision of continental plates, the Himalayas, is also the highest in the world. This is because the crust there is abnormally thick, reaching as much as 72 km in southern Tibet (Figure 16.26). The collision occurred when the Indian continent, part of the mainly oceanic Indian plate, impinged upon the Asian continent. The northward rate of motion of India away from the East African coast was about 100 mm per year until about 40 Ma ago, when it slowed to about 50 mm per year. Now it is still moving northwards but at only about 10 mm per year. The slow-down about 40 Ma ago coincides with geological evidence that sediment deposited at that time along the northern edge of the Indian plate was for the first time derived from the north. This is taken to mark the time of collision. After collision India continued to move northwards at least a further 2000 km, and it is this movement that thickened the crust underneath the Himalayas.

There are several interpretations to explain this remarkable occurrence. Tectonic thickening by the packing of thrust slices under the overriding plate can account for only about 500 km of the 2000 km of post-collision movement. Thickening in the Transhimalaya may be more plausibly explained by Andean-type addition of magma from below. The obvious proposal that the Indian plate was simply pushed underneath the Asian one, so doubling its thickness, is not supported by the structural evidence. Another possibility is that the post-collision movement was taken up by the sideways expulsion, along huge strike-slip faults, of wedges of the crust, as a result of the impaction of the Indian continent into Asia (Figure 16.27). The structures in the thick Himalayan crust become younger and more complex southwards, and earthquakes show that movement is now mostly concentrated on the main boundary thrust at the south side of the Himalayas.

The position of the suture between the Indian and Asian continents is uncertain. One suture, marked by a huge belt of ophiolites, follows the upper Indus and Tsangpo valleys. The Indian continental plate, with sediments and faunas of Gondwana facies, certainly extends this far north. North of this suture is the Transhimalaya, with very large calc-alkaline granodiorite batholiths which clearly indicate that an underlying north-dipping subduction zone was active in the Cretaceous period. Further north, in Tibet, there is probably another suture and this, rather than the Indus–Tsangpo suture, may mark the true southern margin of pre-collision Asia. If so, the Transhimalaya represents a microplate or island arc related to the Indian plate. The Kohistan block in the Karakoram Mountains exposes the whole thickness of the crust and part of the upper mantle of an island arc between the two sutures.

Because the crust has doubled in thickness the Himalayas are rising isostatically, and they have risen about 3 km in the past 3 Ma. The resulting erosion has shed vast volumes of sediment into the Indo-Gangetic plain which is thickly floored with this molasse. Earlier Cenozoic sediments derived from the earlier stages of uplift have themselves been overridden and thrust under the advancing Himalayan nappes. The thickness of the crust will inevitably result in its heating up and recrystallizing at depth into granulites. Isostatic uplift will eventually expose these granulites at the surface, supporting deductions of a collisional origin for granulites in old fold systems.

Effects of the collision between the African and Eurasian plates are seen throughout the zone from the Himalayas to the Alps. In Iran, Oman and westwards into Turkey, innumerable ophiolite fragments, trench melanges and subduction generated calc-alkaline volcanics show the effect of successive collisions of arcs and continents. In the Mediterranean the effects are complicated by the presence of a number of microplates which have moved in different directions. Palaeomagnetic observations show that several microplates, including the Iberian peninsula, Corsica–Sardinia, Italy and Cyprus were rotated anticlockwise during the Mesozoic collision process.

The Alps, perhaps geologically the most thoroughly known mountain range in the world, consist of slices derived from the African plate to the south, separated by slivers of ophiolites on a south-dipping suture from the main structures of the western Alps. These are a series of enormous recumbent folds in which both Palaeozoic basement and Mesozoic cover have been plastically deformed (Figures 16.26, 16.27), overriding sediments derived from the uplifting mountains. Towards the southern side of the Alps high-pressure, low-temperature blueschists are developed.

Conclusion

Since a major goal of innumerable geologists, geophysicists and geochemists, past and present, has been the elucidation of the history

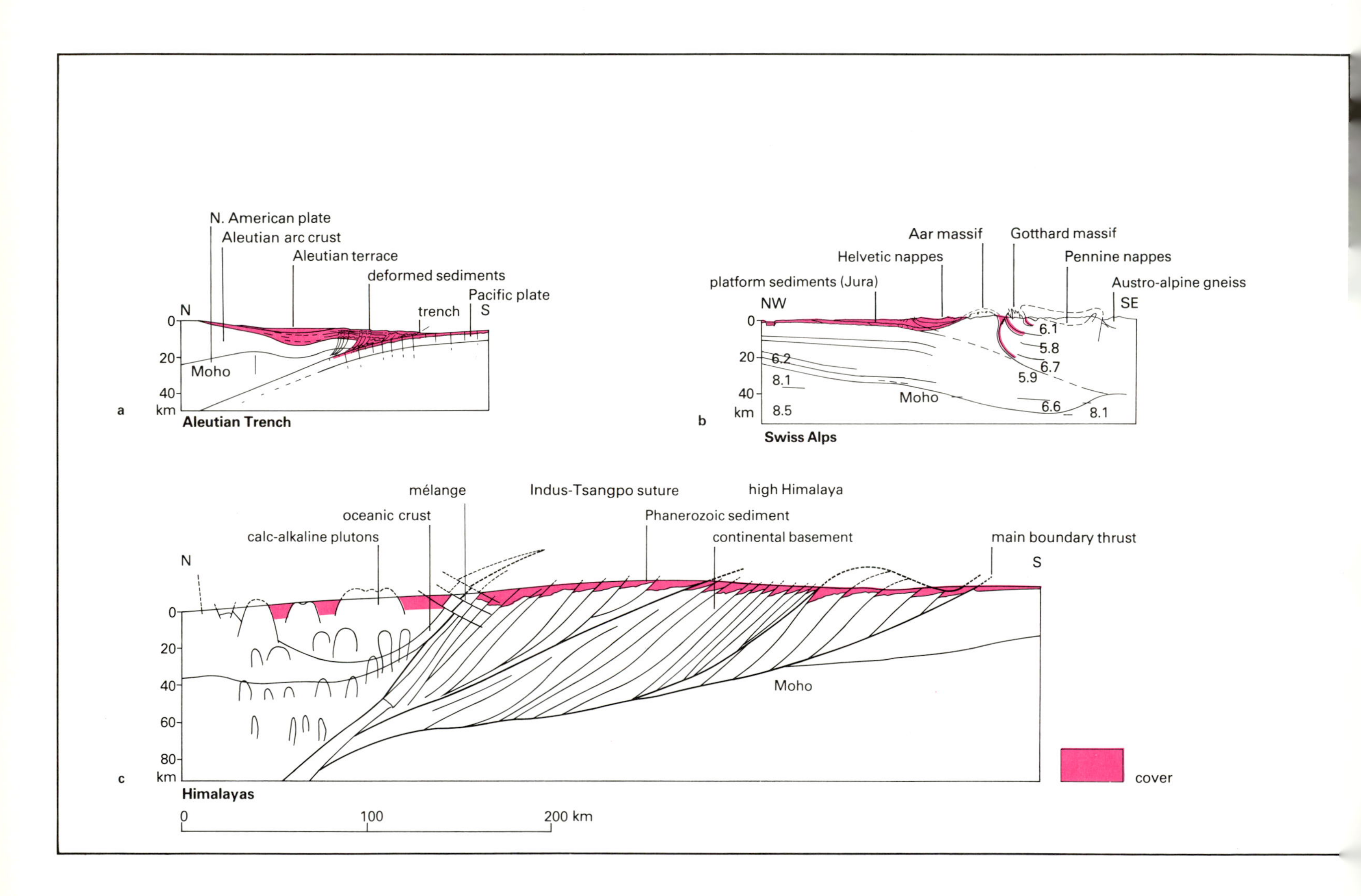

16.26: Diagrammatic cross-sections of major fold belts to same scale as Figure 16.25. (a) The central Aleutian trench—an active island arc subduction zone. (b) The Swiss Alps to the north of the root zone (compare Figure 11.14), showing the thickening of the continental crust. The geological history of the Alps goes back to Palaeozoic times but the nappe structures characteristic of the main continent–continent collision orogeny date from the Palaeogene period. Some seismic data are shown. (c) In the Himalayas, crustal thickening as a result of continent–continent collision is particularly pronounced and is represented here by stacking of numerous thrust slices. The additional effects of lateral faulting appear in Figure 16.27

of the Earth's crust, it is clearly quite impossible even to summarize the major events in a chapter of this length. The inevitable bias of scientific endeavour towards certain geographical areas also makes a global selection of examples to illustrate the story quite impossible as yet. This chapter has therefore taken a number of important events and themes, and omitted a great many more. Nevertheless if one had to collapse the entire story of the evolution of the crust and the development of present-day tectonic processes into a few lines, the result might be as follows. Pre-geological history, unrecorded by any extant rocks, encompassed the accretion of the Earth, the separation of the core and formation of a scum of crust at the surface. Archaean history saw the accretion of many thin and highly mobile fragments into recognizably continental masses. During the Proterozoic tectonic processes were largely confined to the margins of stable cratons which may have aggregated into the first supercontinent. Latest Proterozoic and Phanerozoic times saw a cycle of fragmentation and reassembly (into Pangaea) of those continents, followed by a repeat performance of fragmentation into the configuration that is depicted in present-day atlases.

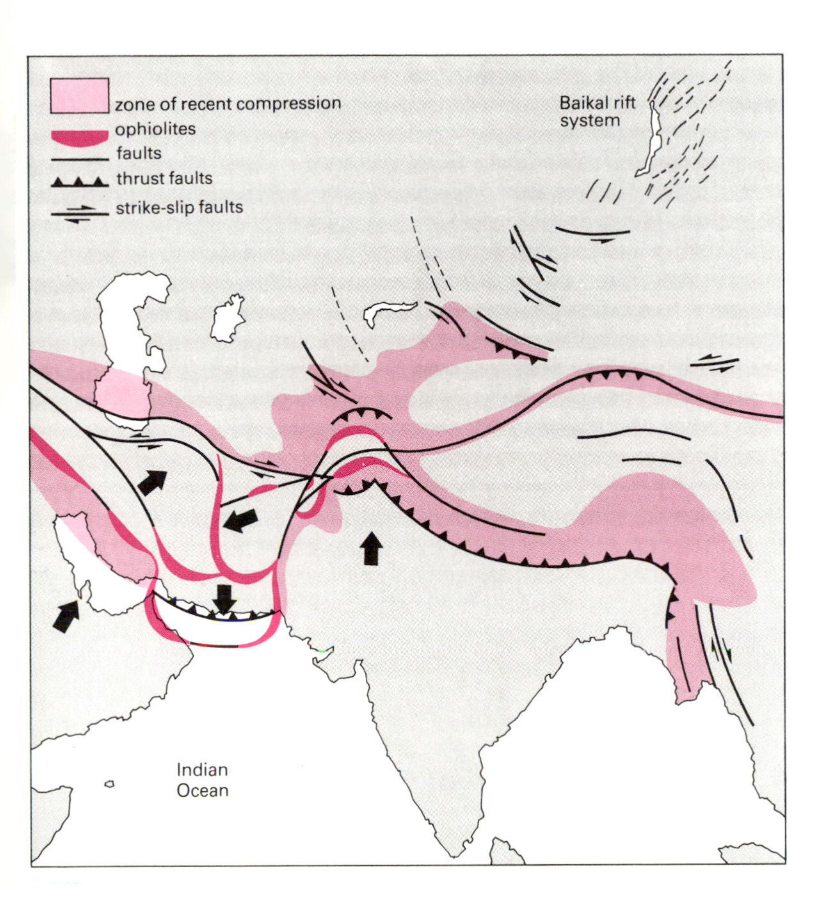

16.27: Outline tectonic map of the Himalayan collision zone between the Indian and Eurasian plates, showing the extent of alpine-type deformation and the various ophiolite bodies representing partly subducted oceanic crust. The complexity of this collision zone is partly attributable to the highly irregular margins of the approaching continents.

16.28: The Dent de Morcles, Rhône valley, Switzerland, exposing the flat-lying fold structures associated with the Morcles nappe of the high calcareous Alps which have travelled north-westwards over the Aiguilles Rouges massif.

Part Four: Surface Processes and Environments

Understanding and description of the present state of the Earth's surface is of direct consequence to us: because it is part of the biosphere we depend on its life-giving qualities. The atmosphere and hydrosphere—those thin and volumetrically insignificant shells of the Earth—are not only vital to life on the planet's surface but also control its continual reshaping by the forces of weathering and erosion, transport and deposition. The surface of Mercury, which lacks an atmosphere, has remained almost completely unaltered for 3000 million years, while for all that time the winds on Earth have never ceased blowing, nor the ocean currents circulating. The combination of solar-powered atmospheric circulation and geothermally-powered crustal movements has resulted in ceaseless readjustments of the surface features of the Earth. The present is like just a single frame in a long sequence of film. Each frame is slightly different from another, but all record a stage in the same story.

Features of the Earth's surface in times past are traditionally interpreted through processes acting at the present. But as we get to know the past better, so the flow of information reverses, and we find ourselves learning more about the present from the past. We are unlikely to witness the disappearance of all the Earth's permanent ice, the erosion of the Himalayas to half their present height, or the evolution of flying molluscs, yet changes of these magnitudes have occurred many times in Earth's history. It is only by studying the versions of such changes recorded, however imperfectly, in the rock strata that we come to see the present-day Earth in its true perspective. Indeed, the more we study the Earth's past, the less representative the present appears to be. To the Earth scientist, study of past and present are both part of the same overall process of understanding the Earth.

Mount Dhaulagiri, 8171 m above sea level, in the Himalayas, Nepal. Below it is the deepest valley in the world, the great gorge of the Kali Gandak River.

17 Atmosphere, water and weather

The Earth's atmosphere

The origin of the atmosphere

In terms of mass the atmosphere represents an insignificant fraction of the Earth, weighing only 5.2×10^{18} kg. However, its importance far exceeds its weight. It is mobile, highly reactive and essential for life. It is entirely accessible, and so knowledge of its composition and chemistry is much more precise than that of regions such as the crust and mantle. The comparative ease with which it can be studied has meant that its composition has been known for many years, as have the compositions of atmospheres of other planets. It has long been evident that atmospheres are highly variable, ranging from the tenuous exospheres of the Moon and Mercury, to the dense atmospheres of Venus and Jupiter. The variation, even among adjacent planets and moons, is so great that it seems certain that atmospheres, regardless of their origin, must undergo considerable evolution. Fortunately the compositions of atmospheres give excellent clues to the processes of both their formation and evolution. The simple question to be asked about the origin of the Earth's atmosphere is: did it merely coalesce about the planet at the time of its formation or was it acquired at a later time? The present consensus is that the primitive atmosphere was lost (Chapter 4) and that the present atmosphere evolved by outgassing, primarily through volcanic activity.

An idea of the composition of the volatiles released during out-

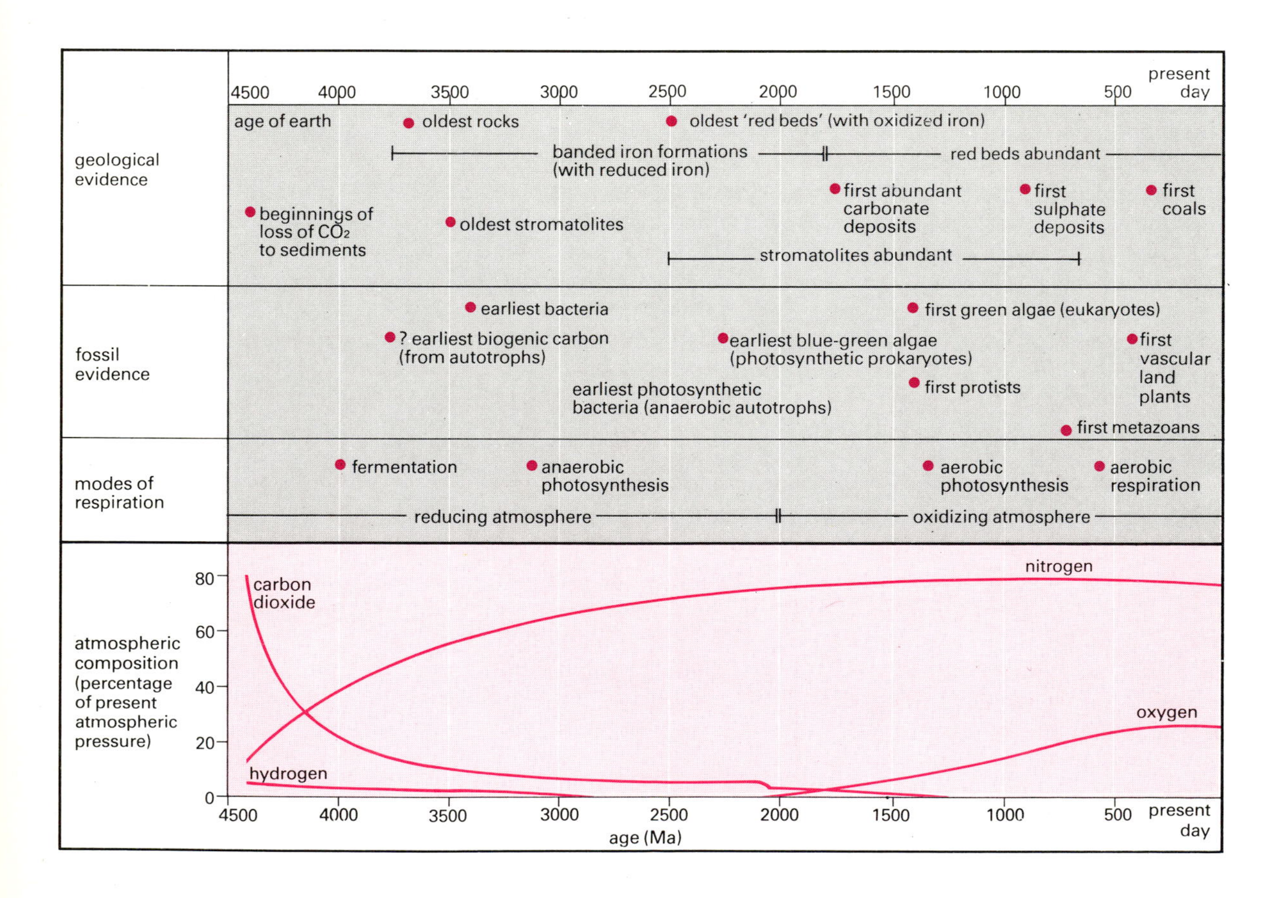

17.1: *A hypothetical scheme of the major changes in the composition of the Earth's atmosphere. Notice how the decline in carbon dioxide was associated with the formation of sedimentary deposits. Autotrophic organisms, and eventually the presence of oxygen, removed hydrogen, which may have been even less abundant in the early atmosphere than the few per cent shown here. The increase in oxygen resulted from photosynthesis, while nitrogen, which resides largely in the atmosphere, shows a rapid rise which reached a plateau as outgassing decreased. The geological and biological evidence referred to are discussed in more detail in Chapters 16 and 21.*

gassing can be gained by examining the composition of modern volcanic fumes, because the composition is determined largely by the equilibrium between the gas and magma and is not dependent on whether the volatiles are truly juvenile or are being recycled. Gas from Hawaiian volcanoes contains 79 per cent water, 0.6 per cent hydrogen, 11.6 per cent carbon dioxide and 0.37 per cent carbon monoxide, with small amounts of sulphur-containing gases, nitrogen, chlorine and argon. The predominance of water is a reminder that most of the outgassed material has gone to make up the oceans.

It is important to note the absence of free oxygen in the outgassed material. The atmosphere would have been in a reduced state, and easily oxidized minerals such as detrital uraninite and pyrite, which are no longer common, would have been stable. Even prior to the origin of life on the Earth the atmosphere would have undergone some evolution. The most interesting chemical change would have been the photolysis of water by ultraviolet radiation: $2H_2O \rightarrow 2H_2 + O_2$. The light hydrogen could then have escaped, leaving some oxygen in the early atmosphere. The magnitude of this oxygen source is not easy to estimate, but it is likely that it was largely consumed by reaction with reduced gases and iron-containing minerals at the Earth's surface.

This chemical evolution was small compared to the changes that occurred after the development of life on the Earth. The earliest organisms would have utilized the adiabatically synthesized molecules present in the environment. A more significant effect on the atmosphere would have come later from chemoautotrophs which derive energy from the reaction between hydrogen and methane. As the hydrogen in the atmosphere became depleted a simple form of photosynthesis developed. The photosynthetic reactions brought about the release of oxygen to change profoundly the atmosphere and the course of biological evolution. The presence of oxygen in the atmosphere about 2000 Ma ago is shown by red beds of that age in which iron is in the oxidized state. Organisms of considerable complexity then evolved to use aerobic respiration as a very efficient source of energy (Chapter 21). Changes in atmospheric composition are still possible, and today human activities are responsible for the largest effects. Carbon dioxide is increasing because of the burning of fossil fuels and the destruction of forests. The ozone layer can be affected by aircraft, aerosol propellants and agricultural production. While the effects on the atmospheric environment are uncertain they may take place very rapidly and have therefore become the subject of much attention.

The composition and structure of the atmosphere

The composition of the atmosphere (Figure 4.12) does not vary greatly with height, except in respect of ozone and water vapour which, together with carbon dioxide, (much increased in concentration over the past two centuries) play very important roles in the atmosphere because of their capacity to absorb radiation. Water vapour is virtually absent above about 8–18 km with an average specific humidity (ratio of the mass of water vapour in a volume to the total mass of air and vapour in the same volume) of 0.008 g/g near the Earth's surface. Unlike water vapour, ozone does not have its source at the Earth's surface and its highest concentrations are at higher levels. Carbon dioxide is produced by living matter and the combustion of fossil fuels at the ground, although large volcanic eruptions can inject relatively substantial amounts into the upper atmosphere. The density of the atmosphere is relatively low at height and so volcanic eruptions may produce considerable changes in composition of the atmosphere at the upper levels. Besides its radiation characteristics, water vapour also has very important thermodynamic properties. It exists in the vapour, liquid and solid phases in the atmosphere, and important heat transfers are associated with the phase changes. Condensation processes lead to the formation of clouds and precipitation which occur almost exclusively in the lowest 8 km of the atmosphere, rarely up to 18 km.

The atmosphere is conveniently regarded as a number of distinctive layers or 'shells' surrounding the Earth. The extremity of the atmosphere lies 30 000–40 000 km from the Earth's surface. The density of the atmosphere decreases with height since it is a compressible fluid and the mean pressure at sea level is 1013 millibars (1 mb = 100 pascal). Figure 17.2 shows the change of atmospheric pressure with height. The total mass of the atmosphere is 5.29×10^{18} kg, compared with 1.35×10^{21} kg for the ocean, and 5.98×10^{24} kg for the solid Earth.

The outermost shell is the magnetosphere which occupies the greatest volume of all the shells. It is the part of the atmosphere which is influenced by the Earth's magnetic field (see Chapter 7). Although

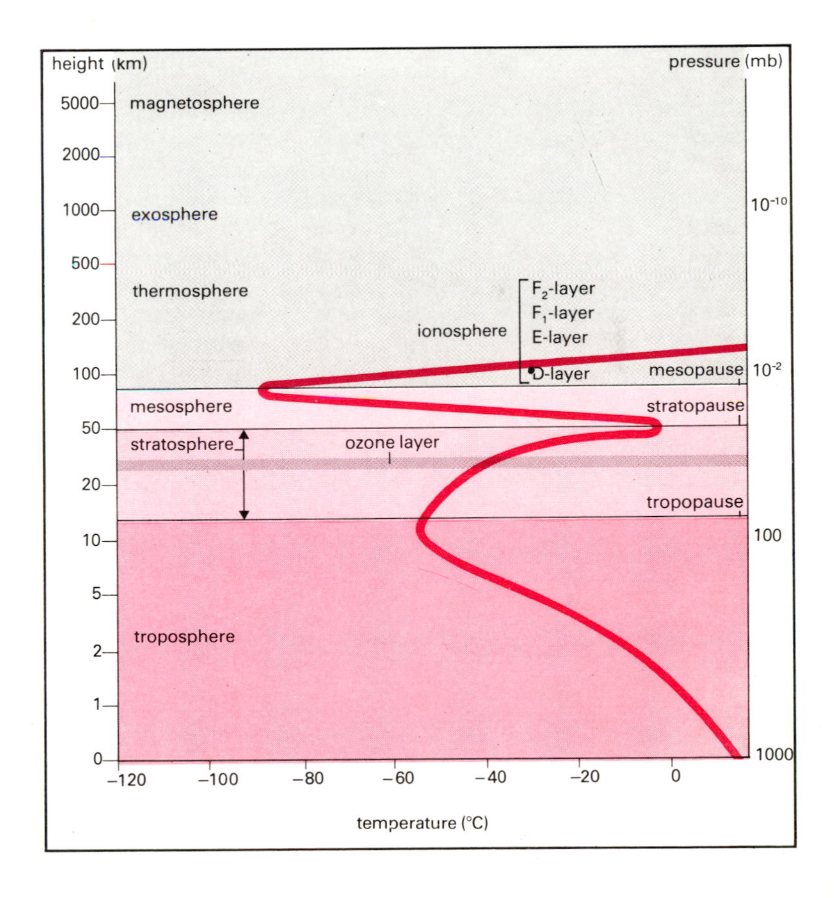

17.2: *The 'shells' of the atmosphere and their temperature characteristics.*

it is thousands of kilometres from the Earth's surface (Figure 17.2) changes in the magnetosphere may directly influence weather. The next shell is the exosphere, from which molecules are continuously escaping into space. Over the lifetime of the Earth most of the light molecules have escaped, leaving behind heavier molecules (nitrogen and oxygen). The thermosphere is characterized by large temperature variations in excess of 300°. At these heights, there are relatively few molecules and so heat retention is low. However, the average temperature of the thermosphere is high because of its inability to re-emit the incoming short-wave radiation from the sun as long-wave radiation.

At heights above about 80 km the absorption of ultraviolet radiation from the sun and the flow of cosmic rays from space separate negatively charged electrons from atoms of oxygen and molecules of nitrogen. This ionization leads to the phenomena of Aurora Borealis and Aurora Australis, the polar lights. The zones of greatest ionization in the ionosphere occur in the so-called D, E, F_1 and F_2 layers (Figure 17.2).

Temperature decreases with height in the mesosphere down to around 183 K at 80 km. This minimum is called the mesopause. Water vapour in small amounts can move up into the mesosphere from lower levels and its condensation results in thin, noctilucent clouds. Chemical reactions have important effects on the temperature structure of the atmosphere and on the climate of the Earth's surface. Nitrogen plays a minor role in chemical reactions but the other major constituent, oxygen, is very active. The electromagnetic radiation from the Sun can be regarded as a stream of photons; those photons with a short wavelength (ν = 0.1–0.2 micrometre (μm)) cause molecular photodissociation of oxygen ($O_2 + h\nu \rightarrow O + O$), where h is Planck's constant. The atomic oxygen produced by this reaction is a major constituent of the atmosphere at around 100 km. At altitudes above about 70 km atomic oxygen recombines to form molecular oxygen in the presence of another molecule (M) which removes excess energy from the reaction ($O + O + M \rightarrow O_2 + M$). At rather lower levels another reaction is very important ($O + O_2 + M \rightarrow O_3 + M$). This reaction is most common between 30 and 60 km where the atmosphere is denser and collisions between O and O_2 are more likely. Ozone (O_3) is unstable and may be destroyed by recombination with atomic oxygen (at higher levels) or by photodissociation ($O_3 + h\nu \rightarrow O_2 + O$). In the lowest levels of the atmosphere it is removed by reactions with dust or at the ground. Maximum ozone density occurs at about 25 km; atmospheric circulations are responsible for its transportation down from its source region at 30–60 km. If all the ozone in the atmosphere were brought down to sea level the layer would be only about 4 mm thick. Nevertheless the trace amounts present in the stratosphere absorb virtually all the ultraviolet radiation between 0.2 and 0.3 μm (the wavelengths at which radiation can cause skin cancer). Maximum absorption in this wave-band occurs at around 50 km and results in the temperature maximum at the stratopause. Recently concern has been shown about the possible effects of pollution on ozone, either injected directly into the stratosphere from supersonic aircraft or via the troposphere from human activities at the ground. Water vapour, nitric oxide and fluorocarbons (released mainly from aerosol propellants) may all affect ozone chemistry and thus bring about changes in the behaviour of the atmosphere.

The quantities of ozone in the atmosphere fluctuate with time and also vary with latitude. Most atmospheric ozone is contained within the stratosphere (the atmospheric shell in which temperature increases with height) and there is a strong ozone concentration gradient across the tropopause into the relatively ozone-free troposphere (lower atmosphere). Below about 25 km ozone is not very reactive and can be used as a 'tracer' of air mass movement since its concentration in any volume of air will tend to remain constant. It is transferred into the troposphere, especially close to frontal zones in the lower layer of the atmosphere. Ozone variations in the troposphere show a relationship with weather, and the gas is eventually destroyed especially by contact with the ground surface.

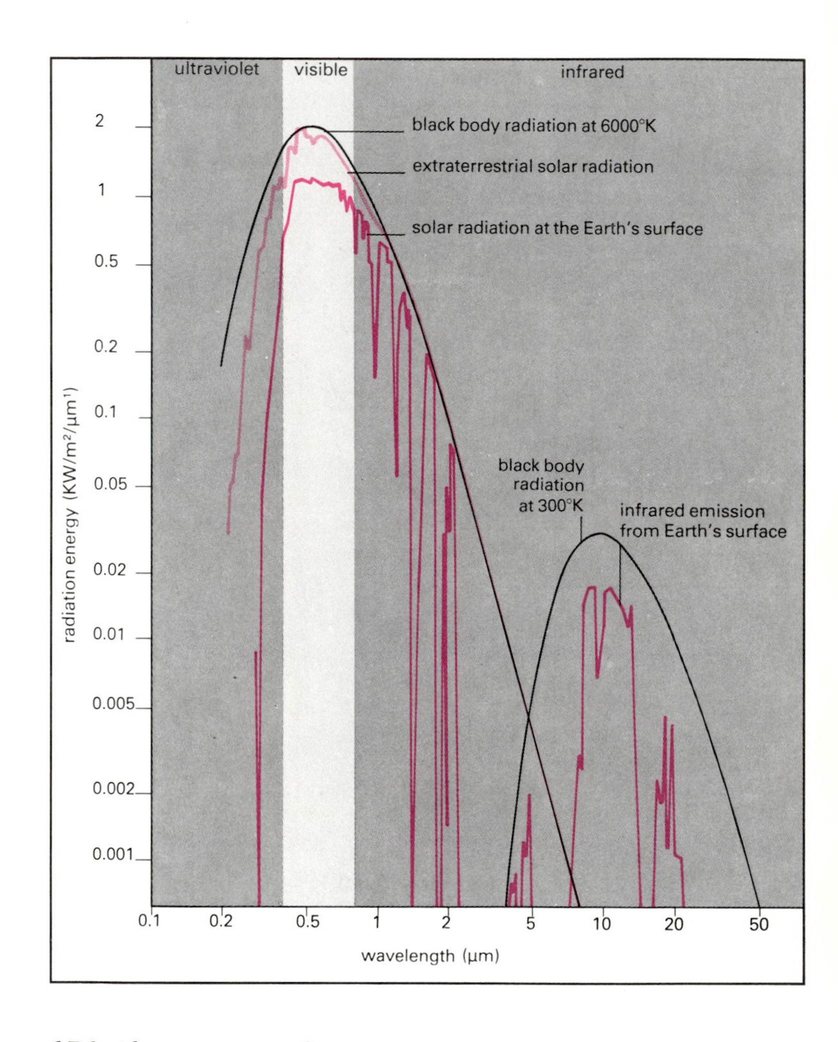

17.3: *Electromagnetic radiation spectra from black bodies with surface temperatures of 6000 K and 300 K (close to the surface temperatures of the Earth and Sun respectively). The 6000 K radiation is reduced by the square of the ratio of the Sun's radius to the average distance between the Sun and Earth to give the radiation incident at the top of the atmosphere. The direct solar radiation incident at the Earth's surface and the long-wave radiation lost to space are modified by reflection and by absorption in the atmosphere.*

Little vertical mixing occurs in the stratosphere (the name means layered sphere). It is inherently stable (ie has an inhibited vertical movement) since warm air overlies colder air. Particles that find their way into the stratosphere (eg from nuclear weapon detonation or volcanic eruptions) stay there for periods of years whereas the residence time in the more turbulent troposphere (where warmer air underlies colder) is of the order of days. For this reason large volcanic eruptions, by altering the radiation balance of the stratosphere, can affect the Earth's climate for a number of years. Within the layer between about 18 to 30 km in the stratosphere is a regular circulation over the lower latitudes which is characterized by a reversal every 12–13 months, from a westerly to an easterly regime and then back again. The nature of this circulation is linked to the changing distribution of ozone in the stratosphere, and the consequent changes in the radiation balance. This stratospheric circulation may well be linked to circulations within the troposphere. A periodicity of about two years is shown by many atmospheric phenomena.

The troposphere is characterized in general by a decrease of temperature upwards. It is possible for smaller particles to travel from the ground to the top of the troposphere in a few minutes within thunderstorm updraughts, although vertical velocities are generally much smaller. Precipitation scavenges particles and some gases from the troposphere and contributes to the relatively short residence time. The troposphere is thicker at the equator than at the poles (ranging from about 18 to 8 km) and contains about 80 per cent of the mass of the atmosphere. It also contains practically all the water vapour, most of which is distributed in the lower layers of the atmosphere. If all the water vapour in the atmosphere were to be removed at once it would represent about 25 mm of rainfall all over the Earth's surface.

The troposphere is the shell of the atmosphere where evaporation, condensation and precipitation occur, pressure systems develop and decay and weather is produced. This region of the atmosphere is of the greatest interest to the meteorologist.

The radiation budget of the atmosphere

The small amounts of water vapour, ozone and carbon dioxide in the atmosphere, and the localized distribution of the first two, belie their importance in the radiation budget of the Earth–atmosphere system. Solar energy drives the motions of the atmosphere, and the Earth intercepts about 0.5×10^{-9} of the total energy emitted by the Sun. Incoming solar energy is sometimes called insolation.

The temperature (T) of the sun is about 6000 K. It can be shown that the wavelength of maximum energy of solar radiation is about 0.5 μm. The total energy emission (E) from the Sun is calculated from the Stefan–Boltzmann law: $E = \sigma T^4$, where σ is the Stefan–Boltzmann constant (5.7×10^{-8} W/m^2/K^4). The solar energy received at the outside of the atmosphere is about 1360 W/m^2. This is the so-called solar constant, although there are small variations. The Earth's orbital geometry gives rise to seasonal and latitudinal variations in the receipt of solar energy, which depends on the angle of the Sun above the horizon and on day length.

The Earth has a mean temperature around 290 K (about 7°C) and

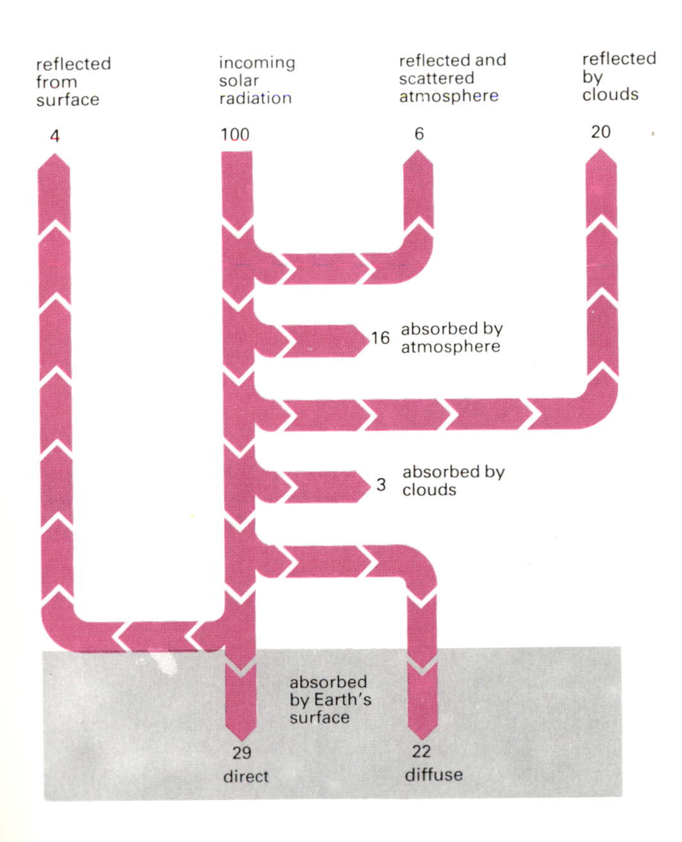

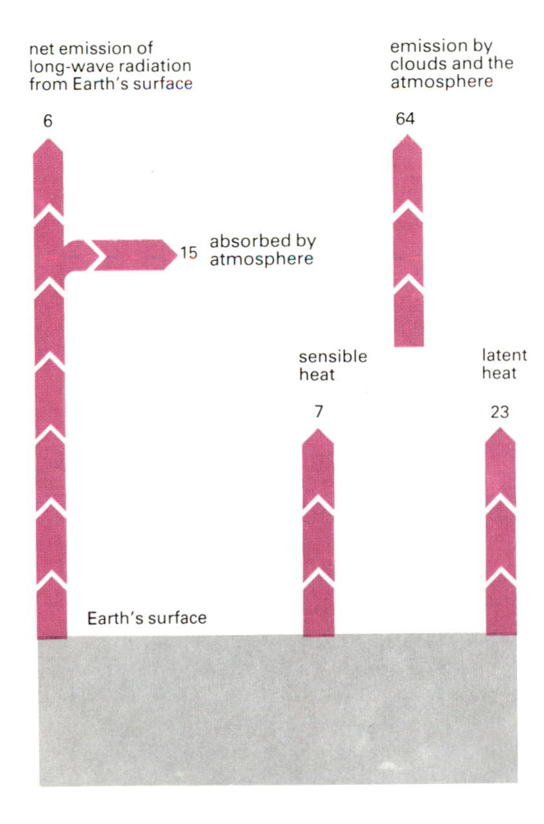

17.4: *Average energy balance for the Earth–atmosphere system. Numbers are all percentages of the total short-wave radiation input at the top of the atmosphere.*

so the emitted terrestrial radiation has its peak intensity at a long wavelength of around 10 μm (Figure 17.3). Figure 17.3 also shows that the spectrum of energy received at the Earth's surface differs from the extraterrestrial spectrum (which is close to the theoretical black body spectrum) because of depletion in the atmosphere. Air molecules, dust and water vapour scatter solar radiation. About 22 per cent of the solar radiation incident at the top of the atmosphere reaches the Earth's surface as 'sky', or diffuse radiation as distinct from the direct beam: the sky is blue because of greater scattering of violets and blues in the visible range. On infrequent occasions—'once in a blue moon'—suspended particles cause scattering of red light from the Sun and Moon which make them appear a bluish white.

Figure 17.4 shows an approximate budget of solar radiation. About 30 per cent of the incoming solar radiation is reflected back to space from the Earth–atmosphere system, around 20 per cent being reflected by clouds, about 6 per cent by the atmosphere and about 4 per cent by the Earth's surface. This reflected short-wave radiation is not available for energy exchanges in the Earth–atmosphere system. The reflectance (or albedo) of the Earth's surface varies spatially and to a lesser extent temporally with, for example, changes in snow cover, growing crops or wind stress on water surfaces.

17.5: *Typical surface albedo values.*

Surface	Albedo (%)
soils	5–10
desert	20–45
grass	16–26
forest	5–20
snow	40–95
water (high Sun elevation)	3–10
water (low Sun elevation)	10–80

Note 0% is nil reflectance; 100% is perfect reflectance.

Albedo values for a selection of surfaces are shown in Figure 17.5. This information illustrates the importance of the geography of the Earth's surface in the radiation balance, although the total albedo of the Earth–atmosphere system includes the albedo of the atmosphere and clouds. Figure 17.5 indicates that typical values of albedo over water with the Sun at a high altitude are 3–10 per cent, whereas albedos measured by satellites over tropical oceans may be greater than 20 per cent because of the additional reflection from clouds. Cloud distribution is a very important factor in the radiation balance.

Only about 19 per cent of the solar radiation input is absorbed by the atmosphere which, therefore, is not heated greatly by short-wave radiation absorption. Figure 17.3 shows that some solar radiation wavelengths are absorbed within the atmosphere before the Earth's surface is reached. About 51 per cent of the solar input is absorbed by the Earth's surface and this is converted into heat energy. Long-wave radiation is emitted from the Earth's surface and most of it is absorbed by the atmosphere (Figure 17.4), especially by carbon dioxide, water vapour and cloud droplets. Figure 17.6 shows the absorption of long-wave radiation at various wavelengths by constituents of the atmosphere. Water vapour is the most important long-wave absorber in the troposphere. Carbon dioxide dominates in the mesosphere. About 6 per cent of the solar input is lost directly to space by net long-wave radiation from the Earth's surface. When cloudless (as liquid water increases absorption), the atmosphere is transparent to long-wave radiation in the 8.0–11.0 μm waveband (except for a narrow band of ozone absorption). This 'atmospheric window' can be partly blocked off by clouds or pollution. Long-wave radiation absorbed by the atmosphere is re-emitted (in all directions) and re-absorbed; this process of radiative exchange in the atmosphere is complicated, but the net result is that the atmosphere has a long-wave radiation deficit that is the equivalent of 49 per cent of the solar input, or 49 units (64 emitted minus 15 absorbed). Short-wave radiation energy absorbed by the atmosphere is equivalent to 19 units and so the atmosphere has a net radiative loss of 30 units (30 per cent of the incoming solar radiation). The Earth's surface has a net radiative surplus of 30 units. If this surface radiation surplus and atmospheric radiation deficit were not balanced the atmosphere would cool at about 0.8°C per day and the

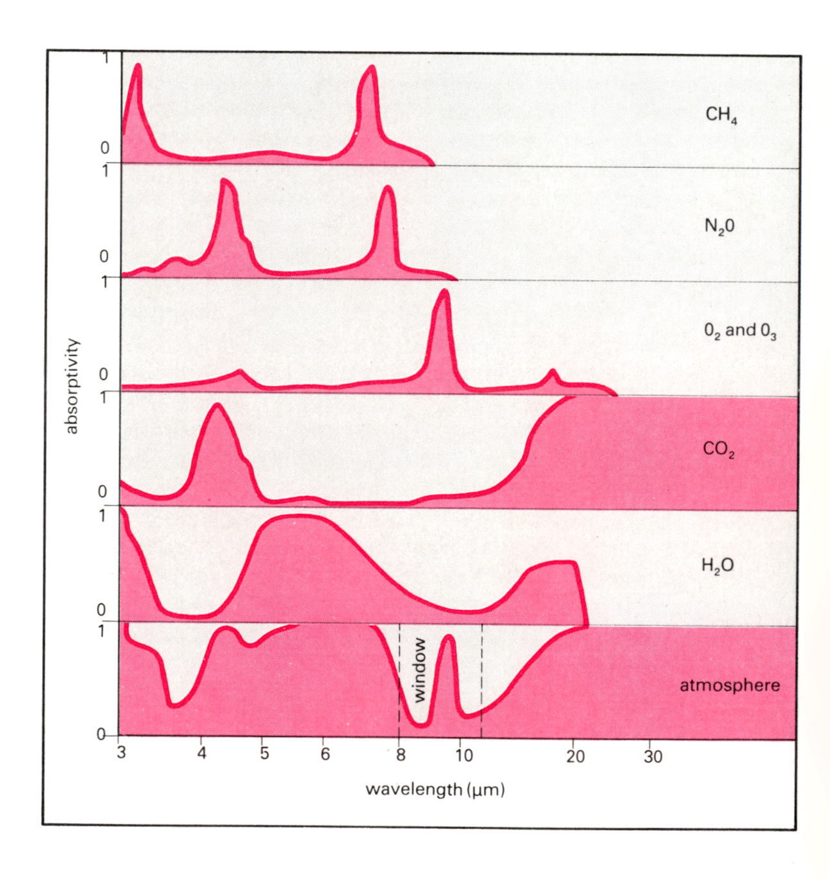

17.6: *Absorption by atmospheric gases of long-wave radiation of various wavelengths (logarithmic scale). 0 = nil absorption, 1 = total absorption.*

Earth's surface would warm at approximately 250°C per day. The required balance is undertaken by transfer of sensible heat and latent heat from the surface into the atmosphere, the latent heat flux being about three times the greater.

Heat transfer

Since the atmospheric radiational cooling is offset largely by transfer of latent heat the hydrologic cycle involves important energy exchanges. The atmosphere contains about 0.001 per cent of the global water resource (or the equivalent of 13×10^3 km^3 of water). Water enters the atmosphere by evaporation and transpiration. When evaporation occurs heat is removed from the evaporating surface (latent heat of vaporization) and when condensation occurs this heat is released (latent heat of condensation). The energy needed to change the water phase from liquid to vapour is 2.45×10^6 J/kg at 20°C. Some cloud droplets will grow to precipitation droplets and the hydrologic cycle will be 'completed' as precipitation falls on the Earth's surface. Precipitation is greater than evaporation polewards of 40° and between 10°N and 10°S. Run-off in these latitudinal zones counters the negative water balance between 20° and 30° where evaporation is greater than precipitation (the subtropical arid zones). There are also important geographical differences (eg the ratio of evaporation to precipitation is much greater for Australia than for Europe). The transfer of water vapour in the atmosphere is polewards at latitudes higher than 20° and equatorwards at lower latitudes—a transfer of water vapour out of the subtropical arid zone where evaporation is greater than precipitation. Although the bulk of the vertical heat transfer in the atmosphere is in the form of latent heat (Figure 17.4), something like 60 per cent of the horizontal heat transfer is in the form of sensible heat. Condensation plays an important part in the atmospheric energy balance.

The Earth–atmosphere system shows a positive radiation balance between 40°N and 40°S, and a deficit outside this zone. The general circulation of the atmosphere (and the oceans) transports heat from the radiation surplus zone of the system to the deficit zone. Most of the latitudinal transfer of sensible and latent heat in the atmosphere is effected within the troposphere (to which, unless otherwise specified, all circulations refer). The transfers maintain the mean temperature structure in the troposphere shown in Figure 17.2 (although changes do occur on time scales of decades or more). The temperature gradient from equator to pole is twice as large in winter as in summer and maximum latitudinal gradients occur in mid-latitudes. The tropopause is broken, usually in more than one location. The break shown in Figure 17.2 represents the statistically averaged position of the break.

About half of the sensible heat that is transported within the atmosphere originates in the latitude zone 0–10°N. Most of this sensible heat is released from the condensation process in the boiler house of the atmosphere—the meteorological equator or the intertropical convergence zone (ITCZ) where north-easterly and south-easterly trade winds converge. The meteorological equator and the geographical equator do not coincide because the atmospheric circulation of the southern hemisphere is stronger, as a result of the different distributions of ocean and land surface which lead to less frictional retardation.

Large-scale motions in the atmosphere

The atmosphere does not function as simply as it might, with warm air ascending at the tropics, travelling polewards in the upper troposphere, sinking in polar latitudes and returning to the equator in the lower troposphere. Complexities arise because all motion in the atmosphere is affected by the Earth's rotation. The effect can be illustrated by rotating a sheet of paper in a horizontal plane, while at the same

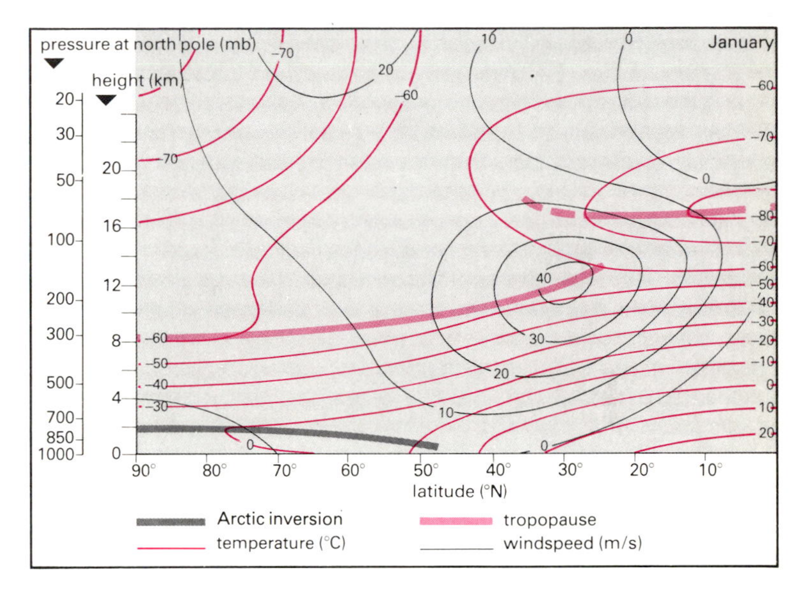

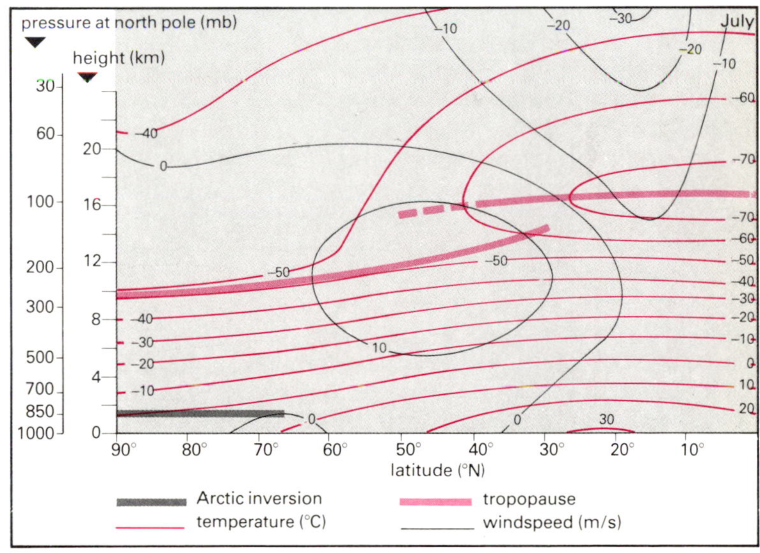

17.7: *Mean meridional (north–south) cross-sections of temperature and of zonal (west–east) wind speeds for the northern hemisphere; a negative wind speed indicates the wind is blowing east to west. The mean height of the tropopause and the top of the Arctic inversion (a layer through which temperature increases with height) are shown.*

time drawing a line at a constant speed in a direction fixed relative to the drawer. The drawn line will be curved, and the curvature is opposite to the direction in which the paper was rotated. The amount of curvature is a function of the relative speeds of rotation and drawing.

In the case of the Earth, the rate of curvature of a particle of air moving above the Earth's surface is proportional to $V\Omega \sin \phi$, where V is the north–south component of the particle's speed relative to the Earth's surface, Ω is the Earth's angular velocity of 7.29×10^{-5} radians/s (one complete revolution being 2π radians), and ϕ is the latitude. To explain this curvature from the point of view of an observer at the Earth's surface, the so-called Coriolis force is introduced, with magnitude $2V\Omega \sin \phi$. The deflection is to the right of the direction of movement in the northern hemisphere and to the left in the southern hemisphere.

Horizontal pressure differences exist within the atmosphere because, for example, of the warming in summer of air over a land mass which is warmer than the surrounding ocean, water having a greater thermal inertia than land. The pressure gradient acts as a force that moves air from zones of high pressure towards those of lower pressure. The pressure gradient force is expressed as $-(1/\rho)\ \partial p/\partial n$, where ρ is air density and $\partial p/\partial n$ is the horizontal pressure gradient (n is measured normal to the isobars or lines joining points of equal atmospheric pressure). The closer the isobars are to one another, the greater the pressure gradient force.

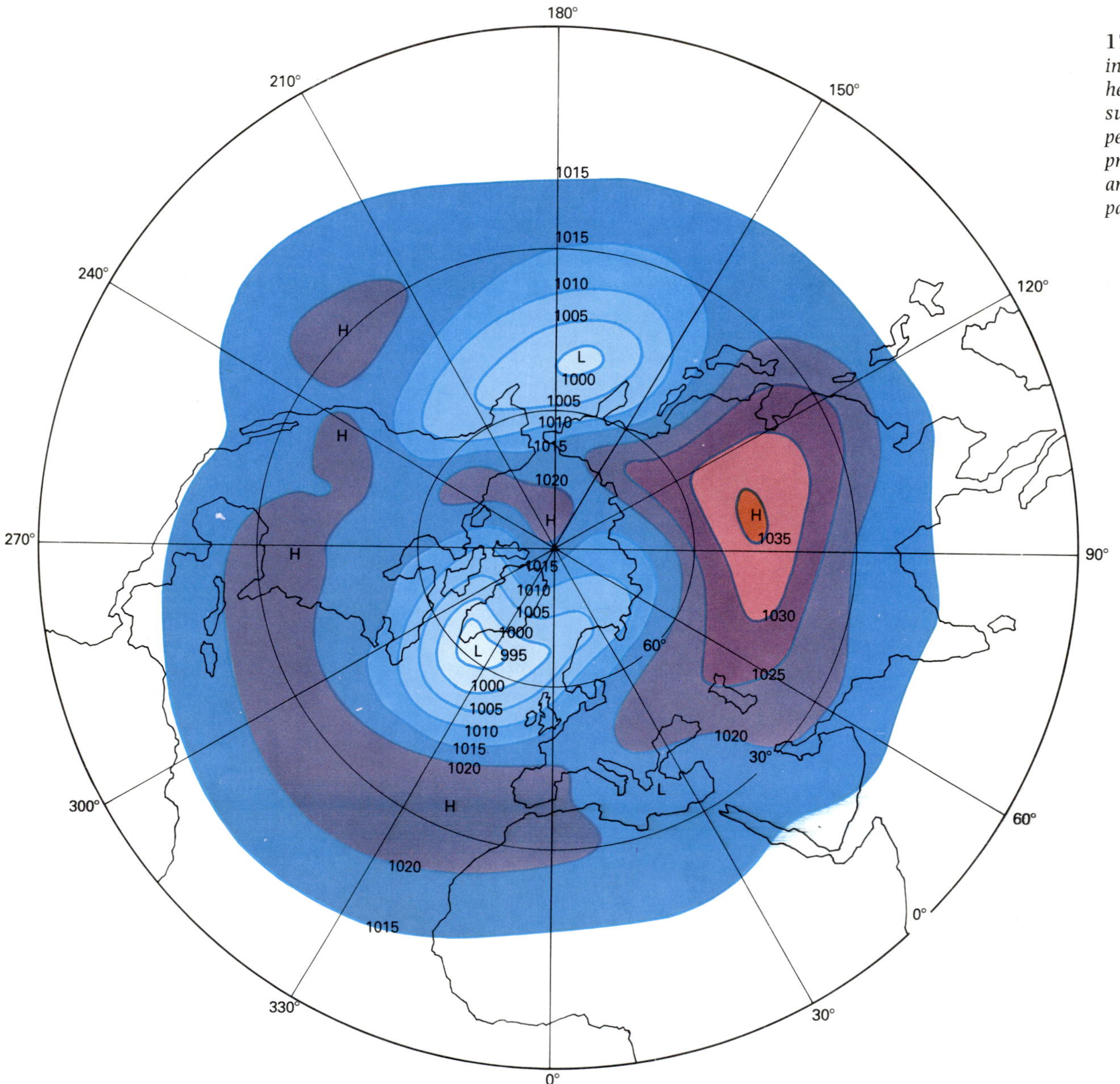

17.8: *Average sea-level pressure in millibars in the northern hemisphere in January. The subtropical high-pressure cells are permanent features, and the low-pressure areas in mid-latitudes are the result of the frequent passage of travelling depressions.*

Geostrophic and gradient wind

Except for the lowest levels of the atmosphere (up to about 1000 m), which are affected by friction, the wind blows more or less parallel to the isobars (at right angles to the pressure gradient). The theoretical wind that results from complete balance between the pressure gradient force and the Coriolis force is called the geostrophic wind. In the northern hemisphere high pressure lies on the right of the geostrophic wind (looking downwind). Over large areas of the atmosphere the wind flow is almost entirely geostrophic although there is cross-isobar flow; without such flow, pressure systems would never fill or deepen. Frictional retardation near the ground has the effect of causing the wind to blow across the isobars, in the direction of the pressure gradient, so that an Ekman spiral develops, as in the oceans (see Chapter 19) with the flow becoming more nearly geostrophic with height. The 'geostrophic approximation' is strictly applicable only to straight isobars. In the case of curved isobars (within a depression for example) another force is required to balance the flow in a curved path, and this force is equated to the inward acceleration. The balanced wind that blows along curved isobars is called the gradient wind. Around a low-pressure area the flow is cyclonic (anticlockwise in the northern hemisphere, clockwise in the southern hemisphere), and around a high-pressure system it is anticyclonic (clockwise in the northern hemisphere, anticlockwise in the southern). Because of friction near to the Earth's surface there is a cross-isobar component to the flow: inwards (or convergent) in the lows, outwards (divergent) in the highs.

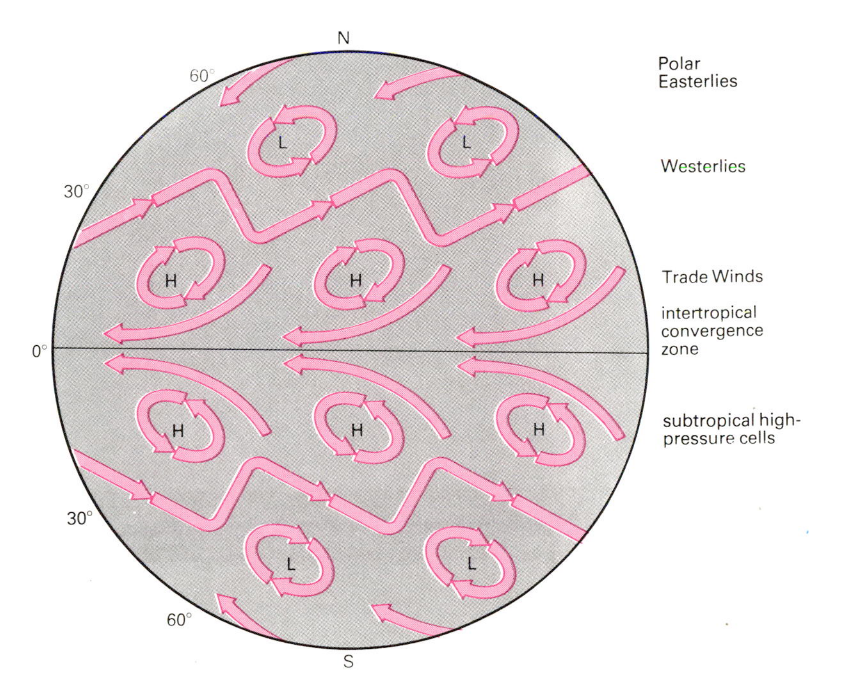

17.9: *Broad features of the overall circulation pattern of the atmosphere. The simple wind belts are modified by cyclonic circulation (anticlockwise in the northern hemisphere, clockwise in the southern hemisphere) around the low pressure cells and anticyclonic circulation around the subtropical high pressure cells.*

Pressure zones and air flows

The mean sea-level pressure distribution for the northern hemisphere confirms that the general circulation does not conform to a simple heat engine. Air blows equatorwards and polewards near the surface from the subtropical high-pressure belt. The high pressure is a result of subsiding air where condensation and precipitation processes are discouraged.

In general terms the circulation of the atmosphere at low level consists of winds converging on the equator from the subtropical high pressure belt (Figure 17.9), with an easterly component because of deflection by the Coriolis force as the air moves from high to low pressures. Air in these easterly trade wind belts is forced to rise in the intertropical convergence zone and large quantities of latent heat are released as the air cools to its condensation temperature. The air, and the sensible heat, then move towards higher latitudes but cannot simply continue polewards because radiative cooling causes the air to descend. In mid-latitudes the Coriolis deflection leads to winds with a westerly component as the air moves to the mid-latitude low pressure zones. In higher latitudes air flowing from the polar high results in the polar easterlies. These winds are shown schematically in Figure 17.13 which also shows the modifications resulting from the permanent and semipermanent pressure features which are in part a function of the geography of the Earth's surface. The subtropical high pressure is strengthened in winter over cold continents and weakened in summer over land because of warming. The high pressure cells over the Bermuda–Azores region in the Atlantic and over the east and north Pacific are the most permanent surface pressure features. The low pressure zones in the northern Pacific (Aleutian low) and northern Atlantic (Icelandic low) in winter are not permanent features but show on mean maps because of the frequent passage of depressions. They practically disappear in summer when the whole pressure distribution moves polewards because of the apparent meridional passage of the Sun. The south–north pressure gradient is greater in winter than in summer because of the larger thermal gradient. Over land, particularly over Asia and northern Africa, winter high pressures are replaced in summer by low pressures. Consequently there are obviously important differences between the summer and winter circulations.

Because the winds persist, and the Earth maintains its rotation rate, there must be zero net transfer of angular momentum between the Earth and the atmosphere. The absolute angular momentum of the system must be conserved. The implications can be appreciated by considering an air parcel which has an initial relative velocity (to the Earth's surface) of zero on the equator and is forced to move to latitude 45°N (Figure 17.10). To conserve the absolute angular momentum of the Earth–atmosphere system wind speeds far in excess of those observed in the real atmosphere are required if the air moves in a simple fashion from low to high latitudes. Theoretical work shows that flow at such great speeds would be very unstable. In lower latitudes, where the poleward rate of decrease in the radius of circles of latitude is quite small, the associated poleward increase in the rate of flow parallel to lines of latitude does not lead to instability, and a band of rapidly moving air persists in the upper troposphere at around 30° in both hemispheres (the subtropical jet streams). Thus, from the intertropical convergence zone to about 30°, the circulation of the atmosphere may be regarded as taking the form of a simple heat engine with one cell (the so-called Hadley cell) in each hemisphere.

Outside the tropics and subtropics is a deep, broad band of westerly winds encircling the globe. Since these winds are rotating west to east faster than is the Earth's surface they lose momentum to the surface through frictional retardation and through the barrier effect created by the great mountain ranges of the world. If the total stress over the Earth's surface is to balance, easterlies (which actually rotate in the same direction as the Earth's surface, but at a slower rate) must exist. In the tropical easterly belt (and in the smaller polar easterly zone) momentum is transferred from the Earth back to the atmosphere. The westerlies are therefore maintained by transfer of momentum from the easterlies. Much north–south transfer of momentum is effected by eddies and waves in the atmosphere. The rotation rate of the Earth favours eddies up to about 2000 km across—the scale on which cyclones occur.

The air subsiding into the subtropics diverges at the Earth's surface. Part of it returns to the equator and continues to the intertropical convergence zone, where the ascending limbs are concentrated into thousands of cumulus clouds (or hot towers) which are warmer than the rest of the tropical air that envelopes them. Above the subtropical high-pressure cells at high level (about 13 km) are the subtropical westerly jet streams with wind speeds typically of 45 m/s. The subtropical jet stream is one of two main ribbons of fast-flowing air in the westerlies. The other is the polar front jet which is embedded in the mid-latitude westerlies, also near the top of the troposphere. Its existence, however, is due to the strong thermal gradient in that part of the atmosphere. Warm air columns have a larger vertical dimension than cold air columns. The pressure surfaces (isobars) are at higher levels in the warm air, which results in a horizontal pressure gradient from the warm air to the cold air. As the thermal gradient increases, so also does the horizontal pressure gradient, and the resulting wind (the thermal wind) becomes greater too. In the northern hemisphere the thermal wind blows with the warmer (thicker) air to the right, looking downwind. Since the direction of the pressure gradient is reversed in the stratosphere the horizontal pressure gradient is at a maximum just below the tropopause (where the polar front jet occurs). Jet streams are defined as cores of very rapidly moving air, but the term is used very loosely. The broad ribbon of upper westerlies in the mid-latitudes is often called the jet stream. Breaks in the tropopause occur near the two jet streams, facilitating exchange between stratosphere and troposphere.

Figure 17.11 shows the height of the 500 mb pressure surface in January. This map gives an idea of the thickness of the lower half of the troposphere, which indicates the distribution of heat, since warm air columns are vertically more extensive than cold ones. Strictly, so that thickness values are directly comparable over the globe, meteorologists use the vertical extent of the layer from 1000 to 500 mb since surface pressure commonly varies in the range 930–1050 mb. However, the 500-mb contours exhibit broadly similar features. The mean wind over both hemispheres blows parallel to these contours, although the flow is stronger in the southern hemisphere.

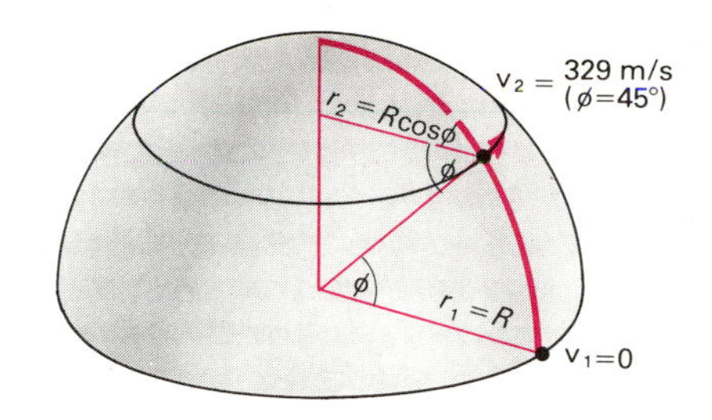

17.10: *For conservation of total angular momentum, the movement of a particle of air from 0° to 45°N, decreasing its radius of rotation (r), must gain a velocity of 329 m/s.*

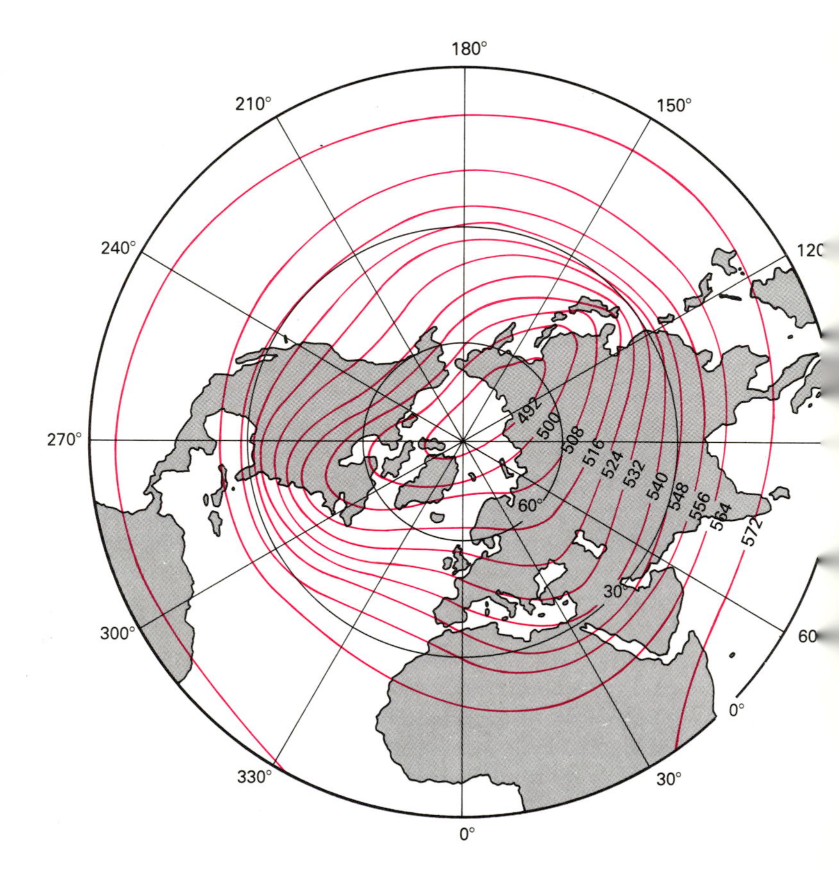

17.11: *A pressure surface map showing contours on the mean 500 mb surface for the northern hemisphere in January. The units of height are geopotential decametres (very close to 10 m). The mean wind flows parallel to the contours such that the great thickness of atmosphere is on the right, and at a speed which is inversely proportional to the contour spacing and to the sine of latitude. The trough over eastern North America is very pronounced; the ridges (eg over the eastern Atlantic) are less pronounced. In summer the thickness gradient is smaller and the mean wind is less strong.*

Westerlies, waves and fronts

The westerly flow of the upper air encircles the globe with a series of large-scale (up to 2000 km) waves (the Rossby (or long) waves). These waves affect the interface between the warmer air masses of low latitudes and the colder air masses of higher latitudes. They may be accentuated by the disturbing effect of mountain barriers, especially the Rockies and the Andes, but they are inherent features of a rotating fluid system which has a thermal gradient applied to it. These and other atmospheric features have been reproduced in the laboratory in rotating dishpans containing a fluid which models the atmosphere. As an air column flows over a mountain it is constricted vertically and its speed increases. Its absolute spin (vorticity) decreases by the same proportion as the vertical shrinking. As the column regains its vertical extent in the lee of the mountain its absolute spin starts increasing again so that a wave extending north–south is formed. The relationships between vertical shrinking and stretching and vorticity have important implications for the weather and are carefully studied by forecasters. In the northern hemisphere, at any one time, the predominant number of Rossby waves is 5, although the map in Figure 17.11 indicates that the short-period detail is lost on mean charts and the seasonal mean wave numbers are 3 in winter and 4 in summer. The major troughs and ridges have preferred regions of occurrence and the locations are important because of the effect on weather at the surface. All the troughs and ridges are closely associated, although some may be regarded as independent anchors (eg the North American trough) in that they are relatively stationary in position and change relatively little in any global adjustment. The trough over eastern Europe is more sensitive to change and so Europe is more susceptible to any variation in the nature of the westerlies. In addition to the Rossby waves there are smaller waves superimposed on the westerly flow, and their shapes and positions may be modified by the underlying surface, as may those of the large-scale waves. The thermal pattern of the Earth's surface affects the patterns of the waves: warm surface areas encourage ridge development and cold areas encourage trough development. Sea surface temperatures have a considerable influence on the atmospheric flow and the weather downstream. The temperatures of the ocean surface waters off Newfoundland in any one month strongly influence the character of the weather in north-west Europe the following month, via their effect on the westerlies.

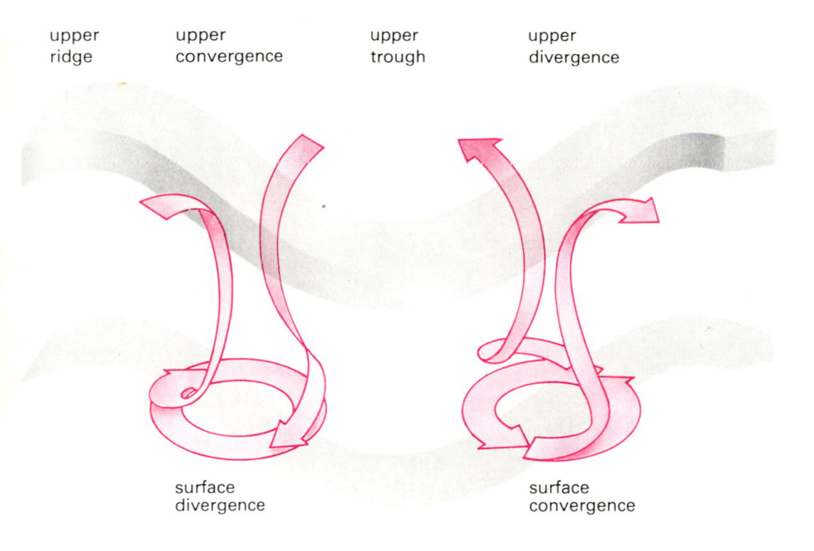

17.12: *The nature of the flow in the waves in the northern hemisphere mid-latitude westerlies leads to convergence in one part of the wave and divergence in another. This pattern of convergence and divergence in the upper westerlies results in compensating divergence and convergence at the surface. The low-level divergence is accompanied by anticyclonic spin; the low-level convergence is accompanied by cyclonic spin. Cyclogenesis (formation of depressions) is favoured downstream of the upper trough; anticyclogenesis is favoured downstream of the upper ridge.*

It is convenient to imagine that the westerly flow in the troposphere flows in the form of waves that exist in the middle and upper troposphere. Because of surface effects the wave pattern is less obvious near the ground and, as Figure 17.12 shows, the flow is anyway much stronger at height. There are important links between the tropospheric flow at all levels. The flow in a trough is cyclonic and the decreasing cyclonic vorticity as air flows out of the trough leads to a divergence. One of the basic theories of fluid flow applied to atmospheric motions relates vorticity to horizontal divergence, such that the fractional increase (or decrease) in vorticity over time is matched by a fractional decrease (or increase) in horizontal area of the column of air whose vorticity is being considered. Downstream of an upper tropospheric trough vorticity is decreasing, leading to horizontal divergence at that level; this encourages convergence below and the development of cyclonic vorticity (Figure 17.12). Cyclogenesis (the formation of the low-pressure cyclones or depressions) is favoured under the polewards-moving limb of an upper westerly wave. By the opposite argument anticyclogenesis is encouraged under the equatorwards-moving arm of the wave. Thus pressure systems at the surface are strongly influenced by the westerly flow above. The motions of cyclones and anticyclones are steered by the westerly flow and they move in association with a migratory wave above with a typical wavelength of about 2000 km.

The effect of cyclonic convergence at the Earth's surface is a confrontation of relatively warm air on the equatorial side and colder air on the poleward side. A boundary between such contrasted and internally homogeneous air masses is known as a front, or frontal zone if the boundary is rarely sharp. The major frontal zone of the northern hemisphere is located around 40–50°N, is discontinuous and is the transition between tropical and polar air masses: the polar front. Across such a frontal zone the large temperature gradient will cause a thermal wind near the tropopause; the air stream will thus be unstable and the wind will meander. The existence of a frontal zone therefore encourages the type of flow that itself aids the development of cyclones or anticyclones. The northwards-moving limb of the North American trough is one of the most important areas of cyclogenesis in the northern hemisphere because of the divergent flow in the upper westerlies and because of the presence of a strong polar front. In winter especially there is a marked contrast between the air mass cooled by the continent and the relatively warm air over the ocean.

The mid-latitude cyclones, and the waves in the westerlies on a similar scale to the cyclones, have been identified as being particularly

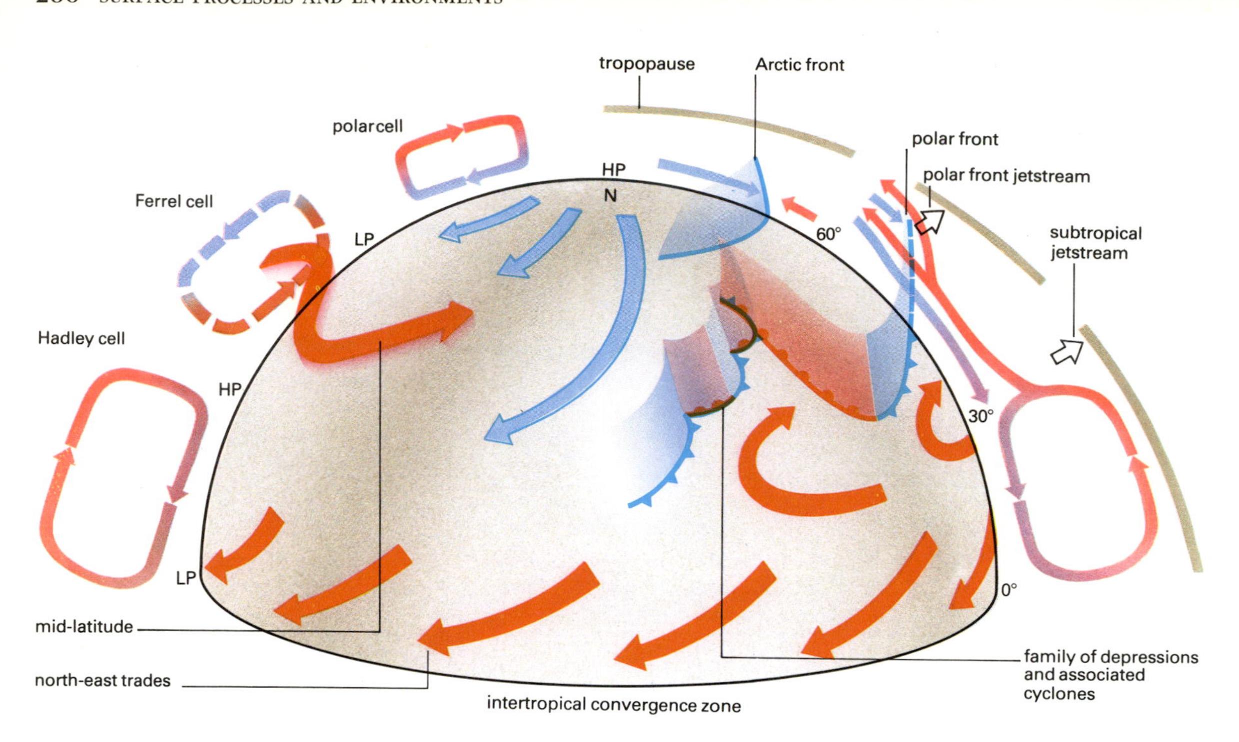

17.13: *A schematic model of general circulation showing surface winds and one family of depressions (and associated travelling anticyclones) on the perturbed polar front.*

important for the transfer of heat and angular momentum. These eddies also maintain the zonal flow by transferring some of their energy to the westerly winds. It has been calculated that five well-developed mid-latitude cyclones can effect the heat transfer needed to balance the Earth–atmosphere heat budget, because of the equatorward component to cold-air flow and the poleward component to the warm-air flow (to be discussed below).

Global air movements

It is now possible to summarize the major components of the global circulation. The meridional circulation can be described as tricellular: the Hadley cell (warm air rising and releasing latent heat as water vapour condenses, colder air sinking); the Ferrel cell in mid-latitudes (thermally indirect with cold air rising and warm air sinking)—driven by frictional coupling between the Hadley cell and the third cell; the weak polar cell (Figure 17.13). The polar front represents the major zone of mixing in each hemisphere, but the mixing is not simply between polar and tropical air. Returning tropical air is also involved. The warmer air slides up over the colder air along a very shallow sloping frontal zone. The tropopause exhibits a tripartite division associated with the cells, the jet streams being located near the breaks. The polar front jet stream is situated close to the top of the polar front zone, where the thermal wind is at its strongest. This model represents only one version of an idealized mean atmospheric circulation. At any one instant eddies disturb the flow. The travelling cyclones and the intervening areas of relatively high pressure are integral parts of the circulation. The large-scale monsoonal circulations can be regarded as perturbations or eddies on the mean flow. The mid-latitude travelling pressure systems, hurricanes, tornadoes and circulations on whatever scale are all eddy motions which play their part in the global system. Their tracks, which are steered by the westerly flow, are influenced by the shape of the wave pattern in that flow. Over a period of weeks the mid-latitude westerlies tend to change from a predominantly west–east zonal (ie latitudinal) flow to a more perturbed, meridional flow accompanying high-amplitude Rossby waves: the so-called index cycle. Under high zonal index conditions the westerlies are strong, the long waves in the westerlies have small amplitude and the pressure systems move quickly west to east in mid-latitudes. Under lower zonal index values greater undulations occur as the westerlies start to meander. The westerlies can become so weak that cut-off circulations occur and blocking anticyclones develop which slow down the passage of travelling depressions. The index cycle plays an important role in the general circulation, for the increased meridional flow and greater length of the westerlies when the index is low improves heat transfer polewards. The deep, blocking (relatively permanent) anticyclones which may exist in association with increased meridional flow act as a barrier to the travelling eddies and have an important influence on the weather. The nature of that influence depends on their position and the time of the year. In England the coldest winter for at least two hundred years in 1963 and the worst drought for an even longer period in 1975–6 were brought about by blocking anticyclones.

Depressions and anticyclones

In mid-latitudes the frontal depressions produce characteristic sequences of weather as they migrate from west to east. The typical weather associated with an anticyclone is usually in marked contrast. Moving pressure systems (depressions or anticyclones) are usually referred to as synoptic-scale systems (500–2000 km). Circulations of around the same size or slightly smaller—the hurricanes—produce

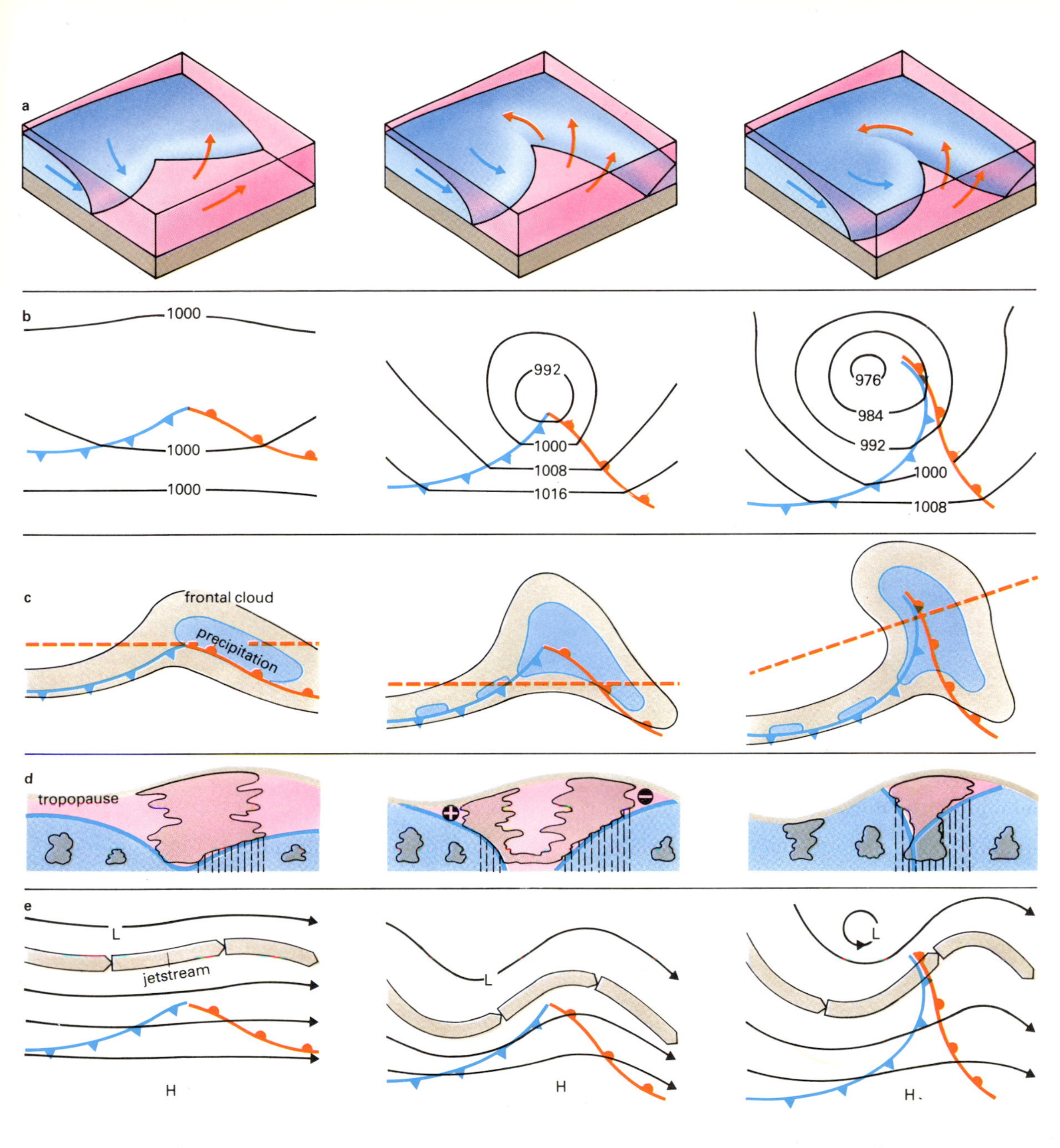

17.14: *The development of a frontal depression.* (**a**) *the view from above shows how the wave on the frontal surface grows and finally occludes over a period of days.* (**b**) *shows the warm and cold fronts on the ground and the associated surface atmospheric pressure distribution.* (**c**) *shows the distribution of frontal cloud and precipitation.* (**d**) *indicates the cloud development and precipitation in relation to the vertical structure of the frontal zones along the cross-sections indicated in (c).* (**e**) *shows the position of the fronts at the ground and the flow aloft including a jet stream axis which is typically associated with a frontal depression. The middle diagram of* (**d**) *shows the jet stream entering the page (+) and leaving the page (−).*

more violent weather sequences in equatorial areas. In common with the travelling cyclones and anticyclones of mid-latitudes hurricanes also play an important role in the general circulation of the atmosphere, as do circulations on smaller scales such as thunderstorms, tornadoes, land and sea breezes or thermally induced breezes in cities.

Certain parts of the mid-latitudes, especially the oceans, experience an almost continuous succession of travelling cyclones with intervening high-pressure systems. The geography of the Earth's surface causes their unequal frequency, and the distribution of depressions is closely related to the distribution of the strongest frontal zones which tend to form over regions of stable, typically high-pressure, conditions. The subtropical anticyclones, continental anticyclones in winter and the polar anticyclones therefore provide favoured source regions. Air masses take on the characteristics of their source region over a period of weeks. Tropical and polar air masses meet along the polar front in the mid-latitudes. This frontal zone is more identifiable in some areas than others and it slopes up towards the polar latitudes with a gradient of about 1 in 200.

The front becomes perturbed on occasions and a wave may develop over a period of two or three days (Figure 17.14). The leading edge of each such wave is a warm front and the trailing edge is a cold front. Ascent, and consequently the formation of clouds and precipitation, occur along the frontal zones. Along the warm front cold air is displaced along a surface with a slope of around 1 in 200; the gentle rate of ascent produces layered cloud and rainfall over a large area ahead of the depression. At the cold front the cold air displaces warm along a steeper surface (about 1 in 50) which leads to greater uplift, the likely development of cumulonimbus clouds, and a narrower belt of heavier precipitation. The wind changes direction at the frontal zones and temperature and humidity change as the fronts pass overhead. The warm and cold fronts are up to 350 km apart at ground level, and the temperature gradient may persist along a frontal region for up to 1000 km. A particular sequence of clouds is often observed as the frontal depression passes overhead at anything up to 1000 km per day.

In the first stages of frontal depression development the tip of the wave at the ground coincides with lowest surface pressure but as occlusion occurs and the warm sector is lifted, the tip of the wave becomes detached from the region of lowest pressure. Uplift of the warm sector at occlusion often produces more rainfall. Frontal depressions often become stationary over the eastern parts of the oceans where the wave decays and may eventually disappear after several days. The Icelandic and Aleutian lows are statistical manifestations of this characteristic.

Frontal depressions tend to occur in families of three or four. Secondary depressions grow on the trailing cold front of the original depression and the polar air pushes further equatorwards. In between the frontal depressions anticyclones often occur producing more settled weather; the horizontal pressure gradients are smaller and so the winds are much lighter. Within these anticyclones the air subsides and warms, so the vertical structure of the air is stable, and the sky tends to be clear. However, when the lowest 1 km or so of air is moist and has been mixed through turbulence a layer of cloud may form. This cloud will often thicken because of cooling radiation at the top of the cloud layer. The winter polar continental highs result from cooling of the land surface and the high density of the shallow layer of cold air in contact with the ground produces high pressure extending upwards to about 2 km. The subtropical anticyclones and the travelling anticyclones of the mid-latitudes are warm and are caused by large-scale subsidence which compresses the air and produces an inversion (a zone in which temperature increases with height). Anticyclones can produce very warm or very cold weather depending on their position and the season.

Clouds and precipitation

The clouds associated with the frontal zones of depressions are caused by wholesale upward gliding of the warm over the cold air. The area over which precipitation is produced on the ground can be up to about 100 km in width and several hundreds of kilometres in length. Precipitation can also be caused by convective uplift in cells up to 15 km high and 1–10 km across. Condensation by this mode leads to cumulus and cumulonimbus clouds, with thunder and lightning a likely outcome. The precipitation resulting from these clouds is much more variable and lasts for a much shorter time than does the frontal zone precipitation. Uplift on a broader scale may occur within hurricanes although the rainfall is distributed in a spiral fashion around the hurricane eye as the air converges into the centre of the circulation with intense cyclonic curvature. Forced ascent over mountains may also lead to precipitation (orographic rainfall).

Air rising sufficiently rapidly will expand and cool adiabatically. Pure air can attain super-saturation of several hundred per cent in laboratory experiments, but in the real atmosphere condensation occurs with very small super-saturations. The atmosphere contains large numbers of particles (condensation nuclei) which range in size from about 0.01 μm radius to a relatively small number with radii of around 10 μm. Maritime air contains about 10^9 condensation nuclei per cubic metre; over an industrial area the concentration can be six times greater. Condensation nuclei are generally natural dust, salt particles and pollution products. Condensation usually occurs first on the larger nuclei, although the solubility of the nucleus is also very important, as the saturation vapour pressure over a solution droplet is less than that over a pure water droplet. Condensation occurs on hygroscopic nuclei well before the air is saturated (at 78 per cent relative humidity in the case of a sodium chloride nucleus).

Water droplets exist in the atmosphere at temperatures down to $-40°$ (super-cooled water droplets) and so water droplet clouds can exist up to very high levels in the troposphere. If ice nuclei are present in the atmosphere freezing will occur at temperatures below about $-12°C$. Ice nuclei are much less common than condensation nuclei, rarely more than about $10^6/m^3$, and consist mainly of natural clay mineral particles. Ice crystal clouds are more wispy and tenuous than water droplet clouds, their edges being more indistinct because ice crystals evaporate very slowly.

Clouds are classified in terms of their shape and structure, their vertical dimension and their height. The ten basic cloud groups are shown in Figure 17.15. Clouds typically contain about 10^6 water droplets per m^3 and have a liquid water content around 1 g/m^3. Cloud droplets are typically 1–100 μm (average 10 μm) in diameter and the typical raindrop has a diameter of 1000 μm, so it would take about one million cloud droplets to form a raindrop. Two processes have been suggested to explain how cloud droplets can grow so large. The first is the ice crystal (or Bergeron) process. When water droplets and ice crystals co-exist in a cloud the water droplets tend to evaporate and the vapour is deposited onto the ice crystals. Snowflakes are formed through aggregation of ice crystals. Water droplets freeze on impact with ice crystals and soon grow large enough to fall out of the cloud, to arrive at the ground as snow, sleet or rain. Hailstones are indicators of strong, vertically extensive updraughts. Larger hailstones consist of alternate concentric spheres of clear and cloudy ice, reflecting either the up-and-down movement repeated within the cloud,

17.15: *Cloud types grouped into families according to height and form. Cirrus (ice-crystal) clouds frequently cause optical phenomena such as haloes appearing to surround the Sun or Moon; they have a fibrous appearance. Cumuliform clouds are bubbly and indicate vertical development. Stratiform clouds indicate more uniform uplift or mixing and are layered. Alto- refers to medium-level clouds and the term nimbus is applied to thick clouds from which continuous precipitation is falling.*

which enables them to grow to very large sizes on occasion, or their passage through differing conditions of growth within the cloud.

In the tropical atmosphere temperatures of less than 0°C may not be reached below a height of 5 km, so there must be another process which leads to the growth of precipitation. This is the coalescence or 'warm rain' process. When some of the water droplets in a cloud are very much larger than the others, they will ascend at a slower rate or descend at a greater rate, catching up the smaller droplets, colliding and coalescing with them. Eventually they grow large enough to precipitate. In clouds in which there are no ice crystals the original large water droplets form on extremely hygroscopic salt particles which originate from sea spray. In many cases, but not in the tropics where ice crystals may be completely absent, precipitation probably results from a combination of the ice crystal and the coalescence processes.

Observation by remote sensing

Technological advances have recently led to marked improvements in atmospheric observation. Automatic weather stations, different types of radar and increasing use of radiosondes (instrument packages attached to balloons) have all contributed. But of all the remote sensing devices, the satellite is becoming the most important (see Chapter 24). Satellites can provide a complete global weather observation service. Polar orbiting satellites ensure twenty-four-hour cover over

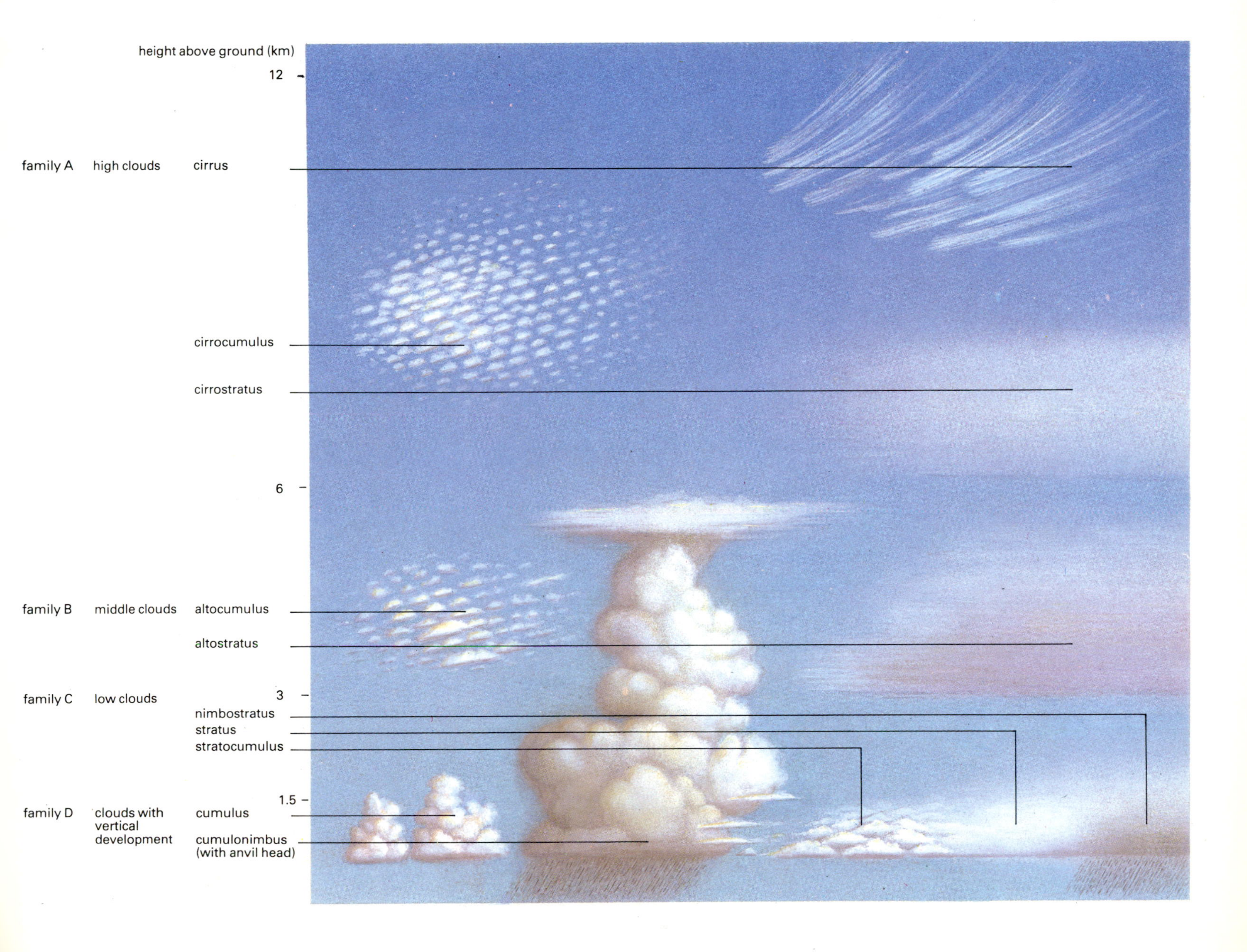

the Earth and geostationary satellites transmit data back at regular intervals for large areas of the Earth's surface. Satellites are still at an early stage of development and their greatest contributions are yet to come. Besides the well-known pictures of clouds that identify weather systems satellites have also been used to measure wind speeds and directions from cloud trajectories. Infrared measurements can be used to produce temperature and moisture profiles for the atmosphere. Because the atmosphere is more transparent to some infrared wavelengths than to others (Figure 17.6) the radiation detected by a satellite sensor can be assumed to have originated from a particular zone in the atmosphere. Radiation of around 11.0 μm wavelength will have originated at or near the Earth's surface since the 'atmospheric window' is transparent to these wavelengths, whereas 15.0 μm radiation is absorbed by carbon dioxide and any that is detected by a satellite must have been re-absorbed and re-emitted, eventually leaving the atmosphere from high levels. Similarly radiation at other wavelengths will characterize intermediate levels in the atmosphere. The intensity of the radiation originating from a particular layer depends strongly on its temperature. Consequently the spatial distribution of temperature at different levels can be mapped.

Geostationary satellites now measure 6.3 μm radiation which is absorbed by water vapour, giving information about water vapour distribution in the upper troposphere (Figure 17.16). Research is now being conducted on a technique that utilizes a microwave pulse from the satellite which is reflected back from the Earth's surface and sensed again by the satellite. At certain microwave wavelengths information is provided on the distribution of water vapour and liquid water throughout the depth of the atmosphere. The distribution of water in a tropical cyclone has already been examined by this technique. Microwave emission may also be used to sense soil moisture, ice (new or old), sea surface temperatures, precipitation and other variables of great interest to meteorologists. Radar and lidar may even be mounted on satellites. It is not inconceivable that satellite instrumentation will all but replace other methods of observation. Already meteorologists are learning more about the components of the radiation budget, the distribution of temperatures and water and other features of the atmosphere than seemed possible a few years ago.

17.16: *An image in the water-vapour spectral band taken by Meteosat-1 on 17 February 1979. White areas show high humidity in the 5–10 km layer of the atmosphere; darker areas correspond to dry regions. The vortex at the top represents a frontal depression close to the UK. The lower vortex is a depression over North Africa. The dry area to the south of the depression is over the Sahara and is caused by subsidence. The intertropical convergence zone is to the south of the subtropical subsidence area.*

Classification of climate

The first serious attempts at climate classification were made by botanists and biologists because of their interest in the effect of the climate in determining the nature of vegetation. Classifications have been developed using monthly values of temperature, precipitation and moisture indices which encompass precipitation, evaporation and characteristics of soil moisture storage. Other classifications are based on the global distribution of net radiation. Another family of classifications (the genetic type) takes account of the prevailing winds as well as conditions of radiation, surface heating and precipitation. Other examples of this genetic type are based on air masses. However, classification is now largely an academic exercise of less significance than forecasting.

Forecasting

The ultimate aim of meteorology is the prediction of the future state of the atmosphere. Weather forecasts are not always accurate. One reason for their imprecision is that the state of the atmosphere at any particular time is not known perfectly. Some parts of the Earth's surface and, especially, the upper atmosphere are poorly observed at present. The observational network is being improved by the application of new technology and through international cooperation programmes such as World Weather Watch and the Global Atmospheric Research Programme. International exchange of data has also been improved.

Many complex interactions in the atmosphere are only partially understood and cannot be perfectly described. The movement of pressure systems, their position and intensity at all levels in the atmosphere, interactions with the Earth's surface, distribution of temperature, humidity etc, the occurrence of clouds and precipitation, all have to be understood and predicted. The accuracy of weather forecasts depends on the element, the area and the period of forecast. Of the elements, temperature, wind speed and wind direction can be most accurately forecast. The occurrence of clouds and precipitation can be predicted with reasonable accuracy, but forecasts of the amounts of precipitation can be much in error. Forecasts for periods of less than one day are most accurate; for periods of two to three days the accuracy is fair; but for greater than three to four days only general indications can be given. Attempts at forecasting for one month or even longer have met with very limited success.

There are four main types of forecasting: synoptic, numerical, statistical and analogue. In the main, weather forecasts are made using a combination of at least the first three methods.

Traditional, or synoptic, forecasting involves the analysis of synoptic weather charts (maps of the weather occurring over a wide area at the same time) which show the distribution of pressure, temperature, fronts etc. The relationships between surface and upper air conditions are paid particular attention. The synoptic forecaster's experience is drawn upon to make the final forecast. The past twenty years have seen great improvements in observation and communications, as well as the development of the high-speed computer. Numerical weather prediction will probably never completely replace human forecasters whose experienced judgment is still needed, especially with smaller-scale features such as thunderstorms. But much routine analysis, such as the prognostication of the pressure pattern, has been taken over by the computer. Sometimes forecasts of individual weather elements are made directly by computers and sometimes they are made by a forecaster using computer analyses.

Numerical weather prediction uses the hydrodynamic equations that pertain to the state of a fluid. The equations are difficult to solve and often there is just not enough information to describe the initial state of the atmosphere. The hydrodynamic equations are arranged so that the rate of change of a variable over time is expressed by the rate of change of the variable over space. The equation is solved by the finite difference method. The rate of change of the variable over space is approximated by the difference between the initial (ie observed) value of the variable at a number of closely spaced points (a grid) and the future spatial distributions are predicted by time integrations of the hydrodynamical equations. The distance between the points on a typical grid is 100 km. In some models computations are done for several levels, and interactions between these levels and the Earth's surface are included. To produce a twenty-four-hour forecast chart by computer typically involves millions of multiplications and tens of millions of program instructions. The consensus opinion among meteorologists is that most of the major steps forward in the field of weather forecasting will be made through the numerical weather prediction method.

Statistical forecasting uses numerical procedures to analyse past data in an attempt to establish predictive relationships. Current data are then used to obtain a forecast. The analogue method assumes that a given synoptic distribution will develop in a way similar to previous cases even though there may be a lengthy intervening period. The existing pressure pattern is matched with past analogues and the forecast made on the basis of the changes that occurred in the past. The statistical and analogue methods are often combined to make long-period forecasts for one or more months ahead.

The world water cycle

The water inventory: storage and transfers

On the global scale the Earth's surface may be viewed as a gigantic reflux distillation unit, powered by the Sun, in which water is continuously evaporated from the oceans and, on land, from the soil, vegetation, lakes and rivers, only to be returned again as precipitation. In this hydrologic cycle the oceans lose more water by evaporation than they receive from precipitation, while the reverse is true on the continents. This transfer of water from the oceans, via the atmosphere, is the source of the flow of rivers.

For any given land area over a long period of time, precipitation = evaporation + riverflow + groundwater outflow. The hydrologic cycle is most vigorous in South America and least in Australia. Europe, North America and Asia have roughly similar water balances. But these similarities mask a great diversity within the continents, from deserts with near-zero precipitation and zero run-off, to mountain areas with more than two metres of run-off each year.

The world water cycle constitutes a transfer of water between reservoirs. The largest store of water is the oceans which contain 97.6 per cent of all water. Of the remainder, the atmosphere contains, at any one time, around 0.035 per cent as vapour. On the land at the present day three-quarters of all non-ocean water occurs as ice in glaciers and ice caps, and almost a quarter consists of groundwater, stored in the voids of rocks beneath the surface. Moisture in the soil accounts for 0.06 per cent while lakes and rivers contain only 0.33 per cent of all non-ocean water. Nevertheless it is in lakes and rivers that water is actually moving fastest, for it is through them that almost the entire terrestrial water cycle is funnelled back to the sea. The relative rapidity of flow processes in the various stores may be seen by comparing their mean residence times, found by dividing their volume by their net throughput. The water cycle is most vigorous at the Earth's surface, with residence times all much less than one year, whereas residence times in groundwaters are from a few years to thousands of years, averaging at a few hundred. The oceans and glaciers have residence times of thousands of years.

A convenient unit of study when considering that part of the hydrologic cycle on land is the drainage basin. It is by means of transfers between stores such as the soil moisture, groundwater and river channels and by retention in these stores that precipitation is converted to stream flow and evapotranspiration. The physical mechanisms involved are the subject of one of the principal areas of scientific study in hydrology.

The two most important stores in the drainage basin are the moisture in the soil and the groundwater filling voids in the rocks. The soil is a most important regulator in the hydrologic cycle. As a store it retains water and partitions it between the run-off part of the drainage basin system and evapotranspiration. The degree of partitioning depends upon the soil moisture content. When the soil is dry, drainage and throughflow are very slow indeed and practically all precipitation is retained in the soil and subsequently evaporated. When the soil is initially moist, addition of further water from precipitation promotes rapid drainage, groundwater recharge and throughflow.

Soil moisture occurs in the pores between grains and crumbs of soil. When these pores are completely full of water the soil is said to be saturated. In most soils for most of the time some pores contain air and the soil is unsaturated. When a soil is saturated, or nearly saturated, water can drain through it fairly easily. Vertical drainage moves downwards through the rocks beneath the soil until it reaches the water table, the level beneath which the rocks are saturated with groundwater. The region between the surface and the water table is called the unsaturated zone. Shallow groundwater flow is normally towards nearby rivers or springs, where the groundwater seeps to the surface to form stream flow, but deeper groundwater may flow beneath drainage basin divides to form regional flow systems. Rates of water movement in the soil and groundwater are much slower than in rivers, ranging from a few millimetres to a few metres per day. Groundwater movement occurs most rapidly in fissures and fractures.

Because groundwater lies beneath the Earth's surface it is difficult to make measurements of it. The level and shape of the water table can be found by measuring the depth to water in wells and boreholes. Contour maps of the water table can be produced, provided enough wells suitable for measurement can be found. Groundwater flow generally follows the line of steepest slope of the water table, and its rate depends upon the permeability of the rocks. Permeable rocks such as sands, gravels and some limestones allow rapid water flow at tens or even hundreds of metres per day. Impermeable strata such as clay form barriers to flow with rates of water movement of less than a metre per year.

Flow of water through the drainage basin hydrologic system

It can now be seen how the soil moisture and groundwater stores regulate the transfer of precipitation input into evapotranspiration or stream-flow output. A single day's rain may form overland flow and run over the ground surface to reach a river. This is comparatively rare and happens only when it rains very hard or when snow melts rapidly over frozen ground. When widespread overland flow does occur, a river flood results. Normally most of the precipitation infiltrates the soil where it is added to the soil moisture store and retained. Most of the stored water may subsequently be evaporated or transpired through vegetation (evapotranspiration), but if the soil moisture content is fairly high then some may drain away through the unsaturated zone to recharge the groundwater. This recharge will raise the level of the water table and cause an increase of seepage to rivers and springs. Thus the flow of water via soil and groundwater is

precipitation
evaporation
from soil and water surface
transpiration

1 water table
2 surface run-off
a 3 infiltration
4 groundwater flow
5 river flow
6 unsaturated zone
7 zone of water table fluctuation
8 saturated zone

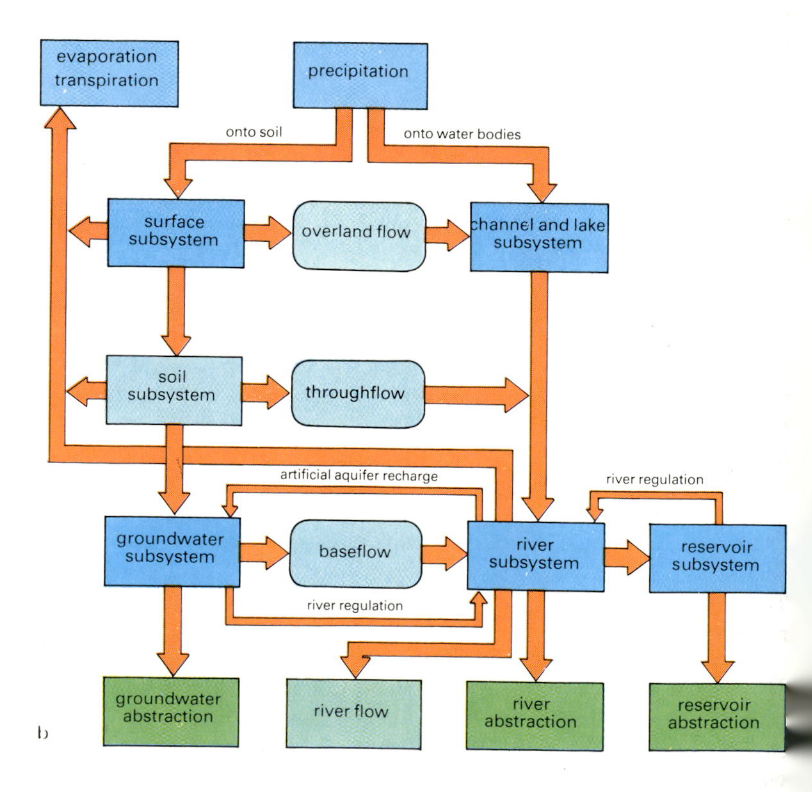

17.17: (a) *The main components of the drainage basin hydrologic system.* (b) *A flow diagram of the drainage basin hydrologic system. Both natural and man-made elements are shown, with darker areas indicating potential for modification of the natural cycle. Boxes with rounded corners indicate transfer processes, square-cornered boxes indicate inputs, outputs and subsystems or stores.*

ultimately expressed as a rise in river flow. A second route from soil to river occurs as throughflow, or downslope seepage of soil moisture on steep slopes.

A further very important regulatory function of the soil is in the infiltration process. Every soil has a particular infiltration capacity which is related to its permeability and is the maximum rate at which water can flow into it. Infiltration capacity for a given soil is greatest when it is dry and least when it is moist. Infiltration capacity forms a regulator for overland flow. Most rain falls at a rate that is less than the soil infiltration capacity and is added in total to the soil moisture store. When overland flow does occur it moves at tens of metres per hour and rapidly increases the flow in river channels, so that floods result.

Groundwater

Groundwater flow takes place on a much wider scale, and much deeper within the crust, than might be imagined. The flow of water within rocks depends upon two properties, their porosity and permeability. Porosity is the fraction of a rock's bulk volume which is made up of voids. In general, clastic rocks are porous, especially if poorly cemented, while crystalline rocks have low porosity. Porosity is usually greatly decreased by diagenesis of sedimentary rocks but is increased by weathering. Permeability is a rather general term used to describe the ease with which a fluid can flow through the voids of a rock. The first experiments on flow through porous media were performed in the mid-nineteenth century by the French engineer Henri Darcy, who studied the flow of water through sand and found that for any given material the flow rate depended upon the hydraulic gradient. Later investigations showed that the permeability depended both upon properties of the fluid (viscosity and density) and upon properties of the rock (pore sizes, tortuosity of flow paths and porosity). For water flow at a constant temperature, each material may be characterized by its coefficient of permeability, sometimes called hydraulic conductivity. Permeability depends strongly upon void sizes, and so it is possible for a rock such as fractured sandstone, with a low porosity of say 1 per cent to have a much higher permeability than clay, even though the latter has a porosity of 20–50 per cent. The reason is that the fractures in the sandstone, while few in number, are relatively large (0.1–1.0 mm across) whereas pores in clay are 0.001 mm or less in diameter.

Rocks with moderate to high permeability allow appreciable flow of groundwater and are called aquifers ('water-bearers'). Rocks such as clays, shales, silts and unfractured crystalline rocks are aquicludes. Their permeability is so low that they form barriers to groundwater movement.

Patterns of groundwater flow

In a uniform, permeable rock mass, patterns of groundwater flow are closely related to surface topography and drainage. The form of the water table closely corresponds to the surface ridges and valleys. Groundwater flows along a path following the steepest slope of the water table and seeps through discharge zones into river beds or from springs on lower slopes. The interfluve areas are recharge zones, where groundwater is replenished by drainage from the soil. The flow patterns can be predicted from mathematical theory and observed in the field. The local circulation of groundwater exerts a partial control on soils and vegetation, with well-drained soils on the interfluve areas and wet-loving plants and waterlogged soils on the discharge zones. The rate of groundwater flow in such a local circulation system depends upon the permeability of the rocks and the hydraulic gradient.

In addition to local groundwater circulation patterns larger-scale regional flow patterns may occur. In general the Earth's land area consists of a series of ridges and valleys superimposed upon a regional slope into a major river valley. Each ridge and valley possesses its own local circulation system, but in addition regional flow occurs beneath these local cells from the highest parts of the land surface to the lowest. The regional groundwater flow constitutes a transfer of water from one local drainage basin to another, across topographic divides. Groundwater flow patterns become even more complex when the rocks do not have uniform permeability. In most real situations the strata consist of alternations of aquicludes and aquifers. The former constitute barriers to flow, which tends to take place in the latter.

Confined and unconfined aquifers

When the upper boundary of the groundwater body in an aquifer is the water table, the aquifer is said to be unconfined. The groundwater is fed by recharge from the unsaturated zone. In some cases an aquifer is overlain by an aquiclude and the water in it is under greater than atmospheric pressure. The aquifer is then confined. Water will rise up a borehole drilled into it until it reaches a level that defines the hydraulic head in the aquifer. At this level the borehole water surface has the same total energy as the water in the aquifer; in the aquifer the water's energy is due to its pressure, and it is located at a low topographic

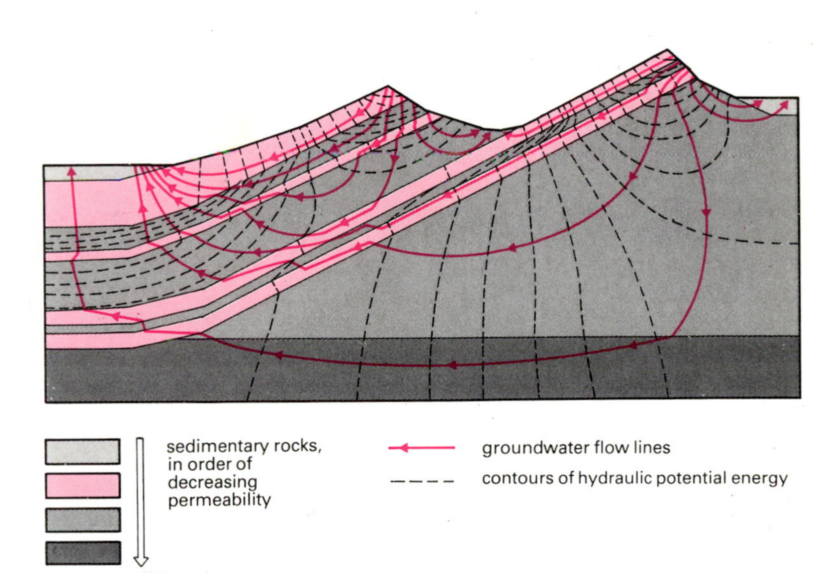

17.18: *Groundwater flow in a complex structure of thrusted sedimentary rocks of varying permeability, showing its relation to hydraulic potential energy contours.*

position, whereas in the borehole the water surface is at atmospheric pressure but is at an elevated topographic position. The imaginary surface defined by water levels in boreholes tapping a confined aquifer is called the piezometric surface. Groundwater flow follows the slope of the piezometric surface just as in an unconfined aquifer it follows the slope of the water table.

An example of a regional aquifer system containing confined and unconfined parts is the Lincolnshire limestone in eastern England. The very permeable limestone aquifer is sandwiched between silt and clay units. Recharge enters it from its outcrop and groundwater flows eastwards beneath two river valleys to a lowland area, the Fens, where upward discharge occurs through springs. Beyond this the aquifer continues but contains saline water, suggesting that very little groundwater movement occurs. In the Fen area the piezometric surface of the aquifer lies above the ground surface, and boreholes drilled there in the nineteenth century gave a natural head of water 12 m above the ground surface. This condition is described as artesian, or overflowing artesian.

River flow regimes

The pattern of storage and release, together with the seasonal patterns of precipitation, temperature and evapotranspiration, gives rise to varying seasonal changes of river flow in different parts of the world. A summary of river regime types and their distribution is shown in Figure 17.19. The distribution of regimes across the globe is similar to the distribution of climatic types. Each regime type arises as follows:

Megathermal Regimes (A) (tropical and subtropical climates): temperature and evapotranspiration high throughout the year; run-off depends upon seasons of excess precipitation.

AF Two precipitation and run-off maximums each year; no low-flow season.

AM A single precipitation maximum and a low-flow season of three months or less.

AW A single short rainy season in savanna-type climate; the low-flow season includes a substantial period with zero run-off.

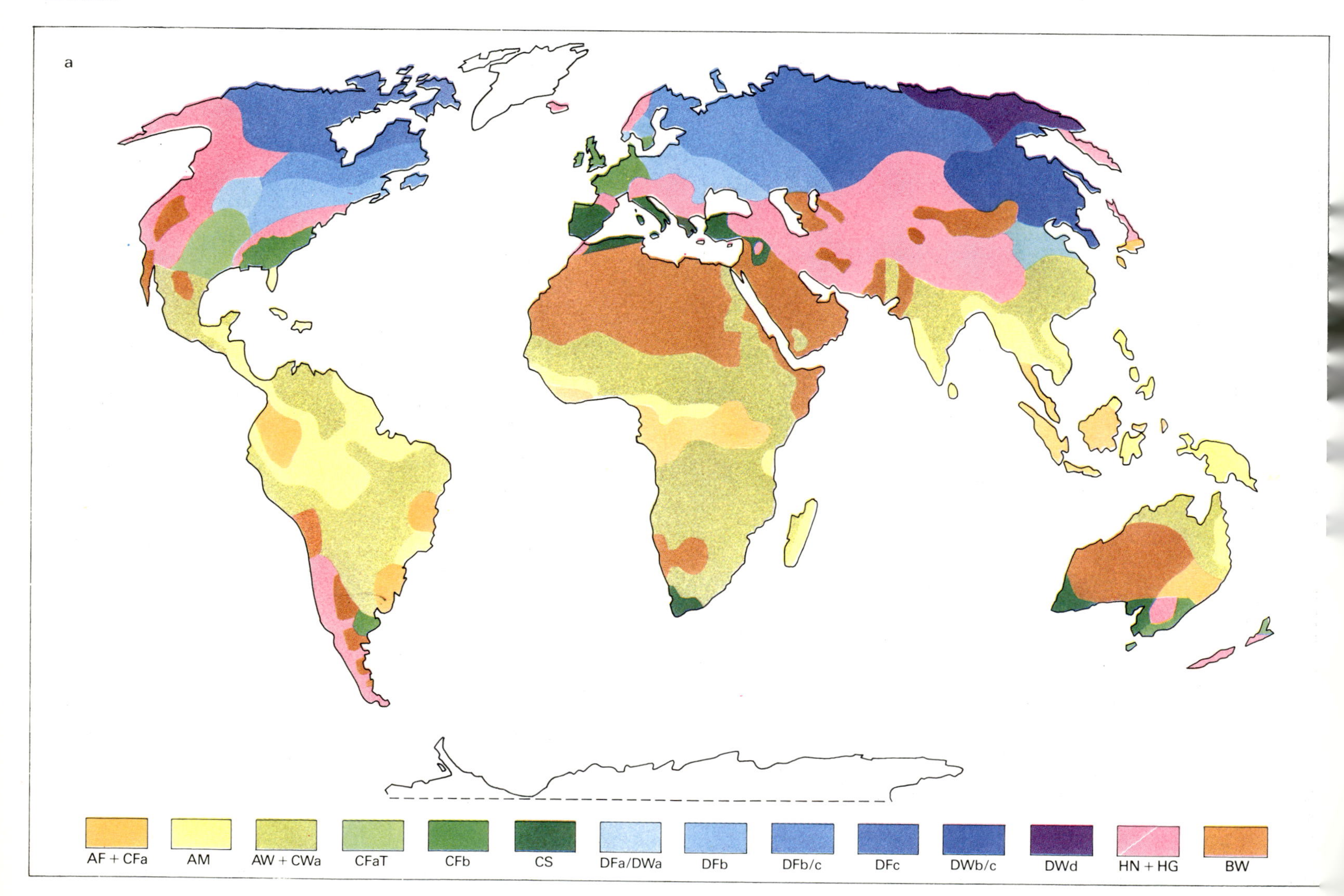

BW A desert regime with run-off confined to flash floods after rare heavy rain.

Mesothermal Regimes (C) (warm temperate climates): temperature and evaporation show marked seasonal variations; the regime depends upon the relative availability of precipitation in different seasons; snow is unimportant.

CFa Warm subtropical climate with a double precipitation maximum; similar to AF.

CWa Warm subtropical climate with strong precipitation in summer; typical of non-tropical monsoon areas.

CS Mediterranean climates; minimum flow in summer; rainfalls and flow maximums are mainly in autumn and spring.

CFb Temperate humid climates; river flow all the year round with the minimum in summer.

CFaT Temperate continental climates; greater precipitation in summer than in winter. Run-off maximums occur in early summer or spring.

Microthermal Regimes (D) (cool temperate climates): in regions where one or more months has a mean temperature less than −3°C substantial storage of snow may occur; annual flow regime will depend upon winter snowfall and warm season rainfall relative to evapotranspiration.

DFa/DWa Temperate continental interiors; summer rainfall produces a peak in flow; winter flows are low because of cold, dry, snowy weather.

DWb/c Continental interiors; strong summer rainfall and lower temperatures prolong the summer peak in flow.

DWd Cold continental interiors; total precipitation is low, mostly summer; most rivers are frozen in winter and the summer flow maximum is very pronounced.

DFb Winter snowfall is heavy and summers are warm; spring snowmelt and early summer rain give maximum run-off.

DFB/c With more pronounced autumn rainfall, snowmelt produces a spring run-off maximum with a subsidiary rainfall peak in autumn.

DFc Cold climates of the continental sub-Arctic produce a violent meltwater flood in early summer and a strong winter minimum of flow.

Mountain Regimes (H): high-altitude regimes characterized by snowfall; warm-season flow contains varying proportions of snow or glacier meltwater.

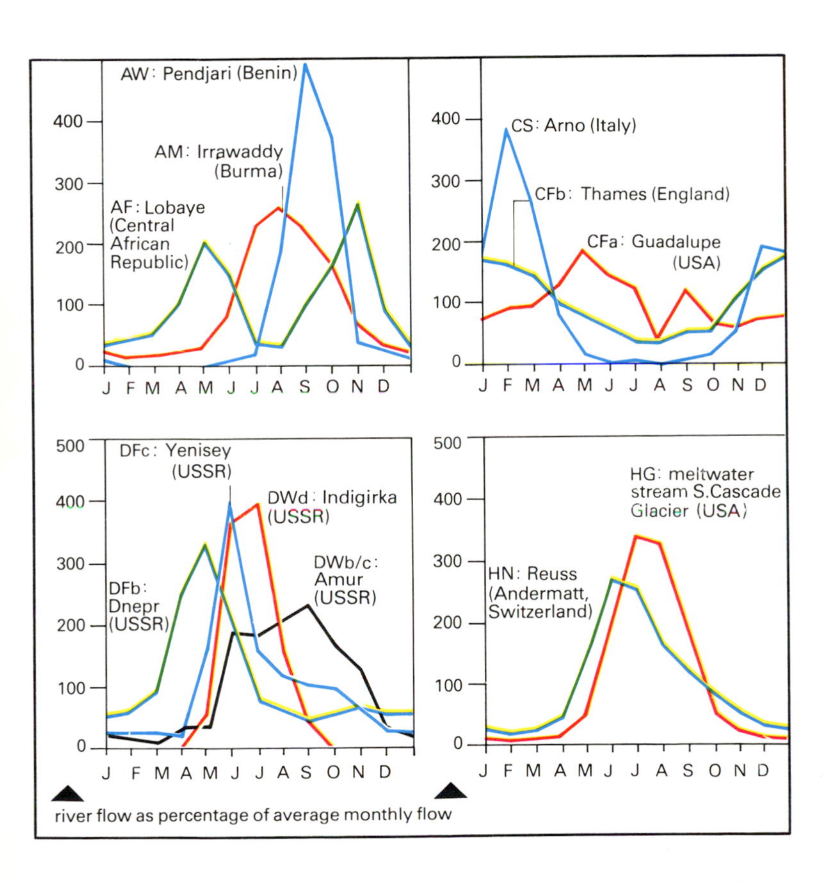

17.19: (**a**) *World distribution of river regimes. The five major groups of river regimes are tropical and subtropical (shown in yellows), warm temperate (greens), cool temperate and arctic (blues), high altitude (pink) and desert (brown). Ice-bound regimes are uncoloured.* (**b**) *Examples of annual flow regimes in each major category of (a). Flows are shown as percentages of the average monthly flow of each river, making it possible to compare rivers with different discharges on the same graph.*

The chemistry of natural waters

The waters flowing along the various pathways of the hydrologic cycle dissolve minerals as they go. The main inorganic constituents of natural waters are hydrogen (H^+), calcium (Ca^{2+}), magnesium (Mg^{2+}), sodium (Na^+), potassium (K^+), carbonate (CO_3^{2-}), bicarbonate (HCO_3^-), chloride (Cl^-), sulphate (SO_4^{2-}), hydroxyl (OH^-), silicate ($HSiO_4^-$) and silica (SiO_2), with smaller quantities of iron (Fe^{2+}, Fe^{3+}), aluminium (Al^{3+}), oxygen (O_2), carbon dioxide (CO_2) and hydrogen sulphide (H_2S). The total concentration varies from 1 mg per litre to 100 g per litre. (Seawater contains around 35 g/l.)

Computer programmes called aqueous models work on the chemical composition of a particular water and can predict whether or not that water will be capable of dissolving or precipitating any particular mineral. The reactions that take place as the water moves through the hydrologic cycle include gas solution, mineral solution, mineral precipitation, ion exchange, oxidation and reduction and reverse osmosis.

Precipitation contains very little in the way of dissolved solids. In sea areas it will carry Na^+ and Cl^- ions with Mg^{2+} and SO_4^{2-} derived from the salt in sea spray. In inland areas there is very much less, the little there is being derived from dust or pollution.

Carbon dioxide (CO_2) dissolved in precipitation forms carbonic acid (H_2CO_3) which reacts with soil minerals by hydrolysis converting the frame- and sheet-silicates into clays. The result on fresh volcanic or alluvial soils is that the soil is leached and converted into an acid type.

Evaporation of soil water in dry seasons leads to seasonal concentrations of the dissolved minerals. These are usually washed away in the wet seasons but in arid climates there may be precipitation of nodular carbonates.

Several hydrologists have identified a general sequence of chemical evolution in groundwater composition. In deep, confined aquifers the composition may evolve from a typical Ca–HCO_3 or Ca–Mg–HCO_3 composition to a Na–Cl-rich brine more concentrated than seawater. In deep zones the typical composition will be affected by SO_4^{2+} and

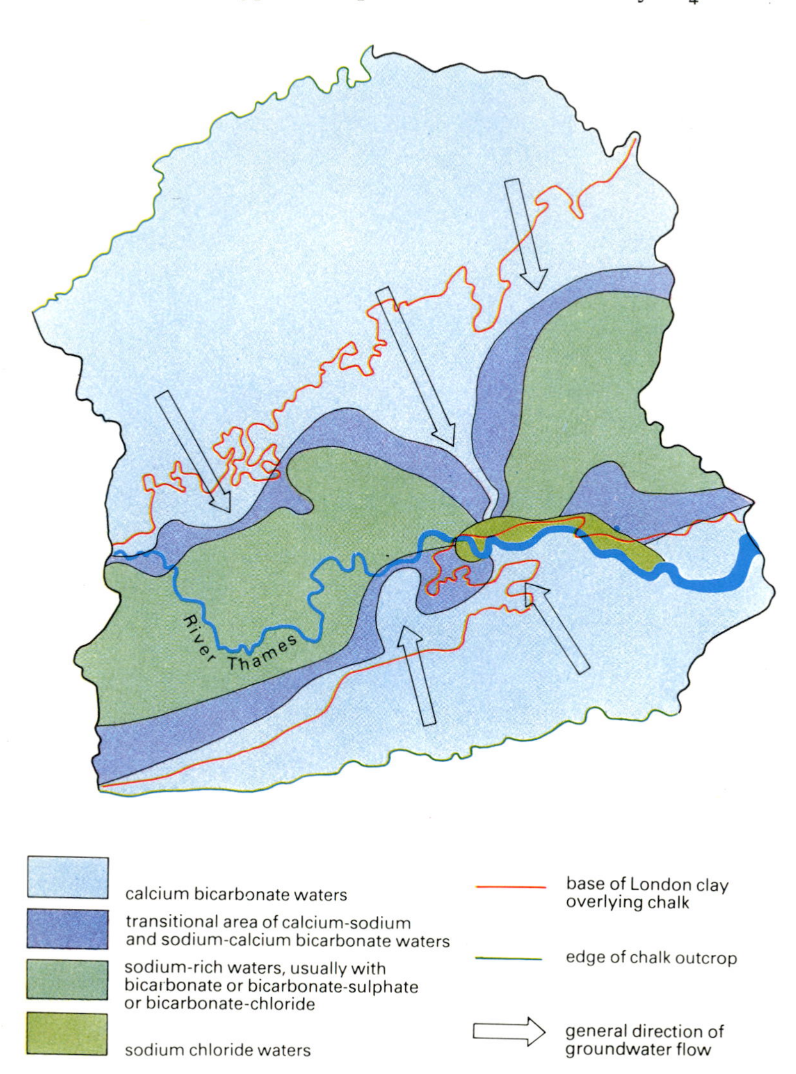

17.20: *Zones of chemical composition in groundwater in the chalk aquifer beneath London, England. The waters in the centre of the basin have been shown by radiocarbon dating to be several thousand years old. Pumping from wells in the past hundred years is causing a gradual flow of water towards the centre.*

cation exchange with Na-rich clay minerals will lead to a decline in Ca and Mg and a rise in Na. Calcite may be precipitated as an intergranular cement here. In deeper bodies Cl^- increases at the expense of SO_4^{2-} and HCO_3^- and sulphide minerals may precipitate. In the deepest parts of regional confined aquifers the resulting brines are sometimes hot.

High concentrations of Cl^- may result from the slow migration of water through confining argillaceous rocks which can act as a semi-permeable membrane. The dissolved solids cannot pass through, and they concentrate as brine in the remaining groundwater. Hypersaline brines are commonly associated with hydrocarbon reservoirs and have been studied extensively by petroleum geologists.

Organic material in rivers is usually broken down by oxidizing bacteria, using atmospheric oxygen, to CO_2 and water. If the organic load is too great, as under conditions of pollution, it will accumulate on the river bed and keep the water anoxic. Sulphate-reducing bacteria may use SO_4^{2-} in the water as an oxygen source and decompose the organic matter, producing foul-smelling hydrogen sulphide (H_2S).

Human interference in the hydrologic cycle

The natural storage and flow of water is interrupted when water is diverted for human use. This may take place in agriculture where the soil is disturbed and the drainage and irrigation balance is altered. Diversion of rivers for irrigation results in a permanent reduction of river flow.

The constant domestic and industrial demand for water is usually satisfied by means of reservoirs that ensure an even supply despite seasonal changes in precipitation. These can be direct-supply reservoirs from which the water is piped directly to the place of use, or river-regulation reservoirs that control the water at a river's headwaters and release it at times of natural low flow. This alters drastically the natural regime of the river. Even more extreme interference occurs when water is pumped out of aquifers to supplement the river at times of natural low flow.

Dams can be used as flood containment structures, to provide water for hydroelectric power or to redistribute water for irrigation. Water supply may involve the transfer of water from one river basin to another and an ambitious scheme in the Soviet Union plans to divert the Arctic rivers to irrigate the dry central Asian steppes.

Quality as well as quantity of water is affected by human use. Sewage or noxious chemicals will overload the natural bacterial cleansing mechanisms of rivers and result in foul-smelling polluted waterways. Such pollution is entirely avoidable but legislation, government agencies and efficient public health engineering are needed.

18 Climate and climatic change

Climate has been defined as the synthesis of weather for a long enough period for the reliable statistical determination of its properties. Weather is essentially the experience of the state of the atmosphere at a particular place and time. It is difficult to attach definite time scales to either; states of weather may be instantaneous or last for weeks, while climate may seem to be relatively constant from year to year or century to century. This chapter looks at evidence that shows that dramatic changes in climate have taken place in the Earth's history, both for particular locations and for the globe as a whole, and at currently acceptable theories for such changes, especially those that could have brought about the repeated but widely separated periods of more or less intense glaciation for which so much evidence has now accumulated.

In recent years concern over the carbon dioxide problem has stimulated much research into climatic change by scientists from a wide range of disciplines and has led to heightened awareness of the need to incorporate climatic factors in planning sensible use of the Earth's depleted resources. A series of short-lived extremes during the 1970s has highlighted the social impact of climatic change (Figure 18.1). New techniques have been developed to reconstruct climates of the past and sophisticated numerical models have been formulated. There are still, however, great uncertainties in our knowledge and understanding of the causes of climatic variations.

Evidence of past climates

The most reliable evidence of climatic change comes from the network of instrumental weather observations collected systematically since the first International Meteorological Congress held in Vienna in 1873. This record has, however, been near global only during the past twenty years. Coverage is still sparse over the oceans, over high latitudes of the southern hemisphere and for the upper levels of the atmosphere. The longest instrumental time series, which are for the European region, extend back only into the late seventeenth century. Satellite data are increasingly used in weather applications, but these records have generally been kept for too short a time to be used in climate studies.

Climatic information from pre-instrumental times is derived from a multiplicity of sources: historical documentary accounts of weather or climatic events and other indirect (or proxy) measures of climate. Indirect climatic data include any type of evidence from which indications of past climate can be gleaned. They cover a range of variables, including geological, chemical and biological, which can be dated with some accuracy and which respond to climate in a determinable way. The climatic evidence derived from any such source, whether instrumental or not, has to be judged critically, and there are major problems of interpretation.

Instrumental data

Reliable surface observations of the major climatic variables (pressure, temperature, precipitation, humidity and wind velocity) began during the seventeenth century. Data collection was systematized during the late nineteenth century, but many records for earlier periods have been preserved in archives and libraries. Observations of the upper air were not taken until much later, and it is only during the past two to three decades that there has been regular coverage in space and time. Even with instrumental data it is difficult to be certain about global climatic trends. For example, while there is consensus about recent variations in northern hemisphere temperature, the record for the southern hemisphere is less firmly established and many details of global patterns of temperature change have yet to be determined. Instrumental records have to be checked carefully for errors, or inhomogeneities, due to changes in, for example, the instruments, observational methods and exact locations, and it must also be ensured that the records are truly representative.

Historical sources

Historical sources are all non-instrumental documentary accounts related directly or indirectly to climate. It is necessary to distinguish between contemporary accounts, so-called primary sources, and compilations of such material made later, or secondary sources. Errors, in dating and translation for example, are frequently found in secondary sources. Many types of material have survived: accounts of severe weather or climatic events such as storms, floods, cold winters and droughts; information about recurring phenomena such as crop yields and harvest dates; and statements concerning weather-related events such as the dates of freezing of rivers and lakes. This kind of information can be found in annals, chronicles, diaries, manorial accounts, state papers, ships' logs and in many other places (Figure 18.2). Documentary accounts are available from Roman times, although dating problems render the earliest virtually unusable. Records are available from Iceland, the Russian plains and most of Europe from about AD 1000. A wealth of material for Asia exists but its potential has yet to be tapped.

Historical accounts are subjective and generally qualitative, and must be interpreted accordingly. The perception of an event as extreme is dependent on a number of factors, in particular the observer's past experience of climate and the reason for recording the event (it may be that an excuse was needed for some failure!). The historical record tends also to be biased towards shorter-term climatic variations, for long-term fluctuations are filtered out as memories fail. The historical record has nevertheless provided much information about the climate of the past millenium.

Indirect evidence

There are many types of indirect climatic data. Any variable that responds in an identifiable manner to climatic change and can be dated with some accuracy can be used. These data vary greatly in terms of the facet of climate that they indicate, the sensitivity of their response to climate (ie the quality of information they contain), the time taken for such a response to occur (from one to thousands of years) and the reliability with which they can be dated.

18.1: *The heatwave and drought in Texas 1980. It has been claimed that such events are signs of a shift towards more variable climatic conditions, but it is likely that much of the apparent increase in the frequency of climatic extremes is due to improved reporting and greater concern about their consequences. The social impact of drought is not restricted to the Third World. The interlinking of national economies means that local climatic extremes can have global repercussions: shortfalls in Soviet harvests during the 1970s resulted in massive grain purchases from the USA and a rise in the world price of wheat.*

Deep-sea sediment cores (Figure 18.3), collected routinely on oceanographic expeditions, have provided information on climates of the distant past. Statistical analyses of fossil flora and fauna provide data on past ocean temperatures. Dating is by isotopic analysis, which can also give information on global ice volume, and by palaeomagnetic stratigraphy. Similarly, in terrestrial sediment cores fossil pollen and soil types are used to identify times of climatic change during postglacial and earlier times. Ice cores, from Antarctica, Greenland and elsewhere, provide indirect climatic data for the past 100 000 years. Sea-level variations can be deduced from dated terraces at continental and island locations.

Climatic information for the Holocene has been gleaned from many sources. Tree rings can provide a reliably dated record of climatic parameters, allowing large-scale reconstructions of climatic variables, such as sea-level pressure, temperature and precipitation. Archaeology and radiocarbon dating, glacier movements, vegetation limits, bog stratigraphy, lake levels, insect and faunal remains, and varves in lake sediments have also contributed useful data.

The climatic history of the Earth

Indirect data have been used to reconstruct the climate of the Earth for various periods since the formation of the earliest dated sedimentary rocks about 3700 million years (Ma) ago. Although rocks from before 2800 Ma ago tend to be so metamorphosed that it is very difficult to get climatic evidence from them, models of the early Earth's atmosphere, the composition of which can only be assumed, have been used to estimate climatic conditions before that time. The climatic record for the first 90 per cent of the Earth's lifetime is indeed extremely fragmentary. Interpretation of the indirect evidence, which is mostly geological, is complicated by the lack of detailed palaeogeographic information. Accurate estimates of the positions of the continents and oceans are available only for post-Palaeozoic times. The climatic history of the Earth is continually under review as additional evidence is collected and as new techniques for extracting information are developed.

The Archaean and Proterozoic eons

The first billion years of the Earth's climatic history can be estimated by use of numerical models. Calculations based on assumed atmospheric compositions give an estimated global mean temperature of about 37°C at 4250 and 25°C at 3500 Ma, values based on estimates of the carbon dioxide content of the Earth's early atmosphere which was much greater than it is today. If the ammonia content was also enhanced, as has been suggested, temperatures may have been some 20°C higher. All these estimates are prone to considerable error because of the assumptions made, but it is likely that the composition of the early atmosphere, particularly the enhanced levels of carbon dioxide, did produce a marked 'greenhouse effect' (see below) and relatively warm climatic conditions. Isotopic data from the Fig Tree

18.2: *A woodcut from a contemporary pamphlet* Lamentable Newes out of Monmouthshire *provides a graphic image of the great floods that affected the coastal regions of the British Channel in January 1607.*

18.3: Swedish scientists preparing to take a sediment core north of Svalbard in the Arctic Ocean in the summer of 1980. The weight stand to which the core tube is being attached weighs 1.4 tonnes and the corer penetrates the sediment by the impact of its own weight. The sea here is 1000 m deep and the core recovered was 5.5 m long. When work is completed on the cores taken during this expedition it will be known if this area of the Arctic Ocean has been continuously covered by ice during the past million years.

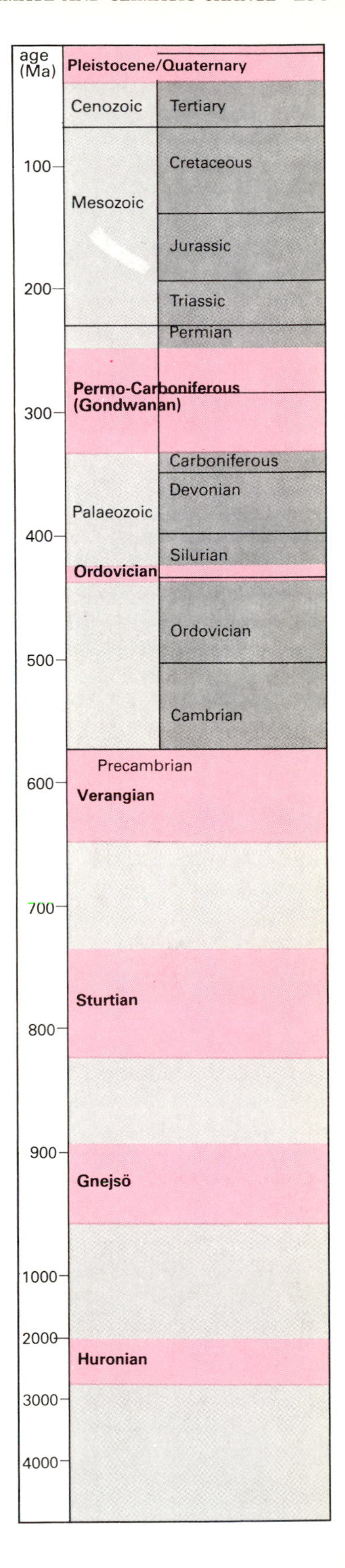

18.4: A schematic presentation of the major glacial periods (pink) in the Earth's history. It is likely that a number of glaciations occurred within each epoch. Note the change in scale on the time axis at 1000 Ma ago.

chert of an age of 3000 Ma suggests a value of 70°C, although this is controversial.

The first known glaciation (Figure 18.4), which occurred around 2700 to 1800 Ma ago in the early Proterozoic, was the Huronian glaciation, best evidenced by sedimentary and morphological data from the Gowganda, Ramsey Lake and Bruce formations in Ontario, Canada. Morphological evidence of glaciation must be interpreted carefully. Often such features can be created by processes other than glaciation or deglaciation. Striated surfaces, roches moutonnées and the like are considered to be the most reliable evidence. The three levels of Huronian strata meeting these criteria suggest that the period was one of recurring glacial events rather than a continuous glaciation. After this, the Earth is believed to have remained warm and free of permanent snow or ice until about 950 Ma ago. During the late Precambrian period at least three distinct glacial epochs occurred, at about 940 (the Gnejsö), 770 (the Sturtian) and 615 (the Varangian) Ma ago, each lasting about 100 Ma. These glaciations were not global and there is some uncertainty as to their precise geographical location, although they were collectively very widespread (see Chapter 23).

The Phanerozoic eon

The indirect record of climate improves towards the end of the Proterozoic eon when animals with shells spread through the oceans and the sedimentary record is better preserved. Palaeozoic climates (570 to 225 Ma ago) were generally warm, similar to today's but possibly drier, although there is evidence of a brief glacial event in the late Ordovician around 430 Ma ago. The best sedimentary and morphological evidence for this period of glaciation has been found in the Sahara: outwash and graded sediments, U-shaped palaeovalleys with polished and striated floors overlain with tills etc. Later in the Palaeozoic, climatic zonation increased and the Carboniferous was a period of cooler and wetter climates ending with the long Permo-Carboniferous glaciation which lasted from about 330 to 250 Ma ago. During this period a single supercontinent, Pangaea, consisting of all the Earth's landmasses, spanned all latitudes on one side of the globe. The centre of glaciation was close to the geographic south pole. Much of the evidence for this well-documented event was discovered in the nineteenth century in the southern hemisphere, and convinced local workers of the concept of continental drift.

During the Mesozoic era climatic conditions were generally warm; ocean-bottom temperatures, calculated by use of deep-sea sediment cores, did not drop below 14°C. Temperatures ranged from 10 to 20°C at the poles to 25 to 30°C at the equator. Variations in humidity may well have occurred. Early Triassic climates were humid but by the middle of this period the climate, particularly in subequatorial latitudes, had become drier and possibly warmer than today's. These conditions did not prevail long past 200 Ma ago. Slight cooling and a decrease in aridity had occurred by the beginning of the Jurassic period, and there is evidence that this was a time of small latitudinal and seasonal temperature contrasts.

That indirect data are more abundant for the Cretaceous period (135 to 65 Ma ago) is due to oxygen isotope analyses of deep-sea cores using benthic and planktonic fauna. These data give information concerning changes in high- and low-latitude ocean temperatures. High-latitude warming occurred during the first half of the Cretaceous period culminating in Albian times (about 100 Ma ago), after which global cooling occurred, particularly towards the end of the Cretaceous.

The pronounced, although brief, cooling that occurred at the end of the Mesozoic era (around 65 Ma ago) was followed by mass extinctions of marine and land organisms, although these events have yet to be reliably linked. A long-term cooling trend commenced at the start of the Eocene epoch (about 55 Ma ago) and was most pronounced at high latitudes. It was occasionally interrupted by rapid, shorter-term cooling events and was accompanied by cooling and then warming in low latitudes. Four marked short-term cooling episodes occurred during the early Cenozoic era; during the middle Palaeocene about 60, the middle Eocene about 45, the Eocene–Oligocene boundary about 38 and the middle Oligocene epoch about 28 Ma ago. The most severe short-term event occurred at the boundary of the Eocene and Oligocene (about 38 Ma ago). Within 100 000 to one million years the whole ocean water column had cooled by 3 to 5°C in high latitudes and at depth in low latitudes. The temperatures of surface waters at low latitudes dropped by about 10°C. It is believed that at this time Antarctic sea-ice began to form and cold bottom water was first produced. North Atlantic deep water production did not begin until about 12 Ma ago.

Rapid cooling also occurred around 25 Ma ago. Mountain glaciers had by now developed in Antarctica and around the time of another rapid cooling about 15 Ma ago mountain glaciers began to form in the northern hemisphere. The Antarctic ice sheet probably began to form about this time also, and by the late Miocene epoch the west Antarctic ice sheet had reached its present size. The east Antarctic sheet took longer to develop, reaching its present size by about 4 Ma ago. Glaciation in the northern hemisphere remained at a low level until 2.5 to 3 Ma ago when sea ice became a semipermanent feature of the Arctic Ocean.

The past two million years have seen a series of cold and warm periods. These changes have been associated with marked fluctuations in the size of continental ice sheets in the northern hemisphere and in sea ice extent in both hemispheres. Fluctuations occurred on a variety of time scales. At least four major cool periods have occurred during this time: at about 1.6 to 1.3, 0.9 to 0.7, 0.55 to 0.4 and 0.08 to 0.01 Ma ago. Rapid short-lived events did, however, occur throughout the period: at least seventeen glacial–interglacial cycles have been identified in Europe. There is some evidence of three distinct phases in the climatic conditions of this period: ice age, warm (as today) and interstadial (intermediate). The shifts between these phases have occurred quite rapidly at times.

The climatic record of this period is again most clearly shown by data from deep-sea sediment cores, which give information on global ice volume and local ocean temperature variations. A cycle about 100 000 years in length and of varying strength is evident from both these and some terrestrial data for the past 500 000 years. On these time scales fluctuations appear to be global in extent, synchronous between continents and in both hemispheres.

The past 125 000 years

The penultimate (Eemian) interglacial occurred some 120 000 years ago. It lasted about 10 000 years, and ended as glacial growth marked the initial stages of the Earth's last major glaciation 85 000 years ago. Large ice sheets developed in less than 10 000 years. Detailed evidence of sea surface temperatures during this time has been extensively studied by scientists in the CLIMAP project. These temperatures were similar to today's, except around Antarctica, where they were 2 to 3°C colder, and in the subpolar north Atlantic Ocean and Norwegian Sea, where they were about 1 to 2°C warmer. Warming of the oceans into the interglacial occurred first around the Antarctic and in the south Atlantic Ocean. The cooling that followed the interglacial was very variable in its timing. Whereas it occurred in the southern oceans some 3000 years before the major ice sheets began to develop in the northern hemisphere, northern waters remained warm during their development, perhaps providing an efficient moisture source for the growth of the ice sheets.

The last major northern glaciation reached its peak about 18 000 years ago (Figures 18.5 and 18.6), when sea level was at least 85 m lower than it is today and ice covered much of the high-latitude oceans. The Gulf Stream flowed eastwards across the north Atlantic towards the Iberian peninsula. The global mean ocean temperature (summer) was about 2.3°C below its present-day value. Ocean temperatures in the north Atlantic around latitudes 40 to 50°N were 12 to 18°C, and in certain areas in the north Pacific 6 to 10°C, below present-day values. In the southern hemisphere, the Antarctic polar front was displaced north and the area of sub-Antarctic water was reduced. Winter sea ice extended to latitude 46°S in the south Atlantic. In equatorial regions greater upwelling of cold bottom water apparently occurred. North America, north-west Europe, the Taimyr peninsula in northern Siberia, Greenland and Antarctica were covered by ice sheets sometimes 3 km thick.

Information like that in Figure 18.6 is fed into numerical models to simulate the general circulation of the atmosphere, and thus 'predict' atmospheric conditions during this glacial maximum. Global mean air temperature has been estimated to have been about 5°C less than at present. High pressure was located over the major ice sheets. The Asian monsoon was weak and shifted southwards so that it did not bring rain to the Indian subcontinent. North Africa, Arabia and Australia were also drier.

The present interglacial

The warming of the north Atlantic Ocean occurred between 16 000 and 7000 years ago, although it was punctuated by short-lived cooling events. The cordilleran ice sheet, which had occupied western North America, had disappeared rapidly by 10 000 years ago, having reached its maximum extent 14 000 years ago. The Scandinavian ice sheet began to collapse about that time and reached its present size by 8500 years ago. The Laurentide ice sheet, in eastern and central North America, had melted by about 7000 years ago. During the post-glacial, minor glacial advances have occurred about every 2500 years. The east Antarctic ice sheet has remained stable, but the west Antarctic ice sheet has continued to collapse during the past 12 000 years. There is some debate as to the precise timing of these events.

Climatic fluctuations have occurred on all time scales within the Holocene. In general the spatial patterns of change have been geographically complex. It appears that the longer-term trends (on the scale of centuries or greater) have been only hemispheric in extent at most during the early postglacial, at the time of the so-called climatic optimum (or Boreal period), and possibly during recent centuries. Summer temperatures have been estimated to have been 2°C greater than at present during the climatic optimum (6000 to 3000 BC) which followed the transition from glacial to interglacial conditions. This period of warmth was followed by cooling during the Sub-Boreal (to 1000 BC) until the cool and moister Sub-Atlantic period which lasted until about 500 BC.

During Roman times, warming occurred in Europe until AD 400 when climatic deterioration again set in. Improvement occurred about AD 800 and the period AD 800 to 1200 was a time of warmer climates —the so-called medieval warm period (or little climatic optimum). This relatively benign climate in higher latitudes occurred as Vikings settled in Greenland and Iceland. Temperatures began to fall at the onset of the Little Ice Age in AD 1200. Although brief warming events interrupted the climatic deterioration, the Little Ice Age was at its most severe during the late seventeenth century. Temperatures then gradually rose over the following two centuries with a temporary setback during the late eighteenth century (Figure 18.7). This warming peaked in the early twentieth century and by the 1950s cooling had again affected many regions of the northern hemisphere.

The evidence suggests that, particularly since Roman times, these generalized climatic trends have not been synchronous, ie occurring

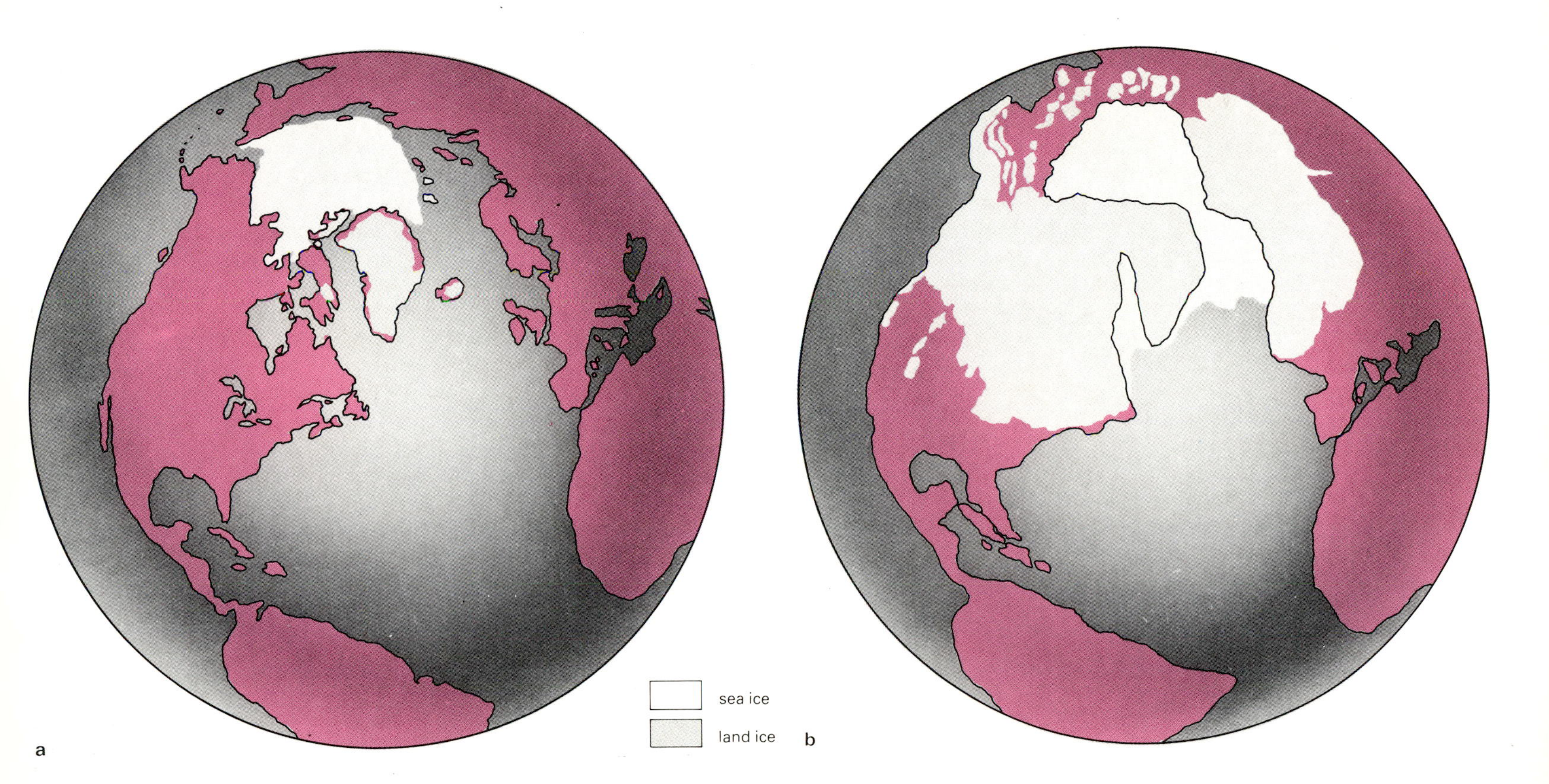

18.5: *The Earth at the present day* (**a**) *and at the time of the last glacial maximum some 20 000 years ago* (**b**). *During this glaciation sea level was about 76 m lower than at present, exposing many parts of the continental shelf. Two great ice sheets covered North America and others occupied northern Europe and Asia. The surface waters of the north Atlantic Ocean were frozen over in high latitudes and sea ice extended well into lower latitudes of the southern hemisphere.*

throughout a hemisphere or the globe simultaneously. The timing of the trends has varied region by region and the peak of the Little Ice Age may have been the only time when the northern hemisphere as a whole experienced cold conditions. The general warming in the twentieth century, which affected practically the whole of the northern hemisphere, may have been the only time of truly large-scale warming. The southern hemisphere has, however, experienced different temperature trends during the present century with warming not affecting certain regions until recent decades.

It is likely that the spatial patterns of climatic change are complex whatever the time scale, fragmentary evidence merely giving the illusion of coherent global trends. The detailed portrait of the last major glaciation now available through the efforts of the CLIMAP (see Chapter 17) team contrasts sharply with the cruder, more generalized portrayals of simultaneous cooling and warming which have been given for earlier glaciations. The finest possible detail of the climatic record is needed to determine the causes of past fluctuations and to predict future climatic developments. Knowledge of the history of the Earth's climates needs further improvement, and there is a continuing effort to enhance the climatic data base.

18.6: *Map of the last major glaciation of the Earth about 18 000 years ago.*

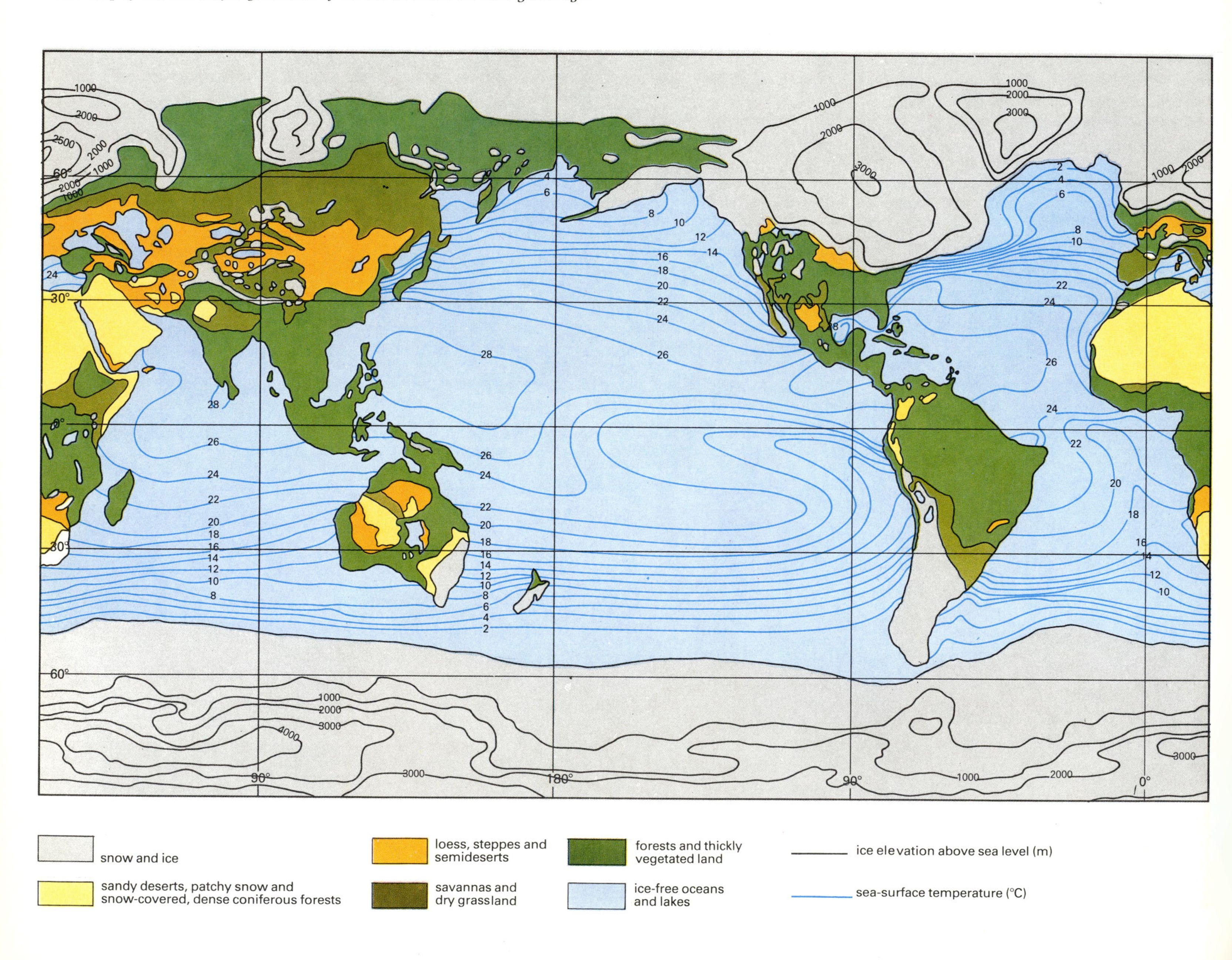

Causes of climatic change

There are many uncertainties in the record of past climatic variations, and even more in current knowledge of their causes. The climatic system is complex, encompassing the atmosphere, cryosphere and hydrosphere, and an understanding of its internal mechanisms and interactions is still rudimentary. The ability to test physically realistic mechanisms for climatic change is severely limited by inadequacies in available models of the climate and by uncertainties in the climatic data base. The search for mechanisms is further complicated by the fact that at any one time many causal factors must determine the climatic regime.

Changes in climate occur as the energy balance of the Earth is disturbed. The main source of energy driving the atmospheric circulation and controlling climate is the Sun. As solar energy enters the atmosphere it is subject to a series of transformations before it reaches the Earth's surface, and becomes available to the atmospheric circulation. The characteristics of the Earth's surface affect energy exchanges, the oceans acting as mobile heat reservoirs, while topography constrains the wind flow. A complex chain of transformations thus links the solar output to the atmospheric circulation and any disturbance in this chain can result in a climatic change. Such disturbances can take the form of changes in the amount of energy entering the Earth's atmosphere from space or reaching the Earth's surface, redistribution of heat within the atmosphere or within the seasonal cycle, redistribution of heat storage within different components of the climate system or major alterations to geography.

A range of mechanisms has been suggested which can produce a significant disturbance in the Earth's energy balance on glacial time scales, but evidence for these is variable in quality. None can be said to have been irrefutably proven to have caused any climatic variation in the past. For example, a reason for the Earth's major glaciations is still being sought.

Changes in solar energy

As the Sun is the main source of the Earth's energy, variations in solar output must be considered a likely cause of climatic change. On time scales of days, years and centuries, much empirical evidence has purported to show links between solar variations (generally in the number of sunspots) and various climatic events. The statistical evidence has, however, been questioned and theoretical models of these links are not convincing. Changes in the solar output, or the solar energy received at the top of the Earth's atmosphere, on the daily to decadal time scales, are small energetically and it is not clear how variations of the magnitude observed could significantly disturb the Earth's energy balance. Trigger mechanisms, by which a small variation in solar output could catalyse a significant climatic change, have been proposed but are difficult to substantiate.

On the century-to-century time scale, evidence from postglacial times suggests a relationship between climate and the ^{14}C content of the atmosphere. The latter is dependent in part on solar output and may therefore act as an index of solar variation. This evidence has, however, been questioned on the grounds that the climatic record is not sufficiently well established to justify the certainty claimed for the correlation. On time scales greater than one hundred years, there is no direct evidence of changes in the solar output. It has been suggested, on theoretical grounds, that thermal relaxation oscillations may exist on time scales of about 100 000 years because of instabilities in the solar convection zone. Periodic disturbance in solar output may also occur because of ^{3}He instability in the solar core every 200 to 300 Ma. According to model calculations a decrease in solar luminosity of about 5 per cent would last for some 4 Ma, followed by a slower luminosity increase of the same amount lasting about 8 Ma. Such fluctuations in solar output are considered speculative, but it is widely accepted that longer-term evolutionary trends in solar luminosity have occurred; the problem lies in detecting their effects and distinguishing them from those brought about by other mechanisms.

18.7: *Two contrasting pictures of the Rhône glacier in Switzerland illustrate the effects of the warming that affected much of the northern hemisphere during the nineteenth and early twentieth centuries. The watercolour, executed about 1770, shows the final phase of the Little Ice Age. The photograph, taken in 1970, shows that an extensive retreat occurred over the intervening two hundred years. Although the response of glaciers to climatic change is complex, evidence such as this is frequently used as an indicator of long-term climatic variation.*

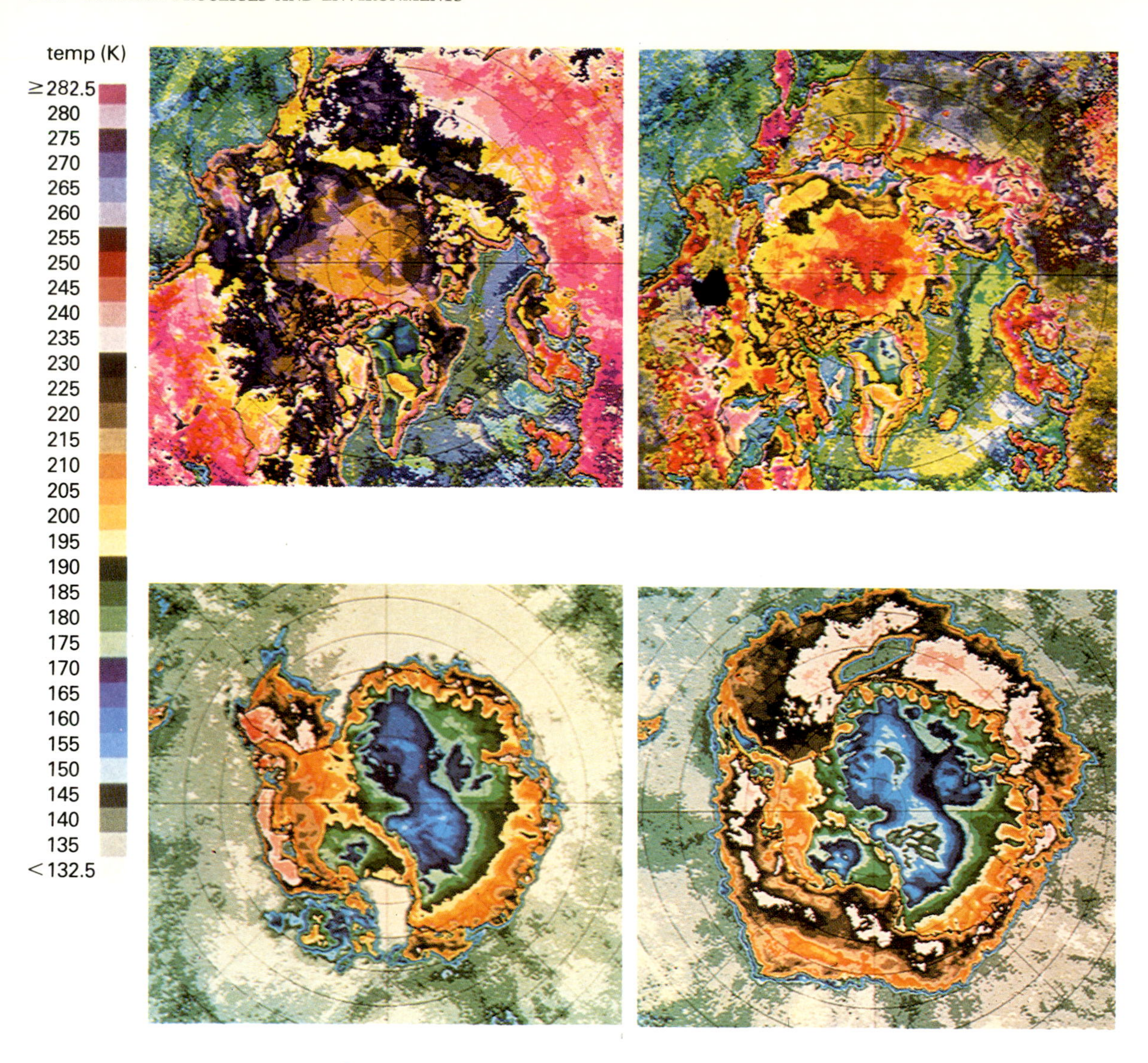

18.8: *The North Pole in January (left) and July (right). The Arctic Ocean is covered by a thin layer of ice at the present day. This ice layer insulates the oceans from the atmosphere, reducing the flow of heat and moisture. Pronounced variations in ice extent occur on all time scales and are associated with significant changes in the local and larger-scale energy balance. Various theories involving variations in the Arctic ice cover as an indicator of glaciation have been proposed, but have yet to be substantiated. Nevertheless the history of the ice cover of the Arctic is important in tracing the evolution of climate.*

18.9: *The South Pole in January (left) and August (right). The Antarctic ice cap contains over 90 per cent of the world's ice at the present day and is much more stable than the ice cover of the Arctic. It is over 3 Ma old and has been growing throughout the past 12 000 years. The stability of the two major ice sheets in west and east Antarctica is in doubt. It has been proposed that major surges could give rise to widespread cooling of the oceans, and hence to glaciation.*

The Sun is thought to be a reasonably well-behaved main-sequence star, and its luminosity is believed to have increased with time as nuclear fusion has increased its mean molecular weight. Accepted stellar theory predicts that solar luminosity 4500 Ma ago was about 75 per cent of its present value and would today result in an ice-covered Earth. Yet it is known that this did not occur. There is firm evidence of unfrozen oceans from the oldest known rocks, and life could not have evolved during the first billion years or so of the Earth's existence had it been ice-covered. This problem is known as the 'faint Sun paradox'. It is necessary to explain how the effects of low solar luminosity during the Earth's early history were counteracted to produce a net warming.

Changes in the Earth's atmosphere

Gases which are transparent to incoming shortwave radiation but relatively opaque to outgoing longwave radiation, such as carbon dioxide, ammonia and water vapour, are said to be radiatively active. Variations in their concentration can alter the thermal structure of the Earth's atmosphere and disturb its energy balance. Significant variations in the composition of the atmosphere are known to have occurred in the past and this mechanism is believed to have been important in determining past climatic change, as well as being a possible influence on future climate (see below).

The carbon dioxide concentration in the early Earth's atmosphere is thought to have been about a thousand times greater than it is today. Calculated models suggest that the consequent enhancement of the greenhouse effect (see below) would have been sufficient to offset the effects of low solar luminosity at that time. Ammonia, which is opaque to infrared radiation, may have also been present in concentrations sufficient to have enhanced the greenhouse effect significantly. There is geological evidence favouring a reducing atmosphere for the early Earth and the evolution of life may have required a considerably greater ammonia concentration than that found at present. Ammonia does, however, have a relatively short life in the atmosphere, since it photodissociates readily, and it is therefore unlikely to have remained in sufficient concentrations for long enough to have produced an appreciable climatic effect.

More recently there is some evidence from Antarctic ice cores of reduced atmospheric carbon dioxide levels (about half today's values) during the last major glaciation, 14 000 to 22 000 years ago. It is likely that more carbon dioxide was stored in the cooler oceans but biospheric changes may also have affected its distribution. This reduction in atmospheric carbon dioxide was a result of the shift to glacial conditions and probably not a major cause of the glaciation. The reverse greenhouse effect may, however, have been important in reinforcing cooling.

Changes in the Earth's geography

Changes due to continental drift and sea floor spreading in the configuration and positions of the landmasses may have played a major role in determining the Earth's climate. As a landmass moved into a polar position, because of continental drift or true polar wandering (ie the Earth's crust moving as a whole relative to the rotational axis), its climate would become colder and glaciation might result. During the Permo-Carboniferous glaciation about 270 Ma ago, which was apparently greatest near the geographic south pole, a single continent, Pangaea, occupied all latitudes on one side of the globe. It is believed that the unusual configuration of the landmasses at that time is sufficient to explain this major glaciation. Necessary preconditions for, though not the prime cause of, the Quaternary glaciations, were the movement of Antarctica to its current polar position and the isolation of the Arctic Ocean by the northern landmasses.

Another geographical effect probably related to plate tectonic processes is the variation in the overall ratio of land and sea areas caused both by changing continental configurations and by eustatic sea-level changes resulting from variations in worldwide activity of mid-ocean ridges. It has been suggested that the land-sea ratio may have varied by as much as 20 per cent in the past 180 Ma, and that this important palaeogeographical effect may have influenced climate in a number of ways, for example by changing the net radiation balance (oceans have a lower albedo, or ability to reflect light, than land), by altering the geographical distribution of albedo, by changing the seasonality of climate, by increasing the likelihood of polar glaciation or by altering the ocean circulation. Such theories are, however, difficult to substantiate without the use of complex numerical models which are not yet available.

Orogeny, the processes of continental uplift and mountain building, became a possible climatic influence in Proterozoic times. Mountain chains affect the dynamics of atmospheric flow and latitudinal heat and moisture transport, and hence climate. The temperature of the atmosphere is lower at height, and so an increase in the mean elevation of the continents can cause relative cooling and greater seasonal contrasts. Secondary effects may arise because of alterations in snow cover or vegetation. Isostatic deformation of land by ice loading can produce local changes in relief and sea level. The build-up and melting of major ice sheets result in global changes in sea level. These processes operate on shorter time scales (10 000 to 100 000 years) than the processes already discussed and may have complex consequences. It has been suggested that the time scale of such variations, coupled with feedback effects, could produce quasi-periodic climatic changes, at least during times of glaciation.

The above effects of continental drift, orogeny and isostasy would all have resulted in alterations in the circulation of the oceans, which may have had greater climatic consequences than the factors that produced them. Oceanographic changes can influence climate by albedo effects, by changes in heat storage or transport as the ocean currents change their paths or by alterations in the distribution of ocean surface temperature which affect the thermal structure of the lower levels of the atmosphere. The increase in bottom water formation during the early Cenozoic, when a long-term cooling trend began, marked a major change in the ocean's capacity to store heat. The opening of the Greenland–Norwegian Sea to the Arctic and the initiation of the Antarctic circumpolar current may have been related to a cooling event of 100 000 years' duration about 38 Ma ago at the Eocene–Oligocene boundary.

Volcanic activity

Large explosive volcanic eruptions disperse solid and liquid particles (aerosols) into the upper layers of the atmosphere (Figure 18.12). These 'dust veils' alter the opacity of the atmosphere and affect the radiative transfer from Earth to space and hence the global energy balance. While the particulates and aerosols have a lifetime in the upper atmosphere of only a few years, changes in the frequency of eruptions may produce climatic changes on longer time scales: a high frequency of volcanic eruptions increases the dust load in the upper atmosphere and leads to surface cooling. This mechanism is well established on theoretical and empirical grounds as a cause of climatic change on time scales of up to a decade during the past couple of hundred years. The early twentieth century warming coincided with a time of significantly low explosive volcanic activity.

Larger eruptions than those experienced in historical times are evidenced by calderas (large volcanic craters) and by ash layers in deep-sea sediment cores. Some were at least ten times as powerful as the eruption of Krakatoa in 1883, and it has been suggested that enhanced volcanic activity may have initiated glaciation. On the million year time scale four periods of enhanced volcanic activity have been identified in deep-sea core records, of which two (in the middle Pliocene and middle Miocene epochs) coincided with the beginning of glaciation in different parts of the world. There is, however, a great deal of uncertainty in the long-term record of volcanic activity which makes it difficult to test this particular theory.

The astronomical theory of climatic change

The astronomical, or Milankovitch, theory of climatic change dates back to the last century. It was originally proposed by John Herschel (1792–1871) in 1830 soon after the discovery of the first evidence that recurrent glaciation had occurred throughout the history of the Earth. Calculations of the effects of orbital variations on the Earth's energy balance were made by J. A. Adhémar in 1842 and later in the century by James Croll. The theory has had a controversial history. It was revived by M. Milankovitch during the present century and, as empirical evidence accumulates in support of the theoretical arguments, astronomical factors are gradually becoming accepted as an explanation for the series of Quaternary glaciations. Changes in the eccentricity of the Earth's solar orbit, in the precession of the Earth's axis of rotation and the axis of the orbital ellipse, and in the obliquity or tilt of the Earth's axis relative to the plane of the ecliptic result in major variations in the latitudinal and seasonal distribution of insolation (Figure 18.10).

The Earth's eccentricity affects the relative intensity and duration of the seasons in the two hemispheres and also the range of insolation received during a particular year. The precession controls changes in length between the astronomical half-year seasons and the Earth–Sun distances at both solstices. Obliquity affects the geographical contrast in annual radiation and the intensity of the seasonal variations in both hemispheres. For glaciation to occur according to this theory, northern summers must be cold in high latitudes to reduce melting of the winter snow. This happens when the eccentricity is high, northern summer is at perihelion (the point of the orbit at which the Earth is nearest to the Sun) and the obliquity is low. This also produces cold winters in the southern hemisphere and therefore the extension of the sea ice limit to lower latitudes, with glaciation in both hemispheres.

These astronomical changes in the seasonal and latitudinal distribution of energy are quasi-periodic, and similar cycles have been detected in deep-sea sediment cores for the past 500 000 years (Figure 18.11) and in some terrestrial records. Obliquity variations may account for earlier glacial epochs in the late Precambrian and climate-related vegetation changes at the end of the Eocene.

While the astronomical theory of glaciation still requires refinement in order to explain the detailed timing (and magnitude) of glaciation during the Quaternary, it is the best established of all theories. The physical processes underlying the theory are well known and have been tested empirically. It is likely that consideration of mechanisms within the climate system will provide the explanation for various inconsistencies in the timing of glaciation predicted by the Milankovitch model, particularly the rapidity of its onset and termination.

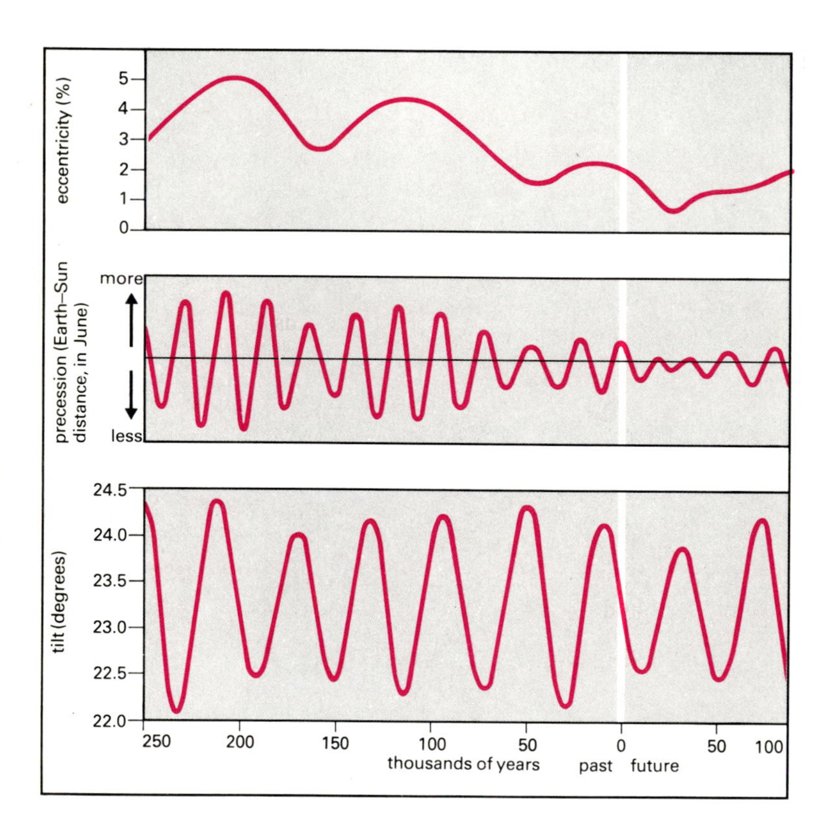

18.10: *Calculated changes in the Earth's eccentricity, precession and tilt. The geometry of the Earth's orbit is affected by variations in the gravitational field due to planetary motion. Analysis of temperature and global ice volume data derived from deep-sea sediment cores reveals evidence of similar cycles (lasting around 100 000, 40 000 and 20 000 years) in the past.*

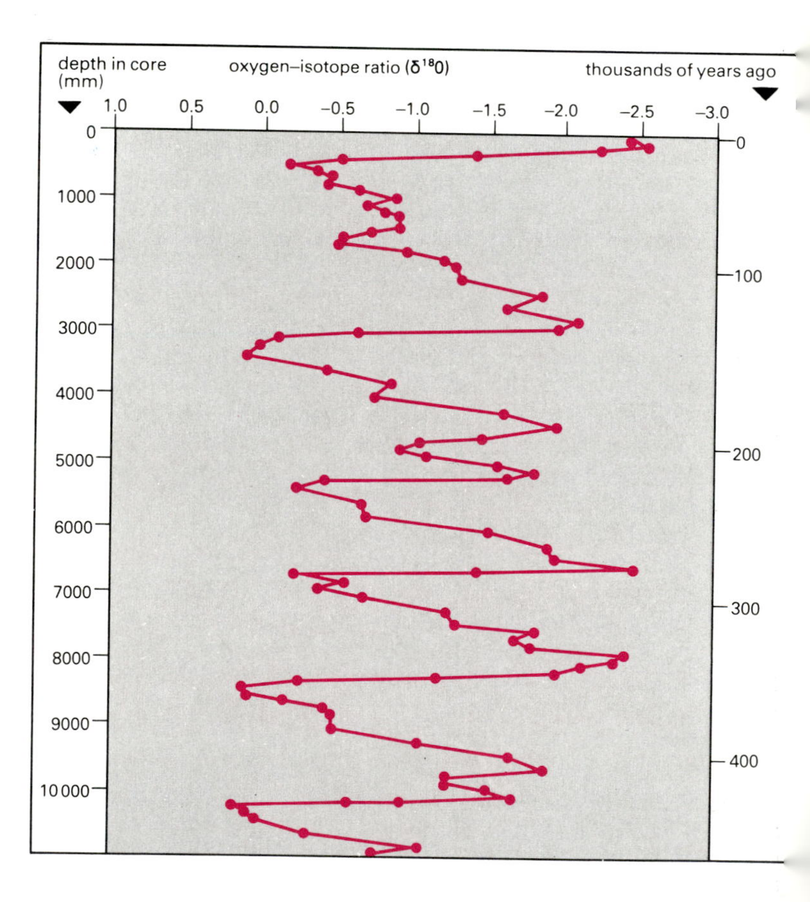

18.11: *Changes in the oxygen isotope ratio in a deep sea sediment core from the Caribbean show a clear 100 000-year cycle. The saw-toothed nature of the curve suggests that de-glaciation occurred much faster than glacial expansion.*

18.12: *The series of eruptions of Mount St Helens in the north-western USA during 1980 enabled scientists to study the climatic impact of a sequence of well-documented volcanic explosions. While the initial eruption was particularly violent much of the blast was directly laterally, and the injection of dust and aerosols into the upper layers of the atmosphere was not great. By late 1980 a veil of volcanic ejecta had covered higher latitudes of the hemisphere probably causing minor cooling of the Earth's surface but not a major climatic change.*

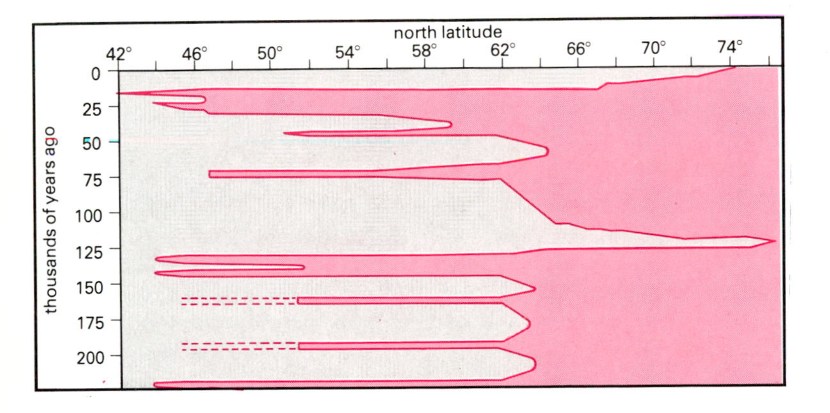

18.13: *Pronounced migrations of polar water (pink) have taken place in the eastern north Atlantic Ocean during the last 225 000 years. The penetration of polar water into southern latitudes is based on analysis of foraminifera and coccoliths found in deep sea sediment cores. The polar water has advanced beyond 51°N seven or eight times during this period. Such changes in the water mass distribution must play an important role in influencing the development and decay of glaciation; they cause major disturbances in the transfer of heat and moisture between the oceans and the atmosphere.*

Internal mechanisms of change

The climate system depends on the complex interplay between several different elements. Variations in one element of the system can influence others and produce, through feedback loops, variations on a range of time scales without any external forcing mechanism (Figure 18.13). Even when a change is initiated by external forcing, internal mechanisms are likely to modify the initial change relatively quickly. It is now thought unlikely that internal mechanisms alone could have initiated major glaciation, but such theories have been popular.

One such mechanism for glaciation is based on positive feedback (ie a secondary change reinforces the initial disturbance) between surface temperature and albedo in high latitudes. An ice-free Arctic Ocean leads to greater snowfall on the adjacent landmasses as high-latitude evaporation is enhanced. The increased surface albedo of the snow-covered land reflects more heat away from the Earth, causing cooling. This, in conjunction with the remaining moisture source, promotes increased snowfall and glaciation. The process continues until the Arctic Ocean freezes over and the moisture source is lost, at which point de-glaciation begins. Unfortunately this theory does not stand up against the indirect climate record. The Arctic Ocean has not been free of ice for at least 700 000 years and possibly 2.5 Ma. It can be argued, however, that a completely ice-free Arctic is not necessary to initiate glaciation in this way, which is often used to explain the rapidity of glaciation.

Another internal mechanism that has been claimed as explaining widespread glaciation involves the east Antarctic ice sheet. Surges in this ice sheet, prompted by decreased precipitation, would lead to a vast increase in the amount of sea ice in the southern hemisphere and higher albedo in the polar region, so causing cooling. The west Antarctic ice sheet, which is grounded below sea level, is thought to be dynamically unstable and there has been some concern over the possibility that future warming induced by increased carbon dioxide levels in the atmosphere might initiate a major surge and a catastrophic rise in sea level. This is, however, unlikely to occur for hundreds of years, if ever.

Other mechanisms of change

Many other mechanisms for long-term climatic change have been suggested. In circling the Galaxy the Earth passes through one of the rotating spiral arms of the Galaxy about once every 300 Ma. Passage through this and the adjacent dust lane might cause climatic effects as interstellar dust affects the radiative transmittancy of interplanetary space or as infalling interstellar material affects the solar luminosity. The Earth's path around the Galaxy is elliptical and its passage nearest to the Galaxy's centre every 270 to 400 Ma might give rise to cyclic changes in climate on the 100 Ma time scale. On a more local scale, there is some suggestion of correlation between the indirect climate record and reversals of the Earth's geomagnetic field, but there is no obvious physical mechanism to account for the link.

It is possible that glaciation may not have been due to deterministic factors such as those discussed and that the climate system may be continually adjusting to the factors that determine the global energy balance. If climate is relatively insensitive to external influences two or more sets of apparently stable climatic conditions could alternate, separated by periods of transition.

It is unlikely that any one mechanism for major changes will ever be conclusively proved to have accounted for the past climatic record because this is determined by a multitude of factors, both internal and external to the climate system. Even if an external factor is shown to have initiated glaciation or de-glaciation, the detailed picture of the last major glaciation has shown that the evolution of the glacial–interglacial cycle is so complex in timing and geographical pattern that much of this fine detail must be due to the internal workings of the climate system itself.

Carbon dioxide and climate

The potential climatic impact of increased levels of atmospheric carbon dioxide is currently considered to be a major environmental problem. The natural balance of the global carbon cycle has been upset by the burning of fossil fuels and, most probably, by deforestation and changing land use. Numerical models predict that carbon dioxide increases will, by the mid-twenty-first century, result in a global warming of 2 to 3°C as the Earth's energy balance alters. A climatic shift of this magnitude and rapidity has not been experienced since the last major glaciation.

The scientific problem is relatively easy to formulate, but determination of the exact effects on climate of the injection of a certain amount of carbon dioxide into the atmosphere is not simple. Knowledge of the global carbon cycle and of the climate system is limited, and projections of future energy strategies and hence carbon dioxide emission are necessarily speculative. The carbon dioxide problem spans a number of interlinked research areas and assessment of the implications of changes in climate induced by carbon dioxide involves a mesh of social, economic and political factors.

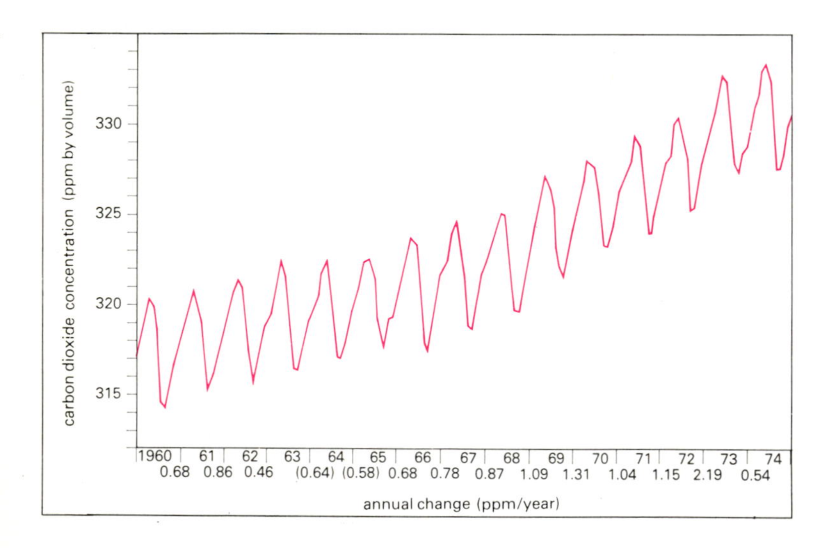

18.14: *The worldwide rise in carbon dioxide concentration, estimated at about 0.33 per cent per year, is shown by data from the Mauna Loa Observatory in Hawaii.*

The rising level of atmospheric carbon dioxide

The burning of fossil fuels (coal, petroleum and natural gas) began to contribute significantly to the level of carbon dioxide in the atmosphere as a whole following the Industrial Revolution of the nineteenth century. About one half of the carbon dioxide produced by the combustion of fossil fuels remains in the atmosphere. The other half is dissolved in the oceans or enters the Earth's biosphere, particularly the forests. Carbon dioxide concentration during the nineteenth century has been estimated at about 280 ppm (parts per million) by volume and is about 340 ppm by volume today. During the 1970s the concentration has been increasing at a rate of 0.5 to 2 ppm by volume per year.

Carbon dioxide is well mixed through the atmosphere and the only regional differences in concentration are thought to be due largely to the fact that the anthropogenic input is greater in the northern hemisphere. The difference of about 1 ppm by volume found between observations taken at the South Pole and at northerly locations implies a time lag of some eighteen months before the South Pole receives the carbon dioxide injected into the northern hemisphere. This is because of the slow exchange of air between the hemispheres.

Figure 18.14 shows the carbon dioxide concentration recorded at the Mauna Loa Observatory in Hawaii. There is evidence of a seasonal cycle, producing an intra-annual swing of some 6 to 7 ppm by volume and marked by a reduction in atmospheric carbon dioxide in spring and summer due to variations in the growth of vegetation (which is mostly in the northern hemisphere). The records also show a long-term rise in carbon dioxide of some 15 ppm by volume over the whole twenty-year period. The rate of increase is not, however, constant. Despite the fact that the worldwide increase in the release of carbon dioxide cannot have changed much from around 5 per cent per year during this time the rate of increase in the level of atmospheric carbon dioxide dropped during the mid-1960s, rose towards the end of that decade and then dropped again in the mid-1970s. It is believed that these changes are due to variations in oceanic uptake which may be the result of large-scale changes in the distribution of sea surface temperatures or alterations in the vertical circulation of the oceans.

Future rises in atmospheric carbon dioxide

While the details of the global carbon cycle are not fully known, the observed increase in atmospheric carbon dioxide concentrations fits well with estimates based on models of storage and fluxes. It is generally assumed that atmospheric carbon dioxide will continue to increase at about one-half the rate of increase of consumption of fossil fuel over the next few decades, but estimation of these levels is, of course, a major problem, involving economic and political as well as technological and scientific factors. There are various possible models of future energy usage. These involve a mix of fossil fuel, nuclear and 'alternative' (for example solar and wind power) energy strategies, the greatest carbon dioxide rise being associated with models involving a large fossil fuel component (Figure 18.15). Despite great uncertainties, both in the models of energy usage and the global carbon cycle, it is now widely accepted that by the twenty-first century the level of atmospheric carbon dioxide will have reached a concentration of around 380 to 390 ppm by volume and that present values will have doubled by the middle of that century. The projection assumes no significant change in the rate of growth of fossil fuel consumption from the present value of about 2 per cent per year.

The greenhouse effect

The variation of the Earth's atmospheric composition through time has already been discussed as a possible cause of past climatic change. The radiative properties of carbon dioxide give rise to the so-called greenhouse effect. The spectrum of solar energy received at the Earth's surface differs from that expected from a perfect radiator at any temperature, partly as a result of absorption within the Sun's atmosphere but mostly from absorption within the Earth's atmosphere. Departures from the expected distribution of solar energy indicate energy absorbed within the Earth's atmosphere, principally by carbon dioxide, ozone, water vapour and liquid water drops or droplets.

Carbon dioxide is virtually transparent to incoming solar radiation (there is some absorption in the wavelength range 4.0 to 4.8 micrometres (μm)) but absorbs outgoing terrestrial infrared radiation in several bands, most notably 12 to 18 μm. Present concentrations of atmospheric carbon dioxide, equivalent to a layer of gas at the surface some 30 mm thick, are sufficient to absorb all radiation in the solar spectrum between wavelengths of 4.0 and 4.8 μm and about 40 per cent of the radiation between 14 and 16 μm. Water vapour is, however, also radiatively active around these wavelengths and is a more effective absorber. Together the present concentrations of carbon dioxide and water vapour absorb nearly all outgoing terrestrial radiation at wavelengths greater than 14 μm.

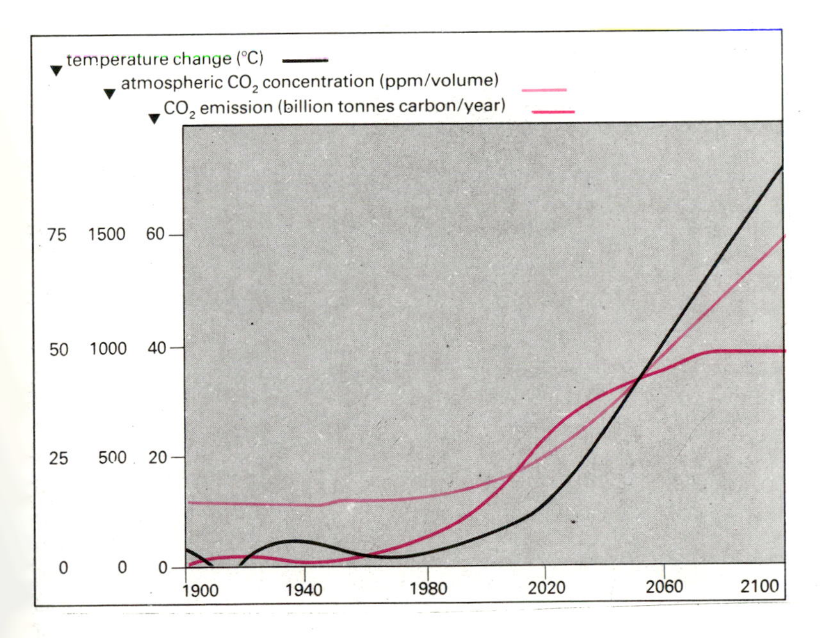

18.15: *A diagram showing the implications for carbon dioxide emission and atmospheric concentration, and hence global climate, of a 'worst possible' energy strategy involving a large fossil fuel component.*

Increases in carbon dioxide content can therefore only increase absorption of radiation at wavelengths of 12 to 14 μm. This has little effect on the amount of incoming solar energy, but outgoing terrestrial radiation which would otherwise escape to space is trapped within the lower levels of the atmosphere. This could lead to a rise in surface temperature and cooling of the upper levels of the atmosphere. This mechanism is known as the greenhouse effect, although in fact a slightly different mechanism operates in a horticultural greenhouse.

Impact on climate

There are two main ways in which quantitative estimates of the climatic impact of atmospheric carbon dioxide increases can be made. The first involves the use of numerical models of the climate system. The second is an analogue approach based on past records of climate from which the possible effects of large-scale warming are sought.

Numerical models of the climate system vary greatly in sophistication, spatial and temporal resolution and the way in which they simulate basic physical processes. They are all based on the fundamental equations of fluid dynamics and thermodynamics which formulate the conservation of energy, mass and momentum. The simplest models of the climate are one-dimensional: they consider only the vertical structure of the atmosphere over the globe. Most exclude simulation of the seasonal cycle. The most sophisticated, developed at the Geophysical Fluid Dynamics Laboratory at Princeton, is a three-dimensional seasonal model of the general circulation of the atmosphere which takes into account coupling of ocean and atmosphere and variations in ocean temperatures and sea ice extent. When this model was applied to the carbon dioxide problem, it predicted that a doubling of present-day carbon dioxide concentrations would lead to a rise in global annual mean surface temperature of 2 to 3°C with greater changes at high latitudes in winter. Evaporation would be enhanced significantly leading to an overall increase in precipitation, with changes again greatest in high latitudes. Although there have been other attempts to model the impact of enhanced levels of atmospheric carbon dioxide these results are considered to be the most reliable. Nevertheless, such numerical models are still at the developmental stage and many problems have yet to be satisfactorily solved. For computational reasons most models only simulate part of the ocean–atmosphere–cryosphere system. Cloud processes have yet to be modelled satisfactorily. These limitations provoke continuing doubts as to the accuracy of even these sophisticated models.

The spatial distribution of changes in temperature and precipitation associated with rising levels of atmospheric carbon dioxide are likely to be complex, and it is the spatial patterns of change that will determine the impact on society. A second approach provides insight into this facet of the carbon dioxide problem: the use of warm periods in the Earth's history as analogues for future prediction. These models have been based on indirect climatic data for the early post-glacial, a period believed to have been significantly warmer than today, and on instrumental data for notably warm and cold years during the present century.

It is still not certain that warming due to enhanced levels of carbon dioxide will in fact occur. It may well be that control systems

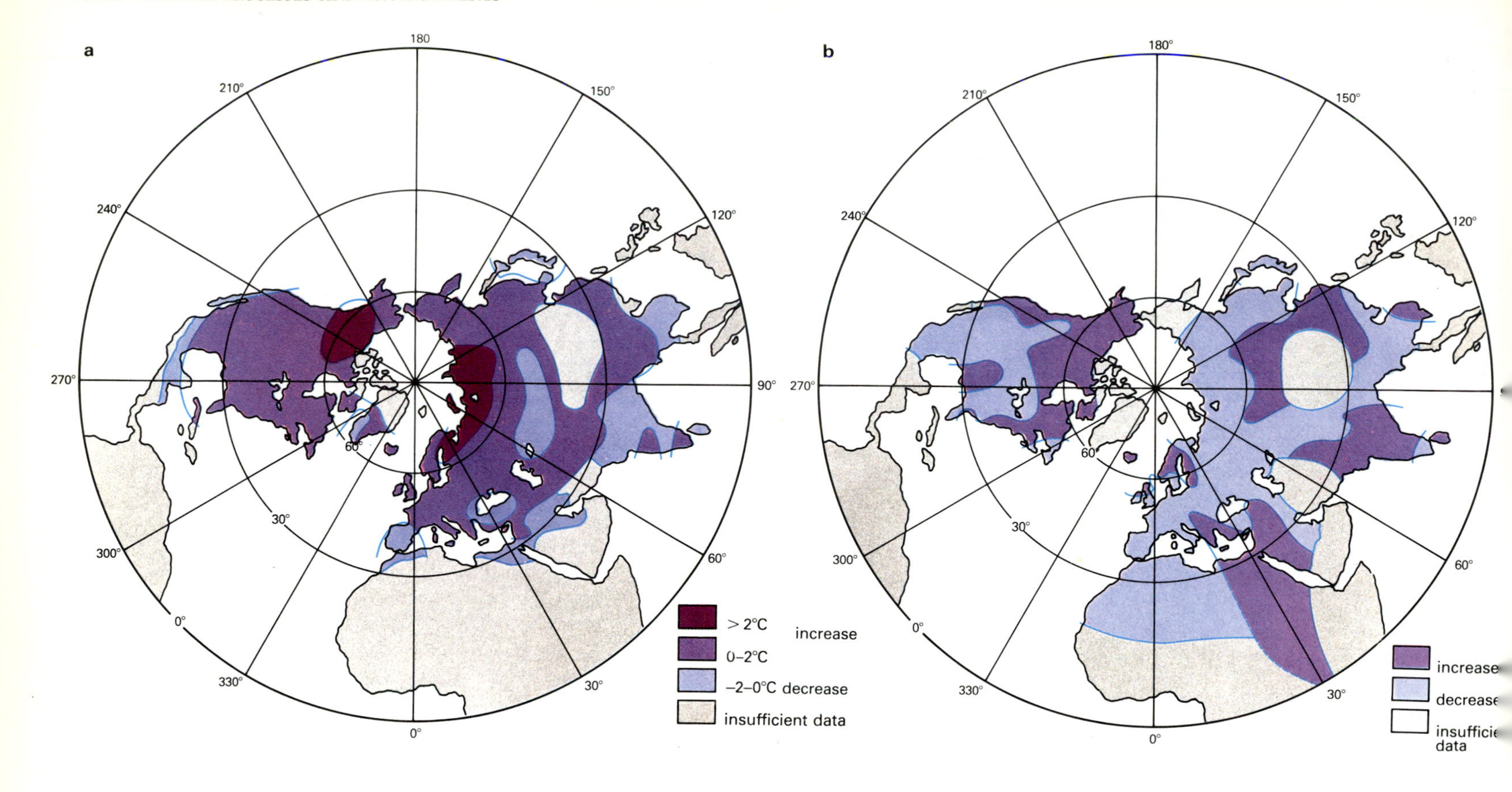

which have yet to be included in the climate models will counteract or amplify the warming induced by carbon dioxide. For example it has been suggested that the warming may initiate changes in cloud cover leading to cooling, thus offsetting the warming (an example of negative feedback). Natural mechanisms, such as changes in the frequency of explosive volcanic activity, may also produce cooling to counteract the anthropogenic warming. The carbon dioxide issue has however presented a challenge to climatologists and palaeoclimatologists, and has stimulated much research in recent years. The climatic history of the Earth is no longer a topic of pure academic interest; the understanding of the climate system of the past is being directly applied to the pressing problem of future climate.

18.16: *Spatial patterns of temperature* (**a**) *and precipitation* (**b**) *changes associated with a hemispheric temperature rise of about 0.6°C, which may take place by the early twenty-first century. Temperature rises are greater over northern Asia (> 2°C), but in some places there is cooling as the atmospheric circulation redistributes available heat. Precipitation increases are evident over India, but decreases over central and south-central USA, Europe and Russia could have important agricultural and social consequences.*

19 The oceans

Considering the size of the oceans and their importance to mankind it is remarkable how recently their study has developed. Oceanography is still a comparatively young science, and one of the most rapidly growing. Compared with the study of the land surface of the Earth, the oceans pose special problems to the scientist, particularly in terms of access to all but their surface layer. This chapter looks at some of the more important aspects of the ocean system, bearing in mind that the subject is a vast one and knowledge is accumulating rapidly, requiring the continuing reappraisal of existing theories.

The ocean basins

The topography of the oceans is more than a description of their limits, and even these are not fixed through time. The form of the ocean basins controls the physical and hence the chemical and biological processes taking place in them. The major topographic features of the ocean floor are the mid-Atlantic ridge, marking the boundary between the American and African plates, and the east Pacific rise with its extension south of Australia and through to the Indian Ocean, marking the southern and eastern boundaries of the Pacific and Indian plates (see Figure 10.17). On each side of these ridges, which are at a depth of about 3 km, the bottom slopes away to the great expanses of the ocean abyssal plains at between 4 and 5 km depth. Only at Iceland does the ridge reach the surface. The ridges form an effective barrier to the motion of the ocean bottom water, except possibly through valleys marking the sites of the numerous transform faults that traverse the ridge systems. The abyssal plains are punctuated by seamounts, some of which are isolated. Others, such as the Emperor seamounts in the northern Pacific Ocean, form chains. Most seamounts are of volcanic origin. Only a small proportion reach the ocean surface. Of the few of these still volcanically active, the Hawaiian Islands are well-known examples. These active undersea volcanoes occur away from plate boundaries in regions that are otherwise tectonically very stable.

The deepest parts of the oceans are at the trenches, where ocean crust is being subducted beneath less dense continental lithosphere, and are associated with regions of earthquake activity and andesite volcanism. The deepest point of the ocean is in the Mindanão trench at 11 515 m, close to the Philippine Islands (Figure 19.1). Although only forming 8.5 per cent of the sea floor the continental shelves (the shallow regions around the edges of the continents) are much more important economically and politically than the deep oceans. They vary in width from a few kilometres off the west coast of the USA to 1500 km in the East Indies. The offshore boundaries of the shelf are difficult to define, although the 100-fathom (~200 m) contour is often taken as the limit. On many continental margins there is a 'shelf-break', with a sharp change in gradient from the shelf (0.1–0.5°) to the continental slope (3–5°), which is cut by numerous canyons

19.1: *Areas, volumes and mean depths of the oceans and seas.*

Oceans and adjacent seas	Area (10^6 km^2)	Volume (10^6 km^3)	Mean depth (m)
Pacific	166 241	696 189	4 188
Asiatic Mediterranean	9 082	11 366	1 252
Bering Sea	2 261	3 373	1 492
Sea of Okhotsk	1 392	1 354	973
Yellow and East China Seas	1 202	327	272
Sea of Japan	1 013	1 690	1 667
Gulf of California	153	111	724
Total	181 344	714 410	3 940
Atlantic	86 557	323 369	3 736
American Mediterranean	4 357	9 427	2 164
Mediterranean	2 510	3 771	1 502
Black Sea	508	605	1 191
Baltic Sea	382	38	101
Total	94 314	337 210	3 575
Indian	73 427	284 340	3 872
Red Sea	453	244	538
Persian Gulf	238	24	100
Total	74 118	284 608	3 840
Arctic	9 485	12 615	1 330
Arctic Mediterranean	2 772	1 087	392
Total	12 257	13 702	1 117
Totals and means depths	362 033	1 349 929	3 729

Note
Mediterranean refers to generally land-locked seas bordering the ocean, particularly where the boundaries of these seas are not well-defined.

formed, it is thought, by infrequent high-speed turbidity currents (see Chapter 20) transporting material from the shelf to the abyssal plain. Large debris fan formations at the bases of some canyons support this idea, although current-meter measurements show that the residual flow of water is often in the up-canyon direction. This has proved an embarrassment: toxic and other undesirable waste has been dumped at the heads of these canyons in the belief that it would rapidly be removed down-canyon and transported to the abyssal plain. Down-canyon flows may be very occasional events initiated by intense surface storms or earthquake activity.

On a geological time scale the ocean basins must have changed their shape considerably because of continental drift, and the ocean current systems must have varied significantly with the juxtaposition of the continental masses, although the ocean currents depend as much on the atmospheric circulation as on the positions of the continents. Assuming that many of the features of atmospheric circulation have changed little with the movement of the continental masses the oceans of earlier times probably had similar major circulating current systems (called gyres) to those existing in the present Atlantic and Pacific Oceans. The critical oceanic width below which a gyre will not form is about 1500 km, one-third of the present width of the Atlantic. The Atlantic is a spreading ocean and would have reached this critical width sixty million years (Ma) ago. The ocean circulations, both deep and surface, did not take on precisely their present configuration until thirty Ma ago with the start of the Antarctic circumpolar current. This occurred when the unrestricted flow of this deep current became possible after the opening of both the Austral passage (between Australia and Antarctica) and the Drake passage (between Antarctica and South America).

Changes in ocean basin geometry can result in global changes in sea level, called eustatic changes. However, global sea-level changes often occur because of changes in seawater volume. Climatic change, with its effects on the volume of polar ice, can also cause variations in sea level. A lowering of global temperature will result in a transfer of freshwater from the oceans to the ice packs, lowering sea level and increasing the concentration of salts in the ocean. A climatic warming will have the opposite effect. Complete melting of the polar ice packs would result in the sea level rising 160 m above its present level. Local, as distinct from global, changes in sea level are also linked with continental drift and climatic change, and are caused by local isostatic adjustments of the Earth's crust to changing conditions. Thus Scandinavia is currently rising by ~5 mm per year as it continues to recover from the weight of the thick ice sheet that overlay the region 20 000 years ago, during the last Ice Age. Local changes, which are due to local rising or sinking of landmasses, and eustatic changes combine to give the observed alterations in sea level, and to make the determination of true mean (eustatic) sea-level changes very difficult. Indirect evidence for eustatic sea-level changes and resulting ocean current changes can be obtained from studies of the abundances in ocean-floor sediments of planktonic foraminifera. The latter live in the ocean surface waters, and their distribution gives evidence of current systems and the extent of the influence of polar ice in the past. This has been done for the north Atlantic at the time of the last Ice Age (the CLIMAP project), using over a hundred deep-sea cores to map out the surface temperature and circulation.

Chemical and physical properties of seawater

Of the ninety-two naturally occurring elements, nearly eighty have been detected as constituents of seawater. The majority are present in only minute concentrations, but the great volume of the oceans is such that the total amounts dissolved in seawater are enormous. Titanium and chromium, elements essential for hardening steel, are present in seawater in concentrations of only 1 and 0.3 nanogram per litre respectively, but their total amounts in the oceans are 1.3 and 0.4 million tonnes. The eleven major constituents (Figure 19.2) comprise 99.9 per cent of the total dissolved material and generally act conservatively—their concentrations are only changed by input or output processes at the ocean boundaries, such as dilution in coastal waters by river run-off, or by rainfall onto the ocean surface. The minor constituents are often non-conservative and may take part in biological reactions; silicon is removed from seawater by some organisms and is therefore not included as a major constituent, although fluorine with a lower concentration is. Neither oxygen nor nitrogen is considered a major element, both existing in seawater partly as dissolved gases and having very variable concentrations.

Most of the elements present in the oceans are in the form of dissolved ions such as sulphate or hydroxide, and a single element may exist in several forms; zinc, for example, is believed to occur as Zn^{2+}, $ZnOH^{+}$ and as $ZnCO_3$. Sodium ions (Na^{+}) and chloride ions (Cl^{-}) are the two commonest species, providing more than 90 per cent by weight of the dissolved constituents. The proportions of the major elements and ions are remarkably uniform throughout the world's seas and oceans. This 'constancy of composition' indicates that, on a geological time scale, the oceans are well mixed and that interchange of waters between oceans must be efficient. Exceptions to the constancy of composition rule occur locally in regions close to hydrothermal plumes on the spreading ridges, or near coasts because of anthropogenic pollution. Whereas the ratios of the major ion concentrations in seawater

Constituent	Total mass (tonnes)	Residence time (Ma)
Chlorine (Cl)	2.5×10^{16}	103
Sodium (Na)	1.4×10^{16}	70
Magnesium (Mg)	1.7×10^{15}	14
Potassium (K)	5.0×10^{14}	6.7
Calcium (Ca)	5.4×10^{14}	1.2
Silicon (Si)	2.6×10^{12}	0.02
Copper (Cu)	2.6×10^{9}	0.02
Zinc (Zn)	6.4×10^{9}	0.02
Cobalt (Co)	6.6×10^{7}	0.014
Manganese (Mn)	2.6×10^{8}	0.014
Iron (Fe)	2.6×10^{9}	0.00006

19.2: *The residence times of some of the dissolved constituents of seawater. The major elements have long residence times while the minor elements are generally removed more quickly. Those constituents with short residence times are generally very reactive in seawater and easily removed from solution*

are largely uniform the total concentration of dissolved material varies considerably. Freshwater from rivers dilutes seawater in estuaries and near coasts, while evaporation, particularly in the tropics beneath the subtropical high pressure cells, increases the concentrations of salts.

Geological evidence indicates that the composition of seawater has not varied significantly over the past 1000 Ma. Dissolved constituents are continually being added through the weathering of rocks and soils by rain water and eventually reach the oceans. To maintain an equilibrium these constituents must be removed from the oceans at a rate equal to that of their addition. If the oceans are, on average, in equilibrium then a residence time can be computed for each chemical constituent, representing the average time a molecule remains in the ocean before being removed:

$$\text{residence time} = \frac{\text{mass of constituent in the ocean}}{\text{rate of input or removal of that constituent}}$$

In general the shorter the residence time of a constituent, the more reactive are its dissolved species. Both sodium (Na) and chlorine (Cl) have very long residence times, indicating that the ionic species Na^+ and Cl^- are unreactive in solution, in contrast with the high reactivity of elemental Na and Cl.

The most important removal mechanism is authigenesis, involving inorganic reactions in seawater. Authigenesis takes several forms. First, seawater can become trapped as pore water in sediments or it can percolate through fissures in the ocean floor basalts in a 'reverse weathering', so that constituents such as Na^+ become parts of new minerals. Second, some reactions take place as sedimentary particles, eg clay minerals, sink from surface waters, adsorbing constituents, mainly potassium, onto their surfaces. Third, manganese nodules on the ocean floor are an authigenic sink for certain metals, particularly manganese and iron. Although manganese nodules consist primarily of the oxides and hydroxides of manganese and iron, their economic importance lies in the presence of copper, nickel and cobalt at concentrations high enough to justify the cost of recovery. They occur only in the surface sediments, and estimates based on deep-sea photographs put their global abundance at 10^{11} tonnes. The precipitation of copper, nickel, cobalt and manganese as nodules is believed to be the major removal mechanism for these metals, accounting for their short residence times, of the order of 15 000 years. The precise reason for the formation of nodules is not understood. Biogenesis, the process by which organic reactions use constituents of seawater, is dominant for a limited number of materials utilized by marine organisms to form skeletal material. The biogenic processes use dissolved SiO_2, Ca^{2+} and bicarbonate ions (HCO_3^-) to form calcium carbonate ($CaCO_3$) and calcium silicate ($CaSiO_3$). Other constituents such as phosphorus and nitrogen are also used by marine organisms, but in the formation of soft parts rather than skeletons. They are quickly recycled and returned to solution.

The amount of dissolved material, irrespective of constituents, is quantified as the 'salinity' and is expressed as the number of parts per thousand by weight of dissolved constituents in seawater. Ocean water has an average salinity of 34.5 parts per thousand ($^0/_{00}$). The accurate routine determination of salinity, a fundamental oceanographic property of seawater, has been a great problem to oceanographers. It cannot be determined by evaporating a weighed sample to dryness because some constituents are volatile, and must therefore be estimated indirectly. This can be done by titrating against chloride with silver nitrate. Salinity (S) is then given by $S = 1.80655 \times$ chlorinity. The constant 1.80655 is based on the principle of constancy of composition of seawater, relating chlorinity to the total amount of dissolved constituents.

This technique, used since the turn of the century, is now being replaced by faster and more reliable electrical conductivity methods. Titration methods have an error of $\pm 0.02^0/_{00}$ while conductivity techniques give accuracy better than $\pm 0.003^0/_{00}$. Instruments capable of rapid *in situ* salinity measurements are now routinely used for obtaining detailed salinity profiles down through thousands of metres of the oceans. Electrical conductivity of seawater is also a function of its temperature, and instrument development followed advances in electronics which allowed compensation to be made for this factor. Salinity is now defined in terms of electrical conductivity.

Relative density determines the depth at which a particular mass of seawater will be found: denser waters sink to the bottom of the oceans while less dense ones 'float' above them. The density of seawater is determined to a great extent by its salinity and temperature. An increase in salinity raises the density, while an increase in temperature decreases it. At the ocean surface meteorological factors are important in determining water densities: evaporation, precipitation and the total radiation balance play their parts. In the Mediterranean low river run-off and high evaporation rates give high surface salinities ($38.4^0/_{00}$), resulting in a high overall density of 1029 kg per cubic metre (although water temperatures are also high). In the Norwegian and Greenland seas the water temperatures are close to 0°C and the salinity is around $34.9^0/_{00}$; these conditions result in a water mass with a density that is less than that of Mediterranean water ($\sim$1028 kg/m^3). The effect of pressure on water density is only small, but must be considered when comparing seawater densities. Oceanographers use densities computed from *in situ* temperature and salinity but reduce all pressures to sea-level pressure (Figure 19.3). Sometimes in the analysis of deep and bottom waters, where temperatures and salinities vary little with depth, it is necessary to allow for the small effect of the adiabatic expansion of water, and a potential temperature (θ) is used in place of the *in situ* temperature.

The dissolved atmospheric gases oxygen (O_2), nitrogen (N_2) and carbon dioxide (CO_2) are important components of seawater. An equilibrium exists between the concentration of each and its partial pressure in the atmosphere. When the equilibrium is upset (if, for example, O_2 is depleted by the respiration of marine animals or the oxidation of decaying organic matter), gas is transferred across the air–sea interface until the balance is restored. Most of this transfer, and the subsequent downward mixing of gases, is accomplished by the action of storm waves, particularly when such waves break, causing 'white caps' and spray.

When the concentration of a gas in the atmosphere changes its equilibrium the concentration in seawater also changes. The concentration of atmospheric CO_2 has been steadily increasing due to anthropogenic factors. Some of this CO_2 is removed by absorption in surface waters, effectively slowing down the rate of atmospheric CO_2 increase. The deep-ocean circulation brings ocean waters to the surface on a time scale of hundreds of years, so the waters now being brought to the surface in regions of upwelling have low CO_2 concentrations. This is because the waters were in equilibrium with the low atmospheric CO_2 concentrations existing at the time they were last at the surface. Considerable amounts of atmospheric CO_2 can be absorbed by these waters. It is not clear for how long this process can continue before the overall CO_2 concentration of the oceans rises, making the latter less effective in removing CO_2.

The vertical structure of the oceans

The oceans can be divided into three layers: an upper wind-mixed layer, the thermocline and the uniform deep waters (Figure 19.4). Below the surface wind-mixed waters there is a rapid decrease in temperature, beginning between 100 and 300 m and continuing for a few hundred metres, from that of the surface waters to the average deep-water value of a few degrees centigrade. This sharp temperature gradient, or main thermocline, should strictly be called a pycnocline (density gradient) because it is often accompanied by a salinity gradient. Transfer of water across the thermocline is difficult because of the sharp density gradient where the water is stably stratified.

The surface layers are influenced by atmospheric conditions such as weather and climate. Weak thermoclines can exist temporarily in these surface waters because of the daily cycle of heating and cooling and seasonal fluctuations in radiation balance, precipitation, evaporation and wave activity at the surface. These weaker thermoclines are transient and may be destroyed by storm activity. The main thermocline represents the maximum depth to which surface-induced mixing of the water can occur. This is largely controlled by the intensity and frequency of the major winter storms that completely mix all the water above the main thermocline.

The vast bulk of the waters of the oceans lie beneath the main thermocline and have uniform temperature and salinity. More than 75 per cent of the ocean waters have potential temperature within the range -1 to 5°C and salinity (S) values between 34.4 and $35.0^{0}/_{00}$. Since the greater part of seawater has highly uniform characteristics, great importance is attached by oceanographers to the accurate measurement of temperature and salinity, in order to show any small change that would indicate the source regions and motions of these deep waters. The highest ocean salinity, $37.5^{0}/_{00}$, occurs in the surface waters of the northern subtropical Atlantic, while the highest surface temperature, 32°C, is found in the Persian Gulf. Even greater extremes exist, particularly at the bottom of the Red Sea where there are 'pools' of dense brine with salinity of $270^{0}/_{00}$ and temperatures in excess of 55°C.

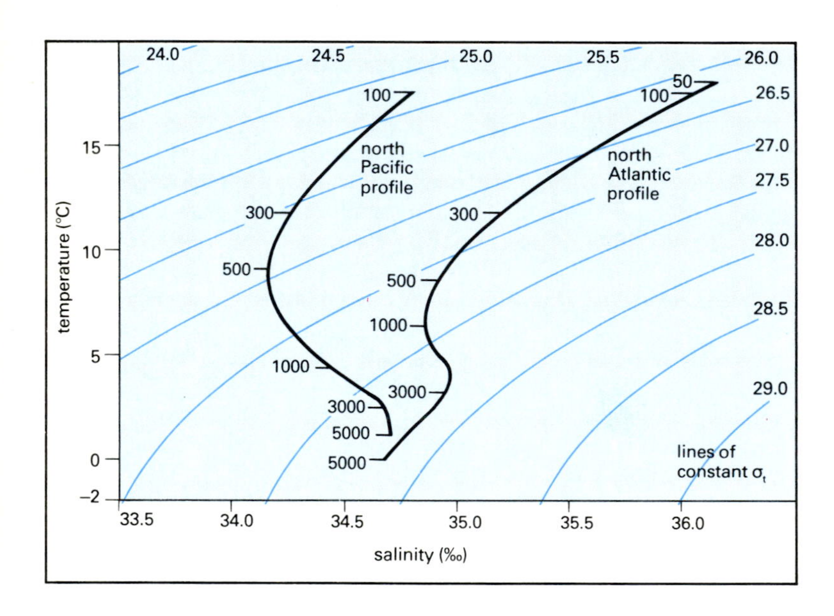

19.3: Both temperature and salinity are important in controlling the density of seawater. Temperature (t) and salinity (S) data are frequently plotted by a diagram of the type shown here. The curved blue lines are isopycnals—lines of constant density (σ_t) expressed as σ_t = density in $kg/m^3 - 1000$. Profiles of water columns from the central north Atlantic and the central north Pacific are shown with depths marked on the individual profiles. The density increases with depth, indicating that the water columns are stable. Towards the bottom of the Atlantic profile the density appears uniform over the bottom 2000 m. However, if adiabatic effects are included, then these deep waters are found to be stable, with the densest waters closest to the ocean bottom.

Light and sound in the oceans

Light penetrates only a relatively short distance into the oceans. As with most other regions of the electromagnetic spectrum, eg heat and radio waves, its energy is rapidly absorbed by water molecules. Even in the clearest of waters found in the subtropical oceans, light penetrates only a few hundred metres, the shorter wavelength blue penetrating further than the red (Figure 19.5). The blue colour of clear ocean water is due to a greater proportion of blue light being scattered and arriving back at the surface without being absorbed. The ocean acquires a greenish hue where the water contains phytoplankton (which possess green chlorophyll), and overall light penetration decreases. The yellowish-brown colours encountered in some coastal waters are produced by industrial and domestic pollutants, or result from the lifting of mud and sands into suspension in shallow waters by tidal currents and wave action. In such waters the turbidity, a measure of the concentration of suspended material, is so high that light penetration at all wavelengths is no more than a metre. Light has only limited use as a surveying tool in oceanography because it is so readily absorbed and scattered. It is used primarily for localized photography of the sea bed from submersibles or remote-controlled cameras lowered to near the sea bed, and also in remote-sensing applications (see Chapter 24) for the examination of surface water turbidity and mixing.

In the oceans sound is used to perform many of the tasks for which light and radio waves are used in the atmosphere. It is used to obtain large-scale 'pictures' of the ocean floor and of the stratigraphy of subsurface sediments, to locate shoals of fish and to communicate with underwater instruments. There are also many military applications concerning submarine detection. Sound waves with frequencies between 10 kHz and 100 kHz (wavelengths of 15 to 1.5 cm) are commonly used for echo-sounding. Sound attenuation in water increases with frequency, and so low-frequency sound penetrates further and less energy is required. However, the angular spread Θ of the sound beam leaving an echo-sounder transducer is related to the diameter of the transducer (d) and to the sound frequency (f) as follows: $\sin \Theta \propto 1/df$. Low-frequency sound waves, while penetrating through water more effectively than high-frequency ones, produce a broader beam and thus give poorer resolution of features on the sea bed.

Sound travels at ~1500 m per second in seawater, but its speed is a function of water temperature, salinity and pressure, and differences in these properties between layers of seawater can cause sound waves to be refracted. In a layer around the main thermocline, called the 'sound channel' (Figure 19.6), sound waves can become trapped and the sound energy can travel thousands of kilometres. The sound channel is used for the transmission of data from long-range floats (called SOFAR floats) which are released to drift freely at predetermined depths in the oceans, tracing the motions of the mid-depth ocean waters. The sound channel also has military importance, allowing submarines to 'hide' by floating below the channel where much of the noise they make is refracted away horizontally and does not reach the ocean surface.

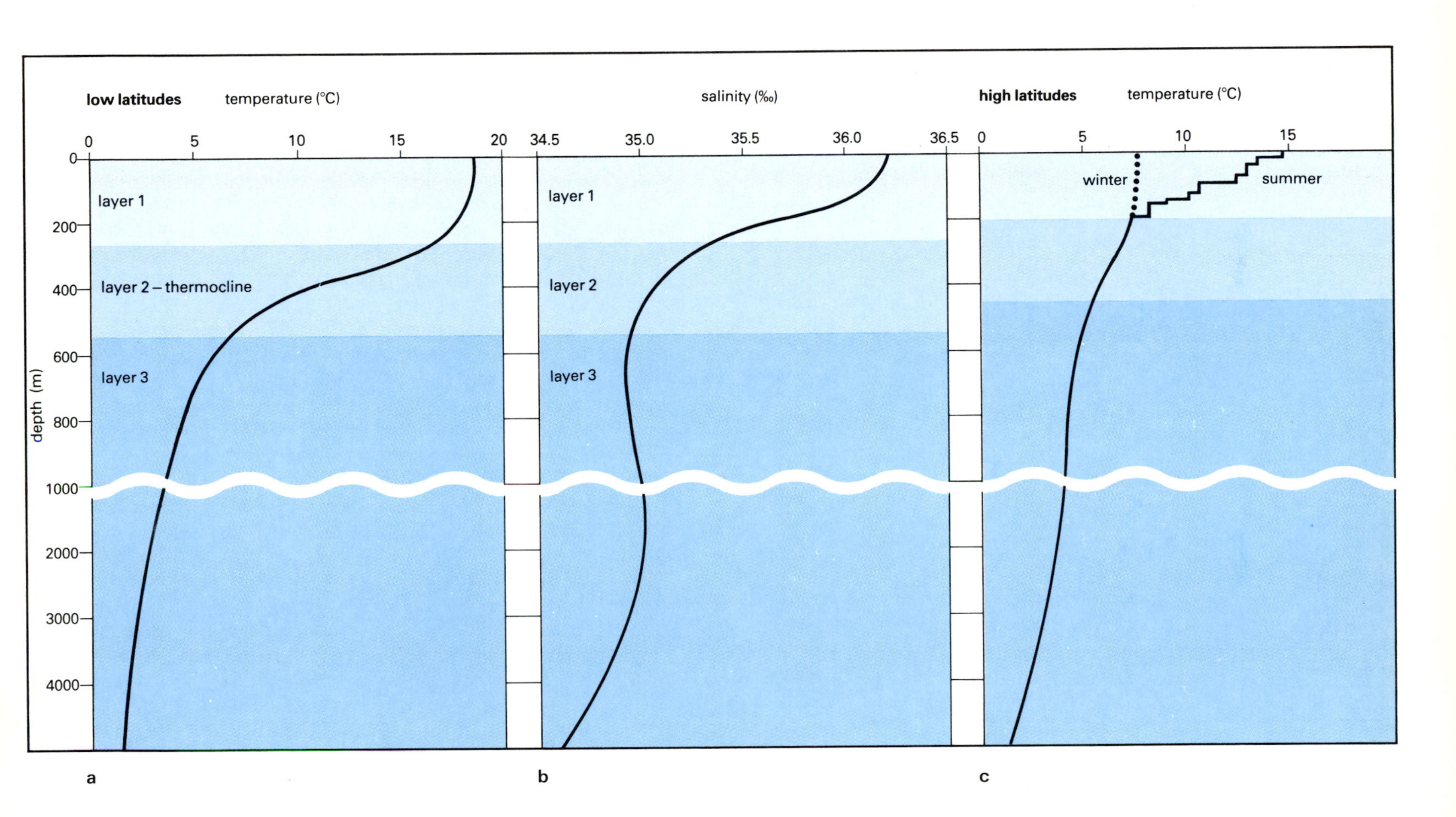

19.4: *The temperature and salinity profiles through a typical low-latitude water mass in* (**a**) *and* (**b**) *show the upper wind-mixed layer, the thermocline (layer 2) and the deep waters of layer 3, where temperature and salinity vary little with depth. The shape of the profiles varies with latitude, the thermocline being sharpest in the tropics, while in polar regions surface cooling can reduce the entire water column to a uniform temperature, allowing mixing throughout the column. In the high latitudes* (**c**), *where summer and winter weather conditions contrast sharply, minor thermoclines can develop during the calmer summer months. These seasonal features are destroyed by the deep mixing events of winter storms.*

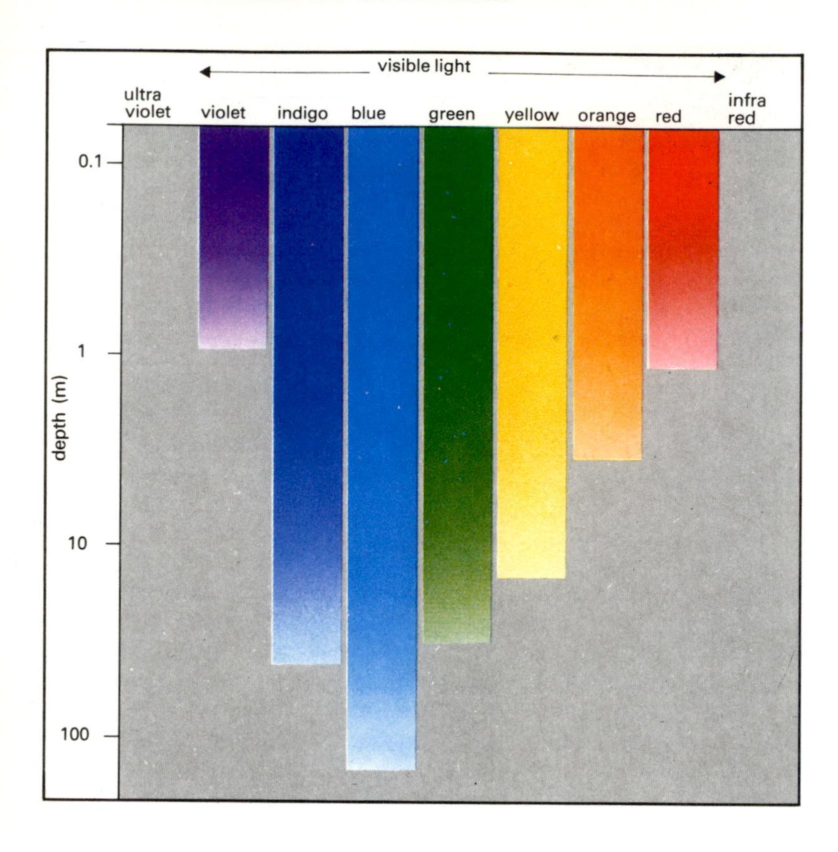

19.5: *Penetration below the ocean surface by visible light. In clear ocean water, blue light can penetrate more than 100 m but red and ultraviolet light are rapidly absorbed close to the surface. In turbid coastal waters light penetration can be very much less than this.*

Motions of the ocean waters

Surface currents

The major motions of ocean waters, including tides and currents, are due directly or indirectly to the Earth's radiation balance and rotation, and to the relative motions of the Earth, Moon and Sun. The mean atmospheric wind circulation patterns, described in Chapter 17, are the main driving force of the oceans' surface currents. There are relatively few features of ocean circulation that respond to wind variations on a time scale of a few days, corresponding to what we would term 'weather'. Some ocean currents respond to seasonal wind changes, eg the generally southwards-flowing Somali current which reverses completely in direction during May and then reverts to normal in August, responding to the monsoon winds over the Indian Ocean. A few currents, such as El Niño (see below), are due to intermittent climatic features. It is, however, the long-term winds that have set up the major surface currents.

It is important to remember that the direction of a wind is that from which it blows, whereas the direction of a current is that towards which the current is flowing. Thus the north-east and south-east trade winds, on the equatorial sides of the subtropical anticyclones, drive the westward-flowing equatorial currents. Poleward of these anticyclones the air tends to turn westwards to produce the mid-latitude westerly winds, which drive the major eastwards-flowing currents, such as the north Atlantic drift, the north Pacific current and the Antarctic circumpolar current. With the exception of the Antarctic circumpolar current in the southern oceans the continents block the flow of the major zonal currents and result in the establishment of a series of oceanic gyres, clockwise in the north Atlantic and north Pacific Oceans and anticlockwise in the south Atlantic, south Pacific and the Indian Oceans (Figure 19.7). These gyres are centred at around 30°N and 30°S, approximately at the same latitudes as the subtropical anticyclones in the atmosphere, and have much more intense currents on their western than on their eastern sides. The intensification of the

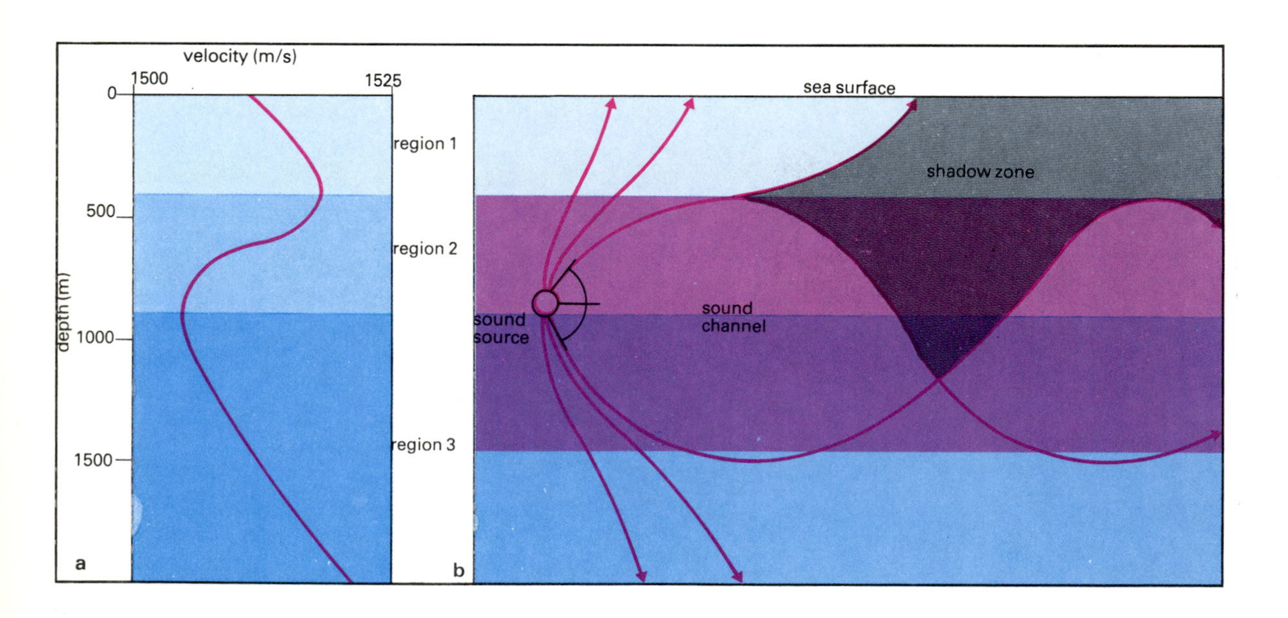

19.6: *A typical profile of the variation in the velocity of sound in seawater as a function of depth* (**a**). *The velocity increases with increasing pressure, but is also affected by changes in salinity and temperatures so that sound waves are refracted as they propagate through seawater* (**b**). *Sound waves from a source in region 2 or the upper portion of region 3 spread out, and some of the energy becomes trapped in a region known as the sound channel.*

western boundary currents of the gyres is due to the rotation of the Earth and the requirement that absolute vorticity (angular momentum) be conserved. The western boundary currents, the Gulf Stream, the Kiro Shio, the Brazil current and the East Australian current, are termed warm currents because they transport much heat towards the poles, helping to restore the imbalance between equatorial and polar regions in the net thermal radiation received from the Sun. Some recent satellite data indicate that these currents may be transporting 40 per cent of the heat energy required to maintain equilibrium, but 25 per cent is a more probable long-term estimate. Western boundary currents are the swiftest and most consistent currents in the oceans, contrasting sharply with the broad sluggish movement towards the equator of the rest of the ocean gyres: the Gulf Stream transports an estimated 150 Sv (1 Sverdrup (Sv) = 10^6 m^3 per second) of seawater northwards with speeds exceeding 1.5 m per second (3 knots).

The currents on the eastern sides of the gyres, flowing towards the equator, are cold currents (cold compared with average ocean water at a given latitude), eg the California, Humboldt, Canaries and Benguela currents. Close to the continental boundaries an additional factor called upwelling decreases the water temperature further. Air and water moving on the surface of the Earth experience an apparent force called the Coriolis force which causes the fluid to turn to the right in the northern hemisphere and to the left in the southern hemisphere. Thus in both northern and southern hemispheres the currents flowing towards the equator are turned away from the coastline. To take the place of the surface water flowing away from the continental boundary cold water flows up from below the thermocline. Upwelling velocities can be as great as 25 m per day. The regions of upwelling, off Peru and along the west coast of Africa, are biologically some of the most productive areas in the world because the cold upwelling water has a high content of nutrients essential for rapid phytoplankton growth.

Away from land the wind has a complex effect on the surface ocean waters. In addition to the formation of waves, the frictional drag of the wind pulls a thin surface layer of water in the direction of the wind. However, as soon as the surface water begins to move, the Coriolis force acts to turn the water to the right (in the northern hemisphere) as much as it can, while the wind still drags it forward. Each

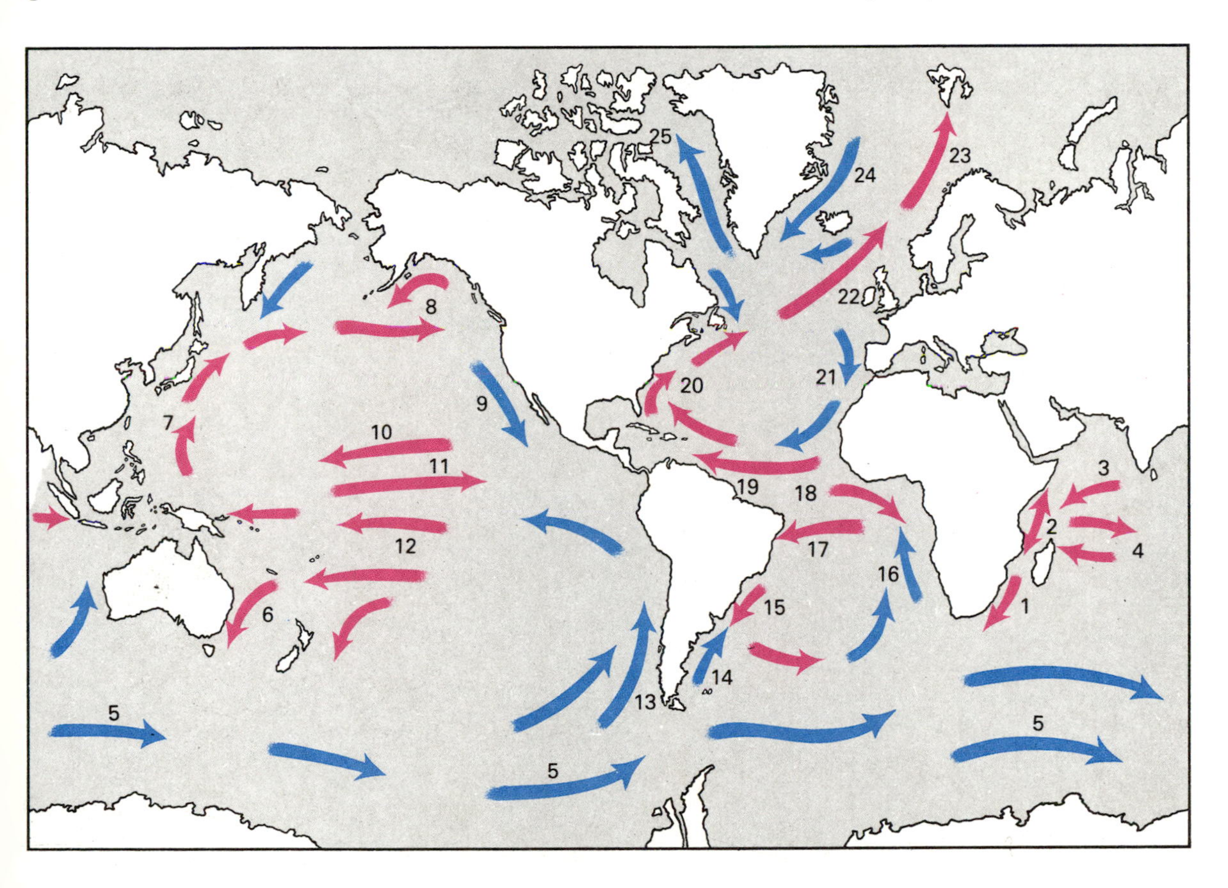

1 Agulhas
2 Somali
3 North Equatorial
4 South Equatorial
5 Antarctic Circumpolar (West Wind Drift
6 West Austrailian
7 Kiro Shio
8 North Pacific/Alaska
9 California
10 North Equatorial
11 Equatorial Counter
12 South Equatorial
13 Peru
14 Falkland
15 Brazil
16 Benguela
17 South Equatorial
18 Equatorial Counter
19 North Equatorial
20 Gulf Stream
21 Canaries
22 North Atlantic Drift
23 Norwegian
24 East Greenland
25 West Greenland

19.7: *In the oceans the surface currents form circulating systems or gyres, centred near the positions of the subtropical high pressure cells. Warm currents flow polewards on the western sides of the oceans in relatively narrow bands of fast-moving water (the Gulf Stream and the Kiro Shio), while the cold currents are generally much broader and slower moving. Near the equator the south equatorial currents cross the equator to several degrees north. The equatorial counter-currents are generally weak. In the Indian Ocean the circulation is more complex because of the restricted area of this ocean and because of the seasonal wind reversal of the Indian monsoon.*

thin sheet of water below also begins to move because it is frictionally coupled to the layer above, and it also tries to turn to the right. The vertical current profile thus generated is known as the Ekman spiral (Figure 19.8), where the surface current moves at 45° to the right (left in the southern hemisphere) of the wind direction at 2–3 per cent of the wind speed. Lower layers move more slowly and flow more to the right than layers above. At a depth called the Ekman depth, the water flows in the opposite direction to that of the surface current, at a speed of 0.043 times the surface current speed. The net mass transport (the average transport over the complete spiral, taking both speed and direction into account) is at 90° to the right of the wind direction. Such wind-driven flows only develop and continue under steady wind conditions in regions free from obstructions such as landmasses, but are difficult to observe even in the open ocean. In the open ocean, this Ekman current system can continue only as long as no pressure gradients build up to oppose the flow. In the case of the major ocean gyres Ekman transport will try to move water inwards towards the gyre centres, causing it to pile up and to develop a pressure gradient force opposing the Ekman flow. The Coriolis force and the pressure gradient balance each other and the flow continues around the gyre.

A pressure gradient force may also be generated by water piling up against a land mass; the current flow then turns along the water slope, the Coriolis force again balancing the pressure gradient force. Such a current is then known as a geostrophic current. The major ocean currents are of this kind. The slope of the water Θ across the current is given by: $\tan \Theta = -2\omega \sin \phi \, V/g$, where ω is the angular velocity of the Earth, ϕ is the latitude, g the acceleration due to gravity, and V is the current speed. Measurement of the ocean surface slope theoretically offers an indirect method of measuring current speeds, but even for a strong current such as the Gulf Stream water slopes are minute, of the order of 10 mm per kilometre width of current.

Close to the equator the ocean current system becomes complex because the south-east trade winds cross the equator. The south equatorial currents extend across the equator to 3–5°N, while the north equatorial currents are found at about 10–15°N. Between the two currents is the region of the doldrums, the equatorial trough, where winds are very light and variable, and where there is a weak equatorial counter-current. The counter-currents are driven by the reverse in direction of the Ekman transport of the south-east trades as they cross the equator and by a pressure gradient force caused by water from the north and south equatorial currents piling up against the continental margins. In the Pacific Ocean, over a period of several months, the irregular and unpredictable relaxation in the strength of the trade winds allows the Pacific equatorial counter-current to increase in strength and to flow back to the coast of South America. The warm current then flows southwards down the Peruvian coast, where it displaces the region of upwelling cold water which is forced offshore. When this occurs the anchovy fishery in the region is destroyed, with damaging economic consequences. The warm inshore coastal current is accompanied by anomalous weather conditions along the Peruvian coast: storms and torrential rains occur in the desert-like coastal region, which is usually in the rain shadow of the Andes. This entire phenomenon is called El Niño (Spanish: the Christ Child), because it occurs around Christmas time. It is one of the clearest indications of the delicate balance between the atmosphere and the ocean.

Deep currents

While most surface currents are directly wind-driven there is little direct linkage between winds and the deep ocean currents because of the thermocline. Deep ocean currents are driven by density differences, denser water flowing beneath less dense waters, and the deep ocean circulation is termed a thermo-haline circulation because the water density is determined mainly by the water temperature and salinity. It is very difficult to measure the speeds and directions of deep currents. Until the development of reliable deep-water internally recording current meters it was nearly impossible, and many indirect techniques have been used to estimate deep-current velocities. One method was to use the concept of water masses, whereby regions of water with approximately the same temperature–salinity characteristics have the same history. On this basis the path of a mass of water can be traced by looking at the temperatures and salinities of ocean waters. A water mass may have a single source region or it may be formed by the mixing of two or more identifiable water masses.

Water masses are classified by their source region (eg the north Atlantic, Mediterranean and Antarctic) and the depth at which they

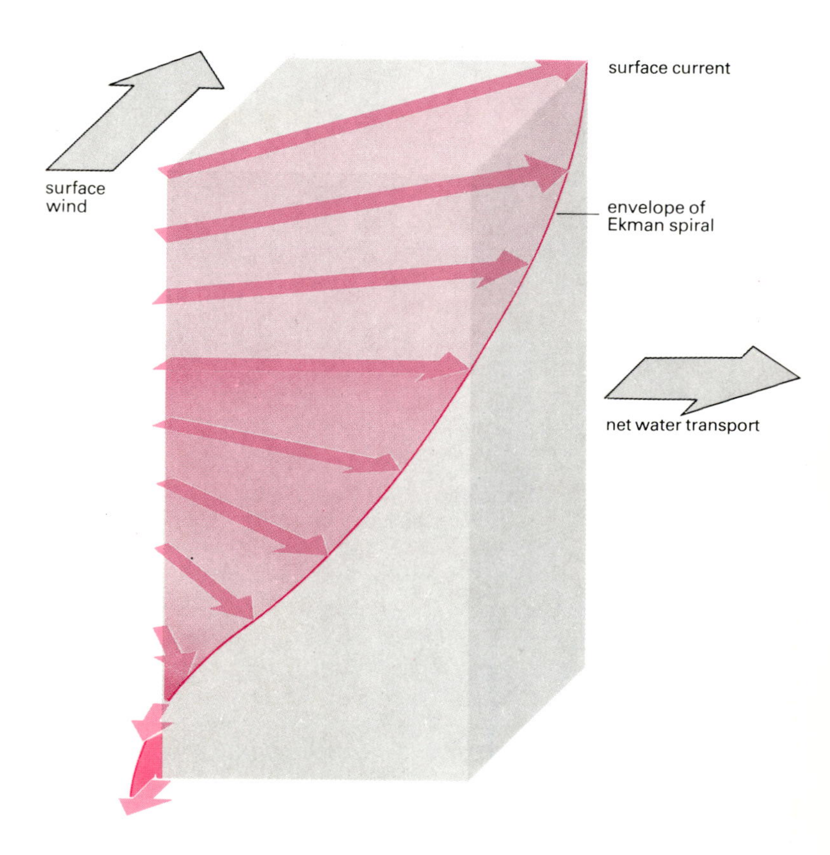

19.8: *A northern hemisphere Ekman spiral in which a surface current moving at 45° to the right of the direction of the surface wind induces subsurface currents successively farther to the right and decreasing in speed with increasing depth. At the Ekman depth the current speed has decreased to less than 5 per cent of that at the surface and the net transport of water due to the wind is at 90° to the right of the wind direction.*

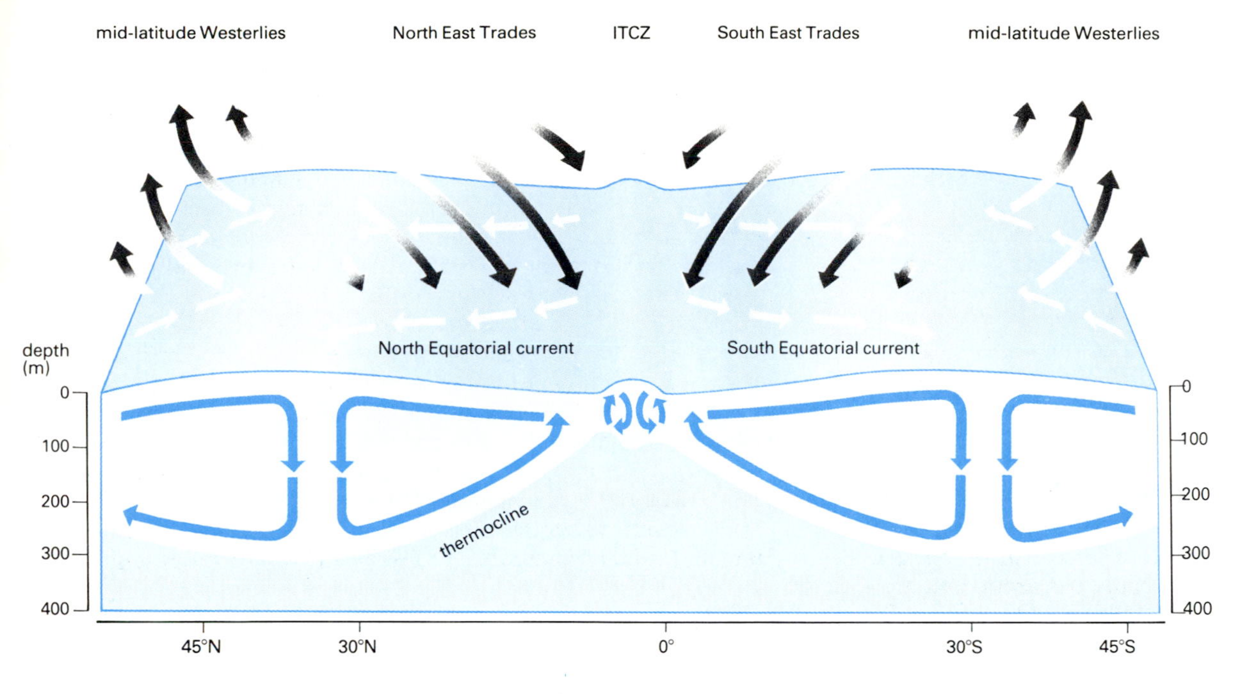

19.9: *The winds blowing over the oceans, the trade winds and the mid-latitude westerlies, cause an Ekman transport of the water in the directions shown by the open arrows attached to the wind symbols. The blue-coloured winds are blowing out of the page, while the red-coloured winds are blowing into the page. The transport of waters into the centre of the gyres causes an upward doming of the water surface, greatly exaggerated in this diagram, and a downward depression of the main thermocline. The slope of the sea surface results in an outward pressure gradient which opposes the Ekman transport of water by the winds. The water transport that does occur sets up a slow circulation as shown by the arrows.*

19.10: *The warm current of the Gulf Stream shown here in a computer-enhanced image from infrared data from the polar orbiting satellite NOAA-5 taken on 23 May 1978. The false colours represent particular temperature ranges, as follows: white, less than 3°C (clouds); blue, 3–13° C (coastal waters and continental shelf); orange to red, 13–17°C (outer Gulf Stream); violet to black, 19–27+°C (core of Gulf Stream and some land areas).*

are predominantly found (surface, intermediate, deep and bottom). In the northern hemisphere the major source regions of water below the thermocline are in the Norwegian and Greenland Seas, where water brought northwards in the north Atlantic drift circulates in a counter-clockwise gyre. This circulating water cools rapidly, by long-wave radiation emission, to -1°C (surface ocean water freezes at -1.9°C), sinks and flows southwards. On its way it spills over irregular barrier sills at depths of 400 to 800 m between Scotland and Iceland and between Iceland and Greenland. During the passage of this very cold water over the barrier sills it mixes with overlying waters to form north Atlantic deep water (NADW), a water mass with temperature and salinity characteristics of 2°C and $34.95^0/_{00}$ which can be identified in the Atlantic Ocean, after some mixing with surrounding water, as far south as 50°S.

Another major source of deep water is around the Antarctic continental shelf, where very dense water forms, again by radiative cooling and increase of salinity caused by the formation of sea ice. This water is not as dense as NADW at formation. However, it has no barrier sill to cross and less mixing with overlying water occurs as it flows down the Antarctic continental slope to become Antarctic bottom water (AABW). It has temperature and salinity characteristics of 0.5°C and $34.7^0/_{00}$. The third source of deep ocean waters is at the convergence of surface waters in the Antarctic circumpolar current, called the Antarctic convergence. When waters converge horizontally at the surface, continuity imposes a downward flow. The Antarctic convergence is the source region for the main water mass, Antarctic intermediate water (AAIW), overlying NADW and AABW in the oceans (4°C, $34.4^0/_{00}$). The largest water mass, 40 per cent of the total ocean volume, is the Pacific and Indian Ocean common water (1.5°C, $34.7^0/_{00}$), formed by the mixing of NADW, AAIW and AABW. The paths taken by the circulating deep ocean waters are not fully known.

Chemical tracers in the oceans

Oceanographers have traditionally used temperature and salinity, together with a few oxygen measurements, to make their deductions concerning the large-scale deep ocean circulation and mixing processes. In this respect oxygen is not an ideal tracer because it is non-conservative, unlike temperature and salinity which, away from the surface and hydrothermal vents, are conserved.

A continuing programme of systematic chemical sampling of the oceans is now being undertaken, mainly along north–south tracks from the Arctic to the Antarctic, using modern techniques to obtain water from different depths. In this way a vast new bank of chemical data for studying many aspects of the oceans is being accumulated. This programme, the Geochemical Ocean Sections Programme (GEOSECS), was designed and initiated to obtain detailed chemical analyses of ocean waters. Each sample collected is subjected to standardized analyses in order to establish the concentrations of twenty-three chemical and fifteen isotopic species, covering most of the major and some of the more important minor constituents. This programme gives a new dimension to studies of ocean movements. Stable chemical species can be used to examine mixing processes and oceanic circulation paths, while radioactive isotopes can be used as clocks to determine current speeds and mixing rates. An example of this is provided by the isotope carbon-14 (^{14}C), which is produced in the upper atmosphere by cosmic bombardment, with the result that the CO_2 in the atmosphere contains a small proportion of ^{14}C. This isotope has a half-life of about 5750 years, and the ratio of ^{14}C to the abundant ^{12}C isotope, measured in the CO_2 dissolved in a deep-sea water sample, can be used to date it. (The 'age' of the sample is the length of time since, as surface water above the thermocline, it has been taking up atmospheric CO_2.) ^{14}C data suggest a time scale of between 200 and 1000 years for deep-ocean circulation. On a shorter time scale tritium (^{3}H), the heaviest isotope of hydrogen, was put into the atmosphere in large quantities by nuclear weapons testing between 1955 and the early 1960s. The GEOSECS Atlantic section for tritium shows the extent to which cold surface waters in the Norwegian and Greenland seas have sunk to form bottom waters. Bomb-tritiated seawater can be detected in bottom waters at 40°N, indicating a time scale of ten years for the sinking of the surface water.

Tides and waves

Tides and waves are the most frequently encountered of ocean phenomena. Both are periodic rises and falls of the sea surface, but they have entirely different origins. Tides are generated by forces at the Earth's surface which occur because of the relative motions of the Earth, Moon and Sun, whereas waves are generated by wind.

Over the Earth as a whole these forces balance (the Earth–Moon–Sun system is in equilibrium), but small horizontal forces, caused by localized imbalance at the Earth's surface, about 10^{-7} times the force of gravity in magnitude, pull the waters to form 'bulges' or regions of high water. These forces are directed such that they tend to produce high waters at positions where the Moon (or the Sun) is overhead (at zenith) or diametrically opposite (nadir). The tides due to the Earth–Sun system (solar tides) are 0.47 times the size of the tides due to the Earth–Moon system (lunar tides). Due to the Earth's rotation the solar tide has a period of exactly twelve hours, while the lunar tide, because of the monthly rotation of the Moon about the Earth, has a period of 12.41 hours. The tides, called the principal lunar and solar semidiurnal (twice daily) tides, move in and out of step causing a cycle of spring tides (above average high waters) and neap tides (below average high waters) every fourteen days. Other tides also occur due to the complex relative motions of the Earth, Sun and Moon, which can modify the simple semidiurnal tides.

The tide-generating forces should always produce high waters when the Moon is overhead, and the top of the tidal wave should follow the Moon's zenith position around the Earth. However, the actual response of the ocean water is somewhat different, as the tidal waves can only move at a speed c defined by the ocean depth d, such that $c^2 = gd$, where g is the acceleration due to gravity. A tidal wave at the equator would need to travel at about 1500 km per hour to remain in a position with the Moon at zenith, and this would require a water depth of 17–18 km, the average water depth being less than 4 km. Additionally, the direct progress of tidal waves is blocked by the continents, and the Coriolis force turns the tidal waves into rotating systems. As a result the tidal response to the causative forces is rather complex.

In the open ocean tides are only a matter of a few hundreds of millimetres in amplitude and tidal currents are small, but as the tidal waves propagate onto the continental shelves their amplitudes and current speeds increase and the effects of tides can be considerable. Bottom topography and coastal shape are important in determining the effects of tides, particularly in estuaries and semi-enclosed seas. Such bodies of water each have their own natural periods of oscillation, and they seiche (slop backwards and forwards) at these periods if they are disturbed. When the natural period of oscillation is close to the period of one of the major tidal cycles, then the sea or estuary will resonate and large tidal ranges will occur. The Mediterranean Sea has only a very small tidal range because its natural period is far from any tidal period, while very large tides occur in the Bay of Fundy, along the Brittany coast of France and in the Bristol Channel, where natural periods of oscillation are around twelve hours. In regions of high tidal range the possibility exists of economic conversion of tidal energy into electrical power by impounding the water behind barrages and using reversible turbines to extract potential energy from it as it flows in or out. There is an operational tidal power station at La Rance in northern France, where the average tidal height is 8.5 m. The disadvantage of this particular type of station is that power is produced in accordance with the tidal cycle, and not necessarily when it is most needed.

The waves generated by the wind blowing over the ocean surface offer another possible source of renewable energy, and several prototype devices capable of harnessing water energy are now operating. Power from waves is intermittent, although greatest during winter months when electrical power requirements are also greatest. Great technical and practical problems still need to be overcome to produce a station capable of generating a significant amount of electricity. Such a station must be able to extract energy from moderate seas with heights of 1 to 2 m, and to survive storms in which waves are ten times higher.

Waves obtain their energy from the wind, and the height of the waves increases with wind strength, the time for which the wind has blown to produce the waves and the fetch (the distance of ocean over which the wind has been blowing). Ocean waves are a complex of many different wave trains of varying wave heights and wave lengths L, each wave train moving at its own speed c such that $c^2 = gL/2\pi$. Longer waves thus move faster than shorter waves. The superposition of these wave trains results in a confused sea, with waves moving generally in the direction of the wind but with no clearly defined wave height or wave length. When the crests of several wave trains come together a very large wave can occur, but such a wave only exists for a short period, the wave trains moving at different speeds and very soon getting out of step. This is the origin of the idea that every seventh wave is higher than the others. However, wave heights are determined by an essentially random process, and a high wave can occur on consecutive waves or after a dozen or so.

Prediction of individual wave heights is not possible, but some statistics describing the sea can be calculated, knowing wind speed, the fetch and the duration for which the wind has been blowing. Of practical interest is the significant wave height H_s, a measure of the total energy E of the waves: $E = \frac{1}{8}\rho\, gH_s^2$, where ρ is the water density. H is important to mariners and engineers where operations can only

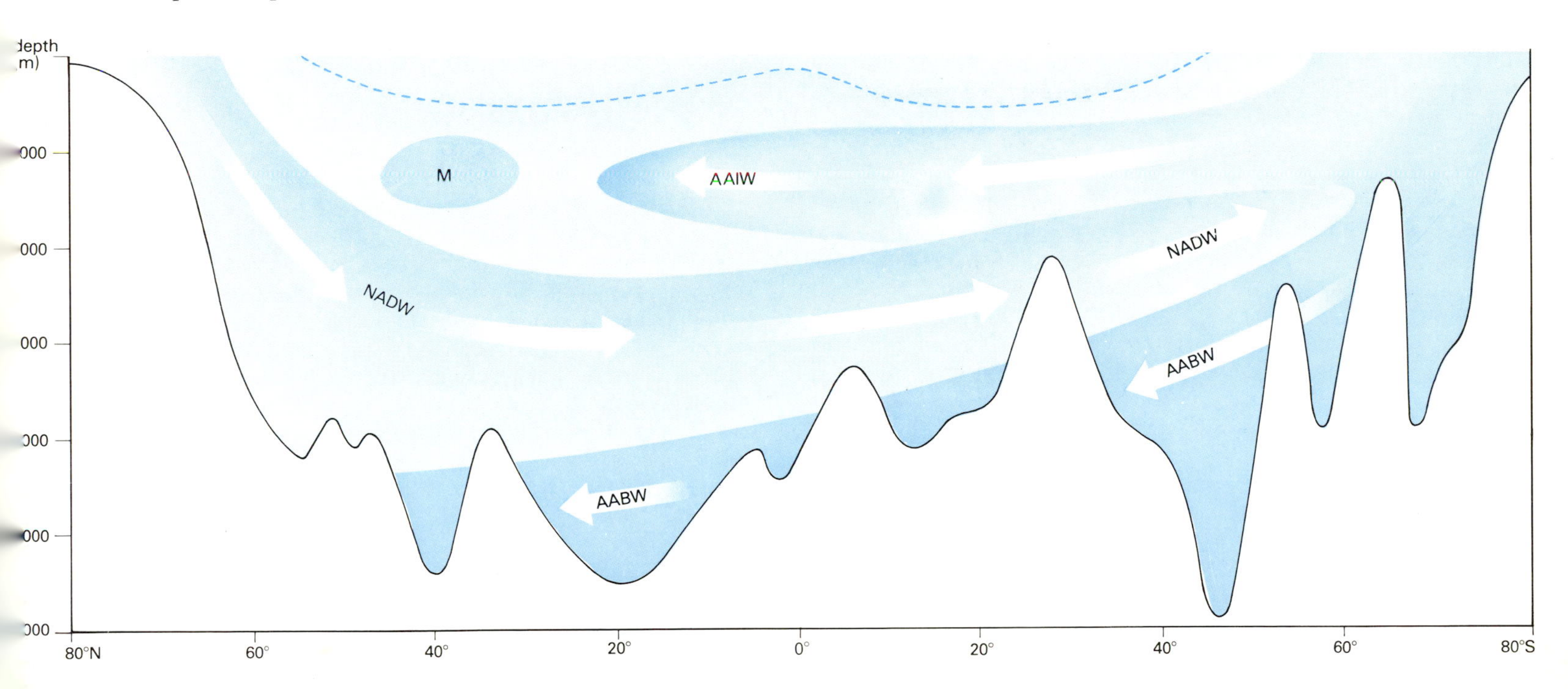

19.11: *The major water masses of the Atlantic Ocean. The densest waters (AABW) flow close to the bottom while above it NADW flows southwards. Mediterranean water enters the Atlantic over a sill near Gibraltar and sinks until it reaches its own density level (~1500 m). Mixing occurs between adjacent water masses, continually modifying their distinct temperature and salinity characteristics. The dotted line is the approximate level of the 10°C isotherm, marking the depth of the main thermocline.*

take place while wave heights remain below a certain threshold, eg for offshore drilling and pipe laying. Also of practical importance is wave height H_{max}, the maximum wave height likely to occur during a given period, eg during the expected production lifetime of an oil rig, which may be as long as twenty-five years. H_{max} is needed to calculate design dimensions and the strengths of the structures. Many oil-producing sites are at inhospitable locations for which few or no wave data are available, and estimations of H_{max} have to be made on the bases of scanty evidence and assuming, with little justification, that the climate will remain constant.

Giant waves in the oceans are rare, and those reported are usually overestimated by human observers. The largest wave recorded by an instrument was 22 m from crest to trough, encountered by an Atlantic Ocean weather ship. Large waves are usually very long. In general the higher a wave, the less steep it is. In some regions of the world's oceans currents can act to focus wave energy and give a high incidence of freak waves. This seems to occur particularly in the southwards-flowing Agulhas current off South Africa, where waves from storms in the southern oceans travel northwards. Several large bulk-carrying ships have been sunk or severely damaged by giant waves while steaming southwards and using the Agulhas current to speed their passage.

The most damaging waves of all are the tsunami waves, often referred to as tidal waves, incorrectly as they have no connection whatever with tides. Tsunami waves are usually caused by submarine earthquakes, eruptions or under-sea landslides. They travel very rapidly across the oceans with speeds of more than 200 km per hour. On the open oceans their height is only a few tenths of a metre and, when this is combined with a wavelength of perhaps 20 km, they are imperceptible to the mariner. Only when the tsunami wave encounters shallow water does it increase in height to tens of metres and surge inland, causing great destruction and loss of life. A tsunami can have several crests, separated by many minutes, and between crests the waters can recede like a very low tide before rushing in again. In the great tsunami that engulfed Lisbon in 1755 a rapid drop in water level, emptying Lisbon harbour, occurred before the arrival of the wave crest. Of the many hundreds of people who ran down into the harbour to salvage lost goods or materials dropped overboard from ships, most were drowned as the ensuing wave crest swamped the harbour and part of the city a few minutes later. In the Pacific Ocean, where the majority of tidal waves occur, there now exist warning services to minimize the impact of a tsunami on coastal regions. When seismographs record a violent under-sea event, tsunami travel times from the epicentre to the nearest coasts or islands are estimated and appropriate warnings are issued.

Productivity in the oceans

Human beings obtain considerable quantities of protein from the oceans, mainly fish (6×10^7 tonnes per year) with smaller but significant contributions from marine plants, crustaceans and molluscs (0.5×10^7 tonnes per year). Excluding fish, most of these organisms are harvested from shallow coastal and estuarine regions. The distribution of world fisheries is not uniform across the oceans. Some areas are highly productive, supporting large fish stocks, while others are sparsely populated. The great variations in productivity can be explained by the structure and dynamics of the oceans, which to a large extent control the availability of nutrients. Excluding human beings, fish constitute the top level in the ocean food chain, ie the path by which proteins and nutrients pass from one group of organisms to another. Fish occur abundantly where their food (and their food's food, etc) is plentiful.

At the base of the food chain are the phytoplankton, small plant-like organisms ranging in size from a few micrometres to several millimetres, which can synthesize proteins by fixation of carbon dioxide, using light as the energy source. Phytoplankton production, known as primary production, is limited to the photic zones of the oceans (where light can penetrate). The vertical distribution of gross primary production (GPP), in the absence of other limiting factors, reflects the light-intensity profile, except close to the ocean surface where light intensities, particularly the ultraviolet component, can kill the phytoplankton. The compensation depth is the depth, averaged over twenty-four hours, at which the GPP rate equals the respiration rate, ie the net primary production (NPP) is zero. The critical depth is that at which total use of oxygen by respiration and oxidation in the water above is equal to total oxygen production by photosynthesis. The oxygen profile usually shows an oxygen minimum below the critical depth, oxygen slowly diffusing downwards from the surface waters and also upwards from the deep oxygen-rich waters. In addition to light and CO_2 phytoplankton require certain nutrients for successful growth. The elements that these nutrients must supply are nitrogen (N, from nitrates), phosphorus (P, from inorganic phosphates) and silicon (Si, from silica), as well as many important trace elements. In regions where GPP is high, N, P and Si can be depleted from the waters, unless constantly replaced, and it is believed that phytoplankton production is limited by the supply of these elements.

Productivity is highest in the regions of upwelling, in coastal waters and in the cold Arctic and Antarctic Oceans. It is lowest in the great ocean gyres, where the stably stratified thermocline acts as a barrier to the vertical exchange of water needed to replenish the nutrients used by phytoplankton in the surface waters. The nutrients can be removed completely from the surface waters when organic and skeletal material settles through the water column (some nutrients are recycled by predators higher in the food chain). In the Arctic and Antarctic very little stable stratification exists and mixing is easily accomplished. In shallow coastal waters on the continental shelves the detrital and skeletal material cannot sink out of the photic zone, since tidal currents are strong and, combined with storm-wave action, mix the coastal waters thoroughly. In regions of upwelling the nutrients are constantly being replenished by the cold rising nutrient-rich waters, and the surface areas can support very high concentrations of phytoplankton. Upwelling also supports the phytoplankton physically, many being non-motile; they would otherwise slowly sink down through the photic zone. The major regions of upwelling are along the eastern boundaries of the oceans and along the equator, where there is a region of surface water divergence (Figure 19.9).

The dominant types of phytoplankton are diatoms, which produce a silica ($SiO_2.nH_2O$) cell wall and link together in chains or rods, possibly to reduce their sinking speeds and thus to remain in the photic zone longer and flagellates, which possess two whiplike flagella that they use for propulsion through the water at speeds of up to 20–30 mm per minute, their movements normally being vertical in response to diurnal variations in light intensity. Species of dinoflagellates are responsible for occasional 'red tides', in which ocean areas turn red because of the presence of enormous numbers of these phytoplankton. As well as simply asphyxiating fish by clogging their gills, dinoflagellates also produce chemicals in concentrations that are toxic to fish. During 1946–7 red tides in the Gulf of Mexico killed an estimated one hundred thousand tonnes of fish, at a time when the world fish catch was twelve million tonnes per year. Fish were washed up on the west coast of Florida in quantities of 150 kg per linear metre of shoreline.

The food chain between phytoplankton and fish can be long or short. In nutrient-rich upwelling waters off Peru, where an enormous anchovy fishery exists, the phytoplankton are so dense that the fish can feed directly on them (mainly colonial diatoms in aggregates). The food chain, being just one or two steps long, results in a very efficient conversion of primary production into edible fish protein, since only a small proportion of the energy of the system is required for respiration. This fishery had a peak catch of twelve million tonnes in 1970, more than 20 per cent of the total world catch for the year. The fishery is particularly susceptible to El Niño (see above) which pushes the cold, upwelling region offshore, and in the El Niño year of 1972 the anchovy catch dropped to four million tonnes. In coastal waters the food chain can be three or four steps long, with zooplankton (small larvae and animals) grazing on the phytoplankton. Some fish feed directly on the plankton, eg the pelagic (midwater) fish such as herring, pilchard, mackerel and anchovy, while the demersal (bottom-feeding) fish, eg plaice, sole and halibut, feed mainly on benthic invertebrates living on the bottom or in the muds and sediments there. The benthic invertebrates are themselves plankton-feeders, or else feed on detritus falling to the sea bed from the surface waters. Other fish are carnivores, preying upon smaller fish and adding a further step to the food chain.

In the open oceans, where primary production is low and the phytoplankton are widely scattered, up to five steps may occur in the food chain, each of which can be of low efficiency, that is each animal in the chain expending considerable energy in hunting for widely scattered prey and so consuming many grams of food for each gram of increase in body weight. Most open ocean waters are well below the photic zone, so there can be no primary production using photosynthesis. Some production does occur—about 5 per cent of what is produced photosynthetically—by planktonic bacteria which fix carbon, using energy from inorganic compounds such as ammonia and methane. This chemosynthesis should be regarded as secondary production, as most of the inorganic compounds used are derived from decaying organic material. The inhabitants of the deep oceans are sea-floor invertebrates and considerable numbers of scavenging fish and crustaceans. These creatures consume the carcasses of dead animals falling from the upper waters, and are adapted so that they can rapidly locate and devour them. Most data on the varieties and densities of the inhabitants of the deep oceans have been gained using bait and a time-lapse camera system dropped to the ocean bottom. The numbers of the various species and the time taken for their arrival provide a measure of their abundance. Some of the densest deep-ocean benthic populations have been found beneath the surface waters with the lowest productivity. Conversely, bait on the sea bed beneath the intensely productive Antarctic surface waters was visited by very few creatures, and beneath the Peruvian upwelling very few fish were photographed although many invertebrates visited the bait.

Some completely new communities of deep-water benthic organisms were discovered in 1977 by a geological expedition that made the first visual observations of hydrothermal plumes at an ocean ridge. The deep submersible *Alvin* made a series of dives on the Galápagos ridge and positively identified hydrothermal vents, initially by the shimmering produced by the mixing of warm water with cold bottom water, and also by a faint milky precipitate consisting of native sulphur formed by the oxidation of hydrogen sulphide. Photographs taken from *Alvin* revealed the area around the vents to be colonized by highly productive communities of densely populated benthic filter-feeding animals, mainly clams, mussels, limpets and tubeworms. The tube-

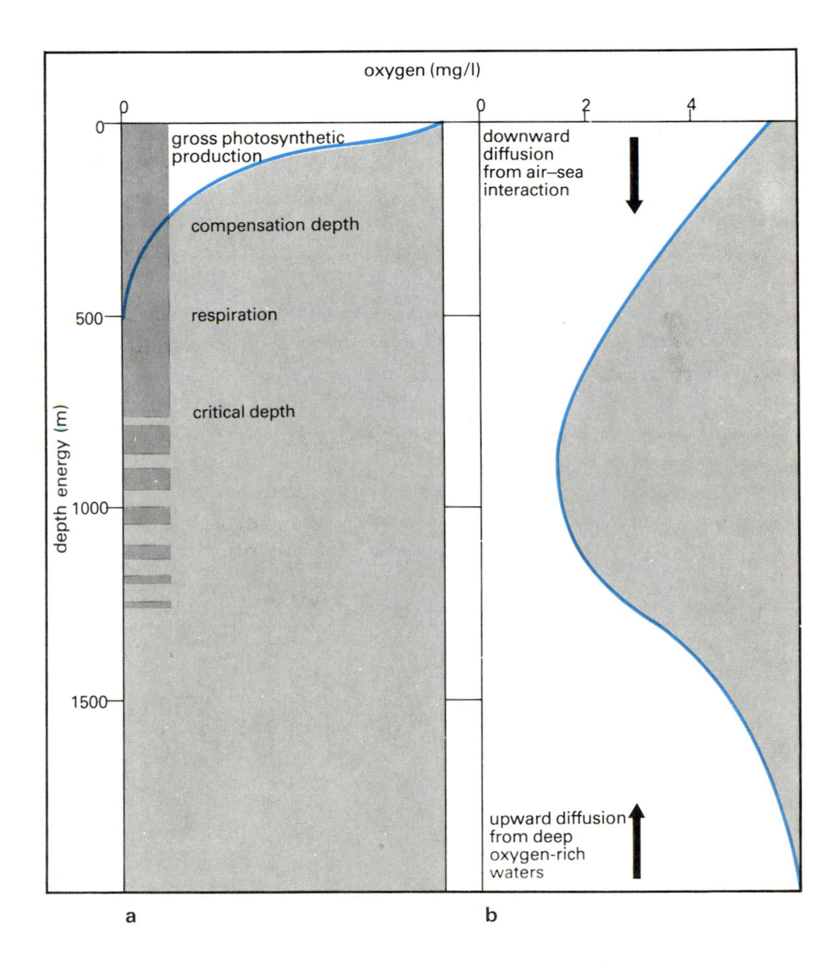

19.12: (**a**) *Changes in gross phytoplankton production with depth in a typical oceanic profile.* (**b**) *Profile of oxygen concentration with depth, showing a minimum some distance below the zone of primary production. At this depth some* O_2 *is used up by decaying organic materials descending from the surface, and it is barely replaced by diffusion from above and below.*

worms, of a new genus, are the largest of the deep-ocean tubeworms, measuring up to 3.5 m in length and 25 mm in diameter. The molluscs were all of unusual or new species, and of large size, the mussels up to 150 mm in length and the clams up to 300 mm. A previously undescribed fish was observed swimming in the hot (6–8°C) waters rising from the vents.

The primary production in the region of the vents, supporting the entire benthic community, is chemosynthesis: high concentrations of a sulphur-oxidizing bacterium (10^{11} bacteria per litre) were found in samples of the water issuing from the vents. A highly productive population of these bacteria is believed to exist within the fissures of the rockmass through which the hydrogen sulphide-laden waters pass on their way upwards. Much remains to be discovered about the ecology of these highly localized communities, particularly with regard to the method by which they locate and colonize new vent areas, as hydrothermal vents are believed to have individual lifetimes of the order of only fifty years.

19.13: *Hydrothermal vent community photographed from the submersible* Alvin *at a depth of 2.5 km on the Galápagos ridge. The vestimentiferan tubeworms, which live at temperatures up to 23°C, do not have either a mouth or a gut but simply absorb the plentiful dissolved food and oxygen. The mussels filter out the abundant micro-organisms in the water and the blind white crabs scavenge among the other organisms.*

20 Sedimentation: environments and processes of deposition

Sediments are composed of materials resulting from reactions between the surface rocks of the Earth's crust, the gases and liquids of the atmosphere and hydrosphere, and the biosphere consisting of organisms and their products. These reactions, called weathering, produce a veneer of broken-down solid detritus, the regolith, which when enriched in organic detritus and attacked by organic agencies ultimately produces soil. Products of weathering are carried away from the site of the reactions—the process of erosion—by various agents influenced by the force of gravity. The most important of these agents are running water, wind and ice (Figure 20.1). The products travel in a solid form as bed load or suspended load and in solution as solution load. Weathering and erosion lead to denudation, the destruction of the upstanding land of the continents. Ultimately, the products of denudation are laid down in a variety of settings, mainly in the marine realm, to form accumulations of sediments in the process known as deposition. After deposition, the unconsolidated sediments are altered to sedimentary rock by diagenesis: a combination of physical processes such as compaction and complex chemical changes.

Production of sediments

The source rock, relief, climate and vegetation control both the amount and the nature of the sediments produced. Today igneous, sedimentary and metamorphic rocks are weathered to produce sedimentary detritus, but early in the Earth's history the surface crust was composed entirely of igneous rocks, as it is on some other planets today. Successive cycles of sedimentation and orogeny followed. Consequently the Earth's surface is now covered with a complex mosaic of igneous, sedimentary and metamorphic rocks which are slowly undergoing destruction to form new sediments.

The type of weathering and which agent of erosion is dominant (downslope mass-movement, running water, wind or ice) depend largely on climate (Figure 20.2). In arid climates the rocks are broken down by mainly mechanical processes. Rapid heating and cooling of rock that contains minerals of different coefficients of expansion leads to rock disintegration. Furthermore disintegration is aided by the forces of crystallization of salts precipitated from waters drawn to the surface by capillary action, and by ice crystallization. In wet climates chemical weathering is dominant. The movement of water through the rocks leads to leaching of the more unstable constituents and their removal in solution or recombination to form new minerals which are stable at the Earth's surface. The most complete chemical weathering is found in lowland areas where gravity does not constantly move the weathered regolith away from its parent rock. Consequently, the broad cratonic or shield areas in the tropics are sites of very extensive chemical weathering and of deeply weathered profiles of altered source rock. Various clay minerals are produced by these processes, and when left undisturbed for long periods the landscape becomes coated with a weathered veneer of iron-rich clay deposits (laterite). If the degree of weathering is sufficient the iron is removed and deposits rich in aluminium (bauxite) are left.

The clay minerals of soils and those carried to the sea by rivers show a distinct global zonation which is related to climate, although local relief and source rock can confuse the picture. Thus kaolinite, a mineral that forms in intensely leached soils, is dominant in tropical and equatorial rivers and seas, illite is more common in middle latitudes and the relatively unstable clay mineral chlorite can only survive in high latitudes where chemical weathering is more subdued.

Vegetation is important in determining the type and amount of sediment produced by a particular source area. A cover of vegetation

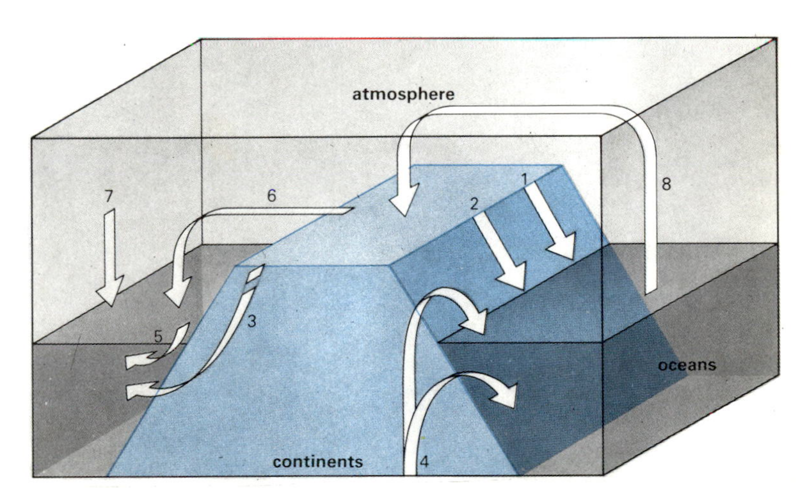

1 rivers 225
2 ice 20
3 groundwater 4.8
4 from subsurface 4.4
5 marine erosion 2.5
6 wind 0.6
7 extraterrestrial source 0.036
8 aerosols 2.6

20.1: *Amounts of material transported to the oceans from the continents (10^{11} kg per year).*

reduces the surface run-off of water, reduces or prevents the surface impact of falling rain, binds together loose weathered material and protects the surface veneer of weathered material from erosion.

The cover of vegetation on the Earth's surface has varied throughout its history. There was virtually no land vegetation such as is known today before the latter part of the Palaeozoic. Gradually plants colonized the lowlands and then invaded the highlands so that, in the Mesozoic, conditions became more similar to those of today. Finally, in the Cenozoic, grasses developed and a modern land flora was established. These changes had dramatic effects on the rate of denudation of the continents. It is claimed that humans, through their widespread alteration of surface vegetation, have increased this denudation by a factor of two.

The interplay of the controlling factors of geology, relief, climate and vegetation has produced a wide variation in the rates of denudation. High relief leads to steep slopes, down which weathered detritus can move relatively quickly to expose more fresh rock to weathering processes. Consequently mountainous areas yield much greater volumes of sediment than adjacent lowlands. As the relief decreases the amount and size of the material being shed decreases, and its stability becomes greater since more time has been made available for the attainment of equilibrium with surface conditions. As the relative relief of various parts of the Earth's surface has changed, there have been periods when the volume of detritus shed from the continents has been greater than normal. The high relief of the relatively young mountains and the very large amounts of rainfall in south-east Asia make it particularly susceptible to denudation, whereas the low relief and aridity of much of Australia results in little denudation (Figure 20.3).

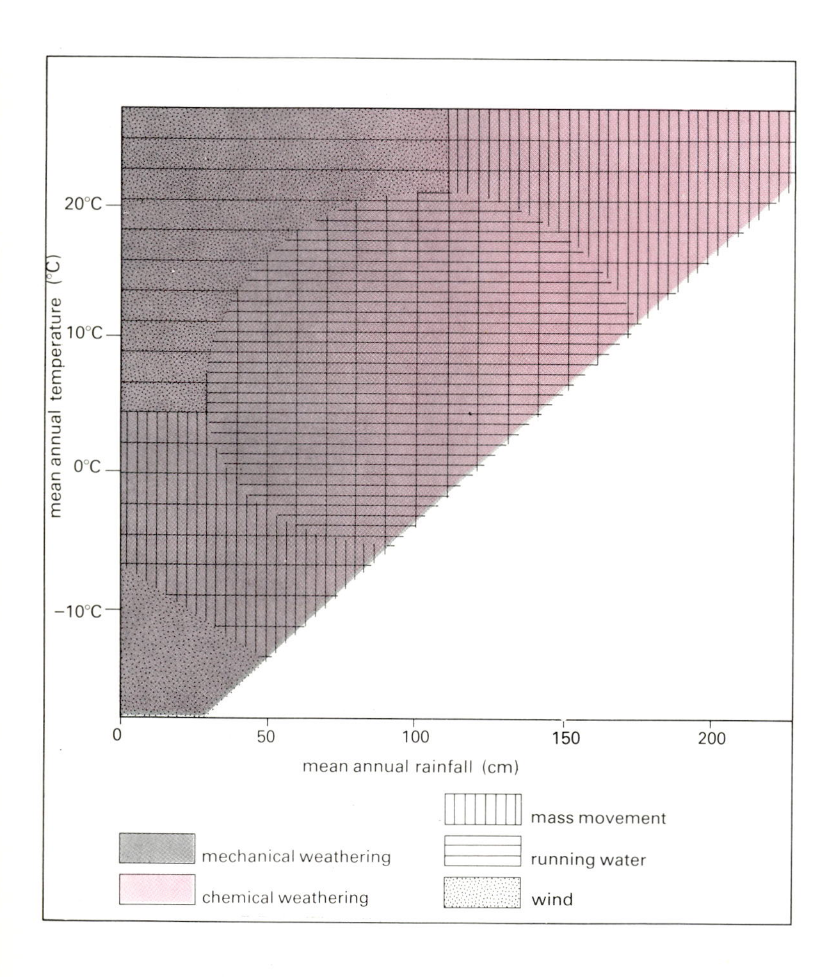

20.2: *Climatic control on the relative importance of various types of weathering and various agents of erosion.*

Transportation of sediments

The weathered solid detritus moves downslope or is transported by water, wind or ice. Material in solution travels in surface waters or seeps down through the rocks of the crust to re-emerge in rivers or on the edges of the continents.

Notwithstanding the mode of transport of the detrital sediment, the released particles undergo broadly similar changes. Minerals are abraded or dissolved; those with good cleavage or fractures break along their planes of weakness to produce smaller grains. Sharp corners are chipped away as grains collide with one another or are ground together, and the particles become more rounded.

Grains are moved with varying ease by the different agents of transportation, depending mainly on their size and density and to a minor extent on their shape, or sphericity (ie the degree of approach of a grain to a sphere) and roundness. Transportation produces a decrease in grain size as well as an increase in roundness in the direction of transport. Sphericity also increases, if rather slowly, in the same direction, because of the greater ease of transport of more spherical grains (Figure 20.4). The general process involves separation of the heterogeneous components of weathered rock debris into a spectrum of grain populations, on the basis of their response to fluid flow. Sediments with a uniform grain size are said to be well sorted; those that are both well sorted and well rounded, indicating a considerable degree of transportation, are said to be texturally mature.

Transportation also produces changes in mineral composition. Many of the minerals of igneous and metamorphic rocks have formed at high temperatures and pressures. Generally, however, it is minerals that formed under low pressures and temperatures that persist when subjected to conditions on the Earth's surface and that are dominant in sediments. The softer and chemically unstable minerals or rocks, the minerals that cleave and fracture, and the rocks that split most easily undergo a gradual process of preferential elimination from the load of transported sediment. Striking changes may occur in the composition of some sediments during transportation, although this is unusual, particularly with sand-sized sediments. Sand samples collected from the upper and lower courses of large rivers show remarkably little change other than a slight reduction in some of the more unstable components. In order to produce a well-sorted, well-rounded sand composed mainly of stable minerals such as quartz, the sediment, unless derived from a source rock of such composition, probably has to go through several cycles of denudation and deposition. Such a

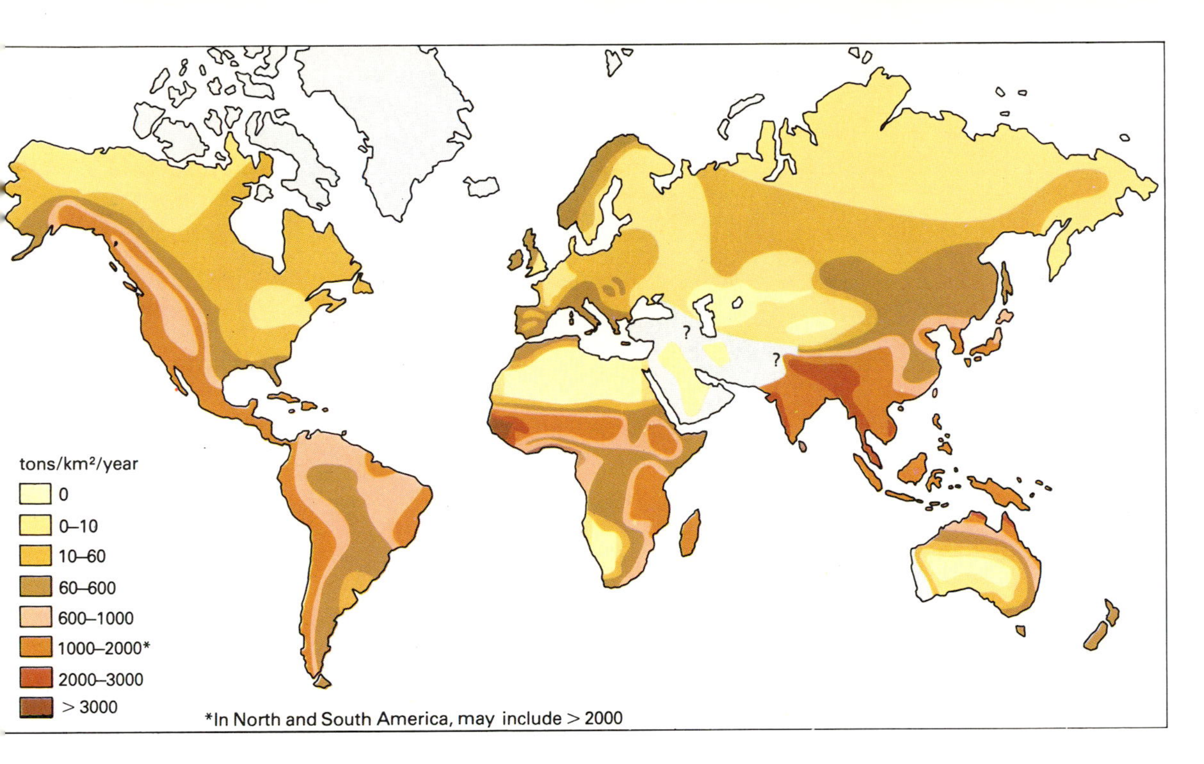

20.3: Rates of denudation of the continental crust.

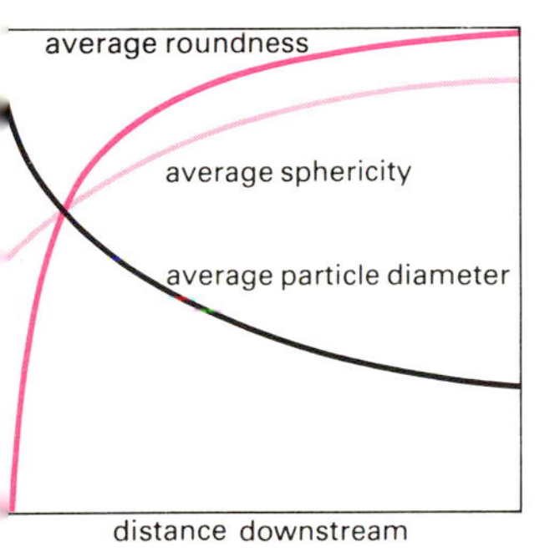

20.4: Changes of sedimentary properties of grains due to selective sorting and abrasion during transportation.

ediment is said to be both texturally and mineralogically mature. onstant recycling of sediments has been a persistent feature of the arth's history. Relatively slight changes in sea level have repeatedly xposed large areas of the continental shelf to subaerial processes and ave brought the sediments of coastal plains into the realm of marine rocesses. On the other hand sediments deposited in deeper water areas re only recycled when orogenic events uplift them to form new nountains, and are often mineralogically and texturally less mature han the sediments of shallow shelf and coastal plain environments.

edimentary environments and facies

Vherever sediments are deposited the local environment of deposition, tself a product of the physical, chemical and biological processes cting at the given locality, imprints itself on the sediments. Accumulations of sediments form coherent units that can be recognized in the geological column by the sum of their lithological and palaeontological characteristics. The term 'facies' has been introduced to describe such distinct entities. A facies is thus a manifestation of a particular sedimentary environment and the processes operating within that environment. Sediments of a particular facies can be recognized in successions of all ages, but show some differences because of the evolution of plants and animals. The recognition and delineation of such facies in space and time allow geologists to build up a series of palaeogeographic pictures of the Earth's surface.

The relationship of the various facies to one another and their geometry and relative thickness are partly dependent upon the agent of deposition and partly on the rates of sedimentation and of change of sea level relative to land. Changes of relative levels of land and sea have produced worldwide transgressions and regressions of the sea over the continents. Superimposed upon these worldwide events, or eustatic changes, there have been local changes caused by local tectonic events and fluctuations in the supply of sediment. The rate of sediment supply relative to the magnitude of land/sea-level changes governs the geometry of the facies (Figure 20.5). Where the rate of sea-level rise is high compared to the rate of sediment supply, the coastline will move landwards and marine sediments will overlie coastal and coastal-plain sediments; the sediments are said to exhibit an onlapping relationship. Where the rate of sediment supply is high compared to the relative sea-level rise, the coastline will advance seawards, or prograde, as in a delta. This also happens if the sea level falls or is constant. Under all these conditions coastal-plain and coastal sediment will overlie marine sediment; they are then said to exhibit an offlapping relationship.

Continental sedimentation

The slopes and lowlands of the continents are covered with the products of denudation en route to the oceans or shallow seas. Much of this sediment cover is transitory but some may stay on the continents for millions of years.

Fluvial sediments

A large part of the surface of the Earth is sculptured by running water. Small streams join to form huge drainage networks which ultimately debouch into the sea. Such river or fluvial systems are the main avenues for the escape of continental detritus to the sea.

Waters flowing down valleys in regions of high relief deposit their load to produce alluvial fans (Figure 20.6). These may vary from tens of metres to hundreds of kilometres in radius. Depending upon the discharge, the whole fan may be covered with water as sheet floods, or water may flow only in a pattern of channels, or stream floods. Where the water is particularly heavily laden with sediment, dense glacier-like mudflows may develop and flow over short distances from the valley mouth.

The sediment in fans is usually coarse-grained and poorly sorted, particularly in the upper reaches; further down the slope of the fan the sediment becomes finer and better sorted. The gravels become better rounded, and there is a decrease in quantity of the less durable rocks and minerals. In the head of the fan the gravels are usually clast-supported, ie the individual gravel particles are in contact, but further downslope they may 'float' in a sandy matrix. The gravels are often imbricate (ie the individual clasts are stacked with long axes inclined upcurrent) and show a gently inclined stratification parallel to the fan surface. Further down the fan the gravels merge into sands. These are frequently stratified, displaying sedimentary structures that can be used in ancient deposits not only to indicate the direction of flow of the transporting medium but also to assess the velocities of the currents and even the depth of water (Figure 20.7). This is possible because, as water passes over a bed of granular material, it begins to move the particles and fashion them into a series of bed forms, determined by the velocity of the current, the depth of water and the grain size of the sediment.

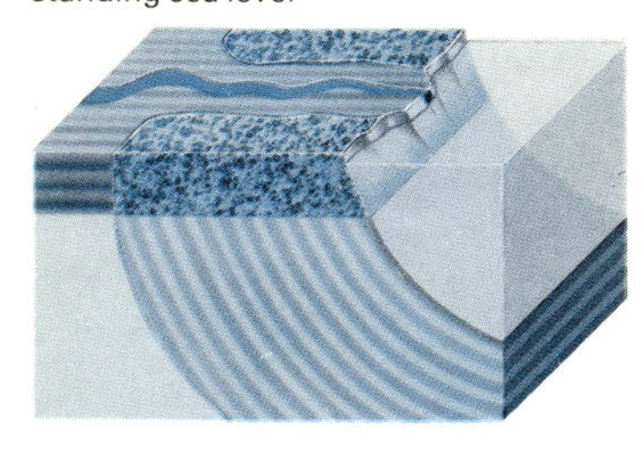

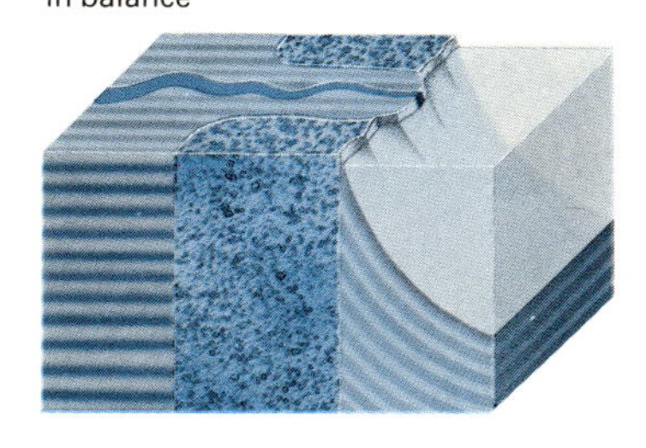

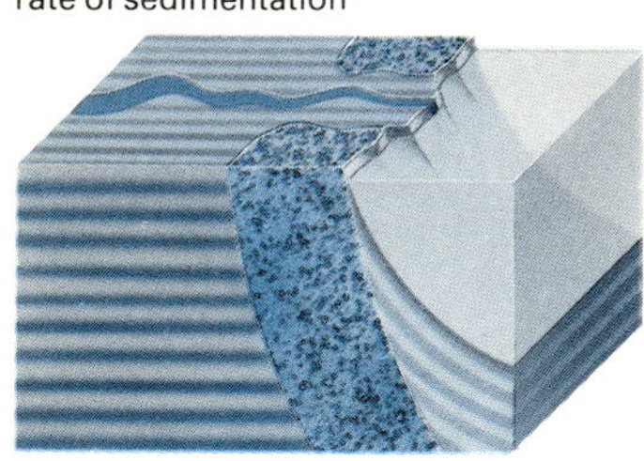

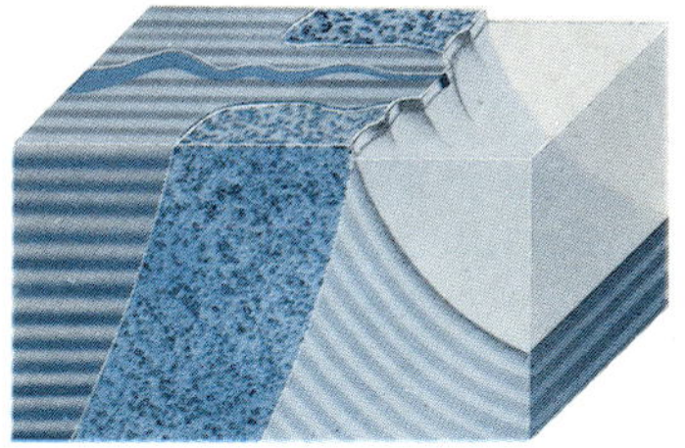

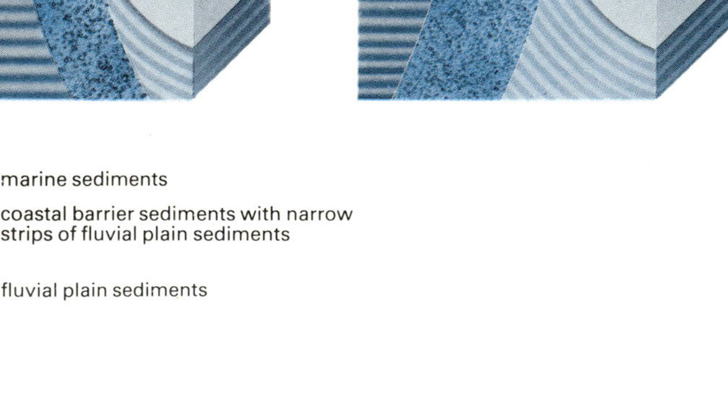

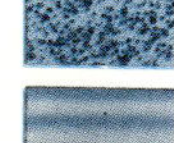

20.5: *The relationship of various facies with varying rates of sedimentation and land/sea-level changes.*

20.6: *An alluvial fan in the Sierra Nevada Mountains, south-east of Granada, in California.*

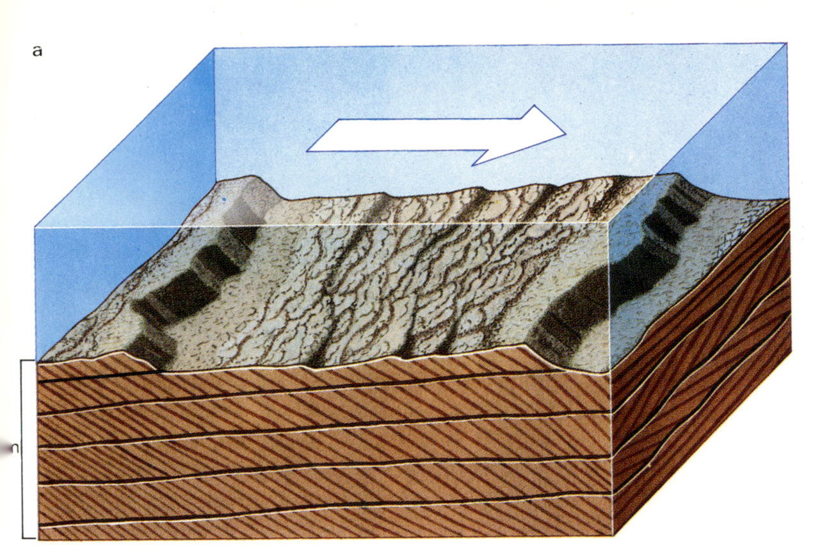

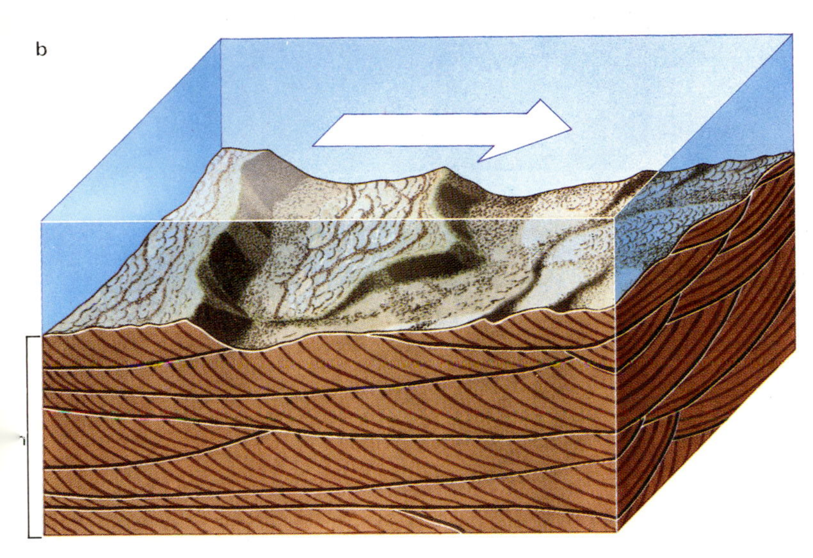

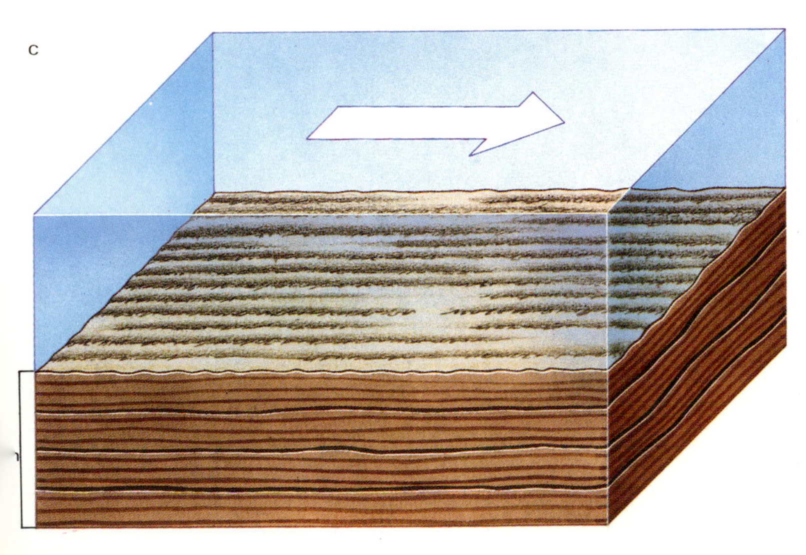

20.7: *Some types of sedimentary structure produced by unidirectional currents:* (**a**) *ripples,* (**b**) *sand waves,* (**c**) *plane bed with current lineation.*

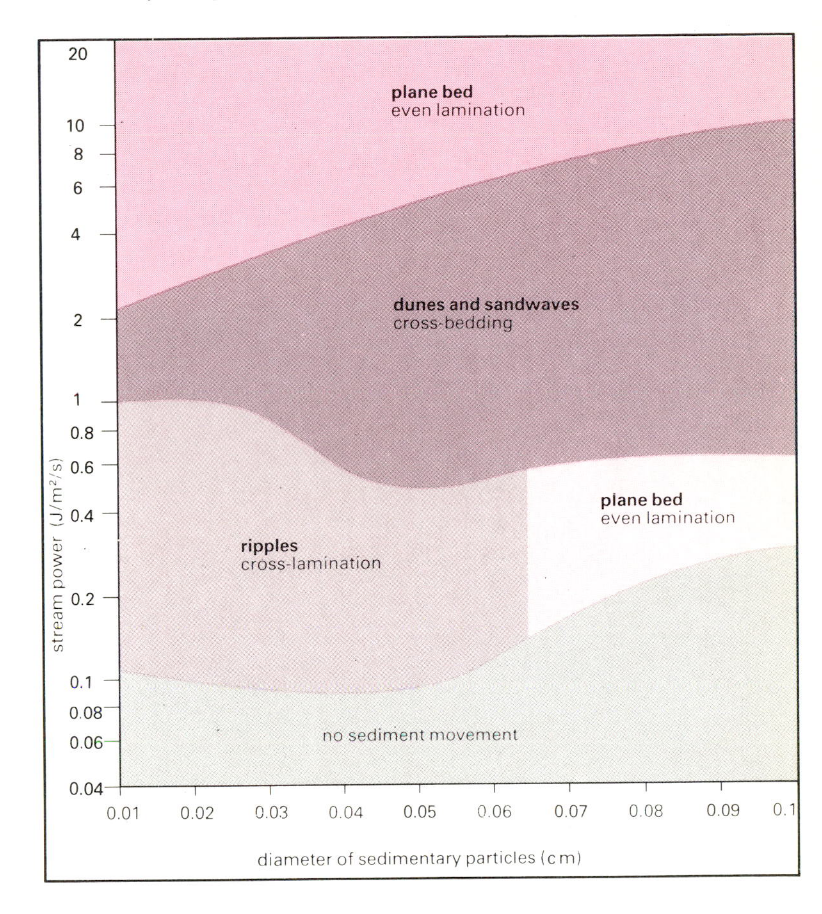

20.8: *Bed forms produced in sediments by unidirectional currents.*

In a unidirectional current, increasing shear stress induced by the fluid produces regular undulations, or ripples, in the sediment, and as the flow increases these develop into larger bed forms—dunes or sand waves (Figure 20.8). Further increase in flow leads to the obliteration of the earlier-formed structures and the production of a flat, or plane, bed with the individual particles aligned in the direction of current flow (current lineation). Finally, more complex bed forms such as antidunes are produced at yet higher velocities, but these are rarely preserved in ancient sediments.

Downslope from the fans, in valley floors or on open lowland areas bordering highlands, sediment is deposited to form a gently sloping braid-plain interlaced by constantly changing channels (Figure 20.9). These have an irregular topography with many gravel bars composed of imbricate, clast-supported gravels arranged in flat beds. There is very little fine-grained sediment, except in abandoned or inactive channels. Such plains may pass downstream into sandy braid-plains characterized by relatively steep slopes and variable discharge (Figure 20.10). Gravel-rich braid-plain channels are sometimes filled with cross-stratified gravels and sand, which pass into rippled sands and then silts. Sand-rich braid-plains also contain channels with bars filled by upward-fining sequences of sediments: basal cross-stratified gravelly sand, cross-stratified sands and a capping sheet of silty mud.

Where the discharge is more regular and the slopes are more gentle, the sediment load is finer grained and rivers develop meanders (Figure 20.11). The river channel migrates across the valley, eroding on the outer banks of the meanders and depositing sediment on the inner banks. The coarsest material remains on the channel floor to form a lag gravel. Sand is piled up on the inner bank where it is fashioned into dunes and ripples, forming cross-stratified sands. During floods the rivers overtop the banks and cover the interchannel area (flood plain or flood basin). They deposit the coarsest parts of their load close to the channel to form slightly raised banks (levées), and distribute the finer material over the flood plain. Crevasse channels are occasionally cut through the levées, and coarser sand is spread over these areas as crevasse splays. In humid areas the flood plain may be colonized with dense woodland to produce backswamps, which will later form peat and ultimately coal. In more arid areas the flood plains may be scrub-covered saline flats where evaporation draws up water to precipitate calcium carbonate (calcrete) or calcium sulphate (gypcrete) and other salts. Lateral migration of channels produces an

20.10: *A braided river on the Sandur Plains of Iceland; a huge area of glacial sediment that is inundated by the glacial meltwaters each spring.*

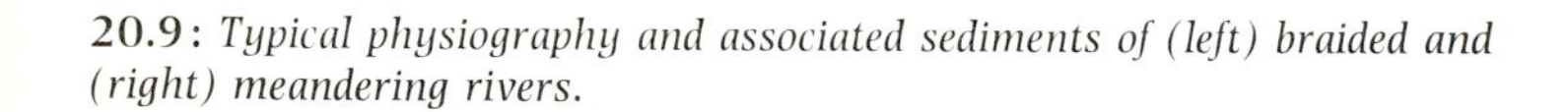

20.9: *Typical physiography and associated sediments of (left) braided and (right) meandering rivers.*

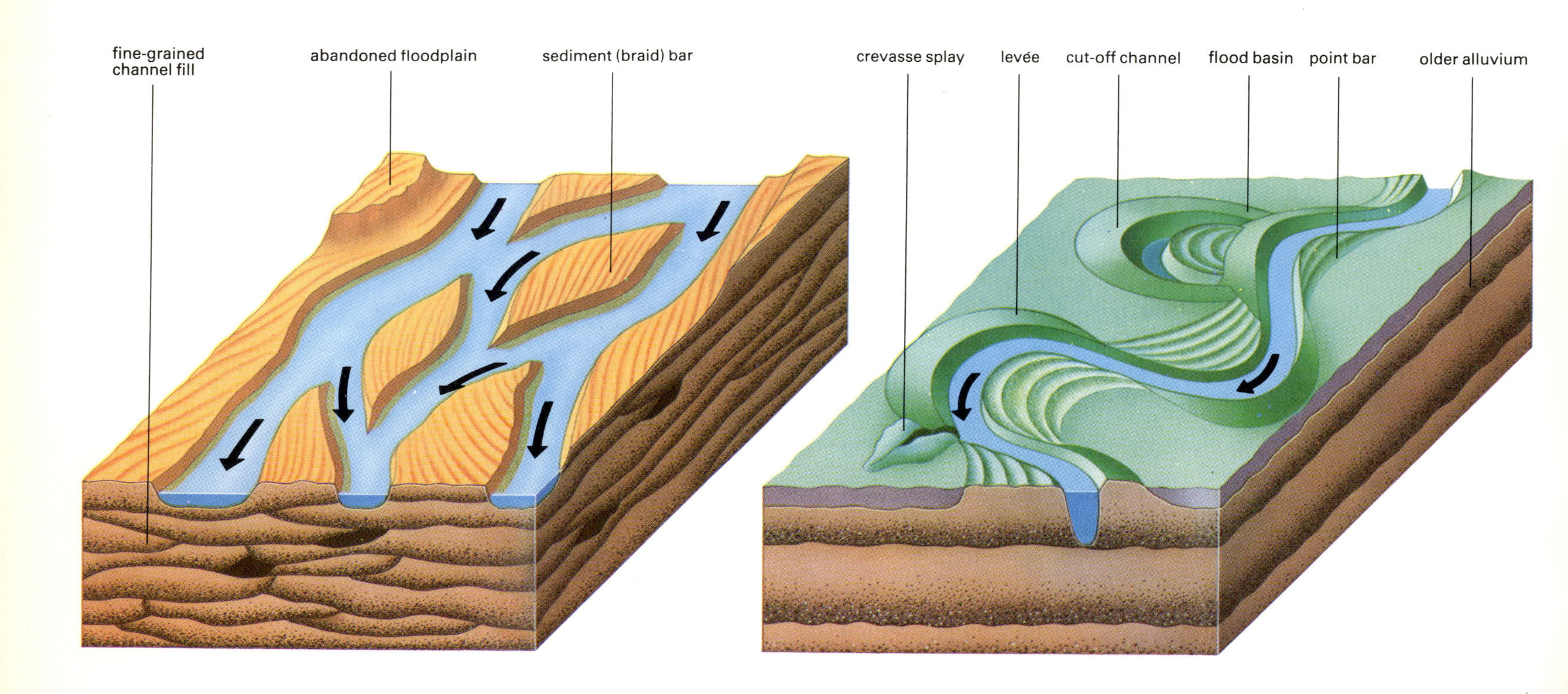

20.11: *A meandering river: the Amazon in Brazil.*

20.12: *Desert dunes in Saudi Arabia.*

upward-fining sequence of deposits: lag gravel, trough cross-stratified sand (deposited on the point bar) and muds (deposited on the flood plain) overlain by peats or salts. Channel migration produces sheet sands. Where the river carries large amounts of fine-grained sediment, the channel may not occupy the whole valley floor but may be confined in a fairly narrow meander belt. Such channels sometimes change course relatively suddenly in a process termed avulsion, and occupy another site on the valley floor. Under such circumstances the sands and coarser sediment formed in the channels are deposited in long, linear sand bodies, approximately parallel to the river course, enclosed in the muddy flood plain deposits. Sands with this shape, when found in the geological column, are known as shoestring sands. Convincing examples of fluvial sediments have been described from the Devonian 'Old Red Sandstone' of the UK, Spitzbergen and Canada, the Carboniferous rocks of the UK and Canada, and from many other parts of the geological column.

Desert sediments

Deserts cover about 30 per cent of the continental surface and are usually thought of as sandy areas covered by dunes, ie sand seas, or ergs. However, only about 20 per cent of deserts have this character; vast areas are covered by bare rock undergoing erosion (hamada), gravel plains (reg or serir), or silt and salts, the latter sometimes forming in shallow lakes called playas or internal sabkhas.

The general aridity and the great diurnal fluctuations in temperature lead to dominantly mechanical weathering. Rock surfaces are veneered with disintegrated rock fragments and mineral grains. Because of the sparse vegetation, this detritus is easily moved by infrequent but sometimes torrential floods. In many fault-bounded areas, eg in the western USA, alluvial fans develop at the feet of mountains and coalesce to form extensive sloping plains, or bajadas. Wind and water carry silt and clay to deposit them with salts in playas towards the centres of such depressions. Water drawn to the surface by capillary action coats the clasts and rock surfaces with skins of iron and manganese, producing a dark surface, or desert varnish. This process also leads to precipitation of calcium carbonate and calcium sulphate, with the formation of hard surface layers of calcrete or gypcrete. Elsewhere, eg in the Chilean deserts, reaction between the saline waters and volcanic debris forms the rare nitrate minerals Chile saltpetre (KNO_3) and saltpetre ($NaNO_3$).

Wind blowing over the surface abrades gravel by the surface blast of transported material, and this, combined with splitting due to heating and cooling, produces very characteristic faceted shapes (ventifacts). Sand creeps along the surface or moves in leaps and bounds over it (saltation) to produce a variety of bed forms analogous to those formed under subaqueous flow. They include small ripples with wavelengths in the centimetre to decimetre range, dunes with wavelengths of hundreds of metres and heights of tens of metres and draas, huge dune complexes with smaller dunes developed over their surfaces, which may extend for many kilometres and are hundreds of metres high. The finer material is separated from the coarser sand and gravels by the wind. It is carried at relatively higher levels than the saltating sand and may be deposited on the playa flats or carried outside the desert to be deposited as a blanket (loess).

The larger dunes have gentle windward slopes and steep lee slopes (Figure 20.12). Sand moving up the windward slope is driven over the crest of the dune to form steeply dipping laminae made of layers of grains that avalanche down the steep slip-slope (grain flow deposits) and others that fall from the air above (grain fall deposits). These processes produce a strikingly laminated, well-sorted sand dipping at up to 34°, markedly steeper than dips in subaqueous sediment. When the surface has been wetted and attains a certain degree of cohesiveness it can slump. The dunes have a wide variety of shapes. Some have fairly regular crests at right angles to the transporting wind; these are transverse dunes. Elsewhere the crests curve to give crescentic and linguoid patterns. In areas with a thin cover of sand the dunes occur as isolated crescent-shaped accumulations of sand (barchans), their horns pointing downwind. Where older sand fields are being eroded parabolic dunes develop with horns facing upwind. Some of the most impressive dunes are the elongate ridges (seif dunes), whose crests parallel the dominant wind direction and are probably formed by spiral vortices flowing parallel to the wind direction. From the limited studies made on such dunes the sands appear to develop a series of slip-faces approximately at right angles to the dominant direction of the wind. The cross-stratification thus produced is bimodal and contrasts in direction with the almost unimodal cross-stratification of sands developed in transverse, barchan and parabolic dunes. Where winds have variable patterns, more complex star-shaped and pyramidal dune forms may develop whose internal structure is poorly known. Spreads of dune sands may develop at the foot of alluvial fans or near bare rock areas producing sand-size detritus and in some cases may climb on to the fan surface. Where the wind direction is relatively constant, the dunes show a persistent trend, mirroring the wind direction. Although there is considerable doubt about the structures of large dunes, sufficient is already known to indicate that cross-stratification produced by advancing dune faces can be used to determine the direction of palaeowinds in ancient sand seas.

Desert sediments are poor in organic matter, so that groundwaters contain free oxygen and oxidize unstable ferromagnesian minerals, releasing iron that is precipitated as a ferric oxide cement surrounding the grains. Red colouration is thus a striking feature of many ancient desert sediments (termed red beds).

The evaporites deposited in desert depressions may show considerable variation, depending upon the surrounding rocks. Some form in bitter lakes as chlorides, mixtures of chlorides or sulphates, or pure sulphates; others form in alkali lakes as carbonates, including calcium carbonate ($CaCO_3$) and dolomite ($CaMg(CO_3)_2$). Gypsum ($CaSO_4.2H_2O$), anhydrite ($CaSO_4$) and halite ($NaCl$) are also found in such environments. Where there is volcanic activity volcanic detritus is attacked by saline water or by waters emanating from volcanic fissures, leading to the formation of zeolites. These mineral deposits and their interstitial brines are extensively exploited in the western USA, Chile, Turkey and other parts of the world as raw materials for the chemical industry.

20.13: *The Baltoro glacier in Karakoram, with a large lateral moraine at the right of the picture.*

Extensive cross-stratified sands of Jurassic age in New Mexico, USA, are among the most spectacular examples of ancient dune sands. Equally famous are the Permian sandstones of northern Europe. These appear to have formed in structural depressions such as the North Sea basin where alluvial fan sediments pass in the direction of the basin to dune sands, which in turn pass into evaporites which were deposited in playa lakes.

Glacial sediments

Although ice now covers only a small part of the continental surface, within the past two million years over 32 per cent of the Earth's surface has been glaciated. Ice moving down to lower latitudes from polar regions or from high altitudes to lower ground has profoundly modified the land surface by erosion in the upland regions and by blanketing lowland areas with a thick cover of glacial sediments. Furthermore considerable volumes of sediments have been carried to offshore locations by icebergs calving off ice sheets when they reach the sea and distributing sediment over thousands of square kilometres of the ocean floor.

Ice is armed with sedimentary detritus in its lower layers. As it moves over the land surface it scrapes off the loose soil and rock and abrades the underlying solid rock floor, producing striae and grooves. The latter may be several metres deep and tens of metres across, and may be followed for kilometres in their extreme development. Melt waters flowing in tunnels in the lower parts of the glaciers wear away the underlying rock as they transport sand and gravel.

Eroded debris accumulates on glaciers when material falls on the ice surface from adjacent slopes and when it is eroded from the base of the ice. Glaciers transport large volumes of sediment, and these are deposited beneath the ice or at its snout (Figure 20.13). This sediment has a very distinctive ill-sorted texture and is known as till. Individual clasts are often grooved or striated and gravel-sized clasts show a marked elongation in the direction of ice flow. These massive sediments are sometimes banded and often intensely deformed by ice moving over them to produce structures very reminiscent of highly folded rocks. Interbedded with such ill-sorted deposits are lenses of better-sorted, cross-stratified sands and gravels representing the deposits of englacial or subglacial tunnels through which water flowed.

Wide braid-plains are developed in front of glaciers. They are covered with gravels and sands and are formed by the seasonal melting of the ice. Aeolian dunes form belts of better-sorted sediment on the margins of the ice-deposited sediment, eg the cover sands overlying parts of Holland. Laminated deposits, consisting of layers of fine and coarse sediment, are found in subaqueous deposits. These often contain scattered pebbles and boulders (dropstones) released from floating ice. The lamination may be produced by cold dense meltwater plunging beneath lake waters to produce a density current that carries sand, silt and clay over the lake floor, or it may be due to seasonal variations in run-off from the glacial streams. The coarser fraction settles quickly during the spring melt while the finer material settles more slowly, thus producing a distinctive pair of laminae known as a varve. Finer-grained sand and silt may be transported by wind many miles away from the outwash plains ultimately to be deposited as a blanket of well-sorted, poorly stratified sediment (loess). Such deposits cover large parts of Europe, North America and Asia.

Ancient glacial deposits are recognized from many ages. The best known are the Precambrian ones of North America, Scandinavia, Scotland and Australia; the Ordovician ones of the Sahara; and those of the Carboniferous and Permian periods in the southern Gondwana continent comprising South America, southern Africa, India and Australia, as well as Antarctica (see Chapter 23).

Coastal plain, nearshore and marine sedimentation

The sediment supplied by rivers, ice, wind and coastal erosion is dispersed along the world's shorelines to produce broad coastal plains with various geomorphological forms (Figure 20.14) on the passive margins of continents and rather narrower fringes on the active margins. Some of the sediment moves seawards to cloak the continental shelves and semi-enclosed shelf seas and is carried even further to the continental slopes and to the floor of the deep ocean.

In the sea, life processes lead to the production of biogenous debris consisting of the hard skeletal parts of organisms inhabiting the sea floor and the waters above. This production of biological sediment depends on the abundance and type of organism dominant in particular areas. This in turn depends on water depth, temperature, salinity, turbidity and the availability of the essential nutrients: silicate, phosphate, nitrate etc. These factors are themselves controlled by nearness to land and oceanic processes. Where there is little supply of terrigenous material from the land such biogenous sediment is dominant. The skeletal debris is broken down to produce particles of a variety of sizes. They are moved in exactly the same way as terrigenous sediment by wave, tidal and wind-driven currents and are fashioned into similar coastal and submarine geomorphological forms. In addition, however, biological processes may develop large wave-resistant structures called reefs or bioherms, which may be of immense size, ie tens to hundreds of metres thick, kilometres wide and hundreds of kilometres long.

Physicochemical or biochemical reactions lead to the production of hydrogenous sediments, which may be in the form of simple precipitates from seawater or from waters emerging on the sea floor, especially along fractures or faults, or may be produced by reaction between seawater and rocks or sediments exposed on the sea floor.

Sediment originating in outer space is a very minor component. However, because of the extremely low rates of sedimentation in some parts of the ocean floor, remote from the continents, such extra-terrestrial sediment is present in recognizable amounts.

Deltas

Vast volumes of sediment are deposited in deltas, lying at the seaward ends of the world's large drainage basins (Figure 20.15). The form and shape of a delta is a function of the supply of sediment from the river and the strength of the waves, tides and wind-induced currents in the sea. Where river flow is dominant the river water flows seawards relatively unhindered to form long finger-like extensions of sediment. Where wave action is important the coarser sand is spread around the delta margin to produce a more rounded outline (Figure 20.16). Where tides are important, the river mouth channels show a distinctive flared pattern formed by reversing tidal currents.

Sand generally remains fairly close to the shoreline, except during periods of very strong river flow or heavy storm action when large waves will suspend it and allow some of it to escape seawards. Occasionally a dense suspension produced by such events can move down the front of the delta as bottom-hugging density currents, so that coarser material is transported into deeper water. In contrast, clay and fine silt are carried seawards with the freshwater as it spreads out over the surface. Thus a wide, shallow, sandy platform surrounds the subaerial margin of the delta, and this is succeeded towards the sea by an apron of fine-grained sediment, composed mainly of silt and clay.

Sedimentation is rapid in deltas. Fine-grained sediment is deposited with very large volumes of water trapped between the particles, and this escapes only slowly. Consequently, when a river changes its course from one part of the delta plain to another, in order to find a more direct route to the sea, the previously active part of the delta continues to compact. As it is no longer receiving much sediment its subaerial parts sink beneath the sea. Waves attack its margins and small beach-dune complexes form which are pushed landwards over the sinking delta surface (Figure 20.17). These events cover the old subaerial delta surface with a veneer of marine sediment. This process can be repeated many times in deltas, such as the Mississippi, and this produces a series of overlapping deltaic lobes.

The rapid deposition of sediment and continuous compaction produce other changes, particularly in deltas with a very high content of silt and clay. Water-saturated sediment deposited on the front of the delta is very unstable and this commonly slumps and slides, producing scars on the upper subaqueous slopes and faults and fractures which continue to grow as sedimentation proceeds. Fine-grained sediment sometimes moves upwards to form vertical columns of mud (mud diapirs), which may emerge on the surface as mud-lump islands.

The net result of sedimentation in deltas is a seawards-building (prograding) embankment of sediment. The sequence in such a body shows an upward coarsening of sediments. Fine silts and clays deposited in the seaward part of the delta are overlain by interbedded sands, silts and clays developed on the upper slopes of the delta. These in turn are overlain by the sands of the shallow-water parts of the delta and those of the cordon of beaches and dunes. The upper part of such a pile is formed of river channel sands and the associated levée, crevasse and backswamp sediments. The upward increase in grain size will be accompanied by an increase in complexity of the sediment units and a decrease in marine organisms.

The subaerial parts of the delta are vast areas of swamp and marsh dotted with lakes and cut by a lacework of major and minor channels. In humid areas, particularly in the subtropics and tropics, where there is rapid plant growth, dense forests develop between the channels. Although sediment can reach these areas when the river waters escape from the confines of the channels, in areas remote from such channels the production of plant debris exceeds the supply of terrigenous sediment. Normally when plants die the vegetable matter is broken down through oxidation and through biological breakdown by aerobic bacteria and fungi. However, where plant debris collects under water the environment soon becomes stagnant and anaerobic, the included oxygen being consumed in the oxidation of organic matter. Anaerobic bacteria continue to degrade the plant debris and the product is peat, a deposit composed of partly decomposed plant debris.

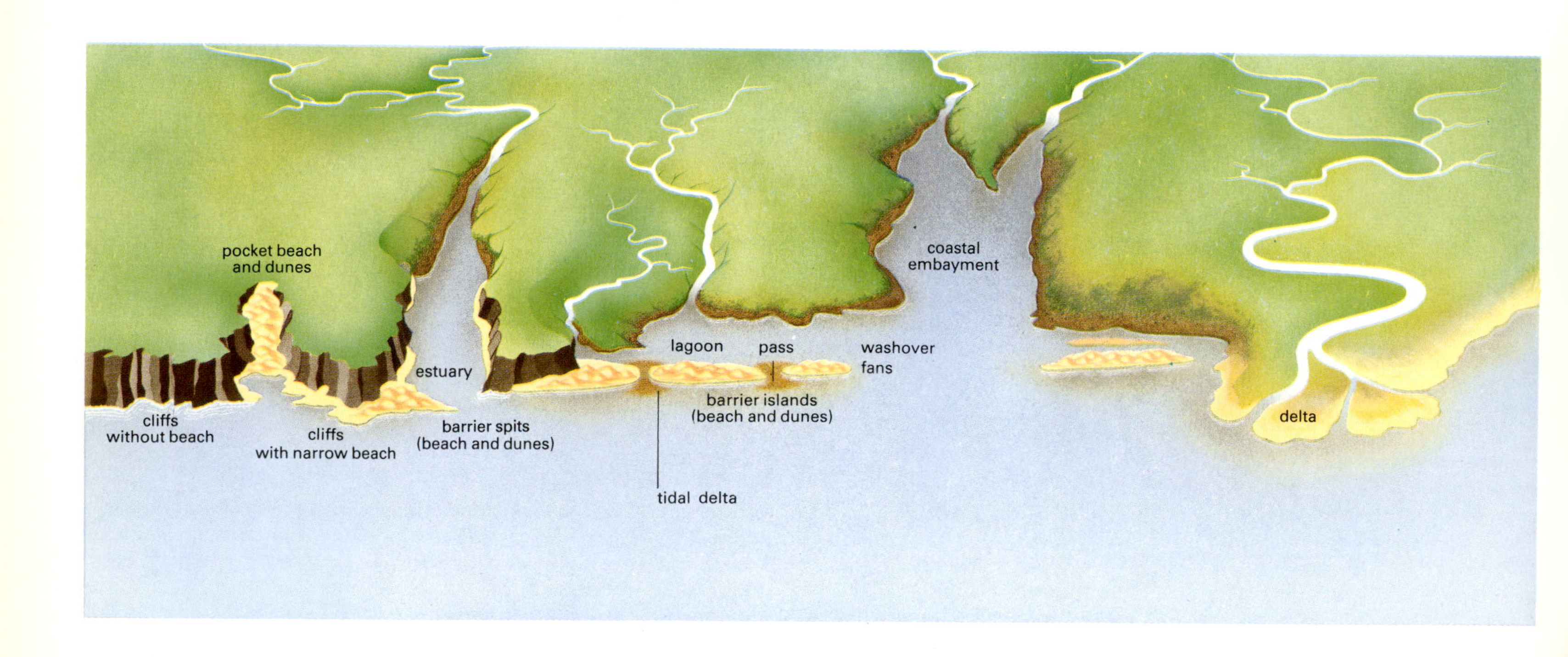

20.14: *A simplified sketch showing features of coastal geomorphology.*

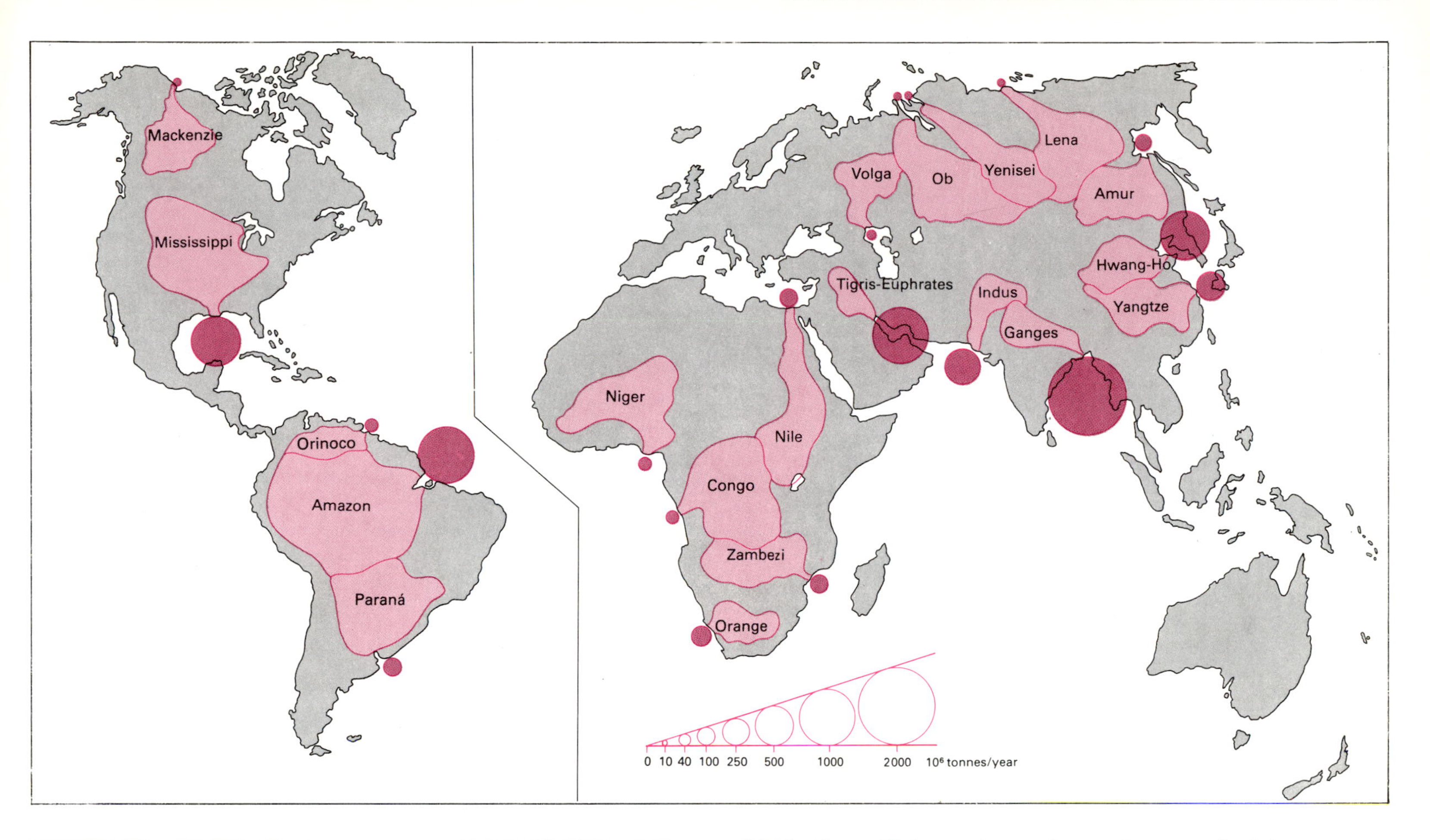

20.15: *The world's largest drainage basins, the magnitude of their river loads and the location of deltas.*

20.16: *The Nile delta, photographed from Gemini 4.*

As the organic matter is buried by later sediment and the temperature rises, conversion to coal takes place.

Seams of peat can be traced for hundreds of square kilometres in some large deltaic plains. Towards the river courses the seams split and interfinger with the terrigenous sediment of the levées. Such seams may be interrupted by crevasse-splays of sediment or by channels that have cut across them and have later become filled with sediment, called 'wash-outs' by miners when found in ancient coal-bearing sequences. In some cases the introduction of the crevasse-splay sediments has entombed trees. These are left standing in their vertical position of growth. Similar trees, now turned into coal, can be seen in ancient coal-bearing sequences. Traces of siderite ($FeCO_3$) can be found in some marsh sediments of deltaic plains, and this mineral is abundant as beds of clay-ironstone in ancient coal-measure sequences. Analogous features, found in both ancient coal sequences and modern deltaic-plain sediments, are the leached soils (seat-earths) of the ancient record, in which the coal-forming vegetation was rooted.

Modern deltaic plains are prone to periodic flooding caused by hurricanes and river floodwaters. Similar events can be recorded in the geological record. In addition more widespread inundations have periodically allowed marine sediments to extend over entire coastal plains. The great extent of the resulting 'marine bands' indicates that they probably record an actual rise of sea level rather than a seasonal storm event. They are invaluable in the correlation of ancient coal seams.

The ancient sequences of coal-bearing strata found in the Carboniferous rocks of northern Europe and America, and similar deposits of other ages, show all the features of some modern deltaic plains. The rhythmic sequence of subaqueous muds passing up into shallow-water sediments and coals is thought to be the result either of repeated switching of delta lobes or repeated rises of sea level causing the spread of shallow-water conditions over vast areas and subsequent prograding of successive deltas.

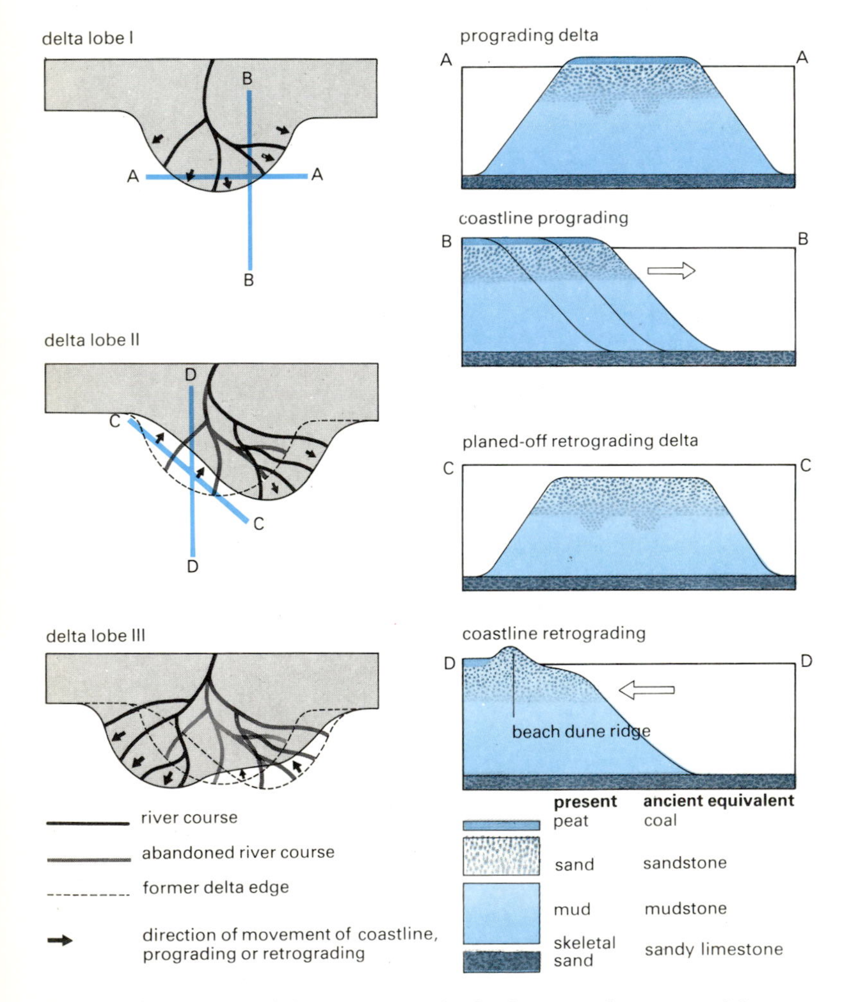

20.17: *The sequence of changes due to the development of successive delta lobes and the resulting sedimentary sequences.*

Ancient deltaic sequences are also of considerable interest to the oil geologist. The presence of abundant marine muds (source rock) and sandstones (reservoir rock), and the fact that the latter are often capped with marine muds (cap rock), makes them potential sites of oil and gas accumulation. Good examples of oil-bearing deltaic sequences are known from Cretaceous sandstones of Colorado and New Mexico, USA, Cenozoic sandstones of Texas and Louisiana, USA, and from many other parts of the world.

Estuaries and intertidal flats

Many rivers do not have deltas at their mouths. Those which cross low-lying topography near the coast, particularly where there has been a eustatic rise in sea level, may meet the sea in a long inlet, or estuary. Estuaries remain unfilled with sediment only if the catchment area of the inflowing river is too small or if the topography of the hinterland is too subdued. The seaward-flowing river water gradually mixes with the seawater which is driven landwards by tidal flow, although the two water bodies may remain separate because of their contrasting densities. Sediment carried from land by the river is thus mixed in the estuary with sediment produced by coastal erosion and may be trapped in the estuary by the pattern of water circulation.

Estuaries are protected from the waves of the open sea, and beaches and dunes are rare. Instead they are usually fringed with salt marshes or swamps in which mud and peat accumulate. In areas of high tidal ranges these are succeeded towards the channels by bare mudflats and sandflats colonized by many burrowing organisms. The subaqueous parts of the estuary are floored with coarse-grained sediments, sands or shell gravels, although these may be covered by mud. In areas of high tidal range the channels are often broken into an interlacing pattern by large sand banks (Figure 20.18) which constantly change position. Large mounds, or bioherms, composed of bottom-living organisms such as mussels and oysters are common features of estuaries.

The margins of estuaries in regions of moderate to high tidal range are one of the areas where intertidal flats, gently sloping areas covered and uncovered by the tides, are well developed. However, they are also found in other coastal embayments such as the Bay of Fundy, Nova Scotia and the Wash, UK; in lagoons, eg the Wadden Sea of Holland and Germany; and even on open coastlines with broad sandbanks or where wave energy is low, eg parts of the German coast and the coast of France around Mont St Michel. In all these areas, marshes or swamps pass seawards into muddy flats which in turn pass into sandy flats. The sediments of these areas are formed of either carbonate or terrigenous material. They are rich in sedimentary structures produced by physical phenomena such as waves and currents, and by burrowing and surface-dwelling organisms. Upward-fining sequences 5–15 m thick develop as the marshes build seawards over the mudflats and sandflats to produce broad coastal plains.

Beach-dune barriers

Many miles of the world's coastlines are bordered by beaches. On rocky shorelines they occur as a narrow fringe where cliffs do not plunge directly into the sea. In small bays they become better developed and are often succeeded landwards by aeolian dunes. The beach and dunes may sometimes form a barrier-spit which partly seals an estuary. On the lowland coasts of the world beach-dune barriers form an almost continuous cordon, broken only occasionally by small rocky headlands or by river outlets; they may become isolated from the coast to form long linear features called barrier islands, enclosing shallow water bodies between the barrier and the land.

Beaches are dominated by breaking wave systems. Tides are of less importance, although beaches of tidal seas are generally broader and are morphologically complex, and inshore tidal currents can impinge on the lower parts of the beach to complicate the normal wave-induced water circulation.

Waves approaching a coastline begin to affect the bottom and are refracted so that they approach the coast at a low angle (wave refraction). When water depth approximately equals wave height, the waves break, water is thrown landwards, surges towards the coast through the surf zone and runs up the beach face as swash. Part of the water sinks into the porous sand or gravel and part returns as backwash. The amounts depend on the degree of saturation of the beach, and this in turn is dependent upon the character of the waves and to a lesser extent in tidal seas on the level of the tide. Returning water is trapped inside the breaker zone, moves alongshore and returns through narrow belts of swiftly flowing waves (rips) cutting across the breaker zones.

The zone extending from just seawards of where the waves break to the landward limit of wave action is highly turbulent. Fine-grained sediment is winnowed from the beach sediments and carried away alongshore and out through the breaker zone into calmer water. Beach sediments are therefore composed of well-sorted sands or gravel. During periods of steep winter waves beaches are cut back and sand is transferred seawards to form distinctive bars under the breaking waves, but during periods of flatter summer waves this material is driven landwards again to build the beach seawards. Large storms may rip sand seawards beyond the breaker zone, from where it either moves landwards again or is picked up by wind-blown or tidal currents and dispersed over the adjacent sea floor. Erosion of the sea floor seawards of the breaker zone also provides some sediment. This moves slowly landwards under the action of the shoaling waves. Where waves approach a coastline obliquely the sand and gravel are driven along parallel to the shoreline, smoothing it by filling estuaries, coastal embayments and harbours. Where there are strong onshore winds, sand is blown landwards from the beaches to form huge dune complexes that extend inland for many kilometres. Sometimes, as on the Californian coast of the USA, sand moving alongshore is lost into

20.18: *A typically intricate pattern of banks and channels with bordering intertidal flats and marshes in estuaries: Cardigan Bay, Wales.*

the upper reaches of submarine canyons. The breaking wave system can thus be regarded as a barrier to the seaward movement of coarse-grained sediment, although some is lost seawards during infrequent but important storms, and some to the heads of submarine canyons.

The sediments of the barrier-dune systems are, in areas close to river mouths or eroding cliffs, composed mainly of terrigenous sediment derived from continental denudation or from erosion of the coast or the adjacent sea floor. Where there is a poor supply of detritus from the continents they are composed of skeletal debris produced by breakdown of marine organisms living on the adjacent sea floor. In warm tropical seas the beach-dune sand may be completely composed of oolitic sands in which sand grains are coated with envelopes of calcium carbonate precipitated from the warm agitated seawater.

Where large quantities of sand are being injected into the system of breaking waves, the beach-dune complex may build seawards to form wide, sandy coastal plains. These may be preserved in the geological column as extensive sheet sands, several tens of metres thick. These show an upward sequence of fine-grained offshore sediments, usually characterized by abundant and scattered ripple-marks or flat-bedding, a series of ripple-bedded sands, which may be burrowed, and coarser sands showing large-scale cross-stratification and representing the deposits of the nearshore breaker bars. The sequence is capped by sands with a gentle seaward dip (deposits of the beach face), followed by some flat-lying sands containing coarse skeletal debris and

20.19: *Cape Hatteras, North Carolina, photographed from the Apollo 9 spacecraft, juts out far into the Atlantic; Cape Lookout is near the bottom of the picture. The current flows from north to south, carrying the sediment with it.*

sometimes gravel clasts. These represent the flat-lying upper part of beaches (the berm). This sequence may also show an uppermost unit of cross-stratified sands with steep dips (aeolian dune deposits). Excellent ancient examples of such sequences have been described from various parts of the world, including the Cretaceous sands of New Mexico, Carboniferous sands of the Appalachian Mountains and the Winn Mountains of the USA. They are of particular importance as potential hydrocarbon reservoirs.

Lagoons

The lagoons enclosed by beach-dune barriers or, in tropical seas, coral reefs are usually very shallow. They connect with the open sea through passes or tidal inlets cut across these barriers (Figure 20.19), and in non-tidal seas may sometimes be completely sealed off for several months. Because of their shallowness and their restricted connection with the sea, the waters of lagoons are subjected to greater extremes of salinity and temperature than the waters of the adjacent marine areas, particularly on coasts with low tidal ranges. In humid areas lagoons may be brackish and in arid areas hypersaline. Both conditions may occur in one lagoon. Water temperature may be very high or so low that the waters freeze. Because of the extreme conditions, lagoons are colonized by a fauna of low diversity, though it may be abundant.

These shallow water bodies are protected from large waves, although their sediment may be vigorously stirred by small waves. Strong winds can initiate seiching and can drive the lagoon waters seawards through the inlets or passes to deposit sand, silt and clay over a wide area of the adjacent sea floor. Similarly strong onshore winds can drive water and sediment into the lagoon either through the inlets or, in extreme cases, by cutting new passes across the barrier islands.

In humid areas, where there is considerable transport of sediment from the land, lagoons are usually covered with muds where the tidal action is low and with muds and sands in tidal lagoons. In the latter, intertidal flats are developed, as in bays and estuaries. Marshes will develop around the lagoonal margins and algal flats may form. In areas where there is little supply of sediment, lagoons are often infilled by sand driven landwards from the barrier either through the inlets or by catastrophic storm events. Skeletal remains of organisms can enrich the sediment with calcium carbonate or build bioherms of salinity-tolerant bivalves and carbonate-secreting worms (serpulids). Where there is no supply of sediment the shallow lagoonal areas may be floored completely with skeletal debris. This is broken down by organisms, such as bottom-feeding fish, and is further attacked by boring algae and fungi to produce structureless sand-size carbonate grains as well as mud. The finer-grained material is ingested by bottom-feeding organisms and secreted as sand-sized faecal pellets. Where temperatures and evaporation are high the waters may become very saline, and calcium carbonate muds may be precipitated. Vigorous plant growth can produce peats in such environments. Blue-green algae build mats, particularly in tropical areas, over the surface of the shallow water and intertidal areas. These mats trap sediments which can accumulate to produce a distinctive laminated structure (stromatolites) known from rocks of all ages.

Reactions can occur between the carbonate muds and sands of the local fringing intertidal areas and the interstitial waters to form dolomite ($CaMg(CO_3)_2$). In very arid areas the high rate of evaporation produces gypsum ($CaSO_4.2H_2O$) in these intertidal sediments. Furthermore, on the supratidal coastal plains (sabkhas), which often border the lagoons on their landward side, and which have been produced by coastal progradation, evaporation becomes very intense: large volumes of gypsum and anhydrite ($CaSO_4$) are precipitated, as well as other evaporitic minerals. The dense saline waters of the lagoons may sink and move seawards through the porous barrier sands or reef, separating the lagoon from the open sea, and react with the calcium carbonate sediment to cause further dolomitization. This process was first inferred from evidence in ancient rocks, and was later found in present-day sediments.

Many examples of barrier islands and their associated lagoonal sediments have been found in the geological column. Some of the most famous are the Oligocene Frio sand and the Jurassic Smackover formation and its associated deposits in the southern USA.

Continental-shelf sedimentation

The continental shelves are important sites of sediment accumulation, and were equally important in the past. They are usually less than 200 m deep but have an average depth that is rather shallower. They are wide on passive continental margins and narrow on active margins. In the past, during periods of higher sea level, these shallow shelf seas extended across much of the present continental surface to form the epeiric seas. Much of the surface of the present shelves was exposed during the low sea levels of the Pleistocene glaciations, and they have only been covered by the present depth of water for a few thousand years.

The surface of the shelves is sometimes very flat, but they may show considerable topographical irregularities and traces of old coastlines and river valleys on their surface. Large valleys called submarine canyons cut the shelf surface and can reach almost to the coast. In tropical areas coral and calcareous algae have sometimes built reefs for hundreds or thousands of kilometres along the shelf margin to produce rimmed shelves.

The shelves are bathed by oceanic waters and are affected by various oceanographical processes. Large waves produced in the open ocean disturb the sediment before ultimately breaking along the shoreline. They affect even the deepest parts of the shelves, although their action is more important in the nearshore areas. The oceanic current systems generally flow parallel to the margin of the continents, but ribbons of these currents flow over the shelf edges and in places produce large gyres.

In areas with a high tidal range, particularly in shallower-water shelf seas such as the North Sea, the sediment on the shelves is stirred by strong tidal currents. Although these currents reverse their flow either diurnally or semidiurnally, they may show strong asymmetry, an important factor in the control of the net direction of sediment transport. Local storms may produce strong winds that give rise to waves and wind-generated currents. Such episodic events have a catastrophic and long-term cumulative effect on shelf sediments.

Coarse-grained sediments are kept close to the coastline by the breaking-wave system, except when storms rip sediment out onto the shelf areas. On the other hand the fine silt and clay (mud) injected into the shelf waters can be transported over vast distances by ocean, tidal and locally produced wind-driven currents before ultimately settling (Figure 20.20). Generally mud is transported parallel to the shoreline and it extends away from the river outlets. It may cover the shelf for hundreds of kilometres parallel to the coastline and shelf margin. Where the shelves are very narrow it may reach the shelf edge. In areas where there is only spasmodic injection of mud, the mud sheet may be only tens of kilometres wide, paralleling and lying seawards of the nearshore belt of sandy sediments constantly stirred by wave action. Where little mud escapes seawards or where particularly stormy conditions or strong tidal currents are common, mud may not settle in appreciable quantities except in sheltered depressions or where the strength of tidal currents decreases. The extensive sheets of mudstones with shallow-marine faunas, such as the Jurassic Oxford and Kimmeridge clays of the UK, are ancient examples of such shelf mud sheets.

Beyond the limits of the encroaching mud sheet, or where no sediment is supplied to the coastline, the only material reaching the shelf is skeletal debris of organisms living on the sea floor or in the overlying water. This is being added to all shelf sediments but dominates only in these areas. The production of calcium carbonate is slow and where, following the post-glacial rise of sea level, the shelves have been covered by the sea for only a few thousand years, the areas away from the blanketing effect of modern terrigenous sediment show the sediment of previous cycles still exposed on the sea floor. The latter are often coarse-grained sands or even gravels; if skeletal debris is present it is old and ecologically out of place. Remains of vertebrates and beds of peat attest to the shallow-water or coastal-plain origin of some of this sediment. These materials are relics of previous conditions and hence they are called 'relict sediments'. However, modern shelf organisms have usually burrowed into these deposits, waves and currents have disturbed them and remains of modern skeletal debris have been incorporated into them. Consequently the term palimpsest sediments is normally used, meaning older sediments that show the imprint of later conditions.

On some shelves, storm and tidal currents produce, in the places where they are strongest, coarse lags of gravel and shell. Sand is transported and fashioned into dunes, large sand-waves and long linear banks, the latter stretching for tens of kilometres and sometimes being several kilometres across and tens of metres high. Mud settles only during quiescent periods and is incorporated as thin wisps in the coarser sediment. Most of the preserved sediment is produced by reworking of older (relict) sediment, although some may be provided by local coastal erosion, as in the case of the east coast of the USA. A sand sheet is produced in such tidal or stormy seas, akin to the mud sheets of other shelves, which covers many thousands of square kilometres. When such a deposit is preserved in the geological column it forms an extensive sandstone with abundant cross-stratification and contains marine fossils. A good example of such a deposit is the Cretaceous Lower Greensand of northern Europe.

In areas where little or no terrigenous mud is reaching the inner shelf and has not done so in the past, the whole shelf is covered with skeletal sands and, in sheltered areas, with carbonate mud. This passes landwards into the skeletal and oolitic sands of the coastal barriers or reefs, and on the shelf edge may pass again into a shelf edge reef. Most of the skeletal sand and mud is produced by biological breakdown of large shells and the complete skeletons of smaller shells, including foraminifera, sponges, molluscs, barnacles, bryozoans, echinoids and calcareous algae. In some areas, eg the Persian Gulf, carbonate mud can be precipitated by the removal of carbon dioxide (CO_2) from seawater by spontaneous blooms of phytoplankton. Off the Yucatan peninsula, large areas of silty mud are produced by the accumulation of the very fine-grained tests of foraminifera. On parts of the shelf of the Caribbean the accumulation of coccoliths has produced large areas of fine-grained sediment. In the past, shelf areas were covered for thousands of square kilometres by such fine-grained carbonate deposits, producing rocks such as the north European Cretaceous chalk.

In the sediments of some carbonate-rich tropical seas the saturated interstitial waters have led to early cementation, and hard rock surfaces are produced on the sea floor. These are colonized and bored into by various organisms and produce very distinctive horizons called 'hard grounds', which are common in ancient carbonate shelf sea deposits (eg the Cretaceous chalk and Jurassic limestones of western Europe).

Upwelling of deep water to replace shelf water driven offshore by

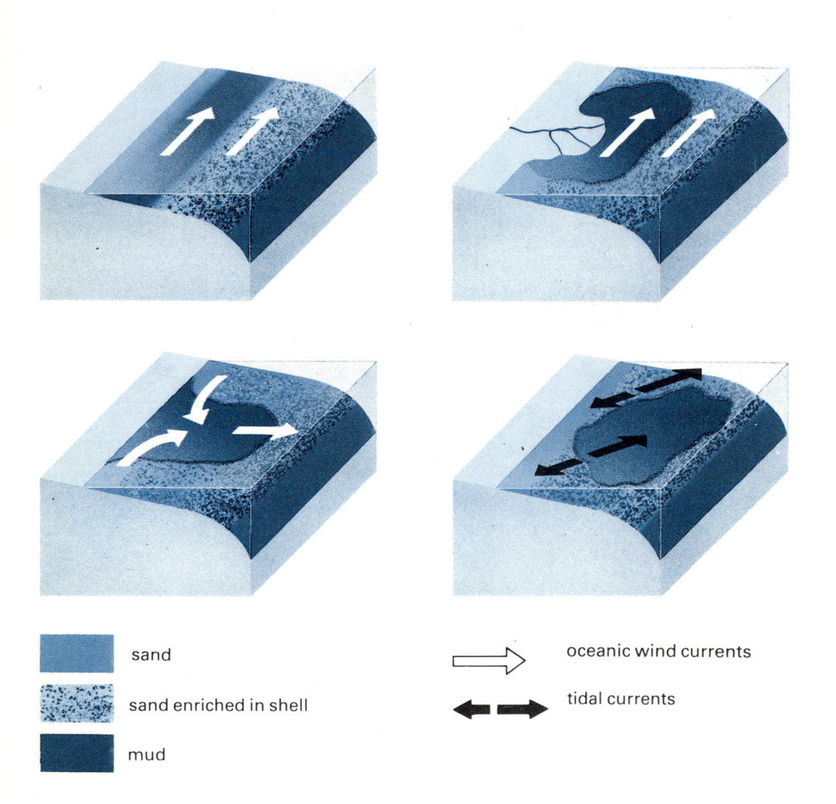

20.20: *Various patterns of mud dispersal and the development of mud sheets on continental shelves.*

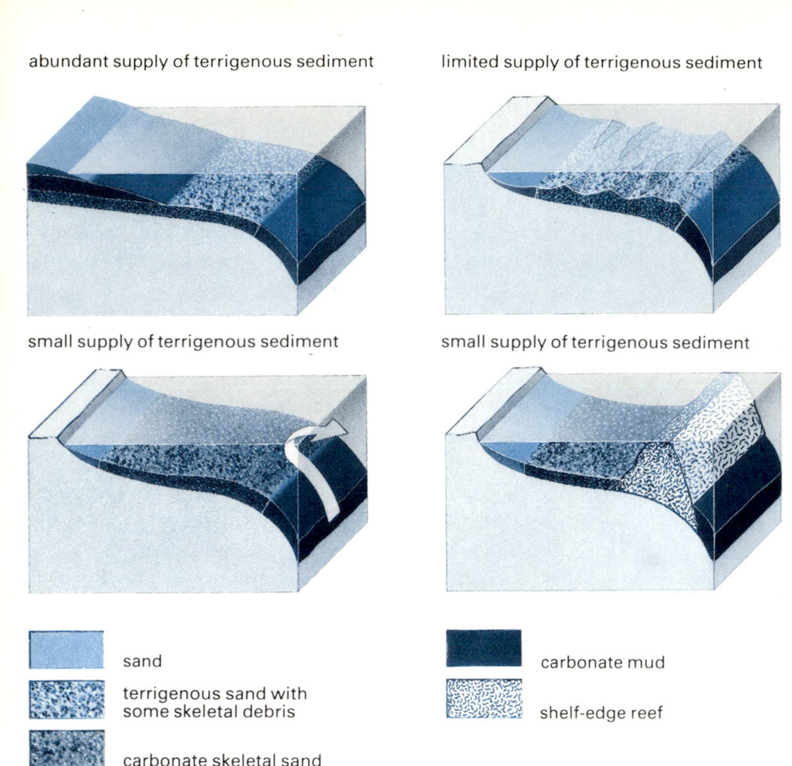

20.21: *Simplified cross-sections of some types of continental shelves and their associated sediments.*

wind action is a particularly common phenomenon on the west coasts of many oceans. The upwelling waters are rich in nutrients such as phosphate, silicate and nitrate. The increased availability of such components leads to the production of enormous amounts of phytoplankton and higher organisms that feed on it. The production of organic matter is thus many times greater than under normal shelf conditions, and bottom sediments become enriched in organic matter. Interstitial waters react with previously deposited calcareous skeletal debris to produce phosphatic deposits as a replacement of skeletal debris and faecal pellets, or as a precipitate between and coating the grains. Such reaction occurs only when the shelf is not being covered by a supply of terrigenous mud. These deposits are common only off tropical desert shorelines or on outer areas remote from a supply of terrigenous mud, such as the shelf off the South African deserts and that off the deserts of Peru. Similar phosphatic deposits are common in ancient shelf limestones showing evidence of slow deposition (condensed sequences) or on surfaces of non-deposition (diastems).

The clay minerals or unstable minerals sometimes react with seawater to produce iron-rich silicate minerals such as glauconite or chamosite. Iron-oxide- and hydroxide-rich clay pellets in nearshore muddy sands of tropical seas pass seawards into chamositic sediments, which in turn pass into glauconitic sediments. Similar lateral changes have been described in Cenozoic sequences found in the eastern USA. The extensively exploited Jurassic ironstones of north-west Europe are thought to have formed in shelf seas.

The huge reef complexes that rim many modern tropical shelves are some of the most impressive effects of marine organic activity. At maximum development, eg in the Great Barrier Reef of Australia, they run for many thousands of kilometres along the shelf edge. The barrier is broken by tidal passes through which oceanic waters can pass onto the shelf area landwards, where small isolated reefs may be developed within the shelter of the main barrier.

Reefs are wave-resistant structures made by organisms such as corals and calcareous algae, forming a rigid framework. This is infilled by broken corals, algal debris as well as shell debris of organisms that live within the shelter of the reef or bore into the coralline skeletons. It is bound together by encrusting calcareous algae, foraminifera and other organisms, and calcium carbonate precipitated in the interstitial spaces. Impressive examples of ancient barrier reefs are the Permian Capitan reef in the western USA and the Devonian Canning Basin reef in Australia. Similar structures have been produced by many other organisms, including rudists (bivalves), and the remains of these are found in Cretaceous rocks of North America, southern Europe and in other parts of the ancient seaway called the Tethys.

Basinal and oceanic sedimentation

Fine-grained sediment is carried in suspension over the edge of the continental shelf by oceanic currents impinging upon the shelf or by currents induced by tides and storms (Figure 20.22). These processes, together with internal waves within the water masses, can produce turbid clouds of water which move downslope under the action of gravity. Such dilute suspensions of fine-grained material lying above the deep ocean floor have been termed nepheloid layers.

The chief avenues for transport of coarse-grained sediment to the deep waters beyond the shelf break are submarine canyons. These huge valleys cut across the continental shelves and lead to large aprons of sediment known as submarine fans. The latter may be quite small where they build into small basins, as in the continental borderlands off California, but where they build onto open ocean floors, as in the case of the Bengal fan in the Indian Ocean, they may be up to 2500 km in length.

On the upper parts of the fans the canyon usually continues as one marked channel bordered by levées, whereas in its middle parts the canyons split into several distributary channels. As the fan profile becomes more gentle the submarine channels become wider and shallower. Some extend seawards beyond the fan onto the flat oceanic, or abyssal, plains. These cover many thousands of square kilometres and are some of the flattest parts of the Earth's surface. Their topography is very different from that of the usual hummocky surface of the deep ocean floor, the bottom having been smoothed by a blanket of sediment.

Submarine fans are best developed oceanwards of a major source of sediment such as a delta (eg the Bengal fan), or below points where shelf sediments are trapped in canyons that cut back onto the shelf. On the west side of the north Atlantic the lower part of the continental margin has a smooth apron at its base with little indication of submarine fans, although it is cut by canyons. This so-called continental rise has been subjected to other phenomena such as contour currents (see below). This relationship of canyons, fans and abyssal plains is analogous to the situation seen on land in arid areas where alluvial fans pass out into playa flats.

The interbedded sediments found in the canyons, in fans and their channels and on abyssal plains are organized into distinctive units with particular sequences of sedimentary structures which, although not exclusive to these deposits, are very characteristic of them. The most striking is graded bedding (Figure 20.23) in which coarse sands occasionally having a shallow-water fauna at the base grade upwards into progressively finer interbedded silt and clay with a pelagic marine fauna through an interval of centimetres to metres. The base of the coarse layer is always sharp, whereas the top grades into the overlying, interbedded, fine sediments. In many cases the coarse layer is structureless at the base and passes up into laminated and then rippled sands. This upward sequence accompanying the grading has been termed the 'Bouma sequence' after the geologist (A. H. Bouma) who drew attention to the common repetitive occurrences of such structures. Each sequence results from the flow of an individual spasmodic current which transported shallow water sediment into the deeper water areas. Normal deep-water pelagic muds then covered the deposit.

Such graded structures had been observed many years ago in ancient sequences. Some geologists, notably Sir Edward Bailey, working in the Highlands of Scotland, had remarked on the difference between well-sorted sands with cross-bedding and ripple structures and muddier sands with graded bedding. He suggested that the latter had been deposited in deep water, unlike the former, which showed normal shallow-water attributes. Furthermore, studies of such sediments in many parts of the geological column, for instance in some of the Lower Palaeozoic sediments of Britain and the Mesozoic–Cenozoic sediments of California and Europe, where they are known as flysch, showed very distinct structures on the underside of the coarse horizons: downward bulges and streaks of sand and other curious markings (sole marks). These are now known to have been produced by currents that scoured a surface of semiconsolidated mud and then, as the current waned, by infilling the scoured surface with sand. Such structures have proved invaluable for determining the flow direction of currents in ancient rocks.

20.22: *A generalized sketch showing the method of dispersal of terrigenous sediment into the deep ocean.*

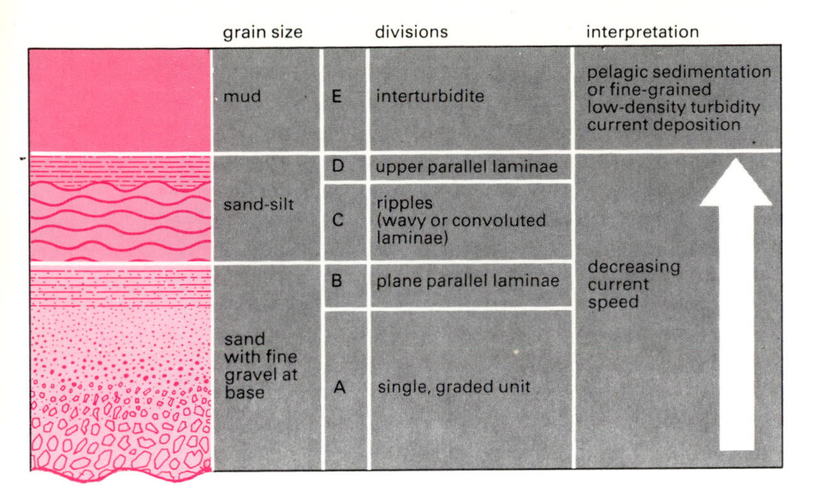

20.23: *An idealized sedimentary sequence deposited by a turbidity current.*

Such sediments are thought to be deposited by turbidity currents, currents made dense by their load of sediment. Because their density is greater than that of seawater they flow down the slope, hugging the bottom and eroding the sea floor as they progress, and finally slowing down and depositing their load. Very dense, turbid water issuing from river mouths or deltas, or driven out from estuaries and lagoons in time of flood, can produce such bottom-hugging currents. Slumping, either due to sediment becoming unstable on steep slopes or being disturbed by seismic shocks, can initiate slides of unconsolidated sediment. As they become more dilute, these can develop into turbidity currents. Where the heads of canyons reach close inshore into the breaker zone they are frequently emptied of sand by sudden down-slope movements. Studies of modern sediments in submarine canyons, fans and on other parts of the steep continental slope, and of sequences in the geological column that contain such sediments, have revealed examples of all gradations; gravity-induced slumped sediments to debris flows consisting of large boulders or rafts of sediment in a muddy matrix and sediments deposited from turbidity currents (turbidites). The debris flows and slumps are more characteristic of the proximal parts of submarine fans where slopes are steeper, or of any steep submarine slope, and the turbidites are found on the lower fan, inter-channel areas and abyssal plain where the slopes are more gentle.

Nowhere have turbidity currents of the size envisaged been observed in modern seas. However, indirect evidence has come from the repeated breaking of submarine telephone cables in various parts of the world. The most celebrated is the sequence of breaks that occurred in 1929 in some cables lying along the continental slope off the Grand Banks of Newfoundland. After an earthquake, cables were broken in sequence down the continental slope. The timing of the breaks and the distances separating them allowed the speed of the destructive agent, thought to be a turbidity current, to be calculated; in places it reached 55 km per hour. Many other breaks have been reported from all over the world, and broken cables which were entwined with shallow-water plant debris that was still green have been retrieved from the deep sea bed.

Investigations of the apron of sediment—the continental rise—at the foot of the continental slope has shown that sediment can be transported parallel to the rises. Deep ocean currents, which are particularly strong on the western sides of oceans, can move mud and fine sand. These contour currents have produced thin layers (contourites) of well-sorted sand and silt, often showing ripple cross-stratification. Unlike the turbidites they have sharp upper and lower contacts with the enclosing fine-grained, normal deep-water sediment and show quite distinctive structures. It is claimed that they can also be recognized in inferred deep-water sediments of Cenozoic age in the Alps and other areas.

Sea-level changes affect deep-water sedimentation. As sea level fell during the Pleistocene ice ages the coastline moved towards the edge of the continental shelf. Rivers poured their load of sediment directly onto the upper slope and into the heads of submarine canyons that are now separated from the large rivers and their deltas. Slumping and the generation of turbidity currents were more common under such conditions, and generally the transfer of continental detritus to the deep sea floor was more important than under present-day conditions. Many of the large submarine fans off large canyons related to major rivers were built in such periods and are receiving virtually no sediment today because of the remoteness of the contemporary coastline.

Wind and ice also distribute fine-grained sediment seawards of the shelf. Fine-grained material transported by winds from arid areas can reach even the remotest parts of the ocean floor. High-level winds and jet streams are particularly effective in transporting continental debris, which is either picked up from the land surface by lower winds or derived from material thrown into the air by volcanic processes. Dusts of obvious continental origin have been collected in the air over many parts of the ocean surface; fine-grained quartz is very characteristic of oceanic areas offshore from desert.

Icebergs, calving off the front of large ice sheets, are laden with sediment of a whole variety of grain sizes, ranging from large boulders to clays. As these icebergs drift from the margins of the ice sheets they melt and drop their sediment load over the deep sea floor. Unlike the bottom-flowing turbidity currents, but similar to the aeolian dusts, this debris is not restricted to submarine channels and drapes the existing submarine topography.

Intermixed with the products of continental denudation along the continental margins, and covering the floors of the oceans, there is biogenous sediment. This is partly produced by organisms living on the sea floor (benthos), but because of the absence of light, the lower temperature and the limited availability of food, the density of bottom-living creatures producing skeletal debris decreases markedly with depth. However, the surface waters abound in floating and swimming forms of such organisms. Pelagic fauna and flora produce a constant rain of skeletal debris which settles to the sea floor, although it is diluted by remains of shallow-water benthonic organisms in shallow areas. The most important producers of deep-sea biogenous sediment are the phytoplankton and zooplankton (Figure 20.24). The main phytoplankton are diatoms that produce siliceous tests composed of poorly crystalline SiO_2 (opal) and the coccolithophores that produce skeletons of calcium carbonate (calcite). The most important zooplankton are radiolarians, producing siliceous tests of opal, and the foraminifera that produce tests of calcium carbonate. In some areas, such as the Mediterranean, planktonic molluscs are very important producers of calcium carbonate, but their remains rarely equal those of the foraminifera in deep-sea deposits.

The importance of the pelagic extraction of calcium carbonate and silica has increased throughout geological history. Whereas radiolaria are known in most post-Cambrian deposits, diatoms did not become abundant until Cretaceous times. Similarly the coccoliths and other nannoplankton appeared and diversified in late Jurassic times. The foraminifera increased enormously in abundance in late Cretaceous times. Calcium carbonate is the dominant biogenic constituent of Mesozoic and younger deep-sea sediments (Figure 20.25). Where there is more than 30 per cent planktonic skeletal material, the term calcareous or siliceous ooze is used.

Away from the diluting effect of terrigenous sediment the abundance of calcareous ooze depends initially on the rate of production of calcareous tests in the upper (photic) water zone of the oceans. Whereas surface oceanic water is supersaturated with respect to calcium carbonate, a large portion of the oceanic waters below about 2500 m is undersaturated. Consequently, as the skeletal remains settle through the water they are dissolved below a particular depth. The level at which a noticeable increase in solution of the tests takes place is known as the lysocline. The depth at which the calcium carbonate is completely removed from the sediments (about 1000 m below the lysocline) is termed the carbonate compensation depth. It does not correspond to the level at which seawater becomes undersaturated because dissolution is not instantaneous and the complete removal of calcium carbonate takes a considerable time. However, in most oceans calcium carbonate has been removed from sediments below 4000–5000 m. The removal of the skeletal remains is partly dependent upon the mineralogy of the skeletons. Hence pteropods (planktonic molluscs), being composed of aragonite, are removed at shallow depths, and their remains usually occur only on the upper parts of seamounts and on the crests of ridges.

In the deep ocean, remote from land areas, and in places too deep for the preservation of calcareous and siliceous oozes, the sea floor is covered with a chocolate-brown clay (called red clay). This is composed of clay minerals and fine quartz derived from continental areas, mixed with feldspar, pyroxenes and other minerals derived from volcanic sources, and contains complex iron-manganese oxides and hydroxides. It also contains minerals that have grown in place on the sea floor by reaction between unstable minerals, including clays, and seawater. The most common of these authigenic minerals are zeolites, eg phillipsite and clinoptilolite. Small cosmic spherules are rather exotic components that must have come from outer space. They are found in all sediments but are particularly noticeable in deep-water deposits because of the slow rate of deposition there.

Ferromanganese nodules are a component of deep-sea deposits that have attracted considerable interest since they were first described. The ferromanganese deposits commonly contain sufficiently high concentrations of nickel, copper, zinc, cobalt etc to allow them to be mined commercially in some areas. Various processes have generated these sediments, including slow precipitation of metals from seawater by chemical precipitation or settling of particulate matter; remobilization of metals due to reduction and solution within the sediment column and its associated organic remains; weathering of submarine volcanic rocks by seawater; and hydrothermal precipitation from solutions emanating from oceanic lavas, particularly at the crests of oceanic ridges, which react with seawater.

Debate has continued for almost a hundred years on the possibility that rocks analogous to present-day deep-sea sediments might be found on land. Some fine-grained limestone, radiolarian cherts and red mudstones containing ferromanganese nodules have all been interpreted as deep-water deposits. When these deposits are associated with complexes of basic and ultrabasic igneous rock they are thought to represent fragments of oceanic floor. The rocks of Timor and Barbados are such deposits, and most geologists would accept them as being of deep-sea origin. Modern concepts of plate tectonics have removed many problems that bedevilled earlier discussion, eg the permanence

20.24: *Typical deep sea deposits. Left, the protozoa* Sarcodina radiolarida, *some of the zooplankton forming a radiolarian ooze. Right, the protozoa* Sarcodina foraminiferida, *some of the phytoplankton in a foramimferan ooze.*

of oceanic and continental areas, but the exact depth of deposition of many ancient sequences is still a subject for heated argument.

Abnormal environments of deposition

Deposition of organic matter

Organic material is a ubiquitous but minor component of almost all sediments. Some accumulations of vegetable matter go to form peat and ultimately coal. Other important accumulations occur in lakes, producing oil shales. However, the most important accumulations occur dispersed in normal marine sediments.

The process that leads to the primary production of organic matter is photosynthesis. This process is the conversion of light energy to chemical energy. The hydrogen of water is combined with carbon dioxide in water and air to form glucose and oxygen. These in turn are converted to cellulose, starch and other organic compounds that make up plant matter.

In the aqueous realm organic matter is produced primarily by phytoplankton, forming the base of the food chain. They produce 550×10^9 tonnes per year in ocean waters, and small amounts in continental water bodies. Mixed with this material is that produced by zooplankton, plant and animal benthos. Bacteria attack and break down this organic matter and are themselves usually considered an important source of it. In coastal areas plant debris and other organic detritus from the land may be a major source of organic matter. This is supplemented by small amounts carried in solution and absorbed on fine-grained clay minerals in river waters.

All marine sediments usually contain some organic matter, but because it is light and fine-grained it accumulates preferentially with fine-grained sediments. A large amount of organic matter may be produced locally or be introduced into a local environment. Chemical and biological attack as it settles through the water column means that only a small amount of that produced reaches the sea floor. In deep parts of the ocean virtually all the organic matter is destroyed during its descent and consequently the sediments are poor in it. Even when organic matter reaches the sea floor it may not survive, as it may be consumed by bottom-living organisms and further oxidized and dissolved. Burrowing by organisms replenishes the oxygen of the interstitial waters of the sediments and allows such chemical changes to continue. However, in undisturbed, fine-grained sediments, the movement of fluids is so slow that oxygen in the interstitial water is soon exhausted.

Breakdown of organic matter under anaerobic conditions becomes important at this stage. Biologically produced organic polymers, proteins and carbohydrates are destroyed and their constituents are recombined in new poly-condensed structures. Oxygen for these metabolic processes is produced by the bacterial breakdown of the sulphate in seawater, hydrogen sulphide forming as a by-product. This precipitates iron from solution as iron sulphide (pyrite or marcasite). Some other cations, eg zinc and copper, also precipitate as sulphides. At this stage most of the carboxyl groups have been removed. The

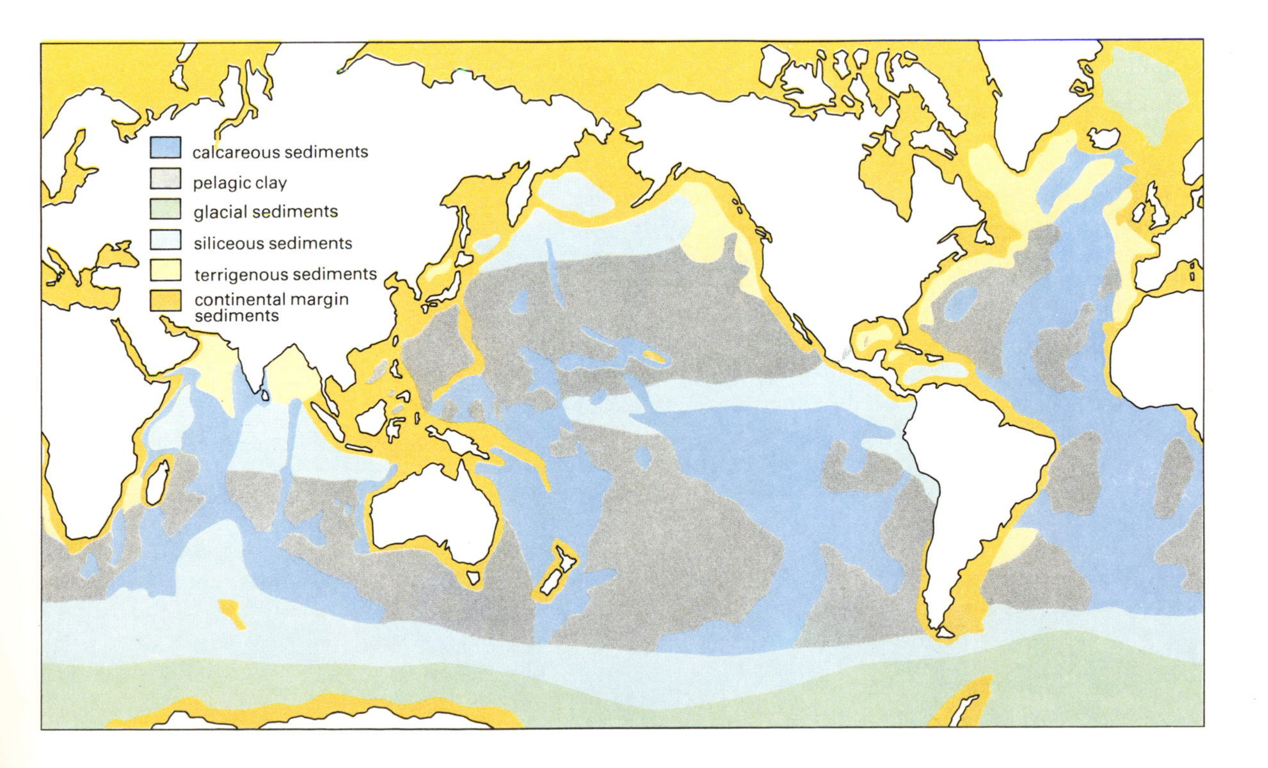

20.25: *The distribution of the major components of present-day sediments on the deep-ocean floor.*

organic matter consists mainly of condensed benzene rings linked together by hydrocarbon chains (kerogen), mixed with some soluble organic material (bitumen) and traces of fluid hydrocarbons. Disseminated kerogen is the most important organic constituent of marine sediments, and the most important form of organic matter on Earth. It is one thousand times more abundant than coal and oil in natural reservoirs, and fifty times more abundant than bitumen and dispersed hydrocarbons in rocks. Kerogen forms 80 to 99 per cent of organic matter in ancient rocks. When the sediments are buried by later deposits the pressure and temperature rise, and kerogen is thermally degraded to produce liquid petroleum and gas, the proportions depending upon the kerogen type.

Under special circumstances the water column is depleted in oxygen and this allows the falling organic matter to reach the underlying sea floor in greater quantities than under normal conditions. Such conditions have recurred throughout the history of the Earth, and good examples can be seen in modern seas (Figure 20.26).

Oxygen depletion may occur locally in lagoons, but these are usually sufficiently stirred by waves or currents for the condition to be ephemeral and for the bottom waters to contain oxygen. The waters of most shelf seas are oxygenated to the sediment–water interface. However, in some shelf seas, such as the Baltic, and deep coastal inlets, such as fjords, the water column is very stratified, and oxygenated conditions do not continue from surface to bottom. This is due to an abundance of freshwater run-off from land. The lighter freshwater lies upon the denser saltwater to produce a stable stratification, and the latter becomes depleted in oxygen because of oxidation of organic material descending from the surface. The oxygen-depleted lower parts of the water column, because of the stable water stratification, are only rarely disturbed by storm events and hence there is an enrichment of the bottom sediments in organic matter. Waters occupying isolated topographical depressions on open shelves develop similar anoxic conditions.

Widespread masses of anoxic waters are found in some large semi-enclosed basins where the water may be up to 2000 m deep. This occurs in the Black Sea, a deep and very large basin, which, because of the large run-off of freshwater from the adjacent land and its separation from the Mediterranean by a very shallow sill, the Bosphorous, has developed a stagnant lower water column below depths of about 200 m.

A similar stratification could of course develop in deep saline basins. The dense bottom water would not be easily disturbed and hence could develop anoxic conditions. Indeed, in many ancient evaporitic basins black shales rich in organic matter and heavy metals (eg the Permian Marl Slate of northern Europe) are found associated with evaporites.

Generally, beyond the shelf break in the open ocean and in other deep basins the waters are well oxygenated. However, in some areas where upwelling occurs the rate of production of organic matter may be so great that the normally developed oxygen minimum layer of the oceans is particularly well developed, as in the Gulf of California, Mexico. Where this intersects the sea floor the bottom sediments are impoverished in respect of bottom-dwelling organisms and enriched in organic matter.

Elsewhere, deep basins in the open sea, without emergent barriers, can develop poorly oxygenated conditions. The Santa Barbara basin off the Californian coast, USA, is one of a series of deep basins of intermediate depth between the shelf break and the ocean floor. Whereas most of the adjacent basins are connected, and normal oxygenated bottom water can reach even their deepest parts, the Santa Barbara basin has no such connection. Consequently the bottom waters below the level of the submerged barrier are rarely disturbed and are the site of the accumulation of sediments rich in organic matter, as the bottom waters are poor in oxygen.

Deep-water sediments beyond the foot of the slope are usually bathed in well-oxygenated water and contain low amounts of organic matter. However, trenches such as the Caraico trench off Venezuela are isolated from such water movements and organic-rich sediments have accumulated in spite of the great depth of water.

Even in normal marine basins and oceans such anoxic events can develop. Thin layers of organic-rich, well-laminated sediment found in the Quaternary sequence on the floor of the Mediterranean indicate that anaerobic conditions sometimes existed for a short period over the whole eastern Mediterranean basin. The widespread presence of ancient dark mudstones and shales suggests that such anoxic events have often been widespread at particular times in the history of the Earth.

The characteristic feature of the organic-rich sediments is their well-laminated structure (Figure 20.27), as the anaerobic conditions prevent bottom organisms from colonizing and disturbing the sediment. Organic-rich laminae alternate with others rich in detrital matter or tests of plankton. Iron (pyrite) and other sulphides are often abundant in these organic-rich sediments which can contain relatively large amounts of metals (zinc, lead and uranium).

Deposition of evaporites

Evaporation of seawater leads to the deposition of marine evaporites. Evaporites are being deposited at the present day in coastal plains and lagoons of arid subtropical areas and in some lakes or inland seas, but these deposits are less widespread than formerly. Nowhere are extensive subaqueous deposits accumulating. Even in the Dead Sea, with its exceptionally high salinities, there is no extensive deposition of subaqueous evaporites. Very special conditions must obviously have occurred to form the huge marine evaporite deposits (salt giants) found in many ancient sedimentary basins.

The essential process for the production of evaporites is the evaporation of a saline solution. A very large amount of seawater needs to be evaporated to produce a significant thickness of evaporite. The resulting sediment consists mainly of halite (NaCl), preceded by some calcite and dolomite ($CaCO_3$ and $CaMg(CO_3)_2$) and by gypsum or anhydrite ($CaSO_4.2H_2O$ and $CaSO_4$), and succeeded by small amounts of other magnesium and potassium salts. Calcium and magnesium carbonates, calcium sulphate and sodium chloride would be in the proportions of 1:10:200 (excluding the insoluble salts of magnesium and potassium as they are of minor volumetric importance). However, this ratio is markedly different from the relative proportions of the various evaporitic sediments found in any one ancient sedimentary basin, and it is clear that simple evaporation of a standing body of water cannot be invoked.

Several simple models for evaporite basins have been proposed (Figure 20.28). The end-product of the simple evaporation of a standing body of water in a closed basin would be a concentric series of belts of evaporitic minerals. No example of such a simple closed basin is known from the geological record. The amount of salts formed would be very thin. Thus if the water was 1000 m deep only 16.5 m of salts would be precipitated. However, if the basin were replenished with water each time it evaporated, a series of such sequences could pile up to give a rhythmic deposit. A sequence approximating to this is found

20.26: *Various situations in which sediments rich in organic matter can accumulate.*

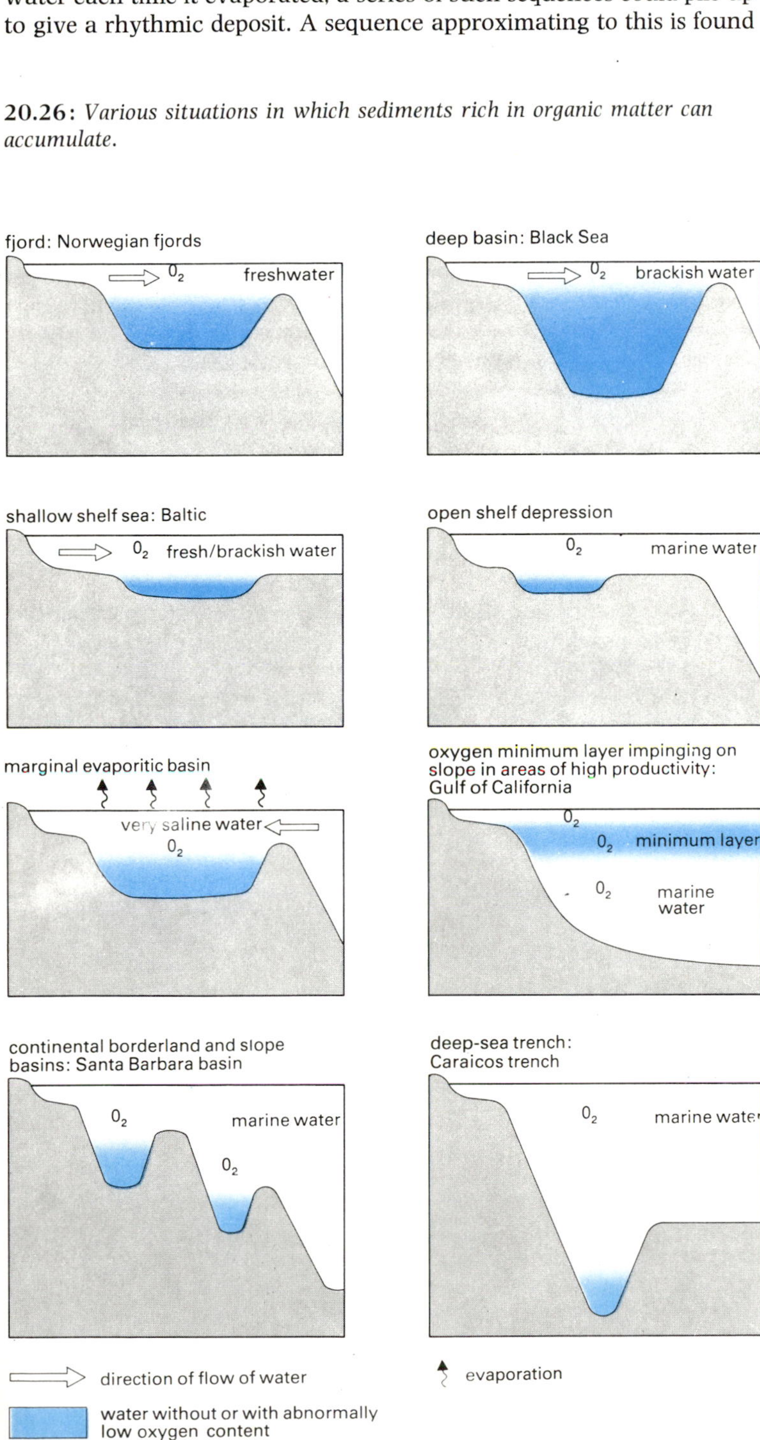

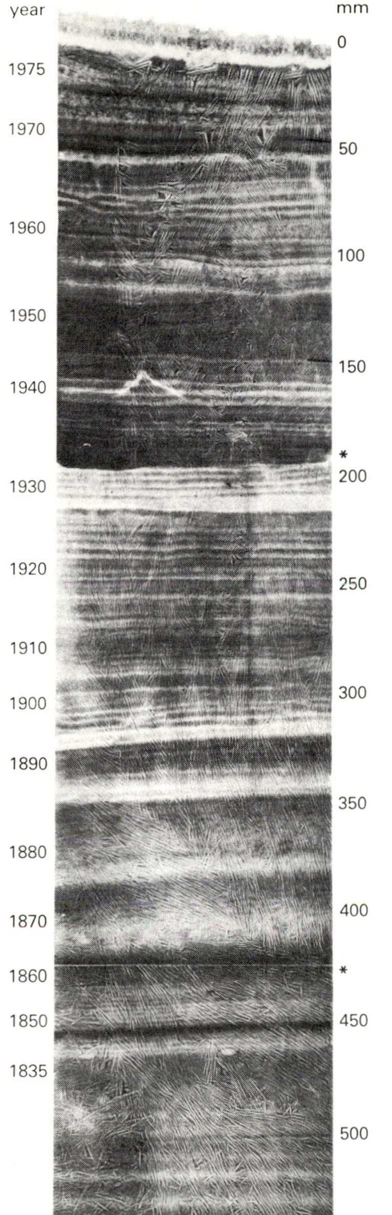

20.27: *Laminated sediments deposited on the floor of the Santa Barbara basin off California, USA.*

in the Permian Zechstein evaporites of the North Sea.

The other simple model is that of the barred basin. In this case water is separated from the main oceanic mass by a barrier. Evaporation increases away from the entrance, producing a series of monomineralic deposits arranged laterally, rather than vertically as in the simple evaporating-pan model. The Silurian evaporites of the Michigan basin, USA, conform in part to this model although there were apparently periods of complete desiccation. A rather similar situation could develop in an open-channel evaporite basin. An example of this is the Carboniferous evaporites of the central Colorado basin, USA.

20.28: *Various possible types of evaporite accumulation.*

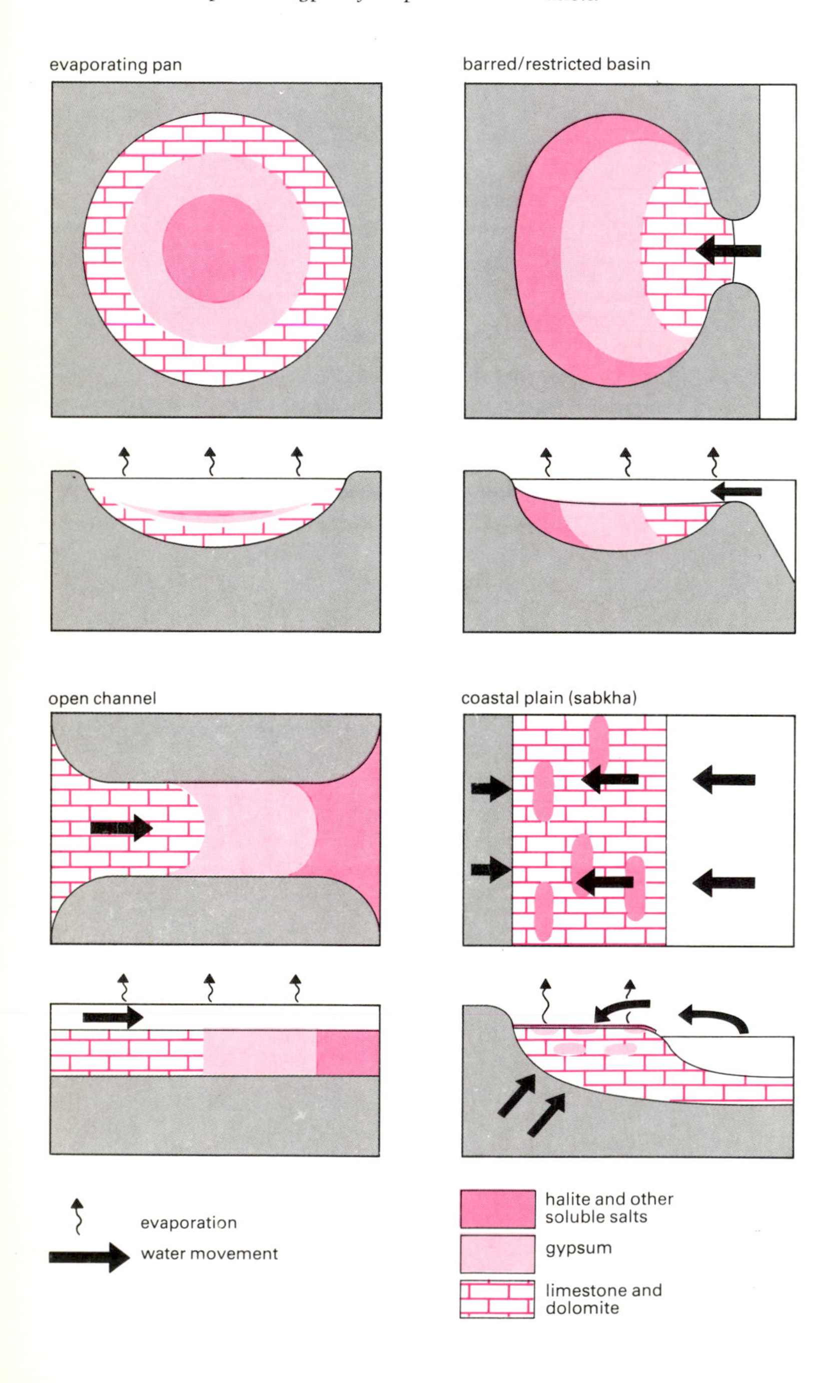

Modern studies of evaporitic sediments have revealed a wide variety of sedimentary sequences and structures. Some evaporites, particularly the gypsum-anhydrite rocks, show irregular bedding and nodular contorted structures. They contain evidence of the former presence of blue-green algal mats (stromatolites) and of complete desiccation (mudcracks). These structures and the associated sediments indicate that they formed in shallow lagoons and on intertidal and supratidal flats, and that they are the ancient analogues of modern arid coastal plain (sabkha) deposits seen forming in the Persian Gulf. Other evaporites show normal sedimentary structures (cross-bedding, rippling, graded-bedding etc), and the grains of gypsum have obviously been transported and deposited by the normal traction currents or by turbidity currents.

Elsewhere, sheaves of coarse crystals, centimetres to several metres in height, are found standing vertically or almost vertically, clearly having grown up subaqueously from the sea floor into the surrounding and overlying brine. In other cases beautifully laminated evaporites, packages of which can be traced over vast distances (hundreds of kilometres), indicate very forcibly that deposition occurred under quiescent conditions where no bottom organisms existed. The laminations often consist of pure anhydrite interlaminated with grey-black layers rich in dolomite and organic matter. A pair of such laminae may be between 0.2 and 2.0 mm or even 10.0 mm thick. Their paired nature and widespread extent have led to suggestions that they are seasonal. Similar laminations in halite rocks are coarser.

Interbedded with such laminated sediments are graded units. These seem to be the result of turbidity currents which transported grains of gypsum from shallow waters to the quieter and presumably deeper waters in which the laminated rocks formed. More puzzling are the occasional layers showing evidence of supratidal–intertidal deposition, ie nodular and contorted structures. These indicate that sometimes the entire basin of deposition dried out.

It thus appears that evaporites can develop in a range of environments from coastal plains to deep-water areas; but in contrast to normal seas the waters of the depositional basins were highly saline. One setting with which they have been particularly associated is the newly rifted juvenile ocean basins such as the Red Sea.

Conclusion

Understanding of the processes and environments of sedimentation is critical to the interpretation of Earth history, since it is largely through the preservation of sediments as sedimentary rocks that that history is recorded (see Chapter 23). It is of course not only history that is locked up in the Earth's great sedimentary basins, and the need to discover ever more reservoirs of oil and gas has been the major stimulus to sedimentological research for many years, leading to new discoveries over the whole range of scales, from the microscopic structure of shales and sandstones to the analysis of the largest basins of deposition. Thus from sand, clay and lime come not only cement, but knowledge of the Earth's past, and energy for its future.

21 The origins and early evolution of life

The problem of the origin of life

Of all the puzzles in the history of life perhaps the most fundamental and perplexing is that of the origin of life. Darwin did not in his published work commit himself to a view on life's origin, but in a letter to Hooker in 1871 he speculated that 'But if (and on what a big if) we could conceive in some warm little pond with all sorts of ammonia and phosphoric salts—light, heat, electricity etc present, that a protein compound was chemically formed, ready to undergo still more complex changes, at the present day such matter would be instantly devoured, or absorbed, which would not have been the case before living creatures were formed'.

There is no simple, straightforward definition of the term 'life'. It is here regarded as a self-replicating, energy-storing system which can produce either similar offspring or mutated offspring capable of producing something different which, in turn, is potentially self-replicating. The ability to grow, together with the potential to produce something different by way of mutation, separates living from non-living systems such as crystals.

Traditionally theologians contend that life is the result of a supernatural event. A related hypothesis is of the spontaneous generation of life from non-living matter, a view widely held in both eastern and western religions and by naturalists well into the nineteenth century. In scientific circles, by the beginning of the twentieth century, two views on the origin of life had come to the fore. According to one view life is coeternal with matter and has no beginning, having propagated from one solar system to another by means of microbial spores. This is the theory of panspermia and is said to have resulted in the Earth receiving the seeds of life early in its history, possibly as a result of the deliberate action of intelligent, extraterrestrial beings (directed panspermia). According to the other hypothesis life arose early on Earth by a series of successive chemical reactions. This chemical evolution, or biopoesis, is a sort of spontaneous generation in the sense that life would in this way have arisen from non-living precursors; once chemical evolution had produced a cell capable of converting energy and matter into an offspring that would grow and resemble the parent, biological evolution could begin.

Chemical evolution and the origin of life itself are tied up with the early physical evolution of the planet. The Earth is some 4550 million years (Ma) old. During its earliest evolution the Earth heated up because of accretion, gravitational compression and radioactive decay. Following the catastrophic formation of the core and its differentiation, convection began to cool the Earth. Critical for the origin of life are surface temperatures which allow water to exist as a liquid. How soon after the formation of the Earth surface temperatures within this critical range occurred is conjectural, but there is little doubt that it was before 3800 Ma ago, the age of the oldest sedimentary rocks.

During the period from its formation to 3800 Ma ago the Earth, in common with the Moon and the other terrestrial planets, was subjected to an intense bombardment of meteoroids and asteroids. This may have affected chemical pathways to the origin of life. The shock waves and heat generated by impact might have increased reaction rates. These extraterrestrial bodies could have introduced carbon-containing compounds, raw materials for monomers or even some of the monomers themselves, and various trace elements necessary for life such as molybdenum. On the other hand this bombardment episode could have been catastrophic, destroying any early life or temporarily disrupting biopoesis.

Experimental evidence for possible mechanisms

Assuming that life originated on Earth, scientific research into chemical evolution has produced an explanation for events leading up to the formation of a living entity, but not yet for its actual creation. As early as 1783, when C. W. Scheele produced hydrogen cyanide (HCN), organic compounds have been synthesized from inorganic precursors. The production of these polyatomic organic molecules forms the basis of contemporary theories and research on the origin of life. As R. E. Dickerson has stated (1978): 'Within one billion years after the formation of the Earth 4.6 billion years ago one-celled organisms had evolved out of organic molecules produced nonbiologically in an atmosphere containing no free oxygen'.

The Earth is of a size and at a distance from its star, the Sun, that enables it to hold gravitationally all the elements except hydrogen. Water is in liquid form and its three phases—solid (in the lithosphere), liquid (hydrosphere) and gaseous (atmosphere)—can coexist. Life must have appeared in this physical setting. In the 1920s A. I. Oparin and J. B. S. Haldane proposed, independently, similar models for the origin of life which in modified forms are widely held today. According to these models organic compounds serving as the building blocks of life were formed from inorganic precursors; life originated on an Earth with a reducing atmosphere (having no free oxygen) composed of methane (CH_4), ammonia (NH_3), water (H_2O) and hydrogen (H_2) and also, possibly, carbon monoxide (CO), carbon dioxide (CO_2) and nitrogen (N_2); and an energy source, such as ultraviolet radiation, synthesized organic compounds from the primitive atmosphere.

In 1953 Stanley Miller, working under Professor Harold Urey at the University of Chicago, reported on the production of amino acids under experimental conditions designed to simulate conditions on the primitive Earth (Figure 21.1). An electric spark was discharged in an apparatus filled with CH_4, NH_3, H_2O (as water vapour) and H_2 (but no O_2). This key experiment, called the Miller–Urey experiment, brought the problem of the origin of life into the laboratory. Other experiments have also produced a variety of organic compounds. Fischer–Tropisch reactions, for example, involve a gas mixture of CO, H_2 and NH_3 at high temperatures (600–900°C) in closed containers with metallic ore catalyzers. Such experiments are believed to simulate reactions between a primitive atmosphere and extrusive igneous lavas. The Miller–Urey apparatus, Fischer–Tropisch reactions and other experiments have yielded several hundred amino acids (including many of the twenty important for biological systems), fatty acids, hydrocarbons, nitrogenous bases, small organic compounds and some polymers.

Other energy sources besides electric discharge and simulated volcanic heat have been used or considered possible in the production of monomers: solar heat; shock waves generated from thunderstorms, meteorite impact or a meteor passing through the atmosphere; beta and gamma rays from radioactive decay; and cosmic rays. Experiments in chemical evolution share common features. All are intended to simulate a primitive atmosphere made up of gases whose constituents are carbon, nitrogen, hydrogen and water vapour; an adequate and sustained energy source is needed of an amount and/or duration sufficient to permit synthetic reactions while not destroying the products; there must be no free oxygen (O_2) in the system. Life on Earth probably could not have originated if molecular oxygen was present either at the sites at which synthetic reactions were taking place, as oxygen destroys the products, or where and at the time when life first appeared. Today, oxygen-mediating enzymes not then available such as catalase, peroxidase and superoxide dismutase prevent cellular oxidation.

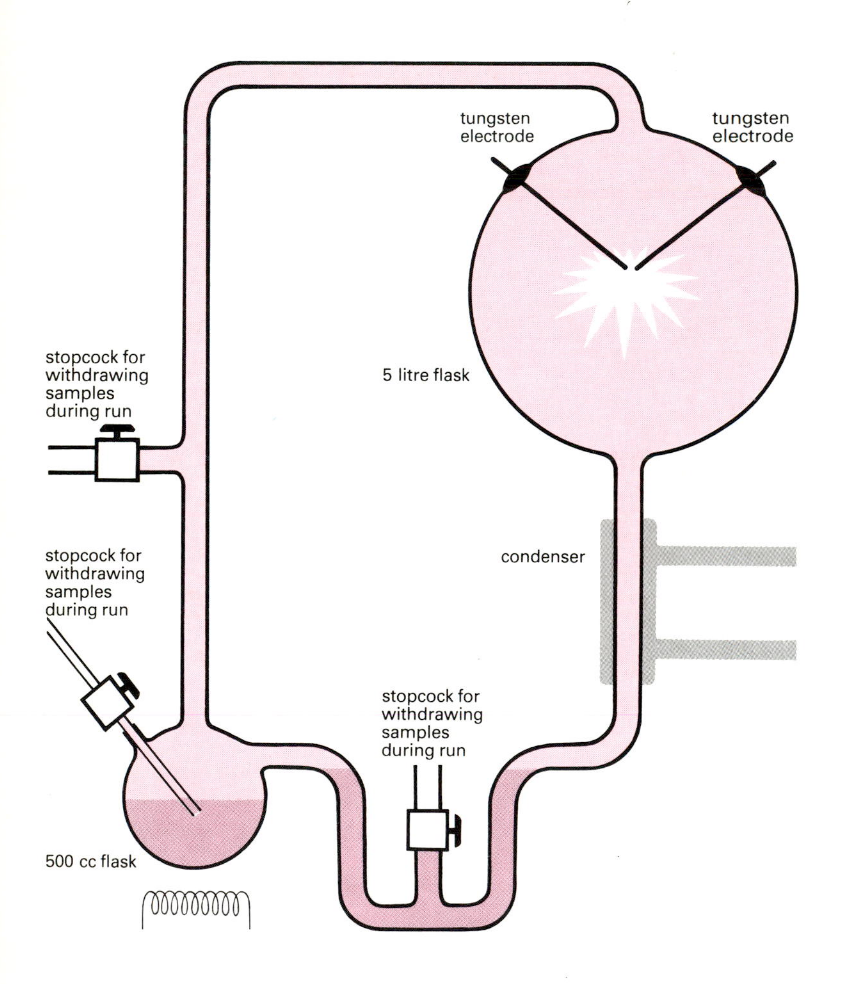

21.1: *The Miller–Urey apparatus.*

Knowledge of extraterrestrial carbon compounds influences current ideas on the origins of life. The presence of organic compounds in meteorites was established in 1834 by Berzelius, and today an extensive literature on these carbonaceous chondrites exists. Radio-astronomy has detected a variety of organic compounds in interstellar space; some fifty-five have so far been detected, including ammonia (NH_3), water (H_2O), formaldehyde (HCHO), carbon monoxide (CO), hydrogen cyanide (HCN), molecular hydrogen (H), methanol (CH_3OH), ethanol (C_2H_5OH) and formic acid (HCOOH). It is thus conceivable that some of the organic compounds that are basic to biopoesis are foreign, but that life itself originated on the Earth.

Experimentally produced amino acids include many that are not found in biological systems. Only twenty amino acids are essential to life. In chemical evolution initial molecular experimentation was a random process, and the protein-building set of twenty amino acids probably represent a combination that happened to survive, without necessarily being the combination that might be predicted theoretically. All life, except for a few bacteria, makes use of only one of two possible configurations of optically active molecules—the L (left-handed) configuration. Experimentally produced amino acids contain an equal mixture of both D (right-handed) and L isomers, and it was probably chance that the L isomer was chosen. Once one configuration occurred similar isomers would be favoured over the other. Mixtures of L and D isomers were not favoured because efficiency increases when only one type is required to bind with an enzyme.

Experiments involving the abiogenic synthesis of organic compounds have concentrated on the production of amino acids (which are the easiest and most straightforward to synthesize). Aldehydes, in common with formaldehyde, and hydrogen cyanide can polymerize into amino acids (Figure 21.2). Proteins can then be produced by the polymerization of amino acids into polypeptides. The production of proteins or 'metabolism' is only half the essence of life, the other half involves reproduction. For this, nucleic acids are needed. Nucleic acids are composed of five nitrogenous bases (adenine, guanine and cytosine (purines) and thyamine and uracil (pyrimidines)), a five-carbon sugar and a phosphoric acid. There are two different kinds of nucleic acid—ribonucleic acid (RNA) and deoxyribonucleic acid (DNA).

The sugar ribose can be synthesized by a multistep reaction involving five molecules of formaldehyde (Figure 21.3). Nitrogenous bases such as adenine are readily synthesized, adenine being simply a pentamer of hydrogen cyanide. Nucleotides—a nitrogenous base plus a sugar—have proved very difficult to synthesize under simulated prebiotic conditions because the proper linkage between the sugar and the base has not been found. A nucleotide is produced (Figure 21.4) by reaction with a phosphate, and the polymerization of nucleosides and phosphates by dehydration give rise to RNA and DNA. The chemistry seems straightforward, and yet no one has been able to make a DNA molecule. The earliest nucleic acid was probably a ribonucleic acid (RNA), which today is found in most viruses and in the cytoplasm of cells. The loss of one atom of oxygen per ribose sugar of every nucleotide unit would give rise to the more stable deoxyribonucleic acid (DNA). It is the DNA molecule that brings about reliable replication and will even pass on mutations.

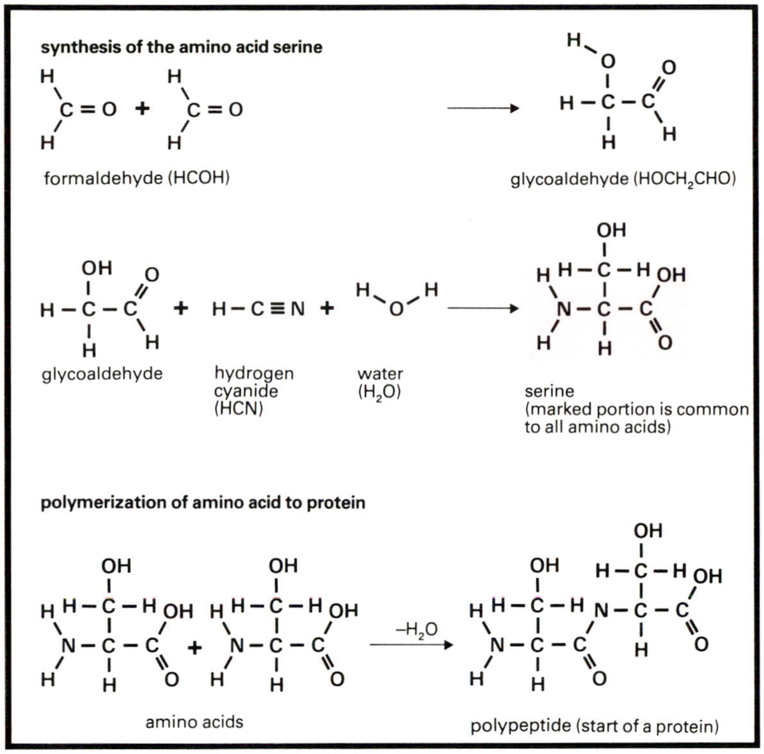

Sugars can be formed by the condensation of formaldehyde though they have not been detected in experiments on chemical evolution. The reaction is not regarded as significant, for sugars produced in aqueous solutions are unstable; in addition they react with experimentally produced intermediates and amino acids. To overcome this problem chemical evolutionists envisage a microniche where sugars could be synthesized, stored and react with nitrogenous bases and in turn with phosphates to form nucleic acids.

How were all these materials concentrated and organized into a living cell? Polymerization can occur spontaneously in highly concentrated solutions. This may have been unlikely on the primitive Earth unless, in Darwin's warm little pond, a prebiotic soup was concentrated by evaporation. Evaporation would also solve the problem of dehydration reactions which accompany polymerization. Oparin envisages the formation of colloidal droplets of organic matter called coacervates. Conceivably these coacervates could dissolve or absorb

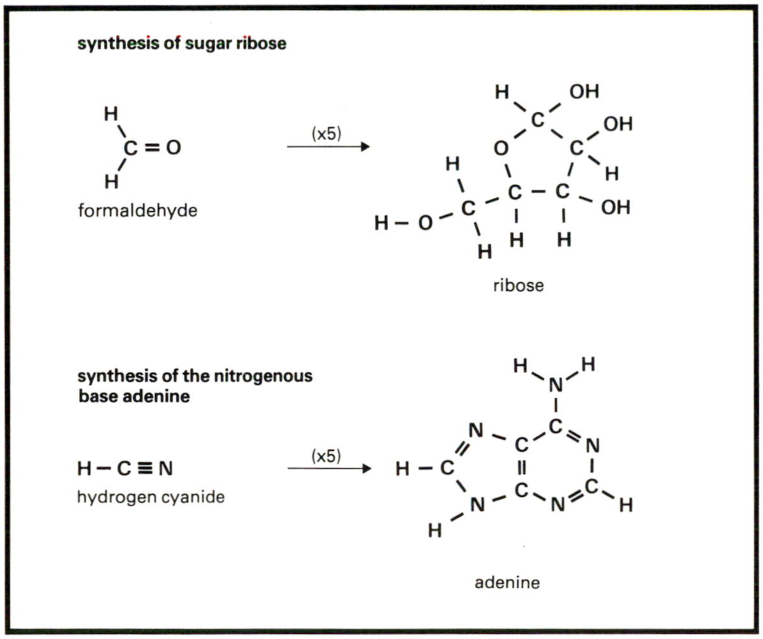

21.3: *Synthesis of a sugar and a nitrogenous base.*

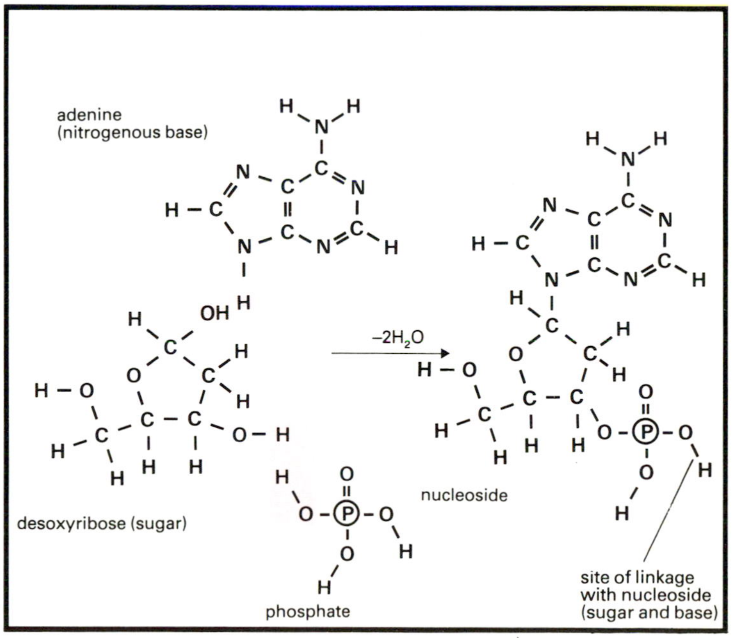

21.4: *Synthesis of a nucleotide (the link in the nucleic acid chain).*

many organic compounds, and a non-aqueous micro-environment could exist in their interiors in which additional prebiotic synthesis might occur.

Spheroidal to globular organic microstructures in organic residues of Miller–Urey experiments, and proteinoid microspheres have been produced from thermally synthesized peptides under almost dry conditions. These microspheres can have layered membranes and are strikingly similar in size and morphology to coccoid prokaryotes such as bacteria and blue-green algae. Some are able to divide themselves, but no nucleic acids are involved. Significantly this shows a tendency for abiotically produced materials to order themselves into complex structures. Random processes, although active, are subordinate to kinetically favoured pathways.

One of the many critical, unresolved problems in understanding the origin of life is the first functional relation between proteins and nucleotides—the origin of the genetic code or, in other words, the transition from biochemistry to genetics. The constituent of genetic material in living cells is nucleic acid. However, the biological synthesis of nucleic acids requires proteins in the form of enzymes. Did enzymes or nucleic acids occur first? One view is that they developed in parallel. Alternatively experiments suggest that ordered amino acid sequences can result from a directive effect of the reacting amino acids themselves. It is conceivable that this self-ordering was the first informational matrix. A protoenzymatic activity could be inherent in the system which ensures the creation of a copy of the original ordered amino acids, and in turn their polymerization to proteins.

Various authors have postulated that a primitive genetic system could have arisen through an association of proteins with mineral crystal sites. Minerals such as clays could concentrate prebiotic material, favour dehydrations and thus drive polymerizations. Because crystals have ordered atomic structures a template would be provided which would favour one path over another—a kind of primitive selection process.

This prebiological evolution would select for specific combinations, polymer colonies composed of proteins being a result. The addition of nucleic acids could have been an example of pre-adaptation; their original functions were mechanical rather than protein-coding. Thus the nucleic acid contributed to colony fitness by coding for proteins that were themselves useful. Once this happened molecular evolution would no longer depend on mineral templates. Once freed of mineral dependency the macromolecules became membrane-bound, possibly through coacervation of microsphere construction, giving rise to the first cell.

Primitive organisms

At this point it is necessary to introduce several key terms with which to refer to successively more advanced grades of complexity among primitive organisms. The vast majority of familiar plants and animals at the present day are eukaryotic, that is, their cells are highly organized and contain a nucleus in which the cell's genetic materials are concentrated. The prokaryotic cell is presumed to be the more primitive, being little more than a bag of protoplasm in which the genetic material is loosely distributed. Prokaryotes include the bacteria and the blue-green algae which are also known as cyanobacteria; all other algae, otherwise the simplest kinds of plant, both unicellular and multicellular, are eukaryotic. Most eukaryotes are capable of sexual reproduction. Prokaryotes, however, rarely exchange genetic material when they reproduce. Hence one surviving mutation can give rise to a new species. The very earliest organisms were also heterotrophic, unable to build their own complex organic molecules and hence dependent on obtaining them from their surroundings. Autotrophy is the ability to construct organic compounds from simple substances such as water and carbon dioxide.

The first living things were small, micrometre-sized, spheroidal, heterotrophic, bacteria-like organisms which survived by fermenting organic molecules of a primitive soup. These heterotrophs were preadapted to the chemical environment that produced them. They would have a distinct advantage over such possibly prebiotic systems as coacervates and proteinoid microspheres in their ability to replicate and to pass on genetic variations upon which natural selection could operate. Primitive living systems might have originated several times during this period of trial and error. But with their dependence on abiologically produced organic matter as a 'food source', these earliest heterotrophs could have brought about one of the first crises in the history of life. At some time the heterotrophic consumption rate of the primitive soup would exceed the rate at which abiologically produced organic compounds were supplied. Selection pressures would favour any mutation or combination of mutations that enabled these heterotrophs to manufacture their own food and become autotrophic. The production of organic compounds would now take place within the cells. This more advanced life form would have a considerable selection advantage over heterotrophs which were competing among themselves for abiological foodstuffs.

Autotrophs today use CO_2 as their principal carbon source with a number of compounds as electron donors (Figure 21.5). The first autotroph might have been a methane-producing bacterium (methanogen) which used CO_2 as a carbon source and hydrogen as an electron donor to produce cellular carbon and CH_4 (methane). It is possible to reconstruct a bacterial phylogeny from the RNA sequences of living groups of bacteria. The differences in RNA between the two groups suggest that methanogens diverged very early from prokaryotes and may have been the first autotrophs.

Before the recent work on methanogens the likeliest candidates for the first autotrophs were photosynthetic bacteria which employ CO_2 as

21.5: *Generalized equations for reactions used by autotrophs.*

Methanogens:

$$CO_2 + H_2 \longrightarrow (CH_2O) + CH_4$$

Photoautotrophs:

purple and green bacteria–Photosystem I

$$CO_2 + 2H_2S \xrightarrow{\text{sunlight}} (CH_2O) + H_2O + 2S$$

blue-green algae (cyanobacteria)—Photosystem II

$$CO_2 + H_2O \xrightarrow{\text{sunlight}} (CH_2O) + O_2$$

the carbon source and H_2 or H_2S as the electron donor, the reaction being catalyzed by light. Modern photosynthetic bacteria (but not blue-green algae) are anaerobes; they do not use water as an electron donor nor do they liberate oxygen.

The attainment of autotrophy, whether first by methanogens or photosynthetic bacteria, probably resulted in the second adaptive microbial radiations. Availability of CO_2, H_2 and H_2S were probably not limiting factors, though the dependence of photosynthetic bacteria on sunlight determined their range of possible environments. Diversification of heterotrophic prokaryotes probably accompanied the appearance of autotrophs.

The next major step in microbial evolution was probably the first photosynthetic autotrophs which released oxygen, an important stage therefore in the evolution of the atmosphere. These autotrophs, represented by the blue-green algae make use of a different photosystem which contains a more advanced type of chlorophyll, chlorophyll-a, which is present in all algae and in higher plants. Many researchers regard the blue-green algae as highly evolved prokaryotes and that their evolution from a photosynthetic bacterium required a great deal of biochemical evolution. This is not a trivial step in microbial evolution. Considering the dependence on oxygen of all higher organisms the appearance of O_2-releasing photosynthetic microbes may have been one of the most important evolutionary events in the history of life.

The nature of Precambrian fossils

The fossil record supplies no direct information on the origin of life, but it does contribute to an understanding of the early evolution of life. Since the 1950s considerable effort has been directed towards discovering, describing and interpreting Precambrian life. Fossils from rocks older than 570 Ma are being found in increasing numbers, and there is now a record that extends back to 3500 Ma or more ago. Scientists are attempting to piece this fossil record together and to integrate it both with models of crustal, atmospheric and hydrospheric evolution and with contemporary ideas on prokaryotic and eukaryotic evolution and phylogeny.

Precambrian fossils may be grouped into four categories: stromatolites, microfossils, algal megafossils and metazoans. Structures now recognized as stromatolites have been known since John Steele's illustration of Cambrian forms from New York state (1825). Stromatolites are a curious type of fossil; characteristically they occur as layered sedimentary structures composed of limestone or dolomite whose laminae are convex away from the substratum and form domes and columns. The study of stromatolites came within the domain of the palaeontologist when James Hall named Steele's structures *Cryptozoön proliferum* (1883). Animal affinities were implied (Greek *kryptos* hidden, *zoon* animal), but suitable modern analogues were unknown at that time. In 1908 Kalkowsky introduced the term *stromatolith* (ie stromatolite), which he used in a descriptive, non-genetic sense. Nevertheless, it was tacitly assumed by that time that stromatolites were biogenic. A blue-green algal origin for stromatolites was suggested in 1914 by Charles D. Walcott who found microfossils in Precambrian carbonate stromatolites from the Belt supergroup in Montana which he compared to modern blue-green algae. He lent further support to

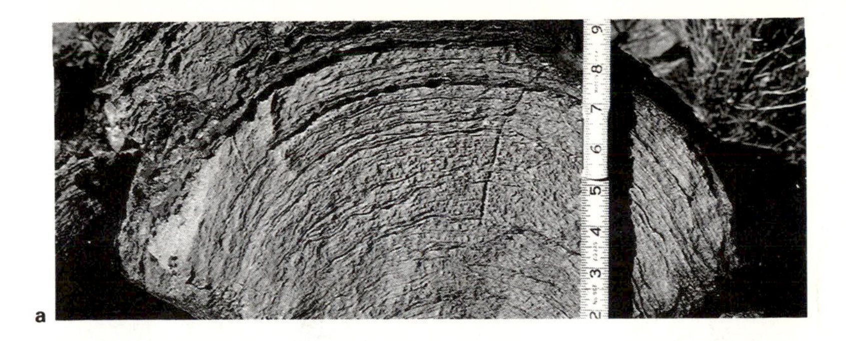
a

b

c

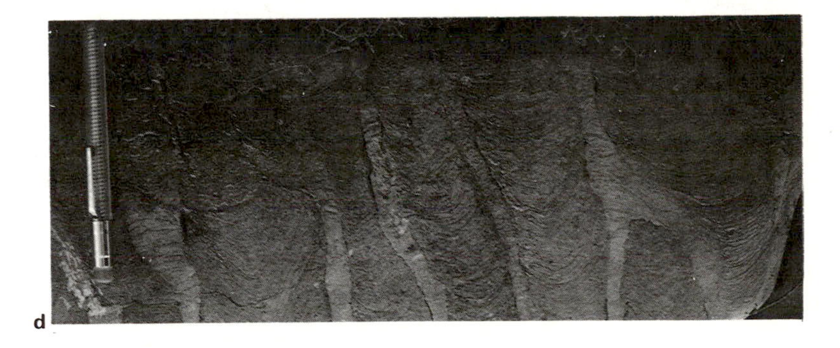
d

21.6: *Fossil stromatolites.* **(a)** *An unnamed dome, South Australia, Lower Cambrian.* **(b)** *A flat to wavy laminated stromatolite, Liaoning, China, Middle to Upper Riphean.* **(c)** Baicalia *cf.* rara, *Liaoning, China.* Baicalia *is common in Middle to Upper Riphean deposits.* **(d)** Chihsienella chihsienensis, *Tieling formation, Chihsien, China, Middle Riphean.*

a

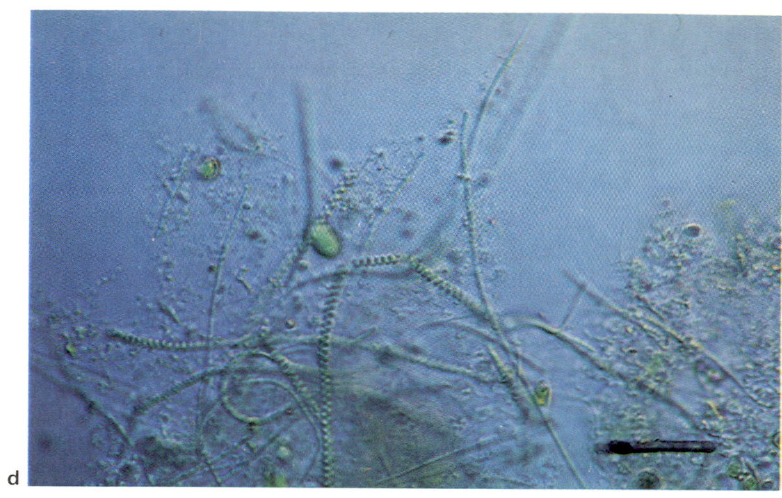
d

b

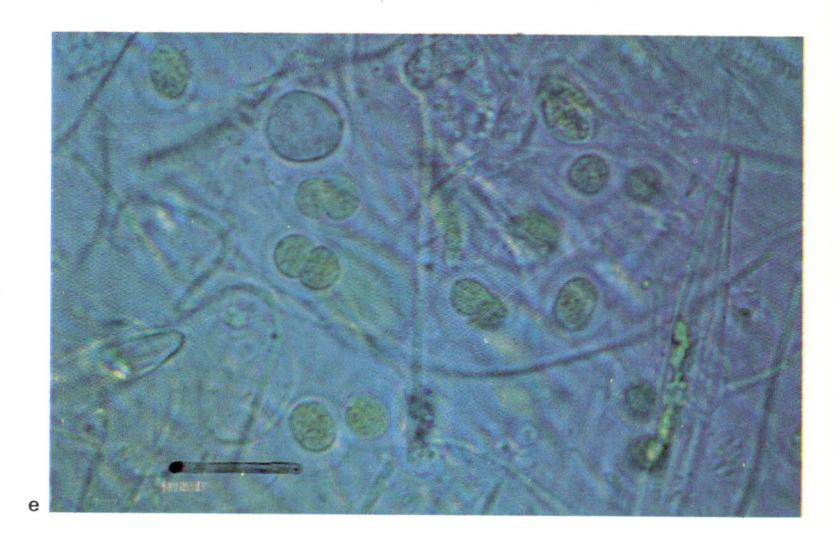
e

c

21.7: *Recent stromatolites, microbial mats and blue-green algae.* (**a**) *Stromatolites exposed at low tide, Hamelin Pool, Shark Bay, Western Australia.* (**b**) *A microbial mat, Yucatan, Mexico.* (**c**) *Section through a microbial mat, Baja California, Mexico, showing the laminated structure.* (**d**) *Coccoids and partially degraded filamentous blue-green algae from the Yucatan microbial mat (bar = 10 μm).* (**e**) *Filamentous blue-green algae from the Yucatan microbial mat (bar = 10 μm).*

this idea by noting the role that modern blue-green algae play in the precipitation of calcium carbonate.

Convincing modern analogues were not reported until 1933, when recent algal mats were described from marine and associated environments in the Bahamas. However, the Bahamian algal mats did not show the columnar or domal growth so common in the fossil record.

The discovery in 1954 of large, nearly metre-sized, stromatolite columns growing in Shark Bay, Western Australia, provided morphological analogues with some of the more conspicuous stromatolites found in the fossil record. The striking similarity of certain ancient forms to recent Shark Bay structures was taken as compelling evidence for similar modes of formation. From these and other discoveries in contemporary stromatolite-forming localities a number of general characteristics can be ascribed to stromatolites. Predominantly built by blue-green algae they accrete by trapping and binding sedimentary particles in marine settings and by mineral precipitation in non-marine settings. Generally, their preferred sedimentary type is $CaCO_3$, their water chemistry hypersaline (if marine) or alkaline (if non-marine) and their environmental setting intertidal (marine) or submerged (non-marine). Their shape is controlled by turbulence and their microstructure by microbiology. Recent stromatolites are not widely distributed and are not diverse morphologically.

Precambrian microfossils are organically preserved prokaryotes and unicellular protists found in a variety of chemical and fine-grained clastic rocks. They apparently lacked mineralized hard parts, hence unusual fossilization conditions must have prevailed to ensure their preservation. After death organisms are quickly decomposed under anaerobic conditions by a succession of bacterial decomposers. Microfossils, therefore, represent the remains of those organisms which eluded complete post-mortem degradation. Some micro-organisms can produce decay-resistant external coverings, enhancing their preservation potential. Bacterial degradation can be arrested by unfavourable environmental conditions such as low pH or high salinity. A particularly effective inhibitor of degradation is the precipitation of silica soon after death.

Microfossils preserved in silica provide the foundation upon which our understanding of early microbial evolution is based. The silica is in the form of chert, a microcrystalline variety of quartz. The significance and potential of chert for the cellular preservation of Precambrian micro-organisms was not appreciated until 1954 when an exquisitely preserved microbiota was discovered in stromatolitic cherts of the Gunflint iron formation, about 2000 Ma old, in Ontario, Canada, by Tyler and Barghoorn. Since 1954 several dozen silicified stromatolitic microbiotas, ranging in age from 3500 to 570 Ma have been reported.

There is another group of microfossils, not associated with stromatolites but found in detrital sedimentary rocks such as shale, siltstone and sandstone. These sedimentary rocks represent a wide variety of depositional environments. The microfossils isolated from these clastic sediments are different from the typically filamentous ones found in stromatolitic cherts; they are usually unicellular and are thought to represent the spores of unicellular algae. Unlike most stromatolitic microfossils many of these clastic facies microfossils cannot be compared with any single taxonomic group of living organisms. For descriptive and classification purposes the term 'acritarch' has been

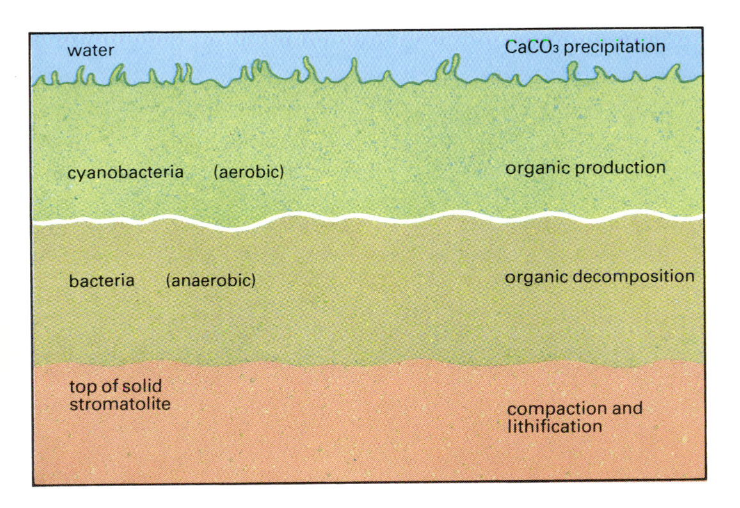

21.8: *Processes involved in the construction of stromatolites. The mat of living blue-green algae traps or precipitates calcium carbonate particles, growing upwards and leaving organic (bacterial) decay and inorganic processes to complete the structure.*

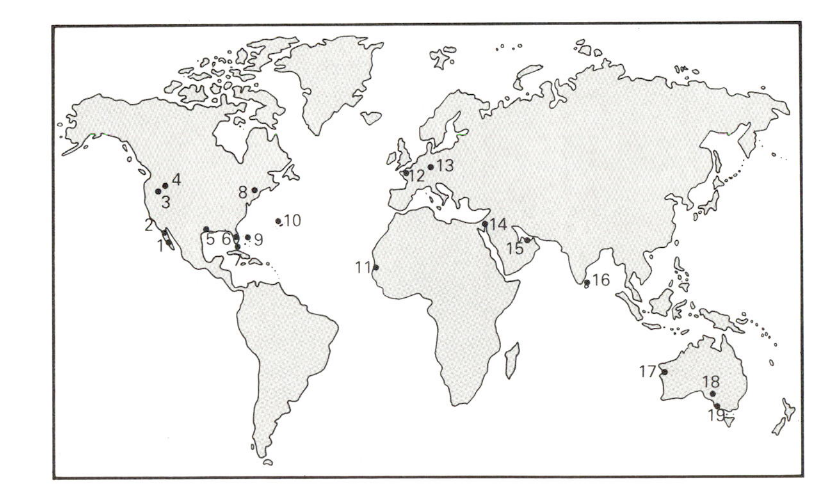

21.9: *World map to show known present-day locations of active stromatolite formation.*

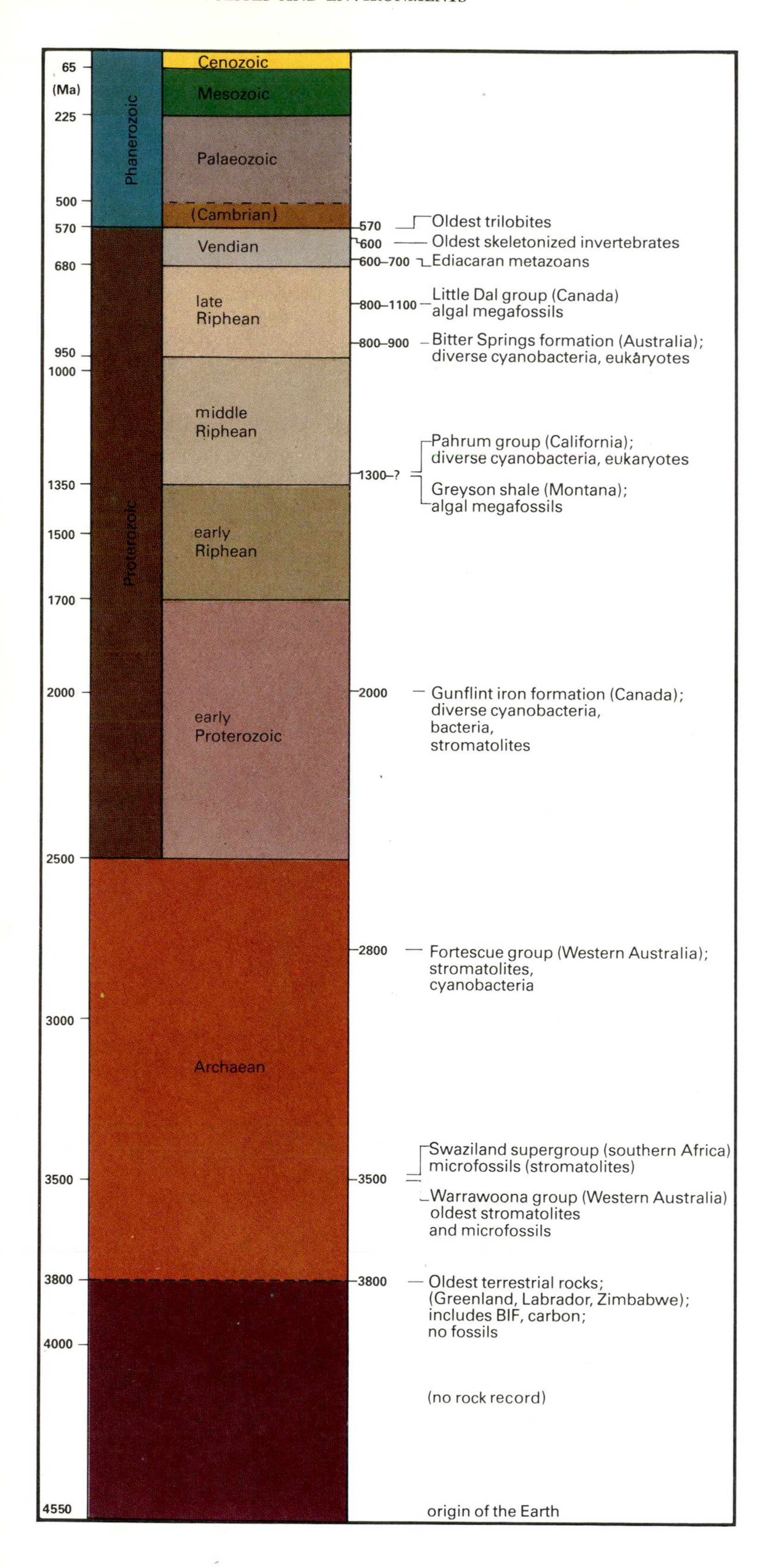

widely adopted to designate these small, organic-walled microfossils of uncertain taxonomic affinities. Although their systematic position is uncertain they show sufficient morphological changes with time in the Precambrian to have been used in biostratigraphy. Because of their association with clastic rocks acritarchs are considered to be the remains of plankton, and because of their generally assumed algal affinities they are attributed to the phytoplankton. Acritarchs, though many may be algal spores, are such a confusing and palaeobiologically poorly understood fossil group that any conclusions as to their eukaryotic nature should be drawn very tentatively. Acritarchs are rare in clastic rocks older than 1600 Ma and show an increase in size, diversity and complexity of organization in progressively younger sediments.

Millimetre- to centimetre-sized carbonaceous compressions of probable algal affinities have been found in rocks of middle and late Proterozoic age on several continents. Few such algal megafossil localities are known but these fossils may represent the oldest reliable evidence for the appearance of eukaryotes in the fossil record.

Finally, in late Precambrian detrital rocks younger than 680 Ma, a distinctive assemblage of soft-bodied animal impressions, casts and moulds have been found on several continents. These late Proterozoic metazoans are discussed in Chapter 23.

The Precambrian fossil record

The chronology of Precambrian time

Precambrian geological time has been subdivided into two eons, the Archaean and Proterozoic. The differences between these two eons are geological rather than palaeontological (see Chapter 16). The Archaean begins at 3800 Ma with the oldest terrestrial rock so far recorded and ends, as currently defined, at 2500 Ma. The Proterozoic begins at 2500 Ma and ends at about the first occurrence of shelly fossils (the base of the Cambrian, about 570 Ma). Various subdivisions of the Proterozoic have been proposed and palaeontologically useful ones are shown in Figure 21.10. Preston Cloud introduced the term 'Hadean' which has been used to designate geological time from the origin of the Earth, some 4550 Ma ago, to the oldest terrestrial rocks now known at 3800 Ma.

Archaean fossils

With the oldest terrains dated at 3800 Ma many geological, biochemical and perhaps biological events remain unrecorded in the rock record. Terrains approaching 3800 Ma old are known from Greenland, Labrador, Minnesota, South Africa and Zimbabwe. The most interesting in terms of interpreting Earth history and yielding possible clues of early life are the Isua supracrustals of south-western Greenland. Though highly metamorphosed the nature of the metasediments indicates deposition under aqueous conditions (therefore surface temperatures were within that narrow range needed for life), and apparently under anaerobic conditions. Carbonates imply the presence of atmospheric CO_2. Sedimentary banded iron formations (BIFs) and reduced

21.10: *A Precambrian time scale.*

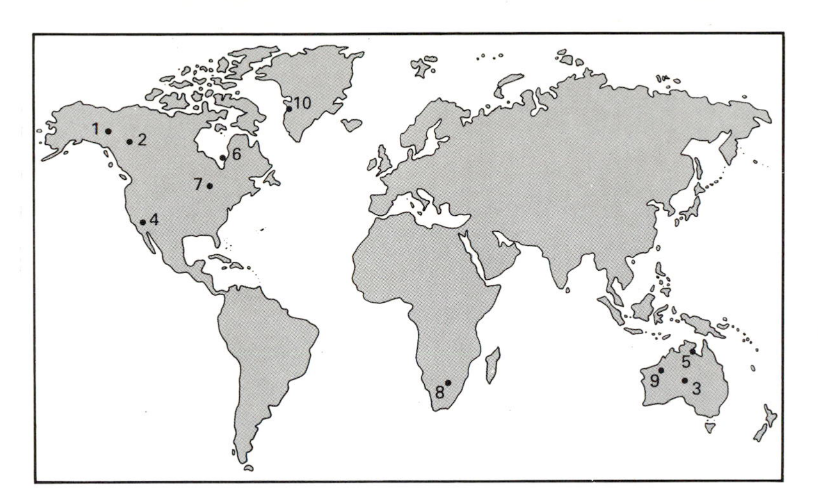

	rock unit	age (Ma)
1	Tindir group	~ 600
2	Little Dal group	1100–800
3	Bitter Springs formation	~ 850
4	Beck Spring dolomite	< 1300
5	Balbarini dolomite	~ 1500
6	Belcher Islands microbiota	~ 2000
7	Gunflint iron formation	~ 2000
8	Swaziland supergroup	~ 3500
9	Warrawoona group	~ 3500
10	Isua supracrustals	~ 3800

21.11: *Location and age of important Precambrian fossil-bearing rock units.*

carbon, in the form of graphite, are also found. This reduced carbon is the subject of considerable debate: is it of abiogenic or biogenic origin? One test that has been applied is to measure the ratios of two stable isotopes of carbon, ^{12}C and ^{13}C. All autotrophs will preferentially reduce the lighter isotope (^{12}C) of carbon. It follows that if the carbon in rocks of a given age is enriched in ^{12}C over ^{13}C, it should indicate the existence of autotrophy at that time. The Isua carbon has been found to be enriched in ^{12}C which suggests that autotrophs were present. Coupled with the presence of iron oxides in the BIF this may even suggest that O_2-releasing photoautotrophs (photosystem II; see Figure 21.5) had evolved by the time of Isua. However, caution is still needed, as it remains possible that a similar isotopic signature could also have been produced by metamorphism.

21.12: *Domal stromatolite from the Warrawoona group, Western Australia, 3500 Ma old.*

The BIFs at Isua present another problem. BIFs are rocks composed of alternating layers of iron-rich and iron-poor silica. This rock type is abundant among, and characteristic of, sedimentary rocks 3800 to about 2000 Ma in age. The most widely held model, proposed by Preston Cloud, for the genesis of BIFs presents them as oxygen sinks. Iron could become widely dispersed in the primitive hydrosphere in the reduced ferrous state (Fe^{2+}), if there was little or no oxygen present. Ferrous iron combines with oxygen and precipitates out of solution as the iron oxide haematite. The most likely source of the oxygen necessary for the formation of the iron oxides in BIFs is biogenic oxygen released by photosynthesis. The BIFs at Isua would thus, like the carbon isotopic ratios, seem to suggest that O_2-releasing autotrophs were present 3800 Ma ago. However, other, non-biogenic, sources for O_2 such as the photodissociation of water in the atmosphere, cannot be ruled out.

Many researchers are reluctant to conclude that the evidence in favour of the existence of life at Isua is compelling. Carbonaceous microstructures from a metaquartzite at Isua in the earliest Archaean have been interpreted recently as the remains of micro-organisms. But microfossils similar to those described would not survive even moderate metamorphism, whereas the Isua terrain has suffered prolonged high-temperature metamorphism. Recent work has shown that the carbonaceous microstructures at Isua are inorganic fluid inclusions which happen to be associated with graphitic material.

The oldest unequivocal remains of life come from slightly metamorphosed rocks about 3500 Ma old of the Warrawoona group at North Pole, Western Australia. Stromatolites have been found here as well as filamentous microfossils in chert. Microscopic carbonaceous spheroids, possibly biogenic, are also known. The Warrawoona stromatolites morphologically resemble other Precambrian and younger stromatolites. These early Archaean structures are interpreted as having formed in shallow water like their younger counterparts.

In laminated cherts a few kilometres from the stromatolite locality filamentous microfossils have been detected. Five distinct morphological categories of filamentous microfossils have been described from the Warrawoona cherts, resembling a number of different prokaryotic organisms. Reduced carbon in these rocks is enriched in ^{12}C, suggestive of autotrophy, and is present in quantities not unlike those obtained from the richly microfossiliferous cherts of the younger Gunflint and Bitter Springs deposits. Interpretations of the Warrawoona fossils suggest that by about 3500 Ma ago life was already diverse and at an autotrophic level. The level of evolution implied by the Warra-woona fossils suggest one of two possibilities: either the beginnings o life appreciably predated Warrawoona deposition or the rates o microbial evolution and the origins of new metabolic pathways were very rapid. Further work on older deposits is needed to settle this.

Cherts from the Swaziland supergroup, in South Africa, abou 3500 Ma old, also contain microfossils. But only spheroidal microfossils are known from these deposits, and no convincing stromatolites have been found.

Younger Archaean rocks contain additional evidence of early life

21.13: *Warrawoona group, Western Australia.* (**a**) *A general view of Warrawoona outcrops, Pilbara region, Western Australia.* (**b**) Eoleptonema australicum, *a degraded filamentous prokaryote (bar = 10 μm).* (**c**) Primaevifilum septatum, *probably a trichome of a prokaryote (bar = 10 μm).* (**d**) Siphonophycus kestron, *a filamentous prokaryote (bar = 10 μm).*

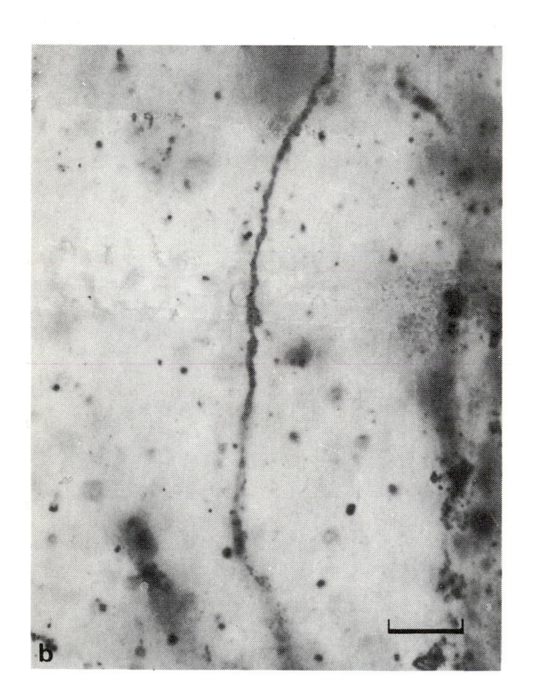

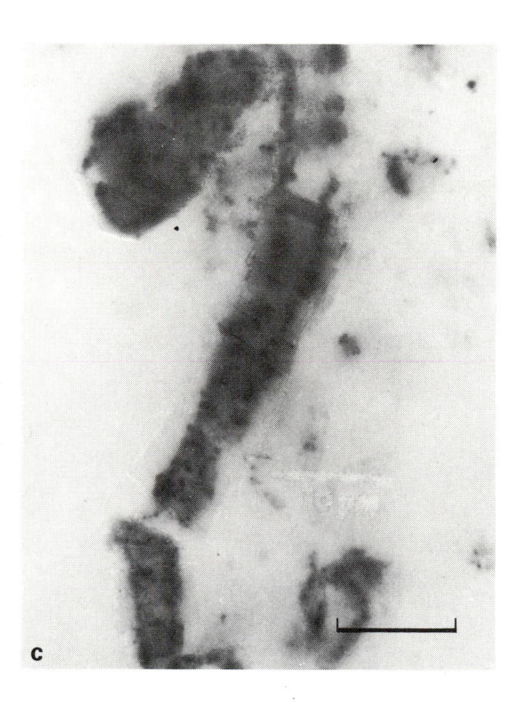

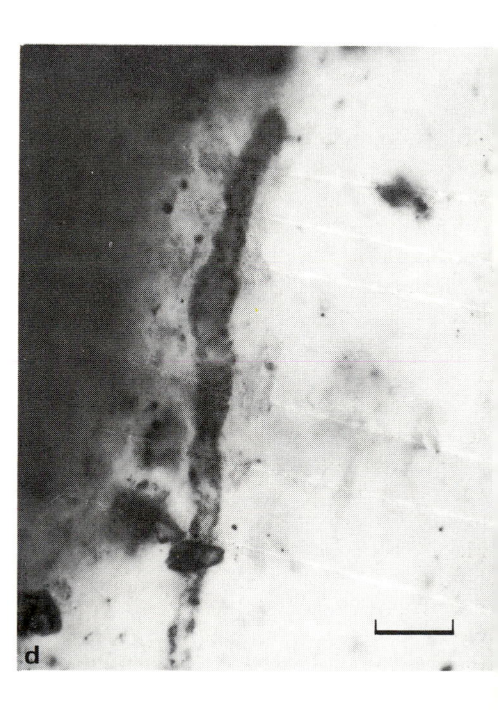

Stromatolites predominate in this record, and the Fortescue group in Western Australia, about 2800 Ma old, is one of the very few Archaean localities in which microfossils have been detected. The Fortescue microfossils, from stromatolitic cherts, morphologically resemble some modern blue-green algae but could also be compared to other types of prokaryotes. Precambrian palaeobiologists are now taking care not to infer or imply the physiology of an ancient microbe solely from its morphological comparison to a modern microbe. The superficial resemblance of a microfossil to a modern blue-green alga does not necessarily mean that it was an O_2-releasing photoautotroph.

The evolutionary origin of photosynthetic blue-green algae probably occurred sometime during the Archaean, but possibly as late as the earliest Proterozoic. Evidence is indirect: photosynthesis is the only mechanism known to produce O_2 in the quantities needed to account for all the BIF in the geological record, and stromatolites can be used to infer the presence of blue-green algae in the Archaean. It is not until about 2000 Ma, however, that fossil microbes with strong resemblances to modern blue-green algae are found.

The paucity of Archaean fossils, both stromatolites and microfossils, may partially result from geological conditions. Archaean shallow-marine environments were of limited extent and were the sites of rapid clastic sediment deposition; this sedimentological regime does not allow for microbial mat development, and hence the small number of stromatolites and stromatolitic microfossils. The lack of Archaean acritarchs is probably a reflection of the fact that eukaryotes had not yet evolved and prokaryotic microbes did not have a sufficiently resistant spore stage to be readily preserved.

Proterozoic fossils

Compared with the Archaean the Proterozoic fossil record is rich and diverse. Not only did stromatolites and their microfossils become more abundant, but acritarchs, algal megafossils and eventually metazoans appeared.

Stromatolites

With the stabilization of the crust and the development of extensive shallow marine environments in the early Proterozoic, stromatolites became abundant, widespread and morphologically diverse, their shapes including columnar, columnar-layered, conical, branching columns, domes and rounded 'balls', and of planar or wavy-laminated construction. Stromatolites dominated shallow marine carbonate environments until the Palaeozoic era when they were replaced in importance by metazoans.

Stromatolites were conspicuously more abundant in the Proterozoic than at any other time in Earth history, and especially from 2000 to 680 Ma. They were more diverse and achieved greater size: columns several metres in diameter are found as well as stromatolitic mounds several tens to hundreds of metres across, and stromatolitic 'reefs' exceed 300 m in height and 5 km across. Branching columnar stromatolites were common in the Proterozoic yet rare in the Phanerozoic. Many (perhaps most) formed by the precipitation of calcium carbonate, while trapping and binding of detrital particles was the dominant process of stromatolite construction in the Phanerozoic. Subtidal regions were the sites of extensive stromatolite growth and development, whereas later stromatolites appear to be restricted to intertidal sites. Most important from a biostratigraphical point of view is that numerous stromatolite morphologies and microstructures have restricted time ranges within the Proterozoic. It is this feature, first utilized by Soviet geologists, that is the most striking characteristic of Proterozoic stromatolites.

Soviet geologists set up a four-fold subdivision of the last billion years of Precambrian time based on the temporal restriction of distinctive stromatolites. Potassium–argon dating of glauconites from unmetamorphosed sedimentary sequences associated with the stromatolite-bearing strata provided the necessary isotopic age control. Though refined and somewhat modified through the years the overall scheme has remained fundamentally unchanged and has been found applicable to regions outside the Soviet Union. Although at first developed for the middle and late Proterozoic, stromatolite biostratigraphy is being applied with some success to rocks older than 1600 Ma, as well as to Cambrian and Ordovician rocks.

The time restriction of columnar stromatolite morphologies in the Proterozoic suggests a strong relation between stromatolite form and biology. Studies on recent stromatolites, however, indicate that the microbes that build stromatolites exercise rather little control on the overall shape assumed by the growing stromatolite, which is thought to be primarily influenced by wave and current action. This has cast doubt on the utility of stromatolites for dating Proterozoic rocks. However, many Proterozoic stromatolites have no modern analogues, and it may be that the controls on their growth and form were more directly linked at that time to the biology of their builders and that their overall form could indeed have evolved through Proterozoic time.

Unfortunately biostratigraphically useful stromatolites are normally preserved as limestone and dolomite and therefore only very

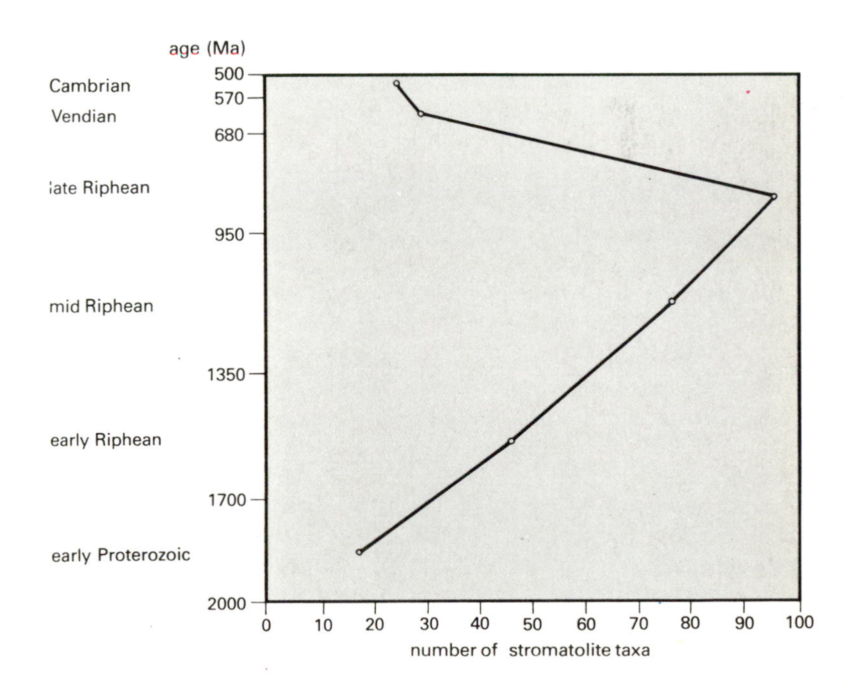

21.14: *A diversity curve of columnar stromatolites.*

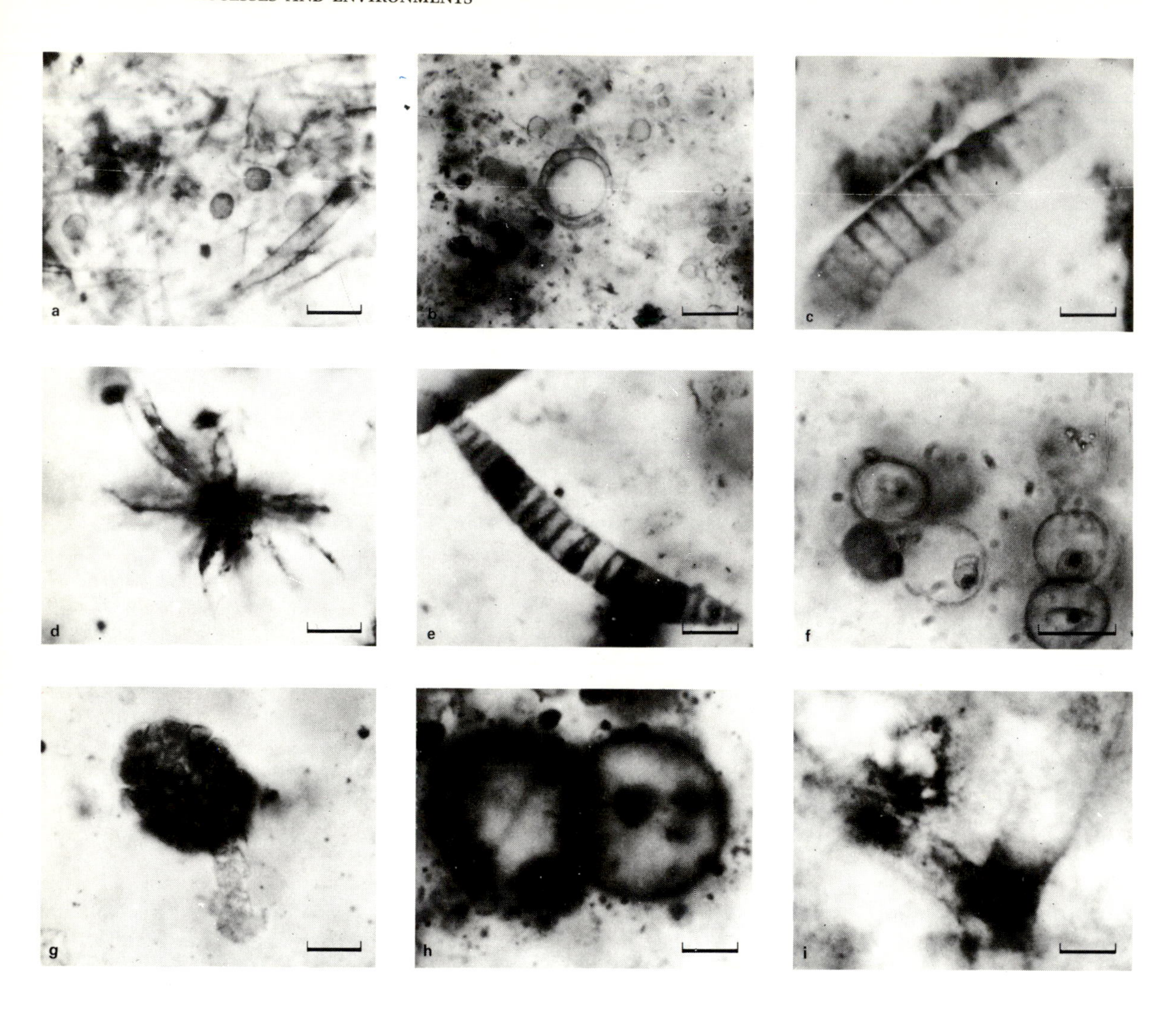

21.15: *Precambrian microfossils; all scale bars are 10 μm long.* (**a**) Gunflintia minuta *filaments, probably a blue-green alga from chert of the 2000-Ma-old Gunflint iron formation, Canada.* (**b**) Eosphaera tyleri, *a form unique to the Gunflint chert.* (**c**) *Blue-green algal trichome in chert. Tindir group (600 Ma old), Yukon Territory, Canada.* (**d**) Eoastrion simplex, *from chert of the Gunflint iron formation.* (**e**) *A tapered, blue-green algal trichome from chert of the Tindir group, Yukon, Canada.* (**f**) *A pair of spheroids from chert of the Bitter Springs formation, central Australia, 800–900 Ma old.* (**g**) *Aggregate of spheroids from the chert of the Tindir group.* (**h**) *Aggregate of spheroids from the Mineral Fork formation, Utah, 800 Ma old.* (**i**) Palaeosiphonella cloudii, *a probable eukaryote from chert of the Kingston Peak formation, California, younger than 1300 Ma old.*

rarely contain microfossils. Within the Proterozoic the apparent diversity of columnar stromatolite forms increased until the late Riphean; during the Vendian period diversity dropped off sharply, perhaps as a result of predation by early herbivorous metazoans.

Stromatolitic microbiotas

The microfossils found in silicified stromatolites provide the most useful palaeobiological data on the early history of life. Microfossils from these rocks are usually studied in petrographic thin sections—extremely thin slices of chert mounted on glass slides. This method of study has several advantages over the alternative procedure of extracting the organic content of the sediment by dissolving away the rock matrix as is done in the study of acritarchs. The microfossil is viewed in its enclosing rock matrix, and it is thus fairly easy to recognize contaminant microbes which have been introduced since deposition of the sediment. The effects of postmortem morphological changes induced by degradation are preserved *in situ*. There is also a possibility that populations and communities can be studied, as the microfossils can be seen in their original relative positions, perhaps forming distinct microbial communities. Most important, the fidelity of preservation in cherts is at times very high, reflecting original morphology and allowing more confident comparisons to modern analogues.

Stromatolitic cherts from the approximately 2000-Ma-old Gunflint iron formation, Ontario, and formations of similar age from the Belcher Islands, Hudson Bay, provide a data base early in the Proterozoic from which to evaluate evolutionary trends in stromatolitic microbiotas. The Belcher Islands microfossils are found in wavy-laminated, planar stromatolites which formed in an intertidal to shallow subtidal setting. It is a diverse assemblage of microbes with members resembling modern blue-green algae of three different families. This fossil microbial mat community is very similar to younger microbiotas from the approximately 1500-Ma-old Balbirini dolomite and the approximately 850-Ma-old Bitter Springs formation of Australia. Both suggest an environmental setting similar to that of the Belcher Islands fossils as well as bearing a striking resemblance to extant blue-green algae from recent microbial mats. Some early Proterozoic fossil blue-green algae were essentially modern in their appearance.

The fossils of the Gunflint iron formation differ from those of the Belcher Islands in taxonomic composition, stromatolite structure and environmental setting. The Gunflint microbiota is dominated by the thin filament *Gunflinta minuta* and the solitary unicell *Huroniospora microreticulata*; both are generally regarded as belonging to the blue-green algae. The Gunflint formation is a BIF deposited in a subtidal setting which may explain some of the palaeontological differences observed. In addition to stromatolites the Gunflint formation contains a non-stromatolitic chert facies with a distinctive microbiota, dominated in places by the presumed fossil bacterium *Eoastrion simplex*. This fossil is morphologically identical to modern bacteria such as *Metallogenium* which concentrates manganese oxides. The nine different types of blue-green algae in the Gunflint formation comprise about 99 per cent of the microbiota. The remainder includes bacteria such as *Eoastrion* and *Eosphaera tyleri*, and others which have no modern analogue and whose systematic positions are not understood. These forms are unique to the Gunflint and are presumed to be prokaryotic.

In an analysis of Proterozoic stromatolitic microbiotas, J. William Schopf found that unicells and cylindrical tubelike sheaths as well as cellular trichome-like filaments show an increase in size through time accompanied by an increase in taxonomic diversity, continuing well into the Vendian. This is consistent with the contention that the restriction in time of particular stromatolites resulted from the evolution of stromatolite-building microbes.

The origin of the eukaryotes

The evolutionary appearance and diversification of eukaryotes occurred sometime in the middle of the Proterozoic. This was perhaps the greatest evolutionary jump in the history of life. Eukaryotes are characterized by membrane-bound organelles, one of which, the nucleus, encloses the genetic material in a distinct membrane-bound package. Respiration and photosynthesis are carried out in separate organelles within the cell. Eukaryotes differ from prokaryotes in their biochemistry, in having a much larger genetic package, and in their capacity for sexual reproduction. The geological record does not unfortunately record the events leading to their appearance and we must look to modern biology for clues to the molecular and biochemical events leading to the origin of the eukaryotes by about 1400 Ma ago. Three competing theories are currently offered. The endosymbiosis theory proposes that eukaryotes originated through the progressive assimilation of different types of prokaryote, performing different functions in symbiotic relationships within a prokaryote that had lost its cell wall. The developmental theory suggests that invaginations developed in the cell wall of a prokaryote, eventually becoming detached as organelles. These developments could have occurred between 2000 and 1400 Ma ago. The progenote theory suggests that eukaryotes did not arise from prokaryotes, but that both were derived from a common ancestor, the progenote, some 3500 Ma ago. Around 1400 to 1500 Ma ago the ancestral eukaryote (prokaryote in grade but eukaryotic in biochemistry) acquired chloroplasts and mitochondria through endosymbiosis. All three theories fail to explain the formation of the eukaryotic nucleus, but all take account of the fossil evidence that organelle-bearing eukaryotes had appeared by about 1400 Ma ago.

Although the fossil record of early eukaryotes is currently controversial, several lines of evidence support the origin of nucleated cells at about 1400 Ma ago. One line of evidence that has been used is the presence of an internal small mass, interpreted as an organelle, within a cell-like microfossil, but there is now experimental evidence that such features are common products of cellular degradation in prokaryotes. More convincing, yet not totally unequivocal, are the large branching tubes (upwards of 100 micrometres (μm) in diameter) described from silicified stromatolites of the Beck Spring dolomite of California which is younger than 1300 Ma old. Such large branching tubes are comparable with those of certain living green algae, which are eukaryotic.

The Soviet micropalaeontologist B. V. Timofeev first noted that acritarchs increased dramatically in both abundance and size at about 1400 Ma ago. This in all probability reflects the appearance of eukaryotes. From their large size, some exceeding 100 μm in diameter, it seems reasonable to conclude that at least some acritarchs represent the remains of eukaryotic algae. Spheroidal prokaryotes are usually less than 100 μm in diameter and more commonly less than 20 μm. *Chuaria*, an acritarch that can reach a few millimetres in diameter, is

21.16: *Three alternative models of eukaryotes.*

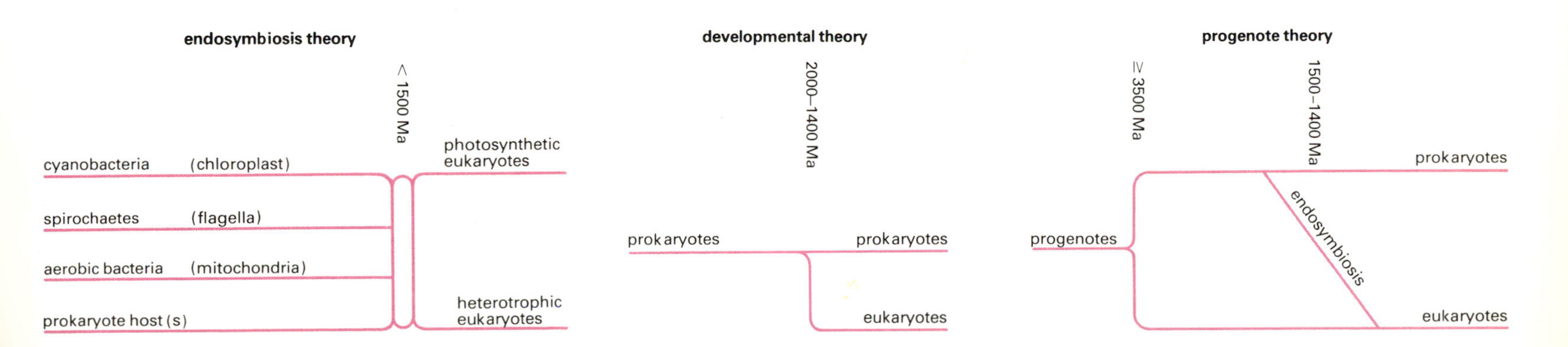

21.17: *Algal growths near hydrothermal springs, Waimongu, New Zealand.*

common in late Proterozoic strata and has recently been found in a rock sequence more than 1600 Ma old in China. Proterozoic acritarchs are predominantly smooth spheroids; spiny acritarchs resembling certain life stages of modern green (eukaryotic) algae appear only in the late Proterozoic.

The most compelling candidates for the earliest eukaryotes are the millimetre-sized carbonaceous ribbons found in several formations 1300 Ma old and slightly younger. The most spectacular such deposit is in the Little Dal group 1100 to 800 Ma old in the Mackenzie Mountains, north-western Canada. Carbonaceous compressions several millimetres wide and a few centimetres long are almost certainly the remains of eukaryotic algae—Proterozoic seaweeds in fact.

Though Proterozoic strata record the remains of eukaryotes, the record is nevertheless dominated by shallow-water stromatolite communities. Eukaryotes, when they appear, are recorded in more offshore facies and rarely are their preserved remains found in or associated with stromatolitic deposits. From their insignificant and even controversial earliest appearances, at some time in the mid- to late Proterozoic, the first unicellular eukaryotes gave rise in late Proterozoic times to multicellular algae and later still to multicellular animals, the metazoans, whose appearance is taken up in the next chapter.

22 Fossils and evolution

The nature and interpretation of fossils

Palaeontology, the study of fossils, has traditionally been a largely descriptive science. Its major concern has been the description and classification of fossils, and its main aim the determination of the relative ages of the rocks that contain them. This descriptive role, as an essential part of historical geology, remains important. The fossil record is, in addition, a unique source of information about the course of evolution. Palaeontologists, however, are no longer satisfied merely to document the sequence of life forms through time, and the discipline is presently undergoing a revolution that makes it one of the most exciting areas in the Earth sciences. This upheaval has been fuelled by discoveries in other fields, such as plate tectonics and ecology, and by the application of new techniques, including computers and the scanning electron microscope. Most importantly, however, it is the result of an approach to the discipline that concentrates on problems in an attempt to determine the processes involved in the evolution of organisms, and their distribution and interrelationships through time.

The term 'fossil' designates any trace of past life—a much wider range of phenomena than is generally appreciated. Fossils are not only the actual remains of organisms, eg shells and bones or the leaves of plants (body fossils), but also the results of their activity, eg burrows and footprints (trace fossils), and the organic compounds they produce by biochemical processes (chemical fossils). Inorganically produced structures may occasionally be confused with traces of life, particularly in the rocks of the Precambrian in which a number of supposed fossils have been shown to be pseudofossils.

22.1: *Types of fossilization. The trilobite produces trace fossils such as the linear walking traces and oval resting traces. When the trilobite dies it may remain intact or fall apart to give a body fossil. Organic molecules will be released from the decaying trilobite into the sediment to give chemical fossils.*

Body fossils

The process of fossilization is complex and involves a number of stages commencing with the death of the organism and culminating in its discovery as a fossil (Figure 22.2). Certain organisms (eg those with mineralized shells) clearly have a much higher 'fossilization potential' than others (eg soft-bodied worms or jellyfish), but many other factors also influence the amount of information preserved. An understanding of the processes of preservation (taphonomy) is critical to an interpretation of the fossil.

Body fossils of animals and plants nearly always consist only of the skeletonized or toughened parts because the soft tissues are destroyed

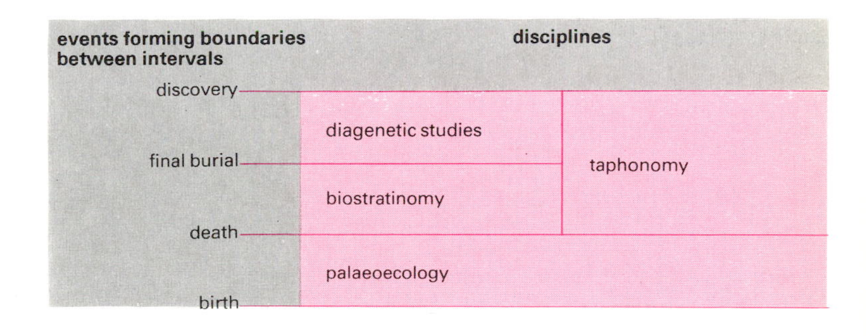

22.2: *The subdisciplines of palaeontology related to stages in the life and fossilization of organisms.*

22.3: A table of the most important exceptional fossil faunas and floras during geological time and their relationship to different environments.

	Terrestial					Marine	
	tundra	freshwater	swamp	lacustrine/fluvial	coastal	shallow	deep
Pleistocene	Permafrost, frozen mammoths, Siberia	La Brea Tar Pits, California, USA					
Miocene		Succor Creek formation, Oregon, USA		Lake Constance, Switzerland/ Germany Calico Mountains, nodules (Barstow formation)		Monterey formation, California, USA	
Oligocene		Amber, Baltic region Quercy phosphorites, Massif Central, France					
Eocene			Geiseltal lignite, Halle, East Germany	Green River formation, Wyoming, Utah, Colorado, USA		Monte Bolca fish beds, Monte Bolca, Italy	
Cretaceous						Sahel Alma fish beds, Lebanon	
Jurassic					Solnhofen limestone Bavaria, West Germany		Posidonia shale, Württemberg, West Germany
Triassic					Grès a Voltzia (Buntsandstein), Alsace, France		
Carboniferous					Mazon Creek nodules (Francis Creek shale) Illinois, USA	Bear Gulch limestone, Montana, USA	
Devonian			Rhynie chert, Aberdeenshire, Scotland				Hünsruck shale, Rhineland, West Germany
Silurian					Eurypterid beds (Rootsikula Stage) Saaremaa, Estonia	Lesmahagow Inlier, Lanarkshire, Scotland	
Ordovician							Beecher's trilobite bed (Frankfort shale), Rome, New York, USA
Cambrian						Anthraconite nodules, Sweden Spence shale, Utah, USA Emu Bay shale, Kangaroo Island, South Australia Kinzers formation, Pennsylvania, USA	Burgess shale (Stephen formation), British Columbia, Canada
Precambrian						Pound quartzite, Flinders Ranges, South Australia Valdai series, Onega Peninsula, USSR Conception group, Avalon Peninsula, Newfoundland, Canada	

by microbial decay or scavengers. Even the hard parts may become so bored by organisms such as sponges, fungi and algae that they are open to destruction. Transportation by waves or currents naturally leads to the fragmentation and abrasion of hard parts so that only the more robust survive (eg the heavily calcified claws of crabs). Many fossil plants and some animals are known only as fragments, and it may be difficult to determine which elements belong together. This often leads to a multiplicity of names for different parts of what may turn out to be the same species. The trunk of a Coal Measure lycopod, for example, may be assigned to the genus *Lepidodendron*, the cones to *Lepidostrobus* and the underground parts to *Stigmaria*. Potential fossils are also occasionally subject to chemical attack, but usually only after burial in sediment. In deep oceanic water, however, aragonite and, at a slightly greater depth, calcite are undersaturated and the sinking shells of pelagic animals may thus be dissolved.

The remains of an organism that survive the agents of biological, physical and chemical attack must next become buried by sediment. It is hardly surprising that organisms living in the sediment (the infauna) have a greatly increased chance of preservation than those living on the sea floor (the epifauna) or swimming or floating above it (pelagic organisms). For example the regular echinoids (sea urchins having pentaradial symmetry), though common today, are epifaunal and have a poor fossil record, in contrast to the sand dollars, which have shifted to an infaunal mode of life. The fossilization potential of the sand dollars has also been increased by strengthening of the skeleton and a change from a globular to a flattened shape which prevents an exhumed test rolling across the sea floor and getting broken up.

'Catastrophic' burial by a rapid influx of sediment is necessary to preserve delicate epifaunal organisms. Field observations have shown that modern crinoids (sea lilies and feather stars), for example, disarticulate within a few days of death; rapid burial prevents this and is thus invoked to explain a number of fossil 'starfish beds' where delicate starfish and brittle stars are preserved. With continued sedimentation the overburden increases, water is driven out of the sediment surrounding a fossil and compaction takes place. The degree of compaction varies but is greatest in shales in which delicate shells may be severely crushed. Experiments with spheres of different composition embedded in different matrices, to simulate the flattening of fossil plant spores, have shown that a variety of artefacts that might be mistaken for original features can be produced by compaction.

Compaction is not the only diagenetic process to which fossils are subjected. Chemical alteration or replacement is very common, particularly among calcareous ($CaCO_3$) skeletons which occur in most marine invertebrates and those algae with hard parts. The two principal forms are calcite (trigonal habit) and aragonite (orthorhombic habit). Aragonite is particularly unstable and normally either inverts to calcite or is dissolved. It generally survives only where it is protected by an impermeable matrix or by the organic sheath surrounding each crystallite. Consequently aragonite becomes progressively rarer in older rocks. If either calcite or aragonite is dissolved completely away the resulting cavity may be infilled, preserving the fossil as a mould or cast. The infilling of original or secondary cavities provides valuable information; for example internal casts of mammal skulls may reveal features of the brain. The original calcite may also be replaced, most commonly by silica (SiO_2) or pyrite (FeS_2).

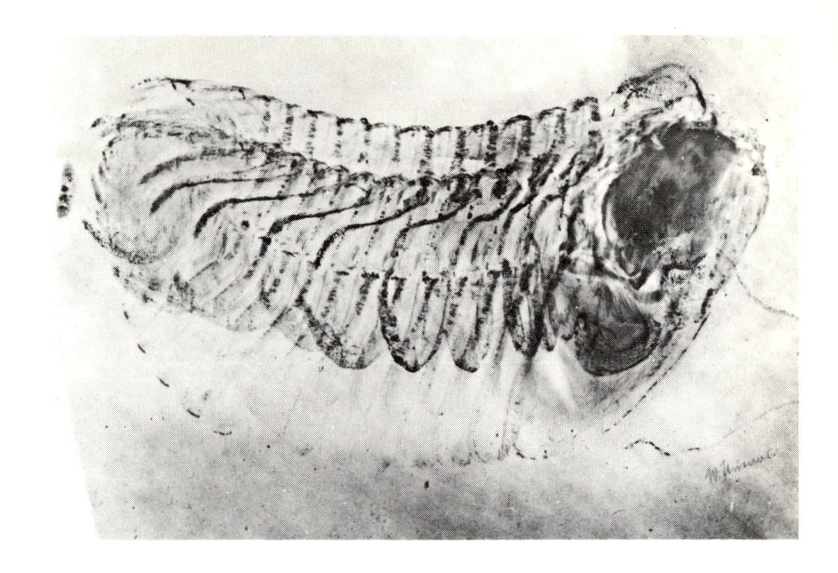

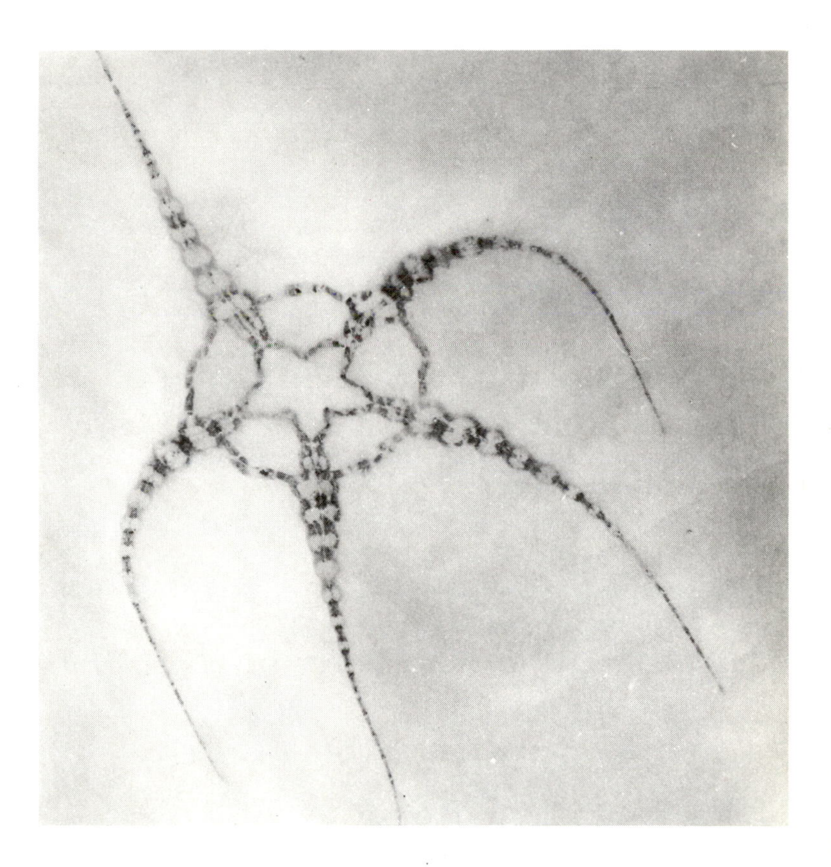

22.4: *X-ray photographs of pyritized Devonian fossils from Germany. Top: a trilobite, showing its appendages and internal anatomy. Above: a starfish.*

Calcium phosphate is a rarer skeletal material, occurring in some arthropods, inarticulate brachiopods and a group of microfossils of uncertain affinity, the conodonts. It usually occurs as a variety of apatite—$Ca_5(PO_4)_3(OH)$—and is also found in the bones and teeth of vertebrates. The only other relatively important skeletal material is silica which forms the delicate skeletons of radiolarians and diatoms and the spicules of some sponges. Where phosphatic or siliceous (either original or replaced) skeletons occur in limestones the rock can be dissolved in dilute acid to release the fossils from the matrix for study. The features of pyritized specimens, on the other hand, may be studied even when concealed in the rock, by means of X-rays. Replacement by phosphate is uncommon but there are remarkably preserved Upper Cambrian ostracodes (small bivalved crustaceans) from Sweden in which the delicate appendages have been coated by a phosphatic crust.

Soft part preservation

It is usually possible to infer a certain amount about the missing soft parts of fossils by comparison with living relatives. The nature of soft parts can also be deduced directly from traces such as muscle scars which they leave on the skeleton. The actual preservation of soft tissues is rare but scattered examples occur throughout the fossil record. The most important of these are listed in Figure 22.3. Two celebrated examples of recently extinct animals, the frozen carcasses of Siberian mammoths (Figure 22.5) and the dehydrated remains, including fur and faeces, of the giant sloth in Patagonia, are due to local climatic circumstances, and such fossils are unlikely to survive for any appreciable interval of geological time. Older soft-bodied faunas also owe their origin, however, to protection from decay and scavenging. Under anaerobic conditions, especially at low temperatures, decay is much slower and is further inhibited when an animal is buried in fine-grained sediment and is thus removed from the attention of scavengers.

The actual details of preservation vary widely. In the Devonian Hunsrück shale, for example, the soft parts are preserved as thin films of pyrite. The Carboniferous Mazon Creek fossils are quite different. Evidently the organisms were rapidly buried and nucleation of an ironstone nodule around the corpse occurred very shortly afterwards, preserving the soft parts. In the Jurassic Posidonia shale the skin of ichthyosaurs survives as a carbonaceous film revealing fins, for example, which would otherwise have been unknown as they were not supported by bone. In the Eocene Geiseltal lignite a process akin to tanning has occurred, and so exquisite is the preservation of the soft tissues of this rich vertebrate fauna that even blood corpuscles, pigment cells and minute parasitic worms have been identified. More widespread cases of soft-part preservation include Cenozoic amber, a fossil resin which when exuded from trees sometimes trapped small creatures, mostly insects, and the Precambrian Ediacaran faunas (see below) which are preserved as impressions cast in fine sediments.

Fossil plants

The preservation of fossil plants often differs in detail from that of animals. The tissues are reduced frequently to coalified material, and in such compressions the internal anatomy is usually obliterated. The external features, however, may be exquisitely preserved as an impression on the sediment. Paradoxically the details in such fossils may be on a finer scale than the grain size of the sediment. The explanation probably lies in the former presence of a mucilaginous film, perhaps of bacterial origin, which covered the dead plant and took the first impression. The internal anatomy of plants, including cell

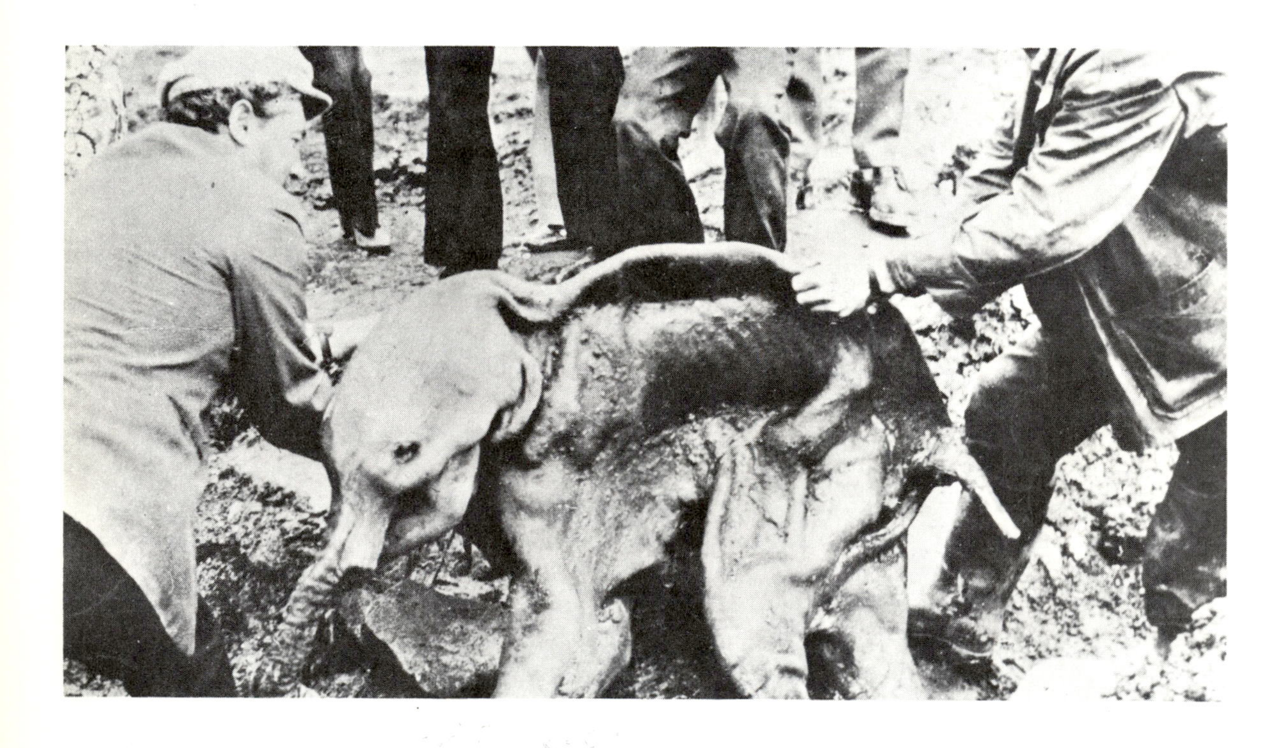

22.5: *The discovery of a frozen baby mammoth from the permafrost of Siberia. The specimen is about 10 000 years old. Traces of the hair, which is reddish-brown, are visible on the legs. The animal is about 1.5 m long.*

walls and even cell contents, may be preserved by permineralization. The tissue is rapidly 'flooded' by mineralizing fluids and the details preserved in calcite, pyrite or, in the finest examples such as the Devonian Rhynie chert of Scotland and the Triassic 'petrified forests' of Arizona, by silica.

Even if the remains of an organism become fossilized they are likely to be destroyed by subsequent exposure and erosion of the sedimentary rock containing them. Some fossils, particularly resistant microfossils, may survive redeposition in younger sediments, providing a source of potential confusion to the stratigrapher. Fossils preserved in subsequently deformed rocks suffer distortion and may be of use to structural geologists in measuring strain. Fossils can survive fairly high-grade metamorphism and the history of an area that has been subjected to thermal metamorphism can now be determined on the basis of the colour of conodonts which shows a progressive and irreversible change with increasing temperature from an original amber-brown to black, finally becoming translucent.

Trace fossils

The conditions conducive to the preservation of the remains of animals differ from those favouring the survival of the traces they produce, and the two are rarely found in association. Thus it is usually impossible to determine the originator of a trace fossil with confidence, particularly as observations of modern animals show that traces produced by widely different animals may be remarkably similar. Trace fossils are therefore classified usually according to the activity they represent rather than the animal that produced them: resting, crawling, feeding, grazing, dwelling and escaping traces are commonly recognized (Figure 22.7). Many types of trace fossil are found throughout geological time, indicating the persistence of a behaviour pattern. The majority of trace fossils were made by infaunal animals, especially deposit feeders. The interpretation of such traces has been hampered by the difficulty of studying modern equivalents which are concealed in unconsolidated sediment. New techniques such as X-raying intact blocks of sediment and injecting quick-setting resins into burrow systems are now overcoming this problem.

22.6: *A fly trapped in Oligocene amber, from the Baltic coast of Poland.*

22.7: *A meandering trace fossil known as* Taphrhelminthopsis *from the Eocene of Spain.*

Apart from indicating the behaviour of fossil organisms trace fossil assemblages can be used in environmental interpretation, particularly in relation to water depth. In shallow turbulent water most traces are made by suspension feeders (those that catch food from the water) which make characteristic U-shaped burrows. Further offshore, in quieter water, deposit feeders (which swallow sediment) are common and leave traces of their searches through the sediment, and in the deep sea the elaborate patterns made by the deposit feeders reflect the need to exploit scarce food supplies efficiently.

Chemical fossils

Large organic molecules do not survive long after the death of an organism, but may break down to specific derivatives which are stable over long periods and therefore useful as an indication of their former presence. Some amino acids, for example, survive as the product of proteins. A wide variety of other organic substances, such as hydrocarbons, carbohydrates and lipids, can be extracted from bulk rock samples or from individual body fossils and identified by such methods as gas chromatography. Chemical fossils are perhaps most significant in the search for early evidence of life in the Precambrian.

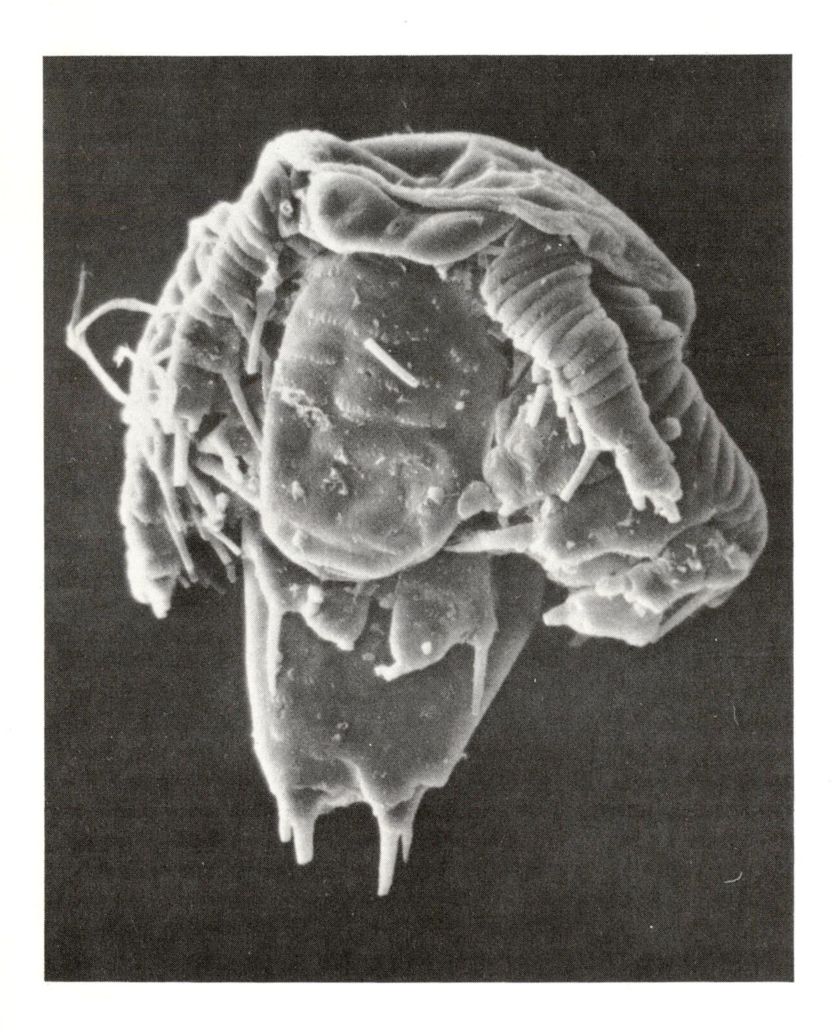

22.8: *Mineral replacement in fossils. Top: silicified specimens of cuplike brachiopods from the Permian. Above: a phosphatized specimen of a juvenile crustacean arthropod from the Cambrian.*

The discovery of phytane and pristane in early Precambrian rocks, for example, indicates the presence of photosynthetic organisms as these compounds are derived from the breakdown of chlorophyll and have no other known natural origin. Extreme care, however, is necessary in this research to avoid contamination from modern sources such as the sweat from the fingers of the investigator.

Palaeoecology

The functional interpretation of fossils

An essential prerequisite to an interpretation of the ecology of fossil animals is an understanding of the functional significance of their morphology and thereby the mode of life of each species. Significant advances have recently been made both in this field (functional morphology) and within the wider framework of interactions between species. The nature of the fossil record, however, limits the information that can be recovered, and it is difficult to apply to fossils many of the latest developments in the study of the ecology of modern organisms. Studies of functional morphology not only allow the mode of life of fossil species to be determined, but can also trace the evolution of the efficiency of an organ or function. Some of the earliest echinoderms, for example, were filter feeders, and this strategy persists until the present day alongside more advanced feeding methods which appeared later. In some of the earliest echinoderms the filtering tube feet arose directly from a spiralling row of plates (the ambulacrum) on the surface of the test. These became extinct in the early Cambrian. Other echinoderms developed several radiating ambulacra which in later forms gave rise to simple arms, thus extending the filtering area. These survived until the end of the Palaeozoic. In the most advanced filtering echinoderms, the crinoids (sea lilies and feather stars), which have modern representatives, the arms are more complex in structure and may branch several times.

The interpretation of function is often straightforward, based on a comparison with modern relatives or analogues. Simple mechanical considerations and a fossil's occurrence may also assist. For example species that have all-round vision, including directly downwards, are unlikely to have lived on a muddy sea bottom and were probably swimmers; infaunal animals are sometimes found *in situ* in their burrows, and pelagic forms usually have a very wide distribution unconstrained by sediment type (reflecting bottom conditions) as they are carried by ocean currents.

In the functional analysis of an extinct organism, particularly if bizarre structures are present with no obvious modern parallel, it may be helpful to postulate a variety of possible functions for a structure and formulate an ideal model, or paradigm, to perform each function. The paradigm that most nearly approaches the structure in question is likely to be the right one. However, the paradigm method of functional analysis is not infallible. An organ may combine two or more functions or may lack a function altogether, remaining only as a vestige of a larval or ancestral form. Elegant functional analyses, sometimes using carefully built models, have recently been reported in areas as diverse as the aerodynamic abilities of flying reptiles, vision in trilobites, the jaw mechanics of Mesozoic reptiles and swimming in Carboniferous horseshoe crabs.

Feeding strategies

Increasing attention is being given to recognizing in fossils the major feeding (or trophic) categories: suspension feeders, deposit feeders, grazers (scraping plant material off surfaces), scavengers (eating corpses) and predators. This is relatively straightforward for groups of organisms with living relatives. Some use only one method (brachiopods are always suspension feeders), while others, such as the gastropods, are more diverse in their habits. Extinct groups present a greater problem. The best evidence for feeding in fossil organisms comes from the mouth parts, which may be hard, preservable jaws or soft-bodied tentacles, and from the gut contents. Fossil faeces, or coprolites, are also abundant and casts of the mud-filled intestines of deposit feeders are known. It is rarely possible to assign a coprolite to a particular species, but those of predators may contain identifiable fragments of their prey.

The most dramatic illustrations of feeding habits come from predators; examples of palaeovoracity include fish swallowed entire (Figure 22.10) and ichthyosaur guts full of the indigestible arm hooks of belemnites. Many of the abundant fossil fish from the Carboniferous Mecca Quarry in Illinois are bitten, torn or otherwise dismembered, apparently as a result of a feeding frenzy. Some predatory habits have evidently persisted for long periods; the association of fossil starfish and bivalves in Palaeozoic rocks suggests that this association between prey and predator was first formed then. Other predatory groups have presumably shifted their habits according to changes in prey. It has been suggested, half seriously, that the holes punched in ammonites by mosasaur teeth are a result of increasing desperation on the part of these Cretaceous marine lizards as their traditional prey disappeared!

Associations between species

Comparatively little is known about interrelationships other than feeding between fossil species. Symbiotic relationships may be divided into several types that include commensalism (one species benefits and the other is not materially affected) and mutualism (both species benefit). The tubicolous animals that encrust the shell margins of some brachiopods and bivalves where the feeding currents entered

22.10: *A fossil example of greed from the Cretaceous of Kansas. The large fish* (Xiphactinus molossus) *has swallowed entire a specimen of a smaller fish* (Gillicus arcuatus).

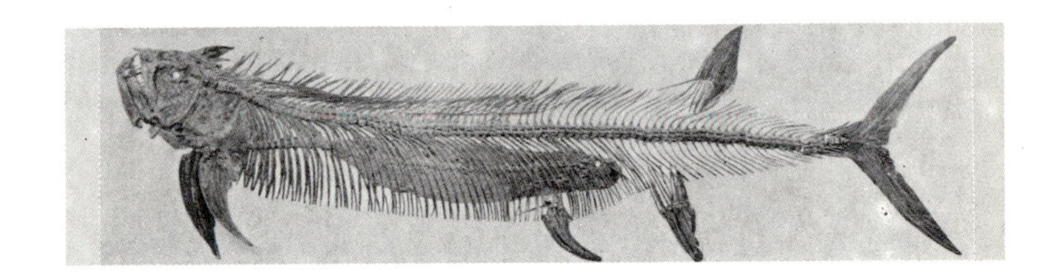

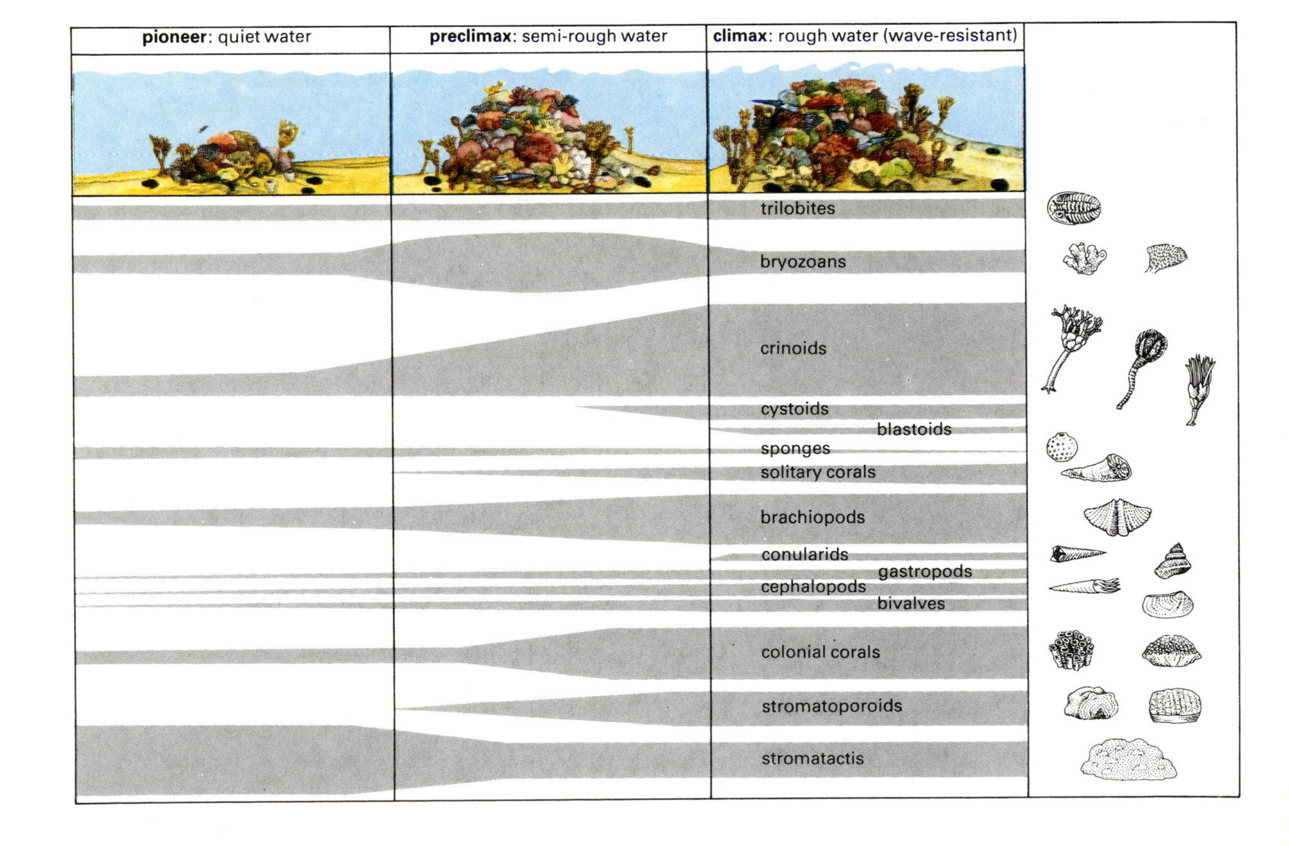

22.9: *The ecological succession in a Silurian fossil reef starting with the pioneer community in quiet water and moving towards the climax community when the reef approaches sea level. The groups that contribute to the reef ecology and structure, and their relative importance at different stages, are also depicted.*

were probably commensals. Known cases of fossil mutualism are very rare. In modern reef-building corals there is a mutualistic association with algae (zooxanthellae). These algae appear to facilitate the secretion of the massive calcareous skeletons that make up the reef, but as they live in the soft parts of the coral their chances of becoming fossilized are minimal. Mesozoic and Cenozoic corals almost certainly possessed zooxanthellae, but there has been debate whether Palaeozoic corals and certain other extinct organisms availed themselves of this association.

Much recent palaeoecological research has concerned the recurrent associations of fossil species, often called communities. However, an association of fossils rarely if ever fully represents a once living community of organisms, except in the rare cases where soft-bodied and lightly skeletonized forms survive. In associations of marine fossils the evidence of trace fossils, many of which were constructed by soft-bodied animals, can be used to estimate the importance of members of the fauna which are otherwise unrepresented.

The study of modern communities shows how they are established by pioneer species that gradually give way to climax forms; initial colonization is usually by opportunistic species with generalized feeding habits which are replaced in due course by more specialized species. Comparatively little is known about this ecological succession in soft sediments in the fossil record. Much better information is available from fossil reefs, where both pioneer and later reef-building organisms are preserved in the framework of the reef.

22.11: *A reconstruction of the Ediacaran fauna from the late Precambrian of Australia. The animals include (1) the sea-pen* Charniodiscus, *(2) the worm* Dickinsonia, *(3) the tiny arthropod* Praecambridium, *(4) the echinoderm* Tribrachidium, *(5) the sea-pen* Glaessnerina, *(6) the jellyfishes* Cyclomedusa *and (7)* Brachina. *The animals lived on sandy sediment in shallow water.*

The colonization of the Palaeozoic seas

Late Precambrian life: the Ediacaran faunas

A hypothetical observer in the late Precambrian would have had few grounds for optimism about the future of life as he was confronted by endless tracts of domelike stromatolites constructed by primitive blue-green algae. Yet over a comparatively short period of about 150 million years the initial diversification of metazoans in the late Precambrian (around 650 million years ago) gave way to a series of dazzling adaptive radiations during the Cambrian, both among the groups that developed hard skeletal parts and those that remained soft-bodied.

The relatively sudden appearance near the base of the Cambrian (570 million years ago) of invertebrates with hard parts is sufficiently clear evidence that there must have been Precambrian metazoans, but it is only comparatively recently that well-documented Precambrian faunas have been described. The first unequivocal Precambrian metazoans come from the Ediacaran faunas (named after the area of their initial discovery in the Ediacaran Ranges of South Australia, although localities are now known in every continent except Antarctica). Dating of the Ediacaran faunas is difficult, but most seem to be between seven and six hundred million years old.

One of the earliest Ediacaran faunas, from the Avalon Peninsula in south-east Newfoundland, is a fairly diverse assemblage of about twenty species. Most of the animals seem to be representative of a low grade of organization, in which the body wall is composed of two layers of cells (diploblastic), ectoderm and endoderm, separated by a layer of gelatinous material, the mesogloea. This is the structure of the cnidarians (the phylum including the hydroids, jellyfish and corals) to which some of the Ediacaran organisms strongly resembling sea pens (pennatulaceans) and jellyfish are believed to belong, although the affinities of others are uncertain.

The Ediacaran fauna from Australia is one of the youngest and is more diverse. Sea pens and jellyfish are still abundant but they are joined by further cnidarians such as probable siphonophores (relatives of the modern Portuguese man-of-war) and more advanced animals believed to be triploblastic. The third layer of cells, the mesoderm, between the ectoderm and endoderm, allows the development of organ systems. These new groups therefore represent an evolutionary advance in the interval since the first appearance of Ediacaran faunas. The worms such as *Dickinsonia* and *Spriggina* have been regarded as annelids (segmented worms) but the former, together with a newly described species from the Valdai Series near the White Sea in the northern Soviet Union, have been reinterpreted as a flatworm. *Spriggina* may be nearer to the annelids, but it has features reminiscent of arthropods. The arthropods appear to be represented by other poorly preserved species, although it seems doubtful that any were directly ancestral to later groups such as the trilobites or crustaceans. *Tribrachidium*, with feeding arms radiating across a disclike body recalls some Palaeozoic echinoderms, but there are only three arms in contrast to the five-fold symmetry characteristic of this phylum.

The recently discovered Ediacaran fauna from the Valdai Series in the Soviet Union has strong similarities with the Australian localities but the presence of new forms shows that a full understanding of these Precambrian metazoans is some way away. A major enigma is the abundance and widespread distribution of the Ediacaran faunas, in marked contrast to the rarity of later soft-bodied faunas which usually owe their existence to unusual combinations of localized circumstances. Some of the Ediacaran faunas include trace fossils, none of which can be confidently attributed to the activities of the known body fossils; thus not all the animals are preserved. It is possible, however, that the soft-bodied preservation was facilitated by an absence of the scavengers or predators which may have appeared later in geological time.

The Ediacaran animals were not the first metazoans. Although, perhaps surprisingly, no early metazoans have been found in Precambrian cherts, among fossil seaweeds 1300 million years old from Montana there is an elongate coiled form which, it has been suggested, could be a worm. Scattered remains interpreted as metazoan trace fossils are found as far back as about two thousand million years. Many such structures have been shown to be pseudofossils, however, and a sceptical attitude is essential towards new claims. A number of traces dated at about a thousand million years from the Grand Canyon do seem to be authentic. Even if these pre-Ediacaran trace fossils are accepted the record is extremely sparse and indicates very low diversity until about 620 million years ago. In younger rocks there is a dramatic increase in the diversity of trace fossils, which reflects more sophisticated activity by a greater variety of animals, and which continues across the Precambrian–Cambrian boundary in many parts of the world.

Cambrian life: the appearance of hard parts

Even the later Ediacaran faunas include only a few major groups of animals (phyla): the cnidarians and forms currently interpreted as platyhelminths (flatworms), annelids, arthropods and echinoderms. They therefore represent only the early stages of metazoan diversification. A much more obvious radiation took place during the latest Precambrian and early Cambrian as is shown in the fossil record by the widespread appearance of fossils with hard skeletal parts.

A period of only thirty million years (about 0.65 per cent of total geological time) saw the appearance of hard parts belonging to groups as diverse as the brachiopods (lamp shells), molluscs, hyolithids (animals with a conical shell closed by an operculum that were once thought to be molluscs), sponges, archaeocyathids (sessile spongelike organisms with a calcareous skeleton), arthropods, echinoderms and conodontophorids (animals bearing minute phosphatic toothlike organs called conodonts). The first hard parts secreted by protistans (foraminiferans) and algae also appeared at about this time. By the mid-Cambrian there were cnidarians with hard parts (the corals); in late Cambrian rocks the first fish scales are found; and by the Ordovician the bryozoans, the last phylum to evolve hard parts, had made their debut.

Hard parts appear in abundance in the Cambrian, but their earliest records are, in fact, Precambrian. Although the Ediacaran faunas are almost entirely soft-bodied, the impressions of some of the sea pens suggest that their soft tissues may have contained spicules of some resistant material. An Ediacaran fauna from Namibia is preserved in sandstone, but in some of the interlayered limestone beds elongate calcareous tubes named *Cloudina* have been found, and it has been suggested that they may have been constructed by some wormlike creature. A younger, but still early, occurrence of hard parts is within the Nemakit–Daldyn horizon of northern Siberia, now well known to palaeontologists. A sparse fauna occurs near the top of this rock unit and includes worm tubes, conodonts and molluscs. The age of this horizon is not entirely certain, but it is generally considered as lying just below the boundary between the Precambrian and Cambrian. It is in the Tommotian stage, the lowest part of the Cambrian system, that abundant hard parts belonging to diverse organisms first appear.

Some palaeontologists have taken the view that mineralized skeletons are simply a grade of evolutionary complexity and that no special explanation is required for their appearance. Others are impressed by the relative rapidity of their development in such a diverse range of organisms (ten or more invertebrate phyla, together with protistans and algae) and continue to search for triggering mechanisms. The apparent absence of predators, or evidence of predation, in the Ediacaran and earliest Cambrian faunas has prompted the inference that hard parts evolved primarily as a protective measure. Correlations have also been drawn with a postulated increase in the concentration of atmospheric oxygen, although estimates for the actual values around this time are very uncertain. Abundant oxygen is necessary for the synthesis of collagen, an essential component of both hard parts and large muscles. The appearance and development of trace fossils, produced by large muscular animals, and the subsequent development of hard parts could be directly related to the initiation of collagen synthesis in an increasingly oxygenated environment.

Important Cambrian phyla

The fossil record of Cambrian life is naturally dominated by marine invertebrates. One of the first groups to rise to prominence were the archaeocyathids (Figure 22.12), simple spongelike organisms which secreted a calcareous skeleton, sometimes sufficiently abundant to form the earliest reeflike structures. The diversity and geographical extent of archaeocyathid faunas increased rapidly during the earliest Cambrian and they are found in all continents. However, they subsequently declined abruptly and became extinct later in the Cambrian. In contrast the molluscs diversified to become one of the most successful groups of invertebrate animals. The oldest mollusc faunas occur in the earliest Cambrian of Siberia and China and show that four major groups were already present at this time, indicating an unknown Precambrian evolutionary history. The monoplacophorans—simple, conical, single-shelled forms with serially repeated muscle scars—apparently gave rise to the other major shelled groups, eg gastropods (snails), rostroconchs (ancestors to the bivalves) and nautiloids (the

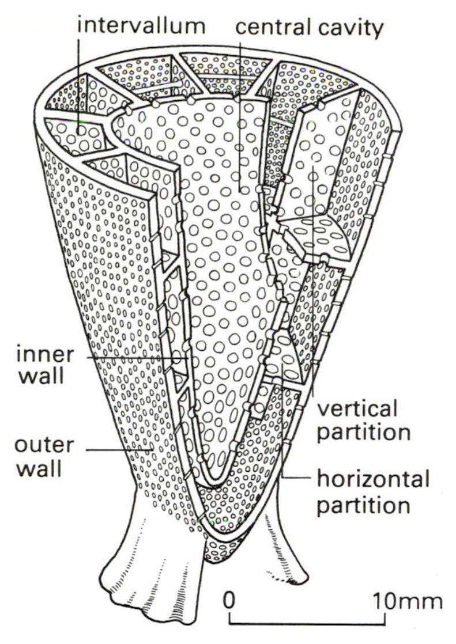

22.12: *An archaeocyathid. Calcareous organisms of uncertain relationships, archaeocyathids flourished during the Lower Cambrian.*

22.13: *A Cambrian trilobite* (Biceratops). *Trilobites are among the most characteristic of Cambrian animals.*

22.14: *A reconstruction of Cambrian marine life. The animals shown include (1) the echinoderm (eocrinoid)* Gogia spiralis, *(2) the brachiopods* Lingulella *and (3)* Nisusia, *(4) the trilobite* Paradoxides, *(5) the shelled animal* Hyolithes, *(6) the echinoderm (eocrinoid)* Lichenoides, *(7) the swimming trilobite* Agnostus. *This reconstruction illustrates characteristic Cambrian animals from several environments, and it is unlikely that any single piece of sea floor exactly resembled this figure.*

first cephalopods), almost all of which appeared during the Cambrian.

Brachiopods and, more locally, echinoderms are also significant elements of Cambrian faunas. The brachiopods are suspension feeders and remain attached to the substratum; their bivalved shell is used both for protection and as a filter chamber. The action of cilia borne by the arms of a tentacular organ called the lophophore actively draws water into the shell where it is filtered for food. The brachiopods are much less important in the modern seas. A rapid early radiation of the echinoderms led to the evolution of fifteen distinct classes within the phylum. This radiation of the echinoderms into such a wide variety of forms may reflect low levels of competition in the early Cambrian seas. The fact that many of the classes disappeared relatively soon suggests that the less efficient became extinct as other organisms, better adapted to their environments, displaced them.

The precise order in which these invertebrate phyla appeared is rather uncertain, but the trilobites were probably latecomers. They apparently appeared after the Tommotian stage but rapidly came to dominate most Cambrian faunas. Evidence for their ecology is limited but most were bottom-dwelling and the evidence of their trace fossils, such as *Cruziana*, suggests that some were deposit-feeders. The trilobites remained relatively simple in morphology during the Cambrian compared to those that evolved later. One exception to this were the tiny bizarre agnostoids which were very widely distributed and were evidently pelagic; for this reason they are invaluable for biostratigraphic correlation.

The evolutionary explosion among Cambrian invertebrates relates to their rapid invasion of new ecological niches. All the major feeding types among the invertebrates were established in the Cambrian. Deposit-feeders were especially important and included arthropods and molluscs. Suspension-feeders included brachiopods, some echinoderms, sponges and probably archaeocyathids. Less is known about grazers, but enormous traces apparently made by a molluscan radula in late Cambrian sediments in Saudi Arabia hint at the size reached by this feeding group. The relative abundance in the Burgess shale in British Columbia of predators and scavengers, especially arthropods and priapulids, indicates that these feeding strategies were fully developed in Cambrian times. The Cambrian adaptive radiation has been particularly well documented in the echinoderms, but the Burgess shale fauna shows a similar pattern in groups as diverse as the arthropods, polychaetes and priapulids. Many species seem to be 'experiments', as various phyla tried out different modes of life, and in consequence they have, to our eyes, bizarre appearances. Some were evidently relatively inefficient and were replaced in due course by more efficient forms.

Later Palaeozoic marine life

By mid-Cambrian times at least eleven phyla which persist to the present day, and about ten others destined to become extinct, had evolved and the initial burst of diversification was over. A state of equilibrium appears to have been reached as rates of extinction and speciation balanced each other. However, during the late Cambrian and continuing into the Ordovician period, a new and dramatic increase in diversity at lower levels in the hierarchy (ie more families, genera and species rather than more phyla) led to a major repatterning of marine life. This set the seal on the style and character of marine faunas until the mass extinctions of the Permian, at the end of the Palaeozoic. This new expansion involved a few new organisms, such as the colonial bryozoans ('moss-animals'), but mainly affected a number of groups that were already present, but in insignificant numbers, such as the corals, articulate brachiopods (those with an articulating hinge), cephalopods, bivalves and crinoids.

The groups that diversified first in the late Precambrian and early Cambrian were essentially generalized, with broad feeding and habitat requirements. They were therefore well suited to colonize what were largely unoccupied habitats. The diversification of more specialized forms was largely postponed until the late Cambrian and Ordovician, for reasons that are still not clear. Once it began, however, the resulting species were often equipped to replace their more generalized predecessors perhaps by direct competition. Furthermore in having narrower requirements a larger number of species were able to be packed into the same habitat space. This explanation of the pattern of diversification in the early Palaeozoic must be considered tentative until more detailed evidence to support it is documented. The equilibrium established at the end of the Ordovician period held until the major extinctions of the Permian period.

Most Cambrian faunas are dominated by trilobites (Figure 22.13) and indeed in terms of numbers of genera their heyday occurred during this period. The wave of metazoan diversification in the Ordovician established the brachiopods, bryozoans and stalked echinoderms as the dominant members of marine faunas in place of trilobites. All these groups represent an increased emphasis on suspension-feeding. They

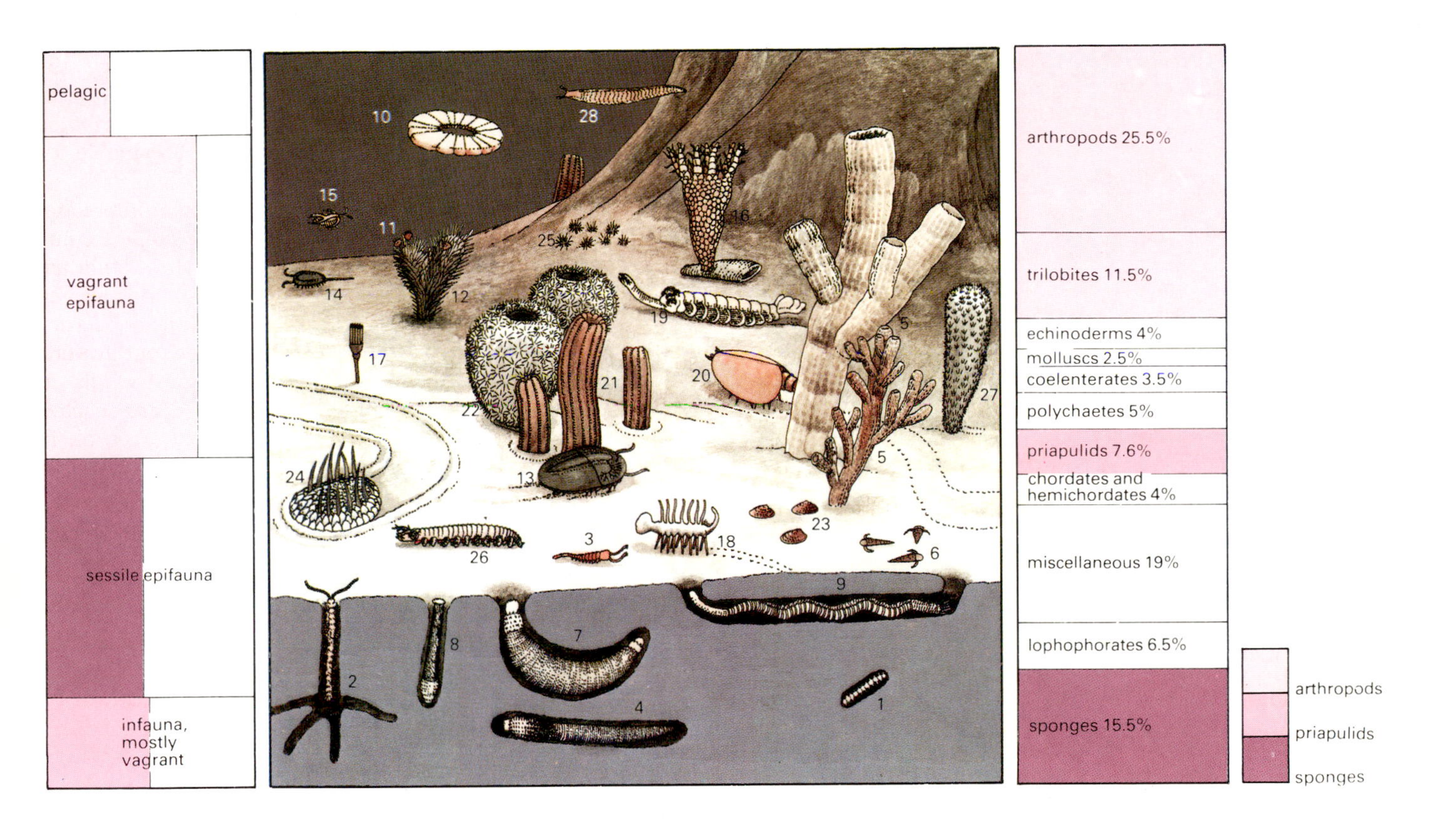

22.15: *A reconstruction of the Burgess shale fauna from the middle Cambrian of Canada. The animals shown are the polychaetes (1)* Peronochaeta *and (2)* Burgessochaeta*; (3) the arthropod* Yohoia*; (4) the priapulid* Ancalagon*; (5) the sponge* Vauxia*; (6) the shelled animal* Hyolithes*; the priapulids (7)* Ottoia*, (8)* Selkirkia *and (9)* Louisella*; (10) the jellyfish* Peytoia*; (11) the brachiopod* Dictyonina *attached to (12) the sponge* Pirania*; (13) the trilobite* Naraoia*; the arthropods (14)* Burgessia *and (15)* Marrella*; (16) the crinoid* Echmatocrinus*; the enigmatic organisms (17)* Dinomischus*, (18)* Hallucigenia *and (19)* Opabinia*; (20) the arthropod* Canadaspis*; (21) the sea-anemone* Mackenzia*; (22) the sponge* Eiffelia*; the molluscs (23)* Scenella *and (24)* Wiwaxia*, (25) the sponge* Choia*; (26) the arthropod* Aysheaia*; (27) the sponge* Chancelloria*; and (28) the chordate* Pikaia*. The two columns show the relative abundances (number of genera) of major groups and habitats.*

probably avoided competition by filtering at different levels, the brachiopods on the sea floor, the bryozoans a few centimetres above it and the crinoids several centimetres higher. Predators are also more evident in post-Cambrian assemblages; from the later Silurian the first fish with jaws, the placoderms, and then in the Devonian the sharks and bony fishes, evolved and rapidly replaced the nautiloids as the dominant marine carnivores. By the Devonian the coiled ammonoid cephalopods appeared and became one of the most successful marine groups of the later Palaeozoic and Mesozoic. In the Ordovician a group of colonial filter-feeding organisms, the graptolites, became planktonic and are an extremely important group for the correlation of Ordovician and Silurian sediments. Corals appeared in significant numbers in the Ordovician, although the first reefs with corals as a primary component occurred in the Silurian. The importance of reefs to the palaeontologist lies in their resistance to destruction, so that the changing tenancy through time of this particular ecosystem is readily available for study.

The development of life on land

The previous chapter has shown that plants evolved more than three thousand million years ago in the Archaean. Until the Silurian period, however, plant life was almost exclusively algal and is preserved mainly in marine rocks. The initial 'greening' of the landscape by green algae and bacteria may have taken place at or before this time, but it was not until the oxygen levels in the atmosphere had built up sufficiently to create an ozone layer shielding the land surfaces from harmful ultraviolet radiation that a large-scale invasion of the land by plants and animals could take place.

The invasion of the land

The vascular land plants have a single common feature in the possession of xylem (for water transport) with characteristically thickened conducting elements (tracheids, or vessels) which also give the plant rigidity—an essential feature for living on land. They have an epidermal covering (the cuticle), which prevents desiccation, with small pores (stomata), which permit gas exchange and control transpiration. Lower groups of vascular land plants such as ferns

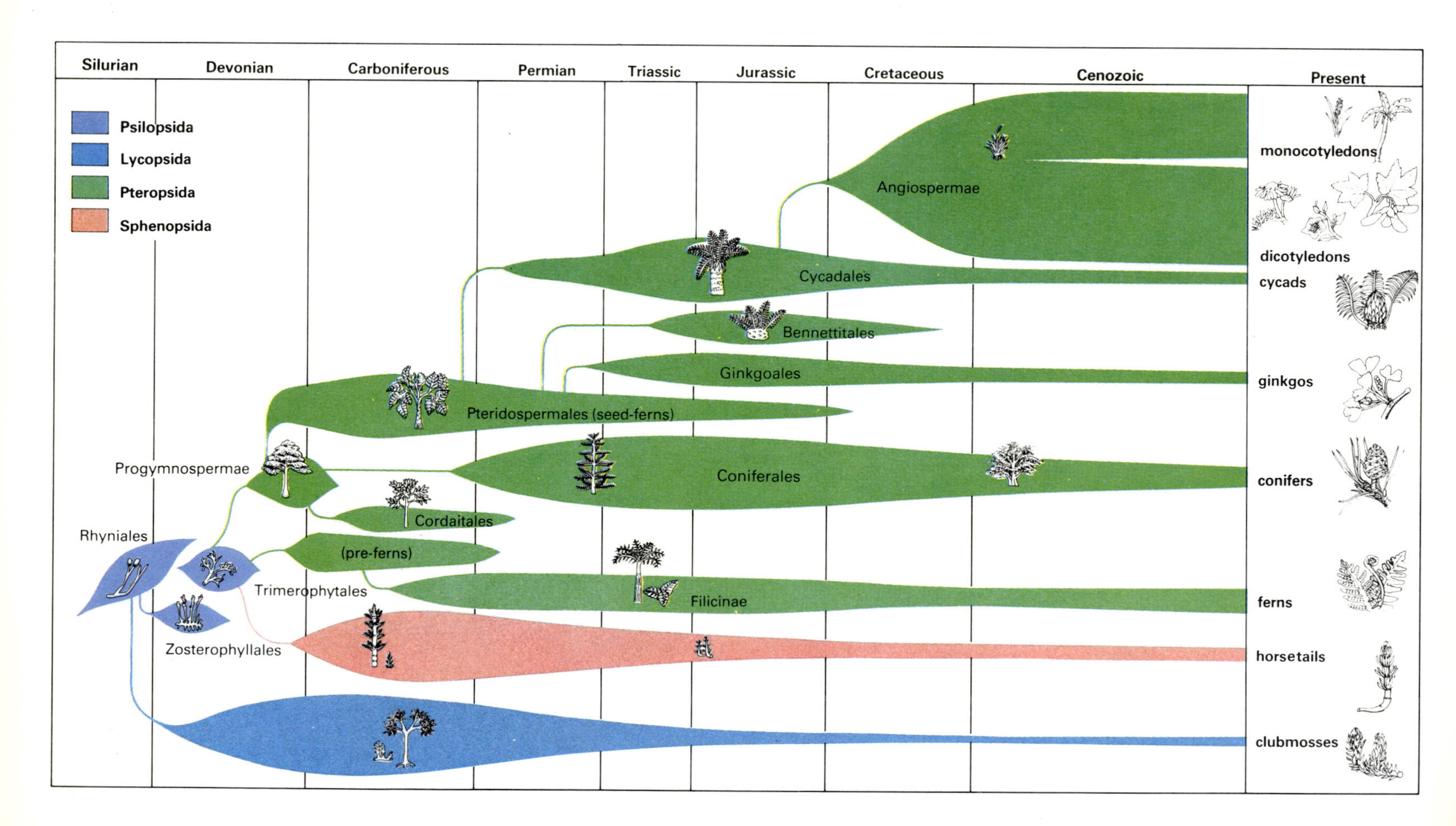

22.16: *The classification of vascular land plants adopted here emphasizes evolutionary rather than simply morphological relationships.*

show an alternation of generations in which the sporophyte is the more significant. It produces spores that possess a very tough coat (exine) to resist desiccation and are dispersed from the parent plant. The earliest-known vascular land plant is *Cooksonia*, a very small plant, no more than 100 mm high and showing no differentiation into stem, leaves and roots.

Once this initial phase of land colonization was accomplished in the late Silurian to early Devonian, a rapid diversification took place. This can be recognized both in the sparse macrofossil record and, even more clearly, in the more extensive microfossil record. The spores produced by these early land plants were about 50 micrometres (μm) in diameter. Many types of plant produced recognizably different types of spore, and the very rapid increase in the number of types of spore that occurred through the Devonian reflects the rapid evolution of land plants at this time.

Limited evidence from late Silurian and early Devonian faunas suggests that arthropods were among the earliest land invertebrates. They included herbivorous myriapods (millipedes) and carnivorous arachnids (spiders). The myriapods, as primary consumers, may have followed plants (primary producers) onto land. The carnivorous arachnids in early Devonian faunas imply the availability of metazoan and protozoan prey which have not been preserved as fossils. The early arthropods probably contributed to building up humus and soil structures, previously non-existent. Some were large, a possible adaptation to prevent water loss, and the late Silurian and early Devonian saw numerous experiments among the arthropods into satisfactory adaptations to a terrestrial life. Those arthropod groups that had major or minor success include the insects, myriapods, arachnids, isopods, decapods, amphipods, eurypterids and the xiphosures.

The Rhynie chert

The most complete knowledge of early land plants, and indeed of early terrestrial ecosystems, comes from the early Devonian Rhynie chert from Aberdeenshire, Scotland. Here both plants and animals are preserved in three dimensions in a petrified peat deposit, showing minute cellular detail. The most common vascular land plant is a simple type known as *Rhynia*, having a simple leafless dichotomizing stem and no true roots. The stem bore terminal sporangia yielding trilete spores. The plant probably grew no more than 300 mm tall. In

22.17: *Generalized relationships among tetrapod vertebrates and their evolution from a fishlike ancestor.*

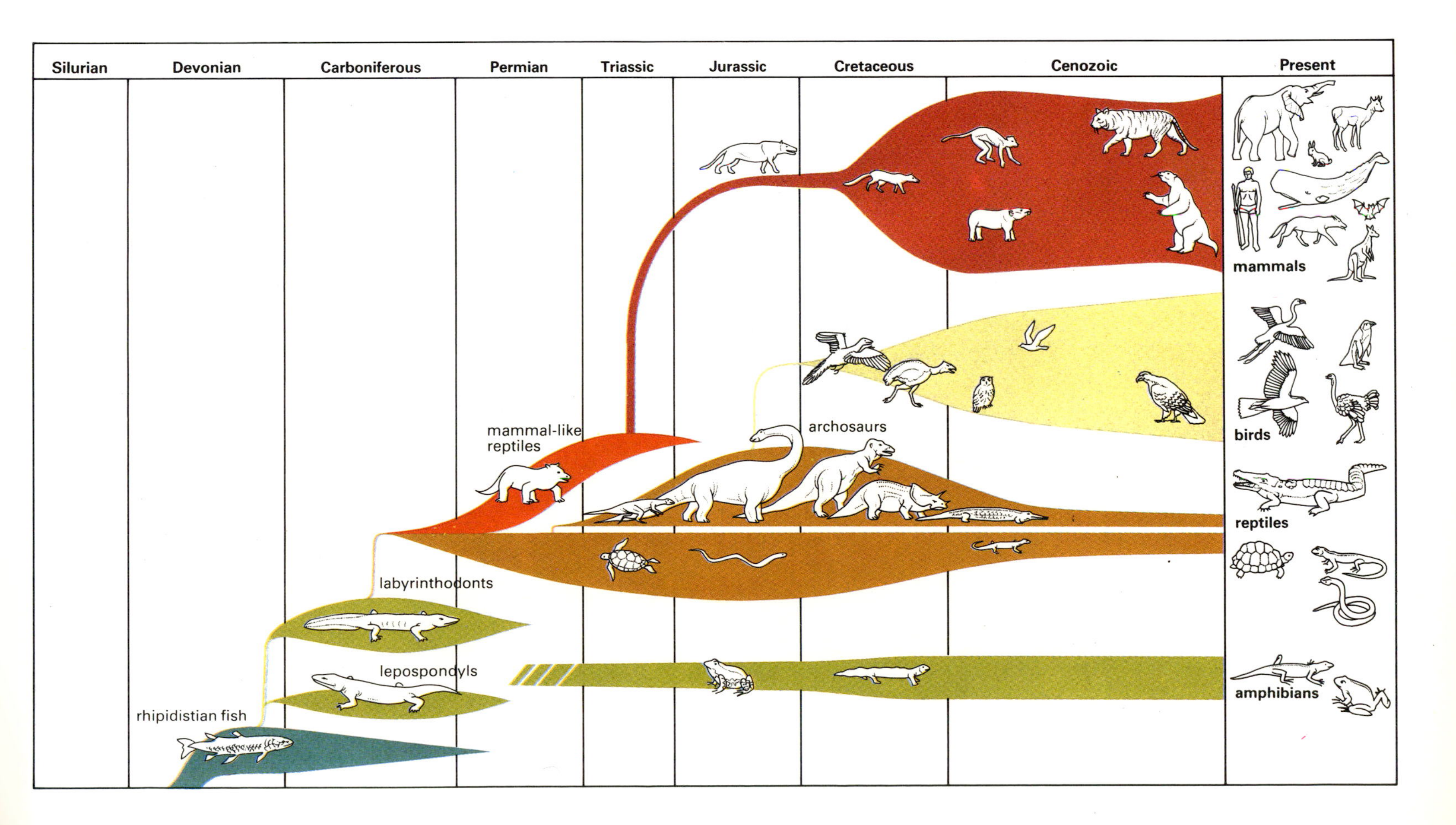

addition to this, and one or two other simple plants, is a rather different type called *Asteroxylon*. Although about the same size as *Rhynia*, the vascular strand of *Asteroxylon* is star-shaped in cross-section, compared with the simple cylindrical strand of *Rhynia*. It also has spines on the stem among which, on fertile specimens, are scattered kidney-shaped sporangia. This plant may be a forerunner of the lycopods which became prominent later in the Palaeozoic. Both algae and fungi are also found in the chert.

The unique fauna of the Rhynie chert consists mainly of micro-arthropods. The terrestrial fauna includes collembolans (springtails), such as *Rhyniella*, mites, spiders and other arachnids. This deposit also provides data concerning interactions between animals and plants and hence a tantalizing glimpse of an early terrestrial ecosystem. Some arachnids have been discovered within empty *Rhynia* sporangia and it has been suggested that these animals might have been feeding on the *Rhynia* spores. Such spore-feeding might also have benefited the plant as well as the animal as it has been shown that some of the spores eaten by modern arthropods can remain viable after they have passed through the gut, thus assisting plant dispersal. Collembolans today are generally herbivorous and it is suggested that *Rhyniella* fed also on soil micro-organisms and spores. Many of the *Rhynia* axes show damage extending into the sap-containing phloem tissue as would be caused by an arthropod feeding on plant juices. The other arthropods appear to have been carnivorous.

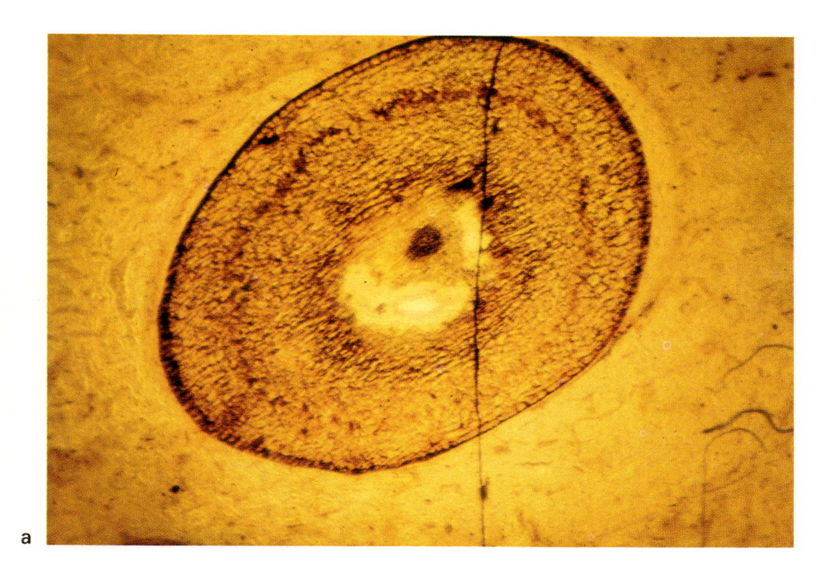

a

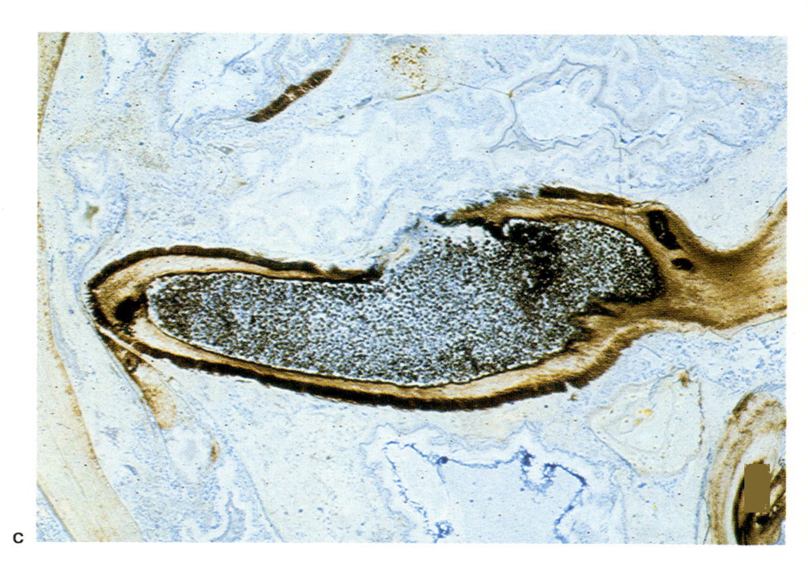

c

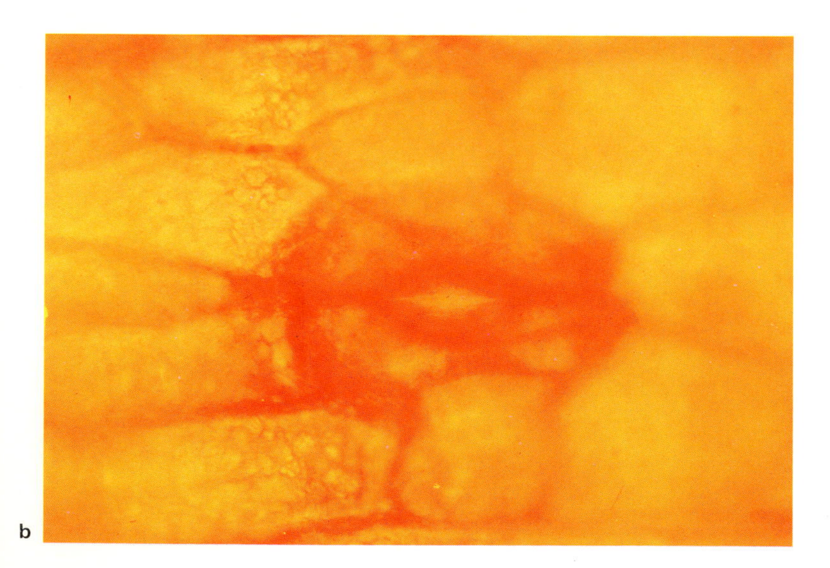

b

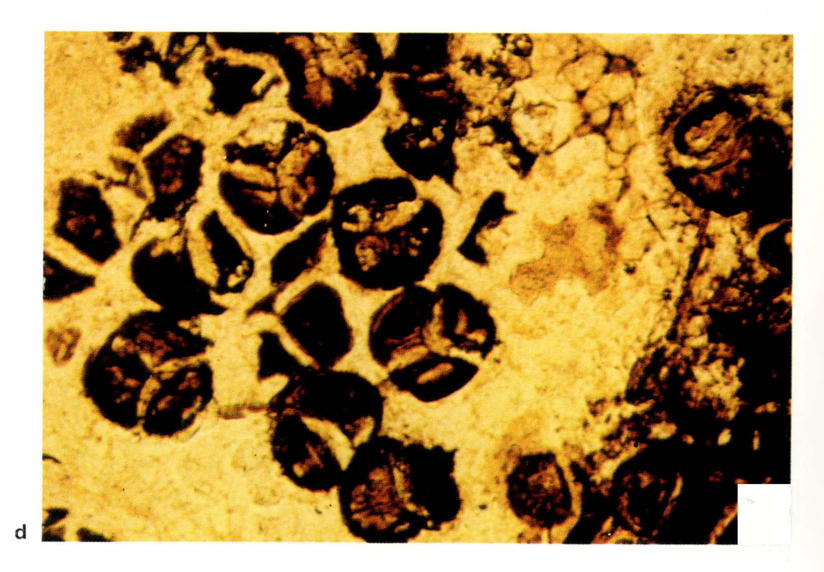

d

22.18: Rhynia *from the Rhynie chert.* (**a**) *A transverse section through the stem, showing the central xylem strand* (×15). (**b**) *Stoma* (×600). (**c**) *A longitudinal section through terminal sporangia containing spores* (×5). (**d**) *Spore tetrads* (×250).

The development of late Devonian ecosystems

Size and strategies of Devonian land plants

By the end of the early Devonian four major groups of land plants had evolved, all believed to have been derived from a *Cooksonia*-type stock. Three of these (Rhyniales, Zosterophyllales and Trimerophytales) have been traditionally included in the Psilopsida which comprises the most primitive and simple vascular plants with no differentiation into stem, leaves or roots. All are homosporous (having spores of the same size) with massive sporangia either terminating or lateral to protostelic axes. The other group is the Lycopsida, in which the plants have either minute enations or leaves derived from this type (microphylls) borne spirally on the stem. The sporangia are massive and borne lateral to the stem and may be either homosporous or heterosporous (having two sizes of spores).

All the late Silurian and early Devonian plants were small (less than 1 m tall), probably because the plants produced only primary strengthening tissue. In addition their life cycle restricted the plants to damp environments. As they could not readily colonize well-drained soil, erosion in the uncolonized upland areas would have remained unchecked. Two main problems faced these early land plant communities: competition for light and competition for space. Two new strategies—reproductive and vegetative—adopted by plants to combat these problems had far-reaching consequences in the later Devonian.

All early land plants were homosporous. This tied the plants to damp environments as the sperm must swim in soil moisture to fertilize the eggs which are produced by the tiny gametophyte generation of the plant. Possibly by the end of the early Devonian and certainly by the mid-Devonian some plants adopted a heterosporous reproductive strategy which brought the process of sexual differentiation to an earlier stage in the reproductive cycle. Two sizes of spore were produced: a microspore, about 50 μm in diameter which gave rise to a motile sperm, and a megaspore (>200 μm), which gave rise to a female gametophyte. These larger spores were produced in smaller numbers but had an increased food reserve which gave the new sporophyte plant an increased chance of survival. Later in the Devonian some plants reduced the number of megaspores in the sporangium to one and then released it only when fertilization had been achieved. This allowed the structure to hold both food and its own water within which the sperm could swim to fertilize the egg, so allowing the plant to live in dryer conditions. This reproductive structure was the seed. Fossil seeds are first found in rocks of latest Devonian age. Initially such seeds were quite small, typical of pioneer plants (for competition in the

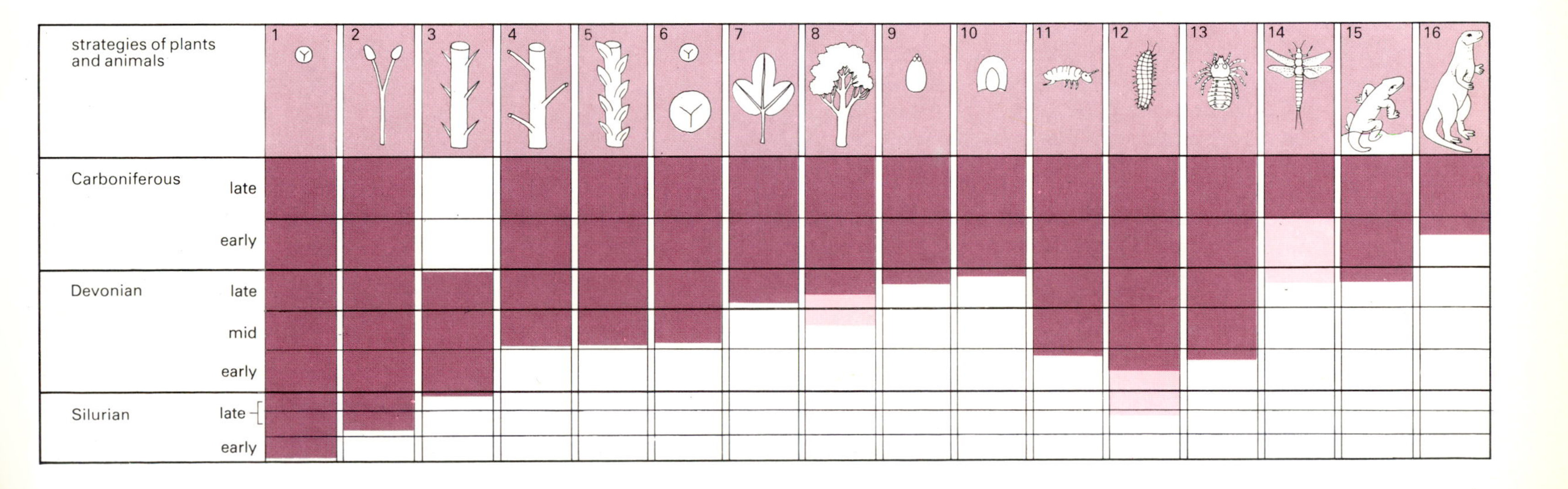

22.19: *The evolution of strategies of animals and plants, showing the innovation of various biological strategies of animals and plants in the late Palaeozoic. (1) Triradiate spores with a resistant organic coat were produced by land plants as reproductive propagules. (2) Vascular land plants. Early forms were small, usually with simple dichotomous branching and terminal sporangia. (3) Spiny axes, possibly an adaptation to discourage invertebrate herbivores. (4) Over-topped branching; the main central stem gave support. (5) Microphyllous leaves evolved from small enations to increase the photosynthetic area. (6) Heterospory: the numerous small microspores give rise to a male gametophyte and the smaller number of megaspores give rise to a female gametophyte giving the resultant new generation an increased survival potential. (7) Megaphyllous leaves, probably derived from a lateral branching system. They increase the plant's photosynthetic area and allow it to shade competing plants. (8) Secondary growth allows plants to increase in girth and hence height and to shade competitors. (9) Radial seeds and seed megaspores. The female propagule is retained and only released when fertilized. Increased independence from damp soil facilitates an increased likelihood of survival in new niches. (10) Bilateral seeds. (11) A millipede arthropod, one of the earliest land animals. (12) A collembolan (springtail; Arthropoda), an early land invertebrate. (13) A carnivorous spider (Insecta). (14) A flying insect, the first conqueror of the air. (15) An amphibian, the first land vertebrate but requiring water for breeding. (16) A reptile, the first land vertebrate to be independent of water for breeding.*

new niches that these plants could inhabit would be low). These earliest gymnosperms could now colonize the vacant upland areas, helping to stabilize some erosional areas for the first time. The evolution of this new reproductive strategy created a second rapid diversification of land plants. At about the same time the earliest leaves were developed by webbing between finely branched stems.

In any plant community there is competition for light. A plant that can shade out its competitor by growing higher will obviously have an advantage. Other consequent advantages might be increased dispersal potential of spores and seeds, particularly by wind, and the elevation of sporangia out of the reach of spore-feeding arthropods. The problem was solved by the production of secondary wood, by continued growth around the circumference of the stem so that the increase in girth could provide support for an increase in height. The appearance of trees in different lineages in the mid to late Devonian and hence the evolution of forests enabled a two-level forest ecosystem to develop.

Arthropods dominated the late Devonian forest-floor ecosystem. It has been suggested that the evolution of trees may have stimulated the evolution of flight in insects although predation by primitive hunting spiders upon early wingless insects might also have been a factor. It seems likely that numerous arthropods climbed the trees in search of food or to escape predation and that a complex tree-top ecosystem evolved during the late Devonian and early Carboniferous.

The evolution of tetrapods

The earliest four-footed animals were amphibians, and they were thus restricted to damp areas in which water was available for their reproductive cycle. The warm, shady and humid environment of the forest floor in equatorial regions during the late Devonian and early Carboniferous periods provided just these conditions. Rather than from freshwater fish it is now believed that the amphibians originated from marine Rhipidistia, a group of crossopterygian fish which were the dominant predators of the late Palaeozoic seas and which reached a length of up to 2 m. The main food source of the early amphibians, of which very little is yet known, may have been fish and arthropods, including some land arthropods. The main responses of amphibians to living on land were the development of lungs, the elaboration of the bone structure, particularly the limbs for support, and the prevention of desiccation. These developments were taken a stage further by the reptiles, which evolved an exclusively terrestrial mode of life during the late Carboniferous. It is now generally believed that their emergence followed rapidly on the development of the amniote egg which freed the animals from dependence on water for reproduction.

The evolution of flight

From the late Silurian until the end of the Devonian all the early land invertebrates were ground- or tree-dwelling. With the evolution of trees and the elevation of nutritious fructifications there was an obvious advantage for an animal to be able to move from tree top to tree top with the minimum of effort, ie by flying. Although on theoretical grounds it is suspected that one group of arthropods, the insects, achieved flight as early as the late Devonian, the earliest fossils are of Namurian (late Carboniferous) age. During the late Palaeozoic the insects rapidly diversified and populated the air. For almost 100 Ma these flying insects had no aerial competitors, for the air was the last major environment to be conquered by the vertebrates. Indeed, no amphibian achieved true flight. The vertebrates needed to solve several basic problems. The first was to be strong and yet have a light skeleton. Reptiles and later birds solved this by developing hollow bones and a keeled breastbone. The second problem was to develop limbs for flying. The limbs were modified either by the addition of feathers or by stretching a membrane between fingers of limbs to produce a wing. Finally birds developed a constant, high metabolic rate which allowed for the strenuous activity of flying over long periods. Gliding lizards are known from late Permian sediments. The Permian was also a time of insect diversification. From the Jurassic to Cretaceous a group of flying reptiles, the Pterosaurs, evolved probably from a thecodont stock; they include the large *Pteranodon* although some believe that

22.20: *The evolution of flight.* (**a**) *The first airborne animal: an insect.* (**b**) *A flying reptile.* (**c**) *A bird.* (**d**) *A flying mammal: a bat. (Limb bones are shown in colour.)*

it was a glider rather than a true flyer. The first true bird, *Archaeopteryx*, from the late Jurassic has many features of a small dinosaur but with the addition of feathers and it probably evolved from a small theropod dinosaur. It has been suggested that the feathers were not originally developed for flying but for catching insects and only later, after further structural and behavioural modifications, was flight probably achieved, as this would assist in capturing potential prey. Generally birds are uncommon as fossils and the record is biased towards aquatic and large flightless genera in the Cenozoic.

Although a number of mammals are capable of gliding, only one group has achieved true conquest of the air: the bats. The bats developed a membrane stretched between the forelimb and the body as a wing, and are known mainly from Cenozoic sediments. In the present day the birds and the insects dominate the air, both groups undergoing a rapid diversification during the Cenozoic.

The diversification of terrestrial environments

During the late Carboniferous a wide variety of ecological niches was inhabited by various, often highly specialized, animals and plants. In the coal-forming equatorial belts (now Europe and North America), in particular, lowland and upland ecosystems are known. Lowland swamp communities were dominated by arborescent lycopods, and other lowland habitats, such as floodplains, were largely occupied by ferns and pteridosperms. Arthropods dominated leaf-litter environments and insects had the skies to themselves. Swamp pools were inhabited by fish and amphibians and insectivorous or carnivorous amphibians lived around the lake margins. Primitive reptiles roamed the lowlands, whereas the uplands, dominated by gymnosperms, probably supported herbivorous mammal-like reptiles. There is evidence from fossil charcoal that forest fire played a major role in some late Carboniferous forests.

A wide variety of evidence suggests the existence of complex food webs in the late Carboniferous swamps. Terrestrial arthropods were possibly the main herbivores while millipedes and mites must have played an important role as decomposers of plant debris. Other arthropods, such as arachnids, preyed upon these invertebrates. The terrestrial vertebrates all appear to have been insectivorous or carnivorous on other tetrapods. Fish dominated the lakes and fed upon plants and other animals and in turn were preyed upon by amphibians with large amphibians at the top of the trophic pyramid.

Late Palaeozoic changes

At the end of the Carboniferous period there was a major climatic change which had a dramatic effect upon terrestrial life, both animals and plants. The climate in equatorial areas became more arid and the drying out of the coal-swamp environments caused a major shift in the ecological balance. The large arborescent lycopods which dominated the coal-forming swamp forests became extinct by the end of the Carboniferous as these environments dried out. When the lake and lakeside environments also dried out in the Euro-American area in the early Permian, almost all the associated animals (mainly amphibians) and plants became extinct. The lowland areas became arid and plants and animals better adapted to these conditions spread down from the upland areas. The floras were dominated by primitive conifers and other gymnosperms, whereas herbivorous reptiles and mammal-like reptiles dominated the vertebrate communities. The late Palaeozoic was also a time for biogeographical separation of floras, with the seed-plant *Glossopteris* dominating those of the southern continents (Gondwana). Land vertebrates are recorded from this area for the first time.

Most of the terrestrial amphibians, common in the Carboniferous, became extinct by the end of the early Permian. The therapsids, a group of mammal-like reptiles, subsequently became the dominant vertebrates. This group diversified and included large and small terrestrial carnivores, semi-aquatic fish-eaters, and large and small herbivores. The insects which probably further diversified at this time were almost certainly a major food source to some reptiles. Of these, the mammal-like reptiles had a more erect posture which it is thought had an influence on metabolic rate, and consequently an internally maintained body temperature was evolved. An excellent fossil record (from the Soviet Union and Gondwanaland) illustrates the changes in the transition from the therapsids to the mammals. The mammals, however, remained only a small and insignificant element of the fauna until the end of the Mesozoic era.

Mesozoic stability

The cycads which evolved during the early Permian together with other seed-bearing plants dominated early Mesozoic land floras with the ferns and sphenopsids (horsetails) as subsidiary elements. The terrestrial faunas were dominated by arthropods, including diverse insect groups, and vertebrates. The mammal-like reptiles which became extinct by the early Jurassic were gradually replaced, from the mid-Triassic onwards, by the archosaurian reptiles, dinosaurs in particular.

The reptiles are capable of reproducing on land, unlike the amphibians from which they evolved. In common with their descendants (birds and mammals) their embryos are surrounded by protective membranes affording water retention and to allow exchange for respiratory gases, waste and nutrients. In order to avoid water loss the reptiles developed a thick leathery skin and also developed a light skeleton, permitting agility. Many of the early reptiles probably fed on insects. The early archosaurian reptiles, the thecodonts, appeared late in the Permian. Some of these may have walked on their hind legs while others remained quadrupedal. The archosaurs, the main group of Mesozoic reptiles, include the dinosaurs and the crocodiles. The diversification of the dinosaurs was spectacular in the late Triassic to early Jurassic and may have been a response to the spread of extensive dry lowland habitats occupied by diverse gymnospermous plants.

Subdivision of the dinosaurs is based on differences in pelvic structure and two main groups can be recognized. The Saurischia (lizard-hipped) were more primitive with a pelvic structure of triradiate form similar to that of the thecodonts. They included both carnivores and herbivores. The Ornithischia (bird-hipped) were herbivorous. The archosaurs evolved to occupy many ecological niches, some even returning to an aquatic mode of life (crocodiles) while another lineage included the flying reptiles.

Mesozoic–Cenozoic changes

The Jurassic and early Cretaceous landscape was dominated by gymnospermous plants, reptiles and insects. The success of the dinosaurs in the Mesozoic restricted the mammals to being small nocturnal insectivores and it was not until the dinosaurs became extinct at the end of the Cretaceous that the mammals rapidly diversified. The major adaptations of the mammals which were to be of such importance later were a coating of hair which may have had a sensory function as well as an insulating one; teeth of only two generations, clearly differentiated into incisors, canines, premolars and molars; reproduction–by definition they give live birth and suckle their young; brain size–larger in relation to body weight and a marked reorganization of the bony brain case.

During the Cretaceous a new group of plants evolved: these were the angiosperms with enclosed female ovules and showy and elaborate modified leaflets around the reproductive structures. These plants with their highly specialized reproductive strategy developed a mutually beneficial interaction with various animals but particularly the insects. Social insects became common and a number of new groups evolved, including the butterflies. The success of the angiosperms was immediate and by the end of the Cretaceous period angiosperms dominated land floras and had evolved into a great diversity of ecological niches. The dinosaurs survived this floral revolution and it has been recently suggested that the origin of the flowering plants may be closely correlated with the spread of low-browsing dinosaurs such as *Triceratops*. The intense, low cropping of vegetation by these animals gave the early angiosperms, probably low shrubs and small trees, a distinctive advantage over cycads and conifers because of their life cycle.

Although their relations, the crocodiles and birds, survived and flourished, the extinction of the dinosaurs at the end of the Cretaceous is an unsolved mystery. Among numerous factors major disruption of the food webs, caused by floral changes, ecological instability and climatic changes may have been contributing causes. Alkaloids and hydrosoluble tannins present in many angiosperms may also have acted as a deterrent to herbivorous reptiles and may have even been directly responsible for interfering with several physiological activities, including the observed thinning of ceratopsian eggshells.

Modern terrestrial ecosystems

The angiosperms went through a second major radiation in the late Cretaceous–early Cenozoic, at which time the mammals underwent a major adaptive radiation and became the dominant land vertebrates. Two groups of carnivorous mammals arose from primitive insectivorous types at this time: an extinct group, the creodonts, and the modern order Carnivora. The diverse modern ungulate groups (hoofed mammals) arose from the primitive condylarths. Regional differences in faunas became more significant, influenced by the separation of the continents, so that the Australian fauna of marsupial and monotreme mammals remained isolated from other faunas and avoided competition from the better-adapted placental mammals.

Co-evolution between angiosperms and insects helped the spread and diversification of these groups. In the late Palaeogene a group of angiosperms evolved, the Gramineae (grasses), which had a major influence on ecosystem development. The grasses are wind-pollinated and with their evolution came extensive grasslands and savanna. This new ecological niche was one which fostered large grazing herds of herbivores with their attendant carnivores. Numerous animal groups adapted from a browsing to a grazing mode of life (eg the horses). Sometime in the Cenozoic primates made a similar transition into an environment where cooperation in organized communities meant survival and success, resulting eventually in the appearance in Africa of bipedal Man. The late Cenozoic glaciations rendered a number of animals and plants extinct, but in general terms modern ecosystems are very similar to mid-Cenozoic ones, apart from the influence exerted by human beings.

Aspects of evolution

While the mechanisms of evolutionary change can be demonstrated by biochemists and geneticists it is only by studying fossils that the long-term effects of evolution can be investigated. Recent palaeontological contributions to evolutionary studies include the testing of speciation models and the recognition of patterns of evolution at higher taxonomic levels. Research into the fossil record provides insights on such fundamental problems as changes in diversity, on whether evolution is directional and how it is influenced by the environment.

The fossil record shows how broad, and at times how rapid, the turnover of life has been. The succession of fossil faunas was initially interpreted as a result of recurrent catastrophes followed by complete restocking after each annihilation. With the demise of this apocalyptic view it was hoped that representatives of ancient faunas would be found lurking in the deep oceans, at a time when that environment was almost completely unexplored. Oceanographic cruises generally failed to dredge up such extinct organisms, but relicts of once major groups are now known from both marine and terrestrial environments. These 'living fossils' include the coelacanth fish (*Latimeria*) from the Indian Ocean, with nearest relatives in the late Cretaceous, a monoplacophoran mollusc (*Neopilina*) from the Pacific Ocean off Central America, thought to have been extinct since the Palaeozoic, and the maidenhair tree (*Ginkgo*) discovered in Chinese temple gardens, but occurring abundantly as fossil leaves in the Mesozoic and Cenozoic.

Species and speciation

Biologists define a species as a potentially interbreeding group of populations reproductively isolated from other such groups. The species, thus defined, is a natural and objective unit. All groupings of species into higher taxa (genus, family, order etc) are, in contrast, subjective, depending on the judgment of the taxonomist who erects the group. They are not, however, arbitrary. Conventionally a classification is designed to reflect evolutionary history, or phylogeny; species are grouped into higher taxa on the basis of a common ancestry although such natural relationships are often difficult to identify. In practice the vast majority of modern species are differentiated on morphological criteria, but some can be separated only by behavioural differences (eg bird song) or trivial morphological differences, such as colour, which would not be preserved in fossil material.

The problem of recognizing a species is compounded when a lineage of fossils evolved continuously and gradually through time. At some point along such a lineage the morphology of a descendant will be sufficiently different from its ancestor to be regarded as a separate species, but if evolution is at a steady rate, where is the line between them to be drawn? Steadily evolving fossil lineages, however, have rarely been recorded. More commonly the appearance of a species in the fossil record is abrupt and its morphology shows little change over its geological range. The sudden appearance of species has traditionally been attributed to gaps in the sedimentary record due either to non-deposition or subsequent erosion, resulting in the removal of intermediate forms in gradually evolving lineages. In contrast the model of 'punctuated equilibrium', which has recently become widely accepted, regards an abrupt appearance in the fossil record as a natural reflection of the very rapid evolution of a new species from a small population that has become isolated from the range of the parent species. If the speciation event is successful the new species may quickly spread over a wide range. The chances of preservation and recovery of the small and rapidly evolving isolated population are remote. On the other hand the migration of the new species, once established, over a wider area is more likely to be preserved in the sedimentary record.

Patterns of evolution

The wealth of information available from the fossil record encourages speculation as to whether the overall patterns of evolutionary change are 'deterministic', the result of particular processes, or 'stochastic', controlled by so many variables that the net result appears random. Persistent trends in fossil lineages were formerly interpreted as the result of orthogenesis, the evolution of organisms in a particular direction. Orthogenesis is now discredited as an evolutionary process. In the evolution of horses, for instance, once thought to show an

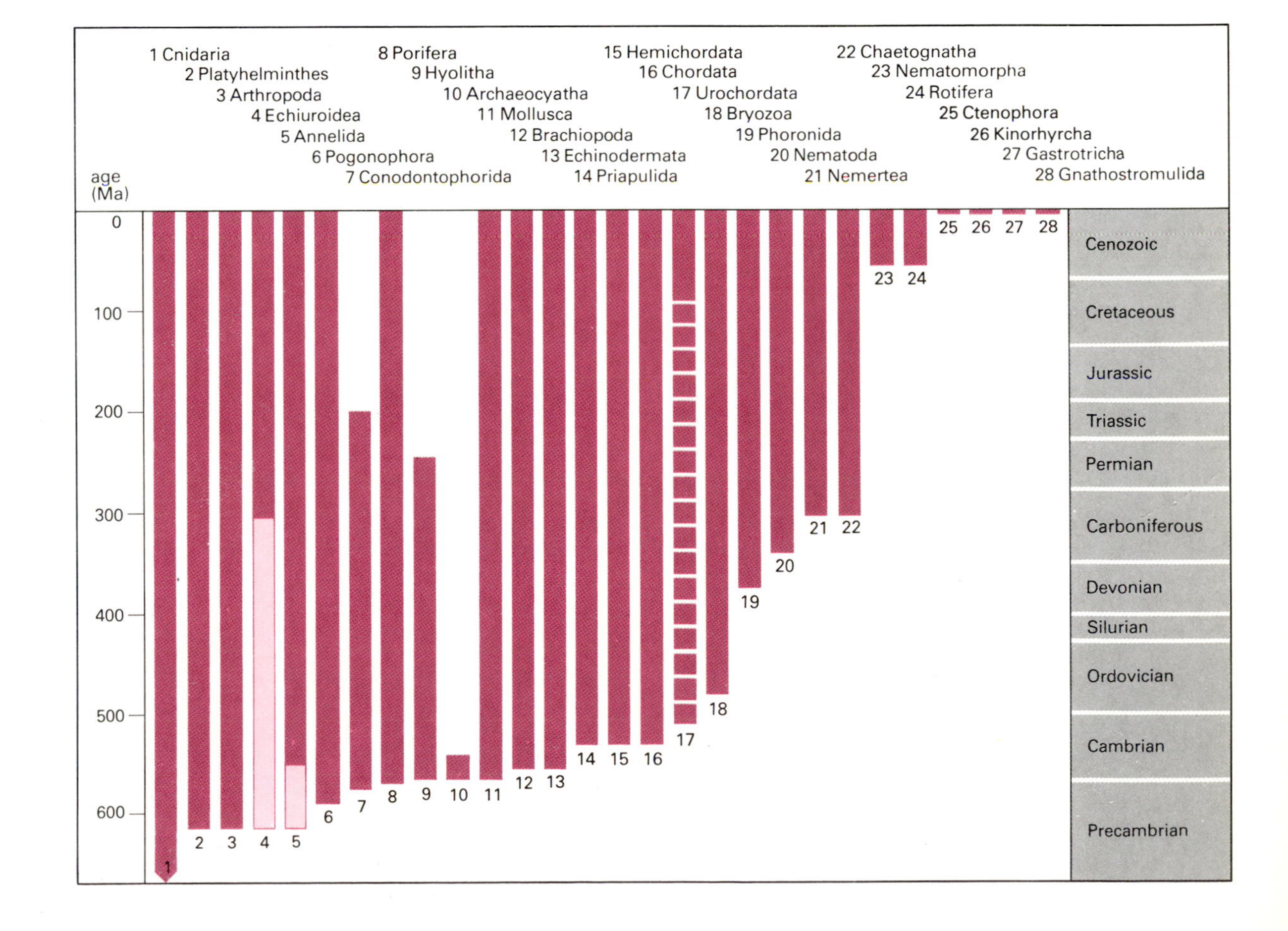

22.21: *The ranges on the geological time scale of all known fossil and living major groups (phyla) of animals. Very few have no fossil record at all, but some are so poorly represented by fossils that new discoveries could alter their known ranges quite significantly.*

inexorable trend from small several-toed browsing animals to the single-hoofed grazers of today, it is now known that there were numerous side branches and diversions. In addition the extreme morphologies evolved by some animals can usually be interpreted as adaptations for specialized functions; the huge antlers of the Giant Irish Elk, for example, were used in sexual display and so conveyed an advantage to outweigh the apparent disadvantage of unwieldiness.

A well-documented evolutionary trend in many groups, known as Cope's rule, is a progressive size increase from small ancestors to larger descendant species; the horse lineage is one of many examples. New groups tend to arise from smaller, unspecialized species and therefore have an inherent potential for size increase. Other more complex trends have also been recognized in the fossil record. Repeated branching from a stock through time to give a succession of similar forms is known as iterative evolution: the evolution of land snails in the Pleistocene of Bermuda is one of many examples documented. A more common trend is convergence, in which unrelated groups end up with similar morphologies, usually as an adaptation to a similar mode of life. Examples include the marsupial mammals of Australia which have evolved striking resemblances to some of the placental mammals, and the convergence between swimming vertebrates.

Changes in diversity

One of the most intriguing of evolutionary questions is whether metazoan diversity, as measured in numbers of species or some higher taxonomic category such as family or order, reached a steady state early in the Palaeozoic or has continued to rise to the present. Inevitably studies of this kind concentrate on marine animals, especially invertebrates, as these are more systematically preserved. The fossil record of species is inadequate and a higher taxon, such as the order, is usually used as a basis. There is a general agreement that diversity climbed exponentially from very low levels during the late Precambrian and earliest Cambrian. A further diversification during the late Cambrian and Ordovician produced about three times as many orders as the more celebrated Cambrian radiation. At the level of order there is evidence that the equilibrium reached in the late Ordovician has been maintained with very little increase for the rest of the Phanerozoic eon (Figure 22.24). Each of the occasional mass extinctions was succeeded by recovery to pre-existing levels of diversity.

However, there is marked disagreement as to how the diversity of species may have changed during the same period of time. There is a close correlation between the recorded species diversity and the estimated volume and surface area of sediments for each period,

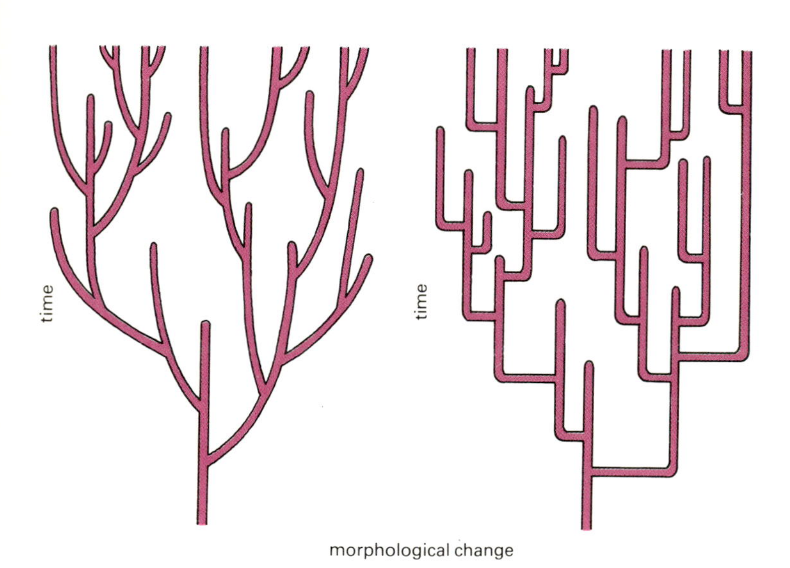

22.22: *The contrast in evolutionary patterns, if the model of evolution is assumed to be gradualistic (left), with steady and continuous shifts from ancestral and descendant species, or if the model is one of punctuated equilibria (right), where each speciation event is extremely rapid and a newly formed species maintains its original morphology until either its extinction or further speciation.*

22.23: *Convergent evolution between three swimming vertebrates: ichthyosaur (top), shark (middle) and marine mammal (bottom). Each animal has evolved from an entirely different ancestor, yet adaptation to rapid swimming has resulted in a pronounced similarity of shape.*

suggesting that the known diversity may be a function of the amount of sedimentary rock from which fossils can be collected rather than a true indication of species diversity. Some have argued from this that species diversity may have been more or less constant during the Phanerozoic. Others, however, believe that the evidence suggests that species diversity has not been constant, although the overall magnitude of increase since the Cambrian is uncertain.

Controls on evolution

Effects of both physical and biological factors in controlling evolution can be recognized in the fossil record. The most significant physical influence is perhaps the movement of the crustal plates. Fragmentation of supercontinents increases the total area of the shallow marine shelf seas while the associated increase in the volume of the spreading ridges results in flooding of the continental margins. An increase in shelf space undoubtedly favours diversification of marine faunas; several such diversifications appear to be linked with major transgressions, such as that following the break-up of the supercontinent of Pangaea in the Mesozoic. Among modern faunas diversity decreases away from the tropical regions and the fossil floras of Carboniferous coal measures deposited in tropical latitudes are likewise very much more diverse than contemporaneous high-latitude floras of Gondwanaland. Similarly if a continent drifts out of contact with its neighbours new endemic faunas will arise as speciation continues in geographic isolation; Madagascar and New Zealand provide numerous examples.

Biological influences on evolution are more difficult to assess as they often leave little trace in the rock record. Repatterning of food supplies, especially at the base of the food chain among the phytoplankton may have had wide-ranging effects on evolution. Competition between groups must also have been a significant factor in evolution, but direct proof is almost impossible to obtain. One exception is that of the modern European acorn barnacles. The more successful, rapidly growing *Semibalanus* has displaced its competitor *Chthamalus* to the highest part of the intertidal zone of rocky shores, a trend that can be traced through the fossil record.

Not all the factors controlling the major patterns of evolution are fully understood. Variations in the diversity of pelagic animals through the Mesozoic and Cenozoic appear to run in cycles of about thirty-two million years. In any cycle there is a swing from complex pelagic ecosystems with a wide variety of protistans and large vertebrates to a much simpler ecosystem dominated by a few species. These cycles are presumably related to changes in climate and oceanic circulation, but their underlying cause is uncertain (see also Chapter 23).

Rates of evolution

Rates of evolutionary change in different groups have clearly varied widely, but quantification, or even comparison, of such rates is difficult. Fossils, of course, normally record evolutionary changes in hard skeletal parts only and advances in, for instance, physiological mechanisms will remain undetected unless they impinge on the hard parts. A crude method of establishing evolutionary rates is by measuring the change in dimensions of morphological characters, such as shell size, through time. However, the connection between genetic change and any corresponding morphological change is poorly understood but probably indirect. An additional problem is that a group poor in fossilizable characters, such as the plants, will necessarily appear to be evolving more slowly than a more complex organism like a trilobite, irrespective of the actual rate.

Despite these obstacles the study of evolutionary rates yields important results. Research on fossil lungfish, for example, has revealed a period of rapid evolution in the Devonian followed by evolutionary stagnation up to the present. It has also been shown that rates of change in the head of lungfish differed from those in the body. This phenomenon, known as mosaic evolution, has now been documented in many groups and it appears to be quite common for some characters to change while others remain relatively static. 'Living fossils' are long-lived groups with very low rates of morphological change. Many inhabit marginal environments in which more specialized species cannot survive and from which they are therefore unlikely to be displaced.

Comparison of evolutionary rates between unrelated groups presents a number of difficulties, but results for Cenozoic bivalves and mammals indicate that bivalve species survive an average of ten million years and mammal species for only one to two million years, showing that the latter group evolved faster. A plausible explanation for this is that the rate is proportional to the intensity of competition—high in mammals, low in bivalves.

Extinction

Most species become extinct a few million years after their origin. In some cases one species evolves directly into another in a continuous lineage; this is termed a pseudo-extinction as the ancestral species does not strictly die out. Pseudo-extinction is probably relatively rare, however, perhaps accounting for about 5 per cent of all extinctions. Although the fossil record is one of constant extinction and replacement this equilibrium was destroyed during a few relatively short periods when much higher rates of extinction prevailed. These mass extinctions usually affected a wide spectrum of organisms. The most important were at the end of the Palaeozoic and Mesozoic eras, but other major extinctions occurred near the end of the Ordovician, the late mid-Devonian, the end of the Triassic and during the Pleistocene.

The magnitude of these extinction events is difficult to judge precisely because the fossil record is so poor at the species level but as many as 96 per cent of all marine species may have become extinct in the late Permian. The Permian mass extinction marked the disappearance of the tabulate and rugose corals, trilobites, fusulinid foraminiferans and many echinoderms, brachiopods and bryozoans. On land a number of reptiles vanished although the plants were apparently less affected. The Permian event was cumulative rather than sudden, the extinctions being spread over millions of years. The mass extinction was related in time to the movement of the continents together to form a single supercontinent (Pangaea). This eliminated shelf areas where the continents joined and the area of shallow seas was further reduced by a major regression due to the subsidence of spreading ridges in the oceans. Many of the marginal marine basins became very saline, and it has been suggested that so much salt was precipitated that the oceans became slightly brackish leading to the extinction of many marine organisms with a limited tolerance to salinity changes.

The mass extinctions at the end of the Triassic saw the demise of many more reptiles and some amphibians. Having just survived the Permian extinctions, the ammonoids almost fell victim again. Only one lineage survived from which a radiation repopulated the Jurassic and Cretaceous seas with an ultimately enormous variety of ammonites.

No mass extinction has received as much attention as the terminal Cretaceous event that marked the disappearance of the ammonites, rudist bivalves and many echinoids, foraminiferans and calcareous phytoplankton. On land the dinosaurs vanished. Overall diversity had been decreasing towards the end of the Cretaceous, but Cretaceous life had already survived periods of low diversity which did not lead to mass extinction. Furthermore, some groups such as most bivalves and all mammals were comparatively unaffected and the belemnites, abundant in the Mesozoic seas, survived into the Eocene. The abrupt disappearance of many planktonic protistans from the oceanic record indicates that the final extinctions may have been rapid. Problems in biostratigraphical correlation between marine and terrestrial life, however, make it difficult to prove that the dinosaur extinctions were exactly contemporaneous with those in the marine realm. If this were the case a global catastrophe may be indicated. The recent discovery of remarkably high concentrations of iridium in deep-sea clays at the Cretaceous–Palaeocene boundary has refocused attention on an extra-terrestrial explanation because this element is much more abundant in extra-terrestrial bodies than on Earth. It is suggested that the impact of an asteroid 10 km in diameter could explain the iridium and other anomalous elements by spreading a pall of dust over the entire Earth which would have severely disrupted marine and terrestrial life. Even if corroborative evidence of this disaster were discovered it seems certain that more gradual effects, especially climatic, were also influencing life in the late Cretaceous. The final extinction of the dinosaurs, for example, seems to have occurred progressively later nearer the equator.

A wealth of ingenious explanations for mass extinctions has been proposed, but the reasons for those of the late Ordovician and late mid-Devonian remain uncertain even though a wide variety of organisms was affected in each case. The discovery of frequent magnetic reversals through time has prompted speculation that the temporary removal of the Earth's magnetic field and influx of dangerous radiation during polarity switches increased extinction rates. Although a few extinctions do coincide with magnetic reversals the overall frequency of reversals far exceeds that of major extinctions. No single recurring factor explains mass extinctions and a combination of causes is usually suggested.

Conclusion

A consideration of the history of life inevitably leads to speculation about its future. The fossil record makes it clear that there is no *a priori* way of identifying successful groups. Many forms, for example the mammals, had long and insignificant histories before an evolutionary

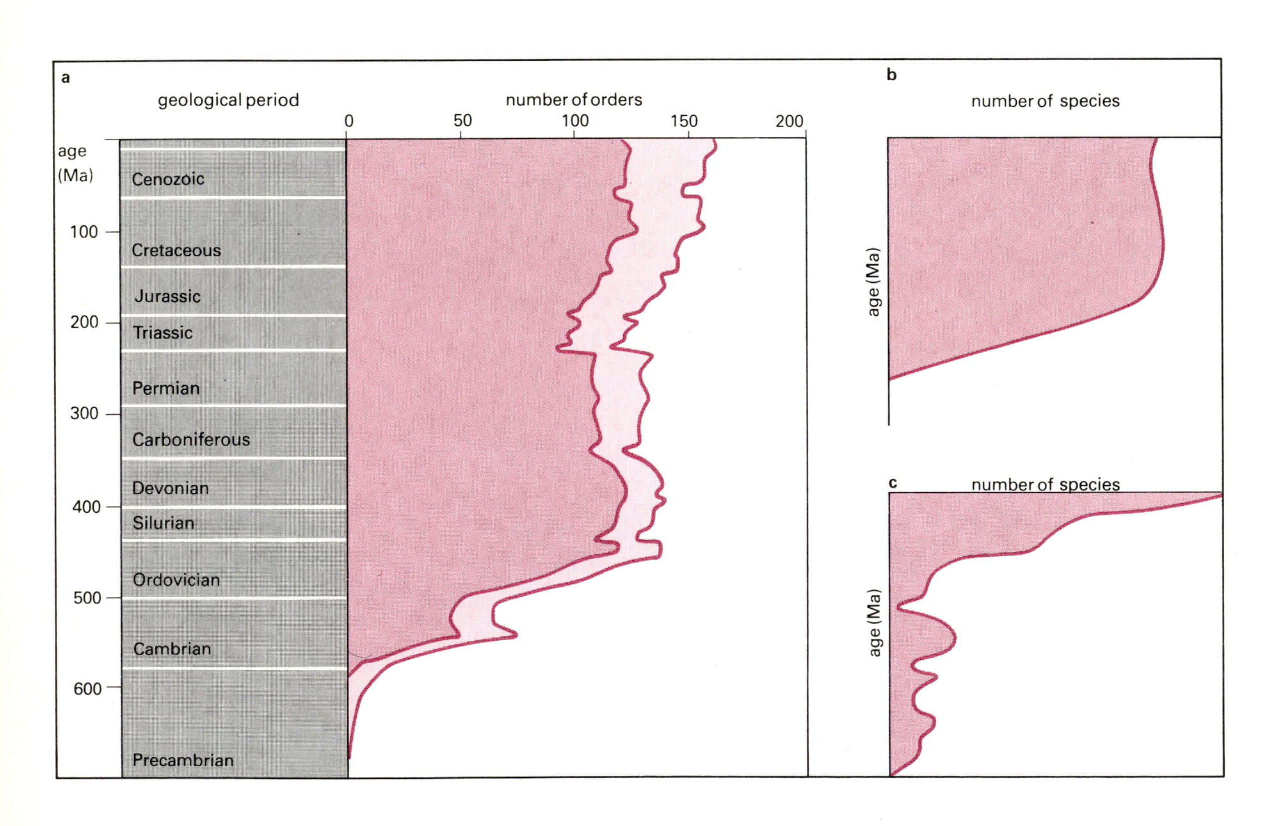

22.24: *Changes in diversity of marine metazoans from late Precambrian to Recent times.* (**a**) *Diversity in terms of taxonomic orders—the left-hand curve ignores the contribution of exceptional faunas, such as the Burgess shale, which are accounted for by the right-hand curve. The number of orders increases steeply to the late Ordovician, since when it has varied much less.* (**b**) *Diversity in terms of known fossil species, showing great fluctuations at times of extinction and a steep rise towards the Recent.* (**c**) *Species diversity corrected for the differing availabilities of rock for the various Systems, showing relatively constant diversity since Cambrian times.*

flowering. Other groups, for example the ammonoids in the Triassic, came within a hair's breadth of extinction before burgeoning once again. Had they become extinct then their ecological niches would presumably have been occupied by some other organisms during the Jurassic and Cretaceous. Thus it is very difficult to predict which of present-day species will give rise to successful evolutionary lineages and which will soon become extinct.

Finally, what of our own species' contribution to the future? It has been argued that concern about extinctions caused by human activities is misplaced (barring a catastrophic emulation of former mass extinctions) in view of the pervasiveness of extinction through geological time. In the sense that *Homo sapiens* is just another species this may be true, but in terms of our immediate moral and ecological responsibility the importance of our commitment to the future well-being of life is inescapable.

22.25: *The effects of the late Cretaceous mass extinctions. Animals and plants that became extinct are uncoloured, whereas groups that survived the mass extinction are coloured. The extinct groups include the reptiles (1)* Plesiosaurus, *(2)* Mosasaurus, *(3)* Deinonychus, *(4)* Tyrannosaurus, *(5)* Edmontosaurus, *(6)* Brachiosaurus, *(7)* Triceratops, *(8)* Pteranodon; *other animals such as (9) ammonites, (10) some types of sea-urchin and (11) peculiar molluscs known as rudists, and plants such as the (12)* Bennetitales.

23 Historical geology: layers of Earth history

An important part of the Earth sciences is the scientific investigation, description and understanding of the Earth's surface as it is today. Exciting as this task is, it hardly compares in magnitude with the exploration of the long and varied history of the planet. These two separate but related aspects of the Earth sciences are sometimes referred to as physical geology (in which must be included oceanography, climatology, geography and so on) and historical geology. Many of the chapters in this book have dealt with particular present-day phenomena and then applied the knowledge to the interpretation of related phenomena in the geological past. This important procedure was developed by Charles Lyell in the early nineteenth century out of James Hutton's conception of the Earth as a constantly self-renewing machine. This 'uniformitarian' approach, by which Earth history could be interpreted in terms of processes identical to those acting on the Earth today, replaced earlier 'creationist' views in which Earth history had essentially finished, having been accomplished in a few short and rather recent cataclysmic events, and was scarcely worth investigating.

One guiding principle of historical geology had already been expounded by Steno in the century before Hutton, albeit in the context of the biblical Flood (see Figure 1.3). This principle is that positional relationships among rocks, minerals and fossils can be interpreted in terms of time relationships. The early emphasis on the investigation of sedimentary strata led to the coining of the word 'stratigraphy' for the study of sequence in layered rocks. The underlying principles are relatively simple to apply at a single locality but problems arise when sequences in different areas are compared and correlated. Methods of correlation are important not only for building up a more complete geological history than is afforded by rocks outcropping in a single region, but also for the practical purposes of tracing economically important units such as coal seams, oil reservoirs and ore bodies.

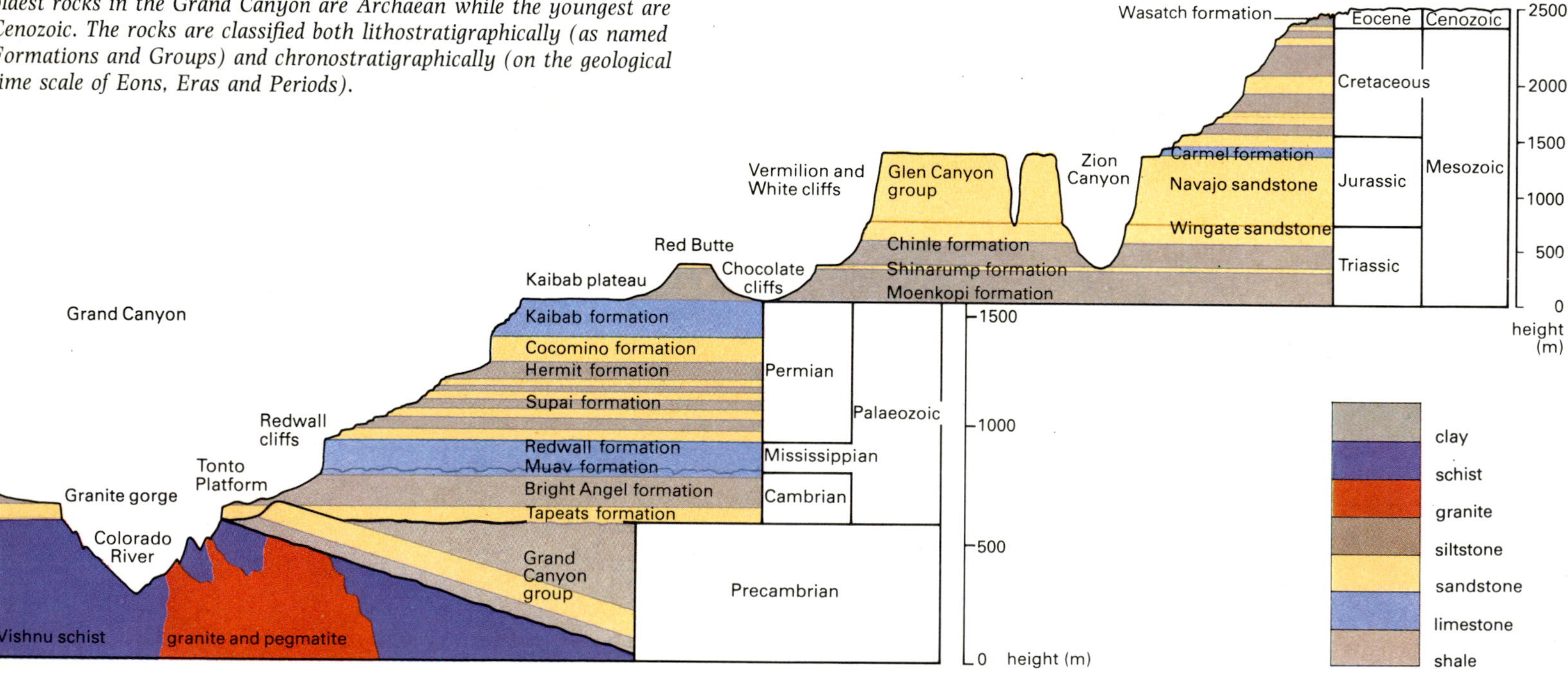

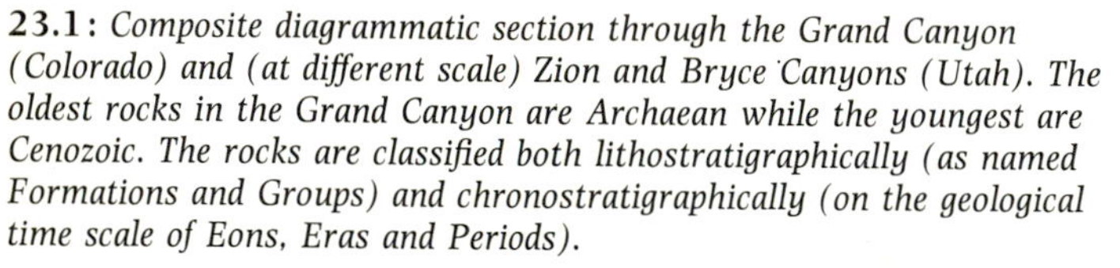

23.1: *Composite diagrammatic section through the Grand Canyon (Colorado) and (at different scale) Zion and Bryce Canyons (Utah). The oldest rocks in the Grand Canyon are Archaean while the youngest are Cenozoic. The rocks are classified both lithostratigraphically (as named Formations and Groups) and chronostratigraphically (on the geological time scale of Eons, Eras and Periods).*

23.2: *The horizontal Permian limestones of Temple Mountain, Spitzbergen, record a period of time when this part of the Earth's crust was in the tropics, 250 million years ago.*

Lithostratigraphy: rocks and time

Lithostratigraphy is the understanding and description of rocks at the local level 'in the field', whether at outcrop or in borehole cores. There are several important premises out of which the geologist builds a story from what looks to anyone else like a pile of rocks. In the simplest case—that of horizontally layered sedimentary rocks—the lowest layer is the oldest and the most recently deposited layer is at the top. Thus the 1600 m of horizontal strata of the Grand Canyon of Arizona record a span of Earth history of some 250 million years. This principle of superposition is sometimes referred to as a law, but like other stratigraphical principles exceptions can readily be found. The principle of superposition presupposes that the strata were deposited in horizontal layers, the principle of original horizontality. The principle of cross-cutting relationships states that Earth movements that produced folds and faults in the rocks must have been of a more recent date than the formation of the rocks themselves. This applies also to intrusive igneous bodies such as dykes and sills. The principle of included fragments states that rock fragments incorporated in some other rock are older than the rock in which they are found. Hence the rocks from which the pebbles in a conglomerate were eroded must already have been in existence at the time the conglomerate was being deposited.

This last principle encapsulates Hutton's vision of the Earth as a recycling machine. The rocks of one uplifted continent are eroded to provide the sediment that will eventually contribute to the next. Where the older denuded landscape meets the new aggradational area Hutton predicted the existence of unconformities; an example he found is shown in Figure 1.7. An unconformity is an example of a cross-cutting relationship, where younger strata overlie older strata which have been folded, uplifted and eroded so that the younger strata rest across their upturned edges. The unconformity surface therefore represents a lapse of some considerable length of time in the formation of the sequence. The angle between the older and younger sets of strata may be sufficiently great to be obvious at a single outcrop, as in Hutton's Scottish examples, or it may be so low that it emerges only from studies of outcrops over a wider area.

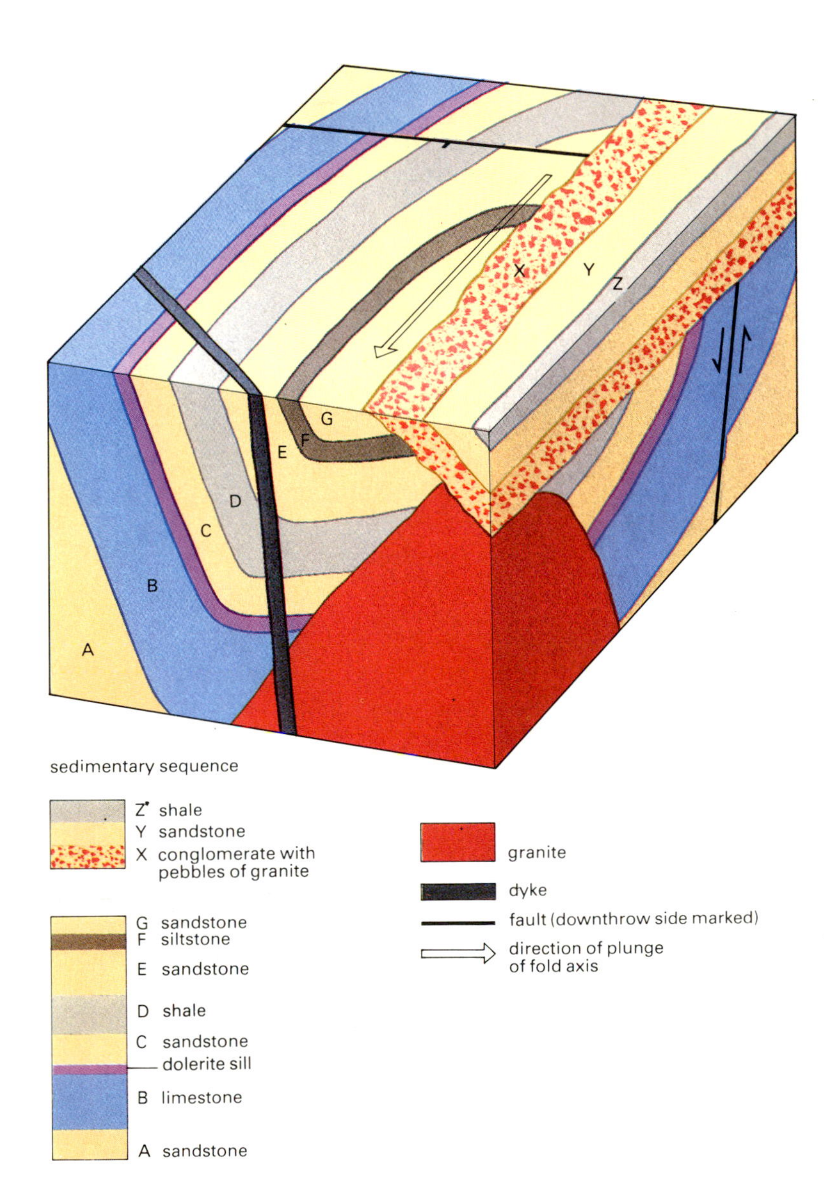

23.3: *Block diagram illustrating the principle of unravelling geological history from positional relationships of rock units. The sedimentary sequence A to G was laid down in that order, intruded by a sill and folded. Further folding (to tilt the axis of the first fold), faulting and intrusion of granite followed in an indeterminate order. After a period of erosion, sedimentary layers X, Y and Z were laid down unconformably over the older rocks and later tilted by another phase of unfolding. The dyke, as it is still vertical, was probably intruded later still.*

A geologist will be exceptionally lucky if he finds a single locality where the entire regional rock sequence is displayed. More usually it is necessary to piece together information from a number of separate outcrops. Such outcrops can be plotted on a map and geological maps usually include an element of interpretation as well as observation. It will be assumed that there is some continuity between the separate outcrops, and dip and strike measurements will help to predict this.

Lithostratigraphical units are bodies of rock sufficiently distinctive in character to be recognizable from adjacent units and to be given names. The basic unit is a formation, which must have sufficient consistency of character over a reasonable area so that it can be represented on a map. Formations can be subdivided at a more local level into members and beds, or combined over wider areas into groups. It is conventional to attach a geographical name to any formally defined rock unit, such as the London Clay Formation, and also to designate a type section—a standard locality or borehole at which the unit is typically developed and can be used for future reference.

Correlation in the lithostratigraphical sense is the business of predicting the geometrical extension of rock units between known occurrences such as outcrops or boreholes. But, while superposition of strata at a single locality implies succession in time, does assignment of a stratum at two distant localities to the same formation necessarily imply any equivalence in time? This is the point at which stratigraphy departs from purely geometrical descriptions of rock bodies and starts to account for the ways in which sedimentary rocks actually form.

Facies and stratigraphy

At any time in Earth history, just as at present, different processes were happening at different places. Erosion in uplifted areas was balanced by deposition in subsiding areas (see Chapter 15). In the areas of deposition a wide variety of sedimentary environments can be expected to exist at the same time. These will be expressed in both the lithological and palaeontological character of the resulting sediment, the overall aspect of which is referred to as its facies. For instance, beach sands, coral reefs, estuarine muds and shallow marine silts may all be represented along a single coastline at the same time, but changes in geography will bring changes in the relative positions of those environments (Figure 23.4). Johannes Walther in 1893 formulated the principle of facies, that facies that succeed one another in vertical succession are normally those found adjacent to one another at any given time. This implies that the vertical sequence reflects the lateral migration of facies with the passing of time. The result of this is a succession of diachronous (time-crossing) strata (see Figure 20.5). Because change is the rule in geological processes individual strata can be expected to be diachronous. This imposes difficulties in trying to reconstruct the geography of a large region at a particular time. However, the succession of facies at one place gives some localized historical information despite the correlation difficulties. Marine sedimentary sequences may trace a record of shallowing or deepening of the sea, implying an approach or recession of the local shoreline. Such sequences are referred to as regressive and transgressive respectively

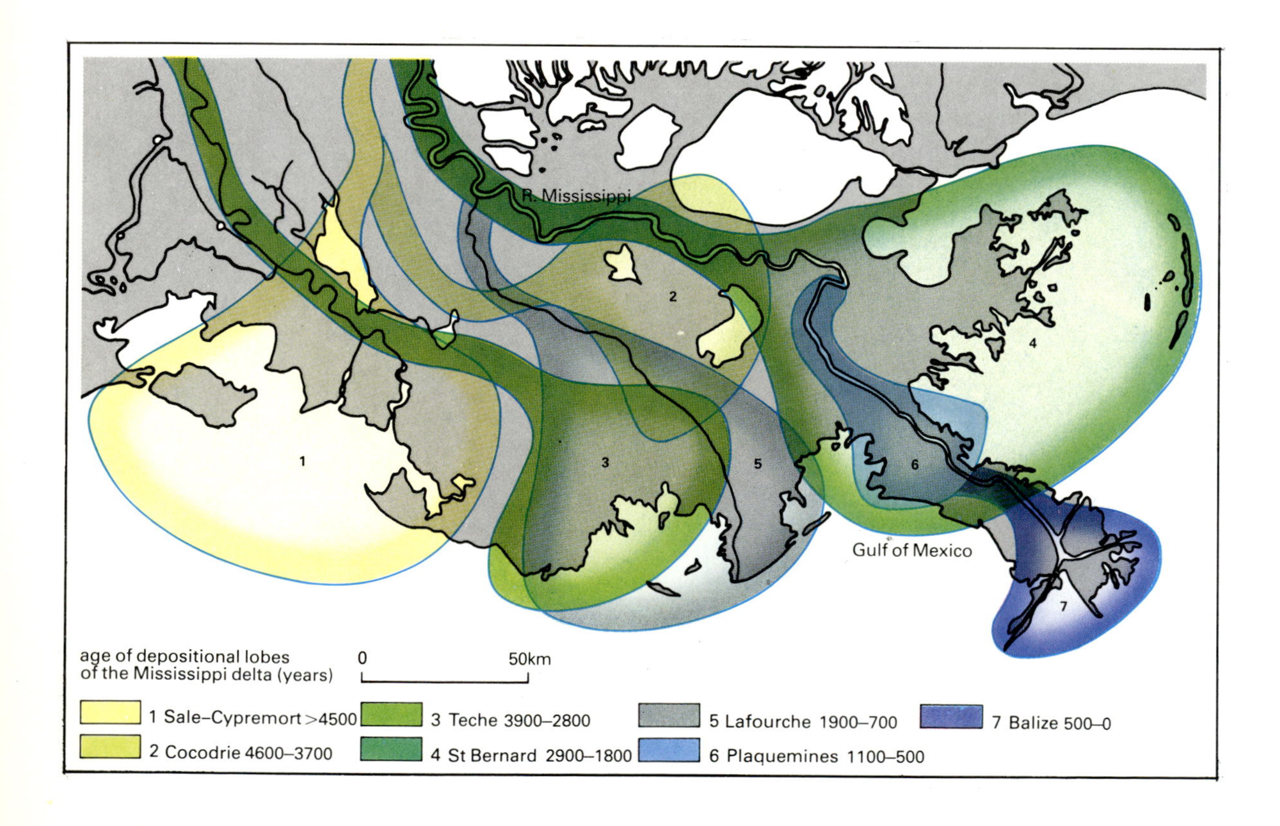

23.4: *Lateral migration of successive depositional lobes of the Mississippi Delta, showing how a sheet of sediment can be built up by lateral (as well as vertical) processes.*

and may imply a local uplift or subsidence of the area or a worldwide change in sea level. Often regressive and transgressive sedimentary sequences follow each other implying cyclical fluctuations in the relative sea level. Such cyclic effects must, however, be superimposed on a net subsidence since subsidence is needed for the accumulation of sedimentary strata (Figure 23.5).

Chronostratigraphy: a standard for geological time

Throughout geological literature reference is made repeatedly to a set of terms—Cambrian, Ordovician, Silurian and so on—that between them describe the entire stratigraphic record from about the point at which abundant fossils appear. These terms may be applied to rocks and fossils, when they are the labels of the geological systems, or to time, when they designate the periods. Additional terms subdivide the systems into series and stages, and the periods into epochs and ages. The periods are each assigned to larger units, the Palaeozoic, Mesozoic and Cenozoic eras, and these in turn compose the Phanerozoic eon.

The science of subdividing and labelling geological time is known as chronostratigraphy. Many of the geological systems were named before the nature of the stratigraphic record was fully understood. They were perceived as natural divisions of the rock record, laid down in distinctive episodes of Earth history. This is exactly what a geologist working in a particular area finds, but it cannot be expected that on a worldwide scale the intervals of deposition and the intervals of Earth

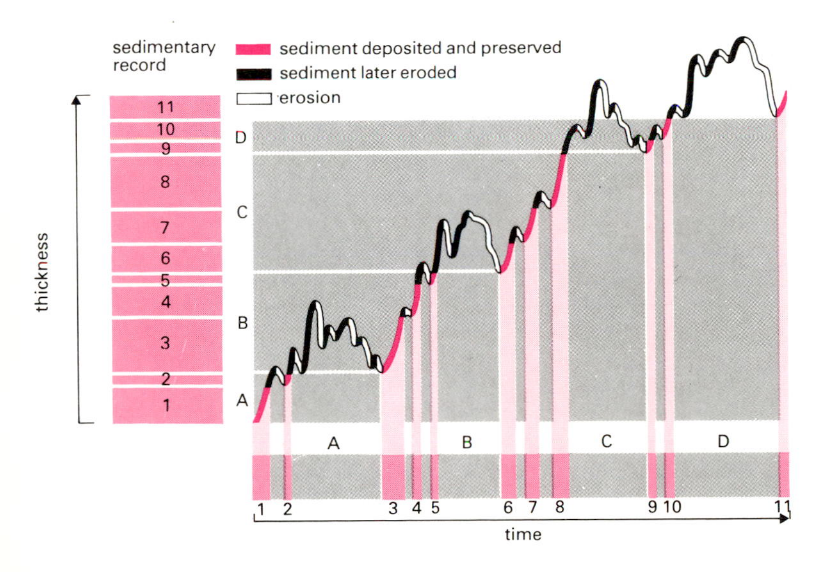

23.5: *The imperfection of the sedimentary record as a 'tape-recording' of palaeo-environmental events. The recording occurs only during those intervals of time when base-level, the level below which deposition can take place at any particular time, is rising. During intervals of stillstand, no net deposition occurs, while when base level is lowered there is a net erosion of sediment and consequent loss of record. In a regime of fluctuating base levels, rather little time (that represented in colour) is recorded in accumulated rock.*

movement would be simultaneous everywhere. However, many of the systems contain fossils unique to themselves and it is the distinctive nature of fossil assemblages that is the principal criterion for identifying strata belonging to each system, series or stage, ie for dating them relative to the geological time scale.

Relative stratigraphic dating uses the accumulated body of knowledge pertaining to each division. But, wherever two adjacent divisions are distinguished by the differences in their overall fossil content, it will necessarily be difficult to identify their boundary since the differences between the two divisions will be minimal at this point. The problem is analogous to the separation of successive species in an evolving lineage. The solution that is currently emerging in chronostratigraphy is to define the boundaries rather than to characterize the divisions themselves.

The only boundary that has yet been established in the standard stratigraphic scale is that between the Silurian and Devonian systems. The top of the Silurian system in its type area on the border between England and Wales is a major facies change from marine to terrestrial rocks at which the stratigraphically important graptolites disappear. The Devonian system in its type area in south-west England is represented by marine rocks, but Silurian rocks do not outcrop in the area. The conventional boundary between the two systems in the Silurian type area is thus difficult to locate in other parts of the world, for instance in Australia and Czechoslovakia where graptolites overlap in range with Devonian land plants. The process of defining an internationally acceptable boundary took a great deal of deliberation among a party of specialists who toured likely areas all over the world and examined all the evidence before having their final choice ratified at an International Geological Congress in Montreal in 1972. The section chosen was in Czechoslovakia, and a 'golden spike' was figuratively hammered into the precise spot at a location aptly named Klonk. This

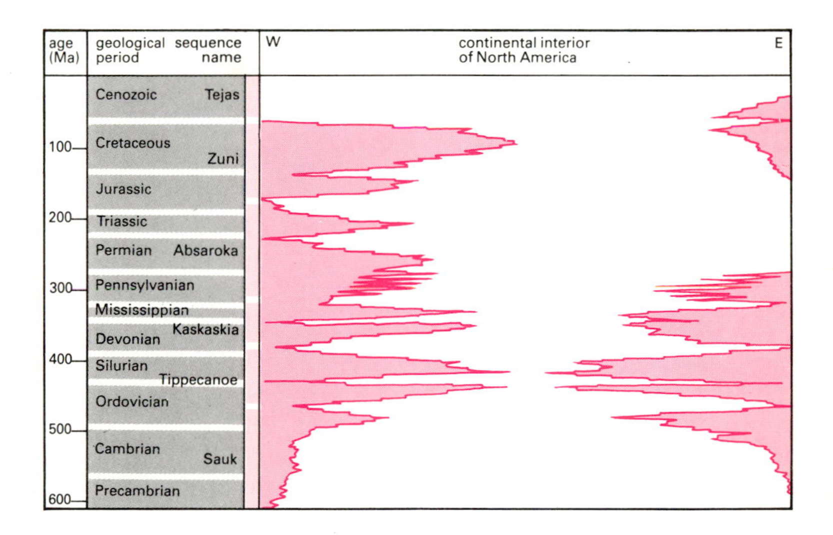

23.6: *Transgressive and regressive cycles have deposited a succession of unconformity-bounded 'sequences' over North America, each being thicker and representing a longer time interval nearer the continental margins.*

point will forever mark the one place in the world where one can actually point to the Silurian–Devonian system boundary. Anywhere else the boundary must be located by careful correlation, via intermediate localities if necessary, with the evidence available from above and below this spot. Unfortunately few further boundaries seem near this stage of international acceptance and it will take many years before the entire stratigraphic scale is finally and irrevocably defined in this way.

Correlation

Correlation is a widely used term covering a variety of rather different operations in stratigraphy. It may be used loosely to suggest that two or more beds were laid down at the same time. It may be used to indicate continuity of lithological units, particularly between occurrences in boreholes. Evidently, if stratigraphy is to be anything more than a description of local successions of rocks and fossils, correlation of one kind or another is an essential process.

Lithological correlation

The lateral tracing of strata between outcrops, mines or boreholes is clearly of economic importance as well as scientific interest. Physical continuity of a stratum between two outcrops does not necessarily imply time equivalence because of the normally diachronous nature of strata. It does nevertheless imply continuity in the existence of a particular environment of deposition, so lithological correlation is restricted to rocks confined to a particular basin of deposition. Certain types of strata, such as coal seams, are more limited in extent than others. Coal is formed from the accumulated vegetation growing on the top of a shifting delta and successive coal seams produced in the same area may not be in physical continuity with each other. Such considerations are important in the search for oil since the most interesting layers are frequently those that lack physical continuity and hence confine the oil.

Lithological correlation may not be possible in subsurface exploration, particularly offshore, where all the information comes from boreholes. The extraction of a cylindrical core of rock with a hollow bit is expensive and time-consuming and the geologist has to make do

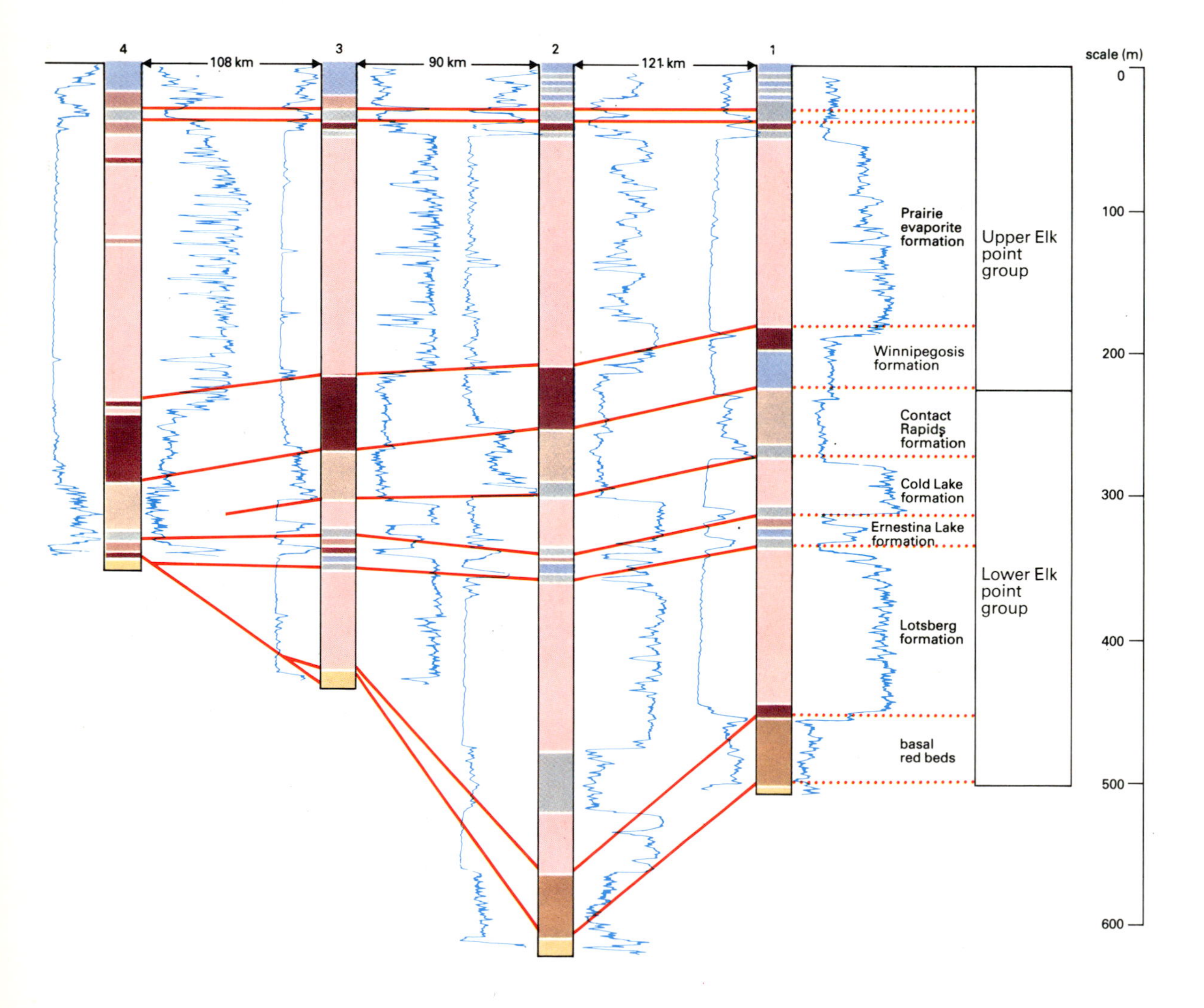

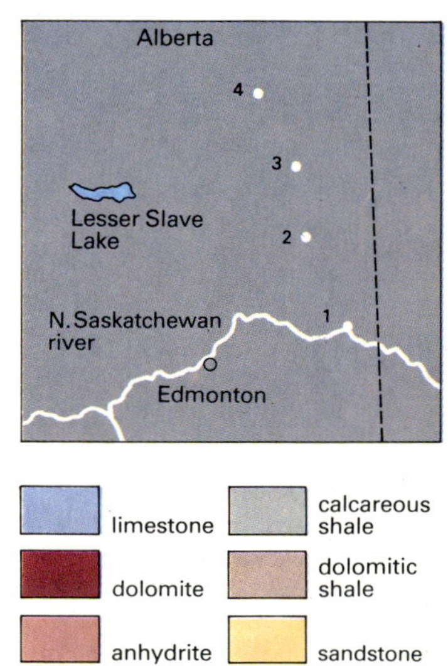

23.7: *An example of the lithological correlation of sedimentary strata by means of borehole logs in the Devonian oilfield of western Canada.*

with the cuttings brought up in the drilling mud. However, data from wireline logs (see Chapter 24) may allow correlation between adjacent boreholes by matching distinctive electrical 'signatures' (Figure 23.7). Problems arise where individual strata or groups of strata are discontinuous and the geologist has to decide whether they were originally discontinuous or have been eroded at an unconformity.

Event correlation

Some strata are less time-transgressive than others. Ash from a volcano gives such a bed and it is easily recognized when interbedded with other sediments. The ash from an individual eruption usually takes a few days or less to reach its maximum extent. Likewise a turbidity current may spread sediment over a huge area of the sea floor in a matter of hours. The importance of these examples is that, although the deposition was certainly time-transgressive, the time involved was geologically insignificant and the bed can be regarded as a time plane embedded in the rock record. Individual volcanic ashes are particularly valuable in correlating between marine and non-marine strata, provided that an individual ash is distinctive enough chemically, mineralogically or radiometrically to be identifiable wherever it occurs. This technique is called tephrochronology and is of particular value in correlating Quaternary deposits in areas such as western North America and New Zealand where volcanic activity is frequent.

A transgression followed by a regression may be regarded as an event that can be used for correlation. The problems arise when the strata are cyclic and show repetition of transgressive and regressive events; examples are common in the Carboniferous rocks of many parts of the world. But which transgressive events at one locality are to be equated with those at another? Extra evidence, from fossils for instance, is needed in such a case. There have, nevertheless, been serious suggestions that cycles of deposition are the 'natural' basis for

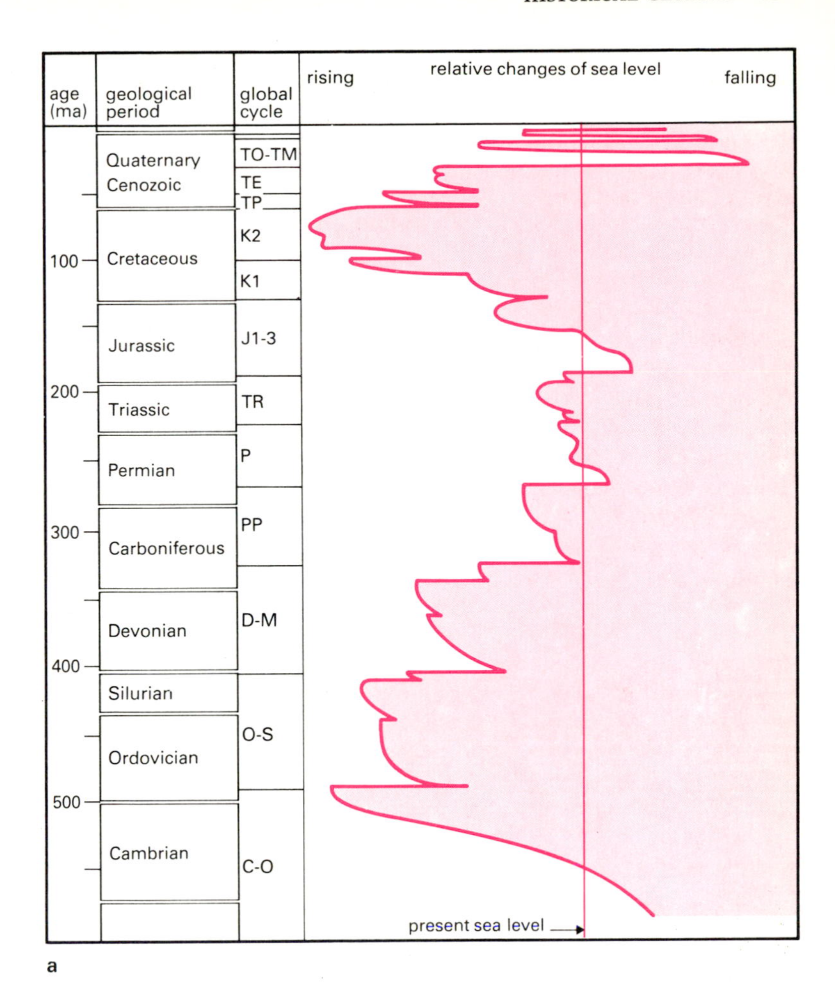

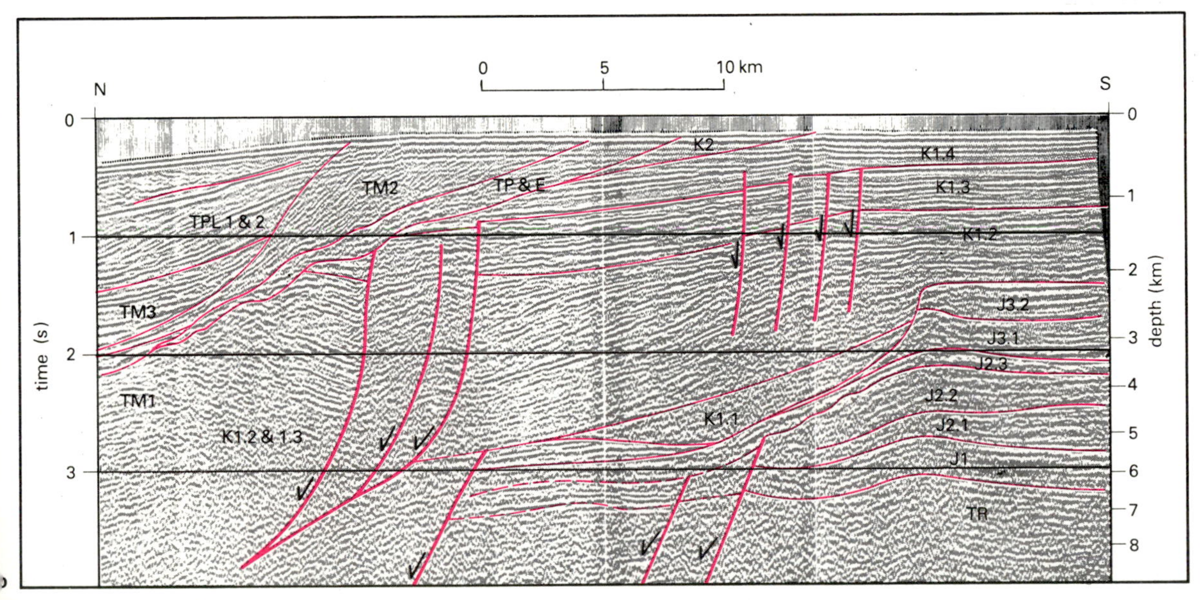

23.8: (**a**) *Proposed composite eustatic sea-level curve for Phanerozoic time, developed by Dr J. R. Vail and associates at Exxon Oil Company. The Mesozoic and Cenozoic curves are based on seismic stratigraphy of continental margins, but the Palaeozoic section is based on the onshore extent of marine sedimentary strata (cf. Figure 23.6).* (**b**) *Seismic reflection profile offshore western Africa, showing identification of eustatic cycles as a means of dating the strata.*

stratigraphy and that their boundaries are the logical means of time correlation.

Worldwide changes in sea level would be valuable in event correlation. Such eustatic changes should be easily recognized and would provide an intercontinental time-correlation tool. These changes certainly took place in the Quaternary as the formation and the melting of the ice caps withdrew water from the sea and returned it, causing sea level fluctuations of 100 m or more. A possible mechanism for slower changes of a similar magnitude is the changing activity of mid-ocean ridges. Changes in sea level recorded in subsurface continental margin rock sequences can be detected as unconformities by seismic stratigraphy. Boreholes enable relative dating of such unconformities to be carried out by palaeontological methods. While some of the unconformities will be due to local upheavals many geologists now believe that others are due to eustatic sea-level changes and will therefore have exact counterparts on other continents—a potentially powerful correlation tool. A global sea-level curve based on such studies has recently been developed for the Cenozoic and Mesozoic eras, and it has been tentatively extended back through the Palaeozoic as well (Figure 23.8). This correlation technique is claimed to have been instrumental in the discovery of a number of oil reservoirs but its validity is not universally accepted. Changes in the Earth's shape could theoretically produce sea-level changes in opposite directions in different parts of the world, and many sea-level changes (for instance in the Quaternary) take place during a much shorter time scale than can be resolved by any known correlation technique.

Climatic change correlation

Correlation on the basis of climatic changes is particularly important for Quaternary stratigraphy where climatic changes have evidently taken place much more rapidly than the organic evolution on which correlation depends for other periods. Many different types of sediment can be interpreted in terms of climate or temperature, such as lake sediments containing drifted pollen, and marine sediments with varying oxygen isotopes. However, as with cyclic sedimentation, the problems come in matching particular sequences with one another.

Magnetostratigraphy

Magnetic minerals in both igneous and sedimentary rocks retain a record of the magnetic field prevailing at the time and place of their formation. Changes in magnetic polarity take place very rapidly and they presumably have a simultaneous effect over the entire Earth. On this basis a standard magnetic polarity time scale is being extended backwards through Cenozoic time (see Figure 10.6) and will eventually include the later Mesozoic. The problems of correlating strata and new localities with the standard are similar to those applying to eustatic and climatic methods—knowledge that a particular stratum or lava is normally or reversely magnetized is not enough on its own to establish its position on the scale. Ideally, closely sampled deep sea sediment cores should be correlatable by starting from the top and matching the reversals in downward succession, but the non-recognition of a single polarity event in one of the cores would throw the correlation out of step. The method is therefore usually used in combination with others.

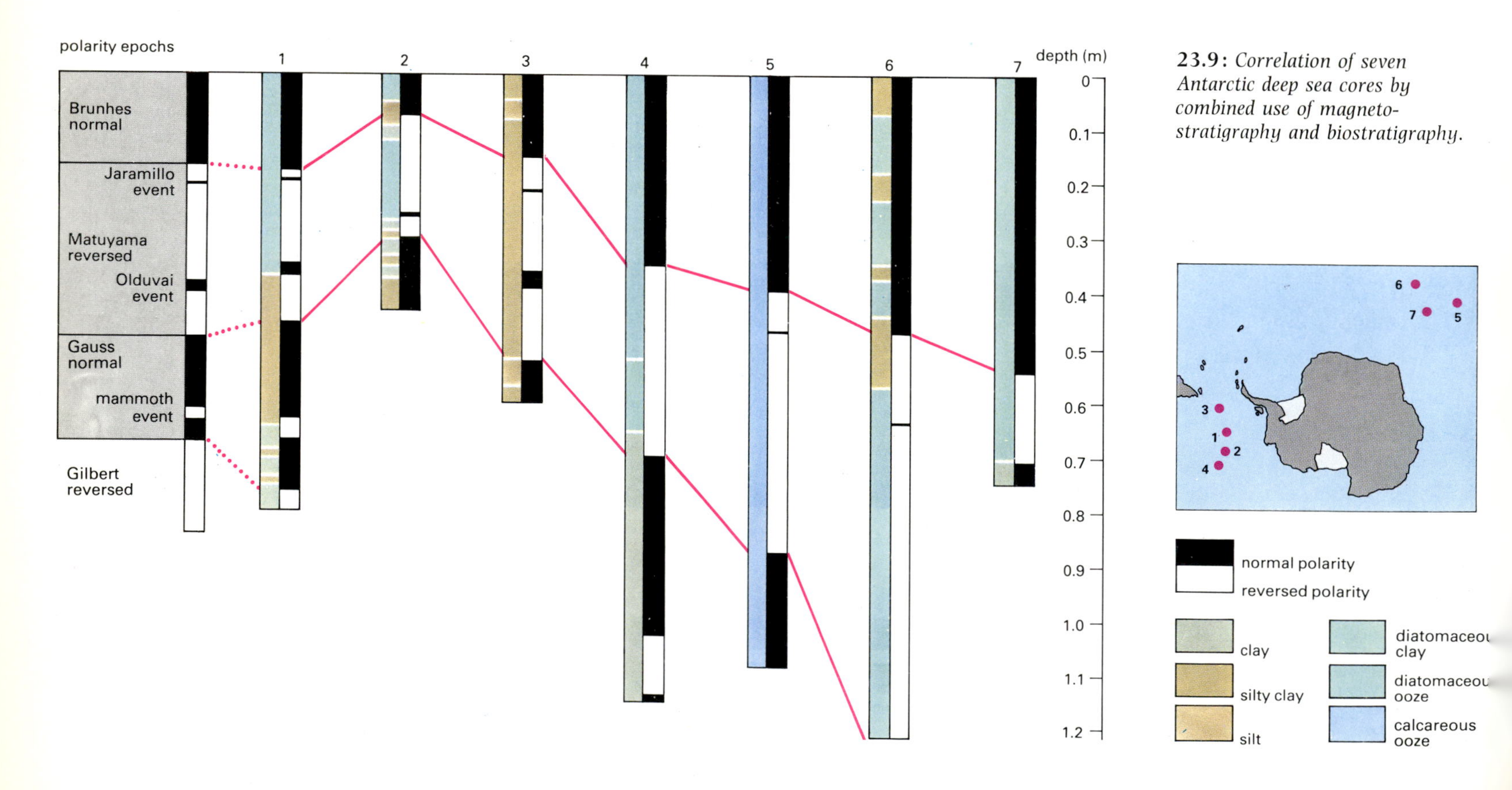

23.9: *Correlation of seven Antarctic deep sea cores by combined use of magnetostratigraphy and biostratigraphy.*

Quantifying the geological time scale

Classical stratigraphy tells us a great deal about the sequence of geological events, but very little about the actual lengths of time involved, the rate of processes and the ages of individual events. To do this some component or characteristic of geological materials that changes regularly and predictably with time is needed. The family of methods used most widely for 'age-dating' is that involving radioactive decay. However, there are other processes that have some in-built periodicity and can, in special cases, be used for measuring geological time. An archaeological analogy is the use of tree rings for dating samples of wood (dendrochronology).

Dating by sedimentation

Processes of sedimentation have frequently been looked to for evidence of the passage of time, on the apparently straightforward lines that the thicker the sediment, the longer it must have taken to accumulate. Darwin's estimate of a cumulative total of 22 km of sediment for the entire geological column of the British Isles clearly indicated a very long period of time. The relative lengths of the periods can be estimated by comparing the maximum known thickness of each system, but the extension of this to measure the passage of geological time by making guesses at the global average rate of sedimentation is more hazardous. The estimated maximum thickness for each period, totalled for the whole of the Phanerozoic has increased steadily from 22 000 m in 1860, to 102 000 m in 1909, to around 138 000 m today. Estimated average rates of deposition have varied between 0.035 mm per year and 3 mm per year. Using an intermediate value of 0.3 mm per year gives 457 Ma for the length of the Phanerozoic eon, quite close to the 570 Ma now reckoned by radiometric methods. However, rates of sedimentation are so enormously variable that this method can give only the crudest estimates of the passage of geological time.

23.10: *Comparison between maximum recorded thickness of sediment deposited and duration of each of the Phanerozoic periods.*

System	Maximum recorded sediment thickness (km)	Length of period (Ma)	Average maximum sedimentation rate (km/Ma)
Neogene and Quaternary	13.0	26	0.50
Palaeogene	20.9	39	0.54
Cretaceous	15.8	71	0.22
Jurassic	13.1	54	0.24
Triassic	8.8	35	0.25
Permian	6.2	55	0.11
Carboniferous	13.8	65	0.21
Devonian	11.7	50	0.23
Silurian	8.9	35	0.25
Ordovician	13.8	70	0.20
Cambrian	11.8	70	0.17
Phanerozoic eon	137.8	570	0.24

23.11: *Varved clay deposited 10 000 years ago in Leppä Koski, Finland, near Hämeenlinna.*

Of more value are those rather rare instances where sedimentation is directly controlled by seasonal processes. In glacial areas the seasonal rhythm is recorded as varves (p. 333). Their thicknesses vary from year to year and the sequences can be diagnostic of a particular time interval. Overlapping sequences of varves from different lakes, any one of which does not generally remain in existence for more than a few tens of thousands of years, can thus be used to build up a detailed chronology of glacial sediments.

Varved sediments in non-glacial areas can be sufficiently rich in organic carbon to be important as oil source rocks. They originate in shallow seas where seasonal 'blooms' of microplankton form thin laminae interlayered with inorganic mud and are preserved from decay by a shortage of oxygen. The Green River Shales of Eocene age in the western USA represent a thick succession of lake deposits, varved throughout, that took between 5 and 8 Ma to deposit.

Dating by evolution

Evolution has provided stratigraphy with one of its most useful tools in relative dating (see below) and attempts have been made to quantify the process. Unfortunately this requires an estimate of rate of evolution —something that has been far from constant through geological time. Early calculations were based on rates of appearances and extinctions of organisms. Charles Lyell in 1867 surmised that about 20 Ma might be needed for a complete turnover of all species of molluscs and on this basis dated the beginning of the Ordovician period at 240 Ma ago.

By an extension of this technique, Lyell subdivided the Cenozoic era according to the average percentages of Holocene species found among the fossil molluscs of groups of strata. His 'newer Pliocene' was characterized by the fact that 90 per cent of its fossil molluscs belong to living species. For the 'older Pliocene' the proportion was 50 per cent, for the Miocene 18 per cent and for the Eocene 3.5 per cent. (The Palaeocene and Oligocene had not then been defined.) Modern investigations suggest that this method could be accurate for dating individual fossil assemblages to about 5 Ma—not particularly good in comparison with other methods. The problems are that rates of evolution are higher for some groups (eg gastropods) than for others (eg bivalves), and that rates of extinction may be different from one area to another.

Now that the mechanisms of heredity and evolution are better understood there may be a possibility of rather more accurate estimates of time, although in a manner that would make them of little value for stratigraphy. It is now possible to quantify the differences in genetic material between distantly related organisms and these differences may be used as a measure of the time elapsed since the evolutionary paths of the two organisms diverged from their common ancestor. The number of amino acid differences in the alpha chain of human and horse haemoglobin (youngest common ancestor in the late Cretaceous, about 80 Ma ago) is eighteen, and the corresponding figure for human and carp (youngest common ancestor in the early Devonian, about 370 Ma ago) is sixty-eight. These figures give the average rate of substitutions per amino acid site per year as about 8.9×10^{-10}. This figure could be used to estimate the age of strata containing the youngest common ancestor of two or more living species for which amino acid differences are known.

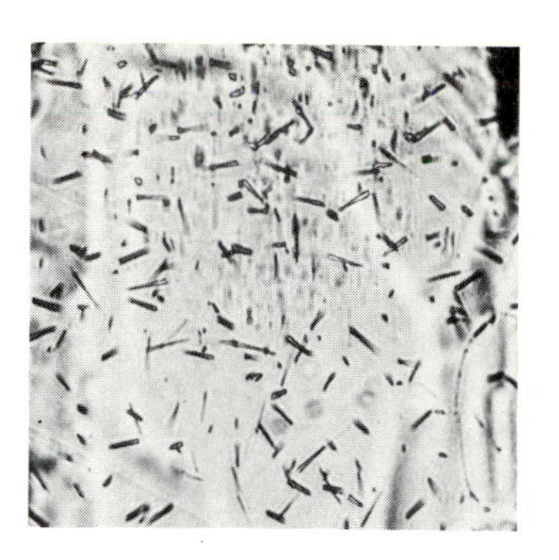

23.12: *Fission tracks about 30–40 μm long in a polished section of a crystal of sphene ($CaTiSiO_5$) from a granite in Scotland, dated by counting the tracks as 401 ± 8 Ma.*

Fission track and radiometric dating

Decay at the atomic level is the basis for two dating techniques. The more recently discovered is fission track dating, first investigated in the 1960s. The method depends on the presence of ^{238}U, the more common isotope of uranium, which has the unlikely property of splitting into two fragments spontaneously. The process is explosive and leaves a tiny trail of damage in the enclosing material. There is a fixed probability of a fission event occurring in a given amount of ^{238}U in a given time interval, so that the number of fission tracks formed is related to the amount of ^{238}U present and to time. Fission tracks are not preserved in materials that have been heated subsequently, so that the number of tracks is related to the time interval since the material was last at a certain critical temperature, which may be the time at which it was formed or the time of some subsequent metamorphic event. Tracks in minerals can be detected under an ordinary microscope if they are first etched in acid to increase their size (Figure 23.12). Their rate of formation is about four tracks per square millimetre per million years for a mineral with one part per million of uranium. The amount of uranium present is measured by irradiating the sample in a nuclear reactor. This induces fission in the much less abundant isotope ^{235}U, and a count of the resulting tracks will give a measure of the quantity of uranium present. Metamorphic events at temperatures as low as 100°C can be dated by this method.

Radiometric dating of rocks is by now as familiar as relative dating by fossils, and is every bit as specialized. Like fission track dating the method depends on the spontaneous changes that take place in certain isotopes. The differences lie in the greater range of possible types of decay and in the method of their detection. The products of the decay process are measured and their relationship to the age of the sample can be calculated as described in Chapter 8. It is important that the products of the decay process, the 'daughter' nuclides, should be retained in the rock. The more daughter nuclide that is lost since the formation of the rock the younger it will appear to be. On the other hand allowance may have to be made for the original presence in the rock of a proportion of daughter nuclide other than that produced by radioactive decay.

The nuclides most commonly used are radiogenic lead, argon, strontium and neodymium, which are the daughter nuclides of uranium, potassium, rubidium and samarium respectively. The uranium–lead method is widely used to date samples of the mineral zircon which occurs in a variety of igneous rocks. The rubidium–strontium method is more commonly used on whole-rock samples. The potassium–argon method has the advantage that potassium, the parent nuclide, is much more common in rocks than uranium or rubidium, and that argon, being an inert gas, has no place in any crystal lattice. The most commonly employed minerals for potassium–argon dating are biotite and muscovite micas and hornblende, all of which contain very little original argon but are excellent at retaining their radiogenic argon. The potassium–argon date may relate to the original formation of the igneous rock or it may relate to a subsequent metamorphic event when the 'clock' was reset. Potassium–argon dating may also be applied, although less reliably, to some sedimentary rocks. Glauconite crystallizes in sedimentary environments and can be used for dating but it may be reworked and incoporated in younger sediments making the dating unreliable.

The decay schemes discussed above all have half-lives of hundreds of millions or thousands of millions of years. They are thus useless for dating rocks less than about 100 000 years old as the amount of radiogenic nuclide becomes too small to measure accurately. The radiocarbon method, developed in Chicago by W. F. Libby in the late 1940s, is used for both geological and archaeological material less than 40 000 years old, as the radioactive isotope carbon-14 has a half-life of only 5730 years. As with other radiometric methods, the experimental accuracy of the age determinations can be calculated and this is usually expressed in the final result as a statistical measure. For example, in the statement that a sample has a radiocarbon age of 10 250 ± 250 years, the figure 250 is the standard deviation estimated for this particular age determination and it means that there is a 68 per cent probability that the true age lies within 250 years of 10 250, and a 95 per cent probability of its lying within 500 years.

Radiocarbon dating has proved to be a key tool in unravelling late Quaternary history since the shortness of this time span means that little detectable organic evolution has taken place. The errors of radiocarbon dating are much smaller than the smallest subdivisions of Quaternary time, and the method thus provides a true correlation tool. Such is not the case with the radiometric dating of much older rocks, since the experimental errors, although proportionately similar to those of radiocarbon dating, become much greater in numbers of years. For example, radiometric determinations of Lower Palaeozoic rocks are commonly quoted with standard deviations of around 10 Ma whereas individual graptolite zones may represent as little as a million years. Nevertheless considerable effort has been expended, since Arthur Holmes first investigated the potential for dating uranium-bearing rocks, on the erection of a numerical geological time scale to stand alongside the traditional relative stratigraphic scale.

Methods of dating Phanerozoic events depend on linking the radiometric dates as closely as possible with biostratigraphic information. This usually means bracketing some igneous event between fossil-bearing horizons.

Biostratigraphy–fossils and strata

Fossils are objects of continuing fascination and frequently of considerable beauty. From being collectors' curios until the eighteenth century, however, they leapt to prominence in the nineteenth century as a vital tool in the emerging discipline of stratigraphy. Their contribution to the documentation of the course of evolution has been described in the previous chapter, but it was long before either Charles Darwin or Alfred Russel Wallace were born that William Smith made his discovery that successive strata in south-west England contained different assemblages of fossils and that, conversely, the fossils could be used to recognize strata that were otherwise poorly distinguishable. Smith had not discovered evolution–he merely made stratigraphic use of his empirical observations. Although Darwin's *Origin of Species* in 1859 provided a theoretical justification for Smith's 'principle of faunal succession', it made surprisingly little difference to the practice of biostratigraphy, which is the use of fossils for the characterization of successive intervals of geological time. Like so much of geology, biostratigraphic practice is based on accumulated experience and published data rather than on any underlying immutable laws. Most of the principles on which present-day biostratigraphy is based were at least partly understood before 1859, and biostratigraphers need not believe in evolution when performing relative dating and correlation.

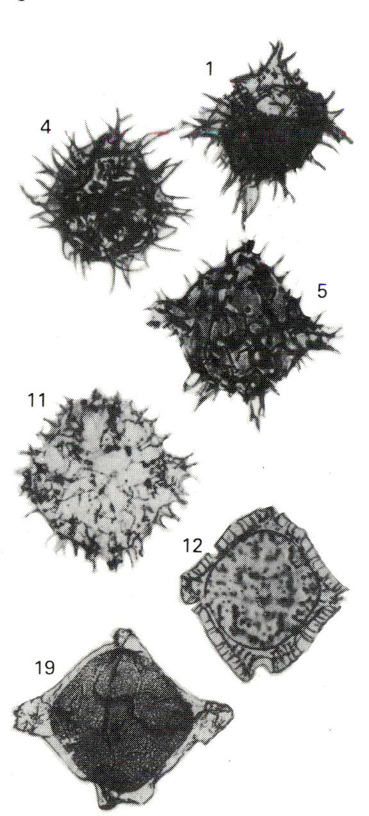

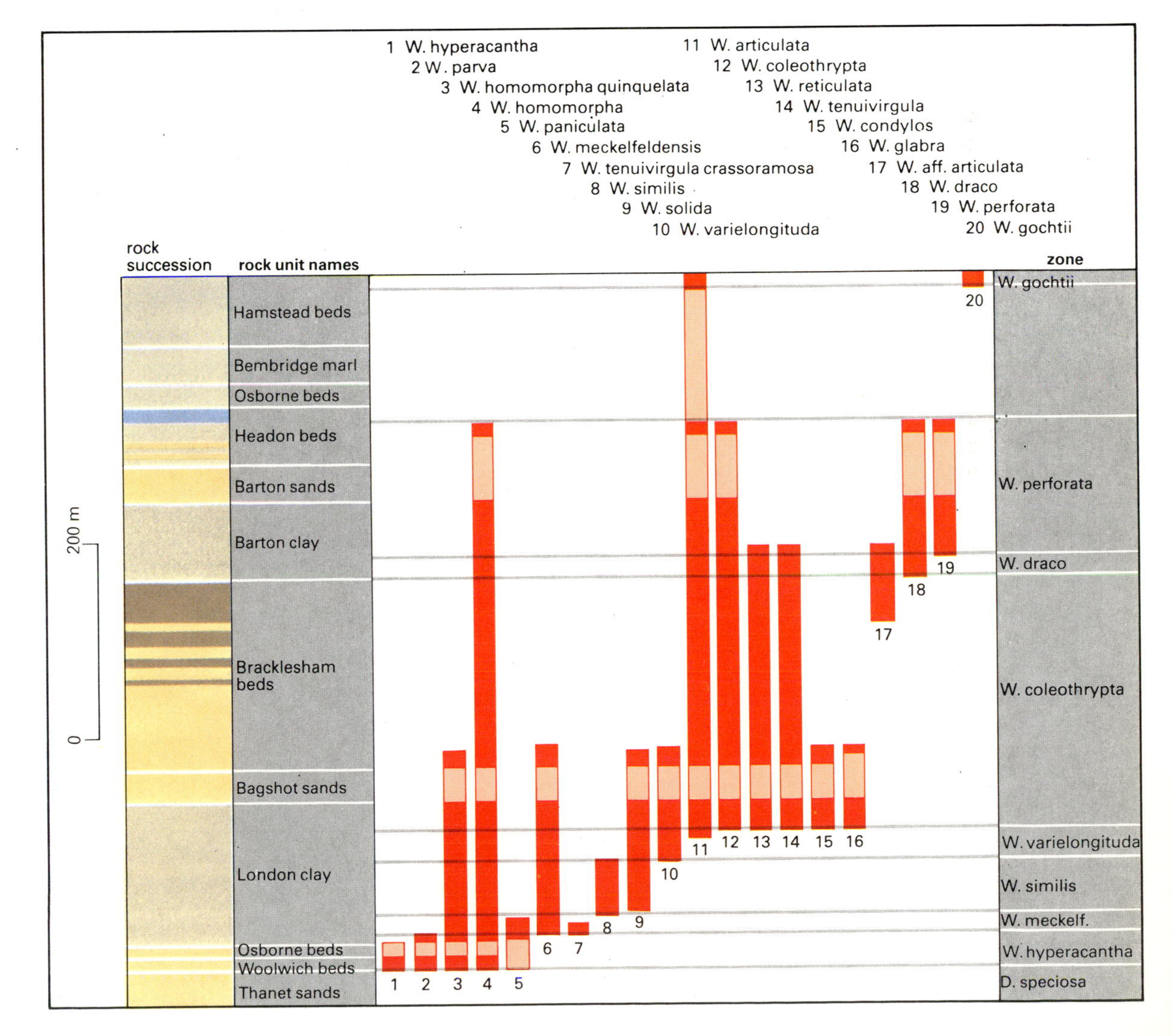

23.13: *Example of Oppel zones:* Wetzeliella *(dinoflagellate) zonation of early Cenozoic, Europe. The relationship between a zone's stratigraphic extension and the range of the species for which it is named is not necessarily one of simple equivalence.*

Stages and zones

The principle of faunal succession enables strata to be traced laterally by means of their distinctive fossil contents. It was soon realized that similar faunal successions can be found in different lithological successions, and that some age equivalence was implied by finding similar fossil assemblages in different areas. This led gradually to the erection of a stratigraphy of fossils which was largely independent of lithology. Successive faunas were seen as representing successive stages in Earth history, which were bounded by events of annihilation of the complete fauna and its replacement by another. The concepts of what constitutes a stage and how it is to be defined have evolved, but many of the Mesozoic stages defined over a century ago have retained much of their intended significance.

The stage concept was put forward by a French geologist, d'Orbigny, but it was a German, Albert Oppel, who founded the biostratigraphically more important zone, fortuitously at just about the time of the announcement by Darwin and Wallace of their theory of evolution. Oppel made very detailed observations of the precise ranges of numerous fossil species in Jurassic strata at many different localities. His zones were carefully selected portions of the faunal succession within which certain species always occurred together. Some of those species, of which there might be a dozen or more, might also occur outside the zone, and so it was the concurrence of all of them that gave the zone its distinctness and recognizability. Oppel divided the Jurassic system into thirty-three such zones and these were immediately found to be applicable elsewhere in Europe outside his area of study.

Oppel zones, or concurrent-range biozones as they are now usually called, have probably not been bettered as a means of classifying strata into units of value in time correlation (Figure 23.13) although it may be that even better time correlation can be achieved by not classifying fossil-bearing strata into zones at all. It is not surprising that they take a great deal of work to define, and so various rather simpler types of zone are sometimes used. The degree of refinement that can ultimately be achieved by this means is an open question, but some examples can

23.14: *Biostratigraphic utility of various fossil groups through geological time.*

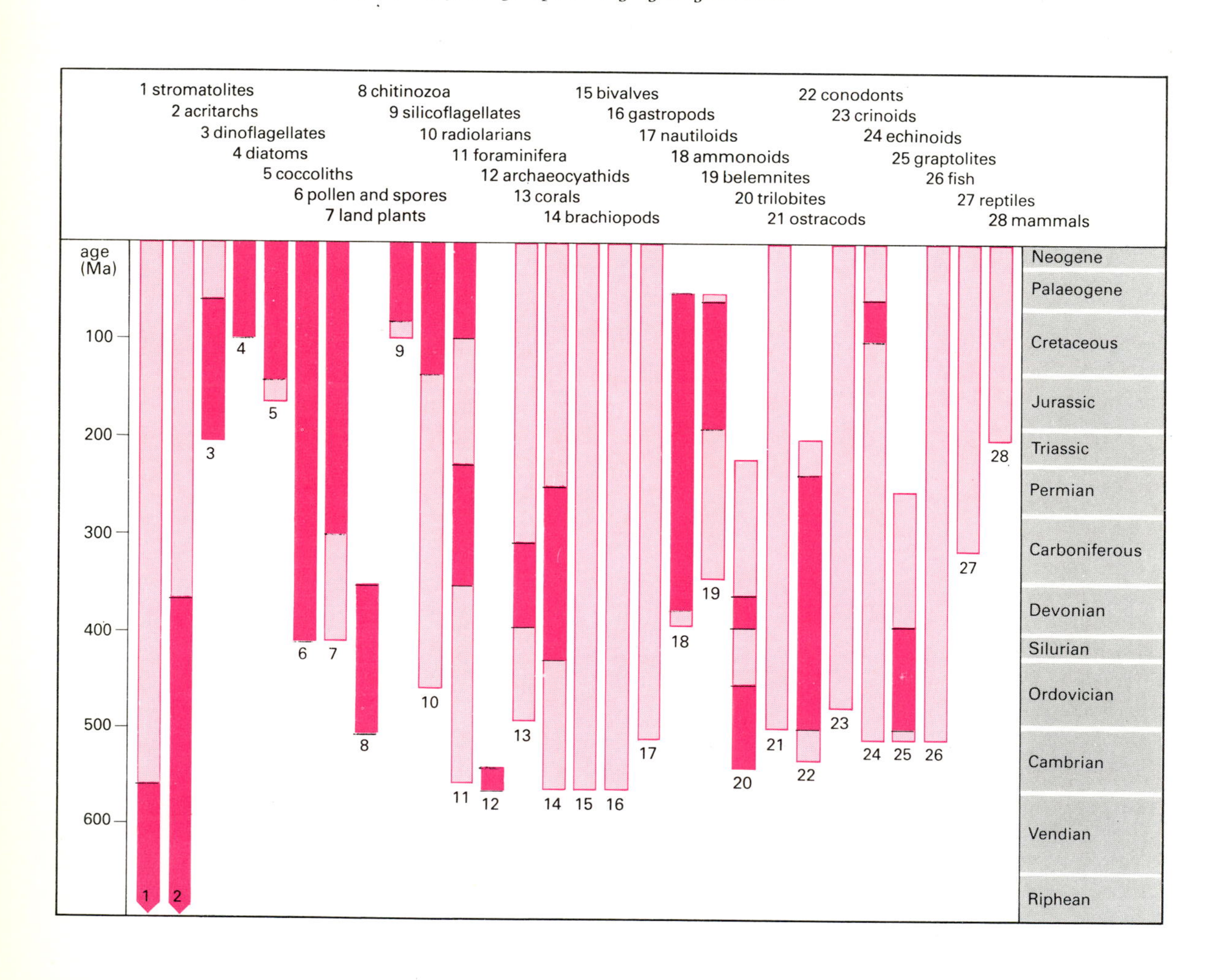

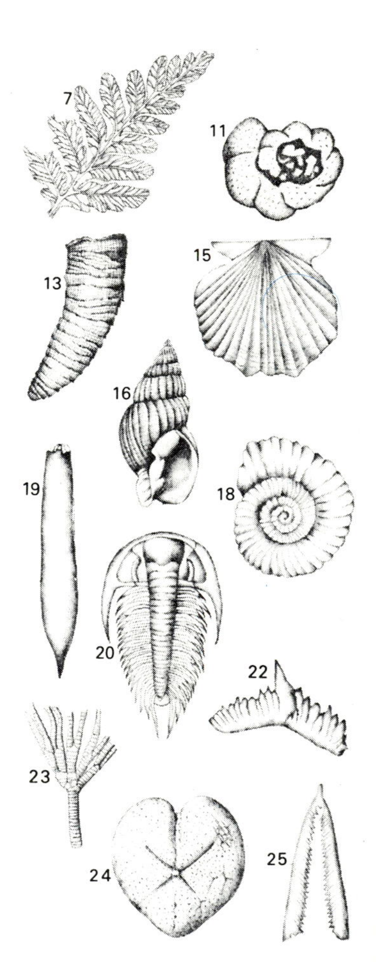

give a rough idea of what has proved possible. The Silurian period is now believed to have spanned some 40 Ma at most, and this is subdivided into about twenty-five graptolite zones, giving an average duration of 1.6 Ma per zone. Jurassic ammonite biozones, many of which can be recognized worldwide, average about 1.2 Ma. Biozonal schemes for more restricted areas of deposition are more precise. The ammonite zones of the Upper Cretaceous succession of the western interior of North America—a largely landlocked sea at the time—average 0.6 Ma and a few may represent a time span as short as 0.2 Ma. Zones based on microplankton of various kinds in deep sea sediments of Cenozoic age also range between a few hundred thousand and a few million years in duration.

Generally the most useful fossils in biostratigraphy are those that have shells or other readily preservable hard parts, are widely distributed geographically (perhaps because they were floating or swimming for at least part of their life cycle) and had relatively short total ranges. Species of ammonites and graptolites fulfil these criteria particularly well, but other groups are useful for various time intervals (Figure 23.14). Of special importance are several groups of microfossils whose small size greatly increases their chances of recovery from boreholes, not only from cores but also from the more commonly produced cuttings. The stimulus for the study of these groups was the need for deeper exploration for oil in the earlier part of this century and the biostratigraphy of foraminifera and, more recently, of spores, pollen and dinoflagellates has been brought to a fine art. A further boost to micropalaeontology was given in the late 1960s by the Deep Sea Drilling Project (DSDP).

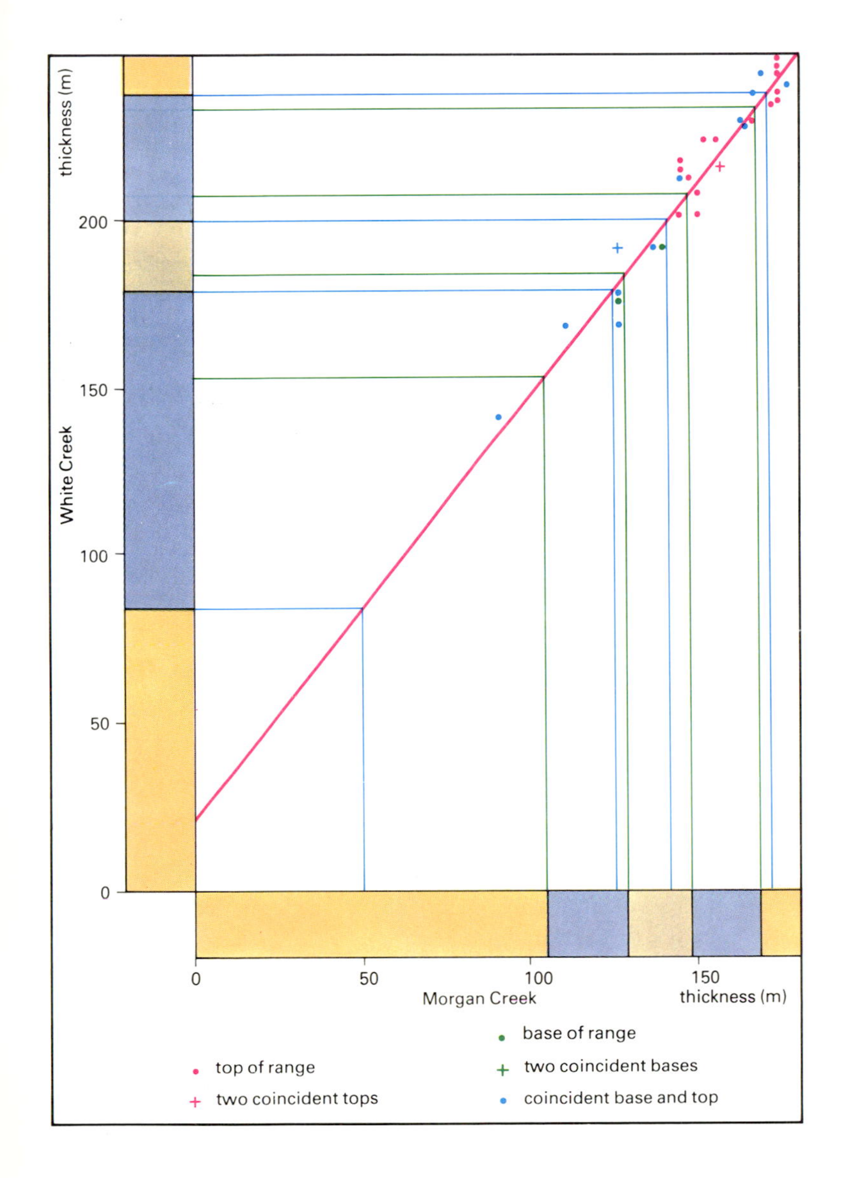

23.15: *Graphical correlation of two Cambrian sections in Texas, based on first and last appearances of 19 trilobite species. The statistical equation of correlation is White Creek = (1.27 × Morgan Creek) + 21.3 metres. The correlation line can be used to show how each of the lithological boundaries is diachronous, each being earlier at White Creek than Morgan Creek.*

Biostratigraphy without zones

Where achievement of the greatest possible refinement in correlation is critical, the use of zones may be inappropriate and some other approaches have more recently been devised.

The graphic correlation technique is so elegantly simple that it is surprising that it was introduced as late as 1964 (by Alan B. Shaw in the USA). It consists simply of treating two measured and collected geological sections as two coordinate axes, and plotting as a graph the coincidence of corresponding palaeontological events such as the first or last appearance of a particular species (Figure 23.15). Given reasonable continuity of sedimentation in both areas, a line of correlation can be fitted through the available points, either by eye or by simple statistics. An extension of the method is to regard one particular section as a standard with which to correlate all others, refining the correlation equation in the process. The equation can then be used to compare the ages of sedimentary events (such as the start of deposition of a particular rock type) from one section to another.

A technique that has been called probabilistic biostratigraphy recognizes that the apparent order of first appearances of a number of species in different areas varies, depending on many factors such as migration and the vagaries of preservation. However, if a number of localities or boreholes are systematically sampled it should be possible to work out a most probable order of appearances and disappearances of the species studied. Information from a new borehole can be compared to this probabilistic standard and correlated with it in a way that can be expressed statistically.

Both of these techniques are continuously refinable with the addition of more raw data, whereas zonal schemes, once published, tend to become entrenched. Information consisting essentially of statements of the presence or absence of fossil species at specified points in a linear sequence are particularly well suited to computer handling, and statistical treatment of biostratigraphic data is likely to increase for applications where accuracy of correlation is essential.

Major events in the surface history of the Earth

A highly idealized goal of stratigraphical studies would be to reconstruct the precise state of every point on the Earth's surface at every moment through its history. It will be all too clear already from this chapter that this is quite out of the question, that there are huge portions of time past in any given area that are not represented by any remaining rock formations, and that there are too many problems of correlating the rock record from one area to another to build up a picture of what was happening at any particular time. What is left is a collection of vignettes from roughly the same time interval scattered over the globe. The world scene that emerges is bound to be highly generalized, like trying to summarize the whole of human history on a single map showing miscellaneous events of unspecified date between 50 000 BC and the present. It is, however, a scene that is being continuously refined as more geological information is documented, and more careful correlation is achieved.

This chapter cannot of course give a complete history of the changing surface environments of the Earth. It aims rather to show how much these environments appear to have changed during the course of Earth history, with some examples of particularly significant events. These events are taken in reverse order, starting with the most recent times, to show how it becomes progressively more difficult to find and interpret the evidence for successively earlier intervals. Some of the information for particular selected time intervals is summarized on a series of palaeocontinental maps. The outlines for the maps were drawn by computer at Cambridge University under the direction of Dr Alan G. Smith. They are based primarily on palaeomagnetic data of palaeolatitude, although for the Mesozoic and Cenozoic maps additional information is available from oceanic magnetic stripes in certain areas. The maps become increasingly speculative with increasing age largely because of the lack of control over relative longitude of the continental masses, and alternative interpretations, some of them very different, will be found in other publications.

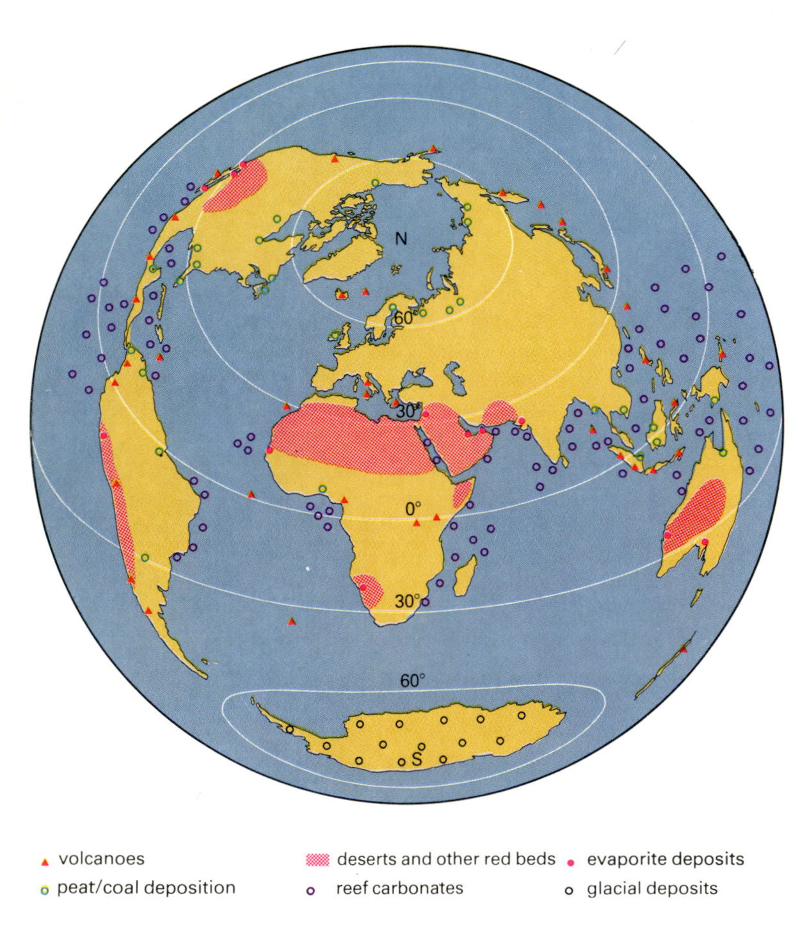

23.16: *Present-day world map on a Lambert equal area projection, showing climate-dependent environments recognizable from their sedimentary deposits, for comparison with the following palaeocontinental maps.*

The Cenozoic era

The Cenozoic era, comprising the Quaternary, Neogene and Palaeogene periods, is that covering the past 65 Ma of Earth history. Cenozoic stratigraphy, traditionally pieced together from incomplete sequences of terrestrial and marginal marine deposits requiring correlation from basin to basin in Europe, has been revolutionized since the late 1960s by the internationally supported DSDP. Ocean floor sediment samples are now available as continuous cores from hundreds of sites all over the world. Oceanic sediments are relatively continuous and sedimentation rates away from the influence of the continents are relatively steady. Furthermore there is a continuous supply of ready-made index fossils with all the right characteristics of small size, durability and rapid evolution. These include the various types of calcareous and siliceous microplankton, such as foraminifera and coccoliths, that now provide the framework for stratigraphic zonation of oceanic sediments. The difficult process of correlating this excellent marine record with the traditional land-based stratigraphy has only just begun.

The Quaternary period

The Quaternary period is that part of the Cenozoic era which includes the present. Its two subdivisions are the Pleistocene and Holocene epochs. The first is roughly the time over which repeated glaciations have been characteristic of both hemispheres. The second is the time since the retreat of the most recent ice sheets in the northern hemisphere. Like any other division of geological time these are arbitrary; northern hemisphere glaciation began within Pliocene time (as currently defined) and ice sheets and glaciers exist at the present day.

The Quaternary, brief as it is in terms of Earth history, requires rather specialized stratigraphic treatment. Deposits of Quaternary age are more widespread than those of any previous period, so there is a wealth of data available. But there has been too little time since the start of the Quaternary, some 1.6 Ma ago, for significant organic evolution. This has two effects: one is that practically all the fossils in Quaternary deposits, with the exception of a few extinct mammals, are familiar to biologists and so reconstruction of the environments in which they lived is relatively straightforward; the other is that stratigraphic correlation cannot depend on evolutionary change between deposits within the Quaternary.

Nineteenth-century European geologists were able to apply the

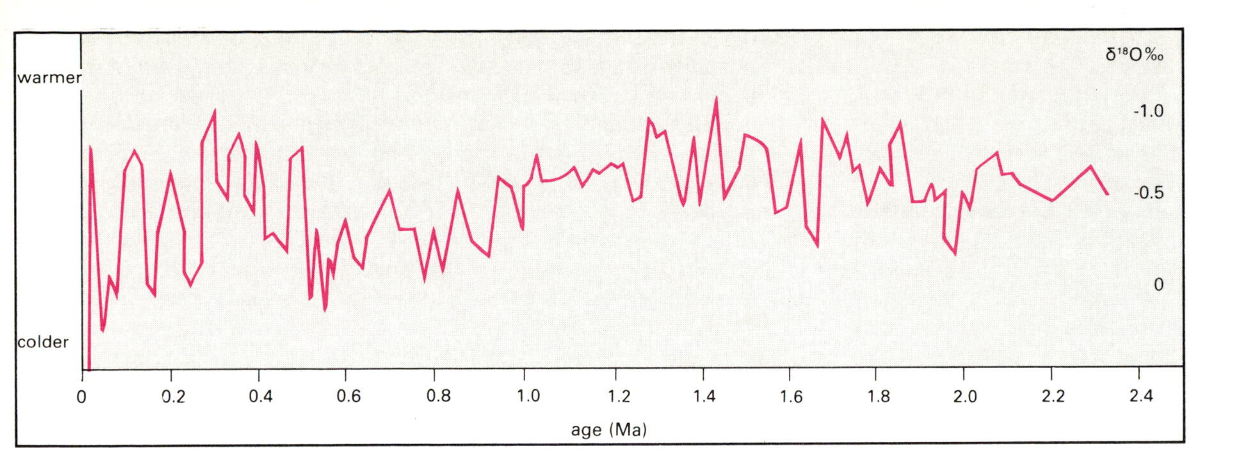

23.17: *Oxygen isotope results in a DSDP core from the Pacific Ocean demonstrate that there have been 18 cold periods since the beginning of the Pleistocene epoch about 1.6 Ma ago.*

basic principle of superposition to the problem of how many times the ice had advanced and retreated. Land-based Quaternary stratigraphy, however, is greatly hampered by the difficulties of correlation (one interglacial deposit is likely to contain a similar selection of fossil remains to any other) and by the fact that successive advances of the ice have tended to sweep away all earlier glacial deposits. The classical Alpine glacial stratigraphy depended on recognition and correlation of deposits of outwash gravels left behind by the retreat of four successive glaciations. The unsatisfactory nature of this scheme as a standard for reference is illustrated by the discovery of part of a bicycle in gravels in northern Czechoslovakia that had been assigned to the most recent glaciation, and thus to a period more than 10 000 years ago.

The standard reference scale for the Quaternary is now becoming more firmly established in the deep sea record, where relatively continuous sedimentation can record changing surface water temperatures of the sea as reflected in the species and oxygen isotope composition of fossil microplankton. It is now clear from these records that there have been no fewer than eighteen complete warm–cold cycles during Quaternary time, more than could ever have been reliably pieced together from land-based studies (Figure 23.17). Both oxygen isotope ratios and changes in the microplankton demonstrate alternations of warmer and colder waters with time. The problem with the deep ocean record is in correlating it with the widely studied land deposits. This is unlikely ever to be finally achievable, since the probability of equating an inferred warmer or colder period with the wrong part of the deep sea record is obviously rather high. Most of the biostratigraphic techniques used for terrestrial deposits are inapplicable to the deep sea record, being based on such fossils as pollen, beetles and mammals. Nevertheless, it seems likely that a standard subdivision of Quaternary time will be based on a selected core or cores of sediment from one of the oceans. Boundary points will be selected to define the subdivisions (probably stages) in the usual way, and the stages will probably each represent either a cold or a warm interval (this would result in a division into about thirty-six stages). Where correlation with this standard is possible, it will represent stratigraphic fine-tuning of the order of 50 000 years.

The beginning of the Pleistocene epoch, which is also the boundary between the Quaternary and Neogene periods, traditionally marks the onset of glacial conditions in Europe, but while this was clearly an important local event, its definition is still a problem. Sediments of this age in glaciated terrain are unknown, and in any case glaciation can be expected to have started later, if only by a short interval, at latitudes successively nearer the equator. The provisional base of the Pleistocene has been taken as the first appearance of a cold-water fauna of shells in a thick sequence of sediments in Calabria in Italy. It has now been proved that the age of this datum approximates to the age of the end of the Olduvai event on the magnetostratigraphic time scale. This could provide the basis for permanent definition of this boundary at a fixed point in a selected deep sea core.

The Palaeogene and Neogene periods

The Palaeogene and Neogene periods have rather recently entered stratigraphical terminology as replacements for the Tertiary, which entered geological nomenclature in 1759. The Palaeogene comprises the Palaeocene, Eocene and Oligocene epochs and the Neogene includes the Miocene and Pliocene epochs.

Climatic deterioration during the Cenozoic era can be documented by reference to plant fossils, for instance in North America where leaf shapes characteristic of temperate climates are found successively farther and farther south in younger Cenozoic deposits. But the most valuable palaeoclimatic evidence now comes from oxygen isotope studies on microfossils from DSDP ocean floor samples. The oxygen isotope record from most parts of the world shows a gradual fall in temperature through the Eocene, with a 4°C drop in temperature at about the Eocene–Oligocene boundary (Figure 23.18). Further fluctuations were then followed by another sharp drop late in the Miocene. Permanent ice is known to have started to form in the northern hemisphere in the late Pliocene, but by then Antarctica had probably been largely ice-covered for millions of years. This curiously asymmetrical arrangement was due to the lack of land at the North Pole while the mountainous continent of Antarctica had been in a south polar position since the early Cretaceous. The opening of the Drake Passage between South America and Antarctica would have allowed polar ocean currents to flow right round Antarctica, isolating it from the effects of warmer currents. Other factors which must have played their part in climatic change in the Cenozoic are the closure of the Tethys

Ocean, with consequent changes in oceanic circulation in low latitudes, falling sea levels from the very high levels of the Cretaceous period and the abundance of newly formed, high mountain ranges.

Now that the world's more obvious locations for petroleum reservoirs have all been exploited it is becoming more and more important to be able to predict, through a knowledge of how, when and where sediments have been deposited in past times, just which sedimentary basins are worth further exploration. This work depends very heavily on stratigraphical skill, in predicting the lateral extension of source and reservoir rocks, and in accurate time-correlation within individual basins. In general the younger the strata the smaller their chances of subsequent tectonic disruption which might allow any stored oil to escape, and many of the world's major oil reservoirs are in Cenozoic strata. Important Cenozoic reservoir rocks are found in Venezuela, the North Sea, the Middle East (where limestones are particularly important), the Caucasus Mountains in the Soviet Union and in Indonesia.

The Cenozoic has been the era of major development among the mammals and the flowering plants. The evolution of the mammals has been clearly influenced by continental drift, most obviously in the geographical separation of the marsupial faunas from the placental mammals (or rather, the ancestors of these groups) before the Cenozoic. At the beginning of the Cenozoic the marsupials were able to migrate between South America and Australia by way of Antarctica, despite the extremes of day length they would have experienced. The separation of Australia from Antarctica in the Eocene completely isolated the Australian faunas and allowed the well-known diversification of marsupials there. Meanwhile South America in turn became separated from Antarctica and when it became linked to North America interchange became possible between the northern placental and the southern marsupial faunas. The numerous extinct marsupials from Cenozoic deposits in South America are evidence of their lack of success in competing with the placental mammals for many of the same ecological niches. The palaeontological evidence in fact helps to establish the date of this land connection more accurately than the palaeomagnetic data alone.

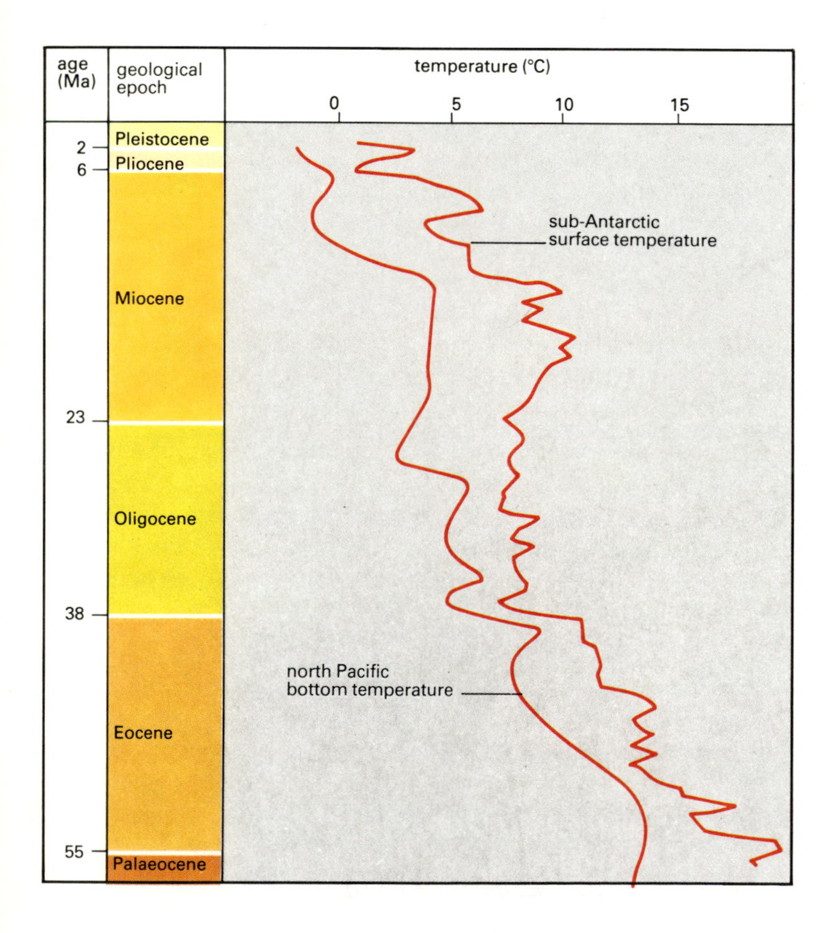

23.18: *Composite $\delta^{18}O$ palaeotemperature curve for the Cenozoic era. Separate curves are shown for temperatures based on bottom-dwelling foraminifera and for surface temperatures from planktonic forms.*

The Mesozoic–Cenozoic boundary

The concept of a division of geological time into major eras based on their typical faunas goes back nearly a century and a half. The changes in fossil content between rocks of Palaeozoic and Mesozoic, and of Mesozoic and Cenozoic ages are no less dramatic even now with so much more detail available. However, major innovations in plant evolution such as the invasion of the land (mid-Silurian), the rise to dominance of the gymnosperms (mid-Permian) and then the angiosperms (mid-Cretaceous) did not occur at the era boundaries that mark similar key events in animal evolution.

The effects and some possible causes of the end-Cretaceous extinction event have been discussed in Chapter 22. Catastrophic agencies such as meteorite or comet impacts have for long been favourites for speculation, but it remains difficult to explain the selectivity shown by the disappearance of the large reptiles but not the smaller ones nor the mammals, by the total disappearance of the ammonites but only the partial extermination of the belemnites, and by the apparent lack of adverse effects on land plants.

It seems that the global cooling that characterized the Cenozoic era actually began before the end of the Cretaceous period, which would make it difficult to relate to a cosmic event at the boundary. The carbonate compensation depth (below which $CaCO_3$ is completely dissolved) seems also to have changed, judging by changes in the relative amounts of calcium carbonate received at the sea floor in different areas. The compensation depth could have risen from its usual depth of several kilometres up to within the surface layers inhabited by most of the microplankton. This could have caused their extinction quite simply by removing the material with which they build their shells and skeletons. The other marine extinctions would follow from the loss of this primary source of food, while the removal of so much $CaCO_3$ to the sea floor could have lowered carbon dioxide levels in the atmosphere, bringing an end to the 'greenhouse effect' that had presumably kept late Cretaceous temperatures so high.

The Mesozoic era

The era of 'middle life' presents many differences from the Cenozoic, not only in its life forms but in its more equitable division of the continental masses between northern and southern hemispheres, and in its climates and circulation patterns. Most important for the geologist trying to unravel Earth history for this era is the rather smaller amount of strata preserved and in particular the complete loss of the deep ocean sedimentary record from before the late Jurassic. With this loss go the ready-made sequences of magnetic polarity reversals and of oxygen isotope ratios. Biostratigraphy, though, was born in the shallow marine limestones and clays of the European Jurassic and Smith, d'Orbigny and Oppel managed without DSDP results. Correlation remains the key to accurate reconstruction of environments in the Mesozoic and it must be remembered that accounts such as this one are ultimately dependent on the accumulated work of thousands of stratigraphers worldwide over the past century and a half.

Continental positions through the Mesozoic (Figure 23.19) were more southerly than at present. Plate movements throughout the Mesozoic were dominated by the gradual break-up of the supercontinent of Pangaea that existed at the beginning of the era, and by the gradual closure of the Tethys Ocean, the great seaway that dominated equatorial palaeogeography from even earlier times. In the mid-Jurassic, the central Atlantic began to form and North and South America became separated. During late Jurassic to early Cretaceous time the southern continents, formerly united as Gondwanaland, all started to break apart, except for Antarctica and Australia, and India began its long northward drift towards Asia (see Chapter 16). Throughout the Mesozoic, then, there was a more or less continual continental mass (though perhaps not always a landmass) stretching from pole to pole. There was an equally impressive ocean, the forerunner of the modern Pacific, of which the Tethys was a mere inlet. The break-up of Pangaea implies a reduction in size of the Pacific, and subduction of an oceanic plate beneath the Americas is implied by the long history of depositional and mountain-building events on the west coast. Evidence of tension in Gondwanaland is provided by the abundant basalt lavas in South Africa, Malagasy, India, South America and Antarctica, much of them of late Jurassic age, coincident with the main period of continental separation.

The beginning of rifting in the southern part of the north Atlantic Ocean was heralded by tensional faulting that allowed elongate basins to subside between faults. As they sank, these basins filled with terrestrial sediments and volcanic rocks of late Triassic age, showing that at this earliest stage the sea was still unable to break in. Earliest Jurassic marine deposits are spread widely around the margins of the north Atlantic showing that by that time the sea had arrived. The south Atlantic separated a little later; sediments along the line that was to split are non-marine right through the Jurassic and early

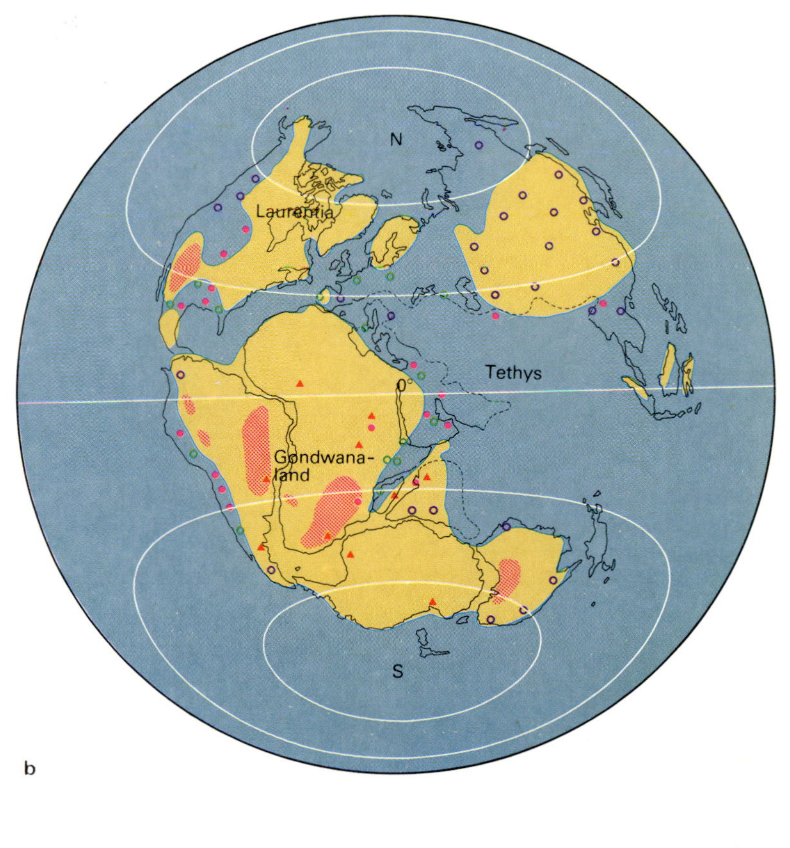

23.19: *Palaeocontinental maps for* (**a**) *80 Ma and* (**b**) *160 Ma ago. Environmental data are for late Cretaceous and late Jurassic sediments respectively.*

▲ volcanoes
○ peat/coal deposition
deserts and other red beds
○ reef carbonates
● evaporite deposits
glacial deposits

Cretaceous. In the mid-Cretaceous a thick succession of evaporite deposits, up to 2000 m thick, was deposited and is now found on the coasts of Brazil and west Africa. This strongly suggests that the initial rifting created a topographic depression in which the only way out for water, given the tropical latitude, was by evaporation. The evaporites are dated to the Aptian stage (roughly 112 Ma ago), while deposits of the following Albian stage (say about 106 Ma ago) are fully marine in character.

The sediments that accumulated in the continental interiors during the Mesozoic are a record of the interplay between changing sea levels and tectonic uplift and subsidence (see Chapter 15). Sea levels must have been unusually low at the start of the Triassic period, which is represented by rather scarce marine sedimentary deposits, and unusually high late in the Cretaceous period, which left widespread marine sediments, including chalks. There is no evidence of glacial deposits of Mesozoic age anywhere in the world, so that regressions and transgressions during this time cannot be ascribed to alternate formation and thawing of major ice caps. It is possible at least that the general rise of sea level was a response to an increase in either the total length of active mid-ocean ridges or the rate of sea floor spreading. Either of these would increase the volume of the world ridge system, displacing more water on to the continents. Changes in ridge activity prolonged over millions of years could result in sea level changes of the order of hundreds of metres—enough to flood the continents to about 40 per cent of their total area as was the case in late Cretaceous time.

Climate is undoubtedly affected by the global extent of the oceans, and the Cretaceous climatic maximum must have been influenced by the major transgression of that time. The greater heat capacity of water compared to rock would result in greater retention of solar radiation and hence a warmer climate; warm and equable climates appear to have been the rule throughout the Mesozoic era. As the palaeolatitudes of most continents are reasonably well known throughout this time there is reasonable certainty that warm dry climates extended further away from the tropics than they do at present. For the Triassic period, red-bed sediments (in which iron is predominantly in the oxidized state) and evaporites were deposited over many continental areas as far north and south as 50° latitude, beyond which coaly deposits of that age are found. This suggests warm and arid conditions in what today would be tropical and temperate zones, and equable but more humid conditions in the equivalent of our cool-temperate and polar regions. This rather even distribution of climates may help to explain the rather low diversity and cosmopolitan character of Triassic marine life. The extent of coral reefs in the Triassic, particularly around the Tethys seaway, was greater than in the succeeding Jurassic period. The long western coast of the Americas experienced desert-like conditions extending well to the north and south of what would be expected today.

There is evidence of stronger climatic zonation in the Jurassic period. Assemblages of plant fossils vary with palaeolatitude in the northern hemisphere, although the southern floras were remarkably widespread throughout Gondwanaland. The marine faunas of the later Jurassic also became progressively more distinct from those of the Tethys and southern seas, possibly due to climatic or oceanographic differences, and correlations between the two regions become more difficult. Oxygen isotope ratios may eventually help to define more precisely any significantly cooler water masses. The rather sparse information so far available is not completely consistent, although a measurement from the calcite of a belemnite of Jurassic age from New Zealand, then at 75°S, gave a sea temperature of 14°C, much warmer than present-day Antarctic waters.

Cretaceous palaeoclimates are better known because of the availability of deep sea data. Results from surface and bottom dwelling foraminifera of Cretaceous age from the Pacific show not only that temperatures increased during the Cretaceous before falling at the end of the period, but also that the difference in temperature between the surface and bottom of the ocean was much less than today, confirming the theory of continuing equability of climate from tropics to poles compared with the strong present-day zonation. For non-marine deposits the Cretaceous shows a pattern similar to the Jurassic and Triassic of an extensive evaporite belt with coal-bearing deposits poleward of about 40° palaeolatitude. Coals are found even in Alaska which was then very near the North Pole, yet the plants fossilized there include trees, shrubs and ferns, suggesting a temperate climate and an unusual adaptation to extremes of summer daylight and winter darkness. In the marine realm the Cretaceous period saw what must have been the most extraordinary explosion of planktonic life since the late Precambrian. The extreme spread of epicontinental seas and the warmth of the climate must have been contributory factors. The widespread distinctive late Cretaceous rock type from which the period derives its name is chalk—pure white limestones consisting largely of the calcareous remains of coccoliths. These chalks are found not only in deep sea sites where they formed as oozes uncontaminated by land-derived sedimentary debris, but they are also abundant on continental margins and interiors, where rather unusual shelf sea conditions must have prevailed.

The Tethyan seaway and its continuation westward between the continental masses of North and South America provided the ideal conditions for the generation of petroleum. This culminated in the mid-Cretaceous climatic optimum, during which it is estimated that 60 per cent of the world's known reserves of oil were generated. The oil fields of the Middle East, North Africa, Venezuela and the Gulf of Mexico are all in this region. This remarkable period, comprising the Aptian, Albian, Cenomanian and Turonian ages dating between about 115 and 85 Ma ago, when the epicontinental shelf seas were five times more extensive than at present, saw the deposition of unprecedented quantities of clays with up to 30 per cent organic carbon. This implies a lack of dissolved oxygen in the sea at the depth at which these 'anoxic' sediments were deposited, suggesting restricted circulation as in the Black Sea today. There was presumably an abundant supply of organic material both from the flourishing microplankton in the oceans and from the rapidly diversifying flowering plants on land which clearly benefited from the warm climatic conditions. The lack of ice caps presumably resulted in less vigorous oceanic circulation than at present.

The late Palaeozoic era

The Palaeozoic era, divided into six or seven periods, is conveniently considered in two parts. Overall it covers the time from the first appearance of abundant fossils with hard parts, through the time of the appearance of abundant land-dwelling life, to the major extinction crisis at which the end of the era is recognized. By comparison with the Mesozoic the late Palaeozoic presents considerable problems of environmental reconstruction because of the major differences in the animals and plants represented in the fossil record and also the lack of surviving oceanic crust. Nevertheless there are still abundant sediments that were deposited on the continental shelves and interiors of the time to provide a partial and perhaps increasingly selective record of events. If the sedimentary record is poorer for Palaeozoic time then the palaeomagnetic evidence for continental positions is nearly destitute. Reconstructions of Permian and earlier continental assemblies rely on far too few data points, with errors of uncertain magnitude, to yield a unique solution for any particular period of time. In any case palaeomagnetic data give little more than the palaeolatitude of the locality sampled. The maps used here for Palaeozoic palaeogeographics are as objective as possible and use the arrangement of continents that seems best to fit all the available palaeomagnetic data. They provide a framework within which to test knowledge of environments as deduced from the record of the rocks and fossils.

A major point of difference between the late Palaeozoic maps in this chapter and those that may be found elsewhere lies in the positioning of Africa and South America relative to North America and Eurasia. There is little doubt about the positions for early Mesozoic (pre-Atlantic) time, but the late Palaeozoic data are rather better satisfied for the reconstruction in Figure 23.20(a), although this implies major dextral transcurrent movement between the late Permian and early Triassic periods (compare with Figure 16.20). Overall the late Palaeozoic was a period of assembly of all the major continents into a single unit, Pangaea. At about the beginning of this time, North America became welded to the western margin of Europe, approximately along the axis that was later to rift again as the north Atlantic. Eastwards, Europe extended only as far as what are now the Urals, then the site of deposition of the rocks that were to form those mountains during the collision with Asia in the Permian period. Asia may well have consisted of more than one plate earlier in the late Palaeozoic, but the details of its assembly are far from clear. Some alternative reconstructions for this period show three plates, comprising the present areas of China, Siberia and part of south-east Asia. Gondwanaland, comprising the present-day southern continents plus India, appears to have remained a single continent throughout this time span. The continental interior of Gondwanaland must have undergone considerable differential uplift and subsidence as thick sequences of sediment accumulated in several major basins. These present problems of correlation with marine successions elsewhere in the world, because Gondwana-

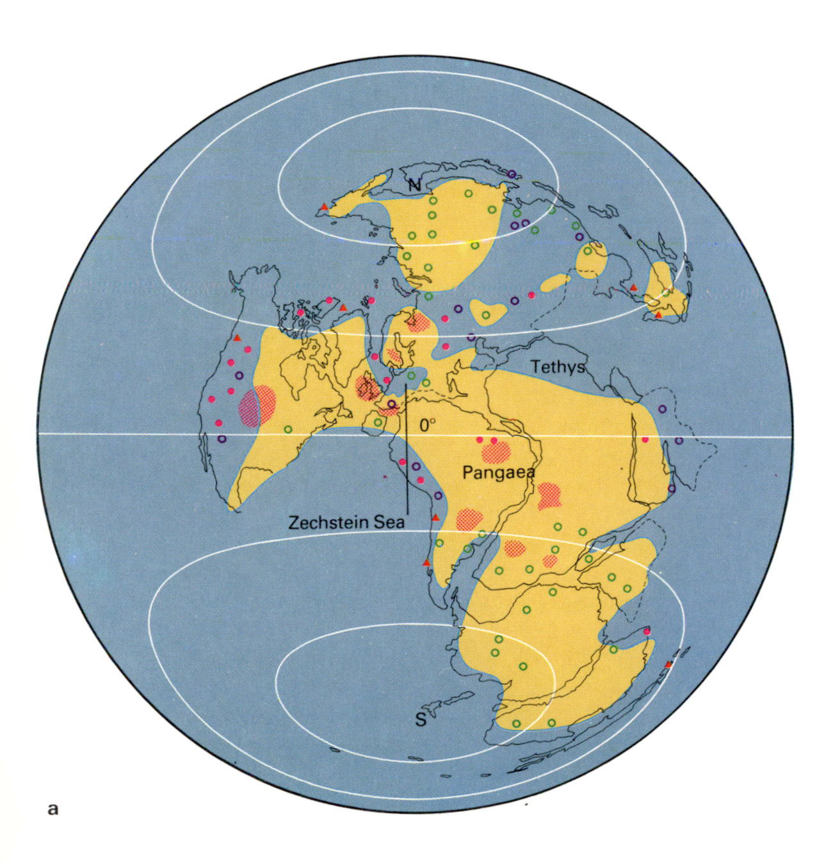

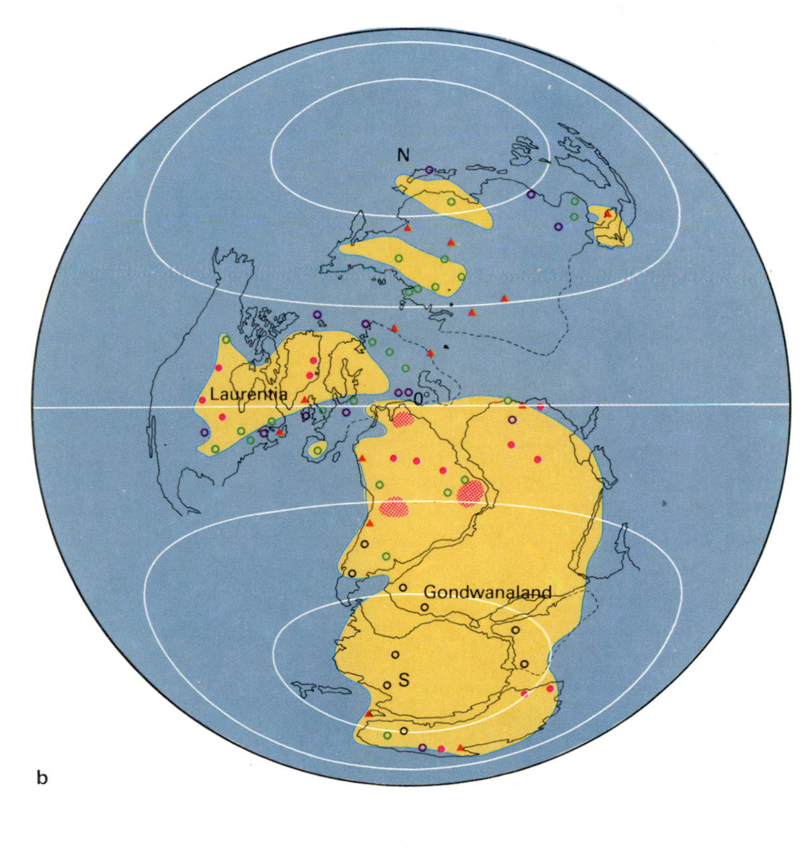

23.20: *Palaeocontinental maps for* (**a**) *240 Ma and* (**b**) *320 Ma ago. Environmental data are for late Permian and mid-Carboniferous sediments respectively.*

23.21: *Devonian barrier reef in the Canning Basin, Australia, now exposed, almost undeformed, in the middle of a desert.*

land tended to have its own flora and fauna evolving in partial isolation from those of the more northerly continents resulting in a lack of fossil species common to both.

Although there are differences of opinion over the precise reconstruction of Gondwanaland in the late Palaeozoic era, there is particularly strong evidence that its constituent parts were all located in high southerly latitudes at the time. All have sedimentary deposits of Carboniferous to Permian age that include all those expected in a glacial environment, including tillites, varved sediments and marine sediments into which pebbles and boulders have been dropped from floating rafts of ice. Their great areal extent allows a certain amount of reconstruction of the size and shape of the ice sheets and the direction in which they moved. These glacial deposits were for long simply referred to as 'Permo-Carboniferous', but increasing precision of correlation is being achieved. The earliest glaciations were in South America and southern Africa, and the latest was in Australia. This is consistent with palaeomagnetic evidence for movement relative to the south pole from early Carboniferous to early Permian times.

Curiously the glacial deposits in parts of Gondwanaland are succeeded by beds with thin layers of coal suggesting temperate climates; close juxtaposition of ice and forest can, however, be seen at the present day in New Zealand and elsewhere. Coal of late Palaeozoic age is much more characteristic of tropical palaeolatitudes for, although the accumulation and preservation of abundant plant material to form coal requires swampy or boggy sites that can be found at any latitude, ideal coal-forming conditions seem to have been unusually persistent in the Carboniferous tropics. Land plants first became truly abundant in Devonian times, and the widespread distribution of many genera over large areas point to rather equable climates. By the late Devonian the coal swamp environment had appeared but it is the Carboniferous period that is named for its remarkable coal deposits which form the bulk of world economic reserves. Combined with the restricted occurrence of evaporites of this age, this evidence is suggestive of unusually high humidity over much of the world. The later Carboniferous also saw the deposition of important oil source rocks in west Texas and elsewhere in the western USA. Devonian source rocks are important in Texas, Canada and the Volga-Ural district of the Soviet Union. The change of climate towards glacial conditions during the Carboniferous period explains both the latitudinal restriction of reefs in the seas, and the increasing provinciality of plants on land. There is a strong tendency for sedimentary sequences of this age, both of terrestrial and shallow marine origin, to show cyclical repetitions, and these have been ascribed to eustatic changes in sea level with advances and retreat of the ice in the southern hemisphere. However, the impossibility of matching the individual interglacials with individual transgressive cycles means that this hypothesis must remain conjectural.

Late Palaeozoic correlation and boundaries

The three (four in North America) systems of the late Palaeozoic era were founded on the differences in overall aspect of the rocks of each in Europe and North America—the Devonian redbeds of the Old Red Sandstone continent that flanked the site of closure of the Iapetus Ocean, the Mississippian or early Carboniferous grey limestones of transgressive shallow seas, the Pennsylvanian or later Carboniferous coal measures of the extensive equatorial swamps, and the Permian marls and limestones of the shallow Zechstein Sea. In other parts of the world these divisions are much less obvious—especially in Gondwanaland—and there is no question of seeking global or cosmic events to explain the boundaries between the late Palaeozoic systems/periods.

On the other hand, the close of the late Palaeozoic era was of considerably wider import, marking as it did the culmination of a period of extinction of even greater proportions than that at the end of the Cretaceous period. It seems that sea level was particularly low at this time, and this combined with the reduction in total coastlines caused by the reassembly of Pangaea might simply have reduced the ecological niches available to organisms in the normally prolific shallow shelf seas. The Permian period presents problems of correlation as marine faunas appear to have become increasingly localized and specialized towards its close. The rarity of very early Triassic deposits also makes precise definition of a boundary difficult. Not only does there always seem to be a pause in the sedimentary record in each of the few localities where there may be a chance of locating a suitable horizon (and the localities tend to be in inaccessible areas such as the boundary between Iran and the Soviet Union, and the Himalayas of Pakistan and Kashmir), but there is a striking change in the whole appearance of the marine fossil faunas. It will probably be some time before it is ascertained that this ecological replacement took place everywhere at the same time so that a site can be selected at which to drive home the 'golden spike' that will finally and permanently separate the Palaeozoic from the Mesozoic era.

The early Palaeozoic era

It has been said that the one evolutionary event on Earth that could have been recognized from space was the colonization of the land by plants. The major environmental difference between early and late Palaeozoic times must surely be that the land surfaces during the Cambrian, Ordovician and most of the Silurian periods were devoid of plant life as we know it, making reconstruction of early Palaeozoic environments a stage more difficult. Most of the stratigraphic evidence concerns the marine environments of that time.

Palaeomagnetic evidence points reasonably clearly to a predominance of subequatorial latitudes for most of the continental masses in the Cambrian period. Gondwanaland, which appears to have remained as a unit throughout the early Palaeozoic, moved to a more southerly position through the Ordovician and Silurian periods. The other continents, shown in Figures 23.22 and 23.24 as three units though they may have been subdivided, remained in tropical latitudes until the late Silurian when Asia drifted rather further north. At about this time North America (including Greenland) collided with Europe, closing an ocean recently named Iapetus from the mythical father of Atlantis.

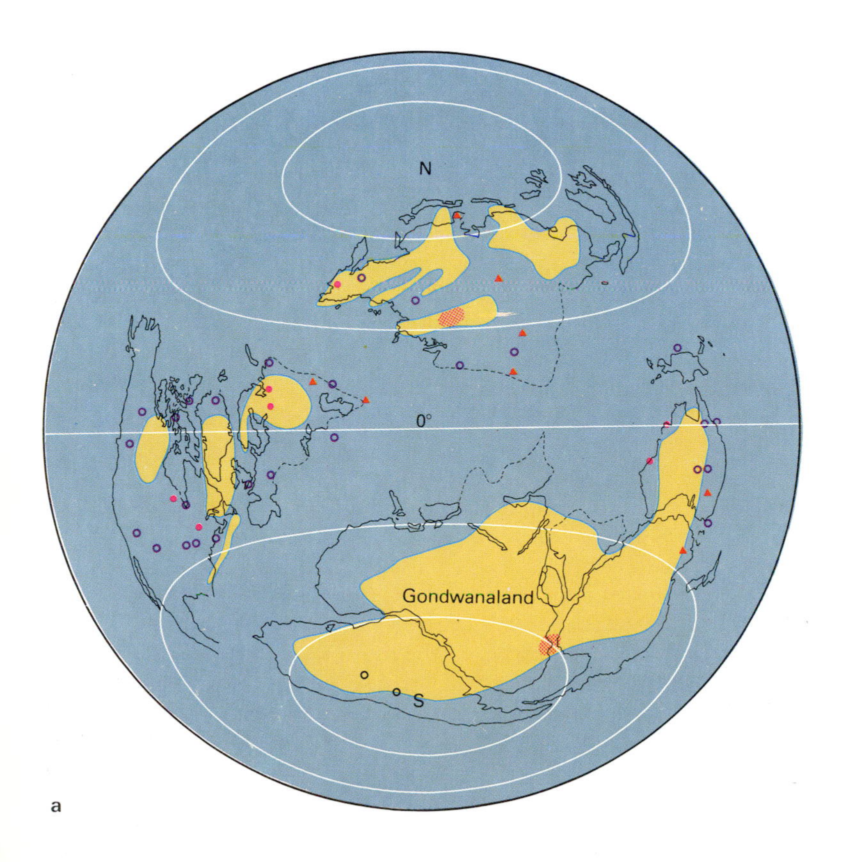

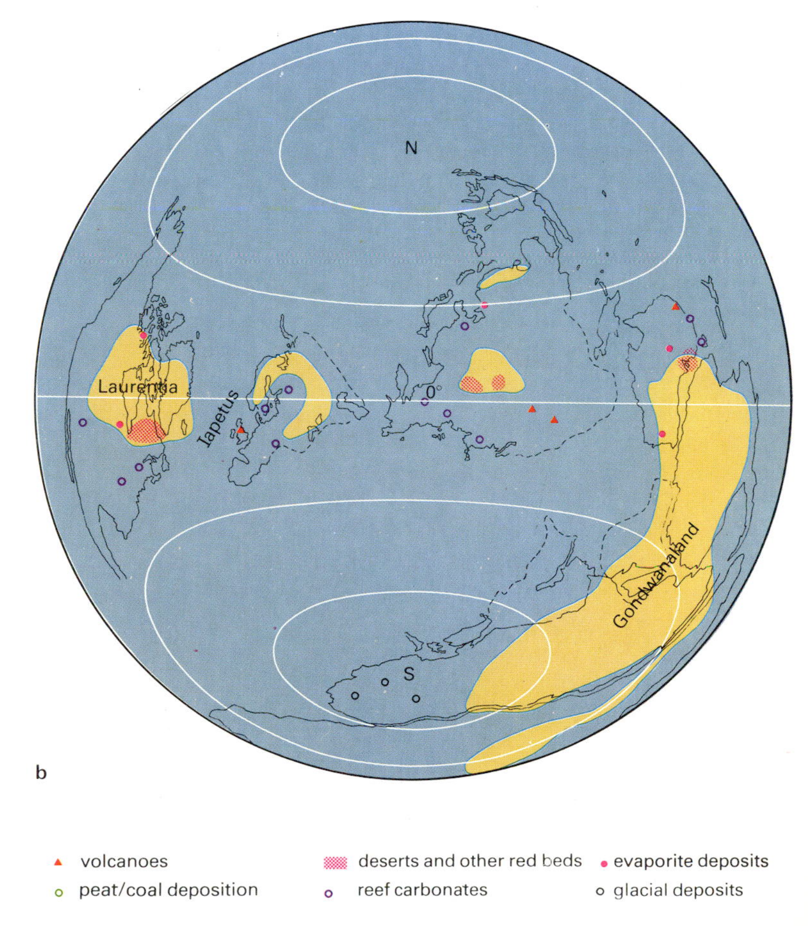

23.22: *Palaeocontinental maps for* (**a**) *400 and* (**b**) *480 Ma ago. Environmental data are from late Silurian and early Ordovician sediments respectively.*

▲ volcanoes
○ peat/coal deposition
deserts and other red beds
○ reef carbonates
● evaporite deposits
○ glacial deposits

23.23: *Ordovician tillites, about 440 Ma old, from the Sahara desert. The striated boulders are clear indicators of a glacial origin.*

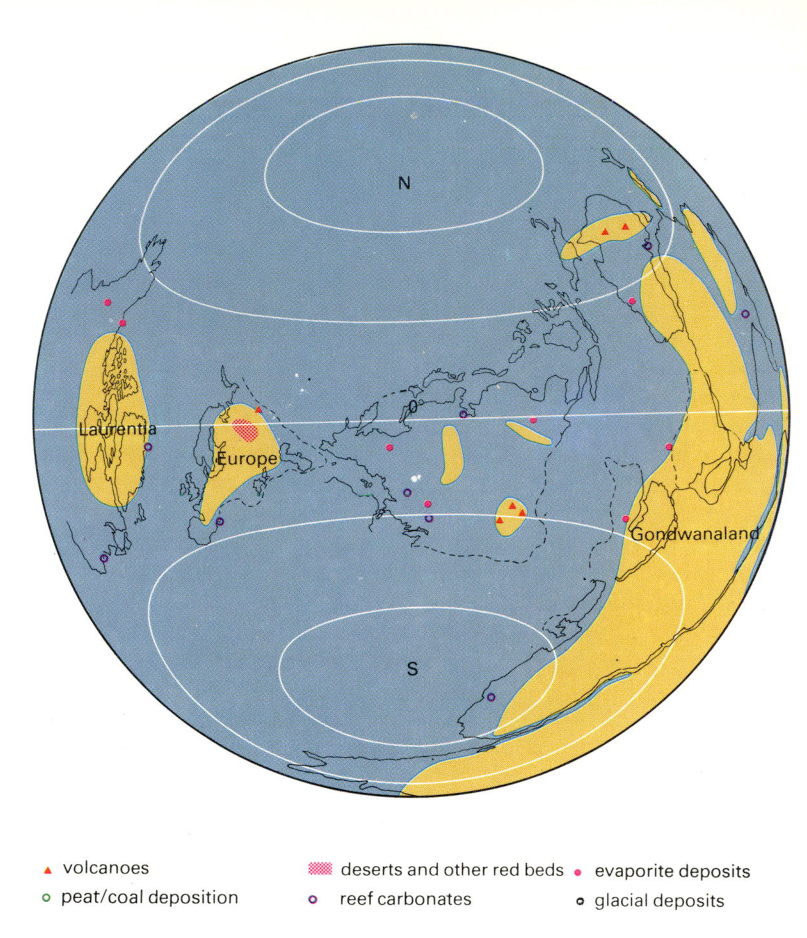

23.24: *Palaeocontinental map for 560 Ma ago. Environmental data are from early Cambrian sediments.*

This event, which threw up the forerunners of the Appalachian, Caledonian and Scandinavian mountains, closed a long period of marine sedimentation during which were laid down the thick deposits of Wales where the Cambrian, Ordovician and Silurian systems were first investigated. The highly deformed rocks of the Iapetus suture zone contrast strongly with the thinner flat-lying strata of the Russian platform. These include unconsolidated blue clays of Cambrian age—surely the oldest sediments anywhere to have remained soft and unlithified. Thick sediments with interbedded volcanic rocks were deposited in many other areas around the continents of both Asia and Gondwanaland, where island arc systems marked subduction zones.

Sea level seems to have been rising through the earlier part of the Cambrian period but it was unusually low at the end of the Ordovician. The Cambrian transgressions mark the first appearance in many parts of the world of fossils with mineralized shells and skeletons, such as brachiopods, trilobites and archaeocycathids. Not all continental shelf areas were flooded simultaneously and the earliest Cambrian fossils found in different areas are therefore of different ages. The biostratigraphy of this time is currently undergoing increasing refinement with the aim of establishing a suitable level, recognizable worldwide, at which to define the beginning of the Cambrian period.

Cambrian climates appear to have been warmer than at present and there are no conclusive signs of glaciation. The extent of the Cambrian seas may have helped to maximize the effects of solar radiation, as did the chalk seas of Cretaceous time. Limestone deposition included the earliest skeletal reef structures although their organisms were very different from those of today. They were formed between 30°N and 30°S latitudes—a little wider than their present-day spread. During the Ordovician period reefs were similarly distributed in shallow tropical seas but some groups of marine organisms, including the trilobites and graptolites, show disjunct distributions suggesting some climatic control. Evidence of glaciation at the end of the period comes from North Africa (Figure 23.23). Melting of the ice early in the Silurian period produced a widespread raising of sea level and climates were restored to their earlier more equable distribution. Marine faunas are notably more cosmopolitan in the Silurian than the Ordovician. Silurian reefs again plot reassuringly within tropical latitudes while evaporites show a strong affinity for west coasts of continental masses, the location that would be expected to have maximum aridity.

An important economic consequence of the Silurian transgression was the spread of an extensive shallow shelf sea over the formerly glaciated area of the Sahara. The organic-rich sediments deposited by this sea form the source of many of the North African oil fields.

Early Palaeozoic periods and correlation

The Cambrian and Silurian systems were originally proposed as essentially lithostratigraphic entities, with inevitable confusion over their limits (see Chapter 1). Lapworth's proposal for an intervening Ordovician system broke new ground in introducing biostratigraphy to the definition of a system. From the outset Ordovician stratigraphy was based on the sequence of graptolite species ranges which Lapworth grouped into zones, and these organisms still provide the key to intercontinental correlation of both Ordovician and Silurian rocks. Other biostratigraphically useful organisms are the trilobites and brachiopods, and also the acritarchs, chitinozoa and conodonts (all microfossils of uncertain biological origin).

Lapworth unfortunately made a mistake among his many positive contributions to Lower Palaeozoic stratigraphy, and the century of controversy that has followed his error remains a fine example of the futility of argument over the definition of chronostratigraphic boundaries. Lapworth clearly stated that the base of the Ordovician system was to be at the base of the Arenig series, but in another publication in the same year (1879) he stated that Lower Arenig graptolites from South Wales were of the same age as Upper Tremadoc graptolites from North Wales, despite the fact that it had already been shown that Sedgwick's Tremadoc rocks were older than the Arenig series (and hence, by Lapworth's definition, Cambrian). The ensuing arguments centred less on what Lapworth had intended than on whether Tremadoc fossils are 'more Cambrian' or 'more Ordovician' in character. The Ordovician–Silurian boundary presents fewer problems, although the withdrawal of the seas from many areas at that time makes selection of a site of 'continuous' deposition that much less easy. The Silurian–Devonian boundary is happily fixed for all time as already described.

The base of the Cambrian system, and the beginning of the Cambrian period, have always had considerable curiosity value. This is the boundary between two eons, the Proterozoic and Phanerozoic. However, the beginning of the Cambrian is rapidly losing some of its mystique, and indeed its significance in Earth history, with increasing research of all kinds on older rocks and earlier times. It seems that no major tectonic or metamorphic events mark this boundary (see Chapter 16) and the appearance of fossilizable skeletal parts among diverse organisms, the event by which the boundary is recognized, was only one stage in a longer process of metazoan evolution.

The Proterozoic eon

The volume of research on Precambrian life and sedimentary environments has grown enormously in recent years, along with the improvements in understanding of overall crustal evolution through this time. Precambrian time is now usually subdivided into the Archaean and Proterozoic eons. Proterozoic time is at present given a geochronometric starting point of 2500 Ma ago based on a clustering of radiometric dates and hence of igneous and metamorphic events at about that time. It should soon acquire a marker point definition of its end when the base of the Cambrian is fixed. A subdivision of Proterozoic time at about 1600 Ma ago is commonly regarded as useful, and the latter half of the eon is divided by Soviet geologists into Riphean and Vendian. Lithostratigraphy and even biostratigraphy of the well-preserved Proterozoic sedimentary sequences of various parts of the Soviet Union (the Baltic Shield, the Urals and eastern Siberia) could perhaps ultimately form the basis for definition of Proterozoic systems. Correlation of these units on other continents is already possible to a limited extent using stromatolites (see Chapter 21). The Vendian unit is widely identifiable by its soft-bodied metazoan faunas and by quite diverse fossil microplankton (acritarchs). It has indeed been suggested that the Cambrian should be extended to include these strata on the grounds that the appearance of metazoans, implying development of sexual reproduction and hence the speeding up of evolutionary processes, was more significant than the development of hard parts.

Evidence for the geographical positions of the continents in Proterozoic time is almost non-existent. It is possibly consistent with the available data that the continents were assembled into a single supercontinent for much of this time (Figure 23.25) but other interpretations are possible (see Chapter 16). Evidence for climates is bound up with the question of the evolution of the atmosphere (see Figure 17.1). Carbonate sediments, in which the minerals calcite and dolomite predominate, are abundant from the early Proterozoic, but are very rare among Archaean rocks. In early Proterozoic rocks evidence of very simple life-forms becomes more abundant, and it is likely that oxygen was then being added to the atmosphere at the expense of carbon dioxide. Furthermore it is in the early Proterozoic that 'red-beds'—sediments containing sufficient iron in the oxidized state to give them a rusty tinge—first become at all common. Sulphate minerals such as gypsum and anhydrite are also first found in rocks of this age—free oxygen would have allowed the oxidation of sulphide ions to sulphate. The chemistry of Proterozoic sediments thus gives important clues to the existence of depositional environments such as shallow shelf seas with precipitation of calcium carbonate and its incorporation into algal frameworks, coastal lagoons with rapid evaporation to form dolomite, gypsum, anhydrite and even rock salt, and flood plains receiving material derived by oxidative weathering from erosional

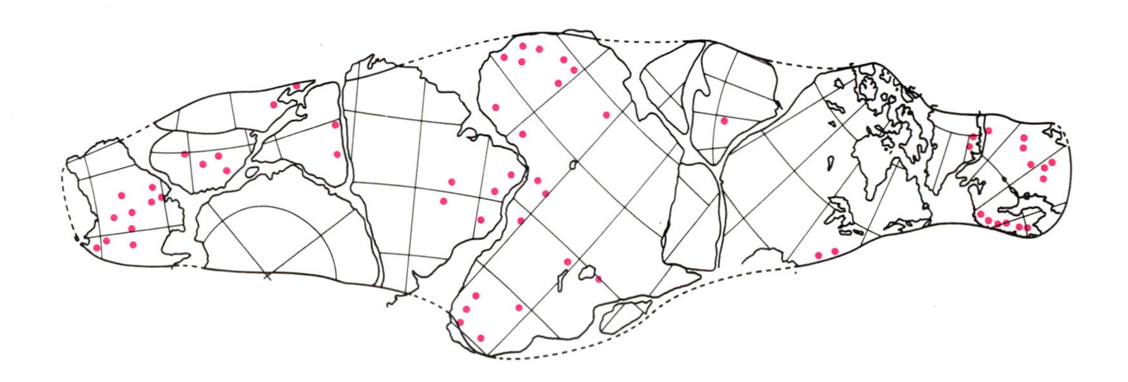

23.25: *Late Proterozoic 'supercontinent', according to one possible reconstruction, showing the wide spread of tillite localities.*

landscapes. These environments all indicate warm and sometimes arid climates.

Warm conditions did not, however, obtain throughout the Proterozoic eon, and there is evidence that repeated glaciation took place. Sediments of probable glacial origin underlie the earliest Cambrian deposits in all continents except Antarctica and the problem is to correlate them. For the youngest occurrences there is a good possibility of relative dating by means of microfossils and stromatolites, though these are not always available. For older occurrences radiometric dating must be used. The dates available suggest that there were three major late Proterozoic glacial events centring on about 615, 770 and 940 Ma ago—similar time intervals separate the late Ordovician, Permo-Carboniferous and late Cenozoic events. Some late Proterozoic tillites were apparently deposited within the tropics, but this could be due to rafting by floating ice. It has been suggested that the Earth's axis had a greater tilt than now giving extreme seasons even in low latitudes. There is also convincing evidence from Canada, South Africa, Australia and India of early Proterozoic glaciation some 2300 Ma ago.

The Archaean eon

There is all too little evidence of the state of the Earth's surface between its formation and about 2500 Ma ago—nearly half its history. The rocks tend to be metamorphosed, sediments are unfamiliar and the tectonic framework seems to have been rather different, perhaps very different (see Chapter 16). It does at least appear that water was present, that land and sea were differentiated and that processes of erosion and sedimentation were in action at the time of the earliest geological record. The continental crust may have been thinner and the global topography therefore less pronounced. However the evidence for the existence of life back to at least 3500 Ma argues for environments of no particularly extreme kind.

The most distinctive sediment type is quite unknown in younger strata, and this is the association of haematite, chert, dolomite and black shale in very thinly layered (perhaps varved) 'banded iron formations' (Figure 23.26). The extensive fine banding in these important sources of iron ore indicates remarkably even conditions of deposition. They were deposited from about 3200 Ma to 1800 Ma ago. Blue-green algae may have been partially responsible for their formation by providing the right chemical environment. Their association with volcanic rocks suggests a volcanic origin for the iron and other metals present, and the closest present-day analogy may be the metalliferous deposits of the Red Sea if the plate tectonic implications are ignored.

Cycles in Earth history

Stratigraphers enjoy looking for evidence of periodicity, or cyclicity, in the events and processes of Earth history. Such cycles are not immediately obvious from the stratigraphic record and the irregular spacing in time of the system/period boundaries is evidently based on random and unique events rather than on truly cyclic revolutions. Nevertheless periodic influences are observable in the record, some predictable, and some for which no explanations are readily available (see Figure 23.28).

Studies of growth bands in bivalve shells and corals (see Figure

23.26: *Banded iron formation, possibly a kind of seasonally varved sediment, from the Archaean of southern India.*

23.27: *Rainprints recorded in sediments of Archaean age from southern India.*

6.11) have succeeded in proving that the number of days in a year has decreased from about 400 in Devonian times, confirming predictions by astronomers that the Earth's rate of rotation is slowing down. Furthermore the lunar month in the Devonian was 30.6 days long, compared with 29.5 days at present. The possible influences on climate of various astronomical cycles are mentioned in Chapter 18. Detailed studies of oxygen isotope (and hence palaeotemperature) variation in Quaternary deep sea sediment cores show a very clear cyclicity with a period of about 100 000 years. Complex filtering techniques suggest that 21 000- and 41 000-year cycles may also be detectable. In older sediments involving the rhythmic interbedding of limestone and shale it has been observed that the relative proportions of the two rock types tend to vary cyclically such as to group the limestone–shale couplets into bundles of five, suggesting a connection with the fact that there are approximately five precessional cycles to every eccentricity cycle.

A. G. Fischer of Princeton University has suggested that Earth history since the late Precambrian shows two cycles of some 300 Ma each, in which alternating low and high sea level and continental topography were caused by variations in sea floor spreading activity, which in turn depended on variable mantle convection. The resultant variations in volcanism affected the balance of carbon dioxide in the atmosphere, and the Earth has thus oscillated between an 'icehouse' and a 'greenhouse' state, the latter being represented by climatic optima of the Ordovician to Devonian and the Jurassic to Cretaceous periods. The transitional intervals include some of the major extinction events. Organic diversity and community structure supposedly show a 32 Ma periodicity between 'polytaxic' diversity peaks with particularly exotic faunas and 'oligotaxic' intervals of reduced diversity.

On a still longer time scale is a hypothesis of major changes in the Earth's obliquity over a 1250 Ma cycle. If the obliquity were increased the seasons would become much more pronounced over the world and the poles would receive more solar radiation than the equator. This might explain the predominantly low palaeolatitudes of late Proterozoic glacial sediments and their frequent association with sediments characteristic of warmer climates. At the other extreme the occurrence of abundant fossil plants at polar palaeolatitudes in the Cretaceous and early Cenozoic is easier to explain if there were no seasonal variation in daylength, due to near zero obliquity.

The problem, and at the same time the joy, of hypotheses of cyclicity as well as of cosmic catastrophes such as cometary impacts, is that the stratigraphic evidence is rarely quite sufficiently refined to refute them while its essentially circumstantial nature allows free rein to the imaginative and comfort to the fundamentalist.

23.28: *Periods and periodicities interpreted from the geological record.*

1 250 Ma	Interval between zero obliquity events
300 Ma	Length of two Phanerozoic 'supercycles'
32 Ma	Cycles of taxonomic diversity
100 000 years	Orbital eccentricity cycle
41 000 years	Obliquity cycle
21 000 years	Precession of equinoxes
300 years	Age of fossil tree
5 years	Lifespan of fossil belemnite
1 year	Varves, tree rings; growth bands of fossils
lunar month	Growth bands of fossils
few days	Deposition of volcanic ash
1 day	Growth bands of corals, bivalves
6 hours	Tidal cycle; growth bands of bivalves
few hours	Lava flow; deposition of turbidite
few minutes	Mudflow
few seconds	Duration of earthquake
fraction of second	Formation of rainprint; lightning strike

Part Five: Evaluation of Earth Resources and Hazards

Man interacts with the surface of the Earth in an immense range of ways, taking the Earth sciences towards the realms of sociology, economics and even politics. This part of the Encyclopedia deals with a selection of topics chosen to represent this diversity of application of the Earth sciences.

The age of electronics has brought some quite astonishing improvements in man's ability to observe things at a distance and to make measurements of hidden objects. Three quite different variations on this theme are remote-sensing (particularly from satellites), seismic reflection profiling, and borehole logging. All three use sophisticated computer techniques and all three are of great importance in the increasingly difficult search for new resources of raw materials for an ever hungrier world.

The activities of the majority of geologists are controlled by the prices of certain raw materials on world markets. A fall in price of a key commodity can put hundreds of scientists out of work. Yet it is geology that helps to determine those prices, by determining the abundance and concentration of minerals, metals and hydrocarbons and their geographical (which can mean political) distribution. Geology is not the only factor in defining a resource, though, and the interplay of need and greed makes a study in itself.

Man's opinion of himself is all too easily brought down to Earth, frequently with more than a bump, when the Earth, active as always, lets us know who is really in charge. A lack of interest in the length of time a seismic wave will take to reach the other side of the Earth is understandable when it has already flattened your house, and few people would bother to find out the chemical composition of volcanic ash that has smothered livestock and crops. But scientific study of these hazards is critical to understanding them, and proper understanding is the only way to minimize the ravages of future episodes.

Climax Mine, Colorado, mining molybdenum for use in steel production.

24 Remote sensing and other techniques

Remote sensing

Collecting data about objects without direct contact is the essence of remote sensing, and the term is most commonly applied to observations of the Earth's surface from aircraft or satellites. By convention remote sensing refers only to techniques that use electromagnetic (EM) radiation, and excludes methods that measure force fields such as magnetism from a distance.

The first experiments in remote sensing were carried out in the 1850s by Aimé Laussedat, a French Army engineer, who used cameras mounted on kites and balloons. Such aerial photography was first applied practically for military intelligence purposes during the American Civil War (1861–5), and every subsequent development in remote sensing has stemmed directly from military applications. Infrared photography was devised to distinguish between live plants, which reflect infrared radiation, and infrared-absorbing camouflage and dead vegetation. Thermal imagery, which measures emitted infrared, stemmed from night-time bombing reconnaissance. Airborne radar scanning was first used to map surface military installations at night or through cloud cover, while satellite remote sensing was developed for rapid monitoring of missile sites. In every case, however, numerous non-military applications were subsequently devised. Remote sensing is now an important synoptic tool for the geologist, geographer, meteorologist and environmental scientist. One of its major advantages is the wide field of view, and the higher the viewing platform, the wider the view. Thus large features of the Earth's surface, which would otherwise have to be mapped painstakingly by field workers, are immediately apparent and require only location and verification by field studies. Moreover, the speed and relative cheapness of remote sensing enables repeated surveys to show up subtle changes in the Earth's surface over a period of time.

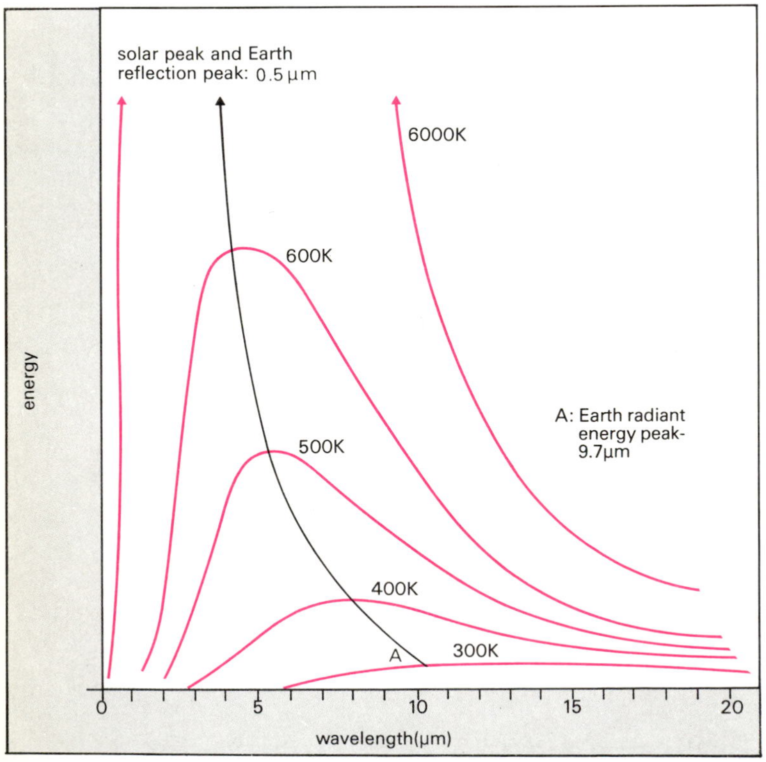

The electromagnetic spectrum and natural materials

The source of EM waves is the movement of charged particles, as in the production of radio waves by alternating currents flowing in a transmitter. More important, however, is the generation of EM radiation by thermal vibration of particles, as expressed by the temperature of a body. The concept of a black body is fundamental to understanding the interaction between temperature, emission of EM radiation and its absorption and reflection by various materials. A black body is a hypothetical material that absorbs all EM radiation falling on it, and also radiates the maximum amount of energy. Its

24.1: *Energy and EM spectrum radiated at different temperatures. According to the Stefan-Boltzmann law the radiant flux of a black body, F, at a temperature T (in K where OK is $-273°C$) is $F = \sigma T^4$, where σ is a constant. The black body and its opposite are abstract concepts. In reality all surfaces fall somewhere between the two such that $F = \varepsilon\sigma T^4$, where ε is the emissivity of the surface. Emissivity is the ratio between the radiant flux of the real body and that of an abstract black body. The effect of the principal law of black body radiation is that, if the temperature of a body is doubled, the energy that it emits as EM radiation goes up by two to the power four, or sixteen times. The emissivity of a surface is different for different EM wavelengths, so that when we measure the energy levels in the spectrum emitted by a heated body the result is a curve.*

converse, a perfect reflector, has the opposite properties (Figures 24.1 and 24.2).

When EM energy strikes matter various interactions are possible, depending on the nature of the surface, the structure and composition of the matter and the EM wavelength. The radiation may be transmitted through the substance. It may be scattered in all directions as diffuse radiation ultimately to be lost by absorption (common in the atmosphere). Finally it may be reflected unchanged from the surface of the substance. These interactions form the basis of remote sensing (Figure 24.3).

The EM radiation that is reflected, emitted or scattered by an object is detected by the relevant remote sensing system and conventionally displayed as an image. The intensity of the radiation is generally portrayed as variations in brightness or tone (shades of grey). Different adjacent tones give different contrasts on the image. Variations in the type of surface give rise to different textures and patterns.

Remote sensors

The waveband between long wavelength ultraviolet and short wavelength infrared is detectable using conventional photography. For visible light ordinary panchromatic black-and-white film or natural colour film is used in conjunction with a filter to eliminate the blue end of the spectrum, which is dominated by the scattering effect of the atmosphere producing haze. Infrared, of short wavelength, can be detected on ordinary film using a red filter or on infrared-sensitive black-and-white film. Coloured photographs can be produced using special colour infrared film.

A more useful technique splits the sensed EM radiation into different wavebands. This can be achieved on film using filters, but is usually done using optical-mechanical systems called multispectral scanners (MSS). These comprise a mirror which sweeps a swathe of ground beneath the aircraft or satellite. The radiation received from the ground is split into discrete wavebands by a system of prisms. Each waveband is detected by photocells, and its intensity is recorded on multichannel magnetic tape. The information is therefore in the form of strips which are reconstructed to a photographic image by computer (Figure 24.4). A similar scanner is the only means of detecting long-wavelength infrared emitted by surfaces.

The systems outlined so far record naturally reflected or emitted EM radiation and are therefore passive systems. An active remote-sensing system supplies its own source of energy, as in flash photography. In this category is radar, where the target is 'illuminated' by microwaves generated aboard the platform. The receiving system displays the returning radiation visually and records it on magnetic tape.

Aerial photography

Photographs taken from aircraft may appear as conventional scenes, taken from the ports or nose and incorporating the horizon. Such photographs show the terrain in perspective and may be visually

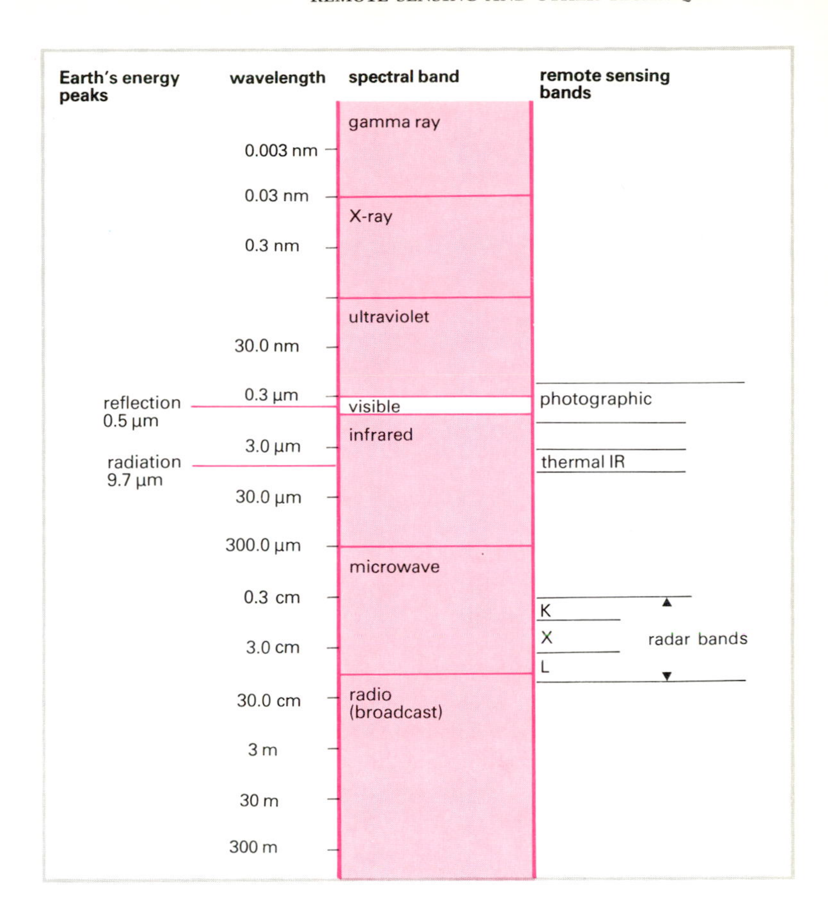

24.2: *The EM spectrum showing wavebands used in remote sensing. The Sun, whose surface temperature is 6000 K, emits most energy at an EM wavelength of 0.5 µm. This corresponds to green light, but it also emits high energies from ultraviolet to infrared. Roughly the same pattern is reflected by the Earth into space during daytime. The energy emitted by the Earth at night (surface temperature about 290 K) has a peak at 9.7 µm in the infrared part of the spectrum.*

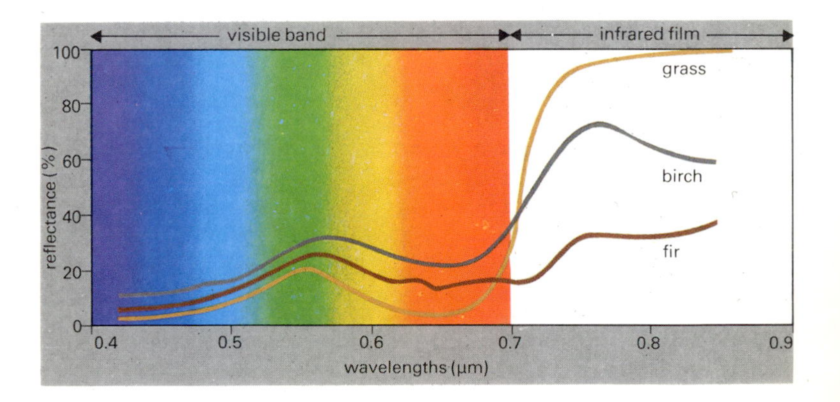

24.3: *Reflectance of different EM wavelengths by typical vegetation.*

appealing, but they are difficult to interpret and convert to map format because of the varying scale. The more usual procedure is to photograph vertically downwards from the aircraft. The cameras used are designed to expose frames at a predetermined rate as the aircraft flies along, so as to give a complete record. If the frames overlap by 50 per cent, so that all points occur on three successive frames, it is possible to view the overlapping photographs in three dimensions, or stereoscopically. A special viewer is used, so that the left eye views the left photograph of an overlapping pair and the right eye views the right photograph. This has the apparent effect of widening the eye base to several hundred metres and considerably exaggerating normal stereoscopic vision. Consequently, relatively insignificant topographic features stand out very boldly on the paired photographs.

On a vertical aerial photograph a point directly below the aircraft is at the centre of the frame and is the only point that appears in true plan view. All other tall features, such as trees, mountains and buildings, are distorted so that their tops are displaced from their bases in a direction radially outwards from the frame centre.

The clearest natural features to show on all remotely sensed images are those relating to topography. They include drainage patterns, ridges, escarpments, lakes and shorelines. As topography is often closely controlled by the underlying geology the first step in geological interpretation is to look for prominent topographic patterns. The most easily seen patterns are those involving simple geometrical shapes such as circles or lines (Figure 24.5). In many cases straight lines joining or displacing topographic features are due to features of geological importance such as vertical strata, igneous dykes or vertical faults that have displaced rocks on either side. Circular features may indicate volcanic calderas or the sites of large meteorite impacts. Sinusoidal features are the signatures of large-scale folds. Other easily recognized shapes relate to constructional landforms such as sand dunes, river terraces, glacial moraines, deltas and lava flows. The shapes of destructional or erosional landforms often tell us much about the underlying geology since they reflect the relative resistance to erosion of the rocks, an important clue to rock type.

It is often quite simple to distinguish between outcrops of rock and their cover of unconsolidated sediments or vegetation. The last two generally have a smooth appearance or texture, whereas surface rock is usually rough because erosion has picked out joints and layering on a small scale. Sedimentary, metamorphic and some volcanic rocks frequently show banded patterns reflecting their structure whereas intrusive rocks appear homogeneous and are transected by numerous joints.

Stereoscopic viewing of aerial photographs means that rock structure can be evaluated, the relative proportions of scarps and dip slopes giving direction and amounts of dip. Mutual relations between rock units, such as evidence of igneous intrusion or unconformity, enable preliminary geological histories to be built as hypothetical models to be tested by detailed ground survey.

The human eye can discriminate many more shades of colour than levels of grey tone, and consequently colour aerial photographs (Figure 24.6) ought to be more readily interpreted than black-and-white. However, for many geological purposes Earth scientists prefer to use black-and-white photographs, partly because they have long been available for most parts of the Earth and their use is familiar, but also because colour photographs may show too much! Remote sensing is most often used as a synoptic tool, and it is easier to construct a simple set of terrain categories with black-and-white than with true colour. With colour, however, it is possible to 'slice' different wavebands using filters during photography and to pick out selected targets for more precise classifications.

By combining black-and-white film with a special filter that cuts out all wavelengths shorter than 0.7 micrometre (μm) it is possible to produce photographs of reflected near infrared radiation. This has some distinct advantages. First, the elimination of atmospheric scattering present in visible light produces much sharper images. Second, vegetation differences show up more clearly. Finally, and most im-

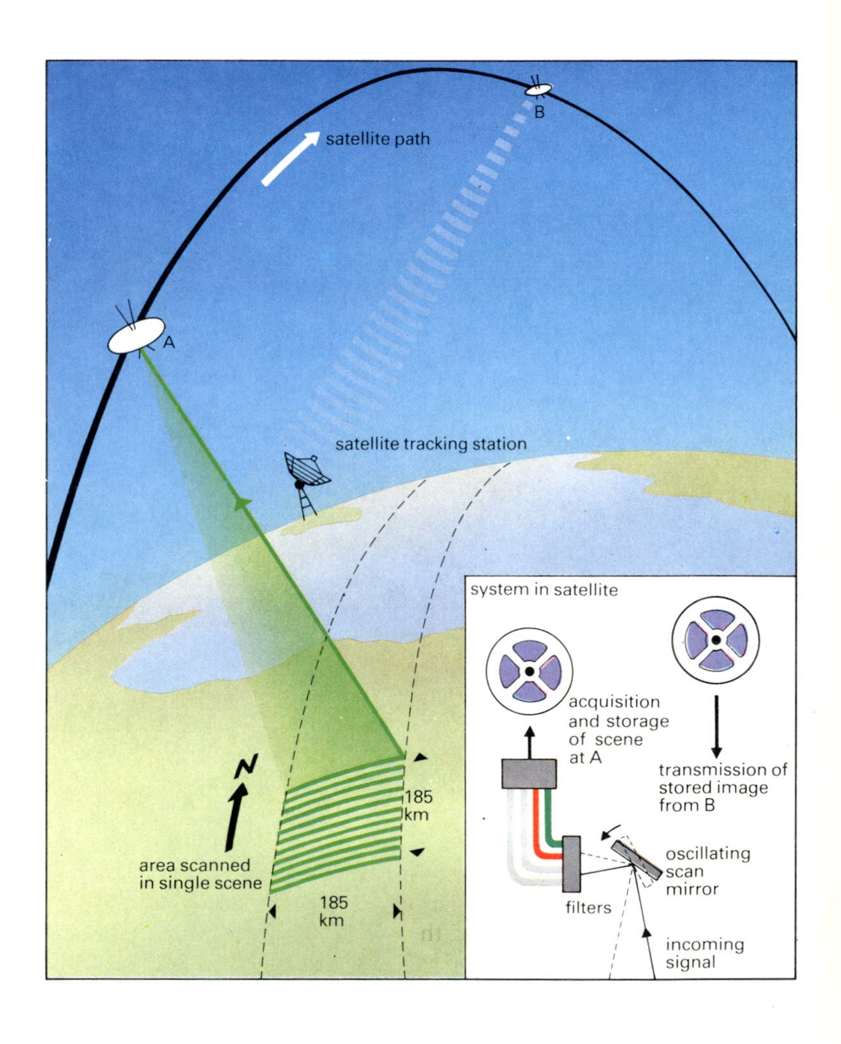

24.4: *The basic components of the Landsat multispectral scanner system. At A the oscillating scan mirror sweeps a patch of ground beneath the satellite. The radiation is filtered into red, green and infrared wavebands, whose intensities are measured and stored on magnetic tape. At B the satellite beams the photographic image reconstructured by the computer to a receiver at a tracking station.*

portantly, water strongly absorbs infrared and appears jet black; in this way water courses and the boundaries between water and land are very clear.

Infrared-sensitive colour film was originally designed for camouflage detection, but is now a very useful addition to the arsenal of remote-sensing techniques. Compared with ordinary colour film the colour-forming layers are sensitive to longer wavelengths, so that blue corresponds to green light, green to red and red to infrared. When developed the photographs show unusual false colours. In particular, different vegetation types are contrasted, so that broad-leaved plants appear red, conifers brown to purple and diseased plants dark red to blue. Soil and rock surfaces appear in various shades of blue if water-saturated and in yellows and browns if dry. Artificial surfaces such as roads and buildings often appear blue. A very important use of false-colour infrared photography is in agricultural and forestry surveys of large areas to check for disease and pest infestation, but the same colour rendering is used in Landsat images (see below) for geological interpretation at much smaller scales.

Thermal infrared images

EM radiation of the wavelengths and energies detected by photography can only be emitted by stars. For the normal range of surface temperatures on the Earth, emitted radiation is largely in the far infrared, at wavelengths longer than 3 μm. Because the atmosphere absorbs some wavelengths, only the 3 to 5 and 8 to 14 μm wavebands can be monitored from aircraft and only the 10.5 to 12.5 μm waveband from satellites. In both cases line-scanning devices must be employed.

Three thermal aspects of the Earth can be monitored by thermal infrared surveys: solar energy absorbed and re-emitted by the surface, geothermal energy emitted in areas of volcanic or hot-spring activity and thermal energy generated by human constructions such as power stations.

Because of the swamping effect of reflected solar radiation (Figure 24.7) thermal infrared surveys are usually conducted at night. When normal terrain is examined thermal images have a markedly different appearance compared with other images. Broadly

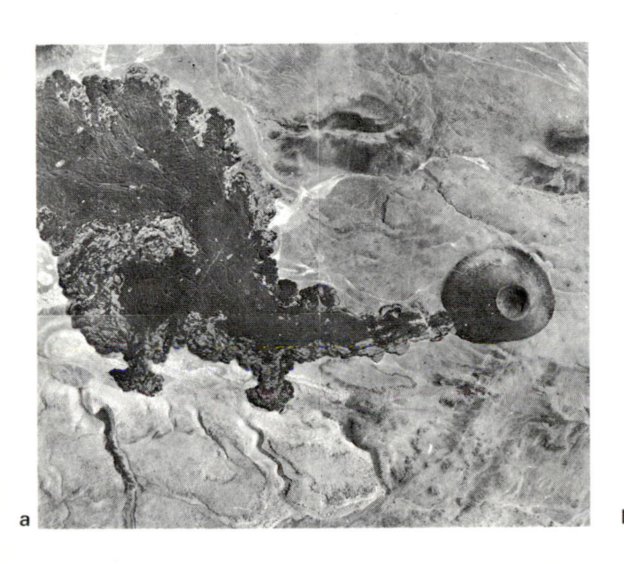
a

b

c

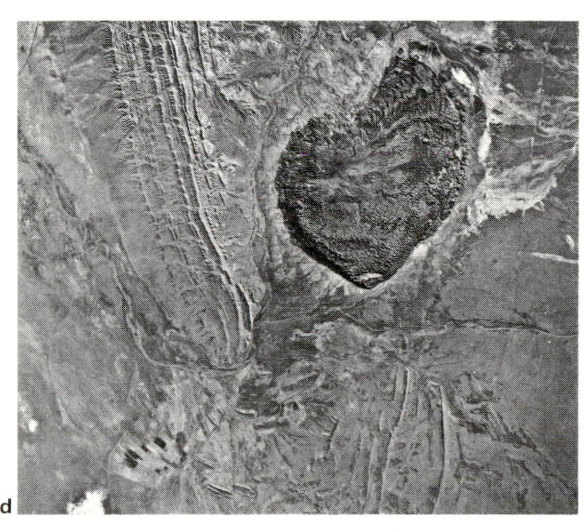
d

24.5: (**a**) *A conical volcano with circular crater from the base of which a lava flow has been erupted on which can be seen flow ridges, levées and secondary flows.* (**b**) *An area of uniformly dipping sedimentary strata in which horizons of different resistances have been picked out by erosion as ridges and valleys. Faulting has been accompanied by folding as shown by the curvature of ridges and valleys.* (**c**) *A very large anticlinal fold in an arid area underlain by well-bedded sediments.* (**d**) *This very clear example of an intrusion is in fact not igneous, the intrusion being low-density rock salt that has risen by gravitational means into more dense sediments, whose bedding it clearly truncates.*

a

b

24.6: (**a**) *A true-colour aerial photograph of farmland and developed forestry in limestone uplands.* (**b**) *A false-colour infrared photograph of the same scene as (a). The bluish tinge of the newly ploughed fields indicates that the soil is wet. Water courses show up strongly as deep blue-black. Vegetation of different types and stages of growth is more clearly distinguished than in the true-colour photograph (a).*

speaking the most absorbent surfaces with the greatest thermal inertia are the strongest night-time emitters of thermal infrared, and vice versa. On a pre-dawn image of a rocky area, shales, mudstones and limestone appear cool whereas basalts, sandstones and granites have warm signatures. The seemingly obvious prediction that dark rocks will absorb and therefore re-emit most energy is not in fact true. A rock's ability to absorb heat is no more significant than its capacity to stay warm, which is a function of density, shale, mudstone and limestone being of low density.

By far the most spectacular thermal images are those associated with volcanoes. Not only do active vents and recent lava flows show up prominently but, more importantly, the thermal structure is brought out very clearly (Figure 24.8).

Side-looking airborne radar (SLAR)

Producing images in the microwave area of the EM spectrum relies on an active system. The platform emits pulses of microwaves, ranging from a metre to a millimetre in wavelength, towards the target area. On an aircraft this is to one side. Since microwaves can penetrate cloud and haze the active mode of the sensor is independent of daylight and weather.

The properties of materials that control their radar signatures are their dielectric constant (their insulating quality), which increases with moisture content, and their surface roughness. The more moist the surface rock or soil, the less penetration by radar, and the more reflection results. The roughness of a surface depends on the height of irregularities in relation to wavelength of the microwaves. Clearly a surface with irregularities smaller than the radar wavelengths will be smooth. Smooth surfaces reflect microwaves and, unless the surface is perpendicular to the wave paths, little reflected radiation returns to the receiver and a dark image results. Rough surfaces randomly scatter the radiation, a proportion of which is picked up by the receiver; thus rough surfaces look bright. Since the SLAR system is active high mountains block transmitted pulses and therefore have quite dark shadows (Figure 24.9). Dense and irregular vegetation has a shadowing effect too and produces speckled textures. Smooth surfaces such as calm water, dried lake beds, highways and buildings are like mirrors to radar and generally look dark. Smooth walls perpendicular to the wave paths show clearly as bright networks of lines.

The highlighting of topography makes SLAR an important tool in mapping, and it tends to pronounce alignments related to features such as faults. However, as there is no direct relationship, as there is with other methods, between SLAR images and rock type it is most commonly used as a means of structure and landscape analysis.

Satellite remote sensing: the Landsat system

Although there are many satellites now in orbit supplying information about the Earth for military, meteorological and other uses, only three are of immediate use to the Earth scientist. One of these is Seasat, which is operated only over oceans and acquires SLAR images for determining the state of the sea and distribution of pack ice and icebergs. It also records over narrow strips of coastline. The second is the currently functioning member of the Landsat series (Landsat C), which gathers multispectral information by the line scan method. Landsat products are of immense importance for Earth scientists, although their usefulness is to some extent fortuitous since they were initially developed for the strategic monitoring of world agriculture. The third satellite was the manned Skylab, now destroyed and unlikely to be replaced for several years. There are plans to launch further Landsat missions, a successor to Skylab, the Space Shuttle and a French system with the acronym SPOT.

Landsat C moves in an almost perfectly circular, near polar orbit 917 km above the Earth. It is Sun-synchronous, crossing the equator in daylight fourteen times per day at approximately 9.30 am local time. The progressive shift in the orbit is so arranged that a completely overlapping view of the entire surface can be obtained every eighteen days. Three imaging systems are aboard Landsat C, one based on television or return beam vidicon (RBV), another for thermal emission images and a third, the most useful, an MSS. The MSS system records the 'view' in 58 × 79 m bits, or pixels, along each line scan, as digital signals on magnetic tape. This is played back to several ground receiving stations where computers reassemble the data in photographic form. Each image corresponds to a parallelogram with 185 km sides on the ground (Figure 24.4), is printed at 1:1 000 000 scale and is as scale-accurate all over as conventional maps at 1:250 000 scale.

Four wavebands, corresponding to green, red and two short-wavelength infrared bands, are recorded. Any of these can be viewed as black-and-white images, or three can be used to produce false colour composite images by printing green through a blue filter, red through green and infrared through red. The result has exactly the same colour signature for different surface materials as colour infrared photography (Figure 24.6(b)). Interpretation combines these signatures with examination of tone, textures and patterns, as with ordinary aerial photographs (Figure 24.5). One great advantage of Landsat images is the synoptic picture of large areas, so that very large geological structures (Figure 24.10) such as lineaments, folds, igneous intrusions and broad geomorphological features can be quickly appreciated and related to ground or aerial observations.

Because each MSS image is contained in digital form on magnetic tape much information can be extracted from it by means of computer techniques. With photographs there is a limit to how much manipula-

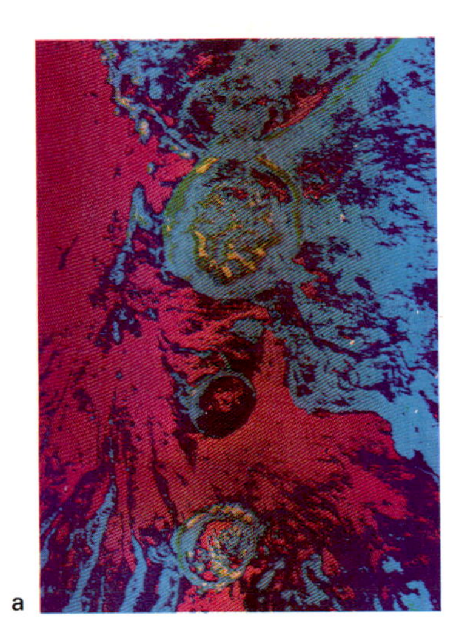

a

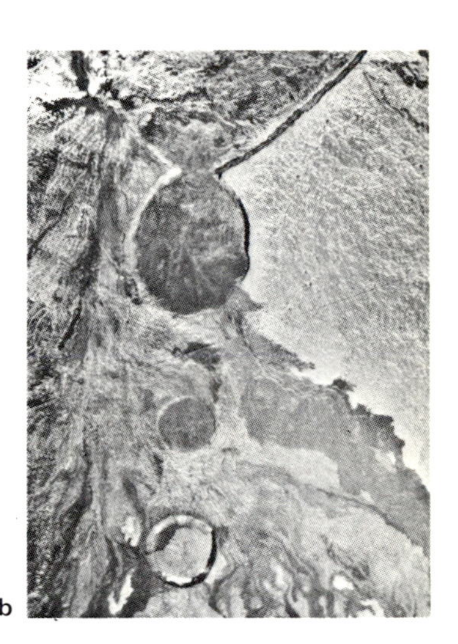

b

24.8: *In this density-sliced thermal infrared linescan image of the volcano Mauna Loa, on Hawaii* (**a**)*, comparison with a black-and-white aerial photomosaic* (**b**) *reveals a complex thermal structure. The crater interiors themselves are relatively cool (blue) except for hot spots (pink and yellow), whereas the main areas of high thermal emission are at the elliptical caldera fault scarps. The caldera margins and lava flows appear hot because they expose dense basalts with high thermal inertia, the craters containing vesicular material. The hot spots in the craters are directly related to volcanic activity.*

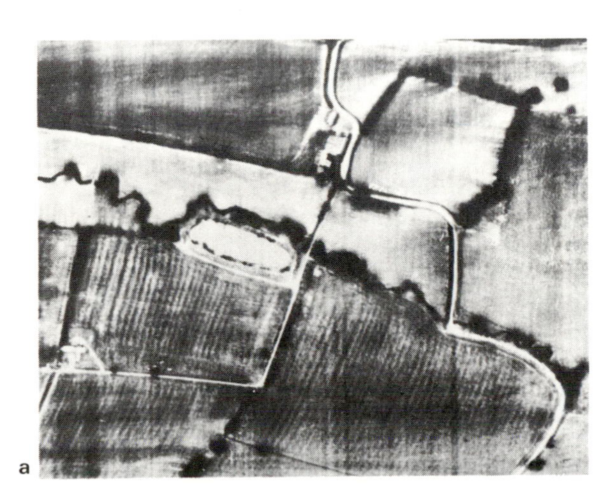

a

b

24.7: *Thermal infrared linescan imagery of agricultural terrain taken in the afternoon* (**a**) *and taken after sunset* (**b**)*. There are some clear differences between the two images. The stream is clearly warmer than its surroundings after sunset, as are trees. A field, which clearly reflects infrared during daylight, has a cool, dark signature at night but is speckled with bright dots that are in fact ruminating cows! (scale 1:1000).*

tion of contrast can be achieved. Computers can produce any kind of contrast imaginable. Subtle contrast changes that are normally difficult to see can be stretched so that they are sharply differentiated. Where broad, continuous variations in tone mask important features, the tone densities can be arbitrarily sliced into precise categories and reprinted as different shades of grey, different colours or with contour lines to make visual examination much easier. Various mathematical filtering techniques can be used to suppress image features of non-geological interest and leave only those that indicate faults, joints and bedding.

As well as the image enhancement described above, computers can be used to extract information from MSS magnetic tapes and use their decision-making capacity to produce entirely artificial images. Each pixel in one waveband is exactly registered with its corresponding pixel on the other three. To each is assigned a number indicating the reflectance of the terrain in a particular waveband. Consequently it is possible to classify terrain not only by spectral reflectance, but also by the ratios between reflectances in different wavebands, and thereby to produce ratio images in twelve different combinations between the four wavebands. These images can then be composited in threes to produce yet another series of false colour images. By combining contrast-stretching, density-slicing, ratioing and compositing, geologically, hydrologically or agriculturally known test areas on the ground can be identified for the computer. In turn, the computer will find all similar areas in the image and produce a thematic map which directly gives preclassified information (Figure 24.11).

Some economic and social applications of remote sensing

It should be made clear that remote sensing has (at the time of writing) neither a mine nor an oil well directly to its credit, nor is it likely to be 'successful' in this respect in the future. However, this apparent failure overlooks the real usefulness of the various techniques outlined here. Remote sensing and its interpretation are a rapid means of assessing the potential of large areas with a low financial outlay per unit area covered. It is a reconnaissance tool that can delimit target areas for follow-up surveys by more detailed and costly ground-based techniques. It can save money by attaching low priority to that vast proportion of the Earth's crust that has little economic potential. A valuable discovery results not from a single exploration method, but from a combination of many methods. In certain economic applications, however, such as geothermal and hydrogeological surveys, remote sensing can be the most important factor in delimiting the extent of a resource.

24.9: *This spectacular radar image of the San Cristobal volcanic complex in Venezuela shows the marked highlighting of topography and the clarity of the SLAR technique. The speckled areas on the lower slopes are forests. Roads, towns and even small walls stand out very sharply.*

In mineral and oil exploration aerial photographs and Landsat images have proved valuable in providing new data and basemaps for surface geological mapping, which will continue to be their primary use. It has also been known for a long time that mineralized areas and oil fields are associated with distinctive, large structures which show up very clearly on small-scale images. Of particular interest is the common association of important mines along, and at the intersections of, local fractures and regional lineaments hundreds of kilometres long. Direct examination and digital processing of Landsat images reveal very rapidly the vast majority of these straight alignments.

Many ore bodies are associated with some form of alteration of local rocks by water-rich solutions. The secondary minerals in such altered zones are often very distinctive in reflective properties and hence appearance, exemplified by the bright red, iron-rich gossans formed by weathering of sulphide minerals. Other distinctive signals occur at wavelengths beyond the sensitivity range of human vision. A combination of contrast stretching, band ratioing and colour compositing of multispectral images, such as Landsat, enables such altered zones to be pinpointed precisely. Unfortunately the present Landsat system, because it was designed for agricultural surveillance, does not incorporate the most useful wavelengths (1.6 and 2.2 μm infrared) for geological exploration, but this will be incorporated in the fourth Landsat payload.

The environmental applications of remote sensing are much more diverse. Infrared and radar images, particularly from satellites, enable the charting of many elusive aspects of lakes, seas and oceans, such as wave patterns, currents, pack ice, oil spills and thermal pollution. Meteorological satellites, using visible and infrared wavebands and relaying data from unmanned sensors, are revolutionizing weather forecasting. The monitoring of vegetation by a variety of imaging systems enables disease, pests and undernourishment to be rapidly identified and remedied, and allows much better resource management than formerly possible. Proper classification and monitoring of land use have only become possible with the development of remote sensing, and will revolutionize national economic planning.

In the context of disasters the wide field of view and the rapidity and repetitiveness of remote sensing provide invaluable assistance to the continual search for means of predicting, minimizing and relieving human catastrophes. In this respect the satellite programme has proved most valuable by identifying, for instance, incipient storm centres and fault lines prone to earthquakes and by rapidly outlining areas of damage. It will be improved by the addition of radar imagery to give it an all-weather capacity. Its ability to relay information from unmanned seismographs, tilt meters, strain gauges, flood and rain gauges and thermal sensors will greatly aid prediction of many types of natural disaster.

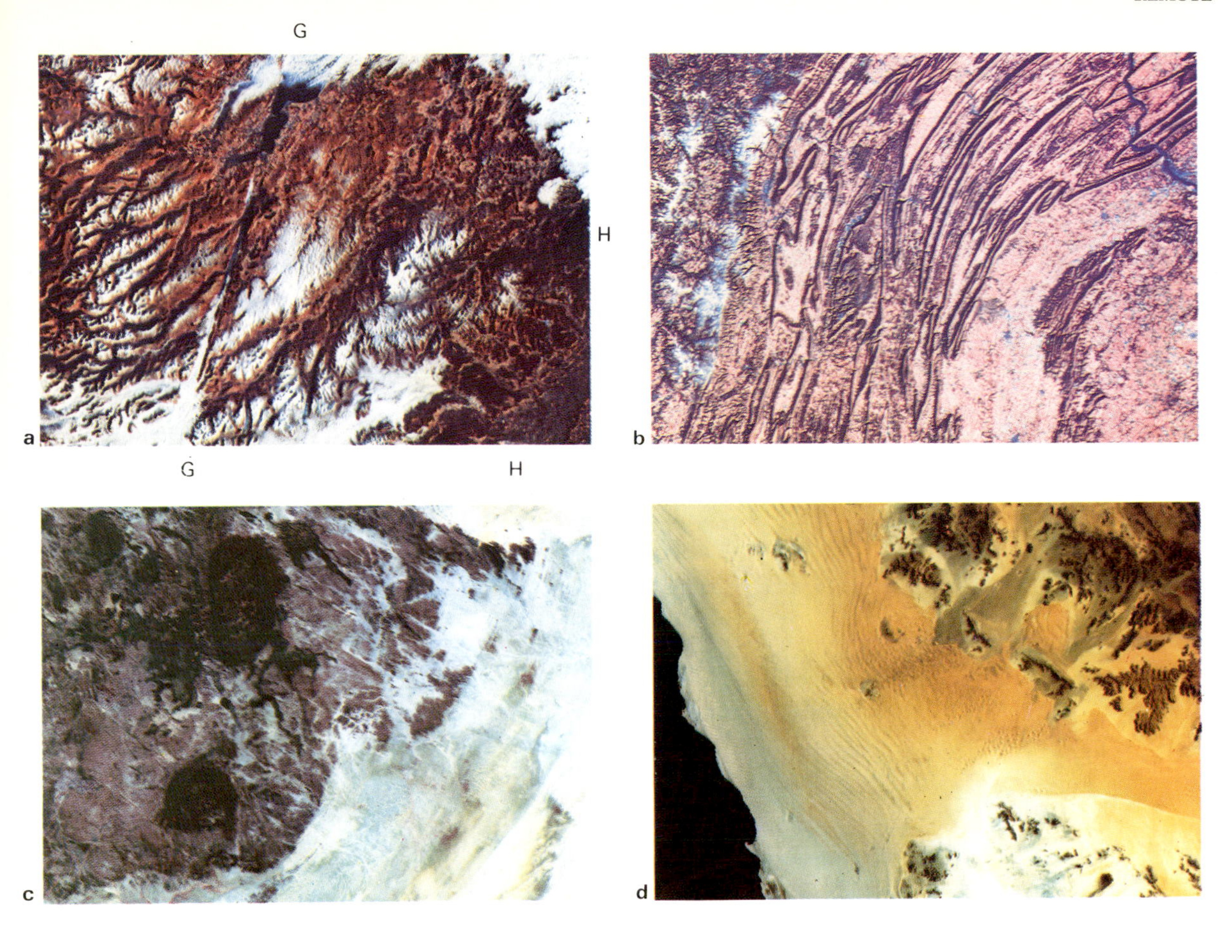

24.10: (**a**) *False-colour Landsat imagery of the Scottish Highlands shows the Great Glen Fault* (G–G)*, a sinistral wrench fault, and the Highland Boundary Fault* (H–H)*, which has a downthrow of several kilometres to the south. The valleys in the south-west are filled by morning mist.* (**b**) *The folded sediments of the Appalachian Valley and Ridge province are sharply defined by sandstone units that resist erosion to form ridges.* (**c**) *The elliptical dark-grey masses are late Palaeozoic to Mesozoic tin-bearing granites in Niger. They intrude a Precambrian igneous and metamorphic complex in which several marked lineaments can be seen. The whole yellow area is a sandy desert virtually devoid of vegetation which would have a red signature.* (**d**) *The Namib Desert of Namibia is characterized by huge linear dunes aligned perpendicular to the south-westerly trade winds. Sediment supply to the desert is from erosion of the rugged mountains in the north-east of the image.*

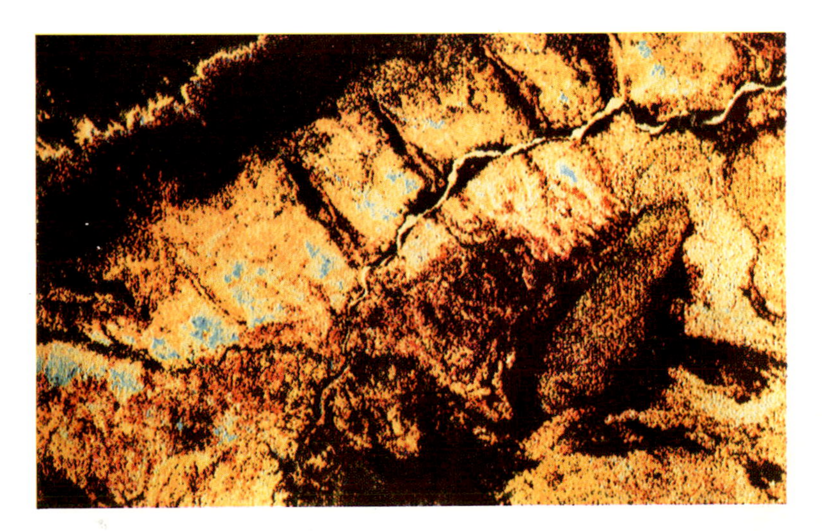

24.11: *Various types of digitally processed Landsat images can be used to highlight selected features. Left, a computer-processed colour composite of north-west Saudi Arabia made from ratio images of bands 4/5, 5/6 and 6/7 has been specially designed to discriminate different rock types such as blue-green granites from multicoloured sediments and volcanics. Right, colour ratio compositing of the uranium district of Arizona has been used to pick out zones of surface altered rock, which relate to uranium enrichment in blue.*

Seismic reflection profiling

A geologist's ideal world would have a transparent Earth so that it would be possible to look down at the rocks below the surface. Unfortunately the Earth is not transparent, at least to light waves, but it is transparent to another kind of wave, ie sound, or seismic, waves. Seismic reflection profiling is a method of remote sensing in which the Earth is probed with seismic waves. It is useful not only as a method for finding out more about the Earth's crust, but also has great economic importance, as it is the main exploration method for oil and gas.

Seismic reflection profiling is an offspring of earthquake seismology, but instead of seismic waves generated by earthquakes other sources of sound energy are used. There is no difference in principle between the seismic waves generated by earthquakes and those produced by the sources used for seismic profiling. Earthquakes, however, release an enormous amount of seismic energy, whereas the seismic energy sources used in profiling are far less powerful. Such seismic waves cannot be detected on the other side of the Earth as can earthquakes, and seismic reflection methods have only a small distance between the energy source and the recording points.

The basic technique of seismic reflection consists of generating seismic waves and recording the arrival of the waves at a series of detectors, usually arranged along a straight line from the source (Figure 24.12). The waves travel downwards from the source, are reflected back to the surface from the boundaries between rock layers and are recorded by the detectors.

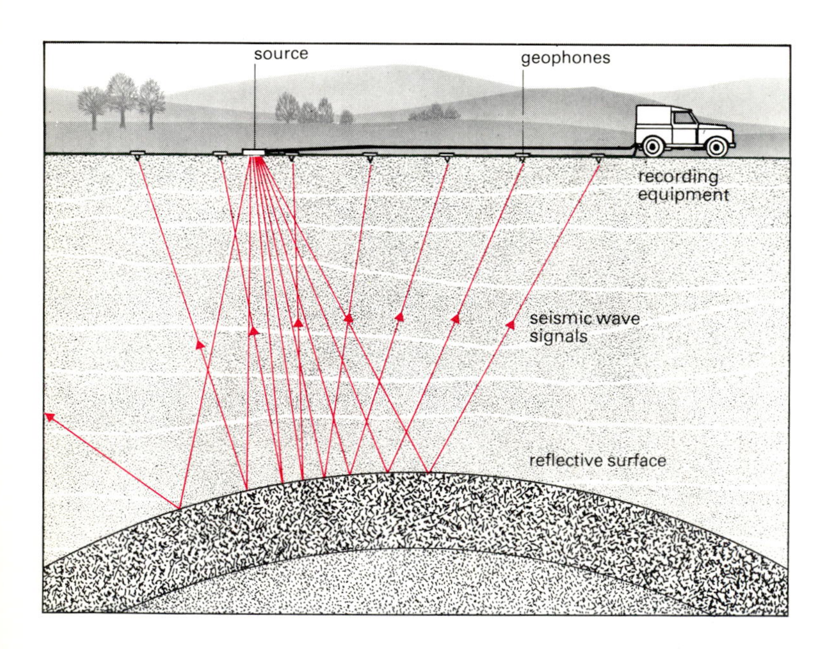

24.12: *The principal components of a seismic reflection profiling system on land.*

Seismic sources

On land, chemical explosives are the most widely used seismic energy source. Charges of between 2 and 25 kg are placed in shot holes drilled to depths of up to 100 m. The hole is filled with water to reduce the blow-out of energy into the air; results are best when the charge is detonated beneath the water table. Explosives give good results but they have disadvantages: they are expensive, time consuming to use (in the drilling of the shot holes) and can be dangerous. Consequently other seismic sources have been developed, including an electromechanical device, which produces seismic waves by vibrating the surface of the ground, and a gas-propelled piston, which thumps the surface.

Over water-covered areas—lakes, rivers and the sea—a charge can be exploded in the water, so no shot holes are necessary. However, most marine surveys use airguns as a sound source; these generate a seismic pulse by suddenly releasing high-pressure compressed air into the water when the gun is fired. Several guns are usually used, in an array. Other marine sources use gas explosions or, for low-penetration surveys, a sparker or a boomer. A sparker produces a seismic pulse by discharging an electric arc in the water, which is thus vaporized, producing a seismic wave; a boomer is an electromechanical source, which produces a seismic pulse when a plate is rapidly repelled by current flowing through a coil.

Detectors

The large seismometers used to record earthquakes are replaced in seismic reflection by the smaller geophones, on land, or hydrophones, in water. A geophone has a permanent magnet, with a gap between the poles. A coil is fixed in the magnetic field in the gap on a spring mounting, which allows it to move vertically. The geophone is placed in the ground, and seismic waves reaching the geophone cause the coil to move in the magnetic field, inducing a current in the coil which is proportional to the seismic signal. A hydrophone is a pressure-sensitive device, a piezoelectric crystal that transforms pressure waves in water into electrical signals. The hydrophones are combined into groups of twenty to a hundred on a line in a flexible plastic tube 30–100 m long whose output is added together. A number of these sections are joined together (typically twenty-four or forty-eight) with 'dead' sections, having no hydrophones between these active sections. This whole arrangement is called a seismic streamer or array, and is towed at depths of 5–15 m behind a ship.

Surveys

There is a variety of different configurations of sources and detectors for seismic reflection surveys. One of the most common is a split-spread configuration, in which geophones are used in a line on either side of a shot point (Figure 24.13). The most usual layout at sea and one used occasionally on land is multiple common depth point (CDP). The shot positions are planned to obtain the maximum number of reflections from the same subsurface point with subsequent shots, so that the data can be processed to give a very detailed picture of the subsurface. The time to fire each shot is calculated by computer from the ship's instantaneous speed. Marine seismic reflection profiling is much faster than land surveys as the source and streamer are moved continuously by the ship, instead of being moved discontinuously between shots. The electrical signal from the geophones or hydrophones is recorded on digital magnetic tape. Digital recording instruments were introduced during the 1960s, and since then they have increased remarkably in their capacity and sophistication. The early instruments were able to record the signal from twenty-four detectors, whereas the more recent ones record forty-eight or ninety-six traces. The electrical signal is sampled every 4 milliseconds or even faster, and it is these numbers that are recorded. Increasing the number of traces recorded makes it possible to increase the quality of the final profile.

Computers are then used to process these traces to produce the best reflection profile. The signals must be added together, with appropriate time delays, so that all signals from a common depth point enhance each other, and the signals add but the noise does not. This is called CDP stacking. The record is also de-convolved to enhance the resolution of the reflected signals and to attenuate multiple reflections.

Figure 24.13 is a typical seismic reflection profile, from the North Sea, and to a first approximation it represents a vertical slice through the Earth. The profile thus shows the rock structure, in this case a dome, a possible site for the accumulation of oil and gas. This profile is not a true section through the Earth, as the vertical scale is in terms of travel time for the seismic waves, not depth. To convert these times to depth it is necessary to know the velocity of seismic waves in the rocks; depth can then be calculated using the equation: depth = travel time × velocity. The velocities of the various rock layers can be obtained from the reflection survey and are calculated during the processing stage.

Migration

There is another problem that prevents the reflection profile from being a true section through the Earth. A boundary between two rock layers that is inclined will give a reflection that is not vertical, so that the reflection is recorded a horizontal distance away from the surface point above this boundary. This means that the reflection profile shows structures distorted from their real shape; anticlines and domes will appear wider than they are, and synclines narrower. The nature of this distortion has been understood for a long time, but before digital processing techniques were used there was no method of moving, or 'migrating', the reflectors back to their true positions. Migration is now possible as a routine stage in processing the profile, to produce a profile that is an ever closer approximation to the real Earth.

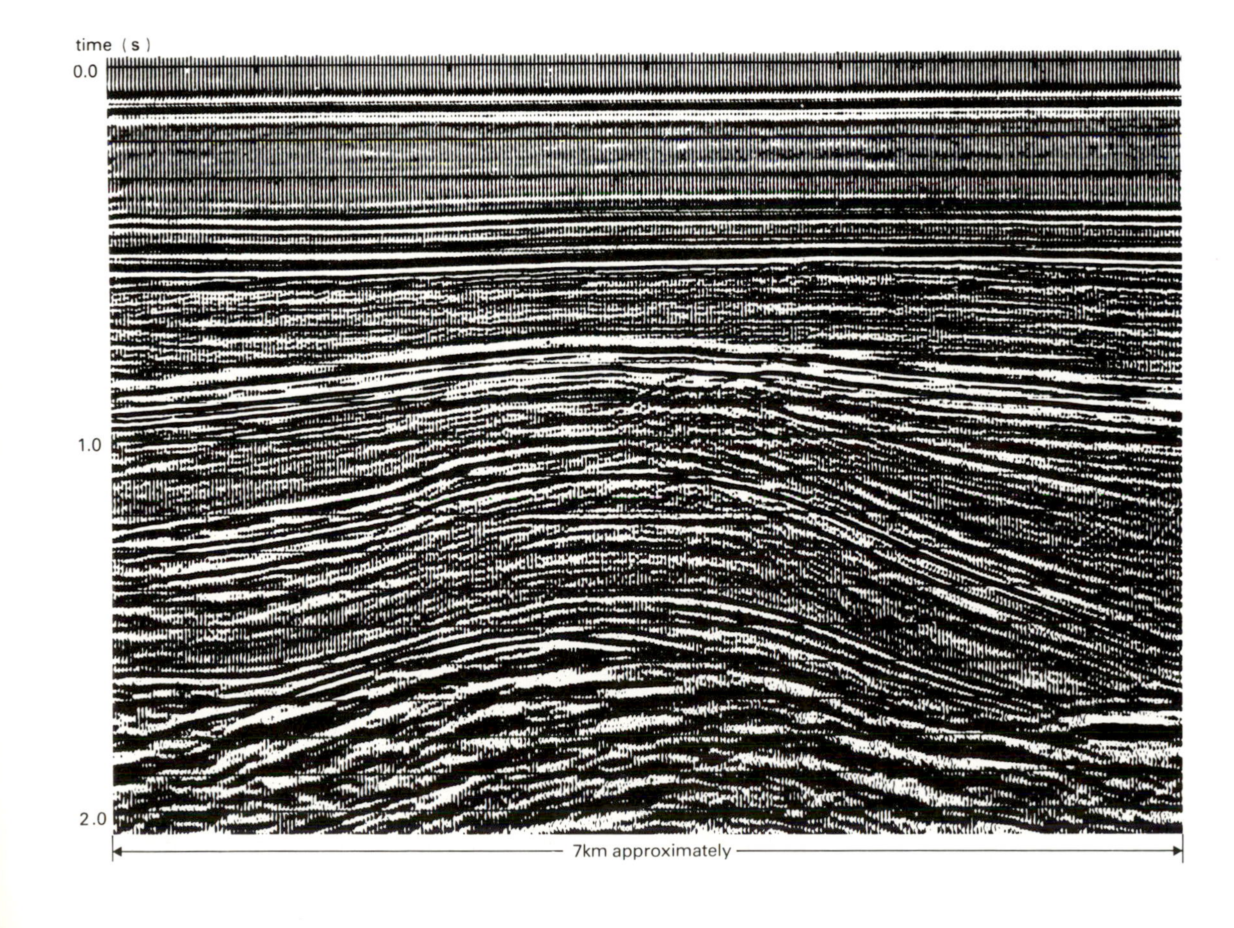

24.13: *A seismic reflection profile from the North Sea. The horizontal extent is about 7 km. The reflector at 0.1 seconds is the sea bed. Strong subsurface reflectors are also seen at 0.3 and 0.8 seconds. The strong horizontal reflector near the centre of the profile, at 1.2 seconds, is the reflection from a gas–water boundary.*

Reflection amplitude

Seismic reflection profiles are used usually to interpret subsurface structure, for example to look for possible oil- or gas-bearing structures. But it is also possible to get more from seismic reflection than this; Figure 24.13 shows that the reflections for different rock layers have different strengths, or amplitudes, and these different amplitudes can be used to tell us more about the rocks. Reflection amplitude depends on the velocity of seismic waves in a rock layer and also on the density of the rock. Amplitudes will be high at a rock boundary when there is a large difference in seismic velocity or density in the rocks on either side of the boundary. This occurs especially when there is oil or gas, as these are much lighter than the surrounding rocks. Figure 24.13 has an example of this: there is a strong horizontal reflection near the middle of the profile 1.2 seconds deep. This is a reflection between gas above and water below. The method of direct detection of oil and gas from reflection amplitudes is called bright-spot interpretation. Until this method was developed many exploration wells were drilled into rock structures that could have contained oil or gas but were found to be dry. Bright-spot interpretation has not totally eliminated drilling of dry wells but it has reduced their number.

Seismic reflection when it is used for oil and gas exploration is only involved with the upper part of the Earth's crust. However, it can be used to look deeper into the Earth. Figure 24.14 is a profile from the surface through to the base of the oceanic crust in the north Atlantic Ocean. Shot in 1975, it was the first profile to show good reflections from the top of the Earth's mantle.

Seismic stratigraphy

A new field of investigation in seismic reflection is seismic stratigraphy, the investigation of what reflectors on a reflection profile mean in geological terms, and their correlation between one area and another. A seismic reflection profile gives a picture through the Earth made up of a number of reflections. It is only possible to interpret these reflections in terms of individual rock layers where boreholes have been drilled to check what rock each reflector represents. But boreholes cannot be drilled everywhere, as they are very expensive, and seismic stratigraphy is attempting to interpret the geology from the character of the reflections (see also Figure 23.8).

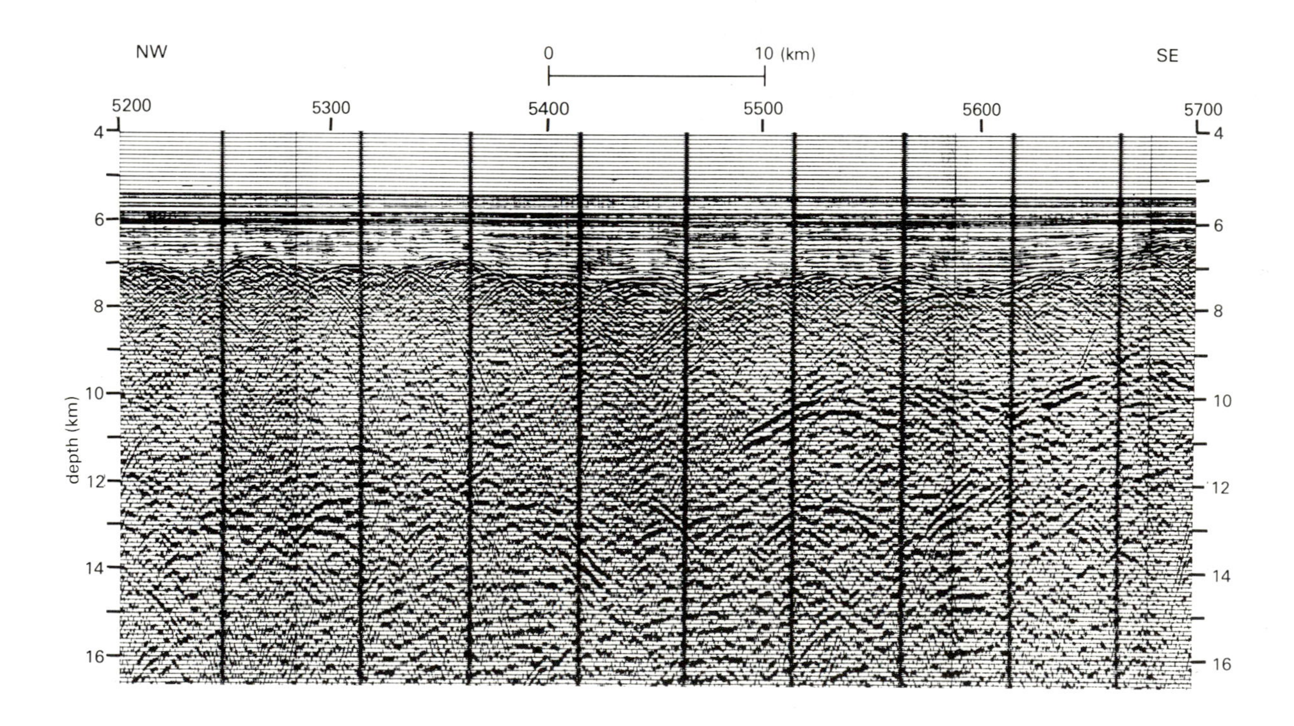

24.14: *A seismic reflection profile in the western north Atlantic Ocean. The profile shows a section through the whole crust, to the top of the mantle. The ocean is about 5.5 km deep, and below this there is a layer of sediment, about 1.5 km thick, in which there are a few horizontal reflections. The bumpy reflector at 7 km is the top of the igneous oceanic crust (oceanic layer 2). The crust–mantle boundary, the Mohorovičić discontinuity, shows as a strong reflector 13 km in depth. The thick vertical lines indicate the shot points.*

Electrical wireline logging

Electrical wireline logging is, nowadays, a well-known and universally applied geophysical technique used in prospecting for hydrocarbons, water and minerals. It is, however, quite young. The first electrical log was run by the French geophysicists Conrad and Marcel Schlumberger in 1927 in an oil well in Alsace. They lowered a sonde device (see below) down the well and took resistivity readings for every metre of depth. The potential of the technique for correlating subsurface rock units was immediately recognized and within a few years logging was put into commercial operation in France, Spain, Yugoslavia, Romania, the Soviet Union and the USA.

As the principles became better understood, more elaborate techniques were applied, culminating in the late 1960s when computers were taken to the well sites. Current work makes full use of modern electronic technology to obtain reliable well site measurements and help with decision-making.

General technique

Most geophysical wireline logging today is applied to oil and gas exploration and production. The primary objectives here are the detection of potential reservoir rocks, the determination of their porosity and permeability and the identification of the fluids present.

Porosity is the proportion of the rock volume that is occupied by pore space. In most reservoirs hydrocarbons fill only a fraction of the pore space, and this fraction is called the hydrocarbon saturation. Where water is the only other fluid present (the usual situation), the water saturation plus the hydrocarbon saturation, both of which are expressed as fractions of the pore space, equals one. The water saturation may determine whether or not a particular well is capable of commercial production. Permeability—the degree to which the pores in a rock are interconnected—is a property of the rock that is often investigated. The permeability determines how easily a fluid will pass through a rock and hence whether or not it will come out of the rock and into the well to be extracted. The unit of permeability is the darcy.

As well as determining the porosity and permeability of a reservoir rock an important function of well-logging is to correlate the rock structures between one well and another, identifying the faults and unconformities and obtaining as full a picture as possible of the reservoir itself.

The logging tool, or sonde, consists of a steel cylinder containing the measurement sensors and the electronics for the power supply and signal transmission. It is lowered down the borehole at the end of a multiconductor steel-armoured cable. The cable is spooled from the surface where the conductors are connected to the instrumentation. The surface instrumentation supplies electrical power for the tools, transmits the command signals to the sonde, and receives and records the measurements sent back. These measurements comprise the well-log.

Measurements are made continuously by pulling the logging tool from the bottom of the hole to the top at a monitored speed. The results are represented as curves plotted against depth. In some special cases stationary measurements are made at particular points and occasionally readings are taken while the tool is descending. The measurements may be recorded on paper, photographic film or magnetic tape for later processing by computer.

A standard logging survey will consist of a number of electrical measurements, several porosity measurements and a natural gamma ray radioactivity measurement. Special readings may also be taken, such as with a dipmeter.

24.15: *A typical wireline logging operation at a wildcat well in the overthrust belt of south-west Wyoming. The truck in the foreground is effectively a mobile laboratory housing a computer for immediate interpretation of the logs.*

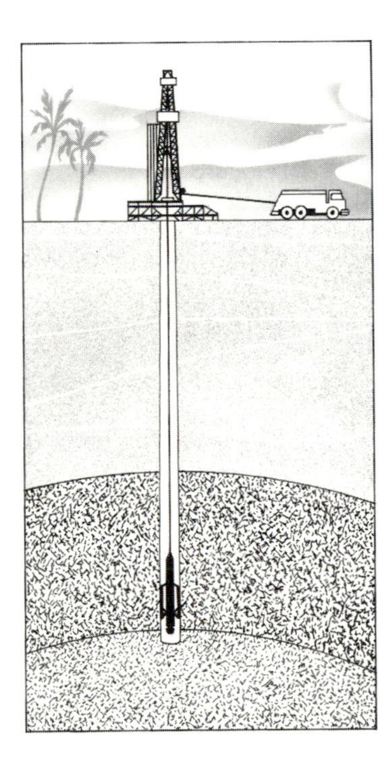

24.16: *A schematic diagram of an electric wireline logging installation. After the sonde has reached the bottom of the borehole it is retrieved at a carefully controlled speed, and formation measurements are made. Measurements are generally stopped when the sonde enters the cased portion of the borehole.*

The problem of fluid invasion

The objective of well-logging is to measure the properties of the undisturbed rocks and the fluids they contain. However, the act of drilling a hole produces disturbances. Pressures within rocks are very great and once these are relieved, as with a borehole, the fluids may be expelled as a blowout. To avoid such an occurrence and its consequences the boreholes are filled with drilling mud—a complex mixture of fluids usually suspended in water. The solids added to the fluid ensure that the mud in the well is kept at a pressure similar to that of the fluids in the rock. The exact balance is rarely achieved and often mud is forced into permeable formations pushing the indigenous fluids back from the borehole—a process called invasion. The solid components of the mud are plastered onto the wall of the well but this cake is continually chipped off and renewed by the flow of the mud. Appreciation of this invasion process is essential in the interpretation of well-logs.

Most readings are taken from the rocks closest to the well—the very rocks that are most affected. Different sensors record the properties of rocks at different distances from the well-tool. The effective depth of investigation, a qualitative term, is the radius from the logging tool that contains the material whose properties dominate the measurements. An instrument with shallow penetration will give readings from the mud in the well or caked on the wall. An instrument with deep penetration should give readings principally from rocks beyond the zone of disturbance. Readings from instruments of intermediate penetration will be from the rocks affected by fluid invasion.

In hydrocarbon exploration boreholes several logs are recorded simultaneously and this is usually done as soon as possible after the drilling. This is necessary if the hole is to be cased but it also avoids the long-term effects of fluid invasion.

Radioactivity logging

All radioactive log measurements are based on the fact that some atomic nuclei emit natural radiations that can be detected and measured, and others can be induced to do so by bombardment from suitable sources. Gamma radiation and neutrons possess appreciable penetrating powers and are both used in radioactivity logging. Well-logging tools that measure the radioactivity of nearby rocks can be considered under three categories: natural radioactivity, density and neutron logs.

The first tool detects gamma rays from the natural radioactivity of uranium, thorium and potassium in the rocks. The gamma ray log is widely used for formation correlation and identification because it is sensitive to the changes between radioactive and nonradioactive beds and these can be pinpointed fairly accurately. In sedimentary rocks the log reflects the shale content since radioactive elements tend to concentrate in clays and shales.

The sonde consists of a detector, such as a scintillation counter, and an amplifier. The calibration is in API units, defined in the API calibration pit at the University of Houston, Texas.

For the density log the sonde contains a concentrated source of gamma rays of a specified energy. These will interact with the surrounding rock and the intensity of the returning rays will be dependent on the density of the material through which they have passed. The detector is located a short distance above the source and the system is designed so that the gamma rays detected are those that have travelled through the rock formation.

The instrument must be calibrated for factors such as mud density and hole diameter and it must be placed hard against the side of the well, cutting through the mud cake beforehand. The readings gained are useful for determining the density and porosity of the rock formation.

Neutrons from a source in the logging tool will lose energy as they pass through the rock, or they may generate gamma rays in the process. The detector may measure either effect. The system responds primarily to the hydrogen content of the rock and is particularly useful for locating porous zones and determining the amount of fluid-filled porosity.

Again corrections must be made to take into account the features of the well and also the nature of the strata. Shales and gypsum have distorting effects because of the water associated with the crystals found in these rocks.

Pulsed neutron logs record the rate of decay of thermal neutrons. Bursts of high-energy neutrons are slowed by certain elements, particularly chlorine, and gamma rays are emitted. The interval between the pulse of neutrons and the arrival of gamma rays at a detector a short distance away is measured and recorded. Since this system is particularly sensitive to chlorine it can be used to detect the presence and the quantity of salt water and can be used to measure the rate of depletion of an oil reservoir by monitoring the rate at which the oil is being replaced by water.

Acoustic logging

Acoustic logs measure elastic or seismic energy in boreholes. The technique was first used in 1954 as an aid in seismic prospecting. It soon became apparent that seismic velocity correlated with porosity, and the estimation of porosity is now its principal application. Other uses include the study of the quality of the cement bond in cased holes and the investigation of fracturing within the rock formation.

A burst of sonic energy from a transmitter will reach a receiver a few metres away by transmission through the borehole fluid, by tube waves and by compressional waves in the rock. The compressional waves in the rock are the fastest and these are recorded first, the time between transmission and reception being logged. The nature of the material whose porosity is being studied has a strong influence on the readings and considerable correction has often to be made to the results.

Resistivity logging

Electrical well logging may measure the electrical resistivity of the formation or its self-potential. The rocks themselves, especially sedimentary rocks, are very poor conductors but their pore spaces contain water in which salts are dissolved. The salts in solution dissociate into positive and negative ions that move in an electric

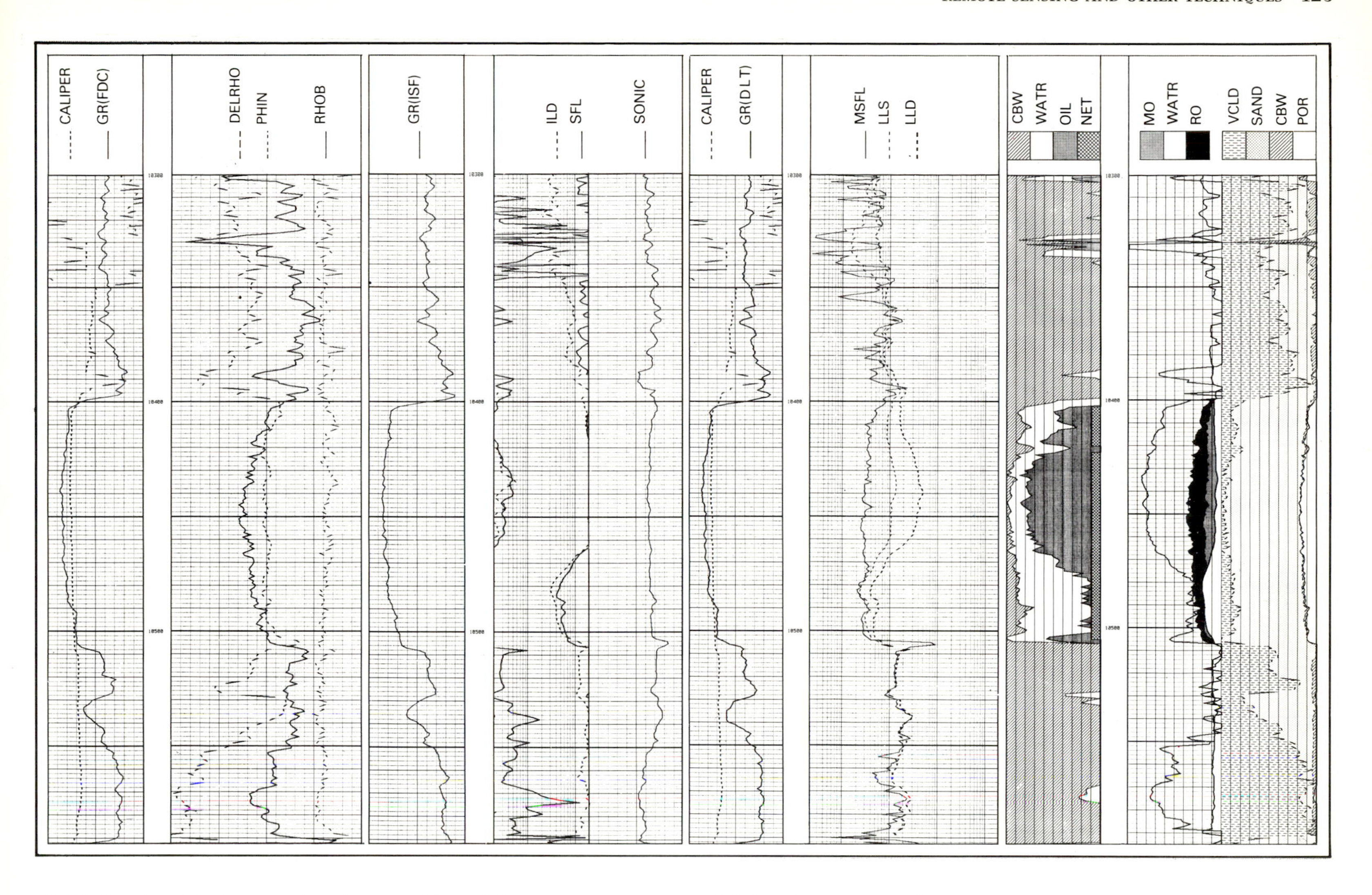

24.17: *Results of the interpretation of a suite of logs, all for the same interval (3756 to 3869 m) in the same borehole. The upper part of the section is a shale formation, with traces of sandstone. The middle part is a sandstone with slight amounts of interstitial clay and is oil bearing, although near the base the freely movable oil (ie the amount which can be recovered) is replaced completely by water in the pore space. The residual oil in this section is oil trapped in the pore space and is in the first instance irrecoverable. Below this is another shale formation. The curves in columns* (**a**) *to* (**c**) *are the results of three separate logging runs. The first run* (**a**) *recorded natural gamma-rays (GR (FDC)), borehole diameter (CALIPER), neutron porosity (PHIN), gamma-ray formation density (RHOB) and a density correction (DELRHO), the usual combination for porosity and lithology determination;* (**b**) *natural gamma-ray (GR (ISF)), deep induction resistivity (ILD), shallow resistivity (SFL) and interval transit time (SONIC);* (**c**) *natural gamma-rays (GR (DLT)), borehole diameter (CALIPER) deep and shallow laterolog resistivity (LLD, LLS), microresistivity (MSFL), this combination giving the invasion profile and hence the true formation resistivity. The gamma-ray log with each run allows cross-correlation between runs.* (**d**) *shows the interpretation from the wireline logs. The different curves and shaded areas identify rock and fluid characteristics as follows: free formation water (WATR), oil (OIL), water bound to the clay/shale particles (CBW), residual and movable oil (RO, MO), clay and quartzite volumes (VCLD, SAND), and porosity (POR). The cross-hatched column marked NET identifies net effective reservoir rock.*

field, thus providing the vehicle for current flow in the rocks. The basic method of resistivity logging consists of applying a direct current or a low-frequency alternating current between current electrodes and measuring the potential difference arising between two or more potential electrodes. The record is then a plot of potential variation, or apparent resistivity, against depth.

A 'normal' log is the term given to a system in which one current electrode and one potential electrode are placed down the hole, while the other current electrode and potential electrode are placed at the surface. The effective penetration into the formation for such a system is about twice the distance between the electrodes. The borehole diameter, the resistivity of the mud and the invasion of the rocks all have an influence on the reading and corrections must be made.

In 'lateral' logs both potential electrodes are placed down the hole along with one current electrode, while the other current electrode is at the surface. The system is now largely obsolete and suffers the same disadvantages as 'normal' logs. If the spacing between the potential electrodes is large the depth of penetration is about equal to the length of the spacing. The microlog is a specialized lateral log in which the potential electrodes are only a few centimetres apart. This gives a sharp definition of thin beds but the shallow depth of investigation—usually less than 100 mm—means that its use is confined to the identification of permeable formations.

Greater resolution and penetration is achieved by focusing the current into a thin horizontal disc that penetrates the formation laterally instead of flowing up the borehole walls. The focusing is achieved by the use of another set of electrodes known as the guard electrodes. Several variations have been developed with different numbers and spacings of electrodes. The depth of investigation is considerable and the focused current log, or laterolog, is a very widely used resistivity tool. A similar electrode arrangement to that in the focused current log but placed against the borehole wall on a rubber pad gives the microlaterolog. Its performance is similar to that of the micrólog but it has the advantage of a greater depth of penetration because of the focusing. It is used to measure resistivity in the invaded zone.

The induction log was developed in 1948 for use in wells with high-resistivity muds. The sonde consists of an arrangement of coils. A transmitting coil produces an electromagnetic field that induces eddy currents in nearby conductive formations. The secondary electromagnetic fields generated by these eddy currents induce voltages in the receiver coil which are amplified and transmitted to the surface. The induction log signal is proportional to the conductivity of the formations rather than to their resistivity. Focusing is achieved by additional coils and this increases the depth of investigation, reducing the effect of nearby conducting material. In this way the influence of mud resistivity, hole diameter and invasion can be reduced. Particularly conductive beds may introduce irregularities and so correction must be made. The best results are obtained in dry holes or in oil-filled holes where other resistivity logging methods fail.

The logging sonde of the dipmeter has four microresistivity pads mounted at 90° from each other. During logging they are applied firmly to the borehole walls. The sequence of readings at each pad will be slightly out of step from the others depending on the dip of the strata because of the difference in depth of corresponding features at each side of the hole. Such information is very valuable in the plotting of oil field structures. The device can also be used to determine the deviation of the borehole and the orientation of the equipment. All the readings are recorded on magnetic tape in digital form at the well site, and sophisticated data processing systems have been developed to interpret and present the results.

Production logging

Once a hydrocarbon well has been confirmed and the reservoir is producing, different wireline logs are needed to monitor production and reservoir behaviour. Such logs are called production logs and usually measure fluid and flow properties. They are invariably run in cased holes with the wells producing normally.

Production rates are measured by the flowmeter log. It consists basically of a propeller immersed in the borehole, and its speed of rotation can be related directly to fluid velocity and hence the production rate of the well. The gradiomanometer tool consists of two pressure sensors mounted a short distance apart. The difference in hydrostatic pressure between the sensors is directly related to fluid density—water has a greater pressure gradient and hence a greater density than oil, which in turn has a greater density than gas. Other reservoir monitoring logs check pressure and temperature.

All these measurements are taken regularly, every six months or so, during the life of the wells in the reservoir. Correlation of well data keeps a check on the decline in pressure, the displacement of the fluid and the encroachment of water. In this way problems are anticipated and the remedies can be put in hand.

Log interpretation

The several dozen measurements taken by different logging tools, either singly or in combination, refer to specific physical properties of the rock, such as radioactivity, acoustic velocity and resistivity. The information that is really required, however, is about the porosity, the fluid saturation and the permeability, and the measurements obtained must be combined and applied to mathematical formulas in order to determine these. This process is called log analysis and interpretation.

Most of the mathematical models are developed from those first formulated by G. E. Archie in 1942. He determined that the relationship between the bulk resistivity of a water-saturated rock sample and the resistivity of the saturating water was a function of the porosity of the rock. He also calculated that the bulk resistivity of the rock increases when the pores are partly filled with gas or oil.

An important equation involves the estimation of the pore volume filled with water and is derived from the resistivity of the water, the true resistivity of the sample and the rock porosity—all of which can be measured from log runs. The water resistivity can be obtained by the resistivity log or a self-potential log, porosity can be read from the density, neutron and acoustic logs, and the true resistivity can be read from resistivity logs. The constants involved in the equation are calculated from a wide-ranging study of rock samples.

To obtain consistent results several log runs of each kind are carried out and their results are combined and compared. The final results of the raw data logs are shown in terms of rock and fluid properties. On the basis of such results the commercial viability of a discovery can be assessed, and development plans prepared.

25 The economics of Earth resources

The material foundations of human society and its economy are those derived by direct intervention in natural systems. We use the term 'resources' here, since it is in common geological usage, for what in strict economic terms are commodities. Resources include food, natural fibres, agricultural products, wood and the gamut of other produce from biological systems, as well as water, metals, fuels, fertilizers and chemical feedstock, which are physical products of the atmosphere, hydrosphere and crust, and the direct concern of geologists.

Physical resources are inseparable from the complex system that forms our planet, and their exploitation is a fundamental aspect of society. The way in which physical resources are exploited affects the surface environment. The effects can be beneficial, as with the use of soil fertilizers, but exploitation also poses the hazard of pollution. Although the environment has been affected by civilization for millennia, the increasing scale and rapidity of modification have suddenly posed hitherto unforeseen problems. The growth of world population and rising material standards have stimulated increased exploitation of physical resources. The tonnages now involved are so great that depletion of resources is conceivable. Since the accessible parts of the Earth are finite there is a limit to continued growth, and the long time scales of the natural cycles in which most physical resources are redistributed suggest that many existing resources are irreplaceable in the foreseeable future. Although this book is primarily about the science of the Earth, the interaction of science with economics and politics in the exploitation of the products of geological activity cannot be ignored.

Physical resources and economics

Resources have two primary economic aspects: they have to be of use to someone and they must be exchangeable for other resources. In a modern economy, no longer dominated by barter, their exchange value is expressed in money. In times of stability paper money expresses this relationship in a general way. However, with inflation paper money loses its value equivalence and is obviously covering up for some more fundamental arbiter of value, some kind of 'real' money.

Gold is a 'money resource' that has never been mined primarily for its usefulness in industry but rather as the most basic form of money. Any metal that is in great demand for its industrial application would be useless as a currency. It would be incorporated in one industrial product or another and it would be incapable of circulation. Gold is intrinsically useful but, by virtue of its unique physical properties, only very small quantities are needed. It takes much effort to mine, thereby expressing a lot of labour in a small volume. It does not corrode and it cannot be debased by dilution with other metals without perceptible change in colour. These features of gold have placed it in its present-day dominant monetary role. Silver once played a monetary role but the discovery of huge South American reserves caused its exchange value to slump disastrously in the seventeenth century.

The usefulness and exchange value of any resource are subject to change. Natural volcanic glass, obsidian, from sources in the Mediterranean area was formerly traded widely as an easily sharpened material for weapons and tools. When it was displaced by metals it became impossible to exchange and therefore valueless. In the nineteenth century aluminium was regarded as a scientific curiosity and a precious metal because it was so difficult to smelt. Since the invention of electrolytic smelting this light, easily worked metal has become the second most commonly used metal after iron.

Chance and necessity both play a part in the process whereby natural components of the Earth's crust, such as metal ores or petroleum, are transformed from 'unusual' Earth materials into widely sought and traded commodities, on which vast amounts of human labour are expended. For example, natural oil seepages have been known for centuries in the Middle East, eastern North America and even in Britain. Until the mid-nineteenth century they were not regarded as useful. A rapidly growing demand for lubricants and lamp oil, hitherto supplied from organic sources or oil shales, led to successful drilling of one such seepage in Pennsylvania. This first underground oil field made all other sources of oil redundant by producing high-quality petroleum for less effort and therefore more cheaply.

Physical resources play a fundamental role in the capitalist economy but they do not determine the course of economic history. The statistics associated with the 1929 Wall Street Crash and the Great Depression of the 1930s reveal the subsidiary role of resources to deeper changes in the capitalist system. Metal prices slumped, the minimum profitable ore grades that could be raised went up, mines closed and metal production and consumption declined only after the crash. Things appeared rosy in mining right up to 29 October 1929. The economics of mining and other modern industries must, therefore, be looked at in the light of wider developments in the world economy, as the postwar experience makes plain.

At the Bretton Woods Conference, held in New Hampshire in 1944, the major capitalist countries other than the Axis powers recognized the threat of revolution if the troops were to return from war to conditions comparable with those of the 1930s. By this time the USA owned a large proportion of the world's gold, the value of which was greater than that expressed by all dollars in circulation. The world's major currencies were temporarily stabilized by fixing the price of gold in US dollars and controlling the dollar exchange rates of all other currencies within fixed limits. Thus, in theory, anyone could exchange paper tokens of money for the real thing, and the result was restored confidence in paper currency. The engineered strength of the dollar was used to print vast amounts of new notes and to support long-term credit, which, through such schemes as Marshall Aid, regenerated capitalist industry in Europe and Japan.

The net result of Bretton Woods was the boom of the 1950s and 1960s from which the mining and oil industries benefited enormously. This period exhibited controlled inflation of the paper money supply, and prices and costs rose slowly but steadily. Since the whole structure relied on gold having a fixed price gold-mining became less and less profitable, except where wages could be fixed at low rates, as in South Africa. A ludicrous situation arose in Australia, where, to make a profit of US$0.5 per ounce, a subsidy of US$13.5 per ounce was required by the gold-mining industry. Despite such subsidies gold production failed to keep pace with the printing of paper money. By the late 1960s, the value of dollars in circulation was five times greater than the value of the gold reserves held by the USA. Eventually there was a run on the dollar, large quantities of gold being bought by European central banks. The only possible response was the decision of the Nixon administration, on 15 August 1971, to allow gold to find its true value relative to the dollar. This broke the Bretton Woods Agreement and the hopes for stability it represented.

Figures 25.1 and 25.2 graphically express the post-1971 orgy of speculation in physical resources, the net result of which was massive, uncontrolled inflation. This affected every aspect of the world economy, in particular drastically reducing the profitability of manufacturing industries and mining. In early 1974 the oil producers had to raise their prices dramatically in order to keep pace with inflation. It is the dollar price of gold that charts this economic collapse most precisely, since gold still maintains its position as the ultimate means of exchange.

The classification of physical resources

A primary classification of physical resources for the purposes of economic planning can be based on use categories, such as energy resources, metallic ores, water, constructional materials and chemical feedstocks. Further classification can distinguish between primary energy sources, such as solar energy and uranium, and secondary energy resources, such as fossil fuels. Metallic ores are commonly classified on a geological basis, subdivided, for instance, into magmatic segregations, hydrothermal replacements and chemical precipitates. The usefulness of these subdivisions lies in planning exploration programmes.

Physical resources are concentrated by the processes in geological

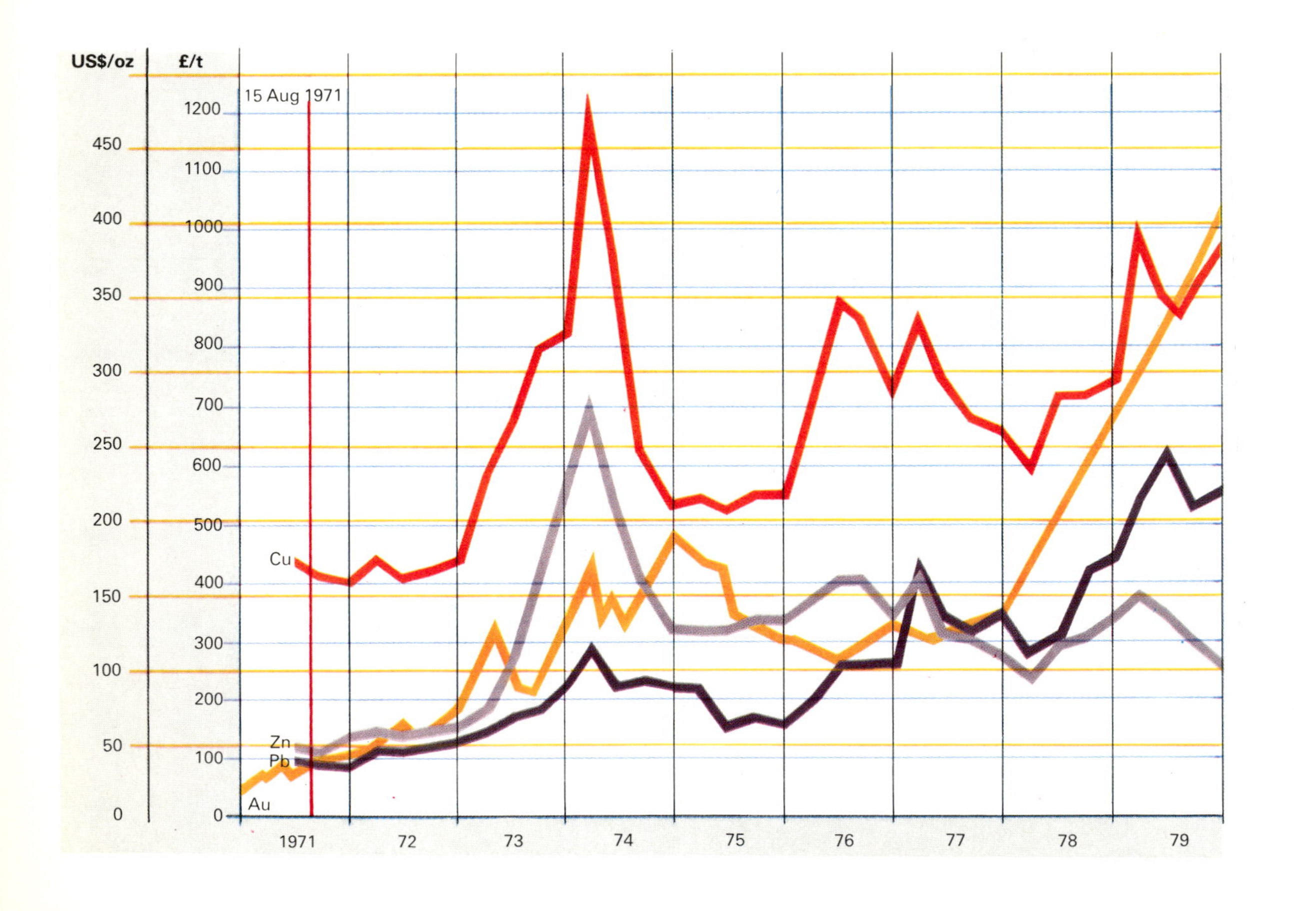

25.1: *The destabilization of the US$ on 15 August 1971 was quickly followed by a dramatic rise in the prices of metals. The metals represented here are gold, copper, zinc and lead, and prices are expressed in the currencies that are used in the metal exchanges. The destabilization inflated the costs of the manufacturing industry and led to drops in consumption, reflected in the decline in prices in late 1974. The influence of speculation, which leads to sudden rises followed by falls reflecting profit-taking, particularly in the case of gold, is also shown here. Such price instability causes grave problems for mining, as production is geared to the smooth regulation of output.*

cycles. These cycles may proceed irregularly and have different periodicities but they are interrelated. For instance, the cycles concerned with soil formation and erosion may be speeded up when crops are grown on them, sometimes with disastrous results, as in the US Midwest in the 1930s. Physical resources are part of the rock cycle, including plate tectonics, erosion and the hydrologic cycle. The formation of coal and oil links the rock cycle to biological processes.

Exploitation of physical resources interrupts geological cycles and often profoundly modifies them. To some extent the cycles may accommodate the results of human activity. For instance, the carbon dioxide released when fossil fuels are burnt returns to the carbon cycle, and waste metals find their way to the oceans, enter sediments and return to igneous and hydrothermal fluids when these sediments are subducted. Although physical resources can be replenished in natural cycles, to be usefully renewable they must be recreated within time scales relevant to human activities. Most physical resources are connected with cycles having periods from thousands to hundreds of millions of years and, therefore, are non-renewable (Figure 25.3). The

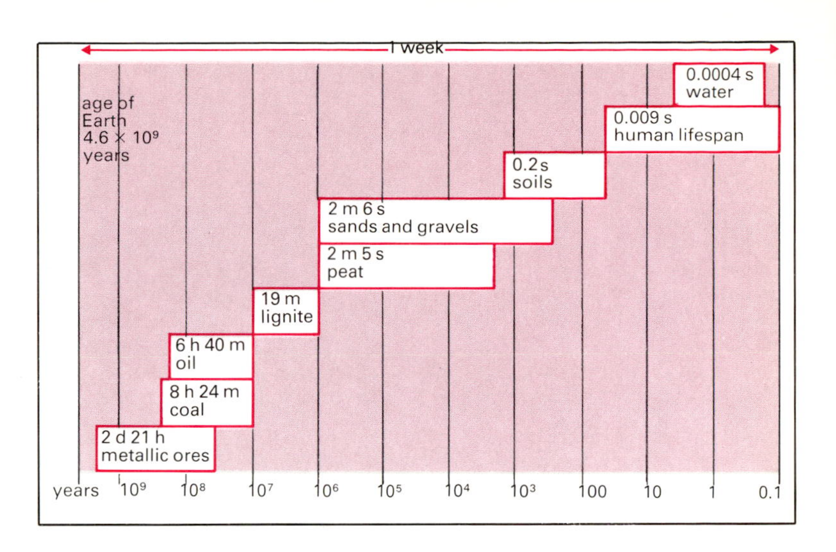

25.3: *The maximum and minimum times estimated to elapse before resources can be replaced by natural cycles are expressed on a logarithmic scale and also on a scale that compresses the Earth's history into a week. Only some constructional materials, soils, water resources and energy resources connected with solar radiation renew themselves in times less than thousands of years and can be considered renewable. All other conventional physical resources, including some not shown, such as inorganic fertilizers and chemical feedstock, are inherently non-renewable. Some newly discovered metal-rich deposits such as manganese nodules may be renewable if rates of replacement always exceed exploitation.*

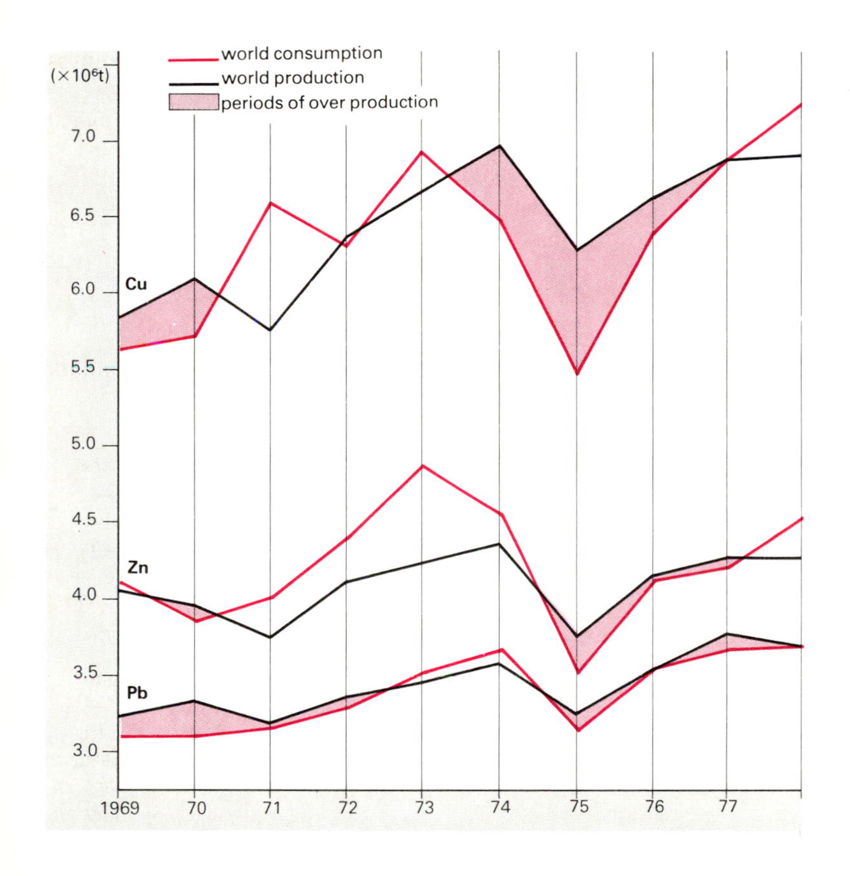

25.2: *World production figures for copper, zinc and lead show how the mining industry is disrupted by economic fluctuations. In particular, a sudden fall in copper consumption in 1973–5 caused huge overproduction and a glut of refined metal which was partly responsible for the rapid decline in copper prices in 1974–5. The graphs show a remarkable similarity, even though copper is not usually mined together with zinc and lead.*

exploitation and depletion of non-renewable resources pose a finite limit to their use. It is significant that natural rates of movement of important elements have been overtaken by human-induced mobilization. For example, it is estimated that each year 375 thousand tonnes of copper are transported from land to the oceans by erosion, water and wind but over five million tonnes are mined.

The term 'total stock' is used when quantifying physical resources to indicate the mass of a potentially useful Earth material within a given area. It merely expresses existence; only a fraction of the total stock can be regarded as a useful resource. In a capitalist economy the use value of a resource is not so important as its exchange value, wherein lies profit. From this standpoint only that fraction of resources that can be acquired at a lower cost than the price they command is exploited. The fraction is referred to as 'reserves' (Figure 25.4). Further categorization of resources and reserves can be based simply on the degree to which it is certain that resources exist and on their likely profitability (Figure 25.5).

The boundary between reserves and conditional resources shifts up or down as exchange value, production costs or technology changes. Advances in technology can have the contradictory effect of removing some reserves to the conditional resource category if, as in the case of some coal seams, they are too small to accommodate the machinery the technology requires. An estimate of resources is not static but depends on many factors. A resource that is abundant in the identified and subeconomic category but for which there are hypothetical resources can have its reserves boosted by a technological breakthrough or a rise in price. Uranium is an example of such a resource, recent rises in fuel prices having rendered new exploration unnecessary. A fall in price or increase in costs in this case leaves little leeway for replenishing reserves by exploration. For resources having small amounts in the identified category there is little incentive to improve techniques, and reserves can be boosted only by exploration. Titanium has low conditional and hypothetical resources, and yet it is fairly certain that a vast amount exists in the form of black-sand beach deposits. Sufficient demand for titanium or depletion of reserves would stimulate new exploration.

In some states—for instance the Soviet Union—more emphasis is laid on the intrinsic usefulness of resources than on their exchange value and profitability, even though the influence of world prices is felt. Consequently many Soviet reserves would fall into the capitalist world's conditional resource category. Finally, there are resources that are new to science, some of which are covered below, and there may be others whose discovery would alter totally the resources picture. They do not yet appear in the world's resource inventory.

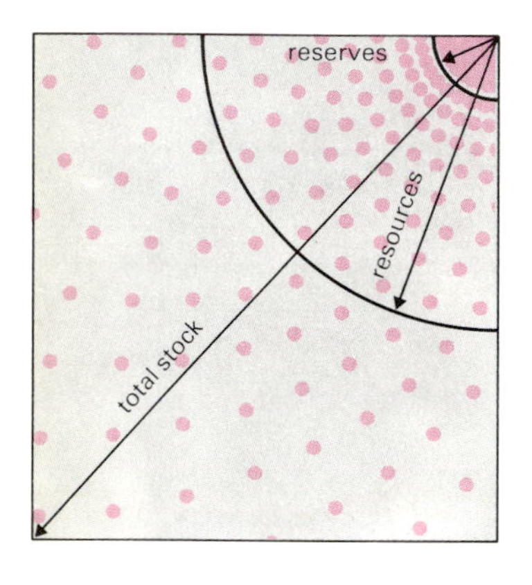

25.4: The relationship between reserves, resources and total stock. The area for each category represents tonnage of available resource. The concentration of a physical resource is expressed by the spacing between the dots.

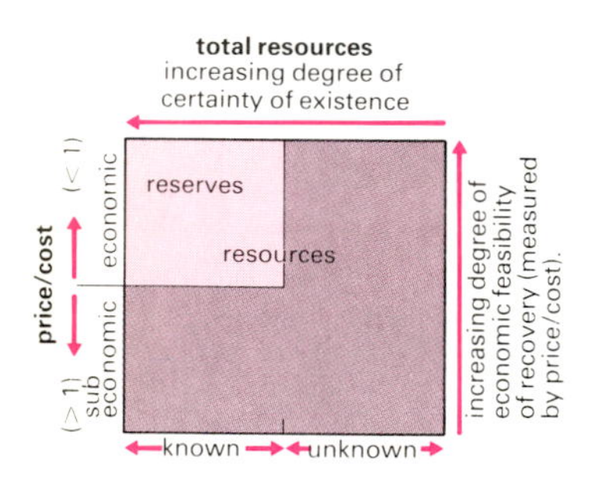

25.5: Left, the way in which the degree of certainty about the existence of a particular resource and the profitability of a deposit separate reserves from the whole field of resources. This is amplified, below, by the US Geological Survey's widely used classification of mineral raw materials.

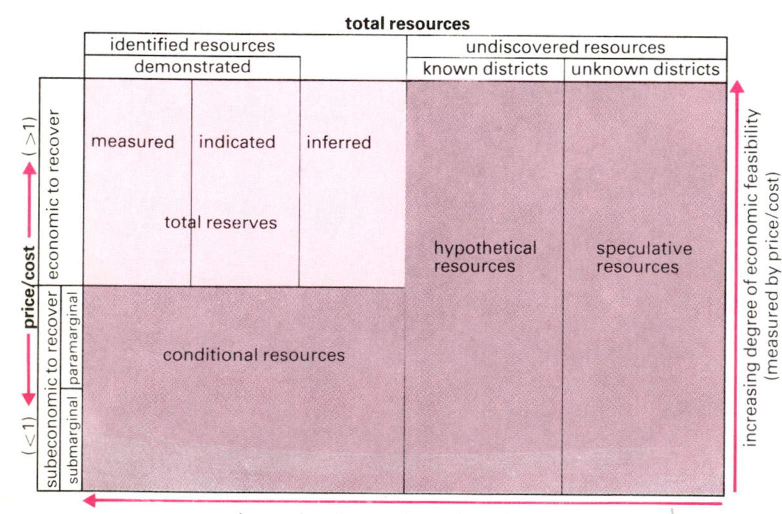

The distribution of resources

The distribution of physical resources is not only a reflection of geological processes but is also intimately related to use and exchange values. Very abundant and widespread resources which require little processing, such as water and constructional materials, do not norally command high prices. Consequently it is unusual for them to be exploited far from the site where they will be used. They are bulky, and transport costs soon inflate their price. When they are in short supply, for geological or climatic reasons, extra effort and cost must be incurred. Such resources are often said to have high place value. Occurrences of such resources remote from industrial conurbations are not regarded as reserves, nor are they usually evaluated until development of remote areas is considered. Water lies in a special category of resources having high place value. Its supply often fluctuates widely from season to season and year to year and its quality varies tremendously, depending on its content of dissolved and suspended solids. Water's ultimate use is consequently the major factor in evaluating supplies.

Relatively rare resources with sophisticated uses, which generally require complex processing, command high prices. Though they are not abundant, high and growing demand, and their potential for making large profits, means that they will be sought and exploited virtually anywhere on the planet, even in very remote areas. They have low place value. However, whether such resources constitute reserves or not depends on many cost factors, some of which relate to location. An example is a vast deposit of iron ore with 60 per cent iron content in the Venezuelan Andes, which is not considered a reserve because of the high cost of providing access. Difficulties in winning a useful resource from its host rock may also limit its classification as a reserve. For instance, extracting oil from some oil shales poses technical difficulties despite their high hydrocarbon content.

In the whole spectrum of resources categorized according to place value, those with low place value are of most concern. They are resources of paramount economic and strategic importance. They include metallic ores, energy resources such as fossil and nuclear fuels, chemical feedstock and inorganic fertilizers, and some specialized resources such as china clay and insulating materials. Each of them is intimately associated with intricate geological processes, and it is these processes that govern the distribution of exploitable reserves rather than economic or political considerations.

The distribution of oil-producing areas is shown in Figure 25.6. They appear to be randomly scattered, but the reserve figures for each illustrate the gross inequalities of distribution. Oil is a product of rapid

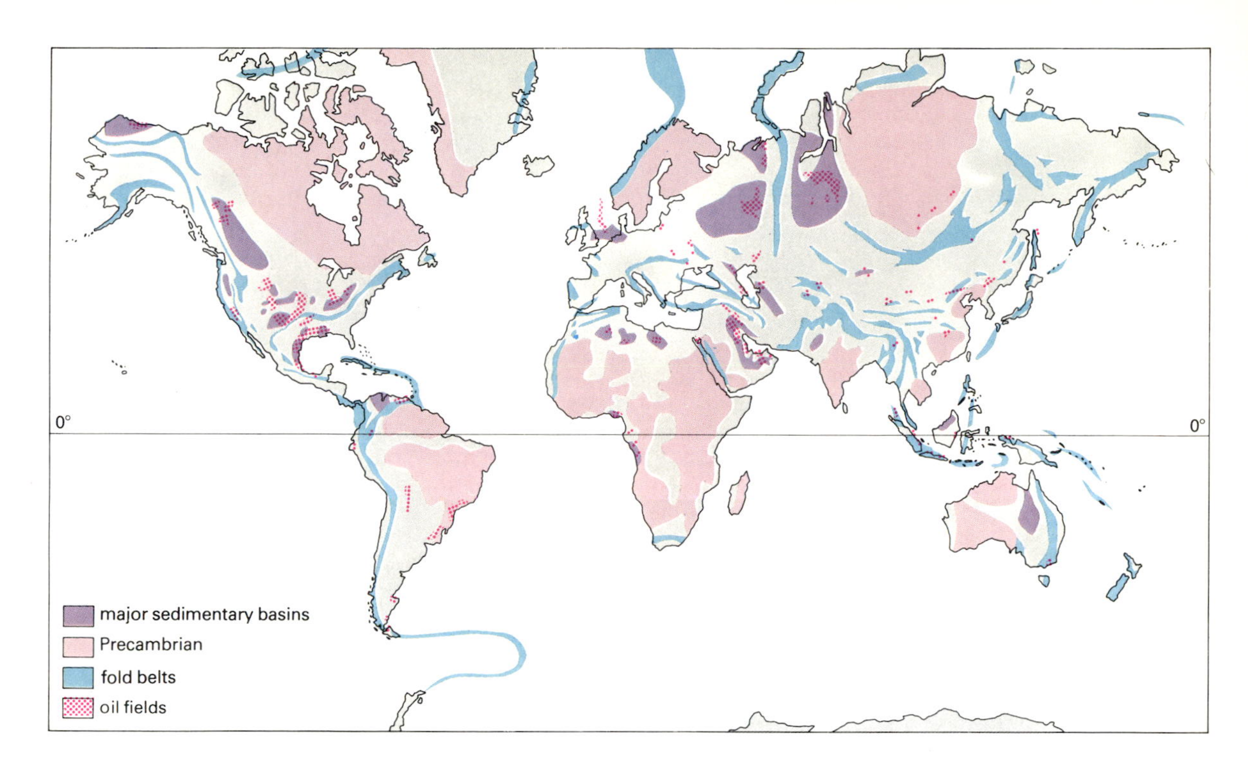

25.6: *World geology severely limits the areas where oil exploration prospects are optimistic. All areas where Precambrian rocks outcrop are barren because of their often crystalline nature and their very age (petroleum degrades to tars and ultimately solid hydrocarbons after a few hundred million years). In areas of great deformation, metamorphism and igneous activity oil cannot survive; intense folding and faulting allows oil to leak to the surface, and temperatures above 300–400°C destroy its molecular structure. Most oil fields are in large sedimentary basins in which conditions during their development were suitable for oil formation, migration and trapping in structures. Such basins formed in close relation to earlier continental crust, sometimes at its margin or where it rifted.*

sedimentary burial of teeming organisms before they can be completely oxidized by decay. Their slow 'fermentation' after burial eventually saturates the containing sediments in organic liquids and gases. Subsequent migration of these liquids and gases into more porous reservoir rocks and accumulation in traps–suitable fold and fault structures–culminate in individual oil fields. Fields could be destroyed by high temperatures associated with deep burial, tectonic thickening of the crust or igneous activity. It is likely, therefore, that productive areas will be deep sedimentary basins which were formerly associated with conditions suitable for organic blooms and have undergone some deformation to produce trap structures. Extreme deformation, metamorphism and igneous activity are incompatible with preservation.

The distribution of coal is more restricted than that of oil. Most occurs in a wide belt in the northern hemisphere, while the southern continents have very few deposits. Coal is a relic of tropical swamp conditions in estuaries and deltas. The greatest bulk of the world's hard coal is late Carboniferous to Permian in age, and the present distribution of deposits relates to the break-up of the Pangaea supercontinent, through which the Carboniferous equator ran.

Metal-ore deposits are frequently grouped together in restricted geographical regions, sometimes in association of elements such as copper–lead–zinc and tin–niobium–tantalum. Such associations occur in metallogenic provinces which can be related to processes connected with sea floor spreading and plate tectonics. Among these are both massive and disseminated coppers, nickel and chromium deposits in obducted ophiolites and the porphyry copper deposits related to deep levels in andesitic volcanoes at destructive plate margins (Figure 25.7). Metallogenic provinces are also known from terrains where plate tectonics have had no direct influence. Although granites are among the most common rocks of the continental crust, relatively few occurrences are associated with tin (Sn) mineralization. Such Sn-granites occur in swarms, in south-east Asia, Nigeria–Niger and western Europe.

From these three examples it is clear that whatever the social, economic or political need for resources of low place value there is little choice about where they may be sought successfully. They relate to geological boundaries rather than political ones. Some resources are remarkably restricted in their occurrence. The world's largest reserves of chromium, tungsten, cobalt, nickel and gold lie respectively in Zimbabwe, China, Zaïre, Cuba and South Africa. Territorial disputes, secessions and major wars have been triggered off by attempts to control the exploitation of physical resources in these areas.

The geological setting of one physical resource is often widely separated from the setting of other physical resources. The setting of one resource may even exclude or destroy certain other resources. It is geologically improbable that a porphyry copper deposit could be found next to an oil field or that obducted oceanic crust with nickel deposits would border a coal field.

The industrial revolution in Britain was, to a large extent, due to the close association of coal, limestone and sedimentary iron ores. At Coalbrookdale in Shropshire, the cradle of the modern steel industry, a single mine provided coal, iron carbonate ore, limestone and even fireclay. On the other hand, Africa, except for its southern parts, has virtually no coal or pure limestone, and its enormous reserves of iron ore, among many other metals, could not be smelted before the development in the twentieth century of appropriate techniques.

The spread of revolutionary industrial techniques closely followed the distribution of the world's major coal fields, and this is reflected today by patterns of development. Even though many underdeveloped states are now known to have enormous reserves of physical resources, such as oil, copper and iron ore, many are unable to foster thriving industry because of their lack of resource diversity. They are often forced to export resources that might otherwise raise their living standards immeasurably.

An important consequence of the unequal distribution of resources has been price and supply manipulation by international organizations formed by underdeveloped states. The defensive actions of the Organization of Petroleum Exporting Countries (OPEC) and the Intergovernmental Council of Copper Exporting Countries (CIPEC) against inflation in the prices of industrial goods exported by developed states have played a crucial role in the world economy.

Some trends in resource exploitation

Rising world population and improvements in the material standard of living, particularly in the developed nations, is naturally reflected by increasing exploitation of physical resources of all kinds. A wealth of literature now exists, dealing with demography, exponential growth and systems analysis of world society and human ecology. It is not profitable to review the associated controversies here. The question of catastrophic depletion of resources has to be qualified by the intricacies of resource inventories and the fact that exponential industrial growth has begun to decline in the face of more fundamental economic disruptions. There are serious problems associated with all physical resources—water supply, energy sources, rare metals and strategic metals—but the development of copper-mining demonstrates several important trends during the twentieth century.

The average minimum profitable grade of copper (Cu) ores has declined from about 1.5 to 2 per cent Cu content to around 0.5 per cent in the 1970s (see Figure 25.8). There are mines today that are productive at grades as low as 0.2 per cent because other metals, such as molybdenum and gold, occur in the same deposit and help pay for copper-mining. This decline is due partly to the depletion of most known rich deposits. More important, the decline in workable grade has been made possible by improvements in the technical efficiency of mining and in separating ore from waste. In the early part of the twentieth century most copper-mining was underground and used labour-intensive methods to extract low-volume, high-grade ore deposits. Today it is dominated by mechanized, capital-intensive surface mining of high-volume, low-grade occurrences, typified by porphyry copper deposits.

An automobile contains about 23 kg of copper, which corresponds to 3 tonnes of ore. Therein lies the major problem associated with low-grade mining. Unlike underground mining, in which waste is usually packed in old workings for stabilization, open-pit waste has to be piled up to avoid blanketing unworked ore. A mine producing ten thousand tonnes of copper metal in a year also produces twelve million tonnes of waste. Sheer bulk causes acute environmental problems, but a more chronic difficulty centres on the nature of the waste. Copper ores, and those of several other metals, are commonly associated with useless iron sulphides. Oxidation produces dilute sulphuric acid and slimy red ochres. Solution of heavy metals by the acid results in highly toxic waters, which may penetrate groundwater and surface run-off, thereby

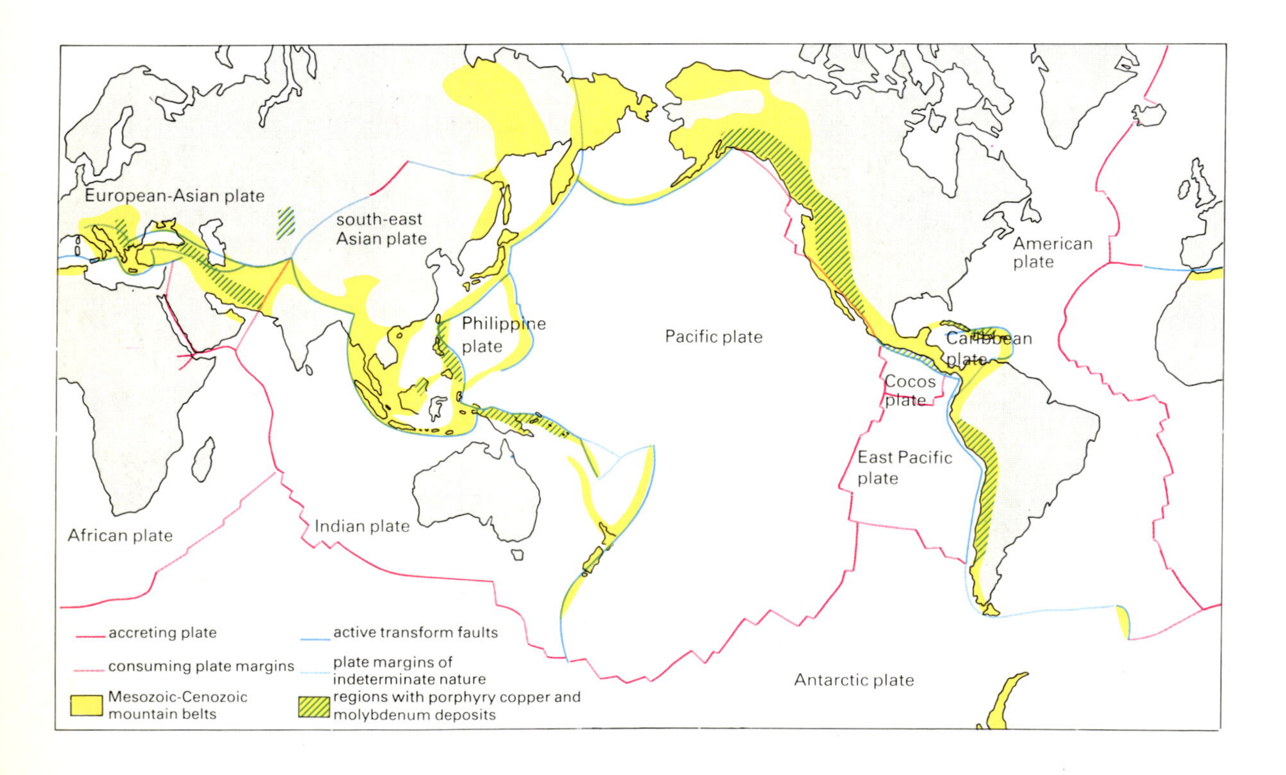

25.7: *Major deposits of copper with molybdenum—the so-called 'porphyry' copper deposits—are found as disseminations within granodioritic intrusions, which are clearly related to Mesozoic and younger orogenic and volcanic belts around the Pacific and in the Alpine–Himalayan chain. However, many other types of copper deposits are also closely and directly controlled by plate tectonic processes.*

adding to pollution caused by dust and fumes from smelters.

As well as the input of capital and human labour, mining requires energy. To produce one tonne of copper from ore in the ground under average conditions requires the use of fifteen megawatt hours of fuel. A high proportion is required to produce ore concentrates, and to mine a lower grade of ore there must be a greater energy input (Figure 25.9). The energy needed to supply present world copper demands by mining common rock would amount to a quarter of the world's current energy consumption. Energy expended in this way ends up heating the atmosphere and modifies climate, as can be seen in heat 'islands' around industrial centres. An industrial society with a higher energy consumption could disturb regional climates.

Shortages are not exclusive to metals and energy. Water resources can be overused, leading to shortfalls in supply, as well as being scarce in desert areas and steppe lands. Shortages can be alleviated by aquifer recharge and ponding, and more bizarre solutions that have been investigated include towing icebergs to desert areas and continent-wide flow diversion. There are plans to divert large rivers flowing north to the Arctic Ocean, such as the Mackenzie and Ob Rivers, across low watersheds to irrigate the North American Midwest and the Russian steppes respectively. Northwards-flowing large rivers have a warming effect on both boreal regions and the Arctic Ocean, but their southward diversion would have a net cooling effect, with potentially disastrous consequences.

Even if population growth is stabilized, to maintain the material standards of developed societies and to improve those of the other two-thirds of the world's population requires that exploitation of all resources must increase. Many resources are squandered through inefficiency, but recycling, though imperative, presents no real solution except in an economically static world economy. At present rates of use, lifetimes of current reserves of non-renewable resources are measurable in tens or, at most, hundreds of years. A conventional approach to resources cannot sustain a growing world economy indefinitely. As yet, extraterrestrial resources cannot be contemplated and an interim solution must be sought on the home planet.

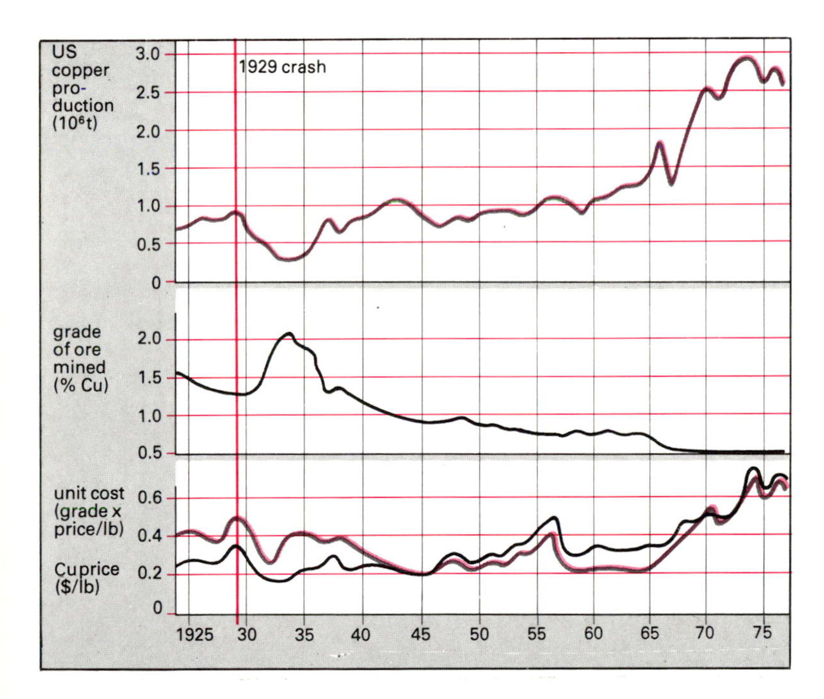

25.8: *The primary role of changes in the whole economy over the fortunes of an individual resource is well expressed by the copper industry. The graph refers to copper mining in the USA and clearly shows the interrelations between metal price, mining costs, production and the relative richness of the ore that could be mined profitably after the worldwide disruption in trade presaged by the stock market crash of October 1929.*

25.9: *Energy requirements for the mining of ores.*

Grade of ore (%)	Energy cost (kWh/t)
2.0	15 000
1.0	22 000
0.5	35 000
0.2	90 000
seawater	500 000
common rock	1 500 000

Note
Each assumes conventional extraction methods.

Future resources

The resource base of the world economy can be broadened by two main thrusts. First, improved techniques of exploration and a fuller understanding of the geology of well-known and as yet unexplored areas may reveal reserves that have not yet been discovered. Second, there may be types of resource and new technologies that have hitherto been overlooked. It is the second category that is the concern here, and alternatives in the field of energy and important metals will now be examined.

Until the 1970s the world's primary fuels were coal and oil, both fossil organic compounds that are definitely non-renewable, having formed slowly in geologically and climatically very special environments. Sooner or later, and in the case of oil and gas probably sooner, they will run out. The statistics of resource depletion have been very carefully studied. Available supplies peak and then tail off so that replacements have by necessity to play a rapidly increasing role in a growing economy.

Literally hundreds of alternative energy resources have been widely publicized, from 'biogas' and ethanol generated by composting agricultural, animal and human waste to natural forces, such as wind, tides and sea waves. No doubt all of them could be locally useful, but even collectively they cannot satisfy the world's needs. This is amply illustrated by geothermal energy; if the Earth's entire heat flow was usable it would just satisfy total projected energy demand by AD 2000. The only suitable geothermal sites for large-scale electricity generation are sparsely associated with active constructive and destructive plate margins, though heating may be widely based upon heat exchange with groundwater in some stable areas of high heat flow.

Various intractable or low-grade fossil fuels, such as tar sands and oil shales, have vast total resources. However, they have an associated energy problem of their own. The energy cost per barrel of oil extracted from them is nearly equivalent to one barrel of cheap conventional oil. They offer no solution though they can be economically profitable.

However, two energy resources, solar and nuclear, offer solutions. As much solar energy falls on a thermal power station per day as the power station generates itself. The main problem lies in the conversion of solar radiation to electricity. This requires very large areas of photoelectric generators, an area equivalent to the state of Arizona being required to satisfy US demand. One solution is the use of generators in stationary orbit beaming the energy down in the form of microwaves. Another is the use of solar furnaces to melt an intermediary, such as common salt, which then 'fires' conventional steam generators. Whatever the technical problems, solar energy is an ideal resource because it is free, non-polluting and does not disturb the Earth's thermal budget.

Nuclear energy from burner-type reactors is an inefficient source since it uses only a tiny proportion of uranium and, therefore, depletes resources. Fast-breeder reactors can use all uranium and also thorium, of which there are huge resources. They generate new fuel in the form of plutonium and require a much smaller input of mined fuel. However, their radioactive waste, which has a long half-life, causes problems of safe disposal. Indeed, the only absolutely final way to dispose of such waste would be to send it by rocket into the sun. The operating conditions for fast-breeder reactors are such that they pose a significantly higher risk of systems failure than burner reactors. One suggestion is to site them in remote areas, perhaps on man-made islands, and to use the energy to electrolyse water and produce hydrogen as an easily transported secondary fuel for electricity generation and industrial feedstock. However, all planned breeder-reactors are sited near areas of high energy demand.

The third form of nuclear energy is controlled thermonuclear fusion of deuterium. The only by-product is helium, itself a very useful material. The total stock of deuterium in seawater and the very high energy produced per fusion reaction offer millennia of supply at levels much higher than required today, but the technical problems associated with producing the most modest controlled fusion have yet to be overcome.

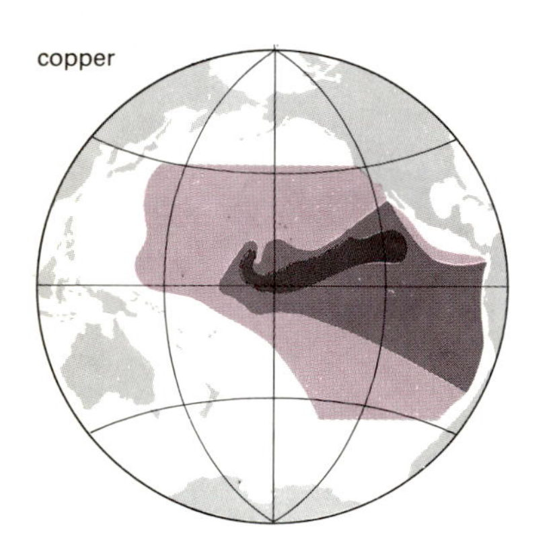

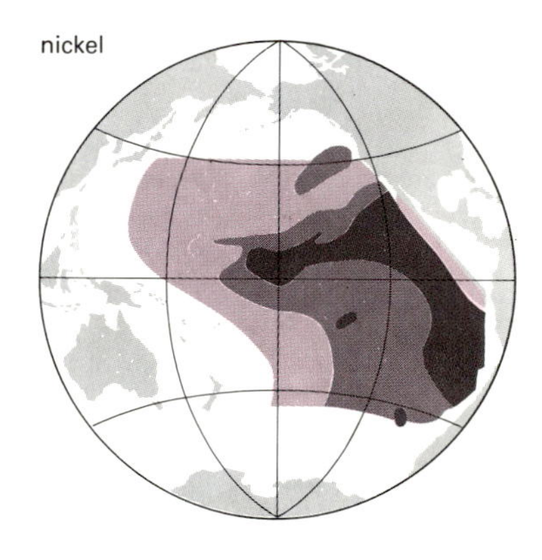

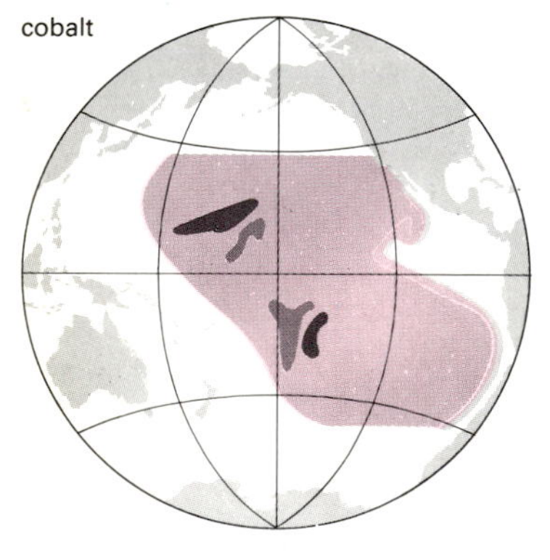

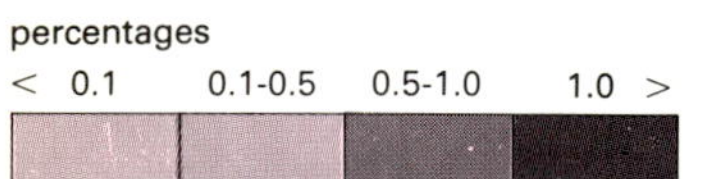

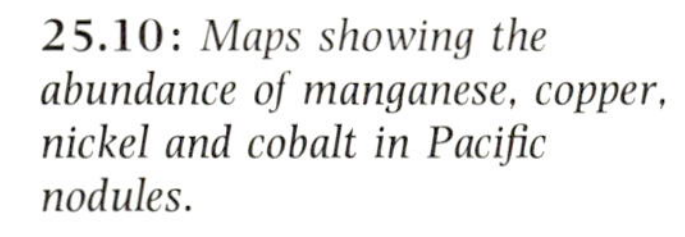

25.10: *Maps showing the abundance of manganese, copper, nickel and cobalt in Pacific nodules.*

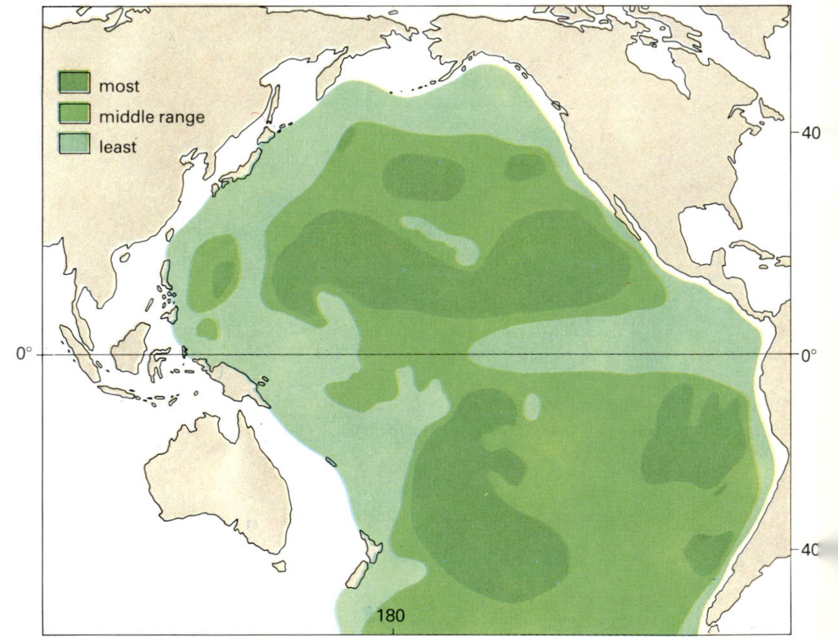

25.11: *Manganese nodules are found in great abundance on some parts of the ocean floor; the concentric structure of nodules is now known to be due to their continual growth, and they may be truly renewable metal resources. Mapping of metal content in Pacific nodules (Figure 25.10) shows that the sites where nodules are most dense and have the greatest content of copper, nickel and cobalt are west of Central America and near the Hawaiian islands.*

25.12: *Metal content of manganese nodules*

Element	Average (Pacific) (%)	Range (Indian and Pacific) Maximum (%)	Minimum (%)	Average minimum exploitable grades of land-based ores (%)
Mn	17.2	34.0	5.4	35.0
Fe	11.8	26.0	4.4	25.0
Ni	0.63	2.0	0.13	1.0
Co	0.36	2.6	0.045	1.0
Cu	0.36	2.5	0.028	0.5
Zn	0.35	0.66	0.010	2.5
Pb	0.047	0.51	0.046	2.0
Ba	0.20	1.6	0.018	
Mo	0.036	0.080	0.0087	0.1
V	0.042	0.093	0.010	0.4
Cr	0.0012	0.012	0.0002	30.0
Cd	0.0010	0.004	0.003	

25.13: *Metal reserves and their lifetimes*

Element	Reserves in land-based deposits (tonnes)	Reserves in Pacific nodules (tonnes)	Lifetime of conventional reserves (years)	Lifetime of Pacific nodule reserves (years)
Mn	8×10^{8}	3×10^{11}	97	60 000
Fe	1×10^{11}	2×10^{11}	240	480
Ni	6×10^{5}	1×10^{10}	150	2 500 000
Co	2×10^{6}	4×10^{9}	110	220 000
Cu	3×10^{8}	6×10^{9}	36	720

25.14: *World metals consumption, and production from manganese nodules*[1].

Element	World metals consumption in 1972 (tonnes)	Annual metals production from nodule-mining (tonnes)	Annual surplus (tonnes)
Fe	6.3×10^{8}	1.9×10^{8}	
Zn	5.5×10^{6}	6.0×10^{6}	5.0×10^{5}
Mn	5.0×10^{6}	2.9×10^{8}	2.8×10^{8}
Cr	5.0×10^{6}	1.7×10^{4}	
Pb	4.4×10^{6}	6.4×10^{5}	
Ni	3.6×10^{5}	1.1×10^{7}	1.1×10^{7}
Mo	7.4×10^{4}	6.0×10^{5}	5.3×10^{5}
Co	2.0×10^{4}	6.0×10^{6}	6.0×10^{6}
Cd	1.0×10^{4}	1.7×10^{4}	7.0×10^{3}
V	5.0×10^{3}	6.3×10^{5}	6.2×10^{5}

Note
1 It is assumed that nodules are mined to supply total world copper requirements.

Mining low-grade resources by conventional means incurs high energy costs. A low-energy approach lies in hydrometallurgical extraction. Instead of separating ore from waste mechanically it is selectively dissolved by some reagent. One interesting development lies in the selective breeding of bacteria that depend on oxidation of sulphides to produce metallic sulphates in solution. They could be introduced into low-grade ore previously pulverized by underground explosion, their waste products being pumped out and electrolysed to win their metal content.

Along the axial rifts of constructive plate margins, such as the Red Sea, pockets of metal-enriched brines and muds have been found. They are thought to result from interaction between seawater and primary magmas (see Chapter 13). They are multimetal resources, the mining of which provides several saleable products. The best known of these pockets contains an estimated two hundred million tonnes of metal-rich mud, and if they are present along all constructive margins their potential is immense.

Another metallic resource is found on the floors of deep oceans. Black potato-sized objects, known as manganese nodules (as it has long been realized that they are rich in that metal), reveal in cross-section concentric growth rings which sophisticated dating techniques show to be added to at rates around a millimetre per million years. Though this may seem to be very slow, there are so many nodules that the yearly growth totals many millions of tonnes. They form by precipitation around small grains from solutions emanating from the oceanic crust and its cover of sediments.

In common with deep-ocean heavy brines, manganese nodules contain a multiplicity of metals, many around conventional ore grades (Figure 25.12). Figure 25.13 shows just how vast a resource those from the Pacific alone constitute (Figure 25.10). As they are growing measurably, they form a renewable resource for any conceivable rate of extraction. If solutions can be found to the technical problems of gathering them from great depths and metallurgically separating their constituent metals, and to the political problems of their location in international waters, they seem to provide the ultimate solution to depletion of many metal resources. However, the present world economy once again poses the major difficulty.

The proportions of metals in manganese nodules do not match the proportions required by world industry (Figure 25.14). An economy in which metal extraction from manganese nodules played a dominant part would produce surpluses of several metals, thereby forcing down prices and bankrupting conventional mining upon which several countries, including Zaïre, Zambia, Peru and Chile, are entirely dependent. Moreover, these surpluses are inevitable and have to be disposed of somehow. This could be polluting to an unimaginable degree.

Scientifically controlled exploitation of nodules and brines combined with conventional production to provide industrially acceptable proportions of metals is possible. However, the only organizations with the requisite expertise to exploit untouched resources are mining companies, which are subject to their investors' requirements for profits; this brings us back to the interaction of economics and Earth sciences. Physical resources are the primary concern of Earth scientists, although their use values, exchange values and ultimate control are outside our sphere of influence.

26 Geological hazards

Earthquakes

Of all geological hazards, earthquakes are among the most frequently occurring and most destructive. While their origins are abrupt movements on faults, which may be deep beneath the ground, their significance lies in the shock waves produced, which may affect large areas of the ground surface. The earthquake regions (seismic zones) of the world lie along the plate boundaries. The most spectacular zone follows the active plate margins that border the Pacific Ocean, and includes Japan, Alaska and California. In contrast, most areas on stable plates are relatively aseismic.

The size of an earthquake may be indicated on the Richter scale of

26.1: *Modified Mercalli Intensity Scale (1956 revision).*

Intensity value	Description
I	Not felt; marginal and long-period effects of large earthquakes.
II	Felt by persons at rest, on upper floors or favourably placed.
III	Felt indoors; hanging objects swing; vibration like passing of light trucks; duration estimated; may not be recognized as an earthquake.
IV	Hanging objects swing; vibration like passing of heavy trucks, or sensation of a jolt like a heavy ball striking the walls; standing cars rock; windows, dishes, doors rattle; glasses clink; crockery clashes; in the upper range of IV, wooden walls and frames creak.
V	Felt outdoors; direction estimated; sleepers wakened; liquids disturbed, some spilled; small unstable objects displaced or upset; doors swing, close, open; shutters, pictures move: pendulum clocks stop, start, change rate.
VI	Felt by all; many frightened and run outdoors; persons walk unsteadily; windows, dishes, glassware break; knickknacks, books, etc, fall off shelves; pictures off walls; furniture moves or overturns; weak plaster and masonry D crack; small bells ring (church, school); trees, bushes shake visibly, or heard to rustle.
VII	Difficult to stand; noticed by drivers; hanging objects quiver; furniture breaks; damage to masonry D, including cracks; weak chimneys broken at roof line; fall of plaster, loose bricks, stones, tiles, cornices, also unbraced parapets and architectural ornaments; some cracks in masonry C; waves on ponds, water turbid with mud; small slides and caving in along sand or gravel banks; large bells ring; concrete irrigation ditches damaged.
VIII	Steering of cars affected; damage to masonry C and partial collapse; some damage to masonry B; none to masonry A; fall of stucco and some masonry walls; twisting, fall of chimneys, factory stacks, monuments, towers, elevated tanks; frame houses move on foundations if not bolted down; loose panel walls thrown out; decayed piling broken off; branches broken from trees; changes in flow or temperature of springs and wells; cracks in wet ground and on steep slopes.
IX	General panic; masonry D destroyed; masonry C heavily damaged, sometimes with complete collapse; masonry B seriously damaged; general damage to foundations; frame structures, if not bolted, shift off foundations; frames racked; serious damage to reservoirs; underground pipes break; conspicuous cracks in ground; in alluviated areas sand and mud ejected, earthquake fountains, sand craters.
X	Most masonry and frame structures destroyed with their foundations; some well-built wooden structures and bridges destroyed; serious damage to dams, dikes, embankments; large landslides; water thrown on banks of canals, rivers, lakes, etc; sand and mud shifted horizontally on beaches and flat land; rails bent slightly.
XI	Rails bent greatly; underground pipelines completely out of service.
XII	Damage nearly total; large rock masses displaced; lines of sight and level distorted; objects thrown into the air.

Note

To avoid ambiguity of language, the quality of masonry, brick or otherwise, is specified by the following lettering.

Masonry A. Good workmanship, mortar and design; reinforced, especially laterally, and bound together by using steel, concrete etc; designed to resist lateral forces.

Masonry B. Good workmanship and mortar; reinforced, but not designed in detail to resist lateral forces.

Masonry C. Ordinary workmanship and mortar; no extreme weakness like failing to tie in at corners, but neither reinforced nor designed against horizontal forces.

Masonry D. Weak materials, such as adobe; poor mortar; low standards of workmanship; weak horizontally.

magnitude (see Chapter 3). More relevant as a measure of earthquake strength is the intensity, for which the modified Mercalli scale is widely used (Figure 26.1). This is a subjective indication of the extent of damage at any point and therefore varies with distance from the epicentre of the earthquake. The scale extends from I to XII and is always expressed in Roman numerals. For each step in Richter magnitude there is roughly a thirty-fold increase in released energy; comparison with intensity can only be approximate (Figure 26.3).

Crucial factors affecting the scale of earthquake hazard include the materials and methods of construction used in buildings. In areas prone to seismic activity, buildings need a stable three-dimensional framework. A disastrous combination, which can result in the pancaking of the whole structure, consists of thin, easily sheared walls and strong, reinforced floor slabs. In the earthquake at San Fernando, California, in 1971 the Sylmar Veteran's Hospital, built forty years before, completely collapsed and forty-four people died. Only 3 km away, the Olive View Hospital, newly constructed with earthquake-resistant design, survived with several hundred people inside safe and untouched. However, all four staircase wings fell away from the main building and an adjacent administration block pancaked—revealing inadequacies in even the most modern designs.

Domestic buildings vary greatly in their response to earthquakes. Wooden-frame houses, so popular in North America, are ideal in their resistance to damage, even surviving actual fault displacement underneath them. In contrast, adobe-type buildings, of rough blocks bonded by clay or lean mortar, are easily destroyed even at intensity VII; their popularity in seismic areas such as Turkey and Iran has accounted for some of the very high earthquake death tolls in these countries.

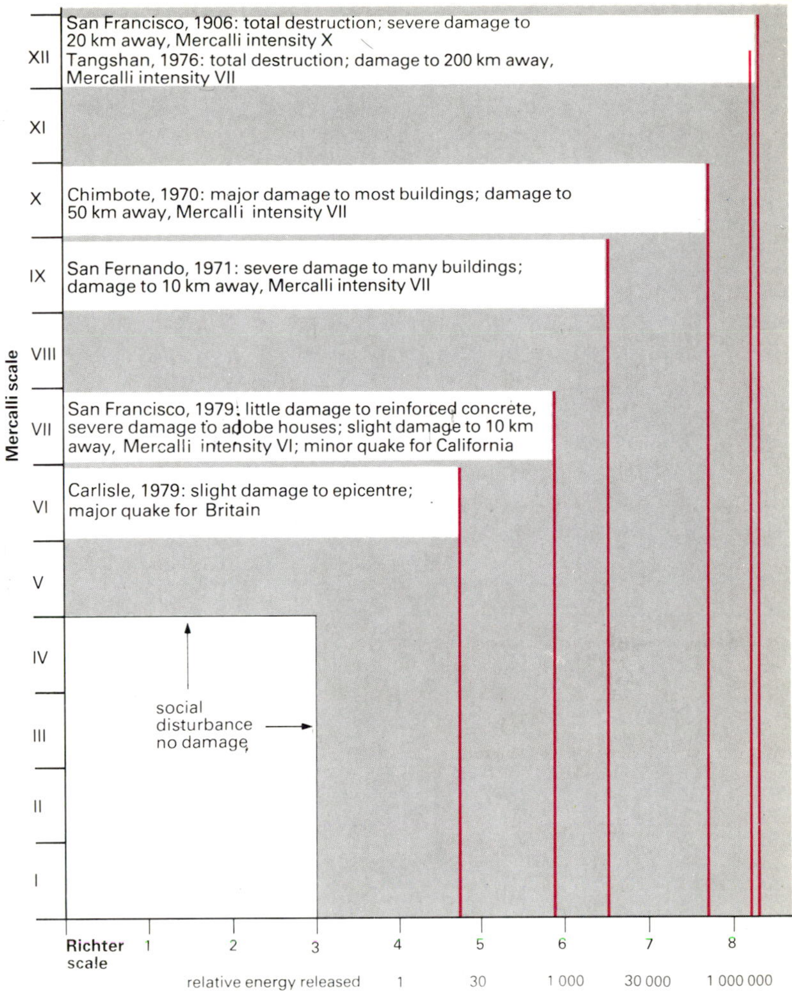

26.3: *The approximate correlation of earthquake damage, magnitude and intensity. Comment on building damage can only be generalized and takes no account of foundation conditions which can be a major influence. Earthquakes of less than magnitude 4 are of only academic interest, but the magnitude 8.2 event at Tangshan killed about 650 000 people.*

26.2: *This building—the Hotel Agava in Bar, Montenegro—crumpled as a result of an earthquake in Yugoslavia and Albania. Some two hundred lives were lost, and Montenegro alone sustained £245 million worth of damage.*

Another major factor in earthquake damage is the nature of the ground on which structures are built. Bedrock is the most stable. Soft ground does not dampen earthquake vibrations but amplifies them, so thick, poorly consolidated sediment saturated with groundwater is the very worst foundation. A simple classification of ground conditions therefore provides one of the best bases for hazard zoning in seismic areas. In such a classification much of the new development in San Francisco and Tokyo, which is on young coastal plains, falls into the high hazard zone.

Development planning that takes account of hazard zoning is one of the best ways of mitigating the potential destruction in earthquake-prone areas. The zoning is not based purely on foundation conditions but must also take into account the many secondary effects of earthquakes. Landslides are the most important. In 1970 a major earthquake in Peru set off a massive slide, which engulfed the town of Yungay and was probably responsible for the deaths of as many as 20 000 people. The slope angle of hillsides is therefore another important element in a hazard zoning map. Of the other secondary hazards, the most important is provided by reservoirs. Dams may be overtopped by seiches (earthquake waves) or may fail because of vibration. In the 1971 San Fernando earthquake the clay-fill dam on the Lower Van Norman Reservoir suffered a major slide on its upstream face. If the water had not been at an abnormally low level at the time, it would have overtopped the dam and swamped the urban development below.

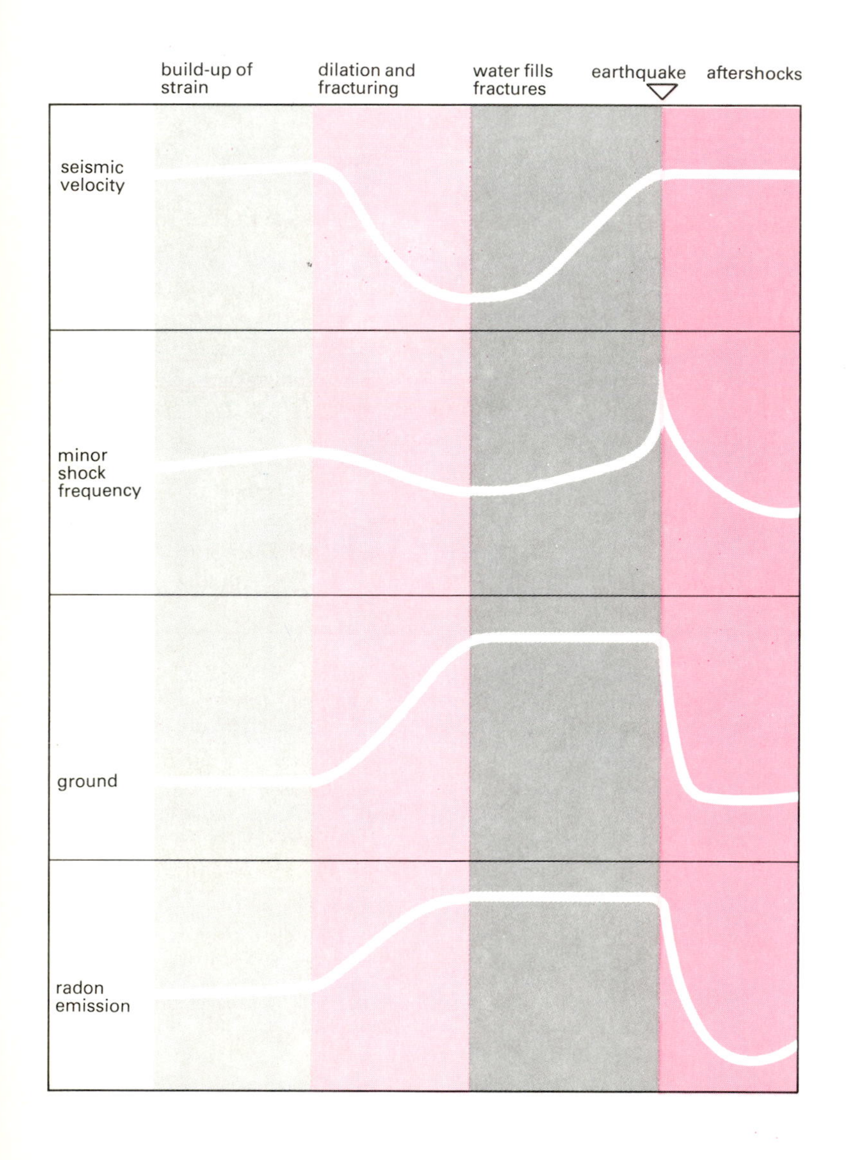

26.4: *Various earthquake precursors of possible value in prediction. The seismic velocity warns of a coming event by recovering to its regional normal as the major shock occurs. Minor shock frequency increases almost as a major event occurs. Increase in ground uplift and radon emission indicate only that an earthquake will probably occur, but cannot be used to predict its time of occurrence.*

Although earthquakes cannot be prevented it may be possible to reduce their scale, for instance, by inducing small non-destructive fault movements in a critical area. Such a critical site is the San Andreas fault zone. Some 200 km south of San Francisco, the two sides of the fault are moving past each other by nearly 20 mm per year, but the movement is slow and smooth with hardly a perceptible earthquake due to it. However, through San Francisco itself the fault has locked solid since 1906. This check to plate movement is building up stresses in the rocks, which one day will give way suddenly in another massive earthquake.

The experiments referred to in Chapter 3, in which water pumped into deep wells induced swarms of minor earthquakes, suggest a means of earthquake control. As increased water pressures in the rock reduce the level of stress required to trigger fault movement, theoretically water pumped into the locked San Andreas fault zone could trigger small quakes before stresses build up enough to create a larger quake on a dry fault. However, the legal consequences for anyone who deliberately caused an earthquake—even if it was in order to prevent a larger quake at a later date—is a deterrent to the application of such a technique.

The significance of water pressures is also shown by reservoir-induced earthquakes. Most of the world's large impounded reservoirs have caused local increases in seismic activity soon after their construction. India's Koyna dam triggered a quake in 1963 that was responsible for the deaths of 177 people. The weight of water in a reservoir, even though immense, is probably not enough to generate earthquake movements on its own. It is more likely that the locally raised groundwater pressures are a cause of this type of quake.

Prediction is, perhaps, a more realistic way of reducing earthquake hazard than is control. In 1975 the Chinese successfully predicted a major earthquake at Haicheng and, by timely evacuation, almost eliminated any loss of life despite the scale of the destruction. There have been other successful predictions of smaller quakes but there have been even more false alarms and earthquakes that were not predicted. The problem is that every earthquake is different, and although a number of warning signs may be available (see Chapter 3) most give no indication of the time when a quake may be expected (Figure 26.4).

Volcanoes

For purposes of hazard assessment, volcanoes can be broadly divided into two groups, according to the composition of their magmas. Basaltic volcanoes, associated with oceanic crust, produce large quantities of lava with minimal explosive activity. In contrast, more acidic volcanoes, normally associated with destructive plate margins, tend to be violent and unpredictable. The viscosity of the magmas can lead to a build-up of gas pressures, resulting in massive explosions such as those of Krakatoa and Santorini.

As a hazard to man basaltic volcanoes are relatively unimportant. They may even represent quite safe tourist attractions, as in Iceland and Hawaii. The front of a lava flow rarely advances faster than walking pace, so it offers little threat to human life, and flow patterns, dictated by topography, are readily predicted. The same volcanoes may produce heavy falls of ash. The 1973 eruption of Iceland's Kirkefell was unfortunately close to the fishing town of Vestmannaeyjar. Lava overran a small part of the town and threatened to block the entrance to the vital harbour. Though ash covered almost the whole town and some houses collapsed under its weight, most was shovelled away after the eruption ceased and a large part of the town has since been reoccupied.

A redeeming feature of the more violent acidic volcanoes, the normally long period between their eruptions, has to be set against their unpredictable patterns of behaviour. As well as proving the hazard of explosions on the scale of Krakatoa and Santorini, though these are rare events, acidic volcanoes can produce flowing gas clouds, *nuées ardentes* or pyroclast surges, which may develop almost without warning during a period of prolonged and less turbulent activity. Hot gases, charged with varying proportions of red-hot ash, can descend a mountainside at well over 100 km per hour bringing instant death and devastation. The destruction of Pompeii by Vesuvius in 79 BC and of St Pierre by Mont Pelée, a disaster in 1902 in which 29 000 lives were lost, were both of this type.

Volcanoes are clearly beyond the control of man, but prediction and risk assessment can reduce the hazards. Planning is an effective defence on many basaltic volcanoes if it prevents development in the radiating valleys that funnel both lava flows and mudflows (lahars) which may develop in saturated ash. Experience in Iceland and Hawaii and on Italy's Mount Etna has proved that lava flows can even be redirected if ground conditions are favourable.

However, zoning is of little use with respect to explosive volcanoes and there the only protection is through prediction. Symptoms of impending eruption include measurable expansion of the volcano; increases in small-scale activity of the volcano; increased earthquake activity at significant depths, indicating magma movement; changes of gas components of fumaroles; and alterations to local magnetic fields, influenced by rock melting and magma generation. Accurate prediction of eruptions from such signals requires study of a particular volcano over long periods. A permanent observatory and many years of monitoring have facilitated reasonably confident predictions concerning Hawaii's Kilauea volcano. But Kilauea is an unusually well-behaved basaltic volcano and few of the world's volcanoes are so extensively monitored, and some are not at all.

Tsunamis

The huge waves, tsunamis, generated by abrupt physical displacement of the sea bed as a side effect of volcanic activity or earthquakes are often incorrectly known as tidal waves despite being unrelated to tidal movements. Massive volumes of water are involved, and the waves travel at up to 800 km per hour. Over ocean depths tsunamis are imperceptibly broad and low, but on approaching land they are slowed down by the shallower water and increase in amplitude. If funnelled by a coastal inlet, a tsunami can arrive at a shore as a destructive breaker over 20 m high. Very large tsunamis occur less than once every ten years and nearly all originate in the Pacific Ocean. Japan has suffered the largest share of tsunami disasters, including one in 1896

26.5: *In 1974 Kirkefell erupted on the Icelandic island of Heimaey. Ash covered most of the town of Vestmannaeyjar (which in this picture is masked by a snow cover), while lavas caught the edge of the town and threatened to block the vital harbour entrance. The steam rising off the lava is from jets of water sprayed onto it to cool it and halt its progress towards the town and harbour.*

that was responsible for the death of 27 000 people. One of the few recorded major volcanic tsunamis was unleashed by the explosion of Krakatoa in 1883 and accounted for nearly all the fatalities from that famous eruption.

The obvious precaution of restricting development on exposed Pacific shores is uneconomic, given the infrequency of tsunamis. As a tsunami takes twenty hours to cross the Pacific a monitoring system can alert vulnerable areas so that coastal evacuations can be organized. However, for the area close to a source earthquake, a tsunami will always be a hazard. The Seismic Sea Wave Warning System, which covers the Pacific and operates from Hawaii, was established after the destruction of the Hawaiian town of Hilo by the tsunami of 1946. Information from seismic stations around the Pacific and a series of sea-level indicators in the northern Pacific are used in conjunction with charts of wave travel time to predict arrival times.

Landslides

One of the most spectacular examples of a landslide is to be found on the floor of the Saidmarreh valley in western Iran. A huge blanket of broken limestone, covering over 200 km^2 and in places as much as 300 m deep, is the debris from a single prehistoric landslide, resulting from the failure of Kabir Kuh, a hillside 1000 m high. Devastating though the impact of such a failure might be on a major town, landslides are rarely on a scale comparable to seismic or volcanic events. Furthermore, many of the larger slides are slow-moving so, while destructive to property, they present little threat to life.

In many respects landslides are very ordinary, natural features of erosion; they merely occupy one end of a spectrum of slope degradation processes. Whether a slope degrades progressively and harmlessly or as a landslide normally depends on geological structure—dipping masses of bedded sedimentary rocks and easily sheared clays provide the most dangerously unstable slope conditions. The hazard to man arises from the near impossibility of quantifying the degree of instability of a given hillside. The weakening effect of irregular networks of joints deep beneath the surface cannot be accurately assessed. Many towns and villages, notably in Canada and Scandinavia, are built on unstable, sensitive clays. These fail even on very low slopes and it is practically impossible to predict just when and where the initial failure will take place.

The mechanisms by which a landslide can be triggered off vary considerably. In 1950 the Swedish village of Surte was destroyed in a slide triggered by vibrations from a pile-driver working on the sensitive clay that underlay the houses. Earthquakes are an obvious source of destabilization. In 1920, in Kansu in China, tremors set off landslides that were responsible for 200 000 deaths. Removal of the toe of a potential landslide is a sure way of disturbing the balance of the area and promoting slope failure; the Saidmarreh slide was caused by the River Karkheh cutting into the foot of the Kabir Kuh hillside. On a smaller scale, quarrying and road cutting excavations have frequently caused significant landslides. A similar effect results from loading the head of a potential slide; in a number of cases housing has provided the critical overload factor on marginally stable hillsides.

Water is a particularly significant factor of hillside stability. Groundwater provides by its own hydrostatic pressure a measure of support for the rock that it permeates and by doing so it lowers the frictional cohesion within the rock mass as a whole. An increase in this critical 'pore water pressure' can therefore reduce the effective strength of a hillside to the point when it becomes unstable and may fail as a slow or rapid landslide. Mountain areas with monsoon climates suffer seasonal landslide damage with monotonous regularity. Continuous monitoring of water pressures is a vital concern of many hillside construction programmes. Drainage is, as the foregoing might suggest, the most effective tool in the control of landslides. The large landslides at Folkestone Warren in England, for instance, have been stabilized by the installation of an effective underground drainage scheme (Figure 26.6).

Subsidence

Generally subsidence only becomes a significant geological hazard when it is accelerated by human interference with the ground. Mining is a straightforward case. Subsidence can also be caused by the pumping of fluids, mainly water, from compressible sediments. The slow sinking of Venice is largely because the city is built on a thick succession of soft sediments which are slowly compacting. The clays and silts within the sediment sequence are partially supported by their pore water pressure, and loss of pressure efficiently hastens compaction. The interbedded sands are productive aquifers, which the industrial hinterland of Venice, with its huge demand for water, has exploited through massive groundwater pumping from beneath the city.

Venice is not the only case of serious subsidence. Mexico City has

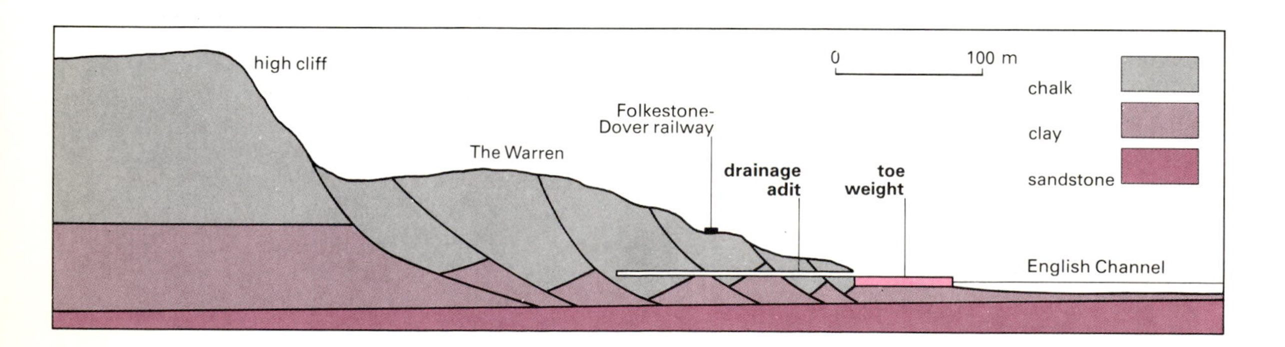

26.6: *Representative cross-section through the complex landslide mass of Folkestone Warren, showing the massive concrete toe weighting and one of the many drainage adits which have resulted in stabilization of the landslip and safe operation of the Folkestone–Dover railway, which traverses its entire length.*

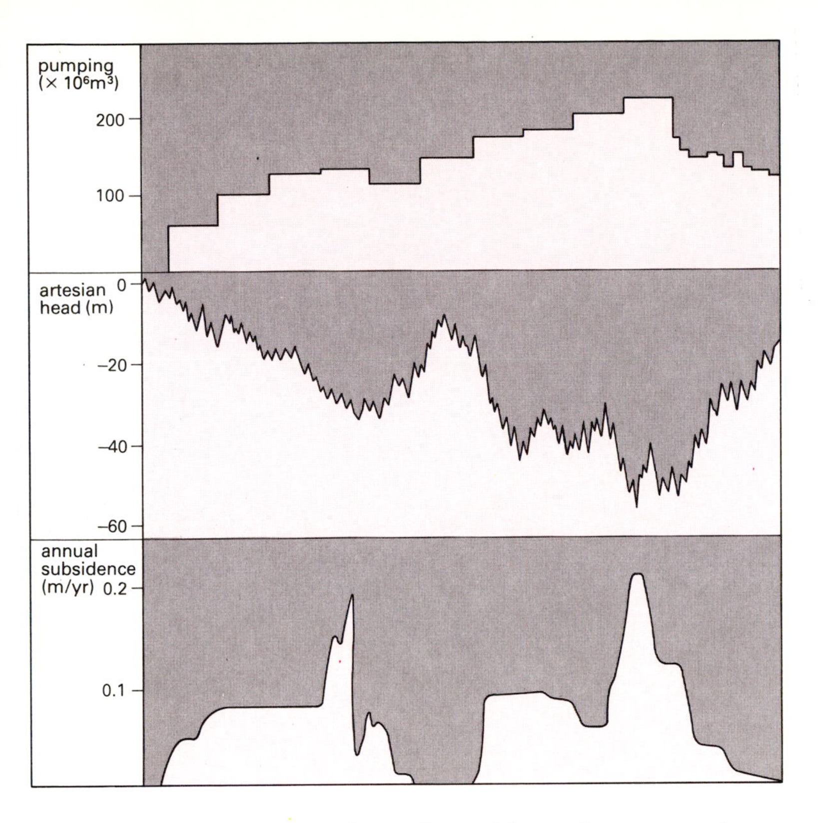

26.7: *Subsidence in the Santa Clara Valley, California, due to compaction of sediments. The rate of subsidence is closely related to the artesian head of the groundwater, which in turn is influenced by artificial pumping. By 1976 a combination of decreased pumping and higher rainfalls allowed sufficient recovery of the artesian head to stop the subsidence.*

26.8: *The motorway bridge near Salzburg in Austria failed in 1959 when heavy rainstorms in a mountainous catchment area provided very high peak flood flows in the river.*

settled in parts by up to 9 m, and subsidence in the central valleys of California (Figure 26.7) has affected an area of over 13 500 km^2. These cases, too, have been due to excessive groundwater pumping. Since the recognition of the cause of subsidence, a major reduction in water extraction has produced a considerable reduction in subsidence rates at all three sites. Venice has stopped sinking since well-pumping was placed under strict controls in 1969, when alternative water supplies from the Alps became available.

Subsidence can also be a hazard where cavernous limestones underlie a blanket of sediments. Collapse of the sediments into voids in the limestone is a rare event, except where local changes in drainage conditions, due to human activities, can act as a trigger. In the Johannesburg region of South Africa, thirty-four lives and hundreds of homes have been lost since 1962 as a result of sinkholes developing over limestones de-watered by gold-mining operations. The drainage factor is critical, and experience in American limestone terrains has shown that sinkholes are most likely to form just after the land has been developed for building, a frustrating aspect of otherwise unpredictable sinkhole subsidence over limestones.

Water

The greatest of all geological phenomena hazardous to man is water. Over the past one hundred years the Hwang Ho (Yellow River) of northern China has been responsible for the deaths of over six million people, either directly by drowning or indirectly by famine caused by flooding of the farmland that covers the extensive floodplain. Throughout the world, cities large and small, sited on similarly active floodplains suffer repeated damage and destruction.

But water is also the key to so many other natural hazards. Most subsidence is related to water movement, and the worst recent landslide disaster was caused by the construction of Italy's Vaiont Reservoir, which precipitated a landslide into itself and drowned two thousand people in its floodwave. Perhaps the greatest threat to life in the next inevitable large-scale earthquake in California lies in the likelihood of reservoirs failing. Earth scientists clearly have a responsibility in all such hazard-prone areas to apply their understanding of geological processes to the minimization of the potential damage and suffering.

Part Six: Extraterrestrial Geology

Some of the most exciting pictures to reach the press and television screens of the world in recent years have been close-ups of other worlds in the Solar System. Scientist and non-scientist alike have gazed in awe at the details of Mercury's surface, at the shining disc of Saturn's rings, and at the sulphurous volcanic outbursts on Io. The space programme has brought the realm of the astronomer within the grasp of the Earth scientist, who can now apply his methods and test his theories on completely different planetary bodies.

There was some discussion in the late 1960s about the propriety of applying the term geology to the study of the Moon and planets, but science has been spared a welter of new sub-disciplines termed 'selenology', 'martology', and 'venerology'. Extraterrestrial geology is not a self-contradiction, since there is more to the science than its subject matter: what counts is the *methods* that the geologist, geophysicist and geochemist are now applying so successfully to the quantities of remote-sensed data that have resulted from recent space missions. The study of the Earth itself benefits in turn from the sharpening of those methods against new and often unexpected planetary phenomena. The earlier chapters of the Earth's own history will only be told when that history, in deep-frozen form, can be read from those other members of the Solar System whose active evolution ceased long ago, at a time when our own was only just beginning.

The lunar module Intrepid *landing on the Moon during the second manned mission, photographed from the Apollo 12 command service module, 14 November 1969.*

27 The geology of the solar system

The Moon and planets have been subjects of absorbing interest to the human race for millennia. Until the seventeenth century human knowledge was limited to what could be gleaned by observation with the naked eye. This did not amount to much, but it was sufficient for Kepler to establish the laws of planetary motion, and the relative distances of the planets from the Sun. The advent of the telescope in 1609, and its subsequent refinement, greatly extended knowledge of the solar system. Not only were three new planets discovered, but so also were a whole swarm of satellites, and a multitude of statistical data on the sizes, masses and densities of the planets was accumulated. The Moon in particular was an easy subject for telescopic observation, and over the years its face was charted in minute detail. The planets proved much more difficult. Almost nothing was learned about Mercury because it is so close to the Sun. Nothing could be learned about the surface of Venus, which is perpetually shrouded in cloud. Mars is so small and so distant that even the best telescopic view of it is akin to observing the Moon with the naked eye, and thus serious misconceptions were easily formed.

Telescopic study of the planets, then, was frustrating and unsatisfactory. While much was learned of their physical characteristics by years of patient observation, it is fair to say that almost nothing was learned about their geological evolution. The age and composition of the surface rocks, their volcanic history, their surface processes and so on all remained inaccessible.

Recent years have seen a transformation. Specifically, the era of spacecraft exploration has broadened the study of the Moon and planets, which are now accessible to investigation not only by astronomers but also by geologists and geophysicists. For the first time men and women have been able to study the history of planets other than their own.

27.1: *Fundamental planetary data*

	Mean distance from the Sun (km × 10^6)	**Period of revolution**	**Period of rotation**	**Equatorial diameter** (km)	**Mass** (Earth = 1)	**Density** (10^3 kg/m^3)	**Surface gravity** (Earth = 1)	**Number of known satellites**	**Oblateness**[1] $\frac{r_e - r_p}{r_e}$
Mercury	57.90	88.00 days	59.00 days	4 880	0.055	5.4	0.37	0	0
Venus	108.20	224.70 days	243.00 days retrograde	12 104	0.815	5.2	0.88	0	0
Earth	149.60	365.26 days	23.93 hours	12 756	1.0	5.5	1.0	1	0.003
Mars	227.90	687.00 days	24.61 hours	6 787	0.108	3.9	0.38	2	0.009
Jupiter	778.30	11.86 years	9.83 hours	142 800	317.9	1.3	2.64	14	0.06
Saturn	1 427.00	29.46 years	10.23 hours	120 000	95.2	0.7	1.15	11	0.1
Uranus	2 869.60	84.01 years	11.00 hours retrograde	51 800	14.6	1.2	1.17	5	0.06
Neptune	4 496.60	164.80 years	16.00 hours	49 500	17.2	1.7	1.18	2	0.02
Pluto	5 900.00	247.70 years	6.37 days	2 700	0.002	1.5	?	1	?

Note A sphere has oblateness = 0 and a disc has oblateness = 1; r_p = polar radius and r_e = equatorial radius.

The Moon

The Moon is one of the smaller bodies in the solar system, but it is the body that has been most closely studied. The data obtained on the Moon, and the techniques developed for its investigation, have been of first importance in the study of the planets. It has a diameter of 3476 km, a mass of 7.4×10^{22} kg, and a density of 3.3×10^{3} kg/m^{3}, and it orbits at a mean distance of 384 000 km from the Earth in a period of 27.32 days.

The surface features of the Moon

The surface of the Moon is dominated by circular crater-like structures ranging from submetre-sized pits to ring basins many hundreds of kilometres in diameter (Figure 27.2). The latter form the smooth, darker-coloured patches known as seas, or maria, which are visible to the naked eye. Although the presence of craters on the Moon has been acknowledged since Galileo made the first systematic observations of it in 1609, their origin was the subject of controversy until the 'Apollo' era. One principal school of thought, propagated by the American mineralogist J. D. Dana in 1846, proposed that the craters were all formed by internal processes, and thus were comparable with terrestrial volcanic craters. The other, contrary, school, initiated by the American geologist G. K. Gilbert, in 1893, considered that they had an external origin in the impact of meteorites and asteroids.

Telescopic observations could not reconcile this dispute. In the period prior to the 'Apollo' landings intensive studies were made of terrestrial impact craters, craters formed during underground tests of nuclear weapons and simulated craters made in laboratory tests. This work led to general agreement that almost all lunar craters of whatever size are of primary impact origin. Craters of purely volcanic origin are exceedingly rare and are difficult to distinguish from small impact craters.

The very large ringed basins, such as the Mare Imbrium, and some of the larger craters, such as Plato, appear to be floored with basalt lavas. This led to the suggestion that the impact events producing the circular structures were so profound that they opened up fractures within the lunar crust and that magmas derived from partial melting of the lunar mantle later welled up along these fractures to be erupted in voluminous lava flows which eventually filled the circular basins. The very thin, extensive nature of the lava flows suggests that they were of low viscosity, and explains why no large volcanic edifices, comparable to terrestrial volcanoes, are found on the Moon. Most of the mare lavas appear to have been erupted from obscure fissures around the rims of the circular basins.

Perhaps the most conspicuous features of the full Moon when it is observed with low-power binoculars are brilliant, pale-coloured rays radiating around craters such as Tycho. These are believed to be thin streaks of ejecta flung out during the impact event that formed the crater; those from Tycho extend for 2000 km radially around it. Rays are found only around young craters because exposure to solar radiation causes the material in the ejecta to fade.

Rilles are essentially linear structures of negative relief. Many different kinds have been recognized since the first, Schroeter's Valley, was discovered by the German, Schroeter, in 1787. Crater chain rilles consist of small craters aligned along graben-like structures or of discontinuous chains of small craters. The former probably have a volcanic origin, analogous to terrestrial volcanic craters aligned along a fissure or rift; the latter were probably formed by secondary impacts, when streams of ejecta were hurled away from neighbouring primary impact sites and fell back to the lunar surface spreading radially around the primary site. Sinuous rilles (Figure 27.4) are the lunar morphological feature most closely equivalent to terrestrial river valleys, but are different in many key respects: many commence at craters, they have a greater depth/width ratio than rivers, they have a smaller meander wavelength, they may be discontinuous for part of their length and the bends are often parallel-sided.

The stratigraphy and geochronology of the Moon

Studies of crater densities on the Moon have revealed that the light-coloured 'highland' areas are much more heavily cratered than the darker-coloured 'seas', and are therefore older. More detailed studies have shown that the highlands have reached saturation, ie no new crater could be formed without obliterating older ones, and that sur-

27.2: *Craters and structures of impact origin are conspicuous in this Apollo 8 view of the lunar surface. The dark circular 'seas' are themselves basalt-filled impact basins and rays of ejected material are visible radiating from the small crater near the top of the picture.*

faces with distinctly different ages could be recognized by crater-counting in the mare areas.

In 1962 Eugene Shoemaker and Robert Hackmann of the US Geological Survey refined these concepts, and presented the first lunar stratigraphy. Their work was based on the same principles as terrestrial stratigraphy, namely the observation of a sequence of rock groups resting on top of each other. However, instead of using sedimentary strata as marker horizons they used ejecta blankets from impact events. They established the following time scale:

Present time
Copernican period
Eratosthenian period
Procellarian period
Imbrian period
Pre-Imbrian period
Beginning of lunar history

This time scale was ultimately adopted, in a slightly revised form, by the US Geological Survey, which commenced geological mapping of the whole of the near side of the Moon at a scale 1:1 000 000.

Perhaps the single most important result of the Apollo project was the recognition of the extreme antiquity of the Moon. Geochronological work on samples returned by the Apollo landings enabled a detailed calibration of Shoemaker and Hackmann's time scale to be established. It was found that all the pale-coloured highland areas were formed before about 4000 million years (Ma) ago, while the younger basalt lavas infilling the ring basins were erupted between 3800 and 3200 Ma ago. The importance of these data is that the mare basalts are unequivocally the youngest lunar rocks, whereas the oldest known terrestrial rocks are 3800 Ma old. Clearly the geological activity of the Moon lasted for only a short time after its formation 4600 Ma ago; it appears to have been more or less inactive for the past 3200 Ma.

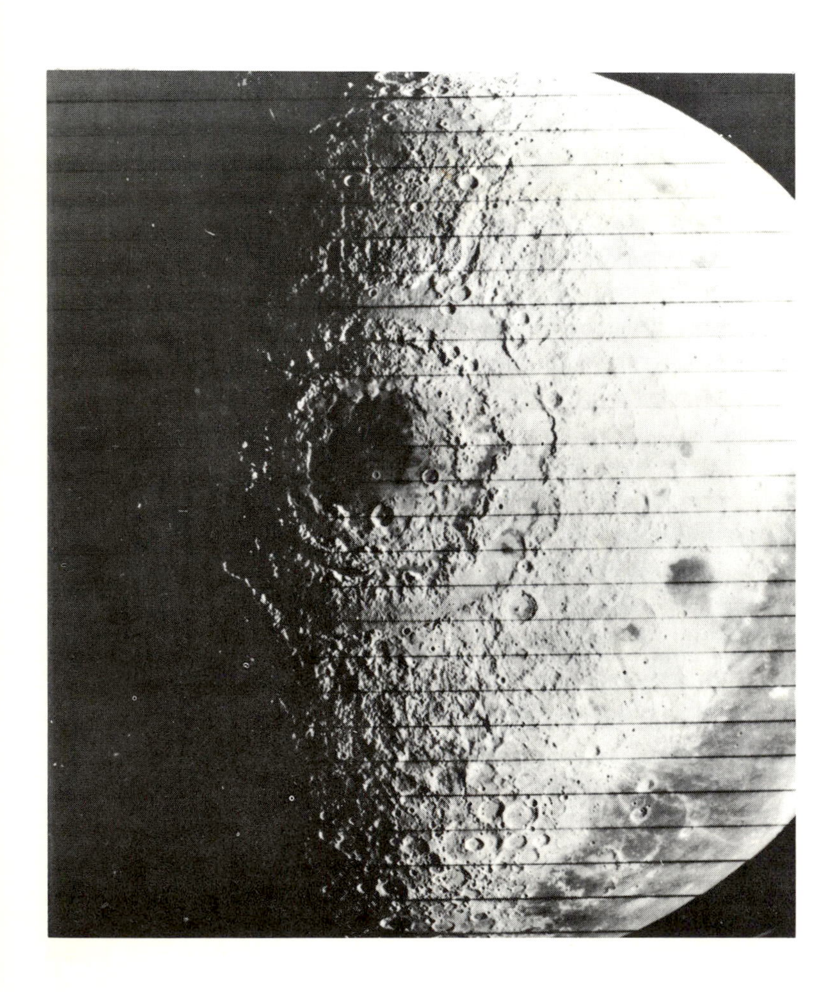

27.3: *The magnificent Orientale basin, more than 1000 km in diameter, demonstrates spectacularly the scale of lunar impact structures. Concentric ridges were thrown up by complex rebounds in the moments after impact. (The dark bands are artefacts of the Lunar Orbiter IV imaging system.)*

27.4: *The origin of sinuous rilles like this one near the Apollo 15 landing site (bottom) is not well understood, despite the results of landing missions. The most likely explanation is that they are collapsed lava tubes, similar to those observed on large terrestrial basaltic volcanoes.*

The petrology, mineralogy and geochemistry of the Moon

A total of 381.69 kg of lunar rocks was obtained by the Apollo missions; the unmanned Russian spacecraft have obtained a further few hundred grams. The specimens consist of regolith material, glasses, breccias and solid rocks.

The regolith material is lunar rock waste; it consists of fine-grained, dusty particles produced by the churning up of the outer surface of the Moon by meteorite impacts. Glasses form an important part of the regolith. They occur either as agglutinate, cementing together fragments of the other rocks, or as discrete, smoothly rounded spheres and dumbbells. These were produced when meteorite impacts took place, the energy of the impact both melting rocks at the impact site and hurling droplets of molten glass huge distances over the Moon. The breccias are angular fragments of other rocks welded together in a glassy matrix. The fragmentation was caused by the shock waves set up by large, crater-forming events, and many breccias show evidence of shock metamorphism, which is often so intense that the original nature of the rock is no longer discernible.

The solid rocks can be loosely divided into two broad groups. The highland rocks consist of gabbroic anorthosites and related basic cumulate rocks. The high plagioclase content of these rocks is responsible for their relatively pale colour. The mare rocks are basalts closely similar in physical appearance to terrestrial basalts, and with similar minerals: feldspars, pyroxenes, olivines and oxides (Figure 27.5). The chief difference petrographically between lunar and terrestrial basalts is that the former are remarkably unaltered. Hydrous minerals are almost entirely absent on the Moon, because of the lack of water, and so also are the weathering and alteration products commonly seen in terrestrial basalts.

The vast majority of lunar feldspars are calcic plagioclases falling within the range bytownite to anorthite (An_{75-100}); sodium and potassium feldspars are rare. Pyroxenes are extremely abundant on the Moon. Interestingly enough, pyroxenes with compositions intermediate between hedenbergite ($CaFeSi_2O_6$) and ferrosilite ($FeSiO_3$) are not known on Earth, but they are found on the Moon and constitute some of the very few new minerals found there. Lunar olivines show broadly the same range in composition as terrestrial ones between magnesium-rich forsterite and iron-rich fayalite, but some have unexpectedly high chromium contents–up to 0.6 per cent. Ilmenite is the most abundant opaque mineral, forming up to 20 per cent of some rocks. Armalcolite $(Fe, Mg)Ti_2O_5$ is a completely new mineral, named after the Apollo 11 astronauts *Arm*strong, *Al*drin and *Col*lins. It closely resembles ilmenite in appearance. Troilite (iron sulphide) and native iron are both observed in lunar basalts, and are rare or absent in terrestrial lavas; their presence reflects the strongly reducing (anoxic) lunar environment.

Chemically the lunar basalts are broadly similar to terrestrial basalts, but the lunar rocks contain more titanium and iron, and less aluminium. Perhaps the most important difference, however, is that the lunar basalts contain much less of the alkali metals sodium and potassium, often only one-tenth as much as terrestrial basalts.

It is thought that the highland rocks represent plagioclase accumulations that formed early in the Moon's history, and that the mare basalts were derived later by partial melting in the lunar mantle. Considerable variations exist in the lunar basalts returned from different landing sites, and between the youngest and oldest rocks sampled. The earliest basalts were probably derived from partial melting of materials in the lunar mantle at relatively shallow levels, while the latest were derived from the deepest levels. The sources at shallow levels probably consisted of plagioclase and pyroxenes; the deeper ones had increasing proportions of olivine and less plagioclase. The progressive deepening of the melting zone with time is attributed to the decreasing concentrations with depth of the radioactive elements potassium, uranium and thorium; these elements were concentrated at a shallow level just beneath the lunar crust. With increasing depth radiogenic heat would take longer to accumulate, and hence melting would be longer delayed. According to this model it is the lack of heat sources below a depth of 300 km, and the thickness and rigidity of the material above, that caused lunar volcanism to fizzle out about 3200 Ma ago.

Seismic studies and the internal structure of the Moon

An essential part of the Apollo lunar science programme was the establishment of seismic stations. Several thousand minor 'moonquakes' were detected, and a small number of events attributed to meteorite impacts. The moonquakes were all extremely feeble, ranking

27.5: *Coarse-grained basalt from the Apollo 17 landing site. Coloured minerals are mostly pyroxenes, grey-striped grains are plagioclase feldspars.*

at 1.3 or less on the Richter scale (earthquakes below about 5 on the Richter scale do not cause perceptible damage; see Chapter 26). The characteristic feature of all moonquakes is that they are extremely deep-seated, taking place at depths of up to 1000 km. Terrestrial earthquake focuses below about 500 km are, in contrast, very rare. Moonquakes are not caused, therefore, by the same kind of shallow-level fault movements that cause earthquakes.

Studies of the frequencies of quakes showed that they exhibit marked periodicities, being most frequent when the Moon's orbit is most highly elliptical and least frequent when it is most nearly circular. When the orbit is at its most elliptical the differences in the gravitational interaction between the Earth and Moon reach their maximum as the distance between the two bodies then varies most. These interactions cause minute distortions of the whole body of the Moon, and it is the energy released in these distortions of shape that is manifested as quakes.

Analysis of the travel times of seismic signals from quakes, meteorite impacts and space vehicles deliberately crashed onto the Moon has enabled the internal structure to be determined. The crust is about 60 km thick and consists of material in which P and S wave velocities are relatively low—probably anorthositic rocks. The giant ring basins are excavated in this crust, and gravity studies show that they may be filled with basalt lavas to a depth of 20 km. Below the crust is the lunar mantle, which probably closely resembles the Earth's and is composed of silicate minerals such as pyroxene and olivines.

The suppression of S waves and the slower propagation of P waves suggest that there is a transition from solid mantle to a core at a depth of about 1000 km. The lack of P wave reflections from the mantle/core boundary has been used to argue that the boundary is diffuse, rather than sharp. The suppression of S waves suggests that the core is probably partly fused, but it may not be completely molten. The composition of the lunar core is very uncertain. The low bulk density of the Moon indicates that an iron-nickel core like that of the Earth cannot be present, but an iron sulphide mixture is a possibility.

The Moon's magnetic field

The absence of a mobile metal core within the Moon has been cited to account for the fact that the Moon's magnetic field is extremely weak. The intensity is about 5 gammas, compared with the Earth's 60 000 gammas at the equator. The Moon's field does resemble the Earth's, however, in that it is dipolar, and the north magnetic pole is located near the Moon's North Pole.

Studies carried out on the Moon and on returned samples show that the natural remanent magnetism (NRM) of lunar rocks is stronger than the present field, and varies from place to place, reaching 313 gammas at the Apollo 16 site. Some attempts have been made to estimate the intensity of the original magnetic field. The results suggest that between about 4000 and 3000 Ma ago the Moon had a field with an intensity of about 2000–3000 gammas, and it is generally agreed that the lunar rocks acquired their NRM by cooling through their Curie points in magnetic fields of at least this strength.

Several hypotheses have been advanced to account for this strong ancestral field. One possibility is that the Moon may have had a molten, convecting core early in its history. This would have been capable of providing the same kind of dynamo that is believed to be responsible for the Earth's magnetic field. A second possibility is that a permanent magnetic field was imprinted on the Moon by some external agency. It is possible that early in the history of the solar system the Sun's magnetic field could have been up to a hundred times stronger than now.

Mercury

Mercury is the smallest and most dense of the inner planets (Figure 27.1). Because it is never more than 69.7 million km from the Sun it is the most difficult of the planets to observe from the Earth. It is only visible for short periods just before sunrise and just after sunset. At these times the planet is necessarily low in the sky, and the effects of turbulence are intensified by the long line of sight through the atmosphere. As a result surface features are extremely difficult to observe telescopically.

For many decades it was universally accepted that Mercury's rotation period was the same as the revolution period, namely 88 days, with the implication that Mercury kept one face pointing always towards the Sun. In 1965 radar studies showed that the rotation period was 58.6 days. This indicates a slightly more complicated relationship between rotation and revolution, in that Mercury rotates on its axis three times in the period during which it completes two revolutions around the Sun. A consequence of this spin–orbit coupling is that solar radiation reaches a maximum at two fixed points on Mercury's equator, known as the 'hot poles'. The temperature at the hot poles is believed to rise as high as 430°C, more than enough to melt lead. Night-time temperatures on Mercury may fall as low as −173°C, since there is no atmosphere to retain heat. The enormous diurnal temperature range must have important implications for surface processes, but these remain to be investigated.

The surface features of Mercury

The surface of Mercury remained effectively unknown until the Mariner 10 mission obtained the first pictures in 1974 (Figure 27.6). Superficially Mercury looks much like the Moon; a heavily cratered world lacking familiar terrestrial landforms. In detail, however, many important differences can be discerned. Three are particularly significant. First, on the Moon there are obvious differences between the pale-coloured highlands and the darker maria. No such simple distinction exists on Mercury. Second, and related to the first difference, while there is abundant evidence of extensive basaltic volcanism on the Moon, the evidence is much less clear on Mercury. Third, the greater acceleration due to gravity on the surface of Mercury produced many subtle differences in crater morphologies. For example, the secondary craters produced by impacts on Mercury are far less widely dispersed than those on the Moon.

The largest single feature on Mercury is the Caloris basin, a large ring basin analogous to the major ring basins on the Moon, and in common with them formed by the impact of a body of asteroidal dimensions. The basin is some 1300 km in diameter and is surrounded by a ring of high mountains rising some 2 km above the plains inside the basin. The mountains are thought to be blocks of the Mercurian crust thrown up by shock waves associated with the impact. The impact also threw out huge quantities of ejected material which fell back

27.6: *One of the first pictures obtained by the Mariner 10 spacecraft of the planet Mercury in 1974. The superficial similarities to the Moon are obvious.*

around the impact site to excavate conspicuous ridges and grooves radiating out around the basin. This terrain, which is similar to that surrounding the Mare Imbrium on the Moon, has been called lineated terrain by the geologists who compiled the first maps of Mercury, N. J. Trask and J. E. Guest.

In the region on the side of Mercury opposite to that of the Caloris basin is an area of chaotic small blocks and hills, known by the Mariner 10 scientists as the 'weird' terrain. It is thought that this jumbled terrain was thrown up by shock waves which were caused by the impact that created the basin and travelled through the whole body of Mercury. So great was the impact that its effects may even have been manifested in the antipodean region of the planet.

Smooth plains resembling the smooth lava plains filling the lunar Mare Imbrium are observed within and surrounding the Caloris basin. The similarities are not close, however, since the Mercurian smooth plains lack the obvious lava-flow features seen on the Moon, and are much paler coloured. While some scientists consider that the smooth plains may indeed be volcanic in origin others think that they may be blankets of impact ejecta. The question whether or not widespread basaltic volcanic landforms exist on Mercury is clearly fundamental to unravelling its geological history, but this problem is unlikely to be finally resolved until spacecraft have landed on the planet and actually sampled its surface.

Whereas many parts of the Mercurian cratered terrain closely resemble the lunar highlands and are saturated with craters, there are other large areas where there appear to be 'gaps' between large craters. These are known as intercrater plains. These plains have been interpreted in two quite different ways. One hypothesis argues that the early Mercurian surface was never wholly saturated, and that the intercrater plains therefore represent the primordial surface of the planet. If this is the case, then the Mercurian intercrater plains must be one of the most ancient surfaces in the solar system, relatively unchanged for about 4600 million years. The alternative suggestion is that the primordial surface of Mercury was saturated with craters, but that a subsequent event—perhaps a planetwide episode of volcanism—smoothed out or erased these earliest craters. Later, continued but less intensive bombardment excavated the craters now exposed, without reaching saturation.

Mercury's scarps appear to be unique in the solar system. They are long, low breaks or steps in the topography which can be traced for hundreds of kilometres, range in height from a few hundred metres to 3 km and have characteristic lobate outlines. They strongly suggest the operation of compressive stresses, which resulted in parts of the Mercurian crust overriding others. A widely discussed hypothesis for their origin suggests that they were formed by bulk contraction of the planet after its interior cooled, causing the brittle outer crust to wrinkle, rather in the way that the skin of an apple wrinkles as it dries out. An alternative explanation is that they were caused by stresses set up in response to 'tidal de-spinning', when Mercury was forced by gravitational interactions into its present resonant rotation period.

The internal structure of Mercury

The absence of seismic data makes it difficult to speculate on the internal structure of Mercury. The high density of the planet, coupled with its small size, indicates that it must have a heavy core which forms a much higher proportion of the whole volume of the planet than the Earth's, and may be some 3600 km in diameter. The Mercurian mantle may form a layer only some 640 km thick, but it probably consists of silicate minerals similar to the Earth's. Little is known of the composition of the crust. Although some early observations suggested that Mercury might possess a thin atmosphere (a pressure of ten millibars was suggested at one time), the ultraviolet spectrometer carried on Mariner 10 showed conclusively that the planet has an atmospheric pressure of only 10^{-12} millibars. It also showed that the most abundant gas is helium, which may be trapped in the vicinity of Mercury by its magnetic field, after having been blown out of the Sun in the solar wind.

Mercury's magnetic field

The Mariner 10 spacecraft revealed that Mercury has a dipole magnetic field with an intensity of only about one-hundredth that of the Earth, with north and south poles closely aligned with the rotational axis. This discovery was not anticipated because it was considered that Mercury's slow period of rotation would preclude the operation of a geomagnetic dynamo similar to the Earth's. As suggested in the case of the Moon's field it is possible that Mercury's magnetic field is essentially a fossil one acquired early in its history and never subsequently lost.

27.7: *Although Venus' dense atmospheric envelope looks featureless to the unaided eye, this Mariner 10 ultraviolet image reveals complex structures, from which details of upper atmosphere circulation can be established.*

Since Mercury is so close to the Sun it seems almost certain that it must have been affected by the Sun's powerful field.

The geological history of Mercury

Crater density studies show that Mercury experienced a history of massive bombardment resembling that of the Moon until, assuming that the meteorite bombardment rate was the same in the region of both bodies, about 4000 Ma ago. Subsequently the rate of bombardment rapidly decreased. Towards the end of the episode of bombardment the Caloris basin was formed on Mercury. This sculpted the 'oceans' around the basin, excavated the lineated terrain and threw up the weird terrain in the antipodean region. The smooth plains areas are slightly younger, but are not younger than about 3000 Ma. There is no evidence of any geological activity after this period.

Venus

With a diameter of 12 104 km and a density of 5.2×10^3 kg/m^3 Venus is the planet that most closely resembles the Earth (Figure 27.8). It is also the Earth's nearest neighbour, coming within 40 million km, and is conspicuously brilliant as a morning and evening 'star', brighter than any celestial object save the Sun and Moon. Notwithstanding its proximity its geology is extremely difficult to study because of the thick blanketing atmosphere.

Although the planet appears featureless when viewed in ordinary light some atmospheric features are visible in ultraviolet light. Observation of these revealed a curious anomaly: a retrograde rotation period was apparently indicated of about four terrestrial days, quite unlike that of any other body in the solar system and difficult to account for in terms of the distribution of angular momentum in the solar system. Since features of the solid surface were invisible beneath the cloud cover the true rotation period could not be established until 1962, when radar studies revealed a retrograde rotation period of 243 days. There is no contradiction between the two observations; the atmospheric features are blown around the planet at speeds of about 100 m/s, comparable to the velocities of terrestrial atmospheric jet streams.

The surface features of Venus

Radar studies made with Earth-based radio telescopes, and radar altimeter data obtained from orbiting spacecraft, are beginning to provide information about large-scale features on Venus. At present it appears that most of Venus consists of monotonous, gently rolling plains, with many large crater-like structures comparable to the large craters and ring basins on the Moon and Mercury. The Venusian craters, however, appear to be extremely flat features. One of the best known is 160 km in diameter, but only about 500 m deep. Such denuded topography could be the result of intensive erosion taking place beneath the thick Venusian atmosphere. The same atmosphere, however, has also shielded the planet from meteorite bombardment even more effectively than the Earth's atmosphere. Small impact craters must be absent, since only the largest objects can penetrate the atmosphere without burning up.

The existence of many large craters on Venus suggests that its surface may be geologically ancient, and therefore that geological processes like those on Earth are not active on Venus. There are some radar data, however, which suggest that this may not be the case. An extremely large, circular feature some 800 km in diameter has been tentatively identified as a massive shield volcano; if it is a volcano it is by far the largest in the solar system. The existence of a giant rift valley some 7 km high and several hundred kilometres long has also been inferred from the radar data. If the existence of volcanoes and large rift valleys is confirmed by later work, it will indicate that plate tectonic processes of a kind may have become established on Venus.

A long series of Russian missions gradually revealed that the Venusian environment is hostile to spacecraft, largely because the temperatures encountered were too high to allow delicate electronic components to continue functioning. In 1970 Venera 7 survived for 15 minutes on the surface, and showed that the surface temperature was 477°C and the atmospheric pressure 95 Earth atmospheres. Under these conditions rocks would just begin to glow dull red. It is therefore very difficult to make simple comparisons between the geological environment of Venus and the Earth.

In 1972 the Venera 8 spacecraft discovered that light levels beneath the Venusian cloud cover were, contrary to earlier speculations, bright enough to allow for surface photography. Venera 9 and 10 were therefore equipped with cameras, and obtained the first pictures from the surface of another planet. Both sets of pictures revealed a barren, rocky terrain, and provided clear evidence that surface processes are active on Venus, eroding and transporting material. The mechanisms responsible are far from understood; although powerful upper atmosphere winds prevail the measured surface winds were extremely gentle.

The Venera spacecraft also carried gamma-ray detectors to study the natural radioactivity of the surface rocks, and these data have important implications in determining the rock compositions. The data from the Venera 8 landing site were comparable with those that are obtained from terrestrial granitic rocks; whereas those from the Venera 9 and 10 sites were both comparable to basaltic rocks. The presence of basaltic rocks was not unexpected: basaltic volcanism seems to have occurred throughout the inner solar system. Granitic rocks, however, indicate that profound geochemical fractionation processes have taken place, and in this respect Venus may resemble the Earth. If large areas of granitic rocks do exist on Venus—and their presence has only been inferred at one place—then it may mean that the Venusian crust is differentiated into continental and oceanic areas in the same way as the Earth's. This in turn would mean that Venus has had a much more 'interesting' geological history than the Moon or Mercury.

The atmosphere of Venus

An understanding of Venus' atmosphere is essential before its geological history can be elucidated, and most spacecraft missions have been concerned with atmospheric studies. They have shown that the atmosphere consists of 95 per cent carbon dioxide, with a few per cent of nitrogen, small traces of oxygen and sulphur dioxide, and a fraction of 1 per cent of water.

Spacecraft data showed that there are three separate 'decks' of cloud in Venus' atmosphere, separated by clear layers. The uppermost deck is at an altitude of about 65 km, and may be several kilometres thick; it is this deck that is visible in telescopic views of the planet. Although the cloud layer is thick the clouds themselves are not dense, and resemble a terrestrial haze. At an altitude of 58 km is the second deck, which is more tenuous but seems to consist of solid particles up to 30 micrometres in diameter. These may be tiny sulphur particles, and may be responsible for the faint yellow tinge in Venus' appearance. The lowermost cloud deck has a well-defined base at an altitude of about 49 km, and is about as dense as ordinary clouds on Earth. The clouds appear to consist of liquid droplets, and occasional showers may rain down from them. The falling droplets probably consist of concentrated sulphuric acid! It is clear that rocks on the surface of Venus are exposed to a violently corrosive atmosphere.

Venus' most striking characteristics are its massive carbon dioxide atmosphere and searing surface temperatures. These features are less anomalous than they may at first appear. There is only about as much carbon dioxide in Venus' atmosphere as there would be in the Earth's if all the carbon dioxide contained in rocks such as limestones were liberated, and the high temperatures are a direct consequence of the thick atmosphere, which acts as a greenhouse. About 80 per cent of the solar radiation incident on the atmosphere is reflected, while the remainder penetrates into the atmosphere and some reaches the surface of Venus where it is re-radiated at longer wavelengths. The clouds are opaque to these wavelengths, which correspond to the infrared, and temperatures therefore rise until an equilibrium is reached. Carbon dioxide, sulphur dioxide and water vapour all block infrared radiation very effectively, and thus provide an effective greenhouse.

It is more difficult to determine how and when the greenhouse effect began to operate. It seems likely that Venus started off 4600 Ma ago with a fairly thin, cool atmosphere. As volcanic de-gassing continued more and more water vapour and carbon dioxide began to

27.8: A preliminary topographic map of Venus compiled from radar data obtained by the Pioneer Venus mission in April and May 1980. The most important feature revealed is the contrast between elevated 'continental' regions and subdued 'oceanic' areas.

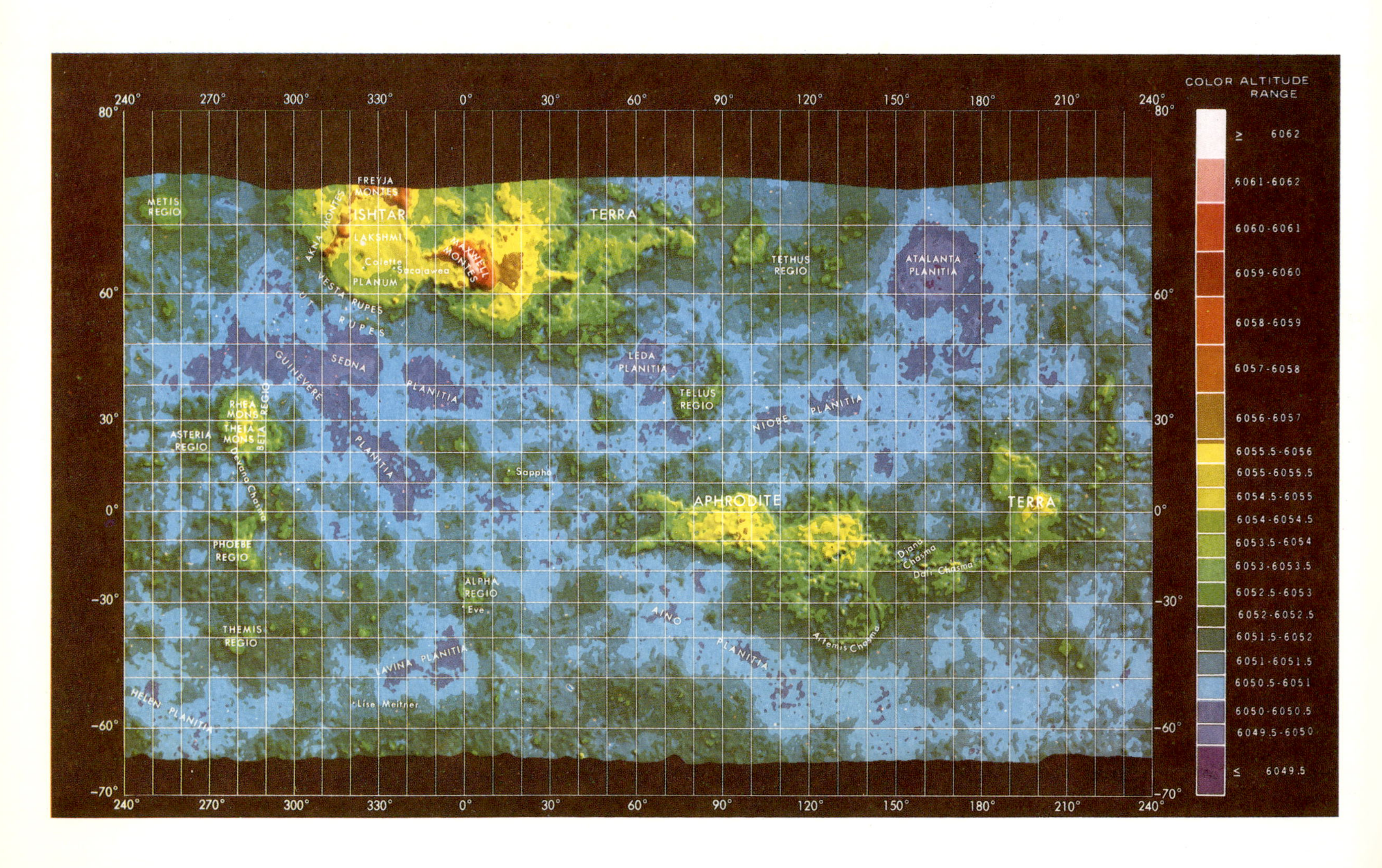

27.9: *An artist's impression of Ishtar Terra, the highest and most spectacular continent-sized region on Venus. The plateau rises several kilometres above the vast expanse of flat, denuded topography, and is larger and higher than the terrestrial Tibetan plateau.*

accumulate on the surface. It is possible that liquid water may have existed at this time. As more and more water, which is much the best radiation blocker, accumulated in the atmosphere the greenhouse effect became established, increasing amounts of water were driven off into the atmosphere and the temperature rose still further until all the combined carbon dioxide in surface material had been driven off, so that Venus resembled an extensive lime kiln. A similar process could conceivably operate on Earth if the temperature were to rise substantially, reaching equilibrium only when all carbonates had been broken down, and the carbon dioxide added to the atmosphere. The principal reason why this has not happened already is that the Earth has so much water that the atmosphere very rapidly becomes saturated with it, and it precipitates out once more. It may be that Venus, which accreted nearer the Sun, simply acquired a smaller total budget of volatiles, or that it lost much of its primary water through photodissociation processes in its upper atmosphere.

The internal structure and composition of Venus

Because of its close physical resemblance to the Earth it is natural to suppose that Venus has a similar structure. However, spacecraft investigations indicated that Venus has only a feeble external magnetic field, which may imply that it lacks a mobile, molten metallic core resembling the Earth's. None the less, its slow rotation rate may mean that it has a similar core, but one in which the same motions have never been generated.

Nothing is known of the thickness of the Venusian crust, except that if exceptionally large volcanoes are present the crust is likely to

27.10: *The first views of the surface of another planet—Venus—provided by the Venera 9 (upper) and Venera 10 (lower) spacecraft. The perspective is distorted, but both terrains are boulder-strewn deserts.*

be much thicker than the Earth's. It is generally accepted that Venus has a silicate mantle broadly comparable with the Earth's, but there must be many important differences of detail. For example, the Australian geochemist A. E. Ringwood has argued that, because the bulk density of Venus is about 2 per cent less than the Earth's, a larger proportion of the total iron in Venus is present as iron oxide in the mantle, and that thus the core to mantle ratio is smaller than the Earth's.

Mars

Mars is a small planet, only about half the size of the Earth. Its orbit never brings it less than 55 million km from the Earth, and consequently it has always been frustrating to observe through telescopes.

Dusky markings are visible on the reddish-coloured surface, but the white polar caps are the only conspicuous features. Seasonal changes in the polar caps, and in some of the dusky markings led to the supposition that large tracts of vegetation might be present (Figure 27.11), and in particular that intelligent life forms might exist on Mars. A prolonged debate on the existence of 'canals' developed, initially as the result of the observations of the Italian astronomer G. V. Schiaparelli (1835–1910), and subsequently through the efforts of Percival Lowell, a wealthy American who established an observatory at Flagstaff, Arizona, specifically with the objective of studying Mars. Although he produced detailed maps showing networks of canals covering the whole planet (Figure 27.12), his ideas were never universally accepted. Even in the second half of the twentieth century, however, when spacecraft missions to Mars were being planned, the possibility of life existing there was taken seriously. Hopes for the existence of more advanced forms of life receded when Earth-based estimates of the Martian atmospheric pressure had to be steadily reduced to a figure of about 10 millibars. They faded entirely when the Mariner 4 spacecraft obtained the first close-up pictures in 1964, revealing a barren, cratered surface, much like the Moon, and a complete absence of any canal-like features.

The surface features of Mars

The first comprehensive view of Mars was obtained in 1971 by the amazingly successful Mariner 9 mission, which obtained photographic cover of the whole planet at a resolution of about 1 km. Mariner 9 showed that the southern hemisphere of Mars is quite different from the northern. The southern consists of heavily cratered terrain, which probably experienced a major episode of bombardment about 4000 Ma ago, much like the lunar highlands. Several large circular ring basins such as Argyre and Hellas, analogous to the lunar ring basins, are present. The northern hemisphere consists mainly of vast expanses of smooth plains relatively free of craters, which are believed

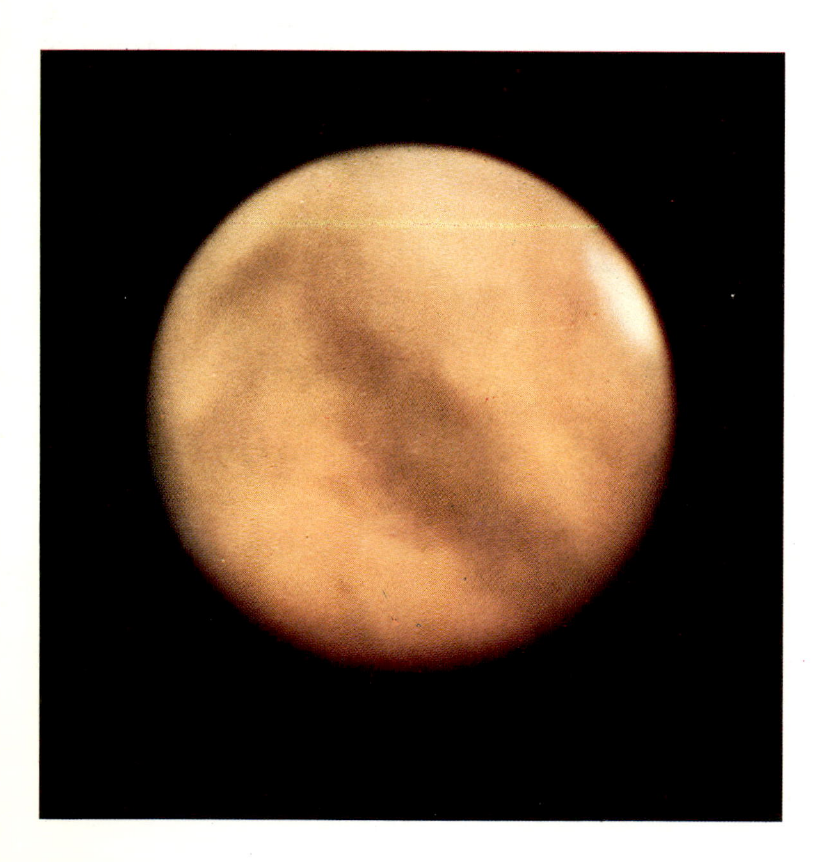

27.11: *This photograph of Mars illustrates the best results obtainable by Earth-based telescopes. The dusky markings were interpreted as patches of vegetation by many observers. One of the icy polar caps is conspicuous.*

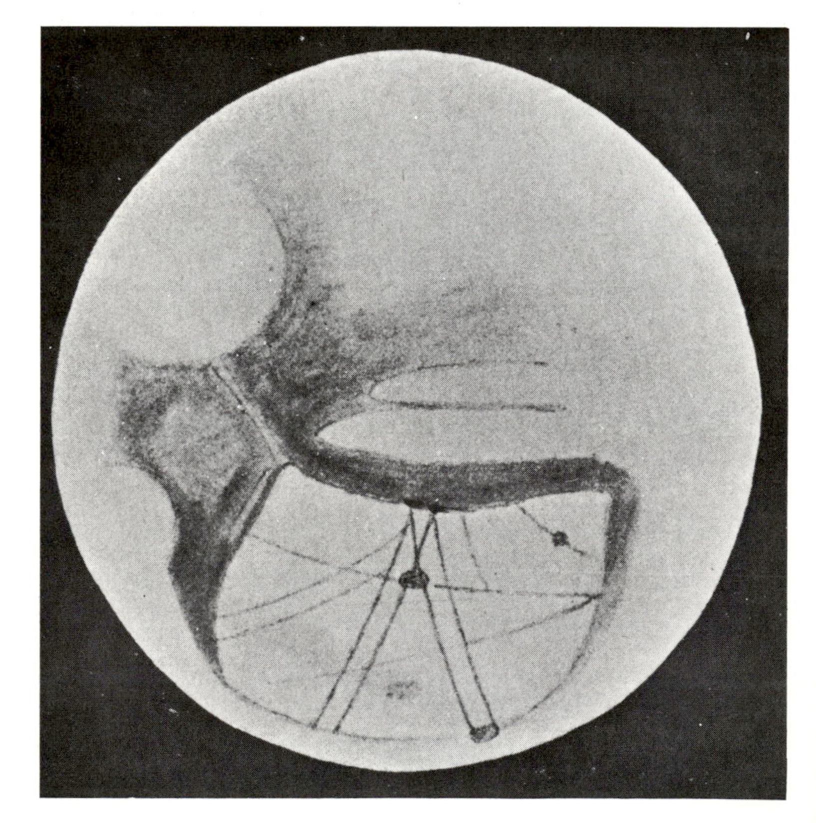

27.12: *One of Percival Lowell's sketches of Mars, showing details of his 'canals'. Although other observers doubted the existence of any canals Lowell claimed that many existed in pairs as seen here.*

to be covered by basaltic lava flows. Crater density studies suggest that these smooth plains may be broadly the same age as the lunar maria, or slightly younger.

Whereas geological activity on the Moon ceased with the eruption of the mare lavas, this was not the case on Mars. Most impressive of the younger geological features are the huge volcanic edifices. Many examples of these have been identified, but the best known are four clustered together in an elevated region of the Martian crust known as the Tharsis dome. The largest of these volcanoes, Olympus Mons (Figure 27.13), is a broad shield some 500 km in diameter, with a summit caldera complex some 23 km above the surrounding plains. Its flanks are covered with innumerable lava flows which spread out for large distances over the surrounding plains. The precise age of the volcanic activity is not known, but some estimates suggest that it may have been active as little as two hundred million years ago.

Detailed studies of lava flow morphology suggest that the lava flows on Olympus Mons are of basaltic or ultrabasic composition. Terrestrial basaltic volcanoes such as the large shields of Hawaii never attain such dimensions. One possible explanation is that the crust of Mars is much thicker than that of the Earth, and far too thick and rigid for plate tectonic movements to take place. Thus, whereas a terrestrial volcano sited above a deep-seated mantle hot spot (such as those in the Pacific) will drift away from it in a matter of a few million years, the volcanoes on Mars remained permanently sited above their hot spots, and thus grow to prodigious sizes.

Further evidence for the great thickness of the Martian crust comes from the discovery of a huge rift valley, Vallis Marineris, some 5000 km long, which cuts across the equatorial regions of the planet, terminating near the Tharsis dome. This valley, far larger than any terrestrial feature, is interpreted as a 'failed' rift, a site that on Earth might have led to the formation of a proto-ocean and to separation of the crustal masses on either side. Geological evidence shows that the Martian crust in the region of Valles Marineris (Figure 27.14) may be as much as 200 km thick, and thus far too rigid to allow for successful spreading to be initiated.

The high-resolution Viking orbiter pictures revealed that Martian crater ejecta blankets possess some remarkable characteristics. Instead of the ejecta falling radially around the impact site to form secondary

27.13: *The great Martian Volcano Olympus Mons, photographed by the Viking 1 orbiter on 31 July 1976. Cliffs, 4 km high, which ring the volcano are visible in the lower part of the picture.*

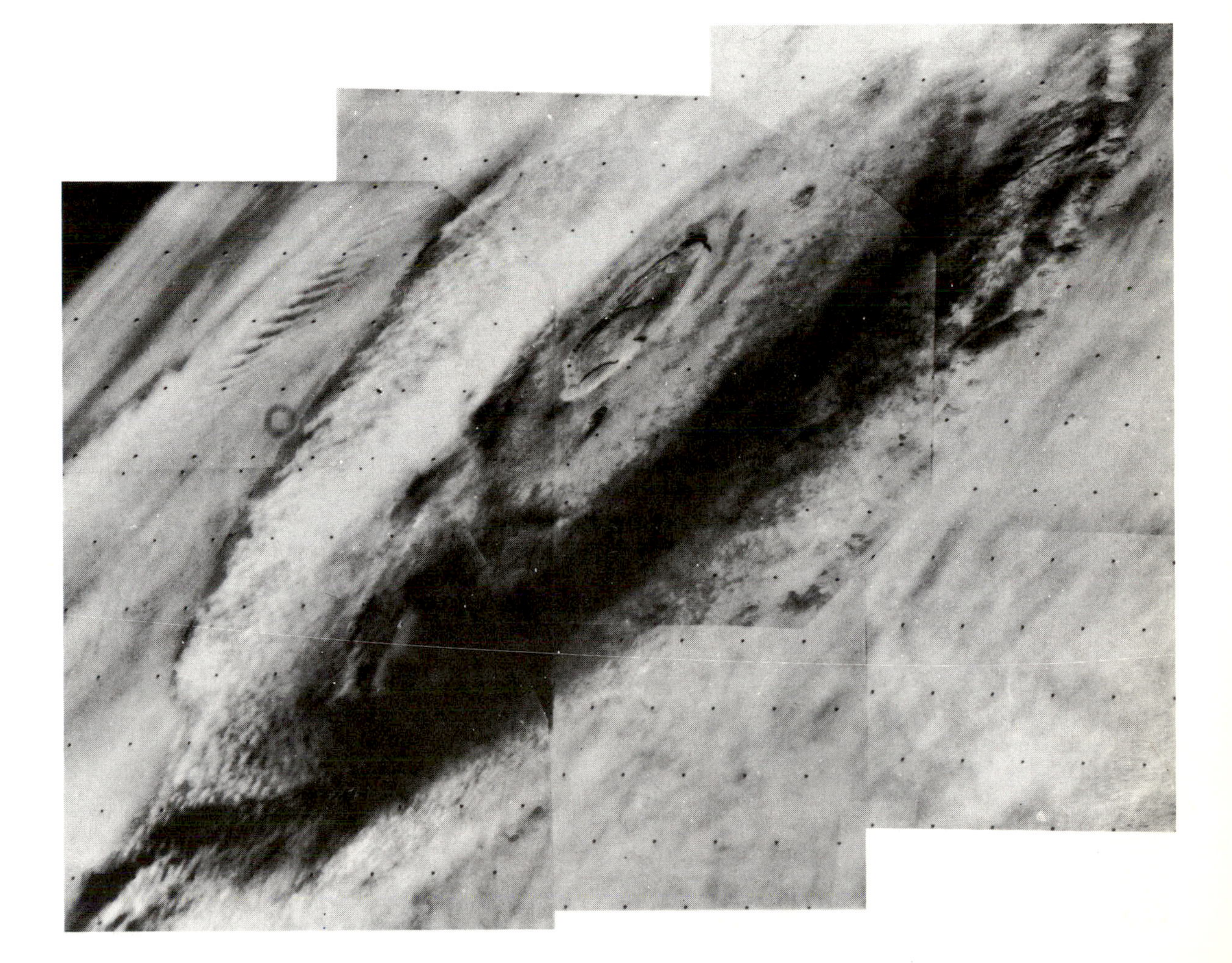

craters, as on the Moon, many Martian craters look as though they have been formed by impacts taking place in a soft, wet surface, and the ejecta have flowed out radially in sludge-like masses to form ramparts around the craters. These unexpected ejecta patterns are now interpreted as the result of impacts taking place into terrain that contained large amounts of subsurface ice in the form of permafrost. This is a condition that has never prevailed on the Moon, but seems not unlikely on Mars, since it possesses polar ice caps at the present day. When the impacts took place the energy released vaporized or melted the ice causing radial flows of pulverized material and water which spread out around the crater to form the distinctive rampart structures.

It is ironic that while the Mariner 9 spacecraft finally disposed of the possibility of canals on Mars they did demonstrate that channels exist there. The channels, which are quite unrelated to the mythical canals, were perhaps the single most important geological discovery made by any of the missions to Mars. Their presence indicates, almost unarguably, that liquid water once flowed on the surface of the planet and, by implication, that conditions might at some point in the past have been favourable for the evolution of life.

Three different types of channel exist. First, there are innumerable small, dendritic channels which drain downwards from elevated regions. These are interpreted as simple run-off channels, of the sort that form after rainfall in arid, hilly terrestrial areas. Second, there are very much larger channels, 200–300 km long, which are cut into the cratered highlands and meander downslope, debouching ultimately on the smooth plains. These often exhibit braided patterns, resembling those formed in shallow terrestrial rivers by flash floods. Third, there are channels that, traced upstream, end abruptly in massive clifflike scarps showing clear evidence of massive slumping and landslipping.

The third type of channel is the most easily explained. It has been suggested that such channels were formed when geothermal heating

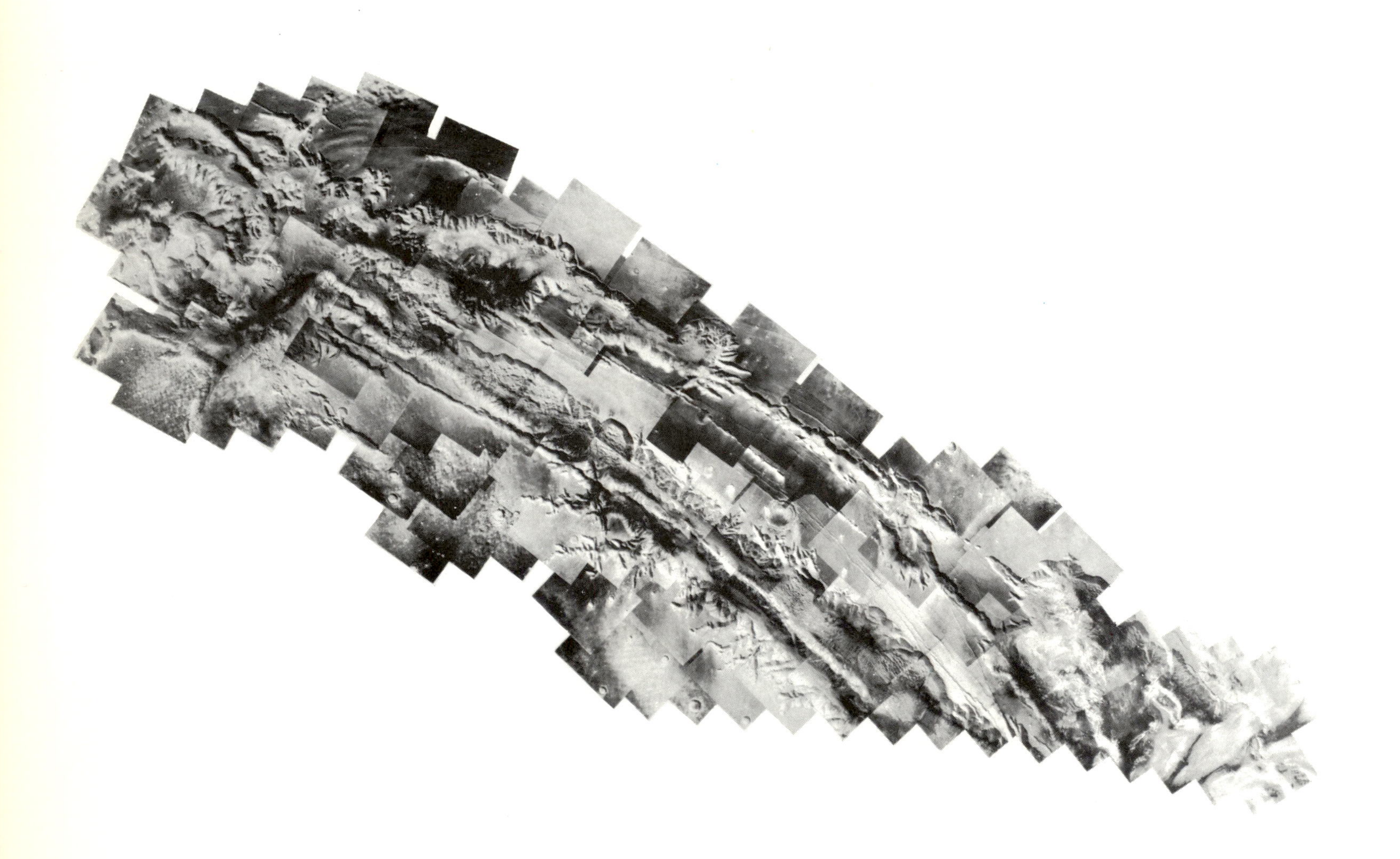

27.14: *A mosaic of Viking orbiter photographs illustrating the scale of Mars' vast equatorial canyon, the Valles Marineris. The canyon would span the entire width of the continental USA.*

caused melting of large volumes of permafrost ice, which resulted in large-scale collapse and an outburst comparable to a terrestrial glacier burst, or *jökullhlaup*. The combination of channel-like features with large collapse scarps strongly supports this suggestion, but unfortunately there is no clear evidence of links between the source regions of the channels and sites of volcanic or geothermal activity. The other two types of channel appear to have been eroded by water, which in turn implies that the Martian climate has experienced massive changes.

Climatic change on Mars

The presence of the channels indicates fairly unequivocally that the Martian climate must at some time in its geological history have been less hostile than it is now, but it is not possible to determine precisely when this was so. Crater density studies, which provide the only clues, indicate that the channels may have been formed over a very long period, between as long ago as 4000 Ma and only 2000 Ma. Three factors can be identified as being partially responsible for climatic changes on Mars.

First, the mass of Jupiter is so great that it periodically perturbs the orbit of Mars from a nearly circular shape to a much more elliptical one, each cycle of perturbation taking about 2 Ma to complete. The mean distance between Mars and the Sun varies significantly during this cycle, and Mars receives more solar radiation when it is nearer the Sun than when it is more distant. Second, superimposed upon this effect is the precession of Mars' spin axis, which takes about one hundred thousand years to complete and causes regular variations in the total amount of solar radiation received. Third, the inclination of the rotation axis to the orbital plane may vary by as much as 30°, unlike the Earth's, which supposedly remains steady at $23\frac{1}{2}°$.

Although these factors must have pronounced climatic effects all seem to operate over far too short a time scale to be responsible for the kind of climatic changes necessary for the cutting of the channel systems, a phenomenon that is far from being fully explained. A major unknown factor is the total mass of water present in the Martian subsurface as ice, since the amount of atmospheric water at present is trivial—it is only enough to form a layer some 40 micrometres thick if spread out over the entire planet.

The atmosphere of Mars

Speculations about causes of climatic variation on Mars are inextricably linked with the evolution of its atmosphere. The Viking spacecraft have provided some important data. The present atmosphere is composed predominantly of carbon dioxide, with small amounts of nitrogen, argon, oxygen and carbon monoxide. The surface atmospheric pressure is somewhat less than 10 millibars, the value varying by as much as 30 per cent when the carbon dioxide ice in the polar ice caps is seasonally melted or frozen.

Mass-spectrometer studies revealed that the Martian atmosphere is depleted in ^{36}Ar but enriched in ^{40}Ar and ^{15}N. The enrichment in ^{15}N has profound implications for the planet's atmospheric evolution. The Martian ^{15}N:^{14}N ratio is about 1.7 times greater than the Earth's and, since the initial ratio was presumably similar to the Earth's, this has been taken to imply that the light isotope of nitrogen has been selectively lost from the Martian atmosphere over geological time. Some estimates suggest that nitrogen alone may have exerted originally a pressure of 30 millibars, and while such estimates are not universally accepted it seems clear that Mars has experienced massive atmospheric loss. The question that remains to be resolved is whether the gases have been entirely lost, to space, or whether they remain in a combined form in surface and subsurface materials. If the latter, then periodic increases in atmospheric pressure and climatic amelioration are possible.

The surface of Mars

The Viking 1 and 2 spacecraft landed thousands of kilometres apart, but the terrain at each site is remarkably similar, a boulder-strewn desert stretching from horizon to horizon. A small crater is visible in the distance in the Viking 1 imagery, and there are numerous small drifts or dunes of dust. The surface closely resembles those of terrestrial deserts where wind erosion has predominated and all fine-grained material has been blown away leaving 'lag' gravels.

Most of the rocks visible in the images have a spongy surface texture, reminiscent of terrestrial vesicular lavas. The vesicular texture is consistent with the view that the plains areas where the craft landed are covered with lava. The material in the dunes appears to be much finer grained than in terrestrial sand dunes, and is probably of silt size. Internal stratification can be seen in the dunes which seem at present

27.15: *Arandas crater, Mars, 25 km in diameter, exhibits a prominent central peak, radial grooves and conspicuous ramparts. The pattern of the ejected material suggests that the crater formed as a result of an impact into wet or icy ground.*

to be in retreat and may be erosional relics of a much larger deposit. Exactly when and how the dunes formed is not known; it is possible that they are extremely ancient features.

The surface chemistry of Mars

Both Viking landers carried miniaturized laboratories capable of carrying out a series of analyses of surface soil samples (Figure 27.17). Most of these analyses were designed to detect the presence of microbial life forms, but an X-ray fluorescence instrument provided the first direct inorganic analysis of surface materials from another planet. The material analysed was all fine grained and was scooped up from the Martian 'soil'. Hence it is not directly representative of any particular rock composition. The results of the analyses were consistent with the weathering products of basic to ultrabasic igneous rocks. The soil appears to consist of silicate minerals coated with a thin film of ferric oxide, less than 2 micrometres thick. It is this film which gives the Martian surface its characteristic red colour.

An interesting discovery was the high proportion of sulphur in the soil. This is believed to be present in the form of sulphate, and is probably a consequence of the very arid Martian environment. Sulphate minerals such as gypsum are only found on Earth in regions characterized by high solar radiation levels, little rainfall and high evaporation rates. Such environments are found in salt flats, and in areas such as the Atacama Desert where prolonged leaching and evaporation of groundwater has led to the accumulation of a thick surface layer rich in sulphates and nitrates. These areas may provide the closest terrestrial equivalents to the Martian surface.

Seismology and the internal structure of Mars

Both Viking landers carried seismographs to detect seismic events and provide data to determine the internal structure of the planet. Unfortunately, the Viking 1 instrument malfunctioned, and the Viking 2 instrument detected only one, rather uncertain, event. Even this result is interesting, because it indicates that Mars must be seismically quiet, implying that its internal processes are either extinct or operate at only a very subdued level.

In the absence of seismic data the only evidence on the internal structure of Mars is provided by its low density relative to the Earth, its marked oblateness (the polar diameter is some 40 km less than the equatorial diameter) and its extremely weak magnetic field (less than that of the Earth). The low bulk density and the oblateness suggest that Mars does not possess a massive metallic core. It is likely that the planet has an outer silicate crust some 200 km thick, overlying a peridotitic mantle which in turn overlies an iron sulphide core some 3600 km in diameter, with a density of about 6×10^3 kg/m^3.

There has been some speculation that the Martian mantle is enriched in iron, most probably in the form of an oxide or sulphide. Partial melting of such iron-enriched mantle materials would lead to the production of ultrabasic lavas, with silica contents around 44 per cent. These are likely to have been erupted at temperatures in excess of 1500°C, with extremely low viscosity. It may be no coincidence that morphological studies of Martian lavas suggest that most were of low viscosity and erupted in huge volumes, flooding the smooth plains.

27.17: *Partial analysis of Martian soil.*

	Per cent (by weight) of total sample
SiO_2	44.7
Al_2O_3	5.7
Fe_2O_3	18.2
MgO	8.3
CaO	5.6
K_2O	less than 0.3
TiO_2	0.8
SO_2	7.7
Cl	0.7
Total	91.9

Note
The identity of the mineral species present could not be deduced with the instruments available; it is likely that they are clay minerals resembling bentonite.

27.16: *A panorama of the Martian surface obtained by the Viking 1 lander in 1976. The long boulder (left) is about 8 m from the spacecraft, and about 3 m long. Sculpting of the low dunes in middle distance is clear evidence of the effectiveness of wind erosion on Mars.*

27.18: *The spongy, vesicular rocks in this detailed Viking 1 view of the Martian surface are likely to be vesicular basalt lavas, probably fragmented on and excavated from the subsurface by impact events.*

The geological history of Mars

M. N. Toksöz and A. T. Hsui, working at the Massachusetts Institute of Technology, have suggested the following history for the planet, based on a consideration of its thermal evolution. Between 4600 and 4000 Ma ago Mars heated up as a result of the accretion process and the decay of short-lived radioactive elements. The core separated out and the crust began to form at the surface, continuing to do so until about 3000 Ma ago. Between 3000 and 2000 Ma ago internal radioactive heating steadily increased, and expansion of the interior caused tensile features to open up on the surface. Between 2000 and 1000 Ma ago volcanic activity reached its peak, and large volumes of lava were erupted and huge quantities of gas were blown out from the mantle into the atmosphere. This may have coincided with one of the periods in which the channels were eroded.

About 1000 Ma ago radioactive heating in the interior waned, and the crust got steadily thicker, reaching its present thickness of 200 km and effectively sealing off the interior. Volcanic activity thereafter diminished. In the past few hundred million years the cooling has continued, and the mobile part of the mantle is now more than 300 km below the surface. Volcanic and tectonic processes are therefore all but terminated on Mars.

Jupiter

Jupiter is much the largest of all the planets in both mass and volume. Its mass is larger than that of all the other planets combined; it is nearly 0.1 per cent that of the Sun and is over three hundred times greater than that of the Earth. Its density, however, is only 1.3×10^3 kg/m^3, indicating that its bulk composition is quite different from that of the Earth. About four-fifths of its mass consists of hydrogen and helium, and thus it is not possible to discuss Jupiter in the same geological terms as the inner, rocky planets. The planet itself is of most interest to atmospheric physicists. The Voyager spacecraft showed that Jupiter's four large satellites—Io, Europa, Ganymede and Callisto—exhibit an extraordinary diversity of surface features, and are among the most interesting bodies in the solar system.

The surface features of Jupiter

Even with a modest telescope surface features can easily be observed on Jupiter, and it was early recognized that Jupiter's rotation period is about nine hours and fifty minutes (measured at the equator)—extremely rapid for such a massive body. The most obvious surface features are broad belts and bands of different hues running parallel to the equator across the planet's disc. These are interpreted as resulting from powerful Coriolis forces acting on convectively rising and descending plumes of gas within the atmosphere. Gases moving towards the equator are swept into a direction opposite to the rotation direction, while gases moving away from the equator towards the poles are swept into a direction parallel to the rotation direction. Similar, but less marked, effects can be seen in the Earth's atmosphere, notably in the distribution of trade wind belts.

The Voyager spacecraft provided an immense amount of new data on the details of Jupiter's atmospheric circulation, and provided the first close-up pictures of persistent features such as the Great Red Spot. First observed telescopically in the seventeenth century, this Spot has remained visible ever since, although varying somewhat in size, and drifting slightly in longitude. Early astronomers thought that the Spot might be a great volcano poking up through the atmosphere, but it was soon realized that it must be a purely atmospheric disturbance—perhaps an enormous cyclone. Although the Voyager pictures provided atmospheric scientists with almost startlingly clear records of circulation around and within the Spot, opinions are still divided on its exact nature.

The atmospheric composition of Jupiter

Our knowledge of Jupiter's atmospheric composition is likely to be much advanced by the US Galileo mission, scheduled for the mid-1980s, which will send a probe deep into the atmosphere. At present all our information is based on remote-sensing studies carried out using spacecraft and Earth-based telescopes. Hydrogen, helium, ammonia, methane and water are all believed to be present in large quantities. It is possible that the clouds visible on the outer surface of the planet consist of tiny suspended particles of ammonia ice; below these may be clouds consisting of water ice particles and, lower still, water droplets.

A major unresolved problem is the nature of the material responsible for the coloration of some of Jupiter's cloud systems, notably the Great Red Spot. Many suggestions have been made, including phosphine, ammonium sulphide, ammonium hydrosulphide, several dif-

27.19: *Jupiter photographed by Voyager 1 in March 1979. Jupiter's Great Red Spot can be seen on the upper right.*

ferent organic compounds and some inorganic polymers. Some planetary scientists have even argued that conditions within the denser parts of Jupiter's atmosphere might be such that life forms could exist there, metabolizing the suspended organic compounds. This seems increasingly unlikely in view of the powerful convective currents operating in the atmosphere, which would subject any organism to intolerable ranges of temperature.

The internal structure of Jupiter

The only guides to Jupiter's internal structure are given by its density, oblateness and magnetic field. The planet's density demonstrates that it must be composed of light elements, dominantly hydrogen and helium. The polar flattening (the polar diameter is 8900 km less than the equatorial diameter), coupled with the very rapid rotation rate, indicates a concentration of mass near the centre of the planet, since an even more marked flattening would otherwise be observed. The existence of a powerful magnetic field, first suspected from Earth-based radio observations and later documented by the 'Pioneer' spacecraft, suggests that there must be moving currents of an electrically conducting material within Jupiter, analogous to the motions within the Earth's core which are held to be responsible for Earth's magnetic field.

These constraints are satisfied by a theoretical model for Jupiter's internal structure in which the outer (observable) atmospheric layer overlies a layer of liquid molecular hydrogen some 24 000 km thick. At the base of this layer pressures are believed to be so great that the hydrogen exists in a form that is liquid and metallic. At the centre of the planet may be a small core some 10 000 km in diameter.

The Galilean satellites of Jupiter

Jupiter has fourteen known satellites, one of which was discovered by the Voyager 2 spacecraft. The planet also possesses a thin ring surrounding it, discovered by the Voyager 1 spacecraft. The presence of this ring was not suspected from terrestrial telescopic observations, but has subsequently been seen. Its discovery, coupled with that of Uranus' ring system, demonstrates that ring systems are not unusual, but rather are to be expected when large amounts of fragmentary material are in close orbit around a parent planet.

Of Jupiter's numerous minor attendants, only the Galilean satellites—Io, Europa, Ganymede and Callisto—are of any geological consequence. These satellites, which are easily observable with the most modest optical aid, were first observed by Galileo Galilei (1564–1642) in 1609. They are so large that they are in effect small planets. Io and Europa are roughly the same size as the Earth's Moon, but Ganymede and Callisto are both bigger than the planet Mercury. There is an interesting decrease in density outwards from Jupiter, such that Io, only 420 000 km from Jupiter, has the greatest density (about 3.5×10^3 kg/m^3), while Callisto, 1 880 000 km from Jupiter has the lowest (about 1.6×10^3 kg/m^3). Io and Europa are therefore thought to be largely rocky, and made of silicate materials broadly similar to the Earth's moon, while Ganymede and Callisto must incorporate a large proportion of icy material.

Io

One of the most remarkable discoveries of the present era of spacecraft exploration of the planets was made by the Voyager 1 spacecraft in March 1979, when a great volcanic eruption was actually observed in progress on Io (Figure 27.20). Subsequently eight active volcanoes were located. In an even more remarkable piece of scientific work US scientists predicted, before the Voyager spacecraft reached Io, that active volcanism would be found there because of the complex gravitational interactions between Jupiter, Io and Europa.

Io is very close to the enormous mass of Jupiter, and so it is not surprising that, in common with the Earth's Moon, it always keeps the same face turned towards the planet. In addition the satellite is compelled to stay in an orbit which causes its period of revolution around Jupiter to be exactly half that of Europa's. A great deal of mechanical work is required to keep Io in this resonant position, and this is manifested in the generation of large amounts of heat in its interior, which is thereby kept almost entirely molten. It has been suggested that electromagnetic induction, due to Jupiter's powerful magnetic field, may also help to heat Io.

The rate of volcanic activity is such that the entire surface of the satellite is renewed at a rate of about 1 mm a year, and as a consequence it shows no evidence of meteorite bombardment—the only craters on Io are volcanic in origin. Many of the present ideas on Io's volcanism are highly speculative, but all point in one direction: that most of Io's internal and surface processes are controlled by sulphur.

In terrestrial volcanic activity the volatile component responsible for explosive eruptions is water, which is liberated in huge volumes in the form of steam during all eruptions. In its primitive state Io possessed a similar, or even greater, proportion of water, but the constant volcanic activity throughout its geological history has led to the loss of all of this to space; Io is too small to retain volatile materials, and thus when volcanic eruptions hurl material into space any water derived from the interior is lost for ever. The other principal volatile component in terrestrial volcanic eruptions is sulphur dioxide, which seems to be the only volatile component now present on Io. Sulphur dioxide is more dense than water, and thus when erupted in volcanic plumes it is retained by the satellite much more easily. Consequently, through geological time sulphur dioxide has replaced water on Io, and almost all its surface features are now being interpreted in terms of sulphur dioxide and sulphur compounds. There may be pools of molten elemental sulphur within some of the large volcanic craters, and it is possible that the 'lava flows' surrounding the volcanoes are composed of sulphur rather than silicate materials. It has even been suggested that while Io's mantle and lower crust may be of 'normal' silicate composition its outer skin may consist of an 'ocean' of liquid sulphur, overlain by a film of solid sulphur and sulphur dioxide. Such hypotheses, which are reminiscent of science fiction, are seriously proposed by planetary scientists studying Io. Its startling surface properties make it certain that Io will be a primary target for future spacecraft exploration.

Europa

Although closely similar in size and density to Io, Europa presents a totally dissimilar appearance. The entire surface of the satellite is sheathed in a layer of ice perhaps 100 km thick. While the tidally induced heating within Io has been sufficiently intense for it to lose all its water through volcanic activity, this is not true of Europa. Europa, however, must have experienced extensive resurfacing of a different kind, because few impact craters can be detected on its surface. It is

27.20: *One of the most striking spacecraft pictures ever obtained, this Voyager 1 photograph shows a violent volcanic eruption in progress on Io, Jupiter's innermost large satellite. The volcanic plume reaches more than 200 km above the surface, indicating an eruption velocity of about 2000 km per hour.*

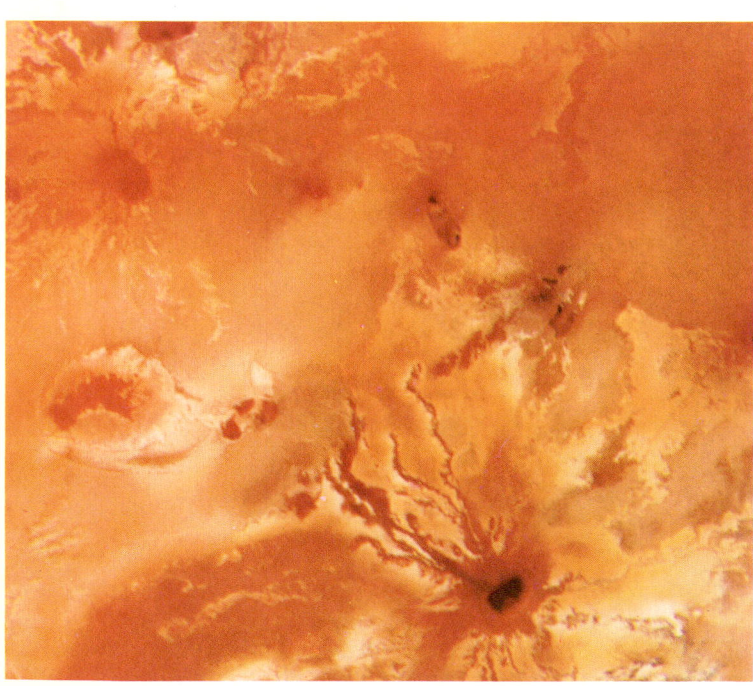

27.21: *A close-up view of one of Io's many volcanic centres. The irregular radiating features are probably lava flows, although there are doubts about their composition. The diffuse reddish and orange colours of the surface are suggestive of sulphur compounds and other volcanic sublimates.*

27.22: *Ganymede, Jupiter's largest satellite. The bright spots dotting the surface are young impact centres. 'Matching' areas of dark-coloured material separated by polar material suggests that 'rifting' has taken place.*

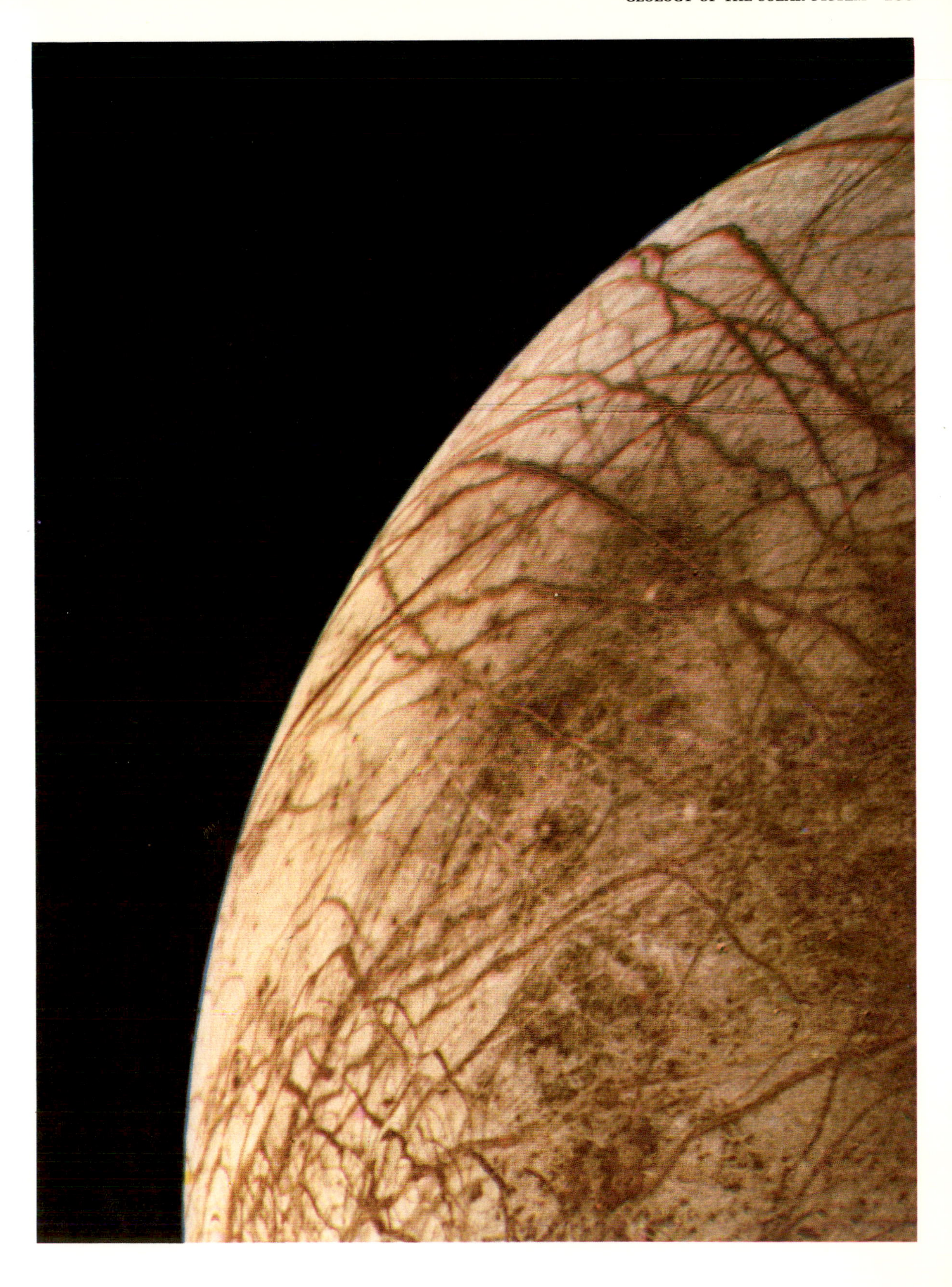

27.23: *Europa, the second Galilean satellite has a smooth, ice-covered surface with few impact centres. The extensive fracture patterns on its surface have been interpreted as being the result of expansion taking place within the satellite.*

likely that it has experienced tidally induced heating of a kind similar to that of Io, but this has never been enough to melt more than some of the subsurface ice, turning it to water. Any water reaching the surface by eruption through 'watery' volcanoes would immediately freeze once more, renewing the smooth, icy surface.

The only surface features visible on Voyager images are extensive networks of dark streaks covering most of the satellite. These features may be crevasse-like fractures in the ice crust, opened up by tidal distortions resulting from its resonant orbit and subsequently filled from beneath with fresh ice. It may be through these fractures that water reaches the surface to renew the exterior ice.

Ganymede

Ganymede is one of the largest satellites in the solar system, and is bigger than the planet Mercury. Its density is so low, however, that it contains much less mass than Mercury, and must be made up largely of ice, with perhaps a small core of rocky silicate materials. Its surface appearance is quite extraordinary. Superficially it resembles the Earth's Moon, and exhibits many impact craters. This indicates that its surface is much more ancient than Io's or Europa's, but it is difficult to assign to it a precise age since nothing is known of the rate of meteorite bombardment in the vicinity of Jupiter. The youngest impact craters are surrounded by bright, whitish ejecta blankets which contrast sharply with the surrounding surface. It is thought that the impacts throw out fresh ice from beneath the surface, and that this falls back around the impact site, covering the older, dust-covered ice surface.

Large areas of the surface are also of a distinctly paler colour than their surroundings. These darker areas resemble 'continental' masses surrounded by 'oceanic' crust, and places are visible where an older, dark 'continent' has been split apart and the gap filled by paler-coloured material. Such observations have been interpreted in terms of a kind of plate tectonic process taking place entirely within the icy outer parts of Ganymede, with the outer surface becoming disrupted and shifting around, perhaps in response to melting of the interior ice in the 'mantle'. Where ancient surface areas are split up younger material wells up from below to fill in the gap, analogous to the formation of oceanic crust in terrestrial sea floor spreading.

27.24: *Callisto is the darkest and most distant of the Galilean satellites. One conspicuous impact basin, 600 km in diameter, is visible, surrounded by numerous concentric rings which were probably thrown up by shock waves on impact.*

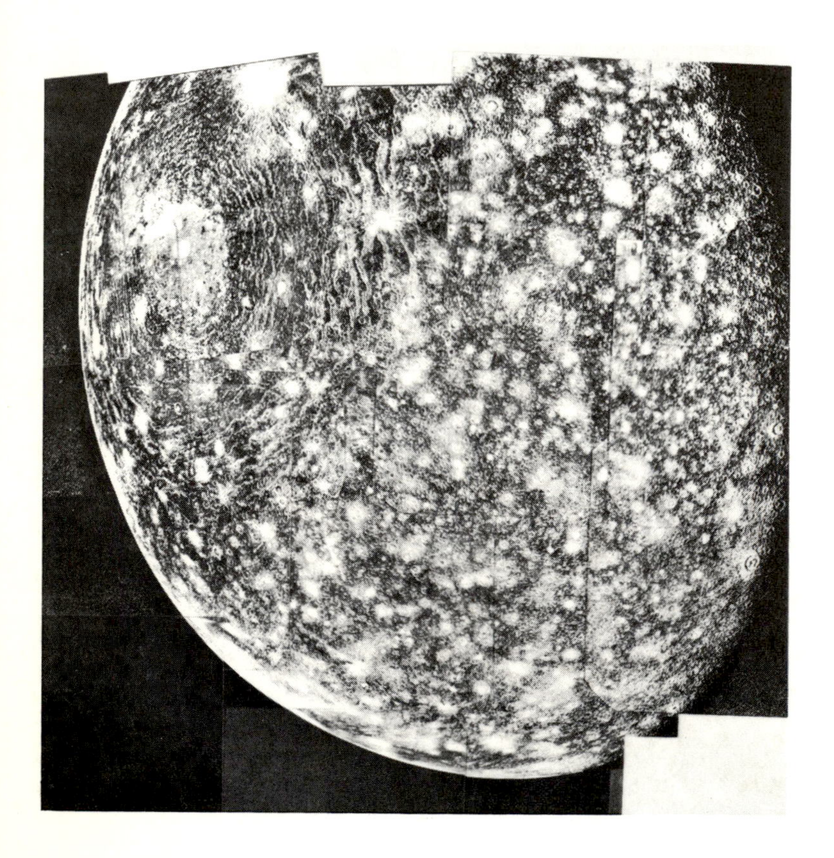

Callisto

Callisto is the Galilean satellite most distant from Jupiter. It shows no evidence of internal processes, probably because tidally induced heating was never sufficiently intense to melt its interior. In common with Ganymede, Callisto is largely composed of ice, but its surface appearance is very different because it is extremely heavily cratered. Its surface is much the most ancient of all the four large satellites, and may be substantially unchanged since its formation.

Callisto is also much darker than the other satellites, reflecting only about 20 per cent of the incident light. This is believed to be because its surface, being ancient, has become covered with meteorite dust collected in the vicinity of Jupiter over the course of perhaps 4000 Ma. Ganymede also shows evidence of a partial dust cover, but on Callisto the dust appears to be much thicker because, in contrast to Ganymede, it is difficult to detect water by spectroscopic means. Yet both satellites must contain a large proportion of ice.

Saturn

Although its magnificent ring system makes it the most spectacular of all the planets Saturn is too distant for telescopic study to be very rewarding, and it is only recently that spacecraft have provided a close-up view. Pioneer 11 made an exploratory fly-past in 1979, and in 1980 Voyager 1 provided the first high-resolution pictures.

The pictures revealed few surprises about the planet itself. The surface appearance is broadly similar to Jupiter's and has a similar pattern of turbulent cloud belts. Although there are some large atmospheric disturbances nothing comparable in magnitude to Jupiter's Great Red Spot exists. The composition of Saturn's atmosphere and the internal structure are also thought to be similar to Jupiter's. Its density is even lower than Jupiter's (only 0.7×10^3 kg/m^3) and it is so highly oblate that the equatorial diameter is 10 per cent greater than the polar diameter. This fact, coupled with the rotation period of ten hours and fourteen minutes, may suggest that there is a small core of dense material at the centre of the planet.

The Voyager mission totally transformed our knowledge of the

27.25: *Saturn's rings, seen by Voyager 1. They proved to be very much more complex structures than had been expected and many details remain to be explained.*

ring system. Previously, it had been thought that there were three separate rings, with a prominent gap, known as Cassini's division, between the outermost and middle rings. The spacecraft pictures showed that each main ring is itself composed of scores of smaller ones, and that there are even rings within Cassini's division. They also revealed the presence of two new satellites, and emphasized the close relationships between the spacing of rings and the orbits of satellites around the planet.

The outermost rings are so thin that stars can from time to time be seen through them. They are probably not more than 1–2 km in thickness. It has been known for many decades that the rings cannot be solid discs but are composed of myriads of solid particles moving around the planet in orbits obeying Kepler's laws. The spacing of the rings appears to be dictated by the orbital periods of Saturn's satellites; there are gaps where the orbital period of a particle would be a simple multiple of that of a more distant satellite. Although the rings contain only a small amount of material—a tiny fraction of Saturn's mass—it is interesting to know what they are made of. Photometric, polarimetric, spectroscopic and radar studies indicate that the particles within the rings are probably mostly ice, in lumps around 300–400 mm across. The surface texture is probably rough and fluffy, like snow rather than solid ice, and the particles may be impregnated with other materials, such as iron sulphide and silicate materials.

Saturn's satellites

Saturn has at least fourteen satellites. Others may exist on the fringes of the ring system. Almost nothing was known of these satellites until Voyager 1 retained pictures of the larger ones, revealing them to have heavily cratered surfaces like those of Ganymede and Callisto. Much the largest and most interesting satellite is Titan, the only known planetary satellite with a dense atmosphere. As a result of telescopic work it was thought that the atmosphere consisted largely of methane, but the Voyager data shows that it is in fact nitrogen, and suggests that the whole satellite might be bathed in a deep ocean of liquid nitrogen.

The perpetual blanket of clouds in Titan's atmosphere prevented the Voyager spacecraft from taking pictures of this frigid ocean, but some of its data did suggest that chemical conditions in the atmosphere may have resembled somewhat those of the early Earth; if temperatures on Titan were a little higher complex organic compounds could be synthesized, the first steps towards the origin of life.

Uranus

Uranus is so distant from the Sun that it is impossible to learn much about it telescopically. No spacecraft has yet visited it, although it is

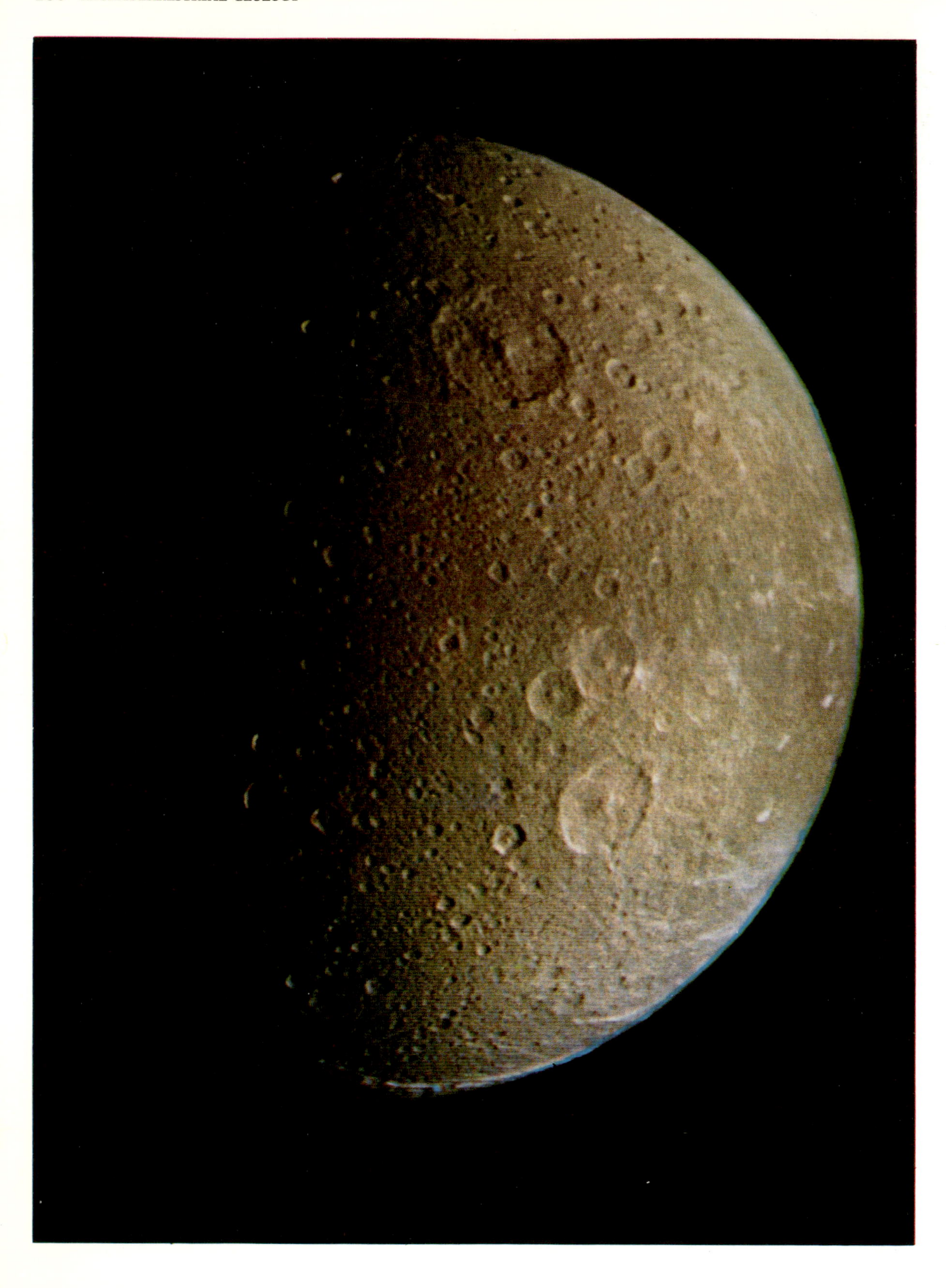

27.26: *Diane, one of Saturn's fifteen satellites, seen in close-up by Voyager 1. The satellite is thought to be composed largely of ice, and has clearly experienced heavy meteorite bombardment. A long irregular fracture is also conspicuous, suggesting internal expansion in the satellite.*

hoped that the Voyager craft will survive to reach the planet in 1986. Uranus is a low-density planet made up largely of condensed gases, although its density is too high for it to be composed predominantly of hydrogen. It has an internal structure broadly similar to Jupiter's. Uranus is interesting in two respects. First, its rotational axis is inclined at at angle of more than 90° to its orbital plane, so that its rotation is effectively retrograde. Second, in 1977 it was discovered that it possessed a system of rings, the second known planet to do so. The rings were only discovered when Uranus passed in front of a star, and the light from the star was seen to be dimmed before the planet itself moved in front of it. Because the rings are not visible telescopically it is inferred that they are rather dark, which suggests that they may not be composed of icy material like Saturn's. An alternative explanation is that they are clouds of gas rather than solid particles.

Neptune

Neptune is closely similar in size, mass and density to Uranus, but it is 'conventional' in that its rotational axis is upright and it is not known to possess rings. It has two satellites, Triton and Nereid. Triton is believed to have a diameter of 6000 km and, if this is correct, is the largest satellite in the solar system. Triton's orbit around Neptune is retrograde whereas the other major satellites move in an anticlockwise orbit.

Pluto

Conventionally regarded as the most distant of the planets, Pluto is at the time of writing nearer than Neptune. Because its orbit is highly elliptical Pluto actually remains nearer the Sun for twenty years of its 247-year revolution period.

Almost nothing was known about Pluto until careful telescopic observations in 1978 showed that it possesses one large satellite, named Charon, which orbits only 17 000 km from its parent. This discovery enabled the mass of Pluto to be estimated for the first time. It proved to be only 1.5×10^{22} kg—remarkably small. It is now known that the diameter of Pluto is only 2700 km, making it much the smallest planet in the solar system. Its density is only about 1.5×10^3 kg/m^3, suggesting that it has a largely icy composition. Pluto may be no more than a large and dirty snowball in space.

Because Pluto is so small it has been suggested that it may be an escaped satellite of Neptune. One suggestion is that immediately after the formation of the solar system Pluto and Triton both revolved around Neptune, until Triton's orbit brought it so close to Pluto that it swept right around it, and went back in the opposite direction; in doing so it caused Pluto to be flung off in the opposite direction, eventually to fall into an orbit around the Sun. Although speculative, this hypothesis has the advantage of simultaneously explaining the retrograde orbit of Triton and Pluto's small size and highly unusual orbit.

Conclusion

The spacecraft exploration of the planets has been richly rewarding. Our knowledge of them and their satellites has advanced more in the past decade than over the previous three millennia, and each has proved to be quite different. In the past 'geology' has meant one thing, and one thing only—the study of the evolution of the Earth. Now there are half a dozen or more distinct 'geologies'. It will take many years of patient study, and many more spacecraft missions, before we can hope to understand them fully, in all their variety.

Perhaps the most valuable result of this wealth of discovery, however, is the dazzling new light that it casts on our view of the Earth. Men and women have always known that the Earth was unique. Now the extent of that uniqueness is appreciated even more. With its active internal processes, its hydrosphere and its oxygen-rich atmosphere and above all its teeming biosphere, the Earth is indeed an extraordinary and wonderful planet.

Glossary

abyssal plain An extensive flat region of the deep ocean floor deeper than the continental shelf but not as deep as the ocean trenches, covered with a thin layer of sediment, often pelagic ooze.
accessory mineral A mineral present in small quantities in an igneous rock, disregarded when classifying the rock. Many are silicates, oxides and sulphides.
acid rocks Igneous rocks containing more than 66 per cent silica, or 10 per cent or more free quartz, eg granite, rhyolite.
adiabatic Denoting physical change occurring without the addition or loss of heat from the system, often applied to changes in temperature in ascending or descending air masses.
advection The transfer of heat in a fluid by horizontal flow, as in the horizontal component of a convection cell in the Earth's mantle or horizontal air movements in the atmosphere.
aeolian Resulting from wind action, eg aeolian erosion, aeolian deposits.
air mass A mass of air in the lower atmosphere having largely uniform properties of temperature and moisture, originating in a specific source area and bounded by fronts.
albedo The ratio of the radiation reflected from a surface to the total radiation received by it. Different surfaces (eg snow, water, forest) have different albedos; the planetary albedo of the Earth is approximately 34 per cent.
albite A sodic plagioclase feldspar, $NaAlSi_3O_8$, a triclinic mineral found mainly in alkali-rich rocks, eg granites.
algae A group of primitive eukaryotic plants, mainly aquatic and without leaves, flowers or vascular system, eg seaweeds and diatoms.
alkaline Denoting igneous rocks and minerals that are rich in sodium and potassium and low in calcium (compare calc-alkaline).
alluvium Unconsolidated sedimentary material transported by a river and deposited on floodplains, estuaries and deltas and on the river bed. Consisting of silts, sands and gravels, it may form fertile alluvial soil.
ammonite A mollusc of the order Ammonoidea which became extinct at the end of the Cretaceous, characterized by a coiled partitioned shell with convoluted suturing. The ammonites are important zone fossils in Mesozoic rocks.
ammonoid An extinct marine mollusc of the class Cephalopoda, usually having a coiled and partitioned external shell. A general term covering members of the ammonite and goniatite groups.
amniote egg A tough-shelled egg in which the embryo is protected by inner (amnion) and outer (chorion) membranes. Its development led to the emergence of reptiles no longer needing water for reproduction.
amorphous Non-crystalline (ie having no regular atomic structure), eg opal.
amphibian A cold-blooded vertebrate of the class Amphibia which characteristically breeds and passes its early stages in water but its adult stages on land, eg the frog and salamander.
amphibole One of a group of common orthorhombic or monoclinic silicate minerals, mostly of magnesium, iron, calcium and sodium with a general composition $-(Mg,Fe^{2+},Ca)_{2-3}(Mg,Fe^{2+},Fe^{3+},Al)_5(Si,Al)_5(Si,Al_4)O_{11}(OH)_2-$occurring in many igneous and metamorphic rocks. Hornblende is a common form.
amphipod A crustacean of the order Amphipoda, with a laterally compressed body, eg the sand hopper.
anaerobe An organism that does not need air or free oxygen for respiration. Certain bacteria are anaerobic.
anatexis Extreme metamorphism in which the structure of the original rock is completely destroyed and partial melting takes place.
andalusite An aluminium silicate, a polymorph of Al_2SiO_5 occurring in prismatic crystals of the orthorhombic system in high grade metamorphic rocks.
andesite A fine-grained igneous rock of intermediate composition, the volcanic equivalent of diorite; it typically contains plagioclase feldspar, amphibole, and sometimes small amounts of quartz and alkali feldspar. Andesitic magma is generated at subduction zones and forms new continental crust.
angiosperm A flowering seed-bearing plant characterized by the production of seeds enclosed in fruits. Angiosperms evolved during the Cretaceous.
angular momentum The product of the angular velocity of a rotating body and its moment of inertia (which is the rotational equivalent of the mass of a body in linear motion).
anhydrite Anhydrous calcium sulphate ($CaSO_4$), an orthorhombic mineral occurring in association with gypsum in evaporite deposits.
anion Ion having a negative charge by virtue of having gained one or more electrons.
annelid A worm of the phylum Annelida, with a segmented (mostly soft) body, eg the earthworm.
anomaly A deviation from the normal or predicted value, conditions etc. A magnetic anomaly is a departure from the predicted value of the Earth's magnetic field at a particular point on its surface; similarly a gravity anomaly, a geothermal anomaly.
anorthite A triclinic feldspar, $CaAl_2Si_2O_8$, the calcium end member of the plagioclase series.
anorthosite A coarse-grained, light-coloured igneous rock occurring as large intrusive bodies including layered basic complexes. It consists of over 90 per cent plagioclase feldspar.
anticline An arch-shaped fold in which the younger strata remain at the top of the succession.
anticyclone A region of high pressure (relative to the surrounding air) in the atmosphere, characterized by descending air; winds, generally light, blow around it in a clockwise direction in the northern hemisphere, anticlockwise in the southern hemisphere.
antidune A small-scale feature of current-bedding produced by the erosion of sediment from the leeside of a ripple and deposition on the other (stoss) side, causing its migration upcurrent.
antiform An arch-shaped fold in the general case in which the direction of younging of the strata is not known; antiforms include inverted synclines.
aquifer A pervious or porous layer of rock that is water-saturated and can therefore supply water to springs or wells.
arachnid An arthropod of the class Arachnida, including spiders, scorpions and mites, characterized by four pairs of legs and simple eyes.
aragonite An orthorhombic carbonate mineral ($CaCO_3$), occurring in sedimentary rocks and in a number of shells. It commonly inverts to the more stable crystalline form calcite in a geologically short period.
archosaur A member of the group known as the ruling reptiles which included the dinosaurs, pterosaurs and modern crocodiles, it was prominent among terrestrial life in the Mesozoic.
arenaceous rock A sedimentary rock composed mainly of cemented grains of sand.
argillite A fine-grained sedimentary rock composed mainly of clay minerals, eg mudstone, shale.
arthropod An animal of the phylum Arthropoda, having jointed limbs, a segmented body and a chitinous exoskeleton, eg crustaceans and insects.
asteroid A minor planet, one of over 1500 orbiting the Sun between Mars and Jupiter; the largest is Ceres (770 km) and their total mass is about 1/5000 that of Earth.
asthenosphere A layer of the Earth's mantle at the base of the lithosphere, between about 100 and 200 km below the surface. Seismic velocities are lower in the asthenosphere than above or below it due to its low rigidity. It is thought to be a region of partial melting.
atmophile Denoting an element occurring in an uncombined state in the atmosphere as a gas: nitrogen and the inert gases.
atmosphere The layer of gases surrounding the Earth, divided into the troposphere (up to 10–15 km, the source of most of the weather), stratosphere (up to about 50 km), mesosphere (up to 80–90 km), ionosphere (up to about 400 km) and the outermost layer, the exosphere. Also a unit of pressure of 101 325 newtons per square metre.
atomic weight The mean mass of the atoms of an element, weighed against a standard of the carbon isotope ^{12}C whose atomic mass is taken to be exactly 12.
authigenesis The formation of minerals *in situ* during or after the deposition of the sediments in which they occur.
autotroph An organism capable of synthesizing its own organic compounds from simple inorganic substances, eg green plants.

barrier island An elongate island lying roughly parallel to the coast from which it is separated by a shallow lagoonal area.
barrier spit A linear accumulation of beach material extending part-way across the mouth of an estuary in the direction of longshore drift along the adjacent coast.
basalt A fine-grained, basic igneous rock consisting mainly of plagioclase feldspar and pyroxene. Basalts occur principally in lava flows and constitute over 90 per cent of volcanic rocks.
basement Igneous or metamorphic rocks, often Precambrian, that unconformably underlie unmetamorphosed sedimentary strata.
basic rocks Igneous rocks containing 45–55 per cent silica. They are quartz-free and contain feldspars that are more calcic than sodic. Also referred to as mafic.

basin 1 The area drained by one river and its tributaries: a drainage basin. 2 An inward-plunging, synclinal fold structure; the opposite of a dome. 3 An area of continued subsidence of the crust that accumulates sediment over a prolonged period: sedimentary basin.
batholith A large intrusive mass of igneous rock, typically granite, outcropping over at least 100 km^2 and extending to unknown depth. Batholiths are particularly characteristic of orogenic belts in subduction zones.
bed See stratum.
belemnite An extinct marine cephalopod mollusc, squidlike and with a characteristic bullet-shaped guard which forms part of the internal chambered shell; a common Mesozoic fossil.
Benioff zone The seismically active zone dipping from the base of an ocean trench into the mantle, marking the path of subduction of an oceanic plate at a destructive plate margin.
Bennettitales An order of plants of the Mesozoic, similar to the cycads but having bisexual flowers.
benthos Plants or animals that live on the sea bed.
biophile An element occurring usually in organic matter: carbon, phosphorus.
biosphere The regions of the Earth in which life exists.
biota The plant and animal life occupying a specific region and/or period of time.
biotite The commonest of the ferromagnesian micas, $K(Mg,Fe^{2+})_3(AlSi_3)O_{10}(OH,F)_2$, occurring in many igneous and metamorphic rocks.
bivalve An aquatic organism of the phylum Mollusca having a bivalved calcareous shell, ranging from early Cambrian to Recent.
blocking Of a more persistent area of high pressure, causing the usual paths of depressions to be diverted round it.
blue-green algae Prokaryotic cellular organisms, lacking membrane-bound organelles such as a nucleus, occurring singly or in groups or filaments. Most are aerobic photoautotrophs, using sunlight as an energy source and producing oxygen. They range from the Archaean to the present day, and are sometimes referred to as the cyanobacteria.
bond The force by which the atoms in a chemical compound are attached to each other, either by sharing of electrons (covalent) or by electrostatic attraction between ions (Van der Waal's bond).
brachiopod A marine invertebrate of the phylum Brachiopoda with two calcareous or chitinous shells. Brachiopods are superficially similar, but unrelated, to the bivalves and distinguished from them by the bilateral symmetry of their shells and by their anatomy.
breccia 1 A coarse sedimentary rock consisting of angular fragments set in a finer groundmass, usually from a localized source, eg scree breccia, formed by the lithification of a scree slope. 2 Fault breccia, formed from fragments broken from country rock by movement along a fault.
brine Water containing a high proportion of a dissolved salt, especially common salt (NaCl).
bryozoan (polyzoan) A small colonial animal of the phylum Bryozoa attached to the sea floor or to other animals or rocks, ranging from Ordovician to Recent.
bulk modulus The ratio of applied stress to resulting change in volume per unit volume; one of the elastic constants.
bytownite A triclinic feldspar of the plagioclase series, near the calcium-rich end of the series, occurring in basic igneous rocks.

calc-alkaline Denoting igneous rocks in which the feldspar is calcium-rich (as opposed to alkaline rocks in which the feldspar is sodic to potassic).
calcareous Containing or consisting of calcium carbonate.
calcite A trigonal mineral consisting of crystalline calcium carbonate ($CaCO_3$), usually white or colourless; the principal constituent of most limestones.
caldera A large volcanic crater caused by repeated explosion, or by the collapse of a volcanic cone after withdrawal of magma from beneath it.
cambium Zone of growth of stems and roots by the production of additional xylem and phloem cells.
canyon 1 A steep-sided river valley, usually deeper than its width, often occurring in arid or semi-arid regions. 2 A deep valley-like incision into the continental slope, thought to be eroded by turbidity currents: a submarine canyon.
carbonaceous Containing carbon; denoting especially a sedimentary rock containing significant organic material and derivatives, eg coal.
carbonates A group of widespread minerals, principally of calcium, magnesium, iron and manganese, or combinations of these. Calcite ($CaCO_3$), dolomite ($CaMg(CO_3)_2$) and siderite ($FeCO_3$) occur as primary sedimentary constituents.
carbonatite An igneous rock consisting chiefly of carbonate minerals, especially calcium carbonate, often containing rare-earth accessory minerals. Carbonatites are of very restricted distribution.
carbon dioxide (CO_2) A gas formed by the oxidation of carbon and carbon compounds and by the reaction of acid with carbonates. It is present in the atmosphere (0.05 per cent, increasing because of the combustion of fossil fuels) and in solution in the hydrosphere. It is absorbed during photosynthesis.
carbon monoxide (CO) A colourless poisonous gas, formed by the incomplete combustion of carbon.
cassiterite (SnO_2) An ore of tin, brown to black in colour and found in hydrothermal veins, acid igneous rocks and alluvial deposits.
cation Ion having a positive charge by virtue of the loss of one or more electrons.
Caytoniales Extinct woody plants having four-lobed, net-veined leaves with ovules and pollen sacs borne in fruitlike bodies. Pollinated directly in gymnospermous fashion. They are regarded as Mesozoic seed ferns.
centrifugal force The outward force that is experienced by a body moving in a curved path; it is directed away from the centre of curvature.
cephalopod A marine mollusc of the class Cephalopoda, including squids, octopuses and the extinct ammonites. The class is characterized by the presence of tentacles and well-developed eyes, and fossils are known from the Cambrian onwards.
chalcophile Denoting an element occurring usually as a sulphide: silver, arsenic, copper, mercury, lead, zinc (and sulphur).
chamosite A greenish monoclinic mineral related to chlorite $(Fe^{2+},Mg,Al,Fe^{3+})_6(AlSi_3)O_{10}(OH)_2$: it is found in sediments where it may form an iron ore.
chert A sedimentary rock consisting of cryptocrystalline chalcedonic silica precipitated chemically or biologically on the sea floor. It is an opaque, dark-coloured massive or stratified rock. It also commonly forms by replacement of limestone (silicification); flint is an example.
china clay A mineral deposit (kaolinite), consisting mainly of kaolin, produced by the weathering or hydrothermal alteration of feldspars in granite.
chlorite A greenish ferromagnesian sheet silicate mineral, $(Mg,Fe^{2+})_{10}Al_2(Si,Al)_8O_{20}(OH,F)_{20}$. Chlorite is most commonly found as an alteration product of other ferromagnesian minerals, eg as a result of weathering in the presence of water, and in low grade regionally metamorphosed rocks.
chondrite A type of stony (ie dominantly silicate) meteorite having chondrules–small globular pockets of pyroxene and olivine.
chromite A cubic mineral ($Fe^{2+}Cr_2O_4$), one of the spinel group. It is the chief ore of chromium, often occurring in basic igneous rocks and in serpentinites.
clast A sedimentary particle, usually a mineral grain or rock fragment.
clastic sediments Sediments derived from pre-existing rock material by weathering and erosion processes; clastic sedimentary rocks include conglomerates, sandstones, siltstones and mudstones.
clay A very fine-grained clastic sediment consisting of particles less than 1/256 mm in diameter, mainly of clay minerals (hydrated silicates, especially of aluminium, which occur as layered platy or fibrous crystals and can take up or lose water). Clay is plastic when wet, cracks on drying out and hardens on heating.
cleavage 1 The tendency of a mineral to split along a plane of weakness in its crystalline structure. 2 The tendency of a rock to split along closely spaced planes, especially as a result of the parallel alignment of minerals due to deformation or metamorphism.
climate The sum of the prevailing long-term weather conditions of an area, determined by factors such as latitude, altitude and location in relation to landmasses and to the main circulation belts of the atmosphere and oceans.
clinopyroxene A monoclinic pyroxene; includes most of the calcium-bearing pyroxenes. Clinopyroxenes formed in rapidly cooled lavas may subsequently invert to orthopyroxene.
coalescence A mechanism by which water droplets or ice crystals increase in size to form raindrops or snowflakes in clouds.
coccolithophore A unicellular planktonic organism secreting coccoliths (microscopic, round, calcareous plates) that may accumulate on the ocean floor as sediment, ultimately contributing to limestones such as chalk.
cnidarian/coelenterate An aquatic invertebrate of the phylum Coelenterata (Cnidaria) which includes the jellyfish, sea anemones and corals. Coelenterates occur as polyps (cylindrical and stationary) or medusae (disc-shaped and free-living) and occur throughout the fossil record from the late Proterozoic.
coacervate A colloidal droplet formed by the spontaneous separation of phases containing high

and low concentrations of certain biological polymers in a solution of such polymers; a possible model for the origins of living cells from abiotically produced organic substances.
coccolith A microscopic calcareous plate a few micrometres in diameter secreted as part of a spherical skeleton by unicellular green algae (coccolithophores). Coccoliths have been important constituents of the marine microplankton since the Jurassic period, contributing in particular to the Cretaceous chalks. Referred to as nanoplankton by biostratigraphers to whom they are important in the correlation of Cretaceous and Cenozoic oceanic sediments.
collembolan (or spring-tail) A small, primitive wingless insect of the order Collembola.
compound A substance consisting of two or more elements that are chemically united in definite proportions.
condensation The change of state from vapour to liquid, eg as a result of cooling or increase of pressure.
conductivity The ability to conduct or transmit heat or electricity.
condylarth A primitive mammal of the early Cenozoic from which modern ungulates evolved.
conglomerate A coarse-grained, clastic sedimentary rock formed largely of rounded, water-worn pebbles.
Coniferales Gymnosperms bearing seed cones, with generally rather small, needle-like leaves first occurring in the late Carboniferous.
connate water Groundwater trapped in sediments at the time of their deposition.
conodont A microscopic fossil consisting of a small, toothlike structure composed of calcium phosphate. Their zoological affinity is unknown and they occur from the Cambrian to the Triassic.
constructive plate margin The boundary between two separating lithospheric plates at which new oceanic crust is generated, typically at the mid-ocean ridges.
continental drift The relative movements of the continents over the Earth's surface as a result of ocean floor spreading.
contour A line on a map or chart joining all points at the same elevation relative to sea level.
convection The transfer of heat through a fluid by movements of the fluid itself; the heated part expands and rises, cooler material flows in to take its place and a convection current is thus established.
coprolite A fossilized faecal pellet, which may be used to determine the diet of an extinct creature.
coral One of a group of marine coelenterate animals of the class Anthozoa (phylum Cnidaria), the hard calcareous skeletons of which accumulate to form the coral reefs of tropical waters.
Cordaitales Large tree-sized plants, bearing strap-shaped leaves and their seeds in cones, common in the late Palaeozoic.
cordierite An orthorhombic silicate mineral, $Al_3(Mg,Fe^{2+})_2(Si_5Al)O_{18}$, found in thermally metamorphosed rocks such as hornfels.
cordillera A chain of mountains, particularly the mountains of western North and South America, but also used generally of mountains in a corresponding tectonic setting.
core That part of the Earth's interior lying below the mantle, separated from it by the Gutenberg discontinuity (about 2900 km below the Earth's surface). It consists of an inner core (believed to be solid) and an outer core (liquid).
core sample A cylindrical section of sediment obtained from the sea floor or of rock extracted from the Earth's crust by drilling with a hollow bit.
Coriolis force A hypothetical force required to explain deflection to the right in the northern hemisphere and to the left in the southern hemisphere of the path of a body or fluid moving north or south.
covalent Denoting a chemical bond in which a pair of electrons are shared by adjacent atoms, one being provided by each.
crater A bowl-shaped depression at the mouth of a volcano or formed by the impact of a meteorite.
craton A region of the Earth's crust that has remained relatively unaffected by orogenic activity for a long period of time.
crevasse 1 A deep crack in glacier ice caused by the differential movement of ice where the containing valley widens (longitudinal crevasse) or steepens (transverse crevasse). 2 A steep-sided channel in a delta or other sedimentary environment.
crinoid (or sea lily) A marine invertebrate of the phylum Echinodermata, having feathery arms radiating from a cup at the end of a stem. Fossils occur from the Cambrian onwards.
crossopterygian A bony fish of the subclass Crossopterygii, characterized by fleshy paired fins. Now almost extinct except for the coelacanth, they were common from the Devonian to the end of the Palaeozoic and gave rise to the amphibians.
crust The outermost shell of the Earth, overlying the mantle; about 35 km in thickness beneath the continents and 10 km in thickness beneath the oceans. It is divided into continental crust, or sial, and heavier oceanic crust, or sima. The Mohorovičić discontinuity marks its lower boundary. The crust forms the outer part of the lithosphere.
crustacean An invertebrate of the class Crustacea, of the phylum Arthropoda, including lobsters, crabs and barnacles. Fossils are known from the Cambrian onwards.
cryosphere The region of the Earth's surface that is permanently frozen.
crystallization The solidification of minerals as crystals of definite geometrical form, either through cooling of molten magma or as a result of metamorphism.
cumulate rock An igneous rock forming the lower ultrabasic layers of an intrusion, where the early-formed crystals settled during slow cooling at the bottom of the magma chamber.
cumulus cloud A type of cloud resulting from convectional cooling of columns of moist air. Fine-weather types are shallow; rain-bearing (cumulonimbus) types are dark and of great vertical extent.
Curie point The critical temperature, specific to each substance, above which it loses its permanent magnetization.
cycad A gymnospermous plant of the order Cycadales which evolved during the Permian, with fern-like leaves at the top of an unbranched stem. Cycads were abundant during the Mesozoic but are much less important today.
cyclogenesis The formation of cyclones or low-pressure systems in the atmosphere, especially the development of intense tropical storms over the oceans.
cyclone A depression or region of low pressure (relative to the surrounding air) in the atmosphere. Winds blow around it in an anticlockwise direction in the northern hemisphere and clockwise in the southern hemisphere. The term cyclone is sometimes used for a small, intense low-pressure system in the tropics (a tropical storm).

dacite A fine-grained acid igneous rock, the volcanic equivalent of quartz-diorite, containing sodic plagioclase and quartz as phenocrysts.
decapod A crustacean of the order Decapoda, characterized by five pairs of walking limbs, eg lobsters and crabs.
deformation Change in shape, volume or structure of some region or feature of the Earth's crust resulting from stress caused by tectonic activity.
delta A generally fan-shaped accumulation of sediment deposited at the mouth of a river into which the river flows, where the lake or sea is relatively free of currents.
dendritic Denoting a branching (treelike) pattern or structure, eg a drainage pattern.
density The mass per unit volume of a substance, expressed in SI units as kilograms per cubic metre.
denudation The wearing down of the land surface by the sum of the processes of weathering, mass movement, transportation and erosion.
destructive plate margin The boundary between two converging lithospheric plates where sea floor is subducted into the mantle, typically at the ocean trenches.
deuterium (D) A rarely occurring natural isotope of hydrogen having twice the mass of the more abundant isotope.
diagenesis The physical and chemical processes (excluding metamorphism) that affect deposited sediment, both causing and subsequent to the formation of solid rock.
diapir A vertical intrusion, bulbous or cylindrical in shape, formed by less dense material rising through country rock. Applied to both igneous and sedimentary rocks.
diatom A microscopic unicellular alga, an important constituent of the marine plankton. Their accumulated siliceous cases often form extensive deposits called diatomite.
diatreme A volcanic vent or pipe cutting through the overlying rocks as a result of the explosive release of magmatic gases and filled with the products of the eruption, eg the kimberlite pipes of South Africa.
differentiation The segregation of constituent minerals. Magmatic differentiation involves the separation of magma into fractions of different compositions, one of the main processes being fractional crystallization. Metamorphic differentiation involves the migration of some rock constituents into preferred zones, resulting in foliation or other inhomogeneities.
diffraction The bending of the direction of propagation of waves of all kinds (seismic, electromagnetic, acoustic) around objects in their path.
dinosaur An extinct terrestrial reptile abundant in the

Mesozoic, the largest reaching about 35 m in length. The two groups are the Saurischia (lizard-hipped) and the Ornithischia (bird-hipped).
diorite A coarse-grained intermediate igneous rock consisting of salic plagioclase feldspar and ferromagnesian minerals, usually hornblende or biotite. Diorite is the plutonic equivalent of andesite.
dipmeter An instrument used to measure the angle of dip of rock strata in a borehole.
discontinuity A boundary within the Earth between zones with different seismic characteristics. The Mohorovičić discontinuity, or Moho, separates the crust from the mantle and the Gutenberg discontinuity separates the mantle from the core. The Conrad discontinuity separates lower and upper continental crust. Seismic discontinuities may be either sharp or gradational.
dolomite 1 A trigonal carbonate mineral ($CaMg(CO_3)_2$). 2 A type of limestone composed of over 15 per cent magnesium carbonate. It occurs either as a primary precipitate from seawater (an evaporite) or by the alteration of calcite rock by magnesium-charged solutions (dolomitization).
dome An anticlinal fold structure that plunges radially, sometimes a topographic feature and sometimes only a structural one. The term is also used generally for any dome-shaped feature or rock formation.
dune A mound of wind-blown sand occurring on beaches above high-tide mark or in sandy deserts.
dyke A sheetlike intrusion of igneous rock, normally of intermediate grain size, that cuts discordantly through the surrounding rock.
dynamo A mechanism that converts mechanical energy into electrical energy by the movement of an electrical conductor across a magnetic field (electromagnetic induction).

earthquake Shock waves, detectable and sometimes causing violent tremors at the Earth's surface, generally originating by movements along deep-seated fault planes releasing an accumulated build-up of stress in the lithosphere.
echinoderm A marine invertebrate of the phylum Echinodermata, including starfish, sea cucumbers, sea urchins and crinoids. They usually have a five-rayed radial symmetry and a skeleton of calcareous plates. Fossils are known from the Cambrian onwards.
echinoid (or sea urchin) A marine animal of the class Echinoidea, of the phylum Echinodermata. Fossils are known from the Cambrian onwards.
eclogite A rare metamorphic rock resembling basalt in chemical composition, formed by regional metamorphism under conditions of very high temperature and pressure. Eclogites consist of pyroxene and a magnesian garnet. They are found as xenoliths in kimberlite pipes.
eddy A rotational flow of fluid within the general flow of the main body. Depressions are large-scale eddies within the general circulation of the atmosphere.
Ekman flow The response of the water beneath the ocean surface to a wind. Because of frictional and Coriolis forces, even surface water flows at about 45° to the direction of the wind, veering increasingly and moving more slowly with depth, until at the **Ekman depth** the water flows at 180° to wind direction. The **Ekman layer** is the surface layer, approximately 100 m thick, in which Ekman flow occurs.
Ekman spiral A representation of the pattern of changing velocity and direction of flow of ocean currents with increasing depth.
elastic waves Seismic waves in which particles are displaced but return to their original position after the passage of the wave.
electrode A conductor through which an electric current is passed into or out of something. The positive electrode is the anode, and the negative the cathode.
electron A negatively charged elementary particle existing in every atom, orbiting the nucleus, in numbers equal to the atomic number of the element.
element A substance consisting wholly of atoms of the same atomic number, listed in the periodic table (see page 53); there are over a hundred known elements, of which the first ninety-two, up to and including uranium, occur naturally.
ensialic Occurring in the sialic layer, ie the continental crust (eg of sedimentary basins occurring away from continental margins).
epeiric sea A comparatively shallow sea lying inland and resulting from large-scale, slow, vertical crustal (epeirogenic) movements.
erosion The breaking down and wearing away of rocks by running water, ice, wind, waves etc.
erratic A boulder or other large rock fragment, transported from its source usually by glacial action.
escarpment (or scarp) A steep slope forming a linear feature, resulting from the differential erosion of gently inclined strata.
essential minerals Those minerals in an igneous or metamorphic rock that determine its mineralogical classification.
estuary The widening tidal portion of a river where it approaches the sea, containing both fresh- and salt-water.
eukaryote A unicellular or multicellular organism in which the cells contain well-defined organelles including a nucleus.
eurypterid A freshwater scorpion-like arthropod of the class Eurypterida, some attaining a length of 2 m, occurring from the Ordovician to the Permian.
eustatic Of world-wide changes in sea level.
evaporation The change of state of liquid to vapour. Evaporation of water into the atmosphere depends on the vapour pressure of the air, its temperature, the amount of wind and the nature of the surface from which evaporation takes place.
evaporite Any sedimentary rock formed by precipitation from saline water, eg rock salt. Evaporites are classified according to chemical composition.
exoskeleton The hard outer structure protecting and supporting the body of certain animals, eg arthropods.
exosphere See atmosphere.
extrusive rock Igneous rock formed from magma that has flowed out at the Earth's surface as lava (ie volcanic rock); it usually has small crystals or a glassy texture.

facies, metamorphic A grouping of rocks that regardless of composition have all been metamorphosed under similar conditions of temperature and/or pressure. The principal metamorphic facies are named from the mineral assemblages typically developed in metamorphosed igneous rocks.
facies, sedimentary The sum of the lithological and palaeontological characteristics of sedimentary strata, with their implications for the environment of deposition of those strata.
fault A fracture along which rocks have been displaced in a horizontal, vertical or oblique sense.
fault scarp A steep topographic slope caused by faulting. A fault line scarp is a slope caused by the differential erosion of rocks juxtaposed by faulting, the original fault scarp having been eroded away.
feldspar The most abundant mineral in igneous rocks, consisting structurally of a complex aluminium silicate framework with varying proportions of potassium, sodium, calcium and (rarely) barium. Feldspars are grouped into alkali feldspars ($KAlSi_3O_8$–$NaAlSi_3O_8$) and plagioclase feldspars ($NaAlSi_3O_8$–$CaAl_2Si_2O_8$) and are monoclinic or triclinic minerals.
felsic minerals Light-coloured minerals, particularly feldspar and quartz.
ferromagnesian minerals Silicate minerals rich in iron and magnesium, usually dark-coloured. Includes amphibole, pyroxene, olivine, biotite and mica.
ferrosilite An iron-bearing orthopyroxene, $Fe^{2+}{}_2Si_2O_6$.
ferruginous Denoting rocks or minerals containing iron, often resulting in a reddish colour.
fissure 1 An extensive crack or fracture in rocks. 2 A linear volcanic vent.
fixation The conversion of a free element (eg nitrogen in the air) to a solid compound, or of any substance to a less volatile form.
flysch Thick sequences of interbedded marine shales and greywacke sandstones, deposited by turbidity currents from rapidly uplifted and eroded fold mountains; the flysch itself is involved in the later stages of orogeny. Strictly an Alpine term, but now more widely applied.
fold A flexure in rocks resulting from compressional or gravitational forces. An upfold is termed an antiform and a downfold a synform, with many variations of different degrees of complexity.
Foraminifera An order of small unicellular marine protistan animals with calcareous shells, accumulations of which have contributed to limestones since Devonian time. Earlier foraminifera built skeletons of pre-existing sediment grains. Foraminifera were exclusively benthic until the mid Jurassic, since when planktonic forms have abounded.
fossil Remains of past life preserved in sedimentary rock. Fossils include skeletal remains, impressions in the form of casts or moulds, trace fossils, coprolites and chemical fossils.
fractionation, magmatic The segregation of minerals in a body of cooling magma due to their crystallization at different temperatures, successive crops of crystals sinking in the magma and accumulating at the base of the intrusion.
fracture A crack (fault or joint) in rocks resulting from deformation rather than cleavage.
fumarole A vent in the ground from which hot

gases, especially steam, are emitted, often found in volcanic areas. There is often a crystalline deposit surrounding the vent.
fusulinid A large foraminifer, up to 20 mm or more in length, often having a spindle-shaped calcareous endoskeleton. They were abundant in the Carboniferous and Permian periods.

gabbro A coarse-grained basic igneous rock, the plutonic equivalent of basalt, and consisting essentially of plagioclase feldspar, pyroxene and olivine.
gametophyte In a plant showing alternation of generations, it is the generation that produces the gametes, or sexual cells; as opposed to the sporophyte —the generation that produces spores.
garnet One of a group of silicate minerals crystallizing in the cubic system, of general formula $(Fe^{2+},Mg,Mn^{2+},Ca)_3(Fe^{3+},Al,Cr)_2Si_3O_{12}$. Garnets are mostly found in metamorphic rocks, particularly in high-grade rocks including eclogite, but they also occur as accessory minerals in a variety of igneous rocks.
gastropod A mollusc of the class Gastropoda, including snails and slugs. They usually have a helically coiled, undivided shell and crawl on a muscular foot. Fossils are known from the Cambrian onwards.
geodesy The scientific study of the exact shape and size of the Earth, with which is associated the study of its rotation and its gravitational field.
geoid The shape that the Earth would have to be for the effect of gravity to be constant at all points, taken as mean sea level extended beneath the continents.
Ginkgoales Tree sized plants with characteristically fan-shaped leaves with dichotomous venation. Ovules and pollen sacs are borne separately and arranged in catkins. Fossils are mostly Mesozoic, although one species is still in existence.
glaciation An increase in the extent and activity of the ice cover of the Earth's surface; a particular period during which this occurred; the geomorphological effects of the presence and movement of ice sheets and glaciers.
glauconite A green, monoclinic, mica-like mineral, a complex hydrated silicate of iron and potassium, occurring in marine sediment as an authigenic mineral, especially in greensands. Formula $K(Fe^{2+},Mg,Al)_2Si_4O_{10}(OH)_2$.
gneiss A coarse-grained crystalline rock formed during high-grade regional metamorphism of igneous (orthogneiss) or sedimentary (paragneiss) rocks, characterized by a banded appearance and linear orientation of minerals.
gossan A mass of hydrated iron oxides and quartz, often indicating the outcrop of a sulphide-bearing vein, the metals and sulphur having been leached from the uppermost part by percolating water.
graben An elongate body of rock downfaulted between parallel normal faults. (It does not necessarily form a topographical feature, but the term is frequently used synonymously with rift valley.)
granite A coarse-grained acid igneous rock consisting mainly of quartz, alkali feldspar and mica, with various accessory minerals. It occurs in intrusive bodies from crystallized magma or the 'granitization' (metasomatic transformation) of pre-existing rocks.
granodiorite A coarse-grained acid igneous rock containing quartz, predominantly plagioclase feldspar, and ferromagnesian minerals (especially biotite or hornblende). Granodiorites are the most abundant plutonic rocks.
graptolite An extinct marine colonial animal of the class Graptolithina, characterized by several minute chitinous cups housing individual polyps arranged along a stem. Fossils occur from the Cambrian to the Carboniferous and are important in biostratigraphy.
gravitational acceleration (g) The acceleration of a body falling in the Earth's gravitational field, equal to 9.80665 m per second per second in a vacuum at the Earth's surface. Measurements are usually made in milligals; one mgal is an acceleration of 0.01 mm per second per second.
greywacke A poorly sorted clastic sedimentary rock, usually with coarse angular rock particles (lithic fragments) set in a much finer cementing material.
groundwater Water present under ground in pores and fissures in rocks and in the soil, derived from rainwater percolating downwards (meteoric water), from water trapped in sediment at the time of its deposition (connate water) and from magmatic sources (juvenile water).
gymnosperm A seed-bearing plant, characterized by the production of seeds not enclosed in an ovary as they are in the angiosperms, eg conifers. The first fossil seeds occur in late Devonian rocks.
gypsum A monoclinic calcium sulphate mineral, $(CaSO_4.2H_2O)$; the hydrated equivalent of anhydrite. Gypsum occurs as an evaporite mineral, and in shales and limestones.

haematite An important iron ore (Fe_2O_3), of trigonal symmetry and often occurring as a cement in sandstones (producing a red tinge), as an accessory mineral in igneous rocks and in hydrothermal veins and replacement deposits.
hedenbergite A calcium–iron clinopyroxene, $CaFe^{2+}Si_2O_6$.
heterotroph An organism unable to synthesize organic substances from inorganic compounds. Includes some bacteria, all fungi and all animals.
horizon 1 A surface or distinctive layer used as a reference in a sequence of sedimentary rocks. 2 A layer of soil forming part of the vertical sequence in a soil profile.
horst An elongate body of rock between parallel normal faults, uplifted relative to surrounding rocks.
humus The organic matter in soil, formed from partial decomposition of plant and animal remains.
hydrocarbons Organic compounds consisting primarily of hydrogen and carbon, including those which occur in petroleum, natural gas and coal.
hydrosphere The sum of the natural waters of the outer Earth, encompassing the oceans (98 per cent), the rivers, water vapour and underground water.
hydrostatic equilibrium The state of a fluid in which the various forces acting on it (including gravity and pressure) balance each other.
hygromagmatophile Of elements which prefer to concentrate in magma rather than in mantle minerals; a form of incompatibility.
hygroscopic nuclei Minute solid particles (dust, salt, pollen etc) in the atmosphere. They attract water vapour and initiate condensation, which may result in precipitation.

iceberg A mass of ice floating in the sea, most of it submerged. In the northern hemisphere icebergs tend to be of castellated form, while in the southern hemisphere they are typically tabular.
ichthyosaur A marine aquatic reptile having a fishlike body and paddle-like limbs. They were prominent especially in the Jurassic and became extinct during the Cretaceous.
igneous Denoting rocks formed by solidification from a molten state, either intrusively below the Earth's surface or extrusively (as lava or pyroclastic fragments). Igneous rocks have an interlocking crystalline texture and vary in mineral content according to the composition of the melt.
ignimbrite (or welded tuff) A type of pyroclastic rock with a streaky appearance, formed by the welding together of hot tuff material deposited from an explosive volcanic eruption.
ilmenite A black trigonal mineral $(FeTiO_3)$, a source of titanium, occurring in basic igneous and metamorphic rocks, in veins, and in detrital sands.
imbricate 1 Denoting a thrust-faulting structure in which a series of high-angle reverse faults occur between parallel thrust planes, causing layers of rock to overlap like tiles on a roof. 2 Denoting an overlapping accumulation of flat pebbles in a stream, with the long axes of the pebbles dipping upstream.
incompatible Of an element with a partition coefficient less than one; ie an element preferring to remain in a magma than to find a place in the lattice of a crystallizing mineral.
induction, electromagnetic The creation of an electromotive force in an electric conductor by a change in the magnetic flux through it.
intermediate rocks Igneous rocks containing 55–66 per cent silica, or less than 10 per cent free quartz, with sodic plagioclase and/or alkali feldspar.
interstadial A relatively warmer stage within a glacial phase during which the ice advance is temporarily halted.
intrusive rock Igneous rock formed from magma that has forced its way among pre-existing rocks. Types of igneous intrusions include batholiths, dykes, sills etc. Crystals are large relative to those of extrusive rocks.
inversion Change from one mineral polymorph to another usually with an increase in ordering of crystal structure, eg aragonite to calcite.
invertebrate Any animal outside the phylum Chordata, having no backbone.
ion An electrically charged atom or group of atoms which has lost or gained one or more electrons: cations are positively charged and anions are negatively charged.
ionosphere See atmosphere.
island arc An arcuate chain of islands in volcanically and seismically active zones at destructive plate margins, often flanked by a deep trench on the convex side which is usually the oceanic side.
isobar A line on a map joining points of equal pressure (eg atmospheric pressure).

isopod A terrestrial or aquatic crustacean of the order Isopoda, with a flattened body, eg woodlice.
isostasy A condition of equilibrium in the Earth's crust. Assuming that the lighter continental masses float on a denser medium, changes in crustal elevation must be compensated in some way at depth.
isotope A form of an element having the same atomic number as other forms of the same element but having a different atomic weight by virtue of having a different number of neutrons in the nucleus.

joint A fracture in a rock with no visible offset of rock on each side.

kimberlite Brecciated peridotite found in pipes or diatremes in ancient metamorphic rocks and sometimes containing diamonds.
kinetic energy The energy of a body resulting from its motion, equal to $\frac{1}{2}mv^2$, when m is the mass of a body moving at velocity v.
kyanite An aluminium silicate, a polymorph of Al_2SiO_5 crystallizing in the triclinic system and found in high grade regionally metamorphosed rocks.

labyrinthodonts Large, late Palaeozoic amphibians having massive skulls and teeth with 'labyrinth' structure in cross-section. All possessed a tail and had well-developed body ribs giving a strong rib cage.
laterite A residual deposit of hydrated iron oxides, formed by weathering under tropical conditions, especially of iron-rich rocks such as basalt.
lava Magma in molten or partly molten state, extruded at the Earth's surface through vents or fissures. Basic lavas are far more abundant and less viscous than acid lavas.
leaching The process by which materials are removed from the upper layers of soil or rock by percolating water and carried away in solution or suspension.
lepospondyls Small, late Palaeozoic amphibians, usually small animals. They may have been ancestral to the reptiles.
levée A bank formed along the sides of a river channel in its floodplain or on a delta top, formed by deposition from occasional floods.
lignite A brown coal with a high moisture content, representing a low grade of coalification, intermediate between peat and bituminous coal.
limestone A sedimentary rock formed principally of calcium carbonate, usually as calcite. The following principal types all intergrade: clastic limestones, which are derived from pre-existing calcareous rocks; organic limestones, including chalk, which result from the accumulation of debris from calcareous plants and animals; and precipitated limestones, which result from chemical processes.
lineament An extensive linear topographic feature which may reflect the trend of some underlying structure, often detectable only from aerial photography.
lithification The formation of solid rock from loose sediment by compaction, cementation and diagenesis.
lithology The general physical (usually visible) characteristics of rocks.
lithophile Denoting an element that occurs mainly in oxygen-bearing compounds, especially in silicate minerals. The eight major lithophile elements (oxygen, silicon, potassium, magnesium, aluminium, iron, copper and sodium) make up nearly 90 per cent of the Earth's crust.
lithosphere The outer, rigid shell of the solid Earth, overlying the less rigid asthenosphere. The lithosphere comprises the crust (both oceanic and continental) and that part of the mantle (the lithospheric mantle) above the asthenosphere to which the crust is mechanically coupled. The base, and thus the thickness, of the lithosphere is difficult to define, but it probably varies between about 50 and 100 km below the Earth's surface.
lodestone A naturally occurring, strongly magnetized form of almost pure magnetite (Fe_3O_4), the magnetization of which can be detected without the use of sensitive instruments.
Love wave A type of seismic wave travelling along the ground surface, in which affected particles move in a horizontal direction at right angles to the direction of propagation.
lutite A sedimentary rock consisting of silt- or clay-sized ($<\frac{1}{16}$ mm) particles.
lycopod A club moss of the class Lycopsida, having a dense covering of spirally arranged simple leaves. They became prominent in the late Palaeozoic when treelike forms inhabited the coal forests.

magma Molten rock material originating in the Earth's mantle and crust and consisting of a liquid silicate phase with or without a solid phase of suspended crystals. Magma cooled at the surface forms volcanic rock; magma cooled at depth forms intrusive rock.
magmagenesis The formation of magma (molten rock) from solid material, either by the partial melting of mantle peridotite or by melting of crustal material above subduction zones.
magnetic field The force field surrounding and associated with a magnetic body. The geomagnetic field resembles a dipole field with subordinate non-dipole fields superimposed upon it.
magnetite A black iron oxide (Fe_3O_4), one of the spinel group of minerals, occurring in metamorphic and igneous rocks and constituting an important iron ore.
mammal A warm-blooded vertebrate of the class Mammalia. The three main groups are the egg-laying mammals, the marsupials, and the placental mammals. Several extinct groups are known from Mesozoic strata.
mammoth An extinct elephant of the Pleistocene. The woolly mammoth (*Mammuthus primigenius*) had a hairy coat and huge curved tusks.
mantle The section of the Earth's interior between the crust and the outer core, bounded at the top by the Mohorovičić discontinuity and at the base by the Gutenberg discontinuity. It is believed to consist of material of the composition of peridotite.
mass spectrometer An instrument for determining the types and amounts of ions present in a sample, by separating them according to their electric charge-to-mass ratio. It is used for isotopic analysis of rocks as well as for geochronology.
matrix The fine-grained material of a rock in which the coarser particles are set. In sedimentary rocks the matrix is often the cementing material, and in igneous rocks it is the groundmass of crystalline or glassy minerals.
meander A pronounced bend or loop in a river channel or valley.
mélange A formation, usually with some degree of lateral continuity, consisting of a chaotic mixture of rock types and showing evidence of intense tectonic deformation; characteristic of the oceanic trench environment in the region of a subduction zone.
melilite A tetragonal silicate mineral group, $(Na,Ca)_2(Mg,Al)(Al,Si)_2O_7$, found in thermally metamorphosed limestones and some undersaturated basic igneous rocks such as nephelinites.
meridian A line of longitude from the North to the South Pole passing through a given point on the Earth's surface. The prime meridian (0°) is taken as that passing through Greenwich (London, UK).
mesosphere See atmosphere.
metamorphism The alteration by heat and/or pressure of the mineralogy and texture of pre-existing rocks without their passing through a liquid phase. Regional metamorphism results from high temperatures and pressures of depth; contact (thermal) metamorphism is more limited in extent and results mainly from the high temperatures around igneous intrusions.
metasomatism Mineralogical and chemical change in rocks resulting from interaction with migrating fluids, commonly associated with contact metamorphism. It is also common in oceanic crust, where seawater penetrates the newly emplaced hot rocks.
metastable Existing in apparent equilibrium but capable of changing to a more stable state, applied to the polymorphs of certain minerals and to naturally occurring glasses.
metazoan An animal with a body formed of more than one cell (ie all animals except the single-celled protozoans and the sponges).
meteorite An extraterrestrial body entering the Earth's gravitational field and falling to the Earth's surface. Meteorites typically comprise silicate minerals and a nickel–iron alloy and are classified as stony, stony–iron or iron meteorites according to their precise composition. Some stony meteorites (chondrites) include small rounded clumps (chondrules) of olivine and pyroxene crystals.
mica A common group of monoclinic minerals consisting of complex hydrous silicates of potassium, aluminium etc, with a general formula $(K,Na,Ca)(Mg,Fe,Li,Al)_{2-8}(Al,Si)_4O_{10}(OH,F)_2$. Micas have a perfect basal cleavage producing a flaky, layered structure and occur particularly in acid igneous rocks, in many metamorphic rocks and in derived sediments. The main varieties are biotite, phlogopite and muscovite.
mid-ocean ridge One of a continuous system of mainly submarine ridges, hundreds to thousands of metres above the abyssal ocean floor, and orientated approximately parallel to the ocean margins. The ridge crest is characterized by a central rift, where seismic activity is concentrated. New oceanic crust is created at the ridges, marking the sites of constructive plate margins.

migmatite A very high-grade metamorphic rock in which extremes of temperature and pressure have induced partial melting so that the rock has taken on some of the characteristics of igneous texture.
mineral A naturally occurring crystalline substance with a definite chemical composition and a regular internal structure. Rocks are aggregates of minerals.
mineralization 1 The concentration of mineral deposits, especially metalliferous ores, in veins, replacement bodies etc in pre-existing rocks. 2 The replacement of the organic constituents of a plant or animal by inorganic materials during fossilization.
Mohorovičić discontinuity The seismic discontinuity between the Earth's crust and the mantle, typically 35 km below the surface of the continents, 10 km below the ocean floor, but up to 70 km below mountain belts.
molasse Thick sequences of sediment deposited during rapid erosion of newly formed and uplifting mountain belts. Molasse is thus post-tectonic, and is typically non-marine, by contrast with flysch, which is syn-tectonic (deposited during folding) and is marine.
molecule Two or more atoms linked by chemical bonding and constituting the smallest unit of a chemical substance that can exist independently while retaining the properties of the substance.
mollusc An invertebrate animal with a soft unsegmented body, a muscular 'foot' and usually a calcareous shell. The phylum Mollusca includes the gastropods, bivalves and cephalopods, all important as fossils.
moraine A deposit of unstratified rock fragments transported in, on or under ice sheets or glaciers and left behind when the ice melts.
mudstone A sedimentary rock consisting of fine-grained (less than 1/256 mm particle size) clay and other minerals. It is distinguished from shale by its inability to split along bedding planes.
muscovite A monoclinic sheet silicate mineral, a mica, $KAl_2(AlSi_3O_{10})(OH,F)_2$. Muscovite occurs most typically in acid plutonic igneous rocks, and in metamorphic rocks including schists and gneisses.
myriapod A terrestrial arthropod of the class Myriapoda having many walking limbs, eg centipedes and millipedes.

nanoplankton (or nannoplankton) Microscopic planktonic organisms, both zooplankton and phytoplankton particularly coccoliths, that can be detected only by means of the highest power of a phase-contrast microscope. Fossil nanoplankton is particularly important in deep ocean (pelagic) sediments.
nappe A recumbent fold, with a roughly horizontal axial plane and horizontal limbs which may be detached from their 'roots' by thrusting up 100 kilometres or more.
nautiloid A mollusc of the class Cephalopoda, having straight, curved or coiled shells with simple suture lines between the shell and the septa. Only one genus *(Nautilus)* still exists, but fossil nautiloids date back to the early Cambrian.
nepheline A feldspathoid mineral crystallizing in the hexagonal system, $NaAlSiO_4$. Nepheline cannot coexist with quartz, and thus is characteristic of undersaturated igneous rocks such as alkali basalts and nepheline syenites.
nephelinite A basic to ultrabasic volcanic rock consisting of nepheline with various ferromagnesian minerals, typically occurring in stable continental areas. The plutonic equivalent is ijolite.
nodule 1 A rounded concretion of minerals within a sedimentary rock. It is usually harder than the enclosing rock. 2 A xenolith of mantle-derived rock in a diatreme.
nucleus The positively charged central mass of an atom, occupying a minute fraction of the volume of the atom; it consists of one or more protons and (except for hydrogen) one or more neutrons.
nuclide An atom of a specific isotope, characterized by the constitution of its nucleus (ie by its atomic number and its mass number).
nuées ardentes Incandescent clouds of hot volcanic gases, made dense by their suspended load of ash and other hot pyroclastic material and thus able to travel at great speed under gravity. The rock deposited from a nuée ardente is an ignimbrite.

oblateness The proportion by which the equatorial diameter of a spheroid is greater than its polar diameter. The Earth is less oblate than Saturn.
occlusion Uplift of a warm front in the late stages of filling of a depression.
ocean floor spreading The separation of the lithospheric plates at mid-ocean ridges (constructive plate margins), with the formation of new oceanic crust along the rift by intrusion and effusion of basalt magma.
offlap The disposition of beds of sedimentary rock, characteristic of a marine regression, where older beds extend further onto the continent than the younger beds.
oligoclase A member of the plagioclase series of feldspars at the soda-rich end of the series.
olivine One of a group of orthorhombic silicate minerals, usually dark green and occurring in basic and ultrabasic igneous rocks such as peridotites; an important constituent of the mantle. Olivines vary in composition from Mg_2SiO_4 (forsterite) to Fe_2SiO_4 (fayalite).
onlap (or overlap) The disposition of beds of sedimentary rock above a plane of unconformity over which the sea has transgressed: progressively younger beds of the upper series extend further over the land surface than the preceding beds.
ophiolite A distinctive association of pelagic sediments and basic and ultrabasic igneous rocks, believed to represent a segment of the oceanic lithosphere, tectonically emplaced onto an island arc or continental margin.
orogeny An episode of tectonic activity (folding, faulting, thrusting) and mountain-building usually related to a destructive plate margin.
orographic Denoting rainfall or clouds produced by air passing over areas of high relief which is forced to rise and cool and then to condense.
orthoclase A potassium-rich alkali feldspar, $KAlSi_3O_8$, crystallizing in the monoclinic system. Orthoclase is an essential constituent of acid igneous rocks, and is also found in many metamorphic rocks.
orthopyroxene An orthorhombic pyroxene mineral, one of a series from enstatite ($MgSiO_3$) to orthoferrosilite ($FeSiO_3$), found in basic and ultrabasic igneous rocks and in strongly metamorphosed rocks where cooling was slower than that where clinopyroxenes of equivalent compositions form.
orthorhombic One of the seven crystal systems, with three twofold axes of symmetry at right angles to each other.
ostracod A small aquatic arthropod of the class Ostracoda, having a bivalved calcareous carapace enclosing the body. Fossils range from the Cambrian to Recent.
outcrop The total area where a particular rock body, bed or vein reaches the Earth's surface, either exposed or covered by surface material.
outgassing Loss of volatile elements and compounds from a planetary body, eg by volcanic activity.
overlap See onlap.
oxide A compound of oxygen with another element.

palaeo- Denoting the geologically ancient, eg palaeontology (the study of fossils), palaeoclimate (the state of the climate at some time in the past), palaeomagnetism (the study of ancient magnetic folds recorded in rocks).
partial melting The melting of some of the minerals of a rock as a result of increased pressure or temperature.
pegmatite Any very coarse-grained igneous rock occurring in minor intrusions related to plutonic bodies and having a related composition, but sometimes with concentrations of certain unusual elements and minerals.
pelagic Denoting the surface layers of the open ocean. Pelagic marine organisms live independently of the shore or sea bed. Pelagic sediments are those which have settled to the ocean floor from near its surface.
peridotite A dense ultrabasic igneous rock consisting mainly of olivine. Peridotites are believed to be a principal constituent of the mantle.
periodicity The length of time between successive occurrences of some cyclic phenomenon.
permafrost Permanently frozen ground extending to depths of up to 500 m; the surface layer sometimes thaws in summer; characteristic of periglacial areas.
permeability The capacity of a rock to allow water or other liquids to pass through it, being either porous or pervious. Permeability is measured in darcies.
perovskite A black or brown cubic mineral ($CaTiO_3$), found in some alkaline igneous rocks and thermally metamorphosed limestones; typical of a particular cubic crystallographic structure.
pervious Denoting a rock that will allow liquids to pass through it along joints, bedding planes etc.
petrogenesis The processes leading to the formation of a particular rock.
phenocryst A large crystal in a finer-grained groundmass in porphyritic igneous rocks. Phenocrysts are crystals of the first minerals to form as magma cools and begins to solidify.
phloem Plant tissue that carries synthesized food substances throughout the plant.
phonolite A fine-grained, undersaturated (silica-deficient) volcanic rock consisting mainly of alkali feldspar and nepheline, with subordinate ferromagnesian minerals. Most phonolites are

porphyritic (ie having a high proportion of phenocrysts set in a finer-grained groundmass). The plutonic equivalent is nepheline syenite.
photodissociation The disintegration of the molecules of a chemical compound (eg water to hydrogen and oxygen) resulting from the absorption of electromagnetic radiation in the visible range of the spectrum.
photosynthesis The process by which plants use carbon dioxide and water to synthesize organic compounds using energy from sunlight absorbed by chlorophyll (green pigment); oxygen is released.
phytobenthos Plants living on the sea floor.
phytophagous Feeding on plants.
phytoplankton The plant component of plankton, eg diatoms, dinoflagellates, coccolithophores.
pillow lava Lava in the form of globular or cylindrical masses, about a metre in diameter, solidified from basaltic magma erupted under water. Pillow lavas occupy much of the top layer of oceanic crust beneath the sediment cover.
placoderm An extinct jawed fish of the class Placodermi, common in Devonian times. Some were armoured and reached a length of 10 m.
plagioclase A triclinic feldspar occurring in most basic and intermediate igneous rocks and many metamorphic rocks, ranging in composition from albite ($NaAlSi_3O_8$) to anorthite ($CaAl_2Si_2O_8$). Plagioclase compositions are conventionally expressed in terms of percentages of the end-members albite (Ab) and anorthite (An), eg a plagioclase with equal proportions of sodium and calcium can be referred to as $Ab_{50}An_{50}$ or simply An_{50}.
planetesimal Small solid body, one of many thought to have constituted the matter of the early Solar System before their accretion into planets.
plankton Aquatic organisms, usually microscopic, that float or drift in the water. Phytoplankton, eg diatoms, are the plant component and zooplankton, eg protozoans and larvae of invertebrates, are the animal component.
plate One of the large rigid sections into which the Earth's lithosphere is divided and which move relative to each other. There are eight major plates and numerous smaller ones.
plateau A relatively flat extensive upland area bounded by steep slopes or mountain ranges.
platyhelminth A flatworm of the phylum Platyhelminthes, some of which are free-living and some parasitic.
plutonic rock Igneous rock formed from magma which has crystallized as an intrusion at depth in the crust and is therefore usually coarsely crystalline.
pneumatolysis Metamorphic alteration of a rock by the introduction of new elements in a gaseous medium.
point-bar The site of deposition of river sediment on the inside of a meander where the current is relatively slow.
pole, geographic Either of the north or south ends of the Earth's axis of rotation, latitude 90°N or 90°S. Geographic north (or south) is also termed true north (or south).
pole, magnetic Either of the two poles of the Earth's magnetic (dipole) field; a magnetic needle swinging freely in a horizontal plane will align itself north–south between them. Magnetic north is at present 79°N, 70°W and magnetic south 79°S, 110°E; these vary over time.
porosity The capacity of a rock such as sandstone to hold water in the spaces (pores) between mineral grains. Porous rocks may or may not be permeable.
porphyry An igneous rock containing a high proportion of phenocrysts set in a finer-grained groundmass.
positron The particle having the same mass as the electron and equal but positive electric charge (+1).
potash Strictly potassium carbonate (K_2CO_3), or potassium hydroxide (KOH), but used loosely of any potassium present in a rock or mineral in any form.
potassium–argon dating A method of radiometric dating based on the rate of decay of the radioactive isotope of potassium ^{40}K to argon, with a half-life of 1.42×10^8 years, used for dating a wide variety of rock types.
precipitation 1 Rain, snow, hail, dew etc deposited on the Earth's surface from moisture in the atmosphere. 2 The deposition of crystalline material from a solution, as in chemically precipitated limestones.
pressure Force per unit area acting on a surface. Atmospheric pressure is the pressure exerted by the weight of the column of air over a point on the Earth's surface, usually expressed in millibars (1 mb = 100 newtons/m^2); average atmospheric pressure at sea level is 1013.25 mb (1 atmosphere). Hydrostatic and geostatic pressures are those exerted at depth below a column of water or rock or both, often measured in bars or kilobars.
profile 1 A vertical section through soil, between the ground surface and parent rock, showing its different layers (horizons); or a section through part of the Earth's crust showing layers of rock. 2 The graphical outline at ground surface of the plane of a section through some feature, eg a valley.
progymnosperms Extinct plants with fernlike reproduction and gymnosperm-like anatomy (ie extensive secondary xylem); ancestors of the seed-bearing gymnosperms.
prokaryote An organism whose cells lack distinct organelles such as a nucleus, and whose hereditary material is not differentiated into chromosomes.
protistans Single-celled organisms forming the kingdom Protista and including the Protozoa, unicellular algae.
proton A positively charged elementary particle in the nucleus of an atom, having an atomic mass of approximately one.
protozoan A unicellular animal of the phylum Protozoa, usually microscopic, eg foraminiferans and radiolarians.
Psilopsida The most primitive order of vascular plants with no differentiation into stem, leaves or roots. Fossils are known from Silurian and Devonian strata.
pteridosperm An extinct seed-producing, fernlike plant belonging to the sub-class Pteridospermae of the gymnosperms; common from the late Palaeozoic to the Mesozoic.
Pteropsida An order of plants with large compound leaves and sporangia borne on the leaves or modified leaves, the ferns; common from the Devonian to the Recent.
P wave A type of seismic wave in which particles are displaced in the direction of propagation of the wave. Also known as a primary wave, pressure wave or compressional wave. P waves pass through solids, liquids and gases.
pyrite A metallic yellow, cubic sulphide mineral (FeS_2), found as an accessory mineral in igneous rocks, and in contact metamorphic rocks, hydrothermal and replacement deposits and anaerobic sediments. It is mined for its sulphur content.
pyroclastic Denoting rocks formed from the debris of explosive volcanic eruptions, such as bombs, pumice, scoriae and ash (tuff) from the magma, and fragments from the wall of the vent (explosion breccia).
pyroxene One of a group of ferromagnesian silicate minerals of orthorhombic (orthopyroxenes) or monoclinic (clinopyroxenes) crystal structure. They occur mostly in basic and ultrabasic igneous rocks and certain metamorphic rocks. Typical compositions are $Mg_2Si_2O_6$ (enstatite), $CaMgSi_2O_6$ (diopside), $NaAlSi_2O_6$ (jadeite).

quartzite A rock consisting almost completely of silica and formed either as a sandstone with purely siliceous cement (orthoquartzite) or from the metamorphism of a pure quartz sandstone (metaquartzite).

radiation The emission or propagation of radiant energy, especially as electromagnetic waves.
radiogenic heat Heat produced naturally in the Earth by the radioactive decay of potassium, uranium and thorium, apparently confined to the higher parts of the continental crust.
Radiolaria An order of small unicellular animals of the phylum Protozoa, one of the main constituents of marine plankton, characterized by an internal siliceous skeleton. They have existed since the Cambrian.
radiometric dating The determination of the age of rocks by identifying the proportions of parent and daughter isotopes of a radioactive element present in a rock, the rate of decay of the isotopes being known.
radionuclide A radioactive nuclide of an element.
Rayleigh wave A type of seismic wave travelling on the ground surface, particles moving in elliptical orbits in the plane of the direction of propagation.
refraction, seismic The change in direction of a seismic wave on passing obliquely from one medium to another in which the speed of propagation differs.
regolith The mantle of rock waste at the Earth's surface; weathered rock and soil overlying the bedrock.
regression The receding of the sea so that former sea floor becomes dry land because of a eustatic fall in sea level or a tectonic uplift of the land; may be recorded stratigraphically as an erosion surface or an unconformity.
relict sediments Sediments laid down under different conditions from those currently existing, eg glacial sediment deposited on the continental shelf around the British Isles.
remanent magnetism The magnetism preserved in a rock or mineral from the time when it cooled through the Curie temperature, providing evidence of the

state of the Earth's magnetic field at that time.
reptile A cold-blooded vertebrate of the class Reptilia, whose young characteristically develop inside a tough-shelled egg, eg crocodiles and snakes.
reservoir 1 A highly porous and permeable mass of rock that is able to hold or transmit fluids. 2 A chamber within the lithosphere beneath a volcano in which molten rock accumulates and is replenished from below.
resistivity The degree of electrical resistance offered by a substance to an electric current.
retrograde 1 Of metamorphism in which lower grade rocks are produced by alteration from higher grade rocks. 2 Opposite of prograde, ie to retreat (eg a delta front) by erosion. 3 Of the orbit or rotation of a planetary body in the reverse of the usual direction.
Rhyniales Extinct primitive plants of the order Psilopsida.
rhyolite A fine-grained or glassy acid igneous rock, mineralogically the volcanic equivalent of granite. It sometimes displays banding resulting from its flow as a viscous lava.
ridge See mid-ocean ridge.
rift valley A steep-sided linear trough in the Earth's surface formed by downfaulting between parallel faults.
rigidity, modulus of The angular strain resulting from applying shear stress to a rigid body, eg rock.
river terrace A near-horizontal, benchlike feature on a valley side, usually paired with another on the opposite side, representing part of the valley floor before further down-cutting by the river, eg as a result of uplift.
Rossby waves (upper air waves) Undulatory disturbances of the uniform flow of the upper-air westerly winds, which develop in a zone of contact between cold polar air and warm tropical air.
rudist A large highly asymmetric bivalved mollusc, one of a group extinct by the end of the Cretaceous, resembling solitary corals and forming reefs in clear, shallow water.
rugose coral One of a group of solitary and colonial corals of the Palaeozoic, having septa in multiples of four.
run-off The proportion of surface water free to flow into streams and rivers, ie not permeating below ground level.

sand bar An elongate accumulation of sand, lying parallel to the shore. It may be permanently submerged or exposed at low tide.
saprophagous Feeding on decaying organic matter.
scarp See escarpment, fault scarp.
scattering The diffusion of electromagnetic radiation by collision with atomic or molecular particles, particularly in the atmosphere.
scheelite A yellowish-white or brownish mineral ($CaWO_4$), an ore of tungsten found in veins and contact metamorphic deposits.
schist A coarse-grained metamorphic rock in which the minerals have crystallized or recrystallized parallel to each other. It can be split along schistosity planes which may be unrelated to the bedding planes of the original rock.
scoriae Angular cinder-like fragments of pyroclastic (volcanic) material containing many cavities and resulting from the rapid surface cooling of a stream of lava.
seamount A mountain-like feature rising from the abyssal plain, sometimes occurring in isolation and sometimes forming a chain. Most are volcanic in origin and only a few reach the surface.
secular Of changes taking place over prolonged periods of time.
sediment Unconsolidated particles of minerals (clastic, organic or chemically precipitated) deposited usually in water, and forming sedimentary rocks on lithification. Sedimentation is the process of sediment deposition.
sedimentary rocks The group of rocks formed from the lithification of deposited particles of pre-existing rocks (clastic sediments), organic remains or chemical precipitates.
seiche Oscillating surface wave in a lake or other body of water of restricted extent.
seismic waves Shock waves generated by earthquakes or underground explosions. The four main types are P waves, S waves, Love waves and Rayleigh waves.
seismograph An instrument for recording the frequency and magnitude of seismic waves; the record produced is a seismogram.
selvedge A sealed margin, such as the chilled 'rind' of pillow lavas, or the chemically altered edge of an ore body.
serpentine A greenish hydrous magnesium silicate mineral, usually greasy to the touch. The two main forms are chrysotile and antigorite. They occur in altered basic and ultrabasic rocks, the result of the breakdown of olivine and pyroxenes by hydrothermal alteration (serpentinization), and have a composition of $Mg_3Si_2O_5(OH)_4$.
shale A laminated fine-grained sedimentary rock composed mainly of clay minerals, which splits easily along bedding planes.
shear A tangential stress in which equal and opposite forces are imposed on either side of a plane and parallel to it. Shear stress tends to deform a body of rock by moving one part of it relative to another.
shield A widely outcropping assemblage of high-grade metamorphic rocks originating at depth in an area of ancient (usually Precambrian) tectonic activity, and unaffected since except by slow uplift and erosion.
shingle Water-worn pebbles or gravel deposited on beaches or offshore bars.
sial The continental crust, composed of granitic rock types rich in silica (65–75 per cent) and aluminium and having a mean density of 2700 kg/m^3.
siderophile Denoting an element soluble in molten iron and often occurring in the native state, including iron, cobalt, nickel and gold. Siderophile elements are found in the metal phases of meteorites and are believed to be components of the Earth's core.
silicates A group of minerals constituting about 95 per cent of the Earth's crust, including the feldspars, quartz, olivines, pyroxenes, amphiboles, micas, clay minerals and garnets. They are all based around the highly stable SiO_4 tetrahedral group of atoms.
sill A sheetlike intrusion of igneous rock concordant with the planar structure of the surrounding rock.
sillimanite An aluminium silicate, a polymorph of Al_2SiO_5 crystallizing in the orthorhombic system as needlelike crystals. Characteristic of high-grade metamorphic rocks rich in aluminia.
silt A fine-grained sediment, of particle size $\frac{1}{16}-\frac{1}{256}$ mm, ie between the clay and sand grades.
siltstone A clastic sedimentary rock formed from lithified silt grade sediments.
sima The crust of oceanic type, composed mainly of basaltic and peridotitic rock types poor in silica (less than 55 per cent) and rich in iron and magnesium. It has a density between 2800 and 3400 kg/m^3.
sinkhole A funnel-shaped hollow in the ground, especially in limestone areas, down which surface water runs into an underground drainage system.
slate A fine-grained metamorphic rock having a perfect (slaty) cleavage due to the parallel alignment of crystals of mica and chlorite. Formed by low-grade regional metamorphism of shale or mudstone.
snout The end of a valley glacier.
soda Strictly sodium carbonate or hydroxide, but used loosely of sodium in any form present in a rock or mineral.
sphene A monoclinic silicate mineral, $CaTiSiO_5$, occurring as an accessory mineral in acid igneous rocks and in some metamorphic rocks.
Sphenopsida (or horse tails) An order of plants in which branches and leaves are borne in whorls. The group was more important in the late Palaeozoic (when it included large trees) than at present.
spicule A tiny needlelike siliceous or calcareous structure, contained in the tissues of corals, sponges etc.
spinel One of a group of oxide minerals of cubic structure. Spinels vary in composition between true spinel ($MgAl_2O_4$) and hercynite ($Fe^{2+}Al_2O_4$), the commonest being magnetite ($Fe^{2+}(Fe^{3+})_2O_4$). They occur mostly in metamorphic rocks, apart from magnetite, which is ubiquitous, and chromite ($Fe^{2+}Cr_2O_4$), which is found in ultrabasic rocks; both are important ore minerals.
sponge A sessile aquatic animal of the phylum Porifera, having a saclike body often containing small skeletal elements (spicules). The most primitive of multicellular organisms, sometimes referred to as Parazoa, ie not truly metazoan.
sporophyte In a plant showing alternation of generations, the sporophyte is the generation that produces the asexual spores (as opposed to the gametophyte—the generation that produces the sexual gametes).
staurolite An orthorhombic aluminium silicate mineral, $FeAl_4Si_2O_{10}(OH)_2$, crystallizing in the orthorhombic system in prismatic crystals, sometimes forming cross-shaped twins; found in metamorphic rocks.
steppe Mid-latitude grasslands consisting of level, generally treeless plains. The term is strictly applied to such areas in Eurasia, those in North America being referred to as prairies and in South America as pampas.
stoss The side of a ripple or dune having the shallower slope.
strain Deformation of a body, as a response to applied forces (stress) and expressed as the ratio of the change in shape (or volume) to the original shape (or volume).
stratification The layering or bedding of sedimentary rocks and of some igneous rocks.

stratigraphy The study of the sequence and correlation of stratified rocks, comprising the description and naming of rocks (lithostratigraphy), their relative dating by means of fossils (biostratigraphy), and the definition of geological time (chronostratigraphy).
stratosphere See atmosphere.
stratum (plural, strata) An individual layer, or bed, of sedimentary rock. Stratified rocks are those in which the original layering can be recognized.
stress A force (tensional, compressional or shear) exerted on a body and producing a corresponding deformation (strain).
striation A groove or scratch on the surface of a rock. Glacial striations are cut by rock debris that is embedded in ice moving over boulders or bed rock; they give an indication of the direction of ice movement.
stromatolite A laminated structure, usually of calcium carbonate, precipitated or accumulated by blue-green algae, occurring particularly in Precambrian strata.
stromatoporoid A colonial organism belonging to the class Sclerospongia of the phylum Porifera (sponges). The laminated calcareous structures built by stromatoporoids were important components of reefs in the Silurian and Devonian periods.
subduction The downward emplacement of a lithospheric plate into the mantle beneath an overriding plate. Ocean floor is thus consumed at subduction zones, which are marked by ocean trenches and earthquake foci.
subsidence 1 The sinking of air, associated with divergence of convection cells at the Earth's surface and normally resulting in dry weather and stable conditions. 2 The depression of part of the Earth's crust, eg in a zone of rifting or in a sedimentary basin.
sulphate minerals A group of minerals including barite ($BaSO_4$), gypsum ($CaSO_4.2H_2O$) and anhydrite ($CaSO_4$).
supernova Short lived 'new star' of intense brightness, formed by the explosive transformation of an old star into a compact neutron star and a cloud of gas (the supernova remnant) which includes elements heavier than hydrogen and helium synthesized in the explosion.
suture 1 The line marking the junction between a septum (partition) and the shell of a cephalopod mollusc such as an ammonite. 2 The line or zone marking the locus at the Earth's surface of the site of closure of an ancient ocean by continent–continent collision.
S wave A type of seismic wave in which particles oscillate at right angles to the direction of propagation. Also known as a secondary wave, shear wave, shake wave or transverse wave. S waves pass through solids but not liquids.
syncline A trough-shaped fold in which the younger strata remain at the top of the sequence.
synform A trough-shaped fold in the general case in which the direction of younging of the strata is not known.

tabulate coral A type of colonial coral existing from the Ordovician to the Permian, with simple septa (or none), and having horizontal calcareous plates (tabulae).
talus (or scree) Rock fragments accumulated on and at the foot of a slope, produced by the mechanical weathering of the rock face above.
tectonics The study of the processes forming the major structural features of the Earth's crust. Tectonism is the activity of the deformational processes involved.
terrigenous Denoting sediments eroded from the land.
tetrapod A four-limbed vertebrate.
thecodont One of a group of early archosaurian reptiles appearing in the late Permian. They had teeth set in sockets and some walked on their hind legs. They probably included the ancestors of the dinosaurs.
therapsid One of a group of mammal-like reptiles important from Permian to Triassic times, considered to be the precursors of mammals.
thermocline A layer of ocean water, about 100–200 m below the surface, through which there is a marked decrease in water temperature with depth.
tholeiite A type of basalt of saturated or oversaturated composition (ie having free silica as quartz). It is characteristic of plateau basalts and the lava erupted at mid-ocean ridges.
thrust A low-angle reverse fault.
titration The addition of incremental amounts of one solution to a known amount of another until the reaction between them is complete: a method of volumetric analysis.
tonalite A coarse-grained igneous rock of intermediate composition, or an acid variety of diorite with up to 10 per cent quartz.
torque A force, or system of forces, tending to produce rotation in a body to which it is applied.
trace element An element occurring in minute quantities in the Earth's crust.
trachyte A fine-grained igneous rock of intermediate composition, the volcanic equivalent of syenite, consisting mainly of alkali feldspar with ferromagnesian minerals and having a characteristic flow structure. It is found mainly in ocean islands and in continental areas subject to rifting and volcanism.
trade winds Winds blowing from the tropical high-pressure belts to the equatorial low-pressure belt, from the north-east in the northern hemisphere and south-east in the southern hemisphere.
transgression (marine) The encroaching of the sea over areas of former dry land because of a eustatic rise in the sea level or a tectonic depression of the land. It may be recorded stratigraphically by an onlapping sedimentary succession.
transition element A metal element having an incomplete inner electron shell. Transition elements are dense, hard and have high melting points and include the technologically important metals such as iron, zinc, copper and nickel.
transportation The process of movement of rock fragments from the site of weathering to the site of deposition. Agents of transportation include gravity, water, ice and wind.
trap A rock formation or structure in which hydrocarbons accumulate, prevented from escaping upwards by impermeable layers.
trench A deep linear depression in the ocean floor, usually associated with a destructive plate margin.
trigonal One of the seven crystal systems, characterized by having one threefold axis of symmetry.
trilobite An extinct marine animal of the phylum Arthropoda, ranging in size from microscopic to over 500 mm in length and having a flattened oval segmented body with many appendages. Fossils are common and range from the early Cambrian to Permian.
trophic Relating to nutrition or food.
troposphere See atmosphere.
trough 1 An elongate area of low pressure extending between two areas of higher pressure in the atmosphere. 2 An elongate depression in the Earth's surface which may also be a sedimentary basin.
tsunami An ocean wave generated by submarine earthquake or volcanic activity, travelling at high speeds over very great distances and reaching a considerable height when slowed down on entering shallow coastal waters; sometimes called a tidal wave.
tufa Calcium carbonate deposited from solution in water, eg stalactites and stalagmites found in caves, and travertine found around hot springs.
tuff A rock composed of consolidated volcanic ash. Varieties include crystal tuffs, lithic tuffs, lapilli tuffs and vitric tuffs.

ultrabasic Denoting igneous rocks containing less than 45 per cent silica. Ultrabasic rocks are generally plutonic and contain virtually no quartz or feldspar, their main constituents being ferromagnesian silicates, oxides and sulphides. Also referred to as ultramafic.
unconformity A planar or irregular surface separating groups of rocks of different ages, the older of which have been folded and eroded, the younger being laid down on the resulting erosion surface. An unconformity thus represents a considerable time interval.

vascular Denoting the conducting tissue of the higher plants, characterized by specialized cells that form a continuous system throughout the plant, circulating water, mineral salts and synthesized food substances.
vector A physical quantity that has direction as well as magnitude; eg velocity represented diagrammatically by an arrow.
vein A thin body of minerals intruded into joints or fissures in rocks, usually of igneous or metasomatic origin.
vertebrate Any animal of the phylum Chordata, characterized by a bony or cartilaginous skeleton including a jointed backbone formed of vertebrae and a cranium for protecting the brain; includes fishes, amphibians, reptiles, birds and mammals. They evolved sometime in the Cambrian.
vesicle A small cavity in a volcanic rock resulting from the presence of a bubble of gas in the melt during solidification. Pumice is a typical vesicular rock.
viscosity The degree to which a liquid is viscous, ie resistant to flow. Viscosity is measured in kg/m/s, or in poise.
volatiles Substances that are easily evaporated.

Volatiles dissolved in magma become gaseous as the pressure is released, the commonest being water and carbon dioxide. They reduce the viscosity of the magma and also lower its temperature of solidification.
vug A cavity in a rock lined with well-formed crystals.

water table The underground level up to which permeable rocks are saturated with water. The water table varies seasonally and tends to follow the contours of the ground surface.
water vapour Water in the gaseous state or dispersed in the form of tiny droplets, present in the atmosphere in varying amounts.
weather The day-to-day atmospheric conditions at a particular place, including temperature, precipitation, wind, cloud cover etc.
wolframite A monoclinic ore mineral of tungsten, $FeWO_4$ to $MnWO_4$, occurring as vein deposits associated with granitic rocks and in placer deposits.

xenolith An inclusion of a 'foreign' rock embedded in an igneous body: either a block of country rock or a block of the igneous rock itself that solidified earlier than the rest and differs in composition.
xiphosure A marine arachnid of the order Xiphosura, including horseshoe crabs and many extinct varieties, characterized by a tough semicircular exoskeleton over the anterior region with a long, spearlike tail-spine or telson.
xylem Plant tissue that conducts water and minerals from the roots throughout the plant and provides mechanical support. Primary xylem is usually a central strand running through a stem and resulting from linear growth. Secondary xylem increases the girth of a stem or trunk and results from growth beneath the bark.

zoobenthos Animals living on the sea floor, including both sessile (non-mobile) and free-living (mobile) forms.
zoophagous Feeding on animals.
zooplankton The animal component of plankton, eg protozoans: minute animals that float or drift passively in the water.

Further Reading

General Books

I. G. Gass, P. J. Smith & R. C. L. Wilson, *Understanding the Earth*, Horsham, 1974
A. Hallam, *Planet Earth*, Oxford, 1977
W. K. Hamblin, *The Earth's Dynamic System* (2nd edn), Minneapolis, 1976
A. Holmes, *Principles of Physical Geology* (3rd edn), London, 1978
C. S. Hurlbut (ed), *The Planet We Live On: An Illustrated Encyclopedia of the Earth Sciences*, New York, 1976
D. N. Lapedes (ed), *McGraw-Hill Encyclopedia of the Geological Sciences*, New York, 1974
H. W. Menard, *Geology, Resources and Society: An Introduction to Earth Science*, San Francisco, 1974
F. Press and R. Siever, *Earth* (2nd edn), Oxford, 1978
F. J. Sawkins, C. G. Chase, D. G. Darby and G. Rapp, *The Evolving Earth*, London, 1974

Chapter 1

Frank D. Adams, *The Birth and Development of the Geological Sciences*, Baltimore, 1938
Gordon L. Davies, *The Earth in Decay: A History of British Geomorphology, 1578–1878*, London, 1969
Ursula B. Marvin, *Continental Drift: The Evolution of a Concept*, Washington, 1973
Roy Porter, *The Making of Geology: Earth Science in Britain, 1660–1815*, Cambridge, 1977
Martin J. S. Rudwick, *The Meaning of Fossils: Episodes in the History of Palaeontology*, London and New York, 1972

Chapter 2

S. Mitton (ed), *The Cambridge Encyclopaedia of Astronomy*, London and New York, 1977
P. Murdin and D. Allen, *Catalogue of the Universe*, New York, 1979
John A. Wood, *The Solar System*, Englewood Cliffs, 1979

Chapter 3

B. A. Bolt, *Earthquakes—A Primer*, San Francisco, 1978
B. A. Bolt, *Nuclear Explosions and Earthquakes: The Parted Veil*, Oxford, 1976

Chapter 4

I. Asimov, *Understanding Physics: The Electron, Proton and Neutron*, New York, 1969
R. W. Fairbridge (ed), *The Encyclopedia of Geochemistry and Environmental Science*, New York, 1972
B. H. Mason, *Principles of Geochemistry* (3rd edn), New York, 1966
K. H. Wedepohl (ed), *Handbook of Geochemistry*, Berlin, 1969

Chapter 5

B. Bayly, *Introduction to Petrology*, Englewood Cliffs, 1968
K. G. Cox, N. B. Price and B. Harte, *An Introduction to the Practical Study of Crystals, Minerals and Rocks*, New York, 1967
W. G. Ernst, *Earth Materials*, Englewood Cliffs, 1968
W. S. Fyfe, *Geochemistry of Solids*, Oxford, 1974
A. Harker, *Petrology for Students*, Cambridge, 1960
W. L. Roberts, G. R. Rapp, J. Weber (eds), *Encyclopedia of Minerals*, New York, 1974

Chapter 6

W. A. Heiskanen and F. A. Vening Meinesz, *The Earth and Its Gravity Field*, New York, 1958
Z. Kopal, *The Moon*, Hingham, 1969
K. Lambeck, *The Earth's Variable Rotations: Geophysical Causes and Consequences*, Cambridge, 1980

Chapter 7

Allen Cox (ed), *Plate Tectonics and Geomagnetic Reversals*, San Francisco, 1973
M. W. McElhinny, *Palaeomagnetism and Plate Tectonics*, Cambridge, 1973
P. J. Smith, *Topics in Geophysics*, Milton Keynes, 1973
David W. Strangway, *History of the Earth's Magnetic Field*, New York, 1970

Chapter 8

A. H. Brownlow, *Geochemistry*, Englewood Cliffs, 1979
G. Faure, *Principles of Isotope Geochemistry*, Chichester, 1977

Chapter 9

R. Bowen, *Geothermal Resources*, Barking, 1979
G. C. Brown and A. E. Mussett, *The Inaccessible Earth*, London, 1981
B. K. Ridley, *The Physical Environment*, Chichester, 1979
F. D. Stacey, *Physics of the Earth*, Chichester, 1969

Chapter 10

X. LePichon, J. Bonnin and J. Francheteau, *Plate Tectonics*, Amsterdam, 1980
M. W. McElhinny (ed), *The Earth: Its Origin, Structure and Evolution*, London, 1979
A. E. Ringwood, *Composition and Petrology of the Earth's Mantle*, New York, 1975
A. E. Ringwood, *Origin of the Earth and Moon*, New York, 1979
D. H. Tarling, *Evolution of the Earth's Crust*, London and New York, 1978
J. T. Wilson (ed), *Continents Adrift: Readings from Scientific American*, San Francisco, 1972
B. F. Windley, *The Early History of the Earth's Crust*, London, 1976

Chapter 11

B. E. Hobbs, W. D. Means and P. F. Williams, *An Outline of Structural Geology*, New York, 1976
L. U. de Sitter, *Structural Geology* (2nd edn), New York, 1964

Chapter 12

I. S. E. Carmichael, F. J. Turner and J. Verhogen, *Igneous Petrology*, New York, 1974
K. G. Cox, J. D. Bell and R. J. Pankhurst, *The Interpretation of Igneous Rocks*, London, 1979
P. W. Francis, *Volcanoes*, London, 1979
H. Williams and A. McBirney, *Volcanoes*, San Francisco, 1979

Chapter 13

R. D. Ballard and J. G. Moore, *Photographic Atlas of the Mid-Atlantic Ridge Rift Valley*, New York, 1977
R. G. Coleman, *Ophiolites*, New York, 1977
The Open University, *The Oceanic Crust (Oceanography, a third level course, unit 2)*, Milton Keynes, 1976
The Open University, *Red Sea Case Study (Crustal and Mantle Processes, a third level course)*, Milton Keynes, 1980

Chapter 14

G. A. Macdonald, *Volcanoes*, Englewood Cliffs, 1972
A. Sugimura and S. Uyeda, *Island Arcs: Japan and its Environs*, Amsterdam, 1973
P. J. Wyllie, *The Way the Earth Works*, Chichester, 1977

Chapter 15
M. H. P. Bott, *The Interior of the Earth*, London, 1971
R. H. Dott and R. H. Shaver (eds), *Modern and Ancient Geosynclinal Sedimentation*, Tulsa, 1974
A. K. Matthews, *Dynamic Stratigraphy*, Englewood Cliffs, 1974

Chapter 16
H. H. Read and J. Watson, *Introduction to Geology*, vol. ii, London, 1975
C. K. Seyfert and L. A. Sirkin, *Earth History and Plate Tectonics*, New York, 1973
B. L. Windley, *The Evolving Continents*, Chichester, 1977

Chapter 17
R. G. Barry and R. J. Chorley, *Atmosphere, Weather and Climate*, London, 1968
R. A. Freeze and J. Cherry, *Groundwater*, Englewood Cliffs, 1979
R. M. Goody and J. C. G. Walker, *Atmospheres*, Englewood Cliffs, 1972
J. G. Lockwood, *Causes of Climate*, Leeds, 1979
D. H. McIntosh and A. S. Thom, *Essentials of Meteorology*, London, 1973
R. C. Pereira, *Water Resources and Land Use*, Cambridge, 1972
J. C. Rodda, R. A. Downing, and F. M. Law, *Systematic Hydrology*, London, 1976
J. C. G. Walker, *Evolution of the Atmosphere*, New York, 1977
R. C. Ward, *Principles of Hydrology*, New York, 1975

Chapter 18
L. A. Frakes, *Climates Through Geologic Time*, Amsterdam, 1979
J. Gribbin, *Climatic Change: What's Wrong With Our Weather?*, Cambridge, 1978
J. Imbrie and K. P. Imbrie, *Ice Ages: Solving the Mystery*, London, 1979
H. H. Lamb, *Climate Present, Past and Future*, 2 vols, London, 1972, 1977
E. LeRoy La Durie, *Times of Feast, Times of Famine: A History of Climate Since the Year 1000*, New York, 1971

Chapter 19
William A. Anikouchire and Richard W. Sterberg, *The World Ocean (An Introduction to Oceanography)*, Englewood Cliffs, 1973
J. G. Harvey, *Atmosphere and Ocean: Our Fluid Environment*, Horsham, 1976
G. L. Pickard, *Descriptive Physical Oceanography* (3rd edn), Oxford, 1979
S. Pond and G. L. Pickard, *Introductory Dynamic Oceanography*, Oxford, 1979

Chapter 20
J. R. L. Allen, *Physical Processes in Sedimentation*, London, 1970
H. Blatt, G. V. Middleton and R. C. Murray, *Origin of Sedimentary Rocks*, Englewood Cliffs, 1972
C. Embleton and J. Thornes, *Process in Geomorphology*, London, 1979
H. G. Reading (ed), *Sedimentary Environments and Facies*, Oxford, 1978
R. C. Selley, *Ancient Sedimentary Environments*, London, 1970

Chapter 21
The Open University, *Evolution, A Third Level Course*, Milton Keynes, 1981
R. E. Dickerson, *Evolution, A Scientific American Book*, San Francisco, 1978
C. E. Folsome, *The Origin of Life*, San Francisco, 1979

Chapter 22
D. Attenborough, *Life on Earth*, London, 1979
R. W. Fairbridge and D. Jablonski (eds), *The Encyclopedia of Paleontology*, London, 1979
A. Hallam (ed), *Patterns of Evolution as Illustrated by the Fossil Record*, Amsterdam, 1977
L. B. Halstead, *The Evolution and Ecology of the Dinosaurs*, London, 1975
L. B. Halstead, *The Evolution of the Mammals*, London, 1978
M. R. Hanse (ed), *The Origin of Major Invertebrate Groups*, London, 1979
R. Leaky and R. Lewis, *People of the Lake*, London, 1979
McAlester, *History of Life*, Englewood Cliffs, 1968
A. Panchen (ed), *The Terrestrial Environment and the Origins of Land Vertebrates*, London, 1980
D. M. Raup and S. M. Stanley, *Principles of Paleontology* (2nd edn), San Francisco, 1978
B. Thomas, *The Evolution of Plants and Flowers*, London, 1980

Chapter 23
D. V. Ager, *The Nature of the Stratigraphical Record*, London, 1973
W. B. N. Berry, *Growth of A Prehistoric Time Scale*, San Francisco, 1968
D. L. Eicher, *Geologic Time*, Englewood Cliffs, 1978
A. Hallam, *Facies Interpretation and the Stratigraphic Record*, San Francisco, 1981
L. F. Laporte, *Ancient Environments*, Englewood Cliffs, 1968
A. L. Shaw, *Time in Stratigraphy*, New York, 1964

Chapter 24
S. J. Pirson, *Geologic Well Log Analysis*, Houston, 1970
F. F. Sabins, *Remote Sensing: Principles and Interpretation*, San Francisco, 1978
Schlumberger Ltd, *Log Interpretation Vol. I: Principles*, and *Log Interpretation Vol. II: Applications*, New York, 1972
N. M. Short, *Mission to Earth: Landsat Views The World*, Washington, DC, 1976
K. K. Waters, *Reflection Seismology*, New York, 1978

Chapter 25
B. J. Skinner, *Earth Resources*, Englewood Cliffs, 1969
K. Warren, *Mineral Resources*, London, 1973

Chapter 26
B. A. Bolt, W. L. Horn, G. A. Macdonald and R. F. Scott, *Geological Hazards*, New York, 1975
G. B. Oakeshott, *Volcanoes and Earthquakes: Geological Violence*, New York, 1976
T. Waltham, *Catastrophe, The Violent Earth*, London, 1978

Chapter 27
P. Cadogan, *The Moon: Our Sister Planet*, Cambridge, 1981
P. W. Francis, *The Planets*, London, 1980
J. A. Wood, *The Solar System*, Englewood Cliffs, 1979

Glossary
R. L. Bates and J. A. Jackson (eds), *Glossary of Geology* (2nd edn), Falls Church, 1980
J. Challinor, *A Dictionary of Geology* (5th edn), Cardiff, 1978
D. Dinely, D. Hawkes, P. Hancock, B. Williams, *Earth Resources: A Dictionary of Terms and Concepts*, London, 1976
S. Stiegeler (ed), *Dictionary of Earth Sciences*, London, 1978
D. G. A. Whitten and J. R. V. Brooks (eds), *A Dictionary of Geology*, London, 1973
Dictionary of Geological Terms, New York, 1957

Index

Notes
Page references in *italics* refer to figures and tables.

Acknowledgements

The publishers gratefully acknowledge the sources listed below for permission to reproduce or adapt illustrations. Every effort has been made to obtain permission to use copyright materials; the publishers trust that their apologies will be accepted for any errors or omissions.

1.2 Royal Geographical Society; **1.4** Ulster Museum; **1.6** Mary Evans Picture Library; **1.7** Travel Photo International; **1.9, 1.12** Ann Ronan Picture Library; **1.13** Institute of Geological Sciences; **2.3** Anglo-Australian Observatory; **2.4** AG Astrofotografie/Space Frontiers; **2.5** S. Mitton, *Cambridge Encyclopedia of Astronomy*, Trewin Copplestone Books Ltd; **2.6** Royal Observatory, Edinburgh; **2.7** Anglo-Australian Observatory; **2.8** Royal Observatory, Edinburgh; **2.10** S. Mitton, *Cambridge Encyclopedia of Astronomy*, Trewin Copplestone Books Ltd; **2.12** NASA/Science Photo Library; **2.13** John S. Shelton; **2.14** David Baker; **3.1** A. Bolt, *Nuclear Explosions and Earthquakes: the Parted Veil*, W. H. Freeman and Co; **3.2** K. E. Bullen, *Seismology*, Methuen and Co. Ltd; **3.3** M. Barazangi and J. Dorman, *Bulletin of the Seismological Society of America*, 59, 1969; **3.5** F. Press and R. Siever, *Earth*, W. H. Freeman and Co; **3.6** Spectrum Colour Library; **3.7** Georg Gerster/John Hillelson Agency; **3.8** Gene Daniels/Black Star; **3.9** F. Press and R. Siever, *Earth*, W. H. Freeman and Co; **3.10** K. E. Bullen, *Seismology*, Methuen and Co. Ltd; **3.13** F. Press and R. Siever, *Earth*, W. H. Freeman and Co; **3.14** F. Birch, *Physical Earth and Planetary Interiors*, Elsevier Scientific Publishing Co; **4.1** Institute of Geological Sciences; **4.8** Paul Brierley; **4.9** Institute of Geological Sciences; **4.16, 4.17** David Bayliss/RIDA; **4.19** J. Veizer and L. Jansen, *Journal of Geology*, 87, 1979; **4.20** P. Thorslund and F. E. Wickman/*Nature*; **5.1** Brian Lloyd; **5.2, 5.3, 5.6** Paul Brierley; **5.7a** B. Booth/Geoscience Features; **5.7b** M. S. Hobbs; **5.7c, 5.8** Paul Brierley; **5.12a, b, c** Courtesy of the Trustees of the British Museum (Natural History); **5.12d, 5.15a, b** Paul Brierley; **5.15c** Institute of Geological Sciences; **5.15d, 5.18a** Paul Brierley; **5.18b** Ian Murphy; **5.18c** Institute of Geological Sciences; **5.18d** Bill March/Mountain Camera; **5.21a, b** Paul Brierley; **5.21c** D. G. Smith; **5.21d** M. P. L. Fogden/Bruce Coleman Ltd; **5.21e** Travel Photo International; **5.21f, g** M. D. Fewtrell; **5.23a** Institute of Geological Sciences; **5.23b** Paul Brierley; **5.23c** J. Myers/The Geological Survey of Greenland; **5.27** S. A. Drury; **6.1** Cooper-Bridgeman Library; **6.7** D. E. Cartwright and R. A. Flather, *Ocean Tides*, Institute of Physics; **6.8** Y. Accad and C. L. Pekeris, *Philosophical Transactions of the Royal Society*, A290, 1978; **6.12** E. M. Gaposchkin, *Journal of Geophysical Research*, 79, 1974; **6.15** Groupe de Recherches de Geodesie Spatiale; **6.11** C. T. Scrutton; **7.1** J. Needham, *Science and Civilisation in China*, Cambridge University Press; **7.8** NASA/Science Photo Library; **7.11** Piers Smith; **7.12** S. E. Haggerty; **7.18** J. R. Heirtzler, *Magnetic Anomalies Measured at Sea*, vol 4 (1), Wiley Interscience; **8.1** Institute of Geological Sciences; **8.5** Pearce and Norry, *Contributions to Mineralogy and Petrology*, Springer-Verlag New York Inc; **8.6** D. Buchanan; **8.7** Institute of Geological Sciences; **8.8** ZEFA and J. M. Edmond/*Nature*; **8.9** Vulcain–Explorer/Vision International; **8.10** C. Hawkesworth; **8.11** Faure, *Principles of Isotope Geology*, John Wiley and Sons Inc; and Hoefs, *Stable Isotope Geochemistry*, Springer-Verlag New York Inc; **9.1** ZEFA; **9.3** *Block 1 Course S237*, Open University Press; **9.7** ENEL Ufficio Stampa e PR; **9.8** *Eurelios Leaflet*, Commission of the European Communities; **9.9** G. R. Roberts/Vision International; **9.10** ZEFA; **9.11** Mats Wibe Lund; **9.13, 9.14** H. N. Pollack and D. S. Chapman, *Scientific American*, 237, 1977; **9.15** J. G. Sclater and J. Crowe, *Journal of Geophysical Research*, 84, 1979; **9.16** H. N. Pollack and D. S. Chapman, *Scientific American*, 237, 1977; **9.17** J. G. Sclater and J. Francheteau, *Geophysics, Journal of the Royal Astronomical Society*, 20, 1970; **9.19** G. R. Roberts/Vision International; **9.20** H. N. Pollack and D. S. Chapman, *Scientific American*, 237, 1977; **9.21** R. Haenel, *Map of Temperature Distribution in the European Community* from the *Atlas of Subsurface Temperatures*, Commission of the European Communities; **9.25, 9.26** G. C. Brown and A. E. Mussett, *The Inaccessible Earth*, George Allen and Unwin Ltd; **9.27, 9.29** R. K. O'Nions, P. J. Hamilton and N. E. Evensen, *Scientific American*, 241, 1980; **10.2** Krafft-Explorer/Vision International; **10.4** V. Vacquier, A. D. Raff and R. E. Warren, *Bulletin of the Geological Society of America*, 72, 1961; **10.6** D. H. Tarling and J. G. Mitchell, *Geology*, 4, 1976; **10.9** D. E. Hayes and M. Ewing, *The Sea*, 4 (2) (ed A. E. Maxwell) Wiley Interscience; **10.11** W. G. Ernst, *Subduction Zone Metamorphism*, Dowden Hutchinson and Ross Inc; **10.12** S. B. Smithson, *Geophysical Research Letters*, 5 (9), 1978; **10.13** H. N. Pollack and D. S. Chapman, *Earth and Planetary Science Letters*, Elsevier Scientific Publishing Co; **10.14** M. Talwani and O. Eldholm, *The Geology of Continental Margins* (eds C. A. Burk and C. L. Drake) Springer-Verlag New York Inc; **10.15** R. E. Sheridan, *The Geology of Continental Margins* (eds C. A. Burk and C. L. Drake) Springer-Verlag New York Inc; **10.16** T. G. Francis et al, *Philosophical Transactions of the Royal Society*, A259, 1966; **11.1** John S. Shelton; **11.4a** Institute of Geological Sciences; **11.4b** R. C. L. Wilson/RIDA; **11.8a** P. J. Hill/RIDA; **11.8b** Paul Brierley; **11.8c** N. H. Woodcock; **11.12** J. H. Illies, *Tectonics and Geophysics of Continental Rifts*, D. Riedel Publishing Co; **11.13** Wilcox, Harding and Seely, *Bulletin of the American Association of Petroleum Geologists*, 57; **11.14** Hobbs, Means and Williams, *An Outline of Structural Geology*, John Wiley and Sons Inc; A. Holmes, *Principles of Physical Geology*, Thomas A. Nelson Ltd; Bailey, *Tectonic Essays*, Oxford University Press; **11.15** Hobbs, Means and Williams, *An Outline of Structural Geology*, John Wiley and Sons Inc; **12.2** H. U. Schmincke; **12.3** Krafft-Explorer/Vision International; **12.4, 12.5, 12.8, 12.9** H. U. Schmincke; **13.1** J. Francheteau/Cyamex Expedition; **13.2** Emory Kristoff © National Geographic Society; **13.4** H. U. Schmincke; **13.6** J. R. Cann, *Geophysics, Journal of the Royal Astronomical Society*, 39, 1974; **13.7** D. A. Wood et al, *Initial Reports of the Deep Sea Drilling Project*, 49, 1978; **13.8** J. M. Hall and P. T. Robinson, *Science*, 204, 1979, American Association for the Advancement of Science; **13.9** Courtesy of the Trustees of the British Museum (Natural History); **13.11** J. M. Hall, *Deep Drilling Results in the Atlantic Ocean*, 1979, Maurice Ewing Series 2, American Geophysical Union; **14.1** NASA/Science Photo Library; **14.2** Nigel Press Associates; **14.6** Dewey, *Tectonophysics*, 40, 1977; **14.11** Miyashiro, *American Journal of Science*, 1972; **14.16** R. Muir Wood/Royal Geographical Society; **14.17, 14.18** R. Muir Wood, *New Scientist*, 29 January 1981; **14.19, 14.20** W. S. Pitcher, *Journal of the Geological Society of London*, 135, 1978; **14.21** Dewey and Burke, 1973; **14.26** A. H. G. Mitchell, *Institution of Mining Metallurgy Congress Paper* 37, 1978; **14.28** Michael Abrams/Science Photo Library; **15.1** Pakiser, *Journal of Geophysical Research*, 68, 1963; **15.2** Talwani, Le Pichon and Ewing, *Journal of Geophysical Research*, 70, 1965; **15.3** P. Molnar and P. Tapponier, *Scientific American*, 263 (4), 1977; **15.4** G. A. Wagner et al, *Mem. Inst. Geol. Univ. Padova*, vol XXX, 1977; **15.5** McConnell, *Journal of Geophysical Research*, 73, 1968; **15.6, 15.7** M. R. Leeder; **15.8** J. L. Worzel, *The Geology of Continental Margins* (eds C. A. Burk and C. L. Drake) Springer-Verlag New York Inc; **15.11** M. R. Leeder; **15.12** Hutchinson and Engels, *Philosophical Transactions of the Royal Society*, 267A, 1970; **15.13** R. E. Sheridan, *The Geology of Continental Margins* (eds C. A. Burk and C. L. Drake) Springer-Verlag New York Inc; **15.14** J. Allen Cash; **15.15** *The Geology of Continental Margins* (eds C. A. Burk and C. L. Drake) Springer-Verlag New York Inc; **16.1** Brian Hawkes/Robert Harding Associates; **16.2** Reproduced with permission (A 55/81) of Geodaetisk Institut, Denmark; **16.3** David Baker; **16.4** J. H. Allaart, *The Early History of the Earth* (ed Windley); **16.5, 16.6** The Geological Survey of Greenland; **16.7** R. D. Davies and H. L. Allsopp, *Geology*, 4, 1976; **16.8** C. H. Antiaeusser, *Geological Society of Australia Special Publication*, 3, 1971; **16.10, 16.11** A. Berthelsen, *Geological Survey of Greenland GGU Bulletin*, 25, 1960; **16.12** J. Myers/The Geological Survey of Greenland; **16.18** McElhinny and McWilliams, *Tectonophysics*, Elsevier Scientific Publishing Co; **16.23** Photri; **16.24** Guy Kieffer/Sygma; **16.25a** *Bulletin of the Geological Society of America*, 84, 1973; **16.25b** Hsü and Nabholz, *American Journal of Science*, 279, 1979; **16.25c** F. Mégard, *Étude Géologique des Andes du Pérou Central*, ORSTOM, Paris; **16.28** J. Ramsay; **17.2** G. M. B. Dobson, *Exploring the Atmosphere*, Oxford University Press; **17.3** W. D. Sellars, *Physical Climatology*, University of Chicago Press; **17.4** J. M. Wallace and P. V. Hobbs, *Atmospheric Science, an Introductory Survey*, Academic Press Inc; **17.6** T. R. Oke, *Boundary Layer Climates*, Methuen and Co. Ltd; **17.8** M. Neiburger, J. G. Edinger and W. D. Bonner, *Understanding Our Atmospheric Environment*, United States Department of Commerce; **17.12** H. Riehl, *Introduction to the Atmosphere*, McGraw-Hill Book Co; **17.13** T. J. Chandler and L. F. Musk, *The Geographical Magazine*, November 1976; **17.14, 17.15** A. N. Strahler, *Physical Geography*, John Wiley and Sons Inc; **17.16** European Space Agency; **17.17b** E. Porter, *Water Management in England and Wales*, Cambridge University Press; **17.18** Freeze and Cherry, *Groundwater*, Prentice-Hall Inc; **17.19** R. J. Chorley, *Water, Earth and Man*, Methuen and Co. Ltd; **17.20** J. C. Rodda, R. A. Downing and F. M. Law,

Systematic Hydrology, Butterworth and Co. (Publishers) Ltd; **18.1** Mark Perlstein/Black Star; **18.3** K. Bostrom/YMER Expedition; **18.5** J. Imbrie and K. P. Imbrie, *Ice Ages: Solving the Mystery*, Ridley Enslow Publishers; **18.6** A. McIntyre, *Science*, 191, 1976, American Association for the Advancement of Science; **18.7** Cooper-Bridgeman Library and Paolo Koch/Vision International; **18.8**, **18.9** Science Photo Library; **18.10** J. Imbrie and K. P. Imbrie, *Ice Ages: Solving the Mystery*, Ridley Enslow Publishers; **18.12** J. Mason/Black Star; **18.15** J. Williams and F. Niehaus, *Studies of Different Energy Strategies in Terms of Their Effects on the Atmospheric CO_2 Concentration*, IIASA; **19.10** R. Legeckis, NOAA; **19.13** J. Frederick Grussle/Woods Hole Oceanographic Institution; **20.1** R. M. Garrells and F. T. Mackenzie, *Evolution of the Sedimentary Rocks*, W. W. Norton and Co. Inc; **20.2** L. B. Leopold, M. Gordon Wolman and J. P. Miller, *Fluvial Processes in Geomorphology*, W. H. Freeman and Co; **20.3** R. J. Chorley, *Introduction to Fluvial Processes*, Methuen and Co. Ltd; **20.4** R. M. Garrells and F. T. Mackenzie, *Evolution of the Sedimentary Rocks*, W. W. Norton and Co. Inc; **20.6** Travel Photo International; **20.7** *Depositional Environment As Interpreted From Primary Sedimentary Structures And Stratification Sequences*, Society of Economic Mineralogists and Palaeonotologists; **20.8**, **20.9** J. R. L. Allen, *Physical Processes of Sedimentation*, George Allen and Unwin Ltd; **20.10** J. Behnke/ZEFA; **20.11** W. Jesco von Puttakamer/Alan Hutchison Library; **20.12** Marc Riboud/John Hillelson Agency; **20.13** John Cleare; **20.15** N. M. Strakhov, *Principles of Lithogenesis*, Oliver and Boyd; **20.16** Photo Research Int; **20.18** Aerofilms; **20.19** Photri; **20.23** G. V. Middleton and M. Hampton, *Turbidities and Deepwater Sedimentation*, Society of Economic Mineralogists and Palaeontologists; **20.24** Peter Parks/Oxford Scientific Films; **20.26** J. P. Riley and R. Chester, *Oceanic Sediments and Sedimentary Processes in Chemical Oceanography*, Academic Press Inc London Ltd; **20.27** A. Souter and P. A. Crill, *Sedimentation and Climatic Patterns in the Santa Barbara Basin During the 19th and 20th Centuries*, Scripps Institution of Oceanography; **21.1** Dickerson, *Scientific American*, 1978; **21.6**, **21.7**, **21.12**, **21.13**, **21.15** S. M. Awramik; **21.17** G. R. Roberts/Vision International; **22.1** Courtesy Paleontologisk Museum, Oslo; **22.4** W. Stürmer; **22.5** Novosti Press Agency; **22.6** Institute of Geological Sciences; **22.7** Travel Photo International; **22.8** K. J. Müller/Smithsonian Institution; **22.10** Sternberg Memorial Museum; **22.13** Institute of Geological Sciences; **22.18** A. C. Scott; **23.1** A. N. Strahler, *Physical Geography*, John Wiley and Sons Inc; **23.2** D. G. Smith; **23.4** R. K. Matthews, *Dynamic Stratigraphy*, Prentice-Hall Inc; **23.5** Barrell, *Bulletin of the Geological Society of America*, 28, 1917; **23.6** L. L. Sloss, *Bulletin of the Geological Society of America*, 74, 1963; **23.7** *Sedimentary Basin Case Study*, third level science course S333 SB, Open University Press; **23.8a** C. E. Payton (ed) *Seismic Stratigraphy–Applications to Hydrocarbon Exploration*, Memoir 26, American Association of Petroleum Geologists; **23.8b** P. R. Vail and R. M. Mitchum, Exxon Production Research Company; **23.9** Opdyke, Glass, Hays and Foster, *Science*, 154, 3747, 1966; **23.11** Edward A. Francis; **23.12** Birkbeck College, London; **23.13** L. Costa and C. Downie, *Palaeontology*, 19, 1976 and W. R. Evitt (ed), *Proceedings of a Forum On Dinoflagellates*, 1975, American Association of Stratigraphic Palynologists; **23.15** A. B. Shaw, *Time in Stratigraphy*, McGraw-Hill Book Co; **23.17** Van Donk, *Memoir 145*, 1976, Geological Society of America; **23.18** L. A. Frakes, *Climates Throughout Geologic Time*, Elsevier Scientific Publishing Co; **23.21** P. Playford; **23.23** P. Allen; **23.25** J. D. A. Piper; **23.26**, **23.27** S. A. Drury; **24.1** Colwell et al, *Photogrammetric Engineering*, 29, 1963; **24.3** *Association of Engineering Geologists Special Publication*, 1978; **24.5a** US Soil Conservation Service; **24.5b** US Soil Survey; **24.5c**, **24.5d** Clyde Surveys; **24.6** Aerofilms; **24.8a** US Dept of Agriculture; **24.8b** Digicolor (R) Image produced by Daedalus Enterprises Inc, using their digicolor (R) date reduction process; **24.9** Aerofilms; **24.10a** David Baker; **24.10b** NASA/Science Photo Library; **24.10c** Nigel Press Associates; **24.10d**, **24.11a** David Baker; **24.11b** USGS Open File Report 75–416; **24.14** J. A. Grow and R. G. Markl, USGS; **24.15** Schlumberger; **24.16** *This is Schlumberger*, Schlumberger Inland Services Inc; **25.3** I. G. Gass et al, *Understanding the Earth*, Artemis Press Ltd; **25.4** R. C. Neary, *The Earth's Physical Resources*, Open University Press; **25.5** V. E. McKelvey, *USGS Professional Paper*, 820, 1972, United States Geological Survey; **25.6** I. G. Gass et al, *The Earth's Physical Resources*, Open University Press; **25.7** R. H. Sillitoe, *Economic Geology*, 67, 1972; **25.10** K. K. Turkeian, *Oceans*, Prentice-Hall Inc; **25.11** S. Calvert and B. Price, *New Scientist*, 1 January 1970; **26.2** Camera Press Ltd; **26.4** Mats Wibe Lund; **26.5** *Predicting Earthquakes*, 1976, National Academy of Sciences, Washington, DC; **26.8** Camera Press Ltd; **27.2** Picturepoint; **27.3**, **27.4** David Baker; **27.6**, **27.7** NASA/Science Photo Library; **27.8** Prepared by Eric Eliason, US Geological Survey, Flagstaff, Arizona for NASA, AMES Research Center and Massachusetts Institute of Technology; **27.9**, **27.11** David Baker; **27.12** Ann Ronan Picture Library; **27.14**, **27.16**, **27.18**, **27.19** NASA/Science Photo Library; **27.20**, **27.21**, **27.22** David Baker; **27.23**, **27.24** NASA/Science Photo Library; **27.25** Science Photo Library; **27.26** David Baker.

Part openers: page 10, Cooper-Bridgeman Library; page 34, J. Myers/The Geological Survey of Greenland; page 162, Vulcain–Explorer/Vision International; page 274, John Cleare; page 410, Robert Harding Picture Library; page 442, J. Allen Cash Ltd.

Artwork· Art Services Ltd: 16.4; Tony Bryan: 3.2, 3.5, 3.9, 3.10, 3.11, 3.13, 15.3, 15.12, 15.15; Robert Burns: 11.2, 11.3, 11.5, 11.6, 11.7, 11.9, 11.10, 11.11, 11.12, 11.13, 11.14, 11.15, 18.4, 18.5, 18.6, 18.10, 18.11, 18.16; Eugene Fleury: 6.7, 6.8, 6.12, 10.17, 23.16, 23.19, 23.20, 23.22, 23.24, 25.6, 25.7, 25.8, 25.10, 25.11; Chris Forsey: 13.3, 13.4, 13.5, 13.6, 13.7, 13.8, 13.11, 13.12, 17.17, 26.2, 26.4, 26.5, 26.6; Vana Haggerty: 22.20, 22.22; Brian Hall: 5.4, 5.11, 5.14, 5.19, 5.20, 5.26, 6.6, 7.17, 7.19, 10.1, 10.5, 10.6, 16.22, 24.2, 24.3; Sidney Roderick: 4.2, 4.4, 14.21, 14.26, 14.27, 16.18, 16.21, 17.2, 17.3, 17.4, 17.6, 17.7, 17.8, 17.9, 17.10, 17.11, 17.12, 17.13, 17.14, 17.18, 17.19, 17.20, 19.3, 19.4, 19.5, 19.6, 23.7, 23.8, 23.9, 23.13, 23.14, 23.15, 23.17, 23.18, 24.12, 25.1, 25.2, 25.3, 25.4, 25.5; Colin Salmon: 3.1, 3.4, 3.12, 4.18, 5.17, 6.14, 10.7, 10.8, 14.3, 14.4, 14.5, 14.6, 14.7, 14.9, 14.11, 14.12, 14.14, 14.15, 14.19, 14.20, 24.4; Hilary Saunders: 10.14, 10.20; Graham Smith: 6.2, 6.3, 6.4, 6.5, 6.10, 6.13, 6.16, 6.17, 6.18, 7.14, 8.3, 8.4, 8.5, 8.11, 8.13, 8.15, 8.16, 8.17, 8.18, 14.17, 14.18, 14.24, 14.25, 15.1, 15.2, 15.8, 15.9, 15.13, 16.25, 16.26, 18.14, 18.15, 23.1, 23.3, 23.4, 23.5; Ed Stuart: 7.3, 7.13, 7.15, 7.16, 10.3, 10.13, 10.18; Charlotte Styles: 17.15; Alan Suttie: 9.3, 9.4, 9.8, 9.12, 9.13, 9.14, 9.15, 9.16, 9.17, 9.20, 9.21, 9.22, 9.23, 9.25, 9.26, 9.27, 9.28, 9.29, 10.10, 12.1, 20.1, 20.2, 20.3, 20.4, 20.5, 20.7, 20.8, 20.9, 20.14, 20.15, 20.17, 20.20, 20.21, 20.22, 20.23, 20.25, 20.26, 20.28, 21.16; Thames Cartographic Ltd: 3.3, 7.4, 7.5, 7.6, 7.9, 7.10, 10.4, 10.11, 10.15, 10.21, 15.5, 16.7; Anne Winterbottom: 22.1, 22.9, 22.11, 22.12, 22.14, 22.15, 22.19, 22.22, 22.25; John Yates: 16.9, 16.10, 16.11, 16.13, 16.14, 16.15, 16.16, 16.17, 19.8, 19.9, 19.11, 19.12, 21.1, 21.8, 21.9, 21.10, 21.11, 21.14.